McGraw-Hill

Dictionary of
Engineering

Second
Edition

McGraw-Hill

New York Chicago San Francisco Lisbon London Madrid
Mexico City Milan New Delhi San Juan Seoul Singapore
Sydney Toronto

The **McGraw·Hill** Companies

 9 0   DSH/DSH   0 1 0

ISBN 0-07-141050-3

 This book is printed on recycled, acid-free paper containing a minimum of 50% recycled, de-inked fiber.

This book was set in Helvetica Bold and Novarese Book by the Clarinda Company, Clarinda, Iowa. It was printed and bound by RR Donnelley, The Lakeside Press.

McGraw-Hill books are available at special quantity discounts to use as premiums and sales promotions, or for use in corporate training programs. For more information, please write to the Director of Special Sales, Professional Publishing, McGraw-Hill, Two Penn Plaza, New York, NY 10121-2298. Or contact your local bookstore.

Library of Congress Cataloging-in-Publication Data

McGraw-Hill dictionary of engineering — 2nd. ed.
    p.   cm.
  "All definitions were drawn from the McGraw-Hill dictionary of scientific and technical terms, 6th ed." — Pref.
   ISBN 0-07-141050-3 (alk. paper)
   1. Engineering—Dictionaries. I. McGraw-Hill dictionary of scientific and technical terms. 6th ed.

TA9.M35   2002
620'.003—dc21   2002033178

Contents

Preface

The *McGraw-Hill Dictionary of Engineering* provides a compendium of more than 18,000 terms that are central to the various branches of engineering and related fields of science. The coverage in this Second Edition is focused on building construction, chemical engineering, civil engineering, control systems, design engineering, electricity and electronics, engineering acoustics, industrial engineering, mechanics and mechanical engineering, systems engineering, and thermodynamics. Many new entries have been added since the previous edition with others revised as necessary. Many of the terms used in engineering are often found in specialized dictionaries and glossaries; this Dictionary, however, aims to provide the user with the convenience of a single, comprehensive reference.

All of the definitions are drawn from the *McGraw-Hill Dictionary of Scientific and Technical Terms*, Sixth Edition (2003). Each definition is classified according to the field with which it is primarily associated; if it is used in more than one area, it is idenfified by the general label [ENGINEERING]. The pronunciation of each term is provided along with synonyms, acronyms, and abbreviations where appropriate. A guide to the use of the Dictionary appears on pages vii and viii, explaining the alphabetical organization of terms, the format of the book, cross referencing, and how synonyms, variant spellings, abbreviations, and similar information are handled. The Pronunciation Key is given on page xi. The Appendix provides conversion tables for commonly used scientific units as well as listings of useful mathematical, engineering, and scientific data.

It is the editors' hope that the Second Edition of the *McGraw-Hill Dictionary of Engineering* will serve the needs of scientists, engineers, students, teachers, librarians, and writers for high-quality information, and that it will contribute to scientific literacy and communication.

Mark D. Licker
Publisher

Staff

Mark D. Licker, Publisher—Science

Elizabeth Geller, Managing Editor
Jonathan Weil, Senior Staff Editor
David Blumel, Staff Editor
Alyssa Rappaport, Staff Editor
Charles Wagner, Digital Content Manager
Renee Taylor, Editorial Assistant

Roger Kasunic, Vice President—Editing, Design, and Production

Joe Faulk, Editing Manager
Frank Kotowski, Jr., Senior Editing Supervisor

Ron Lane, Art Director

Thomas G. Kowalczyk, Production Manager
Pamela A. Pelton, Senior Production Supervisor

Henry F. Beechhold, Pronunciation Editor
Professor Emeritus of English
Former Chairman, Linguistics Program
The College of New Jersey
Trenton, New Jersey

How to Use the Dictionary

ALPHABETIZATION. The terms in the McGraw-Hill Dictionary of Engineering, Second Edition, are alphabetized on a letter-by-letter basis; word spacing, hyphen, comma, solidus, and apostrophe in a term are ignored in the sequencing. For example, an ordering of terms would be:

abat-vent	**ADP**
A block	**air band**
Abney level	**airblasting**

FORMAT. The basic format for a defining entry provides the term in boldface, the field is small capitals, and the single definition in lightface:

    term [FIELD] Definition.

A field may be followed by multiple definitions, each introduced by a boldface number:

    term [FIELD] **1.** Definition. **2.** Definition. **3.** Definition.

A term may have definitions in two or more fields:

    term [CIV ENG] Definition. [ENG ACOUS] Definition.

A simple cross-reference entry appears as:

    term *See* another term.

A cross reference may also appear in combination with definitions:

    term [CIV ENG] Definition. [ENG ACOUS] Definition.

CROSS REFERENCING. A cross-reference entry directs the user to the defining entry. For example, the user looking up "access flooring" finds:

    access flooring *See* raised flooring.

The user then turns to the "R" terms for the definition. Cross references are also made from variant spellings, acronyms, abbreviations, and symbols.

    ARL *See* acceptable reliability level.
    arriswise *See* arrisways.
    at *See* technical atmosphere.

ALSO KNOWN AS . . . , etc. A definition may conclude with a mention of a synonym of the term, a variant spelling, an abbreviation for the term, or other such information, introduced by "Also known as . . . ," "Also spelled . . . ," "Abbreviated . . . ," "Symbolized . . . ," "Derived from . . . ." When a term has more than one definition, the positioning of any of these phrases conveys the extent of applicability. For example:

> **term** |CIV ENG| **1.** Definition. Also known as synonym. **2.** Definition. Symbolized T.

In the above arrangement, "Also known as . . ." applies only to the first definition; "Symbolized . . ." applies only to the second definition.

> **term** |CIV ENG| **1.** Definition. **2.** Definition. |ENG ACOUS| Definition. Also known as synonym.

In the above arrangement, "Also known as . . ." applies only to the second field.

> **term** |CIV ENG| Also known as synonym. **1.** Definition. **2.** Definition. |ENG ACOUS| Definition.

In the above arrangement, "Also known as . . ." applies to both definitions in the first field.

> **term** Also known as synonym. |CIV ENG| **1.** Definition. **2.** Definition. |ENG ACOUS| Definition.

In the above arrangement, "Also known as . . ." applies to all definitions in both fields.

Fields and Their Scope

building construction—The technology of assembling materials into a structure, especially one designated for occupancy.

chemical engineering—A branch of engineering which involves the design and operation of chemical plants.

civil engineering—The planning, design, construction, and maintenance of fixed structures and ground facilities for industry, for transportation, for use and control of water, for occupancy, and for harbor facilities.

control systems—The study of those systems in which one or more outputs are forced to change in a desired manner as time progresses.

design engineering—The branch of engineering concerned with the design of a product or facility according to generally accepted uniform standards and procedures, such as the specification of a linear dimension, or a manufacturing practice, such as the consistent use of a particular size of screw to fasten covers.

electricity—The science of physical phenomena involving electric charges and their effects when at rest and when in motion.

electronics—The technological area involving the manipulation of voltages and electric currents through the use of various devices for the purpose of performing some useful action with the currents and voltages; this field is generally divided into analog electronics, in which the signals to be manipulated take the form of continuous currents or voltages, and digital electronics, in which signals are represented by a finite set of states.

engineering—The science by which the properties of matter and the sources of power in nature are made useful to humans in structures, machines, and products.

engineering acoustics—The field of acoustics that deals with the production, detection, and control of sound by electrical devices, including the study, design, and construction of such things as microphones, loudspeakers, sound recorders and reproducers, and public address sytems.

industrial engineering—A branch of engineering dealing with the design, development, and implementation of integrated systems of humans, machines, and information resources to provide products and services.

mechanical engineering—The branch of engineering concerned with energy conversion, mechanics, and mechanisms and devices for diverse applications, ranging from automotive parts through nanomachines.

mechanics—The branch of physics which seeks to formulate general rules for predicting the behavior of a physical system under the influence of any type of interaction with its environment.

systems engineering—The branch of engineering dealing with the design of a complex interconnection of many elements (a system) to maximize an agreed-upon measure of system performance.

thermodynamics—The branch of physics which seeks to derive, from a few basic postulates, relations between properties of substances, especially those which are affected by changes in temperature, and a description of the conversion of energy from one form to another.

Pronunciation Key

Vowels

a	as in bat, that
ā	as in bait, crate
ä	as in bother, father
e	as in bet, net
ē	as in beet, treat
i	as in bit, skit
ī	as in bite, light
ō	as in boat, note
ò	as in bought, taut
ù	as in book, pull
ü	as in boot, pool
ə	as in but, sofa
aù	as in crowd, power
ói	as in boil, spoil
yə	as in formula, spectacular
yü	as in fuel, mule

Semivowels/Semiconsonants

w	as in wind, twin
y	as in yet, onion

Stress (Accent)

ˈ   precedes syllable with primary stress

ˌ   precedes syllable with secondary stress

| precedes syllable with variable or indeterminate primary/secondary stress

Consonants

b	as in bib, dribble
ch	as in charge, stretch
d	as in dog, bad
f	as in fix, safe
g	as in good, signal
h	as in hand, behind
j	as in joint, digit
k	as in cast, brick
k	as in Bach (used rarely)
l	as in loud, bell
m	as in mild, summer
n	as in new, dent
n	indicates nasalization of preceding vowel
ŋ	as in ring, single
p	as in pier, slip
r	as in red, scar
s	as in sign, post
sh	as in sugar, shoe
t	as in timid, cat
th	as in thin, breath
th	as in then, breathe
v	as in veil, weave
z	as in zoo, cruise
zh	as in beige, treasure

Syllabication

·   Indicates syllable boundary when following syllable is unstressed

xi

a *See* ampere.

Å *See* ampere; angstrom.

Å *See* angstrom.

a axis |MECH ENG| The angle that specifies the rotation of a machine tool about the *x* axis. { 'ā 'ak,sis }

abandon |ENG| To stop drilling and remove the drill rig from the site of a borehole before the intended depth or target is reached. { ə'ban·dən }

abate |ENG| **1.** To remove material, for example, in carving stone. **2.** In metalwork, to excise or beat down the surface in order to create a pattern or figure in low relief. { ə'bāt }

abatement |ENG| **1.** The waste produced in cutting a timber, stone, or metal piece to a desired size and shape. **2.** A decrease in the amount of a substance or other quantity, such as atmospheric pollution. { ə'bāt·mənt }

abat-jour |BUILD| A device that is used to deflect daylight downward as it streams through a window. { ä·bä'zhùr }

abattoir |IND ENG| A building in which cattle or other animals are slaughtered. { ,ab·ə'twär }

abat-vent |BUILD| A series of sloping boards or metal strips, or some similar contrivance, to break the force of wind without being an obstruction to the passage of air or sound, as in a louver or chimney cowl. { ,ä,bä'vän }

ablatograph |ENG| An instrument that records ablation by measuring the distance a snow or ice surface falls during the observation period. { ə'blā·də,graf }

A block |CIV ENG| A hollow concrete masonry block with one end closed and the other open and with a web between, so that when the block is laid in a wall two cells are produced. { 'ā ,bläk }

Abney level *See* clinometer. { 'ab·nē 'lev·əl }

abnormal reading *See* abnormal time. { ab'nòr·məl 'rēd·iŋ }

abnormal time |IND ENG| During a time study, an elapsed time for any element which is excessively longer or shorter than the median of the elapsed times. Also known as abnormal reading. { ,ab,nòr·məl 'tīm }

abort branch |CONT SYS| A branching instruction in the program controlling a robot that causes a test to be performed on whether the tool-center point is properly positioned, and to

reposition it if it drifts out of the acceptable range. { ə'bòrt ,branch }

Abrams' law |CIV ENG| In concrete materials, for a mixture of workable consistency the strength of concrete is determined by the ratio of water to cement. { 'ā·brəmz 'lò }

abrasion |ENG| **1.** The removal of surface material from any solid through the frictional action of another solid, a liquid, or a gas or combination thereof. **2.** A surface discontinuity brought about by roughening or scratching. { ə'brā·zhən }

abrasion test |MECH ENG| The measurement of abrasion resistance, usually by the weighing of a material sample before and after subjecting it to a known abrasive stress throughout a known time period, or by reflectance or surface finish comparisons, or by dimensional comparisons. { ə'brā·zhən test }

abrasive belt |MECH ENG| A cloth, leather, or paper band impregnated with grit and rotated as an endless loop to abrade materials through continuous friction. { ə'brās·əv belt }

abrasive blasting |MECH ENG| The cleaning or finishing of surfaces by the use of an abrasive entrained in a blast of air. { ə'brās·əv 'blast·iŋ }

abrasive cloth |MECH ENG| Tough cloth to whose surface an abrasive such as sand or emery has been bonded for use in grinding or polishing. { ə'brās·əv 'klòth }

abrasive cone |MECH ENG| An abrasive sintered or shaped into a solid cone to be rotated by an arbor for abrasive machining. { ə'brās·əv 'kōn }

abrasive disk |MECH ENG| An abrasive sintered or shaped into a disk to be rotated by an arbor for abrasive machining. { ə'brās·əv 'disk }

abrasive jet cleaning |ENG| The removal of dirt from a solid by a gas or liquid jet carrying abrasives to ablate the surface. { ə'brās·əv 'jet 'klēn·iŋ }

abrasive machining |MECH ENG| Grinding, drilling, shaping, or polishing by abrasion. { ə'brās·əv mə'shēn·iŋ }

abreast milling |MECH ENG| A milling method in which parts are placed in a row parallel to the axis of the cutting tool and are milled simultaneously. { ə'brest 'mil·iŋ }

abreuvoir |CIV ENG| A space between stones in masonry to be filled with mortar. { ab·rü'vwär }

ABS *See* antilock braking system.

absolute altimeter |ENG| An instrument which employs radio, sonic, or capacitive technology to produce on its indicator the measurement of distance from the aircraft to the terrain below. Also known as terrain-clearance indicator. { 'ab·sə,lüt al'tim·ə·dər }

absolute altitude |ENG| Altitude above the actual surface, either land or water, of a planet or natural satellite. { 'ab·sə,lüt 'al·tə·tüd }

absolute blocking |CIV ENG| A control arrangement for rail traffic in which a track is divided into sections or blocks upon which a train may not enter until the preceding train has left. { 'ab·sə,lüt 'bläk·iŋ }

absolute block system |CIV ENG| A block system in which only a single railroad train is permitted within a block section during a given period of time. { 'ab·sə,lüt 'bläk ,sis·təm }

absolute efficiency |ENG ACOUS| The ratio of the power output of an electroacoustic transducer, under specified conditions, to the power output of an ideal electroacoustic transducer. { 'ab·sə,lüt ə'fish·ən·sē }

absolute expansion |THERMO| The true expansion of a liquid with temperature, as calculated when the expansion of the container in which the volume of the liquid is measured is taken into account; in contrast with apparent expansion. { 'ab·sə,lüt ik'span·shən }

absolute instrument |ENG| An instrument which measures a quantity (such as pressure or temperature) in absolute units by means of simple physical measurements on the instrument. { 'ab·sə,lüt 'in·strə·mənt }

absolute magnetometer |ENG| An instrument used to measure the intensity of a magnetic field without reference to other magnetic instruments. { 'ab·sə,lüt mag·nə'täm·ə·dər }

absolute manometer |ENG| **1.** A gas manometer whose calibration, which is the same for all ideal gases, can be calculated from the measurable physical constants of the instrument. **2.** A manometer that measures absolute pressure. { 'ab·sə,lüt mə'näm·ə·dər }

absolute pressure gage |ENG| A device that measures the pressure exerted by a fluid relative to a perfect vacuum; used to measure pressures very close to a perfect vacuum. { 'ab·sə,lüt 'presh·ər ,gāj }

absolute pressure transducer |ENG| A device that responds to absolute pressure as the input and provides a measurable output of a nature different than but proportional to absolute pressure. { 'ab·sə,lüt 'presh·ər tranz'dü·sər }

absolute scale *See* absolute temperature scale. { 'ab·sə,lüt ,skāl }

absolute specific gravity |MECH| The ratio of the weight of a given volume of a substance in a vacuum at a given temperature to the weight of an equal volume of water in a vacuum at a given temperature. { 'ab·sə,lüt spə'sif·ək 'grav·əd·ē }

absolute stop |CIV ENG| A railway signal which indicates that the train must make a full stop and not proceed until there is a change in the signal. Also known as stop and stay. { 'ab·sə,lüt 'stäp }

absolute temperature |THERMO| **1.** The temperature measurable in theory on the thermodynamic temperature scale. **2.** The temperature in Celsius degrees relative to the absolute zero at −273.16°C (the Kelvin scale) or in Fahrenheit degrees relative to the absolute zero at −459.69°F (the Rankine scale). { 'ab·sə,lüt 'tem·prə·chür }

absolute temperature scale |THERMO| A scale with which temperatures are measured relative to absolute zero. Also known as absolute scale. { 'ab·sə,lüt 'tem·prə·chür ,skāl }

absolute volume |ENG| The total volume of the particles in a granular material, including both permeable and impermeable voids but excluding spaces between particles. { 'ab·sə,lüt 'väl·yüm }

absolute weighing |ENG| Determination of the mass of a sample and expressing its value in units, fractions, and multiples of the mass of the prototype of the international kilogram. { 'ab·sə,lüt 'wā·iŋ }

absolute zero |THERMO| The temperature of −273.16°C, or −459.69°F, or 0 K, thought to be the temperature at which molecular motion vanishes and a body would have no heat energy. { 'ab·sə,lüt 'zir·ō }

absorber |CHEM ENG| Equipment in which a gas is absorbed by contact with a liquid. |ELECTR| A material or device that takes up and dissipates radiated energy; may be used to shield an object from the energy, prevent reflection of the energy, determine the nature of the radiation, or selectively transmit one or more components of the radiation. |ENG| The surface on a solar collector that absorbs the solar radiation. |MECH ENG| **1.** A device which holds liquid for the absorption of refrigerant vapor or other vapors. **2.** That part of the low-pressure side of an absorption system used for absorbing refrigerant vapor. { əb'sȯr·bər }

absorber capacity |CHEM ENG| During natural gas processing, the maximum volume of the gas that can be processed through an absorber without alteration of specified operating conditions. { əb'sȯr·bər kə,pas·əd·ē }

absorber plate |ENG| A part of a flat-plate solar collector that provides a surface for absorbing incident solar radiation. { əb'sȯr·bər ,plāt }

absorbing boom |CIV ENG| A device that floats on the water and is used to stop the spread of an oil spill and aid in its removal. { əb'sȯrb·iŋ ,büm }

absorbing well |CIV ENG| A shaft that permits water to drain through an impermeable stratum to a permeable stratum. { əb'sȯrb·iŋ ,wel }

absorption bed |CIV ENG| A sizable pit containing coarse aggregate about a distribution pipe system; absorbs the effluent of a septic tank. { əb'sȯrp·shən ,bed }

absorption column *See* absorption tower. { əb'sȯrp·shən ,käl·əm }

absorption cycle |MECH ENG| In refrigeration, the process whereby a circulating refrigerant, for example, ammonia, is evaporated by heat from an aqueous solution at elevated pressure and subsequently reabsorbed at low pressure, displacing the need for a compressor. { əb'sȯrp·shən ,sī·kəl }

absorption dynamometer |ENG| A device for measuring mechanical forces or power in which the mechanical energy input is absorbed by friction or electrical resistance. { əb'sȯrp·shən dīn·ə'mäm·əd·ər }

absorption-emission pyrometer |ENG| A thermometer for determining gas temperature from measurement of the radiation emitted by a calibrated reference source before and after this radiation has passed through and been partially absorbed by the gas. { əb'sȯrp·shən ə'mish·ən pī'räm·əd·ər }

absorption field |CIV ENG| Trenches containing coarse aggregate about distribution pipes permitting septic-tank effluent to seep into surrounding soil. Also known as disposal field. { əb'sȯrp·shən ,fēld }

absorption hygrometer Also known as chemical hygrometer. |ENG| An instrument with which the water vapor content of the atmosphere is measured by means of the absorption of vapor by a hygroscopic chemical. { əb'sȯrp·shən hī'gräm·əd·ər }

absorption loss |CIV ENG| The quantity of water that is lost during the initial filling of a reservoir because of absorption by soil and rocks. { əb'sȯrp·shən ,lȯs }

absorption meter |ENG| An instrument designed to measure the amount of light transmitted through a transparent substance, using a photocell or other light detector. { əb'sȯrp·shən 'mēd·ər }

absorption number |ENG| A dimensionless group used in the field of gas absorption in a wetted-wall column; represents the liquid side mass-transfer coefficient. { əb'sȯrp·shən ,nəm·bər }

absorption plant |CHEM ENG| A facility to recover the condensable portion of natural or refinery gas. { əb'sȯrp·shən ,plant }

absorption process |CHEM ENG| A method in which absorption oil is introduced into an absorption tower so that it absorbs the gasoline in the rising wet gas; the light oil is then distilled to separate the gasoline. { əb'sȯrp·shən ,präs·əs }

absorption refrigeration |MECH ENG| Refrigeration in which cooling is effected by the expansion of liquid ammonia into gas and absorption of the gas by water; the ammonia is reused after the water evaporates. { əb'sȯrp·shən rə,frij·ə'rā·shən }

absorption system |MECH ENG| A refrigeration system in which the refrigerant gas in the evaporator is taken up by an absorber and is then, with the application of heat, released in a generator. { əb'sȯrp·shən ,sis·təm }

absorption tower |ENG| A vertical tube in which a rising gas is partially absorbed by a liquid in the form of falling droplets. Also known as absorption column. { əb'sȯrp·shən ,taù·ər }

absorption trench |CIV ENG| A trench containing coarse aggregate about a distribution tile pipe through which septic-tank effluent may move beneath earth. { əb'sȯrp·shən ,trench }

absorptivity |THERMO| The ratio of the radiation absorbed by a surface to the total radiation incident on the surface. { əb,sȯrp'tiv·əd·ē }

Abt track |CIV ENG| One of the cogged rails used for railroad tracking in mountains and so arranged that the cogs are not opposite one another on any pair of rails. { 'apt ,trak }

abutment |CIV ENG| A surface or mass provided to withstand thrust; for example, end supports of an arch or a bridge. { ə'bət·mənt }

abutting joint |DES ENG| A joint which connects two pieces of wood in such a way that the direction of the grain in one piece is angled (usually at 90°) with respect to the grain in the other. { ə'bət·iŋ ,jȯint }

abutting tenons |DES ENG| Two tenons inserted into a common mortise from opposite sides so that they contact. { ə'bət·iŋ 'ten·ənz }

ac See alternating current.

accelerated aging |ENG| Hastening the deterioration of a product by a laboratory procedure in order to determine long-range storage and use characteristics. { ak'sel·ə,rād·əd 'āj·iŋ }

accelerated life test |ENG| Operation of a device, circuit, or system above maximum ratings to produce premature failure; used to estimate normal operating life. { ak'sel·ər,ā·dəd 'līf ,test }

accelerated weathering |ENG| A laboratory test used to determine, in a short period of time, the resistance of a paint film or other exposed surface to weathering. { ak'sel·ər,ā·dəd 'weth·ər·iŋ }

accelerating incentive See differential piece-rate system. { ak'sel·ər,ād·iŋ in'sen·tiv }

accelerating potential |ELECTR| The energy potential in electron-beam equipment that imparts additional speed and energy to the electrons. { ak'sel·ər,ād·iŋ pə'ten·shəl }

acceleration |MECH| The rate of change of velocity with respect to time. { ak,sel·ə'rā·shən }

acceleration analysis |MECH ENG| A mathematical technique, often done graphically, by which accelerations of parts of a mechanism are determined. { ak,sel·ə'rā·shən ə,nal·ə·səs }

acceleration-error constant |CONT SYS| The ratio of the acceleration of a controlled variable of a servomechanism to the actuating error when the actuating error is constant. { ak,sel·ə'rā·shən 'er·ər 'kän·stənt }

acceleration measurement |MECH| The technique of determining the magnitude and direction of acceleration, including translational and angular acceleration. { ak,sel·ə'rā·shən 'mezh·ər·mənt }

acceleration of free fall See acceleration of gravity. { ak,sel·ə'rā·shən əv 'frē ,fȯl }

acceleration of gravity |MECH| The acceleration imparted to bodies by the attractive force of the earth; has an international standard value of 980.665 cm/s² but varies with latitude and elevation. Also known as acceleration of free fall; apparent gravity. { ak,sel·ə'rā·shən əv 'grav·ə·dē }

acceleration signature |IND ENG| A printed record that shows the pattern of acceleration and deceleration of an anatomical reference point in the performance of a task. { ak,sel·ə'rā·shən 'sig·nə·chər }

acceleration tolerance |ENG| The degree to which personnel or equipment withstands acceleration. { ak,sel·ə'rā·shən 'täl·ər·əns }

acceleration voltage |ELECTR| The voltage between a cathode and accelerating electrode of an electron tube. { ak,sel·ə'rā·shən 'vōl·təj }

accelerator |MECH ENG| A device for varying the speed of an automotive vehicle by varying the supply of fuel. { ak'sel·ə,rād·ər }

accelerator jet |MECH ENG| The jet through which the fuel is injected into the incoming air in the carburetor of an automotive vehicle with rapid demand for increased power output. { ak 'sel·ə,rād·ər ,jet }

accelerator linkage |MECH ENG| The linkage connecting the accelerator pedal of an automotive vehicle to the carburetor throttle valve or fuel injection control. { ak'sel·ə,rād·ər ,liŋ·kij }

accelerator pedal |MECH ENG| A pedal that operates the carburetor throttle valve or fuel injection control of an automotive vehicle. { ak'sel·ə,rād·ər ,ped·əl }

accelerator pump |MECH ENG| A small cylinder and piston controlled by the throttle of an automotive vehicle so as to provide an enriched air-fuel mixture during acceleration. { ak'sel·ə,rād·ər ,pəmp }

accelerogram |ENG| A record made by an accelerograph. { ak'sel·ə·rə,gram }

accelerograph |ENG| An accelerometer having provisions for recording the acceleration of a point on the earth during an earthquake or for recording any other type of acceleration. { ak 'sel·ə·rə,graf }

accelerometer |ENG| An instrument which measures acceleration or gravitational force capable of imparting acceleration. { ak,sel·ə'räm·əd·ər }

accelerometry |IND ENG| The quantitative determination of acceleration and deceleration in the entire human body or a part of the body in the performance of a task. { ak,sel·ə'räm·ə·drē }

accent lighting |CIV ENG| Directional lighting which highlights an object or attracts attention to a particular area. { 'ak·sent ,līd·iŋ }

acceptability |ENG| State or condition of meeting minimum standards for use, as applied to methods, equipment, or consumable products. { ak,sep·tə'bil·ə·dē }

acceptable quality level |IND ENG| The maximum percentage of defects that has been determined tolerable as a process average for a sampling plan during inspection or test of a product with respect to economic and functional requirements of the item. Abbreviated AQL. { ak 'sep·tə·bəl 'kwäl·ə·dē ,lev·əl }

acceptable reliability level |IND ENG| The required level of reliability for a part, system, device, and so forth; may be expressed in a variety of terms, for example, number of failures allowable in 1000 hours of operating life. Abbreviated ARL. { ak'sep·tə·bəl rə,lī·ə'bil·ə·dē ,lev·əl }

acceptance criteria |IND ENG| Standards of judging the acceptability of manufactured items. { ak'sep·təns krī'tēr·ē·ə }

acceptance number |IND ENG| The maximum allowable number of defective pieces in a sample of specified size. { ak'sep·təns ,nəm·bər }

acceptance sampling |IND ENG| Taking a sample from a batch of material to inspect for determining whether the entire lot will be accepted or rejected. { ak'sep·təns ,sam·pliŋ }

acceptance test |IND ENG| A test used to determine conformance of a product to design specifications, as a basis for its acceptance. { ak'sep·təns ,test }

acceptor |CHEM ENG| A calcined carbonate used to absorb the carbon dioxide evolved during a coal gasification process. { ak'sep·tər }

access |CIV ENG| Freedom, ability, or the legal right to pass without obstruction from a given point on earth to some other objective, such as the sea or a public highway. { 'ak,ses }

access door |BUILD| A provision for access to concealed plumbing or other equipment without disturbing the wall or fixtures. { 'ak,ses ,dór }

access eye |CIV ENG| A threaded plug fitted into bends and junctions of drain, waste, or soil pipes to provide access when a blockage occurs. See cleanout. { 'ak,ses ,ī }

access flooring See raised flooring. { 'ak,ses ,flor·iŋ }

access hole See manhole. { 'ak,ses ,hōl }

accessory |MECH ENG| A part, subassembly, or assembly that contributes to the effectiveness of a piece of equipment without changing its basic function; may be used for testing, adjusting, calibrating, recording, or other purposes. { ak'ses·ə·rē }

access road |CIV ENG| A route, usually paved, that enables vehicles to reach a designated facility expeditiously. { 'ak·ses ,rōd }

access tunnel |CIV ENG| A tunnel provided for an access road. { 'ak·ses ,tən·əl }

accident-cause code |IND ENG| Sponsored by the American Standards Association, the code that classifies accidents under eight defective working conditions and nine improper working practices. { 'ak·sə,dent ,kóz ,kōd }

accident frequency rate |IND ENG| The number of all disabling injuries per million worker-hours of exposure. { 'ak·sə,dent 'fre·kwən·sē ,rāt }

accident severity rate |IND ENG| The number of

worker-days lost as a result of disabling injuries per thousand worker-hours of exposure. { 'ak·sə,dent sə'ver·əd·ē ,rāt }

accommodation |CONT SYS| Any alteration in a robot's motion in response to the robot's environment; it may be active or passive. { ə,käm·ə'dā·shən }

accordion door |BUILD| A door that folds and unfolds like an accordion when it is opened and closed. { ə'kȯrd·ē·ən ,dȯr }

accordion partition |BUILD| A movable, fabric-faced partition which is fitted into an overhead track and folds like an accordion. { ə'kȯrd·ē·ən pər'tish·ən }

accordion roller conveyor |MECH ENG| A conveyor with a flexible latticed frame which permits variation in length. { ə'kȯrd·ē·ən 'rōl·ər kən 'vā·ər }

accretion |CIV ENG| Artificial buildup of land due to the construction of a groin, breakwater, dam, or beach fill. { ə'krē·shən }

accumulated discrepancy |ENG| The sum of the separate discrepancies which occur in the various steps of making a survey. { ə'kyü·myə ,lād·əd də'skrep·ən·sē }

accumulative timing |IND ENG| A time-study method that allows direct reading of the time for each element of an operation by the use of two stopwatches which operate alternately. { ə'kyü·myə,lād·iv 'tīm·iŋ }

accumulator |CHEM ENG| An auxiliary ram extruder on blow-molding equipment used to store melted material between deliveries. |ENG| See air vessel. |MECH ENG| **1.** A device, such as a bag containing pressurized gas, which acts upon hydraulic fluid in a vessel, discharging it rapidly to give high hydraulic power, after which the fluid is returned to the vessel with the use of low hydraulic power. **2.** A device connected to a steam boiler to enable a uniform boiler output to meet an irregular steam demand. **3.** A chamber for storing low-side liquid refrigerant in a refrigeration system. Also known as surge drum; surge header. { ə'kyü·myə,lād·ər }

accustomization |ENG| The process of learning the techniques of living with a minimum of discomfort in an extreme or new environment. { ə,kəs·tə·mə'zā·shən }

acetate process |CHEM ENG| Acetylation of cellulose (wood pulp or cotton linters) with acetic acid or acetic anhydride and sulfuric acid catalyst to make cellulose acetate resin or fiber. { 'as·ə,tāt 'präs·əs }

acetone-benzol process |CHEM ENG| A dewaxing process in petroleum refining, with acetone and benzol used as solvents. { 'as·ə,tōn 'ben·zȯl ,präs·əs }

acetylene cutting See oxyacetylene cutting. { ə'sed·əl,ēn 'kət·iŋ }

acetylene generator |ENG| A steel cylinder or tank that provides for controlled mixing of calcium carbide and water to generate acetylene. { ə'sed·əl,ēn 'jen·ə,rād·ər }

acetylene torch See oxyacetylene torch. { ə'sed·əl,ēn ,tȯrch }

acfm See actual cubic feet per minute.

acid blowcase See blowcase. { 'as·əd 'blō·kās }

acid cleaning |ENG| The use of circulating acid to remove dirt, scale, or other foreign matter from the interior of a pipe. { 'as·əd 'klēn·iŋ }

acid conductor |CHEM ENG| A vessel designed for refortification of hydrolyzed acid by heating and evaporation of water, or sometimes by distillation of water under partial vacuum. { 'as·əd kən'dək·tər }

acid egg See blowcase. { 'as·əd ,eg }

acid gases |CHEM ENG| The hydrogen sulfide and carbon dioxide found in natural and refinery gases which, when combined with moisture, form corrosive acids; known as sour gases when hydrogen sulfide and mercaptans are present. { 'as·əd 'gas·əz }

aciding |ENG| A light etching of a building surface of cast stone. { 'as·əd·iŋ }

acid lining |ENG| In steel production, a silica-brick lining used in furnaces. { 'as·əd 'līn·iŋ }

acid number |ENG| A number derived from a standard test indicating the acid or base composition of lubricating oils; it in no way indicates the corrosive attack of the used oil in service. Also known as corrosion number. { 'as·əd 'nəm·bər }

acid polishing |ENG| The use of acids to polish a glass surface. { 'as·əd 'päl·ish·iŋ }

acid process |CHEM ENG| In paper manufacture, a pulp digestion process that uses an acidic reagent, for example, a bisulfite solution containing free sulfur dioxide. { 'as·əd ,prä·səs }

acid recovery plant |CHEM ENG| In some refineries, a facility for separating sludge acid into acid oil, tar, and weak sulfuric acid, with provision for later reconcentration. { 'as·əd rə'kəv·ə·rē ,plant }

acid sludge |CHEM ENG| The residue left after treating petroleum oil with sulfuric acid for the removal of impurities. { 'as·əd ,sləj }

acid soot |ENG| Carbon particles that have absorbed acid fumes as a by-product of combustion; hydrochloric acid absorbed on carbon particulates is frequently the cause of metal corrosion in incineration. { 'as·əd ,süt }

acid treatment |CHEM ENG| A refining process in which unfinished petroleum products, such as gasoline, kerosine, and diesel oil, are contacted with sulfuric acid to improve their color, odor, and other properties. { 'as·əd 'trēt·mənt }

acid-water pollution |ENG| Industrial wastewaters that are acidic; usually appears in effluent from the manufacture of chemicals, batteries, artificial and natural fiber, fermentation processes (beer), and mining. { 'as·əd 'wȯd·ər pə'lü·shən }

Ackerman linkage See Ackerman steering gear. { 'ak·ər·mən ,liŋ·kij }

acme screw thread |DES ENG| A standard thread having a profile angle of 29° and a flat crest; used on power screws in such devices as automobile jacks, presses, and lead screws on lathes. Also known as acme thread. { 'ak·mē 'skrü ,thred }

acme thread See acme screw thread. { 'ak·mē ,thred }

acoubuoy |ENG| An acoustic listening device similar to a sonobuoy, used on land to form an electronic fence that will pick up sounds of enemy movements and transmit them to orbiting aircraft or land stations. { ə'kü,bȯi }

acoustical ceiling |BUILD| A ceiling covered with or built of material with special acoustical properties. { ə'küs·tə·kəl 'sēl·iŋ }

acoustical ceiling system |BUILD| A system for the structural support of an acoustical ceiling; lighting and air diffusers may be included as part of the system. { ə'küs·tə·kəl 'sēl·iŋ 'sis·təm }

acoustical door |BUILD| A solid door with gasketing along the top and sides, and usually an automatic door bottom, designed to reduce noise transmission. { ə'küs·tə·kəl 'dȯr }

acoustical model |CIV ENG| A model used to investigate certain acoustical properties of an auditorium or room such as sound pressure distribution, sound-ray paths, and focusing effects. { ə'küs·tə·kəl 'mäd·əl }

acoustical treatment |CIV ENG| That part of building planning that is designed to provide a proper acoustical environment; includes the use of acoustical material. { ə'küs·tə·kəl 'trēt·mənt }

acoustic array |ENG ACOUS| A sound-transmitting or sound-receiving system whose elements are arranged to give desired directional characteristics. { ə'küs·tik ə'rā }

acoustic center |ENG ACOUS| The center of the spherical sound waves radiating outward from an acoustic transducer. { ə'küs·tik 'sen·tər }

acoustic clarifier |ENG ACOUS| System of cones loosely attached to the baffle of a loudspeaker and designed to vibrate and absorb energy during sudden loud sounds to suppress these sounds. { ə'küs·tik 'klar·ə,fī·ər }

acoustic coupler |ENG ACOUS| A device used between the modem of a computer terminal and a standard telephone line to permit transmission of digital data in either direction without making direct connections. { ə'küs·tik 'kəp·lər }

acoustic delay |ENG ACOUS| A delay which is deliberately introduced in sound reproduction by having the sound travel a certain distance along a pipe before conversion into electric signals. { ə'küs·tik di'lā }

acoustic detection |ENG| Determination of the profile of a geologic formation, an ocean layer, or some object in the ocean by measuring the reflection of sound waves off the object. { ə'küs·tik di'tek·shən }

acoustic fatigue |MECH| The tendency of a material, such as a metal, to lose strength after acoustic stress. { ə'küs·tik fə'tēg }

acoustic feedback |ENG ACOUS| The reverberation of sound waves from a loudspeaker to a preceding part of an audio system, such as to the microphone, in such a manner as to reinforce, and distort, the original input. Also known as acoustic regeneration. { ə'küs·tik 'fēd,bak }

acoustic generator |ENG ACOUS| A transducer which converts electrical, mechanical, or other forms of energy into sound. { ə'küs·tik 'jen·ə,rād·ər }

acoustic heat engine |ENG| A device that transforms heat energy first into sound energy and then into electrical power, without the use of moving mechanical parts. { ə'küs·tik ¦hēt ,en·jən }

acoustic hologram |ENG| The phase interference pattern, formed by acoustic beams, that is used in acoustical holography; when light is made to interact with this pattern, it forms an image of an object placed in one of the beams. { ə'küs·tik 'häl·ə,gram }

acoustic horn See horn. { ə'küs·tik 'hȯrn }

acoustic jamming |ENG ACOUS| The deliberate radiation or reradiation of mechanical or electroacoustic signals with the objectives of obliterating or obscuring signals which the enemy is attempting to receive and of deterring enemy weapons systems. { ə'küs·tik 'jam·iŋ }

acoustic labyrinth |ENG ACOUS| Special baffle arrangement used with a loudspeaker to prevent cavity resonance and to reinforce bass response. { ə'küs·tik 'lab·ə,rinth }

acoustic line |ENG ACOUS| The acoustic equivalent of an electrical transmission line, involving baffles, labyrinths, or resonators placed at the rear of a loudspeaker and arranged to help reproduce the very low audio frequencies. { ə'küs·tik 'līn }

acoustic ocean-current meter |ENG| An instrument that measures current flow in rivers and oceans by transmitting acoustic pulses in opposite directions parallel to the flow and measuring the difference in pulse travel times between transmitter-receiver pairs. { ə'küs·tik 'ō·shən ,kər·ənt 'mēd·ər }

acoustic position reference system |ENG| An acoustic system used in offshore oil drilling to provide continuous information on ship position with respect to an ocean-floor acoustic beacon transmitting an ultrasonic signal to three hydrophones on the bottom of the drilling ship. { ə'küs·tik pə'zish·ən ¦ref·rəns ,sis·təm }

acoustic radar |ENG| Use of sound waves with radar techniques for remote probing of the lower atmosphere, up to heights of about 5000 feet (1500 meters), for measuring wind speed and direction, humidity, temperature inversions, and turbulence. { ə'küs·tik 'rā,där }

acoustic radiator |ENG ACOUS| A vibrating surface that produces sound waves, such as a loudspeaker cone or a headphone diaphragm. { ə'küs·tik 'rād·ē,ād·ər }

acoustic radiometer |ENG| An instrument for measuring sound intensity by determining the unidirectional steady-state pressure caused by the reflection or absorption of a sound wave at a boundary. { ə'küs·tik ,rād·ə'ä·məd·ər }

acoustic ratio |ENG ACOUS| The ratio of the intensity of sound radiated directly from a source to the intensity of sound reverberating from the walls of an enclosure, at a given point in the enclosure. { ə'küs·tik 'rä·shō }

acoustic reflex enclosure |ENG ACOUS| A loudspeaker cabinet designed with a port to allow a low-frequency contribution from the rear of the speaker cone to be radiated forward. { ə'küs·tik 'rē,fleks in,klō·zhər }

acoustic regeneration See acoustic feedback. { ə'küs·tik rē,jen·ə'rā·shən }

acoustic seal |ENG ACOUS| A joint between two parts to provide acoustical coupling with low losses of energy, such as between an earphone and the human ear. { ə'küs·tik 'sēl }

acoustic signature |ENG| In acoustic detection, the profile characteristic of a particular object or class of objects, such as a school of fish or a specific ocean-bottom formation. { ə'küs·tik 'sig·nə·chər }

acoustic spectrograph |ENG| A spectrograph used with sound waves of various frequencies to study the transmission and reflection properties of ocean thermal layers and marine life. { ə'küs·tik 'spek·trə,graf }

acoustic spectrometer |ENG ACOUS| An instrument that measures the intensities of the various frequency components of a complex sound wave. Also known as audio spectrometer. { ə'küs·tik spek'träm·əd·ər }

acoustic strain gage |ENG| An instrument used for measuring structural strains; consists of a length of fine wire mounted so its tension varies with strain; the wire is plucked with an electromagnetic device, and the resulting frequency of vibration is measured to determine the amount of strain. { ə'küs·tik ,strān ,gāj }

acoustic theodolite |ENG| An instrument that uses sound waves to provide a continuous vertical profile of ocean currents at a specific location. { ə'küs·tik thē'äd·əl,īt }

acoustic transducer |ENG ACOUS| A device that converts acoustic energy to electrical or mechanical energy, such as a microphone or phonograph pickup. { ə'küs·tik tranz'dü·sər }

acoustic transformer |ENG ACOUS| A device, such as a horn or megaphone, for increasing the efficiency of sound radiation. { ə'küs·tik tranz 'fòr·mər }

acoustic treatment |BUILD| The use of sound-absorbing materials to give a room a desired degree of freedom from echo and reverberation. { ə'küs·tik 'trēt·mənt }

acoustic-wave-based sensor |ENG| A device that employs a surface acoustic wave, a thickness-shear-mode resonance (a resonant oscillation of a thin plate of material), or other type of acoustic wave to measure the physical properties of a thin film or liquid layer or, in combination with chemically sensitive thin films, to detect the presence and concentration of chemical analytes. { ə,kü·stik 'wāv,bäst ,sen·sər }

acoustic well logging |ENG| A ground exploration method that uses a high-energy sound source and a receiver, both underground. { ə'küs·tik 'wel ,läg·iŋ }

acoustoelectronics |ENG ACOUS| The branch of electronics that involves use of acoustic waves at microwave frequencies (above 500 megahertz), traveling on or in piezoelectric or other solid substrates. Also known as pretersonics. { ə,küs·tō·ə,lek'trän·iks }

acquisition |ENG| The process of pointing an antenna or a telescope so that it is properly oriented to allow gathering of tracking or telemetry data from a satellite or space probe. { ,ak·wə'zish·ən }

acquisition and tracking radar |ENG| A radar set capable of locking onto a received signal and tracking the object emitting the signal; the radar may be airborne or on the ground. { ,ak·wə'zish·ən ən 'trak·iŋ ,rā,där }

acre |MECH| A unit of area, equal to 43,560 square feet, or to 4046.8564224 square meters. { 'ā·kər }

acrometer |ENG| An instrument to measure the density of oils. { ə'kräm·əd·ər }

actinogram |ENG| The record of heat from a source, such as the sun, as detected by a recording actinometer. { ,ak'tin·ə,gram }

actinograph |ENG| A recording actinometer. { ,ak'tin·ə,graf }

actinometer |ENG| Any instrument used to measure the intensity of radiant energy, particularly that of the sun. { ,ak·tə'näm·əd·ər }

action |MECH| An integral associated with the trajectory of a system in configuration space, equal to the sum of the integrals of the generalized momenta of the system over their canonically conjugate coordinates. Also known as phase integral. { 'ak·shən }

activate |ELEC| To make a cell or battery operative by addition of a liquid. |ELECTR| To treat the filament, cathode, or target of a vacuum tube to increase electron emission. |ENG| To set up conditions so that the object will function as designed or required. { 'ak·tə,vāt }

activated sludge |CIV ENG| A semiliquid mass removed from the liquid flow of sewage and subjected to aeration and aerobic microbial action; the end product is dark to golden brown, partially decomposed, granular, and flocculent, and has an earthy odor when fresh. { 'ak·tə,vād·əd 'sləj }

activated-sludge effluent |CIV ENG| The liquid from the activated-sludge treatment that is further processed by chlorination or by oxidation. { 'ak·tə,vād·əd ,sləj 'ef,lü·ənt }

activated-sludge process |CIV ENG| A sewage treatment process in which the sludge in the secondary stage is put into aeration tanks to facilitate aerobic decomposition by microorganisms; the sludge and supernatant liquor are separated in a settling tank; the supernatant liquor or effluent is further treated by chlorination or oxidation. { 'ak·tə,vād·əd ,sləj ,səs }

active accommodation |CONT SYS| The alteration of preprogrammed robotic motions by the integrated effects of sensors, controllers, and the robotic motion itself. { 'ak·tiv ə,käm·ə'dā·shən }

7

active area |ELECTR| The area of a metallic rectifier that acts as the rectifying junction and conducts current in the forward direction. { 'ak·tiv 'er·ē·ə }

active-cord mechanism |MECH ENG| A slender, chainlike grouping of joints and links that makes active and flexible winding motions under the control of actuators attached along its body. { 'ak·tiv 'kȯrd 'mek·ə,niz·əm }

active detection system |ENG| A guidance system which emits energy as a means of detection; for example, sonar and radar. { 'ak·tiv di'tek·shən ,sis·təm }

active earth pressure |CIV ENG| The horizontal pressure that an earth mass exerts on a wall. { 'ak·tiv 'ərth 'presh·ər }

active illumination |ENG| Lighting whose direction, intensity, and pattern are controlled by commands or signals. { 'ak·tiv ə,lüm·ə'nā·shən }

active infrared detection system |ENG| An infrared detection system in which a beam of infrared rays is transmitted toward possible targets, and rays reflected from a target are detected. { 'ak·tiv 'in·frə,red di'tek·shən ,sis·təm }

active leaf |BUILD| In a door with two leaves, the leaf which carries the latching or locking mechanism. Also known as active door. { 'ak·tiv 'lēf }

active material |ELEC| 1. A fluorescent material used in screens for cathode-ray tubes. 2. An energy-storing material, such as lead oxide, used in the plates of a storage battery. 3. A material, such as the iron of a core or the copper of a winding, that is involved in energy conversion in a circuit. 4. In a battery, the chemically reactive material in either of the electrodes that participates in the charge and discharge reactions. |ELECTR| The material of the cathode of an electron tube that emits electrons when heated. { 'ak·tiv mə'tir·ē·əl }

active sludge |CIV ENG| A sludge rich in destructive bacteria used to break down raw sewage. { 'ak·tiv 'sləj }

active solar system |MECH ENG| A solar heating or cooling system that operates by mechanical means, such as motors, pumps, or valves. { 'ak·tiv 'sō·lər ,sis·təm }

active sonar |ENG| A system consisting of one or more transducers to send and receive sound, equipment for the generation and detection of the electrical impulses to and from the transducer, and a display or recorder system for the observation of the received signals. { 'ak·tiv 'sō,när }

active system |ENG| In radio and radar, a system that requires transmitting equipment, such as a beacon or transponder. { 'ak·tiv 'sis·təm }

active vibration suppression |MECH ENG| The prevention of undesirable vibration by techniques involving feedback control of the vibratory motion, whereby the forces designed to reduce the vibration depend on the system displacements and velocities. { 'ak·tiv vī'brā·shən sə,presh·ən }

activity |SYS ENG| The representation in a PERT or critical-path-method network of a task that takes up both time and resources and whose performance is necessary for the system to move from one event to the next. { ,ak'tiv·əd·ē }

activity chart |IND ENG| A tabular presentation of a series of operations of a process plotted against a time scale. { ,ak'tiv·əd·ē ,chärt }

activity duration |SYS ENG| In critical-path-method terminology, the estimated amount of time required to complete an activity. { ,ak'tiv·əd·ē də'rā·shən }

activity sampling See work sampling. { ,ak'tiv·əd·ē ,sam·pliŋ }

actual cost |IND ENG| Cost determined by an allocation of cost factors recorded during production. { 'ak·chə·wəl 'kȯst }

actual cubic feet per minute |CHEM ENG| A measure of the volume of gas at operating temperature and pressure, as distinct from volume of gas at standard temperature and pressure. Abbreviated acfm. { 'ak·chə·wəl 'kyü·bik ,fēt pər 'min·ət }

actual horsepower See actual power. { 'ak·chə·wəl 'hȯrs,paủ·ər }

actual power |MECH ENG| The power delivered at the output shaft of a source of power. Also known as actual horsepower. { 'ak·chə·wəl 'paủ·ər }

actual time |IND ENG| Time taken by a worker to perform a given task. { 'ak·chə·wəl tīm }

actuate |MECH ENG| To put into motion or mechanical action, as by an actuator. { 'ak·chə·wāt }

actuated roller switch |MECH ENG| A centrifugal sequence-control switch that is placed in contact with a belt conveyor, immediately preceding the conveyor which it controls. { 'ak·chə,wād·əd 'rō·lər 'swich }

actuating system |CONT SYS| An electric, hydraulic, or other system that supplies and transmits energy for the operation of other mechanisms or systems. { 'ak·chə,wād·iŋ ,sis·təm }

actuator |CONT SYS| A mechanism to activate process control equipment by use of pneumatic, hydraulic, or electronic signals; for example, a valve actuator for opening or closing a valve to control the rate of fluid flow. |ENG ACOUS| An auxiliary external electrode used to apply a known electrostatic force to the diaphragm of a microphone for calibration purposes. Also known as electrostatic actuator. |MECH ENG| A device that produces mechanical force by means of pressurized fluid. { 'ak·chə,wād·ər }

adamantine drill |MECH ENG| A core drill with hardened steel shot pellets that revolve under the rim of the rotating tube; employed in rotary drilling in very hard ground. { ,ad·ə'man,tēn 'dril }

Adam's catalyst |CHEM ENG| Finely divided plantinum(IV) oxide, made by fusing hexachloroplatinic(IV) acid with $NaNO_3$. { 'a·dəmz 'kad·əl·əst }

ada mud |ENG| A conditioning material added to drilling mud to obtain satisfactory cores and samples of formations. { 'ä·də ‚məd }

adapter |ENG| A device used to make electrical or mechanical connections between items not originally intended for use together. { ə'dap·tər }

adaptive branch |CONT SYS| A branch instruction in the computer program controlling a robot that may lead the robot to execute a series of instructions, depending on external conditions. { ə'dap·tiv 'branch }

adaptive control |CONT SYS| A control method in which one or more parameters are sensed and used to vary the feedback control signals in order to satisfy the performance criteria. { ə'dap·tiv kən'trōl }

adaptive-control function |CONT SYS| That level in the functional decomposition of a large-scale control system which updates parameters of the optimizing control function to achieve a best fit to current plant behavior, and updates parameters of the direct control function to achieve good dynamic response of the closed-loop system. { ə'dap·tiv kən'trōl ‚faŋk·shən }

adaptive robot |CONT SYS| A robot that can alter its responses according to changes in the environment. { ə'dap·tiv 'rō‚bät }

adaptive structure |ENG| A structure whose geometric and inherent structural characteristics can be changed beneficially in response to external stimulation by either remote commands or automatic means. { ə‚dap·tiv 'strək·chər }

adaptive system |SYS ENG| A system that can change itself in response to changes in its environment in such a way that its performance improves through a continuing interaction with its surroundings. { ə'dap·tiv 'sis·təm }

adaptometer |ENG| An instrument that measures the lowest brightness of an extended area that can barely be detected by the eye. { ‚a‚dap'tä·məd·ər }

addendum |DES ENG| The radial distance between two concentric circles on a gear, one being that whose radius extends to the top of a gear tooth (addendum circle) and the other being that which will roll without slipping on a circle on a mating gear (pitch line). { ə'den·dəm }

addendum circle |DES ENG| The circle on a gear passing through the tops of the teeth. { ə'den·dəm ‚sər·kəl }

adder |ELECTR| A circuit in which two or more signals are combined to give an output-signal amplitude that is proportional to the sum of the input-signal amplitudes. Also known as adder circuit. { 'ad·ər }

adding tape |ENG| A surveyor's tape that is calibrated from 0 to 100 by full feet (or meters) in one direction, and has 1 additional foot (or meter) beyond the zero end which is subdivided in tenths or hundredths. { 'ad·iŋ ‚tāp }

additive synthesis |ENG ACOUS| A method of synthesizing complex tones by adding together an appropriate number of simple sine waves at

harmonically related frequencies. { ¦ad·ə·div 'sin·thə‚səs }

adhesion |ENG| Intimate sticking together of metal surfaces under compressive stresses by formation of metallic bonds. |MECH| The force of static friction between two bodies, or the effects of this force. { ad'hē·zhən }

adhesional work |THERMO| The work required to separate a unit area of a surface at which two substances are in contact. Also known as work of adhesion. { ad'hē·zhən·əl ‚wərk }

adhesive bond |MECH| The forces such as dipole bonds which attract adhesives and base materials to each other. { ad'hēz·iv 'bänd }

adhesive bonding |ENG| The fastening together of two or more solids by the use of glue, cement, or other adhesive. { ad'hēz·iv 'bänd·iŋ }

adhesive strength |ENG| The strength of an adhesive bond, usually measured as a force required to separate two objects of standard bonded area, by either shear or tensile stress. { ad'hēz·iv 'streŋkth }

adiabatic |THERMO| Referring to any change in which there is no gain or loss of heat. { ¦ad·ē·ə¦bad·ik }

adiabatic compression |THERMO| A reduction in volume of a substance without heat flow, in or out. { ¦ad·ē·ə¦bad·ik kəm'presh·ən }

adiabatic cooling |THERMO| A process in which the temperature of a system is reduced without any heat being exchanged between the system and its surroundings. { ¦ad·ē·ə¦bad·ik 'kül·iŋ }

adiabatic curing |ENG| The curing of concrete or mortar under conditions in which there is no loss or gain of heat. { ¦ad·ē·ə¦bad·ik 'kyúr·iŋ }

adiabatic engine |MECH ENG| A heat engine or thermodynamic system in which there is no gain or loss of heat. { ¦ad·ē·ə¦bad·ik 'en·jən }

adiabatic envelope |THERMO| A surface enclosing a thermodynamic system in an equilibrium which can be disturbed only by long-range forces or by motion of part of the envelope; intuitively, this means that no heat can flow through the surface. { ¦ad·ē·ə¦bad·ik 'en·və‚lōp }

adiabatic expansion |THERMO| Increase in volume without heat flow, in or out. { ¦ad·ē·ə¦bad·ik ik'span·chən }

adiabatic extrusion |ENG| Forming plastic objects by energy produced by driving the plastic mass through an extruder without heat flow. { ¦ad·ē·ə¦bad·ik ik'strü·zhən }

adiabatic process |THERMO| Any thermodynamic procedure which takes place in a system without the exchange of heat with the surroundings. { ¦ad·ē·ə¦bad·ik prä·səs }

adiabatic vaporization |THERMO| Vaporization of a liquid with virtually no heat exchange between it and its surroundings. { ¦ad·ē·ə¦bad·ik ‚vā·pər·ə'zā·shən }

adit |CIV ENG| An access tunnel used for excavation of the main tunnel. { 'ad·ət }

adjustable base anchor |BUILD| An item which holds a doorframe above a finished floor. { ə'jəs·tə·bəl ¦bās 'aŋ·kər }

adjustable parallels [ENG] Wedge-shaped iron bars placed with the thin end of one on the thick end of the other, so that the top face of the upper and the bottom face of the lower remain parallel, but the distance between the two faces is adjustable; the bars can be locked in position by a screw to prevent shifting. { ə'jəs·tə·bəl 'par·ə,lelz }

adjustable square [ENG] A try square with an arm that is at right angles to the ruler; the position of the arm can be changed to form an L or a T. Also known as double square. { ə'jəs·tə·bəl 'skwer }

adjustable wrench [ENG] A wrench with one jaw which is fixed and another which is adjustable; the size is adjusted by a knurled screw. { ə'jəs·tə·bəl 'rench }

adjusting [ENG] In measurement technology, setting or compensating a measuring instrument or a weight in such a way that the indicated value deviates as little as possible from the actual value. { ə'jəst·iŋ }

adjutage [ENG] A tube attached to a container of liquid at an orifice to facilitate or regulate outflow. { 'aj·ə,tazh }

admittance [ELEC] A measure of how readily alternating current will flow in a circuit; the reciprocal of impedance, it is expressed in siemens. { əd'mit·əns }

adobe construction [BUILD] Wall construction with sun-dried blocks of adobe soil. { ə'dō·bē kən'strək·shən }

ADP See automatic data processing.

ADR studio [ENG ACOUS] A sound-recording studio used in motion-picture and television production to allow an actor who did not intelligibly record his or her speech during the original filming or video recording to do so by watching himself or herself on the screen and repeating the original speech with lip synchronism; it is equipped with facilities for recreating the acoustical liveness and background sound of the environment of the original dialog. Derived from automatic dialog replacement studio. Also known as postsynchronizing studio. { !ā¦dē'är ,stüd·ē·ō }

adsorption system [MECH ENG] A device that dehumidifies air by bringing it into contact with a solid adsorbing substance. { ad'sórp·shən ,sis·təm }

advance [CIV ENG] In railway engineering, a length of track that extends beyond the signal that controls it. [MECH ENG] To effect the earlier occurrence of an event, for example, spark advance or injection advance. { əd'vans }

advanced programmatic risk analysis [IND ENG] A method for managing engineering programs with multiple projects and strict resource constraints which balances both technical and management risks. { əd¦vanst ,prō·grə¦mad·ik 'risk ə,nal·əsəs }

advanced sewage treatment See tertiary sewage treatment. { əd¦vanst 'sü·ij ,trēt·mənt }

advance signal [CIV ENG] A signal in a block system up to which a train may proceed within a block that is not completely cleared. { əd'vans 'sig·nəl }

advance slope grouting [ENG] A grouting technique in which the front of the mass of grout is forced to move horizontally through preplaced aggregate. { əd'vans 'slōp 'graúd·iŋ }

advance slope method [ENG] A method of concrete placement in which the face of the fresh concrete, which is not vertical, moves forward as the concrete is placed. { əd'vans 'slōp ,meth·əd }

adz [DES ENG] A cutting tool with a thin arched blade, sharpened on the concave side, at right angles on the handle; used for rough dressing of timber. { adz }

adz block [MECH ENG] The part of a machine for wood planing that carries the cutters. { 'adz ,bläk }

aerated flow [ENG] Flowing liquid in which gas is dispersed as fine bubbles throughout the liquid. { 'e,rād·əd 'flō }

aeration [ENG] **1.** Exposing to the action of air. **2.** Causing air to bubble through. **3.** Introducing air into a solution by spraying, stirring, or similar method. **4.** Supplying or infusing with air, as in sand or soil. { e'rā·shən }

aeration tank [ENG] A fluid-holding tank with provisions to aerate its contents by bubbling air or another gas through the liquid or by spraying the liquid into the air. { e'rā·shən ,taŋk }

aerator [DES ENG] A tool having a roller equipped with hollow fins; used to remove cores of soil from turf. [ENG] **1.** One who aerates. **2.** Equipment used for aeration. **3.** Any device for supplying air or gas under pressure, as for fumigating, welding, or ventilating. [MECH ENG] Equipment used to inject compressed air into sewage in the treatment process. { 'e,rād·ər }

aerial cableway See aerial tramway. { 'e·rē·əl 'kā·bəl,wā }

aerial photogrammetry [ENG] Use of aerial photographs to make accurate measurements in surveying and mapmaking. { 'e·rē·əl ,fōt·ə'gram·ə·trē }

aerial photographic reconnaissance See aerial photoreconnaissance. { 'e·rē·əl ,fōd·ə¦graf·ik ri'kän·ə·səns }

aerial photography [ENG] The making of photographs of the ground surface from an aircraft, spacecraft, or rocket. Also known as aerophotography. { 'e·rē·əl fə'täg·rə·fē }

aerial photoreconnaissance [ENG] The obtaining of information by air photography; the three types are strategic, tactical, and survey-cartographic photoreconnaissance. Also known as aerial photographic reconnaissance. { 'e·rē·əl ,fōd·ō,ri'kän·ə·səns }

aerial reconnaissance [ENG] The collection of information by visual, electronic, or photographic means while aloft. { 'e·rē·əl ,ri'kän·ə·səns }

aerial ropeway See aerial tramway. { 'e·rē·əl 'rōp,wā }

aerial spud [MECH ENG] A cable for moving and anchoring a dredge. { 'e·rē·əl 'spəd }

aerial survey [ENG] A survey utilizing photographic, electronic, or other data obtained from an airborne station. Also known as aerosurvey; air survey. { 'e·rē·əl 'sər·vā }

aerial tramway [MECH ENG] A system for transporting bulk materials that consists of one or more cables supported by steel towers and is capable of carrying a traveling carriage from which loaded buckets can be lowered or raised. Also known as aerial cableway; aerial ropeway. { 'e·rē·əl 'tram,wā }

aeroballistics [MECH] The study of the interaction of projectiles or high-speed vehicles with the atmosphere. { ,e·rō·bə'lis·tiks }

aerobic-anaerobic interface [CIV ENG] That point in bacterial action in the body of a sewage sludge or compost heap where both aerobic and anaerobic microorganisms participate, and the decomposition of the material goes no further. { e'rōb·ik 'an·ə,rōb·ik 'in·tər,fās }

aerobic-anaerobic lagoon [CIV ENG] A pond in which the solids from a sewage plant are placed in the lower layer; the solids are partially decomposed by anaerobic bacteria, while air or oxygen is bubbled through the upper layer to create an aerobic condition. { e'rōb·ik 'an·ə,rōb·ik lə'gün }

aerobic digestion [CHEM ENG] Digestion of matter suspended or dissolved in waste by microorganisms under favorable conditions of oxygenation. { e'rōb·ik də'jes·chən }

aerobic lagoon [CIV ENG] An aerated pond in which sewage solids are placed, and are decomposed by aerobic bacteria. Also known as aerobic pond. { e'rō·bik lə'gün }

aerobic pond See aerobic lagoon. { e¦rō·bik 'pand }

aerochlorination [CIV ENG] Treatment of sewage with compressed air and chlorine gas to remove fatty substances. { ,e·rō,klȯr·ə'nā·shən }

aerodrome See airport. { 'e·rō,drōm }

aerodynamic balance [ENG] A balance used for the measurement of the forces exerted on the surfaces of instruments exposed to flowing air; frequently used in tests made on models in wind tunnels. { ,e·rō·dī'nam·ik 'bal·əns }

aerodynamic trajectory [MECH] A trajectory or part of a trajectory in which the missile or vehicle encounters sufficient air resistance to stabilize its flight or to modify its course significantly. { ,e·rō·dī'nam·ik trə'jek·trē }

aeroelasticity [MECH] The deformation of structurally elastic bodies in response to aerodynamic loads. { ,e·rō·i,las'tis·əd·ē }

aerofall mill [MECH ENG] A grinding mill of large diameter with either lumps of ore, pebbles, or steel balls as crushing bodies; the dry load is airswept to remove mesh material. { 'e·rō,fȯl ,mil }

aerofilter [CIV ENG] A filter bed for sewage treatment consisting of coarse material and operated at high speed, often with recirculation. { 'e·rō,fil·tər }

aerograph [ENG] Any self-recording instrument carried aloft by any means to obtain meteorological data. { 'e·rō,graf }

aerometeorograph [ENG] A self-recording instrument used on aircraft for the simultaneous recording of atmospheric pressure, temperature, and humidity. { ,e·ro,mēd·ē'ȯr·ə,graf }

aerometer [ENG] An instrument to ascertain the weight or density of air or other gases. { e'rä·məd·ər }

aerophotography See aerial photography. { ,e·rō·fə'täg·rə'fē }

aerosol generator [MECH ENG] A mechanical means of producing a system of dispersed phase and dispersing medium, that is, an aerosol. { 'e·rə,sȯl 'jen·ə,rād·ər }

aerospace engineering [ENG] Engineering pertaining to the design and construction of aircraft and space vehicles and of power units, and to the special problems of flight in both the earth's atmosphere and space, as in the flight of air vehicles and in the launching, guidance, and control of missiles, earth satellites, and space vehicles and probes. { ¦e·rō¦spās ,en·jə'nir·iŋ }

aerospace industry [ENG] Industry concerned with the use of vehicles in both the earth's atmosphere and space. { ¦e·rō¦spās 'in·dəs·trē }

aerostatic balance [ENG] An instrument for weighing air. { ¦e·rō¦stad·ik 'bal·əns }

aerosurvey See aerial survey. { ¦e·rō¦sər,vā }

aerotrain [ENG] A train that is propelled by a fan jet engine and floats on a cushion of low-pressure air, traveling at speeds up to 267 miles (430 kilometers) per hour. { 'e·rō,trān }

aesthesiometer See esthesiometer. { es,thē·zē'äm·əd·ər }

affreightment [IND ENG] The lease of a vessel for the transportation of goods. { ə'frāt·mənt }

A frame [BUILD] A dwelling whose main frames are in the shape of the letter A. [ENG] Two poles supported in an upright position by braces or guys and used for lifting equipment. Also known as double mast. { 'ā ,frām }

afterboil [MECH ENG] In an automotive engine, coolant boiling after the engine has stopped because of the inability of the engine at rest to dissipate excess heat. { 'af·tər,bȯil }

afterburning [MECH ENG] Combustion in an internal combustion engine following the maximum pressure of explosion. { 'af·tər,bərn·iŋ }

aftercondenser [MECH ENG] A condenser in the second stage of a two-stage ejector; used in steam power plants, refrigeration systems, and air conditioning systems. { 'af·tər·kən'dens·ər }

aftercooler [MECH ENG] A heat exchanger which cools air that has been compressed; used on turbocharged engines. { 'af·tər,kül·ər }

aftercooling [MECH ENG] The cooling of a gas after its compression. { 'af·tər,kül·iŋ }

afterfilter [MECH ENG] In an air-conditioning system, a high-efficiency filter located near a terminal unit. Also known as final filter. { 'af·tər,fil·tər }

afterrunning |MECH ENG| In an automotive engine, continued operation of the engine after the ignition switch is turned off. Also known as dieseling; run-on. { 'af·tər,rən·iŋ }

after top dead center |MECH ENG| The position of the piston after reaching the top of its stroke in an automotive engine. { 'af·tər 'täp 'ded 'sen·tər }

agger |CIV ENG| A material used for road fill over low ground. { 'a·jər }

aggregate bin |ENG| A structure designed for storing and dispensing dry granular construction materials such as sand, crushed stone, and gravel; usually has a hopperlike bottom that funnels the material to a gate under the structure. { 'ag·rə·gət ,bin }

aggregate interlock |ENG| The projection of aggregate particles or portions thereof from one side of a joint or crack in concrete into recesses in the other side so as to effect load transfer in compression and shear, and to maintain mutual alignment. { 'ag·rə·gət 'in·tər,läk }

aggregate production scheduling |IND ENG| A type of planning at a broad level without consideration of individual products and activities in order to develop a program of output that will meet future demand under given constraints. { ¦ag·ri·gət prə¦dək·shən ,ske¦·ə·liŋ }

aggressive carbon dioxide |CHEM ENG| The carbon dioxide dissolved in water in excess of the amount required to precipitate a specified concentration of calcium ions as calcium carbonate; used as a measure of the corrosivity and scaling properties of water. { ə'gres·iv 'kär·bən dī'äk,sīd }

agile manufacturing |IND ENG| Operations that can be rapidly reconfigured to satisfy changing market demands. { ¦a·jəl ,man·yü'fak·chə·riŋ }

aging |ELEC| Allowing a permanent magnet, capacitor, meter, or other device to remain in storage for a period of time, sometimes with a voltage applied, until the characteristics of the device become essentially constant. |ENG| **1.** The changing of the characteristics of a device due to its use. **2.** Operation of a product before shipment to stabilize characteristics or detect early failures. { 'āj·iŋ }

agitating speed |MECH ENG| The rate of rotation of the drum or blades of a truck mixer or other device used for agitation of mixed concrete. { 'aj·ə,tād·iŋ ,spēd }

agitating truck |MECH ENG| A vehicle carrying a drum or agitator body, in which freshly mixed concrete can be conveyed from the point of mixing to that of placing, the drum being rotated continuously to agitate the contents. { 'aj·ə,tād·iŋ ,trak }

agitator |MECH ENG| A device for keeping liquids and solids in liquids in motion by mixing, stirring, or shaking. { 'aj·ə,tād·ər }

agitator body |MECH ENG| A truck-mounted drum for transporting freshly mixed concrete; rotation of internal paddles or of the drum prevents the setting of the mixture prior to delivery. { 'aj·ə,tād·ər 'bäd·ē }

agricultural pipe drain |CIV ENG| A system of porous or perforated pipes laid in a trench filled with gravel or the like; used for draining subsoil. { ¦ag·rə¦kəl·chə·rəl ,pīp ,drān }

agricultural robot |CONT SYS| A robot used to pick and harvest farm products and fruits. { ¦ag·rə¦kəl·chə·rəl 'rō,bät }

AGV See automated guided vehicle.

aided tracking |ENG| A system of radar-tracking a target signal in bearing, elevation, or range, or any combination of these variables, in which the rate of motion of the tracking equipment is machine-controlled in collaboration with an operator so as to minimize tracking error. { 'ād·əd 'trak·iŋ }

aided-tracking mechanism |ENG| A device consisting of a motor and variable-speed drive which provides a means of setting a desired tracking rate into a director or other fire-control instrument, so that the process of tracking is carried out automatically at the set rate until it is changed manually. { 'ād·əd 'trak·iŋ ,mek·ə,niz·əm }

aided-tracking ratio |ENG| The ratio between the constant velocity of the aided-tracking mechanism and the velocity of the moving target. { 'ād·əd 'trak·iŋ ,rā·shō }

aiguille |ENG| A slender form of drill used for boring or drilling a blasthole in rock. { ā'gwēl }

aiming circle |ENG| An instrument for measuring angles in azimuth and elevation in connection with artillery firing and general topographic work; equipped with fine and coarse azimuth micrometers and a magnetic needle. { 'ām·iŋ ,sər·kəl }

aiming screws |MECH ENG| On an automotive vehicle, spring-loaded screws designed to secure headlights to a support frame and permit aiming of the headlights in horizontal and vertical planes. { 'aim·iŋ ,skrüz }

AIR See air-injection reactor. { er }

air-actuated |ENG| Powered by compressed air. { 'er 'ak·chə,wād·əd }

air-arc furnace |ENG| An arc furnace designed to power wind tunnels, the air being superheated to 20,000 K and expanded to emerge at supersonic speeds. { 'er ,ärk 'fər·nəs }

air aspirator valve |MECH ENG| On certain automotive engines, a one-way valve installed on the exhaust manifold to allow air to enter the exhaust system; provides extra oxygen to convert carbon monoxide to carbon dioxide. Also known as gulp valve. { 'as·pə,rād·ər ,valv }

air-assist forming |ENG| A plastics thermoforming method in which air pressure is used to partially preform a sheet before it enters the mold. { 'er ə'sist 'fòrm·iŋ }

air-atomizing oil burner |ENG| An oil burner in which a stream of fuel oil is broken into very fine droplets through the action of compressed air. { 'er 'at·ə'mīz·iŋ ,òil 'bərn·ər }

air bag |MECH ENG| An automotive vehicle passenger safety device consisting of a passive restraint in the form of a bag which is automatically inflated with gas to provide cushioned protection against the impact of a collision. { 'er ‚bag }

air belt |MECH ENG| The chamber which equalizes the pressure that is blasted into the cupola at the tuyeres. { 'er ‚belt }

air bind |ENG| The presence of air in a conduit or pump which impedes passage of the liquid. { 'er ‚bīnd }

airblasting |ENG| A blasting technique in which air at very high pressure is piped to a steel shell in a shot hole and discharged. Also known as air breaking. { 'er‚blast·iŋ }

air bleeder |MECH ENG| A device, such as a needle valve, for removing air from a hydraulic system. { 'er 'blēd·ər }

airborne collision warning system |ENG| A system such as a radar set or radio receiver carried by an aircraft to warn of the danger of possible collision. { 'er‚bórn kə'lizh·ən 'wórn·iŋ ‚sis·təm }

airborne detector |ENG| A device, transported by an aircraft, whose function is to locate or identify an air or surface object. { 'er‚bórn di 'tek·tər }

airborne electronic survey control |ENG| The airborne portion of very accurate positioning systems used in controlling surveys from aircraft. { 'er‚bórn i‚lek'trän·ik 'sər·vā kən'trōl }

airborne intercept radar |ENG| Airborne radar used to track and "lock on" to another aircraft to be intercepted or followed. { 'er‚bórn 'in·tər‚sept ‚rā‚där }

airborne magnetometer |ENG| An airborne instrument used to measure the magnetic field of the earth. { 'er‚bórn ‚mag·na'täm·əd·ər }

airborne profile recorder |ENG| An electronic instrument that emits a pulsed-type radar signal from an aircraft to measure vertical distances between the aircraft and the earth's surface. Abbreviated APR. Also known as terrain profile recorder (TPR). { 'er‚bórn 'prō‚fīl ri‚kórd·ər }

airborne radar |ENG| Radar equipment carried by aircraft to assist in navigation by pilotage, to determine drift, and to locate weather disturbances; a very important use is locating other aircraft either for avoidance or attack. { 'er ‚bórn 'rā‚där }

airborne waste |ENG| Vapors, gases, or particulates introduced into the atmosphere by evaporation, chemical, or combustion processes; a frequent cause of smog and an irritant to eyes and breathing passages. { 'er‚bórn 'wäst }

air-bound |ENG| Of a pipe or apparatus, containing a pocket of air that prevents or reduces the desired liquid flow. { 'er ‚baúnd }

air brake |MECH ENG| An energy-conversion mechanism activated by air pressure and used to retard, stop, or hold a vehicle or, generally, any moving element. { 'er ‚brāk }

air breaking See airblasting. { 'er ‚brāk·iŋ }

air-breathing |MECH ENG| Of an engine or aerodynamic vehicle, required to take in air for the purpose of combustion. { 'er 'brēth·iŋ }

air cap |MECH ENG| A device used in thermal spraying which directs the air pattern for purposes of atomization. { 'er ‚kap }

air casing |ENG| A metal casing surrounding a pipe or reservoir and having a space between to prevent heat transmission. { 'er ‚kās·iŋ }

air cell |ELECTR| A cell in which depolarization at the positive electrode is accomplished chemically by reduction of the oxygen in the air. |MECH ENG| A small auxiliary combustion chamber used to promote turbulence and improve combustion in certain types of diesel engines. { 'er ‚sel }

air chamber |MECH ENG| A pressure vessel, partially filled with air, for converting pulsating flow to steady flow of water in a pipeline, as with a reciprocating pump. { 'er ‚chām·bər }

air change |ENG| A measure of the movement of a given volume of air in or out of a building or room in a specified time period; usually expressed in cubic feet per minute. { 'er ‚chānj }

air check |ENG ACOUS| A recording made of a live radio broadcast for filing purposes at the broadcasting facility. { 'er ‚chek }

air classifier |MECH ENG| A device to separate particles by size through the action of a stream of air. Also known as air elutriator. { 'er 'klas·ə‚fī·ər }

air cleaner |ENG| Any of various devices designed to remove particles and aerosols of specific sizes from air; examples are screens, settling chambers, filters, wet collectors, and electrostatic precipitators. { 'er ‚klēn·ər }

Airco-Hoover sweetening |CHEM ENG| Removal of mercaptans from gasoline by caustic and water washes, then heating the dried gasoline and passing it with some oxygen through a reactor containing a slurry of diatomaceous earth impregnated with copper chloride; the oxygen regenerates the catalyst. { 'er‚kō 'hüv·ər 'swēt·niŋ }

air compressor |MECH ENG| A machine that increases the pressure of air by increasing its density and delivering the fluid against the connected system resistance on the discharge side. { 'er ‚kəm'pres·ər }

air-compressor unloader |MECH ENG| A device for control of air volume flowing through an air compressor. { 'er ‚kəm'pres·ər ən'lōd·ər }

air-compressor valve |MECH ENG| A device for controlling the flow into or out of the cylinder of a compressor. { 'er ‚kəm'pres·ər ‚valv }

air condenser |MECH ENG| **1.** A steam condenser in which the heat exchange occurs through metal walls separating the steam from cooling air. Also known as air-cooled condenser. **2.** A device that removes vapors, such as of oil or water, from the airstream in a compressed-air line. { 'er ‚kən'dens·ər }

air conditioner |MECH ENG| A mechanism primarily for comfort cooling that lowers the temperature and reduces the humidity of air in buildings. { 'er ‚kən'dish·ən·ər }

air conditioning |MECH ENG| The maintenance of certain aspects of the environment within a defined space to facilitate the function of that space; aspects controlled include air temperature and motion, radiant heat level, moisture, and concentration of pollutants such as dust, microorganisms, and gases. Also known as climate control. { 'er ˌkən'dish·ən·iŋ }

air conveyor See pneumatic conveyor. { 'er kən,vā·ər }

air-cooled engine |MECH ENG| An engine cooled directly by a stream of air without the interposition of a liquid medium. { 'er ˌküld 'en·jən }

air-cooled heat exchanger |MECH ENG| A finned-tube (extended-surface) heat exchanger with hot fluids inside the tubes, and cooling air that is fan-blown (forced draft) or fan-pulled (induced draft) across the tube bank. { 'er ˌküld ˌhēt ˌiks'chānj·ər }

air cooling |MECH ENG| Lowering of air temperature for comfort, process control, or food preservation. { 'er ˌkül·iŋ }

air course See airway. { 'er ˌkórs }

aircraft detection |ENG| The sensing and discovery of the presence of aircraft; major techniques include radar, acoustical, and optical methods. { 'er,kraft di'tek·shən }

aircraft impactor |ENG| An instrument carried by an aircraft for the purpose of obtaining samples of airborne particles. { 'er,kraft im'pak·tər }

air-cure |CHEM ENG| To vulcanize at ordinary room temperatures, or without the aid of heat. { 'er ˌkyür }

air curtain |MECH ENG| A stream of high-velocity temperature-controlled air which is directed downward across an opening; it excludes insects, exterior drafts, and so forth, prevents the transfer of heat across it, and permits air-conditioning of a space with an open entrance. { 'er,kərt·ən }

air cushion |MECH ENG| A mechanical device using trapped air to arrest motion without shock. { 'er ˌkúsh·ən }

air-cushion vehicle |MECH ENG| A transportation device supported by low-pressure, low-velocity air capable of traveling equally well over water, ice, marsh, or relatively level land. Also known as ground-effect machine (GEM); hovercraft. { 'er ˌkúsh·ən ˌvē·ə·kəl }

air-cut |ENG| Referring to the inadvertent mechanical incorporation of air into a liquid system. { 'er ˌkət }

air cycle |MECH ENG| A refrigeration cycle characterized by the working fluid, air, remaining as a gas throughout the cycle rather than being condensed to a liquid; used primarily in airplane air conditioning. { 'er ˌsī·kəl }

air cylinder |MECH ENG| A cylinder in which air is compressed by a piston, compressed air is stored, or air drives a piston. { 'er ˌsil·ən·dər }

air density |MECH| The mass per unit volume of air. { 'er ˌden·səd·ē }

air diffuser |BUILD| An air distribution outlet, usually located in the ceiling and consisting of deflecting vanes discharging supply air in various directions and planes, and arranged to promote mixing of the supplied air with the air already in the room. { 'er di,fyüz·ər }

air-distributing acoustical ceiling |BUILD| A suspended acoustical ceiling in which the board or tile is provided with small, evenly distributed mechanical perforations; designed to provide a desired flow of air from a pressurized plenum above. { 'er di'strib·yəd·iŋ ə'kü·sti·kəl 'sēl·iŋ }

air diving |ENG| A type of diving in which the diver's breathing medium is a normal atmospheric mixture of oxygen and nitrogen; limited to depths of 190 feet (58 meters). { 'er ˌdīv·iŋ }

air drain |CIV ENG| An empty space left around the external foundation wall of a building to prevent the earth from lying against it and causing dampness. { 'er ˌdrān }

airdraulic |MECH ENG| Combining pneumatic and hydraulic action for operation. { ˌer'dról·ik }

air drill |MECH ENG| A drill powered by compressed air. { ¦er ˌdril }

air drying |ENG| Removing moisture from a material by exposure to air to the extent that no further moisture is released on contact with air; important in lumber manufacture. { ¦er 'drī·iŋ }

air duct See airflow duct. { ¦er ˌdəkt }

air ejector |MECH ENG| A device that uses a fluid jet to remove air or other gases, as from a steam condenser. { ¦er i'jek·tər }

air eliminator |MECH ENG| In a piping system, a device used to remove air from water, steam, or refrigerant. { ¦er i'lim·ə,nād·ər }

air elutriator See air classifier. { ¦er ē'lü·trē,ād·ər }

air engine |MECH ENG| An engine in which compressed air is the actuating fluid. { ¦er 'en·jən }

air entrainment |ENG| The inclusion of minute bubbles of air in cement or concrete through the addition of some material during grinding or mixing to reduce the surface tension of the water, giving improved properties for the end product. { ¦er in'trān·mənt }

air escape |DES ENG| A device that is fitted to a pipe carrying a liquid for releasing excess air; it contains a valve that controls air release while preventing loss of liquid. { 'er ə,skāp }

air-exhaust ventilator |MECH ENG| Any air-exhaust unit used to carry away dirt particles, odors, or fumes. { 'er ig'zóst 'ven·tə,lād·ər }

airfield |CIV ENG| The area of an airport for the takeoff and landing of airplanes. { 'er,fēld }

air filter |ENG| A device that reduces the concentration of solid particles in an airstream to a level that can be tolerated in a process or space occupancy; a component of most systems in which air is used for industrial processes, ventilation, or comfort air conditioning. { 'er ,fil·tər }

air flotation See dissolved air flotation. { 'er flō'tā·shən }

airflow duct |ENG| A pipe, tube, or channel through which air moves into or out of an enclosed space. Also known as air duct. { 'er ,flō ,dəkt }

airflow orifice |ENG| An opening through which air moves out of an enclosed space. { 'er,flō 'ȯr·ə·fəs }

airflow pipe |ENG| A tube through which air is conveyed from one location to another. { 'er ,flō ,pīp }

air-fuel mixture |MECH ENG| In a carbureted gasoline engine, the charge of air and fuel that is mixed in the appropriate ratio in the carburetor and subsequently fed into the combustion chamber. { 'er 'fyül ,miks·chər }

air gage |ENG| **1.** A device that measures air pressure. **2.** A device that compares the shape of a machined surface to that of a reference surface by measuring the rate of passage of air between the surfaces. { 'er ,gāj }

air gap |ELECTR| **1.** A gap or an equivalent filler of nonmagnetic material across the core of a choke, transformer, or other magnetic device. **2.** A spark gap consisting of two electrodes separated by air. **3.** The space between the stator and rotor in a motor or generator. |ENG| **1.** The distance between two components or parts. **2.** In plastic extrusion coating, the distance from the opening of the extrusion die to the nip formed by the pressure and chill rolls. **3.** The unobstructed vertical distance between the lowest opening of a faucet (or the like) which supplies a plumbing fixture (such as a tank or washbowl) and the level at which the fixture will overflow. { 'er ,gap }

air grating |BUILD| A fixed metal grille on the exterior of a building through which air is brought into or discharged from the building for purposes of ventilation. { 'er ,grād·iŋ }

air hammer See pneumatic hammer. { 'er ,ham·ər }

air-handling system |MECH ENG| An air-conditioning system in which an air-handling unit provides part of the treatment of the air. { 'er 'hand·liŋ ,sis·təm }

air-handling unit |MECH ENG| A packaged assembly of air-conditioning components (coils, filters, fan humidifier, and so forth) which provides for the treatment of air before it is distributed. { 'er 'hand·liŋ ,yü·nət }

air heater See air preheater. { 'er ,hēd·ər }

air-heating system See air preheater. { 'er ,hēd·iŋ 'sis·təm }

air hoist |MECH ENG| A lifting tackle or tugger constructed with cylinders and pistons for reciprocating motion and air motors for rotary motion, all powered by compressed air. Also known as pneumatic hoist. { 'er ,hȯist }

air horn |MECH ENG| In an automotive engine, the upper portion of the carburetor barrel through which entering air passes in quantities controlled by the choke plate and the throttle plate. { 'er ,hȯrn }

air horsepower |MECH ENG| The theoretical (minimum) power required to deliver the specified quantity of air under the specified pressure conditions in a fan, blower, compressor, or vacuum pump. Abbreviated air hp. { 'er 'hȯrs ,pau̇·ər }

air hp See air horsepower.

air-injection reactor |MECH ENG| A unit installed in an automotive engine which mixes fresh air with hot exhaust gases in the exhaust manifold to react with any gasoline that has escaped unburned from the cylinders. Abbreviated AIR. { 'er in'jek·shən rē'ak·tər }

air-injection system |MECH ENG| A device that uses compressed air to inject the fuel into the cylinder of an internal combustion engine. Also known as thermactor. { 'er in'jek·shən ,sis·təm }

air inlet |MECH ENG| In an air-conditioning system, a device through which air is exhausted from a room or building. { 'er 'in,let }

air-inlet valve |MECH ENG| In a heating/air-conditioning system of a motor vehicle, a valve in the plenum blower assembly that permits selection of either inside or outside air. { 'er 'in,let ,valv }

air knife |ENG| A device that uses a thin, flat jet of air to remove the excess coating from freshly coated paper. { 'er ,nīf }

air-knife coating |ENG| An even film of coating left on paper after treatment with an air knife. { 'er 'nīf ,kōd·iŋ }

air-lance |ENG| To direct a pressurized-air stream to remove unwanted accumulations, as in boiler-wall cleaning. { 'er ,lans }

air leakage |MECH ENG| **1.** In ductwork, air which escapes from a joint, coupling, and such. **2.** The undesired leakage or uncontrolled passage of air from a ventilation system. { 'er ,lēk·əj }

airless spraying |ENG| The spraying of paint by means of high fluid pressure and special equipment. Also known as hydraulic spraying. { 'er·ləs 'sprā·iŋ }

air lift |MECH ENG| **1.** Equipment for lifting slurry or dry powder through pipes by means of compressed air. **2.** See air-lift pump. { 'er ,lift }

air-lift hammer |MECH ENG| A gravity drop hammer used in closed die forging in which the ram is raised to its starting point by means of an air cylinder. { 'er,lift 'ham·ər }

air-lift pump |MECH ENG| A device composed of two pipes, one inside the other, used to extract water from a well; the lower end of the pipes is submerged, and air is delivered through the inner pipe to form a mixture of air and water which rises in the outer pipe above the water in the well; also used to move corrosive liquids, mill tailings, and sand. Also known as air lift. { 'er ,ləs 'pəmp }

air line |ENG| A fault, in the form of an elongated bubble, in glass tubing. Also known as hairline. |MECH ENG| A duct, hose, or pipe that supplies compressed air to a pneumatic tool or piece of equipment. { 'er ,līn }

air-line lubricator See line oiler. { 'er ,līn 'lü·brə,kād·ər }

air lock |ENG| **1.** A chamber capable of being hermetically sealed that provides for passage between two places of different pressure, such as between an altitude chamber and the outside atmosphere, or between the outside atmosphere and the work area in a tunnel or shaft being excavated through soil subjected to water pressure higher than atmospheric. Also known as lock. **2.** An air bubble in a pipeline which impedes liquid flow. **3.** A depression on the surface of a molded plastic part that results from air trapped between the surface of the mold and the plastic. { 'er ,läk }

air-lock strip |BUILD| The weather stripping which is fastened to the edges of each wing of a revolving door. { 'er ,läk ,strip }

air meter |ENG| A device that measures the flow of air, or gas, expressed in volumetric or weight units per unit time. Also known as airometer. { 'er ,mēd·ər }

air mileage indicator |ENG| An instrument on an airplane which continuously indicates mileage through the air. { ¦er ,mī·lij 'in·də'kād·ər }

air mileage unit |ENG| A device which derives continuously and automatically the air distance flown, and feeds this information into other units, such as an air mileage indicator. { 'er ,mī·lij ,yü·nət }

air-mixing plenum |MECH ENG| In an air-conditioning system, a chamber in which the recirculating air is mixed with air from outdoors. { 'er ,miks·iŋ 'plēn·əm }

air monitoring |CIV ENG| A practice of continuous air sampling by various levels of government or particular industries. { 'er ,män·ə·triŋ }

air motor |MECH ENG| A device in which the pressure of confined air causes the rotation of a rotor or the movement of a piston. { 'er ,mōd·ər }

air nozzle |MECH ENG| In an automotive engine, a device for supplying air to the air-injection reactor. { 'er ,näz·əl }

airometer |ENG| **1.** An apparatus for both holding air and measuring the quantity of air admitted into it. **2.** *See* air meter. { ,er'ä·məd·ər }

air outlet |MECH ENG| In an air-conditioning system, a device at the end of a duct through which air is supplied to a space. { 'er ,aút·lət }

air-permeability test |ENG| A test for the measurement of the fineness of powdered materials, such as portland cement. { ¦er ,pər·mē·ə'bil·ə·dē ,test }

airplane flare |ENG| A flare, often magnesium, that is dropped from an airplane to illuminate a ground area; a small parachute decreases the rate of descent. { 'er,plān ,fler }

air pocket |ENG| An air-filled space that is normally occupied by a liquid. Also known as air trap. { 'er ,päk·ət }

air-pollution control |ENG| A practical means of treating polluting sources to maintain a desired degree of air cleanliness. { ¦er pə'lü·shən kən ,trōl }

airport |CIV ENG| A terminal facility used for aircraft takeoff and landing and including facilities for handling passengers and cargo and for servicing aircraft. Also known as aerodrome. { 'er,pórt }

airport engineering |CIV ENG| The planning, design, construction, and operation and maintenance of facilities providing for the landing and takeoff, loading and unloading, servicing, maintenance, and storage of aircraft. { 'er,pórt en·jə'nir·iŋ }

air preheater |MECH ENG| A device used in steam boilers to transfer heat from the flue gases to the combustion air before the latter enters the furnace. Also known as air heater; air-heating system. { 'er ,prē'hēd·ər }

airproof *See* airtight. { 'er,prüf }

air propeller |MECH ENG| A rotating fan for moving air. { 'er prə,pel·ər }

air pump |MECH ENG| A device for removing air from an enclosed space or for adding air to an enclosed space. { 'er ,pəmp }

air puncher |MECH ENG| A machine consisting essentially of a reciprocating chisel or pick, driven by air. { 'er ,pən·chər }

air purge |MECH ENG| Removal of particulate matter from air within an enclosed vessel by means of air displacement. { 'er ,pərj }

air-raid shelter |CIV ENG| A chamber, often underground, provided with living facilities and food, for sheltering people against air attacks. { ¦er ,rād ¦shel·tər }

air receiver |MECH ENG| A vessel designed for compressed-air installations that is used both to store the compressed air and to permit pressure to be equalized in the system. { 'er ri,sē·vər }

air register |ENG| A device attached to an air-distributing duct for the purpose of controlling the discharge of air into the space to be heated, cooled, or ventilated. { 'er ,rej·ə·stər }

air regulator |MECH ENG| A device for regulating airflow, as in the burner of a furnace. { ¦er ¦reg·yə,lād·ər }

air reheater |MECH ENG| In a heating system, any device used to add heat to the air circulating in the system. { ¦er ,rē¦hēd·ər }

air release valve |MECH ENG| A valve, usually manually operated, which is used to release air from a water pipe or fitting. { 'er ri¦lēs ,valv }

air resistance |MECH| Wind drag giving rise to forces and wear on buildings and other structures. { 'er ri'zis·təns }

air ring |ENG| In plastics forming, a circular manifold which distributes an even flow of cool air into a hollow tubular form passing through the manifold. { 'er ,riŋ }

air sampling |ENG| The collection and analysis of samples of air to measure the amounts of various pollutants or other substances in the air, or the air's radioactivity. { 'er ,sam·pliŋ }

air scoop |DES ENG| An air-duct cowl projecting from the outer surface of an aircraft or automobile, which is designed to utilize the dynamic pressure of the airstream to maintain a flow of air. { 'er ,sküp }

air screw |MECH ENG| A screw propeller that operates in air. { 'er ,skrü }

air-seasoned |ENG| Treated by exposure to air to give a desired quality. { 'er ,sēz·ənd }

air separator |MECH ENG| A device that uses an air current to separate a material from another of greater density or particles from others of greater size. { ¦er ¦sep·ə¦rād·ər }

air shaft |BUILD| An open space surrounded by the walls of a building or buildings to provide ventilation for windows. Also known as air well. { 'er ¦shaft }

air shot |ENG| A shot prepared by loading (charging) so that an air space is left in contact with the explosive for the purpose of lessening its shattering effect. { 'er ¦shät }

Airslide conveyor |MECH ENG| An air-activated gravity-type conveyor, of the Fuller Company, using low-pressure air to aerate or fluidize pulverized material to a degree which will permit it to flow on a slight incline by the force of gravity. { ¦er¦slīd kən¦vā·ər }

air space |ENG| An enclosed space containing air in a wall for thermal insulation. { ¦er ¦spās }

airspeed head |ENG| Any instrument or device, usually a pitot tube, mounted on an aircraft for receiving the static and dynamic pressures of the air used by the airspeed indicator. { 'er¦spēd ¦hed }

airspeed indicator |ENG| A device that computes and displays the speed of an aircraft relative to the air mass in which the aircraft is flying. { ¦er¦spēd ¦in·də¦kād·ər }

air spring |MECH ENG| A spring in which the energy storage element is air confined in a container that includes an elastomeric bellows or diaphragm. { 'er ¦spriŋ }

air-standard cycle |THERMO| A thermodynamic cycle in which the working fluid is considered to be a perfect gas with such properties of air as a volume of 12.4 cubic feet per pound at 14.7 pounds per square inch (approximately 0.7756 cubic meter per kilogram at 101.36 kilopascals) and 492°R and a ratio of specific heats of 1:4. { ¦er ¦stan·dərd 'sī·kəl }

air-standard engine |MECH ENG| A heat engine operated in an air-standard cycle. { ¦er ¦stan·dərd 'en·jən }

air starting valve |MECH ENG| A device that admits compressed air to an air starter. { 'er ¦stärd·iŋ ¦valv }

air stripping |CHEM ENG| The process of bubbling air through water to remove volatile organic substances from the water. { 'er ¦strip·iŋ }

air-supply mask See air-tube breathing apparatus. { 'er sə¦plī ¦mask }

air surveillance |ENG| Systematic observation of the airspace by visual, electronic, or other means, primarily for identifying all aircraft in that airspace, and determining their movements. { 'er sər'vā·ləns }

air surveillance radar |ENG| Radar of moderate range providing position of aircraft by azimuth and range data without elevation data; used for air-traffic control. { ¦er sər¦vā·ləns ¦rā,där }

air survey See aerial survey. { ¦er ¦sər,vā }

air-suspension encapsulation |CHEM ENG| A technique for microencapsulation of various types of solid particles; the particles undergo a series of cycles in which they are first suspended by a vertical current of air while they are sprayed with a solution of coating material, and are then moved by the airstream into a region where they undergo a drying treatment. Also known as Wurster process. { 'ər sə¦spen·shən in,kap·sə'lā·shən }

air-suspension system |MECH ENG| Parts of an automotive vehicle that are intermediate between the wheels and the frame, and support the car body and frame by means of a cushion of air to absorb road shock caused by passage of the wheels over irregularities. { 'er sə¦spen·shən ¦sis·təm }

air sweetening |CHEM ENG| A process in which air or oxygen is used to oxidize lead mercaptides to disulfides instead of using elemental sulfur. { 'er ¦swēt·ən·iŋ }

air system |MECH ENG| A mechanical refrigeration system in which air serves as the refrigerant in a cycle comprising compressor, heat exchanger, expander, and refrigerating core. { 'er ¦sis·təm }

air terminal |CIV ENG| A facility providing a place of assembly and amenities for airline passengers and space for administrative functions. |ELEC| A structure, such as a tower, that serves as a lightning arrester. { 'er ¦tərm·ən·əl }

air thermometer |ENG| A device that measures the temperature of an enclosed space by means of variations in the pressure or volume of air contained in a bulb placed in the space. { ¦er thə¦mäm·əd·ər }

airtight |ENG| Not permitting the passage of air. Also known as airproof. { 'er,tīt }

air-to-air resistance |CIV ENG| The resistance provided by the wall of a building to the flow of heat. { ¦er tü ¦er ri'sis·təns }

air toxics See hazardous air pollutants. { 'er ¦täk·siks }

air trap |CIV ENG| A U-shaped pipe filled with water that prevents the escape of foul air or gas from such systems as drains and sewers. See air pocket. { 'er ¦trap }

air-tube breathing apparatus |ENG| A device consisting of a smoke helmet, mask, or mouthpiece supplied with fresh air by means of a flexible tube. Also known as air-supply mask. { ¦er ¦tüb 'brēth·iŋ ¦a·pə¦rad·əs }

air-tube clutch |MECH ENG| A clutch fitted with a tube whose inflation causes the clutch to engage, and deflation, to disengage. { 'er ¦tüb ¦kləch }

air valve |MECH ENG| A valve that automatically lets air out of or into a liquid-carrying pipe when the internal pressure drops below atmospheric. { 'er ¦valv }

air vessel |ENG| **1.** An enclosed volume of air which uses the compressibility of air to minimize water hammer. Also known as accumulator. **2.** An enclosed chamber using the compressibility of air to promote a more uniform flow of water in a piping system. { 'er ¦ves·əl }

air washer |MECH ENG| **1.** A device for cooling and cleaning air in which the entering warm, moist air is cooled below its dew point by refrigerated water so that although the air leaves close to saturation with water, it has less moisture per unit volume than when it entered. **2.** Apparatus to wash particulates and soluble impurities from air by passing the airstream through a liquid bath or spray. { 'er ,wash·ər }

air-water jet |ENG| A jet of mixed air and water which leaves a nozzle at high velocity; used in cleaning the surfaces of concrete or rock. { ¦er ¦wód·ər 'jet }

air-water storage tank |ENG| A water storage tank in which the air above the water is compressed. { ¦er ¦wód·ər 'stór·ij ,taŋk }

airway |BUILD| A passage for ventilation between thermal insulation and roof boards. { 'er,wā }

air well See air shaft. { 'er ,wel }

Airy points |ENG| The points at which a horizontal rod is optionally supported to avoid its bending. { ¦er·ē ,póins }

Airy stress function |MECH| A biharmonic function of two variables whose second partial derivatives give the stress components of a body subject to a plane strain. { ¦er·ē 'stres ,faŋk·shən }

aisleway |CIV ENG| A passage or walkway within a factory, storage building, or shop permitting the flow of inside traffic. { 'īl,wā }

Aitken dust counter |ENG| An instrument for determining the dust content of the atmosphere. Also known as Aitken nucleus counter. { ¦āt·kən 'dəst ,kaúnt·ər }

Aitken nucleus counter See Aitken dust counter. { ¦āt·kən 'nü·klē·əs ,kaúnt·ər }

alarm gage |ENG| A device that actuates a signal either when the steam pressure in a boiler is too high or when the water level in a boiler is too low. { ə'lärm ,gaj }

alarm system |ENG| A system which operates a warning device after the occurrence of a dangerous or undesirable condition. { ə'lärm ,sis·təm }

alarm valve |ENG| A device that sounds an alarm when water flows in an automatic sprinkler system. { ə'lärm ,valv }

albedometer |ENG| An instrument used for the measurement of the reflecting power, that is, the albedo, of a surface. { al·bə'dä·məd·ər }

Alberger process |CHEM ENG| A method of manufacturing salt by heating brine at high pressure and passing it to a graveler which removes calcium sulfate; the salt crystallizes as the pressure is reduced and thus is separated from the brine. { 'äl·bər·gər 'präs·əs }

alcoholimeter See alcoholometer. { ,al·kə,hó'lim·əd·ər }

alcoholmeter See alcoholometer. { 'al·kə,hól ,mēd·ər }

alcoholometer |ENG| A device, such as a form of hydrometer, that measures the quantity of alcohol contained in a liquid. Also known as alcoholimeter; alcoholmeter. { ,al·kə,hó'lä·məd·ər }

alcohol thermometer |ENG| A liquid-in-glass thermometer that uses ethyl alcohol as its working substance. { 'al·kə,hól thər'mäm·əd·ər }

alidade |ENG| **1.** An instrument for topographic surveying and mapping by the plane-table method. **2.** Any sighting device employed for angular measurement. { 'al·ə,dād }

aligning drift |MECH ENG| A rod or bar that is used for aligning parts during assembly. { ə'līn·iŋ ,drift }

alignment |CIV ENG| In a survey for a highway, railroad, or similar installation, a ground plan that shows the horizontal direction of the route. |ELECTR| The process of adjusting components of a system for proper interrelationship, including the adjustment of tuned circuits for proper frequency response and the time synchronization of the components of a system. |ENG| Placing of surveying points along a straight line. { ə'līn·mənt }

alignment correction |ENG| A correction applied to the measured length of a line to allow for not holding the tape exactly in a vertical plane of the line. { ə'līn·mənt kə'rek·shən }

alignment pin |DES ENG| Pin in the center of the base of an octal, loctal, or other tube having a single vertical projecting rib that aids in correctly inserting the tube in its socket. { ə'līn·mənt ,pin }

alignment wire See ground wire. { ə'līn·mənt ,wīr }

alkali ion diode |ENG| In testing for leaks, a device which senses the presence of halogen gases by the use of positive ions of alkali metal on the heated diode surfaces. { 'al·kə,lī 'ī·ən 'dī,ōd }

alkaline wash |CHEM ENG| The removal of impurities from kerosine,used for illuminating purposes, by caustic soda solution. { 'al·kə,līn ,wäsh }

Alkar process |CHEM ENG| Catalytic alkylation of aromatic hydrocarbons with olefins to produce alkylaromatics; for example, production of ethylbenzene from benzene and ethylene. { 'al,kär 'präs·əs }

alkylate bottom |CHEM ENG| Residue from fractionation of total alkylate which boils at a higher temperature than aviation gasolines. { 'al·kə,lāt 'bäd·əm }

alkylation |CHEM ENG| A refinery process for chemically combining isoparaffin with olefin hydrocarbons. { ,al·kə'lā·shən }

allège |BUILD| A part of a wall which is thinner than the rest, especially the spandrel under a window. { a'lezh }

Allen screw |DES ENG| A screw or bolt which has an axial hexagonal socket in its head. { 'al·ən ,skrü }

Allen wrench |DES ENG| A wrench made from a straight or bent hexagonal rod, used to turn an Allen screw. { 'al·ən ,rench }

alligator shears |ENG| A cutting tool with a fixed lower blade and a movable upper blade (shearing arm) that moves in an arc around a fulcrum pin; used mainly for shearing applications that do not require great accuracy. { 'al·ə,gād·ər ,shirz }

alligator wrench [DES ENG] A wrench having fixed jaws forming a V, with teeth on one or both jaws. { 'al·ə,gād·ər ,rench }

allocate [IND ENG] To assign a portion of a resource to an activity. { 'a·lō,kāt }

allowable bearing value [CIV ENG] The maximum permissible pressure on foundation soil that provides adequate safety against rupture of the soil mass or movement of the foundation of such magnitude as to impair the structure imposing the pressure. Also known as allowable soil pressure. { ə'laú·ə·bəl 'ber·iŋ ,val·yü }

allowable load [MECH] The maximum force that may be safely applied to a solid, or is permitted by applicable regulators. { ə'laú·ə·bəl 'lōd }

allowable soil pressure See allowable bearing value. { ə'laú·ə·bəl 'sȯil ,presh·ər }

allowable stress [MECH] The maximum force per unit area that may be safely applied to a solid. { ə'laú·ə·bəl 'stres }

allowance [DES ENG] An intentional difference in sizes of two mating parts, allowing clearance usually for a film of oil, for running or sliding fits. { ə'laú·əns }

allowed hours See standard hour. { ə'laúd 'aú·ərz }

allowed time [IND ENG] Amount of time allowed each employee for personal needs during a work cycle. { ə'laúd 'tīm }

alloy junction [ELECTR] A junction produced by alloying one or more impurity metals to a semiconductor to form a p or n region, depending on the impurity used. Also known as fused junction. { 'a,lȯi ,jəŋk·shən }

alloy-junction diode [ELECTR] A junction diode made by placing a pill of doped alloying material on a semiconductor material and heating until the molten alloy melts a portion of the semiconductor, resulting in a pn junction when the dissolved semiconductor recrystallizes. Also known as fused-junction diode. { 'a,lȯi ,jəŋk·shən 'dī,ōd }

all-translational system [CONT SYS] A simple robotic system in which there is no rotation of the robot or its components during movements of the robot's body. { 'ȯl ,tranz'lā·shən·əl 'sis·təm }

all-weather airport [CIV ENG] An airport with facilities to permit the landing of qualified aircraft and aircrewmen without regard to operational weather limits. { 'ȯl ,weth·ər 'er,pȯrt }

alpha [ELECTR] The ratio between the change in collector current and the change in emitter current of a transistor. { 'al·fə }

alpha cutoff frequency [ELECTR] The frequency at the high end of a transistor's range at which current amplification drops 3 decibels below its low-frequency value. { 'al·fə 'kəd,ȯf ,frē·kwən·sē }

alpha-ray vacuum gage [ENG] An ionization gage in which the ionization is produced by alpha particles emitted by a radioactive source, instead of by electrons emitted from a hot filament; used chiefly for pressures from 10⁻³ to 10 torrs. Also known as alphatron. { 'al·fə ,rā 'vak·yüm ,gāj }

alphatron See alpha-ray vacuum gage. { 'al·fə,trän }

alt See altitude.

altazimuth [ENG] An instrument equipped with both horizontal and vertical graduated circles, for the simultaneous observation of horizontal and vertical directions or angles. Also known as astronomical theodolite; universal instrument. { al'taz·ə·məth }

alt-azimuth mounting See altitude-azimuth mounting. { ,alt 'az·ə·məth ,maúnt·iŋ }

alternate energy [ENG] Any source of energy other than fossil fuels that is used for constructive purposes. { 'ȯl·tər·nət 'en·ər·jē }

alternating current [ELEC] Electric current that reverses direction periodically, usually many times per second. Abbreviated ac. { 'ȯl·tər,nād·iŋ 'kər·ənt }

alternating-current welder [ENG] A welding machine utilizing alternating current for welding purposes. { 'ȯl·tər,nād·iŋ 'kər·ənt 'weld·ər }

alternating stress [MECH] A stress produced in a material by forces which are such that each force alternately acts in opposite directions. { 'ȯl·tər,nād·iŋ 'stres }

altigraph [ENG] A pressure altimeter that has a recording mechanism to show the changes in altitude. { 'al·tə,graf }

altimeter [ENG] An instrument which determines the altitude of an object with respect to a fixed level, such as sea level; there are two common types: the aneroid altimeter and the radio altimeter. { al'tim·əd·ər }

altimeter corrections [ENG] Corrections which must be made to the readings of a pressure altimeter to obtain true altitudes; involve horizontal pressure gradient error and air temperature error. { al'tim·əd·ər kə'rek·shənz }

altimeter setting [ENG] The value of atmospheric pressure to which the scale of an aneroid altimeter is set; after United States practice, the pressure that will indicate airport elevation when the altimeter is 10 feet (3 meters) above the runway (approximately cockpit height). { al 'tim·əd·ər ,sed·iŋ }

altimeter-setting indicator [ENG] A precision aneroid barometer calibrated to indicate directly the local altimeter setting. { al'tim·əd·ər ,sed·iŋ 'in·də,kād·ər }

altimetry [ENG] The measurement of heights in the atmosphere (altitude), generally by an altimeter. { al'tim·ə·trē }

altitude Abbreviated alt. [ENG] **1.** Height, measured as distance along the extended earth's radius above a given datum, such as average sea level. **2.** Angular displacement above the horizon measured by an altitude curve. { 'al·tə,tüd }

altitude azimuth [ENG] An azimuth determined by solution of the navigational triangle with altitude, declination, and latitude given. { 'al·tə,tüd 'az·ə·məth }

altitude-azimuth mounting [ENG] A two-axis telescope mounting in which the azimuth of the direction in which the telescope is pointed is

19

determined by rotation about a vertical axis and the corresponding altitude is determined by rotation about a horizontal axis; computer-controlled motors must move the telescope in both altitude and azimuth to compensate for the earth's rotation. Also known as alt-azimuth mounting. { ¦al·tə,tüd 'az·ə·məth ,maúnt·iŋ }

altitude chamber |ENG| A chamber within which the air pressure, temperature, and so on can be adjusted to simulate conditions at different altitudes; used for experimentation and testing. { 'al·tə,tüd ,chām·bər }

altitude curve |ENG| The arc of a vertical circle between the horizon and a point on the celestial sphere, measured upward from the horizon. { 'al·tə,tüd ,kərv }

altitude datum |ENG| The arbitrary level from which heights are reckoned. { 'al·tə,tüd ,dad·əm }

altitude difference |ENG| The difference between computed and observed altitudes, or between precomputed and sextant altitudes. Also known as altitude intercept; intercept. { 'al·tə,tüd 'dif·rəns }

altitude intercept See altitude difference. { 'al·tə,tüd 'in·tər,sept }

aluminize |ENG| To apply a film of aluminum to a material, such as glass. { ə'lüm·ə,nīz }

AM See amplitude modulation.

A-mast |ENG| An A-shaped arrangement of upright poles for supporting a mechanism designed to lift heavy loads. { 'ā ,mast }

ambient |ENG| Surrounding; especially, of or pertaining to the environment about a flying aircraft or other body but undisturbed or unaffected by it, as in ambient air or ambient temperature. { 'am·bē·ənt }

American basement |BUILD| A basement located above ground level and containing the building's main entrance. { ə'mer·ə·kən 'bās·mənt }

American bond |CIV ENG| A bond in which every fifth, sixth, or seventh course of a wall consists of headers and the other courses consist of stretchers. Also known as common bond; Scotch bond. { ə'mer·ə·kən 'bänd }

American caisson See box caisson. { ə'mer·ə·kən 'kā,sän }

American filter See disk filter. { ə'mer·ə·kən 'fil·tər }

American melting point |CHEM ENG| A temperature 3°F (1.7°C) higher than the American Society for Testing and Materials Method D87 paraffin-wax melting point. { ə'mer·ə·kən 'melt·iŋ ,póint }

American standard beam |CIV ENG| A type of I beam made of hot-rolled structural steel. { ə'mer·ə·kən 'stan·dərd 'bēm }

American standard channel |CIV ENG| A C-shaped structural member made of hot-rolled structural steel. { ə'mer·ə·kən 'stan·dərd 'chan·əl }

American standard pipe thread |DES ENG| Taper, straight, or dryseal pipe thread whose dimensions conform to those of a particular series

of specified sizes established as a standard in the United States. Also known as Briggs pipe thread. { ə'mer·ə·kən 'stan·dərd 'pīp ,thred }

American standard screw thread |DES ENG| Screw thread whose dimensions conform to those of a particular series of specified sizes established as a standard in the United States; used for bolts, nuts, and machine screws. { ə'mer·ə·kən 'stan·dərd 'skrü ,thred }

American system drill See churn drill. { ə'mer·ə·kən ¦sis·təm ,dril }

American Table of Distances |ENG| Published data concerning the safe storage of explosives and ammunition. { ə'mer·ə·kən ,tā·bəl əv 'dis·təns·əz }

ammeter |ENG| An instrument for measuring the magnitude of electric current flow. Also known as electric current meter. { 'a,mēd·ər }

ammonia absorption refrigerator |MECH ENG| An absorption-cycle refrigerator which uses ammonia as the circulating refrigerant. { ə'mōn·yə əb¦sorp·shən ri'frij·ə,rād·ər }

ammonia compressor |MECH ENG| A device that decreases the volume of a quantity of gaseous ammonia by the amplification of pressure; used in refrigeration systems. { ə'mōn·yə kəm'pres·ər }

ammonia condenser |MECH ENG| A device in an ammonia refrigerating system that raises the pressure of the ammonia gas in the evaporating coil, conditions the ammonia, and delivers it to the condensing system. { ə'mōn·yə kən'dens·ər }

ammonia liquor |CHEM ENG| Water solution of ammonia, ammonium compounds, and impurities, obtained from destructive distillation of bituminous coal. { ə'mōn·yə 'lik·ər }

ammonia meter |ENG| A hydrometer designed specifically to determine the density of aqueous ammonia solutions. { ə'mō·nyə ,mēd·ər }

ammonia synthesis |CHEM ENG| Chemical combination of nitrogen and hydrogen gases at high temperature and pressure in the presence of a catalyst to form ammonia. { ə'mōn·yə 'sin·thə·səs }

ammonia valve |ENG| A valve that is resistant to corrosion by ammonia. { ə'mōn·yə ,valv }

ammonoxidation See ammoxidation. { ,a·mən,äk·sə'dā·shən }

ammoxidation |CHEM ENG| A process in which mixtures of propylene, ammonia, and oxygen are converted in the presence of a catalyst, with acrylonitrile as the primary product. Also known as ammonoxidation; oxyamination. { ,am,äk·sə'dā·shən }

amortize |IND ENG| To reduce gradually an obligation, such as a mortgage, by periodically paying a part of the principal as well as the interest. { 'am·ər,tīz }

amount limit |IND ENG| In a test for a fixed quantity of work, the time required to complete the work or the total amount of work that can be completed in an unlimited time. { ə'maúnt ,lim·ət }

amp See amperage; ampere. { amp }

ampacity |ELEC| Current-carrying capacity in amperes; used as a rating for power cables. { am'pas·əd·ē }

amperage |ELEC| The amount of electric current in amperes. Abbreviated amp. { 'am·prij }

ampere |ELEC| The unit of electric current in the rationalized meter-kilogram-second system of units; defined in terms of the force of attraction between two parallel current-carrying conductors. Abbreviated a; A; amp. { 'am,pir }

ampere-hour meter |ENG| A device that measures the total electric charge that passes a given point during a given period of time. { 'am,pir ¦aú·ər ,mēd·ər }

amperometric transducer |ENG| A transducer in which the concentration of a dissolved substance is determined from the electric current produced between two electrodes immersed in the test solution when one of the electrodes is kept at a selected electric potential with respect to the solution. { am,pir·ə¦me·trik tranz'dü·sər }

amphibious |MECH ENG| Said of vehicles or equipment designed to be operated or used on either land or water. { am'fib·ē·əs }

amplification factor |ELECTR| In a vacuum tube, the ratio of the incremental change in plate voltage to a given small change in grid voltage, under the conditions that the plate current and all other electrode voltages are held constant. { ,am·plə·fə'kā·shən ,fak·tər }

amplification noise |ELECTR| Noise generated in the vacuum tubes, transistors, or integrated circuits of an amplifier. { ,am·plə·fə'kā·shən ,nóiz }

amplifier |ENG| A device capable of increasing the magnitude or power level of a physical quantity, such as an electric current or a hydraulic mechanical force, that is varying with time, without distorting the wave shape of the quantity. { 'am·plə,fī·ər }

amplifier-type meter |ENG| An electric meter whose characteristics have been enhanced by the use of preamplification for the signal input eventually used to actuate the meter. { 'am·plə,fī·ər ¦tīp 'mēd·ər }

amplify |ENG ACOUS| To strengthen a signal by increasing its amplitude or by raising its level. { 'am·plə,fī }

amplitude-frequency response See frequency response. { 'am·plə,tüd 'frē·kwən·sē ri'späns }

amplitude-modulated indicator |ENG| A general class of radar indicators, in which the sweep of the electron beam is deflected vertically or horizontally from a base line to indicate the existence of an echo from a target. Also known as deflection-modulated indicator; intensity-modulated indicator. { 'am·plə,tüd ¦mäj·ə,lād·əd ¦in·də,kād·ər }

amplitude modulation |ELECTR| Abbreviated AM. **1.** Modulation in which the aplitude of a wave is the characteristic varied in accordance with the intelligence to be transmitted. **2.** In telemetry, those systems of modulation in which each component frequency f of the transmitted intelligence produces a pair of sideband frequencies at carrier frequency plus f and carrier minus f. { 'am·plə,tüd ,maj·ə'lā·shən }

amylograph |ENG| An instrument used to measure and record the viscosity of starch and flour pastes and the temperature at which they gelatinize. { ə'mīl·ə,graf }

analemma [CIV ENG| Any raised construction which serves as a support or rest. { ,an·ə'lem·ə }

analog |ELECTR| **1.** A physical variable which remains similar to another variable insofar as the proportional relationships are the same over some specified range; for example, a temperature may be represented by a voltage which is its analog. **2.** Pertaining to devices, data, circuits, or systems that operate with variables which are represented by continuously measured voltages or other quantities. { 'an·əl,äg }

analog output |CONT SYS| Transducer output in which the amplitude is continuously proportional to a function of the stimulus. { 'an·əl,äg 'aút,pút }

analog readout |ENG| A scale on a balance that continuously indicates measurement values by the position of an index mark, either a line or a pointer, opposite a graduated scale which is usually marked with numbers. { 'an·əl,äg 'rēd,aút }

analog signal |ELECTR| A nominally continuous electrical signal that varies in amplitude or frequency in response to changes in sound, light, heat, position, or pressure. { 'an·əl,äg 'sig·nəl }

analog switch |ELECTR| **1.** A device that either transmits an analog signal without distortion or completely blocks it. **2.** Any solid-state device, with or without a driver, capable of bilaterally switching voltages or current. { 'an·əl,äg ,swich }

analog-to-digital converter |ELECTR| A device which translates continuous analog signals into proportional discrete digital signals. { ¦an·əl,äg tə ¦dij·ət·əl kən'vərd·ər }

analog-to-frequency converter |ELECTR| A converter in which an analog input in some form other than frequency is converted to a proportional change in frequency. { ¦an·əl,äg tə ¦frē·kwən·sē kən'vərd·ər }

analog voltage |ELECTR| A voltage that varies in a continuous fashion in accordance with the magnitude of a measured variable. { 'an·əl,äg 'vōl·tij }

analytical aerotriangulation |ENG| Analytical phototriangulation, performed with aerial photographs. { ,an·əl'id·ə·kəl ¦er·ō,trī,aŋ·gyə'lā·shən }

analytical balance |ENG| A balance with a sensitivity of 0.1–0.01 milligram. { ,an·əl'id·ə·kəl 'bal·əns }

analytical centrifugation |ENG| Centrifugation following precipitation to separate solids from solid-liquid suspensions; faster than filtration. { ,an·əl'id·ə·kəl sen,trif·ə'gā·shən }

21

analytical nadir-point triangulation |ENG| Radial triangulation performed by computational routines in which nadir points are utilized as radial centers. (‚an·əl'id·ə·kəl ¦nə‚dir ¦pȯint ‚trī‚aŋ·gyə'lā·shən)

analytical orientation |ENG| The computational steps required to determine tilt, direction of principal line, flight height, angular elements, and linear elements in preparing aerial photographs for rectification. (‚an·əl'id·ə·kəl ‚ȯr·ē·ən'tā·shən)

analytical photogrammetry |ENG| A method of photogrammetry in which solutions are obtained by mathematical methods. (‚an·əl'id·ə·kəl ‚fōd·ə'gram·ə·trē)

analytical photography |ENG| Photography, either motion picture or still, accomplished to determine (by qualitative, quantitative, or any other means) whether a particular phenomenon does or does not occur. (‚an·əl'id·ə·kəl fə'täg·rə·fē)

analytical phototriangulation |ENG| A phototriangulation procedure in which the spatial solution is obtained by computational routines. (‚an·əl'id·ə·kəl ‚fōd·ō‚trī‚aŋ·gyə'lā·shən)

analytical radar prediction |ENG| Prediction based on proven formulas, power tables, or graphs; considers surface height, structural and terrain information, and criteria for radar reflectivity together with the aspect angle and range to the target. (‚an·əl'id·ə·kəl 'rā‚där prə'dik·shən)

analytical radial triangulation |ENG| Radial triangulation performed by computational routines. (‚an·əl'id·ə·kəl 'rād·ē·əl ‚tri‚aŋ·gyə'lā·shən)

analytical three-point resection radial triangulation |ENG| A method of computing the coordinates of the ground principal points of overlapping aerial photographs by resecting on three horizontal control points appearing in the overlap area. (‚an·əl'id·ə·kəl ¦thrē ¦pȯint rē'sek·shən 'rād·ē·əl ‚tri‚aŋ·gyə'lā·shən)

analytical ultracentrifuge |ENG| An ultracentrifuge that uses one of three optical systems (schlieren, Rayleigh, or absorption) for the accurate determination of sedimentation velocity or equilibrium. (‚an·əl'id·ə·kəl ¦əl·trə¦sen·trə‚fyüj)

analytic mechanics |MECH| The application of differential and integral calculus to classical (nonquantum) mechanics. (‚an·əl'id·ik mi'kan·iks)

analyzer |ENG| A multifunction test meter, measuring volts, ohms, and amperes. Also known as set analyzer. |MECH ENG| The component of an absorption refrigeration system where the mixture of water vapor and ammonia vapor leaving the generator meets the relatively cool solution of ammonia in water entering the generator and loses some of its vapor content. ('an·ə‚līz·ər)

anchor |CIV ENG| A device connecting a structure to a heavy masonry or concrete object to a metal plate or to the ground to hold the structure in place. |ENG| A device, such as a metal rod, wire,or strap, for fixing one object to another, such as specially formed metal connectors used to fasten together timbers, masonry, or trusses. |MECH ENG| 1. In steam plowing, a vehicle located on the side of the field opposite that of the engine and maintaining the tension on the endless wire by means of a pulley. 2. A device for a piping system that maintains the correct position and direction of the pipes and controls pipe movement occurring as a result of thermal expansion. ('aŋ·kər)

anchorage |CIV ENG| 1. An area where a vessel anchors or may anchor because of either suitability or designation. Also known as anchor station. 2. A device which anchors tendons to the posttensioned concrete member. 3. In pretensioning, a device used to anchor tendons temporarily during the hardening of the concrete. 4. See deadman. ('aŋ·kə·rij)

anchorage deformation |CIV ENG| The shortening of tendons due to their modification or slippage when the prestressing force is transferred to the anchorage device. Also known as anchorage slip. ('aŋ·kə·rij dē‚fȯr'mā·shən)

anchorage slip See anchorage deformation. ('aŋ·kə·rij ‚slip)

anchorage zone |CIV ENG| 1. In posttensioning, the region adjacent to the anchorage for the tendon which is subjected to secondary stresses as a result of the distribution of the prestressing force. 2. In pretensioning, the region in which transfer bond stresses are developed. ('aŋ·kə·rij ‚zōn)

anchor and collar |DES ENG| A door or gate hinge whose socket is attached to an anchor embedded in the masonry. ('aŋ·kər ən 'käl·ər)

anchor block |BUILD| A block of wood, replacing a brick in a wall to provide a nailing or fastening surface. |CIV ENG| See deadman. ('aŋ·kər ‚bläk)

anchor bolt |CIV ENG| A bolt used with its head embedded in masonry or concrete and its threaded part protruding to hold a structure or machinery in place. Also known as anchor rod. ('aŋ·kər ‚bōlt)

anchor buoy |ENG| One of a series of buoys marking the limits of an anchorage. ('aŋ·kər ‚bȯi)

anchor charge |ENG| A procedure that allows several charges to be preloaded in a seismic shot hole; the bottom charges are fired first, and the upper charges are held down by anchors. ('aŋ·kər ‚chärj)

anchored bulkhead |CIV ENG| A bulkhead secured to anchor piles. ('aŋ·kərd 'bəlk‚hed)

anchor log |CIV ENG| A log, beam, or concrete block buried in the earth and used to hold a guy rope firmly. Also known as deadman; ground anchor. ('aŋ·kər ‚läg)

anchor nut |DES ENG| A nut in the form of a tapped insert forced under steady pressure into a hole in sheet metal. ('aŋ·kər ‚nət)

anchor pile |CIV ENG| A pile that is located on the land side of a bulkhead or pier and anchors it

22

through such devices as rods, cables, and chains. { 'aŋ·kər ,pīl }

anchor plate |CIV ENG| A metal or wooden plate fastened to or embedded in a support, such as a floor, and used to hold a supporting cable firmly. { 'aŋ·kər ,plāt }

anchor rod *See* anchor bolt. { 'aŋ·kər ,räd }

anchor station *See* anchorage. { 'aŋ·kər ,stā·shən }

anchor tower |CIV ENG| **1.** A tower which is a part of a crane staging or stiffleg derrick and serves as an anchor. **2.** A tower that supports and anchors an overhead transmission line. { 'aŋ·kər ,taů·ər }

anchor wall *See* deadman. { 'aŋ·kər ,wȯl }

AND circuit *See* AND gate. { 'and ,sər·kət }

AND gate |ELECTR| A circuit which has two or more input-signal ports and which delivers an output only if and when every input signal port is simultaneously energized. Also known as AND circuit; passive AND gate. { 'and ,gāt }

AND/NOR gate |ELECTR| A single logic element whose operation is equivalent to that of two AND gates with outputs feeding into a NOR gate. { ¦and ¦nȯr ,gāt }

AND NOT gate |ELECTR| A coincidence circuit that performs the logic operation AND NOT, under which a result is true only if statement A is true and statement B is not. Also known as A AND NOT B gate. { ¦and ¦nät ,gāt }

AND-OR circuit |ELECTR| Gating circuit that produces a prescribed output condition when several possible combined input signals are applied; exhibits the characteristics of the AND gate and the OR gate. { ¦and ¦ȯr ,sər·kət }

AND-OR-INVERT gate |ELECTR| A logic circuit with four inputs, a_1, a_2, b_1, and b_2, whose output is 0 only if either a_1 and a_2 or b_1 and b_2 are 1. Abbreviated A-O-I gate. { ¦and ¦ȯr in'vərt ,gāt }

Andrade's creep law |MECH| A law which states that creep exhibits a transient state in which strain is proportional to the cube root of time and then a steady state in which strain is proportional to time. { 'an,dräd·z 'krēp ,lȯ }

Andrews's curves |THERMO| A series of isotherms for carbon dioxide, showing the dependence of pressure on volume at various temperatures. { 'an,drüz ,kərvz }

anechoic chamber |ENG| **1.** A test room in which all surfaces are lined with a sound-absorbing material to reduce reflections of sound to a minimum. Also known as dead room; free-field room. **2.** A room completely lined with a material that absorbs radio waves at a particular frequency or over a range of frequencies; used principally at microwave frequencies, such as for measuring radar beam cross sections. { ¦an·ə¦kō·ik 'chām·bər }

anelasticity |MECH| Deviation from a proportional relationship between stress and strain. { ¦an·ə·las¦tis·əd·ē }

anemoblagraph |ENG| A recording pressure-tube anemometer in which the wind scale of the float manometer is linear through the use of

springs; an example is the Dines anemometer. { ¦a·nə·mə'bī·ə,graf }

anemoclinometer |ENG| A type of instrument which measures the inclination of the wind to the horizontal plane. { ¦a·nə,mō·klə¦näm·əd·ər }

anemogram |ENG| A record made by an anemograph. { ə'nēm·ə,gram }

anemograph |ENG| **1.** An instrument which records wind velocities. **2.** A recording anemometer. { ə'nēm·ə,graf }

anemometer |ENG| A device which measures air speed. { ,an·ə'mäm·əd·ər }

anemoscope |ENG| An instrument for indicating the direction of the wind. { ə'nēm·ə,skōp }

anemovane |ENG| A combined contact anemometer and wind vane used in the Canadian Meteorological Service. { ¦an·ə·mō¦vān }

aneroid |ENG| **1.** Containing no liquid or using no liquid. **2.** *See* aneroid barometer. { 'an·ə ,rȯid }

aneroid altimeter |ENG| An altimeter containing an aneroid barometer that actuates the indicator. { 'an·ə,rȯid al'tim·əd·ər }

aneroid barograph |ENG| An aneroid barometer arranged so that the deflection of the aneroid capsule actuates a pen which graphs a record on a rotating drum. Also known as aneroidograph; barograph; barometrograph. { 'an·ə,rȯid 'bar·ə,graf }

aneroid barometer |ENG| A barometer which utilizes an aneroid capsule. Also known as aneroid. { 'an·ə,rȯid bə'räm·əd·ər }

aneroid calorimeter |ENG| A calorimeter that uses a metal of high thermal conductivity as a heat reservoir. { 'an·ə,rȯid ,kal·ə'rim·əd·ər }

aneroid capsule |ENG| A thin, disk-shaped box or capsule, usually metallic, partially evacuated and sealed, held extended by a spring, which expands and contracts with changes in atmospheric or gas pressure. Also known as bellows. { 'an·ə,rȯid 'kap·səl }

aneroid diaphragm |ENG| A thin plate, usually metal, covering the end of an aneroid capsule and moving axially as the ambient gas pressure increases or decreases. { 'an·ə,rȯid 'dī·ə,fram }

aneroid flowmeter |ENG| A mechanism to measure fluid flow rate by pressure of the fluid against a bellows counterbalanced by a calibrated spring. { 'an·ə,rȯid 'flō,mēd·ər }

aneroid liquid-level meter |ENG| A mechanism to measure fluid depth by pressure of the fluid against a bellows which in turn acts on a manometer or signal transmitter. { 'an·ə,rȯid ¦lik·wəd ¦lev·əl ,med·ər }

aneroidograph *See* aneroid barograph. { 'an·ə,rȯid·ə·graf }

aneroid valve |MECH ENG| A valve actuated or controlled by an aneroid capsule. { 'an·ə,rȯid 'valv }

angel echo |ENG| A radar echo from a region where there are no visible targets; may be caused by insects, birds, or refractive index variations in the atmosphere. { 'ān·jəl ,ek·ō }

angle back-pressure valve |MECH ENG| A back-pressure valve with its outlet opening at right angles to its inlet opening. { 'aŋ·gəl 'bak ,presh·ər ,valv }

angle bar |BUILD| An upright bar at the meeting of two faces of a polygonal window, bay window, or bow window. { 'aŋ·gəl ,bär }

angle bead |BUILD| A strip, usually of metal or wood, set at the corner of a plaster wall to protect the corner or serve as a guide to float the plaster flush with it. { 'aŋ·gəl ,bēd }

angle beam |ENG| Ultrasonic waves transmitted for the inspection of a metallic surface at an angle measured from the beam center line to a normal to the test surface. { 'aŋ·gəl ,bēm }

angle blasting |ENG| Sandblasting, or the like, at an angle of less that 90° { 'aŋ·gəl ,blast·iŋ }

angle block |ENG| A small block of wood used to fasten adjacent pieces, usually at right angles, or glued into the corner of a wooden frame to stiffen it. Also known as glue block. { 'aŋ· gəl ,bläk }

angle board |DES ENG| A board whose surface is cut at a desired angle; serves as a guide for cutting or planing other boards at the same angle. { 'aŋ·gəl ,bȯrd }

angle bond |CIV ENG| A tie used to bond masonry work at wall corners. { 'aŋ·gəl ,bänd }

angle brace |ENG| A brace across the interior angle of two members that meet at an angle. Also known as angle tie. { 'aŋ·gəl ,brās }

angle brick |ENG| Any brick having an oblique shape to fit an oblique, salient corner. { 'aŋ· gəl ,brik }

angle clip |CIV ENG| A short strip of angle iron used to secure structural elements at right angles. { 'aŋ·gəl ,klip }

angle closer |ENG| A specially shaped brick used to close the bond at the corner of a wall. { 'aŋ·gəl ,klōz·ər }

angle collar |DES ENG| A cast-iron pipe fitting which has a socket at each end for joining with the spigot ends of two pipes that are not in alignment. { 'aŋ·gəl ,käl·ər }

angle-control section See crossover. { 'aŋ·gəl kən¦trōl ,sek·shən }

angle divider |DES ENG| A square for setting or bisecting angles; one side is an adjustable hinged blade. { 'aŋ·gəl də'vīd·ər }

angle dozer |MECH ENG| A power-operated machine fitted with a blade, adjustable in height and angle, for pushing, sidecasting, and spreading loose excavated material as for opencast pits, clearing land, or leveling runways. Also known as angling dozer. { 'aŋ·gəl,dōz·ər }

angle equation |ENG| A condition equation which expresses the relationship between the sum of the measured angles of a closed figure and the theoretical value of that sum, the unknowns being the corrections to the observed directions or angles, depending on which are used in the adjustment. Also known as triangle equation. { 'aŋ·gəl i,kwā·zhən }

angle fillet |ENG| A wooden strip, triangular in cross section, which is used to cover the internal joint between two surfaces meeting at an angle of less than 180° { 'aŋ·gəl ,fil·ət }

angle fishplates |CIV ENG| Plates which join the rails and prevent the rail joint from sagging where heavy cars and locomotives are used. Also known as angle; angle bar. { 'aŋ·gəl 'fish,plāts }

angle float |ENG| A trowel having two edge surfaces bent at 90°; used to finish corners in freshly poured concrete and in plastering. { 'aŋ·gəl ,flōt }

angle gauge |CIV ENG| A template used to set or check angles in building construction. { 'aŋ· gəl ,gāj }

angle gear See angular gear. { 'aŋ·gəl ,gēr }

angle globe valve |ENG| A globe valve having an angular configuration that permits it to be fitted at bends in pipework. { ¦aŋ·gəl ¦glōb ¦valv }

angle hip tile See arris hip tile. { 'aŋ·gəl 'hip ,tīl }

angle iron |CIV ENG| **1.** An L-shaped cleat or brace. **2.** A length of steel having a cross section resembling the letter L. { 'aŋ·gəl ,T·ərn }

angle joint |ENG| A joint between two pieces of lumber which results in a change in direction. { 'aŋ·gəl ,jȯint }

angle lacing |CIV ENG| A system of lacing in which angle irons are used in place of bars. { 'aŋ·gəl ,lās·iŋ }

angle method of adjustment |ENG| A method of adjustment of observations which determines corrections to observed angles. { 'aŋ·gəl ,meth·əd əv ə'jəs·mənt }

angle of action |MECH ENG| The angle of revolution of either of two wheels in gear during which any particular tooth remains in contact. { 'aŋ·gəl əv 'ak·shən }

angle of advance See angular advance. { 'aŋ·gəl əv əd'vans }

angle of approach |CIV ENG| The maximum angle of an incline onto which a vehicle can move from a horizontal plane without interference. |MECH ENG| The angle that is turned through by either of paired wheels in gear from the first contact between a pair of teeth until the pitch points of these teeth fall together. { 'aŋ·gəl əv ə'prōch }

angle of bite See angle of nip. { 'aŋ·gəl əv 'bīt }

angle of departure |CIV ENG| The maximum angle of an incline from which a vehicle can move onto a horizontal plane without interference, such as from rear bumpers. |ELECTR| See angle of radiation. { 'aŋ·gəl əv di'pär·chər }

angle of depression |ENG| The angle in a vertical plane between the horizontal and a descending line. Also known as depression angle; descending vertical angle; minus angle. { 'aŋ·gəl əv di'presh·ən }

angle of elevation |ENG| The angle in a vertical plane between the local horizontal and an ascending line, as from an observer to an object; used in astronomy, surveying, and so on. Also known as ascending vertical angle; elevation angle. { 'aŋ·gəl əv ,el·ə'vā·shən }

angle of external friction | ENG | The angle between the abscissa and the tangent of the curve representing the relationship of shearing resistance to normal stress acting between soil and the surface of another material. Also known as angle of wall friction. { 'aŋ·gəl əv ek'stərn·əl 'frik·shən }

angle of fall | MECH | The vertical angle at the level point, between the line of fall and the base of the trajectory. { 'aŋ·gəl əv 'fȯl }

angle of impact | MECH | The acute angle between the tangent to the trajectory at the point of impact of a projectile and the plane tangent to the surface of the ground or target at the point of impact. { 'aŋ·gəl əv 'im,pakt }

angle of nip | MECH ENG | The largest angle that will just grip a lump between the jaws, rolls, or mantle and ring of a crusher. Also known as angle of bite; nip. { 'aŋ·gəl əv 'nip }

angle of obliquity See angle of pressure. { 'aŋ·gəl əv ō'blik·wəd·ē }

angle of orientation | MECH | Of a projectile in flight, the angle between the plane determined by the axis of the projectile and the tangent to the trajectory (direction of motion), and the vertical plane including the tangent to the trajectory. { 'aŋ·gəl əv ,ȯr·ē·ən'tā·shən }

angle of pressure | DES ENG | The angle between the profile of a gear tooth and a radial line at its pitch point. Also known as angle of obliquity. { 'aŋ·gəl əv 'presh·ər }

angle of recess | MECH ENG | The angle that is turned through by either of two wheels in gear, from the coincidence of the pitch points of a pair of teeth until the last point of contact of the teeth. { 'aŋ·gəl əv 'rē,ses }

angle of repose | ENG | See angle of rest. | MECH | The angle between the horizontal and the plane of contact between two bodies when the upper body is just about to slide over the lower. Also known as angle of friction. { 'aŋ·gəl əv ri'pōz }

angle of rest | ENG | The maximum slope at which a heap of any loose or fragmented solid material will stand without sliding, or will come to rest when poured or dumped in a pile or on a slope. Also known as angle of repose. { 'aŋ·gəl əv 'rest }

angle of thread | DES ENG | The angle occurring between the sides of a screw thread, measured in an axial plane. { 'aŋ·gəl əv 'thred }

angle of torsion | MECH | The angle through which a part of an object such as a shaft or wire is rotated from its normal position when a torque is applied. Also known as angle of twist. { 'aŋ·gəl əv 'tȯr·shən }

angle of twist See angle of torsion. { 'aŋ·gəl əv 'twist }

angle of wall friction See angle of external friction. { 'aŋ·gəl əv ¦wȯl ,frik·shən }

angle of wrap | DES ENG | On a band brake mechanism, the distance, expressed in degrees, that the brake band wraps around the brake flange. { 'aŋ·gəl əv 'rap }

angle paddle | ENG | A hand tool used to finish a plastered surface. { 'aŋ·gəl ,pad·əl }

angle plate | DES ENG | An L-shaped plate or a plate having an angular section. { 'aŋ·gəl ,plāt }

angle post | BUILD | A railing support used at a landing or other break in the stairs. { 'aŋ·gəl ,pōst }

angle press | MECH ENG | A hydraulic plastics-molding press with both horizontal and vertical rams; used to produce complex moldings with deep undercuts. { 'aŋ·gəl ,press }

angle rafter | BUILD | A rafter, such as a hip rafter, at the angle of the roof. { 'aŋ·gəl ,raf·tər }

angle section | CIV ENG | A structural steel member having an L-shaped cross section. { 'aŋ·gəl ,sek·shən }

angle-stem thermometer | ENG | A device used to measure temperatures in oil-custody tanks; the angle of the calibrated stem may be 90° or greater to the sensitive portion of the thermometer, as needed to fit the tank shell contour. { 'aŋ·gəl ¦stem thər'mäm·əd·ər }

angle stile | BUILD | A narrow strip of wood used to conceal the joint between a wall and a vertical wood surface which makes an angle with the wall, as at the edge of a corner cabinet. { 'aŋ·gəl ,stīl }

angle structure | CIV ENG | A method of building a tower for mechanical strength in which braces are placed at angles with respect to the vertical support rods. { 'aŋ·gəl ,strək·chər }

angle strut | CIV ENG | An angle-shaped structural member which is designed to carry a compression load. { 'aŋ·gəl ,strət }

angle valve | DES ENG | A manually operated valve with its outlet opening oriented at right angles to its inlet opening; used for regulating the flow of a fluid in a pipe. { 'aŋ·gəl ,valv }

angle variable | MECH | The dynamical variable w conjugate to the action variable J, defined only for periodic motion. { 'aŋ·gəl 'ver·ē·ə·bal }

angling dozer See angle dozer. { 'aŋ·gliŋ ,dōz·ər }

angstrom | MECH | A unit of length, 10^{-10} meter, used primarily to express wavelengths of optical spectra. Abbreviated A; Å. Also known as tenthmeter. { 'aŋ·strəm }

Ångström compensation pyrheliometer | ENG | A pyrheliometer consisting of two identical Manganin strips, one shaded, the other exposed to sunlight; an electrical current is passed through the shaded strip to raise its temperature to that of the exposed strip, and the electric power required to accomplish this is a measure of the solar radiation. { 'ȯŋ·strəm käm·pən'sā·shən ¦pīr,hē,lē'äm·əd·ər }

angular acceleration | MECH | The time rate of change of angular velocity. { 'aŋ·gyə·lər ak,sel·ə'rā·shən }

angular accelerometer | ENG | An accelerometer that measures the rate of change of angular velocity between two objects under observation. { 'aŋ·gyə·lər ak,sel·ə'räm·əd·ər }

angular advance | MECH ENG | The amount by which the angle between the crank of a steam engine and the virtual crank radius of the eccentric exceeds a right angle. Also known as angle of advance; angular lead. { 'aŋ·gyə·lər əd'vans }

angular bitstalk See angular bitstock. { 'aŋ·gyə·lər 'bit,stòk }

angular bitstock |MECH ENG| A bitstock whose handles are positioned to permit its use in corners and other cramped areas. Also known as angular bitstalk. { 'aŋ·gyə·lər 'bit,stäk }

angular clearance |DES ENG| The relieved space located below the straight of a die, to permit passage of blanks or slugs. { 'aŋ·gyə·lər 'klir·əns }

angular-contact bearing |MECH ENG| A rolling-contact antifriction bearing designed to carry heavy thrust loads and also radial loads. { 'aŋ·gyə·lər 'kän,takt ,ber·iŋ }

angular cutter |MECH ENG| A tool-steel cutter used for finishing surfaces at angles greater or less than 90° with its axis of rotation. { 'an·gyə·lər 'kəd·ər }

angular error of closure See error of closure. { 'an·gyə·lər 'er·ər əv 'klōzh·ər }

angular gear |MECH ENG| A gear that transmits motion between two rotating shafts that are not parallel. Also known as angle gear. { 'an·gyə·lər 'gēr }

angular impulse |MECH| The integral of the torque applied to a body over time. { 'an·gyə·lər 'im,pəls }

angular lead See angular advance. { 'aŋ·gyə·lər 'lēd }

angular length |MECH| A length expressed in the unit of the length per radian or degree of a specified wave. { 'aŋ·gyə·lər 'leŋkth }

angular milling |MECH ENG| Milling surfaces that are flat and at an angle to the axis of the spindle of the milling machine. { 'aŋ·gyə·lər 'mil·iŋ }

angular momentum |MECH| **1.** The cross product of a vector from a specified reference point to a particle, with the particle's linear momentum. Also known as moment of momentum. **2.** For a system of particles, the vector sum of the angular momenta (first definition) of the particles. { 'aŋ·gyə·lər mə'ment·əm }

angular pitch |DES ENG| The angle determined by the length along the pitch circle of a gear between successive teeth. { 'aŋ·gyə·lər 'pich }

angular rate See angular speed. { 'aŋ·gyə·lər ,rāt }

angular shear |MECH ENG| A shear effected by two cutting edges inclined to each other to reduce the force needed for shearing. { 'aŋ·gyə·lər 'shēr }

angular speed |MECH| Change of direction per unit time, as of a target on a radar screen, without regard to the direction of the rotation axis; in other words, the magnitude of the angular velocity vector. Also known as angular rate. { 'aŋ·gyə·lər 'spēd }

angular travel error |MECH| The error which is introduced into a predicted angle obtained by multiplying an instantaneous angular velocity by a time of flight. { 'aŋ·gyə·lər 'trav·əl ,er·ər }

angular velocity |MECH| The time rate of

change of angular displacement. { 'aŋ·gyə·lər və'läs·əd·ē }

angulator |ENG| An instrument for converting angles measured on an oblique plane to their corresponding projections on a horizontal plane; the rectoblique plotter and the photoangulator are types. { 'aŋ·gyə,lād·ər }

aniline point |CHEM ENG| The minimum temperature for a complete mixing of aniline and materials such as gasoline; used in some specifications to indicate the aromatic content of oils and to calculate approximate heat of combustion. { 'an·əl·ən ,pòint }

animal balance |ENG| A balance designed to weigh living animals, with a readout or display relatively unaffected by the pulse or movements of the animal. { 'an·ə·məl ,bal·əns }

animal power |MECH ENG| The time rate at which muscular work is done by a work animal, such as a horse, bullock, or elephant. { 'an·ə·məl ,paú·ər }

anisotropic membrane |CHEM ENG| An ultrafiltration membrane which has a thin skin at the separating surface and is supported by a spongy sublayer of membrane material. { ¦a,nī·sə¦träp·ik 'mem,brān }

anker |MECH| A unit of capacity equal to 10 U.S. gallons (37.854 liters); used to measure liquids, especially honey, oil, vinegar, spirits, and wine. { 'aŋ·kər }

anneal |ENG| To treat a metal, alloy, or glass with heat and then cool to remove internal stresses and to make the material less brittle. Also known as temper. { ə'nēl }

annealing furnace |ENG| A furnace for annealing metals or glass. Also known as annealing oven. { ə'nēl·iŋ ,fər·nəs }

annealing oven See annealing furnace. { ə'nēl·iŋ ,əv·ən }

annealing point |THERMO| The temperature at which the viscosity of a glass is $10^{13.0}$ poises. Also known as annealing temperature; 13.0 temperature. { ə'nēl·iŋ ,pòint }

annealing temperature See annealing point. { ə'nēl·iŋ ,tem·prə·chər }

annual cost comparison |IND ENG| A method of selecting from among several alternative projects or courses of action on the basis of their annual costs, including depreciation. { 'an·yə·wəl 'kòst kəm,par·ə·sən }

annular auger |DES ENG| A ring-shaped boring tool which cuts an annular channel, leaving the core intact. { 'an·yə·lər 'òg·ər }

annular gear |DES ENG| A gear having a cylindrical form. { 'an·yə·lər 'gir }

annular nozzle |DES ENG| A nozzle with a ring-shaped orifice. { 'an·yə·lər 'näz·əl }

annular section |ENG| The open space between two concentric tubes, pipes, or vessels. { 'an·yə·lər 'sek·shən }

annunciator |ENG| A signaling apparatus which operates electromagnetically and serves to indicate visually, or visually and audibly, whether a current is flowing, has flowed, or has changed

direction of flow in one or more circuits.
{ ə'nən·sē·ăd·ər }
anode |ELEC| The terminal at which current enters a primary cell or storage battery; it is positive with respect to the device, and negative with respect to the external circuit. |ELECTR| **1.** The collector of electrons in an electron tube. Also known as plate; positive electrode. **2.** In a semiconductor diode, the terminal toward which forward current flows from the external circuit. { 'a‚nōd }
anode current |ELECTR| The electron current flowing through an electron tube from the cathode to the anode. Also known as plate current. { 'a‚nōd ‚kər·ənt }
anomalous expansion |THERMO| An increase in the volume of a substance that results from a decrease in its temperature, such as is displayed by water at temperatures between 0 and 4°C (32 and 39°F). { ə'näm·ə·ləs ik'span·shən }
anomaly finder |ENG| A computer-controlled data-plotting system used on ships to measure and record seismic, gravity, magnetic, and other geophysical data and water depth, time, course, and speed. { ə'näm·ə·lē ‚fīn·dər }
anonymous dimensionless group 1–4 |CHEM ENG| Four of the dimensionless groups, used to solve problems in transfer processes, gas absorption in wetted-wall columns, and laminar boundary-layer flow. { ə'nän·ə·məs di¦men·shən·ləs 'grüp ‚wən tə ¦fôr }
antenna circuit |ELECTR| A complete electric circuit which includes an antenna. { an'ten·ə ‚sər·kət }
antenna tilt error |ENG| Angular difference between the tilt angle of a radar antenna shown on a mechanical indicator, and the electrical center of the radar beam. { an'ten·ə 'tilt ‚er·ər }
anticathode |ELECTR| The anode or target of an x-ray tube, on which the stream of electrons from the cathode is focused and from which x-rays are emitted. { ¦an·tē'kath‚ōd }
antichlor |CHEM ENG| A chemical used in the manufacture of paper or textiles to remove excess chlorine or bleaching solution. { ¦an·ti'klòr }
anticollision radar |ENG| A radar set designed to give warning of possible collisions during movements of ships or aircraft. { ‚an·tē·kə'li·zhən ‚rä‚där }
anticreeper |CIV ENG| A device attached to a railroad rail to prevent it from moving in the direction of its length. { 'an·tē‚krēp·ər }
antidieseling solenoid See idle-stop solenoid. { ‚ant·i¦dēz·əl·iŋ 'sō·lə‚nòid }
antifriction |MECH| Making friction smaller in magnitude. |MECH ENG| Employing a rolling contact instead of a sliding contact. { ‚an·tē 'frik·shən }
antifriction bearing |MECH ENG| Any bearing having the capability of reducing friction effectively. { ‚an·tē'frik·shən ‚ber·iŋ }
antifriction material |ENG| A machine element made of Babbitt metal, lignum vitae, rubber, or a combination of a soft, easily deformable metal

overlaid on a hard, resistant one. { ‚an·tē'frik·shən mə'tir·ē·əl }
anti-g suit See g suit. { ¦an·tē¦jē ‚süt }
antiknock blending value |ENG| The numerical improvement by an antiknock additive to gasoline octane, often a greater amount than the additive's own octane value. { 'an·tē‚näk 'blend·iŋ ‚val·yü }
antiknock rating |ENG| Measurement of the ability of an automotive gasoline to resist detonation or pinging in spark-ignited engines. { 'an·tē‚näk 'rād·iŋ }
antilock braking system |MECH ENG| For vehicles, a sensor-control system found in braking systems which prevents wheel lockup while allowing the brakes to continue slowing the wheel. Abbreviated ABS. { ¦an·tē‚läk 'brāk·iŋ ‚sis·təm }
antimagnetic |ENG| Constructed so as to avoid the influence of magnetic fields, usually by the use of nonmagnetic materials and by magnetic shielding. { ‚an·tē‚mag'ned·ik }
antinoise microphone |ENG ACOUS| Microphone with characteristics which discriminate against acoustic noise. { ¦an·tē¦nòiz 'mi·krə ‚fōn }
antiozonant |CHEM ENG| A protective agent which can be added to rubber during processing to diminish the deteriorating effects of ozone. { ‚an·tē'ō·zə·nənt }
antipercolator |MECH ENG| In an automotive engine, a valve in the carburetor that is designed to vent vapor when the throttle plate is closed; prevents fuel from dropping into the carburetor due to unvented pressure. { ‚an·tē'pər·kə ‚lād·ər }
antiquing |ENG| **1.** Producing a rich glow on the surface of a leather by applying stain, wax, or oil, allowing it to set, and rubbing or brushing the leather. **2.** A technique of handling wet paint to expose parts of the undercoat, by combing, graining, or marbling. Also known as broken-color work. { an'tēk·iŋ }
antirad |CHEM ENG| An inhibitor incorporated into rubber during manufacturing to reduce the degrading effects of radiation. { ¦an·te¦rad }
antiradar coating |ENG| A surface treatment used to reduce the reflection of electromagnetic waves so as to avoid radar detection. { ‚an·tē'rā ‚där ¦kōd·iŋ }
antirattle spring |MECH ENG| In an automotive vehicle, a spring installed to hold parts in the clutches and the disk brakes together; prevents rattling. { 'an·tē'rad·əl 'spriŋ }
anti-redeposition agent |CHEM ENG| An additive used in a detergent to help prevent soil from resettling on a fabric after it has been removed during washing. { ¦an·tē‚rē‚dep·ə'zish·ən ‚ā· jənt }
antireflection coating |ENG| The application of a thin film of dielectric material to a surface to reduce its reflection and to increase its transmission of light or other electromagnetic radiation. { ‚an·tē·ri'flek·shən ‚kōd·iŋ }
antiresonance |ELEC| See parallel resonance.

|ENG| The condition for which the impedance of a given electric, acoustic, or dynamic system is very high, approaching infinity. { ‚an·tē'rez·ən·əns }

antiskid plate |ENG| A sheet of metal roughed on both sides and placed between piled objects, such as boxes in a freight car, to prevent sliding. { ¦an·tē¦skid ¦plāt }

antismudge ring |BUILD| A frame attached around a ceiling-mounted air diffuser, to minimize the formation of rings of dirt on the ceiling. { ¦an·tē¦sməj 'riŋ }

antitheft device |MECH ENG| A piece of equipment installed on an automotive vehicle in order to prevent or slow down theft; designs include mechanical locks on the steering wheel and ignition switch as well as other means of shutting off the ignition system, shutting off fuel flow, or sounding an alarm. { ‚an·tē'theft di‚vīs }

anvil |ENG| **1.** The part of a machine that absorbs the energy delivered by a sharp force or blow. **2.** The stationary end of a micrometer caliper. { 'an·vəl }

AOQL See average outgoing quality limit.

aperiodic waves |ELEC| The transient current wave in a series circuit with resistance R, inductance L, and capacitance C when $R^2C = 4L$. { ¦a‚pir·ē¦äd·ik 'wāvz }

aperture |ELECTR| An opening through which electrons, light, radio waves, or other radiation can pass. { 'ap·ə‚chər }

aperture disk |ENG| A disk with a small round opening used in a densitometer to vary the amount of light or the area to be measured. { 'ap·ə‚chər ‚disk }

apex |ENG| In architecture or construction, the highest point, peak, or tip of any structure. { 'ā‚peks }

apical angle |MECH| The angle between the tangents to the curve outlining the contour of a projectile at its tip. { 'ap·i·kəl 'aŋ·gəl }

API scale |CHEM ENG| The American Petroleum Institute hydrometer scale for the measurement of the specific gravity of liquids; used primarily in the American petroleum industry. { ¦ā¦pē¦ī ‚skāl }

apophorometer |ENG| An apparatus used to identify minerals by sublimation. { ‚ap·ə·fə'räm·əd·ər }

apothecaries' dram See dram. { ə'päth·ə‚ker·ēz 'dram }

apothecaries' ounce See ounce. { ə'päth·ə‚ker·ēz 'aùns }

apothecaries' pound See pound. { ə'päth·ə‚ker·ēz 'paùnd }

apparent expansion |THERMO| The expansion of a liquid with temperature, as measured in a graduated container without taking into account the container's expansion. { ə'pa·rənt ik'span·shən }

apparent force |MECH| A force introduced in a relative coordinate system in order that Newton's laws be satisfied in the system; examples are the Coriolis force and the centrifugal force incorporated in gravity. { ə'pa·rənt 'fòrs }

apparent gravity See acceleration of gravity. { ə'pa·rənt 'grav·əd·ē }

apparent motion See relative motion. { ə'pa·rənt 'mō·shən }

apparent source See effective center. { ə'pa·rənt 'sòrs }

apparent weight |MECH| For a body immersed in a fluid (such as air), the resultant of the gravitational force and the buoyant force of the fluid acting on the body; equal in magnitude to the true weight minus the weight of the displaced fluid. { ə'pa·rənt 'wāt }

appliance |ENG| A piece of equipment that draws electric or other energy and produces a desired work-saving or other result, such as an electric heater, a radio, or an electronic range. { ə'plī·əns }

appliance panel |ENG| In electric systems, a metal housing containing two or more devices (such as fuses) for protection against excessive current in circuits which supply portable electric appliances. { ə'plī·əns ‚pan·əl }

applied research |ENG| Research directed toward using knowledge gained by basic research to make things or to create situations that will serve a practical or utilitarian purpose. { ə'plīd ri‚sərch }

applied strategic research |ENG| Research done to provide a basic understanding of a current applied project. { ə'plīd strə'tē·jik ri 'sərch }

applied trim |BUILD| Supplementary and separate decorative strips of wood or moldings applied to the face or sides of a frame, such as a doorframe. { ə'plīd 'trim }

approach |MECH ENG| The difference between the temperature of the water leaving a cooling tower and the wet-bulb temperature of the surrounding air. { ə'prōch }

approach signal |CIV ENG| A railway signal warning an engineer of a signal ahead that displays a restrictive indication. { ə'prōch ‚sig·nəl }

approach vector |CONT SYS| A vector that describes the orientation of a robot gripper and points in the direction from which the gripper approaches a workpiece. { ə'prōch ‚vek·tər }

apron |BUILD| **1.** A board on an interior wall beneath a windowsill. **2.** The vertical rear panel of a sink attached to a wall. **3.** A section of a concrete slab extending beyond the face of a building on adjacent ground. Also known as skirt; skirting. **4.** A vertical panel installed behind a sink or lavatory. |CIV ENG| **1.** A hard-surfaced area, usually paved, adjacent to a ship or the like, used to park, load, unload, or service vehicles. **2.** A covering of a material such as concrete or timber over soil to prevent erosion by flowing water, as at the bottom of a dam. **3.** A concrete or wooden shield that is situated along the bank of a river, along a sea wall, or below a dam. **4.** In a railroad system, a bridge structure that carries tracks and is hinged to land for connecting the deck of a railroad-car ferry to the shore. |MECH ENG| A plate serving to protect or cover a machine. { 'ā·prən }

apron conveyor |MECH ENG| A conveyor used for carrying granular or lumpy material and consisting of two strands of roller chain separated by overlapping plates, forming the carrying surface, with sides 2–6 inches (5–15 centimeters) high. { 'ā·prən kən‚vā·ər }

apron feeder |MECH ENG| A limited-length version of apron conveyor used for controlled-rate feeding of pulverized materials to a process or packaging unit. Also known as plate-belt feeder; plate feeder. { 'ā·prən ‚fēd·ər }

apron flashing |BUILD| 1. The flashing that covers the joint between a vertical surface and a sloping roof, as at the lower edge of a chimney. 2. The flashing that diverts water from a vertical surface into a gutter. { 'ā·prən ‚flash·iŋ }

apron lining |BUILD| The piece of boarding which covers the rough apron piece of a staircase. { 'ā·prən ‚līn·iŋ }

apron piece |BUILD| A beam that supports a landing or a series of winders in a staircase. { 'ā·prən ‚pēs }

apron rail |BUILD| A lock rail having a raised ornamental molding. { 'ā·prən ‚rāl }

apron wall |BUILD| In an exterior wall, a panel which extends downward from a windowsill to the top of a window below. { 'ā·prən ‚wȯl }

AQL See acceptable quality level.

aqualung |ENG| A self-contained underwater breathing apparatus (scuba) of the demand or open-circuit type developed by J.Y. Cousteau. { 'ak·wə‚ləŋ }

aqueduct |CIV ENG| An artificial tube or channel for conveying water. { 'ak·wə‚dəkt }

arbitration |IND ENG| A semijudicial means of settling labor-management disputes in which both sides agree to be bound by the decision of one or more neutral persons selected by some method mutually agreed upon. { ‚är·bə'trā·shən }

arbor |MECH ENG| 1. A cylindrical device positioned between the spindle and outer bearing of a milling machine and designed to hold a milling cutter. 2. A shaft or spindle used to hold a revolving cutting tool or the work to be cut. { 'är·bər }

arbor collar |ENG| A cylindrical spacer that positions and secures a revolving cutter on an arbor. { 'är·bər ‚käl·ər }

arbor hole |DES ENG| A hole in a revolving cutter or grinding wheel for mounting it on an arbor. { 'är·bər ‚hōl }

arbor press |MECH ENG| A machine used for forcing an arbor or a mandrel into drilled or bored parts preparatory to turning or grinding. Also known as mandrel press. { 'är·bər ‚pres }

arbor support |ENG| A device to support the outer end or intermediate point of an arbor. { 'är·bər sə‚pȯrt }

arc See electric arc.The graduated scale of an instrument for measuring angles, as a marine sextant;

readings obtained on that part of the arc beginning at zero and extending in the direction usually considered positive are popularly said to be on the arc, and those beginning at zero and extending in the opposite direction are said to be off the arc. { ärk }

arc force |MECH| The force of a plasma arc through a nozzle or opening. { 'ärk ‚fȯrs }

arch |CIV ENG| A structure curved and so designed that when it is subjected to vertical loads, its two end supports exert reaction forces with inwardly directed horizontal components; common uses for the arch are as a bridge, support for a roadway or railroad track, or part of a building. { ärch }

arch band |CIV ENG| Any narrow elongated surface forming part of or connected with an arch. { 'ärch ‚band }

arch bar |BUILD| 1. A curved chimney bar. 2. A curved bar in a window sash. { 'ärch ‚bar }

arch beam |CIV ENG| A curved beam, used in construction, with a longitudinal section bounded by two arcs having different radii and centers of curvature so that the beam cross section is larger at either end than at the center. { 'ärch ‚bēm }

arch brace |BUILD| A curved brace, usually used in pairs to support a roof frame and give the effect of an arch. { 'ärch ‚brās }

arch bridge |CIV ENG| A bridge having arches as the main supports. { 'ärch ‚brij }

arch center |CIV ENG| A temporary structure for support of the parts of a masonry or concrete arch during its construction. { 'ärch ‚sen·tər }

arch corner bead |BUILD| A corner bead which is cut on the job; used to form and reinforce the curved portion of arch openings. { 'ärch ‚kȯr·nər ‚bēd }

arch dam |CIV ENG| A dam having a curved face on the downstream side, the curve being roughly a portion of a cylinder whose axis is vertical. { 'ärch ‚dam }

arched construction |BUILD| A method of construction relying on arches and vaults to support walls and floors. { ‚ärcht kən'strək·shən }

arch girder |CIV ENG| A normal H-section steel girder bent to a circular shape. { 'ärch ‚gər·dər }

arch-gravity dam |CIV ENG| An arch dam stabilized by gravity due to great mass and breadth of the base. { 'ärch ‚grav·əd·ē ‚dam }

Archimedes' screw |MECH ENG| A device for raising water by means of a rotating broad-threaded screw or spirally bent tube within an inclined hollow cylinder. { ‚är·kə‚mēd‚ēz 'skrü }

arching |CIV ENG| 1. The transfer of stress from a yielding part of a soil mass to adjoining less-yielding or restrained parts of the mass. 2. A system of arches. 3. The arched part of a structure. { 'ärch·iŋ }

architectural acoustics |CIV ENG| The science of planning and building a structure to ensure the most advantageous flow of sound to all listeners. { ‚är·kə‚tek·chər·əl ə'kü·stiks }

architectural engineering |CIV ENG| The branch of engineering dealing primarily with building materials and components and with the design of structural systems for buildings, in contrast to heavy construction such as bridges. { ¦är·kə¦tek·chər·əl ,en·jə'nir·iŋ }

architectural millwork |CIV ENG| Ready-made millwork especially fabricated to meet the specifications for a particular job, as distinguished from standard or stock items or sizes. Also known as custom millwork. { ¦är·kə¦tek·chər·əl 'mil,wərk }

architectural volume |CIV ENG| The cubic content of a building calculated by multiplying the floor area by the height. { ¦är·kə¦tek·chər·əl 'väl·yəm }

architecture |ENG| 1. The art and science of designing buildings. 2. The product of this art and science. { 'är·kə,tek·chər }

arch press |MECH ENG| A punch press having an arch-shaped frame to permit operations on wide work. { 'ärch ,pres }

arch rib |CIV ENG| One of a set of projecting molded members subdividing the undersurface of an arch. { 'ärch ,rib }

arch ring |CIV ENG| A curved member that provides the main support of an arched structure. { 'ärch ,riŋ }

arch truss |CIV ENG| A truss having the form of an arch or arches. { 'ärch ,trəs }

arc of action See arc of contact. { ¦ärk əv 'ak·shən }

arc of approach |DES ENG| In toothed gearing, the part of the arc of contact along which the flank of the driving wheel contacts the face of the driven wheel. { ¦ärk əv ə'prōch }

arc of contact |MECH ENG| 1. The angular distance over which a gear tooth travels while it is in contact with its mating tooth. Also known as arc of action. 2. The angular distance a pulley travels while in contact with a belt or rope. { ¦ärk əv 'kän,takt }

arc of recess |DES ENG| In toothed gearing, the part of the arc of contact wherein the face of the driving wheel touches the flank of the driven wheel. { ¦ärk əv 'rē,ses }

arcometer |ENG| A device for determining the density of a liquid by measuring the apparent weight loss of a solid of known mass and volume when it is immersed in the liquid. { är'käm·əd·ər }

arc process |CHEM ENG| A former process that used electric arcs for fixation (oxidation) of atmospheric nitrogen to manufacture nitric acid. { ¦ärk ¦präs·əs }

arcticization |ENG| The preparation of equipment for operation in an environment of extremely low temperatures. { ¦ärd·ik,ī'zā·shən }

arc triangulation |ENG| A system of triangulation in which an arc of a great circle on the surface of the earth is followed in order to tie in two distant points. { ¦ärk ,tri,aŋ·gyə'lā·shən }

are |MECH| A unit of area, used mainly in agriculture, equal to 100 square meters. { är }

area coverage |ENG| Complete coverage of an area by aerial photography having parallel overlapping flight lines and stereoscopic overlap between exposures in the line of flight. { 'er·ē·ə 'kəv·rij }

area drain |CIV ENG| A receptacle designed to collect surface or rain water from an open area. { 'er·ē·ə ¦drān }

area landfill |CIV ENG| A sanitary landfill operation that takes care of the solid waste of more than one municipality in a region. { 'er·ē·ə 'land,fil }

area light |CIV ENG| 1. A source of light with significant dimensions in two directions, such as a window or luminous ceiling. 2. A light used to illuminate large areas. { 'er·ē·ə ,līt }

area meter |ENG| A mechanism to measure fluid flow rate through a fixed-area conduit by the movement of a weighted piston or float supported by the flowing fluid; includes rotameters and piston-type meters. { 'er·ē·ə ,mēd·ər }

area of use |ENG| For a balance depending on gravitational acceleration, an area that includes a sufficient number of locations providing a mean value for the gravitational acceleration of the given balance. { 'er·ē·ə əv 'yüs }

area survey |ENG| A survey of areas large enough to require loops of control. { 'er·ē·ə ¦sər,vā }

area triangulation |ENG| A system of triangulation designed to progress in every direction from a control point. { 'er·ē·ə ,tri,aŋ·gyə'lā·shən }

area wall |CIV ENG| A retaining wall around an areaway. { 'er·ē·ə ,wȯl }

areaway |CIV ENG| An open space at subsurface level adjacent to a building, providing access to and utilities for a basement. { 'er·ē·ə,wā }

Argand lamp |ENG| A gas lamp having a tube-shaped wick, allowing a current of air inside as well as outside the flame. { 'är,gän 'lamp }

argentometer |ENG| A hydrometer used to find the amount of silver salt in a solution. { ,är·jən'täm·əd·ər }

Arkansas stone |ENG| A whetstone made of Arkansas stone, for sharpening edged tools. { 'är·kən,sȯ ,stōn }

ARL See acceptable reliability level.

arm |CONT SYS| A robot component consiting of an interconnected set of links and powered joints that move and support the wrist socket and end effector. See branch. |ELEC| |ENG ACOUS| See tone arm. { ärm }

arm conveyor |MECH ENG| A conveyor in the form of an endless belt or chain to which are attached projecting arms or shelves which carry the materials. { ¦ärm kən'vā·ər }

arm elevator |MECH ENG| A chain elevator with protruding arms to cradle fixed-shape objects, such as drums or barrels, as they are moved upward { ¦ärm ,el·ə'vād·ər }

armored faceplate |DES ENG| A tamper-proof faceplate or lock front, mortised in the edge of a door to cover the lock mechanism. { 'är·mərd 'fās,plāt }

armored front |DES ENG| A lock front used on mortise locks that consists of two plates, the underplate and the finish plate. { 'är·mərd 'frənt }

armor plate |BUILD| A metal plate which protects the lower part of a door from kicks and scratches, covering the door to a height usually 39 inches (1 meter) or more. { 'är·mər ¦plāt }

arm solution |CONT SYS| The computation performed by a robot controller to calculate the joint positions required to achieve desired tool positions. { 'ärm sə,lü·shən }

arm-tool aggregate |IND ENG| A biomechanical unit comprising the arm and the tool that it holds and manipulates. { ¦ärm ¦tül 'ag·rə·gət }

aromatization |CHEM ENG| Conversion of any nonaromatic hydrocarbon structure to aromatic hydrocarbon, particularly petroleum. { ə,rō·məd·ə'zā·shən }

arostat process |CHEM ENG| A process in which aromatic molecules are saturated by catalytic hydrogenation to produce high-quality jet fuels, low-aromatic-content solvents, and high-purity cyclohexane from benzene. { 'ar·ə,stat ,präs·əs }

array |ELECTR| A group of components such as antennas, reflectors, or directors arranged to provide a desired variation of radiation transmission or reception with direction. { ə'rā }

array radar |ENG| A radar incorporating a multiplicity of phased antenna elements. { ə'rā 'rā,där }

array sonar |ENG| A sonar system incorporating a phased array of radiating and receiving transducers. { ə'rā 'sō,när }

arrester |ELEC| See lightning arrester. |ENG| A wire screen at the top of an incinerator or chimney which prevents sparks or burning material from leaving the stack. { ə'res·tər }

arrestment device |ENG| A locking mechanism installed on a balance for holding one of several levers in place; serves to protect the balance. { ə'rest·mənt di,vīs }

arrière-voussure |BUILD| 1. An arch or vault in a thick wall carrying the thickness of the wall, especially one over a door or window frame. 2. A relieving arch behind the face of a wall. { 'ar·ē,er,vü'sür }

arris fillet |BUILD| A triangular wooden piece that raises the slates of a roof against a chimney or wall so that rain runs off. { 'ar·əs ¦fil·ət }

arris gutter |BUILD| A V-shaped wooden gutter fixed to the eaves of a building. { 'ar·əs ¦gəd·ər }

arris hip tile |BUILD| A special roof tile having an L-shaped cross section, made to fit over the hip of a roof. Also known as hip tile. { 'ar·əs 'hip ,tīl }

arris rail |CIV ENG| A rail of triangular section, usually formed by slitting diagonally a strip of square section. { 'ar·əs ,rāl }

arrissing tool |ENG| A tool similar to a float, but having a form suitable for rounding an edge of freshly placed concrete. { 'ar·əs·iŋ ,tül }

arris tile |BUILD| Any angularly shaped tile. { 'ar·əs ,tīl }

arrisways |CIV ENG| Diagonally, in respect to

the manner of laying tiles, slates, bricks, or timber. Also known as arriswise. { 'ar·əs,wāz }

arriswise See arrisways. { 'ar·əs,wīz }

arrival rate |IND ENG| The mean number of new calling units arriving at a service facility per unit time. { ə'rī·vəl ,rāt }

articulated drop chute |ENG| A drop chute, for a falling stream of concrete, which consists of a vertical succession of tapered metal cylinders, so designed that the lower end of each cylinder fits into the upper end of the one below. { är 'tik·yə,lād·əd 'dräp ,shüt }

articulated leader |MECH ENG| A wheel-mounted transport unit with a pivotal loading element used in earth moving. { är'tik·yə,lād·əd 'lēd·ər }

articulated structure |CIV ENG| A structure in which relative motion is allowed to occur between parts, usually by means of a hinged or sliding joint or joints. { är'tik·yə,lād·əd 'strək·chər }

articulated train |ENG| A railroad train whose cars are permanently or semipermanently connected. { är'tik·yə,lād·əd 'trān }

articulation |CONT SYS| The manner and actions of joining components of a robot with connecting parts or links that allow motion. { är ,tik·yə'lā·shən }

articulation point See cut point. { är,tik·yə'lā·shən ,póint }

artificial atmosphere |CHEM ENG| A mixture of gases used in industrial operations in place of air; classified as an active, or process, atmosphere, or an inactive, or protective, atmosphere. { ¦ärd·ə¦fish·əl 'at·mə,sfir }

artificial ear |ENG ACOUS| A device designed to duplicate the frequency response, acoustic impedance, threshold sensitivity, and relative perception of loudness, consisting of a special microphone enclosed in a box with properties similar to those of the human ear. { ¦ärd·ə¦fish·əl 'ir }

artificial ground |ELEC| A common correction for a radio-frequency electrical or electronic circuit that is not directly connected to the earth. { ¦ärd·ə¦fish·əl 'graund }

artificial harbor |CIV ENG| 1. A harbor protected by breakwaters. 2. A harbor formed by dredging. { ¦ärd·ə¦fish·əl 'här·bər }

artificial monument |ENG| A relatively permanent object made by humans, such as an abutment or stone marker, used to identify the location of a survey station or corner. { ¦ärd·ə¦fish·əl 'män·yə·mənt }

artificial nourishment |CIV ENG| The process of replenishing a beach by artificial means, such as the deposition of dredged material. { ¦ärd·ə¦fish·əl 'nər·ish·mənt }

artificial recharge |CIV ENG| The recharge of an aquifier depleted by abnormally large withdrawals, by the use of injection wells and other techniques. { ¦ärd·ə¦fish·əl 'rē,chärj }

artificial variable |IND ENG| One type of variable introduced in a linear program model in order to find an initial basic feasible solution; an artificial variable is used for equality constraints and for greater-than or equal inequality constraints.   { ¦ärd·ə¦fish·əl 'ver·ē·ə·bəl }

artificial voice |ENG ACOUS| **1.** Small loudspeaker mounted in a shaped baffle which is proportioned to simulate the acoustical constants of the human head; used for calibrating and testing close-talking microphones.   **2.** Synthetic speech produced by a multiple tone generator; used to produce a voice reply in some real-time computer applications.   { ¦ärd·ə¦fish·əl 'vȯis }

artificial weathering |ENG| The controlled production of changes in materials under laboratory conditions to simulate actual outdoor exposure.   { ¦ärd·ə¦fish·əl 'weth·ə·riŋ }

asbestos-cement cladding |BUILD| Asbestos board and component wall systems, directly supported by wall framing, forming a wall or wall facing.   { as'bes·təs si¦ment 'klad·iŋ }

as-built drawing See as-fitted drawing.   { ¦az ¦bilt 'drȯ·iŋ }

as-built schedule |IND ENG| The final schedule for a project, reflecting the actual scope, actual completion dates, actual duration of the specified activities, and start dates.   { ¦az ¦bilt ¦skej·əl }

ascending branch |MECH| The portion of the trajectory between the origin and the summit on which a projectile climbs and its altitude constantly increases.   { ə'send·iŋ 'branch }

ascending vertical angle See angle of elevation.   { ə'send·iŋ ¦vərd·i·kəl 'aŋ·gəl }

as-fitted drawing |ENG| A drawing as amended after completion of an industrial facility in order to provide an accurate record of the details of the entire installation in their final form.   Also known as as-built drawing; as-made drawing.   { ¦az ¦fid·əd 'drȯ·iŋ }

ash |ENG| An undesirable constituent of diesel fuel whose quantitative measurement indicates degree of fuel cleanliness and freedom from abrasive material.   { ash }

ash collector See dust chamber.   { 'ash kə'lek·tər }

ash conveyor |MECH ENG| A device that transports refuse from a furnace by fluid or mechanical means.   { 'ash kən'vā·ər }

ash dump |ENG| An opening in the floor of a fireplace or firebox through which ashes are swept to an ash pit below.   { 'ash ,dəmp }

ash furnace |ENG| A furnace in which materials are fritted for glassmaking.   { 'ash 'fər·nəs }

ashlar |CIV ENG| Masonry with an exposed side of square or rectangular stones.   { 'ash·lər }

ashlar line |BUILD| The outer line of a wall above any projecting base.   { 'ash·lər ,līn }

ash pan |ENG| A metal receptacle beneath a fireplace or furnace grating for collection and removal of ashes.   { 'ash ,pan }

ash pit |BUILD| The ash-collecting area beneath a fireplace hearth.   { 'ash ,pit }

ash pit door |ENG| A cast-iron door providing access to an ash pit for ash removal.   { 'ash ,pit ,dȯr }

A size |ENG| **1.** One of a series of sizes to which trimmed paper and board are manufactured; for size AN, with N equal to any integer from 0 to 10, the length of the longer side is $2^{-(2N-1)/4}$ meters, while the length of the shorter side is $2^{-(2N+1)/4}$ meters, with both lengths rounded off to the nearest millimeter.   **2.** Of a sheet of paper, the dimensions 8.5 inches by 11 inches (216 millimeters by 279 millimeters).   { 'ā ,sīz }

as-made drawing See as-fitted drawing.   { ¦az ¦mād 'drȯ·iŋ }

aspect |CIV ENG| Of railway signals, what the engineer sees when viewing the blades or lights in their relative positions or colors.   { 'a,spekt }

aspect angle |ENG| The angle formed between the longitudinal axis of a projectile in flight and the axis of a radar beam.   { 'a,spekt ,aŋ·gəl }

aspect ratio |DES ENG| **1.** The ratio of frame width to frame height in television; it is 4:3 in the United States and Britain.   **2.** In any rectangular configuration (such as the cross section of a rectangular duct), the ratio of the longer dimension to the shorter. |MECH ENG| In an automotive vehicle, the ratio of the height of a tire to its width.   Also known as tire profile.   { 'a,spekt ,rā·shō }

asphalt cutter |MECH ENG| A powered machine having a rotating abrasive blade; used to saw through bituminous surfacing material.   { 'a ,sfȯlt ,kəd·ər }

asphalt heater |ENG| A piece of equipment for raising the temperature of bitumen used in paving.   { 'a,sfȯlt 'hēd·ər }

asphalt leveling course |CIV ENG| A layer of an asphalt-aggregate mixture of variable thickness, used to eliminate irregularities in contour of an existing surface, prior to the placement of a superimposed layer.   { 'a,sfȯlt 'lev·əl·iŋ ,kȯrs }

asphalt overlay |CIV ENG| One or more layers of asphalt construction on an existing pavement.   { 'a,sfȯlt 'ov·ər,lā }

asphalt pavement |CIV ENG| A pavement consisting of a surface layer of mineral aggregate, coated and cemented together with asphalt cement on supporting layers.   { 'a,sfȯlt 'pāv·mənt }

asphalt soil stabilization |CIV ENG| The treatment of naturally occurring nonplastic or moderately plastic soil with liquid asphalt at normal temperatures to improve the load-bearing qualities of the soil.   { 'a,sfȯlt ¦sȯil ,stāb·ə·lə'zā·shən }

aspirating burner |ENG| A burner in which combustion air at high velocity is drawn over an orifice, creating a negative static pressure and thereby sucking fuel into the stream of air; the mixture of air and fuel is conducted into a combustion chamber, where the fuel is burned in suspension.   { 'as·pə,rād·iŋ 'bər·nər }

aspiration meteorograph |ENG| An instrument for the continuous recording of two or more meteorological parameters, with the ventilation being provided by a suction fan.   { ,as·pə'rā·shən ,mēd·ē'ȯr·ə,graf }

aspiration psychrometer |ENG| A psychrometer in which the ventilation is provided by a suction fan. { ,as·pə'rā·shən ,si'kräm·əd·ər }

aspiration thermograph |ENG| A thermograph in which ventilation is provided by a suction fan. { ,as·pə'rā·shən 'thərm·ə,graf }

aspirator |ENG| Any instrument or apparatus that utilizes a vacuum to draw up gases or granular materials. { 'as·pə,rād·ər }

assay balance |ENG| A sensitive balance used in the assaying of gold, silver, and other precious metals. { 'a,sā ,bal·əns }

assembling bolt |CIV ENG| A threaded bolt for holding together temporarily the several parts of a structure during riveting. { ə'sem·bliŋ ,bōlt }

assembly |MECH ENG| A unit containing the component parts of a mechanism, machine, or similar device. { ə'sem·blē }

assembly line |IND ENG| A mass-production arrangement whereby the work in process is progressively transferred from one operation to the next until the product is assembled. { ə'sem·blē ,līn }

assembly-line balancing |IND ENG| Assigning numbers of operators or machines to each operation of an assembly line so as to meet the required production rate with a minimum of idle time. { ə'sem·blē ,līn 'bal·əns·iŋ }

assembly machine |MECH ENG| A machine in a manufacturing facility that produces a configuration of some practical value from discrete components. { ə'sem·blē mə,shēn }

assembly method |IND ENG| The technique used to assemble a manufactured product, such as hand assembly, progressive line assembly, and automatic assembly. { ə'sem·blē ,meth·əd }

assembly time |ENG| 1. The elapsed time after the application of an adhesive until its strength becomes effective. 2. The time elapsed in performing an assembly or subassembly operation. { ə'sem·blē ,tīm }

assets |IND ENG| All the resources, rights, and property owned by a person or a company; the book value of these items as shown on the balance sheet. { 'a,sets }

assignable cause |IND ENG| Any identifiable factor which causes variation in a process outside the predicted limits, thereby altering quality. { ə'sīn·ə·bəl 'kȯz }

assize |CIV ENG| 1. A cylindrical block of stone forming one unit in a column. 2. A layer of stonework. { ə'sīz }

Assmann psychrometer |ENG| A special form of the aspiration psychrometer in which the thermometric elements are well shielded from radiation. { 'äs,män ,sī'kräm·əd·ər }

assumed plane coordinates |ENG| A local plane-coordinate system set up at the convenience of the surveyor. { ə'sümd ¦plān ,kō'ȯrd·nəts }

astatic galvanometer |ENG| A sensitive galvanometer designed to be independent of the earth's magnetic field. { ā'stad·ik ,gal·və'näm·əd·ər }

astatic governor See isochronous governor. { ā 'stad·ik gəv·ə·nər }

astatic gravimeter |ENG| A sensitive gravimeter designed to measure small changes in gravity. { ā'stad·ik grə'vim·əd·ər }

astatic magnetometer |ENG| A magnetometer for determining the gradient of a magnetic field by measuring the difference in reading from two magnetometers placed at different positions. { ā'stad·ik ,mag·nə'täm·əd·ər }

astatic wattmeter |ENG| An electrodynamic wattmeter designed to be insensitive to uniform external magnetic fields. { ā'stad·ik 'wät,mēd·ər }

astatized gravimeter |ENG| A gravimeter, sometimes referred to as unstable, where the force of gravity is maintained in an unstable equilibrium with the restoring force. { 'as·tə,tīzd grə'vim·əd·ər }

astern |ENG| To the rear of an aircraft, vehicle, or vessel; behind; from the back. { ə'stərn }

astragal |BUILD| 1. A small convex molding decorated with a string of beads or bead-and-reel shapes. 2. A plain bead molding. 3. A member, or combination of members, fixed to one of a pair of doors or casement windows to cover the joint between the meeting stiles and to close the clearance gap. { 'as·trə·gəl }

astragal front |DES ENG| A lock front which is shaped to fit the edge of a door with an astragal molding. { 'as·trə·gəl ¦frənt }

astral lamp |ENG| An Argand lamp designed so that its light is not prevented from reaching a table beneath it by the flattened annular reservoir holding the oil. { 'as·trəl ,lamp }

astroballistics |MECH| The study of phenomena arising out of the motion of a solid through a gas at speeds high enough to cause ablation; for example, the interaction of a meteoroid with the atmosphere. { ¦as·trō·bə'lis·tiks }

astrolabe |ENG| An instrument designed to observe the positions and measure the altitudes of celestial bodies. { 'as·trə,lāb }

astronomical instruments |ENG| Specific kinds of telescopes and ancillary equipment used by astronomers to study the positions, motions, and composition of stars and members of the solar system. { ,as·trə'näm·ə·kəl 'in·strə·məns }

astronomical theodolite See altazimuth. { ,as·trə'näm·ə·kəl thē'äd·əl,īt }

astronomical traverse |ENG| A survey traverse in which the geographic positions of the stations are obtained from astronomical observations, and lengths and azimuths of lines are obtained by computation. { ,as·trə'näm·ə·kəl trə'vərs }

asymmetric rotor |MECH ENG| A rotating element for which the axis (center of rotation) is not centered in the element. { ¦ā·sə¦me·trik 'rōd·ər }

asymmetric top |MECH| A system in which all three principal moments of inertia are different. { ¦ā·sə¦me·trik 'täp }

asynchronous control |CONT SYS| A method of control in which the time allotted for performing an operation depends on the time actually required for the operation, rather than on a predetermined fraction of a fixed machine cycle. { ā'siŋ·krə·nəs kən'trōl }

asynchronous device [CONT SYS] A device in which the speed of operation is not related to any frequency in the system to which it is connected. { ā'siŋ·krə·nəs di'vīs }

asynchronous operation [ELECTR] An operation that is started by a completion signal from a previous operation, proceeds at the maximum speed of the circuits until finished, and then generates its own completion signal. { ā'siŋ·krə·nəs ‚äp·ə'rā·shən }

asynchronous timing [IND ENG] A simulation method for queues in which the system model is updated at each arrival or departure, resulting in the master clock being increased by a variable amount. { ā'siŋ·krə·nəs 'tīm·iŋ }

at See technical atmosphere.

ata [MECH] A unit of absolute pressure in the metric technical system equal to 1 technical atmosphere. { 'a·tə }

athermalize [ENG] To make independent of temperature or of thermal effects. { ā'thər·mə‚līz }

atm See atmosphere.

atmidometer See atmometer. { ‚at·mə'däm·əd·ər }

atmometer [ENG] The general name for an instrument which measures the evaporation rate of water into the atmosphere. Also known as atmidometer; evaporation gage; evaporimeter. { ət'mäm·əd·ər }

atmosphere [MECH] A unit of pressure equal to 101.325 kilopascals, which is the air pressure measured at mean sea level. Abbreviated atm. Also known as standard atmosphere. { 'at·mə‚sfir }

atmospheric cooler [MECH ENG] A fluids cooler that utilizes the cooling effect of ambient air surrounding the hot, fluids-filled tubes. { ‚at·mə‚sfir·ik 'kül·ər }

atmospheric distillation [CHEM ENG] Distillation operation conducted at atmospheric pressure, in contrast to vacuum distillation or pressure distillation. { ‚at·mə‚sfir·ik ‚dis·tə'lā·shən }

atmospheric impurity [ENG] An extraneous substance that is mixed as a contaminant with the air of the atmosphere. { ‚at·mə‚sfir·ik im 'pyür·əd·ē }

atmospheric noise [ELECTR] Noise heard during radio reception due to atmospheric interference. { ‚at·mə‚sfir·ik 'nóiz }

atmospheric steam curing [ENG] The steam curing of concrete or cement products at atmospheric pressure, usually at a maximum ambient temperature between 100 and 200°F (40 and 95°C). { ‚at·mə‚sfir·ik 'stēm 'kyür·iŋ }

atomic force microscope [ENG] A device for mapping surface atomic structure by measuring the force acting on the tip of a sharply pointed wire or other object that is moved over the surface. { ə‚täm·ik 'fórs 'mī‚krə‚skōp }

atomic moisture meter [ENG] An instrument that measures the moisture content of coal instantaneously and continuously by bombarding it with neutrons and measuring the neutrons which bounce back to a detector tube after striking hydrogen atoms of water. { ə'täm·ik 'móis·chər ‚med·ər }

atomic power plant See nuclear power plant. { ə'täm·ik 'paú·ər ‚plant }

atomization [MECH ENG] The mechanical subdivision of a bulk liquid or meltable solid, such as certain metals, to produce drops, which vary in diameter depending on the process from under 10 to over 1000 micrometers. { ‚ad·ə·mə'zā·shən }

atomizer [MECH ENG] A device that produces a mechanical subdivision of a bulk liquid, as by spraying, sprinkling, misting, or nebulizing. { 'ad·ə‚mīz·ər }

atomizer burner [MECH ENG] A liquid-fuel burner that atomizes the unignited fuel into a fine spray as it enters the combustion zone. { 'ad·ə‚mīz·ər ‚bər·nər }

atomizer mill [MECH ENG] A solids grinder, the product from which is a fine powder. { 'ad·ə‚mīz·ər ‚mil }

atomizing humidifier [MECH ENG] A humidifier in which tiny particles of water are introduced into a stream of air. { 'ad·ə‚mīz·iŋ ‚hyü'mid·ə‚fī·ər }

atom probe [ENG] An instrument for identifying a single atom or molecule on a metal surface; it consists of a field ion microscope with a probe hole in its screen opening into a mass spectrometer; atoms that are removed from the specimen by pulsed field evaporation fly through the probe hole and are detected in the mass spectrometer. { 'ad·əm ‚prōb }

attached thermometer [ENG] A thermometer which is attached to an instrument to determine its operating temperature. { ə'tacht thər'mäm·əd·ər }

attemperation [ENG] The regulation of the temperature of a substance. { ə‚tem·pə'rā·shən }

attemperation of steam [MECH ENG] The controlled cooling, in a steam boiler, of steam at the superheater outlet or between the primary and secondary stages of the superheater to regulate the final steam temperature. { ə‚tem·pə'rā·shən əv 'stēm }

attenuate [ENG ACOUS] To weaken a signal by reducing its level. { ə'ten·yə‚wāt }

attenuation [ELEC] The exponential decrease with distance in the amplitude of an electrical signal traveling along a very long uniform transmission line, due to conductor and dielectric losses. [ENG] A process by which a material is fabricated into a thin, slender configuration, such as forming a fiber from molten glass. { ə‚ten·yə'wā·shən }

attic [BUILD] The part of a building immediately below the roof and entirely or partly within the roof framing. { 'ad·ik }

attic tank [BUILD] An open tank which is installed above the highest plumbing fixture in a building and which supplies water to the fixtures by gravity. { 'ad·ik ,taŋk }

atticurge |BUILD| Of a doorway, having jambs which are inclined slightly inward, so that the opening is wider at the threshold than at the top. { 'ad·ə,kərj }

attic ventilator |BUILD| A mechanical fan located in the attic space of a residence; usually moves large quantities of air at a relatively low velocity. { 'ad·ik 'vent·əl,ād·ər }

attraction gripper |CONT SYS| A robot component that uses adhesion, suction, or magnetic forces to grasp a workpiece. { ə'trak·shən ,grip·ər }

attribute sampling |IND ENG| A quality-control inspection method in which the sampled articles are classified only as defective or nondefective. { 'a·trə,byüt ,sam·pliŋ }

attributes testing |ENG| A reliability test procedure in which the items under test are classified according to qualitative characteristics. { 'a·trə,byüts ,test·iŋ }

attrition mill |MECH ENG| A machine in which materials are pulverized between two toothed metal disks rotating in opposite directions. { ə'trish·ən ,mil }

Atwood machine |MECH ENG| A device comprising a pulley over which is passed a stretch-free cord with a weight hanging on each end. { 'at,wúd mə'shēn }

audible leak detector |ENG| A device used as an auxiliary to the main leak detector for conversion of the output signal into audible sound. { ¦òd·ə·bəl 'lēk di,tek·tər }

audio-frequency meter |ENG| One of a number of types of frequency meters usable in the audio range; for example, a resonant-reed frequency meter. { 'òd·ē·ō ¦frē·kwən·sē ,mēd·ər }

audiometer |ENG| An instrument composed of an oscillator, amplifier, and attenuator and used to measure hearing acuity for pure tones, speech, and bone conduction. { ,òd·ē'äm·əd·ər }

audio-modulated radiosonde |ENG| A radiosonde with a carrier wave modulated by audio-frequency signals whose frequency is controlled by the sensing elements of the instrument. { ¦òd·ē·o¦mäj·ə·lād·əd ¦rād·ē·ō,sänd }

audio patch bay |ENG ACOUS| Specific patch panels provided to terminate all audio circuits and equipment used in a channel and technical control facility; this equipment can also be found in transmitting and receiving stations. { 'òd·ē·ō ¦pach ,bā }

audio spectrometer See acoustic spectrometer. { 'òd·ē·ō spek'träm·əd·ər }

audio system See sound-reproducing system. { 'òd·ē·ō ,sis·təm }

audio taper |ENG ACOUS| A special type of potentiometer used in a volume-control apparatus to compensate for the nonlinearity of human hearing and give the impression of a linear increase in audibility as volume is raised. Also known as linear taper. { 'òd·ē·ō ,tā·pər }

audiphone |ENG ACOUS| A device that enables persons with certain types of deafness to hear,

consisting of a plate or diaphragm that is placed against the teeth and transmits sound vibrations to the inner ear. { 'òd·ə,fōn }

auger |DES ENG| **1.** A wood-boring tool that consists of a shank with spiral channels ending in two spurs, a central tapered feed screw, and a pair of cutting lips. **2.** A large augerlike tool for boring into soil. { 'ò·gər }

auger bit |DES ENG| a A bit shaped like an auger but without a handle; used for wood boring and for earth drilling. { 'ò·gər ,bit }

auger boring |ENG| **1.** The hole drilled by the use of auger equipment. **2.** See auger drilling. { 'ò·gər ,bòr·iŋ }

auger conveyor See screw conveyor. { 'ò·gər kən'vā·ər }

auger drilling |ENG| A method of drilling in which penetration is accomplished by the cutting or gouging action of chisel-type cutting edges forced into the substance by rotation of the auger bit. Also known as auger boring. { 'ò·gər ,dril·iŋ }

auger packer |MECH ENG| A feed mechanism that uses a continuous auger or screw inside a cylindrical sleeve to feed hard-to-flow granulated solids into shipping containers, such as bags or drums. { 'ò·gər 'pak·ər }

auget |ENG| A priming tube, used in blasting. Also spelled augette. { ò'zhet }

augette See auget. { ò'zhet }

auralization See virtual acoustics. { ,òr·əl·ə'zā·shən }

autoadaptivity |CONT SYS| The ability of an advanced robot to sense the environment, accept commands, and analyze and execute operations. { ¦òd·ò·,ə,dap'tiv·əd·ē }

autoclave |ENG| An airtight vessel for heating and sometimes agitating its contents under high steam pressure; used for industrial processing, sterilizing, and cooking with moist or dry heat at high temperatures. { 'òd·ō,klāv }

autoclave curing |ENG| Steam curing of concrete products, sand-lime brick, asbestos cement products, hydrous calcium silicate insulation products, or cement in an autoclave at maximum ambient temperatures generally between 340 and 420°F (170 and 215°C). { 'òd·ō,klāv 'kyúr·iŋ }

autoclave molding |ENG| A method of curing reinforced plastics that uses an autoclave with 50–100 pounds per square inch (345–690 kilopascals) steam pressure to set the resin. { 'òd·ō,klāv ¦mōld·iŋ }

autocorrelation |ELECTR| A technique used to detect cyclic activity in a complex signal. { ¦òd·ō,kär·ə'lā·shən }

autofrettage |ENG| A process for manufacturing gun barrels; prestressing the metal increases the load at which its permanent deformation occurs. { 'òd·ō,fred·ij }

autogenous grinding |MECH ENG| The secondary grinding of material by tumbling the material in a revolving cylinder, without balls or bars taking part in the operation. { ò'täj·ə·nəs 'grīnd·iŋ }

autogenous healing |ENG| A natural process of closing and filling cracks in concrete or mortar while it is kept damp. { ȯ'täj·ə·nəs 'hēl·iŋ }

autogenous mill *See* autogenous tumbling mill. { ȯ'täj·ə·nəs 'mil }

autogenous tumbling mill |MECH ENG| A type of ball-mill grinder utilizing as the grinding medium the coarse feed (incoming) material. Also known as autogenous mill. { ȯ'täj·ə·nəs 'təm·bliŋ ‚mil }

autoignition |MECH ENG| Spontaneous ignition of some or all of the fuel-air mixture in the combustion chamber of an internal combustion engine. Also known as spontaneous combustion. { ‚ȯd·ō·ig'nish·ən }

automanual system |CIV ENG| A railroad signal system in which signals are set manually but are activated automatically to return to the danger position by a passing train. { ‚ȯd·ō¦man· yə·wəl 'sis·təm }

automated guided vehicle |IND ENG| In a flexible manufacturing system, a driverless computer-controlled vehicle equipped with guidance and collision-avoidance systems and used to transport workpieces and tools between work stations. Abbreviated AGV. { ‚ȯd·ə¦mād·əd ¦gīd·əd 've·ə·kəl }

automated guided vehicle system |CONT SYS| A computer-controlled system that uses pallets and other interface equipment to transport workpieces to numerically controlled machine tools and other equipment in a flexible manufacturing system, moving in a predetermined pattern to ensure automatic, accurate, and rapid work-machine contact. { 'ȯd·ə‚mād·əd ¦gīd·əd 've·ə·kəl ‚sis·təm }

automatic |ENG| Having a self-acting mechanism that performs a required act at a predetermined time or in response to certain conditions. { ‚ȯd·ə¦mad·ik }

automatic balance |ENG| A balance capable of performing weighing procedures without the intervention of an operator. { ‚ȯd·ə¦mad·ik 'ba·ləns }

automatic batcher |MECH ENG| A batcher for concrete which is actuated by a single starter switch, opens automatically at the start of the weighing operations of each material, and closes automatically when the designated weight of each material has been reached. { ‚ȯd·ə¦mad·ik 'bach·ər }

automatic calibration |ENG| A process in which an electronic device automatically performs the recalibration of a measuring range of a weighing instrument, for example an electronic balance. { ‚ȯd·ə¦mad·ik ‚kal·ə'brā·shən }

automatic check-out system |CONT SYS| A system utilizing test equipment capable of automatically and simultaneously providing actions and information which will ultimately result in the efficient operation of tested equipment while keeping time to a minimum. { ‚ȯd·ə¦mad·ik 'chek‚aut ‚sis·təm }

automatic choke |MECH ENG| A system for enriching the air-fuel mixture in a cold automotive engine when the accelerator is first depressed; the choke plate opens automatically when the engine achieves normal operating temperature. { ‚ȯd·ə¦mad·ik 'chōk }

automatic control |CONT SYS| Control in which regulating and switching operations are performed automatically in response to predetermined conditions. Also known as automatic regulation. { ‚ȯd·ə¦mad·ik kən‚trōl }

automatic control balance |ENG| An automatic balance fitted with an accessory which determines whether a package has been filled within preselected limits. Also known as checkweigher. { ‚ȯd·ə¦mad·ik kən‚trōl ‚bal·əns }

automatic-control block diagram |CONT SYS| A diagrammatic representation of the mathematical relationships defining the flow of information and energy through the automatic control system, in which the components of the control system are represented as functional blocks in series and parallel arrangements according to their position in the actual control system. { ‚ȯd·ə¦mad·ik kən'trōl 'bläk ‚dī·ə‚gram }

automatic-control error coefficient |CONT SYS| Three numerical quantities that are used as a measure of the steady-state errors of an automatic control system when the system is subjected to constant, ramp, or parabolic inputs. { ‚ȯd·ə¦mad·ik kən'trōl 'er·ər ‚kō·ə'fish·ənt }

automatic-control frequency response |CONT SYS| The steady-state output of an automatic control system for sinusoidal inputs of varying frequency. { ‚ȯd·ə¦mad·ik 'frē·kwən·sē ri ‚späns }

automatic controller |CONT SYS| An instrument that continuously measures the value of a variable quantity or condition and then automatically acts on the controlled equipment to correct any deviation from a desired preset value. Also known as automatic regulator; controller. { ‚ȯd·ə¦mad·ik kən‚trōl·ər }

automatic-control servo valve |CONT SYS| A mechanically or electrically actuated servo valve controlling the direction and volume of fluid flow in a hydraulic automatic control system. { ‚ȯd·ə¦mad·ik kən'trōl 'sər·vō ‚valv }

automatic-control stability |CONT SYS| The property of an automatic control system whose performance is such that the amplitude of transient oscillations decreases with time and the system reaches a steady state. { ‚ȯd·ə¦mad·ik kən'trōl stə‚bil·ə·dē }

automatic control system |CONT SYS| A control system having one or more automatic controllers connected in closed loops with one or more processes. Also known as regulating system. { ‚ȯd·ə¦mad·ik kən'trōl ‚sis·təm }

automatic-control transient analysis |CONT SYS| The analysis of the behavior of the output variable of an automatic control system as the system changes from one steady-state condition to another in terms of such quantities as maximum overshoot, rise time, and response time. { ‚ȯd·ə¦mad·ik kən'trōl 'tran·zhənt ə‚nal·ə·səs }

automatic coupling |MECH ENG| A device

which couples rail cars when they are bumped together. { ¦ȯd·ə¦mad·ik 'kəp·liŋ }

automatic data processing |ENG| The machine performance, with little or no human assistance, of any of a variety of tasks involving informational data; examples include automatic and responsive reading, computation, writing, speaking, directing artillery, and the running of an entire factory. Abbreviated ADP. { ¦ȯd·ə¦mad·ik ¦dad·ə 'präs,əs·iŋ }

automatic dialog replacement studio See ADR studio. { ¦ȯd·ə¦mad·ik ,dī·ə,läg ri'pläs·mənt ,stüd·ē,ō }

automatic door bottom |ENG| A movable plunger, in the form of a horizontal bar at the bottom of a door, which drops automatically when the door is closed, sealing the threshold and reducing noise transmission. Also known as automatic threshold closer. { ¦ȯd·ə¦mad·ik 'dȯr ,bäd·əm }

automatic drill |DES ENG| A straight brace for bits whose shank comprises a coarse-pitch screw sliding in a threaded tube with a handle at the end; the device is operated by pushing the handle. { ¦ȯd·ə¦mad·ik 'dril }

automatic fire pump |MECH ENG| A pump which provides the required water pressure in a fire standpipe or sprinkler system; when the water pressure in the system drops below a preselected value, a sensor causes the pump to start. { ¦ȯd·ə¦mad·ik ,fīr ,pəmp }

automatic flushing system |CIV ENG| A water tank system which provides automatically for the periodic flushing of urinals or other plumbing fixtures, or of pipes having too small a slope to drain effectively. { ¦ȯd·ə¦mad·ik 'fləsh·iŋ ,sis·təm }

automatic ignition |ENG| A device that lights the fuel in a gas burner when the gas-control valve is turned on. { ¦ȯd·ə¦mad·ik ig'nish·ən }

automatic indexing |CONT SYS| The procedure for determining the orientation and position of a workpiece with respect to an automatically controlled machine, such as a robot manipulator, that is to perform an operation on it. { ¦ȯd·ə¦mad·ik 'in,deks·iŋ }

automatic level control |ELECTR| A circuit that keeps the output of a radio transmitter, tape recorder, or other device essentially constant, even in the presence of large changes in the input amplitude. Abbreviated ALC. |MECH ENG| In an automotive vehicle, a system in which two air-chamber shock absorbers in the rear are fed compressed air by an electric compressor; pressure in the air chambers is determined automatically by sensors to maintain the vehicle at a predetermined height regardless of load. { ¦ȯd·ə¦mad·ik 'lev·əl kən,trōl }

automatic microfilmer |ENG| A device used to place microfilm in jackets at relatively high speeds. { ¦ȯd·ə¦mad·ik 'mi·krō,fil·ər }

automatic mold |ENG| A mold used in injection or compression molding of plastic objects so

that repeated molding cycles are possible, including ejection, without manual assistance. { ¦ȯd·ə¦mad·ik 'mōld }

automatic press |MECH ENG| A press in which mechanical feeding of the work is synchronized with the press action. { ¦ȯd·ə¦mad·ik 'pres }

automatic pumping station |CHEM ENG| An installation on a pipeline that automatically provides the proper pressure when a fluid is being transported. { ¦ȯd·ə¦mad·ik 'pəmp·iŋ ,stā·shən }

automatic ranging See autoranging. { ¦ȯd·ə¦mad·ik 'rānj·iŋ }

automatic record changer |ENG ACOUS| An electric phonograph that automatically plays a number of records one after another. { ¦ȯd·ə¦mad·ik 'rek·ərd ,chānj·ər }

automatic regulation See automatic control. { ¦ȯd·ə¦mad·ik ,reg·yə'lā·shən }

automatic regulator See automatic controller. { ¦ȯd·ə¦mad·ik 'reg·yə,lād·ər }

automatic sampler |MECH ENG| A mechanical device to sample process streams (gas, liquid, or solid) either continuously or at preset time intervals. { ¦ȯd·ə¦mad·ik 'sam·plər }

automatic screw machine |MECH ENG| A machine designed to automatically produce finished parts from bar stock at high production rates; the term is not an exact, specific machine-tool classification. { ¦ȯd·ə¦mad·ik 'skrü mə ,shēn }

automatic shut-off |ENG ACOUS| A switch in some tape recorders which automatically stops the machine when the tape ends or breaks. { ¦ȯd·ə¦mad·ik 'shəd,ȯf }

automatic slips |ENG| A pneumatic or hydraulic device for setting and removing slips automatically. Also known as power slips. { ¦ȯd·ə¦mad·ik 'slips }

automatic stoker |MECH ENG| A device that supplies fuel to a boiler furnace by mechanical means. Also known as mechanical stoker. { ¦ȯd·ə¦mad·ik 'stōk·ər }

automatic test equipment |ENG| Test equipment that makes two or more tests in sequence without manual intervention; it usually stops when the first out-of-tolerance value is detected. { ¦ȯd·ə¦mad·ik 'test i,kwip·mənt }

automatic threshold closer See automatic door bottom. { ¦ȯd·ə¦mad·ik 'thresh,hōld ,klōz·ər }

automatic time switch |ENG| Combination of a switch with an electric or spring-wound clock, arranged to turn an apparatus on and off at predetermined times. { ¦ȯd·ə¦mad·ik ,tīm ,swich }

automatic track shift |ENG ACOUS| A system used with multiple-track magnetic tape recorders to index the tape head, after one track is played, to the correct position for the start of the next track. { ¦ȯd·ə¦mad·ik 'trak ,shift }

automatic tuning system |CONT SYS| An electrical, mechanical, or electromechanical system that tunes a radio receiver or transmitter automatically to a predetermined frequency when a button or lever is pressed, a knob turned, or a telephone-type dial operated. { ¦òd·ə¦mad·ik 'tün·iŋ ˌsis·təm }

automatic-type belt-tensioning device |MECH ENG| Any device which maintains a predetermined tension in a conveyor belt. { ¦òd·ə¦mad·ik ˌtīp 'belt ¦ten·shən·iŋ di,vīs }

automatic volume compressor See volume compressor. { ¦òd·ə¦mad·ik 'väl·yəm kəm,pres·ər }

automatic volume expander See volume expander. { ¦òd·ə¦mad·ik 'väl·yəm ik,spand·ər }

automatic wet-pipe sprinkler system |ENG| A sprinkler system, all of whose parts are filled with water at sufficient pressure to provide an immediate continuous discharge if the system is activated. { ¦òd·ə¦mad·ik ¦wet ¦pīp 'spriŋk·lər ˌsis·təm }

automatic zero setting |ENG| A system for automatic correction of zero-point drifts or for compensation of soiling of load receivers on a balance by means of a special accessory component. { ¦òd·ə¦mad·ik 'zir·ō ˌsed·iŋ }

automation |ENG| **1.** The use of technology to ease human labor or extend the mental or physical capabilities of humans. **2.** The mechanisms, machines, and systems that save or eliminate labor, or imitate actions typically associated with human beings. { ˌòd·ə'mā·shən }

automechanism |CONT SYS| A machine or other device that operates automatically or under control of a servomechanism. { ¦òd·ō'mek·ə,niz·əm }

automobile |MECH ENG| A four-wheeled, trackless, self-propelled vehicle for land transportation of as many as eight people. Also known as car. { ˌòd·ə·mə'bēl }

automobile chassis |MECH ENG| The automobile frame, together with the wheels, power train, brakes, engine, and steering system. { ˌòd·ə·mə'bēl 'chas·ē }

automotive air conditioning |MECH ENG| A system for maintaining comfort of occupants of automobiles, buses, and trucks, limited to air cooling, air heating, ventilation, and occasionally dehumidification. { ¦òd·ə¦mōd·iv 'er kən ˌdish·ən·iŋ }

automotive body |ENG| An enclosure mounted on and attached to the frame of an automotive vehicle, to contain passengers and luggage, or in the case of commercial vehicles the commodities being carried. { ¦òd·ə¦mōd·iv 'bäd·ē }

automotive brake |MECH ENG| A friction mechanism that slows or stops the rotation of the wheels of an automotive vehicle, so that tire traction slows or stops the vehicle. { ¦òd·ə'mōd·iv 'brāk }

automotive engine |MECH ENG| The fuel-consuming machine that provides the motive power for automobiles, airplanes, tractors, buses, and motorcycles and is carried in the vehicle. { ¦òd·ə¦mōd·iv 'en·jən }

automotive engineering |MECH ENG| The branch of mechanical engineering concerned primarily with the special problems of land transportation by a four-wheeled, trackless, automotive vehicle. { ¦òd·ə¦mōd·iv ˌen·jə'nir·iŋ }

automotive frame |ENG| The basic structure of all automotive vehicles, except tractors, which is supported by the suspension and upon which or attached to which are the power plant, transmission, clutch, and body or seat for the driver. { ¦òd·ə¦mōd·iv 'frām }

automotive ignition system |MECH ENG| A device in an automotive vehicle which initiates the chemical reaction between fuel and air in the cylinder charge. { ¦òd·ə¦mōd·iv ig'nish·ən ˌsis·təm }

automotive steering |MECH ENG| Mechanical means by which a driver controls the course of a moving automobile, bus, truck, or tractor. { ¦òd·ə¦mōd·iv 'stir·iŋ }

automotive suspension |MECH ENG| The springs and related parts intermediate between the wheels and frame of an automotive vehicle that support the frame on the wheels and absorb road shock caused by passage of the wheels over irregularities. { ¦òd·ə¦mōd·iv səs'pen·chən }

automotive transmission |MECH ENG| A device for providing different gear or drive ratios between the engine and drive wheels of an automotive vehicle, a principal function being to enable the vehicle to accelerate from rest through a wide speed range while the engine operates within its most effective range. { ¦òd·ə¦mōd·iv tranz 'mish·ən }

automotive vehicle |MECH ENG| A self-propelled vehicle or machine for land transportation of people or commodities or for moving materials, such as a passenger car, bus, truck, motorcycle, tractor, airplane, motorboat, or earthmover. { ¦òd·ə¦mōd·iv 'vē·ə·kəl }

autonomous robot |ENG| A robot that not only can maintain its own stability as it moves, but also can plan its movements. { ò¦tän·ə·məs 'rō,bät }

autonomous vehicle |ENG| A vehicle that is able to plan its path and to execute its plan without human intervention. { ò¦tän·ə·məs 'vē·ə·kəl }

autopatrol |MECH ENG| A self-powered blade grader. Also known as motor grader. { 'òd·ō·pə,trōl }

autoradar plot See chart comparison unit. { ¦òd·ō¦rä,där ˌplät }

autoradiography |ENG| A technique for detecting radioactivity in a specimen by producing an image on a photographic film or plate. Also known as radioautography. { ¦òd·ō,rād·ē'äg·rə·fē }

autorail |MECH ENG| A self-propelled vehicle having both flange wheels and pneumatic tires to permit operation on both rails and roadways. { 'òd·ō,rāl }

autoranging |ENG| Automatic switching of a multirange meter from its lowest to the next higher range, with the switching process repeated until a range is reached for which the full-scale value is not exceeded. Also known as automatic ranging. { 'òd·ō,rānj·iŋ }

autoreducing tachymeter [ENG] A class of tachymeter by which horizontal and height distances are read simultaneously. { ¦ȯd·ō·ri¦düs· iŋ tə'kim·əd·ər }

autorotation [MECH] **1.** Rotation about any axis of a body that is symmetrical and exposed to a uniform airstream and maintained only by aerodynamic moments. **2.** Rotation of a stalled symmetrical airfoil parallel to the direction of the wind. { ¦ȯd·ō,rō'tā·shən }

autosled [MECH ENG] A propeller-driven machine equipped with runners and wheels and adaptable to use on snow, ice, or bare roads. { 'ȯd·ō'sled }

autostability [CONT SYS] The ability of a device (such as a servomechanism) to hold a steady position, either by virtue of its shape and proportions, or by control by a servomechanism. { ¦ȯd·ō·stə'bil·əd·ē }

auxanometer [ENG] An instrument used to detect and measure plant growth rate. { ˌȯg·zə'näm·əd·ər }

auxiliary dead latch [DES ENG] A supplementary latch in a lock which automatically deadlocks the main latch bolt when the door is closed. Also known as auxiliary latch bolt; deadlocking latch bolt; trigger bolt. { ȯg'zil·yə·rē 'ded ˌlach }

auxiliary latch bolt See auxiliary dead latch. { ȯg 'zil·yə·rē 'lach ˌbōlt }

auxiliary power plant [MECH ENG] Ancillary equipment, such as pumps, fans, and soot blowers, used with the main boiler, turbine, engine, waterwheel, or generator of a power-generating station. { ȯg'zil·yə·rē 'paù·ər ˌplant }

auxiliary rafter [BUILD] A member strengthening the principal rafter in a truss. { ȯg'zil·yə·rē 'raf·tər }

auxiliary reinforcement [CIV ENG] In a prestressed structural member, any reinforcement in addition to that whose function is prestressing. { ȯg'zil·yə·rē ˌrē·ən'fȯrs·mənt }

auxiliary rim lock [DES ENG] A secondary or extra lock that is surface-mounted on a door to provide additional security. { ȯg'zil·yə·rē 'rim ˌläk }

auxiliary rope-fastening device [MECH ENG] A device attached to an elevator car, to a counterweight, or to the overhead dead-end rope-hitch support, that automatically supports the car or counterweight in case the fastening for the wire rope (cable) fails. { ȯg'zil·yə·rē 'rōp ˌfas·ən·iŋ di,vīs }

auxiliary thermometer [ENG] A mercury-in-glass thermometer attached to the stem of a reversing thermometer and read at the same time as the reversing thermometer so that the correction to the reading of the latter, resulting from change in temperature since reversal, can be computed. { ȯg'zil·yə·rē thər'mäm·əd·ər }

auxograph [ENG] An automatic device that records changes in the volume of a body. { 'ȯk·sə,graf }

auxometer [ENG] An instrument that measures the magnification of a lens system. { ˌȯk'säm·əd·ər }

availability [SYS ENG] The probability that a system is operating satisfactorily at any point in time, excluding times when the system is under repair. { ə,vāl·ə'bil·ə·dē }

availability ratio [IND ENG] The ratio of the amount of time a system is actually available for use to the amount of time it is supposed to be available. { ə,vāl·ə'bil·əd·ē 'rā·shō }

available draft [MECH ENG] The usable differential pressure in the combustion air in a furnace, used to sustain combustion of fuel or to transport products of combustion. { ə'vāl·ə·bəl 'draft }

available energy [MECH ENG] Energy which can in principle be converted to mechanical work. { ə'vāl·ə·bəl 'en·ər·jē }

available heat [MECH ENG] The heat per unit mass of a working substance that could be transformed into work in an engine under ideal conditions for a given amount of heat per unit mass furnished to the working substance. { ə'vāl·ə·bəl 'hēt }

available motions inventory [IND ENG] A list of all motions available to a human for performing a specific task. { ə¦vāl·ə·bəl ˌmō·shənz 'in·ven ˌtȯr·ē }

avalanche [ELECTR] **1.** The cumulative process in which an electron or other charged particle accelerated by a strong electric field collides with and ionizes gas molecules, thereby releasing new electrons which in turn have more collisions, so that the discharge is thus self-maintained. Also known as avalanche effect; cascade; cumulative ionization; electron avalanche; Townsend avalanche; Townsend ionization. **2.** Cumulative multiplication of carriers in a semiconductor as a result of avalanche breakdown. Also known as avalanche effect. { 'av·ə,lanch }

avalanche breakdown [ELECTR] Nondestructive breakdown in a semiconductor diode when the electric field across the barrier region is strong enough so that current carriers collide with valence electrons to produce ionization and cumulative multiplication of carriers. { 'av· ə,lanch 'brāk,daùn }

avalanche diode [ELECTR] A semiconductor breakdown diode, usually made of silicon, in which avalanche breakdown occurs across the entire *pn* junction and voltage drop is then essentially constant and independent of current; the two most important types are IMPATT and TRAPATT diodes. { 'av·ə,lanch 'dī,ōd }

avalanche effect See avalanche. { 'av·ə,lanch i,fekt }

avalanche impedance [ELECTR] The complex ratio of the reverse voltage of a device that undergoes avalanche breakdown to the reverse current. { 'av·ə,lanch im'pēd·əns }

avalanche-induced migration [ELECTR] A technique of forming interconnections in a field-programmable logic array by applying appropriate voltages for shorting selected base-emitter junctions. { 'av·ə,lanch in¦düsd ˌmī'grā·shən }

avalanche noise [ELECTR] **1.** A junction phenomenon in a semiconductor in which carriers in a high-voltage gradient develop sufficient energy to dislodge additional carriers through physical impact; this agitation creates ragged current flows which are indicated by noise. **2.** The noise produced when a junction diode is operated at the onset of avalanche breakdown. { 'av·ə‚lanch ‚nȯiz }

avalanche oscillator |ELECTR| An oscillator that uses an avalanche diode as a negative resistance to achieve one-step conversion from direct-current to microwave outputs in the gigahertz range. { 'av·ə‚lanch ¦äs·ə‚läd·ər }

avalanche photodiode |ELECTR| A photodiode operated in the avalanche breakdown region to achieve internal photocurrent multiplication, thereby providing rapid light-controlled switching operation. { 'av·ə‚lanch ‚fōd·ō'dī‚ōd }

avalanche protector |MECH ENG| Guard plates installed on an excavator to prevent loose material from sliding into the wheels or tracks. { 'av·ə‚lanch prə‚tek·tər }

avalanche transistor |ELECTR| A transistor that utilizes avalanche breakdown to produce chain generation of charge-carrying hole-electron pairs. { 'av·ə‚lanch tran'zis·tər }

avalanche voltage |ELECTR| The reverse voltage required to cause avalanche breakdown in a *pn* semiconductor junction. { 'av·ə‚lanch ‚vōl·tij }

average acoustic output |ENG ACOUS| Vibratory energy output of a transducer measured by a radiation pressure balance; expressed in terms of watts per unit area of the transducer face. { 'av·rij ə'kü·stik 'aút‚pút }

average noise figure |ELECTR| Ratio in a transducer of total output noise power to the portion thereof attributable to thermal noise in the input termination, the total noise being summed over frequencies from zero to infinity, and the noise temperature of the input termination being standard (290 K). { 'av·rij 'nȯiz ‚fig·yər }

average outgoing quality limit |IND ENG| The average quality of all lots that pass quality inspection, expressed in terms of percent defective. Abbreviated AOQL. { 'av·rij 'aút‚gō·iŋ 'kwäl·əd·ē ‚lim·ət }

average power output |ELECTR| Radio-frequency power, in an audio-modulation transmitter, delivered to the transmitter output terminals, averaged over a modulation cycle. { 'av·rij 'paú·ər 'aút‚pút }

average sample number |IND ENG| An anticipated number of pieces that must be inspected to determine the acceptability of a particular lot. { 'av·rij ¦sam·pəl ‚nəm·bər }

averaging |CONT SYS| The reduction of noise received by a robot sensor by screening it over a period of time. { 'av· rij·iŋ }

averaging device |ENG| A device for obtaining the arithmetic mean of a number of readings, as on a bubble sextant. { 'av·rij·iŋ di'vīs }

averaging pitot tube |ENG| A flowmeter that consists of a rod extending across a pipe with several interconnected upstream holes, which simulate an array of pitot tubes across the pipe,

and a downstream hole for the static pressure reference. { 'av·rij·iŋ ‚pē‚tō ‚tüb }

aviation method |ENG| Determination of knock-limiting power, under lean-mixture conditions, of fuels used in spark-ignition aircraft engines. { ‚ā·vē'ā·shən 'meth·əd }

avionics |ENG| The design and production of airborne electrical and electronic devices; term is derived from aviation electronics. { ‚ā·vē'än·iks }

avogram |MECH| A unit of mass, equal to 1 gram divided by the Avogadro number. { 'a·və‚gram }

avoidable delay |IND ENG| An interruption under the control of the operator during the normal operating time. { ə'vȯid·ə·bəl di'lā }

avoirdupois pound *See* pound. { ‚av·ərd·ə'pȯiz 'paúnd }

avoirdupois weight |MECH| The system of units which has been commonly used in English-speaking countries for measurement of the mass of any substance except precious stones, precious metals, and drugs; it is based on the pound (approximately 453.6 grams) and includes the short ton (2000 pounds), long ton (2240 pounds), ounce (one-sixteenth pound), and dram (one-sixteenth ounce). { ‚av·ərd·ə'pȯiz 'wāt }

awl |DES ENG| A point tool with a short wooden handle used to mark surfaces and to make small holes, as in leather or wood. { ȯl }

awning window |BUILD| A window consisting of a series of vertically arranged, top-hinged rectangular sections; designed to admit air while excluding rain. { 'ȯn·iŋ ‚win·dō }

ax |DES ENG| An implement consisting of a heavy metal wedge-shaped head with one or two cutting edges and a relatively long wooden handle; used for chopping wood and felling trees. { aks }

axed brick |ENG| A brick, shaped with an ax, that has not been trimmed. Also known as rough-axed brick. { ¦akst ¦brik }

axhammer |DES ENG| An ax having one cutting edge and one hammer face. { 'aks‚ham·ər }

axial fan |MECH ENG| A fan whose housing confines the gas flow to the direction along the rotating shaft at both the inlet and outlet. { 'ak·sē·əl 'fan }

axial-flow compressor |MECH ENG| A fluid compressor that accelerates the fluid in a direction generally parallel to the rotating shaft. { 'ak·sē·əl 'flō kəm'pres·ər }

axial-flow pump |MECH ENG| A pump having an axial-flow or propeller-type impeller; used when maximum capacity and minimum head are desired. Also known as propeller pump. { 'ak·sē·əl 'flō ‚pəmp }

axial force diagram |CIV ENG| In statics, a graphical representation of the axial load acting at each section of a structural member, plotted to scale and with proper sign as an ordinate at each point of the member and along a reference line representing the length of the member. { 'ak·sē·əl ¦fȯrs ‚di·ə‚gram }

axial hydraulic thrust [MECH ENG] In single-stage and multistage pumps, the summation of unbalanced impeller forces acting in the axial direction. { 'ak·sē·əl hī'drȯ·lik 'thrəst }

axial lead [ELEC] A wire lead extending from the end along the axis of a resistor, capacitor, or other component. { 'ak·sē·əl 'lēd }

axial load [MECH] A force with its resultant passing through the centroid of a particular section and being perpendicular to the plane of the section. { 'ak·sē·əl 'lōd }

axial modulus [MECH] The ratio of a simple tension stress applied to a material to the resulting strain parallel to the tension when the sides of the sample are restricted so that there is no lateral deformation. Also known as modulus of simple longitudinal extension. { ¦ak·sē·əl 'mäj·ə·ləs }

axial moment of inertia [MECH] For any object rotating about an axis, the sum of its component masses times the square of the distance to the axis. { 'ak·sē·əl 'mō·mənt əv in'ər·shə }

axial nozzle [MECH ENG] An inlet or outlet connection installed in the head of a shell-and-tube exchanger and aligned normal to the plane in which the tube lies. { ¦ak·sē·əl 'näz·əl }

axial rake [MECH ENG] The angle between the face of a blade of a milling cutter or reamer and a line parallel to its axis of rotation. { 'ak·sē·əl 'rāk }

axial relief [MECH ENG] The relief behind the end cutting edge of a milling cutter. { 'ak·sē·əl ri'lēf }

axial runout [MECH ENG] The total amount, along the axis of rotation, by which the rotation of a cutting tool deviates from a plane. { 'ak·sē·əl 'rən,aút }

axial-type mass flowmeter [ENG] An instrument in which fluid in a pipe is made to rotate at a constant speed by a motor-driven impeller, and the torque required by a second, stationary impeller to straighten the flow again is a direct measurement of mass flow. { 'ak·sē·əl ¦tīp 'mas 'flō,med·ər }

axis [MECH] A line about which a body rotates. { 'ak·səs }

axis of freedom [DES ENG] An axis in a gyro about which a gimbal provides a degree of freedom. { 'ak·səs əv frēd·əm }

axis of rotation [MECH] A straight line passing through the points of a rotating rigid body that remain stationary, while the other points of the body move in circles about the axis. { 'ak·səs əv rō'tā·shən }

axis of sighting [ENG] A line taken through the sights of a gun, or through the optical center and centers of curvature of lenses in any telescopic instrument. { 'ak·səs əv 'sīd·iŋ }

axis of symmetry [MECH] An imaginary line about which a geometrical figure is symmetric. Also known as symmetry axis. { 'ak·səs əv 'sim·ə·trē }

axis of torsion [MECH] An axis parallel to the generators of a cylinder undergoing torsion, located so that the displacement of any point on the axis lies along the axis. Also known as axis of twist. { ¦ak·səs əv 'tȯr·shən }

axis of twist See axis of torsion. { ¦ak·səs əv 'twist }

axle [MECH ENG] A supporting member that carries a wheel and either rotates with the wheel to transmit mechanical power to or from it, or allows the wheel to rotate freely on it. { 'ak·səl }

axle box [ENG] A bushing through which an axle passes in the hub of a wheel. { 'ak·səl ,bäks }

axle ratio [MECH ENG] In an automotive vehicle, the ratio of the speed in revolutions per minute of the drive shaft to that of the drive wheels. { 'ak·səl 'rā·shō }

axometer [ENG] An instrument that locates the optical axis of a lens, particularly a lens used in eyeglasses. { ak'säm·əd·ər }

azel mounting See altazimuth mounting. { 'az·əl ,maúnt·iŋ }

azeotropic distillation [CHEM ENG] A process by which a liquid mixture is separated into pure components with the help of an additional substance or solvent. { ¦āz·ē·ə,trō·pik ,dis·tə'lā·shən }

azimuth [ENG] In directional drilling, the direction of the face of the deviation tool with respect to magnetic north. { 'az·ə·məth }

azimuth-adjustment slide rule [ENG] A circular slide rule by which a known angular correction for fire at one elevation can be changed to the proper correction for any other elevation. { 'az·ə·məth ə¦jəs·mənt 'slīd ,rül }

azimuth alignment [ENG ACOUS] The condition whereby the center lines of the playback- and recording-head gaps are exactly perpendicular to the magnetic tape and parallel to each other. { 'az·ə·məth a'līn·mənt }

azimuth angle [ENG] An angle in triangulation or in traverse through which the computation of azimuth is carried. { 'az·ə·məth 'aŋ·gəl }

azimuth bar See azimuth instrument. { 'az·ə·məth ,bär }

azimuth circle [DES ENG] A ring calibrated from 0 to 360° over a compass, compass repeater, radar plan position indicator, direction finder, and so on, which provides means for observing compass bearings and azimuths. { 'az·ə·məth ,sər·kəl }

azimuth dial [ENG] Any horizontal circle dial that reads azimuth. { 'az·ə·məth ,dīl }

azimuth error [ENG] An error in the indicated azimuth of a target detected by radar. { 'az·ə·məth ,er·ər }

azimuth indicator [ENG] An approach-radar scope which displays azimuth information. { 'az·ə·məth ,in·də,kād·ər }

azimuth instrument [ENG] An instrument for measuring azimuths, particularly a device which fits over a central pivot in the glass cover of a magnetic compass. Also known as azimuth bar; bearing bar. { 'az·ə·məth ,in·strə·mənt }

azimuth line [ENG] A radial line from the principal point, isocenter, or nadir point of a photograph, representing the direction to a similar point of an adjacent photograph in the same flight line; used extensively in radial triangulation. { 'az·ə·məth ‚līn }

azimuth marker |ENG| **1.** A scale encircling the plan position indicator scope of a radar on which the azimuth of a target from the radar may be measured. **2.** Any of the reference limits inserted electronically at 10 or 15° intervals which extend radially from the relative position of the radar on an off-center plan position indicator scope. { 'az·ə·məth ‚mär·kər }

azimuth scale |ENG| A graduated angle-measuring device on instruments, gun carriages, and so forth that indicates azimuth. { 'az·ə·məth ‚skāl }

azimuth-stabilized plan position indicator |ENG| A north-upward plan position indicator (PPI), a radarscope, which is stabilized by a gyrocompass so that either true or magnetic north is always at the top of the scope regardless of vehicle orientation. { 'az·ə·məth ¦sta·bə‚līzd 'plan pə'zish·ən 'in·də‚kād·ər }

azimuth transfer |ENG| Connecting, with a straight line, the nadir points of two vertical photographs selected from overlapping flights. { 'az·ə·məth 'tranz‚fər }

azimuth traverse |ENG| A survey traverse in which the direction of the measured course is determined by azimuth and verified by back azimuth. { 'az·ə·məth trə'vərs }

Azusa |ENG| A continuous-wave, high-accuracy, phase-comparison, single-station tracking system operating at C-band and giving two direction cosines and slant range which can be used to determine space position and velocity of a vehicle (usually a rocket or a missile). { ə'züs·ə }

B

backacter *See* backhoe. { 'bak,ak·tər }

backband |BUILD| A piece of millwork used around a rectangular window or door casing as a cover for the gap between the casing and the wall or as a decorative feature. Also known as backbend. { 'bak,band }

backbend |BUILD| **1.** At the outer edge of a metal door or window frame, the face which returns to the wall surface. **2.** *See* backband. { 'bak,bend }

back bias |ELECTR| **1.** Degenerative or regenerative voltage which is fed back to circuits before its originating point; usually applied to a control anode of a tube or other device. **2.** Voltage applied to a grid of a tube (or tubes) or electrode of another device to reduce a condition which has been upset by some external cause. { 'bak ,bī·əs }

back boxing *See* backlining. { 'bak ¦bäk·siŋ }

backbreak *See* overbreak. { 'bak,brāk }

back check |DES ENG| In a hydraulic door closer, a mechanism that slows the speed with which a door may be opened. { 'bak ,chek }

backdigger *See* backhoe. { 'bak¦dig·ər }

back-draft damper |MECH ENG| A damper with blades actuated by gravity, permitting air to pass through them in one direction only. { 'bak ,draft 'dam·pər }

back edging |ENG| Cutting through a glazed ceramic pipe by first chipping through the glaze around the outside and then chipping the pipe itself. { 'bak ,ej·iŋ }

back end *See* thrust yoke. { 'bak ,end }

backfill |CIV ENG| Earth refilling a trench or an excavation around a building, bridge abutment, and the like. { 'bak,fil }

back fillet |BUILD| The return of the margin of a groin, doorjamb, or window jamb when it projects beyond a wall. { 'bak ,fil·ət }

backfire |CIV ENG| A fire that is started in order to burn against and cut off a spreading fire. |ELECTR| *See* arcback. |ENG| Momentary backward burning of flame into the tip of a torch. Also known as flashback. |MECH ENG| In an internal combustion engine, an improperly timed explosion of the fuel mixture in a cylinder, especially one occurring during the period that the exhaust or intake valve is open and resulting in a loud detonation. { 'bak,fīr }

backflap hinge |DES ENG| A hinge having a flat plate or strap which is screwed to the face of a shutter or door. Also known as flap hinge. { 'bak,flap ,hinj }

backflow |CIV ENG| The flow of water or other liquids, mixtures, or substances into the distributing pipes of a potable supply of water from any other than its intended source. { 'bak,flō }

backflow connection |CIV ENG| Any arrangement of pipes, plumbing fixtures, drains, and so forth, in which backflow can occur. { 'bak,flō kə'nek·shən }

backflow preventer *See* vacuum breaker. { 'bak ,flō pri'ven·tər }

backflow valve *See* backwater valve. { 'bak,flō ,valv }

backfurrow |CIV ENG| In an excavation procedure, the first cut made on undisturbed land. { 'bak,fər·ō }

back gearing |MECH ENG| The technique of using gears on machine tools to obtain an increase in the number of speed changes that can be gotten with cone belt drives. { 'bak ,gir·iŋ }

background discrimination |ENG| The ability of a measuring instrument, circuit, or other device to distinguish signal from background noise. { 'bak,graúnd dis,krim·ə'nā·shən }

background noise |ENG| The undesired signals that are always present in an electronic or other system, independent of whether or not the desired signal is present. { 'bak,graúnd ,nóiz }

background returns |ENG| **1.** Signals on a radar screen from objects which are of no interest. **2.** *See* clutter. { 'bak,graúnd ri'tərnz }

background signal |ENG| The output of a leak detector caused by residual gas to which the detector element reacts. { 'bak,graúnd ,sig·nəl }

back gutter |BUILD| A gutter installed on the uphill side of a chimney on a sloping roof to divert water around the chimney. { 'bak ,gəd·ər }

back hearth |BUILD| That part of the hearth (or floor) which is contained within the fireplace itself. Also known as inner hearth. { 'bak ,härth }

backhoe |MECH ENG| An excavator fitted with a hinged arm to which is rigidly attached a bucket that is drawn toward the machine in operation. Also known as backacter; backdigger; dragshovel; pullshovel. { 'bak ,hō }

backing |CIV ENG| **1.** The unexposed, rough masonry surface of a wall that is faced with finer work. **2.** The earth backfill of a retaining wall. |ELECTR| Flexible material, usually cellulose acetate or polyester, used on magnetic tape as the carrier for the oxide coating. { 'bak·iŋ }

backing board |BUILD| In a suspended acoustical ceiling, a flat sheet of gypsum board to which acoustical tile is attached by adhesive or mechanical means. { 'bak·iŋ ,bȯrd }

backing brick |CIV ENG| A relatively low-quality brick used behind face brick or other masonry. { 'bak·iŋ ,brik }

backing off |ENG| Removing excessive body metal from badly worn bits. { 'bak·iŋ ,ȯf }

backing plate |ENG| A plate used to support the hardware for the cavity used in plastics injection molding. { 'bak·iŋ ,plāt }

backing pump |MECH ENG| A vacuum pump, in a vacuum system using two pumps in tandem, which works directly to the atmosphere and reduces the pressure to an intermediate value, usually between 100 and 0.1 pascals. Also known as fore pump. { 'bak·iŋ ,pəmp }

backing ring |ENG| A strip of metal attached at a pipe joint at the root of a weld to prevent spatter and to ensure the integrity of the weld. { 'bak·iŋ ,riŋ }

backing space |ENG| Space between a fore pump and a diffusion pump in a leak-testing system. { 'bak·iŋ ,spās }

backing-space technique |ENG| Testing for leaks by connecting a leak detector to the backing space. { 'bak·iŋ ,spās ,tek'nēk }

backing up |CIV ENG| In masonry, the laying of backing brick. { 'bak·iŋ |əp }

back jamb See backlining. { 'bak ,jam }

backjoint |CIV ENG| In masonry, a rabbet such as that made on the inner side of a chimneypiece to receive a slip. { 'bak,jȯint }

backlash |DES ENG| The amount by which the tooth space of a gear exceeds the tooth thickness of the mating gear along the pitch circles. |ELECTR| A small reverse current in a rectifier tube caused by the motion of positive ions produced in the gas by the impact of thermoelectrons. |ENG| **1.** Relative motion of mechanical parts caused by looseness. **2.** The difference between the actual values of a quantity when a dial controlling this quantity is brought to a given position by a clockwise rotation and when it is brought to the same position by a counterclockwise rotation. { 'bak,lash }

backlining |BUILD| **1.** A thin strip which lines a window casing, next to the wall and opposite the pulley stile, and provides a smooth surface for the working of the weighted sash. Also known as back boxing; back jamb. **2.** That piece of framing forming the back recess for boxing shutters. { 'bak,līn·iŋ }

back lintel |BUILD| A lintel which supports the backing of a masonry wall, as opposed to the lintel supporting the facing material. { 'bak ,lin·təl }

backlog |IND ENG| **1.** An accumulation of orders promising future work and profit. **2.** An accumulation of unprocessed materials or unperformed tasks. { 'bak,läg }

back mixing |CHEM ENG| The tendency of reacted chemicals to intermingle with unreacted feed in reactors, such as stirred tanks, packed towers, and baffled tanks. { 'bak ,mik·siŋ }

back nailing |BUILD| Nailing the plies of a built-up roof to the substrate to prevent slippage. { 'bak ,nāl·iŋ }

back nut |DES ENG| **1.** A threaded nut, one side of which is dished to retain a grommet; used in forming a watertight pipe joint. **2.** A locking nut on the shank of a pipe fitting, tap, or valve. { 'bak ,nət }

back off |ENG| **1.** To unscrew or disconnect. **2.** To withdraw the drill bit from a borehole. **3.** To withdraw a cutting tool or grinding wheel from contact with the workpiece. { 'bak ,ȯf }

back order |IND ENG| **1.** An order held for future completion. **2.** A new order placed for previously unavailable materials of an old order. { 'bak ,ȯrd·ər }

backplastering |BUILD| A coat of plaster applied to the back side of lath, opposite the finished surface. { 'bak,plas·triŋ }

backplate |BUILD| A plate, usually metal or wood, which serves as a backing for a structural member. { 'bak,plāt }

backplate lamp holder |DES ENG| A lamp holder, integrally mounted on a plate, which is designed for screwing to a flat surface. { 'bak ,plāt 'lamp ,hōl·dər }

back pressure |MECH| Pressure due to a force that is operating in a direction opposite to that being considered, such as that of a fluid flow. |MECH ENG| Resistance transferred from rock into the drill stem when the bit is being fed at a faster rate than the bit can cut. { 'bak ,presh·ər }

back-pressure-relief port |ENG| In a plastics extrusion die, an opening for the release of excess material. { 'bak ,presh·ər ri'lēf ,pȯrt }

back rake |DES ENG| An angle on a single-point turning tool measured between the plane of the tool face and the reference plane. { 'bak ,rāk }

back-run process |CHEM ENG| A process for manufacturing water gas in which part of the run is made down, by passing steam through the superheater, thence up through the carburetor, down through the generator, and direct to the scrubbers. { 'bak ,rən 'präs·əs }

backsaw |DES ENG| A fine-tooth saw with its upper edge stiffened by a metal rib to ensure straight cuts. { 'bak,sȯ }

backscatter gage |ENG| A radar instrument used to measure the radiation scattered at 180° to the direction of the incident wave. { 'bak ,skad·ər ,gaj }

backscattering thickness gage |ENG| A device that uses a radioactive source for measuring the thickness of materials, such as coatings, in which the source and the instrument measuring the radiation are mounted on the same side of the material, the backscattered radiation thus being measured. { 'bak¦skad·ə·riŋ 'thik·nəs ,gāj }

backset |BUILD| The horizontal distance from the face of a lock or latch to the center of the keyhole, knob, or lock cylinder. { 'bak,set }

backsight |ENG| **1.** A sight on a previously established survey point or line. **2.** Reading a leveling rod in its unchanged position after moving the leveling instrument to a different location. { 'bak,sīt }

backsight method |ENG| **1.** A plane-table traversing method in which the table orientation produces the alignment of the alidade on an established map line, the table being rotated until the line of sight is coincident with the corresponding ground line. **2.** Sighting two pieces of equipment directly at each other in order to orient and synchronize one with the other in azimuth and elevation. { 'bak,sīt 'meth·əd }

back siphonage |CIV ENG| The flowing back of used, contaminated, or polluted water from a plumbing fixture or vessel into the pipe which feeds it; caused by reduced pressure in the pipe. { 'bak ¦sī·fən·ij }

back solution |CONT SYS| The calculation of the tool-coordinated positions that correspond to specified robotic joint positions. { 'bak sə,lü·shən }

backspace |MECH ENG| To move a typewriter carriage back one space by depressing a backspace key. { 'bak,spās }

backstay |ENG| **1.** A supporting cable that prevents a more or less vertical object from falling forward. **2.** A spring used to keep together the cutting edges of purchase shears. **3.** A rod that runs from either end of a carriage's rear axle to the reach. **4.** A leather strip that covers and strengthens a shoe's back seam. { 'bak,stā }

back sweetening |CHEM ENG| The controlled addition of commercial-grade mercaptans to a petroleum stock having excess free sulfur in order to reduce free sulfur by forming a disulfide. { 'bak ,swēt·ən·iŋ }

backup |BUILD| That part of a masonry wall behind the exterior facing. |CIV ENG| Overflow in a drain or piping system, due to stoppage. |ENG| **1.** An item under development intended to perform the same general functions that another item also under development performs. **2.** A compressible material used behind a sealant to reduce its depth and to support the sealant against sag or indentation. { 'bak,əp }

backup strip |BUILD| A wood strip which is fixed at the corner of a partition or wall to provide a nailing surface for ends of lath. Also known as lathing board. { 'bak,əp ,strip }

backup system |SYS ENG| A system, normally redundant but kept available to replace a system which may fail in operation. { 'bak,əp ,sis·təm }

backup tong |ENG| A heavy device used on a drill pipe to loosen the tool joints. { 'bak,əp ,täŋ }

back vent |CIV ENG| An individual vent for a plumbing fixture located on the downstream (sewer) side of a trap to protect the trap against siphonage. { 'bak ,vent }

backward-bladed aerodynamic fan |MECH ENG| A fan that consists of several streamlined blades mounted in a revolving casing. { 'bak·wərd ,blād·əd ,er·ō·dī'nam·ik ,fan }

backward pass |IND ENG| The calculation of late finish times (dates) for all uncompleted network activities for a specific project by subtracting durations of uncompleted activities from the scheduled finish time of the final activity. { 'bak·wərd 'pas }

backwash |CHEM ENG| **1.** In an ion-exchange resin system, an upward flow of water through a resin bed that cleans and reclassifies the resin particles after exhaustion. **2.** See blowback. { 'bak,wäsh }

backwater valve |ENG| A type of check valve in a drainage pipe; reversal of flow causes the valve to close, thereby cutting off flow. Also known as backflow valve. { 'bak,wȯd·ər ,valv }

badger |DES ENG| See badger plane. |ENG| A tool used inside a pipe or culvert to remove any excess mortar or deposits. { 'baj·ər }

badger plane |DES ENG| A hand plane whose mouth is cut obliquely from side to side, so that the plane can work close up to a corner. Also known as badger. { 'baj·ər ,plān }

baffle |ELEC| Device for deflecting oil or gas in a circuit breaker. |ELECTR| An auxiliary member in a gas tube used, for example, to control the flow of mercury particles or deionize the mercury following conduction. |ENG| A plate that regulates the flow of a fluid, as in a steam-boiler flue or a gasoline muffler. |ENG ACOUS| A cabinet or partition used with a loudspeaker to reduce interaction between sound waves produced simultaneously by the two surfaces of the diaphragm. { 'baf·əl }

bag |ENG| **1.** A flexible cover used in bag molding. **2.** A container made of paper, plastic, or cloth without rigid walls to transport or store material. { 'bag }

bag filter |ENG| Filtering apparatus with porous cloth or felt bags through which dust-laden gases are sent, leaving the dust on the inner surfaces of the bags. { 'bag ,fil·tər }

baghouse |ENG| The large chamber or room for holding bag filters used to filter gas streams from a furnace. { 'bag,haús }

bag molding |ENG| A method of molding plastic or plywood-plastic combinations into curved shapes, in which fluid pressure acting through a flexible cover, or bag, presses the material to be molded against a rigid die. { 'bag ,mōld·iŋ }

Bagnold number |ENG| A dimensionless number used in saltation studies. { 'bag·nəld ,nəm·bər }

bag plug |ENG| An inflatable drain stopper, located at the lowest point of a piping system, that acts to seal a pipe when inflated. { 'bag ,pləg }

bag trap |ENG| An S-shaped trap in which the vertical inlet and outlet pipes are in alignment. { 'bag ,trap }

baguette *See* bead molding. { ba'get }

bail |ENG| A loop of heavy wire snap-fitted around two or more parts of a connector or other device to hold the parts together. { bāl }

bailer |ENG| A long, cylindrical vessel fitted with a bail at the upper end and a flap or tongue valve at the lower extremity; used to remove water, sand, and mud- or cuttings-laden fluids from a borehole. Also known as bailing bucket. { 'bāl·ər }

Bailey bridge |CIV ENG| A lattice bridge built of interchangeable panels connected at the corners with steel pins, permitting rapid construction; developed in Britain about 1942 as a military bridge. { 'bāl·ē ,brij }

Bailey meter |ENG| A flowmeter consisting of a helical quarter-turn vane which operates a counter to record the total weight of granular material flowing through vertical or near-vertical ducts, spouts, or pipes. { 'bāl·ē ,mēd·ər }

bailing |ENG| Removal of the cuttings from a well during cable-tool drilling, or of the liquid from a well, by means of a bailer. { 'bāl·iŋ }

bailing bucket *See* bailer. { 'bāl·iŋ ,bak·ət }

bailing drum |ENG| A reel for winding bailing line. { 'bāl·iŋ ,drəm }

bailing line |ENG| A cable attached to the bailer of a derrick; it is passed over a sheave at the top of the derrick and spooled on a reel. { 'bāl·iŋ ,līn }

baked finish |ENG| A paint or varnish finish obtained by baking, usually at temperatures above 150°F (65°C), thereby developing a tough, durable film. { 'bākt 'fin·ish }

bakeout |ENG| The degassing of surfaces of a vacuum system by heating during the pumping process. { 'bāk,aút }

baker bell dolphin |CIV ENG| A dolphin consisting of a heavy bell-shaped cap pivoted on a group of piles; a blow from a ship will tilt the bell, thus absorbing energy. { 'bāk·ər ¦bel ,däl·fən }

baking |ENG| The use of heat on fresh paint films to speed the evaporation of thinners and to promote the reaction of binder components so as to form a hard polymeric film. Also known as stoving. { 'bāk·iŋ }

balance |ELEC| The state of an electrical network when it is adjusted so that voltage in one branch induces or causes no current in another branch. |ENG| An instrument for measuring mass or weight. { 'bal·əns }

balance arm |BUILD| On a projected window, a side supporting arm which is constructed so that the center of gravity of the sash is not changed appreciably when the window is opened. { 'bal·əns ,ärm }

balance bar *See* balance beam. { 'bal·əns ,bär }

balance beam |CIV ENG| A long beam, attached to a gate (or drawbridge, and such) so as to counterbalance the weight of the gate during opening or closing. Also known as balance bar. { 'bal·əns ,bēm }

balanced armature unit |ENG ACOUS| Driving unit used in magnetic loudspeakers, consisting of an iron armature pivoted between the poles of a permanent magnet and surrounded by coils carrying the audio-frequency current; variations in audio-frequency current cause corresponding changes in armature magnetism and corresponding movements of the armature with respect to the poles of the permanent magnet. { 'bal·ənst 'ärm·ə·chər ,yü·nət }

balanced construction |BUILD| A plywood or sandwich-panel construction which has an odd number of plies laminated together so that the construction is identical on both sides of a plane through the center of the panel. { 'bal·ənst kən'strək·shən }

balanced design |ENG| A winding pattern used in fabricating filament-wound reinforced plastics that renders the stresses in all the filaments equal. { 'bal·ənst di'zīn }

balanced door |BUILD| A door equipped with double-pivoted hardware which is partially counterbalanced to provide easier operation. { 'bal·ənst 'dór }

balanced draft |ENG| The maintenance of a constant draft in a furnace by monitoring both the incoming air and products of combustion. { 'bal·ənst 'draft }

balanced earthwork |CIV ENG| Cut-and-fill work in which the amount of fill equals the amount of material excavated. { 'bal·ənst 'ərth,wərk }

balanced line |ELEC| A transmission line consisting of two conductors capable of being operated so that the voltages of the two conductors at any transverse plane are equal in magnitude and opposite in polarity with respect to ground. |IND ENG| A production line for which the time cycles of the operators are made approximately equal so that the work flows at a desired steady rate from one operator to the next. { 'bal·ənst ,līn }

balanced method |ENG| Method of measurement in which the reading is taken at zero; it may be a visual or audible reading, and in the latter case the null is the no-sound setting. { 'bal·ənst ¦meth·əd }

balanced reinforcement |CIV ENG| An amount and distribution of steel reinforcement in a flexural reinforced concrete member such that the allowable tensile stress in the steel and the allowable compressive stress in the concrete are attained simultaneously. { 'bal·ənst ,rē·ən 'fór·smənt }

balanced sash |BUILD| In a double-hung window, a sash which opens by being raised or lowered and which is balanced with counterweights or pretensioned springs so that little force is required to move the sash. { 'bal·ənst ,sash }

balanced step |BUILD| One of a series of winders arranged so that the width of each winder tread (at the narrow end) is almost equal to the tread width in the straight portion of the adjacent stair flight. Also known as dancing step; dancing winder. { 'bal·ənst ,step }

balanced valve |ENG| A valve having equal fluid pressure in both the opening and closing directions. { 'bal·ənst ,valv }

balance method See null method. { 'bal·əns ‚meth·əd }

balance pipe [ENG] A pipe in a compressed-air piping system that is used to displace trapped air so that the condensate can flow freely into the trap. { 'bal·əns ‚pīpe }

balance tool [MECH ENG] A tool designed for taking the first cuts when the external surface of a piece in a lathe is being machined; it is supported in the tool holder at an unvarying angle. { 'bal·əns ‚tül }

balance wheel [MECH ENG] **1.** A wheel which governs or stabilizes the movement of a mechanism. **2.** See flywheel. { 'bal·əns ‚wēl }

balancing a survey [ENG] Distributing corrections through any traverse to eliminate the error of closure and to obtain an adjusted position for each traverse station. Also known as traverse adjustment. { 'bal·əns·iŋ ə 'sər‚vā }

balancing delay [IND ENG] In motion study, idleness of one hand while the other is active to catch up. { 'bal·əns·iŋ di‚lā }

balancing plug cock See balancing valve. { 'bal·əns·iŋ 'pləg ‚käk }

balancing valve [ENG] A valve used in a pipe for controlling fluid flow; not usually used to shut off the flow. Also known as balancing plug cock. { 'bal·əns·iŋ ‚valv }

balconet [BUILD] A pseudobalcony; a low ornamental railing at a window, projecting only slightly beyond the threshold or sill. { 'bal·kə‚net }

balcony [BUILD] A deck which projects from a building wall above ground level. { 'bal·kə·nē }

balcony outlet [BUILD] In a vertical rainwater pipe that passes through an exterior balcony, a fitting which provides an inlet for the drainage of rainwater from the balcony. { 'bal·kə·nē ‚aut‚let }

bale [IND ENG] **1.** A large package of material, pressed tightly together, tied with rope, wire, or hoops and usually covered with wrapping. **2.** The amount of material in a bale; sometimes used as a unit of measure, as 500 pounds (227 kilograms) of cotton in the United States. { bāl }

baler [MECH ENG] A machine which takes large quantities of raw or finished materials and binds them with rope or metal straps or wires into a large package. { 'bāl·ər }

baling [CIV ENG] A technique used to convert loose refuse into heavy blocks by compaction; the blocks are then burned and are buried in sanitary landfill. { 'bāl·iŋ }

balk [BUILD] A squared timber used in building construction. [CIV ENG] A low ridge of earth that marks a boundary line. { bök }

balking [IND ENG] The refusal of a customer to enter a queue for some reason, such as insufficient waiting room. { 'bök·iŋ }

ball [MECH ENG] In fine grinding, one of the crushing bodies used in a ball mill. { böl }

ball-and-race-type pulverizer [MECH ENG] A grinding machine in which balls rotate under an applied force between two races to crush materials, such as coal, to fine consistency. Also known as ball-bearing pulverizer. { ‚böl ən ‚rās ‚tīp 'pəl·və‚rīz·ər }

ball-and-ring method See ring-and-ball test. { ‚böl ən 'riŋ ‚meth·əd }

ball-and-socket joint [MECH ENG] A joint in which a member ending in a ball is joined in a member ending in a socket so that relative movement is permitted within a certain angle in all planes passing through a line. Also known as ball joint. { ‚böl ən 'säk·ət ‚jöint }

ball-and-trunnion joint [MECH ENG] A joint in which a universal joint and a slip joint are combined in a single assembly. { ‚böl ən 'trən·yən ‚jöint }

ballast [CIV ENG] Crushed stone used in a railroad bed to support the ties, hold the track in line, and help drainage. [ELEC] A circuit element that serves to limit an electric current or to provide a starting voltage, as in certain types of lamps, such as in fluorescent ceiling fixtures. { 'bal·əst }

ball bearing [MECH ENG] An antifriction bearing permitting free motion between moving and fixed parts by means of balls confined between outer and inner rings. { ‚böl 'ber·iŋ }

ball-bearing hinge [MECH ENG] A hinge which is equipped with ball bearings between the hinge knuckles in order to reduce friction. { ‚böl 'ber·iŋ ‚hinj }

ball-bearing pulverizer See ball-and-race-type pulverizer. { ‚böl 'ber·iŋ 'pəl·və‚rīz·ər }

ball bonding [ENG] The making of electrical connections in which a flame is used to cut a wire, the molten end of which solidifies as a ball, which is pressed against the bonding pad on an integrated circuit. { 'böl ‚bänd·iŋ }

ball breaker [ENG] **1.** A steel or iron ball that is hoisted by a derrick and allowed to fall on blocks of waste stone to break them or to swing against old buildings to demolish them. Also known as skull cracker; wrecking ball. **2.** A coring and sampling device consisting of a hollow glass ball, 3 to 5 inches (7.5 to 12.5 centimeters) in diameter, held in a frame attached to the trigger line above the triggering weight of the corer; used to indicate contact between corer and bottom. { 'böl ‚brāk·ər }

ball bushing [MECH ENG] A type of ball bearing that allows motion of the shaft in its axial direction. { 'böl ‚bush·iŋ }

ball catch [DES ENG] A door fastener having a contained metal ball which is under pressure from a spring; the ball engages a striking plate and keeps the door from opening until force is applied. { 'böl ‚kach }

ball check valve [ENG] A valve having a ball held by a spring against a seat; used to permit flow in one direction only. { 'böl 'chek ‚valv }

ball float [MECH ENG] A floating device, usually approximately spherical, which is used to operate a ball valve. { 'böl ‚flōt }

ball-float liquid-level meter [ENG] A float which rises and falls with liquid level, actuating a pointer adjacent to a calibrated scale in order to measure the level of a liquid in a tank or other container. { 'bȯl ‚flōt ‖lik·wəd ‖lev·əl ‚mēd·ər }

ball grinder See ball mill. { 'bȯl ‚grind·ər }

ballhead |MECH ENG| That part of the governor which contains flyweights whose force is balanced, at least in part, by the force of compression of a speeder spring. { 'bȯl‚hed }

Balling hydrometer |ENG| A type of saccharometer used to determine the density of sugar solutions. { 'bȯl·iŋ hī'dräm·əd·ər }

ballistic body |ENG| A body free to move, behave, and be modified in appearance, contour, or texture by ambient conditions, substances, or forces, such as by the pressure of gases in a gun, by rifling in a barrel, by gravity, by temperature, or by air particles. { bə'lis·tik ‚bäd·ē }

ballistic coefficient |MECH| The numerical measure of the ability of a missile to overcome air resistance; dependent upon the mass, diameter, and form factor. { bə'lis·tik ‚kō·ə'fish·ənt }

ballistic conditions |MECH| Conditions which affect the motion of a projectile in the bore and through the atmosphere, including muzzle velocity, weight of projectile, size and shape of projectile, rotation of the earth, density of the air, temperature or elasticity of the air, and the wind. { bə'lis·tik kən'dish·əns }

ballistic curve |MECH| The curve described by the path of a bullet, a bomb, or other projectile as determined by the ballistic conditions, by the propulsive force, and by gravity. { bə'lis·tik 'kәrv }

ballistic deflection |MECH| The deflection of a missile due to its ballistic characteristics. { bə'lis·tik di'flek·shən }

ballistic density |MECH| A representation of the atmospheric density encountered by a projectile in flight, expressed as a percentage of the density according to the standard artillery atmosphere. { bə'lis·tik 'den·səd·ē }

ballistic efficiency |MECH| **1.** The ability of a projectile to overcome the resistance of the air; depends chiefly on the weight, diameter, and shape of the projectile. **2.** The external efficiency of a rocket or other jet engine of a missile. { bə'lis·tik i'fish·ən·sē }

ballistic entry |MECH| Movement of a ballistic body from without to within a planetary atmosphere. { bə'lis·tik 'en·trē }

ballistic instrument |ENG| Any instrument, such as a ballistic galvanometer or a ballistic pendulum, that measures an impact or sudden pulse of energy. { bə'lis·tik 'in·strə·mənt }

ballistic limit |MECH| The minimum velocity at which a particular armor-piercing projectile is expected to consistently and completely penetrate armor plate of given thickness and physical properties at a specified angle of obliquity. { bə'lis·tik 'lim·ət }

ballistic magnetometer |ENG| A magnetometer designed to employ the transient voltage induced in a coil when either the magnetized sample or coil are moved relative to each other. { bə'lis·tik ‚mag·nə'täm·əd·ər }

ballistic measurement |MECH| Any measurement in which an impulse is applied to a device such as the bob of a ballistic pendulum, or the moving part of a ballistic galvanometer, and the subsequent motion of the device is used to determine the magnitude of the impulse, and, from this magnitude, the quantity to be measured. { bə'lis·tik 'mezh·ər·mənt }

ballistic pendulum |ENG| A device which uses the deflection of a suspended weight to determine the momentum of a projectile. { bə'lis·tik 'pen·jə·ləm }

ballistics |MECH| Branch of applied mechanics which deals with the motion and behavior characteristics of missiles, that is, projectiles, bombs, rockets, guided missiles, and so forth, and of accompanying phenomena. { bə'lis·tiks }

ballistic separator |CIV ENG| A device that takes out noncompostable material like stones, glass, metal, and rubber, from solid waste by passing the waste over a rotor that has impellers to fling the material in the air; the lighter organic (compostable) material travels a shorter distance than the heavier (noncompostable) material. { bə'lis·tik 'sep·ə‚rād·ər }

ballistics of penetration |MECH| That part of terminal ballistics which treats of the motion of a projectile as it forces its way into targets of solid or semisolid substances, such as earth, concrete, or steel. { bə'lis·tiks əv pen·ə'trā·shən }

ballistic table |MECH| Compilation of ballistic data from which trajectory elements such as angle of fall, range to summit, time of flight, and ordinate at any time, can be obtained. { bə'lis·tik 'tā·bəl }

ballistic temperature |MECH| That temperature (in °F) which, when regarded as a surface temperature and used in conjunction with the lapse rate of the standard artillery atmosphere, would produce the same effect on a projectile as the actual temperature distribution encountered by the projectile in flight. { bə'lis·tik 'tem·prə·chər }

ballistic trajectory |MECH| The trajectory followed by a body being acted upon only by gravitational forces and resistance of the medium through which it passes. { bə'lis·tik trə'jek·tə·rē }

ballistic uniformity |MECH| The capability of a propellant, when fired under identical conditions from round to round, to impart uniform muzzle velocity and produce similar interior ballistic results. { bə'lis·tik ‚yü·nə'för·məd·ē }

ballistic vehicle |ENG| A nonlifting vehicle; a vehicle that follows a ballistic trajectory. { bə'lis·tik 'vē·ə·kəl }

ballistic wave |MECH| An audible disturbance caused by compression of air ahead of a missile in flight. { bə'lis·tik ‚wāv }

ballistic wind |MECH| That constant wind which would produce the same effect upon the trajectory of a projectile as the actual wind encountered in flight. { bə'lis·tik 'wind }

ball mill |MECH ENG| A pulverizer that consists of a horizontal rotating cylinder, up to three diameters in length, containing a charge of tumbling or cascading steel balls, pebbles, or rods. Also known as ball grinder. { 'bȯl ,mil }

balloon framing |CIV ENG| Framing for a building in which each stud is one piece from roof to foundation. { bə'lün ,fram·iŋ }

balloting |MECH| A tossing or bounding movement of a projectile, within the limits of the bore diameter, while moving through the bore under the influence of the propellant gases. { 'bal·əd·iŋ }

ball-peen hammer |ENG| A hammer with a ball at one end of the head; used in riveting and forming metal. { 'bȯl,pēn 'ham·ər }

ball pendulum test |ENG| A test for measuring the strength of explosives; consists of measuring the swing of a pendulum produced by the explosion of a weighed charge of material. { 'bȯl 'pen·jə·ləm ,test }

ball race |DES ENG| A track, channel, or groove in which ball bearings turn. { 'bȯl ,rās }

ball screw |MECH ENG| An element used to convert rotation to longitudinal motion, consisting of a threaded rod linked to a threaded nut by ball bearings constrained to roll in the space formed by the threads, in order to reduce friction. { 'bȯl ,skrü }

ball test |CIV ENG| In a drain, a test for freedom from obstruction and for circularity in which a ball (less than the diameter of the drain by a specified amount) is rolled through the drain. { 'bȯl ,test }

ball-up |ENG| **1.** During a drilling operation, collection by a portion of the drilling equipment of a mass of viscous consolidated material. **2.** Failure of an anchor to hold on a soft bottom, by pulling out with a large ball of mud attached. { 'bȯl ,əp }

ball valve |MECH ENG| A valve in which the fluid flow is regulated by a ball moving relative to a spherical socket as a result of fluid pressure and the weight of the ball. { 'bȯl ,valv }

baluster |BUILD| A post which supports a handrail and encloses the open sections of a stairway. { 'bal·ə·stər }

balustrade |BUILD| The railing assembly of a stairway consisting of the handrail, balusters, and usually a bottom rail. { 'bal·ə,strād }

band |BUILD| Any horizontal flat member or molding or group of moldings projecting slightly from a wall plane and usually marking a division in the wall. Also known as band course; band molding. |DES ENG| A strip or cord crossing the back of a book to which the sections are sewn. { band }

bandage |BUILD| A strap, band, ring, or chain placed around a structure to secure and hold its parts together, as around the springing of a dome. |ELEC| Rubber ribbon about 4 inches (10 centimeters) wide used for temporarily protecting a telephone or coaxial splice from moisture. { 'ban·dij }

band brake |MECH ENG| A brake in which the frictional force is applied by increasing the tension in a flexible band to tighten it around the drum. { 'band ,brāk }

band chain |ENG| A steel or invar tape, graduated in feet and at least 100 feet (30.5 meters) long, used for accurate surveying. { 'band ,chān }

band clamp |DES ENG| A two-piece metal clamp, secured by bolts at both ends; used to hold riser pipes. { 'band ,klamp }

band clutch |MECH ENG| A friction clutch in which a steel band, lined with fabric, contracts onto the clutch rim. { 'band ,kləch }

band course See band. { 'band ,kȯrs }

banding |DES ENG| A strip of fabric which is used for bands. hydln a glacier, a structure of alternate ice layers of different textures and appearance. { 'band·iŋ }

band molding See band. { 'band ,mōld·iŋ }

band-pass |ELECTR| A range, in hertz or kilohertz, expressing the difference between the limiting frequencies at which a desired fraction (usually half power) of the maximum output is obtained. { 'band ,pas }

band-pass amplifier |ELECTR| An amplifier designed to pass a definite band of frequencies with essentially uniform response. { 'band ,pas ¦am·plə,fī·ər }

band-pass filter |ELECTR| An electric filter which transmits more or less uniformly in a certain band, outside of which the frequency components are attenuated. { 'band ,pas ,fil·tər }

band-pass response |ELECTR| Response characteristics in which a definite band of frequencies is transmitted uniformly. Also known as flat top response. { 'band ,pas ri'späns }

band-pass system |ENG ACOUS| A loudspeaker system, often used for subwoofers, in which the speaker is mounted inside an enclosure on a shelf that divides the enclosure into two parts, and one or both parts are coupled to the outside by a vent; the frequency response of the system is that of a fourth-order band-pass filter (one vent) or an asymmetrical sixth-order band-pass filter (two vents). { 'band,pas ,sis·təm }

band-rejection filter See band-stop filter. { 'band ri'jek·shən ,fil·tər }

band saw |MECH ENG| A power-operated woodworking saw consisting basically of a flexible band of steel having teeth on one edge, running over two vertical pulleys, and operated under tension. { 'band ,sȯ }

band selector |ELECTR| A switch that selects any of the bands in which a receiver, signal generator, or transmitter is designed to operate and usually has two or more sections to make the required changes in all tuning circuits simultaneously. Also known as band switch. { 'band sə'lek·tər }

band wheel |MECH ENG| In a drilling operation, a large wheel that transmits power from the engine to the walking beam. { 'band ,wēl }

49

bang-bang control |CONT SYS| A type of automatic control system in which the applied control signals assume either their maximum or minimum values. { ¦baŋ ¦baŋ kən'trōl }

bang-bang-off control See bang-zero-bang control. { ¦baŋ ¦baŋ 'óf kən,trōl }

bang-bang robot |CONT SYS| A simple robot that can make only two types of motions. { ¦baŋ ¦baŋ 'rō,bät }

bang-zero-bang control |CONT SYS| A type of control in which the control values are at their maximum, zero, or minimum. Also known as bang-bang-off control. { ¦baŋ ,zir·ō 'baŋ kən,trōl }

banister |BUILD| A handrail for a staircase. { 'ban·ə·stər }

bank |CIV ENG| See embankment. |ELEC| **1.** A number of similar electrical devices, such as resistors, connected together for use as a single device. **2.** An assemblage of fixed contacts over which one or more wipers or brushes move in order to establish electrical connections in automatic switching. |ENG| A pipework installation in which the pipes are set parallel to each other, in proximity. |IND ENG| The amount of material allowed to accumulate at a point on a production line where it is not employed or worked upon, to permit reasonable fluctuations in line speed before and after the point. Also known as float. { baŋk }

banker |ENG| The bench or table upon which bricklayers and stonemasons prepare and shape their material. { 'baŋ·kər }

bank material |CIV ENG| Soil or rock in place before excavation or blasting. { 'baŋk mə'tir·ē·əl }

bank measure |CIV ENG| The volume of a given portion of soil or rock as measured in its original position before excavation. { 'baŋk ,mezh·ər }

bar |MECH| A unit of pressure equal to 10^5 pascals, or 10^5 newtons per square meter, or 10^6 dynes per square centimeter. { bär }

Bárány chair |ENG| A chair in which a person is revolved to test his susceptibility to vertigo. { bə'rän·ē ,cher }

barb bolt |DES ENG| A bolt having jagged edges to prevent its being withdrawn from the object into which it is driven. Also known as rag bolt. { 'bärb ,bōlt }

bar bending |CIV ENG| In reinforced concrete construction, the process of bending reinforcing bars to various shapes. { 'bär ,ben·diŋ }

bar chair See bar support. { 'bär ,cher }

bar clamp |DES ENG| A clamping device consisting of a long bar with adjustable clamping jaws; used in carpentry. { 'bär ,klamp }

bare board |ELECTR| A printed circuit board with conductors but no electronic components. { ¦ber 'bórd }

bareboat charter |IND ENG| An agreement to charter a ship without its crew or stores; the fee for its use for a predetermined period of time is based on the price per ton of cargo handled. { 'ber,bōt ,chärd·ər }

barefaced tenon |ENG| A tenon having a shoulder cut on one side only. { 'ber,fāst ¦ten·ən }

bare tube |ENG| In a heat exchanger, a tube whose inner and outer surfaces are both smooth. { ¦ber 'tüb }

bargeboard See vergeboard. { 'bärj,bórd }

barge couple |BUILD| **1.** One of two rafters that support that part of a gable roof which projects beyond the gable wall. **2.** One of the rafters (under the barge course) which serve as grounds for the vergeboards and carry the plastering or boarding of the soffits. Also known as barge rafter. { 'bärj ,kəp·əl }

barge course |BUILD| **1.** The coping of a wall, formed by a course of bricks set on edge. **2.** In a tiled roof, the part of the tiling which projects beyond the principal rafters where there is a gable. { 'bärj ,kórs }

barge rafter See barge couple. { 'bärj ,raf·tər }

barge spike See boat spike. { 'bärj ,spīk }

barge stone |BUILD| One of the stones, generally projecting, which form the sloping top of a gable built of masonry. { 'bärj ,stōn }

bar hole |ENG| A small-diameter hole made in the ground along the route of a gas pipe in a bar test survey. { 'bär ,hōl }

Bari-Sol process |CHEM ENG| Removal of waxes from liquid hydrocarbons by extraction of the wax with a mixed ethylene dichloride-benzene solvent, followed by separation from the hydrocarbon in a centrifuge. { ¦bär·ē ¦säl ,präs·əs }

bar joist |BUILD| A small steel truss with wire or rod web lacing used for roof and floor supports. { 'bär ,jóist }

barker |DES ENG| See bark spud. |ENG| A machine, used mainly in pulp mills, which removes the bark from logs. { 'bär·kər }

barkometer |CHEM ENG| A hydrometer calibrated to test the strength of tanning liquors used in tanning leather. { bär'käm·əd·ər }

bark spud |DES ENG| A tool which peels off bark. Also known as barker. { 'bark ,spəd }

bar linkage |MECH ENG| A set of bars joined together at pivots by means of pins or equivalent devices; used to transmit power and information. { 'bär ,liŋ·kij }

Barlow's equation |MECH| A formula, $t = DP/2S$, used in computing the strength of cylinders subject to internal pressures, where t is the thickness of the cylinder in inches, D the outside diameter in inches, P the pressure in pounds per square inch, and S the allowable tensile strength in pounds per square inch. { 'bär,lōz i'kwā·zhən }

barnacle |ENG| A nodelike deposit that occurs on the surface of a heat exchanger tube or an evaporating device and has a semigranular outer shell bonded to the fouled surface, enclosing a slurry of putrefying organisms. { 'bär·nə·kəl }

barodynamics |MECH| The mechanics of heavy structures which may collapse under their own weight. { ,bar·ə·dī'nam·iks }

barogram |ENG| The record of an aneroid barograph. { 'bar·ə,gram }

barograph See aneroid barograph. { 'bar·ə₁graf }
barometer |ENG| An absolute pressure gage specifically designed to measure atmospheric pressure. { bə'räm·əd·ər }
barometric |ENG| Pertaining to a barometer or to the results obtained by using a barometer. { bar·ə'me·trik }
barometric altimeter See pressure altimeter. { bar·ə'met·rik al'tim·əd·ər }
barometric condenser |MECH ENG| A contact condenser that uses a long, vertical pipe into which the condensate and cooling liquid flow to accomplish their removal by the pressure created at the lower end of the pipe. { bar·ə'met·rik kən'den·sər }
barometric draft regulator |MECH ENG| A damper usually installed in the breeching between a boiler and chimney; permits air to enter the breeching automatically as required, to maintain a constant overfire draft in the combustion chamber. { bar·ə'met·rik 'draft reg·yə'lād·ər }
barometric elevation |ENG| An elevation above mean sea level estimated from the difference in atmospheric pressure between the point in question and an elevation of known value. { bar·ə'met·rik el·ə'vā·shən }
barometric fuse |ENG| A fuse that functions as a result of change in the pressure exerted by the surrounding air. { bar·ə'met·rik 'fyüz }
barometric hypsometry |ENG| The determination of elevations by means of either mercurial or aneroid barometers. { bar·ə'met·rik hip'säm·ə·trē }
barometric leveling |ENG| The measurement of approximate elevation differences in surveying with the aid of a barometer; used especially for large areas. { bar·ə'met·rik 'lev·əl·iŋ }
barometric switch See baroswitch. { bar·ə'met·rik 'swich }
barometrograph See aneroid barograph. { bar·ə'me·tra₁graf }
barometry |ENG| The study of the measurement of atmospheric pressure, with particular reference to ascertaining and correcting the errors of the different types of barometer. { bə'räm·ə·trē }
baromil |MECH| The unit of length used in graduating a mercury barometer in the centimeter-gram-second system. { 'bar·ə₁mil }
baroscope |ENG| An apparatus which demonstrates the equality of the weight of air displaced by an object and its loss of weight in air. { 'bar·ə₁skōp }
barostat |ENG| A mechanism which maintains constant pressure inside a chamber. { 'bar·ə₁stat }
baroswitch |ENG| **1.** A pressure-operated switching device used in a radiosonde which determines whether temperature, humidity, or reference signals will be transmitted. **2.** Any switch operated by a change in barometric pressure. Also known as barometric switch. { 'bar·ə₁swich }

barothermogram |ENG| The record made by a barothermograph. { ¦bar·ō'thər·mə₁gram }
barothermograph |ENG| An instrument which automatically records pressure and temperature. { ¦bar·ō'thər·mə₁graf }
barothermohygrogram |ENG| The record made by a barothermohygrograph. { ¦bar·ō¦thər·mō'hī·grə, gram }
barothermohygrograph |ENG| An instrument that produces graphs of atmospheric pressure, temperature, and humidity on a single sheet of paper. { ¦bar·ō¦thər·mō'hī·grə₁graf }
barotropic phenomenon |THERMO| The sinking of a vapor beneath the surface of a liquid when the vapor phase has the greater density. { ¦bar·ə'träp·ik fə'näm·ə₁nän }
bar post |CIV ENG| One of the posts driven into the ground to form the sides of a field gate. { 'bär ₁pōst }
barrage |CIV ENG| An artificial dam which increases the depth of water of a river or watercourse, or diverts it into a channel for navigation or irrigation. { bə'räzh }
barrage-type spillway |CIV ENG| A passage for surplus water with sluice gates across the width of the entrance. { bə'räzh ₁tīp 'spil₁wā }
barred-and-braced gate |CIV ENG| A gate with a diagonal brace to reinforce the horizontal timbers. { ¦bärd ən ¦brāst 'gāt }
barred gate |CIV ENG| A gate with one or more horizontal timber rails. { ¦bärd 'gāt }
barrel |DES ENG| **1.** A container having a circular lateral cross section that is largest in the middle, and ends that are flat; often made of staves held together by hoops. **2.** A piece of small pipe inserted in the end of a cartridge to carry the squib to the powder. **3.** That portion of a pipe having a constant bore and wall thickness. |MECH| Abbreviated bbl. **1.** The unit of liquid volume equal to 31.5 gallons (approximately 119 liters). **2.** The unit of liquid volume for petroleum equal to 42 gallons (approximately 158 liters). **3.** The unit of dry volume equal to 105 quarts (approximately 116 liters). **4.** A unit of weight that varies in size according to the commodity being weighed. { 'bar·əl }
barrel bolt |DES ENG| A door bolt which moves in a cylindrical casing; not driven by a key. Also known as tower bolt. { 'bar·əl ₁bōlt }
barrel compressor |MECH ENG| A centrifugal compressor having a barrel-shaped housing. { 'bar·əl kəm₁pres·ər }
barrel drain |CIV ENG| Any drain which is cylindrical. { 'bar·əl ₁drān }
barrel-etch reactor |ENG| A type of plasma reactor in which the specimens to be etched are placed in a quartz support stand and a plasma is generated that diffuses and contacts them. { ¦bar·əl ¦ech rē'ak·tər }
barrel fitting |DES ENG| A short length of threaded connecting pipe. { 'bar·əl ₁fid·iŋ }
barrelhead |DES ENG| The flat end of a barrel. { 'bar·əl₁hed }
barrel roof |BUILD| **1.** A roof of semicylindrical

section; capable of spanning long distances parallel to the axis of the cylinder. **2.** See barrel vault. { 'bar·əl ,rüf }

barrels per calendar day |CHEM ENG| A unit measuring the average rate of oil processing in a petroleum refinery, with allowances for downtime over a period of time. Abbreviated BCD. { 'bar·əlz pər ¦kal·ən·dər ,dā }

barrels per day |CHEM ENG| A unit measuring the rate at which petroleum is produced at the refinery. Abbreviated BD; bpd. { 'bar·əlz pər 'dā }

barrels per month |CHEM ENG| A unit measuring the rate at which petroleum is produced at the refinery. Abbreviated BM; bpm. { 'bar·əlz pər 'mənth }

barrels per stream day |CHEM ENG| A measurement used to denote rate of oil or oil-product flow while a fluid-processing unit is in continuous operation. Abbreviated BSD. { 'bar·əlz pər ¦strēm ,dā }

barren liquor |CHEM ENG| Liquid (liquor) from filter-cake washing in which there is little or no recovery value; for example, barren cyanide liquor from washing of gold cake slimes. { 'bar·ən ¦lik·ər }

barricade |ENG| Structure composed essentially of concrete, earth, metal, or wood, or any combination thereof, and so constructed as to reduce or confine the blast effect and fragmentation of an explosion. { 'bar·ə,kād }

barricade shield |ENG| A type of movable shield made of a material designed to absorb ionizing radiation, for protection from radiation. { 'bar·ə,kād ,shēld }

barrier capacitance |ELECTR| The capacitance that exists between the *p*-type and *n*-type semiconductor materials in a semiconductor *pn* junction that is reverse-biased so that it does not conduct. Also known as depletion-layer capacitance; junction capacitance. { 'bar·ē·ər kə,pas·əd·əns }

barrier curb |CIV ENG| A curb with vertical sides high enough to keep vehicles from crossing it. { 'bar·ē·ər ,kərb }

barrier layer See depletion layer. { 'bar·ē·ər ,lā·ər }

barrier separation |CHEM ENG| The separation of a two-component gaseous mixture by selective diffusion of one component through a separative barrier (microporous metal or nonporous polymeric). { 'bar·ē·ər sep·ə'rā·shən }

barrier shield |ENG| A wall or enclosure made of a material designed to absorb ionizing radiation, shielding the operator from an area where radioactive material is being used or processed by remote-control equipment. { 'bar·ē·ər ,shēld }

barrow See handbarrow; wheelbarrow. { 'ba·rō }

barrow run |CIV ENG| A temporary pathway of wood planks or sheets to provide a smooth access for wheeled materials-handling carriers on a building site. { 'ba·rō ,rən }

bar sash lift |BUILD| A type of handle, attached to the bottom rail of a sash, for raising or lowering it. { 'bär 'sash ,lift }

bar screen |MECH ENG| A sieve with parallel steel bars for separating small from large pieces of crushed rock. { 'bär ,skrēn }

bar strainer |DES ENG| A screening device consisting of a bar or a number or parallel bars; used to prevent objects from entering a drain. { 'bär ,strān·ər }

bar support |CIV ENG| A device used to support or hold steel reinforcing bars in proper position before or during the placement of concrete. Also known as bar chair. { 'bär sə'pòrt }

bar test survey |ENG| A leakage survey in which bar holes are driven or bored at regular intervals along the way of an underground gas pipe and the atmosphere in the holes is tested with a combustible gas detector or such. { 'bär ,test 'sər,vā }

Barth plan |IND ENG| A wage incentive plan intended for a low task and for all efficiency points and defined as: earning = rate per hour × square root of the product (hours standard × hours actual). { 'bärth ,plan }

bar turret lathe |MECH ENG| A turret lathe in which the bar stock is slid through the headstock and collet on line with the turning axis of the lathe and held firmly by the closed collet. { 'bär 'tər·ət ,lāth }

bar-type grating |CIV ENG| An open grid assembly of metal bars in which the bearing bars (running in one direction) are spaced by rigid attachment to crossbars. { 'bär ,tīp 'grād·iŋ }

barycentric energy |MECH| The energy of a system in its center-of-mass frame. { ,bar·ə'sen·trik 'en·ər·jē }

barye |MECH| The pressure unit of the centimeter-gram-second system of physical units; equal to 1 dyne per square centimeter (0.001 millibar). Also known as microbar. { 'ba·rē }

basal tunnel |ENG| A water supply tunnel constructed along the basal water table. { 'bā·səl 'tən·əl }

bascule |ENG| A structure that rotates about an axis, as a seesaw, with a counterbalance (for the weight of the structure) at one end. { 'ba,skül }

bascule bridge |CIV ENG| A movable bridge consisting primarily of a cantilever span extending across a channel; it rotates about a horizontal axis parallel with the waterway. { 'ba,skül ,brij }

bascule leaf |CIV ENG| The span of a bascule bridge. { 'ba,skül ,lēf }

base |CHEM ENG| The primary substance in solution in crude oil, and remaining after distillation. |ELECTR| **1.** The region that lies between an emitter and a collector of a transistor and into which minority carriers are injected. **2.** The part of an electron tube that has the pins, leads, or other terminals to which external connections are made either directly or through a socket. **3.** The plastic, ceramic, or other insulating board that supports a printed wiring pattern. |ENG| Foundation or part upon which an object or instrument rests. { bās }

base anchor |BUILD| The metal piece attached to the base of a doorframe for the purpose of securing the frame to the floor. { 'bās ‚aŋ·kər }

base apparatus |ENG| Any apparatus designed for use in measuring with accuracy and precision the length of a base line in triangulation, or the length of a line in first- or second-order traverse. { 'bās ‚ap·ə'rad·əs }

base bias |ELECTR| The direct voltage that is applied to the majority-carrier contact (base) of a transistor. { 'bās ‚bī·əs }

base block |BUILD| **1.** A block of any material, generally with little or no ornament, forming the lowest member of a base, or itself fulfilling the functions of a base, as a member applied to the foot of a door or to window trim. **2.** A rectangular block at the base of a casing or column which the baseboard abuts. **3.** *See* skirting block. { 'bās ‚bläk }

baseboard |BUILD| A finish board covering the interior wall at the junction of the wall and the floor. Also known as skirt; skirting. { 'bās‚bȯrd }

baseboard heater |BUILD| Heating elements installed in panels along the baseboard of a wall. { 'bās‚bȯrd 'hēd·ər }

baseboard radiator |CIV ENG| A heating unit which is located at the lower portion of a wall and to which heat is supplied by hot water, warm air, steam, or electricity. { 'bās‚bȯrd 'rād·ē ‚ād·ər }

base cap *See* base molding. { 'bās ‚kap }

base circle |DES ENG| The circle on a gear such that each tooth-profile curve is an involute of it. { 'bās ‚sər·kəl }

base correction |ENG| The adjustment made to reduce measurements taken in field exploration to express them with reference to the base station values. { 'bās kə'rek·shən }

base course |BUILD| The lowest course or first course of a wall. |CIV ENG| The first layer of material laid down in construction of a pavement. { 'bās ‚kȯrs }

base elbow |DES ENG| A cast-iron pipe elbow having a baseplate or flange which is cast on it and by which it is supported. { 'bās ‚el‚bō }

base electrode |ELECTR| An ohmic or majority carrier contact to the base region of a transistor. { 'bās i'lek‚trōd }

base flashing |BUILD| **1.** The flashing provided by upturned edges of a watertight membrane on a roof. **2.** Any metal or composition flashing at the joint between a roofing surface and a vertical surface, such as a wall or parapet. { 'bās ‚flash·iŋ }

base isolators |CIV ENG| Components placed within a building (not always at the base) which are relatively flexible in the lateral direction, yet can sustain the vertical load. When an earthquake causes ground motions, base isolators allow the structure to respond much more slowly than it would without them, resulting in lower seismic demand on the structure. Isolators may be laminated steel with high-quality rubber pads, sometimes incorporating lead or other energy-absorbing materials. { 'bās ‚ī·sə‚lād· ərz }

base line Abbreviated BL. |ELECTR| The line traced on amplitude-modulated indicators which corresponds to the power level of the weakest echo detected by the radar; it is retraced with every pulse transmitted by the radar but appears as a nearly continuous display on the scope. |ENG| **1.** A surveyed line, established with more than usual care, to which surveys are referred for coordination and correlation. **2.** A cardinal line extending east and west along the astronomic parallel passing through the initial point, along which standard township, section, and quarter-section corners are established. { 'bās ‚līn }

base-line check *See* ground check. { 'bās ‚līn ‚chek }

basement |BUILD| A building story which is wholly or less than half below ground; it is generally used for living space. { 'bās·mənt }

basement wall |BUILD| A foundation wall which encloses a usable area under a building. { 'bās· mənt ‚wȯl }

base molding |BUILD| Molding used to trim the upper edge of interior baseboard. Also known as base cap. { 'bās ‚mōld·iŋ }

base net |ENG| A system, in surveying, of quadrilaterals and triangles that include and are quite close to a base line in a triangulation system. { 'bās ‚net }

base pin *See* pin. { 'bās ‚pin }

base plate |DES ENG| The part of a theodolite which carries the lower ends of the three foot screws and attaches the theodolite to the tripod for surveying. |ENG| A metal plate that provides support or a foundation. { 'bās ‚plāt }

base pressure |MECH| A pressure used as a reference base, for example, atmospheric pressure. { 'bās 'presh·ər }

base screed |ENG| A metal screed with expanded or short perforated flanges that serves as a dividing strip between plaster and cement and acts as a guide to indicate proper thickness of cement or plaster. { 'bās ‚skrēd }

base sheet |BUILD| Saturated or coated felt sheeting which is laid as the first ply in a built-up roofing membrane. { 'bās ‚shēt }

base shoe |BUILD| A molding at the base of a baseboard. { 'bās ‚shü }

base shoe corner |BUILD| A molding piece or block applied in the corner of a room to eliminate the need for mitering the base shoe. { 'bās ‚shü ‚kȯr·nər }

base station |ENG| The point from which a survey begins. { 'bās ‚stā·shən }

base tee |DES ENG| A pipe tee with a connected baseplate for supporting it. { 'bās ‚tē }

base tile |BUILD| The lowest course of tiles in a tiled wall. { 'bās ‚tīl }

base time *See* normal element time; normal time. { 'bās ‚tīm }

basic element *See* elemental motion. { 'bā·sik 'el· ə·mənt }

basic feasible solution |IND ENG| A basic solution to a linear program model in which all the variables are nonnegative. { 'bā·sik ¦fēz·ə·bəl sə'lü·shən }

basic grasp |IND ENG| Any one of the fundamental means of taking hold of an object. { 'bā·sik ¦grasp }

basic motion |IND ENG| A single, complete movement of a body member; determined by motion studies. { 'bā·sik 'mō·shən }

basic motion-time study |IND ENG| A system of predetermined motion-time standards for basic motions. Abbreviated BMT study. { 'bā·sik 'mō·shən 'tīm ‚stəd·ē }

basic solution |IND ENG| A solution to a linear program model, consisting of *m* equations in *n* variables, obtained by solving for *m* variables in terms of the remaining ($n - m$) variables and setting the ($n - m$) variables equal to zero. { 'bā·sik sə'lü·shən }

basic truss |MECH| A framework of bars arranged so that for any given loading of the bars the forces on the bars are uniquely determined by the laws of statics. { ¦bās·ik 'trəs }

basin |CIV ENG| **1.** A dock employing floodgates to keep water level constant during tidal variations. **2.** A harbor for small craft. |DES ENG| An open-top vessel with relatively low sloping sides for holding liquids. { 'bās·ən }

basket |DES ENG| A lightweight container with perforations. |MECH ENG| A type of single-tube core barrel made from thin-wall tubing with the lower end notched into points, which is intended to pick up a sample of granular or plastic rock material by bending in on striking the bottom of the borehole or solid layer; may be used to recover an article dropped into a borehole. Also known as basket barrel; basket tube; sawtooth barrel. { 'bas·kət }

basket strainer |CHEM ENG| A porous-sided or screen-covered vessel used to screen solid particles out of liquid or gas streams. { 'bas·kət ‚strān·ər }

basket sub |ENG| A fishing tool run above a bit or a mill to recover small nondrillable pieces of metal or debris in the well. { 'bas·kət ‚səb }

basket-weave |BUILD| A checkerboard pattern of bricks, flat or on edge. { 'bas·kət ‚wēv }

bass reflex baffle |ENG ACOUS| A loudspeaker baffle having an opening of such size that bass frequencies from the rear of the loudspeaker emerge to reinforce those radiated directly forward. { ¦bas 'rē‚fleks ‚baf·əl }

bass trap |ENG ACOUS| Any device used in a sound-recording studio to absorb sound at frequencies less than about 100 hertz. { 'bās ‚trap }

bassy |ENG ACOUS| Pertaining to sound reproduction that overemphasizes low-frequency notes. { 'bās·ē }

bastard-cut file |DES ENG| A file that has coarser teeth than a rough-cut file. { 'bas·tərd ¦kət ‚fīl }

bastard pointing *See* bastard tuck pointing. { 'bas·tərd ‚póint·iŋ }

bastard thread |DES ENG| A screw thread that does not match any standard threads. { 'bas·tərd ‚thred }

bastard tuck pointing |BUILD| An imitation tuck pointing in which the external face is parallel to the wall, but projects slightly and casts a shadow. Also known as bastard pointing. { 'bas·tərd ¦tək ‚póint·iŋ }

bat bolt |DES ENG| A bolt whose butt or tang is bashed or jagged. { 'bat ‚bōlt }

batch |ENG| **1.** The quantity of material required for or produced by one operation. **2.** An amount of material subjected to some unit chemical process or physical mixing process to make the final product substantially uniform. { bach }

batch box |ENG| A container of known volume used to measure and mix the constituents of a batch of concrete, plaster, or mortar, to ensure proper proportions. { 'bach ‚bäks }

batch distillation |CHEM ENG| Distillation where the entire batch of liquid feed is placed into the still at the beginning of the operation, in contrast to continuous distillation, where liquid is fed continuously into the still. { 'bach dis·tə'lā·shən }

batched water |ENG| The mixing water added to a concrete or mortar mixture before or during the initial stages of mixing. { 'bacht ‚wòd·ər }

batcher |MECH ENG| A machine in which the ingredients of concrete are measured and combined into batches before being discharged to the concrete mixer. { 'bach·ər }

batching |ENG| Weighing or measuring the volume of the ingredients of a batch of concrete or mortar, and then introducing these ingredients into a mixer. { 'bach·iŋ }

batch manufacturing |IND ENG| The manufacture of parts in discrete runs or lots, generally interspersed with other production procedures. { 'bach ‚man·ə'fak·chər·iŋ }

batch mixer |MECH ENG| A machine which mixes concrete or mortar in batches, as opposed to a continuous mixer. { 'bach ‚mik·sər }

batch plant |ENG| An operating installation of equipment including batchers and mixers as required for batching or for batching and mixing concrete materials. { 'bach ‚plant }

batch process |ENG| A process that is not in continuous or mass production; operations are carried out with discrete quantities of material or a limited number of items. { 'bach ‚präs·əs }

batch production *See* series production. { 'bach prə'dək·shən }

batch reactor |CHEM ENG| A chemical reactor in which the reactants and catalyst are introduced in the desired quantities and the vessel is then closed to the delivery of additional material. { 'bach rē‚ak·tər }

batch rectification |CHEM ENG| Batch distillation in which the boiled-off vapor is re-condensed into liquid form and refluxed back into the still to make contact with the rising vapors. { 'bach ‚rek·tə·fə'kā·shən }

batch treatment |CHEM ENG| A corrosion control procedure in which chemical corrosion inhibitors are injected into the lines of a production system. { 'bach ,trēt·mənt }

batch-type furnace |MECH ENG| A furnace used for heat treatment of materials, with or without direct firing; loading and unloading operations are carried out through a single door or slot. { 'bach ,tīp 'fər·nəs }

bathometer |ENG| A mechanism which measures depths in water. { bə'thäm·əd·ər }

bathtub curve |IND ENG| An equipment failure-rate curve with an initial sharply declining failure rate, followed by a prolonged constant-average failure rate, after which the failure rate again increases sharply. { 'bath,təb ,kərv }

bathyclinograph |ENG| A mechanism which measures vertical currents in the deep sea. { ¦bath·ə¦klīn·ə,graf }

bathyconductograph |ENG| A device to measure the electrical conductivity of sea water at various depths from a moving ship. { ¦bath·ə· kən'dək·tə,graf }

bathygram |ENG| A graph recording the measurements of sonic sounding instruments. { 'bath·ə,gram }

bathymetry |ENG| The science of measuring ocean depths in order to determine the sea floor topography. { bə'thim·ə·trē }

bathythermogram |ENG| The record that is made by a bathythermograph. { ¦bath·ə'thər· mə,gram }

bathythermograph |ENG| A device for obtaining a record of temperature against depth (actually, pressure) in the ocean from a ship underway. Abbreviated BT. Also known as bathythermosphere. { ¦bath·ə'thər·mə,graf }

bathythermosphere See bathythermograph. { ¦bath·ə'thər·mə,sfir }

bating |CHEM ENG| Cleaning of depilated leather hides by the action of tryptic enzymes. { 'bād·iŋ }

batted work |ENG| A hand-dressed stone surface scored from top to bottom in narrow parallel strokes (usually 8–10 per inch or 20–25 per centimeter) by use of a batting tool. { 'bad·əd ,wərk }

batten |BUILD| **1.** A sawed timber strip of specific dimension-usually 7 inches (18 centimeters) broad, less than 4 inches (10 centimeters) thick, and more than 6 feet (1.8 meters) long-used for outside walls of houses, flooring, and such. **2.** A strip of wood nailed across a door or other structure made of parallel boards to strengthen it and prevent warping. **3.** See furring. { 'bat·ən }

batten door |BUILD| A wood door without stiles which is constructed of vertical boards held together by horizontal battens on the back side. Also known as ledged door. { 'bat·ən ,dòr }

battened column |CIV ENG| A column consisting of two longitudinal shafts, rigidly connected to each other by batten plates. { 'bat· ənd 'käl·əm }

battened wall |BUILD| A wall to which battens have been affixed. Also known as strapped wall. { 'bat·ənd 'wòl }

batten plate |CIV ENG| A rectangular plate used to connect two parallel structural steel members by riveting or welding. { 'bat·ən ,plāt }

batten roll |BUILD| In metal roofing, a roll joint formed over a triangular-shaped wood piece. Also known as conical roll. { 'bat·ən ,rōl }

batten seam |BUILD| A seam in metal roofing which is formed around a wood strip. { 'bat· ən ,sēm }

batter |CIV ENG| A uniformly steep slope in a retaining wall or pier; inclination is expressed as 1 foot horizontally per vertical unit (in feet). { 'bad·ər }

batter board |CIV ENG| Horizontal boards nailed to corner posts located just outside the corners of a proposed building to assist in the accurate layout of foundation and excavation lines. { 'bad·ər ,bòrd }

batter brace |CIV ENG| A diagonal brace which reinforces one end of a truss. Also known as batter post. { 'bad·ər ,brās }

batter level |ENG| A device for measuring the inclination of a slope. { 'bad·ər ,lev·əl }

batter pile |CIV ENG| A pile driven at an inclination to the vertical to provide resistance to horizontal forces. Also known as brace pile; spur pile. { 'bad·ər ,pīl }

batter post |CIV ENG| **1.** A post at one side of a gateway or at a corner of a building for protection against vehicles. **2.** See batter brace. { 'bad· ər ,pōst }

batter stick |CIV ENG| A tapered board which is hung vertically and used to test the batter of a wall surface. { 'bad·ər ,stik }

battery |CHEM ENG| A series of distillation columns or other processing equipment operated as a single unit. |ELEC| A direct-current voltage source made up of one or more units that convert chemical, thermal, nuclear, or solar energy into electrical energy. { 'bad·ə·rē }

battery limits |CHEM ENG| An area in a refinery or chemical plant encompassing a processing unit or battery of units along with their related utilities and services. { 'bad·ə·rē ,lim·əts }

batting tool |ENG| A mason's chisel usually 3–4½ inches (7.6–11.4 centimeters) wide, used to dress stone to a striated surface. { 'bad·iŋ ,tül }

bauxite treating |CHEM ENG| A catalytic petroleum process in which a vaporized petroleum fraction is passed through beds of bauxite; conversion of many different sulfur compounds, particularly mercaptans into hydrogen sulfide, takes place. { 'bòk,sīt ,trēd·iŋ }

b axis |MECH ENG| The angle that specifies the rotation of a machine tool about the *y* axis. { 'bē ,ak·səs }

bay |ENG| A housing used for equipment. { bā }

bayonet coupling |DES ENG| A coupling in which two or more pins extend out from a plug and engage in grooves in the side of a socket. { ¦bā·ə'net ¦kəp·liŋ }

bayonet socket |DES ENG| A socket, having J-shaped slots on opposite sides, into which a bayonet base or coupling is inserted against a spring and rotated until its pins are seated firmly in the slots. { 'bā·ə'net 'säk·ət }

bayonet-tube exchanger |MECH ENG| A dual-tube apparatus with heating (or cooling) fluid flowing into the inner tube and out of the annular space between the inner and outer tubes; can be inserted into tanks or other process vessels to heat or cool the liquid contents. { 'bā·ə'net ,tüb iks'chānj·ər }

B-B fraction |CHEM ENG| A mixture of butanes and butenes distilled from a solution of light liquid hydrocarbons. { 'bē¦bē 'frak·shən }

bbl See barrel.

BCD See barrels per calendar day.

BD See barrels per day.

BDC See bottom dead center.

bdft See board-foot.

beacon tracking |ENG| The tracking of a moving object by means of signals emitted from a transmitter or transponder within or attached to the object. { 'bē·kən ,trak·iŋ }

bead |DES ENG| A projecting rim or band. { bēd }

bead and butt |BUILD| Framed work in which the panel is flush with the framing and has a bead run on two edges in the direction of the grain; the ends are left plain. Also known as bead butt; bead butt work. { 'bēd ən 'bət }

bead-and-flush panel See beadflush panel. { 'bēd ən 'fləsh ,pan·əl }

bead and quirk See quirk bead. { 'bēd ən 'kwərk }

bead and reel |BUILD| A semiround convex molding decorated with a pattern of disks alternating with round or elongated beads. Also known as reel and bead. { 'bēd ən 'rēl }

bead butt See bead and butt. { 'bēd ,bət }

bead, butt, and square |BUILD| Framed work similar to bead and butt but having the panels flush on the beaded face only, and showing square reveals on the other. { 'bēd ,bət ən 'skwär }

bead butt work See bead and butt. { 'bēd ,bət ,wərk }

beaded molding |BUILD| A molding or cornice bearing a cast plaster string of beads. { 'bēd·əd 'mōl·diŋ }

beaded tube end |MECH ENG| The exposed portion of a rolled tube which is rounded back against the sheet in which the tube is rolled. { 'bēd·əd 'tüb ,end }

beadflush panel |BUILD| A panel which is flush with the surrounding framing and finished with a flush bead on all edges of the panel. Also known as bead-and-flush panel. { 'bēd,fləsh ,pan·əl }

beading |BUILD| Collectively, the bead moldings used in ornamenting a given surface. { 'bēd·iŋ }

beading plane |DES ENG| A plane having a curved cutting edge for shaping beads in wood. Also known as bead plane. { 'bēd·iŋ ,plān }

bead-jointed |ENG| Of a carpentry joint, having a bead along the edge of one piece to make the joint less conspicuous. { 'bēd ,jöin·təd }

bead molding |BUILD| A small, convex molding of semicircular or greater profile. Also known as baguette. { 'bēd ,mōl·diŋ }

bead plane See beading plane. { 'bēd ,plān }

beaking joint |BUILD| A joint formed by several heading joints occurring in one continuous line; especially used in connection with the laying of floor planks. { 'bēk·iŋ ,jöint }

beam |CIV ENG| A body, with one dimension large compared with the other dimensions, whose function is to carry lateral loads (perpendicular to the large dimension) and bending movements. { bēm }

beam-and-girder construction |BUILD| A system of floor construction in which the load is distributed by slabs to spaced beams and girders. { 'bēm ən 'gər·dər kən'strək·shən }

beam-and-slab floor |BUILD| A floor system in which a concrete floor slab is supported by reinforced concrete beams. { 'bēm ən 'slab ,flór }

Beaman stadia arc |ENG| An attachment to an alidade consisting of a stadia arc on the outer edge of the visual vertical arc; enables the observer to determine the difference in elevation of the instrument and stadia rod without employing vertical angles. { 'bē·man 'stād·ē·ə ,ärk }

beam bearing plate |CIV ENG| A foundation plate (usually of metal) placed beneath the end of a beam, at its point of support, to distribute the end load at the point. { 'bēm ,ber·iŋ ,plāt }

beam blocking |BUILD| **1.** Boxing-in or covering a joist, beam, or girder to give the appearance of a larger beam. **2.** Strips of wood used to create a false beam. { 'bēm ,bläk·iŋ }

beam bolster |CIV ENG| A rod which provides support for steel reinforcement in formwork for a reinforced concrete beam. { 'bēm ,bōl·stər }

beam box See wall box. { 'bēm ,bäks }

beam brick |BUILD| A face brick which is used to bond to a poured-in-place concrete lintel. { 'bēm ,brik }

beam bridge |CIV ENG| A fixed structure consisting of a series of steel or concrete beams placed parallel to traffic and supporting the roadway directly on their top flanges. { 'bēm ,brij }

beam clip |ENG| A device for attaching a pipe hanger to its associated structural beam when it is undesirable to weld the pipe hanger to supporting structural steelwork. Also known as girder clamp; girder clip. { 'bēm ,klip }

beam column |CIV ENG| A structural member subjected simultaneously to axial load and bending moments produced by lateral forces or eccentricity of the longitudinal load. { 'bēm ,kál·əm }

beam-deflection amplifier |MECH ENG| A jet-interaction fluidic device in which the direction of a supply jet is varied by flow from one or more control jets which are oriented at approximately 90° to the supply jet. { 'bēm di'flek·shən 'am·plə,fī·ər }

beam fill |BUILD| Masonry, brickwork, or cement fill, usually between joists or horizontal beams at their supports; provides increased fire resistance. { 'bēm ,fil }

beam form |CIV ENG| A form which gives the necessary shape, suppport, and finish to a concrete beam. { 'bēm ,fòrm }

beamhouse |CHEM ENG| A place where the initial wet operations of tanning, involving soaking in water and solutions of alkali, are carried out. { 'bēm,haùs }

beam pattern See directivity pattern. { 'bēm ,pad·ərn }

beam pocket |CIV ENG| **1.** In a vertical structural member, an opening to receive a beam. **2.** An opening in the form for a column or girder where the form for an intersecting beam is framed. { 'bēm ,pàk·ət }

beam splice |CIV ENG| A connection between two lengths of a beam or girder; may be shear or moment connections. { 'bēm ,splīs }

beam spread |ENG| The angle of divergence from the central axis of an electromagnetic or acoustic beam as it travels through a material. { 'bēm ,spred }

Beams servoed rotational method |ENG| A method of measuring the gravitational constant by determining the inertial reaction of a torsional pendulum to the angular acceleration of a rotating table that is required to cancel the attraction of the pendulum to two large masses. { 'bēmz 'sər,vōd rō'tā·shən·əl ,meth·əd }

beam test |CIV ENG| A test of the flexural strength (modulus of rupture) of concrete from measurements on a standard reinforced concrete beam. { 'bēm ,test }

bean |ENG| A restriction, such as a nipple, which is placed in a pipe to reduce the rate of fluid flow. { bēn }

bearer |CIV ENG| Any horizontal beam, joist, or member which supports a load. { 'ber·ər }

bearing |CIV ENG| That portion of a beam, truss, or other structural member which rests on the supports. |MECH ENG| A machine part that supports another part which rotates, slides, or oscillates in or on it. { 'ber·iŋ }

bearing bar |BUILD| A wrought-iron bar placed on masonry to provide a level support for floor joists. |CIV ENG| A load-carrying bar which supports a grating and which extends in the direction of the grating span. |ENG| See azimuth instrument. { 'ber·iŋ ,bär }

bearing cap |DES ENG| A device designed to fit around a bearing to support or immobilize it. { 'ber·iŋ ,kap }

bearing capacity |MECH| Load per unit area which can be safely supported by the ground. { 'ber·iŋ kə'pas·əd·ē }

bearing circle |ENG| A ring designed to fit snugly over a compass or compass repeater, and provided with vanes for observing compass bearings. { 'ber·iŋ ,sər·kəl }

bearing cursor |ENG| Of a radar set, the radial line inscribed on a transparent disk which can be rotated manually about an axis coincident with the center of the plan position indicator;

used for bearing determination. Also known as mechanical bearing cursor. { 'ber·iŋ ,kər·sər }

bearing distance |CIV ENG| The length of a beam between its bearing supports. { 'ber·iŋ ,dis·təns }

bearing partition |BUILD| A partition which supports a vertical load. { 'ber·iŋ pər'tish·ən }

bearing pile |ENG| A vertical post or pile which carries the weight of a foundation, transmitting the load of a structure to the bedrock or subsoil without detrimental settlement. { 'ber·iŋ ,pīl }

bearing plate |CIV ENG| A flat steel plate used under the end of a wall-bearing beam to distribute the load over a broader area. { 'ber·iŋ ,plāt }

bearing pressure |MECH| Load on a bearing surface divided by its area. Also known as bearing stress. { 'ber·iŋ ,presh·ər }

bearing strain |MECH| The deformation of bearing parts subjected to a load. { 'ber·iŋ ,strān }

bearing strength |MECH| The maximum load that a column, wall, footing, or joint will sustain at failure, divided by the effective bearing area. { 'ber·iŋ ,strenkth }

bearing stress See bearing pressure. { 'ber·iŋ ,stres }

bearing test |ENG| A test of the bearing capacities of pile foundations, such as a field loading test of an individual pile; a laboratory test of soil samples for bearing capacities. { 'ber·iŋ ,test }

bearing wall |CIV ENG| A wall capable of supporting an imposed load. Also known as structural wall. { 'ber·iŋ ,wòl }

bear trap gate |CIV ENG| A type of crest gate with an upstream leaf and a downstream leaf which rest in a horizontal position, one leaf overlapping the other, when the gate is lowered. { 'ber ,trap ,gāt }

beater |ENG| **1.** A tool for packing in material to fill a blasthole containing a charge of powder. **2.** A laborer who shovels or dumps asbestos fibers and sprays them with water in order to prepare them for the beating. |MECH ENG| A machine that cuts or beats paper stock. { 'bēd·ər }

beater mill See hammer mill. { 'bēd·ər ,mil }

beating |ENG| A process that reduces asbestos fibers to pulp for making asbestos paper. { 'bēd·iŋ }

Beattie and Bridgman equation |THERMO| An equation that relates the pressure, volume, and temperature of a real gas to the gas constant. { ¦bēd·ē ən ¦brij·mən i'kwā·zhən }

beat tone |ENG ACOUS| Musical tone due to beats, produced by the heterodyning of two high-frequency wave trains. { 'bēt ,tōn }

bêche |MECH ENG| A pneumatic forge hammer having an air-operated ram and an air-compressing cylinder integral with the frame. { besh }

Beckmann thermometer |ENG| A sensitive thermometer with an adjustable range so that small differences in temperature can be measured. { 'bek·mən ther'mäm·əd·ər }

bed |CIV ENG| **1.** In masonry and bricklaying, the side of a masonry unit on which the unit lies in the course of the wall; the underside when

Bedaux plan

the unit is placed horizontally. **2.** The layer of mortar on which a masonry unit is set. |MECH ENG| The part of a machine having precisely machined ways or bearing surfaces which support or align other machine parts. { bed }

Bedaux plan |IND ENG| A wage incentive plan in which work is standardized into man-minute units called bedaux (B); 60 B per hour is 100% productivity, and earnings are based on work units per length of time. { bə'dō ,plan }

bedding |CIV ENG| **1.** Mortar, putty, or other substance used to secure a firm and even bearing, such as putty laid in the rabbet of a window frame, or mortar used to lay bricks. **2.** A base which is prepared in soil or concrete for laying masonry or concrete. { 'bed·iŋ }

bedding course |CIV ENG| The first layer of mortar at the bottom of masonry. { 'bed·iŋ ,kȯrs }

bedding dot |BUILD| A small spot of plaster built out to the face of a finished wall or ceiling; serves as a screed for leveling and plumbing in the application of plaster. { 'bed·iŋ ,dät }

bed joint |CIV ENG| **1.** A horizontal layer of mortar on which masonry units are laid. **2.** One of the radial joints in an arch. { 'bed ,jȯint }

bed molding |BUILD| **1.** The lowest member of a band of moldings. **2.** Any molding under a projection, such as between eaves and sidewalls. { 'bed ,mōl·diŋ }

beehive oven |ENG| An arched oven that carbonizes coal into coke by using the heat of combustion of gases that are formed, and of a small part of the coke that is formed, with no recovery of by-products. { 'bē,hīv ,əv·ən }

beetle See rammer. { 'bēd·əl }

behavioral dynamics |IND ENG| **1.** The behavioral operating characteristics of individuals and groups in terms of how these people are conditioned by their working environments. **2.** The interactions between individuals or groups in the workplace. { bi'hā·vyə·rəl dī'nam·iks }

Belfast truss |CIV ENG| A bowstring beam for large spans, having the upper member bent and the lower member horizontal; constructed entirely of timber components. { 'bel,fast 'trəs }

bell |ENG| **1.** A hollow metallic cylinder closed at one end and flared at the other; it is used as a fixed-pitch musical instrument or signaling device and is set vibrating by a clapper or tongue which strikes the lip. **2.** See bell tap. { bel }

bell-and-spigot joint |ENG| A pipe joint in which a pipe ending in a bell-like shape is joined to a pipe ending in a spigotlike shape. { 'bel ən 'spik·ət ,jȯint }

bell cap |CHEM ENG| A hemispherical or triangular metal casting used on distillation-column trays to force upflowing vapors to bubble through layers of downcoming liquid. { 'bel ,kap }

belled caisson |CIV ENG| A type of drilled caisson with a flared bottom. { 'beld 'kā,sän }

bell glass See bell jar. { 'bel ,glas }

bell jar |ENG| A bell-shaped vessel, usually made of glass, which is used for enclosing a vacuum, holding gases, or covering objects. Also known as bell glass. { 'bel ,jär }

bell-jar testing |ENG| A leak testing method in which a vessel is filled with tracer gas and placed in a vacuum chamber; leaks are evidenced by gas drawn into the vacuum chamber. { 'bel ,jär ,tes·tiŋ }

bell-joint clamp |ENG| A clamp applied to a bell-and-spigot joint to prevent leakage. { 'bel ,jȯint ,klamp }

Bellman's principle of optimality |IND ENG| The principle that an optimal sequence of decisions in a multistage decision process problem has the property that whatever the initial state and decisions are, the remaining decisions must constitute an optimal policy with regard to the state resulting from the first decisions. { 'bel·mənz 'prin·sə·pəl əv ,äp·tə'mal·əd·ē }

bell mouth |DES ENG| A flared mouth on a pipe opening or other orifice. |ENG| A defect which occurs during metal drilling in which a twist drill produces a hole that is not a perfect circle. { 'bel ,mau̇th }

bellows |ENG| **1.** A mechanism that expands and contracts, or has a rising and falling top, to suck in air through a valve and blow it out through a tube. **2.** Any of several types of enclosures which have accordionlike walls, allowing one to vary the volume. **3.** See aneroid capsule. { 'bel·ōz }

bellows expansion joint |DES ENG| In a run of piping, a joint formed with a flexible metal bellows which compress or stretch to compensate for linear expansion or contraction of the run of piping. { 'bel·ōz ik'span·shən ,jȯint }

bellows gage |ENG| A device for measuring pressure in which the pressure on a bellows, with the end plate attached to a spring, causes a measurable movement of the plate. { 'bel·ōz ,gāj }

bellows gas meter |ENG| A device for measuring the total volume of a continuous gas flow stream in which the motion of two bellows, alternately filled with and exhausted of the gas, actuates a register. { 'bel·ōz ,gas ,mēd·ər }

bellows seal |MECH ENG| A boiler seal in the form of a bellows which prevents leakage of air or gas. { 'bel·ōz ,sēl }

bell-type manometer |ENG| A differential pressure gage in which one pressure input is fed into an inverted cuplike container floating in liquid, and the other pressure input presses down upon the top of the container so that its level in the liquid is the measure of differential pressure. { 'bel,tīp mə'näm·əd·ər }

belt |CIV ENG| In brickwork, a projecting row (or rows) of bricks, or an inserted row made of a different kind of brick. |MECH ENG| A flexible band used to connect pulleys or to convey materials by transmitting motion and power. { belt }

belt conveyor |MECH ENG| A heavy-duty conveyor consisting essentially of a head or drive pulley, a take-up pulley, a level or inclined endless belt made of canvas, rubber, or metal, and carrying and return idlers. { 'belt kən'vā·ər }

belt course See string course. { 'belt ‚kȯrs }

belt drive [MECH ENG] The transmission of power between shafts by means of a belt connecting pulleys on the shafts. { 'belt ‚drīv }

belted-bias tire See bias-belted tire. { ¦bel·təd ‚bī·əs 'tīr }

belt feeder [MECH ENG] A short belt conveyor used to transfer granulated or powdered solids from a storage or supply point to an end-use point; for example, from a bin hopper to a chemical reactor. { 'belt ‚fēd·ər }

belt guard [MECH ENG] A cover designed to protect a belt as well as the pulleys it connects. { 'belt ‚gärd }

belt highway See beltway. { ¦belt 'hī‚wā }

belt sander [MECH ENG] A portable sanding tool having a power-driven abrasive-coated continuous belt. { 'belt ‚san·dər }

belt shifter [MECH ENG] A device with fingerlike projections used to shift a belt from one pulley to another or to replace a belt which has slipped off a pulley. { 'belt ‚shif·tər }

belt slip [MECH ENG] The difference in speed between the driving drum and belt conveyor. { 'belt ‚slip }

belt tightener [MECH ENG] In a belt drive, a device that takes up the slack in a belt that has become stretched and permanently lengthened. { 'belt ‚tīt·nər }

beltway [CIV ENG] A highway that encircles an urban area along its perimeter. Also known as belt highway; ring road. { 'belt‚wā }

bench assembly [ENG] A technique of fitting and joining parts using a bench as a work surface. { 'bench ə'sem·blē }

bench check [IND ENG] A workshop or servicing bay check which includes the typical check or actual functional test of an item to ascertain what is to be done to return the item to a serviceable condition or ascertain the item's temporary or permanent disposition. { 'bench ‚chek }

bench dog [ENG] A wood or metal peg, placed in a slot or hole at the end of a bench; used to keep a workpiece from slipping. { 'bench ‚dȯg }

bench hook [ENG] Any device used on a carpenter's bench to keep work from moving toward the rear of the bench. Also known as side hook. { 'bench ‚hůk }

benching [CIV ENG] **1.** Concrete laid on the side slopes of drainage channels where the slopes are interrupted by manholes, and so forth. **2.** Concrete laid on sloping sites as a safeguard against sliding. **3.** Concrete laid along the sides of a pipeline to provide additional support. { 'bench·iŋ }

bench lathe [MECH ENG] A small engine or toolroom lathe suitable for attachment to a workbench; bed length usually does not exceed 6 feet (1.8 meters) and workpieces are generally small. { 'bench ‚lāth }

benchmark [ENG] A relatively permanent natural or artificial object bearing a marked point whose elevation above or below an adopted datum—for example, sea level—is known. Abbreviated BM. [IND ENG] A standard of measurement possessing sufficient identifiable characteristics common to the individual units of a population to facilitate economical and efficient comparison of attributes for units selected from a sample. { 'bench‚märk }

benchmark index [IND ENG] In manufacturing and mining, an index designed to reflect changes in output occurring between census years. { 'bench‚märk 'in‚deks }

benchmark job [IND ENG] A job that can be related or compared to other jobs in terms of common characteristics and considered an acceptable gauge for other jobs without the need of direct measurements. { 'bench‚märk ‚jäb }

bench photometer [ENG] A device which uses an optical bench with the two light sources to be compared mounted one at each end; the comparison between the two illuminations is made by a device moved along the bench until matching brightnesses appear. { 'bench fə'täm·əd·ər }

bench plane [DES ENG] A plane used primarily in benchwork on flat surfaces, such as a block plane or jack plane. { 'bench ‚plān }

bench sander [MECH ENG] A stationary power sander, usually mounted on a table or stand, which is equipped with a rotating abrasive disk or belt. { 'bench ‚san·dər }

bench-scale testing [ENG] Testing of materials, methods, or chemical processes on a small scale, such as on a laboratory worktable. { 'bench ‚skāl 'tes·tiŋ }

bench stop [ENG] A bench hook which is used to fasten work in place, often by means of a screw. { 'bench ‚stäp }

bench table [BUILD] A projecting course of masonry at the foot of an interior wall or around a column; generally wide enough to form a seat. { 'bench ‚tā·bəl }

bench vise [ENG] An ordinary vise fixed to a workbench. { 'bench ‚vīs }

benchwork [ENG] Any work performed at a workbench rather than on machines or in the field. { 'bench‚wərk }

bend [DES ENG] **1.** The characteristic of an object, such as a machine part, that is curved. **2.** A section of pipe that is curved. **3.** A knot formed by a rope fastened to an object or another rope. { bend }

bend allowance [DES ENG] Length of the arc of the neutral axis between the tangent points of a bend in any material. { 'bend ə'lau·əns }

bender See bending machine. { 'ben·dər }

bending [ENG] **1.** The forming of a metal part, by pressure, into a curved or angular shape, or the stretching or flanging of it along a curved path. **2.** The forming of a wooden member to a desired shape by pressure after it has been softened or plasticized by heat and moisture. { 'ben·diŋ }

bending brake [MECH ENG] A press brake for

making sharply angular linear bends in sheet metal. { 'ben·diŋ ˌbrāk }

bending iron [ENG] A tool used to straighten or to expand flexible pipe, especially lead pipe. { 'ben·diŋ ˌī·ərn }

bending machine [MECH ENG] A machine for bending a metal or wooden part by pressure. Also known as bender. { 'ben·diŋ mə,shēn }

bending moment [MECH] Algebraic sum of all moments located between a cross section and one end of a structural member; a bending moment that bends the beam convex downward is positive, and one that bends it convex upward is negative. { 'ben·diŋ ˌmō·mənt }

bending-moment diagram [MECH] A diagram showing the bending moment at every point along the length of a beam plotted as an ordinate. { 'ben·diŋ ˌmō·mənt ˌdī·ə,gram }

bending schedule [CIV ENG] A chart showing the shapes and dimensions of every reinforcing bar and the number of bars required on a particular job for the construction of a reinforced concrete structure. { 'ben·diŋ ˌskej·əl }

bending stress [MECH] An internal tensile or compressive longitudinal stress developed in a beam in response to curvature induced by an external load. { 'ben·diŋ ˌstres }

Bendix-Weiss universal joint [MECH ENG] A universal joint that provides for constant angular velocity of the driven shaft by transmitting the torque through a set of four balls lying in the plane that contains the bisector of, and is perpendicular to, the plane of the angle between the shafts. { ˌben,diks ˈwīs ˌyü·nə'vər·səl ˌjóint }

bend radius [DES ENG] The radius corresponding to the curvature of a bent specimen or part, as measured at the inside surface of the bend. { 'bend ˌrād·ē·əs }

bend wheel [MECH ENG] A wheel used to interrupt and change the normal path of travel of the conveying or driving medium; most generally used to effect a change in direction of conveyor travel from inclined to horizontal or a similar change. { 'bend ˌwēl }

Benioff extensometer [ENG] A linear strainmeter for measuring the change in distance between two reference points separated by 60–90 feet (20–30 meters) or more; used to observe earth tides. { 'ben·ē·óf ˌek,sten'säm·əd·ər }

bent [CIV ENG] A framework support transverse to the length of a structure. { bent }

bent bar [CIV ENG] A longitudinal reinforcing bar which is bent to pass from one face of a structural member to the other face. { 'bent ˌbär }

bent-tube boiler [MECH ENG] A water-tube steam boiler in which the tubes terminate in upper and lower steam-and-water drums Also known as drum-type boiler. { 'bent ˌtüb 'bóil·ər }

bentwood [ENG] Wood formed to shape by bending, rather than by carving or machining. { 'bent,wùd }

benzol-acetone process [CHEM ENG] A solvent

dewaxing process in which a mixture of the solvent and oil containing wax is cooled until the wax solidifies and is then removed by filtration. { 'ben,zól 'as·ə,tōn ˌpräs·əs }

Bergius process [CHEM ENG] Treatment of carbonaceous matter, such as coal or cellulosic materials, with hydrogen at elevated pressures and temperatures in the presence of a catalyst, to form an oil similar to crude petroleum. Also known as coal hydrogenation. { 'ber·gē·əs 'präs·əs }

Berl saddle [CHEM ENG] A type of column packing used in distillation columns. { 'bərl ˌsad·əl }

berm [CIV ENG] A horizontal ledge cut between the foot and top of an embankment to stabilize the slope by intercepting sliding earth. { bərm }

Bernoulli-Euler law [MECH] A law stating that the curvature of a beam is proportional to the bending moment. { ber,nü·lē ˌóil·ər ˌló }

Berthelot method [THERMO] A method of measuring the latent heat of vaporization of a liquid that involves determining the temperature rise of a water bath that encloses a tube in which a given amount of vapor is condensed. { 'ber·tə,ló ˌmeth·əd }

Berthon dynamometer [ENG] An instrument for measuring the diameters of small objects, consisting of two metal straightedges inclined at a small angle and rigidly joined together; a scale on one of the straightedges is used to read the diameters of objects inserted between them. { 'bər,thän ˌdī·nə'mäm·əd·ər }

beryllium detector [ENG] An instrument designed to detect and analyze for beryllium by gamma-ray activation analysis. Also known as berylometer. { bə'ril·ē·əm di'tek·tər }

berylometer See beryllium detector. { ˌber·ə'läm·əd·ər }

best commercial practice [ENG] A manufacturing standard for a process vessel which has not been designed according to standard codes, such as the American Society of Mechanical Engineers Boiler Code. { ˌbest kə'mər·shəl 'prak·təs }

beta [ELECTR] The current gain of a transistor that is connected as a grounded-emitter amplifier, expressed as the ratio of change in collector current to resulting change in base current, the collector voltage being constant. { 'bād·ə }

beta-cutoff frequency [ELECTR] The frequency at which the current amplification of an amplifier transistor drops to 3 decibels below its value at 1 kilohertz. { 'bād·ə ˌkəd,óf ˌfrē·kwən·sē }

Bethell process See full-cell process. { 'beth·əl 'präs·əs }

Betterton-Kroll process [CHEM ENG] A method for obtaining pure bismuth from softened and desilverized lead. { ˌbed·ər·tən ˌkról ˌpräs·əs }

Betti reciprocal theorem [MECH] A theorem in the mathematical theory of elasticity which states that if an elastic body is subjected to two systems of surface and body forces, then the work that would be done by the first system acting through the displacements resulting from the second system equals the work that would be done by the second system acting through the displacements resulting from the first system. { 'bät·tē ri'sip·rə·kəl ,thir·əm }

Betti's method [MECH] A method of finding the solution of the equations of equilibrium of an elastic body whose surface displacements are specified; it uses the fact that the dilatation is a harmonic function to reduce the problem to the Dirichlet problem. { 'bät·tēz ,meth·əd }

Betz momentum theory [MECH ENG] A theory of windmill performance that considers the deceleration in the air traversing the windmill disk. { 'bets mə'ment·əm ,thē·ə·rē }

bevel [DES ENG] **1.** The angle between one line or surface and another line or surface, or the horizontal, when this angle is not a right angle. **2.** A sloping surface or line. { 'bev·əl }

beveled closer See king closer. { 'bev·əld 'klō·zər }

bevel gear [MECH ENG] One of a pair of gears used to connect two shafts whose axes intersect. { 'bev·əl ,gir }

beveling See chamfering. { 'bev·ə·liŋ }

bezel [DES ENG] **1.** A grooved rim used to hold a transparent glass or plastic window or lens for a meter, tuning dial, or some other indicating device. **2.** A sloping face on a cutting tool. { 'bez·əl }

B-H meter [ENG] A device used to measure the intrinsic hysteresis loop of a sample of magnetic material. { 'bē¦āch ,mēd·ər }

bhp See boiler horsepower; brake horsepower.

bias [ELEC] **1.** A direct-current voltage used on signaling or telegraph relays or electromagnets to secure desired time spacing of transitions from marking to spacing. **2.** The restraint of a relay armature by spring tension to secure a desired time spacing of transitions from marking to spacing. **3.** The effect on teleprinter signals produced by the electrical characteristics of the line and equipment. **4.** The force applied to a relay to hold it in a given position. [ELECTR] **1.** A direct-current voltage applied to a transistor control electrode to establish the desired operating point. **2.** See grid bias. { 'bī·əs }

bias-belted tire [ENG] A motor-vehicle pneumatic tire constructed with a belt of textile cord, steel, or fiber glass around the tire underneath the tread and on top of the ply cords, and laid at an acute angle to the center line of the tread. Also known as belted-bias tire. { '¦bī·əs ,bel·təd 'tīr }

bias compensation [ENG ACOUS] The application of an outward-directed tension to the pickup arm of a record player to counteract the tendency of the arm to slide toward the center. { 'bī·əs ,käm·pən,sā·shən }

bias current [ELECTR] **1.** An alternating electric current above about 40,000 hertz added to the audio current being recorded on magnetic tape to reduce distortion. **2.** An electric current flowing through the base-emitter junction of a transistor and adjusted to set the operating point of the transistor. { 'bī·əs ,kər·ənt }

bias distortion [ELECTR] Distortion resulting from the operation on a nonlinear portion of the characteristic curve of a vacuum tube or other device, due to improper biasing. { 'bī·əs dis 'tòr·shən }

bias-ply tire [ENG] A motor-vehicle pneumatic tire that has crossed layers of ply cord set diagonally to the center line of the tread. { 'bī·əs ,plī 'tīr }

bias voltage [ELECTR] A voltage applied or developed between two electrodes as a bias. { 'bī·əs ,vōl·tij }

biaxial stress [MECH] The condition in which there are three mutually perpendicular principal stresses; two act in the same plane and one is zero. { bī'ak·sē·əl ,stress }

Biazzi process [CHEM ENG] A continuous-flow process for the nitration of glycerin to nitroglycerin; also used to produce glycol dinitrate and diethylene glycol nitrate. { bē'at·sē ,präs·əs }

bibb cock See bibcock. { 'bib ,käk }

bibcock [DES ENG] A faucet or stopcock whose nozzle is bent downward. Also spelled bibb cock. { 'bib,käk }

bicable tramway [MECH ENG] A tramway consisting of two stationary cables on which the wheeled carriages travel, and an endless rope, which propels the carriages. { 'bī,kā·bəl 'tram,wā }

BICMOS technology [ELECTR] An integrated circuit technology that combines bipolar transistors and CMOS devices on the same chip. { '¦bī'sē,mós tek,näl·ə·jē }

bicycle [MECH ENG] A human-powered land vehicle with two wheels, one behind the other, usually propelled by the action of the rider's feet on the pedals. { 'bī,sik·əl }

bid [ENG] An estimate of costs for specified construction, equipment, or services proposed to a customer company by one or more supplier or contractor companies. { bid }

bidirectional [ENG] Being directionally responsive to inputs in opposite directions. { ,bī·də'rek·shən·əl }

bidirectional microphone [ENG ACOUS] A microphone that responds equally well to sounds reaching it from the front and rear, corresponding to sound incidences of 0 and 180°. { ,bī·də'rek·shən·əl 'mī·krə,fōn }

Bierbaum scratch hardness test [ENG] A test for the hardness of a solid sample by microscopic measurement of the width of scratch made by a diamond point under preset pressure. { 'bir ,baúm ¦skrach 'härd·nəs ,test }

biface tool [DES ENG] A tool, as an ax, made from a coil flattened on both sides to form a V-shaped cutting edge. { 'bī,fās 'tül }

bifacial [DES ENG] Of a tool, having both sides alike. { bī'fā·shəl }

bifilar electrometer [ENG] An electrostatic voltmeter in which two conducting quartz fibers, stretched by a small weight or spring, are separated by their attraction in opposite directions toward two plate electrodes carrying the voltage to be measured. { bī'fi·lər i·lek'träm·əd·ər }

61

bifilar micrometer See filar micrometer. { bī'fi·lər mī'kräm·əd·ər }

bifilar suspension |ENG| The suspension of a body from two parallel threads, wires, or strips. { bī'fi·lər səs'pen·shən }

bilateral tolerance |DES ENG| The amount that the size of a machine part is allowed to vary above or below a basic dimension; for example, 3.650 ± 0.003 centimeters indicates a tolerance of ± 0.003 centimeter. { bī'lad·ə·rəl 'täl·ə·rəns }

bilge block |CIV ENG| A wooden support under the turn of a ship's bilge in dry dock. { 'bilj ,bläk }

bill |DES ENG| One blade of a pair of scissors. { bil }

billet |ENG| In a hydraulic extrusion press, a large cylindrical cake of plastic material placed within the pressing chamber. { 'bil·ət }

bimetallic strip |ENG| A strip formed of two dissimilar metals welded together; different temperature coefficients of expansion of the metals cause the strip to bend or curl when the temperature changes. { ¦bī·mə'tal·ik ,strip }

bimetallic thermometer |ENG| A temperature-measuring instrument in which the differential thermal expansion of thin, dissimilar metals, bonded together into a narrow strip and coiled into the shape of a helix or spiral, is used to actuate a pointer. Also known as differential thermometer. { ¦bī·mə'tal·ik thər'mäm·əd·ər }

bin |ENG| An enclosed space, box, or frame for the storage of bulk substance. { bin }

binary component |ELECTR| An electronic component that can be in either of two conditions at any given time. Also known as binary device. { 'bīn·ə·rē kəm'pō·nənt }

binary counter See binary scaler. { 'bīn·ə·rē 'kaúnt·ər }

binary device See binary component. { 'bīn·ə·rē di'vīs }

binary encoder |ELECTR| An encoder that changes angular, linear, or other forms of input data into binary coded output characters. { 'bīn·ə·rē en'kōd·ər }

binary logic |ELECTR| An assembly of digital logic elements which operate with two distinct states. { 'bīn·ə·rē 'läj·ik }

binary scaler |ELECTR| A scaler that produces one output pulse for every two input pulses. Also known as binary counter; scale-of-two circuit. { 'bīn·ə·rē ¦skā·lər }

binary separation |CHEM ENG| Separation by distillation or solvent extraction of a fully miscible liquid mixture of two chemical compounds. { 'bīn·ə·rē sep·ə'rā·shən }

binary signal |ELECTR| A voltage or current which carries information by varying between two possible values, corresponding to 0 and 1 in the binary system. { 'bīn·ə·rē 'sig·nəl }

binary system |ENG| Any system containing two principal components. { 'bīn·ə·rē 'sis·təm }

binder course |CIV ENG| Coarse aggregate with a bituminous binder between the foundation course and the wearing course of a pavement. { 'bīn·dər ,kórs }

binderless briquetting |ENG| The briquetting of coal by the application of pressure without the addition of a binder. { 'bīn·dər·ləs bri'ked·iŋ }

binding post |ELEC| A manually turned screw terminal used for making electrical connections. { 'bīn·diŋ ,pōst }

bind-seize See freeze. { ¦bīnd ¦sēz }

biochemical profile |IND ENG| Data recorded by both electromyographic and biomechanical means during the performance of a task to evaluate changes in the functional capacity of a worker resulting from modifications in human-equipment interfaces. { ¦bī·ō'kem·ə·kəl 'prō,fīl }

biocontrol system |CONT SYS| A mechanical system that is controlled by biological signals, for example, a prosthesis controlled by muscle activity. { ,bī·ō·kən'trōl ,sis·təm }

bioengineering |ENG| The application of engineering knowledge to the fields of medicine and biology. { ,bī·ō,en·jə'nir·iŋ }

biofilter |ENG| An emission control device that uses microorganisms to destroy volatile organic compounds and hazardous air pollutants. { 'bī·ō,fil·tər }

bioinstrumentation |ENG| The use of instruments attached to animals and man to record biological parameters such as breathing rate, pulse rate, body temperature, or oxygen in the blood. { ¦bī·ō,in·strə·mən'tā·shən }

biomedical engineering |ENG| The application of engineering technology to the solution of medical problems; examples are the development of prostheses such as artificial valves for the heart, various types of sensors for the blind, and automated artificial limbs. { ,bī·ō'med·ə·kəl ,en·jə'nir·iŋ }

bionics |ENG| The study of systems, particularly electronic systems, which function after the manner of living systems. { bī'än·iks }

biopak |ENG| A container for housing a living organism in a habitable environment and for recording biological functions during space flight. { 'bī·ō,pak }

biosolid |CIV ENG| A recyclable, primarily organic solid material produced by wastewater treatment processes. { ¦bī·ō,säl·əd }

biostabilizer |CIV ENG| A component in mechanized composting systems; consists of a drum in which moistened solid waste is comminuted and tumbled for about 5 days until the aeration and biodegradation turns the waste into a fine dark compost. { ,bī·ō'stāb·əl,īz·ər }

biotechnical robot |CONT SYS| A robot that requires the presence of a human operator in order to function. { ¦bī·ō¦tek·nə·kəl 'rō,bät }

biotelemetry |ENG| The use of telemetry techniques, especially radio waves, to study behavior and physiology of living things. { ¦bī·ō·tə'lem·ə·trē }

Biot-Fourier equation |THERMO| An equation for heat conduction which states that the rate of change of temperature at any point divided by the thermal diffusivity equals the Laplacian of the temperature. { ¦byō ¦für·yā i'kwä·zhən }

biotron |ENG| A test chamber used for biological research within which the environmental conditions can be completely controlled, thus allowing observations of the effect of variations in environment on living organisms. { 'bī·ə,trän }

bipolar amplifier |ELECTR| An amplifier capable of supplying a pair of output signals corresponding to the positive or negative polarity of the input signal. { bī'pō·lər 'am·plə,fī·ər }

bipolar circuit |ELECTR| A logic circuit in which zeros and ones are treated in a symmetric or bipolar manner, rather than by the presence or absence of a signal; for example, a balanced arrangement in a square-loop-ferrite magnetic circuit. { bī'pō·lər 'sər·kət }

bipolar electrode |ELEC| Electrode, without metallic connection with the current supply, one face of which acts as anode surface and the opposite face as a cathode surface when an electric current is passed through a cell. { bī'pō·lər i'lek,trōd }

bipolar integrated circuit |ELECTR| An integrated circuit in which the principal element is the bipolar junction transistor. { bī'pō·lər 'in·tə,grād·əd 'sər·kət }

bipolar junction transistor |ELECTR| A bipolar transistor that is composed entirely of one type of semiconductor, silicon. Abbreviated BJT. Also known as silicon homojunction. { ¦bī,pōl·ər ,jəŋk·shən tran'zis·tər }

bipolar magnetic driving unit |ENG ACOUS| Headphone or loudspeaker unit having two magnetic poles acting directly on a flexible iron diaphragm. { bī'pō·lər mag'ned·ik 'driv·iŋ ,yü·nət }

bipolar spin device See magnetic switch. { ¦bī,pō·lər 'spin di,vīs }

bipolar spin switch See magnetic switch. { ¦bī,pō·lər 'spin ,swich }

bipolar transistor |ELECTR| A transistor that uses both positive and negative charge carriers. { bī'pō·lər tranz'is·tər }

birdcaged wire |ENG| Wire rope whose strands have been distorted into the shape of a birdcage by a sudden release of a load during a hoisting operation. { 'bərd,kājd ,wīr }

Birkeland-Eyde process |CHEM ENG| An arc process of nitrogen fixation in which air passes through an alternating-current arc flattened by a magnetic field to form about 1% nitric oxide. { ¦bərk·lənd ¦ī·də 'präs·əs }

Birmingham wire gage |DES ENG| A system of standard sizes of brass wire, telegraph wire, steel tubing, seamless tubing, sheet spring steel, strip steel, and steel plates, bands, and hoops. Abbreviated BWG. { 'bər·miŋ·əm 'wīr ,gāj }

birth-death process |IND ENG| A simple queuing model in which units to be served arrive (birth) and depart (death) in a completely random manner. { ¦bərth ¦deth ,prä,səs }

biscuit See preform. { 'bis·kət }

bistable circuit |ELECTR| A circuit with two stable states such that the transition between the states cannot be accomplished by self-triggering. { ¦bī¦stā·bəl ,sər·kət }

bistable unit |ENG| A physical element that can be made to assume either of two stable states; a binary cell is an example. { ¦bī¦stā·bəl 'yü·nət }

bistatic radar |ENG| Radar system in which the receiver is some distance from the transmitter, with separate antennas for each. { 'bī,stad·ik 'rā,där }

bit |DES ENG| **1.** A machine part for drilling or boring. **2.** The cutting plate of a plane. **3.** The blade of a cutting tool such as an ax. **4.** A removable tooth of a saw. **5.** Any cutting device which is attached to or part of a drill rod or drill string to bore or penetrate rocks. { bit }

bit blank |DES ENG| A steel bit in which diamonds or other cutting media may be inset by hand peening or attached by a mechanical process such as casting, sintering, or brazing. Also known as bit shank; blank; blank bit; shank. { 'bit ,blaŋk }

bit breaker |DES ENG| A heavy plate that fits in a rotary table for holding the drill bit while it is being inserted or broken out of the drill stem. { 'bit ,brāk·ər }

bit cone See roller cone bit. { 'bit ,kōn }

bit drag |DES ENG| A rotary-drilling bit that has serrated teeth. Also known as drag bit. { 'bit ,drag }

bite |ENG| In glazing, the length of overlap of the inner edge of a frame over the edge of the glass. { bīt }

bit matrix |ENG| The material, usually powdered and fused tungsten carbide, into which diamonds are set in the manufacture of diamond bits. { 'bit ,mā·triks }

bitrochanteric width |IND ENG| A measurement corresponding to hip breadth that is used in seating design. { ,bī·trə,kan¦ter·ik 'width }

bit shank See bit blank. { 'bit ,shaŋk }

bittern |CHEM ENG| Concentrated sea water or brine containing the bromides and magnesium and calcium salts left in solution after sodium chloride has been removed by crystallization. { 'bid·ərn }

bituminous distributor |MECH ENG| A tank truck having a perforated spray bar and used for pumping hot bituminous material onto the surface of a road or driveway. { bī¦tüm·ə·nəs dis'trib·yəd·ər }

bivane |ENG| A double-jointed vane which measures vertical as well as horizontal wind direction. { 'bī,vān }

blackbody |THERMO| An ideal body which would absorb all incident radiation and reflect none. Also known as hohlraum; ideal radiator. { 'blak¦bäd·ē }

blackbody radiation |THERMO| The emission of radiant energy which would take place from a blackbody at a fixed temperature; it takes place at a rate expressed by the Stefan-Boltzmann law, with a spectral energy distribution described by Planck's equation. { 'blak¦bäd·ē ,rā·dē'ā·shən }

blackbody temperature |THERMO| The temperature of a blackbody that emits the same amount of heat radiation per unit area as a given object; measured by a total radiation pyrometer. Also known as brightness temperature. { 'blak¦bäd·ē ,tem·prə·chər }

black box |ENG| Any component, usually electronic and having known input and output, that can be readily inserted into or removed from a specific place in a larger system without knowledge of the component's detailed internal structure. { 'blak ,bäks }

black-bulb thermometer |ENG| A thermometer whose sensitive element has been made to approximate a blackbody by covering it with lampblack. { 'blak ,bəlb thər'mäm·əd·ər }

black smoke |ENG| A smoke that has many particulates in it from inefficient combustion; comes from burning fossil fuel, either coal or oil. { ¦blak 'smōk }

black-surface enclosure |THERMO| An enclosure for which the interior surfaces of the walls possess the radiation characteristics of a blackbody. { 'blak ,sər·fəs in'klozh·ər }

blacktop paver |MECH ENG| A construction vehicle that spreads a specified thickness of bituminous mixture over a prepared surface. { 'blak,täp ,pāv·ər }

bladder press |MECH ENG| A machine which simultaneously molds and cures (vulcanizes) a pneumatic tire. { 'blad·ər ,pres }

blade |ELEC| A flat moving conductor in a switch. |ENG| **1.** A broad, flat arm of a fan, turbine, or propeller. **2.** The broad, flat surface of a bulldozer or snowplow by which the material is moved. **3.** The part of a cutting tool, such as a saw, that cuts. { blād }

bladed-surface aerator |CIV ENG| A bladed, rotating component of a water treatment plant; used to infuse air into the water. { 'blad·əd ,sər·fəs 'er,ād·ər }

Blake jaw crusher |MECH ENG| A crusher with one fixed jaw plate and one pivoted at the top so as to give the greatest movement on the smallest lump. { 'blāk ¦jò ,krəsh·ər }

blank |DES ENG| See bit blank. |ELECTR| To cut off the electron beam of a television picture tube, camera tube, or cathode-ray oscilloscope tube during the process of retrace by applying a rectangular pulse voltage to the grid or cathode during each retrace interval. Also known as beam blank. |ENG| **1.** The result of the final cutting operation on a natural crystal. **2.** See blind. { blaŋk }

blank bit See bit blank. { 'blaŋk ,bit }

blanket gas |CHEM ENG| A gas phase introduced into a vessel above a liquid phase to prevent contamination of the liquid, reduce hazard of detonation, or to exert pressure on the liquid. Also known as cushion gas. { 'blaŋ·kət ,gas }

blank flange |DES ENG| A solid disk used to close off or seal a companion flange. { 'blaŋk 'flanj }

blankholder slide |MECH ENG| The outer slide of a double-action power press; it is usually operated by toggles or cams. { 'blaŋk,hōl·dər ,slīd }

blanking |ENG| **1.** The closing off of flow through a liquid-containing process pipe by the insertion of solid disks at joints or unions; used during maintenance and repair work as a safety precaution. Also known as blinding. **2.** Cutting of plastic or metal sheets into shapes by striking with a punch. Also known as die cutting. { 'blaŋk·iŋ }

blast |ENG| The setting off of a heavy explosive charge. { blast }

blast burner |ENG| A burner in which a controlled burst of air or oxygen under pressure is supplied to the illuminating gas used. Also known as blast lamp. { 'blast ,bər·nər }

blast cleaning |ENG| Any cleaning process in which an abrasive is directed at high velocity toward the surface being cleaned, for example, sand blasting. { 'blast ,klēn·iŋ }

blast ditching |CIV ENG| The use of explosives to aid in ditch excavation, such as for laying pipelines. { 'blast ,dich·iŋ }

blaster |ENG| A device for detonating an explosive charge; usually consists of a machine by which an operator, by pressing downward or otherwise moving a handle of the device, may generate a powerful transient electric current which is transmitted to an electric blasting cap. Also known as blasting machine. { 'blas·tər }

blast freezer |ENG| An upright freezer in which very cold air circulated by blowers is used for rapid freezing of food. { 'blast ,frē·zər }

blast heater |MECH ENG| A heater that has a set of heat-transfer coils through which air is forced by a fan operating at a relatively high velocity. { 'blast ,hēd·ər }

blasthole |ENG| **1.** A hole that takes a heavy charge of explosive. **2.** The hole through which water enters in the bottom of a pump stock. { 'blast,hōl }

blasthole drilling |ENG| Drilling to produce a series of holes for placement of blasting charges. { 'blast,hōl ,dril·iŋ }

blasting |ENG| **1.** Cleaning materials by a blast of air that blows small abrasive particles against the surface. **2.** The act of detonating an explosive. { 'blas·tiŋ }

blasting cap |ENG| A copper shell closed at one end and containing a charge of detonating compound, which is ignited by electric current or the spark of a fuse; used for detonating high explosives. { 'blas·tiŋ ,kap }

blasting fuse |ENG| A core of gunpowder in the center of jute, yarn, and so on for igniting an explosive charge in a shothole. { 'blas·tiŋ ,fyüz }

blasting machine See blaster. { 'blas·tiŋ ma'shēn }

blasting mat |ENG| A heavy, flexible, tear-resistant covering that is spread over the surface during blasting to contain earth fragments. { 'blast·iŋ ,mat }

blast lamp See blast burner; blowtorch. { 'blast ,lamp }

blast wall |ENG| A heavy wall used to isolate buildings or areas which contain highly combustible or explosive materials or to protect a building or area from blast damage when exposed to explosions. { 'blast ,wȯl }

Blears effect |ENG| The dependence of the signal from an ionization gage on the geometry of the system being measured when an organic vapor is present in the vacuum; the effect can falsify measurement results by up to an order of magnitude. { 'blirz i,fekt }

bleed |ENG| To let a fluid, such as air or liquid oxygen, escape under controlled conditions from a pipe, tank, or the like through a valve or outlet. { blēd }

bleeder |ELECTR| A high resistance connected across the dc output of a high-voltage power supply which serves to discharge the filter capacitors after the power supply has been turned off, and to provide a stabilizing load. |ENG| A connection located at a low place in an air line or a gasoline container so that, by means of a small valve, the condensed water or other liquid can be drained or bled off from the line or container without discharging the air or gas. { 'blēd·ər }

bleeder turbine |MECH ENG| A multistage turbine where steam is extracted (bled) at pressures intermediate between throttle and exhaust, for process or feedwater heating purposes. { 'blēd·ər ,tər·bən }

bleeding |CHEM ENG| The undesirable movement of certain components of a plastic material to the surface of a finished article. Also known as migration. |ENG| Natural separation of a liquid from a liquid-solid or semisolid mixture; for example, separation of oil from a stored lubricating grease, or water from freshly poured concrete. Also known as bleedout. { 'blēd·iŋ }

bleeding cycle |MECH ENG| A steam cycle in which steam is drawn from the turbine at one or more stages and used to heat the feedwater. Also known as regenerative cycle. { 'blēd·iŋ ,sī·kəl }

bleedout See bleeding. { 'blēd,aut }

bleed valve |ENG| A small-flow valve connected to a fluid process vessel or line for the purpose of bleeding off small quantities of contained fluid. { 'blēd ,valv }

blended data |ENG| Q point that is the combination of scan data and track data to form a vector. { 'blen·dəd 'dad·ə }

blending problem |IND ENG| A linear programming problem in which it is required to find the least costly mix of ingredients which yields the desired product characteristics. { 'blen·diŋ ,präb·ləm }

blending stock |CHEM ENG| Any substance used for compounding gasoline, including natural gasoline, catalytically reformed products, and additives. Also known as blendstock. { 'blen·diŋ ,stäk }

blending value |ENG| Measure of the ability of

an added component (for example, tetraethyllead, isooctane, and aromatics) to affect the octane rating of a base gasoline stock. { 'blen·diŋ ,val·yü }

blendstock See blending stock. { 'blend,stäk }

blend stop |BUILD| A thin wood strip fastened to the exterior vertical edge of the pulley stile or jamb to hold the sash in position. { 'blend ,stäp }

blind |ENG| A solid disk inserted at a pipe joint or union to prevent the flow of fluids through the pipe; used during maintenance and repair work as a safety precaution. Also known as blank. { blīnd }

blind controller system |CONT SYS| A process control arrangement that separates the in-plant measuring points (for example, pressure, temperature, and flow rate) and control points (for example, a valve actuator) from the recorder or indicator at the central control panel. { ¦blīnd kən'trōl·ər ,sis·təm }

blind drilling |ENG| Drilling in which the drilling fluid is not returned to the surface. { 'blīnd 'dril·iŋ }

blind flange |DES ENG| A flange used to close the end of a pipe. { ¦blīnd 'flanj }

blind floor See subfloor. { ¦blīnd 'flȯr }

blind hole |DES ENG| A hole which does not pass completely through a workpiece. |ENG| A type of borehole that does not have the drilling mud or other circulating medium carry the cuttings to the surface. { 'blīnd 'hōl }

blinding |ENG| **1.** A thin layer of lean concrete, fine gravel, or sand that is applied to a surface to smooth over voids in order to provide a cleaner, drier, or more durable finish. **2.** A layer of small rock chips applied over the surface of a freshly tarred road. **3.** See blanking. { 'blin·diŋ }

blind joint |ENG| A joint which is not visible from any angle. { ¦blīnd 'joint }

blind nipple |MECH ENG| A short piece of piping or tubing having one end closed off; commonly used in boiler construction. { ¦blīnd 'nip·əl }

blind spot |ENG| An area on a filter screen where no filtering occurs. Also known as dead area. { 'blīnd ,spät }

blink |MECH| A unit of time equal to 10^{-5} day or to 0.864 second. { bliŋk }

blister |ENG| A raised area on the surface of a metallic or plastic object caused by the pressure of gases developed while the surface was in a partly molten state, or by diffusion of high-pressure gases from an inner surface. { 'blis·tər }

blistering |ENG| The appearance of enclosed or broken macroscopic cavities in a body or in a glaze or other coating during firing. { 'blis·tə·riŋ }

block |DES ENG| **1.** A metal or wood case enclosing one or more pulleys; has a hook with which it can be attached to an object. **2.** See cylinder block. { bläk }

block and fall See block and tackle. { ¦bläk ən 'fȯl }

block and tackle |MECH ENG| Combination of

a rope or other flexible material and independently rotating frictionless pulleys. Also known as block and fall. { ,bläk ən 'tak·əl }

block brake [MECH ENG] A brake which consists of a block or shoe of wood bearing upon an iron or steel wheel. { 'bläk ,bräk }

block diagram [ENG] A diagram in which the essential units of any system are drawn in the form of rectangles or blocks and their relation to each other is indicated by appropriate connecting lines. { 'bläk ,dī·ə,gram }

blocked operation [CHEM ENG] The use of a single chemical or refinery process unit alternately in more than one operation; for example, a catalytic reactor will first produce a chemical product and then will be blocked from the main process stream during catalyst regeneration. { 'bläkt äp·ə'rā·shən }

blocked resistance [ENG ACOUS] Resistance of an audio-frequency transducer when its moving elements are blocked so they cannot move; represents the resistance due only to electrical losses. { 'bläkt ri'zis·təns }

blocker-type forging [ENG] A type of forging for designs involving the use of large radii and draft angles, smooth contours, and generous allowances. { 'bläk·ər ,tīp 'fór·jiŋ }

block hole [ENG] A small hole drilled into a rock or boulder into which an anchor bolt or a small charge or explosive may be placed; used in quarries for breaking large blocks of stone or boulders. { 'bläk ,hōl }

blockhouse [ENG] **1.** A reinforced concrete structure, often built underground or half-underground, and sometimes dome-shaped, to provide protection against blast, heat, or explosion during rocket launchings or related activities, and usually housing electronic equipment used in launching the rocket. **2.** The activity that goes on in such a structure. { 'bläk,haús }

blocking [ELECTR] **1.** Applying a high negative bias to the grid of an electron tube to reduce its anode current to zero. **2.** Overloading a receiver by an unwanted signal so that the automatic gain control reduces the response to a desired signal. **3.** Distortion occurring in a resistance-capacitance-coupled electron tube amplifier stage when grid current flows in the following tube. [ENG] Undesired adhesion between layers of plastic materials in contact during storage or use. { 'bläk·iŋ }

blocking capacitor See coupling capacitor. { 'bläk·iŋ kə'pas·əd·ər }

blocking layer See depletion layer. { 'bläk·iŋ ,lā·ər }

block plane [DES ENG] A small type of hand plane, designed for cutting across the grain of the wood and for planing end grains. { 'bläk ,plān }

block section [CIV ENG] In a railroad system, a specific length of track that is controlled by stop signals. { 'bläk ,sek·shən }

block signal system [CONT SYS] An automatic railroad traffic control system in which the track is sectionalized into electrical circuits to detect the presence of trains, engines, or cars. { 'bläk 'sig·nəl ,sis·təm }

block system [CIV ENG] A railroad system for controlling train movements by using signals between block posts, that is, the structures that contain the instruments indicating the positions of trains, conditions within block sections, and control levers for signals and other functions. { 'bläk ,sis·təm }

blood bank [ENG] A place for storing whole blood or plasma under refrigeration. { 'bləd ,baŋk }

bloom [ENG] **1.** Fluorescence in lubricating oils or a cloudy surface on varnished or enameled surfaces. **2.** To apply an antireflection coating to glass. { blüm }

blotter [ENG] A disk of compressible material used between a grinding wheel and its flanges to avoid concentrated stress. { 'bläd·ər }

blotter press [CHEM ENG] A plate-and-frame filter in which the filter medium is blotting paper. { 'bläd·ər ,press }

blowback [CHEM ENG] **1.** A continuous stream of liquid or gas bled through air lines from instruments and to the process line being monitored; prevents process fluid from backing up and contacting the instrument. **2.** Reverse flow of fluid through a filter medium to remove caked solids. Also known as backwash. [MECH ENG] See blowdown. { 'blō,bak }

blowby [MECH ENG] Leaking of fluid between a cylinder and its piston during operation. { 'blō,bī }

blowcase [CHEM ENG] A cylindrical or spherical corrosion- and pressure-resistant container from which acid is forced by compressed air to the agitator; used in manufacture of acids but largely superseded by centrifugal pumps. Also known as acid blowcase; acid egg. { 'blō,kās }

blowdown [CHEM ENG] Removal of liquids or solids from a process vessel or storage vessel or a line by the use of pressure. [MECH ENG] The difference between the pressure at which safety valve opens and the closing pressure. Also known as blowback. { 'blō,daún }

blowdown line [CHEM ENG] A large conduit to receive and confine fluids forced by pressure from process vessels. { 'blō,daún ,līn }

blowdown stack [CHEM ENG] A vertical stack or chimney into which the contents of a chemical or petroleum process unit are emptied in case of an operational emergency. { 'blō,daún ,stak }

blower [MECH ENG] A fan which operates where the resistance to gas flow is predominantly downstream of the fan. { 'blō·ər }

blowing [CHEM ENG] The introduction of compressed air near the bottom of a tank or other container in order to agitate the liquid therein. [ENG] See blow molding. { 'blō·iŋ }

blowing pressure [ENG] Pressure of the air or other gases used to inflate the parison in blow molding. { 'blō·iŋ ,presh·ər }

blowing still [CHEM ENG] A still or process column in which blown or oxidized asphalt is made. { 'blō·iŋ ,stil }

blow-lifting gripper |CONT SYS| A robot component that uses compressed air to lift objects. { 'blō ¦lift·iŋ ¸grip·ər }

blow molding |ENG| A method of fabricating hollow plastic objects, such as bottles, by forcing a parison into a mold cavity and shaping by internal air pressure. Also known as blowing. { 'blō ¸mōl·diŋ }

blown glass |ENG| Glassware formed by blowing air into a ball of liquefied glass until it reaches the desired shape. { ¦blōn 'glas }

blown tubing |ENG| A flexible thermoplastic film tube made by applying pressure inside a molten extruded plastic tube to expand it prior to cooling and winding flat onto rolls. { ¦blōn 'tü·biŋ }

blowoff valves |MECH ENG| Valves in boiler piping which facilitate removal of solid matter present in the boiler water. { 'blō¸óf ¸valvz }

blowout |ELEC| The melting of an electric fuse because of excessive current. |ENG| **1.** The bursting of a container (such as a tube pipe, pneumatic tire, or dam) by the pressure of the contained fluid. **2.** The rupture left by such bursting. **3.** The abrupt escape of air from the working chamber of a pneumatic caisson. { 'blō¸aút }

blowpipe |ENG| **1.** A long, straight tube, used in glass blowing, on which molten glass is gathered and worked. **2.** A small, tapered, and frequently curved tube that leads a jet, usually of air, into a flame to concentrate and direct it; used in flame tests in analytical chemistry and in brazing and soldering of fine work. **3.** *See* blowtorch. { 'blō¸pīp }

blowpit *See* blowtank. { 'blō¸pit }

blow pressure |ENG| Air pressure required for plastics blow molding. { 'blō ¸presh·ər }

blow rate |ENG| The speed of the cycle at which air or an inert gas is applied intermittently during the forming procedure of blow molding. { 'blō ¸rāt }

blowtank |CHEM ENG| A tank or pit, used in papermaking, into which the contents of a digester are blown upon completion of a cook. Also known as blowpit. { 'blō¸taŋk }

blowtorch |ENG| A small, portable blast burner which operates either by having air or oxygen and gaseous fuel delivered through tubes or by having a fuel tank which is pressured by a hand pump. Also known as blast lamp; blowpipe. { 'blō¸tórch }

blowup |CIV ENG| The localized buckling or breaking of a rigid pavement caused by excess pressure along its length. { 'blō¸əp }

blowup ratio |ENG| **1.** In blow molding of plastics, the ratio of the diameter of the mold cavity to the diameter of the parison. **2.** In blown tubing, the ratio of the diameter of the finished product to the diameter of the die. { 'blō¸əp ¸rā·shō }

blunger |ENG| **1.** A large spatula-shaped wooden implement used to mix clay with water. **2.** A vat, containing a rotating shaft with fixed knives, for mixing clay and water into slip. { 'blən·jər }

blunging |ENG| The mixing or suspending of ceramic material in liquid by agitation, to form slip. { 'blən·iŋ }

blunt file |DES ENG| A file whose edges are parallel. { ¦blənt ¦fīl }

blunting |DES ENG| Slightly rounding a cutting edge to reduce the probability of edge chipping. { 'blən·tiŋ }

BM *See* barrels per month; benchmark.

BMT *See* basic motion-time study.

BMX bicycle |MECH ENG| A small, extremely strong, type of bicycle, having generally 20-inch (500-millimeter) wheels, large-cleat (knobbly) tires, upright but not high-rise handlebars, and a seat positioned more towards the rear wheel than on a conventional bicycle, and used for stunt riding and tricks. { ¸bē¸em¸eks 'bī¸sik·əl }

board drop hammer |MECH ENG| A type of drop hammer in which the ram is attached to wooden boards which slide between two rollers; after the ram falls freely on the forging, it is raised by friction between the rotating rollers. Also known as board hammer. { 'bórd 'dräp ¸ham·ər }

board-foot |ENG| Unit of volume in measuring lumber; equals 144 cubic inches (2360 cubic centimeters), or the volume of a board 1 foot square and 1 inch thick. Abbreviated bd-ft. { 'bórd'fút }

board hammer *See* board drop hammer. { 'bórd ¸ham·ər }

boarding |ENG| **1.** A batch of boards. **2.** Covering with boards. { 'bor·diŋ }

board measure |ENG| Measurement of lumber in board-feet. Abbreviated bm. { 'bórd ¸mezh·ər }

boast |ENG| **1.** To shape stone or curve furniture roughly in preparation for finer work later on. **2.** To finish the face of a building stone by cutting a series of parallel grooves. { bōst }

boaster *See* boasting chisel. { 'bō·stər }

boasting chisel |DES ENG| A broad chisel used in boasting stone. Also known as boaster. { 'bōs·tiŋ ¸chiz·əl }

boat spike |DES ENG| A long, square spike used in construction with heavy timbers. Also known as barge spike. { 'bōt ¸spīk }

Bobillier's law |MECH| The law that, in general plane body motion, when *a* and *b* are the respective centers of curvature of points A and B, the angle between A*a* and the tangent to the centrode of rotation (pole tangent) and the angle between B*b* and a line from the centrode to the intersection of AB and *ab* (collineation axis) are equal and opposite. { bō'bil·yāz ¸lò }

body |MECH ENG| The part of a drill which runs from the outer corners of the cutting lips to the shank or neck. { 'bäd·ē }

body centrode |MECH| The path traced by the instantaneous center of a rotating body relative to the body. { 'bäd·ē 'sen¸tröd }

body cone |MECH| The cone in a rigid body that is swept out by the body's instantaneous axis during Poinsot motion. Also known as polhode cone. { 'bäd·ē ,kōn }

body force [MECH] An external force, such as gravity, which acts on all parts of a body. { 'bäd·ē ,fòrs }

body-load aggregate [IND ENG] A biomechanical unit that comprises the combined weight of the load being manipulated and the body segments involved in the task. { ¦bäd·ē ¦lōd 'a·grə·gət }

body motion [IND ENG] Motion of parts of a human body requiring a change of posture or weight distribution. { 'bäd·ē ,mō·shən }

body rotation [CONT SYS] An axis of motion of a pick-and-place robot. { 'bäd·e rō,tā·shən }

bogie Also spelled bogey; bogy. [ENG] **1.** A supporting and aligning wheel or roller on the inside of an endless track. **2.** A low truck or cart of solid build. **3.** A truck or axle to which wheels are fixed, which supports a railroad car, the leading end of a locomotive, or the end of a vehicle (such as a gun carriage) and which is allowed to swivel under it. **4.** A railroad car or locomotive supported by a bogie. [MECH ENG] The drive-wheel assembly and supporting frame comprising the four rear wheels of a six-wheel truck, mounted so that they can self-adjust to sharp curves and irregularities in the road. { 'bō·gē }

boiler [MECH ENG] A water heater for generating steam. { 'bòil·ər }

boiler air heater [MECH ENG] A component of a steam-generating unit that transfers heat from the products of combustion after they have passed through the steam-generating and super-heating sections to combustion air, which recycles heat to the furnace. { 'bòil·ər 'er ,hēd·ər }

boiler casing [MECH ENG] The gas-tight structure surrounding the component parts of a steam generator. { 'bòil·ər ,kās·iŋ }

boiler circulation [MECH ENG] Circulation of water and steam in a boiler, which is required to prevent overheating of the heat-absorbing surfaces; may be provided naturally by gravitational forces, mechanically by pumps, or by a combination of both methods. { 'bòil·ər sər·kyə'lā·shən }

boiler cleaning [ENG] A mechanical or chemical process for removal of grease, scale, and other deposits from steam boiler surfaces. { 'bòil·ər ,klēn·iŋ }

boiler code [MECH ENG] A code, established by professional societies and administrative units, which contains the basic rules for the safe design, construction, and materials for steam-generating units, such as the American Society of Mechanical Engineers code. { 'bòil·ər ,kōd }

boiler controls [MECH ENG] Either manual or automatic devices which maintain desired boiler operating conditions with respect to variables such as feedwater flow, firing rate, and steam temperature. { 'bòil·ər kən'trōlz }

boiler draft [MECH ENG] The difference between atmospheric pressure and some lower pressure existing in the furnace or gas passages of a steam-generating unit. { 'bòil·ər ,draft }

boiler economizer [MECH ENG] A component of a steam-generating unit that transfers heat from the products of combustion after they have passed through the steam-generating and super-heating sections to the feedwater, which it receives from the boiler feed pump and delivers to the steam-generating section of the boiler. { 'bòil·ər i'kän·ə,miz·ər }

boiler efficiency [MECH ENG] The ratio of heat absorbed in steam to the heat supplied in fuel, usually measured in percent. { 'bòil·ər i'fish·ən·sē }

boiler feedwater [MECH ENG] Water supplied to a steam-generating unit. { 'bòil·ər 'fēd ,wòd·ər }

boiler feedwater regulation [MECH ENG] Addition of water to the steam-generating unit at a rate commensurate with the removal of steam from the unit. { 'bòil·ər 'fēd,wòd·ər reg·yə'lā·shən }

boiler furnace [MECH ENG] An enclosed space provided for the combustion of fuel to generate steam in a boiler. Also known as steam-generating furnace. { 'bòil·ər ,fər·nəs }

boiler heat balance [MECH ENG] A means of accounting for the thermal energy entering a steam-generating system in terms of its ultimate useful heat absorption or thermal loss. { 'bòil·ər 'hēt ,bal·əns }

boiler horsepower [MECH ENG] A measurement of water evaporation rate; 1 boiler horsepower equals the evaporation per hour of $34^{1}/_{2}$ pounds (15.7 kilograms) of water at 212°F (100°C) into steam at 212°F. Abbreviated bhp. { 'bòil·ər 'hòrs,paù·ər }

boiler hydrostatic test [MECH ENG] A procedure that employs water under pressure, in a new boiler before use or in old equipment after major alterations and repairs, to test the boiler's ability to withstand about $1^{1}/_{2}$ times the design pressure. { 'bòil·ər hī·drə'stad·ik 'test }

boiler layup [MECH ENG] A significant length of time during which a boiler is inoperative in order to allow for repairs or preventive maintenance. { 'bòil·ər 'lā·əp }

boiler setting [MECH ENG] The supporting steel and gastight enclosure for a steam generator. { 'bòil·ər ,sed·iŋ }

boiler storage [MECH ENG] A steam-generating unit that, when out of service, may be stored wet (filled with water) or dry (filled with protective gas). { 'bòil·ər ,stòr·ij }

boiler superheater [MECH ENG] A boiler component, consisting of tubular elements, in which heat is added to high-pressure steam to increase its temperature and enthalpy. { 'bòil·ər ¦sü·pər,hēd·ər }

boiler trim [MECH ENG] Piping or tubing close to or attached to a boiler for connecting controls, gages, or other instrumentation. { 'bòil·ər ,trim }

boiler tube [MECH ENG] One of the tubes in a boiler that carry water (water-tube boiler) to be heated by the high-temperature gaseous products of combustion or that carry combustion products (fire-tube boiler) to heat the boiler water that surrounds them. { 'bȯil·ər ˌtüb }

boiler walls [MECH ENG] The refractory walls of the boiler furnace, usually cooled by circulating water and capable of withstanding high temperatures and pressures. { 'bȯil·ər ˌwȯlz }

boiler water [MECH ENG] Water in the steam-generating section of a boiler unit. { 'bȯil·ər ˌwȯd·ər }

boil-off [THERMO] The vaporization of a liquid, such as liquid oxygen or liquid hydrogen, as its temperature reaches its boiling point under conditions of exposure, as in the tank of a rocket being readied for launch. { 'bȯil,ȯf }

bollard [CIV ENG] A heavy post on a dock or ship used in mooring ships. { 'bäl·ərd }

bolograph [ENG] Any graphical record made by a bolometer; in particular, a graph formed by directing a pencil of light reflected from the galvanometer of the bolometer at a moving photographic film. { 'bōl·əˌgraf }

bolometer [ENG] An instrument that measures the energy of electromagnetic radiation in certain wavelength regions by utilizing the change in resistance of a thin conductor caused by the heating effect of the radiation. Also known as thermal detector. { bə'läm·əd·ər }

bolster [ENG] A plate for maintaining a fixed space between stacked heat exchangers or heat-exchanger shells. { 'bōl·stər }

bolster plate [MECH ENG] A plate fixed on the bed of a power press to locate and support the die assembly. { 'bōl·stər ˌplāt }

bolt [DES ENG] A rod, usually of metal, with a square, round, or hexagonal head at one end and a screw thread on the other, used to fasten objects together. { bōlt }

bolt blank [DES ENG] A threadless bolt with a head that can be threaded for specific applications. Also known as screw blank. { 'bōlt ˌblaŋk }

bolted joint [ENG] The assembly of two or more parts by a threaded bolt and nut or by a screw that passes through one member and threads into another. { ¦bōl·təd 'jȯint }

bolted rail crossing [CIV ENG] A crossing whose running surfaces are made of rolled rail and whose parts are joined with bolts. { ¦bōl·təd ˌrāl 'krȯs·iŋ }

bolting [ENG] A fastening system using screw-threaded devices such as nuts, bolts, or studs. { 'bōl·tiŋ }

bolt sleeve [DES ENG] A tube designed to surround a bolt in a concrete wall to prevent the concrete from adhering to the bolt. { 'bōlt ˌslēv }

Boltzmann engine [THERMO] An ideal thermodynamic engine that utilizes blackbody radiation; used to derive the Stefan-Boltzmann law. { 'bōlts·mən ˌen·jən }

bomb ballistics [MECH] The special branch of ballistics concerned with bombs dropped from aircraft. { 'bäm bə'lis·tiks }

bomb calorimeter [ENG] A calorimeter designed with a strong-walled container constructed of a corrosion-resistant alloy, called the bomb, immersed in about 2.5 liters of water in a metal container; the sample, usually an organic compound, is ignited by electricity, and the heat generated is measured. { 'bäm kal·ə'rim·əd·ər }

bombproof [ENG] Referring to shelter, building, or other installation resistant or impervious to the effects of bomb explosions. { 'bäm,prüf }

bomb shelter [CIV ENG] A bomb-proof structure for protection of people. { 'bäm ˌshel·tər }

bomb test [ENG] A leak-testing technique in which the vessel to be tested is immersed in a pressurized fluid which will be driven through any leaks present. { 'bäm ˌtest }

bond [CIV ENG] A piece of building material that serves to unite or bond, such as an arrangement of masonry units. [ELEC] The connection made by bonding electrically. [ENG] **1.** A wire rope that fixes loads to a crane hook. **2.** Adhesion between cement or concrete and masonry or reinforcement. { bänd }

Bond and Wang theory [MECH ENG] A theory of crushing and grinding from which the energy, in horsepower-hours, required to crush a short ton of material is derived. { ¦bänd ən 'waŋ ˌthē·ə·rē }

bond course [BUILD] A course of headers to bond the facing masonry to the backing masonry. { 'bänd ˌkȯrs }

bonded strain gage [ENG] A strain gage in which the resistance element is a fine wire, usually in zigzag form, embedded in an insulating backing material, such as impregnated paper or plastic, which is cemented to the pressure-sensing element. { ¦bän·dəd 'strān ˌgāj }

bonded transducer [ENG] A transducer which employs a bonded strain gage for sensing pressure. { ¦bän·dəd tranz'dü·sər }

bonder See bondstone. { 'bän·dər }

bond header [BUILD] In masonry, a stone that extends the full thickness of the wall. Also known as throughstone. { 'bänd ˌhed·ər }

bonding [ELEC] The use of low-resistance material to connect electrically a chassis, metal shield cans, cable shielding braid, and other supposedly equipotential points to eliminate undesirable electrical interaction resulting from high-impedance paths between them. [ENG] **1.** The fastening together of two components of a device by means of adhesives, as in anchoring the copper foil of printed wiring to an insulating baseboard. **2.** See cladding. { 'bän·diŋ }

bonding strength [MECH] Structural effectiveness of adhesives, welds, solders, glues, or of the chemical bond formed between the metallic and ceramic components of a cermet, when subjected to stress loading, for example, shear, tension, or compression. { 'bän·diŋ ˌstreŋkth }

Bond's law [MECH ENG] A statement that relates the work required for the crushing of solid materials (for example, rocks and ore) to the product size and surface area and the lengths of cracks formed. Also known as Bond's third theory. { 'bänz 'lò }

Bond's third theory See Bond's law. { 'bänz ˌthərd 'thē·ə·rē }

bondstone |BUILD| A stone joining the coping above a gable to the wall. |CIV ENG| A masonry stone set with its longest dimension perpendicular to the wall face to bind the wall together. Also known as bonder. { 'bänd,stōn }

bond strength |ENG| The amount of adhesion between bonded surfaces measured in terms of the stress required to separate a layer of material from the base to which it is bonded. { 'bänd ˌstreŋkth }

bond timber |BUILD| A section of wood built horizontally into a brick or stone wall in order to strengthen it or to hold it together during construction. { 'bänd ˌtim·bər }

boom |ENG| 1. A row of joined floating timbers that extend across a river or enclose an area of water for the purpose of keeping saw logs together. 2. A temporary floating barrier launched on a body of water to contain material, for example, an oil spill. 3. A structure consisting of joined floating logs placed in a stream to retard the flow. |MECH ENG| A movable steel arm installed on certain types of cranes or derricks to support hoisting lines that must carry loads. { büm }

boom cat |MECH ENG| A tractor supporting a boom and used in laying pipe. { büm ˌkat }

boom dog |MECH ENG| A ratchet device installed on a crane to prevent the boom of the crane from being lowered but permitting it to be raised. Also known as boom ratchet. { 'büm ˌdòg }

boomer |ENG| A device used to tighten chains on pipe or other equipment loaded on a truck to make the cargo secure. { 'büm·ər }

boomerang sediment corer |ENG| A device, designed for nighttime recovery of a sediment core, which automatically returns to the surface after taking the sample. { 'bü·mə,raŋ 'sed·ə·mənt ˌkòr·ər }

boom ratchet See boom dog. { 'büm ˌrach·ət }

boom stop |MECH ENG| A steel projection on a crane that will be struck by the boom if it is raised or lowered too great a distance. { 'büm ˌstäp }

Boord synthesis |CHEM ENG| A method of producing alpha olefins by the reduction of alpha bromo ethers with zinc. { 'bòrd ˌsin·thə·səs }

boost |ELECTR| To augment in relative intensity, as to boost the bass response in an audio system. |ENG| To bring about a more potent explosion of the main charge of an explosive by using an additional charge to set it off. { büst }

booster |ELEC| A small generator inserted in series or parallel with a larger generator to maintain normal voltage output under heavy loads. |ELECTR| 1. A separate radio-frequency amplifier connected between an antenna and a television receiver to amplify weak signals. 2. A radio-frequency amplifier that amplifies and rebroadcasts a received television or communication radio carrier frequency for reception by the general public. |MECH ENG| A compressor that is used as the first stage in a cascade refrigerating system. { 'büs·tər }

booster brake |MECH ENG| An auxiliary air chamber, operated from the intake manifold vacuum, and connected to the regular brake pedal, so that less pedal pressure is required for braking. { 'büs·tər ˌbrāk }

booster ejector |MECH ENG| A nozzle-shaped apparatus from which a high-velocity jet of steam is discharged to produce a continuous-flow vacuum for process equipment. { 'büs·tər ē'jek·tər }

booster fan |MECH ENG| A fan used to increase either the total pressure or the volume of flow. { 'büs·tər ˌfan }

booster pump |MECH ENG| A machine used to increase pressure in a water or compressed-air pipe. { 'büs·tər ˌpəmp }

booster stations |ENG| Booster pumps or compressors located at intervals along a liquid-products or gas pipeline to boost the pressure of the flowing fluid to keep it moving toward its destination. { 'büs·tər ˌstā·shənz }

bootjack |ENG| A fishing tool used in drilling wells. { 'büt,jak }

bootstrap |ENG| A technique or device designed to bring itself into a desired state by means of its own action. { 'büt,strap }

bootstrap circuit |ELECTR| A single-stage amplifier in which the output load is connected between the negative end of the anode supply and the cathode, while signal voltage is applied between grid and cathode; a change in grid voltage changes the input signal voltage with respect to ground by an amount equal to the output signal voltage. { 'büt,strap ˌsər·kət }

bootstrap driver |ELECTR| Electronic circuit used to produce a square pulse to drive the modulator tube; the duration of the square pulse is determined by a pulse-forming line. { 'büt ˌstrap ˌdrīv·ər }

bootstrap integrator |ELECTR| A bootstrap sawtooth generator in which an integrating amplifier is used in the circuit. Also known as Miller generator. { 'büt,strap 'in·tə,grād·ər }

bootstrapping |ELECTR| A technique for lifting a generator circuit above ground by a voltage value derived from its own output signal. { 'büt ˌstrap·iŋ }

bootstrap sawtooth generator |ELECTR| A circuit capable of generating a highly linear positive sawtooth waveform through the use of bootstrapping. { ¦büt,strap ¦sò,tüth 'jen·ə,rād·ər }

bore |DES ENG| Inside diameter of a pipe or tube. |MECH ENG| 1. The diameter of a piston-cylinder mechanism as found in reciprocating engines, pumps, and compressors. 2. To penetrate or pierce with a rotary tool. 3. To machine a workpiece to increase the size of an existing hole in it. { bòr }

borehole See drill hole. { 'bòr,hōl }

borehole bit See noncoring bit. { 'bȯr,hōl ,bit }

borehole logging [ENG] The technique of investigating and recording the character of the formation penetrated by a drill hole in mineral exploration and exploitation work. Also known as drill-hole logging. { 'bȯr,hōl ,läg·iŋ }

borehole survey [ENG] Also known as drill-hole survey. **1.** Determining the course of and the target point reached by a borehole, using an azimuth-and-dip recording apparatus small enough to be lowered into a borehole. **2.** The record of the information thereby obtained. { 'bȯr,hōl ,sər·vā }

borer [MECH ENG] An apparatus used to bore openings into the earth up to about 8 feet (2.4 meters) in diameter. { 'bȯr·ər }

borescope [ENG] A straight-tube telescope using a mirror or prism, used to visually inspect a cylindrical cavity, such as the cannon bore of artillery weapons for defects of manufacture and erosion caused by firing. { 'bȯr,skōp }

boresighting [ENG] Initial alignment of a directional microwave or radar antenna system by using an optical procedure or a fixed target at a known location. { 'bȯr,sīd·iŋ }

boring bar [MECH ENG] A rigid tool holder used to machine internal surfaces. { 'bȯr·iŋ ,bär }

boring log See drill log. { 'bȯr·iŋ ,läg }

boring machine [MECH ENG] A machine tool designed to machine internal work such as cylinders, holes in castings, and dies; types are horizontal, vertical, jig, and single. { 'bȯr·iŋ mə 'shēn }

boring mill [MECH ENG] A boring machine tool used particularly for large workpieces; types are horizontal and vertical. { 'bȯr·iŋ ,mil }

borrow [CIV ENG] Earth material such as sand and gravel that is taken from one location to be used as fill at another. { 'bä·rō }

borrow pit [CIV ENG] An excavation dug to provide material (borrow) for fill elsewhere. { 'bä·rō ,pit }

bort bit See diamond bit. { 'bȯrt ,pit }

Bosch fuel injection pump [MECH ENG] A pump in the fuel injection system of an internal combustion engine, whose pump plunger and barrel are a very close lapped fit to minimize leakage. { ¦bȯsh 'fyül in¦jek·shən ,pəmp }

Bosch metering system [MECH ENG] A system having a helical groove in the plunger which covers or uncovers openings in the barrel of the pump; most usually applied in diesel engine fuel-injection systems. { ¦bȯsh 'mēd·ə·riŋ ,sis·təm }

boss [DES ENG] Protuberance on a cast metal or plastic part to add strength, facilitate assembly, provide for fastenings, or so forth. { bȯs }

Boston ridge [BUILD] A method of applying shingles to the ridge of a house by which the shingles alternate in overlap from one side of the ridge to the other. { 'bȯs·tən ,rij }

bottle [ENG] A container made from pipe or plate with drawn, forged, or spun end closures, and used for storing or transporting gas. { 'bäd·əl }

bottle centrifuge [ENG] A centrifuge in which the mixture to be separated is poured into small bottles or test tubes; they are then placed in a rotor assembly which is spun rapidly. { 'bäd·əl 'sen·trə,fyüj }

bottleneck assignment problem [IND ENG] A linear programming problem in which it is required to assign machines to jobs (or vice versa) so that the efficiency of the least efficient operation is maximized. { 'bäd·əl,nek ə'sīn·mənt ,präb·ləm }

bottle thermometer [ENG] A thermoelectric thermometer used for measuring air temperature; the name is derived from the fact that the reference thermocouple is placed in an insulated bottle. { 'bäd·əl thər'mäm·əd·ər }

bottom blow [ENG] A type of plastics blow molding machine in which air is injected into the parison from the bottom of the mold. { 'bäd·əm ,blō }

bottom chord [CIV ENG] Any of the bottom series of truss members parallel to the roadway of a bridge. { 'bäd·əm ,kȯrd }

bottom dead center [MECH ENG] The position of the crank of a vertical reciprocating engine, compressor, or pump when the piston is at the end of its downstroke. Abbreviated BDC. { 'bäd·əm ,ded 'sen·tər }

bottom dump [ENG] A construction wagon with movable gates in the bottom to allow vertical discharge of its contents. { 'bäd·əm ,dəmp }

bottomed hole [ENG] A completed borehole, or a borehole in which drilling operations have been discontinued. { ¦bäd·əmd 'hōl }

bottom flow [ENG] A molding apparatus that forms hollow plastic articles by injecting the blowing air at the bottom of the mold. { 'bäd·əm ,flō }

bottoming drill [DES ENG] A flat-ended twist drill designed to convert a cone at the bottom of a drilled hole into a cylinder. { 'bäd·əm·iŋ ,dril }

bottoms [CHEM ENG] Residual fractions that remain at the bottom of a fractionating tower following distillation of the lighter components. { 'bäd·əmz }

bottom sampler [ENG] Any instrument used to obtain a sample from the bottom of a body of water. { 'bäd·əm ,sam·plər }

bottom tap [DES ENG] A tap with a chamfer 1 to 1½ threads in length. { 'bäd·əm ,tap }

boulder buster [ENG] A heavy, pyramidal- or conical-point steel tool which may be attached to the bottom end of a string of drill rods and used to break, by impact, a boulder encountered in a borehole. Also known as boulder cracker. { 'bōl·dər ,bəs·tər }

boulder cracker See boulder buster. { 'bōl·dər ,krak·ər }

bounce table [MECH ENG] A testing device which subjects devices and components to impacts such as might be encountered in accidental dropping. { 'baúns ,tā·bəl }

boundary [ELECTR] An interface between *p*- and *n*-type semiconductor materials, at which donor and acceptor concentrations are equal. { 'baún·drē }

boundary friction |MECH| Friction between surfaces that are neither completely dry nor completely separated by a lubricant. { 'baún·drē ‚frik·shən }

boundary lubrication |ENG| A lubricating condition that is a combination of solid-to-solid surface contact and liquid-film shear. { 'baún·drē ‚lü·brə'kā·shən }

boundary monument |ENG| A material object placed on or near a boundary line to preserve and identify the location of the boundary line on the ground. { 'baún·drē ‚män·yə·mənt }

boundary survey |ENG| A survey made to establish or to reestablish a boundary line on the ground or to obtain data for constructing a map or plat showing a boundary line. { 'baún·drē ‚sər·vā }

bound vector |MECH| A vector whose line of application and point of application are both prescribed, in addition to its direction. { ‚baúnd 'vek·tər }

Bourdon pressure gage |ENG| A mechanical pressure-measuring instrument employing as its sensing element a curved or twisted metal tube, flattened in cross section and closed. Also known as Bourdon tube. { ‚búr·dən 'presh·ər ‚gāj }

Bourdon tube See Bourdon pressure gage. { 'búr·dən 'tüb }

Boussinesq equation |ENG| A relation used to calculate the influence of a concentrated load on the backfill behind a retaining wall. { 'bü·si'nesk i'kwā·shən }

Boussinesq's problem |MECH| The problem of determining the stresses and strains in an infinite elastic body, initially occupying all the space on one side of an infinite plane, and indented by a rigid punch having the form of a surface of revolution with axis of revolution perpendicular to the plane. Also known as Cerruti's problem. { 'bü·si'nesks ‚präb·ləm }

Bowden cable |MECH ENG| A wire made of spring steel which is enclosed in a helical casing and used to transmit longitudinal motions over distances, particularly around corners. { 'bōd·ən ‚kā·bəl }

bowl classifier |CHEM ENG| A shallow bowl with a concave bottom so that a liquid-solid suspension can be fed to the center; coarse particles fall to the bottom, where they are raked to a central discharge point, and liquid and fine particles overflow the edges and are collected. { ‚bōl ‚klas·ə‚fī·ər }

bowl mill See bowl-mill pulverizer. { 'bōl ‚mil }

bowl-mill pulverizer |MECH ENG| A type of pulverizer which directly feeds a coal-fired furnace, in which springs press pivoted stationary rolls against a rotating bowl grinding ring, crushing the coal between them. Also known as a bowl mill. { 'bōl ‚mil 'pəl·və‚riz·ər }

bowl scraper |MECH ENG| A towed steel bowl hung within a fabricated steel frame, running on four or two wheels; transports soil, in addition to spreading and leveling it. { 'bōl ‚skrāp·ər }

Bow's notation |MECH| A graphical method of representing coplanar forces and stresses, using alphabetical letters, in the solution of stresses or in determining the resultant of a system of concurrent forces. { 'bōz nō'tā·shən }

bowstring beam |CIV ENG| A steel, concrete, or timber beam or girder shaped in the form of a bow and string; the string resists the horizontal forces caused by loads on the arch. { 'bō ‚striŋ ‚bēm }

box |DES ENG| See boxing. |ENG| A protective covering or housing. { bäks }

box beam See box girder. { 'bäks ‚bēm }

box caisson |CIV ENG| A floating steel or concrete box with an open top which will be filled and sunk at a foundation site in a river or seaway. Also known as American caisson; stranded caisson. { 'bäks 'kā‚sän }

boxcar |ENG| A railroad car with a flat roof and vertical sides, usually with sliding doors, which carries freight that needs to be protected from weather and theft. { 'bäks‚kär }

box-coking test |ENG| A laboratory test which forecasts the quality of coke producible in commercial practice; uses a specially designed sheet-steel box containing about 60 pounds (27 kilograms) of coal in a commercial coke oven. { 'bäks ‚kōk·iŋ ‚test }

box girder |CIV ENG| A hollow girder or beam with a square or rectangular cross section. Also known as box beam. { 'bäks ‚gər·dər }

box-girder bridge |CIV ENG| A fixed bridge consisting of steel girders fabricated by welding four plates into a box section. { 'bäks ‚gər·dər ‚brij }

box header boiler |MECH ENG| A horizontal boiler with a front header and rear inclined rectangular header connected by tubes. { 'bäks ‚hed·ər ‚bóil·ər }

boxing |DES ENG| The threaded nut for the screw of a mounted auger drill. Also known as box. |ENG| A method of securing shafts solely by slabs and wooden pegs. { 'bäks·iŋ }

boxing shutter |BUILD| A window shutter which can be folded into a boxlike enclosure or recess at the side of the window frame. { 'bäks·iŋ ‚shəd·ər }

box piles |CIV ENG| Pile foundations made by welding together two sections of steel sheet piling or combinations of beams, channels, and plates. { 'bäks ‚pīlz }

boxplot |IND ENG| In quality control, a graph summarizing the distribution, central value, and variability of a set of data values; used to identify problems (or potential problems) that affect the quality of processes and products. { 'bäks‚plät }

box wrench |ENG| A closed-end wrench designed to fit a variety of sizes and shapes of bolt heads and nuts. { 'bäks ‚rench }

Boyle's temperature |THERMO| For a given gas, the temperature at which the virial coefficient B in the equation of state $Pv = RT[1 + (B/v) + (C/v^2) + \cdots]$ vanishes. { 'bóilz 'tem·prə·chər }

bpd See barrels per day.

bpm See barrels per month.

brace |DES ENG| A cranklike device used for turning a bit. |ENG| A diagonally placed structural member that withstands tension and compression, and often stiffens a structure against wind. { brās }

brace and bit |DES ENG| A small hand tool to which is attached a metal- or wood-boring bit. { ¦brās ən 'bit }

braced framing |CIV ENG| Framing a building with post and braces for stiffness. { ¦brāst 'frām·iŋ }

braced-rib arch |CIV ENG| A type of steel arch, usually used in bridge construction, which has a system of diagonal bracing. { ¦brāst ¦rib 'ärch }

brace head |ENG| A cross handle attached at the top of a column of drill rods by means of which the rods and attached bit are turned after each drop in chop-and-wash operations while sinking a borehole through overburden. Also known as brace key. { 'brās ,hed }

brace key See brace head. { 'brās ,kē }

brace pile See batter pile. { 'brās ,pīl }

brachiating motion |CONT SYS| A type of robotic motion that employs legs or other equipment to help the manipulator move in its working environment. { ¦brā·kē'ād·iŋ 'mō·shən }

brachiating robot |CONT SYS| A robot that is capable of moving over the surface of an object. { ¦brā·kē'ād·iŋ 'rō,bät }

brachistochrone |MECH| The curve along which a smooth-sliding particle, under the influence of gravity alone, will fall from one point to another in the minimum time. { brə'kis·tə ,krōn }

bracing |ENG| The act or process of strengthening or making rigid. { 'brās·iŋ }

bracket |BUILD| A vertical board to support the tread of a stair. |CIV ENG| A projecting support. { 'brak·ət }

brad |DES ENG| A small finishing nail whose body either is of uniform thickness or is tapered. { brad }

bradding |ENG| A distortion of a bit tooth caused by the application of excessive weight, causing the tooth to become dull so that its softer inner portion caves over the harder case area. { 'brad·iŋ }

Bragg spectrometer |ENG| An instrument for x-ray analysis of crystal structure and measuring wavelengths of x-rays and gamma rays, in which a homogeneous beam of x-rays is directed on the known face of a crystal and the reflected beam is detected in a suitably placed ionization chamber. Also known as crystal spectrometer; crystal-diffraction spectrometer; ionization spectrometer. { 'brag spek'träm·əd·ər }

braiding |ENG| Weaving fibers into a hollow cylindrical shape. { 'brād·iŋ }

brainstorming |IND ENG| A procedure used to find a solution for a problem by collecting all the ideas, without regard for feasibility, which occur from a group of people meeting together. { 'brān ,stôrm·iŋ }

brake |MECH ENG| A machine element for applying friction to a moving surface to slow it (and often, the containing vehicle or device) down or bring it to rest. { brāk }

brake band |MECH ENG| The contracting element of the band brake. { 'brāk ,band }

brake block |MECH ENG| A portion of the band brake lining, shaped to conform to the curvature of the band and attached to it with countersunk screws. { 'brāk ,bläk }

brake drum |MECH ENG| A rotating cylinder attached to a rotating part of machinery, which the brake band or brake shoe presses against. { 'brāk ,drəm }

brake horsepower |MECH ENG| The power developed by an engine as measured by the force applied to a friction brake or by an absorption dynamometer applied to the shaft or flywheel. Abbreviated bhp. { 'brāk 'hôrs,paú·ər }

brake line |MECH ENG| One of the pipes or hoses that connect the master cylinder and the wheel cylinders in a hydraulic brake system. { 'brāk ,līn }

brake lining |MECH ENG| A covering, riveted or molded to the brake shoe or brake band, which presses against the rotating brake drum; made of either fabric or molded asbestos material. { 'brāk ,lin·iŋ }

brake mean-effective pressure |MECH ENG| Applied to reciprocating piston machinery, the average pressure on the piston during the power stroke, derived from the measurement of brake power output. { 'brāk ¦mēn i'fek·tiv 'presh·ər }

brake shoe |MECH ENG| The renewable friction element of a shoe brake. Also known as shoe. { 'brāk ,shü }

brake thermal efficiency |MECH ENG| The ratio of brake power output to power input. { 'brāk 'thər·məl ə'fish·ən·sē }

branch |ELEC| A portion of a network consisting of one or more two-terminal elements in series. Also known as arm. |ENG| In a piping system, a pipe that originates in or discharges into another pipe. Also known as branch line. { branch }

branch-and-bound technique |IND ENG| A technique in nonlinear programming in which all sets of feasible solutions are divided into subsets, and those having bounds inferior to others are rejected. { ¦branch ən ¦baúnd tek'nēk }

branch gain See branch transmittance. { 'branch ,gān }

branch line |CIV ENG| A secondary line in a railroad system that connects to the main line. |ENG| See branch. { 'branch ,līn }

branch sewer |CIV ENG| A part of a sewer system that is larger in diameter than the lateral sewer system; receives sewage from both house connections and lateral sewers. { ¦branch ¦sü· ər }

branch transmittance |CONT SYS| The amplification of current or voltage in a branch of an electrical network; used in the representation of such a network by a signal-flow graph. Also known as branch gain. (¦branch trans'mit·əns)

brandy |CHEM ENG| A potable alcoholic beverage distilled from wine or fermented fruit juice, usually after the aging of the wine in wooden casks; cognac is a brandy distilled from wines made from grapes from the Cognac region of France. ('bran·dē)

Brayton cycle |THERMO| A thermodynamic cycle consisting of two constant-pressure processes interspersed with two constant-entropy processes. Also known as complete-expansion diesel cycle; Joule cycle. ('brāt·ən ‚sī·kəl)

brazed shank tool |MECH ENG| A metal cutting tool made of a material different from the shank to which it is brazed. (¦brāzd 'shaŋk ‚tül)

breaching |MECH ENG| The space between the end of the tubing and the jacket of a hot-water or steam boiler. ('brēch·iŋ)

breadboard model |ENG| Uncased assembly of an instrument or other piece of equipment, such as a radio set, having its parts laid out on a flat surface and connected together to permit a check or demonstration of its operation. ('bred‚bórd ‚mäd·əl)

breakaway wrist |CONT SYS| A robotic wrist that has a safety feature that guarantees its protection from damage if too much force is exerted on the wrist or end effector. ('brāk·ə‚wā ‚rist)

break-bulk cargo |IND ENG| Miscellaneous goods packed in boxes, bales, crates, cases, bags, cartons, barrels, or drums; may also include lumber, motor vehicles, pipe, steel, and machinery. (¦brāk ¦bəlk 'kär·gō)

breakdown |ELEC| A large, usually abrupt rise in electric current in the presence of a small increase in voltage; can occur in a confined gas between two electrodes, a gas tube, the atmosphere (as lightning), an electrical insulator, and a reverse-biased semiconductor diode. Also known as electrical breakdown. ('brāk‚daún)

breakdown diode |ELEC| A semiconductor diode in which the reverse-voltage breakdown mechanism is based either on the Zener effect or the avalanche effect. ('brāk‚daún ¦dī‚ōd)

breakdown impedance |ELECTR| Of a semiconductor, the small-signal impedance at a specified direct current in the breakdown region. ('brāk‚daún im'pēd·əns)

breakdown potential See breakdown voltage. ('brāk‚daún pə'ten·shəl)

breakdown region |ELECTR| Of a semiconductor diode, the entire region of the volt-ampere characteristic beyond the initiation of breakdown for increasing magnitude of bias. ('brāk‚daún ‚rē·jən)

breakdown voltage |ELEC| **1.** The voltage measured at a specified current in the electrical breakdown region of a semiconductor diode. Also known as Zener voltage. **2.** The voltage at which an electrical breakdown occurs in a dielectric. **3.** The voltage at which an electrical breakdown occurs in a gas. Also known as breakdown potential; sparking potential; sparking voltage. ('brāk‚daún ‚vól·tij)

breaker cam |MECH ENG| A rotating, engine-driven device in the ignition system of an internal combustion engine which causes the breaker points to open, leading to a rapid fall in the primary current. ('brā·kər ‚kam)

breaker plate |ENG| In plastics die forming, a perforated plate at the end of an extruder head; often used to support a screen to keep foreign particles out of the die. ('brā·kər ‚plāt)

break-even analysis |IND ENG| Determination of the break-even point. (brā'kē·vən ə'nal·ə·səs)

break-even point |IND ENG| The point at which a company neither makes a profit nor suffers a loss from the operations of the business, and at which total costs are equal to total sales volume. (brā'kē·vən ‚póint)

break frequency |CONT SYS| The frequency at which a graph of the logarithm of the amplitude of the frequency response versus the logarithm of the frequency has an abrupt change in slope. Also known as corner frequency; knee frequency. ('brāk ‚frē·kwən·sē)

breaking load |MECH| The stress which, when steadily applied to a structural member, is just sufficient to break or rupture it. Also known as ultimate load. ('brāk·iŋ ‚lōd)

breaking pin device |ENG| A device designed to relieve pressure resulting from inlet static pressure by the fracture of a loaded part of a pin. ('brāk·iŋ ‚pin di'vīs)

breaking strength |MECH| The ability of a material to resist breaking or rupture from a tension force. ('brāk·iŋ ‚streŋkth)

breaking stress |MECH| The stress required to fracture a material whether by compression, tension, or shear. ('brāk·iŋ ‚stres)

breakout |ELEC| A joint at which one or more conductors are brought out from a multiconductor cable. |ENG| Failure or collapse of a borehole wall due to stress anisotropy. ('brā‚kaút)

breakout schedule |IND ENG| A schedule for a construction job site, generally in the form of a bar chart, that communicates detailed day-to-day activities to all working levels on the project. ('brāk‚aút ‚skej·əl)

breakover |ELECTR| In a silicon controlled rectifier or related device, a transition into forward conduction caused by the application of an excessively high anode voltage. ('brā‚kō·vər)

breakover voltage |ELECTR| The positive anode voltage at which a silicon controlled rectifier switches into the conductive state with gate circuit open. ('brā‚kō·vər ‚vól·tij)

breakpoint |CHEM ENG| See breakthrough. |IND ENG| In a time study, the end of an element in a work cycle and the point at which a reading is made. Also known as end point; reading point. ('brāk‚póint)

breakthrough |CHEM ENG| **1.** A localized break in a filter cake or precoat that permits fluid to pass through without being filtered. Also known as breakpoint. **2.** In an ion-exchange system, the first appearance of unadsorbed ions of the type which deplete the activity of the resin bed; this indicates that the bed must be regenerated. { 'brāk,thrü }

breakwater [CIV ENG] A wall built into the sea to protect a shore area, harbor, anchorage, or basin from the action of waves. { 'brāk,wód·ər }

breast boards [CIV ENG] Timber planks used to support the tunnel face when excavation is in loose soil. { 'brest ,bórdz }

breast drill [DES ENG] A small, portable hand drill customarily used by handsetters to drill the holes in bit blanks in which diamonds are to be set; it includes a plate that is pressed against the worker's breast. { 'brest ,dril }

breasting dolphin [CIV ENG] A pile or other structure against which a moored ship rests. { ¦brest·iŋ ¦däl·fən }

breast wall [CIV ENG] A low wall built to retain the face of a natural bank of earth. { 'brest ,wól }

breather pipe [MECH ENG] A pipe that opens into a container for ventilation, as in a crankcase or oil tank. Also known as crankcase breather. { 'brē·thər ,pīp }

breath-hold diving [ENG] A form of diving without the use of any artificial breathing mixtures. { 'breth ,hōld ,div·iŋ }

breathing [ENG] **1.** Opening and closing of a plastics mold in order to let gases escape during molding. Also known as degassing. **2.** Movement of gas, vapors, or air in and out of a storage-tank vent line as a result of liquid expansions and contractions induced by temperature changes. { 'brēth·iŋ }

breathing apparatus [ENG] An appliance that enables a person to function in irrespirable or poisonous gases or fluids; contains a supply of oxygen and a regenerator which removes the carbon dioxide exhaled. { 'brēth·iŋ ap·ə'rad·əs }

breathing bag [ENG] A component of a semi-closed-circuit breathing apparatus that mixes the gases to provide low breathing resistance. { 'brēth·iŋ ,bag }

breathing line [CIV ENG] A level of 5 feet (1.5 meters) above the floor; suggested temperatures for various occupancies of rooms and other chambers are usually given at this level. { 'brēth·iŋ ,līn }

breeching [MECH ENG] A duct through which the products of combustion are transported from the furnace to the stack; usually applied in steam boilers. { 'brē·chiŋ }

Brennan monorail car [MECH ENG] A type of car balanced on a single rail so that when the car starts to tip, a force automatically applied at the axle end is converted gyroscopically into a strong righting moment which forces the car back into a position of lateral equilibrium. { ¦bren·ən 'män·ə,rāl ,kär }

Brewster process [CHEM ENG] Concentration of dilute acetic acid by use of an extraction solvent (for example, isopropyl ether), followed by distillation. { 'brü·stər ,präs·əs }

brick molding [BUILD] A wooden molding applied to the gap between the frame of a door or window and the masonry into which the frame has been set. { 'brik ¦mōld·iŋ }

brick seat [BUILD] A ledge on a footing or a wall for supporting a course of masonry. { 'brik ,sēt }

bridge [CIV ENG] A structure erected to span natural or artificial obstacles, such as rivers, highways, or railroads, and supporting a footpath or roadway for pedestrian, highway, or railroad traffic. [ELEC] **1.** An electrical instrument having four or more branches, by means of which one or more of the electrical constants of an unknown component may be measured. **2.** An electrical shunt path. { brij }

bridge abutment [CIV ENG] The end foundation upon which the bridge superstructure rests. { 'brij ə,bət·mənt }

bridge bearing [CIV ENG] The support at a bridge pier carrying the weight of the bridge; may be fixed or seated on expansion rollers. { 'brij ,ber·iŋ }

bridge cable [CIV ENG] Cable from which a roadway or truss is suspended in a suspension bridge; may be of pencil-thick wires laid parallel or strands of wire wound spirally. { 'brij ,kabəl }

bridge crane [MECH ENG] A hoisting machine in which the hoisting apparatus is carried by a bridgelike structure spanning the area over which the crane operates. { 'brij ,krān }

bridge foundation [CIV ENG] The piers and abutments of a bridge, on which the superstructure rests. { 'brij faùn'dā·shən }

bridge hybrid *See* hybrid junction. { 'brij 'hī·brəd }

bridge limiter [ELECTR] A device employed in analog computers to keep the value of a variable within specified limits. { ¦brij ¦lim·əd·ər }

bridge magnetic amplifier [ELECTR] A magnetic amplifier in which each of the gate windings is connected in series with an arm of a bridge rectifier; the rectifiers provide self-saturation and direct-current output. { ¦brij mag'ned·ik 'am·plə,fī·ər }

bridge oscillator [ELECTR] An oscillator using a balanced bridge circuit as the feedback network. { 'brij äs·ə'lād·ər }

bridge pier [CIV ENG] The main support for a bridge, upon which the bridge superstructure rests; constructed of masonry, steel, timber, or concrete founded on firm ground below river mud. { 'brij ,pir }

bridge rectifier [ELECTR] A full-wave rectifier with four elements connected as a bridge circuit with direct voltage obtained from one pair of opposite junctions when alternating voltage is applied to the other pair. { ¦brij ,rek·tə,fī·ər }

bridge trolley [MECH ENG] Either of the wheeled attachments at the ends of the bridge of an overhead traveling crane, permitting the bridge to move backward and forward on elevated tracks. { 'brij ,träl·ē }

bridge vibration [MECH] Mechanical vibration of a bridge superstructure due to natural and human-produced excitations. { 'brij vī'brā·shən }

bridgewall |MECH ENG| A wall in a furnace over which the products of combustion flow. { 'brij,wȯl }

bridging amplifier |ELECTR| Amplifier with an input impedance sufficiently high so that its input may be bridged across a circuit without substantially affecting the signal level of the circuit across which it is bridged. { 'brij·iŋ ,am·plə,fī·ər }

bridging connection |ELECTR| Parallel connection by means of which some of the signal energy in a circuit may be withdrawn frequently, with imperceptible effect on the normal operation of the circuit. { 'brij·iŋ kə,nek·shən }

bridging loss |ELECTR| Loss resulting from bridging an impedance across a transmission system; quantitatively, the ratio of the signal power delivered to that part of the system following the bridging point, and measured before the bridging, to the signal power delivered to the same part after the bridging. { 'brij·iŋ ,lȯs }

bridle |ENG| A pumping unit cable that is looped over the horse head and then connected to the carrier bar; supports the polished-rod clamp. { 'brīd·əl }

bridled-cup anemometer |ENG| A combination cup anemometer and pressure-plate anemometer, consisting of an array of cups about a vertical axis of rotation, the free rotation of which is restricted by a spring arrangement; by adjustment of the force constant of the spring, an angular displacement can be obtained which is proportional to wind velocity. { ¦brīd·əld ¦kəp an·ə'mäm·əd·ər }

Briggs equalizer |ENG| A breathing device consisting of head harness, mouthpiece, nose clip, corrugated breathing tube, an equalizing device, 120 feet (37 meters) of reinforced air tubes, and a strainer and spike. { ¦brigz 'ē·kwə,līz·ər }

Briggs pipe thread See American standard pipe thread. { ¦brigz 'pīp ,thred }

brightness temperature See blackbody temperature. { 'brīt·nəs ,tem·prə·chər }

brine cooler |MECH ENG| The unit for cooling brine in a refrigeration system; the brine usually flows through tubes or pipes surrounded by evaporating refrigerant. { 'brīn ,kül·ər }

Brinell number |ENG| A hardness rating obtained from the Brinell test; expressed in kilograms per square millimeter. { brə'nel ,nəm·bər }

Brinell test |ENG| A test to determine the hardness of a material, in which a steel ball 1 centimeter in diameter is pressed into the material with a standard force (usually 3000 kilograms); the spherical surface area of indentation is measured and divided into the load; the results are expressed as Brinell number. { brə'nel ,test }

briquetting |ENG| **1.** The process of binding together pulverized minerals, such as coal dust, into briquets under pressure, often with the aid of a binder, such as asphalt. **2.** A process or method of mounting mineral ore, rock, or metal fragments in an embedding or casting material, such as natural or artificial resins, waxes, metals, or alloys, to facilitate handling during grinding, polishing, and microscopic examination. { bri'ked·iŋ }

brisance index |ENG| The ratio of an explosive's power to shatter a weight of graded sand as compared to the weight of sand shattered by TNT. { brə'zäns ¦in,deks }

British imperial pound |MECH| The British standard of mass, of which a standard is preserved by the government. { 'brid·ish im'pir·ē·əl 'paund }

British thermal unit |THERMO| Abbreviated Btu. **1.** A unit of heat energy equal to the heat needed to raise the temperature of 1 pound of air-free water from 60° to 61°F at a constant pressure of 1 standard atmosphere; it is found experimentally to be equal to 1054.5 joules. Also known as sixty degrees Fahrenheit British thermal unit ($Btu_{60/61}$). **2.** A unit of heat energy that is equal to 1/180 of the heat needed to raise 1 pound of air-free water from 32°F (0°C) to 212°F (100°C) at a constant pressure of 1 standard atmosphere; it is found experimentally to be equal to 1055.79 joules. Also known as mean British thermal unit (Btu_{mean}). **3.** A unit of heat energy whose magnitude is such that 1 British thermal unit per pound equals 2326 joules per kilogram; it is equal to exactly 1055.05585262 joules. Also known as international table British thermal unit (Btu_{IT}). { 'brid·ish 'thər·məl ,yü·nət }

brittleness |MECH| That property of a material manifested by fracture without appreciable prior plastic deformation. { 'brid·əl·nəs }

brittle temperature |THERMO| The temperature point below which a material, especially metal, is brittle; that is, the critical normal stress for fracture is reached before the critical shear stress for plastic deformation. { 'brid·əl ,tem·prə·chər }

Brix degree |CHEM ENG| A unit of the Brix scale. { 'briks də,grē }

Brix scale |CHEM ENG| A hydrometer scale for sugar solutions indicating the percentage by weight of sugar in the solution at a specified temperature. { 'briks ,skāl }

broach |MECH ENG| A multiple-tooth, barlike cutting tool; the teeth are shaped to give a desired surface or contour, and cutting results from each tooth projecting farther than the preceding one. { brōch }

broaching |ENG| **1.** The restoration of the diameter of a borehole by reaming. **2.** The breaking down of the walls between two contiguous drill holes. |MECH ENG| The machine-shaping of metal or plastic by pushing or pulling a broach across a surface or through an existing hole in a workpiece. { 'brōch·iŋ }

broaching bit See reaming bit. { 'brōch·iŋ bit }

broken-color work See antiquing. { ¦brō·kən ¦kal·ər ,wərk }

bromine test |CHEM ENG| A laboratory test in which the unsaturated hydrocarbons present in a crude oil are determined by mixing a sample with bromine; the lower the rate of bromine absorption, the more paraffinic the test sample. { 'brō,mēn ,test }

bromine value [CHEM ENG] An expression representing the number of centigrams of bromine absorbed by 1 gram of oil under test conditions; an indication of the degree of unsaturation of a given oil. { 'brō,mēn ,val·yü }

brooming [CIV ENG] A method of finishing uniform concrete surfaces, such as the tops of pavement slabs or floor slabs, by dragging a broom over the surface to produce a grooved texture. { 'brü·miŋ }

brown acid [CHEM ENG] Oil-soluble petroleum sulfonate found in sludge following sulfuric acid treatment of petroleum products. { ¦braún ¦as·əd }

brown smoke [ENG] Smoke with less particulates than black smoke; comes from burning fossil fuel, usually fuel oil. { ¦braún ¦smōk }

Brunton See Brunton compass. { 'brənt·ən }

Brunton compass [ENG] A compact field compass, with sights and reflector attached, used for geological mapping and surveying. Also known as Brunton; Brunton pocket transit. { 'brənt·ən ,käm·pəs }

Brunton pocket transit See Brunton compass. { 'brənt·ən ,päk·ət 'tran·zət }

brush [ELEC] A conductive metal or carbon block used to make sliding electrical contact with a moving part. { brəsh }

brush hopper [IND ENG] A rotating brush that wipes quantities of eyelets, rivets, and other small special parts past shaped openings in a chute. { 'brəsh ,häp·ər }

brush rake [MECH ENG] A device with heavy-duty tines that is fixed to the front of a tractor or other prime mover for use in land clearing. { 'brəsh ,rāk }

brush-shifting motor [ENG] A category of alternating-current motor in which the brush contacts shift to modify operating speed and power factor. { 'brəsh ,shif·tiŋ ,mōd·ər }

BSD See barrels per stream day.

B size [ENG] **1.** One of a series of sizes to which trimmed paper and board are manufactured; for size BN, with N equal to any integer from 0 to 10, the length of the shorter side is $2^{-N/2}$ meters, and the length of the longer side is $2^{(1-N)/2}$ meters, with both lengths rounded off to the nearest millimeter. **2.** Of a sheet of paper, the dimensions 11 inches by 17 inches (279 millimeters by 432 millimeters). { 'bē ,sīz }

BT See bathythermograph.

Btu See British thermal unit.

bu See bushel.

bubble cap [CHEM ENG] A metal cap covering a hole in the plate within a distillation tower; designed to permit vapors to rise from below the plate, pass through the cap, and make contact with liquid on the plate. { 'bəb·əl ,kap }

bubble-cap plate [CHEM ENG] One of the devices in large-diameter fractional distillation columns that are designed to produce a bubbling action to exchange the vapor bubbles flowing up the column. { 'bəb·əl ,kap ,plāt }

bubble-cap tray See bubble tray. { 'bəb·əl ¦kap ,trā }

bubble mold cooling [ENG] In plastics injection molding, cooling by means of a continuous liquid stream flowing into a cavity equipped with an outlet at the end opposite the inlet. { 'bəb·əl ,mōld ,kü·liŋ }

bubble test [ENG] Measurement of the largest opening in the mesh of a filter screen; determined by the pressure needed to force air or gas through the screen while it is submerged in a liquid. { 'bəb·əl ,test }

bubble tower [CHEM ENG] A plate tower used in distillation, with plates containing bubble caps. { 'bəb·əl ,taú·ər }

bubble tray [CHEM ENG] A perforated, circular plate placed within a distillation tower at specific places to collect the fractions of petroleum produced in fractional distillation. Also known as bubble-cap tray. { 'bəb·əl ,trā }

bubble-tray column [CHEM ENG] A fractionating column whose plates are formed from bubble caps. { 'bəb·əl ,trā ,käl·əm }

bubble tube [ENG] The glass tube in a spirit level containing the liquid and bubble. { 'bəb·əl ,tüb }

buck [BUILD] The frame into which the finished door fits. { bək }

bucket [ENG] **1.** A cup on the rim of a Pelton wheel against which water impinges. **2.** A reversed curve at the toe of a spillway to deflect the water horizontally and reduce erosiveness. **3.** A container on a lift pump or chain pump. **4.** A container on some bulk-handling equipment, such as a bucket elevator, bucket dredge, or bucket conveyor. **5.** A water outlet in a turbine. **6.** See calyx. { 'bək·ət }

bucket carrier See bucket conveyor. { 'bək·ət ,kar·ē·ər }

bucket conveyor [MECH ENG] A continuous bulk conveyor constructed of a series of buckets attached to one or two strands of chain or in some instances to a belt. Also called bucket carrier. { 'bək·ət kən'vā·ər }

bucket dredge [MECH ENG] A floating mechanical excavator equipped with a bucket elevator. { 'bək·ət ,drej }

bucket elevator [MECH ENG] A bucket conveyor operating on a steep incline or vertical path. Also known as elevating conveyor. { 'bək·ət ¦el·ə,vād·ər }

bucket excavator [MECH ENG] An elevating scraper, that is, one that does the work of a conventional scraper but has a bucket elevator mounted in front of the bowl. { 'bək·ət ¦ek·skə,vād·ər }

bucket ladder See bucket-ladder dredge. { 'bək·ət ,lad·ər }

bucket-ladder dredge [MECH ENG] A dredge whose digging mechanism consists of a ladderlike truss on the periphery of which is attached an endless chain riding on sprocket wheels and carrying attached buckets. Also known as bucket ladder; bucket-line dredge; ladder-bucket dredge; ladder dredge. { 'bək·ət ˌlad·ər ˌdrej }

bucket-ladder excavator See trench excavator. { 'bək·ət ˌlad·ər 'ek·skə·vād·ər }

bucket-line dredge See bucket-ladder dredge. { 'bək·ət ˌlīn ˌdrej }

bucket loader [MECH ENG] A form of portable, self-feeding, inclined bucket elevator for loading bulk materials into cars, trucks, or other conveyors. { 'bək·ət ˌlōd·ər }

bucket temperature [ENG] The surface temperature of ocean water as measured by a bucket thermometer. { 'bək·ət ˌtem·prə·chər }

bucket thermometer [ENG] A thermometer mounted in a bucket and used to measure the temperature of water drawn into the bucket from the surface of the ocean. { 'bək·ət thər'mäm·əd·ər }

bucket-wheel excavator [MECH ENG] A continuous digging machine used extensively in large-scale stripping and mining. Abbreviated BWE. Also known as rotary excavator. { 'bək·ət ˌwēl 'ek·skə,vād·ər }

Buckingham's equations [MECH ENG] Equations which give the durability of gears and the dynamic loads to which they are subjected in terms of their dimensions, hardness, surface endurance, and composition. { 'bək·iŋ·əmz i'kwä·zhənz }

buckle plate [CIV ENG] A steel floor plate which is slightly arched to increase rigidity. { 'bək·əl ˌplāt }

Buckley gage [ENG] A device that measures very low gas pressures by sensing the amount of ionization produced in the gas by a predetermined electric current. { 'bək·lē ˌgāj }

buckling [ENG] Wrinkling or warping of fibers in a composite material. [MECH] Bending of a sheet, plate, or column supporting a compressive load. { 'bək·liŋ }

buckling stress [MECH] Force exerted by the crippling load. { 'bək·liŋ ˌstres }

buckstay [MECH ENG] A structural support for a furnace wall. { 'bək,stā }

buffer [ELEC] An electric circuit or component that prevents undesirable electrical interaction between two circuits or components. [ELECTR] 1. An isolating circuit in an electronic computer used to prevent the action of a driven circuit from affecting the corresponding driving circuit. 2. See buffer amplifier. [ENG] A device, apparatus, or piece of material designed to reduce mechanical shock due to impact. { 'bəf·ər }

buffered FET logic [ELECTR] A logic gate configuration used with gallium-arsenide field-effect transistors operating in the depletion mode, in which the level shifting required to make the input and output voltage levels compatible is achieved with Schottky barrier diodes. Abbreviated BFL. { 'bəf·ərd ˌef¦e¦tē 'läj·ik }

buffing [ENG] The smoothing and brightening of a surface by an abrasive compound pressed against it by a soft wheel or belt. { 'bəf·iŋ }

buffing wheel [DES ENG] A flexible wheel with a surface of fine abrasive particles for buffing operations. { 'bəf·iŋ ˌwēl }

bug [ELECTR] 1. A semiautomatic code-sending telegraph key in which movement of a lever to one side produces a series of correctly spaced dots and movement to the other side produces a single dash. 2. An electronic listening device, generally concealed, used for commercial or military espionage. [ENG] 1. A defect or imperfection present in a piece of equipment. 2. See bullet. { bəg }

buggy See concrete buggy. { 'bəg·ē }

buhrstone mill [MECH ENG] A mill for grinding or pulverizing grain in which a flat siliceous rock (buhrstone), generally of cellular quartz, rotates against a stationary stone of the same material. { 'bər,stōn ˌmil }

build [ELECTR] To increase in received signal strength. { bild }

building [CIV ENG] A fixed structure for human occupancy and use. { 'bil·diŋ }

building-block approach [IND ENG] A technique for development of a set of standard data by creating fixed groups or modules of work elements that may be added together to obtain time values for elements and entire operations. { 'bild·iŋ ˌbläk ə,prōch }

building code [CIV ENG] Local building laws to promote safe practices in the design and construction of a building. { 'bil·diŋ ˌkōd }

building dock [CIV ENG] A type of graving dock or basin, usually built of concrete, in which ships are constructed and then floated out through a caisson gate after flooding the dock. { 'bil·diŋ ˌdäk }

building envelope [CIV ENG] The interior, enclosed space of a building. { 'bil·diŋ 'en·və,lōp }

building footprint See footprint. { 'bil·diŋ ˌfut,print }

building line [CIV ENG] A designated line beyond which a building cannot extend. { 'bil·diŋ ˌlīn }

buildup index See fire-danger meter. { 'bil,dəp ˌin,deks }

built-in beam See fixed-end beam. { ¦bilt,in 'bēm }

built-up beam [ENG] A structural steel member that is fabricated by welding or riveting rather than being rolled. { 'bilt,əp 'bēm }

built-up edge [ENG] Chip material adhering to the tool face adjacent to a cutting edge during cutting. { 'bilt,əp 'ej }

built-up roof [BUILD] A roof constructed of several layers of felt and asphalt. { 'bilt,əp 'rüf }

bulb angle [DES ENG] A steel angle iron enlarged to a bulbous thickening at one end. { 'bəlb ˌaŋ·gəl }

bulge forming [ENG] A process by which contours are formed on the sides of tubular workpieces by exerting pressure inside the tube to force expansion into a die clamped around the exterior. { 'bəlj ˌfôrm·iŋ }

bulk cargo [IND ENG] Cargo which is loaded into a ship's hold without being boxed, bagged, or hand stowed, or is transported in large tank spaces. { 'bəlk 'kär‚gō }

bulk density [ENG] The mass of powdered or granulated solid material per unit of volume. { 'bəlk 'den·səd·ē }

bulk diode [ELECTR] A semiconductor microwave diode that uses the bulk effect, such as Gunn diodes and diodes operating in limited space-charge-accumulation modes. { 'bəlk 'dī ‚ōd }

bulk effect [ELECTR] An effect that occurs within the entire bulk of a semiconductor material rather than in a localized region or junction. { 'bəlk i'fekt }

bulk-effect device [ELECTR] A semiconductor device that depends on a bulk effect, as in Gunn and avalanche devices. { 'bəlk i'fekt di'vīs }

bulk factor [ENG] The ratio of the volume of loose powdered or granulated solids to the volume of an equal weight of the material after consolidation into a voidless solid. { 'bəlk ‚fak·tər }

bulk-handling machine [MECH ENG] Any of a diversified group of materials-handling machines designed for handling unpackaged, divided materials. { 'bəlk ‚hand·liŋ mə'shēn }

bulkhead line [CIV ENG] The farthest offshore line to which a structure may be constructed without interfering with navigation. { 'bəlk ‚hed ‚līn }

bulkhead wharf [CIV ENG] A bulkhead that may be used as a wharf by addition of mooring appurtenances, paving, and cargo-handling facilities. { 'bəlk‚hed ‚wòrf }

bulking value [CHEM ENG] The relative ability of a pigment or other substance to increase the volume of paint. { 'bəl·kiŋ ‚val·yü }

bulk insulation [ENG] A type of insulation that retards the flow of heat by the interposition of many air spaces and, in most cases, by opacity to radiant heat. { 'bəlk in·sə'lā·shən }

bulk material [IND ENG] Material purchased in uniform lots and in quantity for distribution as required for a project. { 'bəlk mə'tir·ē·əl }

bulk micromachining [ENG] A set of processes that enable the three-dimensional sculpting of single-crystal silicon to make small structures that serve as components of microsensors. { 'bəlk ‚mī·krō·mə'shēn·iŋ }

bulk modulus See bulk modulus of elasticity. { 'bəlk 'mäj·ə·ləs }

bulk modulus of elasticity [MECH] The ratio of the compressive or tensile force applied to a substance per unit surface area to the change in volume of the substance per unit volume. Also known as bulk modulus; compression modulus; hydrostatic modulus; modulus of compression; modulus of volume elasticity. { 'bəlk 'mäj·‚ləs əv i‚las'tis·əd·ē }

bulk rheology [MECH] The branch of rheology wherein study of the behavior of matter neglects effects due to the surface of a system. { 'bəlk rē'äl·ə·jē }

bulk photoconductor [ELECTR] A photoconductor having high power-handling capability

and other unique properties that depend on the semiconductor and doping materials used. { 'bəlk ‚fō·dō·kən'dək·tər }

bulk resistor [ELECTR] An integrated-circuit resistor in which the n-type epitaxial layer of a semiconducting substrate is used as a noncritical high-value resistor; the spacing between the attached terminals and the sheet resistivity of the material together determine the resistance value. { 'bəlk ri'zis·tər }

bulk strain [MECH] The ratio of the change in the volume of a body that occurs when the body is placed under pressure, to the original volume of the body. { 'balk ‚strān }

bulk strength [MECH] The strength per unit volume of a solid. { 'bəlk 'streŋkth }

bulk transport [MECH ENG] Conveying, hoisting, or elevating systems for movement of solids such as grain, sand, gravel, coal, or wood chips. { 'bəlk 'tranz‚pórt }

bulldozer [MECH ENG] A wheeled or crawler tractor equipped with a reinforced, curved steel plate mounted in front, perpendicular to the ground, for pushing excavated materials. { 'bul ‚dōz·ər }

bullet [ENG] **1.** A conical-nosed cylindrical weight, attached to a wire rope or line, either notched or seated to engage and attach itself to the upper end of a wire line core barrel or other retrievable or retractable device that has been placed in a borehole. Also known as bug; godevil; overshot. **2.** A scraper with self-adjusting spring blades, inserted in a pipeline and carried forward by the fluid pressure, clearing away accumulations or debris from the walls of a pipe. Also known as go-devil. **3.** A bullet-shaped weight or small explosive charge dropped to explode a charge of nitroglycerin placed in a borehole. Also known as go-devil. **4.** An electric lamp covered by a conical metal case, usually at the end of a flexible metal shaft. **5.** See torpedo. { 'bul·ət }

bullet drop [MECH] The vertical drop of a bullet. { 'bul·ət ‚dräp }

bull gear [DES ENG] A bull wheel with gear teeth. { 'bul ‚gir }

bulling bar [ENG] A bar for ramming clay into cracks containing blasting charges which are about to be exploded. { 'bul·iŋ ‚bär }

bull nose [BUILD] A rounded external angle, as one used at window returns and doorframes. { 'bul ‚nōz }

bull-nose bit See wedge bit. { 'bul ‚nōz ‚bit }

bull-nose plane [DES ENG] A small rabbet plane used to smooth or shape joints or other places that cannot be reached by larger planes. { 'bul ‚nōz ‚plān }

bull wheel [MECH ENG] **1.** The main wheel or gear of a machine, which is usually the largest and strongest. **2.** A cylinder which has a rope wound about it for lifting or hauling. **3.** A wheel attached to the base of a derrick boom which swings the derrick in a vertical plane. { 'bul ‚wēl }

Bulygen number [THERMO] A dimensionless

number used in the study of heat transfer during evaporation. { 'bül·ə·jən ˌnəm·bər }

bump contact [ELECTR] A large-area contact used for alloying directly to the substrate of a transistor for mounting or interconnecting purposes. { 'bəmp ˌkän,takt }

bumper [ENG] **1.** A metal bar attached to one or both ends of a powered transportation vehicle, especially an automobile, to prevent damage to the body. **2.** In a drilling operation, the supporting stay between the main foundation sill and the engine block. **3.** In drilling, a fishing tool for loosening jammed cable tools. { 'bəm·pər }

bumping See chugging. { 'bəm·piŋ }

bund [CIV ENG] An embankment or embanked thoroughfare along a body of water; the term is used particularly for such structures in the Far East. { bənd }

bundling machine [MECH ENG] A device that automatically accumulates cans, cartons, or glass containers for semiautomatic or automatic loading or for shipping cartons by assembling the packages into units of predetermined count and pattern which are then machine-wrapped in paper, film paperboard, or corrugated board. { 'bənd·liŋ mə'shēn }

bund wall [ENG] A retaining wall designed to contain the contents of a tank or a storage vessel in the event of a rupture or other emergency. { 'bənd ˌwól }

bunker [CIV ENG] A bin, often elevated, that is divided into compartments for storing material such as coal or sand. [MECH ENG] A space in a refrigerator designed to hold a cooling element. { 'bəŋ·kər }

bunkering [ENG] Storage of solid or liquid fuel in containers from which the fuel can be continuously or intermittently withdrawn to feed a furnace, internal combustion engine, or fuel tank, for example, coal bunkering and fuel-oil bunkering. { 'bəŋ·kər·iŋ }

bunny suit [ENG] Protective clothing worn by an individual who works in a clean room to prevent contamination of equipment and materials. { 'bən·ē ˌsüt }

Bunsen burner [ENG] A type of gas burner with an adjustable air supply. { 'bən·sən 'bər·nər }

Bunsen ice calorimeter [ENG] Apparatus to gage heat released during the melting of a compound by measuring the increase in volume of the surrounding ice-water solution caused by the melting of the ice. Also known as ice calorimeter. { 'bən·sən 'īs kal·ə'rim·əd·ər }

buoy [ENG] An anchored or moored floating object, other than a lightship, intended as an aid to navigation, to attach or suspend measuring instruments, or to mark the position of something beneath the water. { bói }

buoyancy-type density transmitter [ENG] An instrument which records the specific gravity of a flowing stream of a liquid or gas, using the principle of hydrostatic weighing. { 'bói·ən·sē ˌtīp 'den·səd·ē tranz'mid·ər }

buoy sensor [ENG ACOUS] A hydrophone used as a sensor in buoy projects; some hydrophone arrays are designed for telemetering. { 'bói ˌsen·sər }

burden [ELEC] The amount of power drawn from the circuit connecting the secondary terminals of an instrument transformer, usually expressed in volt-amperes. [ENG] **1.** The distance from a drill hole to the more or less vertical surface of rock that has already been exposed by blasting or excavating. **2.** The volume of the rock to be removed by blasting in a drill hole. { 'bərd·ən }

burglar alarm [ENG] An alarm in which interruption of electric current to a relay, caused, for example, by the breaking of a metallic tape placed at an entrance to a building, deenergizes the relay and causes the relay contacts to operate the alarm indicator. Also known as intrusion alarm. { 'bər·glər ə¦lärm }

buried set-point method [CONT SYS] A procedure for guiding a robot manipulator along a template, in which low-gain servomechanisms apply a force along the edge of the template, while the manipulator's tool is parallel to, and buried below, the template surface. { 'ber·ēd 'set,póint ˌmeth·əd }

burn [ENG] To consume fuel. { bərn }

burn cut See parallel cut. { 'bərn ˌkət }

burner [CHEM ENG] A furnace where sulfur or sulfide ore are burned to produce sulfur dioxide and other gases. [ENG] **1.** The part of a fluid-burning device at which the flame is produced. **2.** Any burning device used to soften old paint to aid in its removal. **3.** A worker who operates a kiln which burns brick or tile. **4.** A worker who alters the properties of a mineral substance by burning. **5.** A worker who uses a flame-cutting torch to cut metals. [MECH ENG] A unit of a steam boiler which mixes and directs the flow of fuel and air so as to ensure rapid ignition and complete combustion. { 'bər·nər }

burner windbox [ENG] A chamber surrounding a burner, under positive air pressure, for proper distribution and discharge of secondary air. { 'bər·nər 'wind,bäks }

burnettize [ENG] To saturate fabric or wood with a solution of zinc chloride under pressure to keep it from decaying. { bər'ned,īz }

burn-in [ELECTR] Operation of electronic components before they are applied in order to stabilize their characteristics and reveal defects. [ENG] See freeze. { 'bərn ˌin }

burning [ENG] The firing of clay products placed in a kiln. { 'bər·niŋ }

burning index See fire-danger meter. { 'bər·niŋ ¦in,deks }

burning point [ENG] The lowest temperature at which a volatile oil in an open vessel will continue to burn when ignited by a flame held close to its surface; used to test safety of kerosine and other illuminating oils. { 'bər·niŋ ˌpóint }

burning quality [ENG] Rated performance for a burning oil as determined by specified ASTM (American Society for Testing and Materials) tests. { 'bər·niŋ ˌkwal· əd·ē }

burning-quality index |ENG| Prediction of burning performance of furnace and heater oils; derived from ASTM (American Society for Testing and Materials) distillation, API (American Petroleum Institute) gravity, paraffinicity, and volatility. { 'bər·niŋ ¦kwäl·əd·ē ‚in‚deks }

burnish |ENG| To polish or make shiny. { 'bər·nish }

burnisher |ENG| A tool with a hard, smooth rounded edge or surface; used for finishing the edges of scraper blades, for smoothing or polishing plastic or metal surfaces, or for other applications requiring manipulation by rubbing. { 'bər·nə·shər }

burnout |ELEC| Failure of a device due to excessive heat produced by excessive current. |ENG| An instance of a device or a part overheating so as to result in destruction or damage. { 'bərn‚aut }

Burnside boring machine |MECH ENG| A machine for boring in all types of ground with the feature of controlling water immediately if it is tapped. { 'bərn‚sīd 'bȯr·iŋ mə'shən }

bursting strength |MECH| A measure of the ability of a material to withstand pressure without rupture; it is the hydraulic pressure required to burst a vessel of given thickness. { 'bər·stiŋ ‚streŋkth }

burst pressure |MECH| The maximum inside pressure that a process vessel can safely withstand. { 'bərst ‚presh·ər }

burton |MECH ENG| A small hoisting tackle with two blocks, usually a single block and a double block, with a hook block in the running part of the rope. { 'bərt·ən }

bus |ELEC| **1.** A set of two or more electric conductors that serve as common connections between load circuits and each of the polarities (in direct-current systems) or phases (in alternating-current systems) of the source of electric power. **2.** See busbar. |ELECTR| One or more conductors in a computer along which information is transmitted from any of several sources to any of several destinations. |ENG| A motor vehicle for carrying a large number of passengers. { bəs }

bus cable |ELECTR| An electrical conductor that can be attached to a bus to extend it outside the computer housing or join it to another bus within the same computer. { 'bəs ‚kā·bəl }

bushel |MECH| Abbreviated bu. **1.** A unit of volume (dry measure) used in the United States, equal to 2150.42 cubic inches or approximately 35.239 liters. **2.** A unit of volume (liquid and dry measure) used in Britain, equal to 2219.36 cubic inches or 8 imperial gallons (approximately 36.369 liters). { 'bush·əl }

bush hammer |MECH ENG| A hand-held or power-driven hammer that has a serrated face containing pyramid-shaped points and is used to dress a concrete or stone surface. { 'bush ‚ham·ər }

bushing |DES ENG| See nipple. |ELEC| See sleeve. |MECH ENG| A removable piece of soft metal or graphite-filled sintered metal, usually

in the form of a bearing, that lines a support for a shaft. { 'bush·iŋ }

Butamer process |CHEM ENG| A method of isomerizing normal butane into isobutane in the presence of hydrogen and a solid, noble-metal catalyst; used to prepare raw material in a gasoline alkylation process. { 'byüd·ə·mər ‚präs·əs }

butane dehydrogenation |CHEM ENG| A process to remove hydrogen from butane to produce butene or butadiene. { 'byü‚tān dē‚hī·drə·jə'nā·shən }

butane vapor-phase isomerization |CHEM ENG| A process to isomerize normal butane into isobutane in the presence of aluminum chloride catalyst and hydrogen chloride promoter. { 'byü‚tān 'vā·pər ‚fāz ī‚säm·ə·rə'zā·shən }

butt |BUILD| The bottom or cover edge of a shingle. |DES ENG| The enlarged and squared-off end of a connecting rod or similar link in a machine. { bət }

butterfly damper See butterfly valve. { 'bəd·ər‚flī ‚dam·pər }

butterfly nut See wing nut. { 'bəd·ər‚flī ‚nət }

butterfly valve |ENG| A valve that utilizes a turnable disk element to regulate flow in a pipe or duct system, such as a hydraulic turbine or a ventilating system. Also known as butterfly damper. { 'bəd·ər‚flī ‚valv }

Butterworth filter |ELECTR| An electric filter whose pass band (graph of transmission versus frequency) has a maximally flat shape. { 'bəd·ər‚wərth 'fil·tər }

Butterworth head |MECH ENG| A mechanical hose head with revolving nozzles; used to wash down shipboard storage tanks. { 'bəd·ər‚wərth ‚hed }

butt fusion |ENG| The joining of two pieces of plastic or metal pipes or sheets by heating the ends until they are molten and then pressing them together to form a homogeneous bond. { 'bət ‚fyü·zhən }

butt gage |ENG| A tool used to mark the outline for the hinges on a door. { 'bət ‚gāj }

butt joint |ELEC| A connection formed by placing the ends of two conductors together and joining them by welding, brazing, or soldering. |ENG| A joint in which the parts to be joined are fastened end to end or edge to edge with one or more cover plates (or other strengthening) generally used to accomplish the joining. { 'bət ‚jȯint }

buttock lines |ENG| The lines of intersection of the surface of an aircraft or its float, or of the hull of a ship, with its longitudinal vertical planes. Also known as buttocks. { 'bəd·ək ‚līnz }

buttocks See buttock lines. { 'bəd·əks }

button |ELECTR| **1.** A small, round piece of metal alloyed to the base wafer of an alloy-junction transistor. Also known as dot. **2.** The container that holds the carbon granules of a carbon microphone. Also known as carbon button. { 'bət·ən }

button bit |DES ENG| A drilling bit made with button-shaped tungsten carbide inserts. { 'bət·ən ,bit }

button die |DES ENG| A mating member, usually replaceable, for a piercing punch. Also known as die bushing. { 'bət·ən ,dī }

buttonhead |DES ENG| A screw, bolt, or rivet with a hemispherical head. { 'bət·ən,hed }

buttress |CIV ENG| A pier constructed at right angles to a restraining wall on the side opposite to the restrained material; increases the strength and thrust resistance of the wall. { 'bə·trəs }

buttress dam |CIV ENG| A concrete dam constructed as a series of buttresses. { 'bə·trəs ,dam }

buttress thread |DES ENG| A screw thread whose forward face is perpendicular to the screw axis and whose back face is at an angle to the axis, so that the thread is both efficient in transmitting power and strong. { 'bə·trəs ,thred }

buzz |CONT SYS| See dither. |ELECTR| The condition of a combinatorial circuit with feedback that has undergone a transition, caused by the inputs, from an unstable state to a new state that is also unstable. { bəz }

BWE See bucket-wheel excavator.

BWG See Birmingham wire gage.

BX cable |ELEC| Insulated wires in flexible metal tubing used for bringing electric power to electronic equipment. { ¦bē¦eks ¦kā·bəl }

bypass |CIV ENG| A road which carries traffic around a congested district or temporary obstruction. |ELEC| A shunt path around some element or elements of a circuit. |ENG| An alternating, usually smaller, diversionary flow path in a fluid dynamic system to avoid some device, fixture, or obstruction. { 'bī,pas }

bypass channel |CIV ENG| **1.** A channel built to carry excess water from a stream. Also known as flood relief channel; floodway. **2.** A channel constructed to divert water from a main channel. { 'bī,pas ,chan·əl }

bypass filter |ELECTR| Filter which provides a low-attenuation path around some other equipment, such as a carrier frequency filter used to bypass a physical telephone repeater station. { 'bī,pas ,fil·tər }

bypass valve |ENG| A valve that opens to direct fluid elsewhere when a pressure limit is exceeded. { 'bī,pas ,valv }

by-product |ENG| A product from a manufacturing process that is not considered the principal material. { 'bī,präd·əkt }

C

c *See* calorie.

C *See* capacitance; capacitor; coulomb.

C² *See* command and control. { 'sē 'tü }

C³ *See* command, control, and communications. { 'sē 'thrē }

cab [ENG] In a locomotive, truck, tractor, or hoisting apparatus, a compartment for the operator. { kab }

cabinet file [DES ENG] A coarse-toothed file with flat and convex faces used for woodworking. { 'kab·ə·nət ‚fīl }

cabinet hardware [DES ENG] Parts for the final trim of a cabinet, such as fastening hinges, drawer pulls, and knobs. { 'kab·ə·nət ‚härd ‚wer }

cabinet saw [DES ENG] A short saw, one edge used for ripping, the other for crosscutting. { 'kab·ə·nət ‚só }

cabinet scraper [DES ENG] A steel tool with a contoured edge used to remove irregularities on a wood surface. { 'kab·ə·nət ‚skrāp·ər }

cable [DES ENG] A stranded, ropelike assembly of wire or fiber. [ELEC] Strands of insulated electrical conductors laid together, usually around a central core, and surrounded by a heavy insulation. { 'kā·bəl }

cable buoy [ENG] A buoy used to mark one end of a submarine underwater cable during time of installation or repair. { 'kā·bəl ‚bói }

cable conveyor [MECH ENG] A powered conveyor in which a trolley runs on a flexible, torque-transmitting cable that has helical threads. { 'kā·bəl kən'vā·ər }

cable drilling [ENG] Rock drilling in which the rock is penetrated by percussion, at the bottom of the hole, of a bit suspended from a wire line and given motion by a beam pivoted at the center. { 'kā·bəl ‚dril·iŋ }

cable duct [ENG] A pipe, either earthenware or concrete, through which prestressing wires or electric cable are pulled. { 'kā·bəl ‚dəkt }

cable-laid [DES ENG] Consisting of three ropes with a left-hand twist, each rope having three twisted strands. { 'kā·bəl ‚lād }

cableman [ENG] A person who installs, repairs, or otherwise works with cables. { 'kā·bəl·mən }

cable railway [MECH ENG] An inclined track on which rail cars travel, with the cars fixed to an endless steel-wire rope at equal spaces; the rope is driven by a stationary engine. { 'kā·bəl ‚rāl‚wā }

cable release [ENG] A wire plunger to actuate the shutter of a camera, thus avoiding undesirable camera movement. { 'kā·bəl ri'lēs }

cable-stayed bridge [CIV ENG] A modification of the cantilever bridge consisting of girders or trusses cantilevered both ways from a central tower and supported by inclined cables attached to the tower at top or sometimes at several levels. { 'kā·bəl ‚stād ‚brij }

cable-system drill *See* churn drill. { 'kā·bəl ‚sis· təm ‚dril }

cable-tool drilling [ENG] A drilling procedure in which a sharply pointed bit attached to a cable is repeatedly picked up and dropped on the bottom of the hole. { 'kā·bəl ‚tül ‚dril·iŋ }

cable vault [CIV ENG] A manhole containing electrical cables. [ELEC] Vault in which the outside plant cables are spliced to the tipping cables. { 'kā·bəl ‚vólt }

cableway [MECH ENG] A transporting system consisting of a cable extended between two or more points on which cars are propelled to transport bulk materials for construction operations. { 'kā·bəl‚wā }

cableway carriage [MECH ENG] A trolley that runs on main load cables stretched between two or more towers. { 'kā·bəl‚wā 'kar·ij }

caboose [ENG] A car on a freight train, often the last car, usually for use by the train crew. { kə'büs }

cab signal [ENG] A signal in a locomotive that informs the engine operator about conditions affecting train movement. { 'kab ‚sig·nəl }

cadastral survey [CIV ENG] A survey made to establish property lines. { kə'das·trəl }

cage [MECH ENG] A frame for maintaining uniform separation between the balls or rollers in a bearing. Also known as separator. { kāj }

cage mill [MECH ENG] Pulverizer used to disintegrate clay, press cake, asbestos, packing-house by-products, and various tough, gummy, high-moisture-content or low-melting-point materials. { 'kāj ‚mil }

cairn [ENG] An artificial mound of rocks, stones, or masonry, usually conical or pyramidal, whose purpose is to designate or to aid in identifying a point of surveying or of cadastral importance. { kern }

caisson [CIV ENG] **1.** A watertight, cylindrical or rectangular chamber used in underwater construction to protect workers from water pressure and soil collapse. **2.** A float used to raise a sunken vessel. **3.** See dry-dock caisson. { 'kā ,sän }

caisson foundation [CIV ENG] A shaft of concrete placed under a building column or wall and extending down to hardpan or rock. Also known as pier foundation. { 'kā,sän foún'dā- shən }

caking [ENG] Changing of a powder into a solid mass by heat, pressure, or water. { 'kāk·iŋ }

cal See calorie.

Cal See kilocalorie.

calandria [CHEM ENG] One of the tubes through which the heating fluid circulates in an evaporator. { kə'lan·drē·ə }

calandria evaporator See short-tubevertical evaporator. { kə'lan·drē·ə i'vap·ə,rād·ər }

calcimeter [ENG] An instrument for estimating the amount of lime in soils. { kal'sim·əd·ər }

calcination [CHEM ENG] A process in which a material is heated to a temperature below its melting point to effect a thermal decomposition or a phase transition other than melting. { ,kal- sə'nā·shən }

calcine [ENG] **1.** To heat to a high temperature without fusing, as to heat unformed ceramic materials in a kiln, or to heat ores, precipitates, concentrates, or residues so that hydrates, carbonates, or other compounds are decomposed and the volatile material is expelled. **2.** To heat under oxidizing conditions. { 'kal,sīn }

calcining furnace [ENG] A heating device, such as a vertical-shaft kiln, that raises the temperature (but not to the melting point) of a substance such as limestone to make lime. Also known as calciner. { 'kal,sin·iŋ ,fər·nəs }

calefaction [ENG] **1.** Warming. **2.** The condition of being warmed. { ¦kal·ə¦fak·shən }

calender [ENG] **1.** To pass a material between rollers or plates to thin it into sheets or to make it smooth and glossy. **2.** The machine which performs this operation. { 'kal·ən·dər }

calibrating tank [ENG] A tank having known capacity used to check the volumetric accuracy of liquid delivery by positive-displacement meters. Also known as meter-proving tank. { 'kal· ə,brād·iŋ ,taŋk }

calibration curve [ENG] A plot of calibration data, giving the correct value for each indicated reading of a meter or control dial. { 'kal·ə,brā- shən ,kərv }

calibration markers [ENG] On a radar display, electronically generated marks which provide numerical values for the navigational parameters such as bearing, distance, height, or time. { 'kal·ə,brā·shən ,mär·kərz }

California polymerization [CHEM ENG] A polymerization process for converting C_3–C_4 olefins to motor fuel by utilizing a catalyst of phosphoric acid on quartz chips. { ¦kal·ə¦fȯr·nyə pə,lim·ə- rə'zā·shən }

caliper [DES ENG] An instrument with two legs or jaws that can be adjusted for measuring linear dimensions, thickness, or diameter. { 'kal·ə- pər }

caliper gage [DES ENG] An instrument, such as a micrometer, of fixed size for calipering. { 'kal· ə·pər ,gāj }

calk See caulk. { kȯk }

Callendar and Barnes' continuous-flow calorimeter [ENG] A calorimeter in which the heat to be measured is absorbed by water flowing through a tube at a constant rate, and the quantity of heat is determined by the rate of flow and the temperature difference between water at ends of the tube. { ¦kal·ən·dər ən 'bärnz kən'tin- yə·wəs ,flō kal·ə'rim·əd·ər }

Callendar's compensated air thermometer [ENG] A type of constant-pressure gas thermometer in which errors resulting from temperature differences between the thermometer bulb and the connecting tubes and manometer used to maintain constant pressure are eliminated by the configuration of the connecting tubes. { ¦kal·ən- dərz ¦käm·pan,sād·əd 'er thər,mäm·əd·ər }

Callendar's equation [THERMO] **1.** An equation of state for steam whose temperature is well above the boiling point at the existing pressure, but is less than the critical temperature: $(V - b) = (RT/p) - (a/T^n)$, where V is the volume, R is the gas constant, T is the temperature, p is the pressure, n equals 10/3, and a and b are constants. **2.** A very accurate equation relating temperature and resistance of platinum, according to which the temperature is the sum of a linear function of the resistance of platinum and a small correction term, which is a quadratic function of temperature. { 'kal·ən·dərz i'kwā- zhən }

Callendar's thermometer See platinum resistance thermometer. { 'kal·ən·dərz thər'mäm·əd·ər }

calorie [THERMO] Abbreviated cal; often designated c. **1.** A unit of heat energy, equal to 4.1868 joules. Also known as International Table calorie (IT calorie). **2.** A unit of energy, equal to the heat required to raise the temperature of 1 gram of water from 14.5° to 15.5°C at a constant pressure of 1 standard atmosphere; equal to 4.1855 ± 0.0005 joules. Also known as fifteen-degrees calorie; gram-calorie (g-cal); small calorie. **3.** A unit of heat energy equal to 4.184 joules; used in thermochemistry. Also known as thermochemical calorie. { 'kal·ə·rē }

calorific value [ENG] Quantity of heat liberated on the complete combustion of a unit weight or unit volume of fuel. { ¦kal·ə¦rif·ik 'val·yü }

calorifier [ENG] A device that heats fluids by circulating them over heating coils. { kə'lȯr· ə,fī·ər }

calorimeter [ENG] An apparatus for measuring heat quantities generated in or emitted by materials in processes such as chemical reactions, changes of state, or formation of solutions. { ,kal·ə'rim·əd·ər }

calorimetric test [ENG] The use of a calorimeter to determine the thermochemical characteristics of propellants and explosives; properties normally determined are heat of combustion, heat of explosion, heat of formation, and heat of reaction. { kə¦lór·ə¦me·trik 'test }

calorimetry |ENG| The measurement of the quantity of heat involved in various processes, such as chemical reactions, changes of state, and formations of solutions, or in the determination of the heat capacities of substances; fundamental unit of measurement is the joule or the calorie (4.184 joules). { kal·ə'rim·ə·trē }

calyx |ENG| A steel tube that is a guide rod and is also used to catch cuttings from a drill rod. Also known as bucket; sludge barrel; sludge bucket. { 'kā¸liks }

calyx drill |ENG| A rotary core drill with hardened steel shot for cutting rock. Also known as shot drill. { 'kā¸liks ¸dril }

cam |MECH ENG| A plate or cylinder which communicates motion to a follower by means of its edge or a groove cut in its surface. { kam }

cam acceleration |MECH ENG| The acceleration of the cam follower. { 'kam ak·sel·ə'rā·shən }

camber |DES ENG| Deviation from a straight line; the term is applied to a convex, edgewise sweep or curve, or to the increase in diameter at the center of rolled materials. { 'kam·bər }

camber angle |MECH ENG| The inclination from the vertical of the steerable wheels of an automobile. { 'kam·bər ¸aŋ·gəl }

cam cutter |MECH ENG| A semiautomatic or automatic machine that produces the cam contour by swinging the work as it revolves; uses a master cam in contact with a roller. { 'kam ¸kəd·ər }

cam dwell |DES ENG| That part of a cam surface between the opening and closing acceleration sections. { 'kam ¸dwel }

cam engine |MECH ENG| A piston engine in which a cam-and-roller mechanism seems to convert reciprocating motion into rotary motion. { 'kam ¸en·jən }

camera study See memomotion study. { 'kam·rə ¸stəd·ē }

cam follower |MECH ENG| The output link of a cam mechanism. { 'kam ¸fäl·ə·wər }

cam mechanism |MECH ENG| A mechanical linkage whose purpose is to produce, by means of a contoured cam surface, a prescribed motion of the output link. { 'kam ¸mek·ə¸niz·əm }

cam nose |MECH ENG| The high point of a cam, which in a reciprocating engine holds valves open or closed. { 'kam ¸nōz }

cam pawl |MECH ENG| A pawl which prevents a wheel from turning in one direction by a wedging action, while permitting it to rotate in the other direction. { 'kam ¸pól }

Campbell-Stokes recorder |ENG| A sunshine recorder in which the time scale is supplied by the motion of the sun and which has a spherical lens that burns an image of the sun upon a specially prepared card. { ¦kam·əl ¦stōks ri 'kórd·ər }

camp ceiling |BUILD| A ceiling that is flat in the center portion and sloping at the sides. { 'kamp ¸sē·liŋ }

cam profile |DES ENG| The shape of the contoured cam surface by means of which motion is communicated to the follower. Also known as pitch line. { 'kam ¸prō¸fīl }

camshaft |MECH ENG| A rotating shaft to which a cam is attached. { 'kam¸shaft }

can |DES ENG| A cylindrical metal vessel or container, usually with an open top or a removable cover. { kan }

canal |CIV ENG| An artificial open waterway used for transportation, waterpower, or irrigation. |DES ENG| A groove on the underside of a corona. { kə'nal }

canalization |ENG| Any system of distribution canals or conduits for water, gas, electricity, or steam { ¸kan·al·ə'zā·shən }

cancellation circuit |ELECTR| A circuit used in providing moving-target indication on a plan position indicator scope; cancels constant-amplitude fixed-target pulses by subtraction of successive pulse trains. { kan·sə'lā·shən ¸sər·kət }

canister See charcoal canister. { 'kan·ə'stər }

canned motor |MECH ENG| A motor enclosed within a casing along with the driven element (that is, a pump) so that the motor bearings are lubricated by the same liquid that is being pumped. { ¦kand 'mōd·ər }

canned pump |MECH ENG| A watertight pump that can operate under water. { ¦kand 'pəmp }

cannibalize |ENG| To remove parts from one piece of equipment and use them to replace like, defective parts in a similar piece of equipment in order to keep the latter operational. { 'kan·ə·bə¸līz }

canonical equations of motion See Hamilton's equations of motion. { kə'nän·ə·kəl i'kwā·zhənz əv 'mō·shən }

canonical form |CONT SYS| A specific type of dynamical system representation in which the associated matrices possess specific row-column structures. { kə'nän·ə·kəl ¸fòrm }

canonically conjugate variables |MECH| A generalized coordinate and its conjugate momentum. { kə'nän·ə·klē ¦kan·jə·gət 'ver·ē·ə·bəlz }

canonical momentum See conjugate momentum. { kə'nän·ə·kəl mə'ment·əm }

canonical transformation |MECH| A transformation which occurs among the coordinates and momenta describing the state of a classical dynamical system and which leaves the form of Hamilton's equations of motion unchanged. Also known as contact transformation. { kə'nän·ə·kəl ¸tranz·fər'mā·shən }

cant file |DES ENG| A fine-tapered file with a triangular cross section, used for sharpening saw teeth. { 'kant ¸fīl }

cant hook |DES ENG| A lever with a hooklike attachment at one end, used in lumbering. { 'kant ¸hük }

cantilever |ENG| **1.** A beam or member securely fixed at one end and hanging free at the other end. **2.** In particular, in an atomic force microscope a very small beam that has a tip attached to its free end; the deflection of the beam is used to measure the force acting on the tip.   { 'kant·əl,ē·vər }

cantilever bridge |CIV ENG| A fixed bridge consisting of two spans projecting toward each other and joined at their ends by a suspended simple span.   { 'kant·əl,ē·vər ¦brij }

cantilever footing |CIV ENG| A footing used to carry a load from two columns, with one column and one end of the footing placed against a building line or exterior wall.   { 'kant·əl,ē·vər 'füd·iŋ }

cantilever retaining wall |CIV ENG| A type of wall formed of three cantilever beams: stem, toe projection, and heel projection.   { 'kant·əl,ē·vər ri'tān·iŋ wól }

cantilever spring |MECH ENG| A flat spring supported at one end and holding a load at or near the other end.   { 'kant·əl,ē·vər ,spriŋ }

cantilever vibration |MECH| Transverse oscillatory motion of a body fixed at one end.   { 'kant·əl,ē·vər vī'brā·shən }

canting |MECH| Displacing the free end of a beam which is fixed at one end by subjecting it to a sideways force which is just short of that required to cause fracture.   { 'kant·iŋ }

canting strip   See water table.   { 'kant·iŋ ,strip }

cant strip |BUILD| **1.** A strip placed along the angle between a wall and a roof so that the roofing will not bend sharply.   **2.** A strip placed under the edge of the lowest row of tiles on a roof to give them the same slope as the other tiles.   { 'kant ,strip }

cap |ENG| A detonating or blasting cap.   { kap }

capacitance |ELEC| The ratio of the charge on one of the conductors of a capacitor (there being an equal and opposite charge on the other conductor) to the potential difference between the conductors. Symbolized C. Formerly known as capacity. |ENG| In a closed feedwater heater, the volume of water required for proper operation of the drain control valve.   { kə'pas·ə·təns }

capacitance altimeter |ENG| An absolute altimeter which determines height of an aircraft aboveground by measuring the variations in capacitance between two conductors on the aircraft when the ground is near enough to act as a third conductor.   { kə'pas·ə·təns al'tim·əd·ər }

capacitance bridge |ELEC| A bridge for comparing two capacitances, such as a Schering bridge.   { kə'pas·ə·təns ,brij }

capacitance level indicator |ENG| A level indicator in which the material being monitored serves as the dielectric of a capacitor formed by a metal tank and an insulated electrode mounted vertically in the tank.   { kə'pas·ə·təns ¦lev·əl 'in·də,kād·ər }

capacitance meter |ENG| An instrument used to measure capacitance values of capacitors or of circuits containing capacitance.   { kə'pas·ə·təns ,mēd·ər }

capacitance-operated intrusion detector |ENG| A boundary alarm system in which the approach of an intruder to an antenna wire encircling the protected area a few feet above ground changes the antenna-ground capacitance and sets off the alarm.   { kə'pas·ə·təns ¦p·ə,rād·əd in'trü·zhən di'tek·tər }

capacitance standard   See standard capacitor.   { kə'pas·ə·təns ,stan·dərd }

capacitive coupling |ELEC| Use of a capacitor to transfer energy from one circuit to another.   { kə'pas·ə·təns ,kəp·liŋ }

capacitive electrometer |ENG| An instrument for measuring small voltages; the voltage is applied to the plates of a capacitor when they are close together, then the voltage source is removed and the plates are separated, increasing the potential difference between them to a measurable value.   Also known as condensing electrometer.   { kə¦pas·əd·iv ,i,lek'träm·əd·ər }

capacitive pressure transducer |ENG| A measurement device in which variations in pressure upon a capacitive element proportionately change the element's capacitive rating and thus the strength of the measured electric signal from the device.   { kə¦pas·əd·iv 'presh·ər tranz,dü·sər }

capacitor |ELEC| A device which consists essentially of two conductors (such as parallel metal plates) insulated from each other by a dielectric and which introduces capacitance into a circuit, stores electrical energy, blocks the flow of direct current, and permits the flow of alternating current to a degree dependent on the capacitor's capacitance and the current frequency. Symbolized C.   Also known as condenser; electric condenser.   { kə'pas·əd·ər }

capacitor bank |ELEC| A number of capacitors connected in series or in parallel.   { kə'pas·əd·ər ,baŋk }

capacitor color code |ELEC| A method of marking the value on a capacitor by means of dots or bands of colors as specified in the Electronic Industry Association color code.   { kə'pas·əd·ər 'kəl·ər ,kōd }

capacitor hydrophone |ENG ACOUS| A capacitor microphone that responds to waterborne sound waves.   { kə'pas·əd·ər 'hī·drə,fōn }

capacitor loudspeaker   See electrostatic loudspeaker.   { kə'pas·əd·ər 'laúd,spēk·ər }

capacitor microphone |ENG ACOUS| A microphone consisting essentially of a flexible metal diaphragm and a rigid metal plate that together form a two-plate air capacitor; sound waves set the diaphragm in vibration, producing capacitance variations that are converted into audiofrequency signals by a suitable amplifier circuit.   Also known as condenser microphone; electrostatic microphone.   { kə'pas·əd·ər 'mī·krə,fōn }

capacitor pickup |ENG ACOUS| A phonograph pickup in which movements of the stylus in a record groove cause variations in the capacitance of the pickup.   { kə'pas·əd·ər 'pik·əp }

capacity   See capacitance.   { kə'pas·əd·ē }

capacity correction |ENG| The correction applied to a mercury barometer with a nonadjustable cistern in order to compensate for the change in the level of the cistern as the atmospheric pressure changes. { kə'pas·əd·ē kə'rek·shən }

capacity factor [IND ENG] The ratio of average actual use to the available capacity of an apparatus or industrial plant to store, process, treat, manufacture, or produce. { kə'pas·əd·ē ,fak·tər }

cap crimper [ENG] A tool resembling a pliers that is used to press the open end of a blasting cap onto the safety fuse before placing the cap in the primer. { 'kap ,krim·pər }

cape chisel [DES ENG] A chisel that tapers to a flat, narrow cutting end; used to cut flat grooves. { 'kāp ,chiz·əl }

cape foot [MECH] A unit of length equal to 1.033 feet or to 0.3148584 meter. { 'kāp ,fût }

capillary correction [ENG] As applied to a mercury barometer, that part of the instrument correction which is required by the shape of the meniscus of the mercury. { ,kap·ə'lar·əd·ē kə,rek·shən }

capillary collector [ENG] An instrument for collecting liquid water from the atmosphere; the collecting head is fabricated of a porous material having a pore size of the order of 30 micrometers; the pressure difference across the water-air interface prevents air from entering the capillary system while allowing free flow of water. { 'kap·ə,ler·ē kə'lek·tər }

capillary drying [ENG] Progressive removal of moisture from a porous solid mass by surface evaporation followed by capillary movement of more moisture to the drying surface from the moist inner region, until the surface and core stabilize at the same moisture concentration. { 'kap·ə,ler·ē 'drī·iŋ }

capillary electrometer [ENG] An electrometer designed to measure a small potential difference between mercury and an electrolytic solution in a capillary tube by measuring the effect of this potential difference on the surface tension between the liquids. Also known as Lippmann electrometer. { 'kap·ə, ler·ē i,lek'träm·əd·ər }

capillary fitting [ENG] A pipe fitting having a socket-type end so that when the fitting is soldered to a pipe end, the solder flows by capillarity along the annular space between the pipe exterior and the socket within it, forming a tight fit. { 'kap·ə,ler·ē ,fid·iŋ }

capillary tube [ENG] A tube sufficiently fine so that capillary attraction of a liquid into the tube is significant. { 'kap·ə,ler·ē ,tüb }

capillary viscometer [ENG] A long, narrow tube that is used to measure the laminar flow of fluids. { 'kap·ə,ler·ē vis'käm·əd·ər }

capital amount factor [IND ENG] Any of 20 common compound interest formulas used to calculate the equivalent uniform annual cost of all cash flows. { 'kap·ət·əl ə'maúnt ,fak·tər }

capital budgeting [IND ENG] Planning the most effective use of resources to obtain the highest possible level of sustained profits. { 'kap·ət·əl 'bəj·əd·iŋ }

capital expenditure [IND ENG] Money spent for long-term additions or improvements and charged to a capital assets account. { 'kap·ət·əl ik'spen·di·chər }

capped fuse [ENG] A length of safety fuse with the cap or detonator crimped on before it is taken to the place of use. { 'kapt 'fyüz }

capping [ENG] Preparation of a capped fuse. { 'kap·iŋ }

cap screw [DES ENG] A screw which passes through a clear hole in the part to be joined, screws into a threaded hole in the other part, and has a head which holds the parts together. { 'kap ,skrü }

capstan [ENG] A shaft which pulls magnetic tape through a machine at constant speed. { 'kap·stən }

capstan nut [DES ENG] A nut whose edge has several holes, in one of which a bar can be inserted for turning it. { 'kap·stən ,nət }

capstan screw [DES ENG] A screw whose head has several radial holes, in one of which a bar can be inserted for turning it. { 'kap·stən ,skrü }

capsule [ENG] A boxlike component or unit, often sealed. { 'kap·səl }

captive fastener [DES ENG] A screw-type fastener that does not drop out after it has been unscrewed. { 'kap·tiv 'fas·ən·ər }

captive test [ENG] A hold-down test of a propulsion subsystem, rocket engine, or motor. { 'kap·tiv 'test }

capture area [ENG ACOUS] The effective area of the receiving surface of a hydrophone, or the available power of the acoustic energy divided by its equivalent plane-wave intensity. { 'kap·chər ,er·ē·ə }

capturing [ENG] The use of a torquer to restrain the spin axis of a gyro to a specified position relative to the spin reference axis. { 'kap·chə·riŋ }

car See automobile. { kär }

Carathéodory's principle [THERMO] An expression of the second law of thermodynamics which says that in the neighborhood of any equilibrium state of a system, there are states which are not accessible by a reversible or irreversible adiabatic process. Also known as principle of inaccessibility. { ,kär·ə,tā·ə'dór·ēz 'prin·sə·pəl }

carbide tool [DES ENG] A cutting tool made of tungsten, titanium, or tantalum carbides, having high heat and wear resistance. { 'kär,bīd ,tül }

carbometer [ENG] An instrument for measuring the carbon content of steel by measuring the magnetic properties of the steel in a known magnetic field. { kär'bäm·əd·ər }

carbonation [CHEM ENG] The process by which a fluid, especially a beverage, is impregnated with carbon dioxide. { ,kär·bə'nā·shən }

carbon bit [DES ENG] A diamond bit in which the cutting medium is inset carbon. { 'kär·bən ¦bit }

carbon burning rate [CHEM ENG] The weight of carbon burned per unit time from the catalytic-cracking catalyst in the regenerator. { ¦kär·bən 'bərn·iŋ ,rāt }

carbon canister See charcoal canister. { ¦kär·bən 'kan·ə·stər }

carbon dioxide fire extinguisher |CHEM ENG| A type of chemical fire extinguisher in which the extinguishing agent is liquid carbon dioxide, stored under 800–900 pounds per square inch (5.5–6.2 megapascals) at normal room temperature. { ¦kär·bən dī'äk,sīd 'fīr ik'stiŋ·gwish·ər }

carbon hydrophone |ENG ACOUS| A carbon microphone that responds to waterborne sound waves. { ¦kär·bən 'hī·drə,fōn }

carbon knock |MECH ENG| Premature ignition resulting in knocking or pinging in an internal combustion engine caused when the accumulation of carbon produces overheating in the cylinder. { 'kär·bən ,näk }

carbon microphone |ENG ACOUS| A microphone in which a flexible diaphragm moves in response to sound waves and applies a varying pressure to a container filled with carbon granules, causing the resistance of the microphone to vary correspondingly. { ¦kär·bən 'mī·krə,fōn }

carbon-pile pressure transducer |ENG| A measurement device in which variations in pressure upon a conductive carbon core proportionately change the core's electrical resistance, and thus the strength of the measured electric signal from the device. { 'kär·bən ,pīl 'presh·ər tranz ,dü·sər }

carbon residue |CHEM ENG| The quantity of carbon produced from a lubricating oil heated in a closed container under standard conditions. { 'kär·bən 'rez·ə,dü }

carbon-residue test |CHEM ENG| A destructive-distillation method for estimation of carbon residues in fuels and lubricating oils. Also known as Conradson carbon test. { 'kär·bən 'rez·ə,dü ,test }

carbon resistance thermometer |ENG| A highly sensitive resistance thermometer for measuring temperatures in the range 0.05–20 K; capable of measuring temperature changes of the order 10^{-5} degree. { 'kär·bən ri¦zis·təns thər,mäm ·əd·ər }

carbon transducer |ENG| A transducer consisting of carbon granules in contact with a fixed electrode and a movable electrode, so that motion of the movable electrode varies the resistance of the granules. { 'kär·bən tranz'dü·sər }

carburetion |CHEM ENG| The process of enriching a gas by adding volatile carbon compounds, such as hydrocarbons, to it, as in the manufacture of carbureted water gas. |MECH ENG| The process of mixing fuel with air in a carburetor. { ,kär·bə'rā·shən }

carburetor |CHEM ENG| An apparatus for vaporizing, cracking, and enriching oils in the manufacture of carbureted water gas. |MECH ENG| A device that makes and controls the proportions and quantity of fuel-air mixture fed to a spark-ignition internal combustion engine. { 'kär·bə,red·ər }

carburetor icing |MECH ENG| The formation of ice in an engine carburetor as a consequence of expansive cooling and evaporation of gasoline. { 'kär·bə,red·ər ,ī·siŋ }

card |ELECTR| A printed circuit board or other arrangement of miniaturized components that can be plugged into a computer or peripheral device. { kärd }

Cardan joint See Hooke's joint. { 'kär,dan ,jóint }

Cardan motion |MECH ENG| The straight-line path followed by a moving centrode in a four-bar centrode linkage. { 'kär,dan 'mō·shən }

Cardan shaft |MECH ENG| A shaft with a universal joint at its end to accommodate a varying shaft angle. { 'kär,dan ,shaft }

Cardan's suspension |DES ENG| An arrangement of rings in which a heavy body is mounted so that the body is fixed at one point; generally used in a gyroscope. { 'kär,danz səs'pen·shən }

card-edge connector |ELEC| A connector that mates with printed-wiring leads running to the edge of a printed circuit board on one or both sides. Also known as edgeboard connector. { 'kärd ,ej kə'nek·tər }

cardioid microphone |ENG ACOUS| A microphone having a heart-shaped, or cardioid, response pattern, so it has nearly uniform response for a range of about 180° in one direction and minimum response in the opposite direction. { 'kärd·ē,óid 'mī·krə,fōn }

cardioid pattern |ENG| Heart-shaped pattern obtained as the response or radiation characteristic of certain directional antennas, or as the response characteristic of certain types of microphones. { 'kärd·ē,óid ,pad·ərn }

card key access |ENG| A physical security system in which doors are unlocked by placing a badge that contains magnetically coded information in proximity to a reading device; some systems also require the typing of this information on a keyboard. { 'kärd ,kē 'ak,ses }

car dump |MECH ENG| Any one of several devices for unloading industrial or railroad cars by rotating or tilting the car. { 'kär ,dəmp }

car-following theory |ENG| A mathematical model of the interactions between motor vehicles in terms of relative speed, absolute speed, and separation. { 'kär,fäl·ə·wiŋ ,thē·ə·rē }

cargo boom |MECH ENG| A long spar extending from the mast of a derrick to support or guide objects lifted or suspended. { 'kär·gō ,büm }

cargo mill |IND ENG| A sawmill equipped with docks so the product can be loaded directly onto ships. { 'kär·gō ,mil }

cargo winch |MECH ENG| A motor-driven hoisting machine for cargo having a drum around which a chain or rope winds as the load is lifted. { 'kär·gō ,winch }

carillon |ENG| A musical instrument played from a keyboard with two or more full chromatic octaves of fine bells shaped for homogeneity of timbre. { 'kär·ə,län }

Carnot-Clausius equation |THERMO| For any system executing a closed cycle of reversible changes, the integral over the cycle of the infinitesimal amount of heat transferred to the system divided by its temperature equals 0. Also known as Clausius theorem. { kär¦nōt 'klóz·ē· əs i‚kwä·zhən }

Carnot cycle |THERMO| A hypothetical cycle consisting of four reversible processes in succession: an isothermal expansion and heat addition, an isentropic expansion, an isothermal compression and heat rejection process, and an isentropic compression. { kär'nō ‚si·kəl }

Carnot efficiency |THERMO| The efficiency of a Carnot engine receiving heat at a temperature absolute T_1 and giving it up at a lower temperature absolute T_2; equal to $(T_1 - T_2)/T_1$. { kär'nō i'fish·ən·sē }

Carnot engine |MECH ENG| An ideal, frictionless engine which operates in a Carnot cycle. { kär'nō 'en·jən }

Carnot number |THERMO| A property of two heat sinks, equal to the Carnot efficiency of an engine operating between them. { kär'nō ‚nəm·bər }

Carnot's theorem |THERMO| **1.** The theorem that all Carnot engines operating between two given temperatures have the same efficiency, and no cyclic heat engine operating between two given temperatures is more efficient than a Carnot engine. **2.** The theorem that any system has two properties, the thermodynamic temperature T and the entropy S, such that the amount of heat exchanged in an infinitesimal reversible process is given by $dQ = TdS$; the thermodynamic temperature is a strictly increasing function of the empirical temperature measured on an arbitrary scale. { kär'noz 'thir·əm }

carousel |MECH ENG| A rotating transport system that transfers and presents workpieces for loading and unloading by a robot or other machine. { ‚kar·ə'sel }

carpenter's level |DES ENG| A bar, usually of aluminum or wood, containing a spirit level. { 'kär·pən·tərz ‚lev·əl }

car retarder |ENG| A device located along the track to reduce or control the velocity of railroad or mine cars. { ¦kär ri'tärd·ər }

carriage |ENG| **1.** A device that moves in a predetermined path in a machine and carries some other part, such as a recorder head. **2.** A mechanism designed to hold a paper in the active portion of a printing or typing device, for example, a typewriter carriage. |MECH ENG| A structure on an industrial truck or stacker that supports forks or other attached equipment and travels vertically within the mast. { 'kar·ij }

carriage bolt |DES ENG| A round-head type of bolt with a square neck, used with a nut as a through bolt. { 'kar·ij ‚bōlt }

carriage stop |MECH ENG| A device added to the outer way of a lathe bed for accurately spacing grooves, turning multiple diameters and lengths, and cutting off pieces of specified thickness. { 'kar·ij ‚stäp }

carrier |MECH ENG| Any machine for transporting materials or people. { 'kar·ē·ər }

carrier line |ELEC| Any transmission line used for multiple-channel carrier communication. { 'kar·ē·ər ‚līn }

carrier pipe |ENG| Pipe used to carry or conduct fluids, as contrasted with an exterior protective or casing pipe. { 'kar·ē·ər ‚pīp }

carrousel |IND ENG| In an assembly-line operation, a conveyor that moves objects in a complete circuit on a horizontal plane. { ‚ka·rə'sel }

carrying capacity |ELEC| The maximum amount of current or power that can be safely handled by a wire or other component { 'kar· ē·iŋ kə'pas·əd·ē }

carry-over |CHEM ENG| Unwanted liquid or solid material carried by the overhead effluent from a fractionating column, absorber, or reaction vessel. { 'kar·ē ‚ō·vər }

car shaker |MECH ENG| A device consisting of a heavy yoke on an open-top car's sides that actively vibrates and rapidly discharges a load, such as coal, gravel, or sand, when an unbalanced pulley attached to the yoke is rotated fast. { 'kär ‚shāk·ər }

car stop |ENG| An appliance used to arrest the movement of a mine or railroad car. { 'kär ‚stäp }

Cartesian-coordinate robot |CONT SYS| A robot having orthogonal, sliding joints and supported by a nonrotary base as the axis. { kär'tē·zhən kō¦órd·ən·ət 'rō‚bät }

Cartesian diver manostat |ENG| Preset, on-off-control manometer arrangement by which a specified low pressure (high vacuum) is maintained via the rise or submergence of a marginally buoyant float within a liquid mercury reservoir. { kär'tē·zhan ¦dīv·ər 'man·ə‚stat }

cartridge |ENG| A cylindrical, waterproof, paper shell filled with high explosive and closed at both ends; used in blasting. |ENG ACOUS| See phonograph pickup; tape cartridge. { 'kär·trij }

cartridge filter |ENG| A filter for the clarification of process liquids containing small amounts of solids; turgid liquid flows between thin metal disks, assembled in a vertical stack, to openings in a central shaft supporting the disks, and solids are trapped between the disks. { 'kär·trij ‚fil·tər }

cartridge starter |MECH ENG| An explosive device which, when placed in an engine and detonated, moves a piston, thereby starting the engine. { 'kär·trij ‚stärd·ər }

car tunnel kiln |ENG| A long kiln with the fire located near the midpoint; ceramic ware is fired by loading it onto cars which are pushed through the kiln. { 'kär ‚tən·əl ‚kil }

Casale process |CHEM ENG| A process that employs promoted iron oxide catalyst for synthesis of ammonia from nitrogen and hydrogen. { kə‚säl·ē ‚präs·əs }

cascade |ELEC| An electric-power circuit arrangement in which circuit breakers of reduced interrupting ratings are used in the branches, the circuit breakers being assisted in their protection function by other circuit breakers which operate almost instantaneously. Also known as backup arrangement. |ELECTR| See avalanche. |ENG| An arrangement of separation devices, such as isotope separators, connected in series

cascade compensation

so that they multiply the effect of each individual device. { ka'skād }

cascade compensation [CONT SYS] Compensation in which the compensator is placed in series with the forward transfer function. Also known as series compensation; tandem compensation. { ka'skād käm·pən'sā·shən }

cascade control [CONT SYS] An automatic control system in which various control units are linked in sequence, each control unit regulating the operation of the next control unit in line. { ka'skād kən,trōl }

cascade cooler [CHEM ENG] Fluid-cooling device through which the fluid flows in a series of horizontal tubes, one above the other; cooling water from a trough drips over each tube, then to a drain. Also known as serpentine cooler; trickle cooler. { ka'skād ,kü·lər }

cascaded [ENG] Of a series of elements or devices, arranged so that the output of one feeds directly into the input of another, as a series of dynodes or a series of airfoils. { ka'skād·əd }

cascade impactor [ENG] A low-speed impaction device for use in sampling both solid and liquid atmospheric suspensoids; consists of four pairs of jets (each of progressively smaller size) and sampling plates working in series and designed so that each plate collects particles of one size range. { ka'skād im'pak·tər }

cascade limiter [ELECTR] A limiter circuit that uses two vacuum tubes in series to give improved limiter operation for both weak and strong signals in a frequency-modulation receiver. Also known as double limiter. { ka 'skād 'lim·əd·ər }

cascade mixer-settler [CHEM ENG] Series of liquid-holding vessels with stirrers, each connected to an unstirred vessel in which solids or heavy immiscible liquids settle out of suspension; light liquid moves through the mixer-settler units, counterflowing to heavy material, in such a manner that fresh liquid contacts treated heavy material, and spent (used) liquid contacts fresh (untreated) heavy material. { ka'skād ¦mik·sər ¦set·lər }

cascade pulverizer [MECH ENG] A form of tumbling pulverizer that uses large lumps to do the pulverizing. { ka'skād 'pəl·və,rīz·ər }

cascade system [MECH ENG] A combination of two or more refrigeration systems connected in series to produce extremely low temperatures, with the evaporator of one machine used to cool the condenser of another. { ka'skād ,sis·təm }

cascade tray [CHEM ENG] A fractionating apparatus that consists of a series of parallel troughs arranged in stairstep fashion. { ka'skād ,trā }

cascading [ELEC] An effect in which a failure of an electrical power system causes this system to draw excessive amounts of power from power systems which are interconnected with it, causing them to fail, and these systems cause adjacent systems to fail in a similar manner, and so forth. [MECH ENG] An effect in ball-mill rotating devices when the upper level of crushing

bodies breaks clear and falls to the top of the crop load. { ka'skād·iŋ }

cascading drain [MECH ENG] A flow of water into the closed shell of a feedwater heater from a water source maintained at a higher pressure. { ka'skād·iŋ 'drān }

case [ENG] An item designed to hold a specific item in a fixed position by virtue of conforming dimensions or attachments; the item which it contains is complete in itself for removal and use outside the container. { kās }

case bay [BUILD] A division of a roof or floor, consisting of two principal rafters and the joists between them. { 'kās ,bā }

casement window [BUILD] A window hinged on the side that opens to the outside. { 'kās·mənt 'win·dō }

casing [BUILD] A finishing member around the opening of a door or window. [DES ENG] The outer portion of a tire assembly consisting of fabric or cord to which rubber is vulcanized. [MECH ENG] A fire-resistant covering used to protect part or all of a steam generating unit. { 'kā,siŋ }

casing nail [DES ENG] A nail about half a gage thinner than a common wire nail of the same length. { 'kā,siŋ ,nāl }

casing shoe [ENG] A ring with a cutting edge on the bottom of a well casing. { 'kā,siŋ ,shü }

cassette [ENG] A light-tight container designed to hold photographic film or plates. [ENG ACOUS] A small, compact container that holds a magnetic tape and can be readily inserted into a matching tape recorder for recording or playback; the tape passes from one hub within the container to the other hub. { kə'set }

cast [ENG] **1.** To form a liquid or plastic substance into a fixed shape by letting it cool in the mold. **2.** Any object which is formed by placing a castable substance in a mold or form and allowing it to solidify. Also known as casting. { kast }

Castaing-Slodzian mass analyzer See direct-imaging mass analyzer. { ¦kas·taŋ ¦slō·zhən ,mas 'an·ə,līz·ər }

castellated bit [DES ENG] **1.** A long-tooth, saw-tooth bit. **2.** A diamond-set coring bit with a few large diamonds or hard metal cutting points set in the face of each of several upstanding prongs separated from each other by deep waterways. Also known as padded bit. { 'kas·tə,lād·əd 'bit }

castellated nut [DES ENG] A type of hexagonal nut with a cylindrical portion above through which slots are cut so that a cotter pin or safety wire can hold it in place. { 'kas·tə,lād·əd 'nət }

caster [ENG] **1.** The inclination of the kingpin or its equivalent in automotive steering, which is positive if the kingpin inclines forward, negative if it inclines backward, and zero if it is vertical as viewed along the axis of the front wheels. **2.** A wheel which is free to swivel about an axis at right angles to the axis of the wheel, used to support trucks, machinery, or furniture. { 'kas·tər }

cast-film extrusion See chill-roll extrusion. { 'kast ¦film ik'strü·zhən }

Castigliano's principle See Castigliano's theorem. { ,kas·til'yä·nōz ,prin·sə·pəl }

Castigliano's theorem |MECH| The theorem that the component in a given direction of the deflection of the point of application of an external force on an elastic body is equal to the partial derivative of the work of deformation with respect to the component of the force in that direction. Also known as Castigliano's principle. { ,kas·til'yä·nōz ,thir·əm }

casting See cast. { 'kast·iŋ }

casting area |ENG| In plastics injection molding, the moldable area of a thermoplastic material for a given thickness and under given conditions of molding. { 'kast·iŋ ,er·ē·ə }

casting strain |MECH| Any strain that results from the cooling of a casting, causing casting stress. { 'kast·iŋ ,strān }

casting stress |MECH| Any stress that develops in a casting due to geometry and casting shrinkage. { 'kast·iŋ ,stres }

Castner cell |CHEM ENG| A type of mercury cell used in the commercial production of chlorine and sodium. { 'kast·nər ,sel }

Castner process |CHEM ENG| A process used industrially to make high-test sodium cyanide by reacting sodium, glowed charcoal, and dry ammonia gas to form sodamide, which is converted to cyanamide immediately; the cyanamide is converted to cyanide with charcoal. { 'kast·nər ,präs·əs }

cast setting See mechanical setting. { 'kast ,sed·iŋ }

catalyst stripping |CHEM ENG| Introduction of steam to remove hydrocarbons retained on the catalyst, the steam is introduced where the spent catalyst leaves the reactor. { 'kad·əl·əst ,strip· iŋ }

catalytic activity |CHEM ENG| The ratio of the space velocity of a catalyst being tested, to the space velocity required for a standard catalyst to give the same conversion as the catalyst under test. { ¦kad·əl¦id·ik ak'tiv·əd·ē }

catalytic converter |CHEM ENG| A device that is fitted to the exhaust system of an automotive vehicle and contains a catalyst capable of converting potentially polluting exhaust gases into harmless or less harmful products. { ¦kad·əl¦id· ik kən'vərd·ər }

catalytic cracker See catalytic cracking unit. { ¦kad·əl¦id·ik 'krak·ər }

catalytic cracking |CHEM ENG| Conversion of high-boiling hydrocarbons into lower-boiling types by a catalyst. { ¦kad·əl¦id·ik 'krak·iŋ }

catalytic cracking unit |CHEM ENG| A unit in a petroleum refinery in which a catalyst is used to carry out cracking of hydrocarbons. Also known as catalytic cracker. { ¦kad·əl¦id·ik 'krak·iŋ ,yü· nət }

catalytic hydrogenation |CHEM ENG| Hydrogenating by means of catalysts such as nickel or palladium. { ¦kad·əl¦id·ik ,hī·drə·jə'nā·shən }

catalytic polymerization |CHEM ENG| Polymerization of monomers to form high-molecular-weight molecules in the presence of catalysts. { ¦kad·əl¦id·ik pə,lim·ə·rə'zā·shən }

catalytic reforming |CHEM ENG| Rearranging of hydrocarbon molecules in a gasoline boiling-range feedstock to form hydrocarbons having a higher antiknock quality. Abbreviated CR. { ¦kad·əl¦id·ik rē'fȯr·miŋ }

cat-and-mouse engine |MECH ENG| A type of rotary engine, typified by the Tschudi engine, which is an analog of the reciprocating piston engine, except that the pistons travel in a circular motion. Also known as scissor engine. { ¦kat ən 'maús ,en·jən }

cataracting |MECH ENG| A motion of the crushed bodies in a ball mill in which some, leaving the top of the crop load, fall with impact to the toe of the load. { 'kad·ə,rak·tiŋ }

catastrophic failure |ENG| **1.** A sudden failure without warning, as opposed to degradation failure. **2.** A failure whose occurrence can prevent the satisfactory performance of an entire assembly or system. { ,kad·ə'sträf·ik 'fāl·yər }

catch |DES ENG| A device used for fastening a door or gate and usually operated manually from only one side, for example, a latch. { kach }

catch basin |CIV ENG| **1.** A basin at the point where a street gutter empties into a sewer, built to catch matter that would not easily pass through the sewer. **2.** A well or reservoir into which surface water may drain off. { 'kach ,bās·ən }

catching diode |ELECTR| Diode connected to act as a short circuit when its anode becomes positive; the diode then prevents the voltage of a circuit terminal from rising above the diode cathode voltage. { 'kach·iŋ ,dī,ōd }

catchwater |CIV ENG| A ditch for catching water on sloping land. { 'kach,wȯd·ər }

cat cracker |CHEM ENG| A refinery unit where catalytic cracking is done. { 'kat ,krak·ər }

catenary suspension |ENG| Holding a flexible wire or chain aloft by its end points; the wire or chain takes the shape of a catenary. { 'kat· ə,ner·ē səs'pen·shən }

caterpillar |MECH ENG| A vehicle, such as a tractor or army tank, which runs on two endless belts, one on each side, consisting of flat treads and kept in motion by toothed driving wheels. { 'kad·ər,pil·ər }

caterpillar chain |DES ENG| A short, endless chain on which dogs (grippers) or teeth are arranged to mesh with a conveyor. { 'kad·ər,pil· ər ,chān }

caterpillar gate |CIV ENG| A steel gate carried on crawler tracks that is used to control water flow through a spillway. { 'kad·ər,pil·ər ,gāt }

catforming |CHEM ENG| A naphtha-reforming process with a catalyst of platinum-silica-alumina which results in very high hydrogen purity. { 'kat,fȯr·miŋ }

cathetometer [ENG] An instrument for measuring small differences in height, for example, between two columns of mercury. { ,kath·ə'täm· əd·ər }

cathode [ELEC] The terminal at which current leaves a primary cell or storage battery; it is negative with respect to the device, and positive with respect to the external circuit. [ELECTR] **1.** The primary source of electrons in an electron tube; in directly heated tubes the filament is the cathode, and in indirectly heated tubes a coated metal cathode surrounds a heater. Designated K. Also known as negative electrode. **2.** The terminal of a semiconductor diode that is negative with respect to the other terminal when the diode is biased in the forward direction. { 'kath,ōd }

cathode efficiency [CHEM ENG] The proportion of current used for completion of a given process at the cathode. { 'kath,ōd i,fish·ən·sē }

cathode-ray tube [ELECTR] An electron tube in which a beam of electrons can be focused to a small area and varied in position and intensity on a surface. Abbreviated CRT. Originally known as Braun tube; also known as electron-ray tube. { 'kath,ōd ¦rā ,tüb }

cathodic inhibitor [CHEM ENG] A compound, such as calcium bicarbonate or sodium phosphate, which is deposited on a metal surface in a thin film that operates at the cathodes to provide physical protection over the entire surface against corrosive attack in a conducting medium. { kə'thäd·ik in'hib·əd·ər }

catwalk [ENG] A narrow, raised platform or pathway used for passage to otherwise inaccessible areas, such as a raised walkway on a ship permitting fore and aft passage when the main deck is awash, a walkway on the roof of a freight car, or a walkway along a vehicular bridge. { 'kat,wȯk }

caul [ENG] A sheet of metal or other material that is heated and used to equalize pressure during fabricating plywood, shaping surface veneer, and hot-pressing composite materials. { kȯl }

caulk [ENG] To make a seam or point airtight, watertight, or steamtight by driving in caulking compound, dry pack, lead wool, or other material. Also spelled calk. { kȯk }

caulking iron [DES ENG] A tool for applying caulking to a seam. { 'kȯk·iŋ ,ī·ərn }

causality [MECH] In classical mechanics, the principle that the specification of the dynamical variables of a system at a given time, and of the external forces acting on the system, completely determines the values of dynamical variables at later times. Also known as determinism. { kȯ 'zal·əd·ē }

causal system [CONT SYS] A system whose response to an input does not depend on values of the input at later times. Also known as nonanticipatory system; physical system. { 'kȯ·zəl ,sis·təm }

causticization [CHEM ENG] A process for converting an alkaline carbonate into lime. { 'kȯs· tə·sə'zā·shən }

caustic treater [CHEM ENG] A vessel containing a strong alkali through which solutions are passed for removal of undesirable substances, for example, sulfides, mercaptans, or acids. { 'kȯ·stik ,trēd·ər }

cautious control [CONT SYS] A control law for a stochastic adaptive control system which hedges and uses lower gain when the estimates are uncertain. { 'kȯ·shəs kən'trōl }

cave [ENG] A pit or tunnel under a glass furnace for collecting ashes or raking the fire. { kāv }

Cavendish balance [ENG] An instrument for determining the constant of gravitation, in which one measures the displacement of two small spheres of mass m, which are connected by a light rod suspended in the middle by a thin wire, caused by bringing two large spheres of mass M near them. { 'kav·ən·dish 'bal·əns }

cavings See slough. { 'kāv·iŋz }

cavitation [ENG] Pitting of a solid surface such as metal or concrete. { ,kav·ə'tā·shən }

cavitation resistance inducer [MECH ENG] In liquid flows through rotating machinery, an axial flow pump with high-solidity blades that is used in front of a main pump in order to increase the inlet head and thereby prevent cavitation in the downstream impeller. { ,kav·ə¦tā·shən ri'sis· təns in,dü·sər }

cavity frequency meter [ENG] A device that employs a cavity resonator to measure microwave frequencies. { 'kav·əd·ē 'frē·kwən·sē ,mēd·ər }

cavity impedance [ELECTR] The impedance of the cavity of a microwave tube which appears across the gap between the cathode and the anode. { 'kav·əd·ē im'pēd·əns }

cavity magnetron [ELECTR] A magnetron having a number of resonant cavities forming the anode; used as a microwave oscillator. { 'kav· əd·ē 'mag·nə,trän }

cavity radiator [THERMO] A heated enclosure with a small opening which allows some radiation to escape or enter; the escaping radiation approximates that of a blackbody. { 'kav·əd·ē 'rād·ē,ād·ər }

cavity resonance [ENG ACOUS] The natural resonant vibration of a loudspeaker baffle; if in the audio range, it is evident as unpleasant emphasis of sounds at that frequency. { 'kav·əd·ē 'rez· ən·əns }

cavity wall [BUILD] A wall constructed in two separate thicknesses with an air space between; provides thermal insulation. Also known as hollow wall. { 'kav·əd·ē ,wȯl }

c axis [MECH ENG] The angle that specifies the rotation of a machine tool about the z axis. { 'sē ,ak·səs }

CCD See charge-coupled device.

C chart [IND ENG] A quality-control chart showing number of defects in subgroups of constant size; gives information concerning quality level, its variability, and evidence of assignable causes of variation. { 'sē ,chärt }

CCR process See cyclic catalytic reforming process. { ,sē,sē'är ,präs·əs }

CD-4 sound See compatible discrete four-channel sound. { ¦sē¦de 'fór ,saúnd }

ceiling [BUILD] The covering made of plaster, boards, or other material that constitutes the overhead surface in a room. { 'sē·liŋ }

ceiling light [ENG] A type of cloud-height indicator which uses a searchlight to project vertically a narrow beam of light onto a cloud base. Also known as ceiling projector. { 'sē·liŋ ,līt }

ceiling projector See ceiling light. { 'sē·liŋ prə'jek·tər }

ceilometer [ENG] An automatic-recording cloud-height indicator. { sē'läm·əd·ər }

cell [ELEC] A single unit of a battery. [IND ENG] A manufacturing unit consisting of a group of work stations and their interconnecting materials-transport mechanisms and storage buffers. { sel }

cellular cofferdam [CIV ENG] A cofferdam consisting of interlocking steel-sheet piling driven as a series of interconnecting cells; cells may be of circular type or of straight-wall diaphragm type; space between lines of pilings is filled with sand. { 'sel·yə·lər 'kóf·ər,dam }

cellular horn See multicellular horn. { 'sel·yə·lər 'hórn }

cellular manufacturing [IND ENG] A type of manufacturing in which equipment is organized into groups or cells according to function and intermachine relationships. { ¦sel·yə·lər ,man·ə'fak·chər·iŋ }

cellular striation [ENG] Stratum of cells inside a cellular-plastic object that differs noticeably from the cell structure of the remainder of the material. { 'sel·yə·lər strī'ā·shən }

celo [MECH] A unit of acceleration equal to the acceleration of a body whose velocity changes uniformly by 1 foot (0.3048 meter) per second in 1 second. { 'se·lō }

Celsius degree [THERMO] Unit of temperature interval or difference equal to the kelvin. { 'sel·sē·əs di'grē }

Celsius temperature scale [THERMO] Temperature scale in which the temperature θ_c in degrees Celsius (°C) is related to the temperature T_k in kelvins by the formula $\theta_c = T_k - 273.15$; the freezing point of water at standard atmospheric pressure is very nearly 0°C and the corresponding boiling point is very nearly 100°C. Formerly known as centigrade temperature scale. { 'sel·se·əs 'tem·prə·chər ,skāl }

cementation [ENG] **1.** Plugging a cavity or drill hole with cement. Also known as dental work. **2.** Consolidation of loose sediments or sand by injection of a chemical agent or binder. { ,sē,men'tā·shən }

cement gun [MECH ENG] **1.** A machine for mixing, wetting, and applying refractory mortars to hot furnace walls. Also known as cement injector. **2.** A mechanical device for the application of cement or mortar to the walls or roofs of mine openings or building walls. { si'ment ,gən }

cement injector See cement gun. { si'ment in 'jek·tər }

cement kiln [ENG] A kiln used to fire cement to less than complete melting. { si'ment ,kil }

cement mill [MECH ENG] A mill for grinding rock to a powder for cement. { si'ment ,mil }

cement pump [MECH ENG] A piston device used to move concrete through pipes. { si'ment ,pəmp }

cement silo [ENG] A silo used to store dry, bulk cement. { si'ment 'sī,lō }

cement valve [MECH ENG] A ball-, flapper-, or clack-type valve placed at the bottom of a string of casing, through which cement is pumped, so that when pumping ceases, the valve closes and prevents return of cement into the casing. { si'ment ,valv }

centare See centiare. { 'sen,tär }

center [IND ENG] A manufacturing unit containing a number of interconnected cells. { 'sen·tər }

center-bearing swing bridge [CIV ENG] A type of swing bridge that has a single large bearing on a pier, called the pivot pier, in the waterway. { sen·tər ,ber·iŋ 'swiŋ ,brij }

center drill [ENG] A two-fluted tool consisting of a twist drill with a 60° countersink; used to drill countersink center holes in a workpiece to be mounted between centers for turning or grinding. { 'sen·tər ,dril }

center gage [DES ENG] A gage used to check angles; for example, the angles of cutting tool points or screw threads, or the angular position of cutting tools. { 'sen·tər ,gāj }

center-gated mold [ENG] A plastics injection mold with the filling orifice interconnected to the nozzle and the center of the cavity area. { 'sen·tər ,gād·əd 'mōld }

centering [CIV ENG] A curved, temporary support for an arch or dome during a casting or laying operations. { 'sen·tə·riŋ }

centering machine [MECH ENG] A machine for drilling and countersinking work to be turned on a lathe. { 'sen·tə·riŋ ma'shēn }

centerless grinder [MECH ENG] A cylindrical metal-grinding machine that carries the work on a support or blade between two abrasive wheels. { 'sen·tər·ləs 'grin·dər }

center line [ENG] A line that represents an axis of symmetry on a plane figure such as a plan for a structure or a machine. { 'sen·tər ,līn }

center of attraction [MECH] A point toward which a force on a body or particle (such as gravitational or electrostatic force) is always directed; the magnitude of the force depends only on the distance of the body or particle from this point. { 'sen·tər əv ə'trak·shən }

center of buoyancy [MECH] The point through which acts the resultant force exerted on a body by a static fluid in which it is submerged or floating; located at the centroid of displaced volume. { 'sen·tər əv 'bói·ən·sē }

center of force [MECH] The point toward or from which a central force acts. { 'sen·tər əv 'fórs }

center of gravity [MECH] A fixed point in a material body through which the resultant force of gravitational attraction acts. { 'sen·tər əv 'grav·əd·ē }

center of inertia See center of mass. { 'sen·tər əv i'nər·shə }

center of mass [MECH] That point of a material body or system of bodies which moves as though the system's total mass existed at the point and all external forces were applied at the point. Also known as center of inertia; centroid. { 'sen·tər əv 'mas }

center-of-mass coordinate system [MECH] A reference frame which moves with the velocity of the center of mass, so that the center of mass is at rest in this system, and the total momentum of the system is zero. Also known as center of momentum coordinate system. { 'sen·tər əv 'mas kō'órd·nət ,sis·təm }

center-of-momentum coordinate system See center-of-mass coordinate system. { 'sen·tər əv mə'men·təm kō'órd·nət ,sis·təm }

center of oscillation [MECH] Point in a physical pendulum, on the line through the point of suspension and the center of mass, which moves as if all the mass of the pendulum were concentrated there. { 'sen·tər əv ,äs·ə'lā·shən }

center of percussion [MECH] If a rigid body, free to move in a plane, is struck a blow at a point O, and the line of force is perpendicular to the line from O to the center of mass, then the initial motion of the body is a rotation about the center of percussion relative to O; it can be shown to coincide with the center of oscillation relative to O. { 'sen·tər əv pər'kəsh·ən }

center of suspension [MECH] The intersection of the axis of rotation of a pendulum with a plane perpendicular to the axis that passes through the center of mass. { 'sen·tər əv sə'spen·shən }

center of twist [MECH] A point on a line parallel to the axis of a beam through which any transverse force must be applied to avoid twisting of the section. Also known as shear center. { 'sen·tər əv 'twist }

center plug [DES ENG] A small diamond-set circular plug, designed to be inserted into the annular opening in a core bit, thus converting it to a noncoring bit. { 'sen·tər ,pləg }

center punch [DES ENG] A tool similar to a prick punch but having the point ground to an angle of about 90°; used to enlarge prick-punch marks or holes. { 'sen·tər ,pənch }

center square [DES ENG] A straight edge with a sliding square; used to locate the center of a circle. { 'sen·tər ,skwer }

centiare [MECH] Unit of area equal to 1 square meter. Also spelled centare. { 'sen·tē,är }

centibar [MECH] A unit of pressure equal to 0.01 bar or to 1000 pascals. { 'sent·ə,bär }

centigrade heat unit [THERMO] A unit of heat energy, equal to 0.01 of the quantity of heat needed to raise 1 pound of air-free water from 0 to 100°C at a constant pressure of 1 standard atmosphere; equal to 1900.44 joules. Symbolized CHU; (more correctly) CHU$_{mean}$. { 'sent·ə,grād 'hēt ,yü·nət }

centigrade temperature scale See Celsius temperature scale. { 'sent·ə,grād 'tem·prə·chər ,skāl }

centigram [MECH] Unit of mass equal to 0.01 gram or 10^{-5} kilogram. Abbreviated cg. { 'sent·ə,gram }

centihg See centimeter of mercury. { 'sen,tig or ¦sent·ē,āch'¦ē }

centiliter [MECH] A unit of volume equal to 0.01 liter or to 10^{-5} cubic meter. { 'sent·ə,lēd·ər }

centimeter [MECH] A unit of length equal to 0.01 meter. Abbreviated cm. { 'sent·ə,mēd·ər }

centimeter of mercury [MECH] A unit of pressure equal to the pressure that would support a column of mercury 1 centimeter high, having a density of 13.5951 grams per cubic centimeter, when the acceleration of gravity is equal to its standard value (980.665 centimeters per second per second); it is equal to 1333.22387415 pascals; it differs from the dekatorr by less than 1 part in 7,000,000. Abbreviated cmHg. Also known as centihg. { 'sent·ə,mēd·ər əv 'mər·kyə·rē }

central control [SYS ENG] Control exercised over an extensive and complicated system from a single center. { 'sen·trəl kən'trōl }

central force [MECH] A force whose line of action is always directed toward a fixed point; the force may attract or repel. { 'sen·trəl 'fórs }

central gear [MECH ENG] The gear on the central axis of a planetary gear train, about which a pinion rotates. Also known as sun gear. { 'sen·trəl 'gir }

central heating [CIV ENG] The use of a single steam or hot-water heating plant to serve a group of buildings, facilities, or even a complete community through a system of distribution pipes. { 'sen·trəl 'hēd·iŋ }

centralized traffic control [CIV ENG] Control of train movements by signal indications given by a train director at a central control point. Abbreviated CTC. { 'sen·trə,līzd 'traf·ik kən'trōl }

central orbit [MECH] The path followed by a body moving under the action of a central force. { 'sen·trəl 'ór·bət }

centrifugal [MECH] Acting or moving in a direction away from the axis of rotation or the center of a circle along which a body is moving. { ,sen 'trif·i·gəl }

centrifugal atomizer [MECH ENG] Device that atomizes liquids with a spinning disk; liquid is fed onto the center of the disk, and the whirling motion (3000 to 50,000 revolutions per minute) forces the liquid outward in thin sheets to cause atomization. { ,sen'trif·i·gəl 'ad·ə,mīz·ər }

centrifugal barrier [MECH] A steep rise, located around the center of force, in the effective potential governing the radial motion of a particle of nonvanishing angular momentum in a central force field, which results from the centrifugal force and prevents the particle from reaching the center of force, or causes its Schrödinger wave function to vanish there in a quantum-mechanical system. { ,sen'trif·i·gəl 'bar·ē·ər }

centrifugal brake [MECH ENG] A safety device on a hoist drum that applies the brake if the drum speed is greater than a set limit. { ,sen'trif·i·gəl 'brāk }

centrifugal casting [ENG] A method for casting metals or forming thermoplastic resins in which the molten material solidifies in and conforms to the shape of the inner surface of a heated, rapidly rotating container. { ,sen'trif·i·gəl 'kast·iŋ }

centrifugal clarification [MECH ENG] The removal of solids from a liquid by centrifugal action which decreases the settling time of the particles from hours to minutes. { ,sen'trif·i·gəl ,klar·i·fə'kā·shən }

centrifugal classification [MECH ENG] A type of centrifugal clarification purposely designed to settle out only the large particles (rather than all particles) in a liquid by reducing the centrifuging time. { ,sen'trif·i·gəl ,klas·ə·fə'kā·shən }

centrifugal classifier [MECH ENG] A machine that separates particles into size groups by centrifugal force. { ,sen'trif·i·gəl 'klas·ə,fī·ər }

centrifugal clutch [MECH ENG] A clutch operated by centrifugal force from the speed of rotation of a shaft, as when heavy expanding friction shoes act on the internal surface of a rim clutch, or a flyball-type mechanism is used to activate clutching surfaces on cones and disks. { ,sen 'trif·i·gəl 'kləch }

centrifugal collector [MECH ENG] Device used to separate particulate matter of 0.1–1000 micrometers from an airstream; some types are simple cyclones, high-efficiency cyclones, and impellers. { ,sen'trif·i·gəl ka'lek·tər }

centrifugal compressor [MECH ENG] A machine in which a gas or vapor is compressed by radial acceleration in an impeller with a surrounding casing, and can be arranged multistage for high ratios of compression. { ,sen'trif·i·gəl kəm'pres·ər }

centrifugal discharge elevator [MECH ENG] A high-speed bucket elevator from which free-flowing materials are discharged by centrifugal force at the top of the loop. { ,sen'trif·i·gəl 'dis,charj ,el·ə,vād·ər }

centrifugal extractor [CHEM ENG] A device for separating components of a liquid solution, consisting of a series of perforated concentric rings in a cylindrical drum that rotates at 2000–5000 revolutions per minute around a cylindrical shaft; liquids enter and leave through the shaft; they flow radially and concurrently in the rotating drum. { ,sen'trif·i·gəl ik'strak·tər }

centrifugal fan [MECH ENG] A machine for moving a gas, such as air, by accelerating it radially outward in an impeller to a surrounding casing, generally of scroll shape. { ,sen'trif·i·gəl 'fan }

centrifugal filter [ENG] An adaptation of the centrifugal settler; centrifugal action of a spinning container segregates heavy and light materials but heavy materials escape through nozzles as a thick slurry. { ,sen'trif·i·gəl 'fil·tər }

centrifugal filtration [MECH ENG] The removal of a liquid from a slurry by introducing the slurry into a rapidly rotating basket, where the solids are retained on a porous screen and the liquid is forced out of the cake by the centrifugal action. { ,sen'trif·i·gəl fil'trā·shən }

centrifugal force [MECH] **1.** An outward pseudo-force, in a reference frame that is rotating with respect to an inertial reference frame, which is equal and opposite to the centripetal force that must act on a particle stationary in the rotating frame. **2.** The reaction force to a centripetal force. { ,sen'trif·i·gəl 'fȯrs }

centrifugal governor [MECH ENG] A governor whose flyweights respond to centrifugal force to sense speed. { ,sen'trif·i·gəl 'gəv·ə·nər }

centrifugal molecular still [CHEM ENG] A device used for molecular distillation; material is fed to the center of a hot, rapidly rotating cone housed in a chamber at a high vacuum; centrifugal force spreads the material rapidly over the hot surface, where the evaporable material goes off as a vapor to the condenser. { ,sen'trif·i·gəl mə'lek·yə·lər 'stil }

centrifugal moment [MECH] The product of the magnitude of centrifugal force acting on a body and the distance to the center of rotation. { ,sen'trif·i·gəl 'mō·mənt }

centrifugal pump [MECH ENG] A machine for moving a liquid, such as water, by accelerating it radially outward in an impeller to a surrounding volute casing. { ,sen'trif·i·gəl 'pəmp }

centrifugal sedimentation [CHEM ENG] Removing solids from liquids by causing particles to settle through the liquid radially toward or away from the center of rotation (depending on the solid-liquid relative densities) by use of a centrifuge. { ,sen'trif·i·gəl ,sed·ə·mən'tā·shən }

centrifugal separation [MECH ENG] The separation of two immiscible liquids in a centrifuge within a much shorter period of time than could be accomplished solely by gravity. { ,sen'trif·i·gəl ,sep·ə'rā·shən }

centrifugal settler [CHEM ENG] Spinning container that separates solid particles from liquids; centrifugal force causes suspended solids to move toward or away from the center of rotation, thus concentrating them in one area for removal. { ,sen'trif·i·gəl 'set·lər }

centrifugal switch [MECH ENG] A switch opened or closed by centrifugal force; used on some induction motors to open the starting winding when the motor has almost reached synchronous speed. { ,sen'trif·i·gəl 'swich }

centrifugal tachometer [MECH ENG] An instrument which measures the instantaneous angular speed of a shaft by measuring the centrifugal force on a mass rotating with it. { ,sen'trif·i·gəl tə'käm·əd·ər }

centrifuge [MECH ENG] **1.** A rotating device for separating liquids of different specific gravities or for separating suspended colloidal particles, such as clay particles in an aqueous suspension, according to particle-size fractions by centrifugal force. **2.** A large motor-driven apparatus with a long arm, at the end of which human and animal subjects or equipment can be revolved and rotated at various speeds to simulate the prolonged accelerations encountered in rockets and spacecraft. { 'sen·trə¦fyüj }

centrifuge refining |CHEM ENG| The use of centrifuges for liquids processing, such as separation of solids or immiscible droplets from liquid carriers, or for liquid-liquid solvent extraction. { 'sen·trə¸fyüj ri'fīn·iŋ }

centripetal |MECH| Acting or moving in a direction toward the axis of rotation or the center of a circle along which a body is moving. { ¸sen'trip·əd·əl }

centripetal acceleration |MECH| The radial component of the acceleration of a particle or object moving around a circle, which can be shown to be directed toward the center of the circle. Also known as radial acceleration. { ¸sen'trip·əd·əl ik¸sel·ə'rā·shən }

centripetal force |MECH| The radial force required to keep a particle or object moving in a circular path, which can be shown to be directed toward the center of the circle. { ¸sen'trip·əd·əl 'fôrs }

centrobaric |MECH| **1.** Pertaining to the center of gravity, or to some method of locating it. **2.** Possessing a center of gravity. { ¦sen·trō ¦bar·ik }

centrode |MECH| The path traced by the instantaneous center of a plane figure when it undergoes plane motion. { 'sen¸trōd }

centroid See center of mass. { 'sen¸trȯid }

centroid of asymptotes |CONT SYS| The intersection of asymptotes in a root-locus diagram. { 'sen¸trȯid əv 'as·əm¸tȯd·ēz }

cepstrum vocoder |ENG ACOUS| A digital device for reproducing speech in which samples of the cepstrum of speech, together with pitch information, are transmitted to the receiver, and are then converted into an impulse response that is convolved with an impulse train generated from the pitch information. { 'sep·trəm 'vō¦kōd·ər }

ceramic capacitor |ELEC| A capacitor whose dielectric is a ceramic material such as steatite or barium titanate, the composition of which can be varied to give a wide range of temperature coefficients. { sə'ram·ik kə'pas·əd·ər }

ceramic cartridge |ENG ACOUS| A device containing a piezoelectric ceramic element, used in phonograph pickups and microphones. { sə'ram·ik 'kär¸trij }

ceramic earphones See crystal headphones. { sə'ram·ik 'ir¸fōnz }

ceramic glaze |ENG| A glossy finish on a clay body obtained by spraying with metallic oxides, chemicals, and clays and firing at high temperature. { sə'ram·ik 'glāz }

ceramic microphone |ENG ACOUS| A microphone using a ceramic cartridge. { sə'ram·ik 'mī·krə¸fōn }

ceramic pickup |ENG ACOUS| A phonograph pickup using a ceramic cartridge. { sə'ram·ik 'pik·əp }

ceramic radiant |ENG| A baked-clay component of a gas heating unit which radiates heat when incandescent from the gas flame. { sə'ram·ik 'rād·ē·ənt }

ceramics |ENG| The art and science of making ceramic products. { sə'ram·iks }

ceramic tool |DES ENG| A cutting tool made from metallic oxides. { sə'ram·ik ¸tül }

ceramic transducer See electrostriction transducer. { sə'ram·ik tranz'dü·sər }

ceraunograph |ENG| An instrument that detects radio waves generated by lightning discharges and records their occurrence. { sə'rȯn·ə¸graf }

Cermak-Spirek furnace |ENG| An automatic reverberatory furnace of rectangular form divided into two sections by a wall; used for roasting zinc and quicksilver ores. { ¦sər¸mak ¦spir·ek ¸fər·nəs }

cermet resistor |ELEC| A metal-glaze resistor, consisting of a mixture of finely powdered precious metals and insulating materials fired onto a ceramic substrate. { 'sər¸met ri'zis·tər }

Cerruti's problem See Boussinesq's problem. { se'rü·dēz ¸präb·ləm }

certainty equivalence control |CONT SYS| An optimal control law for a stochastic adaptive control system which is obtained by solving the control problem in the case of known parameters and substituting the known parameters with their estimates. { 'sərt·ən·tē i'kwiv·ə·ləns kən'trōl }

cesium magnetometer |ENG| A magnetometer that uses a cesium atomic-beam resonator as a frequency standard in a circuit that detects very small variations in magnetic fields. { 'sē·zē·əm ¸mag·nə'täm·əd·ər }

cesspit See cesspool. { 'ses¸pit }

cesspool |CIV ENG| An underground tank for raw sewage collection; used where there is no sewage system. Also known as cesspit. { 'ses¸pül }

cetane index |CHEM ENG| An empirical method for finding the cetane number of a fuel based on API gravity and the mid boiling point. { 'sē¸tān ¸in¸deks }

cetane number |CHEM ENG| The percentage by volume of cetane (cetane number 100) in a blend with α-methylnaphthalene (cetane number 0); indicates the ability of a fuel to ignite quickly after being injected into the cylinder of an engine. { 'sē¸tān ¸nəm·bər }

CFIA See component-failure-impact analysis.

cfs See cusec.

cg See centigram.

chain |CIV ENG| See engineer's chain; Gunter's chain. |DES ENG| **1.** A flexible series of metal links or rings fitted into one another; used for supporting, restraining, dragging, or lifting objects or transmitting power. **2.** A mesh of rods or plates connected together, used to convey objects or transmit power. { chān }

chain belt |DES ENG| Belt of flat links to transmit power. { 'chān ¸belt }

chain block |MECH ENG| A tackle which uses an endless chain rather than a rope, often operated from an overhead track to lift heavy weights especially in workshops. Also known as chain fall; chain hoist. { 'chān ,bläk }

chain bond |CIV ENG| A masonry bond formed with a chain or bar. { 'chān ,bänd }

chain conveyor |MECH ENG| A machine for moving materials that carries the product on one or two endless linked chains with crossbars; allows smaller parts to be added as the work passes. { 'chān kən'vā·ər }

chain course |CIV ENG| A course of stone held together by iron cramps. { 'chān ,kórs }

chain drive |MECH ENG| A flexible device for power transmission, hoisting, or conveying, consisting of an endless chain whose links mesh with toothed wheels fastened to the driving and driven shafts. { 'chān ,drīv }

chain fall See chain block. { 'chān ,fól }

chain-float liquid-level gage |ENG| Float device to measure the level of liquid in a vessel; the float, suspended from a counterweighted chain draped over a toothed sprocket, rises or falls with the liquid level, and the chain movement turns the sprocket to position a calibrated depth-indicator. { 'chān ¦flōt 'lik·wəd ¦lev·əl ,gāj }

chain gear |MECH ENG| A gear that transmits motion from one wheel to another by means of a chain. { 'chān ,gir }

chain grate stoker |MECH ENG| A wide, endless chain used to feed, carry, and burn a noncoking coal in a furnace, control the air for combustion, and discharge the ash. { 'chān ,grāt ,stōk·ər }

chain hoist See chain block. { 'chān ,hóist }

chaining |CIV ENG| In land surveying, measuring distance by means of a chain or tape. { 'chān·iŋ }

chain pump |MECH ENG| A pump containing an endless chain that is fitted at intervals with disks and moves through a pipe and raises sludge. { 'chān ,pəmp }

chain radar system |ENG| A number of radar stations located at various sites on a missile range to enable complete radar coverage during a missile flight; the stations are linked by data and communication lines for target acquisition, target positioning, or data-recording purposes. { ¦chān 'rā,där ,sis·təm }

chain riveting |ENG| Riveting consisting of rivets one behind the other in rows along the seam. { 'chān ,riv·əd·iŋ }

chain saw |MECH ENG| A gasoline-powered saw for felling and bucking timber, operated by one person; has cutting teeth inserted in a sprocket chain that moves rapidly around the edge of an oval-shaped blade. { 'chān ,só }

chain tongs |DES ENG| A tool for turning pipe, using a chain to encircle and grasp the pipe. { 'chān ,täŋz }

chain vise |DES ENG| A vise in which the work is encircled and held tightly by a chain. { 'chān ,vīs }

chaldron |MECH| **1.** A unit of volume in common use in the United Kingdom, equal to 36 bushels, or 288 gallons, or approximately

1.30927 cubic meters. **2.** A unit of volume, formerly used for measuring solid substances in the United States, equal to 36 bushels, or approximately 1.26861 cubic meters. { 'chól·drən }

chamber |CIV ENG| The space in a canal lock between the upper and lower gates. { 'chām·bər }

chamber kiln |ENG| A kiln consisting of a series of adjacent chambers in a ring or oval through which the fire moves, taking several days to make a circuit; waste gas from the fire preheats ware in chambers toward which the fire is moving, while combustion air is preheated by ware in chambers already fired. { 'chām·bər ,kil }

chamber process |CHEM ENG| An obsolete method of manufacturing sulfuric acid in which sulfur dioxide, air, and steam are reacted in a lead chamber with oxides of nitrogen as the catalyst. { 'chām·bər ,präs·əs }

chamber test |ENG| A fire test developed specifically for floor coverings that measures the speed and distance of the spread of flames under specified conditions. { 'chām·bər ,test }

chamfer |ENG| To bevel a sharp edge on a machined part. { 'cham·fər }

chamfer angle |DES ENG| The angle that a beveled surface makes with one of the original surfaces. { 'cham·fər ,aŋ·gəl }

chamfering |MECH ENG| Machining operations to produce a beveled edge. Also known as beveling. { 'cham·fə·riŋ }

chamfer plane |DES ENG| A plane for chamfering edges of woodwork. { 'cham·fər ,plān }

change gear |MECH ENG| A gear used to change the speed of a driven shaft while the speed of the driving remains constant. { 'chānj ,gir }

changing bag |ENG| An enclosure of lightproof material used for operations such as loading of film holders in daylight. { 'chānj·iŋ ,bag }

channel |CHEM ENG| In percolation filtration, a portion of the clay bed where there is a preponderance of flow. |CIV ENG| A natural or artificial waterway connecting two bodies of water or containing moving water. |ELECTR| **1.** A path for a signal, as an audio amplifier may have several input channels. **2.** The main current path between the source and drain electrodes in a field-effect transistor or other semiconductor device. |ENG| The forming of cavities in a gear lubricant at low temperatures because of congealing. { 'chan·əl }

channeler See channeling machine. { 'chan·əl·ər }

channel FET microphone |ENG ACOUS| A microphone in which a membrane is used as the gate to a field-effect transistor (FET) located just below it, and motion of the membrane modulates the current between the source and drain of the transistor. { ¦chan·əl ¦fet 'mī·krə,fōn or ¦ef¦e¦tē }

channeling machine |MECH ENG| An electrically powered machine that operates by a chipping action of three to five chisels while traveling back and forth on a track; used for primary separation from the rock ledge in marble, limestone, and soft sandstone quarries. Also known as channeler. { 'chan·əl·iŋ mə'shēn }

channel iron |DES ENG| A metal strip or beam with a U-shape. { 'chan·əl ˌī·ərn }

channel process |CHEM ENG| A carbon-black process in which iron channel beams are used as depositing surfaces for carbon black. { 'chan·əl ˌpräs·əs }

chaos *See* chaotic behavior. { 'kā·äs }

chaotic behavior |MECH| The behavior of a system whose final state depends so sensitively on the system's precise initial state that the behavior is in effect unpredictable and cannot be distinguished from a random process, even though it is strictly determinate in a mathematical sense. Also known as chaos. { kā'äd·ik bi'hā·vyər }

Chapman-Jouguet plane |MECH| A hypothetical, infinite plane, behind the initial shock front, in which it is variously assumed that reaction (and energy release) has effectively been completed, that reaction product gases have reached thermodynamic equilibrium, and that reaction gases, streaming backward out of the detonation, have reached such a condition that a forward-moving sound wave located at this precise plane would remain a fixed distance behind the initial shock. { ¦chap·mən zhü¦gwā ˌplān }

characteristic |ELECTR| A graph showing how the voltage or current between two terminals of an electronic device varies with the voltage or current between two other terminals. { ˌkar·ik·tə'ris·tik }

characteristic length |MECH| A convenient reference length (usually constant) of a given configuration, such as overall length of an aircraft, the maximum diameter or radius of a body of revolution, or a chord or span of a lifting surface. { ˌkar·ik·tə'ris·tik 'leŋkth }

characterization factor |CHEM ENG| A number which expresses the variations in physical properties with change in character of the paraffinic stock; ranges from 12.5 for paraffinic stocks to 10.0 for the highly aromatic stocks. Also known as Watson factor. { ˌkar·ik·tə·rə'zā·shən 'fak·tər }

charcoal canister |MECH ENG| In an evaporative control system, a container filled with activated charcoal that traps gasoline vapors emitted by the fuel system. Also known as canister; carbon canister. { 'chär·kōl 'kan·ə'stər }

charcoal test |CHEM ENG| A determination of the natural gasoline content of natural gas by adsorbing the gasoline on activated charcoal and then recovering it by distillation. { 'chär·kōl ˌtest }

charge |ELEC| **1.** A basic property of elementary particles of matter; the charge of an object may be a positive or negative number or zero; only integral multiples of the proton charge occur, and the charge of a body is the algebraic sum of the charges of its constituents; the value of the charge may be inferred from the Coulomb force between charged objects. Also known as

electric charge, quantity of electricity. **2.** To convert electrical energy to chemical energy in a secondary battery. **3.** To feed electrical energy to a capacitor or other device that can store it. |ENG| **1.** A unit of an explosive, either by itself or contained in a bomb, projectile, mine, or the like, or used as the propellant for a bullet or projectile. **2.** To load a borehole with an explosive. **3.** The material or part to be heated by induction or dielectric heating. **4.** The measurement or weight of material, either liquid, preformed, or powder, used to load a mold at one time during one cycle in the manufacture of plastics or metal. |MECH ENG| **1.** In refrigeration, the quantity of refrigerant contained in a system. **2.** To introduce the refrigerant into a refrigeration system. { chärj }

charge collector |ELEC| The structure within a battery electrode that provides a path for the electric current to or from the active material. Also known as current collector. { 'chärj kəˌlek·tər }

charge conservation *See* conservation of charge. { 'chärj ˌkän·sər'vā·shən }

charge-coupled device |ELECTR| A semiconductor device wherein minority charge is stored in a spatially defined depletion region (potential well) at the surface of a semiconductor and is moved about the surface by transferring this charge to similar adjacent wells. Abbreviated CCD. { 'chärj ¦kəp·əld di'vīs }

charge-coupled image sensor |ELECTR| A device in which charges are introduced when light from a scene is focused on the surface of the device; image points are accessed sequentially to produce a television-type output signal. Also known as solid-state image sensor. { 'chärj ¦kəp·əld 'im·ij ˌsen·sər }

charge density |ELEC| The charge per unit area on a surface or per unit volume in space. { 'chärj ˌden·səd·ē }

charge-mass ratio |ELEC| The ratio of the electric charge of a particle to its mass. { ˌchärj ˌmas 'rā·shō }

charge quantization |ELEC| The principle that the electric charge of an object must equal an integral multiple of a universal basic charge. { 'chärj ˌkwan·tə'zā·shən }

charge-transfer device |ELECTR| A semiconductor device that depends upon movements of stored charges between predetermined locations, as in charge-coupled and charge-injection devices. { 'chärj ˌtranz·fər di'vīs }

charging current |ELEC| The current that flows into a capacitor when a voltage is first applied. { 'chär·jiŋ ˌkər·ənt }

charging pump |CHEM ENG| Pump that provides pressurized fluid flow for the input of another unit, such as to a triplex pump that requires positive pressure. { 'chär·jiŋ ˌpəmp }

chart comparison unit |ENG| A device that permits simultaneous viewing of a radar plan position indicator display and a navigation chart so that one appears superimposed on the other.

Also known as autoradar plot. { 'chärt kəm'par·ə·sən ,yü·nət }

chart datum *See* datum plane. { 'chärt ,dad·əm }

chart desk [ENG] A flat surface on which charts are spread out, usually with storage space for charts and other navigating equipment below the plotting surface. { 'chärt ,desk }

chart recorder [ENG] A recorder in which a dependent variable is plotted against an independent variable by an ink-filled pen moving on plain paper, a heated stylus on heat-sensitive paper, a light beam or electron beam on photosensitive paper, or an electrode on electrosensitive paper. The plot may be linear or curvilinear on a strip chart recorder, or polar on a circular chart recorder. { 'chärt ri'kord·ər }

chart table [ENG] A flat surface on which charts are spread out, particularly one without storage space below the plotting surface, as in aircraft and VPR (virtual PPI reflectoscope) equipment. { 'chärt ,tā·bəl }

chase [BUILD] A vertical passage for ducts, pipes, or wires in a building. [DES ENG] A series of cuts, each having a path that follows the path of the cut before it; an example is a screw thread. [ENG] **1.** The main body of the mold which contains the molding cavity or cavities. **2.** The enclosure used to shrink-fit parts of a mold cavity in place to prevent spreading or distortion, or to enclose an assembly of two or more parts of a split-cavity block. **3.** To straighten and clean threads on screws or pipes. { chās }

chase mortise [DES ENG] A mortise with a sloping edge from bottom to surface so that a tenon can be inserted when the outside clearance is small. { 'chās ,mord·əs }

chaser [ENG] A thread-cutting tool with many teeth. { 'chās·ər }

chase ring [MECH ENG] In hobbing, the ring which restrains the blank from spreading during hob sinking. { 'chās ,riŋ }

chasing tool [DES ENG] A hammer or chisel used to decorate metal surfaces. { 'chās·iŋ ,tül }

chassis [ENG] **1.** A frame on which the body of an automobile or airplane is mounted. **2.** A frame for mounting the working parts of a radio or other electronic device. { 'chas·ē }

chassis ground [ELEC] A connection made to the metal chassis on which the components of a circuit are mounted, to serve as a common return path to the power source. { 'chas·ē ,graünd }

chassis punch [DES ENG] A hand tool used to make round or square holes in sheet metal. { 'chas·ē ,pənch }

chatter [ELEC] Prolonged undesirable opening and closing of electric contacts, as on a relay. Also known as contact chatter. [ENG] An irregular alternating motion of the parts of a relief valve due to the application of pressure where contact is made between the valve disk and the seat. [ENG ACOUS] Vibration of a disk-recorder cutting stylus in a direction other than that in which it is driven. { 'chad·ər }

chattering [CONT SYS] A mode of operation of a relay-type control system in which the relay switches back and forth infinitely fast. { 'chad·ə·riŋ }

Chattock gage [ENG] A form of micromanometer in which observation of the interface between two immiscible liquids is used to determine when the pressure to be measured has been balanced by the pressure head resulting from tilting of the entire apparatus. { 'chad·ək ,gāj }

check [ENG] A device attached to something in order to limit the movement, such as a door check. { chek }

check dam [CIV ENG] A low, fixed structure, constructed of timber, loose rock, masonry, or concrete, to control water flow in an erodable channel or irrigation canal. { 'chek ,dam }

checkerboard regenerator [ENG] An open-checkerwork arrangement of firebrick in a high-temperature chamber that absorbs heat during a batch processing cycle, then releases it to preheat fresh combustion air during the down cycle; used, for example, in the steel industry with open-hearth and heat-treating furnaces. { 'chek·ər,bord ri'jen·ə,rād·ər }

checker plate [ENG] A type of slip-resistant floor plate with a distinctive raised pattern that is used for walkways and platforms. { 'chek·ər ,plāt }

checkers [ENG] Open brickwork in a checkerboard regenerator allowing for the passage of hot, spent gases. { 'chek·ərz }

check fillet [BUILD] A curb set into a roof to divert or control the flow of rainwater. { 'chek ,fil·ət }

checkout [ENG] A sequence of actions to test or examine a thing as to its readiness for incorporation into a new phase of use or as to the performance of its intended function. { 'chek,aüt }

check rail [BUILD] A rail, thicker than the window, that spans the opening between the top and bottom sash; usually beveled and rabbeted. *See* guardrail. { 'chek ,rāl }

check stop [BUILD] A narrow length of wood or metal that is installed to hold a sliding element in place, such as the lower part of a sash of a double-hung window. { 'chek ,stäp }

check study [IND ENG] A review of a job or operation in part or in its entirety to evaluate the validity of a standard time. { 'chek ,stəd·ē }

check valve [MECH ENG] A device for automatically limiting flow in a piping system to a single direction. Also known as nonreturn valve. { 'chek ,valv }

cheesebox still [CHEM ENG] One of the first types of vertical cylindrical stills designed with a vapor dome. { 'chēz,bäks ,stil }

cheese head [DES ENG] A raised cylindrical head on a screw or bolt. { 'chēz ,hed }

chemical engineering [ENG] That branch of engineering serving those industries that chemically convert basic raw materials into a variety of products, and dealing with the design and operation of plants and equipment to perform such work; all products are formed in chemical processes involving chemical reactions carried out under a wide range of conditions and frequently accompanied by changes in physical state or form. { 'kem·i·kəl ,en·jə'nir·iŋ }

chemical film dielectric |ELEC| An extremely thin layer of material on one or both electrodes of an electrolytic capacitor, which conducts electricity in only one direction and thereby constitutes the insulating element of the capacitor. { 'kem·i·kəl ,film ,dī·ə'lek·trik }

chemical fire extinguisher [CHEM ENG] Any of three types of fire extinguishers (vaporizing liquid, carbon dioxide, and dry chemical) which expel chemicals in solid, liquid, or gaseous form to blanket or smother a fire. { 'kem·i·kəl 'fīr ik'stiŋ·gwish·ər }

chemical force microscope [ENG] A modification of the atomic force microscope in which an organic monolayer on the probe tip that terminates with specific chemical functional groups is sensitive to specific molecular interactions between these groups and those on the sample surface. { ¦kem·ə·kəl ,fòrs 'mī·krə,skōp }

chemical hygrometer See absorption hygrometer. { 'kem·i·kəl hī'gräm·əd·ər }

chemical ion pump [CHEM ENG] A vacuum pump whose pumping action is based on evaporation of a metal whose vapor then reacts with the chemically active molecules in the gas to be evacuated. { 'kem·i·kəl 'ī·ən ,pəmp }

chemically sensitive field-effect transistor |ELECTR| A field-effect transistor in which the ordinary gate electrode is replaced by a chemically sensitive membrane so that the gain of the transistor depends on the concentration of chemical substances. { 'kem·ik·lē ¦sen·səd·iv 'fēld i¦fekt tran,zis·tər }

chemical process industry [CHEM ENG] An industry in which the raw materials undergo chemical conversion during their processing into finished products, as well as (or instead of) the physical conversions common to industry in general; includes the traditional chemical, petroleum, and petrochemical industries. { ¦kem·i·kəl 'prä·səs ,in·də·strē }

chemical pulping [CHEM ENG] Separation of wood fiber for paper pulp by chemical treatment of wood chips to dissolve the lignin that cements the fibers together. { 'kem·i·kəl 'pəlp·iŋ }

chemical reactor [CHEM ENG] Vessel, tube, pipe, or other container within which a chemical reaction is made to take place; may be batch or continuous, open or packed, and can use thermal, catalytic, or irradiation actuation. { 'kem·i·kəl rē'ak·tər }

chemical similitude [CHEM ENG] A procedure used to ensure satisfactory operation of a full-scale chemical process by comparison with pilot plant data. { 'kem·i·kəl sə'mil·ə,tüd }

chemical sterilization [ENG] The use of bactericidal chemicals to sterilize solutions, air, or solid surfaces. { 'kem·i·kəl ,ster·ə·lə'zā·shən }

chemical thermometer [ENG] A filled-system temperature-measurement device in which gas or liquid enclosed within the device responds to heat by a volume change (rising or falling of mercury column) or by a pressure change (opening or closing of spiral coil). { 'kem·i·kəl thər'mäm·əd·ər }

chemurgy [CHEM ENG] A branch of chemistry concerned with the profitable utilization of organic raw materials, especially agricultural products, for nonfood purposes such as for paints and varnishes. { 'ke·mər,jē }

cherry picker [MECH ENG] Any of several small traveling cranes, especially one used to hoist a passenger on the end of a boom. { 'cher·ē ,pik·ər }

Chicago boom [MECH ENG] A hoisting device that is supported on the structure being erected. { shə'kä·gō ,büm }

Chicago caisson [CIV ENG] A cofferdam about 4 feet (1.2 meters) in diameter lined with planks and sunk in medium-stiff clays to hard ground for pier foundations. Also known as open-well caisson. { shə'kä·gō 'kā,sän }

Child-Langmuir equation See Child's law.

Child-Langmuir-Schottky equation See Child's law. { ¦chīld ¦laŋ·myür 'shät,kē i'kwā·zhən }

Child's law [ELECTR] A law stating that the current in a thermionic diode varies directly with the three-halves power of anode voltage and inversely with the square of the distance between the electrodes, provided the operating conditions are such that the current is limited only by the space charge. Also known as Child-Langmuir equation; Child-Langmuir-Schottky equation; Langmuir-Child equation. { 'chīldz ,lò }

Chile mill [MECH ENG] A crushing mill having vertical rollers running in a circular enclosure with a stone or iron base or die. Also known as edge runner. { 'chil·ē ,mil }

chiller [CHEM ENG] Oil-refining apparatus in which the temperature of paraffin distillates is lowered preparatory to filtering out the solid wax components. { 'chil·ər }

chill roll [ENG] A cored roll used in chill-roll extrusion of plastics. { 'chil ,rōl }

chill-roll extrusion [ENG] Method of extruding plastic film in which the film is cooled while being drawn around two or more highly polished chill rolls, inside of which there is cooling water. Also known as cast-film extrusion. { 'chil ,rōl ek'strü·zhən }

chimney [BUILD] A vertical, hollow structure of masonry, steel, or concrete, built to convey gaseous products of combustion from a building. |ELECTR| A pipelike enclosure that is placed over a heat sink to improve natural upward convection of heat and thereby increase the dissipating ability of the sink. { 'chim,nē }

chimney apron [BUILD] A flashing made of a nonferrous metal, such as copper, that is built into the masonry of the chimney and the roofing material at the place where the roof is penetrated by the chimney. { 'chim·nē ,ā·prən }

chimney bar |BUILD| A wrought-iron or steel lintel which is supported by the sidewalls and carries the masonry above the fireplace opening. Also known as turning bar. { 'chim,nē ,bär }

chimney cap [CIV ENG] A rotary device fitted to a chimney and moved by the wind so that the chimney is turned away from the wind to permit the escape of smoke while rain or snow is prevented from entering the chimney. { 'chim·nē ,kap }

chimney core [MECH ENG] The inner section of a double-walled chimney which is separated from the outer section by an air space. { 'chim,nē ,kór }

chip [ELECTR] **1.** The shaped and processed semiconductor die that is mounted on a substrate to form a transistor, diode, or other semiconductor device. **2.** An integrated microcircuit performing a significant number of functions and constituting a subsystem. Also known as microchip. { chip }

chip breaker [DES ENG] An irregularity or channel cut into the face of a lathe tool behind the cutting edge to cause removed stock to break into small chips or curls. { 'chip ,brāk·ər }

chip cap [DES ENG] A plate or cap on the upper part of the cutting iron of a carpenter's plane designed to give the tool rigidity and also to break up the wood shavings. { 'chip ,kap }

chip capacitor [ELECTR] A single-layer or multilayer monolithic capacitor constructed in chip form, with metallized terminations to facilitate direct bonding on hybrid integrated circuits. { 'chip kə'pas·əd·ər }

chip log [ENG] A line, marked at intervals (commonly 50 feet or 15 meters), that is paid out over the stern of a moving ship and is pulled out by a drag (the chip), to determine the ship's speed. { 'chip ,läg }

chipper [ENG] A tool such as a chipping hammer used for chipping. [MECH ENG] A machine with revolving knives for reducing large pieces of wood to chips. { 'chip·ər }

chipping hammer [ENG] A hand or pneumatic hammer with chisel-shaped or pointed faces used to remove rust and scale from metal surfaces. { 'chip·iŋ ,ham·ər }

chip resistor [ELECTR] A thick-film resistor constructed in chip form, with metallized terminations to facilitate direct bonding on hybrid integrated circuits. { 'chip ri'zis·tər }

chirp radar [ENG] Radar in which a swept-frequency signal is transmitted, received from a target, then compressed in time to give a narrow pulse called the chirp signal. { 'chərp ,rā,där }

chisel [DES ENG] A tool for working the surface of various materials, consisting of a metal bar with a sharp edge at one end and often driven by a mallet. { 'chiz·əl }

chisel bit See chopping bit. { 'chiz·əl ,bit }

chisel bond [ENG] A thermocompression bond in which a contact wire is attached to a contact pad on a semiconductor chip by applying pressure with a chisel-shaped tool. { 'chiz·əl ,bänd }

chisel-edge angle [DES ENG] The angle included between the chisel edge and the cutting edge, as seen from the end of the drill. Also known as web angle. { 'chiz·əl ,ej 'aŋ·gəl }

chisel-tooth saw [DES ENG] A circular saw with chisel-shaped cutting edges. { 'chiz·əl 'tüth 'só }

Chladni's figures [MECH] Figures produced by sprinkling sand or similar material on a horizontal plate and then vibrating the plate while holding it rigid at its center or along its periphery; indicate the nodal lines of vibration. { 'klad,nēz ,fig·yərz }

chloralkali [CHEM ENG] Either of the products of the industrial electrolysis of sodium chloride, that is, sodium hydroxide or chlorine. { klór'al·kə,lī }

chloralkali process [CHEM ENG] An industrial chemical process based on the electrolysis of sodium chloride for the production of sodium hydroxide and chlorine. { ,klór'al·kə,lī ,prä·səs }

chlorinator [CHEM ENG] The apparatus used in chlorinating. { 'klór·ə,nād·ər }

choke [ELEC] An inductance used in a circuit to present a high impedance to frequencies above a specified frequency range without appreciably limiting the flow of direct current. Also known as choke coil. [MECH ENG] To increase the fuel feed to an internal combustion engine through the action of a choke valve. See choke valve. { chōk }

choke coil See choke. { 'chōk ,kóil }

choked neck [DES ENG] Container neck which has a narrowed or constricted opening. { ¦chōkt 'nek }

choke valve [MECH ENG] A valve which supplies the higher suction necessary to give the excess fuel feed required for starting a cold internal combustion engine. Also known as choke. { 'chōk ,valv }

chopper [ENG] Any knife, axe, or mechanical device for chopping or cutting an object into segments. { 'chäp·ər }

chopper amplifier [ELECTR] A carrier amplifier in which the direct-current input is filtered by a low-pass filter, then converted into a square-wave alternating-current signal by either one or two choppers. { 'chäp·ər 'am·plə,fī·ər }

chopper-stabilized amplifier [ELECTR] A direct-current amplifier in which a direct-coupled amplifier is in parallel with a chopper amplifier. { ¦chäp·ər ¦stā·bə,līzd 'am·plə,fī·ər }

chopper transistor [ELECTR] A bipolar or field-effect transistor operated as a repetitive "on/off" switch to produce square-wave modulation of an input signal. { 'chäp·ər tran'zis·tər }

chopping [ELECTR] The removal, by electronic means, of one or both extremities of a wave at a predetermined level. { 'chäp·iŋ }

chopping bit [MECH ENG] A steel bit with a chisel-shaped cutting edge, attached to a string of drill rods to break up, by impact, boulders, hardpan, and a lost core in a drill hole. Also known as chisel bit. { 'chäp·iŋ ,bit }

chop-type feeder [MECH ENG] Device for semicontinuous feed of solid materials to a process unit, with intermittent opening and closing of a hopper gate (bottom closure) by a control arm actuated by an eccentric cam. { 'chäp,tīp ,fēd·ər }

chord [CIV ENG] The top or bottom, generally horizontal member of a truss. { kȯrd }

chordal thickness [DES ENG] The tangential thickness of a tooth on a circular gear, as measured along a chord of the pitch circle. { 'kȯrd·əl 'thik·nəs }

chrome tanning [CHEM ENG] Tanning treatment of animal skin with chromium salts. { ¦krōm 'tan·iŋ }

chromoradiometer [ENG] A radiation meter that uses a substance whose color changes with x-ray dosage. { ¦krō·mō·,rād·ē'äm·əd·ər }

chronocyclegraph [IND ENG] A device used in micromotion studies to record a complete work cycle by taking still pictures with long exposures, the motion paths being traced by small electric lamps fastened to the worker's hands or fingers; time is obtained by interrupting the light circuits with a controlled frequency which produces dots on the film. { ,krän·ō'sī·klə,graf }

chronograph [ENG] An instrument used to register the time of an event or graphically record time intervals such as the duration of an event. { 'krän·ə,graf }

chronometric data [ENG] Data in which the desired quantity is the time of occurrence of an event or the time interval between two or more events. { 'krän·ə,me·trik 'dad·ə }

chronometric radiosonde [ENG] A radiosonde whose carrier wave is switched on and off in such a manner that the interval of time between the transmission of signals is a function of the magnitude of the meteorological elements being measured. { ¦krän·ə¦me·trik 'räd·ē·ō,sänd }

chronometric tachometer [ENG] A tachometer which repeatedly counts the revolutions during a fixed interval of time and presents the average speed during the last timed interval. { ¦krän·ə¦me·trik tə'käm·əd·ər }

chronothermometer [ENG] A thermometer consisting of a clock mechanism whose speed is a function of temperature; automatically calculates the mean temperature. { ¦krän·ō·thər'mäm·əd·ər }

CHU See centigrade heat unit.

CHU$_{mean}$ See centigrade heat unit.

chuck [DES ENG] A device for holding a component of an instrument rigid, usually by means of adjustable jaws or set screws, such as the workpiece in a metalworking or woodworking machine, or the stylus or needle of a phonograph pickup. { chək }

chucking [MECH ENG] The grasping of an outsize workpiece in a chuck or jawed device in a lathe. { 'chək·iŋ }

chucking machine [MECH ENG] A lathe or grinder in which the outsize workpiece is grasped in a chuck or jawed device. { 'chək·iŋ mə'shēn }

churn drill [MECH ENG] Portable drilling equipment, with drilling performed by a heavy string

of tools tipped with a blunt-edge chisel bit suspended from a flexible cable, to which a reciprocating motion is imparted by its suspension from an oscillating beam or sheave, causing the bit to be raised and dropped. Also known as American system drill; cable-system drill. { 'chərn ,dril }

churn shot drill [MECH ENG] A boring rig with both churn and shot drillings. { 'chərn ,shät ,dril }

chute [ENG] A conduit for conveying free-flowing materials at high velocity to lower levels. { shüt }

chute spillway [CIV ENG] A spillway in which the water flow passes over a crest into a sloping, lined, open channel; used for earth and rock-fill dams. { ¦shüt 'spil,wā }

C³I See command, control, communications, and intelligence. { 'sē 'thrē 'ī }

cinetheodolite [ENG] A surveying theodolite in which 35-millimeter motion picture cameras with lenses of 60- to 240-inch (1.5- to 6.1-meter) focal length are substituted for the surveyor's eye and telescope; used for precise time-correlated observation of distant airplanes, missiles, and artificial satellites. { ¦sin·ə·thē'äd·ə,līt }

Cipolletti weir [CIV ENG] Trapezoidal weir in which the sides of the notch slope are one horizontal to four vertical; used to measure water flow in open channels, especially streams and rivers. { chip·ə'led·ē 'wer }

circle shear [MECH ENG] A shearing machine that cuts circular disks from a metal sheet rolling between the cutting wheels. { 'sər·kəl ,shēr }

circuit See electric circuit. { 'sər·kət }

circuit analyzer See volt-ohm-milliammeter. { 'sər·kət ,an·ə,līz·ər }

circuit board See printed circuit board. { 'sər·kət ,bȯrd }

circuit breaker [ELEC] An electromagnetic device that opens a circuit automatically when the current exceeds a predetermined value. { 'sər·kət ,brāk·ər }

circuit conditioning [ELECTR] Test, analysis, engineering, and installation actions to upgrade a communications circuit to meet an operational requirement; includes the reduction of noise, the equalization of phase and level stability and frequency response, and the correction of impedance discontinuities, but does not include normal maintenance and repair activities. { 'sər·kət kən'dish·ə·niŋ }

circuit diagram [ELEC] A drawing, using standardized symbols, of the arrangement and interconnections of the conductors and components of an electrical or electronic device or installation. Also known as schematic circuit diagram; wiring diagram. { 'sər·kət ,dī·ə,gram }

circuit element See component. { 'sər·kət ¦el·ə·mənt }

circuit interrupter [ELEC] A device in a circuit breaker to remove energy from an arc in order to extinguish it. { 'sər·kət ,in·tə,rəp·tər }

circuit loading [ELEC] Power drawn from a circuit by an electric measuring instrument, which may alter appreciably the quantity being measured. { 'sər·kət ,lōd·iŋ }

circuit protection |ELECTR| Provision for automatically preventing excess or dangerous temperatures in a conductor and limiting the amount of energy liberated when an electrical failure occurs. { 'sər·kət prə'tek·shən }

circuitry |ELEC| The complete combination of circuits used in an electrical or electronic system or piece of equipment. { 'sər·kə·trē }

circuit testing |ELEC| The testing of electric circuits to determine and locate an open circuit, or a short circuit or leakage. { 'sər·kət ,tes·tiŋ }

circuit theory |ELEC| The mathematical analysis of conditions and relationships in an electric circuit. Also known as electric circuit theory. { 'sər·kət ,thē·ə·rē }

circular burner |ENG| A fuel burner having a round opening. { 'sər·kyə·lər 'bərn·ər }

circular channel |ENG| Continuous-length opening with circular cross section through which liquid or gas can be made to flow. { 'sər·kyə·lər 'chan·əl }

circular-chart recorder |ENG| Graphic pen-and-ink recorder where measured values are drawn onto a rotating circular chart by the backward and forward movement of a pivoted pen actuated by the input signal (such as temperature, pressure, flow, or force) from an instrument transmitter. { 'sər·kyə·lər ,chärt ri'kórd·ər }

circular cutter |MECH ENG| A rotating blade with a square or knife edge used to slit or shear metal. { 'sər·kyə·lər 'kəd·ər }

circular form tool |DES ENG| A round or disk-shaped tool with the cutting edge on the periphery. { 'sər·kyə·lər ,fórm ,tül }

circular inch |MECH| The area of a circle 1 inch (25.4 millimeters) in diameter. { 'sər·kyə·lər 'inch }

circular mil |MECH| A unit equal to the area of a circle whose diameter is 1 mil (0.001 inch); used chiefly in specifying cross-sectional areas of round conductors. Abbreviated cir mil. { 'sər·kyə·lər 'mil }

circular motion |MECH| **1.** Motion of a particle in a circular path. **2.** Motion of a rigid body in which all its particles move in circles about a common axis, fixed with respect to the body, with a common angular velocity. { 'sər·kyə·lər 'mō·shən }

circular pitch |DES ENG| The linear measure in inches along the pitch circle of a gear between corresponding points of adjacent teeth. { 'sər·kyə·lər 'pich }

circular plane |DES ENG| A plane that can be adjusted for convex or concave surfaces. { 'sər·kyə·lər 'plān }

circular saw |MECH ENG| Any of several power tools for cutting wood or metal, having a thin steel disk with a toothed edge that rotates on a spindle. { 'sər·kyə·lər 'só }

circular scanning |ENG| Radar scanning in which the direction of maximum radiation describes a right circular cone. { 'sər·kyə·lər 'skan·iŋ }

circular spike |ENG| A metal timber connector fitted with a circular series of sharp teeth that dig into the wood, preventing lateral motion, as a bolt is tightened through the wood and the spike. { 'sər·kyə·lər 'spīk }

circular velocity |MECH| At any specific distance from the primary, the orbital velocity required to maintain a constant-radius orbit. { 'sər·kyə·lər və'läs·əd·ē }

circulating fluid |ENG| A fluid pumped into a borehole through the drill stem, the flow of which cools the bit and transports the cuttings out of the borehole. { 'sər·kyə,lād·iŋ 'flü·əd }

circulating pump |CHEM ENG| Pump used to circulate process liquid out of and back into a process system, as in the circulation of distillation column bottoms through an external heater, or the circulation of storage tank bottoms to mix tank contents. { 'sər·kyə,lād·iŋ 'pəmp }

circulating system |CHEM ENG| Fluid system in which the process fluid is taken from and pumped back into the system, as in the circulation of distillation column bottoms through an external heater. { 'sər·kyə,lād·iŋ 'sis·təm }

circulation area |BUILD| The area required for human traffic in a building, including permanent corridors, stairways, elevators, escalators, and lobbies. { ,sər·kyə·'lā·shən ,er·ē·ə }

circumferentor |ENG| A horizontal compass used in surveying that has arms diametrically placed with vertical slit sights in them. { sər'kəm·fə,ren·tər }

cir mil See circular mil.

cistern |CIV ENG| A tank for storing water or other liquid. { 'sis·tərn }

cistern barometer |ENG| A pressure-measuring device in which pressure is read by the liquid rise in a vertical, closed-top tube as a result of system pressure on a liquid reservoir (cistern) into which the bottom, open end of the tube is immersed. { 'sis·tərn bə'ram·əd·ər }

civil engineering |ENG| The planning, design, construction, and maintenance of fixed structures and ground facilities for industry, transportation, use and control of water, or occupancy. { 'siv·əl en·jə'nir·iŋ }

cladding |ENG| Process of covering one material with another and bonding them together under high pressure and temperature. Also known as bonding. { 'klad·iŋ }

clamp |DES ENG| A tool for binding or pressing two or more parts together, by holding them firmly in their relative positions. See clamping circuit. { klamp }

clamping coupling |MECH ENG| A coupling with a split cylindrical element which clamps the shaft ends together by direct compression, through bolts or rings, and by the wedge action of conical sections; not considered a permanent part of the shaft. { 'klamp·iŋ ,kəp·liŋ }

clamping gripper |CONT SYS| A robot element that uses two-link movements, parallel-jaw movements, and combination movements to grasp and handle objects. { 'klamp·iŋ 'grip·ər }

clamping plate |ENG| A plate on a mold which

attaches the mold to a machine. { 'klamp·iŋ ,plāt }

clamping pressure |ENG| In injection and transfer-molding of plastics, the pressure applied to keep the mold closed in opposition to the fluid pressure of the molding material. { 'klamp·iŋ ,presh·ər }

clamp screw |DES ENG| A screw that holds a part by forcing it against another part. { 'klamp ,skrü }

clamp-screw sextant |ENG| A marine sextant having a clamp screw for controlling the position of the tangent screw. { 'klamp ,skrü ,seks·tənt }

clamshell bucket |MECH ENG| A two-sided bucket used in a type of excavator to dig in a vertical direction; the bucket is dropped while its leaves are open and digs as they close. Also known as clamshell grab. { 'klam,shel ,bək·ət }

clamshell grab See clamshell bucket. { 'klam ,shel ,grab }

clamshell snapper |MECH ENG| A marine sediment sampler consisting of snapper jaws and a footlike projection which, upon striking the bottom, causes a spring mechanism to close the jaws, thus trapping a sediment sample. { 'klam,shel ,snap·ər }

Clapeyron-Clausius equation See Clausius-Clapeyron equation. { kla·pā·rōn ¦klōz·ē·əs i,kwā·zhən }

Clapeyron equation See Clausius-Clapeyron equation. { kla·pā·rōn i'kwā·zhən }

Clapeyron's theorem |MECH| The theorem that the strain energy of a deformed body is equal to one-half the sum over three perpendicular directions of the displacement component times the corresponding force component, including deforming loads and body forces, but not the six constraining forces required to hold the body in equilibrium. { kla·pā·rōnz ,thir·əm }

clapper box |MECH ENG| A hinged device that permits a reciprocating cutting tool (as in a planer or shaper) to clear the work on the return stroke. { 'klap·ər ,bäks }

clarification |CHEM ENG| The removal of small amounts (usually less than 0.2%) of fine particulate solids from liquids (such as drinking water) by methods such as gravity sedimentation, centrifugal sedimentation, filtration, and magnetic separation. { ,klar·ə·fə'kā·shən }

clarifier |ENG| A device for filtering a liquid. { 'klar·ə,fī·ər }

clarifying agent See fining. { 'klar·ə,fī·iŋ ,ā·jənt }

clarifying centrifuge |MECH ENG| A device that clears liquid of foreign matter by centrifugation. { 'klar·ə,fī·iŋ 'sen·trə,fyüj }

clarifying filter |ENG| Any filter, such as a sand filter or a cartridge filter, used to purify liquids with a low solid-liquid ratio; in some instances color may be removed as well. { 'klar·ə,fī·iŋ ,fil·tər }

clarity |CHEM ENG| Measure of the amount of opaque suspended solids in a liquid, determined by visual or optical methods. { 'klar·əd·ē }

Clark process |CHEM ENG| Softening of water by adding alkaline solutions of calcium hydroxide so that the acid carbonates are converted to normal carbonates. { 'klärk ,präs·əs }

clasp |DES ENG| A releasable catch which holds two or more objects together. { klasp }

clasp lock |DES ENG| A spring lock with a self-locking feature. { 'klasp ,läk }

clasp nut |DES ENG| A split nut that clasps a screw when closed around it. { 'klasp ,nət }

class A push-pull sound track |ENG ACOUS| Two single photographic sound tracks side by side, the transmission of one being 180° out of phase with the transmission of the other; both positive and negative halves of the sound wave are linearly recorded on each of the two tracks. { ,klas 'ā ¦push ¦pul 'saun ,trak }

class B push-pull sound track |ENG ACOUS| Two photographic sound tracks side by side, one of which carries the positive half of the signal only, and the other the negative half; during the inoperative half-cycle, each track transmits little or no light. { ,klas 'bē ¦push ¦pul 'saun ,trak }

classical mechanics |MECH| Mechanics based on Newton's laws of motion. { 'klas·ə·kəl mə'kan·iks }

classification |ENG| **1.** Sorting out or categorizing of particles or objects by established criteria, such as size, function, or color. **2.** Stratification of a mixture of various-sized particles (that is, sand and gravel), with the larger particles migrating to the bottom. See grading. { ,klas·ə·fə'kā·shən }

classification track |CIV ENG| A railroad track used to separate cars from a train according to destination. { ,klas·ə·fə'kā·shən ,trak }

classification yard |CIV ENG| A railroad yard for separating trains according to car destination. { ,klas·ə·fə'kā·shən ,yärd }

classifier |MECH ENG| Any apparatus for separating mixtures of materials into their constituents according to size and density. { 'klas·ə,fī·ər }

Claude process |CHEM ENG| A process of ammonia synthesis which uses high operating pressures and a train of converters. { 'klōd ,präs·əs }

clausius |THERMO| A unit of entropy equal to the increase in entropy associated with the absorption of 1000 international table calories of heat at a temperature of 1 K, or to 4186.8 joules per kelvin. { 'klōz·ē·əs }

Clausius-Clapeyron equation |THERMO| An equation governing phase transitions of a substance, $dp/dT = \Delta H/(T\Delta V)$, in which p is the pressure, T is the temperature at which the phase transition occurs, ΔH is the change in heat content (enthalpy), and ΔV is the change in volume during the transition. Also known as Clapeyron-Clausius equation; Clapeyron equation. { klōz·ē·əs kla·pā,rōn i,kwā·zhən }

Clausius-Dickel column See thermogravitational column. { ¦klōz·ē·əs ¦dik·əl ¦käl·əm }

Clausius equation |THERMO| An equation of state in reference to gases which applies a correction to the van der Waals equation:

$\{P + (n^2a/|T(V + c)^2|)\} (V - nb) = nRT,$

where P is the pressure, T the temperature, V the volume of the gas, n the number of moles in the gas, R the gas constant, a depends only on temperature, b is a constant, and c is a function of a and b. { 'klōz·ē·əs i'kwä·zhən }

Clausius inequality |THERMO| The principle that for any system executing a cyclical process, the integral over the cycle of the infinitesimal amount of heat transferred to the system divided by its temperature is equal to or less than zero. Also known as Clausius theorem; inequality of Clausius. { 'klōz·ē·əs in·i'kwäl·əd·ē }

Clausius law |THERMO| The law that an ideal gas's specific heat at constant volume does not depend on the temperature. { 'klōz·ē·əs ,lō }

Clausius number |THERMO| A dimensionless number used in the study of heat conduction in forced fluid flow, equal to $V^3L\rho/k\Delta T$, where V is the fluid velocity, ρ is its density, L is a characteristic dimension, k is the thermal conductivity, and ΔT is the temperature difference. { 'klōz·ē·əs ,nəm·bər }

Clausius' statement |THERMO| A formulation of the second law of thermodynamics, stating it is not possible that, at the end of a cycle of changes, heat has been transferred from a colder to a hotter body without producing some other effect. { 'klōz·ē·əs 'stāt·mənt }

Clausius theorem See Clausius inequality. { 'klōz·ē·əs 'thir·əm }

Claus method |CHEM ENG| Industrial method of obtaining sulfur by a partial oxidation of gaseous hydrogen sulfide in the air to give water and sulfur. { 'klaús ,meth·əd }

claw |DES ENG| A fork for removing nails or spikes. { klò }

claw bar See ripping bar. { 'klò ,bär }

claw clutch |MECH ENG| A clutch consisting of claws that interlock when pushed together. { 'klò ,kləch }

claw coupling |MECH ENG| A loose coupling having projections or claws cast on each face which engage in corresponding notches in the opposite faces; used in situations in which shafts require instant connection. { 'klò ,kəp·liŋ }

claw hammer |DES ENG| A woodworking hammer with a flat working surface and a claw to pull nails. { 'klò ,ham·ər }

clay atmometer |ENG| An atmometer consisting of a porous porcelain container connected to a calibrated reservoir filled with distilled water; evaporation is determined by the depletion of water. { 'klā at'mäm·əd·ər }

clay bit A bit designed for use on a clay barrel. See mud bit. { 'klā ,bit }

clay digger |MECH ENG| A power-driven, hand-held spade for digging hard soil or soft rock. { 'klā ,dig·ər }

clay press |ENG| A press used to remove excess water from a pottery-clay slurry. { 'klā ,pres }

clay refining |CHEM ENG| A treating process for vaporized gasoline or other light petroleum product; the material is passed through a bed of granular clay, and certain olefins are polymerized to gums and absorbed by the clay. { 'klā rə'fin·iŋ }

clay regeneration |CHEM ENG| Cleaning coarse-grained absorbent clays for reuse in percolation processes by deoiling them with naphtha, steaming out excess naphtha, and roasting in a stream of air to remove carbonaceous matter. { 'klā ri·jen·ə'rā·shən }

cleaning eye See cleanout. { 'klēn·iŋ ,ī }

cleaning lane |ENG| A space that is located between adjacent rows of tubes in a heat exchanger and allows passage of a cleaning device. { 'klēn·iŋ ,lān }

cleaning turbine |MECH ENG| A tool for cleaning the interior surfaces of heat exchangers and boiler tubes; consists of a drive motor, a flexible drive cable or hose, and a head that is an arrangement of blades, modified drill bits, or brushes. { 'klēn·iŋ ,tər·bən }

cleanout |ENG| A pipe fitting containing a removable plug that provides access for inspection or cleaning of the pipe run. Also known as access eye; cleaning eye. { 'klēn,aút }

cleanout auger See cleanout jet auger. { 'klēn,aút ,óg·ər }

cleanout door |ENG| An opening in the side of a tank usually at ground level and covered by a plate to provide access for removal of sediments from the bottom of the tank. { 'klēn,aút ,dór }

cleanout jet auger |ENG| An auger equipped with water-jet orifices designed to clean out collected material inside a driven pipe or casing before taking soil samples from strata below the bottom of the casing. Also known as cleanout auger. { 'klēn,aút 'jet ,óg·ər }

clean room |ENG| A room in which elaborate precautions are employed to reduce dust particles and other contaminants in the air, as required for assembly of delicate equipment. { 'klēn ,rüm }

clean track |ENG ACOUS| A sound track having no leakage from other tracks. { ¦klēn ¦trak }

cleanup |ELECTR| Gradual disappearance of gases from an electron tube during operation, due to absorption by getter material or the tube structure. |ENG| The time required for a leak-testing system to reduce its signal output to 37% of the signal transmitted at the instant when tracer gases enter the system. { 'klē,nəp }

clearance |ENG| Unobstructed space required for occasional removal of parts of equipment. |MECH ENG| **1.** In a piston-and-cylinder mechanism, the space at the end of the cylinder when the piston is at dead-center position toward the end of the cylinder. **2.** The ratio of the volume of this space to the piston displacement during a stroke. { 'klir·əns }

clearance angle |MECH ENG| The angle between a plane containing the end surface of a cutting tool and a plane passing through the cutting edge in the direction of cutting motion. { 'klir·əns ,aŋ·gəl }

clearance volume |MECH ENG| The volume remaining between piston and cylinder when the piston is at top dead center. { 'klir·əns ,väl·yəm }

clear octane [ENG] The octane number of a particular gasoline before it has been blended with antiknock additives. { klir 'äk,tān }

cleat [CIV ENG] A strip of wood, metal, or other material fastened across something to serve as a batten or to provide strength or support. [DES ENG] A fitting having two horizontally projecting horns around which a rope may be made fast. { klēt }

cleet See cleat. { klēt }

clevis [DES ENG] A U-shaped metal fitting with holes in the open ends to receive a bolt or pin; used for attaching or suspending parts. { 'klev·əs }

clevis pin [DES ENG] A fastener with a head at one end, used to join the ends of a clevis. { 'klev·əs ,pin }

click [ENG ACOUS] A perforation in a sound track which produces a clicking sound when passed over the projector sound head. { klik }

click filter [ELECTR] A capacitor connected across a switch, relay, or key to lengthen the decay time from the closed to the open condition when the device is opened or closed. { 'klik ,fil·tər }

click track [ENG ACOUS] A sound track containing a series of clicks, which may be spaced regularly (uniform click track) or irregularly (variable click track). { 'klik ,trak }

climate control See air conditioning. { 'klī·mət kən'trōl }

climbing crane [MECH ENG] A crane used on top of a high-rise construction that ascends with the building as work progresses. { 'klīm·iŋ 'krān }

climbing irons [DES ENG] Spikes attached to a steel framework worn on shoes to climb wooden utility poles and trees. { 'klīm·iŋ ,ī·ərnz }

clinical thermometer [ENG] A thermometer used to accurately determine the temperature of the human body; the most common type is a mercury-in-glass thermometer, in which the mercury expands from a bulb into a capillary tube past a constriction that prevents the mercury from receding back into the bulb, so that the thermometer registers the maximum temperature attained. { 'klin·ə·kəl thər'mäm·əd·ər }

clinker building [DES ENG] A method of building ships and boilers in which the edge of the wooden planks or steel plates used for the outside covering overlap the edge of the plank or plate next to it; clinched nails fasten the planks together, and rivets fasten the steel plates. { 'kliŋ·kər ,bil·diŋ }

clinograph [ENG] A type of directional surveying instrument that records photographically the direction and magnitude of deviations from the vertical of a borehole, well, or shaft; the information is obtained by the instrument in one trip into and out of the well. { 'klī·nə,graf }

clinometer [ENG] **1.** A hand-held surveying device for measuring vertical angles; consists of a

sighting tube surmounted by a graduated vertical arc with an attached level bubble; used in meteorology to measure cloud height at night, in conjunction with a ceiling light, and in ordnance for boresighting. Also known as Abney level. **2.** A device for measuring the amount of roll aboard ship. { klə'näm·əd·ər }

clip [DES ENG] A device that fastens by gripping, clasping, or hooking one part to another. { klip }

clip bond [CIV ENG] A bond in which the inner edge of face brick is cut off so that bricks laid diagonal to a wall can be joined to those laid parallel to it. { 'klip ,bänd }

clip lead [ELEC] A short piece of flexible wire with an alligator clip or similar temporary connector at one or both ends. { 'klip ,lēd }

clipper See limiter. { 'klip·ər }

clipper diode [ELECTR] A bidirectional breakdown diode that clips signal voltage peaks of either polarity when they exceed a predetermined amplitude. { 'klip·ər ,dī,ōd }

clipper-limiter [ELECTR] A device whose output is a function of the instantaneous input amplitude for a range of values lying between two predetermined limits but is approximately constant, at another level, for input values above the range. { 'klip·ər ,lim·əd·ər }

clivvy See clevis. { 'kliv·ē }

clo [ENG] The amount of insulation which will maintain normal skin temperature of the human body when heat production is 50 kilogram-calories per meter squared per hour, air temperature is 70°F (21°C), and the air is still. { klō }

clock [ELECTR] A source of accurately timed pulses, used for synchronization in a digital computer or as a time base in a transmission system. { kläk }

clock control system [CONT SYS] A system in which a timing device is used to generate the control function. Also known as time-controlled system. { 'kläk kən'trōl ,sis·təm }

clock drive [ENG] The mechanism that causes an equatorial telescope to revolve about its polar axis so that it keeps the same star in its field of view. { 'kläk ,drīv }

clocked flip-flop [ELECTR] A flip-flop circuit that is set and reset at specific times by adding clock pulses to the input so that the circuit is triggered only if both trigger and clock pulses are present simultaneously. { 'kläkt 'flip,fläp }

clocked logic [ELECTR] A logic circuit in which the switching action is controlled by repetitive pulses from a clock. { 'kläkt ,läj·ik }

clock frequency [ELECTR] The master frequency of the periodic pulses that schedule the operation of a digital computer. Also known as clock rate; clock speed. { 'kläk ,frē·kwən·sē }

clock motor See timing motor. { 'kläk ,mōd·ər }

clock oscillator [ELECTR] An oscillator that controls an electronic clock. { 'kläk 'äs·ə,läd·ər }

clock rate See clock frequency. { 'kläk ,rāt }

clock speed See clock frequency. { 'kläk ,spēd }

close-control radar [ENG] Ground radar used

with radio to position an aircraft over a target that is normally difficult to locate or is invisible to the pilot. { ¦klōs kən'trōl 'rā,där }

close-coupled pump [MECH ENG] Pump with built-in electric motor (sometimes a steam turbine), with the motor drive and pump impeller on the same shaft. { ¦klōs ¦kəp·əld 'pəmp }

closed-belt conveyor [MECH ENG] Solids-conveying device with zipperlike teeth that mesh to form a closed tube wrapped snugly around the conveyed material; used with fragile materials. { 'klōzd ¦belt kən'vā·ər }

closed cycle [THERMO] A thermodynamic cycle in which the thermodynamic fluid does not enter or leave the system, but is used over and over again. { ‚klōzd 'sī·kəl }

closed-cycle turbine [MECH ENG] A gas turbine in which essentially all the working medium is continuously recycled, and heat is transferred through the walls of a closed heater to the cycle. { ¦klōzd ¦sī·kəl 'tər‚bīn }

closed fireroom system [MECH ENG] A fireroom system in which combustion air is supplied via forced draft resulting from positive air pressure in the fireroom. { ¦klōzd 'fīr‚rüm ‚sis·təm }

closed loop [CONT SYS] A family of automatic control units linked together with a process to form an endless chain; the effects of control action are constantly measured so that if the controlled quantity departs from the norm, the control units act to bring it back. { ¦klōzd 'lüp }

closed-loop control system See feedback control system. { ¦klōzd ¦lüp kən'trōl ‚sis·təm }

closed-loop telemetry system [ENG] **1.** A telemetry system which is also used as the display portion of a remote-control system. **2.** A system used to check out test vehicle or telemetry performance without radiation of radio-frequency energy. { ¦klōzd ¦lüp tə'lem·ə·trē ‚sis·təm }

closed nozzle [MECH ENG] A fuel nozzle having a built-in valve interposed between the fuel supply and combustion chamber. { ¦klōzd 'näz·əl }

closed pair [MECH] A pair of bodies that are subject to constraints which prevent any relative motion between them. { ¦klōzd 'per }

closed respiratory gas system [ENG] A self-contained system within a sealed cabin, capsule, or spacecraft that will provide adequate oxygen for breathing, maintain adequate cabin pressure, and absorb the exhaled carbon dioxide and water vapor. { ¦klōzd 'res·prə‚tȯr·ē 'gas ‚sis·təm }

closed shop [IND ENG] An establishment permitting only union members to be employed. { ¦klōzd 'shäp }

closed steam [ENG] Steam that flows through a heating coil or annulus so that there is no direct contact between the steam and the material being heated. { ¦klōzd 'stēm }

closed system [ENG] A system for water handling that does not permit air to enter. [THERMO] A system which is isolated so that it cannot exchange matter or energy with its surroundings

and can therefore attain a state of thermodynamic equilibrium. Also known as isolated system. { ¦klōzd 'sis·təm }

close nipple [ENG] A short length of pipe that is completely threaded. { 'klōs 'nip·əl }

close-off rating [MECH ENG] **1.** The maximum allowable pressure drop to which a valve can be subjected at commercial shutoff. **2.** The maximum allowable pressure drop between the out let of a three-way valve and either of the two inlets, or between the inlet and either of the two outlets. { 'klōs ‚ȯf ‚rād·iŋ }

closer [CIV ENG] **1.** In masonry work, the last brick or other masonry component that is laid in a horizontal course. Also known as closure. **2.** A stone course that extends from one windowsill to another. { 'klō·zər }

close-talking microphone [ENG ACOUS] A microphone designed for use close to the mouth, so noise from more distant points is suppressed. Also known as noise-canceling microphone. { 'klōs ‚tȯk·iŋ 'mī·krə‚fōn }

closing line [MECH] The vector required to complete a polygon consisting of a set of vectors whose sum is zero (such as the forces acting on a body in equilibrium). { 'klōz·iŋ ‚līn }

closing machine [ENG] A machine for manufacturing wire rope by braiding wire into strands, and strands into rope. Also known as stranding machine. { 'klōz·iŋ mə‚shēn }

closing pressure [MECH ENG] The amount of static inlet pressure in a safety relief valve when the valve disk has a zero lift above the seat. { 'klōz·iŋ ‚presh·ər }

closure See closer. { 'klō·zhər }

cloth wheel [DES ENG] A polishing wheel made of sections of cloth glued or sewn together. { 'klȯth ‚wēl }

cloud-detection radar [ENG] A type of weather radar designed specifically for the detection of clouds (rather than precipitation). { 'klaud di'tek·shən ‚rā,där }

cloud-drop sampler [ENG] An instrument for collecting cloud particles, consisting of a sampling plate or cylinder and a shutter, which is so arranged that the sampling surface is exposed to the cloud for a predetermined length of time; the sampling surface is covered with a material which either captures the cloud particles or leaves an impression characteristic of the impinging elements. { 'klaud ‚dräp 'sam·plər }

cloud-height indicator [ENG] General term for an instrument which measures the height of cloud bases. { 'klaud ‚hīt ‚in·də‚kād·ər }

cloud mirror See mirror nephoscope. { 'klaud ‚mir·ər }

cloud point [CHEM ENG] The temperature at which paraffin wax or other solid substance begins to separate from a solution of petroleum oil; a cloudy appearance is seen in the oil at this point. { 'klaud ‚pȯint }

cloud test [CHEM ENG] An American Society for Testing and Materials method for determining the cloud point of petroleum oil. { 'klaud ‚test }

clout nail |DES ENG| A nail with a large, thin, flat head used in building. { 'klaút ,nāl }

cloverleaf |CIV ENG| A highway intersection resembling a clover leaf and designed to allow movement and interchange of traffic without direct crossings and left turns. { 'klō·vər,lēf }

clusec |MECH ENG| A unit of power used to measure the power of evacuation of a vacuum pump, equal to the power associated with a leak rate of 1 centiliter per second at a pressure of 1 millitorr, or to approximately 1.33322×10^{-6} watt. { 'klü¦sek }

cluster |ENG| **1.** A pyrotechnic signal consisting of a group of stars or fireballs. **2.** A grouping of rocket motors fastened together. { 'kləs·tər }

clutch |MECH ENG| A machine element for the connection and disconnection of shafts in equipment drives, especially while running. { kləch }

cm See centimeter.

cmHg See centimeter of mercury.

CMOS device |ELECTR| A device formed by the combination of a PMOS (*p*-type-channel metal oxide semiconductor device) with an NMOS (*n*-type-channel metal oxide semiconductor device). Derived from complementary metal oxide semiconductor device. { se,mós di'vīs }

CNC See computer numerical control.

coach screw |DES ENG| A large, square-headed, wooden screw used to join heavy timbers. Also known as lag bolt; lag screw. { 'kōch ,skrü }

coak |DES ENG| **1.** A projection from the end of a piece of wood or timber that is designed to fit into a hole in another piece so that they can be joined to form a continuous unit. **2.** A dowel or hardwood pin that joins overlapping timbers. { kōk }

coalescent pack |CHEM ENG| High-surface-area packing to consolidate liquid droplets for gravity separation from a second phase (for example, gas or immiscible liquid); packing must be wettable by the droplet phase; Berl saddles, Raschig rings, knitted wire mesh, excelsior, and similar materials are used. { ,kō·ə'les·ənt 'pak }

coalescer |CHEM ENG| Mechanical process vessel with wettable, high-surface area packing on which liquid droplets consolidate for gravity separation from a second phase (for example, gas or immiscible liquid). { ,kō·ə'les·ər }

coal gasification |CHEM ENG| The conversion of coal, char, or coke to a gaseous product by reaction with air, oxygen, steam, carbon dioxide, or mixtures of these. { 'kōl ,gas·ə·fə'kā·shən }

coal hydrogenation See Bergius process. { 'kōl ,hī·drə·jə'nā·shən }

coal liquefaction |CHEM ENG| The conversion of coal (with the exception of anthracite) to petroleum-like hydrocarbon liquids, which are used as refinery feedstocks for the manufacture of gasoline, heating oil, diesel fuel, jet fuel, turbine fuel, fuel oil, and petrochemicals. { ¦kōl lik·wə'fak·shən }

coast |ENG| A memory feature on a radar which, when activated, causes the range and angle systems to continue to move in the same direction and at the same speed as that required to track an original target. { kōst }

coastal berm See berm. { 'kōs·təl 'bərm }

coastal engineering |CIV ENG| A branch of civil engineering pertaining to the study of the action of the seas on shorelines and to the design of structures to protect against this action. { 'kōs·təl en·jə'nir·iŋ }

coat hanger die |ENG| A plastics-sheet slot die shaped like a coat hanger on the inside. { 'kōt ,haŋ·ər ,dī }

coaxial |MECH| Sharing the same axes. |MECH ENG| Mounted on independent concentric shafts. { kō'ak·sē·əl }

coaxial speaker |ENG ACOUS| A loudspeaker system comprising two, or less commonly three, speaker units mounted on substantially the same axis in an integrated mechanical assembly, with an acoustic-radiation-controlling structure. { kō'ak·sē·əl 'spēk·ər }

coaxial wavemeter |ENG| A device for measuring frequencies above about 100 megahertz, consisting of a rigid metal cylinder that has an inner conductor along its central axis, and a sliding disk that shorts the inner conductor and the cylinder. { kō'ak·sē·əl 'wāv,mēd·ər }

cobalt glance See cobaltite. { 'kō,bȯlt 'glans }

cobalt-molybdate desulfurization |CHEM ENG| A process for desulfurization of petroleum by using cobalt molybdate as a catalyst. { ¦kō ,bȯlt mə¦lib,dāt dē,səl·fə·ri'zā·shən }

cock |ENG| Any mechanism which starts, stops, or regulates the flow of liquid, such as a valve, faucet, or tap. { käk }

Coddington shape factor See shape factor. { 'käd·iŋ·tən 'shāp ,fak·tər }

coded mask |ENG| A pattern of tungsten blocks that absorb gamma-ray photons in a gamma-ray telescope, and are arranged so that an astronomical gamma-ray source projects on a position-sensitive detector a pattern that is characteristic of the direction of arrival of the photons. { 'kōd·əd ,mask }

code-sending radiosonde |ENG| A radiosonde which transmits the indications of the meteorological sensing elements in the form of a code consisting of combinations of dots and dashes. Also known as code-type radiosonde; contracted code sonde. { 'kōd ,send·iŋ 'rād·ē·ō,sänd }

code-type radiosonde See code-sending radiosonde. { 'kōd ,tīp 'rād·ē·ō,sänd }

codistor |ELECTR| A multijunction semiconductor device which provides noise rejection and voltage regulation functions. { kō'dis·tər }

coefficient of capacitance |ELEC| One of the coefficients which appears in the linear equations giving the charges on a set of conductors in terms of the potentials of the conductors; a coefficient is equal to the ratio of the charge on a given conductor to the potential of the same conductor when the potentials of all the other conductors are 0. { ¦kō·ə'fish·ənt əv kə'pas·ə·təns }

coefficient of compressibility |MECH| The decrease in volume per unit volume of a substance resulting from a unit increase in pressure; it is the reciprocal of the bulk modulus. { ¦kō·ə'fish·ənt əv kəm,pres·ə'bil·əd·ē }

coefficient of conductivity See thermal conductivity. { ¦kō·ə'fish·ənt əv ,kän·dək'tiv·əd·ē }

coefficient of cubical expansion |THERMO| The increment in volume of a unit volume of solid, liquid, or gas for a rise of temperature of 1° at constant pressure. Also known as coefficient of expansion; coefficient of thermal expansion; coefficient of volumetric expansion; expansion coefficient; expansivity. { ¦kō·ə'fish·ənt əv 'kyüb·ə·kəl ik'span·shən }

coefficient of elasticity See modulus of elasticity. { ¦kō·ə'fish·ənt əv i,las'tis·əd·ē }

coefficient of expansion See coefficient of cubical expansion. { ¦kō·ə'fish·ənt əv ik'span·shən }

coefficient of friction |MECH| The ratio of the frictional force between two bodies in contact, parallel to the surface of contact, to the force, normal to the surface of contact, with which the bodies press against each other. Also known as friction coefficient. { ¦kō·ə'fish·ənt əv 'frik·shən }

coefficient of friction of rest See coefficient of static friction. { ¦kō·ə'fish·ənt əv 'frik·shən əv 'rest }

coefficient of induction |ELEC| One of the coefficients which appears in the linear equations giving the charges on a set of conductors in terms of the potentials of the conductors; a coefficient is equal to the ratio of the charge on a given conductor to the potential on another conductor, when the potentials of all the other conductors equal 0. { ¦kō·ə'fish·ənt əv in'dək·shən }

coefficient of kinetic friction |MECH| The ratio of the frictional force, parallel to the surface of contact, that opposes the motion of a body which is sliding or rolling over another, to the force, normal to the surface of contact, with which the bodies press against each other. { ¦kō·ə'fish·ənt əv kə'ned·ik 'frik·shən }

coefficient of linear expansion |THERMO| The increment of length of a solid in a unit of length for a rise in temperature of 1° at constant pressure. Also known as linear expansivity. { ¦kō·ə'fish·ənt əv 'lin·ē·ər ik'span·shən }

coefficient of performance |THERMO| In a refrigeration cycle, the ratio of the heat energy extracted by the heat engine at the low temperature to the work supplied to operate the cycle; when used as a heating device, it is the ratio of the heat delivered in the high-temperature coils to the work supplied. { ¦kō·ə'fish·ənt əv pər'fór·məns }

coefficient of potential |ELEC| One of the coefficients which appears in the linear equations giving the potentials of a set of conductors in terms of the charges on the conductors. { ¦kō·ə'fish·ənt əv pə'ten·chəl }

coefficient of restitution |MECH| The constant e, which is the ratio of the relative velocity of two elastic spheres after direct impact to that before impact; e can vary from 0 to 1, with 1 equivalent to an elastic collision and 0 equivalent to a perfectly elastic collision. Also known

as restitution coefficient. { ¦kō·ə'fish·ənt əv ,res·tə'tü·shən }

coefficient of rigidity See modulus of elasticity in shear. { ¦kō·ə'fish·ənt əv rə'jid·əd·ē }

coefficient of rolling friction |MECH| The ratio of the frictional force, parallel to the surface of contact, opposing the motion of a body rolling over another, to the force, normal to the surface of contact, with which the bodies press against each other. { ¦kō·ə'fish·ənt əv 'rōl·iŋ 'frik·shən }

coefficient of sliding friction |MECH| The ratio of the frictional force, parallel to the surface of contact, opposing the motion of a body sliding over another, to the force, normal to the surface of contact, with which the bodies press against each other. { ¦kō·ə'fish·ənt əv 'slīd·iŋ 'frik·shən }

coefficient of static friction |MECH| The ratio of the maximum possible frictional force, parallel to the surface of contact, which acts to prevent two bodies in contact, and at rest with respect to each other, from sliding or rolling over each other, to the force, normal to the surface of contact, with which the bodies press against each other. Also known as coefficient of friction of rest. { ¦kō·ə'fish·ənt əv 'stad·ik 'frik·shən }

coefficient of strain |MECH| For a substance undergoing a one-dimensional strain, the ratio of the distance along the strain axis between two points in the body, to the distance between the same points when the body is undeformed. { ¦kō·ə'fish·ənt əv 'strān }

coefficient of superficial expansion |THERMO| The increment in area of a solid surface per unit of area for a rise in temperature of 1° at constant pressure. Also known as superficial expansivity. { ¦kō·ə'fish·ənt əv ,sü·pər'fish·əl ik'span·chən }

coefficient of thermal expansion See coefficient of cubical expansion. { ¦kō·ə'fish·ənt əv 'thər·məl ik'span·shən }

coefficient of volumetric expansion See coefficient of cubical expansion. { ¦kō·ə'fish·ənt əv ¦väl·yə¦me·trik ik'span·chən }

coelostat |ENG| A device consisting of a clockwork-driven mirror that enables a fixed telescope to continuously keep the same region of the sky in its field of view. { 'sē·lə,stat }

coercimeter |ENG| An instrument that measures the magnetic intensity of a natural magnet or electromagnet. { ,kō,ər'sim·əd·ər }

coextrusion |ENG| Extrusion-forming of plastic or metal products in which two or more compatible feed materials are used in physical admixture through the same extrusion die. { ¦kō,ik'strü·zhən }

cofferdam |CIV ENG| A temporary damlike structure constructed around an excavation to exclude water. { 'kò·fər,dam }

coffered ceiling |BUILD| An ornamental ceiling constructed of panels that are sunken or recessed. { 'kò·fərd 'sēl·iŋ }

cog |DES ENG| A tooth on the edge of a wheel. |ELEC| A fluctuation in the torque delivered by a motor when it runs at low speed, due to electro-mechanical effects. Also known as torque ripple. { käg }

cog belt [MECH ENG] A flexible device used for timing and for slip-free power transmission. { 'käg ,belt }

cogeneration [MECH ENG] The simultaneous on-site generation of electric energy and process steam or heat from the same plant. { ,kō,jen·ə'rā·shən }

cogged belt See timing belt. { 'kägd ,belt }

cog railway [CIV ENG] A steep railway that employs a cograil that meshes with a cogwheel on the locomotive to ensure traction. { 'käg 'rāl,wā }

cogwheel [DES ENG] A wheel with teeth around its edge. { 'käg,wēl }

coherent moving-target indicator [ENG] A radar system in which the Doppler frequency of the target echo is compared to a local reference frequency generated by a coherent oscillator. { kō'hir·ənt 'müv·iŋ 'tär·gət ,in·də,kād·ər }

coherent noise [ENG] Noise that affects all tracks across a magnetic tape equally and simultaneously. { kō'hir·ənt 'nóiz }

cohesive strength [MECH] 1. Strength corresponding to cohesive forces between atoms. 2. Hypothetically, the stress causing tensile fracture without plastic deformation. { kō'hē·siv 'streŋkth }

coil [CONT SYS] Any discrete and logical result that can be transmitted as output by a programmable controller. { kóil }

coil spring [DES ENG] A helical or spiral spring, such as one of the helical springs used over the front wheels in an automotive suspension. { 'kóil ,spriŋ }

coil winder [ENG] A manual or motor-driven mechanism for winding coils individually or in groups. { 'kóil ,wīn·dər }

coincidence amplifier [ELECTR] An electronic circuit that amplifies only that portion of a signal present when an enabling or controlling signal is simultaneously applied. { kō'in·sə·dəns ,am·plə,fī·ər }

coincidence circuit [ELECTR] A circuit that produces a specified output pulse only when a specified number or combination of two or more input terminals receives pulses within an assigned time interval. Also known as coincidence counter; coincidence gate. { kō'in·sə·dəns ,sər·kət }

coincidence correction See dead-time correction. { kō'in·sə·dəns kə'rek·shən }

coincidence counter See coincidence circuit. { kō'in·sə·dəns ,kaúnt·ər }

coincidence gate See coincidence circuit. { kō'in·sə·dəns ,gāt }

coinjection molding [ENG] A technique used in polymer processing whereby two or more materials are simultaneously injected into the cavity of a mold. Also known as sandwich molding. { ,kō·in'jek·shən ,mōld·iŋ }

coke breeze [MECH ENG] Undersized coke screenings passing through a screen opening of approximately 5/8 inch (16 millimeters). { 'kōk ,brēz }

coke drum [CHEM ENG] A vessel in which coke is produced. { 'kōk ,drəm }

coke knocker [MECH ENG] A mechanical device used to break loose coke within a drum or tower. { 'kōk ,näk·ər }

coke number [CHEM ENG] A number used to report the results of the Ramsbottom carbon residue test. { 'kōk ,nəm·bər }

coke oven [CHEM ENG] A retort in which coal is converted to coke by carbonization. { 'kōk ,əv·ən }

coke-oven regenerator [CHEM ENG] Arrangement of refractory blocks in the flue system of a coke oven to recover waste heat from hot, exiting combustion gases; the blocks, in turn, release heat to warm, incoming fuel gas. { 'kōk ,əv·ən ri'jen·ə,rād·ər }

coker [CHEM ENG] The processing unit in which coking occurs. { 'kōk·ər }

coking [CHEM ENG] 1. Destructive distillation of coal to make coke 2. A process for thermally converting the heavy residual bottoms of crude oil entirely to lower-boiling petroleum products and by-product petroleum coke. { 'kok·iŋ }

coking still [CHEM ENG] A still in which coking is done; usually, it is a batch still. { 'kok·iŋ ,stil }

Colburn j factor equation [THERMO] Dimensionless heat-transfer equation to calculate the natural convection movement of heat from vertical surfaces or horizontal cylinders to fluids (gases or liquids) flowing past these surfaces. { 'kol·bərn 'jā ,fak·tər i'kwā·zhən }

Colburn method [CHEM ENG] Graphical method, and equations to calculate the theoretical number of plates (trays) needed to separate light and heavy liquids in a distillation column. { 'kōl·bərn ,meth·əd }

cold-air machine [MECH ENG] A refrigeration system in which air serves as the refrigerant in a cycle of adiabatic compression, cooling to ambient temperature, and adiabatic expansion to refrigeration temperature; the air is customarily reused in a closed superatmospheric pressure system. Also known as dense-air system. { 'kōld ¦er mə,shēn }

cold-chamber die casting [ENG] A die-casting process in which molten metal is ladled either manually or mechanically into a relatively cold cylinder from which it is forced into the die cavity. { 'kōld ,chām·bər 'dī ,kast·iŋ }

cold chisel [DES ENG] A chisel specifically designed to cut or chip cold metal; made of specially tempered tool steel machined into various cutting edges. Also known as cold cutter. { 'kōld ,chiz·əl }

cold cure [CHEM ENG] Vulcanization of rubber at nonelevated temperatures with a solution of a sulfur compound. { 'kōld ,kyúr }

cold cutter See cold chisel. { 'kōld ,kəd·ər }

cold differential test pressure [ENG] The inlet pressure of a pressure-relief valve at which the valve is set to open during testing. { 'kōld ,dif·ə'ren·chəl 'test ,presh·ər }

cold flow |MECH| Creep in polymer plastics. { 'kōld ,flō }

cold joint |ENG| A soldered connection which was inadequately heated, with the result that the wire is held in place by rosin flux, not solder. { 'kōld 'jóint }

cold lime-soda process |CHEM ENG| A water-softening process in which water is treated with hydrated lime (sometimes in combination with soda ash), which reacts with dissolved calcium and magnesium compounds to form precipitates that can be removed as sludge. { 'kōld ¦līm ¦sō·də ,präs·əs }

cold molding |ENG| Shaping of an unheated compound in a mold under pressure, followed by heating the article to cure it. { 'kōld ,mōld·iŋ }

cold plasma |CHEM ENG| Low-energy ionized gas. { ¦kōld 'plaz·mə }

cold plate |MECH ENG| An aluminum or other plate containing internal tubing through which a liquid coolant is forced, to absorb heat transferred to the plate by transistors and other components mounted on it. Also known as liquid-cooled dissipator. { 'kōld ,plāt }

cold saw |MECH ENG| **1.** Any saw for cutting cold metal, as opposed to a hot saw. **2.** A disk made of soft steel or iron which rotates at a speed such that a point on its edge has a tangential velocity of about 15,000 feet per minute (75 meters per second), and which grinds metal by friction. { 'kōld ,só }

cold settling |CHEM ENG| A process that removes wax from high-viscosity stocks. { 'kōld ,set·liŋ }

cold slug |ENG| The first material to enter an injection mold in plastics manufacturing. { 'kōld ,sləg }

cold-slug well |ENG| The area in a plastic injection mold which receives the cold slug from the sprue opening. { 'kōld ,sləg 'wel }

cold-spot hygrometer *See* dew-point hygrometer. { 'kōld ,spät hī'gräm·əd·ər }

cold storage |ENG| The storage of perishables at low temperatures produced by refrigeration, usually above freezing, to increase storage life. { ¦kōld 'stòr·ij }

cold-storage locker plant |ENG| A plant with many rental steel lockers, each with a capacity of about 6 cubic feet (0.17 cubic meter) and generally for food storage by an individual family, placed in refrigerated rooms, at about 0°F (−18°C). { ¦kōld 'stòr·ij 'läk·ər ,plant }

cold stress |MECH| Forces tending to deform steel, cement, and other materials, resulting from low temperatures. { 'kōld ,stres }

cold stretch |ENG| A pulling operation on extruded plastic filaments in which little or no heat is used; improves tensile properties. { 'kōld ,strech }

cold test |CHEM ENG| A test to determine the temperature at which clouding or coagulation is first visible in a sample of oil, as the temperature of the sample is reduced. { 'kōld ,test }

cold trap |MECH ENG| A tube whose walls are cooled with liquid nitrogen or some other liquid to condense vapors passing through it; used with diffusion pumps and to keep vapors from entering a McLeod gage. { 'kōld ,trap }

collapse |ENG| Contraction of plastic container walls during cooling; produces permanent indentation. { kə'laps }

collapse properties |MECH| Strength and dimensional attributes of piping, tubing, or process vessels, related to the ability to resist collapse from exterior pressure or internal vacuum. { kə'laps ,präp·ərd·ēz }

collapsing pressure |MECH| The minimum external pressure which causes a thin-walled body or structure to collapse. { kə'lap·siŋ ,presh·ər }

collar |DES ENG| A ring placed around an object to restrict its motion, hold it in place, or cover an opening. { 'käl·ər }

collar beam |BUILD| A tie beam in a roof truss connecting the rafters well above the wall plate. { 'käl·ər ,bēm }

collar bearing |MECH ENG| A bearing that resists the axial force of a collar on a rotating shaft. { 'käl·ər ,ber·iŋ }

collared hole |ENG| A started hole drilled sufficiently deep to confine the drill bit and prevent slippage of the bit from normal position. { 'käl·ərd ,hōl }

collet |DES ENG| A sleeve or flange that can be tightened about a rotating shaft to halt motion. { kə'lekt }

collective bargaining |IND ENG| The negotiation for mutual agreement in the settlement of a labor contract between an employer or his representatives and a labor union or its representatives. { kə'lek·tiv 'bär·gən·iŋ }

collector |ELECTR| **1.** A semiconductive region through which a primary flow of charge carriers leaves the base of a transistor; the electrode or terminal connected to this region is also called the collector. **2.** An electrode that collects electrons or ions which have completed their functions within an electron tube; a collector receives electrons after they have done useful work, whereas an anode receives electrons whose useful work is to be done outside the tube. Also known as electron collector. |ENG| A class of instruments employed to determine the electric potential at a point in the atmosphere, and ultimately the atmospheric electric field; all collectors consist of some device for rapidly bringing a conductor to the same potential as the air immediately surrounding it, plus some form of electrometer for measuring the difference in potential between the equilibrated collector and the earth itself; collectors differ widely in their speed of response to atmospheric potential changes. { kə'lek·tər }

collector capacitance |ELECTR| The depletion-layer capacitance associated with the collector junction of a transistor. { kə'lek·tər kə'pas·əd·əns }

collector current |ELECTR| The direct current that passes through the collector of a transistor. { kə'lek·tər ,kər·ənt }

collector cutoff [ELECTR] The reverse saturation current of the collector-base junction. { kə'lek·tər 'kəd,óf }

collector junction [ELECTR] A semiconductor junction located between the base and collector electrodes of a transistor. { kə'lek·tər ,jəŋk·shən }

collector modulation [ELECTR] Amplitude modulation in which the modulator varies the collector voltage of a transistor. { kə'lek·tər ,mäj·ə'lā·shən }

collector resistance [ELECTR] The back resistance of the collector-base diode of a transistor. { kə'lek·tər ri'zis·təns }

collector voltage [ELECTR] The direct-current voltage, obtained from a power supply, that is applied between the base and collector of a transistor. { kə'lek·tər ,vól·tij }

collet [DES ENG] A split, coned sleeve to hold small, circular tools or work in the nose of a lathe or other type of machine. [ENG] **1.** The glass neck remaining on a bottle after it is taken off the glass-blowing iron. **2.** Pieces of glass, ordinarily discarded, that are added to a batch of glass. Also spelled cullet. { käl·ət }

collimation error [ENG] **1.** Angular error in magnitude and direction between two nominally parallel lines of sight. **2.** Specifically, the angle by which the line of sight of a radar differs from what it should be. { ,käl·ə'mā·shən ,er·ər }

collimation tower [ENG] Tower on which a visual and a radio target are mounted to check the electrical axis of an antenna. { ,käl·ə'mā·shən ,taú·ər }

collision-avoidance radar [ENG] Radar equipment utilized in a collision-avoidance system. { kə'lizh·ən ə'vòid·əns ,rā,där }

collision-avoidance system [ENG] Electronic devices and equipment used by a pilot to perform the functions of conflict detection and avoidance. { kə'lizh·ən ə'vòid·əns ,sis·təm }

collision blasting [ENG] The blasting out of different sections of rocks against each other. { kə'lizh·ən ,blast·iŋ }

colloider [CIV ENG] A device that removes colloids from sewage. { kə'lòid·ər }

colloid mill [MECH ENG] A grinding mill for the making of very fine dispersions of liquids or solids by breaking down particles in an emulsion or paste. { 'käl,óid ,mil }

color-bar code [IND ENG] A code that uses one or more different colors of bars in combination with black bars and white spaces, to increase the density of binary coding of data printed on merchandise tags or directly on products for inventory control and other purposes. { 'kəl·ər ,bär ,kōd }

color code [ELEC] A system of colors used to indicate the electrical value of a component or to identify terminals and leads. [ENG] **1.** Any system of colors used for purposes of identification, such as to identify dangerous areas of a factory. **2.** A system of colors used to identify the type of material carried by a pipe; for example, dangerous materials, protective materials, extra valuable materials. { 'kəl·ər ,kōd }

color coder See matrix. { 'kəl·ər ,kōd·ər }

color decoder See matrix. { 'kəl·ər dē'kod·ər }

color Doppler flow imaging scanner [ENG] A device that obtains B-mode images and Doppler blood flow data simultaneously, and superimposes a color Doppler image on the gray-scale B-mode image. { 'kəl·ər 'däp·lər ,flō 'im·ij·iŋ ,skan·ər }

color emissivity See monochromatic emissivity. { 'kəl·ər ,e·mi'siv·əd·ē }

color encoder See matrix. { 'kəl·ər en'kōd·ər }

column See tower. [ENG] A vertical shaft designed to bear axial loads in compression. { 'käl·əm }

column crane [MECH ENG] A jib crane whose boom pivots about a post attached to a building column. { 'käl·əm ,krān }

column drill [MECH ENG] A tunnel rock drill supported by a vertical steel column. { 'käl·əm ,dril }

column splice [CIV ENG] A connection between two lengths of a compression member (column); an erection device rather than a stress-carrying element. { 'käl·əm ,splīs }

comb See drag. { kōm }

combination chuck [DES ENG] A chuck used in a lathe whose jaws either move independently or simultaneously. { ,käm·bə'nā·shən 'chək }

combination collar [DES ENG] A collar that has left-hand threads at one end and right-hand threads at the other. { ,käm·bə'nā·shən 'käl·ər }

combination cycle See mixed cycle. { ,käm·bə'nā·shən 'sik·əl }

combination lock [ENG] A lock that can be opened only when its dial has been set to the proper combination of symbols, in the proper sequence. { ,käm·bə'nā·shən 'läk }

combination pliers [DES ENG] Pliers that can be used either for holding objects or for cutting and bending wire. { ,käm·bə'nā·shən 'plī·ərz }

combination saw [MECH ENG] A saw made in various tooth arrangement combinations suitable for ripping and crosscut mitering. { ,käm·bə'nā·shən 'só }

combination square [DES ENG] A square head and steel rule that when used together have both a 45° and 90° face to allow the testing of the accuracy of two surfaces intended to have these angles. { ,käm·bə'nā·shən 'skwer }

combination unit [CHEM ENG] A processing unit that combines more than one process, such as straight-run distillation together with selective cracking. { ,käm·bə'nā·shən 'yü·nət }

combination wrench [DES ENG] A wrench that is an open-end wrench at one end and a socket wrench at the other. { ,käm·bə'nā·shən 'rench }

combined flexure [MECH] The flexure of a beam under a combination of transverse and longitudinal loads. { kəm'bīnd 'flek·shər }

combined footing [CIV ENG] A footing, either rectangular or trapezoidal, that supports two columns. { kəm'bīnd 'fúd·iŋ }

combined sewers |CIV ENG| A drainage system that receives both surface runoff and sewage. { kəm'bīnd 'sü·ərz }

combined stresses |MECH| Bending or twisting stresses in a structural member combined with direct tension or compression. { kəm'bīnd 'stres·əz }

combing |BUILD| In roofing, the topmost row of shingles which project above the ridge line. |ENG| **1.** Using a comb or stiff bristle brush to create a pattern by pulling through freshly applied paint. **2.** Scraping or smoothing a soft stone surface. { 'kōm·iŋ }

comb nephoscope |ENG| A direct-vision nephoscope constructed with a comb (a crosspiece containing equispaced vertical rods) attached to the end of a column 8–10 feet (2.4–3 meters) long and supported on a mounting that is free to rotate about its vertical axis; in use, the comb is turned so that the cloud appears to move parallel to the tips of the vertical rods. { ¦kōm ¦nef·ə,skōp }

combplate |MECH ENG| The toothed portion of the stationary threshold plate that is set into both ends of an escalator or moving sidewalk and meshes with the grooved surface of the moving steps or treadway. { 'kōm,plāt }

combustible loss |ENG| Thermal loss resulting from incomplete combustion of fuel. { kəm 'bəs·tə·bəl ,lós }

combustion chamber |ENG| Any chamber in which a fuel such as oil, coal, or kerosine is burned to provide heat. |MECH ENG| The space at the head end of an internal combustion engine cylinder where most of the combustion takes place. { kəm'bəs·chən ,chām·bər }

combustion-chamber volume |MECH ENG| The volume of the combustion chamber when the piston is at top dead center. { kəm'bəs·chən ,chām·bər ,väl·yəm }

combustion deposit |ENG| A layer of ash on the heat-exchange surfaces of a combustion chamber, resulting from the burning of a fuel. { kəm'bəs·chən də'päz·ət }

combustion engine |MECH ENG| An engine that operates by the energy of combustion of a fuel. { kəm'bəs·chən ,en·jən }

combustion engineering |MECH ENG| The design of combustion furnaces for a given performance and thermal efficiency, involving study of the heat liberated in the combustion process, the amount of heat absorbed by heat elements, and heat-transfer rates. { kəm'bəs·chən en·jə'nir·iŋ }

combustion furnace |ENG| A furnace whose source of heat is the energy released in the oxidation of fossil fuel. { kəm'bəs·chən ,fər·nəs }

combustion knock See engine knock. { kəm'bəs·chən ,näk }

combustion shock |ENG| Shock resulting from abnormal burning of fuel in an internal combustion engine, caused by preignition or fuel-air detonation; or in a diesel engine, the uncontrolled burning of fuel accumulated in the combustion chamber. { kəm'bəs·chən ,shäk }

combustion turbine See gas turbine. { kəm'bəs·chən 'tər,bīn }

combustor |MECH ENG| The combustion chamber together with burners, igniters, and injection devices in a gas turbine or jet engine. { kəm 'bəs·tər }

come-along |DES ENG| A device for gripping and effectively shortening a length of cable, wire rope, or chain by means of two jaws which close when one pulls on a ring. See puller. { 'kəm ə,lóŋ }

comfort chart |ENG| A diagram showing curves of relative humidity and effective temperature superimposed upon rectangular coordinates of wet-bulb temperature and dry-bulb temperature. { 'kəm·fərt ,chärt }

comfort control |ENG| Control of temperature, humidity, flow, and composition of air by using heating and air-conditioning systems, ventilators, or other systems to increase the comfort of people in an enclosure. { 'kəm·fərt kən'trōl }

comfort curve |ENG| A line drawn on a graph of air temperature versus some function of humidity (usually wet-bulb temperature or relative humidity) to show the varying conditions under which the average sedentary person feels the same degree of comfort; a curve of constant comfort. { 'kəm·fərt ,kərv }

comfort standard See comfort zone. { 'kəm·fərt ,stan·dərd }

comfort temperature |MECH ENG| Any one of the indexes in which air temperatures have been adjusted to represent human comfort or discomfort under prevailing conditions of temperature, humidity, radiation, and wind. { 'kəm·fərt ,tem·prə·chər }

comfort zone |ENG| The ranges of indoor temperature, humidity, and air movement, under which most persons enjoy mental and physical well-being. Also known as comfort standard. { 'kəm·fərt ,zōn }

command |CONT SYS| An independent signal in a feedback control system, from which the dependent signals are controlled in a predetermined manner. { kə'mand }

command and control |SYS ENG| The process of military commanders and civilian managers identifying, prioritizing, and achieving strategic and tactical objectives by exercising authority and direction over human and material resources by utilizing a variety of computer-based and computer-controlled systems, many driven by decision-theoretic methods, tools, and techniques. Abbreviated C². { kə'mand ən kən 'trōl }

command, control, and communications |SYS ENG| A version of command and control in which the role of communications equipment is emphasized. Abbreviated C³. { kə'mand kən'trōl ən kə,myü·ne'kā·shənz }

command, control, communications, and intelligence |SYS ENG| A version of command and control in which the roles of communications equipment and intelligence are emphasized.

Abbreviated C³I. { kə'mand kən'trōl kə,myü·nə'kā·shənz ən in'tel·ə·jəns }

command destruct |CONT SYS| A command control system that destroys a flightborne test rocket or a guided missile, actuated by the safety officer whenever the vehicle's performance indicates a safety hazard. { kə'mand di'strəkt }

command guidance |ENG| A type of electronic guidance of guided missiles or other guided aircraft wherein signals or pulses sent out by an operator cause the guided object to fly a directed path. Also known as command control. { kə'mand ,gīd·əns }

commercial diesel cycle See mixed cycle. { kə'mər·shəl 'dē·zəl ,sī·kəl }

commercial harbor |CIV ENG| A harbor in which docks are provided with cargo-handling facilities. { kə'mər·shəl 'här·bər }

comminution |MECH ENG| Breaking up or grinding into small fragments. Also known as pulverization. { ,käm·ə'nü·shən }

comminutor |MECH ENG| A machine that breaks up solids. { 'käm·ə,nüd·ər }

common-base connection See grounded-base connection. { ¦käm·ən 'bās kə'nek·shən }

common-base feedback oscillator |ELECTR| A bipolar transistor amplifier with a common-base connection and a positive feedback network between the collector (output) and the emitter (input). { ¦käm·ən 'bās 'fēd,bak ,äs·ə,lād·ər }

common bond See American bond. { ¦käm·ən ¦bänd }

common carrier |IND ENG| A company recognized by an appropriate regulatory agency as having a vested interest in furnishing communications services or in transporting commodities or people. { ¦käm·ən 'kar·ē·ər }

common-collector connection See grounded-collector connection. { ¦käm·ən kə'lek·tər kə'nek·shən }

common-drain amplifier |ELECTR| An amplifier using a field-effect transistor so that the input signal is injected between gate and drain, while the output is taken between the source and drain. Also known as source-follower amplifier. { ¦käm·ən 'drān 'am·plə,fī·ər }

common-emitter connection See grounded-emitter connection. { ¦käm·ən i'mid·ər kə'nek·shən }

common-gate amplifier |ELECTR| An amplifier using a field-effect transistor in which the gate is common to both the input circuit and the output circuit. { ¦käm·ən 'gāt 'am·plə,fī·ər }

common joist |BUILD| An ordinary floor beam to which floor boards are attached. { ¦käm·ən ¦jȯist }

common labor |IND ENG| Unskilled workers. { ¦käm·ən ¦lā·bər }

common mode |ELECTR| Having signals that are identical in amplitude and phase at both inputs, as in a differential operational amplifier. { ¦käm·ən ,mōd }

common-mode error |ELECTR| The error voltage that exists at the output terminals of an operational amplifier due to the common-mode voltage at the input. { ¦käm·ən ,mōd 'er·ər }

common-mode gain |ELECTR| The ratio of the output voltage of a differential amplifier to the common-mode input voltage. { ¦käm·ən ,mōd 'gān }

common-mode input capacitance |ELECTR| The equivalent capacitance of both inverting and noninverting inputs of an operational amplifier with respect to ground. { ¦käm·ən ,mōd 'in,pút kə'pas·əd·əns }

common-mode input impedance |ELECTR| The open-loop input impedance of both inverting and noninverting inputs of an operational amplifier with respect to ground. { ¦käm·ən ,mōd 'in ,pút im'ped·əns }

common-mode input resistance |ELECTR| The equivalent resistance of both inverting and noninverting inputs of an operational amplifier with respect to ground or reference. { ¦käm·ən ,mōd 'in,pút ri'zis·təns }

common-mode rejection |ELECTR| The ability of an amplifier to cancel a common-mode signal while responding to an out-of-phase signal. Also known as in-phase rejection. { ¦käm·ən ,mōd ri'jek·shən }

common-mode rejection ratio |ELECTR| The ratio of the gain of an amplifier for difference signals between the input terminals, to the gain for the average or common-mode signal component. Abbreviated CMRR. { 'käm·ən ,mōd ri 'jek·shən 'rā·shō }

common-mode signal |ELECTR| A signal applied equally to both ungrounded inputs of a balanced amplifier stage or other differential device. Also known as in-phase signal. { ¦käm·ən ,mōd 'sig·nəl }

common-mode voltage |ELECTR| A voltage that appears in common at both input terminals of a device with respect to the output reference (usually ground). { ¦käm·ən ,mōd 'vōl·tij }

common rafter |BUILD| A rafter which extends from the plate of the roof to the ridge board at right angles to both members, and to which roofing is attached. { ¦käm·ən 'raf·tər }

common-rail injection |MECH ENG| A type of diesel engine fuel-injection system in which one rail maintains the fuel at a specified pressure while feed lines run from the rail to each fuel injector. { 'käm·ən ¦rāl in'jek·shən }

common return |ELECTR| A return conductor that serves two or more circuits. { ¦käm·ən ri'tərn }

common wall |BUILD| A wall that is shared by two dwelling units. { ¦käm·ən ¦wȯl }

communications |ENG| The science and technology by which information is collected from an originating source, transformed into electric currents or fields, transmitted over electrical networks or space to another point, and reconverted into a form suitable for interpretation by a receiver. { kə,myü·nə'kā·shənz }

compaction |ENG| Increasing the dry density of a granular material, particularly soil, by means such as impact or by rolling the surface layers. { kəm'pak·shən }

compactor |MECH ENG| 1. Machine designed

to consolidate earth and paving materials by kneading, weight, vibration, or impact, to sustain loads greater than those sustained in an uncompacted state. **2.** A machine that compresses solid waste material for convenience in disposal. { kəm'pak·tər }

companion flange [DES ENG] A pipe flange that can be bolted to a similar flange on another pipe. { kəm'pan·yən ,flanj }

comparative rabal [ENG] A rabal observation (that is, a radiosonde balloon tracked by theodolite) taken simultaneously with the usual rawin observation (tracking by radar or radio direction-finder), to provide a rough check on the alignment and operating accuracy of the electronic tracking equipment. { kəm'par·əd·iv 'rā,bal }

comparator [CONT SYS] A device which detects the value of the quantity to be controlled by a feedback control system and compares it continuously with the desired value of that quantity. [ENG] A device used to inspect a gaged part for deviation from a specified dimension, by mechanical, electrical, pneumatic, or optical means. { kəm'par·əd·ər }

comparator circuit [ELECTR] An electronic circuit that produces an output voltage or current whenever two input levels simultaneously satisfy predetermined amplitude requirements; may be linear (continuous) or digital (discrete). { kəm 'par·əd·ər ,sər·kət }

comparator method [THERMO] A method of determining the coefficient of linear expansion of a substance in which one measures the distance that each of two traveling microscopes must be moved in order to remain centered on scratches on a rod-shaped specimen when the temperature of the specimen is raised by a measured amount. { kəm'par·əd·ər ,meth·əd }

compartment mill [MECH ENG] A multisection pulverizing device divided by perforated partitions, with preliminary grinding at one end in a short ball-mill operation, and finish grinding at the discharge end in a longer tube-mill operation. { kəm'pärt·mənt ,mil }

compass [ENG] An instrument for indicating a horizontal reference direction relative to the earth. { 'käm·pəs }

compass bowl [ENG] That part of a compass in which the compass card is mounted. { 'käm·pəs ,bōl }

compass card [DES ENG] The part of a compass on which the direction graduations are placed; it is usually in the form of a thin disk or annulus graduated in degrees, clockwise from 0° at the reference direction to 360°, and sometimes also in compass points. { 'käm·pəs ,kärd }

compass card axis [DES ENG] The line joining 0° and 180° on a compass card. { 'käm·pəs ,kärd ,ak·səs }

compass declinometer [ENG] An instrument used for magnetic distribution surveys; employs a thin compass needle 6 inches (15 centimeters) long, supported on a sapphire bearing and steel pivot of high quality; peep sights serve for

aligning the compass box on an azimuth mark. { 'käm·pəs ,dek·lə'näm·əd·ər }

compass roof [BUILD] A roof in which each truss is in the form of an arch. { |käm·pəs |rüf }

compass saw [DES ENG] A handsaw which has a handle with several attachable thin, tapering blades of varying widths, making it suitable for a variety of work, such as cutting circles and curves. { 'käm·pəs ,sȯ }

compatibility [SYS ENG] The ability of a new system to serve users of an old system. { kəm,pad·ə'bil·ə·dē }

compatibility conditions [MECH] A set of six differential relations between the strain components of an elastic solid which must be satisfied in order for these components to correspond to a continuous and single-valued displacement of the solid. { kəm,pad·ə'bil·əd·ē kən,dish·ənz }

compatible discrete four-channel sound [ENG ACOUS] A sound system in which a separate channel is maintained from each of the four sets of microphones at the recording studio or other input location to the four sets of loudspeakers that serve as the output of the system. Abbreviated CD-4 sound. { kəm'pad·ə·bəl dis'krēt |fȯr |chan·əl 'saȯnd }

compatible monolithic integrated circuit [ELECTR] Device in which passive components are deposited by thin-film techniques on top of a basic silicon-substrate circuit containing the active components and some passive parts. { kəm'pad·ə·bəl ,män·ə'lith·ik 'in·tə,grād·əd 'sər·kət }

compensated neutron logging [ENG] Neutron well logging using one source and two detectors; the apparent limestone porosity is calculated by computer from the ratio of the count rate of one detector to that of the other. { 'käm·pən,sād· əd |nü,trän ,läg·iŋ }

compensated pendulum [DES ENG] A pendulum made of two materials with different coefficients of expansion so that the distance between the point of suspension and center of oscillation remains nearly constant when the temperature changes. { 'käm·pən,sād·əd 'pen·jə·ləm }

compensated semiconductor [ELECTR] Semiconductor in which one type of impurity or imperfection (for example, donor) partially cancels the electrical effects on the other type of impurity or imperfection (for example, acceptor). { 'käm·pən,sād·əd 'sem·i·kən'dək·tər }

compensated volume control See loudness control. { 'käm·pən,sād·əd 'väl·yəm kən'trōl }

compensating leads [ENG] A pair of wires, similar to the working leads of a resistance thermometer or thermocouple, which are run alongside the working leads and are connected in such a way that they balance the effects of temperature changes in the working leads. { 'käm· pən,sād·iŋ 'lēdz }

compensating network [CONT SYS] A network used in a low-energy-level method for suppression of excessive oscillations in a control system. { 'käm·pən,sād·iŋ 'net,wərk }

compensation |CONT SYS| Introduction of additional equipment into a control system in order to reshape its root locus so as to improve system performance. Also known as stabilization. |ELECTR| The modification of the amplitude-frequency response of an amplifier to broaden the bandwidth or to make the response more nearly uniform over the existing bandwidth. Also known as frequency compensation. { ˌkäm·pən'sā·shən }

compensation signals |ENG| In telemetry, signals recorded on a tape, along with the data and in the same track as the data, used during the playback of data to correct electrically the effects of tape-speed errors. { ˌkäm·pən'sā·shən ˌsig·nəlz }

compensator |CONT SYS| A device introduced into a feedback control system to improve performance and achieve stability. Also known as filter. |ELECTR| A component that offsets an error or other undesired effect. { 'käm·pən ˌsād·ər }

complementary |ELECTR| Having *pnp* and *npn* or *p*- and *n*- channel semiconductor elements on or within the same integrated-circuit substrate or working together in the same functional amplifier state. { ˌkäm·plə'men·trē }

complementary constant-current logic |ELECTR| A type of large-scale integration used in digital integrated circuits and characterized by high density and very fast switching times. Abbreviated CCCL; C³L. { ˌkäm·plə¦men·trē ¦kän·stənt ¦kə·rənt 'läj·ik }

complementary logic switch |ELECTR| A complementary transistor pair which has a common input and interconnections such that one transistor is on when the other is off, and vice versa. { ˌkäm·plə'men·trē 'läj·ik ˌswich }

complementary metal oxide semiconductor device See CMOS device. { ˌkäm·plə¦men·trē ¦med·əl ¦äk ˌsīd 'sem·i·kən ˌdək·tər di'vīs }

complementary symmetry |ELECTR| A circuit using both *pnp* and *npn* transistors in a symmetrical arrangement that permits push-pull operation without an input transformer or other form of phase inverter. { ˌkäm·plə'men·trē 'sim·ə·trē }

complementary transistors |ELECTR| Two transistors of opposite conductivity (*pnp* and *npn*) in the same functional unit. { ˌkäm·plə'men·trē tran'zis·tərs }

complete-expansion diesel cycle See Brayton cycle. { kəm'plēt ik'span·shən 'dē·zəl ˌsi·kəl }

complete lubrication |ENG| Lubrication taking place when rubbing surfaces are separated by a fluid film, and frictional losses are due solely to the internal fluid friction in the film. Also known as viscous lubrication. { kəm'plēt 'lüb·rə'kā·shən }

complex frequency |ENG| A complex number used to characterize exponential and damped sinusoidal motion in the same way that an ordinary frequency characterizes simple harmonic motion; designated by the constant *s* corresponding to a motion whose amplitude is given by Ae^{st}, where A is a constant and *t* is time. { 'käm ˌpleks 'frē·kwən·sē }

complex impedance See electrical impedance; impedance. { 'käm ˌpleks im'pēd·əns }

complex permittivity |ELEC| A property of a dielectric, equal to $\epsilon_0(C/C_0)$, where C is the complex capacitance of a capacitor in which the dielectric is the insulating material when the capacitor is connected to a sinusoidal voltage source, and C_0 is the vacuum capacitance of the capacitor. { 'käm ˌpleks ˌpər·mə'tiv·əd·ē }

complex reflector |ENG| A structure or group of structures having many radar-reflecting surfaces facing in different directions. { 'käm ˌpleks ri'flek·tər }

complex relative attenuation |ELECTR| The ratio of the peak output voltage, in complex notation, of an electric filter to the output voltage at the frequency being considered. { 'käm ˌpleks ¦rel·əd·iv ə ˌten·yə'wā·shən }

complex target |ENG| A radar target composed of a number of reflecting surfaces that, in the aggregate, are smaller in all dimensions than the resolution capabilities of the radar. { 'käm ˌpleks 'tär·gət }

compliance |MECH| The displacement of a linear mechanical system under a unit force. { kəm'plī·əns }

compliance constant |MECH| Any one of the coefficients of the relations in the generalized Hooke's law used to express strain components as linear functions of the stress components. Also known as elastic constant. { kəm'plī·əns ˌkän·stənt }

compliant substrate |ELECTR| A semiconductor substrate into which an artificially formed interface is introduced near the surface which makes the substrate more readily deformable and allows it to support a defect-free semiconductor film of essentially any lattice constant, with dislocations forming in the substrate instead of in the film. Also known as sacrificial compliant substrate. { kəm¦plī·ənt 'səb ˌstrāt }

component |ELEC| Any electric device, such as a coil, resistor, capacitor, generator, line, or electron tube, having distinct electrical characteristics and having terminals at which it may be connected to other components to form a circuit. Also known as circuit element; element. { kəm'pō·nənt }

component distillation |CHEM ENG| A distillation process in which a fraction that cannot normally be separated by distillation is removed by forming an azeotropic mixture. { kəm'pō·nənt dis·tə'lā·shən }

component-failure-impact analysis |SYS ENG| A study that attempts to predict the consequences of failures of the major components of a system. Abbreviated CFIA. { kəm'pō·nənt ¦fāl·yər 'im ˌpakt ə ˌnal·ə·səs }

composite |ENG ACOUS| A re-recording consisting of at least two elements. { kəm'päz·ət }

composite beam |CIV ENG| A structural member composed of two or more dissimilar materials joined together to act as a unit in which the resulting system is stronger than the sum of its parts. An example in civil structures is the steel-concrete composite beam in which a steel wide-flange shape (I or W shape) is attached to a concrete floor slab. { kəm'päz·ət 'bēm }

composite column [CIV ENG] A concrete column having a structural-steel or cast-iron core with a maximum core area of 20. { kəm'päz·ət 'käl·əm }

composite filter [ELECTR] A filter constructed by linking filters of different kinds in series. { kəm'päz·ət 'fil·tər }

composite I-beam bridge [CIV ENG] A beam bridge in which the concrete roadway is mechanically bonded to the I beams by means of shear connectors. { kəm'päz·ət 'ī ,bēm ,brij }

composite macromechanics [ENG] The study of composite material behavior wherein the material is presumed homogeneous and the effects of the constituent materials are detected only as averaged apparent properties of the composite. { kəm'päz·ət ¦mak·rō·mə'kan·iks }

composite material See composite. { kəm¦päz·ət mə¦tir·ē·əl }

composite micromechanics [ENG] The study of composite material behavior wherein the constituent materials are studied on a microscopic scale with specific properties being assigned to each constituent; the interaction of the constituent materials is used to determine the properties of the composite. { kəm'päz·ət ¦mik·rō·mə'kan·iks }

composite pile [CIV ENG] A pile in which the upper and lower portions consist of different types of piles. { kəm'päz·ət 'pīl }

composite sampler [ENG] A hydrometer cylinder equipped with sample cocks at regular intervals along its vertical height; used to take representative (vertical composite) samples of oil from storage tanks. { kəm'päz·ət 'sam·plər }

composite truss [CIV ENG] A truss having compressive members and tension members. { kəm'päz·ət 'trəs }

composition [MECH] The determination of a force whose effect is the same as that of two or more given forces acting simultaneously; all forces are considered acting at the same point. { ,käm·pə'zish·ən }

composition diagram [CHEM ENG] Graphical plots to show the solvent-solute concentration relationships during various stages of extraction operations (leaching, or solid-liquid extraction; and liquid-liquid extraction). { ,käm·pə'zish·ən ,dī·ə,gram }

composition-of-velocities law [MECH] A law relating the velocities of an object in two references frames which are moving relative to each other with a specified velocity. { ,käm·pə'zish·ən əv və'läs·əd·ēz ,lò }

compound angle [ENG] The angle formed by two mitered angles. { 'käm,paùnd 'aŋ·gəl }

compound engine [MECH ENG] A multicylinder-type displacement engine, using steam, air, or hot gas, where expansion proceeds successively (sequentially). { 'käm,paùnd 'en·jən }

compounding [MECH ENG] The series placing of cylinders in an engine (such as steam) for greater ratios of expansion and consequent improved engine economy. { 'käm,paùnd·iŋ }

compound lever [MECH ENG] A train of levers in which motion or force is transmitted from the arm of one lever to that of the next. { 'käm ,paùnd 'lev·ər }

compound rest [MECH ENG] A principal component of a lathe consisting of a base and an upper part dovetailed together; the base is graduated in degrees and can be swiveled to any angle; the upper part includes the tool post and tool holder. { 'käm,paùnd 'rest }

compound screw [DES ENG] A screw having different or opposite pitches on opposite ends of the shank. { 'käm,paùnd 'skrü }

compregnate [ENG] Compression of materials into a dense, hard substance with the aid of heat. { kəm'preg,nāt }

compressadensity function [MECH] A function used in the acoustic levitation technique to determine either the density or the adiabatic compressibility of a submicroliter droplet suspended in another liquid, if the other property is known. { kəm,pres·ə'den·səd·ē ,fəŋk·shən }

compressed air [MECH] Air whose density is increased by subjecting it to a pressure greater than atmospheric pressure. { kəm'prest 'er }

compressed-air diving [ENG] Any form of diving in which air is supplied under high pressure to prevent lung collapse. { kəm¦prest ¦er 'dīv·iŋ }

compressed-air loudspeaker [ENG ACOUS] A loudspeaker having an electrically actuated valve that modulates a stream of compressed air. { kəm¦prest ¦er 'laùd,spēk·ər }

compressed-air power [MECH ENG] The power delivered by the pressure of compressed air as it expands, utilized in tools such as drills, in hoists, grinders, riveters, diggers, pile drivers, motors, locomotives, and in mine ventilating systems. { kəm¦prest ¦er 'paùr·ər }

compressibility [MECH] The property of a substance capable of being reduced in volume by application of pressure; quantitively, the reciprocal of the bulk modulus. { kəm,pres·ə'bil·əd·ē }

compressibility factor [THERMO] The product of the pressure and the volume of a gas, divided by the product of the temperature of the gas and the gas constant; this factor may be inserted in the ideal gas law to take into account the departure of true gases from ideal gas behavior. Also known as deviation factor; gas-deviation factor; supercompressibility factor. { kəm,pres·ə'bil·əd·ē ,fak·tər }

compressible fluid flow [CHEM ENG] Gas flow when the pressure drop due to the flow of a gas through a system is large enough, compared with the inlet pressure, to cause a 10% or greater decrease in gas density. { kəm'pres·ə·bəl 'flü·əd ,flō }

compression [ELECTR] **1.** Reduction of the effective gain of a device at one level of signal with respect to the gain at a lower level of signal, so that weak signal components will not be lost in background and strong signals will not overload the system. **2.** *See* compression ratio. |MECH| Reduction in the volume of a substance due to pressure; for example in building, the type of stress which causes shortening of the fibers of a wooden member. |MECH| *See* compression ratio. { kəm'presh·ən }

compression coupling |MECH ENG| **1.** A means of connecting two perfectly aligned shafts in which a slotted tapered sleeve is placed over the junction and two flanges are drawn over the sleeve so that they automatically center the shafts and provide sufficient contact pressure to transmit medium loads. **2.** A type of tubing fitting. { kəm'presh·ən ,kəp·liŋ }

compression cup |ENG| A cup from which lubricant is forced to a bearing by compression. { kəm'presh·ən ,kəp }

compression failure |ENG| Buckling or collapse caused by compression, as of a steel or concrete column or of wood fibers. { kəm'presh·ən ,fāl·yər }

compression fitting |ENG| A leak-resistant pipe joint designed with a tight-fitting sleeve that exerts a large inward pressure on the exterior of the pipe. { kəm'presh·ən ,fid·iŋ }

compression gage |ENG| An instrument that measures pressures greater than atmospheric pressure. { kəm'presh·ən ,gāj }

compression ignition |MECH ENG| Ignition produced by compression of the air in a cylinder of an internal combustion engine before fuel is admitted. { kəm'presh·ən ig'nish·ən }

compression-ignition engine *See* diesel engine. { kəm¦presh·ən ig¦nish·ən 'en·jən }

compression member |ENG| A beam or other structural member which is subject to compressive stress. { kəm'presh·ən ,mem·bər }

compression modulus *See* bulk modulus of elasticity. { kəm'presh·ən ,mäj·ə·ləs }

compression mold |ENG| A mold for plastics which is open when the material is introduced and which shapes the material by heat and by the pressure of closing. { kəm'presh·ən ,mōld }

compression pressure |MECH ENG| That pressure developed in a reciprocating piston engine at the end of the compression stroke without combustion of fuel. { kəm'presh·ən ,presh·ər }

compression process |CHEM ENG| The recovery of natural gasoline from gas containing a high proportion of hydrocarbons. { kəm'presh·ən ,prä·səs }

compression ratio |ELECTR| The ratio of the gain of a device at a low power level to the gain at some higher level, usually expressed in decibels. Also known as compression. |MECH ENG| The ratio in internal combustion engines between the volume displaced by the piston plus the clearance space, to the volume of the clearance space. Also known as compression. { kəm'presh·ən ,rā·shō }

compression refrigeration |MECH ENG| The cooling of a gaseous refrigerant by first compressing it to liquid form (with resultant heat

buildup), cooling the liquid by heat exchange, then releasing pressure to allow the liquid to vaporize (with resultant absorption of latent heat of vaporization and a refrigerative effect). { kəm'presh·ən ri,frij·ə'rā·shən }

compression release |MECH ENG| Release of compressed gas resulting from incomplete closure of intake or exhaust valves. { kəm'presh·ən ri'lēs }

compression ring |MECH ENG| A ring located at the upper part of a piston to hold the burning fuel charge above the piston in the combustion chamber, thus preventing blowby. { kəm'presh·ən ,riŋ }

compression spring |ENG| A spring, usually a coil spring, which resists a force tending to compress it. { kəm'presh·ən ,spriŋ }

compression strength |MECH| Property of a material to resist rupture under compression. { kəm'presh·ən ,streŋkth }

compression stroke |MECH ENG| The phase of a positive displacement engine or compressor in which the motion of the piston compresses the fluid trapped in the cylinder. { kəm'presh·ən ,strōk }

compression test |ENG| A test to determine compression strength, usually applied to materials of high compression but low tensile strength, in which the specimen is subjected to increasing compressive forces until failure occurs. { kəm'presh·ən ,test }

compressive member |CIV ENG| A structural member subject to tension. { kəm'pres·iv 'mem·bər }

compressive strength |MECH| The maximum compressive stress a material can withstand without failure. { kəm'pres·iv 'streŋkth }

compressive stress |MECH| A stress which causes an elastic body to shorten in the direction of the applied force. { kəm'pres·iv 'stres }

compressor |ELECTR| The part of a compandor that is used to compress the intensity range of signals at the transmitting or recording end of a circuit. |MECH ENG| A machine used for increasing the pressure of a gas or vapor. Also known as compression machine. { kəm'pres·ər }

compressor blade |MECH ENG| The vane components of a centrifugal or axial-flow, air or gas compressor. { kəm'pres·ər ,blād }

compressor station |MECH ENG| A permanent facility which increases the pressure on gas to move it in transmission lines or into storage. { kəm'pres·ər ,stā·shən }

compressor valve |MECH ENG| A valve in a compressor, usually automatic, which operates by pressure difference (less than 5 pounds per square inch or 35 kilopascals) on the two sides of a movable, single-loaded member and which has no mechanical linkage with the moving parts of the compressor mechanism. { kəm'pres·ər ,valv }

compromise joint |CIV ENG| **1.** A joint bar used for joining rails of different height or section.

2. A rail that has different joint drillings from that of the same section. { 'käm·prə‚mīz ‚jȯint }

compromise rail |CIV ENG| A short rail having different sections at the ends to correspond with the rail ends to be joined, thus providing a transition between rails of different sections. { 'käm‚prə‚mīz ‚rāl }

computational numerical control See computer numerical control. { ‚käm·pyə'tā·shən·əl nü'mer·ə·kəl kən'trōl }

computed path control |CONT SYS| A control system designed to follow a path calculated to be the optimal one to achieve a desired result. { kəm'pyüd·əd ¦path kən'trōl }

computer-aided design |CONT SYS| The use of computers in converting the initial idea for a product into a detailed engineering design. Computer models and graphics replace the sketches and engineering drawings traditionally used to visualize products and communicate design information. Abbreviated CAD. { kəm'pyüd·ər ‚ād·əd də'zīn }

computer-aided engineering |ENG| The use of computer-based tools to assist in solution of engineering problems. { kəm'pyüd·ər ‚ād·əd ‚en·jə'nir·iŋ }

computer-aided manufacturing |CONT SYS| The use of computers in converting engineering designs into finished products. Computers assist managers, manufacturing engineers, and production workers by automating many production tasks, such as developing process plans, ordering and tracking materials, and monitoring production schedules, as well as controlling the machines, industrial robots, test equipment, and systems that move and store materials in the factory. Abbreviated CAM. { kəm'pyüd·ər ‚ād·əd ‚man·ə'fak·chə·riŋ }

computer control |CONT SYS| Process control in which the process variables are fed into a computer and the output of the computer is used to control the process. { kəm'pyüd·ər kən'trōl }

computer-controlled system |CONT SYS| A feedback control system in which a computer operates on both the input signal and the feedback signal to effect control. { kəm'pyüd·ər kən'trōld ‚sis·təm }

computer-integrated manufacturing |IND ENG| A computer-automated system in which individual engineering, production, marketing, and support functions of a manufacturing enterprise are organized; functional areas such as design, analysis, planning, purchasing, cost accounting, inventory control, and distribution are linked through the computer with factory floor functions such as materials handling and management, providing direct control and monitoring of all process operations. Abbreviated CIM. { kəm'pyüd·ər ¦int·ə‚grād·əd ‚man·ə'fak·chər·iŋ }

computer numerical control |CONT SYS| A control system in which numerical values corresponding to desired tool or control positions are generated by a computer. Abbreviated CNC. Also known as computational numerical control; soft-wired numerical control; stored-program

numerical control. { kəm'pyüd·ər nü'mer·i·kəl kən'trōl }

computer part programming |CONT SYS| The use of computers to program numerical control systems. { kəm'pyüd·ər 'pärt 'prō‚gram·iŋ }

concatenation |ELEC| A method of speed control of induction motors in which the rotors of two wound-rotor motors are mechanically coupled together and the stator of the second motor is supplied with power from the rotor slip rings of the first motor. |ENG ACOUS| The linking together of phonemes to produce meaningful sounds. { kən‚kat·ən'ā·shən }

concave bit |DES ENG| A type of tungsten carbide drill bit having a concave cutting edge; used for percussive boring. { 'kän‚kāv ‚bit }

concentrated load |MECH| A force that is negligible because of a small contact area; a beam supported on a girder represents a concentrated load on the girder. { 'kän·sən‚träd·əd 'lōd }

concentrator |ELECTR| Buffer switch (analog or digital) which reduces the number of trunks required. |ENG| **1.** An apparatus used to concentrate materials. **2.** A plant where materials are concentrated. { 'kän·sən‚träd·ər }

concentric groove See locked groove. { kən'sen·trik 'grüv }

concentric locating |DES ENG| The process of making the axis of a tooling device coincide with the axis of the workpiece. { kən'sen·trik 'lō‚kād·iŋ }

concentric orifice plate |DES ENG| A fluid-meter orifice plate whose edges have a circular shape and whose center coincides with the center of the pipe. { kən'sen·trik 'ȯr·ə·fəs ‚plāt }

concentric reducer |ENG| A threaded or butt-welded pipe fitting whose ends are of different sizes but are concentric about a common axis. { kən'sen·trik ri'dü·sər }

concentric tube column |CHEM ENG| A carefully insulated distillation apparatus which is capable of very high separating power, and in which the outer vapor-rising annulus of the column is concentric around an inner, bottom-discharging reflux return. { kən'sen·trik ¦tüb ¦käl·əm }

concrete beam |CIV ENG| A structural member of reinforced concrete, placed horizontally over openings to carry loads. { 'käŋ‚krēt 'bēm }

concrete bridge |CIV ENG| A bridge constructed of prestressed or reinforced concrete. { 'käŋ‚krēt 'brij }

concrete bucket |ENG| A container with movable gates at the bottom that is attached to power cranes or cables to transport concrete. { 'käŋ‚krēt ‚bək·ət }

concrete buggy |ENG| A cart which carries up to 6 cubic feet (0.17 cubic meter) of concrete from the mixer or hopper to the forms. Also known as buggy; concrete cart. { 'käŋ‚krēt ‚bəg·ē }

concrete caisson sinking |CIV ENG| A shaft-sinking method similar to caisson sinking except that reinforced concrete rings are used and an airtight working chamber is not adopted. { 'käŋ‚krēt 'kā‚sän ‚siŋk·iŋ }

concrete cart *See* concrete buggy. { 'käŋ‚krēt ‚kärt }

concrete chute |ENG| A long metal trough with rounded bottom and open ends used for conveying concrete to a lower elevation. { 'käŋ ‚krēt ‚shüt }

concrete column |CIV ENG| A vertical structural member made of reinforced or unreinforced concrete. { 'käŋ‚krēt 'käl·əm }

concrete dam |CIV ENG| A dam that is built of concrete. { 'käŋ‚krēt 'dam }

concrete mixer |MECH ENG| A machine with a rotating drum in which the components of concrete are mixed. { 'käŋ‚krēt ‚mik·sər }

concrete nail |DES ENG| A hardened-steel nail that has a flat countersunk head and a tapered point and is used for nailing various materials to concrete or masonry. { 'käŋ‚krēt 'nāl }

concrete pile |CIV ENG| A reinforced pile made of concrete, either precast and driven into the ground, or cast in place in a hole bored into the ground. { 'käŋ‚krēt 'pīl }

concrete pipe |CIV ENG| A porous pipe made of concrete and used principally for subsoil drainage; diameters over 15 inches (38 centimeters) are usually reinforced. { 'käŋ‚krēt 'pīp }

concrete pump |MECH ENG| A device which drives concrete to the placing position through a pipeline of 6-inch (15-centimeter) diameter or more, using a special type of reciprocating pump. { 'käŋ‚krēt ‚pəmp }

concrete slab |CIV ENG| A flat, reinforced-concrete structural member, relatively sizable in length and width, but shallow in depth; used for floors, roofs, and bridge decks. { 'käŋ‚krēt 'slab }

concrete vibrator |MECH ENG| Vibrating device used to achieve proper consolidation of concrete; the three types are internal, surface, and form vibrators. { 'käŋ‚krēt ‚vī‚brād·ər }

concurrent engineering |ENG| The simultaneous design of products and related processes, including all product life-cycle aspects such as manufacturing, assembly, test, support, disposal, and recycling. { kən¦kər·ənt ‚en·jə'nir·iŋ }

concussion |ENG| Shock waves in the air caused by an explosion underground or at the surface or by a heavy blow directly to the ground surface during excavation, quarrying, or blasting operations. { kən'kəsh·ən }

condensate flash |CHEM ENG| Partial evaporation (flash) of hot condensed liquid by a stepwise reduction in system pressure, the hot vapor supplying heat to a cooler evaporator step (stage). { 'kän·dən‚sat ‚flash }

condensate strainer |MECH ENG| A screen used to remove solid particles from the condensate prior to its being pumped back to the boiler. { 'kän·dən‚sat ‚strān·ər }

condensate well |MECH ENG| A chamber into which condensed vapor falls for convenient accumulation prior to removal. { 'kän·dən‚sat ‚wel }

condensation |ELEC| An increase of electric charge on a capacitor conductor. |MECH| An increase in density. { ‚kän·dən'sā·shən }

condenser |ELEC| *See* capacitor. |MECH ENG| A heat-transfer device that reduces a thermodynamic fluid from its vapor phase to its liquid phase, such as in a vapor-compression refrigeration plant or in a condensing steam power plant. { kən'den·sər }

condenser-discharge anemometer |ENG| A contact anemometer connected to an electrical circuit which is so arranged that the average wind speed is indicated. { kən¦den·sər'dis‚chärj an·ə'mäm·əd·ər }

condenser microphone *See* capacitor microphone. { kən'den·sər 'mī·krə‚fōn }

condenser transducer *See* electrostatic transducer. { kən'den·sər ‚tranz'dü·sər }

condenser tubes |MECH ENG| Metal tubes used in a heat-transfer device, with condenser vapor as the heat source and flowing liquid such as water as the receiver. { kən'den·sər ‚tübs }

condensing electrometer *See* capacitive electrometer. { kən¦dens·iŋ ə‚lek'träm·əd·ər }

condensing engine |MECH ENG| A steam engine in which the steam exhausts from the cylinder to a vacuum space, where the steam is liquefied. { kən¦dens·iŋ ¦en·jən }

conditionally periodic motion |MECH| Motion of a system in which each of the coordinates undergoes simple periodic motion, but the associated frequencies are not all rational fractions of each other so that the complete motion is not simply periodic. { kən'dish·ən·əl·ē ‚pir·ē¦ad·ik ‚mō·shən }

conditionally stable circuit |ELECTR| A circuit which is stable for certain values of input signal and gain, and unstable for other values. { kən'dish·ən·əl·ē ¦stā·bəl ‚sər·kət }

conductance |ELEC| The real part of the admittance of a circuit; when the impedance contains no reactance, as in a direct-current circuit, it is the reciprocal of resistance, and is thus a measure of the ability of the circuit to conduct electricity. Also known as electrical conductance. Designated G. |THERMO| *See* thermal conductance. { kən'dək·təns }

conduction |ELEC| The passage of electric charge, which can occur by a variety of processes, such as the passage of electrons or ionized atoms. Also known as electrical conduction. { kən'dək·shən }

conduction cooling |ELECTR| Cooling of electronic components by carrying heat from the device through a thermally conducting material to a large piece of metal with cooling fins. { kən'dək·shən ‚kül·iŋ }

conduction pump |ENG| A pump in which liquid metal or some other conductive liquid is moved through a pipe by sending a current across the liquid and applying a magnetic field at right angles to current flow. { kən'dək·shən ‚pəmp }

conductive coupling |ELEC| Electric connection of two electric circuits by their sharing the same resistor. { kən'dək·tiv 'kəp·liŋ }

conductive interference |ELECTR| Interference to electronic equipment that originates in power lines supplying the equipment, and is conducted to the equipment and coupled through the power supply transformer. { kən'dək·tiv ,in·tər'fir·əns }

conductivity |ELEC| The ratio of the electric current density to the electric field in a material. Also known as electrical conductivity; specific conductance. { ,kän,dək'tiv·əd·ē }

conductivity bridge |ELEC| A modified Kelvin bridge for measuring very low resistances. { ,kän,dək'tiv·əd·ē ,brij }

conductivity cell |ELEC| A glass vessel with two electrodes at a definite distance apart and filled with a solution whose conductivity is to be measured. { ,kän,dək'tiv·əd·ē ,sel }

conductivity modulation |ELECTR| Of a semiconductor, the variation of the conductivity of a semiconductor through variation of the charge carrier density. { ,kän,dək'tiv·əd·ē ,mäj·ə'lā·shən }

conductivity modulation transistor |ELECTR| Transistor in which the active properties are derived from minority carrier modulation of the bulk resistivity of the semiconductor. { ,kän,dək 'tiv·əd·ē ,mäj·ə'lā·shən tran'zis·tər }

conductometer |ENG| An instrument designed to measure thermal conductivity; in particular, one that compares the rates at which different rods transmit heat. { ,kän,dək'täm·əd·ər }

conductor |ELEC| A wire, cable, or other body or medium that is suitable for carrying electric current. Also known as electric conductor. { kən'dək·tər }

conductor pipe |BUILD| A metal pipe through which water is drained from the roof. { kən'dək·tər ,pīp }

conduit |ELEC| Solid or flexible metal or other tubing through which insulated electric wires are run. |ENG| Any channel or pipe for conducting the flow of water or other fluid. { 'kän·də·wət }

cone |ENG ACOUS| The cone-shaped paper or fiber diaphragm of a loudspeaker. { kōn }

cone bearing |MECH ENG| A cone-shaped journal bearing running in a correspondingly tapered sleeve. { 'kōn ,ber·iŋ }

cone-bottom tank |ENG| Liquids-storage tank with downward-pointing conical bottom to facilitate drainage of bottom, as of water or sludge. { 'kōn ,bäd·əm ,taŋk }

cone brake |MECH ENG| A type of friction brake whose rubbing parts are cone-shaped. { 'kōn ,brāk }

cone classifier |MECH ENG| Inverted-cone device for the separation of heavy particulates (such as sand, ore, or other mineral matter) from a liquid stream; feed enters the top of the cone, heavy particles settle to the bottom where they can be withdrawn, and liquid overflows the top edge, carrying the smaller particles or those of lower gravity over the rim; used in the mining and chemical industries. { 'kōn 'klas·ə,fī·ər }

cone clutch |MECH ENG| A clutch which uses the wedging action of mating conical surfaces to transmit friction torque. { 'kōn ,kləch }

cone crusher |MECH ENG| A machine that reduces the size of materials such as rock by crushing in the tapered space between a truncated revolving cone and an outer chamber. { 'kōn ,krəsh·ər }

conehead rivet |DES ENG| A rivet with a head shaped like a truncated cone. { 'kōn,hed 'riv·ət }

cone key |DES ENG| A taper saddle key placed on a shaft to adapt it to a pulley with a too-large hole. { 'kōn ,kē }

cone loudspeaker |ENG ACOUS| A loudspeaker employing a magnetic driving unit that is mechanically coupled to a paper or fiber cone. Also known as cone speaker. { 'kōn 'laùd ,spēk·ər }

cone mandrel |DES ENG| A mandrel in which the diameter can be changed by moving conical sleeves. { 'kōn ,man·drəl }

cone nozzle |DES ENG| A cone-shaped nozzle that disperses fluid in an atomized mist. { 'kōn ,näz·əl }

cone of friction |MECH| A cone in which the resultant force exerted by one flat horizontal surface on another must be located when both surfaces are at rest, as determined by the coefficient of static friction. { 'kōn əv 'frik·shən }

cone pulley See step pulley. { 'kōn ,púl·ē }

cone rock bit |MECH ENG| A rotary drill with two hardened knurled cones which cut the rock as they roll. Also known as roller bit. { 'kōn 'räk ,bit }

cone-roof tank |ENG| Liquids-storage tank with flattened conical roof to allow a vapor reservoir at the top for filling operations. { 'kōn ,rüf ,taŋk }

cone speaker See cone loudspeaker. { 'kōn ,spēk·ər }

cone valve |CIV ENG| A divergent valve whose cone-shaped head in a fixed cylinder spreads water around the wide, downstream end of the cone in spillways of dams or hydroelectric facilities. Also known as Howell-Bunger valve. { 'kōn ,valv }

confidence level |IND ENG| The probability in acceptance sampling that the quality of accepted lots manufactured will be better than the rejectable quality level (RQL); 90% level indicates that accepted lots will be better than the RQL 90 times in 100. { 'kän·fə·dəns ,lev·əl }

configuration |ELEC| A group of components interconnected to perform a desired circuit function. |MECH| The positions of all the particles in a system. |SYS ENG| A group of machines interconnected and programmed to operate as a system. { kən,fig·yə'rā·shən }

confined flow |ENG| The flow of any fluid (liquid or gas) through a continuous container (process vessel) or conduit (piping or tubing). { kən'fīnd 'flō }

confinement |ENG| Physical restriction, or degree of such restriction, to passage of detonation wave or reaction zone, for example, that of a

121

resistant container which holds an explosive charge. { kən'fīn·mənt }

confining liquid [CHEM ENG] A liquid seal (most often mercury or sodium sulfate brine) that is displaced during the no-loss transfer of a gas sample from one container to another. { kən'fīn·iŋ ,lik·wəd }

congruent melting point [THERMO] A point on a temperature composition plot of a nonstoichiometric compound at which the one solid phase and one liquid phase are adjacent. { kən'grü·ənt 'melt·iŋ ,point }

conical ball mill [MECH ENG] A cone-shaped tumbling pulverizer in which the steel balls are classified, with the larger balls at the feed end where larger lumps are crushed, and the smaller balls at the discharge end where the material is finer. { 'kän·ə·kəl 'bȯl ,mil }

conical bearing [MECH ENG] An antifriction bearing employing tapered rollers. { 'kän·ə·kəl 'ber·iŋ }

conical pendulum [MECH] A weight suspended from a cord or light rod and made to rotate in a horizontal circle about a vertical axis with a constant angular velocity. { 'kän·ə·kəl 'pen·jə·ləm }

conical refiner [MECH ENG] In paper manufacture, a cone-shaped continuous refiner having two sets of bars mounted on the rotating plug and fixed shell for beating unmodified cellulose fibers. { 'kän·ə·kəl ri'fīn·ər }

conical roll See batten roll. { ¦kän·ə·kəl ¦rōl }

coniscope See koniscope. { 'kän·ə,skōp }

conjugate momentum [MECH] If q_i ($i = 1,2, . .$) are generalized coordinates of a classical dynamical system, and L is its Lagrangian, the momentum conjugate to q_i is $p_i = \partial L/\partial q_i$. Also known as canonical momentum; generalized momentum. { 'kän·jə·gət mə'men·təm }

connecting rod [MECH ENG] Any straight link that transmits motion or power from one linkage to another within a mechanism, especially linear to rotary motion, as in a reciprocating engine or compressor. { kə'nekt·iŋ ,räd }

connector [ELECTR] A switch, or relay group system, which finds the telephone line being called as a result of digits being dialed; it also causes interrupted ringing voltage to be placed on the called line or of returning a busy tone to the calling party if the line is busy. [ENG] **1.** A detachable device for connecting electrical conductors. **2.** A metal part for joining timbers. **3.** A symbol on a flowchart indicating that the flow jumps to a different location on the chart. { kə'nek·tər }

Conradson carbon test See carbon-residue test. { 'kän·rəd·sən 'kär·bən ,test }

conservation of angular momentum [MECH] The principle that, when a physical system is subject only to internal forces that bodies in the system exert on each other, the total angular momentum of the system remains constant, provided that both spin and orbital angular momentum are taken into account. { ,kän·sər'vā·shən əv 'aŋ·gyə·lər mə'men·təm }

conservation of areas [MECH] A principle governing the motion of a body moving under the action of a central force, according to which a line joining the body with the center of force sweeps out equal areas in equal times. { ,kän·sər'vā·shən əv 'er·ē·əz }

conservation of charge [ELEC] A law which states that the total charge of an isolated system is constant; no violation of this law has been discovered. Also known as charge conservation. { ,kän·sər'vā·shən əv 'chärj }

conservation of momentum [MECH] The principle that, when a system of masses is subject only to internal forces that masses of the system exert on one another, the total vector momentum of the system is constant; no violation of this principle has been found. Also known as momentum conservation. { ,kän·sər'vā·shən əv mə'men·təm }

conservative force field [MECH] A field of force in which the work done on a particle in moving it from one point to another depends only on the particle's initial and final positions. { kən'sər·və·tiv 'fȯrs ,fēld }

conservative property [THERMO] A property of a system whose value remains constant during a series of events. { kən'sər·və·tiv 'präp·ərd·ē }

console [ENG] **1.** A main control desk for electronic equipment, as at a radar station, radio or television station, or airport control tower. Also known as control desk. **2.** A large cabinet for a radio or television receiver, standing on the floor rather than on a table. **3.** A grouping of controls, indicators, and similar items contained in a specially designed model cabinet for floor mounting; constitutes an operator's permanent working position. { 'kän,sōl }

consolute temperature [THERMO] The upper temperature of immiscibility for a two-component liquid system. Also known as upper consolute temperature; upper critical solution temperature. { 'kan·sə,lüt 'tem·prə·chər }

constant-amplitude recording [ENG ACOUS] A sound-recording method in which all frequencies having the same intensity are recorded at the same amplitude. { ¦kän·stənt 'am·plə,tüd ri,kȯrd·iŋ }

constant-distance sphere [ENG ACOUS] The relative response of a sonar projector to variations in acoustic intensity, or intensity per unit band, over the surface of a sphere concentric with its center. { 'kän·stənt 'dis·təns ,sfir }

constant element [IND ENG] Under a specified set of conditions, an element for which the standard time allowance should always be the same. { 'kän·stənt 'el·ə·mənt }

constant-force spring [MECH ENG] A spring which has a constant restoring force, regardless of displacement. { ¦kän·stənt ¦fȯrs ,spriŋ }

constant-head meter [ENG] A flow meter which maintains a constant pressure differential but varies the orifice area with flow, such as a rotameter or piston meter. { 'kän·stənt ,hed ,mēd·ər }

constant-load balance [ENG] An instrument for measuring weight or mass which consists of a single pan (together with a set of weights that can be suspended from a counterpoised beam) that has a constant load (200 grams for the microbalance). { ¦kän·stənt¦lŏd 'bal·əns }

constant-load support |ENG| A spring-loaded support designed to maintain a constant and balanced load on a pipe in the event of vertical movement. { ¦kän·stənt ¦lŏd sə'pórt }

constant of gravitation *See* gravitational constant. { 'kän·stənt əv grav·ə'tā·shən }

constant of motion |MECH| A dynamical variable of a system which remains constant in time. { 'kän·stənt əv 'mō·shən }

constant-pressure combustion |MECH ENG| Combustion occurring without a pressure change. { ¦kän·stənt ¦presh·ər kəm'bəs·chən }

constant-pressure gas thermometer |ENG| A thermometer in which the volume occupied by a given mass of gas at a constant pressure is used to determine the temperature. { ¦kän·stənt ¦presh·ər 'gas thər,mäm·əd·ər }

constant-speed drive |MECH ENG| A mechanism transmitting motion from one shaft to another that does not allow the velocity ratio of the shafts to be varied, or allows it to be varied only in steps. { ¦kän·stənt ¦spēd 'drīv }

constant-velocity recording |ENG ACOUS| A sound-recording method in which, for input signals of a given amplitude, the resulting recorded amplitude is inversely proportional to the frequency; the velocity of the cutting stylus is then constant for all input frequencies having that given amplitude. { ¦kän·stənt və'läs·əd·ē ri ,kòrd·iŋ }

constant-velocity universal joint |MECH ENG| A universal joint that transmits constant angular velocity from the driving to the driven shaft, such as the Bendix-Weiss universal joint. { ¦kän·stənt və'läs·əd·ē ,yü·nə,vər·səl 'jóint }

constant-volume gas thermometer *See* gas thermometer. { ¦kän·stənt 'väl·yəm 'gas thər,mäm·əd·ər }

constrained mechanism |MECH ENG| A mechanism in which all members move only in prescribed paths. { kən'stränd 'mek·ə,niz·əm }

constraint |ENG| Anything that restricts the transverse contraction which normally occurs in a solid under longitudinal tension. |MECH| A restriction on the natural degrees of freedom of a system; the number of constraints is the difference between the number of natural degrees of freedom and the number of actual degrees of freedom. { kən'stränt }

construction |DES ENG| The number of strands in a wire rope and the number of wires in a strand; expressed as two numbers separated by a multiplication sign. |ENG| **1.** Putting parts together to form an integrated object. **2.** The manner in which something is put together. { kən'strək·shən }

construction area |BUILD| The area of exterior walls and permanent interior walls and partitions. { kən'strək·shən ,er·ē·ə }

construction cost |IND ENG| The total costs, direct and indirect, associated with transforming

a design plan for material and equipment into a project ready for operation. { kən'strək·shən ,kòst }

construction engineering |CIV ENG| A specialized branch of civil engineering concerned with the planning, execution, and control of construction operations for projects such as highways, dams, utility lines, and buildings. { kən'strək·shən ,en·jə'nir·iŋ }

construction equipment |MECH ENG| Heavy power machines which perform specific construction or demolition functions. { kən'strək·shən i'kwip·mənt }

construction joint |CIV ENG| A vertical or horizontal surface in reinforced concrete where concreting was stopped and continued later. { kən'strək·shən ,jóint }

construction survey |CIV ENG| A survey that gives locations for construction work. { kən'strək·shən ,sər,vā }

construction wrench |DES ENG| An open-end wrench with a long handle; the handle is used to align matching rivet or bolt holes. { kən'strək·shən ,rench }

consumer's risk |IND ENG| The probability that a lot whose quality equals the poorest quality that a consumer is willing to tolerate in an individual lot will be accepted by a sampling plan. { kən'süm·ərz 'risk }

contact |ELEC| *See* electric contact. |ENG| Initial detection of an aircraft, ship, submarine, or other object on a radarscope or other detecting equipment. { 'kän,takt }

contact adsorption |CHEM ENG| Process for removal of minor constituents from fluids by stirring in direct contact with powdered or granulated adsorbents, or by passing the fluid through fixed-position adsorbent beds (activated carbon or ion-exchange resin); used to decolorize petroleum lubricating oils and to remove solvent vapors from air. { 'kän,takt ad'sórp·shən }

contact aerator |CIV ENG| A tank in which sewage that is settled on a bed of stone, cement-asbestos, or other surfaces is treated by aeration with compressed air. { 'kän,takt 'er,ād·ər }

contact anemometer |ENG| An anemometer which actuates an electrical contact at a rate dependent upon the wind speed. Also known as contact-cup anemometer. { 'kän,takt an·ə'mäm·əd·ər }

contact bed |CIV ENG| A bed of coarse material such as coke, used to purify sewage. { 'kän ,takt ,bed }

contact catalysis |CHEM ENG| Process of change in the structure of gas molecules adsorbed onto solid surfaces; the basis of many industrial processes. { 'kän,takt kə'tal·ə·səs }

contact ceiling |BUILD| A ceiling in which the lath and construction are in direct contact, without use of furring or runner channels. { 'kän ,takt ,sēl·iŋ }

contact condenser |MECH ENG| A device in which a vapor, such as steam, is brought into direct contact with a cooling liquid, such as water, and is condensed by giving up its latent heat to the liquid. Also known as direct-contact condenser. { 'kän‚takt kən'den·sər }

contact-cup anemometer *See* contact anemometer. { 'kän‚takt ‚kəp an·ə'mäm·əd·ər }

contact electricity [ELEC] An electric charge at the surface of contact of two different materials. { 'kän‚takt i‚lek'tris·əd·ē }

contact electromotive force *See* contact potential difference. { 'kän‚takt i‚lek·trə'mōd·iv 'förs }

contact filtration [CHEM ENG] A process in which finely divided adsorbent clay is mixed with oil to remove color bodies and to improve the oil's stability. { 'kän‚takt fil'trā·shən }

contact gear ratio *See* contact ratio. { 'kän‚takt ‚gir ‚rā·shō }

contact grasp [IND ENG] A basic grasp that is used to push an object over a surface, such as using the index finger to push a coin over a flat surface. { 'kän‚takt ‚grasp }

contact-initiated discharge machining [MECH ENG] An electromachining process in which the discharge is initiated by allowing the tool and workpiece to come into contact, after which the tool is withdrawn and an arc forms. { 'kän‚takt ə‚nish·ē‚ād·əd ‚dis‚chärj mə‚shēn·iŋ }

contact inspection [ENG] A method by which an ultrasonic search unit scans a test piece in direct contact with a thin layer of couplant for transmission between the search unit and entry surface. { 'kän‚takt in'spek·shən }

contact microphone [ENG ACOUS] A microphone designed to pick up mechanical vibrations directly and convert them into corresponding electric currents or voltages. { 'kän‚takt 'mī·krə‚fōn }

contactor [CHEM ENG] A vessel designed to bring two or more substances into contact. [ELEC] A heavy-duty relay used to control electric power circuits. Also known as electric contactor. { 'kän‚tak·tər }

contactor control system [CONT SYS] A feedback control system in which the control signal is a discontinuous function of the sensed error and may therefore assume one of a limited number of discrete values. { 'kän‚tak·tər kən'trōl ‚sis·təm }

contact potential *See* contact potential difference. { 'kän‚takt pə'ten·chəl }

contact potential difference [ELEC] The potential difference that exists across the space between two electrically connected materials. Also known as contact electromotive force; contact potential; Volta effect. { 'kän‚takt pə'ten·chəl 'dif·rəns }

contact process [CHEM ENG] Catalytic manufacture of sulfuric acid from sulfur dioxide and oxygen. { 'kän‚takt ‚präs·əs }

contact ratio [DES ENG] The ratio of the length of the path of contact of two gears to the base pitch, equal to approximately the average number of pairs of teeth in contact. Also known as contact gear ratio. { 'kän‚takt ‚rā·shō }

contact rectifier *See* metallic rectifier. { 'kän‚takt 'rek·tə ‚fī·ər }

contact resistance [ELEC] The resistance in

ohms between the contacts of a relay, switch, or other device when the contacts are touching each other. { 'kän‚takt ri'zis·təns }

contact sensor [ENG] A device that senses mechanical contact and gives out signals when it does so. { 'kän‚takt 'sen·sər }

contact thermography [ENG] A method of measuring surface temperature in which a thin layer of luminescent material is spread on the surface of an object and is excited by ultraviolet radiation in a darkened room; the brightness of the coating indicates the surface temperature. { 'kän‚takt thər'mäg·rə·fē }

contact time [ENG] The length of time a substance is held in direct contact with a treating agent. { 'kän‚takt ‚tīm }

container [IND ENG] A portable compartment of standard, uniform size, used to hold cargo for air, sea, or ground transport. { kən'tā·nər }

container car [ENG] A railroad car designed specifically to hold containers. { kən'tā·nər ‚kär }

containerization [IND ENG] The practice of placing cargo in large containers such as truck trailers to facilitate loading on and off ships and railroad flat cars. { kən‚tā·nə·rə'zā·shən }

containment [ENG] An enclosed space or facility to contain and prevent the escape of hazardous material. { kən'tān·mənt }

continous-type furnace [MECH ENG] A furnace used for heat treatment of materials, with or without direct firing; pieces are loaded through one door, progress continuously through the furnace, and are discharged from another door. { kən‚tin·yə·wəs ‚tīp 'fər·nəs }

continuity [CIV ENG] Joining of structural members to each other, such as floors to beams, and beams to beams and to columns, so they bend together and strengthen each other when loaded. Also known as fixity. [ELEC] Continuous effective contact of all components of an electric circuit to give it high conductance by providing low resistance. { ‚känt·ən'ü·əd·ē }

continuity of state [THERMO] Property of a transition between two states of matter, as between gas and liquid, during which there are no abrupt changes in physical properties. { ‚känt·ən'ü·əd·ē əv 'stāt }

continuity test [ELEC] An electrical test used to determine the presence and location of a broken connection. { ‚känt·ən'ü·əd·ē ‚test }

continuous beam [CIV ENG] **1.** A beam resting upon several supports, which may be in the same horizontal plane. **2.** A beam having several spans in one straight line; generally has at least three supports. { kən‚tin·yə·wəs 'bēm }

continuous brake [MECH ENG] A train brake that operates on all cars but is controlled from a single point. { kən‚tin·yə·wəs 'brāk }

continuous bridge [CIV ENG] A fixed bridge supported at three or more points and capable of resisting bending and shearing forces at all sections throughout its length. { kən‚tin·yə·wəs 'brij }

continuous bucket elevator [MECH ENG] A

bucket elevator on an endless chain or belt. { kən¦tin·yə·wəs ¦bək·ət 'el·ə,vād·ər }

continuous bucket excavator [MECH ENG] A bucket excavator with a continuous bucket elevator mounted in front of the bowl. { kən¦tin·yə·wəs ¦bək·ət 'ek·skə,vād·ər }

continuous contact coking [CHEM ENG] A thermal conversion process using the mass-flow lift principle to give continuous coke circulation; oil-wetted particles of coke move downward into the reactor in which cracking, coking, and drying take place; pelleted coke, gas, gasoline, and gas oil are products of the process. { kən¦tin·yə·wəs ¦kän,takt 'kōk·iŋ }

continuous control [CONT SYS] Automatic control in which the controlled quantity is measured continuously and corrections are a continuous function of the deviation. { kən¦tin·yə·wəs kən'trōl }

continuous countercurrent leaching [CHEM ENG] Process of leaching by the use of continuous equipment in which the solid and liquid are both moved mechanically, and by the use of a series of leach tanks and the countercurrent flow of solvent through the tanks in reverse order to the flow of solid. { kən¦tin·yə·wəs ¦kaúnt·ər¦kər·ənt 'lēch·iŋ }

continuous distillation [CHEM ENG] Separation by boiling of a liquid mixture with different component boiling points; feed is introduced continuously, with continuous removal of overhead ǀvapors and high-boiling bottoms liquids. { kən¦tin·yə·wəs ,dis·tə'lā·shən }

continuous dryer [ENG] An apparatus in which drying is accomplished by passing wet material through without interruption. { kən¦tin·yə·wəs 'drī·ər }

continuous equilibrium vaporization See equilibrium flash vaporization. { kən¦tin·yə·wəs ,ē·kwə¦lib·rē·əm vā·pə·rə'zā·shən }

continuous-flow conveyor [MECH ENG] A totally enclosed, continuous-belt conveyor pulled transversely through a mass of granular, powdered or small-lump material fed from an overhead hopper. { kən¦tin·yə·wəs ¦flō kən'vā·ər }

continuous footing [CIV ENG] A footing that supports a wall. { kən¦tin·yə·wəs 'fúd·iŋ }

continuous industry [IND ENG] An industry in which raw material is subjected to successive operations, turning it into a finished product. { kən¦tin·yə·wəs 'in·dəs·trē }

continuous kiln [ENG] **1.** A long kiln through which ware travels on a moving device, such as a conveyor. **2.** A kiln through which the fire travels progressively. { kən¦tin·yə·wəs 'kiln }

continuous mixer [MECH ENG] A mixer in which materials are introduced, mixed, and discharged in a continuous flow. { kən¦tin·yə·wəs 'mik·sər }

continuous operation [ENG] A process that operates on a continuous flow (materials or time) basis, in contrast to batch, intermittent, or sequenced operations. { kən¦tin·yə·wəs äp·ə'rā·shən }

continuous production [IND ENG] Manufacture

of products, such as chemicals or paper, involving a sequence of processes performed by a series of machines receiving the materials through a closed channel of flow. { kən¦tin·yə·wəs prə'dək·shən }

continuous-rail frog [ENG] A metal fitting that holds continuous welded rail sections to railroad ties. { kən¦tin·yə·wəs ¦rāl 'fräg }

continuous rating [ENG] The rating of a component or equipment which defines the substantially constant conditions which can be tolerated for an indefinite time without significant reduction of service life. { kən¦tin·yə·wəs 'rād·iŋ }

continuous recorder [ENG] A recorder whose record sheet is a continuous strip or web rather than individual sheets. { kən¦tin·yə·wəs ri 'kórd·ər }

continuous system [CONT SYS] A system whose inputs and outputs are capable of changing at any instant of time. Also known as continuous-time signal system. { kən¦tin·yə·wəs 'sis·təm }

continuous task [IND ENG] A task that requires a continuously changing response by a worker to a continuously changing stimulus. { kən¦tin·yə·wəs 'task }

continuous-time signal system See continuous system. { kən¦tin·yə·wəs ¦tīm 'sig·nəl ,sis·təm }

continuous tube process [ENG] Plastics blow-molding process that uses a continuous extrusion of plastic tubing as feed to a series of blow molds as they clamp in sequence. { kən¦tin·yə·wəs ¦tüb ,präs·əs }

continuous-wave Doppler radar See continuous-wave radar. { kən¦tin·yə·wəs ¦wäv 'däp·lər ,rā ,där }

continuous-wave radar [ENG] A radar system in which a transmitter sends out a continuous flow of radio energy; the target reradiates a small fraction of this energy to a separate receiving antenna. Also known as continuous-wave Doppler radar. { kən¦tin·yə·wəs ¦wäv 'rā,där }

continuous work [IND ENG] A sustained and uninterrupted work activity, for example, exertion of a muscular force. { kən¦tin·yə·wəs 'wərk }

contouring temperature recorder [ENG] A device that records data from temperature sensors towed behind a ship and then plots the vertical distribution of isotherms on a continuous basis. { 'kän,túr·iŋ 'tem·prə·chər ri,kórd·ər }

contour machining [MECH ENG] Machining of an irregular surface. { 'kän,túr mə'shēn·iŋ }

contour turning [MECH ENG] Making a three-dimensional reproduction of the shape of a template by controlling the cutting tool with a follower that moves over the surface of a template. { 'kän,túr ,tərn·iŋ }

contracted code sonde See code-sending radiosonde. { kən'trak·təd ¦kōd ,sänd }

contraction [MECH] The action or process of becoming smaller or pressed together, as a gas on cooling. { kən'trak·shən }

contraction crack [ENG] A crack resulting from restriction of metal in a mold while contracting. { kən'trak·shən ,krak }

contraction joint [CIV ENG] A break designed in a structure to allow for drying and temperature shrinkage of concrete, brickwork, or masonry, thereby preventing the formation of cracks. { kən'trak·shən ˌjȯint }

contraflexure point [CIV ENG] The point in a structure where bending occurs in opposite directions. { ¦kän·trə'flek·shər ˌpȯint }

contrapropagating ultrasonic flowmeter [ENG] An instrument for determining the velocity of a fluid flow from the difference between the times required for high-frequency sound to travel between two transducers in opposite directions along a path having a component parallel to the flow. { ¦kän·trə'prä·pə,gād·iŋ 'əl·trə,sän·ik 'flō,mēd·ər }

contrarotating propellers [MECH ENG] A pair of propellers on concentric shafts, turning in opposite directions. { ¦kän·trə'rō,tād·iŋ prə'pel·ərz }

contrarotation [ENG] Rotation in the direction opposite to another rotation. { ¦kän·trə·rō'tā·shən }

control [CONT SYS] A means or device to direct and regulate a process or sequence of events. [ELECTR] An input element of a cryotron. { kən'trōl }

control accuracy [CONT SYS] The degree of correspondence between the ultimately controlled variable and the ideal value in a feedback control system. { kən'trōl ˌak·yə·rə·sē }

control agent [CHEM ENG] In process automatic-control work, material or energy within a process system of which the manipulated (controlled) variable is a condition or characteristic. { kən'trōl ˌā·jənt }

control board [ELEC] A panel at which one can make circuit changes, as in lighting a theater. [ENG] A panel in which meters and other indicating instruments display the condition of a system, and dials, switches, and other devices are used to modify circuits to control the system. Also known as control panel; panel board. { kən'trōl ˌbȯrd }

control chart [IND ENG] A statistical tool used to detect excessive process variability due to specific assignable causes that can be corrected. It serves to determine whether a process is in a state of statistical control, that is, the extent of variation of the output of the process does not exceed that which is expected based on the natural statistical variability of the process. { kən'trōl ˌchärt }

control circuit [ELEC] A circuit that controls some function of a machine, device, or piece of equipment. [ELECTR] The circuit that feeds the control winding of a magnetic amplifier. { kən'trōl ˌsər·kət }

control diagram See flow chart. { kən'trōl ˌdī·ə,gram }

control echo [ENG] In an ultrasonic inspection system, consistent reflection from a surface, such as a back reflection, which provides a reference signal. { kən'trōl ˌek·ō }

control element [CONT SYS] The portion of a feedback control system that acts on the process

or machine being controlled. { kən'trōl ˌel·ə·mənt }

control hierarchy See hierarchical control. { kən'trōl 'hī·ər,är·kē }

control joint [CIV ENG] An expansion joint in masonry to allow movement due to expansion and contraction. { kən'trōl ˌjȯint }

controllability [CONT SYS] Property of a system for which, given any initial state and any desired state, there exists a time interval and an input signal which brings the system from the initial state to the desired state during the time interval. { kən,trōl·ə'bil· əd·ē }

controllable-pitch propeller [MECH ENG] An aircraft or ship propeller in which the pitch of the blades can be changed while the propeller is in motion; five types used for aircraft are two-position, variable-pitch, constant-speed, feathering, and reversible-pitch. Abbreviated CP propeller. { kən'trōl·ə·bəl 'pich prə'pel·ər }

controlled avalanche device [ELECTR] A semiconductor device that has rigidly specified maximum and minimum avalanche voltage characteristics and is able to operate and absorb momentary power surges in this avalanche region indefinitely without damage. { kən¦trōld 'av·ə,lanch di'vīs }

controlled avalanche rectifier [ELECTR] A silicon rectifier in which carefully controlled, nondestructive internal avalanche breakdown across the entire junction area protects the junction surface, thereby eliminating local heating that would impair or destroy the reverse blocking ability of the rectifier. { kən¦trōld 'av·ə,lanch 'rek·tə,fī·ər }

controlled avalanche transit-time triode [ELECTR] A solid-state microwave device that uses a combination of IMPATT diode and npn bipolar transistor technologies; avalanche and drift zones are located between the base and collector regions. Abbreviated CATT. { kən¦trōld 'av·ə,lanch ¦tranz·ət ˌtīm 'trī,ōd }

controlled medium [CHEM ENG] In process automatic-control work, material within a process system in which a variable (for example, concentration) is controlled. { kən¦trōld 'mēd·ē·əm }

controlled parameter [ENG] In the formulation of an optimization problem, one of the parameters whose values determine the value of the criterion parameter. { kən¦trōld pə'ram·əd·ər }

controlled variable [CONT SYS] In process automatic-control work, that quantity or condition of a controlled system that is directly measured or controlled. { kən¦trōld 'ver·ē·ə·bəl }

controller See automatic controller. { kən'trōl·ər }

controller-structure interaction [CONT SYS] Feedback of an active control algorithm in the process of model reduction; this occurs through observation spillover and control spillover. { kən'trōl·ər ,strək·chər in·tər'ak·shən }

control limits [ELECTR] In radar evaluation, upper and lower control limits are established at those performance figures within which it is expected that 95% of quality-control samples will fall when the radar is performing normally.

[IND ENG] In statistical quality control, the limits of acceptability placed on control charts; parts outside the limits are defective. { kən'trōl ‚lim·əts }

controlling magnet [ENG] An auxiliary magnet used with a galvanometer to cancel the effect of the earth's magnetic field. { kən'trōl· iŋ ‚mag·nət }

control panel [ENG] See control board; panel. { kən'trōl ‚pan·əl }

control room [ENG] A room from which space flights are directed. { kən'trōl ‚rüm }

control signal [CONT SYS] The signal applied to the device that makes corrective changes in a controlled process or machine. { kən'trōl ‚sig·nəl }

control spillover [CONT SYS] The excitation by an active control system of modes of motion that have been omitted from the control algorithm in the process of model reduction. { kən'trōl 'spil‚ō·vər }

control spring [DES ENG] A spring designed so that its torque cancels that of the instrument of which it is a part, for all deflections of the pointer. { kən'trōl ‚spriŋ }

control system [ENG] A system in which one or more outputs are forced to change in a desired manner as time progresses. { kən'trōl ‚sis·təm }

control-system feedback [CONT SYS] A signal obtained by comparing the output of a control system with the input, which is used to diminish the difference between them. { kən'trōl ‚sis·təm 'fēd‚bak }

control track [ENG ACOUS] A supplementary sound track, usually containing tone signals that control the reproduction of the sound track, such as by changing feed levels to loudspeakers in a theater to achieve stereophonic effects. { kən'trōl ‚trak }

control valve [ENG] A valve which controls pressure, volume, or flow direction in a fluid transmission system. { kən'trōl ‚valv }

control variable [CONT SYS] One of the input variables of a control system, such as motor torque or the opening of a valve, which can be varied directly by the operator to maximize some measure of performance of the system. { kən'trōl ‚ver·ē·ə·bəl }

convection coefficient See film coefficient. { kən'vek·shən ‚kō·i'fish·ənt }

convection cooling [ENG] Heat transfer by natural, upward flow of hot air from the device being cooled. { kən'vek·shən ‚kül·iŋ }

convection current [ELECTR] The time rate at which the electric charges of an electron stream are transported through a given surface. { kən'vek·shən ‚kər·ənt }

convection oven [ENG] An oven containing a fan that continuously circulates hot air around the food being prepared. { kən'vek·shən ‚əv·ən }

convection section [ENG] That portion of the furnace in which tubes receive heat from the flue gases by convection. { kən'vek·shən ‚sek·shən }

convective current See convection current. { kən'vek·div ‚kər·ənt }

convector [ENG] A heat-emitting unit for the heating of room air; it has a heating element surrounded by a cabinet-type enclosure with openings below and above for entrance and egress of air. { kən'vek·tər }

convectron [ENG] An instrument for indicating deviation from the vertical which is based on the principle that the convection from a heated wire depends strongly on its inclination; it consists of a Y-shaped tube, each of whose arms contains a wire forming part of a bridge circuit. { kən'vek‚trän }

conventional current [ELEC] The concept of current as the transfer of positive charge, so that its direction of flow is opposite to that of electrons which are negatively charged. { kən'ven·chən·əl 'kər·ənt }

convergent die [ENG] A die having internal channels which converge. { kən'vər·jənt ‚dī }

convergent-divergent nozzle [DES ENG] A nozzle in which supersonic velocities are attained; has a divergent portion downstream of the contracting section. Also known as supersonic nozzle. { kən‚vər·jənt də‚vər·jənt 'näz·əl }

conversion [CHEM ENG] The chemical change from reactants to products in an industrial chemical process. Also known as chemical conversion. { kən'vər·zhən }

converted water See product water. { kən'vərd·əd 'wòd·ər }

conveyor [MECH ENG] Any materials-handling machine designed to move individual articles such as solids or free-flowing bulk materials over a horizontal, inclined, declined, or vertical path of travel with continuous motion. { kən'vā·ər }

conveyor belt balance [ENG] A balance used for weighing unpackaged, loose, continuously transported material on a conveyor belt by weighing the load being moved and measuring the belt speed. { kən'vā·ər ‚belt ‚bal·əns }

cooled-tube pyrometer [ENG] A thermometer for high-temperature flowing gases that uses a liquid-cooled tube inserted in the flowing gas; gas temperature is deduced from the law of convective heat transfer to the outside of the tube and from measurement of the mass flow rate and temperature rise of the cooling liquid. { 'küld ‚tüb pī'räm·əd·ər }

cooler nail [DES ENG] A thin, cement-coated wire nail. { 'kül·ər ‚nāl }

cooling channel [ENG] A channel in the body of mold through which a cooling liquid is circulated. { 'kül·iŋ ‚chan·əl }

cooling coil [MECH ENG] A coiled arrangement of pipe or tubing for the transfer of heat between two fluids. { 'kül·iŋ ‚kòil }

cooling correction [THERMO] A correction that must be employed in calorimetry to allow for heat transfer between a body and its surroundings. Also known as radiation correction. { 'kül·iŋ kə'rek·shən }

cooling curve |THERMO| A curve obtained by plotting time against temperature for a solid-liquid mixture cooling under constant conditions. { 'kül·iŋ ˌkərv }

cooling degree day |MECH ENG| A unit for estimating the energy needed for cooling a building; one unit is given for each degree Fahrenheit that the daily mean temperature exceeds 75°F (24°C). { 'kül·iŋ di'grē ˌdā }

cooling fin |MECH ENG| The extended element of a heat-transfer device that effectively increases the surface area. { 'kül·iŋ ˌfin }

cooling fixture |ENG| A wooden or metal block used to hold the shape or dimensional accuracy of a molding until it cools enough to retain its shape. { 'kül·iŋ ˌfiks·chər }

cooling load |MECH ENG| The total amount of heat energy that must be removed from a system by a cooling mechanism in a unit time, equal to the rate at which heat is generated by people, machinery, and processes, plus the net flow of heat into the system not associated with the cooling machinery. { 'kül·iŋ ˌlōd }

cooling method |THERMO| A method of determining the specific heat of a liquid in which the times taken by the liquid and an equal volume of water in an identical vessel to cool through the same range of temperature are compared. { 'kül·iŋ ˌmeth·əd }

cooling pond |CHEM ENG| Outdoor depression into which hot process water is pumped for purposes of cooling by evaporation, convection, and radiation. { 'kül·iŋ ˌpänd }

cooling power |MECH ENG| A parameter devised to measure the air's cooling effect upon a human body; it is determined by the amount of heat required by a device to maintain the device at a constant temperature (usually 34°C); the entire system should be made to correspond, as closely as possible, to the external heat exchange mechanism of the human body. { 'kül·iŋ ˌpau̇·ər }

cooling-power anemometer |ENG| Any anemometer operating on the principle that the heat transfer to air from an object at an elevated temperature is a function of airspeed. { 'kül·iŋ ˌpau̇r an·ə'mäm·əd·ər }

cooling process |ENG| Physical operation in which heat is removed from process fluids or solids; may be by evaporation of liquids, expansion of gases, radiation or heat exchange to a cooler fluid stream, and so on. { 'kül·iŋ ˌpräs·əs }

cooling range |MECH ENG| The difference in temperature between the hot water entering and the cold water leaving a cooling tower. { 'kül·iŋ ˌrānj }

cooling stress |MECH| Stress resulting from uneven contraction during cooling of metals and ceramics due to uneven temperature distribution. { 'kül·iŋ ˌstres }

cooling tower |ENG| A towerlike device in which atmospheric air circulates and cools warm water, generally by direct contact (evaporation). { 'kül·iŋ ˌtau̇·ər }

coolometer |ENG| An instrument which measures the cooling power of the air, consisting of a metal cylinder electrically heated to maintain a constant temperature; the electrical heating power required is taken as a measure of the air's cooling power. { kü'läm·əd·ər }

cooperative system |ENG| A missile guidance system that requires transmission of information from a remote ground station to a missile in flight, processing of the information by the missile-borne equipment, and retransmission of the processed data to the originating or other remote ground stations, as in azusa and dovap. { kō'äp·rəd·iv ˌsis·təm }

coordinated-axis control |CONT SYS| Robotic control in which the robot axes reach their end points simultaneously, thus giving the robot's motion a smooth appearance. { kō'ȯrd·ən,ăd·əd ¦ak·səs kən,trōl }

coordinating holes |DES ENG| Holes in two parts of an assembly which form a single continuous hole when the parts are joined. { kō'ȯrd·ən,ăd·iŋ ˌhōlz }

cope chisel |DES ENG| A chisel used to cut grooves in metal. { 'kōp ˌchiz·əl }

coping |BUILD| A covering course on a wall. |MECH ENG| Shaping stone or other nonmetallic substance with a grinding wheel. { 'kōp·iŋ }

coping saw |DES ENG| A type of handsaw that has a narrow blade, usually about 1/8 inch (3 millimeters) wide, held taut by a U-shaped frame equipped with a handle; used for shaping and cutout work. { 'kōp·iŋ ˌsȯ }

coplanar forces |MECH| Forces that act in a single plane; thus the forces are parallel to the plane and their points of application are in the plane. { kō'plān·ər ˌfȯrs·əz }

copper dish gum |CHEM ENG| The milligrams of gum found in 100 milliliters of gasoline when evaporated under controlled conditions in a polished copper dish. { 'käp·ər ˌdish 'gəm }

copper loss |ELEC| Power loss in a winding due to current flow through the resistance of the copper conductors. Also known as I^2R loss. { 'käp·ər ˌlȯs }

copper-strip corrosion |ENG| A qualitative method of determining the corrosivity of a petroleum product by observing its effect on a strip of polished copper suspended or placed in the product. Also known as copper strip test. { 'käp·ər ˌstrip ki'rō·zhən }

copper-strip test See copper-strip corrosion. { 'käp·ər 'strip ˌtest }

copper sweetening |CHEM ENG| Those refining processes using cupric chloride to oxidize mercaptans in petroleum. { 'käp·ər ˌswēt·ən·iŋ }

corbinotron |ENG| The combination of a corbino disk, made of high-mobility semiconductor material, and a coil arranged to produce a magnetic field perpendicular to the disk. { 'kȯr'bē·na,trän }

cordage |ENG| Number of cords of lumber per given area. { 'kȯrd·ij }

cord foot |ENG| A stack of wood measuring 16

cubic feet (approximately 0.45307 cubic meter). { 'kòrd ¦füt }

cord tire |DES ENG| A pneumatic tire made with cords running parallel to the tread. { 'kòrd ‚tīr }

core |ELECTR| See magnetic core. |ENG| The inner material of a wall, column, veneered door, or similar structure. { kòr }

core array |ELECTR| A rectangular grid arrangement of magnetic cores. { 'kòr ə'rā }

core bank |ELECTR| A stack of core arrays and associated electronics, the stack containing a specific number of core arrays. { 'kòr ‚baŋk }

core barrel |DES ENG| A hollow cylinder attached to a specially designed bit; used to obtain a continuous section of the rocks penetrated in drilling. { 'kòr ‚bar·əl }

core bit |DES ENG| The hollow, cylindrical cutting part of a core drill. { 'kòr ‚bit }

core catcher See split-ring core lifter. { 'kòr ‚kach·ər }

core cutterhead |ENG| The cutting element in a core barrel unit. { 'kòr 'kəd·ər‚hed }

core drill |MECH ENG| A mechanism designed to rotate and to cause an annular-shaped rock-cutting bit to penetrate rock formations, produce cylindrical cores of the formations penetrated, and lift such cores to the surface, where they may be collected and examined. { 'kòr ‚dril }

core flow |ENG| A pattern of powder flow occurring in hoppers that is characterized by a central core of flowing powder with the powder near the hopper walls remaining stationary. { 'kòr ‚flō }

core gripper See split-ring core lifter. { 'kòr ‚grip·ər }

coreless-type induction heater |ENG| A device in which a charge is heated directly by induction, with no magnetic core material linking the charge. Also known as coreless-type induction furnace. { 'kòr·ləs ‚tīp in'dək·shən ‚hēd·ər }

core lifter See split-ring core lifter. { 'kòr ‚lif·tər }

core logic |ELECTR| Logic performed in ferrite cores that serve as inputs to diode and transistor circuits. { 'kòr ‚läj·ik }

corer |ENG| An instrument used to obtain cylindrical samples of geological materials or ocean sediments. { 'kòr·ər }

core stack |ELECTR| A number of core arrays, next to one another and treated as a unit. { 'kòr ‚stak }

core wall See cutoff wall. { 'kòr ‚wòl }

coring reel See sand reel. { 'kòr·iŋ ‚rēl }

Coriolis acceleration |MECH| **1.** An acceleration which, when added to the acceleration of an object relative to a rotating coordinate system and to its centripetal acceleration, gives the acceleration of the object relative to a fixed coordinate system. **2.** A vector which is equal in magnitude and opposite in direction to that of the first definition. { kòr·ē'ō·ləs ik‚sel·ə'rā·shən }

Coriolis deflection See Coriolis effect. { kòr·ē'ō·ləs di'flek·shən }

Coriolis effect |MECH| Also known as Coriolis deflection. **1.** The deflection relative to the earth's surface of any object moving above the

earth, caused by the Coriolis force; an object moving horizontally is deflected to the right in the Northern Hemisphere, to the left in the Southern. **2.** The effect of the Coriolis force in any rotating system. { kòr·ē'ō·ləs i'fekt }

Coriolis force |MECH| A velocity-dependent pseudoforce in a reference frame which is rotating with respect to an inertial reference frame; it is equal and opposite to the product of the mass of the particle on which the force acts and its Coriolis acceleration. { kòr·ē'ō·ləs ‚fòrs }

Coriolis-type mass flowmeter |ENG| An instrument which determines mass flow rate from the torque on a ribbed disk that is rotated at constant speed when fluid is made to enter at the center of the disk and is accelerated radially. { kòr·ē'ō·ləs ‚tīp ¦mas 'flō‚med·ər }

Corliss valve |MECH ENG| An oscillating type of valve gear with a trip mechanism for the admission and exhaust of steam to and from an engine cylinder. { 'kòr·ləs ‚valv }

corner bead |BUILD| **1.** Any vertical molding used to protect the external angle of the intersecting surfaces. **2.** A strip of formed galvanized iron, sometimes combined with a strip of metal lath, placed on corners to reinforce them before plastering. { 'kòr·nər ‚bēd }

corner chisel |DES ENG| A chisel with two cutting edges at right angles. { 'kòr·nər ‚chiz·əl }

corner effect |ELECTR| The departure of the frequency-response curve of a band-pass filter from a perfect rectangular shape, so that the corners of the rectangle are rounded. |ENG| In ultrasonic testing, reflection of an ultrasonic beam directed perpendicular to the intersection of two surfaces 90° apart. { 'kòr·nər i'fekt }

corner frequency See break frequency. { 'kòr·nər ‚frē·kwən·sē }

corner head |BUILD| A metal molding that is built into plaster in corners to prevent plaster from accidentally breaking off. { 'kòr·nər ‚hed }

cornering tool |DES ENG| A cutting tool with a curved edge, used to round off sharp corners. { 'kòr·nər·iŋ ‚tül }

cornerite |BUILD| A corner reinforcement for interior plastering. { 'kòr·nə‚rīt }

corner joint |ENG| An L-shaped joint formed by two members positioned perpendicular to each other. { 'kòr·nər ‚jòint }

cornerload test |ENG| A test to determine whether the display of an analytical balance is affected by the load distribution on the weighing pan. { 'kòr·nər‚lōd ‚test }

cornerstone |BUILD| An inscribed stone laid at the corner of a building, usually at a ceremony. { 'kòr·nər‚stōn }

cornice brake |MECH ENG| A machine used to bend sheet metal into different forms. { 'kòr·nəs ‚brāk }

corona See corona discharge. { kə'rō·nə }

corona current |ELEC| The current of electricity equivalent to the rate of charge transferred to the air from an object experiencing corona discharge. { kə'rō·nə ¦kər·ənt }

corona discharge |ELEC| A discharge of electricity appearing as a bluish-purple glow on the surface of and adjacent to a conductor when the voltage gradient exceeds a certain critical value; due to ionization of the surrounding air by the high voltage. Also known as aurora; corona; electric corona. { kə'rō·nə 'dis,chärj }

correction chamber |ENG| A closable cavity in a weight on an analytical balance; holds material to adjust weight to nominal value. { kə'rek·shən ˌchām·bər }

correction time |CONT SYS| The time required for the controlled variable to reach and stay within a predetermined band about the control point following any change of the independent variable or operating condition in a control system. Also known as settling time. { kə'rek·shən ˌtīm }

corrective action |CONT SYS| The act of varying the manipulated process variable by the controlling means in order to modify overall process operating conditions. { kə'rek·tiv 'ak·shən }

corrective maintenance |ENG| A procedure of repairing components or equipment as necessary either by on-site repair or by replacing individual elements in order to keep the system in proper operating condition. { kə'rek·tiv mānt·ən·ans }

corrective operation See remedial operation. { kə'rek·tiv ˌrē·ə'rā·shən }

corrector |ENG| A magnet, piece of soft iron, or device used in the adjustment or compensation of a magnetic compass. { kə'rek·tər }

correlated orientation tracking and range See cotar. { 'kär·ə,lād·əd ˌȯr·ē·ən'tā·shən 'trak·iŋ ən 'rānj }

correlation detection |ENG| A method of detection of aircraft or space vehicles in which a signal is compared, point to point, with an internally generated reference. Also known as cross-correlation detection. { ˌkär·ə'lā·shən di'tek·shən }

correlation direction finder |ENG| Satellite station separated from a radar to receive jamming signals; by correlating the signals received from several such stations, range and azimuth of many jammers may be obtained. { ˌkär·ə'lā·shən də'rek·shən ˌfīnd·ər }

correlation tracking and triangulation See cotat. { ˌkär·ə'lā·shən 'trak·iŋ ən trī,aŋ·gyə'lā·shən }

correlation tracking system |ENG| A trajectory-measuring system utilizing correlation techniques where signals derived from the same source are correlated to derive the phase difference between the signals. { ˌkär·ə'lā·shən 'trak·iŋ ˌsis·təm }

correlation ultrasonic flowmeter |ENG| An instrument for determining the velocity of a fluid flow from the time required for discontinuities in the fluid stream to pass between two pairs of transducers that generate and detect high-frequency sound. { ˌkär·ə'lā·shən əl·trə'sän·ik 'flō,mēd·ər }

correlative kinesiology |IND ENG| A field that involves determination of the quantitative relationship between the electrical potential generated by muscular activity and the resultant movement; used in developing a design for a workplace that minimizes fatigue. { kə¦rel·əd·iv kə,nēz·ē'äl·ə·jē }

corrosion coupon See coupon. { kə'rō·zhən ˌkü,pän }

corrosion number See acid number. { kə¦rō·zhən ˌnəm·bər }

corrosive product |CHEM ENG| In petroleum refining, a product that contains a quantity of corrosion-inducing compounds in excess of the limits specified for products classified as sweet. { kə'rō·siv 'präd·əkt }

corrugated bar |DES ENG| Steel bar with transverse ridges; used in reinforced concrete. { 'kär·ə,gād·əd 'bär }

corrugated fastener |DES ENG| A thin corrugated strip of steel that can be hammered into a wood joint to fasten it. { 'kär·ə,gād·əd 'fas·nər }

corrugating |DES ENG| Forming straight, parallel, alternate ridges and grooves in sheet metal, cardboard, or other material. { 'kär·ə,gād·iŋ }

cosmic-ray telescope |ENG| Any device for detecting and determining the directions of either cosmic-ray primary protons and heavier-element nuclei, or the products produced when these particles interact with the atmosphere. { 'käz·mik ˌrā 'tel·ə,skōp }

cosolvent |CHEM ENG| During chemical processing, a second solvent added to the original solvent, generally in small concentrations, to form a mixture that has greatly enhanced solvent powers due to synergism. { kō'säl·vənt }

cost accounting |IND ENG| The branch of accounting in which one records, analyzes, and summarizes costs of material, labor, and burden, and compares these actual costs with predetermined budgets and standards. { 'kȯst ə'kaunt·iŋ }

cost analysis |IND ENG| Analysis of the factors contributing to the costs of operating a business and of the costs which will result from alternative procedures, and of their effects on profits. { 'kȯst ə'nal·ə·səs }

cost control See industrial cost control. { 'kȯst kən'trōl }

cost engineering |IND ENG| A branch of industrial engineering concerned with cost estimation, cost control, business planning and management, profitability analysis, and project management, planning, and scheduling. { 'kȯst ˌen·jə,nir·iŋ }

cost function |SYS ENG| In decision theory, a loss function which does not depend upon the decision rule. { 'kȯst ˌfəŋk·shən }

cost-plus contract |ENG| A contract under which a contractor furnishes all material, construction equipment, and labor at actual cost, plus an agreed-upon fee for his services. { ¦kȯst 'pləs ˌkän,trakt }

cotar |ENG| A passive system used for tracking a vehicle in space by determining the line of direction between a remote ground-based receiving antenna and a telemetering transmitter in the missile, using phase-comparison techniques. Derived from correlated orientation tracking and range. { 'kō,tär }

cotat |ENG| A trajectory-measuring system using several antenna base lines, each separated by large distances, to measure direction cosines to an object; then the object's space position is computed by triangulation. Derived from correlation tracking and triangulation. { 'kō,tat }

cotter |DES ENG| A tapered piece that can be driven in a tapered hole to hold together an assembly of machine or structural parts. { 'käd·ər }

cottered joint |MECH ENG| A joint in which a cotter, usually a flat bar tapered on one side to ensure a tight fit, transmits power by shear on an area at right angles to its length. { 'käd·ərd ,jóint }

cotter pin |DES ENG| A split pin, inserted into a hole, to hold a nut or cotter securely to a bolt or shaft, or to hold a pair of hinge plates together. { 'käd·ər ,pin }

Cotton balance |ENG| A device which employs a current-carrying conductor of special shape to determine the strength of a magnetic field. { 'kät·ən 'bal·əns }

Cottrell precipitator |ENG| A machine for removing dusts and mists from gases, in which the gas passes through a grounded pipe with a fine axial wire at a high negative voltage, and particles are ionized by the corona discharge of the wire and migrate to the pipe. { 'kä·trəl prə'sip·ə,tād·ər }

Couette viscometer |ENG| A viscometer in which the liquid whose viscosity is to be measured fills the space between two vertical coaxial cylinders, the inner one suspended by a torsion wire; the outer cylinder is rotated at a constant rate, and the resulting torque on the inner cylinder is measured by the twist of the wire. Also known as rotational viscometer. { kü'et vis 'käm·əd·ər }

coul See coulomb.

coulisse |ENG| A piece of wood that has a groove cut in it to enable another piece of wood to slide in it. Also known as cullis. { kü'lēs }

coulomb |ELEC| A unit of electric charge, defined as the amount of electric charge that crosses a surface in 1 second when a steady current of 1 absolute ampere is flowing across the surface; this is the absolute coulomb and has been the legal standard of quantity of electricity since 1950; the previous standard was the international coulomb, equal to 0.999835 absolute coulomb. Abbreviated coul. Symbolized C. { 'kü,läm }

Coulomb attraction |ELEC| The electrostatic force of attraction exerted by one charged particle on another charged particle of opposite sign. Also known as electrostatic attraction. { 'kü ,läm ə'trak·shən }

Coulomb field |ELEC| The electric field created by a stationary charged particle. { 'kü,läm ,fēld }

Coulomb force |ELEC| The electrostatic force of attraction or repulsion exerted by one charged particle on another, in accordance with Coulomb's law. { 'kü,läm ,fórs }

Coulomb friction |MECH| Friction occurring between dry surfaces. { 'kü,läm ,frik·shən }

Coulomb interactions |ELEC| Interactions of charged particles associated with the Coulomb forces they exert on one another. Also known as electrostatic interactions. { 'kü,läm in·tər'ak·shənz }

coulombmeter |ENG| An instrument that measures quantity of electricity in coulombs by integrating a stored charge in a circuit which has very high input impedance. { 'kü,läm,mēd·ər }

Coulomb potential |ELEC| A scalar point function equal to the work per unit charge done against the Coulomb force in transferring a particle bearing an infinitesimal positive charge from infinity to a point in the field of a specific charge distribution. { kü'läm pə'ten·chəl }

Coulomb repulsion |ELEC| The electrostatic force of repulsion exerted by one charged particle on another charged particle of the same sign. Also known as electrostatic repulsion. { kü'läm ri'pəl·shən }

Coulomb's law |ELEC| The law that the attraction or repulsion between two electric charges acts along the line between them, is proportional to the product of their magnitudes, and is inversely proportional to the square of the distance between them. Also known as law of electrostatic attraction. { 'kü'lämz ,lö }

Coulomb's theorem |ELEC| The proposition that the intensity of an electric field near the surface of a conductor is equal to the surface charge density on the nearby conductor surface divided by the absolute permittivity of the surrounding medium. { 'kü,lämz ,thir·əm }

count |DES ENG| The number of openings per linear inch in a wire cloth. { kaunt }

countdown |ENG| A step-by-step process that culminates in a climatic event, each step being performed in accordance with a schedule marked by a count in inverse numerical order. { 'kaunt,daun }

counter |ELECTR| See scaler. |ENG| A complete instrument for detecting, totalizing, and indicating a sequence of events. { 'kaunt·ər }

counterbalance See counterweight. { ¦kaunt·ər¦bal·əns }

counterbalanced truck |MECH ENG| An industrial truck configured so that all of its load during a normal transporting operation is external to the polygon formed by the points where the wheels contact the surface. { ¦kaun·tər¦bal·ənst 'trək }

counterbalance system See two-step grooving system. { ¦kaunt·ər¦bal·əns ,sis·təm }

counterblow hammer |MECH ENG| A forging hammer in which the ram and anvil are driven toward each other by compressed air or steam. { 'kaunt·ər,blō ,ham·ər }

131

counterbore |DES ENG| A flat-bottom enlargement of the mouth of a cylindrical bore to enlarge a borehole and give it a flat bottom. |ENG| To enlarge a borehole by means of a counterbore. { 'kaůnt·ər,bór }

counter circuit *See* counting circuit. { 'kaůnt·ər ,sər·kət }

countercurrent distribution |CHEM ENG| A profile of a compound's concentration in different ratios of two immiscible liquids. { 'kaůnt· ər,kər·ənt dis·trə'byü·shən }

countercurrent extraction |CHEM ENG| A liquid-liquid extraction process in which the solvent and the process stream in contact with each other flow in opposite directions. Also known as countercurrent separation. { 'kaůnt·ər,kər· ənt ,ek'strak·shən }

countercurrent flow |MECH ENG| A sensible heat-transfer system in which the two fluids flow in opposite directions. { 'kaůnt·ər,kər·ənt 'flō }

countercurrent leaching |CHEM ENG| A process utilizing a series of leach tanks and countercurrent flow of solvent through them in reverse order to the flow of solid. { 'kaůnt·ər,kər·ənt 'lēch·iŋ }

countercurrent separation *See* countercurrent extraction. { 'kaůnt·ər,kər·ənt ,sep·ə'rā·shən }

countercurrent spray dryer |ENG| A dryer in which drying gases flow in a direction opposite to that of the spray. { 'kaůnt·ər,kər·ənt 'sprā ,drī·ər }

counterfloor *See* subfloor. { 'kaůn·tər,flór }

counterflow |ENG| Fluid flow in opposite directions in adjacent parts of an apparatus, as in a heat exchanger. { 'kaůnt·ər,flō }

counterfort |CIV ENG| A strengthening pier perpendicular and bonded to a retaining wall. { 'kaůnt·ər,fórt }

counterfort wall |CIV ENG| A type of retaining wall that resembles a cantilever wall but has braces at the back; the toe slab is a cantilever and the main steel is placed horizontally. { 'kaůnt· ər,fórt ,wól }

counter/frequency meter |ENG| An instrument that contains a frequency standard and can be used to measure the number of events or the number of cycles of a periodic quantity that occurs in a specified time, or the time between two events. { 'kaůnt·ər 'frē·kwən·sē ,mēd·ər }

counterlath |BUILD| **1.** A strip placed between two rafters to support crosswise laths. **2.** A lath placed between a timber and a sheet lath. **3.** A lath nailed at a more or less random spacing between two precisely spaced laths. **4.** A lath put on one side of a partition after the other side has been finished. { 'kaůnt·ər,lath }

counterpoise |ELEC| A system of wires or other conductors that is elevated above and insulated from the ground to form a lower system of conductors for an antenna. Also known as antenna counterpoise. |MECH ENG| *See* counterweight. { 'kaůnt·ər,póiz }

counterpoise method *See* substitution weighing. { 'kaůn·tər,póiz ,meth·əd }

countershaft |MECH ENG| A secondary shaft that is driven by a main shaft and from which power is supplied to a machine part. { 'kaůnt·ər,shaft }

countersink |DES ENG| The tapered and relieved cutting portion in a twist drill, situated between the pilot drill and the body. { 'kaůnt·ər,siŋk }

countersinking |MECH ENG| Drilling operation to form a flaring depression around the rim of a hole. { 'kaůnt·ər,siŋk·iŋ }

countersunk bolt |DES ENG| A bolt that has a circular head, a flat top, and a conical bearing surface tapering in from the top; in place, the head is flush-mounted. { ,kaůn·tər,səŋk 'bōlt }

counterweight |MECH ENG| **1.** A device which counterbalances the original load in elevators and skip and mine hoists, going up when the load goes down, so that the engine must only drive against the unbalanced load and overcome friction. **2.** Any weight placed on a mechanism which is out of balance so as to maintain static equilibrium. Also known as counterbalance; counterpoise. { 'kaůnt·ər,wāt }

counting circuit |ELECTR| A circuit that counts pulses by frequency-dividing techniques, by charging a capacitor in such a way as to produce a voltage proportional to the pulse count, or by other means. Also known as counter circuit. { 'kaůnt·iŋ ,sər·kət }

couplant |ENG| A substance such as water, oil, grease, or paste used to avoid the retarding of sound transmission by air between the transducer and the test piece during ultrasonic examination. { 'kəp·lənt }

couple |ELEC| To connect two circuits so signals are transferred from one to the other. |ELECTR| Two metals placed in contact, as in a thermocouple. |ENG| To connect with a coupling, such as two belts or two pipes. |MECH| A system of two parallel forces of equal magnitude and opposite sense. { 'kəp·əl }

coupled circuits |ELEC| Two or more electric circuits so arranged that energy can transfer electrically or magnetically from one to another. { 'kəp·əld 'sər·kəts }

coupled engine |MECH ENG| A locomotive engine having the driving wheels connected by a rod. { 'kəp·əld 'en·jən }

coupled oscillators |MECH| A set of particles subject to elastic restoring forces and also to elastic interactions with each other. { 'kəp·əld 'äs·ə,läd·ərz }

coupler |ELEC| A component used to transfer energy from one circuit to another. |ENG| A device that connects two railroad cars. { 'kəp· lər }

coupling |ELEC| **1.** A mutual relation between two circuits that permits energy transfer from one to another, through a wire, resistor, transformer, capacitor, or other device. **2.** A hardware device used to make a temporary connection between two wires. |ENG| **1.** Any device that serves to connect the ends of adjacent parts, as railroad cars. **2.** A metal collar with internal threads used to connect two sections of threaded pipe. |MECH ENG| The mechanical fastening that connects shafts together for power transmission. Also known as shaft coupling. { 'kəp·liŋ }

coupling capacitor |ELECTR| A capacitor used to block the flow of direct current while allowing alternating or signal current to pass; widely used for joining two circuits or stages. Also known as blocking capacitor; stopping capacitor. { 'kəp·liŋ kə'pas·əd·ər }

coupon |CHEM ENG| Polished metal strip of specified size and weight used to detect the corrosive action of liquid or gas products or to test the efficiency of corrosion-inhibitor additives. Also known as corrosion coupon. { 'kü,pän }

course |CIV ENG| A row of stone, block, or brick of uniform height. { kórs }

coursed rubble |CIV ENG| Masonry in which rough stones are fitted into approximately level courses. { 'kórsd 'rəb·əl }

course programmer |CONT SYS| An item which initiates and processes signals in a manner to establish a vehicle in which it is installed along one or more projected courses. { 'kórs 'prō ,gram·ər }

coursing joint |CIV ENG| A mortar joint connecting two courses of brick or pebble. { 'kórs·iŋ ,jóint }

covering power |ENG| The degree to which a coating obscures the underlying material. { 'kəv·riŋ ,paü·ər }

cover plate |ENG| A pane of glass in a welding helmet or goggles which protects the colored lens excluding harmful light rays from damage by weld spatter. { 'kəv·ər ,plāt }

cowling |ENG| A metal cover that houses an engine. { 'kaü·liŋ }

coyote hole *See* gopher hole. { 'kī,ōd·ē ,hōl }

CPM *See* critical path method.

CP propeller *See* controllable-pitch propeller. { ¦sē¦pē prə'pel·ər }

CR *See* catalytic reforming.

crack |ENG| To open something slightly, for instance, a valve. { krak }

cracked residue |CHEM ENG| The residue of fuel resulting from decomposition of hydrocarbons during thermal or catalytic cracking. { 'krakt 'rez·ə,dü }

cracking |CHEM ENG| A process that is used to reduce the molecular weight of hydrocarbons by breaking the molecular bonds by various thermal, catalytic, or hydrocracking methods. |ENG| Presence of relatively large cracks extending into the interior of a structure, usually produced by overstressing the structural material. { 'krak·iŋ }

cracking coil |CHEM ENG| A coil used for cracking heavy petroleum products. { 'krak·iŋ ,kóil }

cracking still |CHEM ENG| The furnace, reaction chamber, and fractionator for thermal conversion of heavier charging stock to gasoline. { 'krak·iŋ ,stil }

cradle |CIV ENG| A structure that moves along an inclined track on a riverbank and is equipped

with a horizontal deck carrying tracks for transferring railroad cars to and from boats at different water elevations. |ENG| A framework or other resting place for supporting or restraining objects. { 'krād·əl }

cramp |DES ENG| A metal plate with bent ends used to hold blocks together. { kramp }

crampon |DES ENG| A device for holding heavy objects such as rock or lumber to be lifted by a crane or hoist; shaped like scissors, with points bent inward for grasping the load. Also spelled crampoon. { 'kram,pän }

crampoon *See* crampon. { 'kram,pün }

crane |MECH ENG| A hoisting machine with a power-operated inclined or horizontal boom and lifting tackle for moving loads vertically and horizontally. { krān }

crane hoist |MECH ENG| A mobile construction machine built principally for lifting loads by means of cables and consisting of an undercarriage on which the unit moves, a cab or house which envelops the main frame and contains the power units and controls, and a movable boom over which the cables run. { 'krān ,hóist }

crane hook |DES ENG| A hoisting fixture designed to engage a ring or link of a lifting chain, or the pin of a shackle or cable socket. { 'krān ,húk }

crane truck |MECH ENG| A crane with a jiblike boom mounted on a truck. Also known as yard crane. { 'krān ,trək }

crank |MECH ENG| A link in a mechanical linkage or mechanism that can turn about a center of rotation. { kraŋk }

crank angle |MECH ENG| 1. The angle between a crank and some reference direction. 2. Specifically, the angle between the crank of a slider crank mechanism and a line from crankshaft to the piston. { 'kraŋk ,aŋ·gəl }

crank arm |MECH ENG| The arm of a crankshaft attached to a connecting rod and piston. { 'kraŋk ,ärm }

crank axle |MECH ENG| 1. An axle containing a crank. 2. An axle bent at both ends so that it can accommodate a large body with large wheels. { 'kraŋk ,ak·səl }

crankcase |MECH ENG| The housing for the crankshaft of an engine, where, in the case of an automobile, oil from hot engine parts is collected and cooled before returning to the engine by a pump. { 'kraŋk,kās }

crankcase breather *See* breather pipe. { 'kraŋ·kās ,brēth·ər }

crankpin |DES ENG| A cylindrical projection on a crank which holds the connecting rod. { 'kraŋk,pin }

crank press |MECH ENG| A punch press that applies power to the slide by means of a crank. { 'kraŋk ,pres }

crankshaft |MECH ENG| The shaft about which a crank rotates. { 'kraŋk,shaft }

crank throw |MECH ENG| 1. The web or arm of a crank. 2. The displacement of a crankpin from the crankshaft. { 'kraŋk ,thrō }

crank web [MECH ENG] The arm of a crank connecting the crankshaft to crankpin, or connecting two adjacent crankpins. { 'kraŋk ,web }

crash bar [ENG] A bar that is installed on a panic exit device located on a door and serves to unlock the door and, sometimes, to activate an alarm. { 'krash ,bär }

crater [MECH ENG] A depression in the face of a cutting tool worn down by chip contact. { 'krād·ər }

crawler [MECH ENG] **1.** One of a pair of an endless chain of plates driven by sprockets and used instead of wheels by certain power shovels, tractors, bulldozers, drilling machines, and such, as a means of propulsion. **2.** Any machine mounted on such tracks. { 'krò·lər }

crawler crane [MECH ENG] A self-propelled crane mounted on two endless tracks that revolve around wheels. { 'krò·lər ,krān }

crawler tractor [MECH ENG] A tractor that propels itself on two endless tracks revolving around wheels. { 'krò·lər ,trak·tər }

crawler wheel [MECH ENG] A wheel that drives a continuous metal belt, as on a crawler tractor. { 'krò·lər ,wēl }

crawl space [BUILD] **1.** A shallow space in a building which workers can enter to gain access to pipes, wires, and equipment. **2.** A shallow space located below the ground floor of a house and surrounded by the foundation wall. { 'kròl ,spās }

crazing [ENG] A network of fine cracks on or under the surface of a material such as enamel, glaze, metal, or plastic. { 'krāz·iŋ }

creep [ELECTR] A slow change in a characteristic with time or usage. [ENG] The tendency of wood to move while it is being cut, particularly when being mitered. [MECH] A time-dependent strain of solids caused by stress. { krēp }

creepage [ELEC] The conduction of electricity across the surface of a dielectric. { 'krē·pij }

creep buckling [MECH] Buckling that may occur when a compressive load is maintained on a member over a long period, leading to creep which eventually reduces the member's bending stiffness. { 'krēp ,bək·liŋ }

creeper [ENG] A low platform on small casters that is used for back support and mobility when a person works under a car. { 'krē·pər }

creep error [ENG] The error that occurs during a mass determination with a digital analytical balance when a value is read, printed, or processed before the display has reached its final position. { 'krēp ,er·ər }

creep-feed grinding See creep grinding. { ¦krēp ¦fēd 'grīnd·iŋ }

creep grinding [MECH ENG] A grinding operation that uses slow feed rates and produces heavy stock removal. Also known as creep-feed grinding. { 'krēp ,grīnd·iŋ }

creep limit [MECH] The maximum stress a given material can withstand in a given time without exceeding a specified quantity of creep. { 'krēp ,lim·ət }

creep recovery [MECH] Strain developed in a period of time after release of load in a creep test. { 'krēp ri'kəv·ə·rē }

creep rupture strength [MECH] The stress which, at a given temperature, will cause a material to rupture in a given time. { 'krēp 'rəp·chər ,streŋkth }

creep strength [MECH] The stress which, at a given temperature, will result in a creep rate of 1% deformation within 100,000 hours. { 'krēp ,streŋkth }

creep test [ENG] Any one of a number of methods of measuring creep, for example, by subjecting a material to a constant stress or deforming it at a constant rate. { 'krēp ,test }

cremone bolt [DES ENG] A fastening for double doors or casement windows; employs vertical rods that move up and down to engage the top and bottom of the frame. { krə'mōn ,bōlt }

crescent beam [ENG] A beam bounded by arcs having different centers of curvature, with the central section the largest. { 'kres·ənt ,bēm }

crest [DES ENG] The top of a screw thread. { krest }

crest clearance [DES ENG] The clearance, in a radial direction, between the crest of the thread of a screw and the root of the thread with which the screw mates. { 'krest ,klir·əns }

crest gate [CIV ENG] A gate in the spillway of a dam which functions to maintain or change the water level. { 'krest ,gāt }

crib [CIV ENG] The space between two successive ties along a railway track. [ENG] **1.** Any structure composed of a layer of timber or steel joists laid on the ground, or two layers across each other, to spread a load. **2.** Any structure composed of frames of timber placed horizontally on top of each other to form a wall. { krib }

cricket [BUILD] A device that is used to divert water at the intersections of roofs or at the intersection of a roof and chimney. { 'krik·ət }

crimp [ENG] **1.** To cause something to become wavy, crinkled, or warped, such as lumber. **2.** To pinch or press together, especially a tubular or cylindrical shape, in order to seal or unite. { krimp }

crimp contact [ELEC] A contact whose back portion is a hollow cylinder that will accept a wire; after a bared wire is inserted, a swaging tool is applied to crimp the contact metal firmly against the wire. Also known as solderless contact. { 'krimp ,kän,takt }

crinal [MECH] A unit of force equal to 0.1 newton. { 'krīn·əl }

cripple [BUILD] A structural member, such as a stud above a window, that is cut less than full length. { 'krip·əl }

crith [MECH] A unit of mass, used for gases, equal to the mass of 1 liter of hydrogen at standard pressure and temperature; it is found experimentally to equal 8.9885×10^{-5} kilogram. { krith }

critical compression ratio [MECH ENG] The lowest compression ratio which allows compression ignition of a specific fuel. { 'krid·ə·kəl kəm'presh·ən ,rā·shō }

critical density |CIV ENG| For a highway, the density of traffic when the volume equals the capacity. |THERMO| The density of a substance at the liquid-vapor critical point. { 'krid·ə·kəl 'den·səd·ē }

critical exponent |THERMO| A parameter *n* that characterizes the temperature dependence of a thermodynamic property of a substance near its critical point; the temperature dependence has the form $|T - T_c|^n$, where T is the temperature and T_c is the critical temperature. { 'krid·ə·kəl ik'spō·nənt }

critical humidity |CHEM ENG| The humidity of a system's atmosphere above which a crystal of a water-soluble salt will always become damp (absorb moisture from the atmosphere) and below which it will always stay dry (release moisture to the atmosphere). { 'krid·ə·kəl yü'mid·əd·ē }

critical isotherm |THERMO| A curve showing the relationship between the pressure and volume of a gas at its critical temperature. { 'krid·ə·kəl 'T·sə,thərm }

critical moisture content |CHEM ENG| The average moisture throughout a solid material being dried, its value being related to drying rate, thickness of material, and the factors that influence the movement of moisture within the solid. { 'krid·ə·kəl 'mȯis·chər ,kän·tent }

critical path method |SYS ENG| A systematic procedure for detailed project planning and control. Abbreviated CPM. { 'krid·ə·kəl 'path ,meth·əd }

critical pressure |THERMO| The pressure of the liquid-vapor critical point. { 'krid·ə·kəl 'presh·ər }

critical slope |CIV ENG| The maximum angle with the horizontal at which a sloped bank of soil of a given height will remain undeformed without some form of support. { 'krid·ə·kəl 'slōp }

critical speed |MECH ENG| The angular speed at which a rotating shaft becomes dynamically unstable with large lateral amplitudes, due to resonance with the natural frequencies of lateral vibration of the shaft. { 'krid·ə·kəl 'spēd }

critical vibration |MECH ENG| A vibration that is significant and harmful to a structure. { 'krid·ə·kəl vī'brā·shən }

critical weight |ENG| In a drilling operation, the weight placed on a bit that will cause the drill string to become resonant with the angular speed at which the rotating shaft is operating. { 'krid·ə·kəl 'wāt }

CR law |ELEC| A law which states that when a constant electromotive force is applied to a circuit consisting of a resistor and capacitor connected in series, the time taken for the potential on the plates of the capacitor to rise to any given fraction of its final value depends only on the product of capacitance and resistance. { ¦sē¦är ¦lȯ }

crochet file |DES ENG| A thin, flat, round-edged file that tapers to a point. { krō'shā ,fīl }

crocodile shears *See* lever shears. { 'kräk·ə,dīl ,shirz }

cross axle |MECH ENG| **1.** A shaft operated by levers at its ends. **2.** An axle with cranks set at 90° { 'krȯs ,ak·səl }

crossbar |CIV ENG| In a grating, one of the connecting bars which extend across bearing bars, usually perpendicular to them. { 'krȯs,bär }

crossbar micrometer |ENG| An instrument consisting of two bars mounted perpendicular to each other in the focal plane of a telescope, and inclined to the east-west path of stars by 45°; used to measure differences in right ascension and declination of celestial objects. { 'krȯs,bär mī'kräm·əd·ər }

crossbeam |BUILD| **1.** Also known as trave. **2.** A horizontal beam. **3.** A beam that runs transversely to the center line of a structure. { 'krȯs,bēm }

cross-belt drive |DES ENG| A belt drive having parallel shafts rotating in opposite directions. { 'krȯs ,belt ,drīv }

crossbolt |DES ENG| A lock bolt with two parts which can be moved in opposite directions. { 'krȯs,bōlt }

cross bond |CIV ENG| A masonry bond in which a course of alternating lengthwise and endwise bricks (Flemish bond) alternates with a course of bricks laid lengthwise. { 'krȯs ,bänd }

cross box |MECH ENG| A boxlike structure for the connection of circulating tubes to the longitudinal drum of a header-type boiler. { 'krȯs ,bäks }

cross bracing |BUILD| Boards which are nailed diagonally across studs or other boards so as to impart rigidity to a framework. { 'krȯs ,brās·iŋ }

cross-correlation detection *See* correlation detection. { 'krȯs kär·ə'lā·shən di'tek·shən }

crosscut |ENG| A cut made through wood across the grain. { 'krȯs,kət }

crosscut file |DES ENG| A file with a rounded edge on one side and a thin edge on the other; used to sharpen straight-sided saw teeth with round gullets. { 'krȯs,kət ,fīl }

crosscut saw |DES ENG| A type of saw for cutting across the grain of the wood; designed with about eight teeth per inch. { 'krȯs,kət ,sȯ }

cross drum boiler |MECH ENG| A sectional header or box header type of boiler in which the axis of the horizontal drum is perpendicular to the axis of the main bank of tubes. { 'krȯs ,drəm ,bȯil·ər }

crossed belt |MECH ENG| A pulley belt arranged so that the sides cross, thereby making the pulleys rotate in opposite directions. { ¦krȯst ¦belt }

crossed-field amplifier |ELECTR| A forward-wave, beam-type microwave amplifier that uses crossed-field interaction to achieve good phase stability, high efficiency, high gain, and wide bandwidth for most of the microwave spectrum. { ¦krȯst ,fēld 'am·plə,fī·ər }

crossed-field device |ELECTR| Any instrument

which uses the motion of electrons in perpendicular electric and magnetic fields to generate microwave radiation, either as an amplifier or oscillator. { 'kròst ,fēld di'vīs }

crossed-needle meter [ENG] A device consisting of two pointer-type analog meters inside a single enclosure with pointer movements centered at different positions so that their point of crossing indicates the value of some function of the two readings. { 'kròst ¦nēd·əl 'mēd·ər }

cross-fade [ENG ACOUS] In dubbing, the overlapping of two sound tracks, wherein the outgoing track fades out while the incoming track fades in. { 'kròs ,fād }

cross-flow baffle [ENG] A type of baffle in a shell-and-tube heat exchanger that directs shellside fluid back and forth or up and down across the tubes. Also known as transverse baffle. { 'kròs ,flō ,baf·əl }

cross furring ceiling [BUILD] A ceiling in which furring members are attached perpendicular to the main runners or other structural members. { 'kròs ,fər·iŋ ,sēl·iŋ }

cross hair [ENG] An inscribed line or a strand of hair, wire, silk, or the like used in an optical sight, transit, or similar instrument for accurate sighting. { 'kròs ,her }

crosshaul [MECH ENG] A device for loading objects onto vehicles, consisting of a chain that is hooked on opposite sides of a vehicle, looped under the object, and connected to a power source and that rolls the object onto the vehicle. { 'kròs,hol }

crosshead [MECH ENG] A block sliding between guides and containing a wrist pin for the conversion of reciprocating to rotary motion, as in an engine or compressor. { 'kròs,hed }

crossing plates [CIV ENG] Plates placed between a crossing and the ties to support the crossing and protect the ties. { 'kròs·iŋ ,plāts }

crosslap joint [BUILD] A joint in which two wood members cross each other; half the thickness of each is removed so that at the joint the thickness is the same as that of the individual members. { 'kròs,lap ,jóint }

cross-level [ENG] To level at an angle perpendicular to the principal line of sight. { 'kròs ,lev·əl }

crossover [CIV ENG] **1.** An S-shaped section of railroad track joining two parallel tracks. **2.** A connection between two pipes in the same water supply system or a connection between two water supply systems. [ELEC] A point at which two conductors cross, with appropriate insulation between them to prevent contact. [ELECTR] The plane at which the cross section of a beam of electrons in an electron gun is a minimum. [ENG] The portion of a draw works' drum containing grooves for angle control so the wire rope can cross over to begin a new wrap. Also known as angle-control section. { 'kròs,ō·vər }

crossover distortion [ELECTR] Amplitude distortion in a class B transistor power amplifier which occurs at low values of current, when input

impedance becomes appreciable compared with driver impedance. { 'kròs,ō·vər dis'tór·shən }

crossover flange [ENG] Intermediate pipe flange used to connect flanges of different working pressures. { 'kròs,ō·vər ,flanj }

crossover frequency [ENG ACOUS] **1.** The frequency at which a dividing network delivers equal power to the upper and lower frequency channels when both are terminated in specified loads. **2.** See transition frequency. { 'kròs,ō·vər ,frē·kwən·sē }

crossover network [ENG ACOUS] A selective network used to divide the audio-frequency output of an amplifier into two or more bands of frequencies. Also known as dividing network; loudspeaker dividing network. { 'kròs,ō·vər ,net,wərk }

crossover spiral See lead-over groove. { 'kròs,ō·vər ,spī·rəl }

crossover voltage [ELECTR] In a cathode-ray storage tube, the voltage of a secondary writing surface, with respect to cathode voltage, on which the secondary emission is unity. { 'kròs,ō·vər ,vōl·tij }

cross-peen hammer [ENG] A hammer with a wedge-shaped surface at one end of the head. { 'kròs ,pēn 'ham·ər }

cross slide [MECH ENG] A part of a machine tool that allows the tool carriage to move at right angles to the main direction of travel. { 'kròs ,slīd }

crosstalk See magnetic printing. { 'kròs,tòk }

cross-thread [ENG] To screw together two threaded pieces without aligning the threads correctly. { 'kròs ,thred }

crosstie [ENG] A timber or metal sill placed transversely under the rails of a railroad, tramway, or mine-car track. { 'kròs,tī }

cross turret [MECH ENG] A turret that moves horizontally and at right angles to the lathe guides. { 'kròs ,tər·ət }

cross ventilation [ENG] The movement of air from one side of a building or room and out the other side or through a monitor. { 'kròs ,vent·əl'ā·shən }

crowbar [DES ENG] An iron or steel bar that is usually bent and has a wedge-shaped working end; used as a lever and for prying. [ELEC] A device or action that in effect places a high overload on the actuating element of a circuit breaker or other protective device, thus triggering it. { 'krō,bär }

crown [CIV ENG] **1.** Center of a roadway elevated above the sides. **2.** In plumbing, that part of a trap where the direction of flow changes from upward to horizontal or downward. [ENG] **1.** The part of a drill bit inset with diamonds. **2.** The vertex of an arch or arched surface. **3.** The top or dome of a furnace or kiln. **4.** A high spot forming on a tool joint shoulder as the result of drill pipe wobbling. { kraun }

crown post [BUILD] Any upright member of a roof truss assembly, such as a king post. { 'kraun ,pōst }

crown saw |DES ENG| A saw consisting of a hollow cylinder with teeth around its edge; used for cutting round holes. Also known as hole saw. { 'kraùn ,sȯ }

crown sheet |MECH ENG| The structural element which forms the top of a furnace in a firetube boiler. { 'kraùn ,shēt }

crown weir |CIV ENG| The highest point on the internal bottom surface of the crown of a plumbing trap. { 'kraùn ,wer }

crown wheel |DES ENG| A gear that is light and crown-shaped. { 'kraùn ,wēl }

crow's nest |ENG| An elevated passageway for personnel located at the top of a derrick, refinery, or similar installation. { 'krōz ,nest }

CRT See cathode-ray tube.

crude assay |CHEM ENG| A procedure for determining the general distillation characteristics and other quality information of crude oil. { ¦krüd 'as·ā }

crude desalting |CHEM ENG| The washing of crude oil with water in order to remove materials such as dirt, silt, and water-soluble minerals. { 'krüd dē'sȯlt·iŋ }

crude material See raw material. { 'krüd me,tir·ē·əl }

crude still |CHEM ENG| The distillation equipment in which crude oil is separated into various products. { 'krüd ,stil }

crusher |MECH ENG| A machine for crushing rock and other bulk materials. { 'krəsh·ər }

crush-forming |ENG| Shaping the face of a grinding wheel by forcing a rotating metal roll into it. { 'krəsh ,fȯr·miŋ }

crushing strain |MECH| Compression which causes the failure of a material. { 'krəsh·iŋ ,strān }

crushing strength |MECH| The compressive stress required to cause a solid to fail by fracture; in essence, it is the resistance of the solid to vertical pressure placed upon it. { 'krəsh·iŋ ,streŋkth }

crushing test |ENG| A test of the suitability of stone that might be mined for roads or building use. { 'krəsh·iŋ ,test }

cryochem process |CHEM ENG| A freeze-drying technique involving conduction heat transfer to the frozen solid held on a metallic surface. { 'krī·ō,kem ,präs·əs }

cryoelectronics |ELECTR| A branch of electronics concerned with the study and application of superconductivity and other low-temperature phenomena to electronic devices and systems. Also known as cryolectronics. { ¦krī·ō·i,lek 'trän·iks }

cryogenic engineering |ENG| A branch of engineering specializing in technical operations at very low temperatures (about 200 to 400°R, or −160 to −50°C). { ,krī·ə'jen·ik en·jə'nir·iŋ }

cryogenic gyroscope |ENG| A gyroscope in which a spherical rotor of superconducting niobium spins while in levitation at cryogenic temperatures. Also known as superconducting gyroscope. { ,krī·ə'jen·ik 'jī·rə,skōp }

cryogenic transformer |ELECTR| A transformer designed to operate in digital cryogenic circuits, such as a controlled-coupling transformer. { ,krī·ə'jen·ik tranz'fȯr·mər }

cryolectronics See cryoelectronics. { ¦krī·ō·i,lek 'trän·iks }

cryology |MECH ENG| The study of low-temperature (approximately 200°R, or −160°C) refrigeration. { krī'äl·ə·jē }

cryometer |ENG| A thermometer for measuring low temperatures. { krī'äm·əd·ər }

cryopreservation |ENG| Preservation of food, biologicals, and other materials at extremely low temperatures. { ¦krī·ō,prez·ər'vā·shən }

cryosar |ELECTR| A cryogenic, two-terminal, negative-resistance semiconductor device, consisting essentially of two contacts on a germanium wafer operating in liquid helium. { 'krī·ō,sär }

cryoscope |ENG| A device to determine the freezing point of a liquid. { 'krī·ə,skōp }

cryosistor |ELECTR| A cryogenic semiconductor device in which a reverse-biased *pn* junction is used to control the ionization between two ohmic contacts. { ¦krī·ə'zis·tər }

cryosorption pump |MECH ENG| A high-vacuum pump that employs a sorbent such as activated charcoal or synthetic zeolite cooled by nitrogen or some other refrigerant; used to reduce pressure from atmospheric pressure to a few millitorr. { ,krī·ə'sȯrp·shən ,pəmp }

cryostat |ENG| An apparatus used to provide low-temperature environments in which operations may be carried out under controlled conditions. { 'krī·ə,stat }

cryotron |ELECTR| A switch that operates at very low temperatures at which its components are superconducting; when current is sent through a control element to produce a magnetic field, a gate element changes from a superconductive zero-resistance state to its normal resistive state. { 'krī·ə,trän }

cryotronics |ELECTR| The branch of electronics that deals with the design, construction, and use of cryogenic devices. { ,krī·ə'trän·iks }

cryptoclimate |ENG| The climate of a confined space, such as inside a house, barn, or greenhouse, or in an artificial or natural cave; a form of microclimate. Also spelled kryptoclimate. { ¦krip·tō'klī·mət }

crystal |ELECTR| A natural or synthetic piezoelectric or semiconductor material whose atoms are arranged with some degree of geometric regularity. { 'krist·əl }

crystal activity |ELECTR| A measure of the amplitude of vibration of a piezoelectric crystal plate under specified conditions. { 'krist·əl ak 'tiv·əd·ē }

crystal calibrator |ELECTR| A crystal-controlled oscillator used as a reference standard to check frequencies. { ¦krist·əl 'kal·ə,brād·ər }

crystal cartridge |ENG ACOUS| A piezoelectric unit used with a stylus in a phonograph pickup to convert disk recordings into audio-frequency signals, or used with a diaphragm in a crystal

137

microphone to convert sound waves into af signals. { ¦krist·əl 'kär,trij }

crystal control [ELECTR] Control of the frequency of an oscillator by means of a quartz crystal unit. { 'krist·əl kən'trōl }

crystal current [ELECTR] The actual alternating current flowing through a crystal unit. { 'krist·əl ,kər·ənt }

crystal cutter [ENG ACOUS] A cutter in which the mechanical displacements of the recording stylus are derived from the deformations of a crystal having piezoelectric properties. { 'krist·əl ,kəd·ər }

crystal-diffraction spectrometer See Bragg spectrometer. { 'krist·əl di'frak·shən spek'träm·əd·ər }

crystal headphones [ENG ACOUS] Headphones using Rochelle salt or other crystal elements to convert audio-frequency signals into sound waves. Also known as ceramic earphones. { 'krist·əl 'hed,fōnz }

crystal holder [DES ENG] A housing designed to provide proper support, mechanical protection, and connections for a quartz crystal plate. { 'krist·əl ,hōl·dər }

crystal hydrophone [ENG ACOUS] A crystal microphone that responds to waterborne sound waves. { 'krist·əl 'hī·drə,fōn }

crystallizer [CHEM ENG] Process vessel within which dissolved solids in a supersaturated solution are forced out of solution by cooling or evaporation, and then recovered as solid crystals. { 'kris·tə,līz·ər }

crystal loudspeaker [ENG ACOUS] A loudspeaker in which movements of the diaphragm are produced by a piezoelectric crystal unit that twists or bends under the influence of the applied audio-frequency signal voltage. Also known as piezoelectric loudspeaker. { 'krist·əl 'laùd,spēk·ər }

crystal microphone [ENG ACOUS] A microphone in which deformation of a piezoelectric bar by the action of sound waves or mechanical vibrations generates the output voltage between the faces of the bar. Also known as piezoelectric microphone. { 'krist·əl 'mī·krə,fōn }

crystal oven [ENG] A temperature-controlled oven in which a crystal unit is operated to stabilize its temperature and thereby minimize frequency drift. { 'krist·əl ,əv·ən }

crystal pickup [ENG ACOUS] A phonograph pickup in which movements of the needle in the record groove cause deformation of a piezoelectric crystal, thereby generating an audio-frequency output voltage between opposite faces of the crystal. Also known as piezoelectric pickup. { ¦krist·əl 'pik,əp }

crystal spectrometer See Bragg spectrometer. { 'krist·əl spek'träm·əd·ər }

C size [ENG] One of a series of sizes to which trimmed paper and board are manufactured; for size CN, with N equal to any integer, the length of the longer side is $2^{3/8-N/2}$ meters, while the length of the shorter side is $2^{1/8-N/2}$ meters, with both lengths rounded off to the nearest millimeter. { 'sē ,sīz }

CTC See centralized traffic control.

CTD recorder See salinity-temperature-depth recorder. { ¦sē¦tē¦dē ri'kórd·ər }

C-tube bourdon element [ENG] Hollow tube of flexible (elastic) metal shaped like the arc of a circle; changes in internal gas or liquid pressure flexes the tube to a degree related to the pressure change; used to measure process-stream pressures. { 'sē ,tüb 'búrd·ən ,el·ə·mənt }

cu See cubic.

cubic [MECH] Denoting a unit of volume, so that if x is a unit of length, a cubic x is the volume of a cube whose sides have length $1x$; for example, a cubic meter, or a meter cubed, is the volume of a cube whose sides have a length of 1 meter. Abbreviated cu. { 'kyü·bik }

cubical dilation [MECH] The isotropic part of the strain tensor describing the deformation of an elastic solid, equal to the fractional increase in volume. { 'kyü·bə·kəl di'lā·shən }

cubic boron nitride [MECH ENG] A synthetic material composed of boron and nitrogen (1:1) that is almost as hard as diamond, used as a superabrasive powder and for cutting and grinding applications. { ¦kyü·bik¦bó,rän 'nī,trīd }

cubic foot per minute [MECH] A unit of volume flow rate, equal to a uniform flow of 1 cubic foot in 1 minute; equal to 1/60 cusec. Abbreviated cfm. { ¦kyü·bik ¦fút pər 'min·ət }

cubic foot per second See cusec. { ¦kyü·bik ¦fút pər 'sek·ənd }

cubicle [BUILD] Any small, approximately square room or compartment. [ENG] An enclosure for high-voltage equipment. { 'kyü·bə·kəl }

cubic measure [MECH] A unit or set of units to measure volume. { 'kyü·bik 'mezh·ər }

cul-de-sac [CIV ENG] A dead-end street with a circular area for turning around. { 'kəl·də,sak }

cull [CHEM ENG] In a plastics molding operation, material remaining in the transfer chamber after the mold has been filled. { kəl }

cullet See collet. { 'kəl·ət }

cullis See coulisse. { 'kəl·əs }

cultellation [ENG] Transferring a surveyed point from a high level (such as on overhang) to a lower level by dropping a marking pin. { kəl·tə'lā·shən }

culvert [ENG] A covered channel or a large-diameter pipe that takes a watercourse below ground level. { 'kəl·vərt }

cumec [MECH] A unit of volume flow rate equal to 1 cubic meter per second. { 'kyü,mek }

cumulative compound motor [MECH ENG] A motor with operating characteristics between those of the constant-speed (shunt-wound) and the variable-speed (series-wound) types. { 'kyü·myə·ləd·iv ,käm,paúnd 'mōd·ər }

cumulative sum chart [IND ENG] A statistical control chart on which the cumulative sum of deviations is plotted over a period of time and which often has a sliding V-shaped mask for comparing the plot with allowable limits. Also known as cusum chart. { 'kyü·myə·ləd·iv 'səm ‚chärt }

cup |DES ENG| A cylindrical part with only one end open. |ENG| A low spot forming on a tool joint shoulder as a result of wobbling. { kəp }

cup anemometer |ENG| A rotation anemometer, usually consisting of three or four hemispherical or conical cups mounted with their diametral planes vertical and distributed symmetrically about the axis of rotation; the rate of rotation of the cups, which is a measure of the wind speed, is determined by a counter. { 'kəp an·ə'mäm·əd·ər }

cup barometer |ENG| A barometer in which one end of a graduated glass tube is immersed in a cup, both cup and tube containing mercury. { 'kəp bə'räm·əd·ər }

cup-case thermometer |ENG| Total-immersion type of thermometer with a cup container at the bulb end to hold a specified amount and depth of the material whose temperature is to be measured. { 'kəp ‚kās thər'mäm·əd·ər }

cup electrometer |ENG| An electrometer that has a metal cup attached to its plate so that a charged body touching the inside of the cup gives up its entire charge to the instrument. { 'kəp i‚lek'träm·əd·ər }

curb |CIV ENG| A border of concrete or row of joined stones forming part of a gutter along a street edge. { kərb }

curb weight |MECH ENG| The weight of a motor vehicle plus fuel and other components or equipment necessary for standard operation; does not include driver weight or payload. { 'kərb ‚wāt }

cure |CHEM ENG| *See* vulcanization. |ENG| A process by which concrete is kept moist for its first week or month to provide enough water for the cement to harden. Also known as mature. { kyür }

cure time |CHEM ENG| The amount of time required for a rubber compound to reach maximum viscosity or modulus at a given temperature. { 'kyür ‚tīm }

Curie balance |ENG| An instrument for determining the susceptibility of weakly magnetic materials, in which the deflection produced by a strong permanent magnet on a suspended tube containing the specimen is measured. { 'kyür·ē ‚bal·əns }

Curie principle |THERMO| The principle that a macroscopic cause never has more elements of symmetry than the effect it produces; for example, a scalar cause cannot produce a vectorial effect. { 'kyür·ē ‚prin·sə·pəl }

Curie scale of temperature |THERMO| A temperature scale based on the susceptibility of a paramagnetic substance, assuming that it obeys Curie's law; used at temperatures below about 1 kelvin. { ‚kyür·ē ‚skāl əv 'tem·prə·chər }

curling |CHEM ENG| A process in which polymers or oligomers are chemically cross-linked to form polymer networks. |CIV ENG| A process for bringing freshly placed concrete to required strength and quality by maintaining the humidity and temperature at specified levels for a given period of time. Also known as seasoning. { 'kyür·iŋ }

curing time |ENG| Time interval between the stopping of moving parts during thermoplastics molding and the release of mold pressure. Also known as molding time. { 'kyür·iŋ ‚tīm }

curling |MECH ENG| A forming process in which the edge of a sheet-metal part is rolled over to produce a hollow tubular rim. { 'kərl·iŋ }

curling dies |MECH ENG| A set of tools that shape the ends of a piece of work into a form with a circular cross section. { 'kərl·iŋ ‚dīz }

curling machine |MECH ENG| A machine with curling dies; used to curl the ends of cans. { 'kərl·iŋ ‚mə'shēn }

current |ELEC| The net transfer of electric charge per unit time; a specialization of the physics definition. Also known as electric current. { 'kər·ənt }

current amplification |ELECTR| The ratio of output-signal current to input-signal current for an electron tube, transistor, or magnetic amplifier, the multiplier section of a multiplier phototube, or any other amplifying device; often expressed in decibels by multiplying the common logarithm of the ratio by 20. { 'kər·ənt am·plə·fə'kā·shən }

current amplifier |ELECTR| An amplifier capable of delivering considerably more signal current than is fed in. { 'kər·ənt ‚am·plə‚fī·ər }

current attenuation |ELECTR| The ratio of input-signal current for a transducer to the current in a specified load impedance connected to the transducer; often expressed in decibels. { 'kər·ənt ə‚ten·yə'wā·shən }

current collector *See* charge collector. { 'kər·ənt kə‚lek·tər }

current-controlled switch |ELECTR| A semiconductor device in which the controlling bias sets the resistance at either a very high or very low value, corresponding to the "off" and "on" conditions of a switch. { 'kər·ənt kən‚trōld 'swich }

current density |ELEC| The current per unit cross-sectional area of a conductor; a specialization of the physics definition. Also known as electric current density. { 'kər·ənt ‚den·səd·ē }

current drain |ELEC| The current taken from a voltage source by a load. Also known as drain. { 'kər·ənt ‚drān }

current drogue |ENG| A current-measuring assembly consisting of a weighted current cross, sail, or parachute, and an attached surface buoy. { 'kər·ənt ‚drōg }

current feedback |ELECTR| Feedback introduced in series with the input circuit of an amplifier. { 'kər·ənt ‚fēd‚bak }

current feedback circuit |ELECTR| A circuit used to eliminate effects of amplifier gain instability in an indirect-acting recording instrument, in which the voltage input (error signal) to an amplifier is the difference between the measured quantity and the voltage drop across a resistor. { 'kər·ənt ‚fēd‚bak ‚sər·kət }

current gain |ELECTR| The fraction of the current flowing into the emitter of a transistor which flows through the base region and out the collector. { 'kər·ənt ˌgān }

current generator [ELECTR] A two-terminal circuit element whose terminal current is independent of the voltage between its terminals. { 'kər·ənt ˌjen·ə,rād·ər }

current intensity [ELEC] The magnitude of an electric current. Also known as current strength. { 'kər·ənt in'ten· səd·ē }

current limiter [ELECTR] A device that restricts the flow of current to a certain amount, regardless of applied voltage. Also known as demand limiter. { 'kər·ənt ˌlim·əd·ər }

current line [ENG] In marine operations, a graduated line attached to a current pole, used to measure the speed of a current; as the pole moves away with the current, the speed of the current is determined by the amount of line paid out in a specified time. Also known as log line. { 'kər·ənt ˌlīn }

current meter See ammeter; velocity-type flowmeter. { 'kər·ənt ˌmēd·ər }

current mirror [ELECTR] An electronic circuit that generates, at a high-impedance output node, an inflowing or outflowing current that is a scaled replica of an input current flowing into or out of a low-impedance input node. { 'kər·ənt ˌmir·ər }

current-mode filter [ELECTR] An integrated-circuit filter in which the signals are represented by current levels rather than voltage levels. { 'kər·ənt,môd ˌfil·tər }

current-mode logic [ELECTR] Integrated-circuit logic in which transistors are paralleled so as to eliminate current hogging. Abbreviated CML. { 'kər·ənt ˌmôd 'läj·ik }

current noise [ELECTR] Electrical noise of uncertain origin which is observed in certain resistances when a direct current is present, and which increases with the square of this current. { 'kər·ənt ˌnoiz }

current pole [ENG] A pole used to determine the direction and speed of a current; the direction is determined by the direction of motion of the pole, and the speed by the amount of an attached current line paid out in a specified time. { 'kər·ənt ˌpōl }

current regulator [ELECTR] A device that maintains the output current of a voltage source at a predetermined, essentially constant value despite changes in load impedance. { 'kər·ənt ˌreg·yə,lād·ər }

current saturation See anode saturation. { 'kər·ənt sach·ə'rā·shən }

current source [ELECTR] An electronic circuit that generates a constant direct current into or out of a high-impedance output node. { 'kər·ənt ˌsòrs }

current strength See current intensity. { 'kər·ənt ˌstreŋkth }

current-type flowmeter [ENG] A mechanical device to measure liquid velocity in open and closed channels; similar to the vane anemometer

(where moving liquid turns a small windmill-type vane), but more rugged. { 'kər·ənt ˌtīp 'flō ˌmēd·ər }

cursor [DES ENG] A clear or amber-colored filter that can be placed over a radar screen and rotated until an etched diameter line on the filter passes through a target echo; the bearing from radar to target can then be read accurately on a stationary 360° scale surrounding the filter. { 'kər·sər }

curtain board [BUILD] A fire-retardant partition applied to a ceiling. { 'kərt·ən ˌbórd }

curtain coating [CHEM ENG] A method in which the substrate to be coated with low-viscosity resins or solutions is passed through, and is perpendicular to, a freely falling liquid curtain. { 'kərt·ən ˌkōd·iŋ }

curtain wall [CIV ENG] An external wall that is not load-bearing. { 'kərt·ən ˌwòl }

curved beam [ENG] A beam bounded by circular arcs. { ˌkərvd 'bēm }

curve resistance [MECH] The force opposing the motion of a railway train along a track due to track curvature. { 'kərv ri'zis·təns }

curve tracer [ENG] An instrument that can produce a display of one voltage or current as a function of another voltage or current, with a third voltage or current as a parameter. { 'kərv ˌtrā·sər }

curvilinear motion [MECH] Motion along a curved path. { 'kər·və'lin·ē·ər 'mō·shən }

cusec [MECH] A unit of volume flow rate, used primarily to describe pumps, equal to a uniform flow of 1 cubic foot in 1 second. Also known as cubic foot per second (cfs). { 'kyü,sek }

cushion gas See blanket gas. { 'kúsh·ən ˌgas }

custodial area [BUILD] Area of a building designated for service and custodial personnel; includes rooms, closets, storage, toilets, and lockers. { kə'stōd·ē·əl ˌer·ē·ə }

custom millwork See architectural millwork. { 'kəs·təm 'mil,wərk }

cusum chart See cumulative sum chart. { ˌkyü ˌsəm ˌchärt }

cut [CHEM ENG] A fraction obtained by a separation process. { kət }

cut and fill [CIV ENG] Construction of a road, a railway, or a canal which is partly embanked and partly below ground. { ˌkət ən 'fil }

cutback [CHEM ENG] Blending of heavier oils with lighter ones to bring the heavier to desired specifications. { 'kət,bak }

cut constraint [SYS ENG] A condition sometimes imposed in an integer programming problem which excludes parts of the feasible solution space without excluding any integer points. { 'kət kən'stränt }

cut-in [CONT SYS] A value of temperature or pressure at which a control circuit closes. [ELEC] An electrical device that allows current to flow through an electric circuit. { 'kət ˌin }

cut methods [SYS ENG] Methods of solving integer programming problems that employ cut constraints derived from the original problem. { 'kət ˌmeth·əds }

140

cut nail |DES ENG| A flat, tapered nail sheared from steel plate; it has greater holding power than a wire nail and is generally used for fastening flooring. { 'kət ,nāl }

cutoff |CIV ENG| **1.** A channel constructed to straighten a stream or to bypass large bends, thereby relieving an area normally subjected to flooding or channel erosion. **2.** An impermeable wall, collar, or other structure placed beneath the base or within the abutments of a dam to prevent or reduce losses by seepage along otherwise smooth surfaces or through porous strata. |ELECTR| **1.** The minimum value of bias voltage, for a given combination of supply voltages, that just stops output current in an electron tube, transistor, or other active device. **2.** *See* cutoff frequency. |ENG| **1.** A misfire in a round of shots because of severance of fuse owing to rock shear as adjacent charges explode. **2.** The line on a plastic object formed by the meeting of the two halves of a compression mold. Also known as flash groove; pinch-off. |MECH ENG| **1.** The shutting off of the working fluid to an engine cylinder. **2.** The time required for this process. { 'kət,óf }

cutoff bias |ELECTR| The direct-current bias voltage that must be applied to the grid of an electron tube to stop the flow of anode current. { 'kət,óf ,bī·əs }

cutoff frequency |ELECTR| A frequency at which the attenuation of a device begins to increase sharply, such as the limiting frequency below which a traveling wave in a given mode cannot be maintained in a waveguide, or the frequency above which an electron tube loses efficiency rapidly. Also known as critical frequency; cutoff. { 'kət,óf ,frē·kwən·sē }

cutoff limiting |ELECTR| Limiting the maximum output voltage of a vacuum tube circuit by driving the grid beyond cutoff. { 'kət,óf ,lim·əd·iŋ }

cutoff point |MECH ENG| **1.** The point at which there is a transition from spiral flow in the housing of a centrifugal fan to straight-line flow in the connected duct. **2.** The point on the stroke of a steam engine where admission of steam is stopped. { 'kət,óf ,póint }

cutoff tool |MECH ENG| A tool used on bar-type lathes to separate the finished piece from the bar stock. { 'kət,óf ,tül }

cutoff trench |CIV ENG| A trench which is below the foundation base line of a dam or other structure and is filled with an impervious material, such as clay or concrete, to form a watertight barrier. { 'kət,óf ,trench }

cutoff valve |MECH ENG| A valve used to stop the flow of steam to the cylinder of a steam engine. { 'kət,óf ,valv }

cutoff voltage |ELECTR| **1.** The electrode voltage value that reduces the dependent variable of an electron-tube characteristic to a specified low value. **2.** *See* critical voltage. { 'kət,óf ,vól·tij }

cutoff wall |CIV ENG| A thin, watertight wall of clay or concrete built up from a cutoff trench to reduce seepage. Also known as core wall. { 'kət,óf ,wól }

cutoff wheel |MECH ENG| A thin wheel impregnated with an abrasive used for severing or cutting slots in a material or part. { 'kət,óf ,wēl }

cut-out |CONT SYS| A value of temperature or pressure at which a control circuit opens. { 'kət ,aút }

cutout angle |ELECTR| The phase angle at which a semiconductor diode ceases to conduct; it is slightly less than 180° because the diode requires some forward bias to conduct. { 'kət,aút ,aŋ·gəl }

cutover |ENG| **1.** To place equipment in active use. **2.** The time when testing of equipment is completed and regular usage begins. { 'kət,ō·vər }

cut point |CHEM ENG| The boiling-temperature division between cuts of a crude oil or base stock. { 'kət ,póint }

cutscore |ENG| A knife used in die-cutting processes, designed to cut just partway into the paper or board so that it can be folded. { 'kət,skór }

cutter |ENG ACOUS| An electromagnetic or piezoelectric device that converts an electric input to a mechanical output, used to drive the stylus that cuts a wavy groove in the highly polished wax surface of a recording disk. Also known as cutting head; head; phonograph cutter; recording head. |MECH ENG| *See* cutting tool. { 'kəd·ər }

cutter bar |MECH ENG| The bar that supports the cutting tool in a lathe or other machine. { 'kəd·ər ,bär }

cutter compensation |CONT SYS| The process of taking into account the difference in radius between a cutting tool and a programmed numerical control operation in order to achieve accuracy. { 'kəd·ər ,käm·pən'sā·shən }

cutterhead |MECH ENG| A device on a machine tool for holding a cutting tool. { 'kəd·ər,hed }

cutter sweep |MECH ENG| The section that is cut off or eradicated by the milling cutter or grinding wheel in entering or leaving the flute. { 'kəd·ər ,swēp }

cutting angle |MECH ENG| The angle that the cutting face of a tool makes with the work surface back of the tool. { 'kəd·iŋ ,aŋ·gəl }

cutting down |MECH ENG| Removing surface roughness or irregularities from metal by the use of an abrasive. { 'kəd·iŋ 'daún }

cutting drilling |MECH ENG| A rotary drilling method in which drilling occurs through the action of the drill steel rotating while pressed against the rock. { 'kəd·iŋ ,dril·iŋ }

cutting edge |DES ENG| **1.** The point or edge of a diamond or other material set in a drill bit. Also known as cutting point. **2.** The edge of a lathe tool in contact with the work during a machining operation. { 'kəd·iŋ 'ej }

cutting head *See* cutter. { 'kəd·iŋ ,hed }

cutting in |MECH ENG| An undesirable action occurring during loose-drum spooling in which a layer of wire rope spreads apart and forms grooves in which the next layer travels. { 'kəd·iŋ 'in }

cutting-off machine [MECH ENG] A machine for cutting off metal bars and shapes; includes the lathe type using single-point cutoff tools, and several types of saws. { 'kəd·iŋ ˌof mə'shēn }

cutting pliers [DES ENG] Pliers with cutting blades on the jaws. { 'kəd·iŋ ˌplī·ərz }

cutting point See cutting edge. { 'kəd·iŋ ˌpȯint }

cutting ratio [ENG] As applied to metal cutting, the ratio of depth of cut to chip thickness for a given shear angle. { 'kəd·iŋ ˌrā·shō }

cutting rule [ENG] A sharp steel rule used in a machine for cutting paper or cardboard. { 'kəd·iŋ ˌrül }

cutting speed [MECH ENG] The speed of relative motion between the tool and workpiece in the main direction of cutting. Also known as feed rate; peripheral speed. { 'kəd·iŋ ˌspēd }

cutting stylus [ENG ACOUS] A recording stylus with a sharpened tip that removes material to produce a groove in the recording medium. { 'kəd·iŋ ˌstī·ləs }

cutting tip [ENG] The end of the snout of a cutting torch from which gas flows. { 'kəd·iŋ ˌtip }

cutting tool [MECH ENG] The part of a machine tool which comes into contact with and removes material from the workpiece by the use of a cutting medium. Also known as cutter. { 'kəd·iŋ ˌtül }

cutting torch [ENG] A torch that preheats metal while the surface is rapidly oxidized by a jet of oxygen issuing through the flame from an additional feed line. { 'kəd·iŋ ˌtȯrch }

cutwater [CIV ENG] A sharp-edged structure built around a bridge pier to protect it from the flow of water and material carried by the water. { 'kət,wȯd·ər }

cybernation [IND ENG] The use of computers in connection with automation. { sī·bər'nā·shən }

cycle [ENG] To run a machine through a single complete operation. { 'sī·kəl }

cyclegraph technique [IND ENG] Recording a brief work cycle by attaching small lights to various parts of a worker and then exposing the work motions on a still-film time plate; motion will appear on the plate as superimposed streaks of light constituting a cyclegraph. { 'sī·klə,graf ˌtek,nēk }

cycle plant [CHEM ENG] A plant in which the liquid hydrocarbons are removed from natural gas and then the gas is put back into the earth to maintain pressure in the oil reservoir. { 'sī·kəl ˌplant }

cycle skip See skip logging. { 'sī·kəl ˌskip }

cycle stock [CHEM ENG] The unfinished product taken from a stage of a refinery process and recharged to the process at an earlier stage in the operation. { 'sī·kəl ˌstäk }

cycle timer [ELECTR] A timer that opens or closes circuits according to a predetermined schedule. { 'sī·kəl ˌtīm·ər }

cyclic catalytic reforming process [CHEM ENG] A method for the production of low-Btu reformed gas consisting of the conversion of carbureted water-gas sets by installing a bed of nickel catalyst in the superheater and using the carburetor as a combustion chamber and process steam superheater. Abbreviated CCR process. { 'sīk·lik ˌkäd·ə¦lid·ik ri'fȯr·miŋ ˌpräs·əs }

cyclic coordinate [MECH] A generalized coordinate on which the Lagrangian of a system does not depend explicitly. Also known as ignorable coordinate. { 'sīk·lik kō'ȯrd·ən·ət }

cyclic element [IND ENG] An element of an operation or process that occurs in each of its cycles. { 'sīk·lik 'el·ə·mənt }

cyclic testing [ENG] The repeated testing of a device or system at regular intervals to be assured of its reliability. { 'sīk·lik 'test·iŋ }

cyclic train [MECH ENG] A set of gears, such as an epicyclic gear system, in which one or more of the gear axes rotates around a fixed axis. { 'sīk·lik 'trān }

cycling [CHEM ENG] A series of operations in petroleum refining or natural-gas processing in which the steps are repeated periodically in the same sequence. [CONT SYS] A periodic change of the controlled variable from one value to another in an automatic control system. { 'sīk·liŋ }

cyclograph [ENG] An electronic instrument that produces on a cathode-ray screen a pattern which changes in shape according to core hardness, carbon content, case depth, and other metallurgical properties of a test sample of steel inserted in a sensing coil. { 'sī·klə,graf }

cycloidal gear teeth [DES ENG] Gear teeth whose profile is formed by the trace of a point on a circle rolling without slippage on the outside or inside of the pitch circle of a gear; now used only for clockwork and timer gears. { sī'klȯid·əl 'gir ˌtēth }

cycloidal pendulum [MECH] A modification of a simple pendulum in which a weight is suspended from a cord which is slung between two pieces of metal shaped in the form of cycloids; as the bob swings, the cord wraps and unwraps on the cycloids; the pendulum has a period that is independent of the amplitude of the swing. { sī'klȯid·əl 'pen·jə·ləm }

cyclone [CHEM ENG] A static reaction vessel in which fluids under pressure form a vortex. [MECH ENG] Any cone-shaped air-cleaning apparatus operated by centrifugal separation that is used in particle collecting and fine grinding operations. { 'sī,klōn }

cyclone cellar [CIV ENG] An underground shelter, often built in areas frequented by tornadoes. Also known as storm cellar; tornado cellar. { 'sī,klōn ˌsel·ər }

cyclone classifier See cyclone separator. { 'sī,klōn ˌklas·ə,fī·ər }

cyclone furnace [ENG] A water-cooled, horizontal cylinder in which fuel is fired cyclonically and heat is released at extremely high rates. { 'sī,klōn ˌfar·nəs }

cyclone separator [MECH ENG] A funnel-shaped device for removing particles from air or other fluids by centrifugal means; used to remove dust from air or other fluids, steam from water, and water from steam, and in certain applications to separate particles into two or more size classes. Also known as cyclone classifier. { 'sī,klōn 'sep·ə,rād·ər }

cylinder |CIV ENG| **1.** A steel tube 10–60 inches (25–152 centimeters) in diameter with a wall at least 1/8 inch (3 millimeters) thick that is driven into bedrock, excavated inside, filled with concrete, and used as a pile foundation. **2.** A domed, closed tank for storing hot water to be drawn off at taps. Also known as storage calorifier. |ENG| **1.** A container used to hold and transport compressed gas for various pressurized applications. **2.** The piston chamber in a pump from which the liquid is expelled. |MECH ENG| See engine cylinder. { 'sil·ən·dər }

cylinder actuator |MECH ENG| A device that converts hydraulic power into useful mechanical work by means of a tight-fitting piston moving in a closed cylinder. { 'sil·ən·dər ,ak·chə,wād·ər }

cylinder block |DES ENG| The metal casting comprising the piston chambers of a multicylinder internal combustion engine. Also known as block; engine block. { 'sil·ən·dər ,bläk }

cylinder bore |DES ENG| The internal diameter of the tube in which the piston of an engine or pump moves. { 'sil·ən·dər ,bȯr }

cylinder head |MECH ENG| The cap that serves to close the end of the piston chamber of a reciprocating engine, pump, or compressor. { 'sil·ən·dər ,hed }

cylinder liner |MECH ENG| A separate cylindrical sleeve inserted in an engine block which serves as the cylinder. { 'sil·ən·dər ,līn·ər }

cylinder machine |ENG| A paper-making machine consisting of one or a series of rotary cylindrical filters on which wet paper sheets are formed. { 'sil·ən·dər mə'shēn }

cylindrical cam |MECH ENG| A cam mechanism in which the cam follower undergoes translational motion parallel to the camshaft as a roller attached to it rolls in a groove in a circular cylinder concentric with the camshaft. { sə'lin·drə·kəl 'kam }

cylindrical-coordinate robot |CONT SYS| A robot in which the degrees of freedom of the manipulator arm are defined chiefly by cylindrical coordinates. { sə'lin·drə·kəl kȯ¦ȯrd·ən·ət 'rō,bät }

cylindrical cutter |DES ENG| Any cutting tool with a cylindrical shape, such as a milling cutter. { sə'lin·drə·kəl 'kəd·ər }

cylindrical grinder |MECH ENG| A machine for doing work on the peripheries or shoulders of workpieces composed of concentric cylindrical or conical shapes, in which a rotating grinding wheel cuts a workpiece rotated from a power headstock and carried past the face of the wheel. { sə'lin·drə·kəl 'grīnd·ər }

D

dac *See* digital-to-analog converter.

dado head [MECH ENG] A machine consisting of two circular saws with one or more chippers in between; used for cutting flat-bottomed grooves in wood. { 'dā·dō ,hed }

dado joint [BUILD] A joint made by fitting the full thickness of the edge or the end of one board into a corresponding groove in another board. Also known as housed joint. { 'dā,dō ,jóint }

dado plane [DES ENG] A narrow plane for cutting flat grooves in woodwork. { 'dā·dō ,plān }

Dahlin's algorithm [CONT SYS] A digital control algorithm in which the requirement of minimum response time used in the deadbeat algorithm is relaxed to reduce ringing in the system response. { 'dä·lənz ,al·gə,rith·əm }

d'Alembert's principle [MECH] The principle that the resultant of the external forces and the kinetic reaction acting on a body equals zero. { ¦dal·əm¦bərz ,prin·sə·pəl }

Dall tube [MECH ENG] Fluid-flow measurement device, similar to a venturi tube, inserted as a section of a fluid-carrying pipe; flow rate is measured by pressure drop across a restricted throat. { 'dól ,tüb }

Dalton's temperature scale [THERMO] A scale for measuring temperature such that the absolute temperature T is given in terms of the temperature on the Dalton scale τ by $T = 273.15(373.15/273.15)^{\tau/100}$. { 'dól·tənz 'tem·prə·chər ,skāl }

dam [CIV ENG] **1.** A barrier constructed to obstruct the flow of a watercourse. **2.** A pair of cast-steel plates with interlocking fingers built over an expansion joint in the road surface of a bridge. { dam }

damage tolerance [ENG] The ability of a structure to maintain its load-carrying capability after exposure to a sudden increase in load. { 'dam·ij ,täl·ə·rəns }

damaging stress [MECH] The minimum unit stress for a given material and use that will cause damage to the member and make it unfit for its expected length of service. { ¦dam·ə·jiŋ 'stres }

damp [ENG] To reduce the fire in a boiler or a furnace by putting a layer of damp coals or ashes on the fire bed. { damp }

damp course [CIV ENG] A layer of impervious material placed horizontally in a wall to keep out water. { ¦damp ,kórs }

dampener [ENG] A device for damping spring oscillations after abrupt removal or application of a load. { 'dam·pə·nər }

damper [ELECTR] A diode used in the horizontal deflection circuit of a television receiver to make the sawtooth deflection current decrease smoothly to zero instead of oscillating at zero; the diode conducts each time the polarity is reversed by a current swing below zero. [MECH ENG] A valve or movable plate for regulating the flow of air or the draft in a stove, furnace, or fireplace. { 'dam·pər }

damper loss [ENG] The reduction in rate of flow or of pressure of gas across a damper. { 'dam·pər ,lós }

damper pedal [ENG] A pedal that controls the damping of piano strings. { 'dam·pər ,ped·əl }

damping [ENG] Reducing or eliminating reverberation in a room by placing sound-absorbing materials on the walls and ceiling. Also known as soundproofing. { 'dam·piŋ }

damping capacity [MECH] A material's capability in absorbing vibrations. { 'dam·piŋ kə'pas·əd·ē }

damping coefficient *See* resistance. { 'dam·piŋ ,kō·i,fish·ənt }

damping constant *See* resistance. { 'dam·piŋ ,kän·stənt }

damping resistor [ELEC] **1.** A resistor that is placed across a parallel resonant circuit or in series with a series resonant circuit to decrease the Q factor and thereby eliminate ringing. **2.** A noninductive resistor placed across an analog meter to increase damping. { 'dam·piŋ ri,zis·tər }

dancing step *See* balanced step. { ¦dan·siŋ ¦step }

dancing winder *See* balanced step. { ¦dan·siŋ ¦wīn·dər }

Danckwerts model [CHEM ENG] Theory applied to liquid flow across packing in a liquid-gas absorption tower; allows for liquid eddies that bring fresh liquid from the interior of the liquid body to the surface, thus contacting the gas in the column. { 'daŋk·verts ,mäd·əl }

dandy roll [MECH ENG] A roll in a Fourdrinier paper-making machine; used to compact the sheet and sometimes to imprint a watermark. { 'dan·dē ,ról }

Daniell hygrometer [ENG] An instrument for measuring dew point; dew forms on the surface of a bulb containing ether which is cooled by evaporation into another bulb, the second bulb being cooled by the evaporation of ether on its outer surface. { 'dan·yəl hī'gräm·əd·ər }

Danjon prismatic astrolabe |ENG| A type of astrolabe in which a Wollaston prism just inside the focus of the telescope converts converging beams of light into parallel beams, permitting a great increase in accuracy. { 'dän·yən priz'mad·ik 'as·trə,lāb }

daraf |ELEC| The unit of elastance, equal to the reciprocal of 1 farad. { 'da,raf }

darby |ENG| A flat-surfaced tool for smoothing plaster. { 'där·bē }

d'Arsonval galvanometer |ENG| A galvanometer in which a light coil of wire, suspended from thin copper or gold ribbons, rotates in the field of a permanent magnet when current is carried to it through the ribbons; the position of the coil is indicated by a mirror carried on it, which reflects a light beam onto a fixed scale. Also known as light-beam galvanometer. { 'dars·ən,vòl gal·və'näm·əd·ər }

dashpot |MECH ENG| A device used to dampen and control a motion, in which an attached piston is loosely fitted to move slowly in a cylinder containing oil. { 'dash,pät }

datum |ENG| **1.** A direction, level, or position from which angles, heights, speeds or distances are conveniently measured. **2.** Any numerical or geometric quantity or value that serves as a base reference for other quantities or values (such as a point, line, or surface in relation to which others are determined). { 'dad·əm, 'däd·əm, or 'däd·əm }

datum level See datum plane. { 'dad·əm ,lev·əl }

datum plane |ENG| A permanently established horizontal plane, surface, or level to which soundings, ground elevations, water surface elevations, and tidal data are referred. Also known as chart datum; datum level; reference level; reference plane. { 'dad·əm ,plän }

daylight See daylight opening. { 'dā,līt }

daylight controls |ENG| Special devices which automatically control the electric power to the lamp, causing the light to operate during hours of darkness and to be extinguished during daylight hours. { 'dā,līt kən'trōlz }

daylighting |CIV ENG| To light an area with daylight. { 'dā,līd·iŋ }

daylight opening |ENG| The space between two press platens when open. Also known as daylight. { 'dā,līt ,ō·pən·iŋ }

day wage |IND ENG| A fixed rate of pay per shift or per daily hours of work, irrespective of the amount of work completed. { 'dā ,wāj }

dc See direct current.

dc-to-ac converter See inverter { ¦dē,sē tü ¦ā,sē kən'vərd·ər }

dc-to-ac inverter See inverter. { ¦dē,sē tü ¦ā,sē in'vərd·ər }

dc-to-dc converter |ELEC| An electronic circuit which converts one direct-current voltage into another, consisting of an inverter followed by a

step-up or step-down transformer and rectifier. { ¦dē,sē tü ¦dē,sē kən'vərd·ər }

Deacon process |CHEM ENG| A method of chlorine production by passing a hot mixture of gaseous hydrochloric acid with oxygen over a cuprous chloride catalyst. { 'dēk·ən ,präs·əs }

dead-air space |BUILD| A sealed air space, such as in a hollow wall. { ¦ded 'er ,spās }

dead area See blind spot. { 'ded ,er·ē·ə }

dead axle |MECH ENG| An axle that carries a wheel but does not drive it. { ¦ded 'ak·səl }

dead band |ELEC| The portion of a potentiometer element that is shortened by a tap; when the wiper traverses this area, there is no change in output. |ENG| The range of values of the measured variable to which an instrument will not effectively respond. Also known as dead zone; neutral zone. { 'ded ,band }

deadbeat |MECH| Coming to rest without vibration or oscillation, as when the pointer of a meter moves to a new position without overshooting. Also known as deadbeat response. { 'ded,bēt }

deadbeat algorithm |CONT SYS| A digital control algorithm which attempts to follow set-point changes in minimum time, assuming that the controlled process can be modeled approximately as a first-order plus dead-time system. { 'ded,bēt 'al·gə,rith·əm }

deadbeat response See deadbeat. { 'ded,bēt ri'späns }

dead block |ENG| A device placed on the ends of railroad passenger cars to absorb the shock of impacts. { 'ded ,bläk }

dead bolt |DES ENG| A lock bolt that is moved directly by the turning of a knob or key, not by spring action. { 'ded ,bōlt }

dead center |MECH ENG| **1.** A position of a crank in which the turning force applied to it by the connecting rod is zero; occurs when the crank and rod are in a straight line. **2.** A support for the work on a lathe which does not turn with the work. { ¦ded 'sen·tər }

dead-end tower |CIV ENG| Antenna or transmission line tower designed to withstand unbalanced mechanical pull from all the conductors in one direction together with the wind strain and vertical loads. { 'ded ,end ,taú·ər }

dead load See static load. { 'ded ,lōd }

deadlocking latch bolt See auxiliary dead latch. { 'ded,läk·iŋ 'lach ,bōlt }

deadman |CIV ENG| **1.** A buried plate, wall, or block attached at some distance from and forming an anchorage for a retaining wall. Also known as anchorage; anchor block; anchor wall. **2.** See anchor log. { 'ded,man }

deadman's brake |MECH ENG| An emergency device that automatically is activated to stop a vehicle when the driver removes his or her foot from the pedal. { ¦ded,manz 'brāk }

deadman's handle |MECH ENG| A handle on a machine designed so that the operator must continuously press on it in order to keep the machine running. { ¦ded,manz 'han·dəl }

dead rail |CIV ENG| One of two rails on a railroad weighing platform that permit an excessive load to leave the platform. { 'ded ,rāl }

dead room *See* anechoic chamber. { 'ded ,rüm }

dead sheave |ENG| A grooved wheel on a crown block over which the deadline is fastened. { 'ded 'shēv }

dead space |THERMO| A space filled with gas whose temperature differs from that of the main body of gas, such as the gas in the capillary tube of a constant-volume gas thermometer. { 'ded ,spās }

dead-stroke |MECH ENG| Having a recoilless or nearly recoilless stroke. { 'ded ,strōk }

dead-stroke hammer |MECH ENG| A power hammer provided with a spring on the hammer head to reduce recoil. { 'ded ,strōk 'ham·ər }

dead time |CONT SYS| The time interval between a change in the input signal to a process control system and the response to the signal. |ENG| The time interval, after a response to one signal or event, during which a system is unable to respond to another. Also known as insensitive time. { 'ded ,tīm }

dead-time compensation |CONT SYS| The modification of a controller to allow for time delays between the input to a control system and the response to the signal. { 'ded ,tīm käm·pən 'sā·shən }

dead-time correction |ENG| A correction applied to an observed counting rate to allow for the probability of the occurrence of events within the dead time. Also known as coincidence correction. { 'ded ,tīm kə'rek·shən }

dead track |CIV ENG| **1.** Railway track that is no longer used. **2.** A section of railway track that is electrically isolated from the track signal circuits. { ¦ded ¦trak }

deadweight gage |ENG| An instrument used as a standard for calibrating pressure gages in which known hydraulic pressures are generated by means of freely balanced (dead) weights loaded on a calibrated piston. { 'ded,wāt ,gāj }

deaeration |ENG| Removal of gas or air from a substance, as from feedwater or food. { dē ,er'ā·shən }

deaerator |MECH ENG| A device in which oxygen, carbon dioxide, or other noncondensable gases are removed from boiler feedwater, steam condensate, or a process stream. { dē'er,ād·ər }

deagglomeration |CHEM ENG| Size-reduction process in which loosely adhered clumps (agglomerates) of powders or crystals are broken apart without further disintegration of the powder or crystal particles themselves. { ,dē· ə,gläm·ə'rā·shən }

deal |DES ENG| **1.** A face on which numbers are registered by means of a pointer. **2.** A disk usually with a series of markings around its border, which can be turned to regulate the operation of a machine or electrical device. { dēl }

deasphalting |CHEM ENG| The process of removing asphalt from petroleum fractions. { dē'as,fȯl·tiŋ }

deblooming |CHEM ENG| The process by which the fluorescence, or bloom, is removed from petroleum oils by exposing them in shallow tanks to the sun and atmospheric conditions or by using chemicals. { dē'blüm·iŋ }

Deborah number |MECH| A dimensionless number used in rheology, equal to the relaxation time for some process divided by the time it is observed. Symbolized D. { də'bȯr·ə ,nəm·bər }

debris dam |CIV ENG| A fixed dam across a stream channel for the retention of sand, gravel, driftwood, or other debris. { də'brē ,dam }

debubblizer |ENG| A worker who removes bubbles from plastic rods and tubing. { dē,bə· bə,līz·ər }

debug |ELECTR| To detect and remove secretly installed listening devices popularly known as bugs. |ENG| To eliminate from a newly designed system the components and circuits that cause early failures. { dē'bəg }

debutanization |CHEM ENG| Removal of butane and lighter components in a natural-gasoline plant. { dē,byüt·ən·ə'zā·shən }

debutanizer |CHEM ENG| The fractionating column in a natural-gasoline plant in which butane and lighter components are removed. { de 'byüt·ən,īz·ər }

debye |ELEC| A unit of electric dipole moment, equal to 10^{-18} Franklin centimeter. { də'bī }

Debye theory |ELEC| The classical theory of the orientation polarization of polar molecules in which the molecules have a single relaxation time, and the plot of the imaginary part of the complex relative permittivity against the real part is a semicircle. { də'bī ,thē·ə·rē }

decade |ELEC| A group or assembly of 10 units; for example, a decade counter counts 10 in one column, and a decade box inserts resistance quantities in multiples of powers of 10. { de'kād }

decade bridge |ELECTR| Electronic apparatus for measurement of unknown values of resistances or capacitances by comparison with known values (bridge); one secondary section of the oscillator-driven transformer is tapped in decade steps, the other in 10 uniform steps. { de'kād ,brij }

decaliter |MECH| A unit of volume, equal to 10 liters, or to 0.01 cubic meter. { 'dek·ə,lēd·ər }

decameter |MECH| A unit of length in the metric system equal to 10 meters. { 'dek·ə,mēd·ər }

decantation |ENG| A method for mechanical dewatering of a wet solid by pouring off the liquid without disturbing underlying sediment or precipitate. { dē,kan'tā·shən }

decanter |ENG| Tank or vessel in which solids or immiscible dispersions in a carrier liquid settle or coalesce, with clear upper liquid withdrawn (decanted) as overflow from the top. { də'kant·ər }

decastere |MECH| A unit of volume, equal to 10 cubic meters. { 'dek·ə,stir }

deceleration |MECH| The rate of decrease of speed of a motion. { dē,sel·ə'rā·shən }

decelerometer [ENG] An instrument that measures the rate at which the speed of a vehicle decreases. { dē,sel·ə'räm·əd·ər }

declare [MECH] A unit of area, equal to 0.1 are or 10 square meters. { 'des·ē,er }

decibar [MECH] A metric unit of pressure equal to one-tenth bar. { 'des·ə,bär }

decibel meter [ENG] An instrument calibrated in logarithmic steps and labeled with decibel units and used for measuring power levels in communication circuits. { 'des·ə,bel ,mēd·ər }

decigram [MECH] A unit of mass, equal to 0.1 gram. { 'des·ə,gram }

deciliter [MECH] A unit of volume, equal to 0.1 liter, or 10^{-4} cubic meter. { 'des·ə,lēd·ər }

decimal balance [ENG] A balance having one arm 10 times the length of the other, so that heavy objects can be weighed by using light weights. { 'des·məl ,bal·əns }

decimal-binary switch [ELEC] A switch that connects a single input lead to appropriate combinations of four output leads (representing 1, 2, 4, and 8) for each of the decimal-numbered settings of its control knob; thus, for position 7, output leads 1, 2, and 4 would be connected to the input. { ¦des·məl ¦bīn·ə·rē 'swich }

decimeter [MECH] A metric unit of length equal to one-tenth meter. { 'des·ə,mēd·ər }

decision calculus [SYS ENG] A guide to the process of decision-making, often outlined in the following steps: analysis of the decision area to discover applicable elements; location or creation of criteria for evaluation; appraisal of the known information pertinent to the applicable elements and correction for bias; isolation of the unknown factors; weighting of the pertinent elements, known and unknown, as to relative importance; and projection of the relative impacts on the objective, and synthesis into a course of action. { di'sizh·ən 'kal·kyə·ləs }

decision rule [SYS ENG] In decision theory, the mathematical representation of a physical system which operates upon the observed data to produce a decision. { di'sizh·ən ,rül }

decision theory [SYS ENG] A broad spectrum of concepts and techniques which have been developed to both describe and rationalize the process of decision making, that is, making a choice among several possible alternatives. { di'sizh·ən ,the·ə·rē }

decision tree [IND ENG] Graphic display of the underlying decision process involved in the introduction of a new product by a manufacturer. { di'sizh·ən ,trē }

deck [CIV ENG] **1.** A floor, usually of wood, without a roof. **2.** The floor or roadway of a bridge. [ENG] A magnetic-tape transport mechanism. { dek }

deck bridge [CIV ENG] A bridge that carries the deck on the very top of the superstructure. { 'dek ,brij }

decking [CIV ENG] Surface material on a deck. [ENG] Separating explosive charges containing primers with layers of inert material to prevent passage of concussion. { 'dek·iŋ }

deckle [ENG] A detachable wood frame fitted around the edges of a papermaking mold. { 'dek·əl }

deckle rod [ENG] A small rod inserted at each end of the extrusion coating die to adjust the die opening length. { 'dek·əl ,räd }

deckle strap [ENG] An endless rubber band which runs longitudinally along the wire edges of a paper machine and determines web width. { 'dek·əl ,strap }

deck roof [BUILD] A roof that is nearly flat and without parapet walls. { 'dek ,rüf }

deck truss [CIV ENG] The frame of a deck. { 'dek ,trəs }

declination axis [ENG] For an equatorial mounting of a telescope, an axis of rotation that is perpendicular to the polar axis and allows the telescope to be pointed at objects of different declinations. { ,dek·lə'nā·shəs ,ak·səs }

declination circle [ENG] For a telescope with an equatorial mounting, a setting circle attached to the declination axis that shows the declination to which the telescope is pointing. { ,dek·lə'nā·shən ,sər·kəl }

declination compass See declinometer. { ,dek·lə'nā·shən ,kəm· pəs }

declination variometer [ENG] An instrument that measures changes in the declination of the earth's magnetic field, consisting of a permanent bar magnet, usually about 0.4 inch (1 centimeter) long, suspended with a plane mirror from a fine quartz fiber 2–6 inches (5–15 centimeters) in length; a lens focuses to a point a beam of light reflected from the mirror to recording paper mounted on a rotating drum. Also known as D variometer. { ,dek·lə'nā·shən ,ver·ē'äm·əd·ər }

declinometer [ENG] A magnetic instrument similar to a surveyor's compass, but arranged so that the line of sight can be rotated to conform with the needle or to any desired setting on the horizontal circle; used in determining magnetic declination. Also known as declination compass. { ,dek·lə'näm·əd·ər }

decoking [CHEM ENG] Removal of petroleum coke from equipment. { dē'kōk·iŋ }

decolorize [CHEM ENG] To remove the color from, as from a liquid. { dē'kəl·ə,rīz }

decolorizer [CHEM ENG] An agent used to decolorize; the removal of color may occur by a chemical reaction or a physical reaction. { dē'kəl·ə,rīz·ər }

decompression [ENG] Any procedure for the relief of pressure or compression. { dē·kəm'presh·ən }

decompression chamber [ENG] **1.** A steel chamber fitted with auxiliary equipment to raise its air pressure to a value two to six times atmospheric pressure; used to relieve a diver who has decompressed too quickly in ascending. **2.** Such a chamber in which conditions of high atmospheric pressure can be simulated for experimental purposes. { dē·kəm'presh·ən ,chām·bər }

decompression table [ENG] A diving guide that

deflectometer

lists ascent rates and breathing mixtures to provide safe pressure reduction to atmospheric pressure after a dive. { dē·kəm'presh·ən ,tā·bəl }

deconcentrator [ENG] An apparatus for removing dissolved or suspended material from feedwater. { dē'käns·ən,trād·ər }

decontamination [ENG] The removing of chemical, biological, or radiological contamination from, or the neutralizing of it on, a person, object, or area. { dē·kən,tam·ə'nā·shən }

decouple [ENG] **1.** To minimize or eliminate airborne shock waves of a nuclear or other explosion by placing the explosives deep under the ground. **2.** To minimize the seismic effect of an underground explosion by setting it off in the center of an underground cavity. { dē'kəp·əl }

decoupler [IND ENG] A materials handling device designed specifically for cellular manufacturing. { dē'kəp·lər }

decrement gage [ENG] A type of molecular gage consisting of a vibrating quartz fiber whose damping is used to determine the viscosity and, thereby, the pressure of a gas. Also known as quartz-fiber manometer. { 'de·krə·mənt ,gāj }

decremeter [ENG] An instrument for measuring the logarithmic decrement (damping) of a train of waves. { 'dek·rə,mēd·ər }

dedendum [DES ENG] The difference between the radius of the pitch circle of a gear and the radius of its root circle. { də'den·dəm }

dedendum circle [DES ENG] A circle tangent to the bottom of the spaces between teeth on a gear wheel. { də'den·dəm ,sər·kəl }

deemphasis [ENG ACOUS] A process for reducing the relative strength of higher audio frequencies before reproduction, to complement and thereby offset the preemphasis that was introduced to help override noise or reduce distortion. Also known as postemphasis; postequalization. { dē'em·fə·səs }

deemphasis network [ENG ACOUS] An RC filter inserted in a system to restore preemphasized signals to their original form. { dē'em·fə·səs ,net,wərk }

deep-draw mold [ENG] A mold for plastic material that is long in relation to the thickness of the mold wall. { ¦dēp ¦drò 'mōld }

deep underwater muon and neutrino detector [ENG] A proposed device for detecting and determining the direction of extraterrestrial neutrinos passing through a volume of approximately 1 cubic kilometer of ocean water, using an array of several thousand Cerenkov counters suspended in the water to sense the showers of charged particles generated by neutrinos. Abbreviated DUMAND. { ¦dēp ,ən·dər'wòd·ər 'myü ,än an nü'trē·nō di,tek·tər }

deep well [CIV ENG] A well that draws its water from beneath shallow impermeable strata, at depths exceeding 22 feet (6.7 meters). { 'dēp ,wel }

deep-well pump [MECH ENG] A multistage centrifugal pump for lifting water from deep, small-diameter wells; a surface electric motor operates

the shaft. Also known as vertical turbine pump. { 'dēp ,wel ,pəmp }

deethanize [CHEM ENG] To separate and remove ethane and sometimes lighter fractions from heavy substances, such as propane, by distillation. { dē'eth·ə,nīz }

deethanizer [CHEM ENG] The equipment used to deethanize. { dē'eth·ə,nīz·ər }

defecation [CHEM ENG] Industrial purification, or clarification, of sugar solutions. { ,def·ə'kā·shən }

defender [IND ENG] A machine or facility which is being considered for replacement. { di'fen·dər }

deferrization [CHEM ENG] Removal of iron, for example, from water in an industrial process. { dē,fer·ə'zā·shən }

deflashing [ENG] Finishing technique to remove excess material (flash) from a plastic or metal molding. { dē'flash·iŋ }

deflected jet fluidic flowmeter See fluidic flow sensor. { di¦flek·təd 'jet flü'id·ik 'flō,mēd·ər }

deflecting torque [MECH] An instrument's moment, resulting from the quantity measured, that acts to cause the pointer's deflection. { di'flek·diŋ ,tòrk }

deflection [ELECTR] The displacement of an electron beam from its straight-line path by an electrostatic or electromagnetic field. [ENG] **1.** Shape change or reduction in diameter of a conduit, produced without fracturing the material. **2.** Elastic movement or sinking of a loaded structural member, particularly of the mid-span of a beam. { di'flek·shən }

deflection bit [DES ENG] A long, cone-shaped, noncoring bit used to drill past a deflection wedge in a borehole. { di'flek·shən ,bit }

deflection curve [MECH] The curve, generally downward, described by a shot deviating from its true course. { di'flek·shən ,kərv }

deflection magnetometer [ENG] A magnetometer in which magnetic fields are determined from the angular deflection of a small bar magnet that is pivoted so that it is free to move in a horizontal plane. { di'flek·shən ,mag·nə'täm·əd·ər }

deflection meter [ENG] A flowmeter that applies the differential pressure generated by a differential-producing primary device across a diaphragm or bellows in such a way as to create a deflection proportional to the differential pressure. { di'flek·shən ,mēd·ər }

deflection-modulated indicator See amplitude-modulated indicator. { di'flek·shən ¦mäj·ə,lād·əd 'in·də,kād·ər }

deflection ultrasonic flowmeter [ENG] A flowmeter for determining velocity from the deflection of a high-frequency sound beam directed across the flow. Also known as drift ultrasonic flowmeter. { di'flek·shən ¦əl·trə¦sän·ik 'flō ,mēd·ər }

deflection wedge [DES ENG] A wedge-shaped tool inserted into a borehole to direct the drill bit. { di'flek·shən ,wej }

deflectometer [ENG] An instrument used for

149

measuring minute deformations in a structure under transverse stress. { ‚dē‚flek'täm·əd·ər }

deflector |ENG| A plate, baffle, or the like that diverts the flow of a forward-moving stream. { di'flek·tər }

deflocculate |CHEM ENG| To break up and disperse agglomerates and form a stable colloid. { dē'fläk·yə‚lāt }

defoaming |CHEM ENG| Reduction or elimination of foam. { dē'fōm·iŋ }

defocus |ENG| To make a beam of x-rays, electrons, light, or other radiation deviate from an accurate focus at the intended viewing or working surface. { dē'fō·kəs }

deformation |MECH| Any alteration of shape or dimensions of a body caused by stresses, thermal expansion or contraction, chemical or metallurgical transformations, or shrinkage and expansions due to moisture change. { ‚def·ər'mā·shən }

deformation curve |MECH| A curve showing the relationship between the stress or load on a structure, structural member, or a specimen and the strain or deformation that results. Also known as stress-strain curve. { ‚def·ər'mā·shən ‚kərv }

deformation ellipsoid See strain ellipsoid. { ‚def·ər'mā·shən ə'lip‚sóid }

deformation thermometer |ENG| A thermometer with transducing elements which deform with temperature; examples are the bimetallic thermometer and the Bourdon-tube type of thermometer. { ‚def·er'mā·shən thər‚mäm·əd·ər }

deformed bar |CIV ENG| A steel bar with projections or indentations to increase mechanical bonding; used to reinforce concrete. { dē 'fórmd ‚bär }

deformeter |ENG| An instrument used to measure minute deformations in materials in structural models. { dē'fór‚mēd·ər }

defrost |ENG| To keep free of ice or to remove ice. |THERMO| To thaw out from a frozen state. { dē'fróst }

degas |ELECTR| To drive out and exhaust the gases occluded in the internal parts of an electron tube or other gastight apparatus, generally by heating during evacuation. |ENG| To remove gas from a liquid or solid. { dē'gas }

degassing See breathing. { dē'gas·iŋ }

degauss |ELECTR| To remove, erase, or clear information from a magnetic tape, disk, drum, or core. { dē'gaús }

degradation |THERMO| The conversion of energy into forms that are increasingly difficult to convert into work, resulting from the general tendency of entropy to increase. { ‚deg·rə'dā·shən }

degradation failure |ENG| Failure of a device because of a shift in a parameter or characteristic which exceeds some previously specified limit. { ‚deg·rə'dā·shən ‚fāl·yər }

degrease |CHEM ENG| **1.** To remove grease from wool with chemicals. **2.** To remove grease from hides or skins in tanning by tumbling them in solvents. { dē'grēs }

degreaser |ENG| A machine designed to clean grease and foreign matter from mechanical parts and like items, usually metallic, by exposing them to vaporized or liquid solvent solutions confined in a tank or vessel. { dē'grēs·ər }

degree |THERMO| One of the units of temperature or temperature difference in any of various temperature scales, such as the Celsius, Fahrenheit, and Kelvin temperature scales (the Kelvin degree is now known as the kelvin). { di'grē }

degree-day |MECH ENG| A measure of the departure of the mean daily temperature from a given standard; one degree-day is recorded for each degree of departure above (or below) the standard during a single day; used to estimate energy requirements for building heating and, to a lesser extent, for cooling. { di'grē ‚dā }

degree of curve |CIV ENG| A measure of the curvature of a railway or highway, equal to the angle subtended by a 100-foot (32.8-meter) chord (railway) or by a 100-foot arc (highway). { di'grē əv 'kərv }

degree of freedom |MECH| **1.** Any one of the number of ways in which the space configuration of a mechanical system may change. **2.** Of a gyro, the number of orthogonal axes about which the spin axis is free to rotate, the spin axis freedom not being counted; this is not a universal convention; for example, the free gyro is frequently referred to as a three-degree-of-freedom gyro, the spin axis being counted. { di'grē əv 'frē·dəm }

degritting |CHEM ENG| Removal of fine solid particles (grit) from a liquid carrier by gravity separation (settling) or centrifugation. { dē 'grid·iŋ }

dehumidification |MECH ENG| The process of reducing the moisture in the air; serves to increase the cooling power of air. { ‚dē·yü‚mid·ə·fə'kā·shən }

dehumidifier |MECH ENG| Equipment designed to reduce the amount of water vapor in the ambient atmosphere. { ‚dē·yü'mid·ə‚fī·ər }

dehydration tank |CHEM ENG| A tank in which warm air is blown through oil to remove moisture. { ‚dē·hī'drā·shən ‚taŋk }

dehydrator |CHEM ENG| Vessel or process system for the removal of liquids from gases or solids by the use of heat, absorbents, or adsorbents. { dē'hī‚drād·ər }

dehydrocyclization |CHEM ENG| Any process involving both dehydrogenation and cyclization, as in petroleum refining. { dē‚hī·drō‚sīk·lə'zā·shən }

deicing |ENG| The removal of ice deposited on any object, especially as applied to aircraft icing, by heating, chemical treatment, and mechanical rupture of the ice deposit. { dē'īs·iŋ }

deinking |CHEM ENG| The process of removing ink from recycled paper so that the fibers can be used again. { dē'iŋk·iŋ }

delamination |ENG| Separation of a laminate into its constituent layers. { dē‚lam·ə'nā·shən }

Delaunay orbit element |MECH| In the *n*-body

problem, certain functions of variable elements of an ellipse with a fixed focus along which one of the bodies travels; these functions have rates of change satisfying simple equations. { də·lō·nā 'ȯr·bət ‚el·ə·mənt }

delay |IND ENG| Interruption of the normal tempo of an operation; may be avoidable or unavoidable. { di'lā }

delay-action detonator See delay blasting cap. { di'lā ‚ak·shən 'det·ən‚ād·ər }

delay allowance |IND ENG| A percentage of the normal operating time added to the normal time to allow for delays. { di'lā ə‚laú·əns }

delay blasting cap |ENG| A blasting cap which explodes at a definite time interval after the firing current has been passed by the exploder. Also known as delay-action detonator. { di'lā 'blast·iŋ ‚kap }

delayed coking |CHEM ENG| A semicontinuous thermal process for converting heavy petroleum stock to lighter material. { di'lād 'kōk·iŋ }

delayed combustion |ENG| Secondary combustion in succeeding gas passes beyond the furnace volume of a boiler. { di'lād kəm'bəs·chən }

delay time |CONT SYS| The amount of time by which the arrival of a signal is retarded after transmission through physical equipment or systems. |ELECTR| The time taken for collector current to start flowing in a transistor that is being turned on from the cutoff condition. |IND ENG| A span of time during which a worker is idle because of factors beyond personal control. { di'lā ‚tīm }

delignification |CHEM ENG| A chemical process for removing lignin from wood. { dē‚lig·nə·fə'kā·shən }

delta |ELECTR| The difference between a partial-select output of a magnetic cell in a one state and a partial-select output of the same cell in a zero state. { 'del·tə }

delta modulation |ELECTR| A pulse-modulation technique in which a continuous signal is converted into a binary pulse pattern, for transmission through low-quality channels. { 'del·tə ‚mäj·ə'lā·shən }

demand See demand factor. { də'mand }

demanded motions inventory |IND ENG| A list of all motions that are required to perform a specific task, including an exact characterization of each. { də‚man·dəd ‚mō·shənz 'in·vən‚tȯr·ē }

demand factor |ELEC| The ratio of the maximum demand of a building for electric power to the total connected load. Also known as demand. { də'mand ‚fak·tər }

demand meter |ENG| Any of several types of instruments used to determine a customer's maximum demand for electric power over an appreciable time interval; generally used for billing industrial users. { də'mand ‚mēd·ər }

demand regulator |ENG| A component of an open-circuit diving system that permits the diver to expel used air directly into the water without rebreathing exhaled carbon dioxide. { də'mand ‚reg·yə‚lād·ər }

demand system |ENG| A system in an airplane that automatically dispenses oxygen according to the demand of the flyer's body. { də'mand ‚sis·təm }

demethanation See demethanization. { dē‚meth·ə'nā·shən }

demethanator |CHEM ENG| The apparatus in which demethanization isconducted. { dē 'meth·ə‚nād·ər }

demethanization |CHEM ENG| The process of distillation in which methane is separated from the heavier components. Also known as demethanation. { dē‚meth·ən·ə'zā·shən }

demineralization |CHEM ENG| Removal of mineral constituents from water. { dē‚min·rə·lə 'zā·shən }

demister |MECH ENG| A series of ducts in automobiles arranged so that hot, dry air directed from the heat source is forced against the interior of the windscreen or windshield to prevent condensation. { dē'mis·tər }

demister blanket |ENG| A section of knitted wire mesh that is placed below the vapor outlet of a vaporizer or an evaporator to separate entrained liquid droplets from the stream of vapor. Also known as demister pad. { dē'mis·tər ‚blaŋ·kət }

demister pad See demister blanket. { dē'mis·tər ‚pad }

demodulator See detector. { dē'mäj·ə‚lad·ər }

demolition |CIV ENG| The act or process of tearing down a building or other structure. { ‚dem·ə'lish·ən }

demon of Maxwell |THERMO| Hypothetical creature who controls a trapdoor over a microscopic hole in an adiabatic wall between two vessels filled with gas at the same temperature, so as to supposedly decrease the entropy of the gas as a whole and thus violate the second law of thermodynamics. Also known as Maxwell's demon. { 'dē·mən əv 'maks‚wel }

demulsification |CHEM ENG| Prevention or breaking of liquid-liquid emulsions by chemical, mechanical or electrical demulsifiers. { də‚məl·sə·fə'kā·shən }

demulsifier |CHEM ENG| A chemical, mechanical, or electrical system that either breaks liquid-liquid emulsions or prevents them from forming. { dē'məl·sə‚fī·ər }

demultiplexer |ELECTR| A device used to separate two or more signals that were previously combined by a compatible multiplexer and transmitted over a single channel. { dē‚məl·tə‚plek·sər }

Denison sampler |ENG| A soil sampler consisting of a central nonrotating barrel which is forced into the soil as friction is removed by a rotating external barrel; the bottom can be closed to retain the sample during withdrawal. { 'den·ə·sən ‚sam·plər }

De Nora cell |CHEM ENG| Mercury-cathode cell used for production of chlorine and caustic soda by electrolysis of sodium chloride brine. { də'nȯr·ə ‚sel }

dense-air refrigeration cycle See reverse Brayton cycle. { ¦dens ¦er ri‚frij·ə'rā·shən ‚sī·kəl }

dense-air system See cold-air machine. { ¦dens 'er ‚sis·təm }

densify |ENG| To increase the density of a material such as wood by subjecting it to pressure or impregnating it with another material. { 'den·sə‚fī }

densimeter |ENG| An instrument which measures the density or specific gravity of a liquid, gas, or solid. Also known as densitometer; density gage; density indicator; gravitometer. { den 'sim·əd·ər }

densitometer |ENG| **1.** An instrument which measures optical density by measuring the intensity of transmitted or reflected light; used to measure photographic density. **2.** See densimeter. { ‚den·sə'täm·əd·ər }

density |MECH| The mass of a given substance per unit volume. { 'den· səd·ē }

density bottle See specific-gravity bottle. { 'den· səd·ē ‚bäd·əl }

density correction |ENG| **1.** The part of the temperature correction of a mercury barometer which is necessitated by the variation of the density of mercury with temperature. **2.** The correction, applied to the indications of a pressure-tube anemometer or pressure-plate anemometer, which is necessitated by the variation of air density with temperature. { 'den·səd·ē kə'rek·shən }

density gage See densimeter. { 'den·səd·ē ‚gāj }

density indicator See densimeter. { 'den·səd·ē ‚in·də‚kād·ər }

density rule |ENG| A grading system for lumber based on the width of annual rings. { 'den·səd·ē ‚rül }

density transmitter |ENG| An instrument used to record the density of a flowing stream of liquid by measuring the buoyant force on an air-filled chamber immersed in the stream. { 'den·səd·ē tranz'mid·ər }

dental coupling |MECH ENG| A type of flexible coupling used to join a steam turbine to a reduction-gear pinion shaft; consists of a short piece of shaft with gear teeth at each end, and mates with internal gears in a flange at the ends of the two shafts to be joined. { 'dent·əl 'kəp·liŋ }

dental work See cementation. { 'dent·əl ‚wərk }

deodorizing |CHEM ENG| A process for removing odor-creating substances from oil or fat, in which the oil or fat is held at high temperatures and low pressure while steam is blown through. { dē'ōd·ə‚rīz·iŋ }

deoil |CHEM ENG| To reduce the amount of liquid oil entrained in solid wax. { dē'óil }

departure track |CIV ENG| A railroad yard track for combining freight cars into outgoing trains. { di'pär·chər ‚trak }

depentanizer |CHEM ENG| A fractionating column for removal of pentane and lighter fractions from a hydrocarbon mixture. { də'pent·ən ‚īz·ər }

deperm See degauss. { dē'pərm }

dephlegmation |CHEM ENG| In a distillation operation, the partial condensation of vapor to form a liquid richer in higher boiling constituents than the original vapor. { dē‚fleg'mā·shən }

dephlegmator |CHEM ENG| An apparatus used in fractional distillation to cool the vapor mixture, thereby condensing higher-boiling fractions. { dē'fleg‚mād·ər }

depilation |ENG| Removal of hair from animal skins in processing leather. { ‚dep·ə'lā·shən }

depletion |ELECTR| Reduction of the charge-carrier density in a semiconductor below the normal value for a given temperature and doping level. { də'plē·shən }

depletion layer |ELECTR| An electric double layer formed at the surface of contact between a metal and a semiconductor having different work functions, because the mobile carrier charge density is insufficient to neutralize the fixed charge density of donors and acceptors. Also known as barrier layer (deprecated); blocking layer (deprecated); space-charge layer. { də'plē·shən ‚lā·ər }

depletion-layer capacitance See barrier capacitance. { di'plē·shən ‚lā·ər kə'pas·əd·əns }

depletion-layer rectification |ELECTR| Rectification at the junction between dissimilar materials, such as a pn junction or a junction between a metal and a semiconductor. Also known as barrier-layer rectification. { də'plē·shən ‚lā·ər ‚rek·tə·fə'kā·shən }

depletion-layer transistor |ELECTR| A transistor that relies directly on motion of carriers through depletion layers, such as spacistor. { də'plē·shən ‚lā·ər tran'zis·tər }

depletion region |ELECTR| The portion of the channel in a metal oxide field-effect transistor in which there are no charge carriers. { də'plē·shən ‚rē·jən }

depolarization |ELEC| The removal or prevention of polarization in a substance (for example, through the use of a depolarizer in an electric cell) or of polarization arising from the field due to the charges induced on the surface of a dielectric when an external field is applied. { dē‚pō·lə·rə'zā·shən }

deposit gage |ENG| The general name for instruments used in air pollution studies for determining the amount of material deposited on a given area during a given time. { də'päz·ət ‚gāj }

depreciation |IND ENG| Loss of value due to physical deterioration. { di‚prē·shē'ā·shən }

depressed center car |ENG| A flat railroad car having a low center section; used to provide adequate tunnel clearance for oversized loads. { di¦prest 'sent·ər ‚kär }

depression angle See angle of depression. { di 'presh·ən ‚aŋ·gəl }

depressor |CHEM ENG| An agent that prevents or retards a chemical reaction or process. { di 'pres·ər }

depropanization |CHEM ENG| In processing of petroleum, the removal of propane and sometimes higher fractions. { dē‚prō·pə·nə'zā·shən }

depropanizer |CHEM ENG| A fractionating column in a gasoline plant for removal of propane and lighter components. { dē'prō·pə,nīz·ər }

depth finder |ENG| A radar or ultrasonic instrument for measuring the depth of the sea. { 'depth ,fīnd·ər }

depth gage |DES ENG| An instrument or tool for measuring the depth of depression to a thousandth inch { 'depth ,gāj }

depth marker |ENG| A thin board or other lightweight substance used as a means of identifying the surface of snow or ice which has been covered by a more recent snowfall. { 'depth ,märk·ər }

depth micrometer |DES ENG| A micrometer used to measure the depths of holes, slots, and distances of shoulders and projections. { 'depth mī'kräm·əd·ər }

depth of engagement |DES ENG| The depth of contact, in a radial direction, between mating threads. { 'depth əv ,en'gāj·mənt }

depth of thread |DES ENG| The distance, in a radial direction, from the crest of a screw thread to the base. { 'depth əv 'thred }

depth sounder |ENG| An instrument for mechanically measuring the depth of the sea beneath a ship. { 'depth ,saúnd·ər }

depth-type filtration |CHEM ENG| Removal of solids by passing the carrier fluid through a mass-filter medium that provides a tortuous path with many entrapments to catch the solids. { 'depth ,tīp fil'trā·shən }

dequeue |ENG| To select an item from a queue. { dē'kyü }

derail |ENG| **1.** To cause a railroad car or engine to run off the rails. **2.** A device to guide railway cars or engines off the tracks to avoid collision or other accident. { dē'rāl }

derating |ELECTR| The reduction of the rating of a device to improve reliability or to permit operation at high ambient temperatures. { dē'rād·iŋ }

derivative action |CONT SYS| Control action in which the speed at which a correction is made depends on how fast the system error is increasing. Also known as derivative compensation; rate action. { də'riv·əd·iv ,ak·shən }

derivative compensation See derivative action. { də'riv·əd·iv ,käm·pən'sā·shən }

derivative network |CONT SYS| A compensating network whose output is proportional to the sum of the input signal and its derivative. Also known as lead network. { də'riv·əd·iv 'net ,wərk }

derived sound system |ENG ACOUS| A four-channel sound system that is artificially synthesized from conventional two-channel stereo sound by an adapter, to provide feeds to four loudspeakers for approximating quadraphonic sound. { də'rīvd 'saúnd ,sis·təm }

derosination |CHEM ENG| Removing excess resins from wood by saponification with alkaline aqueous solutions or organic solvents. { dē ,räz·ən'ā·shən }

derrick |MECH ENG| A hoisting machine consisting usually of a vertical mast, a slanted boom, and associated tackle; may be operated mechanically or by hand. { 'der·ik }

derrick crane See stiffleg derrick. { 'der·ik ,krān }

derrick post See king post. { 'der·ik ,pōst }

desalination |CHEM ENG| Removal of salt, as from water or soil. Also known as desalting. { dē,sal·ə'nā·shən }

desalinization See desalination. { dē,sal·ə· nə'zā·shən }

desalting |CHEM ENG| **1.** The process of extracting inorganic salts from oil. **2.** See desalination. { dē'sól·tiŋ }

desander |ENG| A centrifuge-type device for removing sand from drilling fluid in order to prevent abrasion damage to pumps. { dē'san·dər }

descaling |ENG| Removing scale, usually oxides, from the surface of a metal or the inner surface of a pipe, boiler, or other object. { dē 'skāl·iŋ }

descending branch |MECH| That portion of a trajectory which is between the summit and the point where the trajectory terminates, either by impact or air burst, and along which the projectile falls, with altitude constantly decreasing. Also known as descent trajectory. { di'sen·diŋ 'branch }

descending vertical angle See angle of depression. { di'sen·diŋ |vərd·i·kəl 'aŋ·gəl }

descent trajectory See descending branch. { di 'sent trə'jek·tə·rē }

describing function |CONT SYS| A function used to represent a nonlinear transfer function by an approximately equivalent linear transfer function; it is the ratio of the phasor representing the fundamental component of the output of the nonlinearity, determined by Fourier analysis, to the phasor representing a sinusoidal input signal. { di'skrīb·iŋ ,faŋk·shən }

desiccator |CHEM ENG| A closed vessel, usually made of glass and having an airtight lid, used for drying solid chemicals by means of a desiccant. { 'des·ə,kād·ər }

design engineering |ENG| A branch of engineering concerned with the creation of systems, devices, and processes useful to and sought by society. { di'zīn ,en·jə'nir·iŋ }

design factor |ENG| A safety factor based on the ratio of ultimate load to maximum permissible load that can be safely placed on a structure. { di'zīn ,fak·tər }

design flood |CIV ENG| The flood, either observed or synthetic, which is chosen as the basis for the design of a hydraulic structure. { di 'zīn ,fləd }

design for environment |SYS ENG| A methodology for the design of products and systems that promotes pollution prevention and resource conservation by including within the design process the systematic consideration of the environmental implications of engineering designs. Abbreviated DFE. { di|zīn fər in'vī·ərn·mənt }

design head |CIV ENG| The planned elevation between the free level of a water supply and the point of free discharge or the level of free discharge surface. { di'zīn ,hed }

design heating load [ENG] The space heating needs of a building or an enclosed area expressed in terms of the probable maximum requirement. { di'zīn 'hēd·iŋ ,lōd }

design load [DES ENG] The most stressful combination of weight or other forces a building, structure, or mechanical system or device is designed to sustain. { di'zīn ,lōd }

design pressure [CIV ENG] 1. The force exerted by a body of still water on a dam. 2. The pressure which the dam can withstand. [DES ENG] The pressure used in the calculation of minimum thickness or design characteristics of a boiler or pressure vessel in recognized code formulas; static head may be added where appropriate for specific parts of the structure. { di'zīn 'presh·ər }

design speed [CIV ENG] The highest continuous safe vehicular speed as governed by the design features of a highway. { di'zīn ,spēd }

design standards [DES ENG] Generally accepted uniform procedures, dimensions, materials, or parts that directly affect the design of a product or facility. { di'zīn ,stan·dərdz }

design storm [CIV ENG] A storm whose magnitude, rate, and intensity do not exceed the design load for a storm drainage system or flood protection project. { di'zīn ,stórm }

design stress [DES ENG] A permissible maximum stress to which a machine part or structural member may be subjected, which is large enough to prevent failure in case the loads exceed expected values, or other uncertainties turn out unfavorably. { di'zīn ,stres }

design thickness [DES ENG] The sum of required thickness and corrosion allowance utilized for individual parts of a boiler or pressure vessel. { di'zīn ,thik·nəs }

desilter [MECH ENG] Wet, mechanical solids classifier (separator) in which silt particles settle as the carrier liquid is slowly stirred by horizontally revolving rakes; solids are plowed outward and removed at the periphery of the container bowl. { dē'sil·tər }

desilting basin [CIV ENG] A space or structure constructed just below a diversion structure of a canal to remove bed, sand, and silt loads. Also known as desilting works. { dē'sil·tiŋ ,bā·sən }

desilting works See desilting basin. { dē'sil·tiŋ ,wərks }

desired track See course. { də'zīrd 'trak }

deslimer [MECH ENG] Apparatus, such as a bowl-type centrifuge, used to remove fine, wet particles (slime) from cement rocks and to size pigments and abrasives. { dē'slīm·ər }

destearinate [CHEM ENG] A process of removing from a fatty oil the lower melting point compounds. { dē'stir·ə,nāt }

destraction [CHEM ENG] A high-pressure technique for separating high-boiling or nonvolatile material by dissolving it with application of supercritical gases. { di'strak·shən }

destructive breakdown [ELECTR] Breakdown of

the barrier between the gate and channel of a field-effect transistor, causing failure of the transistor. { di'strək·tiv 'brāk,daún }

destructive testing [ENG] 1. Intentional operation of equipment until it fails, to reveal design weaknesses. 2. A method of testing a material that degrades the sample under investigation. { di'strək·tiv 'test·iŋ }

desulfurization [CHEM ENG] The removal of sulfur, as from molten metals or petroleum oil. { dē,səl·fə·rə'zā·shən }

desulfurization unit [CHEM ENG] A unit in petroleum refining for removal of sulfur compounds or sulfur. { dē·səl·fə·rə'zā·shən ,yü·nət }

detachable bit [ENG] An all-steel drill bit that can be removed from the drill steel, and can be resharpened. Also known as knock-off bit; rip bit. { di'tach·ə·bəl 'bit }

detailing See screening. { 'dē,tāl·iŋ }

det drill See fusion-piercing drill. { 'det ,dril }

detector bar [CIV ENG] A device that keeps a railroad switch locked while a train is passing over it. { di'tek·tər ,bär }

detector car [ENG] A railroad car used to detect flaws in rails. { di'tek·tər ,kär }

detent [MECH ENG] A catch or lever in a mechanism which initiates or locks movement of a part, especially in escapement mechanisms. { 'dē ,tent }

detention basin [CIV ENG] A reservoir without control gates for storing water over brief periods of time until the stream has the capacity for ordinary flow plus released water; used for flood regulation. { di'ten·chən ,bā·sən }

deterioration [ENG] Decline in the quality of equipment or structures over a period of time due to the chemical or physical action of the environment. { di,tir·ē·ə'rā·shən }

determinant [CONT SYS] The product of the partial return differences associated with the nodes of a signal-flow graph. { də'tər·mə·nənt }

determinate structure [MECH] A structure in which the equations of statics alone are sufficient to determine the stresses and reactions. { də'tər·mə·nət 'strək·chər }

determinism See causality. { də'tər·mə,niz·əm }

detonating fuse [ENG] A device consisting of a core of high explosive within a waterproof textile covering and set off by an electrical blasting cap fired from a distance by means of a fuse line; used in large, deep boreholes. { 'det,ən,ād·iŋ 'fyüz }

detonating rate [MECH] The velocity at which the explosion wave passes through a cylindrical charge. { 'det·ən,ād·iŋ ,rāt }

detonating relay [ENG] A device used in conjunction with the detonating fuse to avoid short-delay blasting. { 'det·ən,ād·iŋ ,rē,lā }

detonation [MECH ENG] Spontaneous combustion of the compressed charge after passage of the spark in an internal combustion engine; it is accompanied by knock. { ,det·ən'ā·shən }

detonation front [ENG] The reaction zone of a detonation. { ,det·ən'ā·shən ,frənt }

detonator |ENG| A device, such as a blasting cap, employing a sensitive primary explosive to detonate a high-explosive charge. { 'det·ən ‚ād·ər }

detonator safety |ENG| A fuse has detonator safety or is detonator safe when the functioning of the detonator cannot initiate subsequent explosive train components. { 'det·ən‚ād·ər ‚sāf·tē }

detonics |ENG| The study of detonating and explosives performance. { de'tän·iks }

detritus tank |CIV ENG| A tank in which heavy suspended matter is removed in sewage treatment. { də'trīd·əs ‚taŋk }

Detroit rocking furnace |ENG| An indirect arc type of rocking furnace having graphite electrodes entering horizontally from opposite ends. { də'trȯit 'räk·iŋ 'fər·nəs }

development |ENG| The exploratory work required to determine the best production techniques to bring a new process or piece of equipment to the production stage. { də'vel·əp·mənt }

deviation |ENG| The difference between the actual value of a controlled variable and the desired value corresponding to the set point. { ‚dēv·ē'ā·shən }

deviation factor See compressibility factor. { ‚dēv·ē'ā·shən ‚fak·tər }

deviatonic stress |MECH| The portion of the total stress that differs from an isostatic hydrostatic pressure; it is equal to the difference between the total stress and the spherical stress. { ‚dev·ē·ə'tän·ik 'stres }

device |ELECTR| An electronic element that cannot be divided without destroying its stated function; commonly applied to active elements such as transistors and transducers. |ENG| A mechanism, tool, or other piece of equipment designed for specific uses. { di'vīs }

devil See devil float. { 'dev·əl }

devil float |ENG| A hand float containing nails projecting at each corner and used to roughen the surface of plaster to provide a key for the next coat. Also known as devil; nail coat. { 'dev·əl ‚flōt }

devil's pitchfork |DES ENG| A tool with flexible prongs used in recovery of a bit, underreamer, cutters, or such lost during drilling. { 'de·vəlz 'pich‚fȯrk }

devolatilize |CHEM ENG| To remove volatile components from a material. { ‚dē'väl·ə·tə‚līz }

Dewar calorimeter |ENG| **1.** Any calorimeter in which the sample is placed inside a Dewar flask to minimize heat losses. **2.** A calorimeter for determining the mean specific heat capacity of a solid between the boiling point of a cryogenic liquid, such as liquid oxygen, and room temperature, by measuring the amount of the liquid that evaporates when the specimen is dropped into the liquid. { ¦dü·ər ‚kal·ə'rim·əd·ər }

dewaterer |MECH ENG| Wet-type mechanical classifier (solids separator) in which solids settle out of the carrier liquid and are concentrated for recovery. { dē'wȯd·ər·ər }

dewatering |ENG| **1.** Removal of water from solid material by wet classification, centrifugation, filtration, or similar solid-liquid separation techniques. **2.** Removing or draining water from an enclosure or a structure, such as a riverbed, caisson, or mine shaft, by pumping or evaporation. { dē'wȯd·ər·iŋ }

dewaxing |CHEM ENG| Removing wax from a material or object, a process used to separate solid hydrocarbons from petroleum. { dē 'waks·iŋ }

dew cell |ENG| An instrument used to determine the dew point, consisting of a pair of spaced, bare electrical wires wound spirally around an insulator and covered with a wicking wetted with a water solution containing an excess of lithium chloride; an electrical potential applied to the wires causes a flow of current through the lithium chloride solution, which raises the temperature of the solution until its vapor pressure is in equilibrium with that of the ambient air. { 'dü ‚sel }

dew-point boundary |CHEM ENG| On a phase diagram for a gas-condensate reservoir (pressure versus temperature with constant gas-oil ratios), the area along which the gas-oil ratio approaches zero. { 'dü ‚pȯint ‚baún·drē }

dew-point composition |CHEM ENG| The water vapor-air composition at saturation, that is, at the temperature at which water exerts a vapor pressure equal to the partial pressure of water vapor in the air-water mixture. { 'dü ‚pȯint ‚käm·pə'zish·ən }

dew-point curve |CHEM ENG| On a PVT phase diagram, the line that separates the two-phase (gas-liquid) region from the one-phase (gas) region, and indicates the point at a given gas temperature or pressure at which the first dew or liquid phase occurs. { 'dü ‚pȯint ‚kərv }

dew-point depression |CHEM ENG| Reduction of the liquid-vapor dew point of a gas by removal of a portion of the liquid (such as water) from the gas (such as air). { 'dü ‚pȯint di'presh·ən }

dew-point hygrometer |CHEM ENG| An instrument for determining the dew point by measuring the temperature at which vapor being cooled in a silver vessel begins to condense. Also known as cold-spot hygrometer. { 'dü ‚pȯint hī'gräm·əd·ər }

dew-point pressure |CHEM ENG| The gas pressure at which a system is at its dew point, that is, the conditions of gas temperature and pressure at which the first dew or liquid phase occurs. { 'dü ‚pȯint ‚presh·ər }

dew-point recorder |ENG| An instrument which gives a continuous recording of the dew point; it alternately cools and heats the target and uses a photocell to observe and record the temperature at which the condensate appears and disappears. Also known as mechanized dew-point meter. { 'dü ‚pȯint ri'kȯrd·ər }

DFE See design for environment.

diabatic |THERMO| A thermodynamic change of state of a system in which there is a transfer of heat across the boundaries of the system. Also known as nonadiabatic. { ¦dī·ə¦bad·ik }

diagnostics [ENG] Information on what tests a device has failed and how they were failed; used to aid in troubleshooting. { ‚dī·əg'näs·tiks }

diagonal [CIV ENG] A sloping structural member, under compression or tension or both, of a truss or bracing system. { dī'ag·ən·əl }

diagonal bond [CIV ENG] A masonry bond with diagonal headers. { dī'ag·ən·əl 'bänd }

diagonal pitch [ENG] In rows of staggered rivets, the distance between the center of a rivet in one row to the center of the adjacent rivet in the next row. { dī'ag·ən·əl 'pich }

diagonal pliers [DES ENG] Pliers with cutting jaws at an angle to the handles to permit cutting off wires close to terminals. { dī'ag·ən·əl 'plī·ərz }

diagonal stay [MECH ENG] A diagonal member between the tube sheet and shell in a fire-tube boiler. { dī'ag·ən·əl 'stā }

diagram factor [MECH ENG] The ratio of the actual mean effective pressure, as determined by an indicator card, to the map of the ideal cycle for a steam engine. { 'dī·ə‚gram ‚fak·tər }

dial [DES ENG] A separate scale or other device for indicating the value to which a control is set. { dīl }

DIAL See differential absorption lidar. { 'dī‚al }

dial cable [DES ENG] Braided cord or flexible wire cable used to make a pointer move over a dial when a separate control knob is rotated, or used to couple two shafts together mechanically. { dīl ‚kā·bəl }

dial cord [DES ENG] A braided cotton, silk, or glass fiber cord used as a dial cable. { 'dīl ‚kórd }

dial feed [MECH ENG] A device that rotates workpieces into position successively so they can be acted on by a machine. { 'dīl ‚fēd }

dial indicator [DES ENG] Meter or gage with a calibrated circular face and a pivoted pointer to give readings. { 'dīl ‚in·də‚kād·ər }

dialing step [ENG] The minimum amount, expressed in units of mass, that can be added or removed on a balance fitted with dial weights. { 'dīl·iŋ ‚step }

dial press [MECH ENG] A punch press with dial feed. { 'dīl ‚pres }

dial weight [ENG] A weight piece that acts on the invariable arm of an analytical balance and is added or removed from outside the case by a weight-lifting dialing system. { 'dīl ‚wāt }

dialyzer [CHEM ENG] **1.** The semipermeable membrane used for dialyzing liquid. **2.** The container used in dialysis; it is separated into compartments by membranes. { 'dī·ə‚līz·ər }

diameter group [MECH ENG] A dimensionless group, used in the study of flow machines such as turbines and pumps, equal to the fourth root of pressure number 2 divided by the square root of the delivery number. { dī'am·əd·ər ‚grüp }

diameter tape [ENG] A tape for measuring the diameter of trees; when wrapped around the circumference of a tree, it reads the diameter directly. { dī'am·əd·ər ‚tāp }

diametral pitch [DES ENG] A gear tooth design factor expressed as the ratio of the number of teeth to the diameter of the pitch circle measured in inches. { dī'am·ə·trəl 'pich }

diamond anvil [ENG] A brilliant-cut diamond of extremely high quality that is modified to have 16 sides and has the culet cut off to create either a flat tip or a flat surface followed by a bevel of 5–10° { dī·mənd 'an·vəl }

diamond-anvil cell [ENG] A device for generating an extremely high pressure in a sample that is sandwiched between two diamond anvils to which forces are applied. { ¦dī·mənd ¦an·vəl ‚sel }

diamond bit [DES ENG] A rotary drilling bit crowned with bort-type diamonds, used for rock boring. Also known as bort bit. { 'dī‚mənd ‚bit }

diamond boring [ENG] Boring with a diamond tool. { 'dī·mənd ‚bór·iŋ }

diamond chisel [DES ENG] A chisel having a V-shaped or diamond-shaped cutting edge. { 'dī·mənd ‚chiz·əl }

diamond circuit [ELECTR] A gate circuit that provides isolation between input and output terminals in its off state, by operating transistors in their cutoff region; in the on state the output voltage follows the input voltage as required for gating both analog and digital signals, while the transistors provide current gain to supply output current on demand. { 'dī·mənd ‚sər·kət }

diamond coring [ENG] Obtaining core samples of rock by using a diamond drill. { 'dī·mənd 'kór·iŋ }

diamond count [DES ENG] The number of diamonds set in a diamond crown bit. { 'dī·mənd ‚kaúnt }

diamond crossing [CIV ENG] An oblique railroad crossing that forms a diamond shape between the tracks. { 'dī·mənd ‚krós·iŋ }

diamond crown [DES ENG] The cutting bit used in diamond drilling; it consists of a steel shell set with black diamonds on the face and cutting edges. { 'dī·mənd ‚kraún }

diamond drill [DES ENG] A drilling machine with a hollow, diamond-set bit for boring rock and yielding continuous and columnar rock samples. { 'dī·mənd ‚dril }

Diamond-Hinman radiosonde [ENG] A variable audio-modulated radiosonde used by United States weather services; the carrier signal from the radiosonde is modulated by audio signals determined by the electrical resistance of the humidity- and temperature-transducing elements and by fixed reference resistors; the modulating signals are transmitted in a fixed sequence at predetermined pressure levels by means of a baroswitch. { ¦dī·mənd ¦hin·mən 'rād·ē·ō‚sänd }

diamond indenter [ENG] An instrument that measures hardness by indenting a material with a diamond point. { 'dī·mənd in'den·tər }

diamond matrix |DES ENG| The metal or alloy in which diamonds are set in a drill crown. { 'dī·mənd 'mā·triks }

diamond orientation |DES ENG| The set of a diamond in a cutting tool so that the crystal face will be in contact with the material being cut. { 'dī·mənd ˌȯr·ē·ən·tā·shən }

diamond-particle bit |DES ENG| A diamond bit set with small fragments of diamonds. { 'dī·mənd͡pärd·ə·kəl ˌbit }

diamond pattern |DES ENG| The arrangement of diamonds set in a diamond crown. { 'dī·mənd ˌpad·ərn }

diamond point |DES ENG| A cutting tool with a diamond tip. { 'dī·mənd ˌpȯint }

diamond-point bit See mud auger. { 'dī·mənd ͡pȯint ˌbit }

diamond reamer |DES ENG| A diamond-inset pipe behind, and larger than, the drill bit and core barrel that is used for enlarging boreholes. { 'dī·mənd ˌrēm·ər }

diamond saw |DES ENG| A circular, band, or frame saw inset with diamonds or diamond dust for cutting sections of rock and other brittle substances. { 'dī·mənd ˌsȯ }

diamond setter |ENG| A person skilled at setting diamonds by hand in a diamond bit or a bit mold. { 'dī·mənd ˌsed·ər }

diamond size |ENG| In the bit-setting and diamond-drilling industries, the number of equal-size diamonds having a total weight of 1 carat; a 10-diamond size means 10 stones weighing 1 carat. { 'dī·mənd ˌsīz }

diamond stylus |ENG ACOUS| A stylus having a ground diamond as its point. { 'dī·mənd 'stī·ləs }

diamond tool |DES ENG| **1.** Any tool using a diamond-set bit to drill a borehole. **2.** A diamond shaped to the contour of a single-pointed cutting tool, used for precision machining. { 'dī·mənd ˌtül }

diamond wheel |DES ENG| A grinding wheel in which synthetic diamond dust is bonded as the abrasive to cut very hard materials such as sintered carbide or quartz. { 'dī·mənd ˌwēl }

diaphragm |ENG| A thin sheet placed between parallel parts of a member of structural steel to increase its rigidity. |ENG ACOUS| A thin, flexible sheet that can be moved by sound waves, as in a microphone, or can produce sound waves when moved, as in a loudspeaker. { 'dī·ə‚fram }

diaphragm cell |CHEM ENG| An electrolytic cell used to produce sodium hydroxide and chlorine from sodium chloride brine; porous diaphragm separates the anode and cathode compartments. { 'dī·ə‚fram ˌsel }

diaphragm compressor |MECH ENG| Device for compression of small volumes of a gas by means of a reciprocally moving diaphragm, in place of pistons or rotors. { 'dī·ə‚fram kəm'pres·ər }

diaphragm gage |ENG| Pressure- or vacuum-sensing instrument in which pressures act against opposite sides of an enclosed diaphragm

that consequently moves in relation to the difference between the two pressures, actuating a mechanical indicator or electric-electronic signal. { 'dī·ə‚fram ˌgāj }

diaphragm horn |ENG ACOUS| A horn that produces sound by means of a diaphragm vibrated by compressed air, steam, or electricity. { 'dī·ə‚fram ˌhȯrn }

diaphragm meter |ENG| A flow meter which uses the movement of a diaphragm in the measurement of a difference in pressure created by the flow, such as a force-balance-type or a deflection-type meter. { 'dī·ə‚fram ˌmēd·ər }

diaphragm pump |MECH ENG| A metering pump which uses a diaphragm to isolate the operating parts from pumped liquid in a mechanically actuated diaphragm pump, or from hydraulic fluid in a hydraulically actuated diaphragm pump. { 'dī·ə‚fram ˌpəmp }

diaphragm valve |ENG| A fluid valve in which the open-close element is a flexible diaphragm; used for fluids containing suspended solids, but limited to low-pressure systems. { 'dī·ə‚fram ˌvalv }

diathermous envelope |THERMO| A surface enclosing a thermodynamic system in equilibrium that is not an adiabatic envelope; intuitively, this means that heat can flow through the surface. { ͡dī·ə͡thər·məs 'en·və‚lōp }

dice See die. { dīs }

dicing |ELECTR| Sawing or otherwise machining a semiconductor wafer into small squares, or dice, from which transistors and diodes can be fabricated. { 'dīs·iŋ }

dicing cutter |MECH ENG| A cutting mill for sheet material; sheet is first slit into horizontal strands by blades, then fed against a rotating knife for dicing. { 'dīs·iŋ ˌkəd·ər }

die |DES ENG| A tool or mold used to impart shapes to, or to form impressions on, materials such as metals and ceramics. |ELECTR| The tiny, sawed or otherwise machined piece of semiconductor material used in the construction of a transistor, diode, or other semiconductor device; plural is dice. { dī }

die adapter |ENG| That part of an extrusion die which holds the die block. { 'dī ə'dap·tər }

die blade |ENG| A deformable member attached to a die body which determines the slot opening and is adjusted to produce uniform thickness across plastic film or sheet. { 'dī ‚blād }

die block |ENG| **1.** A tool-steel block which is bolted to the bed of a punch press and into which the desired impressions are machined. **2.** The part of an extrusion mold die holding the forming bushing and core. { 'dī ‚bläk }

die body |ENG| The stationary part of an extrusion die, used to separate and form material. { 'dī ‚bäd·ē }

die bushing See button die. { 'dī ‚bùsh·iŋ }

die casting |ENG| A metal casting process in which molten metal is forced under pressure into a permanent mold; the two types are hot-chamber and cold-chamber. { 'dī ‚kast·iŋ }

die chaser |ENG| One of the cutting parts of a composite die or a die used to cut threads. { 'dī ,chās·ər }

Dieckman condensation |CHEM ENG| Any condensation of esters of dicarboxylic acids which produce cyclic β-ketoesters. { 'dēk·män ,kän ,den'sā·shən }

die clearance |ENG| The distance between die members that meet during an operation. { 'dī ,klir·əns }

die cushion |ENG| A device located in or under a die block or bolster to provide additional pressure or motion for stamping. { 'dī ,kùsh·ən }

die cutting See blanking. { 'dī ,kəd·iŋ }

die gap |ENG| In plastics and metals forming, the distance between the two opposing metal faces forming the opening of a die. { 'dī ,gap }

die holder |ENG| A plate or block on which the die block is mounted; it is fastened to the bolster or press bed. { 'dī ,hōld·ər }

dieing machine |MECH ENG| A vertical press with the slide activated by pull rods attached to the drive mechanism below the bed of the press. { 'dī·iŋ mə'shēn }

die insert |ENG| A removable part or the liner of a die body or punch. { 'dī ,in·sərt }

dielectric breakdown |ELECTR| Breakdown which occurs in an alkali halide crystal at field strengths on the order of 10^6 volts per centimeter. { ,dī·ə'lek·trik 'brāk,daùn }

dielectric constant |ELEC| **1.** For an isotropic medium, the ratio of the capacitance of a capacitor filled with a given dielectric to that of the same capacitor having only a vacuum as dielectric. **2.** More generally, $1 + \gamma\chi$, where γ is 4π in Gaussian and cgs electrostatic units or 1 in rationalized mks units, and χ is the electric susceptibility tensor. Also known as relative dielectric constant; relative permittivity; specific inductive capacity (SIC). { ,dī·ə'lek·trik 'kän·stənt }

dielectric curing |ENG| A process for curing a thermosetting resin by subjecting it to a high-frequency electric charge. { ,dī·ə'lek·trik 'kyùr·iŋ }

dielectric fatigue |ELECTR| The property of some dielectrics in which resistance to breakdown decreases after a voltage has been applied for a considerable time. { ,dī·ə'lek·trik fə'tēg }

dielectric field |ELEC| The average total electric field acting upon a molecule or group of molecules inside a dielectric. Also known as internal dielectric field. { ,dī·ə'lek·trik 'fēld }

dielectric film |ELEC| A film possessing dielectric properties; used as the central layer of a capacitor. { ,dī·ə'lek·trik 'film }

dielectric leakage |ELEC| A very small steady current that flows through a dielectric subject to a steady electric field. { ,dī·ə'lek·trik 'lēk·ij }

dielectric loss factor |ELEC| Product of the dielectric constant of a material and the tangent of its dielectric loss angle. { ,dī·ə'lek·trik ¦los ,fak·tər }

dielectric shielding |ELEC| The reduction of an electric field in some region by interposing a

dielectric substance, such as polystyrene, glass, or mica. { ,dī·ə'lek·trik 'shēld·iŋ }

dielectric strength |ELEC| The maximum electrical potential gradient that a material can withstand without rupture; usually specified in volts per millimeter of thickness. Also known as electric strength. { ,dī·ə'lek·trik 'streŋkth }

dielectric susceptibility See electric susceptibility. { ,dī·ə'lek·trik sə,sep·tə'bil·əd·ē }

die lines |ENG| Lines or markings on the surface of a drawn, formed, or extruded product due to imperfections in the surface of the die. { 'dī ,līnz }

diesel cycle |THERMO| An internal combustion engine cycle in which the heat of compression ignites the fuel. { 'dē·zəl' ,sī·kəl }

diesel electric locomotive |MECH ENG| A locomotive with a diesel engine driving an electric generator which supplies electric power to traction motors for propelling the vehicle. Also known as diesel locomotive. { ¦dē·zəl ə¦lek·trik ,lō·kə'mōd·iv }

diesel electric power generation |MECH ENG| Electric power generation in which the generator is driven by a diesel engine. { ¦dē·zəl ə¦lek·trik 'paù·ər ,jen·ə,rā·shən }

diesel engine |MECH ENG| An internal combustion engine operating on a thermodynamic cycle in which the ratio of compression of the air charge is sufficiently high to ignite the fuel subsequently injected into the combustion chamber. Also known as compression-ignition engine. { ¦dē·zəl 'en·jən }

diesel index |CHEM ENG| An empirical expression for the correlation between the aniline number of a diesel fuel and its ignitability. |MECH ENG| Diesel fuel rating based on ignition qualities; high-quality fuel has a high index number. { 'dē·zəl ,in,deks }

dieseling |MECH ENG| **1.** Explosions of mixtures of air and lubricating oil in the compression chambers or in other parts of the air system of a compressor. **2.** Continuation of running by a gasoline spark-ignition engine after the ignition is turned off. Also known as run-on. { 'dē·zəl·iŋ }

diesel knock |MECH ENG| A combustion knock caused when the delayed period of ignition is long so that a large quantity of atomized fuel accumulates in the combustion chamber; when combustion occurs, the sudden high pressure resulting from the accumulated fuel causes diesel knock. { 'dē·zəl ,näk }

diesel locomotive See diesel electric locomotive. { 'dē·zəl ,lō·kə'mōd·iv }

diesel rig |MECH ENG| Any diesel engine apparatus or machinery. { 'dē·zəl ,rig }

die set |ENG| A tool or tool holder consisting of a die base for the attachment of a die and a punch plate for the attachment of a punch. { 'dī ,set }

die shoe |MECH ENG| A block placed beneath the lower part of a die upon which the die holder is mounted; spreads the impact over the die bed, thereby reducing wear. { 'dī ,shü }

diesinking [ENG] Making a depressed pattern in a die by forming or machining. { 'dī,siŋk·iŋ }

die slide [MECH ENG] A device in which the lower die of a power press is mounted; it slides in and out of the press for easy access and safety in feeding the parts. { 'dī ,slīd }

die swell ratio [ENG] The ratio of the outer parison diameter (or parison thickness) to the outer diameter of the die (or die gap). { 'dī ,swel ,rā·shō }

Dieterici equation of state [THERMO] An empirical equation of state for gases, $pe^{a/RT}(v - b) =$ RT, where p is the pressure, T is the absolute temperature, v is the molar volume, R is the gas constant, and a and b are constants characteristic of the substance under consideration. { dē·də're·chē i'kwā·zhen əv 'stāt }

difference channel [ENG ACOUS] An audio channel that handles the difference between the signals in the left and right channels of a stereophonic sound system. { 'dif·rəns ,chan·əl }

differential [CONT SYS] The difference between levels for turn-on and turn-off operation in a control system. [MECH ENG] Any arrangement of gears forming an epicyclic train in which the angular speed of one shaft is proportional to the sum or difference of the angular speeds of two other gears which lie on the same axis; allows one shaft to revolve faster than the other, the speed of the main driving member being equal to the algebraic mean of the speeds of the two shafts. Also known as differential gear. { ,dif·ə'ren·chəl }

differential absorption lidar [ENG] A technique for the remote sensing of atmospheric gases, in which lasers transmit pulses of radiation into the atmosphere at two wavelengths, one of which is absorbed by the gas to be measured and one is not, and the difference between the return signals from atmospheric backscattering on the absorbed and nonabsorbed wavelengths is used as a direct measure of the concentration of the absorbing species. Abbreviated DIAL. { ,dif·ə'ren·chəl əb'sórp·shən 'lī,där }

differential air thermometer [ENG] A device for detecting radiant heat, consisting of a U-tube manometer with a closed bulb at each end, one clear and the other blackened. { ,dif·ə'ren·chəl 'er thər'mäm·əd·ər }

differential brake [MECH ENG] A brake in which operation depends on a difference between two motions. { ,dif·ə'ren·chəl 'brāk }

differential calorimetry [THERMO] Technique for measurement of and comparison (differential) of process heats (reaction, absorption, hydrolysis, and so on) for a specimen and a reference material. { ,dif·ə'ren·chəl ,kal·ə'rim·ə·trē }

differential chemical reactor [CHEM ENG] A flow reactor operated at constant temperature and very low concentrations (resulting from very short residence times), with product and reactant concentrations essentially constant at the levels in the feed. { ,dif·ə'ren·chəl 'kem·i·kəl rē'ak·tər }

differential effects [MECH] The effects upon the elements of the trajectory due to variations from standard conditions. { ,dif·ə'ren·chəl i'feks }

differential extraction [CHEM ENG] Theoretical limiting case of crosscurrent extraction in a single vessel where feed is continuously extracted with infinitesimal amounts of fresh solvent; true differential extraction cannot be achieved. { ,dif·ə'ren·chəl ik'ctrak·shon }

differential frequency meter [ENG] A circuit that converts the absolute frequency difference between two input signals to a linearly proportional direct-current output voltage that can be used to drive a meter, recorder, oscilloscope, or other device. { ,dif·ə'ren·chəl 'frē·kwən·sē ,mēd·ər }

differential game [CONT SYS] A two-sided optimal control problem. { ,dif·ə'ren·chəl 'gām }

differential gap controller [CONT SYS] A two-position (on-off) controller that actuates when the manipulated variable reaches the high or low value of its range (differential gap). { ,dif·ə'ren·chəl 'gap kən,trōl·ər }

differential gear See differential. { ,dif·ə'ren·chəl 'gir }

differential heat of solution [THERMO] The partial derivative of the total heat of solution with respect to the molal concentration of one component of the solution, when the concentration of the other component or components, the pressure, and the temperature are held constant. { ,dif·ə'ren·chəl 'hēt əv sə'lü·shən }

differential indexing [MECH ENG] A method of subdividing a circle based on the difference between movements of the index plate and index crank of a dividing engine. { ,dif·ə'ren·chəl 'in ,deks·iŋ }

differential instrument [ENG] Galvanometer or other measuring instrument having two circuits or coils, usually identical, through which currents flow in opposite directions; the difference or differential effect of these currents actuates the indicating pointer. { ,dif·ə'ren·chəl 'in·strə·mənt }

differential leak detector [ENG] A leak detector consisting of two tubes and a trap which directs the tracer gas from the system into the desired tube. { ,dif·ə'ren·chəl 'lēk di'tek·tər }

differential leveling [ENG] A surveying process in which a horizontal line of sight of known elevation is intercepted by a graduated standard, or rod, held vertically on the point being checked. { ,dif·ə'ren·chəl 'lev·əl·iŋ }

differential manometer [ENG] An instrument in which the difference in pressure between two sources is determined from the vertical distance between the surfaces of a liquid in two legs of an erect or inverted U-shaped tube when each of the legs is connected to one of the sources. { ,dif·ə'ren·chəl mə'näm·əd·ər }

differential microphone See double-button microphone. { ,dif·ə'ren·chəl 'mī·krə,fōn }

differential motion [MECH ENG] A mechanism in which the follower has two driving elements; the net motion of the follower is the difference between the motions that would result from either driver acting alone. { ¦dif·ə'ren·chəl 'mō·shən }

differential piece-rate system |IND ENG| A wage plan based on a standard task time whereby the worker receives increased or decreased piece rates as his or her production varies from that expected for the standard time. Also known as accelerating incentive. { ¦dif·ə'ren·chəl 'pēs ¦rāt ¦sis·təm }

differential-pressure fuel valve |MECH ENG| A needle or spindle normally closed, with seats at the back side of the valve orifice. { ¦dif·ə¦ren·chəl ¦presh·ər 'fyül ¦valv }

differential-pressure gage |ENG| Apparatus to measure pressure differences between two points in a system; it can be a pressured liquid column balanced by a pressured liquid reservoir, a formed metallic pressure element with opposing force, or an electrical-electronic gage (such as strain, thermal-conductivity, or ionization). { ¦dif·ə¦ren·chəl 'presh·ər ¦gāj }

differential process |CHEM ENG| A process in which a system is caused to move through a bubble point and as a result to form two phases, the minor phase being removed from further contact with the major phase; thus the system continuously changes in quantity and composition. { ¦dif·ə¦ren·chəl 'präs·əs }

differential-producing primary device |ENG| An instrument that modifies the flow pattern of a fluid passing through a pipe, duct, or open channel, and thereby produces a difference in pressure between two points, which can then be measured to determine the rate of flow. { ¦dif·ə'ren·chəl prə¦düs·iŋ ¦prī¸mer·ē di'vīs }

differential pulley |MECH ENG| A tackle in which an endless cable passes through a movable lower pulley, which carries the load, and two fixed coaxial upper pulleys having different diameters; yields a high mechanical advantage. { ¦dif·ə'ren·chəl 'pul·ē }

differential scanning calorimeter |CHEM ENG| An instrument for studying overall chemical reactions by measuring the associated exothermic and endothermic reactions that occur over a specified temperature cycle. { ¦dif·ə¦ren·chəl ¦skan·iŋ ¸kal·ə'rim·əd·ər }

differential scatter |ENG| A technique for the remote sensing of atmospheric particles in which the ackscattering from laser beams at a number of infrared wavelengths is measured and correlated with scattering signatures that are uniquely related to particle composition. Abbreviated DISC. { ¦dif·ə'ren·chəl 'skad·ər }

differential screw |MECH ENG| A type of compound screw which produces a motion equal to the difference in motion between the two component screws. { ¦dif·ə'ren·chəl 'skrü }

differential separation |CHEM ENG| Release of gas (vapor) from liquids by a reduction in pressure that allows the vapor to come out of the solution, so that the vapor can be removed from the system; differs from flash separation, in which the vapor and liquid are kept in contact

following pressure reduction. { ¦dif·ə'ren·chəl ¸sep·ə'rā·shən }

differential steam calorimeter |ENG| An instrument for measuring small specific-heat capacities, such as those of gases, in which the amount of steam condensing on a body containing the substance whose heat capacity is to be measured is compared with the amount condensing on a similar body which is evacuated or contains a substance of known heat capacity. { ¦dif·ə'ren·chəl 'stēm kal·ə'rim·əd·ər }

differential thermal analysis |THERMO| A method of determining the temperature at which thermal reactions occur in a material undergoing continuous heating to elevated temperatures; also involves a determination of the nature and intensity of such reactions. { ¦dif·ə'ren·chəl 'thər·məl ə'nal·ə·səs }

differential thermogravimetric analysis |THERMO| Thermal analysis in which the rate of material weight change upon heating versus temperature is plotted; used to simplify reading of weight-versus-temperature thermogram peaks that occur close together. { ¦dif·ə'ren·chəl ¦thər·mō¸grav·ə¦me·trik ə'nal·ə·səs }

differential thermometer See bimetallic thermometer. { ¦dif·ə'ren·chəl thər'mäm·əd·ər }

differential timing |IND ENG| A time-study technique in which the time value of an element of extremely short duration is determined by various calculations involving cycle values that first include and then exclude the element under consideration. { ¦dif·ə'ren·chəl 'tīm·iŋ }

differential windlass |MECH ENG| A windlass in which the barrel has two sections, each having a different diameter; the rope winds around one section, passes through a pulley (which carries the load), then winds around the other section of the barrel. { ¦dif·ə'ren·chəl 'wind·ləs }

diffuser |ENG| A duct, chamber, or section in which a high-velocity, low-pressure stream of fluid (usually air) is converted into a high-velocity, high-pressure flow { də'fyüz·er }

diffusion |ELECTR| A method of producing a junction by difusing an impurity metal into a semiconductor at a high temperature. |MECH ENG| The conversion of air velocity into static pressure in the diffuser casing of a centrifugal fan, resulting from increases in the radius of the air spin and in area. { də'fyü·zhən }

diffusion barrier |CHEM ENG| Porous barrier through which gaseous mixtures are passed for enrichment of the lighter-molecular-weight constituent of the diffusate; used as a many-stage cascade system for the recovery of $^{235}UF_6$ isotopes from a $^{238}UF_6$ stream. { də'fyü·zhən ¸bar·ē·ər }

diffusion hygrometer |ENG| A hygrometer based upon the diffusion of water vapor through a porous membrane; essentially, it consists of a closed chamber having porous walls and containing a hygroscopic compound, whose absorption of water vapor causes a pressure drop within the chamber that is measured by a manometer. { də'fyü·zhən hī'gräm·əd·ər }

diffusion pump |ENG| A vacuum pump in which a stream of heavy molecules, such as mercury vapor, carries gas molecules out of the volume being evacuated; also used for separating isotopes according to weight, the lighter molecules being pumped preferentially by the vapor stream. { də'fyü·zhən ˌpəmp }

diffusiophoresis |CHEM ENG| A process in a scrubber whereby water vapor moving toward the cold water surface carries particulates with it. { də'fyü·zē·ō·fə'rē·səs }

diffusivity |THERMO| The quantity of heat passing normally through a unit area per unit time divided by the product of specific heat, density, and temperature gradient. Also known as thermal diffusivity; thermometric conductivity. { dif·yü'ziv·əd·ē }

digested sludge |CIV ENG| Sludge or thickened mixture of sewage solids with water that has been decomposed by anaerobic bacteria. { də'jes·təd 'sləj }

digester |CHEM ENG| A vessel used to produce cellulose pulp from wood chips by cooking under pressure. |CIV ENG| A sludge-digestion tank containing a system of hot water or steam pipes for heating the sludge. { də'jes·tər }

digestion |CHEM ENG| **1.** Preferential dissolving of mineral constituents in concentrations of ore. **2.** Liquefaction of organic waste materials by action of microbes. **3.** Separation of fabric from tires by the use of hot sodium hydroxide. **4.** Removing lignin from wood in manufacture of chemical cellulose paper pulp. |CIV ENG| The process of sewage treatment by the anaerobic decomposition of organic matter. { də'jes·chən }

digger |ENG| A tool or apparatus for digging in the ground. { 'dig·ər }

digging |ENG| A sudden increase in cutting depth of a cutting tool due to an erratic change in load. { 'dig·iŋ }

digging line See inhaul cable. { 'dig·iŋ ˌlīn }

digital circuit |ELECTR| A circuit designed to respond at input voltages at one of a finite number of levels and, similarly, to produce output voltages at one of a finite number of levels. { 'dij·əd·əl 'sər·kət }

digital control |CONT SYS| The use of digital or discrete technology to maintain conditions in operating systems as close as possible to desired values despite changes in the operating environment. { 'dij·əd·əl kən'trōl }

digital delayer |ENG ACOUS| A device for introducing delay in the audio signal in a sound-reproducing system, which converts the audio signal to digital format and stores it in a digital shift register before converting it back to analog form. { 'dij·əd·əl di'lā·ər }

digital log |ENG| A well log that has undergone discrete sampling and recording on a magnetic tape preparatory to use in computerized interpretation and plotting. { 'dij·əd·əl 'läg }

digital-to-analog converter |ELECTR| A converter in which digital input signals are changed to essentially proportional analog signals. { 'dij·əd·əl tü ¦an·ə,läg kən'vərd·ər }

dike |CIV ENG| An embankment constructed on dry ground along a riverbank to prevent overflow of lowlands and to retain floodwater. { dīk }

dilatometer |ENG| An instrument for measuring thermal expansion and dilation of liquids or solids. { ,dil·ə'täm·əd·ər }

dilute phase |CHEM ENG| In liquid-liquid extraction, the liquid phase that is dilute with respect to the material being extracted. { də'lüt ˌfāz }

dimpling |ENG| Forming a conical depression in a metal surface in order to countersink a rivet head. { 'dim·pliŋ }

Dines anemometer |ENG| A pressure-tube anemometer in which the pressure head on a weather vane is kept facing into the wind, and the suction head, near the bearing which supports the vane, develops a suction independent of wind direction; the pressure difference between the heads is proportional to the square of the wind speed and is measured by a float manometer with a linear wind scale. { 'dīnz an·ə'mäm·əd·ər }

Dings magnetic separator |MECH ENG| A device which is suspended above a belt conveyor to pull out and separate magnetic material from burden as thick as 40 inches (1 meter) and at belt speeds up to 750 feet (229 meters) per minute. { 'diŋz mag'ned·ik ,sep·ə,rād·ər }

dinking |MECH ENG| Using a sharp, hollow punch for cutting light-gage soft metals or nonmetallic materials. { 'diŋk·iŋ }

dioctyl phthalate test |ENG| A method used to evaluate air filters to be used in critical air-cleaning applications; a light-scattering technique counts the number of particles of controlled size (0.3 micrometer) entering and emerging from the test filter. Abbreviated DOP test. { dī¦äkt·əl ¦tha,lāt ,test }

diode |ELECTR| **1.** A two-electrode electron tube containing an anode and a cathode. **2.** See semiconductor diode. { 'dī,ōd }

diode alternating-current switch See trigger diode. { 'dī,ōd ¦ȯl·tər,näd·iŋ ¦kər·ənt ,swich }

diode amplifier |ELECTR| A microwave amplifier using an IMPATT, TRAPATT, or transferred-electron diode in a cavity, with a microwave circulator providing the input/output isolation required for amplification; center frequencies are in the gigahertz range, from about 1 to 100 gigahertz, and power outputs are up to 20 watts continuous-wave or more than 200 watts pulsed, depending on the diode used. { 'dī,ōd 'am·plə,fī·ər }

diode bridge |ELECTR| A series-parallel configuration of four diodes, whose output polarity remains unchanged whatever the input polarity. { 'dī,ōd ,brij }

diode-capacitor transistor logic |ELECTR| A circuit that uses diodes, capacitors, and transistors to provide logic functions. { 'dī,ōd kə¦pas·əd·ər tran'zis·tər ,läj·ik }

diode characteristic |ELECTR| The composite

Abbreviated dac. { 'dij·əd·əl tü ¦an·ə,läg kən'vərd·ər }

161

electrode characteristic of an electron tube when all electrodes except the cathode are connected together. { 'dī,ōd ,kar·ik·tə·'ris·tik }

diode clamp See diode clamping circuit. { 'dī,ōd ,klamp }

diode clamping circuit |ELECTR| A clamping circuit in which a diode provides a very low resistance whenever the potential at a certain point rises above a certain value in some circuits or falls below a certain value in others. Also known as diode clamp. { ¦dī,ōd 'klamp·iŋ ,sər·kət }

diode clipping circuit |ELECTR| A clipping circuit in which a diode is used as a switch to perform the clipping action. { ¦dī,ōd 'klip·iŋ ,sər·kət }

diode-connected transistor |ELECTR| A bipolar transistor in which two terminals are shorted to give diode action. { 'dī,ōd kə¦nek·təd tran'zis·tər }

diode demodulator |ELECTR| A demodulator using one or more diodes to provide a rectified output whose average value is proportional to the original modulation. Also known as diode detector. { 'dī,ōd dē'mäj·ə,lād·ər }

diode detector See diode demodulator. { 'dī,ōd di'tek·tər }

diode drop See diode forward voltage. { 'dī,ōd ,dräp }

diode forward voltage |ELECTR| The voltage across a semiconductor diode that is carrying current in the forward direction; it is usually approximately constant over the range of currents commonly used. Also known as diode drop; diode voltage; forward voltage drop. { 'dī,ōd ¦fór·wərd 'vōl·tij }

diode function generator |ELECTR| A function generator that uses the transfer characteristics of resistive networks containing biased diodes; the desired function is approximated by linear segments. { 'dī,ōd 'feŋk·shən ,jen·ə,rād·ər }

diode gate |ELECTR| An AND gate that uses diodes as switching elements. { 'dī,ōd ,gāt }

diode limiter |ELECTR| A peak-limiting circuit employing a diode that becomes conductive when signal peaks exceed a predetermined value. { ,dī,ōd 'lim·əd·ər }

diode logic |ELECTR| An electronic circuit using current-steering diodes, such that the relations between input and output voltages correspond to AND or OR logic functions. { 'dī,ōd ,läj·ik }

diode matrix |ELECTR| A two-dimensional array of diodes used for a variety of purposes such as decoding and read-only memory. { 'dī,ōd ,mā·triks }

diode mixer |ELECTR| A mixer that uses a crystal or electron tube diode; it is generally small enough to fit directly into a radio-frequency transmission line. { 'dī,ōd ,mik·sər }

diode switch |ELECTR| Diode which is made to act as a switch by the successive application of positive and negative biasing voltages to the anode (relative to the cathode), thereby allowing or preventing, respectively, the passage of other applied waveforms within certain limits of voltage. { 'dī,ōd ,swich }

diode transistor logic |ELECTR| A circuit that uses diodes, transistors, and resistors to provide logic functions. Abbreviated DTL. { ¦dī,ōd tran'zis·tər ,läj·ik }

diode-triode |ELECTR| Vacuum tube having a diode and a triode in the same envelope. { ¦dī,ōd 'trī,ōd }

diode voltage See diode forward voltage. { 'dī,ōd ,vōl·tij }

diode voltage regulator |ELECTR| A voltage regulator with a Zener diode, making use of its almost constant voltage over a range of currents. Also known as Zener diode voltage regulator. { ¦dī,ōd 'vōl·tij ,reg·yə,lād·ər }

diolefin hydrogenation |CHEM ENG| A fixed-bed catalytic process used to hydrogenate diolefins in C_4 and C_5 fractions to mono-olefin in alkylation feedstocks. { dī'ō·lə,fən ,hī·drə·jə'nā·shən }

dip |ENG| The vertical angle between the sensible horizon and a line to the visible horizon at sea, due to the elevation of the observer and to the convexity of the earth's surface. Also known as dip of horizon. { dip }

DIP See dual in-line package. { dip }

dip circle See inclinometer. { 'dip ,sər·kəl }

dip coating |ENG| A coating applied to ceramic ware or metal by immersion into a tank of melted nonmetallic material, such as resin or plastic, then chilling the adhering melt. { 'dip ,kōd·iŋ }

dip inductor See earth inductor. { 'dip in,dək·tər }

dipmeter |ENG| **1.** An instrument used to measure the direction and angle of dip of geologic formations. **2.** An absorption wavemeter in which bipolar or field-effect transistors replace the electron tubes used in older grid-dip meters. { 'dip,mēd·ər }

dip mold |ENG| A one-piece glassmaking mold with an open top; used to mold patterns. { 'dip ,mōld }

dip needle |ENG| An obsolete type of magnetometer consisting of a magnetized needle that rotates freely in the vertical plane, with an adjustable weight on one side of the pivot. { 'dip ,nēd·əl }

dip of horizon See dip. { 'dip əv hə'rīz·ən }

dipole moment See electric dipole moment. { 'dī,pōl ,mō·mənt }

dipper dredge |MECH ENG| A power shovel resembling a grab crane mounted on a flat-bottom boat for dredging under water. Also known as dipper shovel. { 'dip·ər ,drej }

dipper stick |MECH ENG| A straight shaft connecting the digging bucket of an excavating machine or power shovel with the boom. { 'dip·ər ,stik }

dipper trip |MECH ENG| A device which releases the door of a shovel bucket. { 'dip·ər ,trip }

dipping sonar |ENG| A sonar transducer that is lowered into the water from a hovering antisubmarine-warfare helicopter and recovered after the search is complete. Also known as dunking sonar. { 'dip·iŋ 'sō,när }

dipstick |ENG| A graduated rod which measures depth when dipped in a liquid, used, for example, to measure the oil in an automobile engine crankcase. { 'dip,stik }

dipstick microscopy |ENG| A technique for mapping the variation of thickness of a thin liquid film by repeatedly dipping the tip of an atomic force microscope into the film at different locations and calculating its thickness at each location. { 'dip,stik mī'kräs·kə·pē }

direct-acting pump |MECH ENG| A displacement reciprocating pump in which the steam or power piston is connected to the pump piston by means of a rod, without crank motion or flywheel. { də¦rekt ¦akt·iŋ 'pəmp }

direct-acting recorder |ENG| A recorder in which the marking device is mechanically connected to or directly operated by the primary detector. { də¦rekt ¦akt·iŋ ri'kòrd·ər }

direct-arc furnace |ENG| A furnace in which a material in a refractory-lined shell is rapidly heated to pour temperature by an electric arc which goes directly from electrodes to the material. { də¦rekt ¦ärk ,fər·nəs }

direct bearing |CIV ENG| A direct vertical support in a structure. { də¦rekt 'ber·iŋ }

direct-bonded bearing |MECH ENG| A bearing formed by pouring molten babbitt metal directly into the bearing housing, allowing it to cool, and then machining the metal to the specified diameter. { də¦rekt ¦bän·dəd 'ber·iŋ }

direct command guidance |ENG| Control of a missile or drone entirely from the launching site by radio or by signals sent over a wire. { də¦rekt kə¦mand 'gīd·əns }

direct-connected |MECH ENG| The connection between a driver and a driven part, as a turbine and an electric generator, without intervening speed-changing devices, such as gears. { də¦rekt kə'nek·təd }

direct-contact condenser See contact condenser. { də¦rekt ¦kän,takt kən,den·sər }

direct control function See regulatory control function. { də¦rekt kən'tröl ,fəŋk·shən }

direct cost |IND ENG| The cost in goods and labor to produce a product which would not be spent if the product were not made. { də¦rekt 'kòst }

direct-coupled |MECH ENG| Joined without intermediate connections. { də¦rekt 'kəp·əld }

direct coupling |ELEC| Coupling of two circuits by means of a non-frequency-sensitive device, such as a wire, resistor, or battery, so both direct and alternating current can flow through the coupling path. |MECH ENG| The direct connection of the shaft of a prime mover (such as a motor) to the shaft of a rotating mechanism (such as a pump or compressor). { də¦rekt 'kəp·liŋ }

direct current |ELEC| Electric current which flows in one direction only, as opposed to alternating current. Abbreviated dc. { də¦rekt 'kə·rənt }

direct-current power supply |ELEC| A power supply that provides one or more dc output voltages, such as a dc generator, rectifier-type power supply, converter, or dynamotor. { də¦rekt ¦kə·rənt 'paù·ər sə,plī }

direct digital control |CONT SYS| The use of a digital computer generally on a time-sharing or multiplexing basis, for process control in petroleum, chemical, and other industries. { də¦rekt ¦dij·əd·əl kən'tröl }

direct drive |MECH ENG| A drive in which the driving part is directly connected to the driven part. { də¦rekt 'drīv }

direct-drive arm |CONT SYS| A robot arm whose joints are directly coupled to high-torque motors. { də'rekt 'drīv ,ärm }

direct-drive vibration machine |MECH ENG| A vibration machine in which the vibration table is forced to undergo a displacement by a positive linkage driven by a direct attachment to eccentrics or camshafts. { də¦rekt ¦drīv vī'brā·shən mə,shēn }

direct energy conversion |ENG| Conversion of thermal or chemical energy into electric power by means of direct-power generators. { də¦rekt 'en·ər·jē kən,vər·zhən }

direct-expansion coil |MECH ENG| A finned coil, used in air cooling, inside of which circulates a cold fluid or evaporating refrigerant. Abbreviated DX coil. { də¦rekt ik'span·chən ,kòil }

direct expert control system |CONT SYS| An expert control system that contains rules that directly associate controller output values with different values of the controller measurements and set points. Also known as rule-based control system. { də¦rekt ,eks·pərt kən'tröl ,sis·təm }

direct extrusion |ENG| Extrusion by movement of ram and product in the same direction against a die orifice. { də¦rekt ik'strü·zhən }

direct-feedback system |CONT SYS| A system in which electrical feedback is used directly, as in a tachometer. { də¦rekt 'fēd,bak ,sis·təm }

direct-fire |ENG| To fire a furnace without preheating the air or gas. { də'rekt ,fīr }

direct-fired evaporator |CHEM ENG| An evaporator in which the flame and combustion gases are separated from the boiling liquid by a metal wall, or other heating surface. { də'rekt ¦fīrd i'vap·ə,rād·ər }

direct-geared |MECH ENG| Joined by a gear on the shaft of one machine meshing with a gear on the shaft of another machine. { də'rekt ¦gird }

direct-imaging mass analyzer |ENG| A type of secondary ion mass spectrometer in which secondary ions pass through an electrostatic immersion lens which forms an image that bears a point-to-point relation to the ion's place of origin on the sample surface, and then traverse magnetic sectors which effect mass separation. Also known as Castaing-Slodzian mass analyzer. { də¦rekt ¦im·ij·iŋ ¦mas 'an·ə,līz·ər }

direction |ENG| The position of one point in space relative to another without reference to the distance between them; may be either three-dimensional or two-dimensional, the horizontal being the usual plane of the latter; usually indicated in terms of its angular distance from a reference direction. { də'rek·shən }

directional control [ENG] Control of motion about the vertical axis; in an aircraft, usually by the rudder. { də'rek·shən·əl kən'trōl }

directional control valve [ENG] A control valve serving primarily to direct hydraulic fluid to the point of application. { də'rek·shən·əl kən'trōl ‚valv }

directional drilling [ENG] A drilling method involving intentional deviation of a wellbore from the vertical. { də'rek·shən·əl 'dril·iŋ }

directional gain See directivity index. { də'rek·shən·əl 'gān }

directional gyro [MECH] A two-degrees-of-freedom gyro with a provision for maintaining its spin axis approximately horizontal. { də'rek·shən·əl 'jī·rō }

directional hydrophone [ENG ACOUS] A hydrophone whose response varies significantly with the direction of sound incidence. { də'rek·shən·əl 'hī·drə‚fōn }

directional microphone [ENG ACOUS] A microphone whose response varies significantly with the direction of sound incidence. { də'rek·shən·əl 'mī·krə‚fōn }

directional response pattern See directivity pattern. { də'rek·shən·əl ri'spāns ‚pad·ərn }

direction cosine [ENG] In tracking, the cosine of the angle between a baseline and the line connecting the center of the baseline with the target. { də'rek·shən 'kō‚sīn }

direction-independent radar [ENG] Doppler radar used in sentry applications. { də'rek·shən ‚in·də‚pen·dənt 'rā‚där }

directivity factor [ENG ACOUS] **1.** The ratio of radiated sound intensity at a remote point on the principal axis of a loudspeaker or other transducer, to the average intensity of the sound transmitted through a sphere passing through the remote point and concentric with the transducer; the frequency must be stated. **2.** The ratio of the square of the voltage produced by sound waves arriving parallel to the principal axis of a microphone or other receiving transducer, to the mean square of the voltage that would be produced if sound waves having the same frequency and mean-square pressure were arriving simultaneously from all directions with random phase; the frequency must be stated. { də‚rek'tiv·əd·ə ‚fak·tər }

directivity index [ENG ACOUS] The directivity factor expressed in decibels; it is 10 times the logarithm to the base 10 of the directivity factor. Also known as directional gain. { də‚rek'tiv·əd·ə ‚in‚deks }

directivity pattern [ENG ACOUS] A graphical or other description of the response of a transducer used for sound emission or reception as a function of the direction of the transmitted or incident sound waves in a specified plane and at a specified frequency. Also known as beam pattern; directional response pattern. { də‚rek'tiv·əd·ə ‚pad·ərn }

direct labor [IND ENG] The labor or effort actually producing goods or services. { də'rekt 'lā·bər }

direct labor standard See standard time. { də‚rekt ‚lā·bər 'stan·dərd }

directly heated cathode See filament. { də‚rect·lē ‚hēd·əd 'kā‚thōd }

direct material [IND ENG] Any raw or semifinished material which will be incorporated into the product. { də‚rekt mə'tir·ē·əl }

direct-power generator [ENG] Any device which converts thermal or chemical energy into electric power by methods more direct than the conventional thermal cycle. { də‚rekt ‚paů·ər 'jen·ə‚rād·ər }

direct-radiator speaker [ENG ACOUS] A loudspeaker in which the radiating element acts directly on the air, without a horn. { də‚rekt ‚rād·ē‚ād·ər ‚spēk·ər }

direct-reading gage [ENG] Gage that records directly (instead of inferentially) measured values, for example, a liquid-level gage pointer actuated by direct linkage with a float. { də‚rekt ‚rēd·iŋ 'gāj }

direct recording [ENG ACOUS] Recording in which a record is produced immediately, without subsequent processing, in response to received signals. { də'rekt ri'kȯrd·iŋ }

direct return system [MECH ENG] In a heating or cooling system, a piping arrangement in which the fluid is returned to its origin (boiler or evaporator) by the shortest direct path after it has passed through each heat exchanger. { di‚rekt ri'tərn ‚sis·təm }

direct-writing galvanometer [ENG] A direct-writing recorder in which the stylus or pen is attached to a moving coil positioned in the field of the permanent magnet of a galvanometer. { də‚rekt ‚wrīd·iŋ ‚gal·və'näm·əd·ər }

direct-writing recorder [ENG] A recorder in which the permanent record of varying electrical quantities or signals is made on paper, directly by a pen attached to the moving coil of a galvanometer or indirectly by a pen moved by some form of motor under control of the galvanometer. Also known as mechanical oscillograph. { də‚rekt ‚wrīd·iŋ ri'kȯrd·ər }

disappearing filament pyrometer See optical pyrometer. { 'dis·ə‚pir·iŋ ‚fil·ə·mənt pī'räm·əd·ər }

disappearing stair [BUILD] A stair that can be swung up into a ceiling space. { 'dis·ə‚pir·iŋ 'ster }

disassemble [ENG] To take apart into constituent parts. { ‚dis·ə'sem·bəl }

disc See disk. { disk }

DISC See differential scatter. { disk }

discharge [ELEC] To remove a charge from a battery, capacitor, or other electric-energy storage device. [ELECTR] The passage of electricity through a gas, usually accompanied by a glow, arc, spark, or corona. Also known as electric discharge. { 'dis‚chärj }

discharge channel [MECH ENG] The passage in a pressure-relief device through which the fluid is released to the outside of the device. { 'dis‚chärj ‚chan·əl }

discharged solids *See* residue. { ¦dis‚chärjd 'säl·ədz }

discharge head [MECH ENG] Vertical distance between the intake level of a water pump and the level at which it discharges water freely to the atmosphere. { 'dis‚chärj ‚hed }

discharge hydrograph [CIV ENG] A graph showing the discharge or flow of a stream or conduit with respect to time. { 'dis‚chärj 'hī·drə‚graf }

discharge line [ENG] The length of pipe through which drilling mud travels from the mud pump through the standpipe on its way to the borehole. { 'dis‚chärj ‚līn }

discharge liquor [CHEM ENG] Liquid that has passed through a processing operation. Also known as effluent; product. { 'dis‚chärj ‚lik·ər }

discharge tube [ELECTR] An evacuated enclosure containing a gas at low pressure, through which current can flow when sufficient voltage is applied between metal electrodes in the tube. Also known as electric-discharge tube. [MECH ENG] A tube through which steam and water are released into a boiler drum. { 'dis‚chärj ‚tüb }

discharge-tube leak indicator [ENG] A device which detects the presence of a tracer gas by using a glass tube attached to a high-voltage source; the presence of leaked gas is indicated by the color of the electric discharge. { 'dis‚chärj ‚tüb 'lēk ‚in·də‚kād·ər }

discharging arch [CIV ENG] A support built over, and not touching, a weak structural member, such as a wooden lintel, to carry the main load. Also known as relieving arch. { 'dis‚chärj·iŋ ‚ärch }

disconnect [ELEC] To open a circuit by removing wires or connections, as distinguished from opening a switch to stop current flow. [ENG] To sever a connection. { ‚dis·kə'nekt }

discontinuous construction [BUILD] A building in which there is no solid connection between the rooms and the building structure or between different sections of the building; the design aims to reduce the transmission of noise. { ‚dis·kən'tin·yə·wəs kən'strək·shən }

discount [IND ENG] A reduction from the gross amount, price, or value. { 'dis‚kaunt }

discrete sound system [ENG ACOUS] A quadraphonic sound system in which the four input channels are preserved as four discrete channels during recording and playback processes; sometimes referred to as a 4-4-4 system. { di'skrēt 'saund ‚sis·təm }

discrete system [CONT SYS] A control system in which signals at one or more points may change only at discrete values of time. Also known as discrete-time system. { di'skrēt 'sis·təm }

discrete-time system *See* discrete system. { di'skrēt ‚tīm 'sis·təm }

discrete transfer function *See* pulsed transfer function. { di¦skrēt 'tranz·fər ‚fəŋk·shən }

disdrometer [ENG] Equipment designed to measure and record the size distribution of raindrops as they occur in the atmosphere. { diz'dräm·əd·ər }

disengage [ENG] To break the contact between two objects. { ‚dis·ən'gāj }

dishing [ENG] In metal-forming or plastics-molding operations, producing a shallow concave surface. { 'dish·iŋ }

disintegrator [MECH ENG] An apparatus used for pulverizing or grinding substances, consisting of two steel cages which rotate in opposite directions. { dis'in·tə‚grād·ər }

disk *See* phonograph record. { disk }

disk-and-doughnut [CHEM ENG] A type of fractionating tower construction of alternating disks and plates that are doughnut-shaped, to provide mixing. { ¦disk ən 'dō·nət }

disk attrition mill *See* disk mill. { ¦disk ə'trish·ən ‚mil }

disk brake [MECH ENG] A type of brake in which disks attached to a fixed frame are pressed against disks attached to a rotating axle or against the inner surfaces of a rotating housing. { ¦disk ‚brāk }

disk cam [MECH ENG] A disk with a contoured edge which rotates about an axis perpendicular to the disk, communicating motion to the cam follower which remains in contact with the edge of the disk. { ¦disk ‚kam }

disk canvas wheel [DES ENG] A polishing wheel made of disks of canvas sewn together with heavy twine or copper wire, and reinforced by steel side plates and side rings with bolts or screws. { ¦disk 'kan·vəs ‚wēl }

disk centrifuge [MECH ENG] A centrifuge with a large bowl having a set of disks that separate the liquid into thin layers to create shallow settling chambers. { ¦disk 'sen·trə‚fyüj }

disk clutch [MECH ENG] A clutch in which torque is transmitted by friction between friction disks with specially prepared friction material riveted to both sides and contact plates keyed to the inner surface of an external hub. { ¦disk ‚kləch }

disk coupling [MECH ENG] A flexible coupling in which the connecting member is a flexible disk. { 'disk ‚kəp·liŋ }

disk engine [MECH ENG] A rotating engine in which the piston is a disk. { 'disk ‚en·jən }

disk filter [ENG] A filter in which the substance to be filtered is drawn through membranes stretched on segments of revolving disks by a vacuum inside each disk; the solids left on the membrane are lifted from the tank and discharged. Also known as American filter. { 'disk ¦fil·tər }

disk grinder [MECH ENG] A grinding machine that employs abrasive disks. { 'disk ‚grīnd·ər }

disk grinding [MECH ENG] Grinding with the flat side of a rigid, bonded abrasive disk or segmental wheel. { 'disk ‚grīnd·iŋ }

disk leather wheel [DES ENG] A polishing wheel made of leather disks glued together. { ¦disk 'leth·ər ‚wēl }

disk meter |ENG| A positive displacement meter to measure flow rate of a fluid; consists of a disk that wobbles or nutates within a chamber so that each time the disk nutates a known volume of fluid passes through the meter. { 'disk ‚mēd·ər }

disk mill |MECH ENG| Size-reduction apparatus in which grinding of feed solids takes place between two disks, either or both of which rotate. Also known as disk attrition mill. { 'disk ‚mil }

disk recording |ENG ACOUS| 1. The process of inscribing suitably transformed acoustical or electrical signals on a phonograph record. 2. See phonograph record. { ‚disk ri'kord·iŋ }

disk sander |MECH ENG| A machine that uses a circular disk coated with abrasive to smooth or shape surfaces. { 'disk ‚sand·ər }

disk signal |CIV ENG| Automatic block signal with colored disks that indicate train movements. { ‚disk 'sig·nəl }

disk spring |MECH ENG| A mechanical spring that consists of a disk or washer supported by one force (distributed by a suitable chuck or holder) at the periphery and by an opposing force on the center or hub of the disk. { 'disk ‚spriŋ }

disk wheel |DES ENG| A wheel in which a solid metal disk, rather than separate spokes, joins the hub to the rim. { 'disk ‚wēl }

dispatching |IND ENG| The selecting and sequencing of tasks to be performed at individual work stations and the assigning of these tasks to the personnel. { dis'pach·iŋ }

dispenser |ENG| Device that automatically dispenses radar chaff from an aircraft. { də'spen·sər }

dispersal |CIV ENG| The practice of building or establishing industrial plants, government offices, or the like, in separated areas, to reduce vulnerability to enemy attack. { də'spər·səl }

dispersion mill |MECH ENG| Size-reduction apparatus that disrupts clusters or agglomerates of solids, rather than breaking down individual particles; used for paint pigments, food products, and cosmetics. { də'spər·zhən ‚mil }

displacement |ELEC| See electric displacement. |MECH| 1. The linear distance from the initial to the final position of an object moved from one place to another, regardless of the length of path followed. 2. The distance of an oscillating particle from its equilibrium position. |MECH ENG| The volume swept out in one stroke by a piston moving in a cylinder as for an engine, pump, or compressor. { dis'plās·mənt }

displacement compressor |MECH ENG| A type of compressor that depends on displacement of a volume of air by a piston moving in a cylinder. { dis'plās·mənt kəm‚pres·ər }

displacement engine See piston engine. { dis'plās·mənt ‚en·jən }

displacement gyroscope |ENG| A gyroscope that senses, measures, and transmits angular displacement data. { dis'plās·mənt 'jī·rə‚skōp }

displacement manometer |ENG| A differential manometer which indicates the pressure difference across a solid or liquid partition which can be displaced against a restoring force. { dis'plās·mənt mə'näm·əd·ər }

displacement meter |ENG| A water meter that measures water flow quantitatively by recording the number of times a vessel of known capacity is filled and emptied. { dis'plās·mənt ‚mēd·ər }

displacement pump |MECH ENG| A pump that develops its action through the alternate filling and emptying of an enclosed volume as in a piston-cylinder construction. { dis'plās·mənt ‚pəmp }

displacer-type meter |ENG| Apparatus to detect liquid level or gas density by measuring the effect of the fluid (gas or liquid) on the buoyancy of a displacer unit immersed within the fluid. { di'splās·ər ‚tīp ‚mēd·ər }

disposable |ENG| Within a manufacturing system, designed to be discarded after use and replaced by an identical item, such as a filter element. { də'spō·zə·bəl }

disposal field See absorption field. { də'spō·zəl ‚fēld }

dissipation factor |ELEC| The inverse of Q, the storage factor. { ‚dis·ə'pā·shən ‚fak·tər }

dissipation function See Rayleigh's dissipation function. { ‚dis·ə'pā·shən ‚faŋk·shən }

dissipation loss |ELEC| A measure of the power loss of a transducer in transmitting signals, expressed as the ratio of its input power to its output power. { ‚dis·ə'pā·shən ‚lós }

dissipative muffler |ENG| A device which absorbs sound energy as the gas passes through it; a duct lined with sound-absorbing material is the most common type. { ‚dis·ə'pād·iv 'məf·lər }

dissolved air flotation |CHEM ENG| A liquid-solid separation process wherein the main mechanism of suspended-solids removal is the change of apparent specific gravity of those suspended solids in relation to that of the suspending liquid by the attachment of small gas bubbles formed by the release of dissolved gas to the solids. Also known as air flotation. { də'zälvd ‚er flō'tā·shən }

distance |MECH| The spatial separation of two points, measured by the length of a hypothetical line joining them. { 'dis·təns }

distance marker |ENG| One of a series of concentric circles, painted or otherwise fixed on the screen of a plan position indicator, from which the distance of a target from the radar antenna can be read directly; used for surveillance and navigation where the relative distances between a number of targets are required simultaneously. Also known as radar range marker; range marker. { 'dis·təns ‚märk·ər }

distance ratio |MECH ENG| The ratio of the distance moved by the effort or input of a machine in a specified time to the distance moved by the load or output. { 'dis·təns ‚rā·shō }

distance resolution |ENG| The minimum radial distance by which targets must be separated to be separately distinguishable by a particular radar. Also known as range discrimination; range resolution. { 'dis·təns ˌrez·ə,lü·shən }

distance/velocity lag |CONT SYS| The delay caused by the amount of time required to transport material or propagate a signal or condition from one point to another. Also known as transportation lag; transport lag. { ¦dis·təns və'läs·əd·ē ˌlag }

distant signal |CIV ENG| A signal placed at a distance from a block of track to give advance warning when the block is closed. { ¦dis·tənt 'sig·nəl }

distillation test |CHEM ENG| A standardized procedure for finding the initial, intermediate, and final boiling points in the boiling range of petroleum products. { ˌdis·tə'lā·shən ˌtest }

distortion |ELECTR| Any undesired change in the waveform of an electric signal passing through a circuit or other transmission medium. |ENG| In general, the extent to which a system fails to accurately reproduce the characteristics of an input signal at its output. |ENG ACOUS| Any undesired change in the waveform of a sound wave. { di'stòr·shən }

distortion meter |ENG| An instrument that provides a visual indication of the harmonic content of an audio-frequency wave. { di'stòr·shən ˌmēd·ər }

distributed collector |ENG| A component of a solar heating system comprising a series of modular focusing collectors that are interconnected with an absorber pipe network to carry the working fluid to a heat exchanger. { di'strib·yəd·əd kə'lek·tər }

distributed control system |CONT SYS| A collection of modules, each with its own specific function, interconnected tightly to carry out an integrated data acquisition and control application. { di'strib·yəd·əd kən'trōl,sis·təm }

distributed numerical control |CONT SYS| The use of central computers to distribute part-classification data to machine tools which themselves are controlled by computers or numerical control tapes. { di'strib·yəd·əd nü'mer·ə·kəl kən'trōl }

distributed-parameter system See distributed system. { di'strib·yəd·əd pə'ram·əd·ər ˌsis·təm }

distributed system |CONT SYS| A collection of modules, each with its own specific function, interconnected to carry out integrated data acquisition and control in a critical environment. |SYS ENG| A system whose behavior is governed by partial differential equations, and not merely ordinary differential equations. Also known as distributed-parameter system. { di'strib·yəd·əd 'sis·təm }

distribution |IND ENG| All activities that involve efficient movement of finished products from the end of the production line to the consumer. { ˌdis·trə'byü·shən }

distribution amplifier |ELECTR| A radio-frequency power amplifier used to feed television or radio signals to a number of receivers, as in an apartment house or a hotel. |ENG ACOUS| An audio-frequency power amplifier used to feed a speech or music distribution system and having sufficiently low output impedance so changes in load do not appreciably affect the output voltage. { ˌdis·trə'byü·shən 'am·plə,fī·ər }

distribution box |CIV ENG| In sanitary engineering, a box in which the flow of effluent from a septic tank is distributed equally into the lines that lead to the absorption field. { ˌdis·trə'byü·shən 'bäks }

distribution reservoir |CIV ENG| A service reservoir connected with the conduits of a primary water supply; used to supply water to consumers according to fluctuations in demand over short time periods and serves for local storage in case of emergency. { ˌdis·trə'byü·shən 'rez·əv,wär }

distributor |ELEC| **1.** Any device which allocates a telegraph line to each of a number of channels, or to each row of holes on a punched tape, in succession. **2.** A rotary switch that directs the high-voltage ignition current in the proper firing sequence to the various cylinders of an internal combustion engine. |ELECTR| The electronic circuitry which acts as an intermediate link between the accumulator and drum storage. |ENG| A device for delivering an exact amount of fuel at the exact time at which it is required. { də'strib·yəd·ər }

distributor gear |MECH ENG| A gear which meshes with the camshaft gear to rotate the distributor shaft. { də'strib·yəd·ər ,gir }

district heating |MECH ENG| The supply of heat, either in the form of steam or hot water, from a central source to a group of buildings. { 'di·strikt 'hēd·iŋ }

disturbance |CONT SYS| An undesired command signal in a control system. { də'stər·bəns }

ditch |CIV ENG| **1.** A small artificial channel cut through earth or rock to carry water for irrigation or drainage. **2.** A long narrow cut made in the earth to bury pipeline, cable, or similar installations. { dich }

ditch check |CIV ENG| A small dam positioned at intervals in a road ditch to prevent erosion. { 'dich ,chek }

ditcher See trench excavator. { 'dich·ər }

ditching |ENG| The digging of ditches, as around storage tanks or process areas to hold liquids in the event of a spill or along the sides of a roadway for drainage. { 'dich·iŋ }

dither |CONT SYS| A force having a controlled amplitude and frequency, applied continuously to a device driven by a servomotor so that the device is constantly in small-amplitude motion and cannot stick at its null position. Also known as buzz. { 'dith·ər }

divariant system |THERMO| A system composed of only one phase, so that two variables, such as pressure and temperature, are sufficient to define its thermodynamic state. { di¦ver·ē·ənt 'sis·təm }

dive |ENG| To submerge into an underwater environment so that it may be studied or utilized; includes the use of specialized equipment such as scuba, diving helmets, diving suits, diving bells, and underwater research vessels. { 'dīv }

divergent die |ENG| A die with the internal channels that lead to the orifice diverging, such as the dies used for manufacture of hollow-body plastic items. { də'vər·jənt 'dī }

divergent nozzle |DES ENG| A nozzle whose cross section becomes larger in the direction of flow. { də'vər·jənt 'näz·əl }

diverging duct |DES ENG| Fluid-flow conduit whose internal cross-sectional area increases in the direction of flow. { də'vərj·iŋ ,dəkt }

diversion canal |CIV ENG| An artificial channel for diverting water from one place to another. { də'vər·zhən kə,nal }

diversion chamber |ENG| A chamber designed to direct a stream into a channel or channels. { də'vər·zhən ,chām·bər }

diversion dam |CIV ENG| A fixed dam for diverting stream water away from its course. { də'vər·zhən ,dam }

diversion gate |CIV ENG| A gate which may be closed to divert water from the main conduit or canal to a lateral or some other channel. { də'vər·zhən ,gāt }

diversion tunnel |CIV ENG| An underground passageway used to divert flowing water around a construction site. { də'vər·zhən ,tən·əl }

diversity radar |ENG| A radar that uses two or more transmitters and receivers, each pair operating at a slightly different frequency but sharing a common antenna and video display, to obtain greater effective range and reduce susceptibility to jamming. { də'vər·səd·ē 'rā,där }

diverter valve See air bypass valve. { də'vərd·ər ,valv }

divided lane |CIV ENG| A highway divided into lanes by a median strip. { də'vīd·əd 'lān }

divided pitch |DES ENG| In a screw with multiple threads, the distance between corresponding points on two adjacent threads measured parallel to the axis. { də'vīd·əd 'pich }

divider |DES ENG| A tool like a compass, used in metalworking to lay out circles or arcs and to space holes or other dimensions. { də'vīd·ər }

dividing network See crossover network. { də'vīd·iŋ ,net,wərk }

diving bell |ENG| An early diving apparatus constructed in the shape of a box or cylinder without a bottom and connected to a compressed-air hose. { 'dīv·iŋ ,bel }

diving suit |ENG| A waterproof outfit designed for diving, especially one with a helmet connected to a compressed-air hose. { 'dīv·iŋ ,süt }

division plate |MECH ENG| A diaphragm which surrounds the piston rod of a crosshead-type engine and separates the crankcase from the lower portion of the cylinder. { də'vizh·ən ,plāt }

division wall |BUILD| A wall used to create major subdivisions in a building. { də'vizh·ən ,wòl }

dock |CIV ENG| 1. The slip or waterway that is between two piers or cut into the land for the berthing of ships. 2. A basin or enclosure for

reception of vessels, provided with means for controlling the water level. { däk }

docking block |CIV ENG| A timber used to support a ship in dry dock. { 'däk·iŋ ,bläk }

dockyard |CIV ENG| A yard utilized for ship construction and repair. { 'däk,yärd }

doctor bar See doctor blade. { 'däk·tər ,bär }

doctor blade |ENG| A device for regulating the amount of liquid material on the rollers of a spreader. Also known as doctor bar; doctor knife; doctor roll. { 'däk·tər ,blād }

doctor knife See doctor blade. { 'däk·tər ,nīf }

doctor roll |CHEM ENG| Roller device used to remove accumulated filter cake from rotary filter drums. See doctor blade. { 'däk·tər ,rōl }

doctor solution |CHEM ENG| Sodium plumbite solution used to remove mercaptan sulfur from gasoline and other light petroleum distillates; used in doctor treatment. { 'däk·tər sə'lü·shən }

doctor test |CHEM ENG| A procedure using doctor solution (sodium plumbite) to detect sulfur compounds in light petroleum distillates which react with the sodium plumbite. { 'däk·tər ,test }

doctor treatment |CHEM ENG| Refining process to sweeten (reduce the odor) of gasoline, solvents, and kerosine; sodium plumbite and sulfur convert the odoriferous mercaptans into disulfides. { 'däk·tər ,trēt·mənt }

dodge chain |DES ENG| A chain with detachable bearing blocks between the links. { 'däj ,chān }

Dodge-Romig tables |IND ENG| Tabular data for acceptance sampling, including lot tolerance and AOQL tables. { ¦däj ¦rō·mig ,tā·bəlz }

dodo |ENG| A rectangular groove cut across the grain of a board. { 'dō,dō }

Doebner-Miller synthesis |CHEM ENG| Synthesis of methylquinoline by heating aniline with paraldehyde in the presence of hydrochloric acid. { ¦deb·nər ¦mil·ər 'sin·thə·səs }

dog |DES ENG| 1. Any of various simple devices for holding, gripping, or fastening, such as a hook, rod, or spike with a ring, claw, or lug at the end. 2. An iron for supporting logs in a fireplace. 3. A drag for the wheel of a vehicle. { dòg }

dog clutch |DES ENG| A clutch in which projections on one part fit into recesses on the other part. { 'dòg ,kləch }

dog iron |DES ENG| 1. A short iron bar with ends bent at right angles. 2. An iron pin that can be inserted in stone or timber in order to lift it. { 'dòg ,ī·ərn }

dog screw |DES ENG| A screw with an eccentric head; used to mount a watch in its case. { 'dòg ,skrü }

dog's tooth |CIV ENG| A masonry string course in which the brick corner projects. { 'dògz ,tüth }

dolly |ENG| Any of several types of industrial hand trucks consisting of a low platform or specially shaped carrier mounted on rollers or combinations of fixed and swivel casters; used to carry such things as furniture, milk cans, paper rolls, machinery weighing up to 80 tons, and television cameras short distances. { 'däl·ē }

dolphin |CIV ENG| **1.** A group of piles driven close and tied together to provide a fixed mooring in the open sea or a guide for ships coming into a narrow harbor entrance. **2.** A mooring post on a wharf. { 'däl·fən }

dome |ENG| The portion of a cylindrical container used in a filament-winding process that forms an integral end of the container. |ENG ACOUS| An enclosure for a sonar transducer, projector, or hydrophone and associated equipment; designed to have minimum effect on sound waves traveling underwater. { dōm }

domestic induction heater |ENG| A cooking utensil heated by current (usually of commercial power line frequency) induced in it by a primary inductor. { də'mes·tik in'dək·shən ,hēd·ər }

domestic refrigerator |MECH ENG| A refrigeration system for household use which typically has a compression machine designed for continuous automatic operation and for conservation of the charges of refrigerant and oil, and is usually motor-driven and air-cooled. Also known as refrigerator. { də'mes·tik ri'frij·ə,rād·ər }

donkey engine |MECH ENG| A small auxiliary engine which is usually portable or semiportable and powered by steam, compressed air, or other means, particularly one used to power a windlass to lift cargo on shipboard or to haul logs. { 'dəŋ·kē ,en·jən }

Donohue equation |THERMO| Equation used to determine the heat-transfer film coefficient for a fluid on the outside of a baffled shell-and-tube heat exchanger. { 'dän·ə·hü i,kwā·zhən }

doodlebug |MECH ENG| **1.** A small tractor. **2.** A motor-driven railcar used for maintenance and repair work. { 'düd·əl,bəg }

door |ENG| A piece of wood, metal, or other firm material pivoted or hinged on one side, sliding along grooves, rolling up and down, revolving, or folding, by means of which an opening into or out of a building, room, or other enclosure is open or closed to passage. { dòr }

door check See door closer. { 'dòr ,chek }

door closer |DES ENG| **1.** A device that makes use of a spring for closing, and a compression chamber from which liquid or air escapes slowly, to close a door at a controlled speed. Also known as door check. **2.** In elevators, a device or assembly of devices which closes an open car or hoistway door by the use of gravity or springs. { 'dòr ,klōz·ər }

doorstop |BUILD| A strip positioned on the doorjamb for the door to close against. { 'dòr,stäp }

dope See doping agent. { dōp }

doped junction |ELECTR| A junction produced by adding an impurity to the melt during growing of a semiconductor crystal. { |dōpt 'jəŋk·shən }

doping |ELECTR| The addition of impurities to a semiconductor to achieve a desired characteristic, as in producing an n-type or p-type material. Also known as semiconductor doping. |ENG|

Coating the mold or mandrel with a substance which will prevent the molded plywood part from sticking to it and will facilitate removal. { 'dōp·iŋ }

doping agent |ELECTR| An impurity element added to semiconductor materials used in crystal diodes and transistors. Also known as dopant; dope. { 'dōp·iŋ ,ā·jənt }

doping compensation |ELECTR| The addition of donor impurities to a p-type semiconductor or of acceptor impurities to an n-type semiconductor. { 'dōp·iŋ käm·pən'sā·shən }

Doppler current meter |ENG| An acoustic current meter in which a collimated ultrasonic signal of known frequency is projected into the water and the reverberation frequency is measured; the difference in frequencies (Doppler shift) is proportional to the speed of water traveling past the meter. { 'däp·lər ,kər·ənt ,mēd·ər }

Doppler radar |ENG| A radar that makes use of the Doppler shift of an echo due to relative motion of target and radar to differentiate between fixed and moving targets and measure target velocities. { 'däp·lər 'rā,där }

Doppler range See doran. { 'däp·lər ,rānj }

Doppler sonar |ENG| Sonar based on Doppler shift measurement technique. Abbreviated DS. { 'däp·lər 'sō,när }

Doppler tracking |ENG| Tracking of a target by using Doppler radar. { 'däp·lər ,trak·iŋ }

Doppler ultrasonic flowmeter |ENG| An instrument for determining the velocity of fluid flow from the Doppler shift of high-frequency sound waves reflected from particles or discontinuities in the flowing fluid. { 'däp·lər əl·trə'sän·ik 'flō ,mēd·ər }

DOP test See dioctyl phthalate test. { 'däp ,test }

doran |ENG| A Doppler ranging system that uses phase comparison of three different modulation frequencies on the carrier wave, such as 0.01, 0.1, and 1 megahertz, to obtain missile range data with high accuracy. Derived from Doppler range. { 'dò,rän }

dormer window |BUILD| An extension of an attic room through a sloping roof to accommodate a vertical window. { 'dòr·mər 'win·dō }

Dorr agitator |MECH ENG| A tank used for batch washing of precipitates which cannot be leached satisfactorily in a tank; equipped with a slowly rotating rake at the bottom, which moves settled solids to the center, and an air lift that lifts slurry to the launders. Also known as Dorr thickener. { 'dòr 'aj·ə,tād·ər }

Dorr classifier |MECH ENG| A horizontal flow classifier consisting of a rectangular tank with a sloping bottom, a rake mechanism for moving sands uphill along the bottom, an inlet for feed, and outlets for sand and slime. { 'dòr 'klas· ə,fī·ər }

Dorr thickener See Dorr agitator. { 'dòr 'thik·ə· nər }

dosing tank |CIV ENG| A holding tank that discharges sewage at a rate required by treatment processes. { 'dōs·iŋ ,taŋk }

dot See button. { dät }

double-acting |MECH ENG| Acting in two directions, as with a reciprocating piston in a cylinder with a working chamber at each end. { ¦dəb·əl 'ak·tiŋ }

double-acting compressor |MECH ENG| A reciprocating compressor in which both ends of the piston act in working chambers to compress the fluid. { ¦dəb·əl ¦ak·tiŋ kəm'pres·ər }

double-acting pawl |MECH ENG| A double pawl which can drive in either direction. { ¦dəb·əl ¦ak·tiŋ 'pȯl }

double-action mechanical press |MECH ENG| A press having two slides which move one within the other in parallel movements. { ¦dəb·əl ¦ak·shən mə¦kan·ə·kəl 'pres }

double-amplitude-modulation multiplier |ELECTR| A multiplier in which one variable is amplitude-modulated by a carrier, and the modulated signal is again amplitude-modulated by the other variable; the resulting double-modulated signal is applied to a balanced demodulator to obtain the product of the two variables. { ¦dəb·əl ¦am·plə,tüd ¦mäj·ə,lā·shən 'məl·tə,plī·ər }

double-barrier resonant tunneling diode |ELECTR| A variant of the tunnel diode with thin layers of aluminum gallium arsenide and gallium arsenide that have sharp interfaces and have widths comparable to the Schrödinger wavelengths of the electrons, permitting resonant behavior. Abbreviated DBRT diode. { ¦dəb·əl ,bar·ē·ər ¦rez·ən·ənt ,tən·əl·iŋ 'dī,ōd }

double-base diode See unijunction transistor. { ¦dəb·əl ¦bās 'dī,ōd }

double-base junction diode See unijunction transistor. { ¦dəb·əl ¦bās 'jəŋk·shən 'dī,ōd }

double-base junction transistor |ELECTR| A tetrode transistor that is essentially a junction triode transistor having two base connections on opposite sides of the central region of the transistor. Also known as tetrode junction transistor. { ¦dəb·əl ¦bās 'jəŋk·shən tran'zis·tər }

double block and bleed system |ENG| A valve system configuration in which a full-flow vent valve is installed in a pipeline between two shutoff valves to provide a means of releasing excess pressure between them. { 'dəb·əl ¦bläk ən 'blēd ,sis·təm }

double-block brake |MECH ENG| Two single-block brakes in symmetrical opposition, where the operating force on one lever is the reaction on the other. { ¦dəb·əl ¦bläk 'brāk }

double bridge See Kelvin bridge. { ¦dəb·əl 'brij }

double-button microphone |ENG ACOUS| A carbon microphone having two carbon-filled buttonlike containers, one on each side of the diaphragm, to give twice the resistance change obtainable with a single button. Also known as differential microphone. { ¦dəb·əl ¦bət·ən 'mī·krə,fōn }

double-cone bit |DES ENG| A type of roller bit having only two cone-shaped cutting members. { ¦dəb·əl ¦kōn 'bit }

double-core barrel drill |DES ENG| A core drill consisting of an inner and an outer tube; the

inner member can remain stationary while the outer one revolves. { ¦dəb·əl ,kȯr 'bar·əl ,dril }

double-coursed |BUILD| Covered with a material such as shingles in such a way that no area is covered with less than two thicknesses. { ¦dəb·əl 'kȯrst }

double-crank press |MECH ENG| A mechanical press with a single wide slide operated by a crankshaft having two crank pins. { ¦dəb·əl ¦kraŋk 'pres }

double crossover See scissors crossover. { ¦dəb·əl 'krȯs,ō·vər }

double-cut file |DES ENG| A file covered with two series of parallel ridges crossing at angles to each other. { ¦dəb·əl ¦kət 'fīl }

double-cut planer |MECH ENG| A planer designed to cut in both the forward and reverse strokes of the table. { ¦dəb·əl ¦kət 'plān·ər }

double-cut saw |DES ENG| A saw with teeth that cut during the forward and return strokes. { ¦dəb·əl ¦kət 'sȯ }

double-diffused transistor |ELECTR| A transistor in which two pn junctions are formed in the semiconductor wafer by gaseous diffusion of both p-type and n-type impurities; an intrinsic region can also be formed. { ¦dəb·əl də¦fyüzd tran'zis·tər }

double diode See binode; duodiode. { ¦dəb·əl 'dī,ōd }

double-diode limiter |ELECTR| Type of limiter which is used to remove all positive signals from a combination of positive and negative pulses, or to remove all the negative signals from such a combination of positive and negative pulses. { ¦dəb·əl ¦dī,ōd 'lim·əd·ər }

double distribution |CHEM ENG| The product distribution resulting from counter double-current extraction, a scheme in which each of the two liquid phases is transferred simultaneously and continuously in opposite directions through an interconnected train of contact vessels. { ¦dəb·əl dis·trə'byü·shən }

double-doped transistor |ELECTR| The original grown-junction transistor, formed by successively adding p-type and n-type impurities to the melt during growing of the crystal. { ¦dəb·əl ,dōpt tran'zis·tər }

double-drum hoist |MECH ENG| A hoisting device consisting of two cable drums which rotate in opposite directions and can be operated separately or together. { ¦dəb·əl ¦drəm 'hȯist }

double floor |BUILD| A floor in which binding joists support the ceiling joists below as well as the floor joists above. { ¦dəb·əl 'flȯr }

doublehand drilling |ENG| A rock-drilling method performed by two men, one striking the rock with a long-handled sledge hammer while a second holds the drill and twists it between strokes. Also known as double jacking. { 'dəb·əl,hand 'dril·iŋ }

double Hooke's joint |MECH ENG| A universal joint which eliminates the variation in angular displacement and angular velocity between driving and driven shafts, consisting of two Hooke's joints with an intermediate shaft. { 'dəb·əl 'hŭks ˌjóint }

double-housing planer |MECH ENG| A planer having two housings to support the cross rail, with two heads on the cross rail and one side-head on each housing. { ¦dəb·əl ˌhaúz·iŋ 'plān·ər }

double-hung |BUILD| Of a window, having top and bottom sashes which are counterweighted or equipped with a spring on each side for easier raising and lowering. { ¦dəb·əl 'həŋ }

double impeller breaker See impact breaker. { ¦dəb·əl im'pel·ər ˌbrāk·ər }

double-integrating gyro |MECH| A single-degree-of-freedom gyro having essentially no restraint of its spin axis about the output axis. { ¦dəb·əl ¦in·tə,grād·iŋ 'jī·rō }

double jack |DES ENG| A heavy hammer, weighing about 10 pounds (4.5 kilograms), requiring the use of both hands. { ¦dəb·əl 'jak }

double jacking See doublehand drilling. { ¦dəb·əl 'jak·iŋ }

double load |ENG| A charge separated by inert material in a borehole. { ¦dəb·əl 'lōd }

double mast See A frame. { ¦dəb·əl 'mast }

double pendulum |MECH| Two masses, one suspended from a fixed point by a weightless string or rod of fixed length, and the other similarly suspended from the first; often the system is constrained to remain in a vertical plane. { ¦dəb·əl 'pen·jə·ləm }

double-pipe exchanger |CHEM ENG| Fluid-fluid heat exchanger made of two concentric pipe sections; one fluid (such as a coolant) flows in the annular space between pipes, and the other fluid (such as hot process stream) flows through the inner pipe. { ¦dəb·əl ˌpīp iks'chān·jər }

double-quirked bead See quirk bead. { ¦dəb·əl ¦kwərkt 'bēd }

double-rivet |ENG| To rivet a lap joint with two rows of rivets or a butt joint with four rows. { ¦dəb·əl 'riv·ət }

double-roll crusher |MECH ENG| A machine which crushes materials between teeth on two roll surfaces; used mainly for coal. { ¦dəb·əl 'rōl 'krəsh·ər }

double sampling |IND ENG| Inspecting one sample and then deciding whether to accept or reject the lot or to defer action until a second sample is inspected. { ¦dəb·əl 'sam·pliŋ }

double-shot molding |ENG| A means of turning out two-color parts in thermoplastic materials by successive molding operations. { ¦dəb·əl ˌshät 'mōld·iŋ }

double-sided board |ELECTR| A printed wiring board that contains circuitry on both external layers. { ¦dəb·əl ˌsīd·əd 'bórd }

double-slider coupling See slider coupling. { ¦dəb·əl ¦slīd·ər 'kəp·liŋ }

double-solvent refining |CHEM ENG| Petroleum-refining process using two solvents to simultaneously deasphalt and solvent-treat lubricating-oil stocks. { ¦dəb·əl 'säl·vənt ra'fīn·iŋ }

double square See adjustable square. { ¦dəb·əl 'skwer }

double-stream amplifier |ELECTR| Microwave traveling-wave amplifier in which amplification occurs through interaction of two electron beams having different average velocities. { ¦dəb·əl ˌstrēm 'am·plə,fī·ər }

double-theodolite observation |ENG| A technique for making winds-aloft observations in which two theodolites located at either end of a base line follow the ascent of a pilot balloon; synchronous measurements of the elevation and azimuth angles of the balloon, taken at periodic intervals, permit computation of the wind vector as a function of height. { ¦dəb·əl thē'äd·əl,īt äb·zər'vā·shən }

double-track tape recorder |ENG ACOUS| A tape recorder with a recording head that covers half the tape width, so two parallel tracks can be recorded on one tape. Also known as dual-track tape recorder; half-track tape recorder. { ¦dəb·əl ˌtrak 'tāp ri,kórd·ər }

double-tuned circuit |ELECTR| A circuit that is resonant to two adjacent frequencies, so that there are two approximately equal values of peak response, with a dip between. { ¦dəb·əl ˌtünd 'sər·kət }

double-tuned detector |ELECTR| A type of frequency-modulation discriminator in which the limiter output transformer has two secondaries, one tuned above the resting frequency and the other tuned an equal amount below. { ¦dəb·əl ˌtünd di'tek·tər }

double-wall cofferdam |CIV ENG| A cofferdam consisting of two lines of steel piles tied to each other, and having the space between filled with sand. { ¦dəb·əl ˌwól 'kóf·ər,dam }

double weighing |MECH| A method of weighing to allow for differences in lengths of the balance arms, in which object and weights are balanced twice, the second time with their positions interchanged. Also known as Gauss method of weighing. { ¦dəb·əl 'wā·iŋ }

dovetail joint |DES ENG| A joint consisting of a flaring tenon in a fitting mortise. { 'dəv,tāl 'jóint }

dovetail saw |DES ENG| A short stiff saw with a thin blade and fine teeth; used for accurate woodwork. { 'dəv,tāl 'só }

dowel |DES ENG| **1.** A headless, cylindrical pin which is sunk into corresponding holes in adjoining parts, to locate the parts relative to each other or to join them together. Also known as dowel pin. **2.** A round wooden stick from which dowel pins are cut. { 'daúl }

dowel pin See dowel. { 'daúl ˌpin }

dowel plate |DES ENG| A hardened steel plate with drilled holes that is used to fashion dowels by driving pegs through the holes to remove excess wood. { 'daúl ˌplāt }

dowel screw |DES ENG| A dowel with threads at both ends. { 'daúl ˌskrü }

down |ENG| Not in operation. { daún }

downcomer |BUILD| See downspout. |CHEM ENG| A method of conveying liquid from one tray to the one below in a bubble-tray column. |ENG| In an air-pollution control system, a pipe that conducts gases downward to a device that removes undesirable substances. |MECH ENG| A tube in a boiler waterwall system wherein the fluid flows downward. { 'dauṅ,kam·ər }

downdraft carburetor |MECH ENG| A carburetor in which the fuel is fed into a downward current of air. { 'dauṅ,draft 'kär·bə,rād·ər }

down-feed system |MECH ENG| In a heating or cooling system, a piping arrangement in which the fluid is circulated through supply mains that are located above the levels of the units they serve. { 'dauṅ ˌfēd ˌsis·təm }

downhole equipment See drill fittings. { 'dauṅ ˌhōl i̩'kwip·mənt }

Downs cell |CHEM ENG| A brick-lined steel vessel with four graphite anodes projecting upward from the bottom, with cathodes in the form of steel cylinders concentric with the anodes, containing an electrolyte which is 40% sodium chloride ($NaCl$) and 60% calcium chloride ($CaCl_2$) at 590°C; used to make sodium. { 'dauṅz ˌsel }

downspout |BUILD| A vertical pipe that leads water from a roof drain or gutter down to the ground or a cistern. Also known as downcomer; leader. { 'dauṅ,spauṫ }

Down's process |CHEM ENG| A method for producing sodium and chlorine from sodium chloride; potassium chloride and fluoride are added to the sodium chloride to reduce the melting point; the fused mixture is electrolyzed, with sodium forming at the cathode and chlorine at the anode. { 'dauṅz ˌpräs·əs }

downstream |CHEM ENG| Portion of a product stream that has already passed through the system; that portion located after a specific process unit. { 'dauṅ,strēm }

downtime |IND ENG| The lost production time during which a piece of equipment is not operating correctly due to a breakdown, maintenance, necessities, or power failure. { 'dauṅ,tīm }

dr See dram.

drachm See dram. { dram }

draft Also spelled draught. |CIV ENG| A line of a traverse survey. |ENG| **1.** In molds, the degree of taper on a side wall or the angle of clearance present to facilitate removal of cured or hardened parts from a mold. **2.** The area of a water discharge opening. { draft }

draft gage |ENG| **1.** A modified U-tube manometer used to measure draft of low gas heads, such as draft pressure in a furnace, or small differential pressures, for example, less than 2 inches (5 centimeters) of water. **2.** A hydrostatic depth indicator, installed in the side of a vessel below the light load line, to indicate amount of submergence. { 'draft ˌgāj }

draft hood |ENG| A device used to facilitate the escape of combustion products from the combustion chamber of an appliance, to prevent a backdraft in the combustion chamber, and to neutralize the effect of stack action of the chimney or gas vent on the efficient operation of the appliance. { 'draft ˌhuḋ }

draft loss |MECH ENG| A decrease in the static pressure of a gas in a furnace or boiler due to flow resistance. { 'draft ˌlȯs }

draftsman |ENG| An individual skilled in drafting, especially of machinery and structures. { 'draf·smən }

draft tube |MECH ENG| The piping system for a reaction-type hydraulic turbine that allows the turbine to be set safely above tail water and yet utilize the full head of the site from head race to tail race. { 'draf ˌtüb }

drag |ENG| **1.** A tool fashioned from sheet steel and having a toothed edge along the long dimension; used to level and scratch plaster to produce a key for the next coat of plaster. Also known as comb. **2.** A tool consisting of a steel plate with a finely serrated edge; dragged over the surface to dress stone. { drag }

drag bit See bit drag. { 'drag ˌbit }

drag-body flowmeter |ENG| Device to meter liquid flow; measures the net force parallel to the direction of flow; the resulting pressure difference is used to solve flow equations. { 'drag ˌbäd·ē 'flō,mēd·ər }

drag chain |ENG| **1.** A chain dragged along the ground from a motor vehicle chassis to prevent the accumulation of static electricity. **2.** A chain for coupling rail cars. { 'drag ˌchān }

drag-chain conveyor |MECH ENG| A conveyor in which the open links of a chain drag material along the bottom of a hard-faced concrete or cast iron trough. Also known as dragline conveyor. { 'drag ˌchān kən'vā·ər }

drag classifier |MECH ENG| A continuous belt containing transverse rakes, used to separate coarse sand from fine; the belt moves up through an inclined trough, and fast-settling sands are dragged along by the rakes. { 'drag 'klas·ə ˌfī·ər }

drag conveyor See flight conveyor. { 'drag kən'vā·ər }

drag-cup generator |ENG| A type of tachometer which uses eddy currents and functions in control systems; it consists of two stationary windings, positioned so as to have zero coupling, and a nonmagnetic metal cup, which is revolved by the source whose speed is to be measured; one of the windings is used for excitation, inducing eddy currents in the rotating cup. Also known as drag-cup tachometer. { 'drag ˌkəp 'jen·ə,rād·ər }

drag-cup tachometer See drag-cup generator. { 'drag ˌkəp tə'käm·əd·ər }

drag cut |ENG| A drill hole pattern for breaking out rock, in which angled holes are drilled along a floor toward a parting, or on a free face and then broken by other holes drilled into them. { 'drag ˌkət }

drag factor |CHEM ENG| Ratio of hindered diffusion rate to unhindered rate through a swollen dialysis membrane. Also known as Faxen drag factor; hindrance factor. { 'drag ˌfak·tər }

dragline |MECH ENG| An excavator operated by pulling a bucket on ropes towards the jib from which it is suspended. Also known as dragline excavator. { 'drag,līn }

dragline conveyor *See* drag-chain conveyor. { 'drag,līn kən'vā·ər }

dragline excavator *See* dragline. { 'drag,līn 'eks·kə,vād·ər }

dragline scraper |MECH ENG| A machine with a flat, plowlike blade or partially open bucket pulled on rope for withdrawing piled material, such as stone or coal, from a stockyard to the loading platform; the empty bucket is subsequently returned to the pile of material by means of a return rope. { 'drag,līn 'skrāp·ər }

drag link |MECH ENG| A four-bar linkage in which both cranks traverse full circles; the fixed member must be the shortest link. { 'drag ,link }

dragsaw |DES ENG| A saw that cuts on the pulling stroke; used in power saws for cutting felled trees. { 'drag,só }

drag-type tachometer *See* eddy-current tachometer. { 'drag ,tīp tə'käm·əd·ər }

drain |CIV ENG| **1.** A channel which carries off surface water. **2.** A pipe which carries off liquid sewage. |ELEC| *See* current drain. |ELECTR| The region into which majority carriers flow in a field-effect transistor; it is comparable to the collector of a bipolar transistor and the anode of an electron tube. { drān }

drainage |CIV ENG| Removal of groundwater or surface water, or of water from structures, by gravity or pumping. { 'drān·ij }

drainage canal |CIV ENG| An artificial canal built to drain water from an area having no natural outlet for precipitation accumulation. { 'drān·ij kə,nal }

drainage gallery |CIV ENG| A gallery in a masonry dam parallel to the top of the dam, to intercept seepage from the upstream face and conduct it away from the downstream face. { 'drān·ij ,gal·rē }

drainage well |CIV ENG| A vertical shaft in a masonry dam to intercept seepage before it reaches the downstream side. { 'drān·ij ,wel }

drain tile |BUILD| A cylindrical tile with holes in the walls used at the base of a building foundation to carry away groundwater. { 'drān ,tīl }

drain valve |CHEM ENG| A valve used to drain off material that has separated from a fluid or gas stream, or one used to empty a process line, vessel, or storage tank. { 'drān ,valv }

dram |MECH| **1.** A unit of mass, used in the apothecaries' system of mass units, equal to 1/8 apothecaries' ounce or 60 grains or 3.8879346 grams. Also known as apothecaries' dram (dram ap); drachm (British). **2.** A unit of mass, formerly used in the United Kingdom, equal to 1/16 ounce (avoirdupois) or approximately 1.77185 grams. Abbreviated dr. { dram }

dram ap *See* dram. { 'dram ,ap }

drape forming |ENG| A method of forming thermoplastic sheet in which the sheet is clamped into a movable frame, heated, and draped over

high points of a male mold; vacuum is then applied to complete the forming operation. { 'drāp ,fór·miŋ }

Draper effect |CHEM ENG| The increase in volume at constant pressure at the start of the reaction of hydrogen and chlorine to form hydrogen chloride; the volume increase is caused by an increase in temperature of the reactants, due to heat released in the reaction. { 'drā·pər i,fekt }

draught *See* draft. { draft }

draught stop *See* fire stop. { 'draf ,stóp }

draw |ENG| To haul a load. { dró }

drawbar |ENG| **1.** A bar used to connect a tender to a steam locomotive. **2.** A beam across the rear of a tractor for coupling machines and other loads. **3.** A clay block submerged in a glass-making furnace to define the point at which sheet glass is drawn. { 'dró,bär }

drawbar horsepower |MECH ENG| The horsepower available at the drawbar in the rear of a locomotive or tractor to pull the vehicles behind it. { 'dró,bär 'hórs,paú·ər }

drawbar pull |MECH ENG| The force with which a locomotive or tractor pulls vehicles on a drawbar behind it. { 'dró,bär ,púl }

drawbridge |CIV ENG| Any bridge that can be raised, lowered, or drawn aside to provide clear passage for ships. { 'dró,brij }

drawdown ratio |ENG| The ratio of die opening thickness to product thickness. { 'dró,daún ,rā·shō }

drawer |ENG| A box or receptacle that slides or rolls on tracks within a cabinet. { 'dró·ər }

draw-filing |ENG| Filing by pushing and pulling a file sideways across the work. { 'dró ,fīl·iŋ }

drawing |CHEM ENG| Removing ceramic ware from a kiln after it has been fired. { 'dró·iŋ }

drawknife |DES ENG| A woodcutting tool with a long, narrow blade and two handles mounted at right angles to the blade. { 'dró,nīf }

drawpoint |ENG| A steel point used to scratch lines or to pierce holes. { 'dró,póint }

dredge |ENG| A cylindrical or rectangular device for collecting samples of bottom sediment and benthic fauna. |MECH ENG| A floating excavator used for widening or deepening channels, building canals, constructing levees, raising material from stream or harbor bottoms to be used elsewhere as fill, or mining. { drej }

dredging |ENG| Removing solid matter from the bottom of a water area. { 'drej·iŋ }

dress |CIV ENG| To smooth the surface of concrete or stone. |ELECTR| The arrangement of connecting wires in a circuit to prevent undesirable coupling and feedback. |MECH ENG| **1.** To shape a tool. **2.** To restore a tool to its original shape and sharpness. { dres }

dresser |ENG| Any tool or apparatus used for dressing something. { 'dres·ər }

dressing |CIV ENG| The process of smoothing or squaring lumber or stone for use in a building. |ENG| The sharpening, repairing, and replacing of parts, notably drilling bits and tool joints, to ready equipment for reuse. { 'dres·iŋ }

Dressler kiln |MECH ENG| The first successful muffle-type tunnel kiln. { 'dres·lər ‚kil }

drier |ENG| A device to remove water. { 'drī·ər }

drift |ENG| **1.** A gradual deviation from a set adjustment, such as frequency or balance current, or from a direction. **2.** The deviation, or the angle of deviation, of a borehole from the vertical or from its intended course. **3.** To measure the size of a pipe opening by passing a mandrel through it. |MECH ENG| The water lost in a cooling tower as mist or droplets entrained by the circulating air, not including the evaporative loss. { drift }

drift bolt |ENG| **1.** A bolt used to force out other bolts or pins. **2.** A metal rod used to secure timbers. { 'drift ‚bōlt }

drifter |MECH ENG| A rock drill, similar to but usually larger than a jack hammer, mounted for drilling holes up to 4¹/₂ inches (11.4 centimeters) in diameter. { 'drif·tər }

drift indicator |ENG| Device used to record directional logs; records only the amount of drift (deviation from the vertical), and not the direction. { 'drift ‚in·də‚kād·ər }

driftpin |DES ENG| A round, tapered metal rod that is driven into matching rivet holes of two metal parts for stretching the parts and bringing them into alignment. { 'drift‚pin }

drift plug |ENG| A plug that can be driven into a pipe to straighten it or to flare its opening. { 'drift ‚pləg }

drift ultrasonic flowmeter See deflection ultrasonic flowmeter. { ‚drift ‚əl·trə‚sän·ik 'flō‚mēd·ər }

drill |ENG| A rotating-end cutting tool for creating or enlarging holes in a solid material. Also known as drill bit. { dril }

drillability |ENG| Fitness for being drilled, denoting ease of penetration. { ‚dril·ə'bil·əd·ē }

drill angle gage See drill grinding gage. { 'dril ‚aŋ·gəl ‚gāj }

drill bit See drill. { 'dril ‚bit }

drill cable |ENG| A cable used to pull up drill rods, casing, and other drilling equipment used in making a borehole. { 'dril ‚kā·bəl }

drill capacity |MECH ENG| The length of drill rod of specified size that the hoist on a diamond or rotary drill can lift or that the brake can hold on a single line. { 'dril kə‚pas·əd·ē }

drill carriage |MECH ENG| A platform or frame on which several rock drills are mounted and which moves along a track, for heavy drilling in large tunnels. Also known as jumbo. { 'dril ‚kar·ij }

drill chuck |DES ENG| A chuck for holding a drill or other cutting tool on a spindle. { 'dril ‚chək }

drill collar |DES ENG| A ring which holds a drill bit and gives it radial location with respect to a bearing. { 'dril ‚käl·ər }

drill cuttings |ENG| Cuttings of rock and other subterranean materials brought to the surface during the drilling of wellholes. { 'dril ‚kəd·iŋz }

drill drift |ENG| A steel wedge used to remove tapered shank tools from spindles, sockets, and sleeves. { 'dril ‚drift }

drilled caisson |CIV ENG| A drilled hole filled with concrete and lined with a cylindrical steel casing if needed. { ‚drild 'kā‚sän }

driller |ENG| A person who operates a drilling machine. |MECH ENG| See drilling machine. { 'dril·ər }

drill extractor |ENG| A tool for recovering broken drill pieces or a detached drill from a borehole. { 'dril ik‚strak·tər }

drill feed |MECH ENG| The mechanism by which the drill bit is fed into the borehole during drilling. { 'dril ‚fēd }

drill fittings |ENG| All equipment used in a borehole during drilling. Also known as downhole equipment. { 'dril ‚fid·iŋz }

drill floor |ENG| A work area covered with planks around the collar of a borehole at the base of a drill tripod or derrick. { 'dril ‚flȯr }

drill footage |ENG| The lineal feet of borehole drilled. { 'dril ‚fu̇d·ij }

drill gage |DES ENG| A thin, flat steel plate that has accurate holes for many sizes of drills; each hole, identified as to drill size, enables the diameter of a drill to be checked. |ENG| Diameter of a borehole. { 'dril ‚gāj }

drill grinding gage |DES ENG| A tool that checks the angle and length of a twist drill while grinding it. Also known as drill angle gage; drill point gage. { 'dril ‚grīnd·iŋ ‚gāj }

drill hole |ENG| A hole created or enlarged by a drill or auger. Also known as borehole. { 'dril ‚hōl }

drill-hole logging See borehole logging. { 'dril ‚hōl 'läg·iŋ }

drill-hole pattern |ENG| The number, position, angle, and depth of the shot holes forming the round in the face of a tunnel or sinking pit. { 'dril ‚hōl ‚pad·ərn }

drill-hole survey See borehole survey. { 'dril ‚hōl ‚sər‚vā }

drilling |ENG| The creation or enlarging of a hole in a solid material with a drill. { 'dril·iŋ }

drilling column |ENG| The column of drill rods, with the drill bit attached to the end. { 'dril·iŋ ‚käl·əm }

drilling machine |MECH ENG| A device, usually motor-driven, fitted with an end cutting tool that is rotated with sufficient power either to create a hole or to enlarge an existing hole in a solid material. Also known as driller. { 'dril·iŋ mə‚shēn }

drilling platform |ENG| The structural base upon which the drill rig and associated equipment is mounted during the drilling operation. { 'dril·iŋ ‚plat‚fȯrm }

drilling rate |MECH ENG| The number of lineal feet drilled per unit of time. { 'dril·iŋ ‚rāt }

drilling time |ENG| **1.** The time required in rotary drilling for the bit to penetrate a specified thickness (usually 1 foot) of rock. **2.** The actual time the drill is operating. { 'dril·iŋ ‚tīm }

drilling time log |ENG| Foot-by-foot record of how fast a formation is drilled. { 'dril·iŋ 'tīm ‚läg }

drill jig |MECH ENG| A device fastened to the

work in repetition drilling to position and guide the drill. { 'dril ,jig }

drill log |ENG| **1.** A record of the events and features of the formations penetrated during boring. Also known as boring log. **2.** A record of all occurrences during drilling that might help in a complete logging of the hole or in determining the cost of the drilling. { 'dril ,läg }

drill out |ENG| **1.** To complete one or more boreholes. **2.** To penetrate or remove a borehole obstruction. **3.** To locate and delineate the area of a subsurface ore body or of petroleum by a series of boreholes. { ¦dril 'aut }

drill-over |ENG| The act or process of drilling around a casing lodged in a borehole. { 'dril ,ō·vər }

drill point gage See drill grinding gage. { 'dril ,póint ,gāj }

drill press |MECH ENG| A drilling machine in which a vertical drill moves into the work, which is stationary. { 'dril ,pres }

drill rod |ENG| The long rod that drives the drill bit in drilling boreholes. { 'dril ,räd }

drill sleeve |ENG| A tapered, hollow steel shaft designed to fit the tapered shank of a cutting tool to adapt it to the drill press spindle. { 'dril ,slēv }

drill socket |ENG| An adapter to fit a tapered shank drill to a taper hole that is larger than that in the drill press spindle. { 'dril ,säk·ət }

drill string |MECH ENG| The assemblage of drill rods, core barrel, and bit, or of drill rods, drill collars, and bit in a borehole, which is connected to and rotated by the drill collar of the borehole. { 'dril ,striŋ }

drip cap |BUILD| A horizontal molding installed over the frame for a door or window to direct water away from the frame. { 'drip ,kap }

drip edge |BUILD| A metal strip that extends beyond the other parts of the roof and is used to direct rainwater off. { 'drip ,ej }

drive |ELECTR| See excitation. |MECH ENG| The means by which a machine is given motion or power (as in steam drive, diesel-electric drive), or by which power is transferred from one part of a machine to another (as in gear drive, belt drive). { drīv }

drive-by-wire |MECH ENG| Electronic throttle control in automobiles. { ¦drīv bī 'wīr }

drive chuck |MECH ENG| A mechanism at the lower end of a diamond-drill drive rod on the swivel head by means of which the motion of the drive rod can be transmitted to the drill string. { drīv ,chək }

drive fit |DES ENG| A fit in which the larger (male) part is pressed into a smaller (female) part; the assembly must be effected through the application of an external force. { ,drīv ,fit }

drivehead |ENG| A cap fitted over the end of a mechanical part to protect it while it is being driven. { 'drīv,hed }

driveline |MECH ENG| In an automotive vehicle, the group of parts, including the universal joint and the drive shaft, that connect the transmission with the driving wheels. { 'drīv,līn }

driven caisson |CIV ENG| A caisson formed by driving a cylindrical steel shell into the ground with a pile-driving hammer and then placing concrete inside; the shell may be removed when concrete sets. { ¦driv·ən 'kā,sän }

driven gear |MECH ENG| The member of a pair of gears to which motion and power are transmitted by the other. { ¦driv·ən 'gir }

drivepipe |ENG| A thick-walled casing pipe that is driven through overburden or into a deep drill hole to prevent caving. { 'drīv,pīp }

drive pulley |MECH ENG| The pulley that drives a conveyor belt. { 'drīv ,púl·ē }

driver |ELECTR| The amplifier stage preceding the output stage in a receiver or transmitter. |ENG ACOUS| The portion of a horn loudspeaker that converts electrical energy into acoustical energy and feeds the acoustical energy to the small end of the horn. { 'drī·vər }

drive rod |ENG| Hollow shaft in the swivel head of a diamond-drill machine through which energy is transmitted from the drill motor to the drill string. Also known as drive spindle. { 'drīv ,räd }

drive sampling |ENG| The act or process of driving a tubular device into soft rock material for obtaining dry samples. { 'drīv ,sam·pliŋ }

drivescrew |DES ENG| A screw that is driven all the way in, or nearly all the way in, with a hammer. { 'drīv,skrü }

drive shaft |MECH ENG| A shaft which transmits power from a motor or engine to the rest of a machine. { 'drīv ,shaft }

drive shoe |DES ENG| A sharp-edged steel sleeve attached to the bottom of a drivepipe or casing to act as a cutting edge and protector. { 'drīv,shü }

drive spindle See drive rod. { 'drīv ,spin·dəl }

drive train See power train. { 'drīv ,trān }

driving clock |ENG| A mechanism for driving an instrument at a required rate. { 'drīv·iŋ ,kläk }

driving pinion |MECH ENG| The input gear in the differential of an automobile. { 'drīv·iŋ ,pin·yən }

driving-point function |CONT SYS| A special type of transfer function in which the input and output variables are voltages or currents measured between the same pair of terminals in an electrical network. { 'drīv·iŋ ,póint, faŋk·shən }

driving resistance |MECH| The force exerted by soil on a pile being driven into it. { 'drīv·iŋ ri'zis·təns }

driving wheel |MECH ENG| A wheel that supplies driving power. { 'drīv·iŋ ,wēl }

drogue |ENG| **1.** A device, such as a sea anchor, usually shaped like a funnel or cone and dragged or towed behind a boat or seaplane for deceleration, stabilization, or speed control. **2.** A current-measuring assembly consisting of a weighted current cross, sail, or parachute and an attached surface buoy. Also known as drag anchor; sea anchor. { drōg }

droop governor |MECH ENG| A governor whose equilibrium speed decreases as the load on the machinery controlled by the governor increases. { 'drüp ˌgə·vər·nər }

drop ball |ENG| A ball, weighing 3000–4000 pounds (1400–1800 kilograms), dropped from a crane through about 20–33 feet (6–10 meters) onto oversize quarry stones left after blasting; this method is used to avoid secondary blasting. { 'dräp ˌból }

drop bar |ELEC| Protective device used to ground a high-voltage capacitor when opening a door. |MECH ENG| A bar that guides sheets of paper into a printing or folding machine. { 'dräp ˌbär }

drop hammer See pile hammer. { 'dräp ˌham·ər }

droplet condensation |THERMO| The formation of numerous discrete droplets of liquid on a wall in contact with a vapor, when the wall is cooled below the local vapor saturation temperature and the liquid does not wet the wall. { ¦dräp·lət ˌkän·dən'sā·shən }

dropout |ELEC| Of a relay, the maximum current, voltage, power, or such, at which it will release from its energized position. |ELECTR| A reduction in output signal level during reproduction of recorded data, sufficient to cause a processing error. { 'dräp,aút }

dropout error |ELECTR| Loss of a recorded bit or any other error occurring in recorded magnetic tape due to foreign particles on or in the magnetic coating or to defects in the backing. { 'dräp,aút ,er·ər }

drop press See punch press. { 'dräp ,pres }

drop repeater |ELECTR| Microwave repeater that is provided with the necessary equipment for local termination of one or more circuits. { 'dräp ri,pēd·ər }

drop siding |BUILD| Building siding with a shiplap joint. { 'dräp ,sīd·iŋ }

dropsonde |ENG| A radiosonde dropped by parachute from a high-flying aircraft to measure weather conditions and report them back to the aircraft. Also known as dropwindsonde; parachute radiosonde. { 'dräp,sänd }

dropsonde dispenser |ENG| A chamber from which dropsonde instruments are released from weather reconnaissance aircraft; used only for some models of equipment, ejection chambers being used for others. { 'dräp,sänd də'spen·sər }

drop spillway |CIV ENG| A spillway usually less than 20 feet (6 meters) high having a vertical downstream face, and water drops over the face without touching the face. { ¦dräp 'spil,wā }

drop vent |ENG| In a plumbing system, a type of vent that is connected to a drain or vent pipe at a point below the fixture it is serving. { 'dräp ,vent }

dropwindsonde See dropsonde { ,dräp'wind ,sänd }

dropwise condensation |THERMO| Condensation of a vapor on a surface in which the condensate forms into drops. { ¦dräp,wīz ,kän·dən'sā·shən }

drosometer |ENG| An instrument used to measure the amount of dew deposited on a given surface. { drō'säm· əd·ər }

drum |CHEM ENG| Tower or vessel in a refinery into which heated products are conducted so that volatile portions can separate. |DES ENG| **1.** A hollow, cylindrical container. **2.** A metal cylindrical shipping container for liquids having a capacity of 12–110 gallons (45–416 liters). |ELECTR| A computer storage device consisting of a rapidly rotating cylinder with a magnetizable external surface on which data can be read or written by many read/write heads floating a few millionths of an inch off the surface. Also known as drum memory; drum storage; magnetic drum; magnetic drum storage. |MECH ENG| **1.** A horizontal cylinder about which rope or wire rope is wound in a hoisting mechanism. **2.** A hollow or solid cylinder or barrel that acts on, or is acted upon by, an exterior entity, such as the drum in a drum brake. Also known as hoisting drum. { drəm }

drum brake |MECH ENG| A brake in which two curved shoes fitted with heat- and wear-resistant linings are forced against the surface of a rotating drum. { 'drəm ,bräk }

drum cam |MECH ENG| A device consisting of a drum with a contoured surface which communicates motion to a cam follower as the drum rotates around an axis. { 'drəm ,kam }

drum dryer |MECH ENG| A machine for removing water from substances such as milk, in which a thin film of the product is moved over a turning steam-heated drum and a knife scrapes it from the drum after moisture has been removed. { 'drəm ,drī·ər }

drum feeder |MECH ENG| A rotating drum with vanes or buckets to lift and carry parts and drop them into various orienting or chute arrangements. Also known as tumbler feeder. { 'drəm ,fēd·ər }

drum filter |MECH ENG| A cylindrical drum that rotates through thickened ore pulp, extracts liquid by a vacuum, and leaves solids, in the form of a cake, on a permeable membrane on the drum end. Also known as rotary filter; rotary vacuum filter. { 'drəm ,fil·tər }

drum gate |CIV ENG| A movable crest gate in the form of an arc hinged at the apex and operated by reservoir pressure to open and close a spillway. { 'drəm ,gāt }

drum memory See drum. { ¦drəm 'mem·rē }

drum meter See liquid-sealed meter. { 'drəm ,mēd·ər }

drum plotter |ENG| A graphics output device that draws lines with a continuously moving pen on a sheet of paper rolled around a rotating drum that moves the paper in a direction perpendicular to the motion of the pen. { 'drəm ,pläd·ər }

drum storage See drum. { 'drəm ,stór·ij }

drum trap |ENG| In plumbing, a trap in the form of a cylinder with a vertical axis that is fitted with a removable cover plate. { 'drəm ,trap }

drum-type boiler See bent-tube boiler. { 'drəm ,tīp ,bóil·ər }

dry abrasive cutting |MECH ENG| Frictional cutting using a rotary abrasive wheel without the use of a liquid coolant. { ¦drī ə,brā·siv 'kəd·iŋ }

dry-back boiler See scotch boiler. { ¦drī ,bak 'bȯil·ər }

dry bed |CHEM ENG| A configuration of solid adsorption materials, for example molecular sieves or charcoal, used to recover liquid from or purify a gas stream. { 'drī 'bed }

dry blast cleaning |ENG| Cleaning of metallic surfaces by blasting with abrasive material traveling at a high velocity; abrasive may be accelerated by an air nozzle or a centrifugal wheel. { ¦drī ,blast 'klēn·iŋ }

dry-box process |CHEM ENG| The passing of coke-oven or other industrial gases through boxes containing trays of iron oxide coated on wood shavings or other supporting material in order to remove hydrogen sulfide. { 'drī ,bäks ,präs·əs }

dry-bulb thermometer |ENG| An ordinary thermometer, especially one with an unmoistened bulb; not dependent upon atmospheric humidity. { ¦drī ,bəlb thər'mäm·əd·ər }

dry cargo |IND ENG| Nonliquid cargo, including minerals, grain, boxes, and drums. { ¦drī 'kär,gō }

dry cell |ELEC| A voltage-generating cell having an immobilized electrolyte. { 'drī ,sel }

dry-chemical fire extinguisher |CHEM ENG| A dry powder, consisting principally of sodium bicarbonate, which is used for extinguishing small fires, especially electrical fires. { 'drī ,kem·i·kəl 'fīr ik,stiŋ·gwə·shər }

dry cleaning |ENG| To utilize dry-cleaning fluid to remove stains from textile. { 'drī klēn·iŋ }

dry coloring |CHEM ENG| A plastics coloring method in which uncolored particles of the plastic material are tumble-blended with selected dyes and pigments. |ENG| A method to color plastics by tumbleblending colorless plastic particles with dyes and pigments. { 'drī ,kəl·ə·riŋ }

dry cooling tower |MECH ENG| A structure in which water is cooled by circulation through finned tubes, transferring heat to air passing over the fins; there is no loss of water by evaporation because the air does not directly contact the water. { 'drī ,kül·iŋ ,tau̇·ər }

dry course |BUILD| An initial roofing course of felt or paper not bedded in tar or asphalt. { 'drī ,kȯrs }

dry-desiccant dehydration |CHEM ENG| Use of silica gel or other solid absorbent to remove liquids from gases, such as water from air, or liquid hydrocarbons from natural gas. { ¦drī ¦des·ə·kənt ,dē·hī'drā·shən }

dry-disk rectifier See metallic rectifier. { ¦drī ,disk 'rek·tə,fī·ər }

dry dock |CIV ENG| A dock providing support for a vessel and a means for removing the water so that the bottom of the vessel can be exposed. { 'drī ,däk }

dry-dock caisson |CIV ENG| The floating gate to a dry dock. Also known as caisson. { 'drī ,däk 'kā,sän }

dry friction |MECH| Resistance between two dry solid surfaces, that is, surfaces free from contaminating films or fluids. { 'drī 'frik·shən }

dry grinding |ENG| Reducing particle sizes without a liquid medium. { ¦drī 'grīnd·iŋ }

dry hole |ENG| A hole driven without the use of water. { ¦drī ¦hōl }

drying oven |ENG| A closed chamber for drying an object by heating at relatively low temperatures. { 'drī·iŋ ,əv·ən }

dry kiln |ENG| A heated room or chamber used to dry and season cut lumber. { 'drī ¦kil }

dry limestone process |CHEM ENG| An air-pollution control method in which sulfur oxides are exposed to limestone to convert them to disposable residues. { 'drī 'līm,stōn ,präs·əs }

dry machining |MECH ENG| Cutting, drilling, and grinding operations in which the use of a cutting fluid (lubricant) has been eliminated. { ¦drī mə'shēn·iŋ }

dry measure |MECH| A measure of volume for commodities that are dry. { ¦drī ¦mezh·ər }

dry mill |MECH ENG| Grinding device used to powder or pulverize solid materials without an associated liquid. { ¦drī ¦mil }

dry permeability |ENG| A property of dried bonded sand to permit passage of gases while molten material is poured into a mold. { ¦drī ,pər·mē·ə'bil·əd·ē }

dry pint See pint. { ¦drī ¦pīnt }

dry pipe |MECH ENG| A perforated metal pipe above the normal water level in the steam space of a boiler which prevents moisture or extraneous matter from entering steam outlet lines. { ¦drī ¦pīp }

dry-pipe system |ENG| A sprinkler system that admits water only when the air it normally contains has been vented; used for systems subjected to freezing temperatures. { 'drī ,pīp ,sis·təm }

dry-pit pump |MECH ENG| A pump operated with the liquid conducted to and from the unit by piping. { 'drī ,pit ,pəmp }

dry plasma etching See plasma etching. { ¦drī 'plaz·mə }

dry pressing |ENG| Molding clayware by compressing moist clay powder in metal dies. { ¦drī 'pres·iŋ }

dry pt See pint.

dry run |ENG| Any practice test or session. { ¦drī 'rən }

Drysdale ac polar potentiometer |ENG| A potentiometer for measuring alternating-current voltages in which the voltage is applied across a slide-wire supplied with current by a phase-shifting transformer; this current is measured by an ammeter and brought into phase with the unknown voltage by adjustment of the transformer rotor, and the unknown voltage is measured by observation of the slide-wire setting for a null indication of a vibration galvanometer. { 'drīz,dāl ¦ā¦sē ¦pō·lər pə,ten·chē'äm·əd·ər }

dry sieving |ENG| Particle-size distribution analysis of powdered solids; the sample is placed on the top sieve screen of a nest (stack), with mesh openings decreasing in size from the top to the bottom of the nest. { ¦drī 'siv·iŋ }

dry sleeve |MECH ENG| A cylinder liner which is not in contact with the coolant. { ¦drī ¦slēv }

dry spot |CHEM ENG| **1.** An open area of an incomplete surface film on laminated plastic. **2.** A section of laminated glass where the interlayer and glass are not bonded. { ¦drī ‚spät }

dry-steam drum |MECH ENG| **1.** Pressurized chamber into which steam flows from the steam space of a boiler drum. **2.** That portion of a two-stage furnace that extends forward of the main combustion chamber; fuel is dried and gasified therein, with combustion of gaseous products accomplished in the main chamber; the refractory walls of the Dutch oven are sometimes water-cooled. { ¦drī ‚stēm 'drəm }

dry-steam energy system |ENG| **1.** A geothermal energy source that produces superheated steam. **2.** A hydrothermal convective system driven by vapor with a temperature in excess of 300°F (150°C). { 'drī ¦stēm 'en·ər·jē ‚sis·təm }

dry storage |MECH ENG| Cold storage in which refrigeration is provided by chilled air. { 'drī ‚stór·ij }

dry strength |ENG| The strength of an adhesive joint determined immediately after drying under specified conditions or after a period of conditioning in the standard laboratory atmosphere. { 'drī ‚streŋkth }

dry test meter |ENG| Gas-flow rate meter with two compartments separated by a movable diaphragm which is connected to a series of gears that actuate a dial; when one chamber is full, a valve switches to the other, empty chamber; used to measure household gas-flow rates and to calibrate flow-measurement instruments. { 'drī ¦test ‚mēd·ər }

dry ticket |IND ENG| Tank inspection form signed by shore and ship inspectors before loading and after discharging the ship. { 'drī ‚tik·ət }

dry wall |BUILD| A wall covered with wallboard, in contrast to plaster. |ENG| A wall constructed of rock without cementing material. { 'drī ‚wól }

dry well |CIV ENG| **1.** A well that has been completely drained. **2.** An excavated well filled with broken stone and used to receive drainage when the water percolates into the soil. **3.** Compartment of a pumping station in which the pumps are housed. { 'drī ‚wel }

Drzewiecki theory |MECH ENG| In theoretical investigations of windmill performance, a theory concerning the air forces produced on an element of the blade. { 'dərz·vē·kē ‚thē·ə·rē }

DS See Doppler sonar.

Dualayer distillate process |CHEM ENG| A process for the removal of mercaptan and oxygenated compounds from distillate fuel oils; treatment is with concentrated caustic Dualayer solution and electrical precipitation of the impurities. { 'dü·ə‚lā·ər 'dis·təl·ət ‚präs·əs }

Dualayer solution |CHEM ENG| A concentrated potassium or sodium hydroxide solution containing a solubilizer; used in the Dualayer distillate process. { 'dü·ə‚lā·ər sə'lü·shən }

dual-bed dehumidifier |MECH ENG| A sorbent dehumidifier with two beds, one bed dehumidifying while the other bed is reactivating, thus providing a continuous flow of air. { ¦dü·əl ¦bed ‚dē·yü'mid·ə‚fī·ər }

dual-channel amplifier |ENG ACOUS| An audio-frequency amplifier having two separate amplifiers for the two channels of a stereophonic sound system, usually operating from a common power supply mounted on the same chassis. { ¦dü·əl ¦chan·əl 'am·plə‚fī·ər }

dual control |CONT SYS| An optimal control law for a stochastic adaptive control system that gives a balance between keeping the control errors and the estimation errors small. { ¦dü·əl kən'trōl }

dual-flow oil burner |MECH ENG| An oil burner with two sets of tangential slots in its atomizer for use at different capacity levels. { ¦dü·əl ¦flō 'óil ‚bər·nər }

dual-fuel engine |MECH ENG| Internal combustion engine that can operate on either of two fuels, such as natural gas or gasoline. { ¦dü·əl ¦fyül 'en·jən }

dual-gravity valve |CHEM ENG| A float-operated valve that operates on the interface between two immiscible liquids of different specific gravities. { ¦dü·əl 'grav·əd·ē ‚valv }

dual in-line package |ELECTR| Microcircuit package with two rows of seven vertical leads that are easily inserted into an etched circuit board. Abbreviated DIP. { ¦dü·əl ¦in ‚līn 'pak·ij }

dual meter |ENG| Meter constructed so that two aspects of an electric circuit may be read simultaneously. { 'dü·əl ‚mēd·ər }

dual-mode control |CONT SYS| A type of control law which consists of two distinct types of operation; in linear systems, these modes usually consist of a linear feedback mode and a bang-bang-type mode. { 'dü·əl ‚mōd kən'trōl }

dual-track tape recorder See double-track tape recorder.

dub |ENG ACOUS| **1.** To transfer recorded material from one recording to another, with or without the addition of new sounds, background music, or sound effects. **2.** To combine two or more sources of sound into one record. **3.** To add a new sound track or new sounds to a motion picture film, or to a recorded radio or television production. { dəb }

Dubbs cracking |CHEM ENG| A continuous, liquid-phase, thermal cracking process. { ¦dəbz 'krak·iŋ }

duckbill |MECH ENG| A shaking type of combination loader and conveyor whose loading end is generally shaped like a duck's bill. { 'dək‚bil }

duckfoot |ENG| In a piping system, a support fitted to the bend of a vertical pipe to permit the direct load of the pipework and fittings to be transferred to the floor, foundation, or associated installations. { 'dək‚fút }

duct |MECH ENG| A fluid flow passage which may range from a few inches in diameter to many feet in rectangular cross section, usually constructed of galvanized steel, aluminum, or copper, through which air flows in a ventilation system or to a compressor, supercharger, or other equipment at speeds ranging to thousands of feet per minute. { dəkt }

ducted fan |MECH ENG| A propeller or multibladed fan inside a coaxial duct or cowling. Also known as ducted propeller; shrouded propeller. { ¦dək·təd 'fan }

ducted propeller See ducted fan. { ¦dək·təd prə'pel·ər }

ductile fracture See fibrous fracture. { ¦dək·təl 'frak·chər }

Dufour effect |THERMO| Energy flux due to a mass gradient occurring as a coupled effect of irreversible processes. { ¦dü·fòr i'fekt }

Dufour number |THERMO| A dimensionless number used in studying thermodiffusion, equal to the increase in enthalpy of a unit mass during isothermal mass transfer divided by the enthalpy of a unit mass of mixture. Symbol Du_2. { ¦dü·fòr ,nəm·bər }

Duhem-Margules equation |THERMO| An equation showing the relationship between the two constituents of a liquid-vapor system and their partial vapor pressures:
$$\frac{d \ln p_A}{d \ln x_A} = \frac{d \ln p_B}{d \ln x_B}$$
where x_A and x_B are the mole fractions of the two constituents, and p_A and p_B are the partial vapor pressures. { dü'em 'mär·gyə·lēz i,kwā·zhən }

Dukler theory |CHEM ENG| Relationship of velocity and temperature distribution in thin films on vertical walls; used to calculate eddy viscosity and thermal conductivity near the solid boundary. { 'dük·lər ,thē·ə·rē }

Dulong-Petit law |THERMO| The law that the product of the specific heat per gram and the atomic weight of many solid elements at room temperature has almost the same value, about 6.3 calories (264 joules) per degree Celsius. { də'lòṅ pə'tē ,lò }

Dulong's formula |ENG| A formula giving the gross heating value of coal in terms of the weight fractions of carbon, hydrogen, oxygen, and sulfur from the ultimate analysis. { də'lòṅz ,fòr·myə·lə }

DUMAND See deep underwater muon and neutrino detector. { 'dü,mand }

dumb iron |ENG| **1.** A rod for opening seams prior to caulking. **2.** A rigid connector between the frame of a motor vehicle and the spring shackle. { 'dəm ,ī·ərn }

dumbwaiter |MECH ENG| An industrial elevator which carries small objects but is not permitted to carry people. { 'dəm,wād·ər }

dummy |ENG| Simulating device with no operating features, as a dummy heat coil. { 'dəm·ē }

dummy joint |ENG| A groove cut into the top half of a concrete slab, sometimes packed with filler, to form a line where the slab can crack with only minimum damage. { ¦dəm·ē ,jòint }

dump bailer |ENG| A cylindrical vessel designed to deliver cement or water into a well which otherwise might cave in if fluid was poured from the top. { 'dəmp ,bāl·ər }

dump bucket |MECH ENG| A large bucket with movable discharge gates at the bottom; used to move soil or other construction materials by a crane or cable. { 'dəmp ,bək·ət }

dump car |MECH ENG| Any of several types of narrow-gage rail cars with bodies which can easily be tipped to dump material. { 'dəmp ,kär }

dump tank See measuring tank. { 'dəmp ,taṅk }

dump truck |ENG| A motor or hand-propelled truck for hauling and dumping loose materials, equipped with a body that discharges its contents by gravity. { 'dəmp ,trək }

dump valve |ENG| A large valve located at the bottom of a tank or container used in emergency situations to empty the tank quickly; for example, to jettison fuel from an airplane fuel tank. { 'dəmp ,valv }

dumpy level |ENG| A surveyor's level which has the telescope with its level tube rigidly attached to a vertical spindle and is capable only of horizontal rotary movement. { ¦dəm·pē 'lev·əl }

dunking sonar See dipping sonar. { 'dəṅk·iṅ ,sō,när }

dunnage |ENG| A configuration of members that forms a structural support for a cooling tower or similar appendage to a building but is not part of the building itself. |IND ENG| **1.** Padding material placed in a container to protect shipped goods from damage. **2.** Loose wood or waste material placed in the ship's hold to protect the cargo from shifting and damage. { 'dən·ij }

duplex |ENG| Consisting of two parts working together or in a similar fashion. { 'dü,pleks }

duplexed system |ENG| A system with two distinct and separate sets of facilities, each of which is capable of assuming the system function while the other assumes a standby status. Also known as redundant system. { 'dü,plekst ,sis·təm }

duplex lock |DES ENG| A lock with two independent pin-tumbler cylinders on the same bolt. { 'dü,pleks 'läk }

duplex operation |ENG| In radar, a condition of operation when two identical and interchangeable equipments are provided, one in an active state and the other immediately available for operation. { ¦dü,pleks äp·ə'rā·shən }

duplex pump |MECH ENG| A reciprocating pump with two parallel pumping cylinders. { 'dü,pleks ,pəmp }

duplex tandem compressor |MECH ENG| A compressor having cylinders on two parallel frames connected through a common crankshaft. { ¦dü,pleks ¦tan·dəm kəm'pres·ər }

duplicate cavity plate |ENG| In plastics molds, the removable plate in which the molding cavities are retained; used in operating where two plates are necessary for insert loading. { ¦düp·lə·kət 'kav·əd·ē ,plāt }

Dupré equation |THERMO| The work W_{LS} done

by adhesion at a gas-solid-liquid interface, expressed in terms of the surface tensions γ of the three phases, is $W_{LS} = \gamma_{GS} + \gamma_{GL} - \gamma_{LS}$. { dü'prä i,kwä·zhən }

durability |ENG| The quality of equipment, structures, or goods of continuing to be useful after an extended period of time and usage. { ,dùr·ə'bil·əd·ē }

durable goods |ENG| Products whose usefulness continues for a number of years and that are not consumed or destroyed in a single usage. Also known as durables; hard goods. { ¦dùr·ə· bəl 'gùdz }

durables See durable goods. { 'dùr·ə·bəlz }

duration |MECH| A basic concept of kinetics which is expressed quantitatively by time measured by a clock or comparable mechanism. { də'rā·shən }

durometer |ENG| An instrument consisting of a small drill or blunt indenter point under pressure; used to measure hardness of metals and other materials. { də'räm·əd·ər }

durometer hardness |ENG| The hardness of a material as measured by a durometer. { də 'räm·əd·ər ,härd·nəs }

dust chamber |ENG| A chamber through which gases pass to permit deposition of solid particles for collection. Also known as ash collector; dust collector. { 'dəst ,chām·bər }

dust collector See dust chamber. { 'dəst kə,lek· tər }

dust control system |ENG| System to capture, settle, or inert dusts produced during handling, drying, or other process operations; considered important for safety and health. { 'dəst kən,trōl ,sis·təm }

dust counter |ENG| A photoelectric apparatus which measures the size and number of dust particles per unit volume of air. Also known as Kern counter. { 'dəst ,kaùnt·ər }

dust-counting microscope |ENG| A microscope equipped for quantitative dust sample analysis; magnification is usually 100X. { 'dəst ,kaùnt·iŋ 'mī·krə,skōp }

dust explosion |ENG| An explosion following the ignition of flammable dust suspended in the air. { 'dəst ik'splō·zhən }

dust filter |ENG| A gas-cleaning device using a dry or viscous-coated fiber or fabric for separation of particulate matter. { 'dəst ,fil·tər }

dust separator |ENG| Device or system to remove dust from a flowing stream of gas; includes electrostatic precipitators, wet scrubbers, bag filters, screens, and cyclones. { 'dəst ,sep·ə,rād· ər }

Dutch door |BUILD| A door with upper and lower parts that can be opened and closed independently. { ¦dəch 'dòr }

dutchman |ENG| A filler piece for closing a gap between two pipes or between a pipe or fitting and a piece of equipment, if the pipe is too short to achieve closure or if the pipe and equipment are not aligned. { 'dəch·mən }

Dutchman's log |ENG| A buoyant object thrown overboard to determine the speed of a vessel;

the time required for a known length of the vessel to pass the object is measured, and the speed can then be computed. { ¦dəch·mənz 'läg }

Dutch process |CHEM ENG| A process for making white lead; metallic lead is placed in vessels containing a dilute acetic acid, and the vessels are stacked in bark or manure. { 'dəch ,präs·əs }

duty cycle |ELECTR| See duty ratio. |ENG| 1. The time intervals devoted to starting, running, stopping, and idling when a device is used for intermittent duty. 2. The ratio of working time to total time for an intermittently operating device, usually expressed as a percent. Also known as duty factor. { 'düd·ē ,sī·kəl }

duty cyclometer |ENG| Test meter which gives direct reading of duty cycle. { 'düd·ē sī'kläm· əd·ər }

D variometer See declination variometer. { 'dē ,ver·ē'äm·əd·ər }

Dvorak keyboard |ENG| A keyboard whose layout is altered from that of the standard qwerty keyboard to speed up typing; more of the frequently used keys are on the home row. { də¦vòr ,ak 'kē,bórd }

dwell |DES ENG| That part of a cam that allows the cam follower to remain at maximum lift for a period of time. |ELEC| The number of degrees through which the distributor cam rotates from the time that the contact points close to the time that they open again. Also known as dwell angle. |ENG| A pause in the application of pressure to a mold. { dwel }

dwell angle See dwell. { 'dwel ,aŋ·gəl }

dwt See pennyweight.

DX coil See direct-expansion coil. { ¦dē¦eks ,kóil }

dyecrete process |ENG| A process of adding permanent color to concrete with organic dyes. { 'dī,krēt ,präs·əs }

dyeing |CHEM ENG| The application of color-producing agents to material, usually fibrous or film, in order to impart a degree of color permanence demanded by the projected end use. { 'dī·iŋ }

dynamical similarity |MECH| Two flow fields are dynamically similar if one can be transformed into the other by a change of length and velocity scales. All dimensionless numbers of the flows must be the same. { dī¦nam·ə·kəl sim·ə'lar· əd·ē }

dynamical variable |MECH| One of the quantities used to describe a system in classical mechanics, such as the coordinates of a particle, the components of its velocity, the momentum, or functions of these quantities. { dī¦nam·ə·kəl 'ver·ē·ə·bəl }

dynamic augment |MECH ENG| Force produced by unbalanced reciprocating parts in a steam locomotive. { dī¦nam·ik 'òg,ment }

dynamic balance |MECH| The condition which exists in a rotating body when the axis about which it is forced to rotate, or to which reference is made, is parallel with a principal axis of inertia; no products of inertia about the center of gravity of the body exist in relation to the selected rotational axis. { dī¦nam·ik 'bal·əns }

dynamic behavior |ENG| A description of how a system or an individual unit functions with respect to time. { dī¦nam·ik bə'hāv·yər }

dynamic braking |MECH| A technique of electric braking in which the retarding force is supplied by the same machine that originally was the driving motor. { dī¦nam·ik 'brāk·iŋ }

dynamic check |ENG| Check used to ascertain the correct performance of some or all components of equipment or a system under dynamic or operating conditions. { dī¦nam·ik 'chek }

dynamic compressor |MECH ENG| A compressor which uses rotating vanes or impellers to impart velocity and pressure to the fluid. { dī¦nam·ik kəm'pres·ər }

dynamic creep |MECH| Creep resulting from fluctuations in a load or temperature. { dī¦nam·ik 'krēp }

dynamic equilibrium |MECH| The condition of any mechanical system when the kinetic reaction is regarded as a force, so that the resultant force on the system is zero according to d'Alembert's principle. Also known as kinetic equilibrium. { dī¦nam·ik ē·kwə'lib·rē·əm }

dynamic holdup |CHEM ENG| Liquid held by a tank or process vessel, with constant introduction of fresh material and counteracting withdrawal of held material to maintain a constant liquid level. { dī¦nam·ik 'hōld,əp }

dynamic leak test |ENG| A type of leak test in which the vessel to be tested is evacuated and an external tracer gas is applied; an internal leak detector will respond if gas is drawn through any leaks. { dī¦nam·ik 'lēk ,test }

dynamic load |CIV ENG| A force exerted by a moving body on a resisting member, usually in a relatively short time interval. Also known as energy load. { dī¦nam·ik 'lōd }

dynamic loudspeaker |ENG ACOUS| A loudspeaker in which the moving diaphragm is attached to a current-carrying voice coil that interacts with a constant magnetic field to give the in-and-out motion required for the production of sound waves. Also known as dynamic speaker; moving-coil loudspeaker. { dī¦nam·ik 'laúd ,spēk·ər }

dynamic microphone |ENG ACOUS| A moving-conductor microphone in which the flexible diaphragm is attached to a coil positioned in the fixed magnetic field of a permanent magnet. Also known as moving-coil microphone. { dī¦nam·ik 'mī·krə,fōn }

dynamic model |ENG| A model of an aircraft or other object which has its linear dimensions and its weight and moments of inertia reproduced in scale in proportion to the original. { dī¦nam·ik 'mäd·əl }

dynamic noise suppressor |ENG ACOUS| An audio-frequency filter circuit that automatically adjusts its band-pass limits according to signal level, generally by means of reactance tubes; at low signal levels, when noise becomes more noticeable, the circuit reduces the low-frequency response and sometimes also reduces the high-frequency response. { dī¦nam·ik 'nóiz sə,pres·ər }

dynamic packing |ENG| Any packing that operates on moving surfaces; in functioning, to retain fluid under pressure, they carry the hydraulic load and therefore operate like bearings. { dī¦nam·ik 'pak·iŋ }

dynamics |MECH| That branch of mechanics which deals with the motion of a system of material particles under the influence of forces, especially those which originate outside the system under consideration. { dī¦nam·iks }

dynamic sensitivity |ENG| The minimum leak rate which a leak detector is capable of sensing. { dī¦nam·ik sen·sə'tiv·əd·ē }

dynamic similarity |MECH ENG| A relation between two mechanical systems (often referred to as model and prototype) such that by proportional alterations of the units of length, mass, and time, measured quantities in the one system go identically (or with a constant multiple for each) into those in the other; in particular, this implies constant ratios of forces in the two systems. { dī¦nam·ik ,sim·ə'lar·əd·ē }

dynamic speaker See dynamic loudspeaker. { dī¦nam·ik 'spēk·ər }

dynamic stability |MECH| The characteristic of a body, such as an aircraft, rocket, or ship, that causes it, when disturbed from an original state of steady motion in an upright position, to damp the oscillations set up by restoring moments and gradually return to its original state. Also known as stability. { dī¦nam·ik stə'bil·əd·ē }

dynamic test |ENG| A test conducted under active or simulated load. { dī¦nam·ik 'test }

dynamic time warping |ENG ACOUS| In speech recognition, the operation of compressing or stretching the temporal pattern of speech signals to take speaker variations into account. { dī,nam·ik 'tīm ,wórp·iŋ }

dynamic unbalance |MECH ENG| Failure of the rotation axis of a piece of rotating equipment to coincide with one of the principal axes of inertia due to forces in a single axial plane and on opposite sides of the rotation axis, or in different axial planes. { dī¦nam·ik ən'bal·əns }

dynamic work |IND ENG| A sustained pattern of work that results in motion around an anatomical joint, for example, a handling or assembly task. { dī¦nam·ik ,wərk }

dynamometer |ENG| **1.** An instrument in which current, voltage, or power is measured by the force between a fixed coil and a moving coil. **2.** A special type of electric rotating machine used to measure the output torque or driving torque of rotating machinery by the elastic deformation produced. { ,dī·nə'mäm·əd·ər }

dyne |MECH| The unit of force in the centimeter-gram-second system of units, equal to the force which imparts an acceleration of 1 cm/s² to a 1 gram mass. { dīn }

E

E See electric-field vector.

earliest finish time |IND ENG| The earliest time for completion of an activity of a project; for the entire project, it equals the earliest start time of the final event included in the schedule. { ¦ər·lē·əst 'fin·ish ˌtīm }

earliest start time |IND ENG| The earliest time at which an activity may begin in the schedule of a project; it equals the earliest time that all predecessor activities can be completed. { ¦ər·lē·əst 'start ˌtīm }

early finish date |IND ENG| The earliest time that an activity can be completed. { ¦ər·lē 'fin·ish ˌdāt }

early start date |IND ENG| The earliest time that an activity may be commenced. { ¦ər·lē 'start ˌdāt }

earned value |IND ENG| The budgeted cost of the work performed for a given project. { ¦ərnd 'val·yü }

earphone |ENG ACOUS| 1. An electroacoustical transducer, such as a telephone receiver or a headphone, actuated by an electrical system and supplying energy to an acoustical system of the ear, the waveform in the acoustical system being substantially the same as in the electrical system. 2. A small, lightweight electroacoustic transducer that fits inside the ear, used chiefly with hearing aids. { 'ir,fōn }

earplug |ENG| A device made of a pliable substance which fits into the ear opening; used to protect the ear from excessive noise or from water. { 'ir,pləg }

ear protector |ENG| A device, such as a plug or ear muff, used to protect the human ear from loud noise that may be injurious to hearing, such as that of jet engines. { 'ir prə,tek·tər }

earth See ground. { ərth }

earth current |ELEC| Return, fault, leakage, or stray current passing through the earth from electrical equipment. Also known as ground current. { 'ərth ,kə·rənt }

earth dam |CIV ENG| A dam having the main section built of earth, sand, or rock, and a core of impervious material such as clay or concrete. { 'ərth ,dam }

earthenware |ENG| Ceramic products of natural clay, fired at 1742–2129°F (950–1165°C), that is slightly porous, opaque, and usually covered with a nonporous glaze. { 'ər·thən,wer }

earth inductor |ENG| A type of inclinometer that has a coil which rotates in the earth's field and in which a voltage is induced when the rotation axis does not coincide with the field direction; used to measure the dip angle of the earth's magnetic field. Also known as dip inductor; earth inductor compass; induction inclinometer. { 'ərth in,dək·tər }

earth inductor compass See earth inductor. { 'ərth in'dək·tər ,käm·pəs }

earthmover |MECH ENG| A machine used to excavate, transport, or push earth. { 'ərth,müv·ər }

earth pressure |CIV ENG| The pressure which exists between earth materials (such as soil or sediments) and a structure (such as a wall). { 'ərth ,presh·ər }

earthquake-resistant |CIV ENG| Of a structure or building, able to withstand lateral seismic stresses at the base. { 'ərth,kwāk ri,zis·tənt }

earth thermometer See soil thermometer. { 'ərth thər,mäm·əd·ər }

earthwork |CIV ENG| 1. Any operation involving the excavation or construction of earth embankments. 2. Any construction made of earth. { 'ərth,wərk }

easement |CIV ENG| The right held by one person over another person's land for a specific use; rights of tenants are excluded. { 'ēz·mənt }

easement curve |CIV ENG| A curve, as on a highway, whose degree of curvature is varied to provide a gradual transition between a tangent and a simple curve, or between two simple curves which it connects. Also known as transition curve. { 'ēz·mənt ,kərv }

eave |BUILD| The border of a roof overhanging a wall. { ēv }

eaves board |BUILD| A strip nailed along the eaves of a building to raise the end of the bottom course of tile or slate on the roof. { 'ēvz ,bórd }

eaves molding |BUILD| A cornicelike molding below the eaves of a building. { 'ēvz ,mōl·diŋ }

Ebert ion counter |ENG| An ion counter of the aspiration condenser type, used for the measurement of the concentration and mobility of small ions in the atmosphere. { 'ā·bərt ī·ən ,kaúnt·ər }

ebullating-bed reactor |CHEM ENG| A type of fluidized bed in which catalyst particles are held in suspension by the upward movement of the liquid reactant and gas flow. Also known as slurry-bed reactor. { ¦eb·yə,lād·iŋ ¦bed rē,ak·tər }

eccentric bit |DES ENG| A modified chisel for drilling purposes having one end of the cutting edge extended further from the center of the bit than the other. { ek¦sen·trik 'bit }

eccentric cam |DES ENG| A cylindrical cam with the shaft displaced from the geometric center. { ek¦sen·trik 'kam }

eccentric gear |DES ENG| A gear whose axis deviates from the geometric center. { ek¦sen·trik 'gir }

eccentricity |MECH| The distance of the geometric center of a revolving body from the axis of rotation. { ,ek·sən'tris·əd·ē }

eccentric load |ENG| A load imposed on a structural member at some point other than the centroid of the section. { ek¦sen·trik 'lōd }

eccentric reducer |ENG| A threaded or butt-welded fitting for pipes whose ends are not the same size and are eccentric to each other. { ek ¦sen·trik ri'düs·ər }

eccentric rotor engine |MECH ENG| A rotary engine, such as the Wankel engine, wherein motion is imparted to a shaft by a rotor eccentric to the shaft. { ek¦sen·trik 'rōd·ər ,en·jən }

eccentric signal |ENG| A survey signal whose position is not in a vertical line with the station it is representing. { ek¦sen·trik 'sig·nəl }

eccentric station |ENG| A survey point over which an instrument is centered and which is not positioned in a vertical line with the station it is representing. { ek¦sen·trik 'stā·shən }

eccentric valve |ENG| A rubber-lined slurry or fluid valve with an eccentric rotary cut-off body to reduce corrosion and wear on mechanical moving valve parts. { ek¦sen·trik 'valv }

ECDIS *See* electronic chart display and information system. { 'ek,dis or ¦ē¦sē¦dē¦ī'es }

echogram |ENG| The graphic presentation of echo soundings recorded as a continuous profile of the sea bottom. { 'ek·ō,gram }

echograph |ENG| An instrument used to record an echogram. { 'ek·ō,graf }

echo matching |ENG| Rotating an antenna to a position in which the pulse indications of an echo-splitting radar are equal. { 'ek·ō ,mach·iŋ }

echo ranging |ENG| Active sonar, in which underwater sound equipment generates bursts of ultrasonic sound and picks up echoes reflected from submarines, fish, and other objects within range, to determine both direction and distance to each target. { 'ek·ō ,rānj·iŋ }

echo-ranging sonar |ENG| Active sonar, in which underwater sound equipment generates bursts of ultrasonic sound and picks up echoes reflected from submarines, fish, and other objects within range, to determine both direction and distance to each target. { 'ek·ō ,rānj·iŋ 'sō,när }

echo recognition |ENG| Identification of a sonar reflection from a target, as distinct from energy returned by other reflectors. { 'ek·ō ,rek·ig,nish·ən }

echo repeater |ENG ACOUS| In sonar calibration and training, an artificial target that returns a synthetic echo by receiving a signal and retransmitting it. { 'ek·ō ri,pēd·ər }

echosonogram |ENG| A graphic display obtained with ultrasound pulse-reflection techniques; for example, an echocardiogram. { ¦ek·ō 'sän·ə,gram }

echo sounder *See* sonic depth finder. { 'ek·ō ,saůnd·ər }

echo sounding |ENG| Determination of the depth of water by measuring the time interval between emission of a sonic or ultrasonic signal and the return of its echo from the sea bottom. { 'ek·ō ,saůnd·iŋ }

echo-splitting radar |ENG| Radar in which the echo is split by special circuits associated with the antenna lobe-switching mechanism, to give two echo indications on the radarscope screen; when the two echo indications are equal in height, the target bearing is read from a calibrated scale. { ¦ek·ō ,splid·iŋ 'rā,där }

econometrics |IND ENG| The application of mathematical and statistical techniques to the estimation of mathematical relationships for testing of economic theories and the solution of economic problems. { ē¦kän·ə¦me·triks }

economic life |IND ENG| The number of years after which a capital good should be replaced in order to minimize the long-run annual cost of operation, repair, depreciation, and capital. Also known as project life. { ,ek·ə'näm·ik 'līf }

economic lot size |IND ENG| The number of units of a product or item to be manufactured at each setup or purchased on each order so as to minimize the cost of purchasing or setup, and the cost of holding the average inventory over a given period, usually annual. Also known as project life. { ,ek·ə'näm·ik 'lät ,sīz }

economic order quantity |IND ENG| The number of orders required to fulfill the economic lot size. { ,ek·ə'näm·ik 'òr·dər ,kwän·ə·dē }

economic purchase quantity |IND ENG| The economic lot size for a purchased quantity. { ,ek·ə'näm·ik 'pər·chəs ,kwän·ə·dē }

economics |IND ENG| A social science that deals with production, distribution, and consumption of commodities, or wealth. { ,ek·ə'näm·iks *or*,ē·kə'näm·iks }

economic tool life |IND ENG| In metal machining, the total time, usually expressed in minutes, during which a given tool performs its required function under the most efficient cutting conditions. { ,ek·ə'näm·ik 'tül ,līf }

economizer |ENG| A reservoir in a continuous-flow oxygen system in which oxygen exhaled by the user is collected for recirculation in the system. |MECH ENG| A forced-flow, once-through, convection-heat-transfer tube bank in which feedwater is raised in temperature on its way to the evaporating section of a steam boiler.

thus lowering flue gas temperature, improving boiler efficiency, and saving fuel. { ē'kän·ə,miz·ər }

economy |CHEM ENG| In a multiple-effect evaporation system, the total weight of water vaporized in an evaporator per unit weight of the original steam supplied. { ē'kän·ə·mē }

ECR See electronic cash register.

ED See electronic dummy.

eddy conduction See eddy heat conduction. { 'ed·ē kən,dək·shən }

eddy conductivity |THERMO| The exchange coefficient for eddy heat conduction. { 'ed·ē ,kän ,dək'tiv·əd·ē }

eddy-current brake |MECH ENG| A control device or dynamometer for regulating rotational speed, as of flywheels, in which energy is converted by eddy currents into heat. { 'ed·ē ,kə·rənt ,brāk }

eddy-current clutch |MECH ENG| A type of electromagnetic clutch in which torque is transmitted by means of eddy currents induced by a magnetic field set up by a coil carrying direct current in one rotating member. { 'ed·ē ,kə·rənt ,kləch }

eddy-current heating See induction heating. { 'ed·ē ,kə·rənt ,hēd·iŋ }

eddy-current sensor |ENG| A proximity sensor which uses an alternating magnetic field to create eddy currents in nearby objects, and then the currents are used to detect the presence of the objects. { 'ed·ē ,kə·rənt 'sen·sər }

eddy-current tachometer |ENG| A type of tachometer in which a rotating permanent magnet induces currents in a spring-mounted metal cylinder; the resulting torque rotates the cylinder and moves its attached pointer in proportion to the speed of the rotating shaft. Also known as drag-type tachometer. { 'ed·ē ,kə·rənt tə'käm·əd·ər }

eddy heat conduction |THERMO| The transfer of heat by means of eddies in turbulent flow, treated analogously to molecular conduction. Also known as eddy heat flux; eddy conduction. { 'ed·ē 'hēt kən'dək·shən }

eddy heat flux See eddy heat conduction. { 'ed·ē 'hēt ,fləks }

Edeleanu process |CHEM ENG| A process for removal of compounds of sulfur from petroleum fractions by an extraction procedure utilizing liquid sulfur dioxide, or liquid sulfur dioxide and benzene. { ə,del·ē'ä·nü ,präs·əs }

EDEL room |ENG ACOUS| A control room in a sound-recording studio in which reflective or diffusive surfaces are placed near the loudspeaker and above the mixing console, while the rear wall behind the mixer is made absorptive. Derived from LEDE room (by reverse spelling). { 'ed·əl ,rüm or |ē|de|ē'el ,rüm }

edge connector |ELECTR| A row of etched lines on the edge of a printed circuit board that is inserted into a slot to establish a connection with another printed circuit board. { 'ej kə,nek·tər }

edge effect |ELEC| An outward-curving distortion of lines of force near the edges of two parallel metal plates that form a capacitor. { 'ej i,fekt }

edge runner See Chile mill. { 'ej ,rən·ər }

Edison effect See thermionic emission. { 'ed·ə·sən i,fekt }

eductor |ENG| **1.** An ejectorlike device for mixing two fluids. **2.** See ejector. { ē'dək·tər }

effective area |CHEM ENG| Absolute or cross-sectional area of process media involved in the process, such as the actual area of filter media through which a fluid passes, or the available surface area of absorbent contacted by a gas or liquid. { ə¦fek·tiv 'er·ē·ə }

effective bandwidth |ELECTR| The bandwidth of an assumed rectangular band-pass having the same transfer ratio at a reference frequency as a given actual band-pass filter, and passing the same mean-square value of a hypothetical current having even distribution of energy throughout that bandwidth. { ə¦fek·tiv 'band,width }

effective center |ENG ACOUS| In a sonar projector, the point where lines coincident with the direction of propagation, as observed at different points some distance from the projector, apparently intersect. Also known as apparent source. { ə¦fek·tiv 'sen·tər }

effective confusion area |ENG| Amount of chaff whose radar cross-sectional area equals the radar cross-sectional area of the particular aircraft at a particular frequency. { ə¦fek·tiv kən'fyü·zhən ,er·ē·ə }

effective discharge area |DES ENG| A nominal or calculated area of flow through a pressure relief valve for use in flow formulas to determine valve capacity. { ə¦fek·tiv 'dis,chärj ,er·ē·ə }

effective force See inertial force. { ə¦fek·tiv 'fórs }

effective gun bore line |MECH| The line which a projectile should follow when the muzzle velocity of the antiaircraft gun is vectorially added to the aircraft velocity. { ə¦fek·tiv 'gən ¦bór ,līn }

effective launcher line |MECH| The line along which the aircraft rocket would go if it were not affected by gravity. { ə¦fek·tiv 'lón·chər ,līn }

effective rake |MECH ENG| The angular relationship between the plane of the tooth face of the cutter and the line through the tooth point measured in the direction of chip flow. { ə¦fek·tiv 'rāk }

effective surface |ENG| In a heat exchanger, a surface that actively transfers heat. { ə¦fek·tiv 'sər·fəs }

effective thermal resistance |ELECTR| Of a semiconductor device, the effective temperature rise per unit power dissipation of a designated junction above the temperature of a stated external reference point under conditions of thermal equilibrium. Also known as thermal resistance. { ə¦fek·tiv ¦thər·məl ri'zis·təns }

effector |CONT SYS| A motor, solenoid, or hydraulic piston that turns commands to a teleoperator into specific manipulatory actions. { ə'fek·tər }

efficiency Abbreviated eff. |ENG| **1.** Measure of

the degree of heat output per unit of fuel when all available oxidizable materials in the fuel have been burned. **2.** Ratio of useful energy provided by a dynamic system to the energy supplied to it during a specific period of operation. |THERMO| The ratio of the work done by a heat engine to the heat energy absorbed by it. Also known as thermal efficiency. { ə'fish·ən·sē }

efficiency expert |IND ENG| An individual who analyzes procedures, productivity, and jobs in order to recommend methods for achieving maximum utilization of resources and equipment. { ə'fish·ən·sē ,ek·spərt }

effluent |CHEM ENG| See discharge liquor. |CIV ENG| The liquid waste of sewage and industrial processing. { ə'flü·ənt }

effluent weir |CIV ENG| A dam at the outflow end of a watercourse. { ə'flü·ənt 'wer }

effluvium |IND ENG| By-products of food and chemical processes, in the form of wastes. { ə'flü·vē·əm }

effort-controlled cycle |IND ENG| A work cycle which is performed entirely by hand or in which the hand time controls the place. Also known as manually controlled work. { 'ef·ərt kən,trōld ,sī·kəl }

effort rating |IND ENG| Assessing the level of manual effort expended by the operator, based on the observer's concept of normal effort, in order to adjust time-study data. Also known as pace rating; performance rating. { 'ef·ərt ,rād·iŋ }

Egerton's effusion method |THERMO| A method of determining vapor pressures of solids at high temperatures, in which one measures the mass lost by effusion from a sample placed in a tightly sealed silica pot with a small hole; the pot rests at the bottom of a tube that is evacuated for several hours, and is maintained at a high temperature by a heated block of metal surrounding it. { 'ej·ər·tənz ə'fyü·zhən ,meth·əd }

Ehrenfest's equations |THERMO| Equations which state that for the phase curve P(T) of a second-order phase transition the derivative of pressure P with respect to temperature T is equal to $(C_{fp} - C_{ip})/TV(\gamma^i - \gamma^f) = (\gamma^i - \gamma^f)/(K^f - K^i)$, where i and f refer to the two phases, γ is the coefficient of volume expansion, K is the compressibility, C_p is the specific heat at constant pressure, and V is the volume. { 'er·ən,fests i,kwā·zhənz }

Einthoven galvanometer See string galvanometer. { 'īnt,hō·vən ,gal·və'näm·əd·ər }

ejection |ENG| The process of removing a molding from a mold impression by mechanical means, by hand, or by compressed air. { ē'jek·shən }

ejector |ENG| **1.** Any of various types of jet pumps used to withdraw fluid materials from a space. Also known as eductor. **2.** A device that ejects the finished casting from a mold. { ē'jek·tər }

ejector condenser |MECH ENG| A type of direct-contact condenser in which vacuum is maintained by high-velocity injection water; condenses steam and discharges water, condensate, and noncondensables to the atmosphere. { ē'jek·tər kən,den·sər }

ejector pin |ENG| A pin driven into the rear of a mold cavity to force the finished piece out. Also known as knockout pin. { ē'jek·tər ,pin }

ejector plate |ENG| The plate backing up the ejector pins and holding the ejector assembly together. { ē'jek·tər ,plāt }

ejector rod |ENG| A rod that activates the ejector assembly of a mold when it is opened. { ē'jek·tər ,räd }

Ekman current meter |ENG| A mechanical device for measuring ocean current velocity which incorporates a propeller and a magnetic compass and can be suspended from a moored ship. { 'ek·mən 'kə·rənt ,mēd·ər }

Ekman dredge |ENG| A special type of dredge for sampling sediment that is fitted with opposable jaws operated by a messenger traveling down a cable to release a spring catch. { 'ek·mən ,drej }

Ekman water bottle |ENG| A cylindrical tube fitted with plates at both ends and used for deep-water samplings; when hit by a messenger it turns 180°, closing the plates and capturing the water sample. { 'ek·mən 'wōd·ər ,bäd·əl }

elastance |ELEC| The reciprocal of capacitance. { i'las·təns }

elastic |MECH| Capable of sustaining deformation without permanent loss of size or shape. { i'las·tik }

elastica |MECH| The elastic curve formed by a uniform rod that is originally straight, then is bent in a principal plane by applying forces, and couples only at its ends. { i'las·tə·kə }

elastic aftereffect |MECH| The delay of certain substances in regaining their original shape after being deformed within their elastic limits. Also known as elastic lag. { i'las·tik 'af·tər·i,fekt }

elastic axis |MECH| The lengthwise line of a beam along which transverse loads must be applied in order to produce bending only, with no torsion of the beam at any section. { i'las·tik 'ak·səs }

elastic body |MECH| A solid body for which the additional deformation produced by an increment of stress completely disappears when the increment is removed. Also known as elastic solid. { i'las·tik 'bäd·ē }

elastic buckling |MECH| An abrupt increase in the lateral deflection of a column at a critical load while the stresses acting on the column are wholly elastic. { i'las·tik 'bək·liŋ }

elastic center |MECH| That point of a beam in the plane of the section lying midway between the flexural center and the center of twist in that section. { i'las·tik 'sen·tər }

elastic collision |MECH| A collision in which the sum of the kinetic energies of translation of the participating systems is the same after the collision as before. { i'las·tik kə'lizh·ən }

elastic constant *See* compliance constant; stiffness constant. { i'las·tik 'kän·stənt }

elastic curve [MECH] The curved shape of the longitudinal centroidal surface of a beam when the transverse loads acting on it produced wholly elastic stresses. { i'las·tik 'kərv }

elastic deformation [MECH] Reversible alteration of the form or dimensions of a solid body under stress or strain. { i'las·tik ˌdē·fər'mā·shən }

elastic design [CIV ENG] In the design of a structural member, a method of analysis based on a linear stress-strain relationship, with the assumption that the working stresses constitute only a fraction of the elastic limit of the material. { i¦las·tik di'zīn }

elastic equilibrium [MECH] The condition of an elastic body in which each volume element of the body is in equilibrium under the combined effect of elastic stresses and externally applied body forces. { i¦las·tik ˌē·kwə'lib·rē·əm }

elastic failure [MECH] Failure of a body to recover its original size and shape after a stress is removed. { i'las·tik 'fāl·yər }

elastic flow [MECH] Return of a material to its original shape following deformation. { i'las·tik 'flō }

elastic force [MECH] A force arising from the deformation of a solid body which depends only on the body's instantaneous deformation and not on its previous history, and which is conservative. { i'las·tik 'fórs }

elastic hysteresis [MECH] Phenomenon exhibited by some solids in which the deformation of the solid depends not only on the stress applied to the solid but also on the previous history of this stress; analogous to magnetic hysteresis, with magnetic field strength and magnetic induction replaced by stress and strain respectively. { i'las·tik ˌhis·tə'rē·səs }

elasticity [MECH] **1.** The property whereby a solid material changes its shape and size under action of opposing forces, but recovers its original configuration when the forces are removed. **2.** The existence of forces which tend to restore to its original position any part of a medium (solid or fluid) which has been displaced. { i,las'tis·əd·ē }

elasticity modulus *See* modulus of elasticity. { i,las'tis·əd·ē ,mäj·ə·ləs }

elastic lag *See* elastic aftereffect. { i'las·tik 'lag }

elastic limit [MECH] The maximum stress a solid can sustain without undergoing permanent deformation. { i,las'tis·tik 'lim·ət }

elastic modulus *See* modulus of elasticity. { i,las·tik 'mäj·ə·ləs }

elastic potential energy [MECH] Capacity that a body has to do work by virtue of its deformation. { i'las·tik pə¦ten·chəl ¦en·ər·jē }

elastic ratio [MECH] The ratio of the elastic limit to the ultimate strength of a solid. { i'las·tik 'rā·shō }

elastic recovery [MECH] That fraction of a given deformation of a solid which behaves elastically. { i'las·tik ri'kəv·ə·rē }

elastic scattering [MECH] Scattering due to an elastic collision. { i'las·tik 'skad·ə·riŋ }

elastic solid *See* elastic body. { i'las·tik 'säl·əd }

elastic strain energy [MECH] The work done in deforming a solid within its elastic limit. { i'las·tik 'strān ,en·ər·jē }

elastic theory [MECH] Theory of the relations between the forces acting on a body and the resulting changes in dimensions. { i'las·tik 'thē·ə·rē }

elastic vibration [MECH] Oscillatory motion of a solid body which is sustained by elastic forces and the inertia of the body. { i'las·tik vī'brā·shən }

elastodynamics [MECH] The study of the mechanical properties of elastic waves. { i¦la·stō·dī¦nam·iks }

elastoplasticity [MECH] State of a substance subjected to a stress greater than its elastic limit but not so great as to cause it to rupture, in which it exhibits both elastic and plastic properties. { i¦las·tō·pla'stis·əd·ē }

elastoresistance [ELEC] The change in a material's electrical resistance as it undergoes a stress within its elastic limit. { i¦las·tō·ri'zis·təns }

elbow [DES ENG] **1.** A fitting that connects two pipes at an angle, often of 90° **2.** A sharp corner in a pipe. { 'el,bō }

elbow meter [ENG] Pipe elbow used as a liquids flowmeter; flow rate is measured by determining the differential pressure developed between the inner and outer radii of the bend by means of two pressure taps located midway on the bend. { 'el,bō ,mēd·ər }

electret [ELEC] A solid dielectric possessing persistent electric polarization, by virtue of a long time constant for decay of a charge instability. { i'lek,tret }

electret headphone [ENG ACOUS] A headphone consisting of an electret transducer, usually in the form of a push-pull transducer. { i'lek,tret 'hed,fōn }

electret microphone [ENG ACOUS] A microphone consisting of an electret transducer in which the foil electret diaphragm is placed next to a perforated, ridged, metal or metal-coated backplate, and output voltage, taken between diaphragm and backplate, is proportional to the displacement of the diaphragm. { i'lek,tret 'mī·krə,fōn }

electret transducer [ELECTR] An electroacoustic or electromechanical transducer in which a foil electret, stretched out to form a diaphragm, is placed next to a metal or metal-coated plate, and motion of the diaphragm is converted to voltage between diaphragm and plate, or vice versa. { i'lek,tret tranz'dü·sər }

electric [ELEC] Containing, producing, arising from, or actuated by electricity; often used interchangeably with electrical. { i'lek·trik }

electrical [ELEC] Related to or associated with electricity, but not containing it or having its properties or characteristics; often used interchangeably with electric. { ə'lek·trə·kəl }

187

electrical blasting cap |ENG| A blasting cap ignited by electric current and not by a spark. { ə'lek·trə·kəl 'blast·iŋ ‚kap }

electrical breakdown See breakdown. { ə'lek·trə·kəl 'brāk‚daûn }

electrical conductance See conductance. { ə'lek·trə·kəl kən'dək·təns }

electrical conduction See conduction. { ə'lek·trə·kəl kən'dək·shən }

electrical conductivity See conductivity. { ə'lek·trə·kəl ‚kän‚dək'tiv·əd·ē }

electrical drainage |ELEC| Diversion of electric currents from subterranean pipes to prevent electrolytic corrosion. { i'lek·trə·kəl 'drān·ij }

electrical engineer |ENG| An engineer whose training includes a degree in electrical engineering from an accredited college or university (or who has comparable knowledge and experience), to prepare him or her for dealing with the generation, transmission, and utilization of electric energy. { i'lek·trə·kəl ‚en·jə'nir }

electrical engineering |ENG| Engineering that deals with practical applications involving current flow through conductors, as in motors and generators. { i'lek·trə·kəl ‚en·jə'nir·iŋ }

electrical fault See fault. { i'lek·trə·kəl 'fȯlt }

electrical image |ENG| An image that is obtained in the course of borehole logging and is based on electrical rather than optical contrasts. { i'lek·trə·kəl 'im·ij }

electrical impedance Also known as impedance. |ELEC| **1.** The total opposition that a circuit presents to an alternating current, equal to the complex ratio of the voltage to the current in complex notation. Also known as complex impedance. **2.** The ratio of the maximum voltage in an alternating-current circuit to the maximum current; equal to the magnitude of the quantity in the first definition. { i'lek·trə·kəl im'pēd·əns }

electrical insulator See insulator. { i'lek·trə·kəl 'in·sə‚lād·ər }

electrical loading See loading. { i'lek·trə·kəl 'lōd·iŋ }

electrical log |ENG| Recorded measurement of the conductivities and resistivities down the length of uncased borehole; gives a complete record of the formations penetrated. { i'lek·trə·kəl 'läg }

electrical logging |ENG| The recording in uncased sections of a borehole of the conductivities and resistivities of the penetrated formations; used for geological correlations of the strata and evaluation of possibly productive horizons. Also known as electrical well logging. { i'lek·trə·kəl 'läg·iŋ }

electrically suspended gyro |ENG| A gyroscope in which the main rotating element is suspended by an electromagnetic or an electrostatic field. { i'lek·trə·klē səs'pen·dəd 'jī·rō }

electrical pressure transducer See pressure transducer. { i'lek·trə·kəl 'presh·ər tranz‚dü·sər }

electrical properties |ELEC| Properties of a substance which determine its response to an electric field, such as its dielectric constant or conductivity. { i'lek·trə·kəl 'präp·ərd·ēz }

electrical prospecting |ENG| The use of downhole electrical logs to obtain subsurface information for geological analysis. { i'lek·trə·kəl 'präs‚pek·tiŋ }

electrical resistance See resistance. { i'lek·trə·kəl ri'zis·təns }

electrical-resistance meter See resistance meter. { i'lek·trə·kəl ri'zis·təns ‚mēd·ər }

electrical-resistance strain gage |ENG| A vibration-measuring device consisting of a grid of fine wire cemented to the vibrating object to measure fluctuating strains. { i'lek·trə·kəl ri'zis·təns 'strān ‚gāj }

electrical-resistance thermometer See resistance thermometer. { i'lek·trə·kəl ri'zis·təns thər 'mäm·əd·ər }

electrical resistivity |ELEC| The electrical resistance offered by a material to the flow of current, times the cross-sectional area of current flow and per unit length of current path; the reciprocal of the conductivity. Also known as resistivity; specific resistance. { i'lek·trə·kəl ‚rē·zis'tiv·əd·ē }

electrical resistor See resistor. { i'lek·trə·kəl ri'zis·tər }

electrical symbol |ELEC| A simple geometrical symbol used to represent a component of a circuit in a schematic circuit diagram. { i'lek·trə·kəl 'sim·bəl }

electrical transcription See transcription. { i'lek·trə·kəl tranz'krip·shən }

electrical unit |ELEC| A standard in terms of which some electrical quantity is evaluated. { i'lek·trə·kəl 'yü·nət }

electrical weighing system |ENG| An instrument which weighs an object by measuring the change in resistance caused by the elastic deformation of a mechanical element loaded with the object. { i'lek·trə·kəl 'wā·iŋ ‚sis·təm }

electrical well logging See electrical logging. { i'lek·trə·kəl 'wel ‚läg·iŋ }

electric arc |ELEC| A discharge of electricity through a gas, normally characterized by a voltage drop approximately equal to the ionization potential of the gas. Also known as arc. { i'lek·trik 'ärk }

electric battery See battery. { i'lek·trik 'bad·ə·rē }

electric boiler |MECH ENG| A steam generator using electric energy, in immersion, resistor, or electrode elements, as the source of heat. { i'lek·trik 'bȯil·ər }

electric brake |MECH ENG| An actuator in which the actuating force is supplied by current flowing through a solenoid, or through an electromagnet which is thereby attracted to disks on the rotating member, actuating the brake shoes; this force is counteracted by the force of a compression spring. Also known as electromagnetic brake. { i'lek·trik 'brāk }

electric bridge See bridge. { i'lek·trik 'brij }

electric car |MECH ENG| An automotive vehicle that is propelled by one or more electric motors powered by a special rechargeable electric battery rather than by an internal combustion engine. { i'lek·trik 'kär }

electric cell [ELEC] **1.** A single unit of a primary or secondary battery that converts chemical energy into electric energy. **2.** A single unit of a device that converts radiant energy into electric energy, such as a nuclear, solar, or photovoltaic cell. { i¦lek·trik 'sel }

electric charge See charge. { i¦lek·trik 'chärj }

electric circuit [ELEC] Also known as circuit. **1.** A path or group of interconnected paths capable of carrying electric currents. **2.** An arrangement of one or more complete, closed paths for electron flow. { i¦lek·trik 'sər·kət }

electric coil See coil. { i¦lek·trik 'kȯil }

electric conductor See conductor. { i¦lek·trik kən'dək·tər }

electric connection [ELEC] A direct wire path for current between two points in a circuit. { i¦lek·trik kə'nek·shən }

electric connector [ELEC] A device that joins electric conductors mechanically and electrically to other conductors and to the terminals of apparatus and equipment. { i¦lek·trik kə'nek·tər }

electric contact [ELEC] A physical contact that permits current flow between conducting parts. Also known as contact. { i¦lek·trik 'kän,takt }

electric contactor See contactor. { i¦lek·trik 'kän ,tak·tər }

electric coupling [MECH ENG] Magnetic-field coupling between the shafts of a driver and a driven machine. { i¦lek·trik 'kəp·liŋ }

electric current density See current density. { i¦lek·trik ¦kə·rənt ,den·səd·ē }

electric current meter See ammeter. { i¦lek·trik ¦kə·rənt ,mēd·ər }

electric desalting [CHEM ENG] A process to remove impurities such as inorganic salts from crude oil by settling out in an electrostatic field. { i¦lek·trik dē'sȯlt·iŋ }

electric detonator [ENG] A detonator ignited by a fuse wire which serves to touch off the primer. { i¦lek·trik 'det·ən,ād·ər }

electric dipole [ELEC] A localized distribution of positive and negative electricity, without net charge, whose mean positions of positive and negative charges do not coincide. { i¦lek·trik 'dī,pōl }

electric dipole moment [ELEC] A quantity characteristic of a charge distribution, equal to the vector sum over the electric charges of the product of the charge and the position vector of the charge. { i¦lek·trik 'dī,pōl ,mō·mənt }

electric discharge See discharge. { i¦lek·trik 'dis,chärj }

electric displacement [ELEC] The electric field intensity multiplied by the permittivity. Symbolized D. Also known as dielectric displacement; dielectric flux density; displacement; electric displacement density; electric flux density; electric induction. { i'lek·trik dis'plās·mənt }

electric drive [MECH ENG] A mechanism which transmits motion from one shaft to another and controls the velocity ratio of the shafts by electrical means. { i¦lek·trik 'drīv }

electric fence [ENG] A fence consisting of one or more lengths of wire energized with high-voltage, low-current pulses, and giving a warning shock when touched. { i¦lek·trik 'fens }

electric field [ELEC] **1.** One of the fundamental fields in nature, causing a charged body to be attracted to or repelled by other charged bodies; associated with an electromagnetic wave or a changing magnetic field. **2.** Specifically, the electric force per unit test charge. { i¦lek·trik 'fēld }

electric-field intensity See electric-field vector. { i¦lek·trik ¦fēld in'ten·səd·ē }

electric-field strength See electric-field vector. { i¦lek·trik ¦fēld 'streŋkth }

electric-field vector [ELEC] The force on a stationary positive charge per unit charge at a point in an electric field. Designated E. Also known as electric-field intensity; electric-field strength; electric vector. { i¦lek·trik ¦fēld 'vek·tər }

electric flowmeter [ELEC] Fluid-flow measurement device relying on an inductance or impedance bridge or on electrical-resistance rod elements to sense flow-rate variations. { i¦lek·trik 'flō,mēd·ər }

electric flux [ELEC] **1.** The integral over a surface of the component of the electric displacement perpendicular to the surface; equal to the number of electric lines of force crossing the surface. **2.** The electric lines of force in a region. { i¦lek·trik 'fləks }

electric flux density See electric displacement. { i¦lek·trik 'fləks ,den·səd·ē }

electric flux line See electric line of force. { i¦lek·trik 'fləks ,līn }

electric furnace [ENG] A furnace which uses electricity as a source of heat. { i¦lek·trik 'fər·nəs }

electric fuse See fuse. { i¦lek·trik 'fyüz }

electric guitar [ENG ACOUS] A guitar in which a contact microphone placed under the strings picks up the acoustic vibrations for amplification and for reproduction by a loudspeaker. { i¦lek·trik gə'tär }

electric hammer [MECH ENG] An electric-powered hammer; often used for riveting or caulking. { i¦lek·trik 'ham·ər }

electric heating [ENG] Any method of converting electric energy to heat energy by resisting the free flow of electric current. { i¦lek·trik 'hēd·iŋ }

electric hygrometer [ENG] An instrument for indicating by electrical means the humidity of the ambient atmosphere; usually based on the relation between the electric conductance of a film of hygroscopic material and its moisture content. { i¦lek·trik hī'gräm·əd·ər }

electric hysteresis See ferroelectric hysteresis. { i¦lek·trik ,his·tə'rē·səs }

electrician [ENG] A skilled worker who installs, repairs, maintains, or operates electric equipment. { i,lek'trish·ən }

electric ignition [MECH ENG] Ignition of a charge of fuel vapor and air in an internal combustion engine by passing a high-voltage electric current between two electrodes in the combustion chamber. { i¦lek·trik ig'nish·ən }

electric image |ELEC| A fictitious charge used in finding the electric field set up by fixed electric charges in the neighborhood of a conductor; the conductor, with its distribution of induced surface charges, is replaced by one or more of these fictitious charges. Also known as image. { i¦lek·trik 'im·ij }

electric induction See electric displacement. { i¦lek·trik in'dək·shən }

electric instrument |ENG| An electricity-measuring device that indicates, such as an ammeter or voltmeter, in contrast to an electric meter that totalizes or records. { i¦lek·trik 'in·strə·mənt }

electric locomotive |MECH ENG| A locomotive operated by electric power picked up from a system of continuous overhead wires, or, sometimes, from a third rail mounted alongside the track. { i¦lek·trik ‚lō·kə'mōd·iv }

electric meter |ENG| An electricity-measuring device that totalizes with time, such as a watthour meter or ampere-hour meter, in contrast to an electric instrument. { i¦lek·trik 'mēd·ər }

electric motor See motor. { i¦lek·trik 'mōd·ər }

electric polarization See polarization. { i¦lek·trik ‚pō·lə·rə'zā·shən }

electric potential |ELEC| The work which must be done against electric forces to bring a unit charge from a reference point to the point in question; the reference point is located at an infinite distance, or, for practical purposes, at the surface of the earth or some other large conductor. Also known as electrostatic potential; potential. Abbreviated V. { i¦lek·trik pə'ten·chəl }

electric power |ELEC| The rate at which electric energy is converted to other forms of energy, equal to the product of the current and the voltage drop. { i¦lek·trik 'paů·ər }

electric power generation |MECH ENG| The large-scale production of electric power for industrial, residential, and rural use, generally in stationary plants designed for that purpose. { i¦lek·trik ‚paů·ər ‚jen·ə'rā·shən }

electric power line See power line. { i¦lek·trik 'paů·ər ‚līn }

electric power meter |ENG| A device that measures electric power consumed, either at an instant, as in a wattmeter, or averaged over a time interval, as in a demand meter. Also known as power meter. { i¦lek·trik 'paů·ər ‚mēd·ər }

electric power plant |MECH ENG| A power plant that converts a form of raw energy into electricity, for example, a hydro, steam, diesel, or nuclear generating station for stationary or transportation service. { i¦lek·trik 'paů·ər ‚plant }

electric power station |ELEC| A generating station or an electric power substation. { i¦lek·trik 'paů·ər ‚stā·shən }

electric power substation |ELEC| An assembly of equipment in an electric power system

through which electric energy is passed for transmission, transformation, distribution, or switching. Also known as substation. { i¦lek·trik ‚paů·ər 'səb‚stā·shən }

electric power system |MECH ENG| A complex assemblage of equipment and circuits for generating, transmitting, transforming, and distributing electric energy. { i¦lek·trik ‚paů·ər ‚sis·təm }

electric power transmission |ELEC| Process of transferring electric energy from one point to another in an electric power system. { i¦lek·trik ‚paů·ər tranz‚mish·ən }

electric precipitation |CHEM ENG| A process that utilizes an electric field to improve the separation of hydrocarbon reagent dispersions. { i¦lek·trik prə‚sip·ə'tā·shən }

electric pressure transducer See pressure transducer. { i¦lek·trik ‚presh·ər tranz‚dü·sər }

electric railroad |MECH ENG| A railroad which has a system of continuous overhead wires or a third rail mounted alongside the track to supply electric power to the locomotive and cars. { i¦lek·trik 'rāl‚rōd }

electric reactor See reactor. { i¦lek·trik rē'ak·tər }

electric resistance See resistance. { i¦lek·trik ri'zis·təns }

electric resistance furnace See resistance furnace. { i¦lek·trik ri'zis·təns ‚fər·nəs }

electric shunt See shunt. { i¦lek·trik 'shənt }

electric stacker |MECH ENG| A stacker whose carriage is raised and lowered by a winch powered by electric storage batteries. { i¦lek·trik 'stak·ər }

electric strength See dielectric strength. { i¦lek·trik 'streŋkth }

electric susceptibility |ELEC| A dimensionless parameter measuring the ease of polarization of a dielectric, equal (in meter-kilogram-second units) to the ratio of the polarization to the product of the electric field strength and the vacuum permittivity. Also known as dielectric susceptibility. { i¦lek·trik sə‚sep·tə'bil·əd·ē }

electric tachometer |ENG| An instrument for measuring rotational speed by measuring the output voltage of a generator driven by the rotating unit. { i¦lek·trik tə'käm·əd·ər }

electric tank See electrolytic tank. { i¦lek·trik 'taŋk }

electric thermometer |ENG| An instrument that utilizes electrical means to measure temperature, such as a thermocouple or resistance thermometer. { i¦lek·trik thər'mäm·əd·ər }

electric typewriter |MECH ENG| A typewriter having an electric motor that provides power for all operations initiated by the touching of the keys. { i¦lek·trik 'tīp‚rīd·ər }

electric vehicle |MECH ENG| A ground vehicle propelled by a motor powered by electrical energy from rechargeable batteries or other source onboard the vehicle, or from an external source in, on, or above the roadway; examples include the electrically powered golf cart, automobile, and trolley bus. { i¦lek·trik 'vē·ə·kəl }

electric wire See wire. { i¦lek·trik 'wīr }

electroacoustic effect See acoustoelectric effect.
{ i¦lek·trō·ə¦kü·stik i'fekt }

electroacoustics |ENG ACOUS| The conversion of acoustic energy and waves into electric energy and waves, or vice versa. { i¦lek·trō·ə'kü·stiks }

electroacoustic transducer |ENG ACOUS| A transducer that receives waves from an electric system and delivers waves to an acoustic system, or vice versa. Also known as sound transducer. { i¦lek·trō·ə¦kü·stik tranz'dü·sər }

electrochemical grinding See electrolytic grinding. { i,lek·trō¦kem·i·kəl 'grīnd·iŋ }

electrochemical power generation |ENG| The direct conversion of chemical energy to electric energy, as in a battery or fuel cell. { i,lek·trō 'kem·ə·kəl 'paů·ər ,jen·ə,rā·shən }

electrochemical recording |ELECTR| Recording by means of a chemical reaction brought about by the passage of signal-controlled current through the sensitized portion of the record sheet. { i,lek·trō'kem·ə·kəl ri'kòrd·iŋ }

electrochemical thermodynamics |THERMO| The application of the laws of thermodynamics to electrochemical systems. { i,lek·trō'kem·ə·kəl ,thərm·ō·dī'nam·iks }

electrochemical transducer |ENG| A device which uses a chemical change to measure the input parameter; the output is a varying electrical signal proportional to the measurand. { i,lek·trō'kem·ə·kəl tranz'dü·sər }

electrochemical valve |ELEC| Electric valve consisting of a metal in contact with a solution or compound, across the boundary of which current flows more readily in one direction than in the other direction, and in which the valve action is accompanied by chemical changes. { i,lek·trō'kem·ə·kəl 'valv }

electrochromic device |ENG| A self-contained, hermetically sealed, two-electrode electrolytic cell that includes one or more electrochromic materials and an electrolyte. { i,lek·trə¦krōm·ik di'vīs }

electrochromic display |ELECTR| A solid-state passive display that uses organic or inorganic insulating solids which change color when injected with positive or negative charges. { i¦lek·trō¦krō·mik di'splā }

electrode |ELEC| **1.** An electric conductor through which an electric current enters or leaves a medium, whether it be an electrolytic solution, solid, molten mass, gas, or vacuum. **2.** One of the terminals used in dielectric heating or diathermy for applying the high-frequency electric field to the material being heated. { i'lek,trōd }

electrode admittance |ELECTR| Quotient of dividing the alternating component of the electrode current by the alternating component of the electrode voltage, all other electrode voltages being maintained constant. { i'lek,trōd ad'mit·əns }

electrode capacitance |ELECTR| Capacitance between one electrode and all the other electrodes connected together. { i'lek,trōd kə'pas·əd·əns }

electrode characteristic |ELECTR| Relation between the electrode voltage and the current to an electrode, all other electrode voltages being maintained constant. { i'lek,trōd ,kar·ik·tə'ris·tik }

electrode conductance |ELECTR| Quotient of the inphase component of the electrode alternating current by the electrode alternating voltage, all other electrode voltage being maintained constant; this is a variational and not a total conductance. Also known as grid conductance. { i'lek,trōd kən'dək·təns }

electrode couple |ELEC| The pair of electrodes in an electric cell, between which there is a potential difference. { i'lek,trōd ,kə·pəl }

electrode current |ELECTR| Current passing to or from an electrode, through the interelectrode space within a vacuum tube. { i'lek,trōd ,kə·rənt }

electrode impedance |ELECTR| Reciprocal of the electrode admittance. { i'lek,trōd im'pēd·əns }

electrode resistance |ELECTR| Reciprocal of the electrode conductance; this is the effective parallel resistance and is not the real component of the electrode impedance. { i'lek,trōd ri'zis·təns }

electrode-type liquid-level meter |ENG| Device that senses liquid level by the effect of the liquid-gas interface on the conductance of an electrode or probe. { i'lek,trōd ,tīp ¦lik·wəd ¦lev·əl 'mēd·ər }

electrode voltage See electrode potential. { i'lek ,trōd ,vōl·tij }

electrodrill |MECH ENG| A drilling machine driven by electric power. { i'lek,trō,dril }

electrodynamic ammeter |ENG| Instrument which measures the current passing through a fixed coil and a movable coil connected in series by balancing the torque on the movable coil (resulting from the magnetic field of the fixed coil) against that of a spiral spring. { i,lek·trō·dī'nam·ik 'a,mēd·ər }

electrodynamic instrument |ENG| An instrument that depends for its operation on the reaction between the current in one or more movable coils and the current in one or more fixed coils. Also known as electrodynamometer. { i,lek·trō·dī'nam·ik 'in·strə·mənt }

electrodynamic loudspeaker |ENG ACOUS| Dynamic loudspeaker in which the magnetic field is produced by an electromagnet, called the field coil, to which a direct current must be furnished. { i,lek·trō·dī'nam·ik 'laůd,spēk·ər }

electrodynamic wattmeter |ENG| An electrodynamic instrument connected as a wattmeter, with the main current flowing through the fixed coil, and a small current proportional to the voltage flowing through the movable coil. Also known as moving-coil wattmeter. { i,lek·trō·dī'nam·ik 'wät,mēd·ər }

electrodynamometer See electrodynamic instrument. { i,lek·trō,dī·nə'mäm·əd·ər }

electroexplosive |ENG| An initiator or a system in which an electric impulse initiates detonation

191

electrograph

or deflagration of an explosive. { i̇‚lek·trō·ik 'splō·siv }

electrograph |ENG| Any plot, graph, or tracing produced by the action of an electric current on prepared sensitized paper (or other chart material) or by means of an electrically controlled stylus or pen. { i'lek·trə‚graf }

electrohydraulic |ENG| Operated or effected by a combination of electric and hydraulic mechanisms. { i̇‚lek·trō·hī'dról·ik }

electrokinetograph |ENG| An instrument used to measure ocean current velocities based on their electrical effects in the magnetic field of the earth. { i̇‚lek·trō·kə'ned·ə‚graf }

electroluminescence |ELECTR| The emission of light, not due to heating effects alone, resulting from application of an electric field to a material, usually solid. { i̇‚lek·trō‚lü·mə'nes·əns }

electrolyte-MOSFET |ENG| A metal oxide semiconductor field-effect transistor (MOSFET) that is immersed in a solution to determine the concentrations of dissolved redox active species; the bulk part of the work function of the gate electrode of the transistor changes when the sensor membrane is oxidized or reduced. Abbreviated EMOSFET. { i̇‚lek·trə‚līt 'mós‚fet }

electrolytic grinding |MECH ENG| A combined grinding and machining operation in which the abrasive, cathodic grinding wheel is in contact with the anodic workpiece beneath the surface of an electrolyte. Also known as electrochemical grinding. { i'lek·trə‚lid·ik 'grīnd·iŋ }

electrolytic mercaptan process |CHEM ENG| A process in which an aqueous caustic solution is used to extract mercaptans from refinery streams. { i'lek·trə‚lid·ik mər'kap·tan ‚prä·səs }

electrolytic refining See electrorefining. { i'lek·trə‚lid·ik rə'fīn·iŋ }

electrolytic strip See humidity strip. { i'lek·trə‚lid·ik 'strip }

electrolytic tank |ENG| A tank in which voltages are applied to an enlarged scale model of an electron-tube system or a reduced scale model of an aerodynamic system immersed in a poorly conducting liquid, and equipotential lines between electrodes are traced; used as an aid to electron-tube design or in computing ideal fluid flow; the latter application is based on the fact that the velocity potential in ideal flow and the stream function in planar flow satisfy the same equation, Laplace's equation, as an electrostatic potential. Also known as electric tank; potential flow analyzer. { i'lek·trə‚lid·ik 'taŋk }

electromachining |MECH ENG| The application of electric or ultrasonic energy to a workpiece to effect removal of material. { i̇‚lek·trō·mə'shēn·iŋ }

electromagnetic brake See electric brake. { i̇‚lek·trō·mag'ned·ik 'brāk }

electromagnetic clutch |MECH ENG| A clutch based on magnetic coupling between conductors, such as a magnetic fluid and powder clutch, an eddy-current clutch, or a hysteresis clutch. { i̇‚lek·trō·mag'ned·ik 'kləch }

electromagnetic flowmeter |ENG| A flowmeter that offers no obstruction to liquid flow; two coils produce an electromagnetic field in the conductive moving fluid; the current induced in the liquid, detected by two electrodes, is directly proportional to the rate of flow. Also known as electromagnetic meter. { i̇‚lek·trō·mag'ned·ik 'flō‚mēd·ər }

electromagnetic interference |ELEC| Interference, generally at radio frequencies, that is generated inside systems, as contrasted to radio-frequency interference coming from sources outside a system. Abbreviated emi. { i̇‚lek·trō·mag'ned·ik ‚in·tər'fir·əns }

electromagnetic log |ENG| A log containing an electromagnetic sensing element extended below the hull of the vessel; this device produces a voltage directly proportional to speed through the water. { i̇‚lek·trō·mag'ned·ik 'läg }

electromagnetic logging |ENG| A method of well logging in which a transmitting coil sets up an alternating electromagnetic field, and a receiver coil, placed in the drill hole above the transmitter coil, measures the secondary electromagnetic field induced by the resulting eddy currents within the formation. Also known as electromagnetic well logging. { i̇‚lek·trō·mag'ned·ik 'läg·iŋ }

electromagnetic meter See electromagnetic flowmeter. { i̇‚lek·trō·mag'ned·ik 'mēd·ər }

electromagnetic noise |ELEC| Noise in a communications system resulting from undesired electromagnetic radiation. Also known as radiation noise. { i̇‚lek·trō·mag'ned·ik 'nóiz }

electromagnetic prospecting See electromagnetic surveying. { i̇‚lek·trō·mag'ned·ik 'prä‚spek·tiŋ }

electromagnetic surveying |ENG| Underground surveying carried out by generating electromagnetic waves at the surface of the earth; the waves penetrate the earth and induce currents in conducting ore bodies, thereby generating new waves that are detected by instruments at the surface or by a receiving coil lowered into a borehole. Also known as electromagnetic prospecting. { i̇‚lek·trō·mag'ned·ik sər'vā·iŋ }

electromagnetic well logging See electromagnetic logging. { i̇‚lek·trō·mag'ned·ik 'wel ‚läg·iŋ }

electromanometer |ENG| An electronic instrument used for measuring pressure of gases or liquids. { i̇‚lek·trō·mə'näm·əd·ər }

electromechanical |MECH ENG| Pertaining to a mechanical device, system, or process which is electrostatically or electromagnetically actuated or controlled. { i̇‚lek·trō·mi'kan·ə·kəl }

electromechanical circuit |ELEC| A circuit containing both electrical and mechanical parameters of consequence in its analysis. { i̇‚lek·trō·mi'kan·ə·kəl 'sər·kə t }

electromechanics |MECH ENG| The technology of mechanical devices, systems, or processes which are electrostatically or electromagnetically actuated or controlled. { i̇‚lek·trō·mi'kan·iks }

electrometer |ENG| An instrument for measuring voltage without drawing appreciable current. { i̇‚lek'träm·əd·ər }

electronic chart display

electron beam |ELECTR| A narrow stream of electrons moving in the same direction, all having about the same velocity. { i'lek,trän ,bēm }

electron-beam channeling |ELECTR| The technique of transporting high-energy, high-current electron beams from an accelerator to a target through a region of high-pressure gas by creating a path through the gas where the gas density may be temporarily reduced; the gas may be ionized; or a current may flow whose magnetic field focuses the electron beam on the target. { i'lek,trän ,bēm 'chan·əl·iŋ }

electron-beam drilling |ELECTR| Drilling of tiny holes in a ferrite, semiconductor, or other material by using a sharply focused electron beam to melt and evaporate or sublimate the material in a vacuum. { i'lek,trän ,bēm 'dril·iŋ }

electron-beam generator |ELECTR| Velocity-modulated generator, such as a klystron tube, used to generate extremely high frequencies. { i'lek,trän ,bēm 'jen·ə,rād·ər }

electron-beam ion source |ELECTR| A source of multiply charged heavy ions which uses an intense electron beam with energies of 5 to 10 kiloelectronvolts to successively ionize injected gas. Abbreviated EBIS. { i'lek,trän ,bēm 'ī,än ,sȯrs }

electron-beam ion trap |ELECTR| A device for producing the highest possible charge states of heavy ions, in which impact ionization or excitation by successive electrons is efficiently achieved by causing the ions to be trapped in a compressed electron beam by the electron beam's space charge. Abbreviated EBIT { i,lek,trän ,bē 'i·ən ,trap }

electron-beam lithography |ELECTR| Lithography in which the radiation-sensitive film or resist is placed in the vacuum chamber of a scanning-beam electron microscope and exposed by an electron beam under digital computer control; after exposure, the film is removed from the vacuum chamber for conventional development and other production processes. { i'lek,trän ,bēm li'thäg·rə·fē }

electron-beam magnetometer |ENG| A magnetometer that depends on the change in intensity or direction of an electron beam that passes through the magnetic field to be measured. { i'lek,trän ,bēm mag·nə'täm·əd·ər }

electron-beam parametric amplifier |ELECTR| A parametric amplifier in which energy is pumped from an electrostatic field into a beam of electrons traveling down the length of the tube, and electron couplers impress the input signal at one end of the tube and translate spiraling electron motion into electric output at the other. { i'lek,trän ,bēm ,par·ə¦me·trik 'am·plə,fī·ər }

electron-beam pumping |ELECTR| The use of an electron beam to produce excitation for population inversion and lasing action in a semiconductor laser. { i'lek,trän ,bēm 'pəmp·iŋ }

electron-beam recorder |ELECTR| A recorder in which a moving electron beam is used to record signals or data on photographic or thermoplastic film in a vacuum chamber. { i'lek,trän ,bēm ri'kȯrd·ər }

electron-beam tube |ELECTR| An electron tube whose performance depends on the formation and control of one or more electron beams. { i'lek,trän ,bēm 'tüb }

electron conduction |ELEC| Conduction of electricity resulting from motion of electrons, rather than from ions in a gas or solution, or holes in a solid. |THERMO| The transport of energy in highly ionized matter primarily by electrons of relatively high temperature moving in one direction and electrons of lower temperature moving in the other. { i'lek,trän kən,dək·shən }

electron cyclotron resonance reactor |ENG| A plasma reactor in which resonant coupling of microwave energy into an electron gas at electron cyclotron resonance accelerates electrons, which in turn ionize and excite the neutral gas, resulting in a low-pressure, almost collisionless plasma. { i¦lek,trän ¦sī·klə,trän 'rez·ə·nəns rē,ak·tər }

electronegative |ELEC| **1.** Carrying a negative electric charge. **2.** Capable of acting as the negative electrode in an electric cell. { i¦lek·trō 'neg·əd·iv }

electron flow |ELEC| A current produced by the movement of free electrons toward a positive terminal; the direction of electron flow is opposite to that of current. { i'lek,trän ,flō }

electron holography |ELECTR| An imaging technique using the wave nature of electrons and light, in which an interference pattern between an object wave and a reference wave is formed using a coherent field-emission electron beam from a sharp tungsten needle, and is recorded on film as a hologram, and the image of the original object is then reconstructed by illuminating a light beam equivalent to the reference wave onto the hologram. { i,lek,trän hō 'läg·rə·fē }

electronically agile radar |ENG| An airborne radar that uses a phased-array antenna which changes radar beam shapes and beam positions at electronic speeds. { i,lek'trän·ik·lē ,a·jəl 'rā,där }

electronic altimeter See radio altimeter. { i,lek 'trän·ik al'tim·əd·ər }

electronic cash register |ENG| A system for automatically checking out goods from retail food stores, consisting of a device that scans packages and reads symbols imprinted on the label, and a computer that converts the symbol information to tell a cash register the price of the item; the computer can also keep records of sales and inventories. Abbreviated ECR. { i,lek'trän·ik 'kash ,rej·ə·stər }

electronic chart display and information system |ENG| A navigation information system with an electronic chart database, as well as navigational and piloting information (typically, vessel-route-monitoring, track-keeping, and track-planning information). Abbreviated ECDIS. { i·lek ¦trän·ik 'chärt di¦splā ən ,in·fər'mā·shən ,sis·təm }

193

electronic dummy [ENG ACOUS] A vocal simulator which is a replica of the head and torso of a person, covered with plastisol flesh that simulates the acoustical and mechanical properties of real flesh, and possessing an artificial voice and two artificial ears. Abbreviated ED. { i,lek'trän·ik 'dəm·ē }

electronic engineering [ENG] Engineering that deals with practical applications of electronics. { i,lek'trän·ik ,en·jə'nir·iŋ }

electronic flame safeguard [MECH ENG] An electrode used in a burner system which detects the main burner flame and interrupts fuel flow if the flame is not detected. { i,lek'trän·ik 'flām 'sāf,gärd }

electronic fuse [ENG] A fuse, such as the radio proximity fuse, set off by an electronic device incorporated in it. { i,lek'trän·ik 'fyüz }

electronic heating [ENG] Heating by means of radio-frequency current produced by an electron-tube oscillator or an equivalent radio-frequency power source. Also known as high-frequency heating; radio-frequency heating. { i,lek'trän·ik 'hēd·iŋ }

electronic humidistat [ENG] A humidistat in which a change in the relative humidity causes a change in the electrical resistance between two sets of alternate metal conductors mounted on a small flat plate with plastic coating, and this change in resistance is measured by a relay amplifier. { i,lek'trän·ik hyü'mid·ə,stat }

electronic logger See Geiger-Müller probe. { i,lek 'trän·ik 'läg·ər }

electronic music [ENG ACOUS] Music consisting of tones originating in electronic sound and noise generators used alone or in conjunction with electroacoustic shaping means and sound-recording equipment. { i,lek'trän·ik 'myü·zik }

electronic musical instrument [ENG ACOUS] A musical instrument in which an audio signal is produced by a pickup or audio oscillator and amplified electronically to feed a loudspeaker, as in an electric guitar, electronic carillon, electronic organ, or electronic piano. { i,lek'trän·ik ¦myü·zə·kəl 'in·strə·mənt }

electronic packaging [ENG] The technology of packaging electronic equipment; in current usage it refers to inserting discrete components, integrated circuits, and MSI and LSI chips (usually attached to a lead frame by beam leads) into plates through holes on multilayer circuit boards (also called cards), where they are soldered in place. { i,lek'trän·ik 'pak·ij·iŋ }

electronic photometer See photoelectric photometer. { i,lek'trän·ik fō'täm·əd·ər }

electronic polarization [ELEC] Polarization arising from the displacement of electrons with respect to the nuclei with which they are associated, upon application of an external electric field. { i,lek'trän·ik ,pō·lə·rə'zā·shən }

electronic robot [CONT SYS] A robot whose motions are powered by a direct-current stepper motor. { i,lek'trän·ik 'rō,bät }

electronic speedometer [ENG] A speedometer in which a transducer sends speed and distance pulses over wires to the speed and mileage indicators, eliminating the need for a mechanical link involving a flexible shaft. { i,lek'trän·ik spē'däm·əd·ər }

electronic thermometer [ENG] A thermometer in which a sensor, usually a thermistor, is placed on or near the object being measured. { i,lek 'trän·ik thər'mäm·əd·ər }

electronic voltmeter [ENG] Voltmeter which uses the rectifying and amplifying properties of electron devices and their associated circuits to secure desired characteristics, such as high-input impedance, wide-frequency range, crest indications, and so on. { i,lek'trän·ik 'vōlt,mēd·ər }

electron injection [ELECTR] 1. The emission of electrons from one solid into another. 2. The process of injecting a beam of electrons with an electron gun into the vacuum chamber of a mass spectrometer, betatron, or other large electron accelerator. { i'lek,trän in'jek·shən }

electron microscope [ELECTR] A device for forming greatly magnified images of objects by means of electrons, usually focused by electron lenses. { i'lek,trän 'mī·krə,skōp }

electron vacuum gage [ENG] An instrument used to measure vacuum by the ionization effect that an electron flow (from an incandescent filament to a charged grid) has on gas molecules. { i'lek,trän 'vak·yüm ,gāj }

electrooptic radar [ENG] Radar system using electrooptic techniques and equipment instead of microwave to perform the acquisition and tracking operation. { i,lek·trō'äp·tik 'rā,där }

electropainting [ENG] Electrolytic deposition of a thin layer of paint on a metal surface which is made an anode. { i'lek·trō,pānt·iŋ }

electrophotoluminescence [ELECTR] Emission of light resulting from application of an electric field to a phosphor which is concurrently, or has been previously, excited by other means. { i¦lek·trō¦fōd·ō,lü·mə'nes·ə ns }

electrorefining [CHEM ENG] Petroleum refinery process for light hydrocarbon streams in which an electrostatic field is used to assist in separation of chemical treating agents (acid, caustic, doctor) from the hydrocarbon phase. { i¦lek·trō·ri'fīn·iŋ }

electroresistive effect [ELECTR] The change in the resistivity of certain materials with changes in applied voltage. { i¦lek·tro·ri'zis·tiv ,fekt }

electroscope [ENG] An instrument for detecting an electric charge by means of the mechanical forces exerted between electrically charged bodies. { i'lek,trə,skōp }

electrostatic [ELEC] Pertaining to electricity at rest, such as an electric charge on an object. { i,lek,trə'stad·ik }

electrostatic actuator See actuator. { i,lek·trə 'stad·ik 'ak·chə,wäd·ər }

electrostatic atomization [MECH ENG] Atomization in which a liquid jet or film is exposed to an electric field, and forces leading to atomization arise from either free charges on the surface or liquid polarization. { i,lek·trə'stad·ik ,ad·ə· mə'zā·shən }

electrostatic attraction See Coulomb attraction. { i,lek·trə'stad·ik ə'trak·shən }

electrostatic energy |ELEC| The potential energy which a collection of electric charges possesses by virtue of their positions relative to each other. { i,lek·trə'stad·ik 'en·ər·jē }

electrostatic field |ELEC| A time-independent electric field, such as that produced by stationary charges. { i,lek·trə'stad·ik 'fēld }

electrostatic force |ELEC| Force on a charged particle due to an electrostatic field, equal to the electric field vector times the charge of the particle. { i,lek·trə'stad·ik ' fȯrs }

electrostatic force microscopy |ENG| The use of an atomic force microscope to measure electrostatic forces from electric charges on a surface. { i¦lek·trə,stad·ik ¦fȯrs mī'krä·skə·pē }

electrostatic generator |ELEC| Any machine which produces electric charges by friction or (more commonly) electrostatic induction. { i,lek·trə'stad·ik 'jen·ə,rād·ər }

electrostatic gyroscope |ENG| A gyroscope in which a small beryllium ball is electrostatically suspended within an array of six electrodes in a vacuum inside a ceramic envelope. { i,lek· trə'stad·ik 'jī·rə,skōp }

electrostatic induction |ELEC| The process of charging an object electrically by bringing it near another charged object, then touching it to ground. Also known as induction. { i,lek· trə'stad·ik in'dək·shən }

electrostatic interactions See Coulomb interactions. { i,lek·trə'stad·ik int·ə'rak·shənz }

electrostatic loudspeaker |ENG ACOUS| A loudspeaker in which the mechanical forces are produced by the action of electrostatic fields; in one type the fields are produced between a thin metal diaphragm and a rigid metal plate. Also known as capacitor loudspeaker. { i,lek· trə'stad·ik 'laud,spēk·ər }

electrostatic microphone See capacitor microphone. { i'lek·trə,stad·ik 'mī·krə,fōn }

electrostatic painting |ENG| A painting process that uses the particle-attracting property of electrostatic charges; direct current of about 100,000 volts is applied to a grid of wires through which the paint is sprayed to charge each particle; the metal objects to be sprayed are connected to the opposite terminal of the high-voltage circuit, so that they attract the particles of paint. { i'lek·trə,stad·ik 'pānt·iŋ }

electrostatic potential See electric potential. { i'lek·trə,stad·ik pə'ten·chəl }

electrostatic precipitator |ENG| A device which removes dust or other finely divided particles from a gas by charging the particles inductively with an electric field, then attracting them to highly charged collector plates. Also known as precipitator. { i'lek·trə,stad·ik prə'sip·ə,tād· ər }

electrostatic repulsion See Coulomb repulsion. { i'lek·trə,stad·ik ri'pəl·shən }

electrostatics |ELEC| The study of electric charges at rest, their electric fields, and potentials. { i,lek·trə'stad·iks }

electrostatic separation |ENG| Separation of finely pulverized materials by placing them in electrostatic separators. Also known as high-tension separation. { i'lek·trə,stad·ik ,sep· ə'rā·shən }

electrostatic separator |ENG| A separator in which a finely pulverized mixture falls through a powerful electric field between two electrodes; materials having different specific inductive capacitances are deflected by varying amounts and fall into different sorting chutes. { i'lek· trə,stad·ik 'sep·ə,rād·ər }

electrostatic shielding |ELEC| The placing of a grounded metal screen, sheet, or enclosure around a device or between two devices to prevent electric fields from interacting. { i'lek· trə,stad·ik 'shēld·iŋ }

electrostatic stress |ELEC| An electrostatic field acting on an insulator, which produces polarization in the insulator and causes electrical breakdown if raised beyond a certain intensity. { i'lek·trə,stad·ik 'stres }

electrostatic transducer |ENG ACOUS| A transducer consisting of a fixed electrode and a movable electrode, charged electrostatically in opposite polarity; motion of the movable electrode changes the capacitance between the electrodes and thereby makes the applied voltage change in proportion to the amplitude of the electrode's motion. Also known as condenser transducer. { i'lek·trə,stad·ik tranz'dü·sər }

electrostatic tweeter |ENG ACOUS| A tweeter loudspeaker in which a flat metal diaphragm is driven directly by a varying high voltage applied between the diaphragm and a fixed metal electrode. { i'lek·trə,stad·ik 'twēd·ər }

electrostatic units |ELEC| A centimeter-gram-second system of electric and magnetic units in which the unit of charge is that charge which exerts a force of 1 dyne on another unit charge when separated from it by a distance of 1 centimeter in vacuum; other units are derived from this definition by assigning unit coefficients in equations relating electric and magnetic quantities. Abbreviated esu. { i'lek·trə,stad·ik 'yü· nəts }

electrostatic voltmeter |ENG| A voltmeter in which the voltage to be measured is applied between fixed and movable metal vanes; the resulting electrostatic force deflects the movable vane against the tension of a spring. { i'lek· trə,stad·ik 'vōlt,mēd·ər }

electrostatic wattmeter |ENG| An adaptation of a quadrant electrometer for power measurements in which two quadrants are charged by the voltage drop across a noninductive shunt resistance through which the load current passes, and the line voltage is applied between one of the quadrants and a moving vane. { i'lek· trə,stad·ik 'wät,mēd·ər }

electrostriction |MECH| A form of elastic deformation of a dielectric induced by an electric field, associated with those components of strain which are independent of reversal of field direction, in contrast to the piezoelectric effect. Also known as electrostrictive strain. { i¦lek·trō 'strik·shən }

electrostriction transducer [ENG ACOUS] A transducer which depends on the production of an elastic strain in certain symmetric crystals when an electric field is applied, or, conversely, which produces a voltage when the crystal is deformed. Also known as ceramic transducer. { i¦lek·trō'strik·shən tranz'dü·sər }

electrostrictive strain See electrostriction. { i¦lek· trō'strik·tiv 'strān }

electrothermal ammeter See thermoammeter. { i¦lek·trō'thər·məl 'a¸med·ər }

electrothermal energy conversion [ENG] The direct conversion of electric energy into heat energy, as in an electric heater. { i¦lek·trō'thər·məl 'en·ər·jē kən¸vər·zhən }

electrothermal process [ENG] Any process which uses an electric current to generate heat, utilizing resistance, arcs, or induction; used to achieve temperatures higher than can be obtained by combustion methods. { i¦lek·trō'thər· məl 'präs·əs }

electrothermal voltmeter [ENG] An electrothermal ammeter employing a series resistor as a multiplier, thus measuring voltage instead of current. { i¦lek·trō'thər·məl 'vōlt¸mēd·ər }

Elektrion process [CHEM ENG] A process of condensation and polymerization in which a mixture of a relatively light mineral oil and a fatty oil is subjected to an electric discharge in an atmosphere of hydrogen; the product is a very viscous oil used for blending with lighter lubricating oils. { i'lek·trē¸än ¸präs·əs }

element [CIV ENG] See member. [ELEC] See component. [IND ENG] A brief, relatively homogeneous part of a work cycle that can be described and identified. { 'el·ə·mənt }

elemental motion [IND ENG] In time-and-motion study, a fundamental subdivision of the hand movements in manipulating an object. Also known as basic element; fundamental motion; therblig. { ¸el·ə'ment¸əl ¸mō·shən }

elementary commodity group [IND ENG] The lowest level of goods or services for which consistent values can be determined. Also known as elementary group. { el·ə¦men·trē kə'mad· əd·ē ¸grüp }

elementary group See elementary commodity group. { ¸el·ə'men·trē 'grüp }

element breakdown [IND ENG] Separation of a work cycle into elemental motions. { 'el·ə· mənt 'brāk¸daún }

elements [MECH] The various features of a trajectory such as the angle of departure, maximum ordinate, angle of fall, and so on. { 'el·ə· mənts }

element time [IND ENG] The time to complete a specific motion element. { 'el·ə·mənt ¸tīm }

elevate [ENG] To increase the angle of elevation of a gun, launcher, optical instrument, or the like. { 'el·ə¸vāt }

elevated flooring See raised flooring. { ¦el·ə¸vād· əd 'flór·iŋ }

elevation [ENG] Vertical distance to a point or object from sea level or some other datum. { ¸el·ə'vā·shən }

elevation angle See angle of elevation. { ¸el·ə'vā· shən ¸aŋ·gəl }

elevation meter [ENG] An instrument that measures the change of elevation of a vehicle. { ¸el·ə'vā·shən ¸mēd·ər }

elevation stop [ENG] Structural unit in a gun or other equipment that prevents it from being elevated or depressed beyond certain fixed limits. { ¸el·ə'vā·shən ¸stäp }

elevator [MECH ENG] Also known as elevating machine. **1.** Vertical, continuous-belt, or chain device with closely spaced buckets, scoops, arms, or trays to lift or elevate powders, granules, or solid objects to a higher level. **2.** Pneumatic device in which air or gas is used to elevate finely powdered materials through a closed conduit. **3.** An enclosed platform or car that moves up and down in a shaft for transporting people or materials. Also known as lift. { 'el·ə¸vād·ər }

elevator dredge [MECH ENG] A dredge which has a chain of buckets, usually flattened across the front and mounted on a nearly vertical ladder; used principally for excavation of sand and gravel beds under bodies of water. { 'el·ə¸vād· ər ¸drej }

Elgin extractor [CHEM ENG] Spray-tower, multistage, counterflow extractor in which the diameter of the base section is expanded to eliminate flow restriction at the light-liquid distribution location. { ¦el·jən ik'strak·tər }

ell [BUILD] A wing built perpendicular to the main section of a building. { el }

elliptical orbit [MECH] The path of a body moving along an ellipse, such as that described by either of two bodies revolving under their mutual gravitational attraction but otherwise undisturbed. { ə'lip·tə·kəl 'ór·bət }

elliptical system [ENG] A tracking or navigation system where ellipsoids of position are determined from time or phase summation relative to two or more fixed stations which are the focuses for the ellipsoids. { ə'lip·tə·kəl 'sis·təm }

elliptic gear [MECH ENG] A change gear composed of two elliptically shaped gears, each rotating about one of its focal points. { ə'lip·tik 'gir }

elliptic spring [DES ENG] A spring made of laminated steel plates, arched to resemble an ellipse. { ə'lip·tik 'spriŋ }

elongation [MECH] The fractional increase in a material's length due to stress in tension or to thermal expansion. { ē¸loŋ'gā·shən }

elutriation [CHEM ENG] The process of removing substances from a mixture through washing and decanting. [ENG] In a mixture, the separation of finer lighter particles from coarser heavier particles through a slow stream of fluid moving upward so that the lighter particles are carried with it. { ē¸lü·trē'ā·shən }

elutriator [ENG] An apparatus used to separate

suspended solid particles according to size by the process of elutriation. { ē'lü·trē,ad·ər }

emagram [THERMO] A graph of the logarithm of the pressure of a substance versus its temperature, when it is held at constant volume; in meteorological investigations, the potential temperature is often the parameter. { 'em·ə,gram }

emanometer [ENG] An instrument for the measurement of the radon content of the atmosphere; radon is removed from a sample of air by condensation or adsorption on a surface, and is then placed in an ionization chamber and its activity determined. { ,em·ə'näm·əd·ər }

embankment [CIV ENG] **1.** A ridge constructed of earth, stone, or other material to carry a roadway or railroad at a level above that of the surrounding terrain. **2.** A ridge of earth or stone to prevent water from passing beyond desirable limits. Also known as bank. { em'baŋk·mənt }

embossing stylus [ENG ACOUS] A recording stylus with a rounded tip that forms a groove by displacing material in the recording medium. { em'bäs·iŋ ,stī·ləs }

embrittlement [MECH] Reduction or loss of ductility or toughness in a metal or plastic with little change in other mechanical properties. { ,em'brid·əl·mənt }

emergency brake [MECH ENG] A brake that can be set by hand and, once set, continues to hold until released; used as a parking brake in an automobile. { ə'mər·jən·sē ,brāk }

Emerson wage incentive plan [IND ENG] A plan comprising time wages to 662/3% of standard performance, empiric bonuses from there to standard performance, ending at 120% time wages, and thereafter a straight-line earning which is 20% above and parallel to basic piece rate. { 'em·ər·sən 'wāj in,sen·tiv ,plan }

Emery-Dietz gravity corer [ENG] A tube, with weights attached, which forces sediment samples into its interior as it is dropped on the ocean bottom. { |em·ə·rē |dēts 'grav·əd·ē ,kòr·ər }

emery wheel [DES ENG] A grinding wheel made of or having a surface of emery powder; used for grinding and polishing. { 'em·ə·rē ,wēl }

emi See electromagnetic interference.

emission standard [ENG] The maximum legal quantity of pollutant permitted to be discharged from a single source. { i'mish·ən ,stan·dərd }

emissive power See emittance. { i|mis·iv 'pau·ər }

emissivity [THERMO] The ratio of the radiation emitted by a surface to the radiation emitted by a perfect blackbody radiator at the same temperature. Also known as thermal emissivity. { ,ē·mə'siv·əd·ē }

emittance [THERMO] The power radiated per unit area of a radiating surface. Also known as emissive power; radiating power. { i'mit·əns }

emitter [ELECTR] A transistor region from which charge carriers that are minority carriers in the base are injected into the base, thus controlling the current flowing through the collector; corresponds to the cathode of an electron tube. Symbolized E. Also known as emitter region. { i'mid·ər }

emitter barrier [ELECTR] One of the regions in which rectification takes place in a transistor, lying between the emitter region and the base region. { i'mid·ər ,bar·ē·ər }

emitter junction [ELECTR] A transistor junction normally biased in the low-resistance direction to inject minority carriers into a base. { i'mid·ər ,jəŋk·shən }

EMOSFET See electrolyte-MOSFET.

employment test [IND ENG] Any of a wide variety of tests to measure intelligence, personality traits, skills, interests, aptitudes, or other characteristics; used to supplement interviews, physical examinations, and background investigations before employment. { em'plòi·mənt ,test }

empty-cell process [ENG] A wood treatment in which the preservative coats the cells without filling them. { 'em·tē ,sel 'präs·əs }

emulsification test [CHEM ENG] Standard laboratory procedure for evaluating the resistance of insulating oils, turbine oils, and other lubricating oils to emulsification. { ə,məl·sə·fə'kā·shən ,test }

emulsion cleaner [CHEM ENG] A cleaner composed of organic solvents dispersed in an aqueous solution with the aid of an emulsifying agent. { ə'məl·shən ,klēn·ər }

enamel See glaze. { i'nam·əl }

enameling [ENG] The application of a vitreous glaze to pottery or metal surfaces, followed by fusing in a kiln or furnace. { i'nam·liŋ }

enamel kiln [ENG] A kiln in which enamel colors are fired. { i'nam·əl ,kil }

encastré beam See fixed-end beam. { än·ka·strā bēm }

encoder See matrix. { en'kōd·ər }

encrustation [ENG] The buildup of slag or other material inside furnaces and kilns. { en·krə'stā·shən }

end-bearing pile [CIV ENG] A bearing pile that is driven down to hard ground so that it carries the full load at its point. Also known as a point-bearing pile. { 'end ,ber·iŋ ,pīl }

end construction [CIV ENG] Structural blocks or tiles laid so that the hollow cells run vertically. { 'end kən,strək·shən }

end effector [CONT SYS] The component of a robot that comes into contact with the workpiece and does the actual work on it. Also known as hand. { 'end i,fek·tər }

end-feed centerless grinding [MECH ENG] Centerless grinding in which the piece is fed through grinding and regulating wheels to an end stop. { 'end ,fēd |sen·tər·ləs 'grīnd·iŋ }

end item [ENG] A final combination of end products, component parts, or materials which is ready for its intended use; for example, ship, tank, mobile machine shop, or aircraft. { 'end ,īd·əm }

end lap [DES ENG] A joint in which two joining members are made to overlap by removal of half the thickness of each. { 'end ,lap }

end loader [MECH ENG] A platform elevator at the rear of a truck. { 'end ,lōd·ər }

end mill [MECH ENG] A machine which has a

rotating shank with cutting teeth at the end and spiral blades on the peripheral surface; used for shaping and cutting metal. { 'end ,mil }

end-milled keyway See profiled keyway. { 'end ,mild 'kē,wā }

end-of-arm speed [CONT SYS] The speed at which an end effector arrives at its desired position. { ¦end əv ¦ärm 'spēd }

endoradiosonde [ENG] A miniature battery-powered radio transmitter encapsulated like a pill, designed to be swallowed for measuring and transmitting physiological data from the gastro-intestinal tract. { ¦en·dō'rād·ē·ō,sänd }

end play [MECH ENG] Axial movement in a shaft-and-bearing assembly resulting from clearances between the components. { 'end ,plā }

end point [CHEM ENG] In the distillation analysis of crude petroleum and its products, the highest reading of a thermometer when a specified proportion of the liquid has boiled off. Also known as final boiling point. [CONT SYS] The point at which a robot stops along its path of motion. See breakpoint. { 'end ,pȯint }

end-point rigidity [CONT SYS] The resistance of a robot to further movement after it has reached its end point. { 'en ,pȯint ri'jid·əd·ē }

end stop [MECH ENG] A limit to the movement of a mechanical system or part, usually brought about by valves or shock absorbers. { 'end ,stäp }

end turning See boxing. { 'end ,tərn·iŋ }

endurance [ENG] The time an aircraft, vehicle, or ship can continue operating under given conditions without refueling. { in'dür·əns }

endurance limit See fatigue limit. { in'dür·əns ,lim·ət }

endurance ratio See fatigue ratio. { in'dür·əns ,rā·shō }

endurance strength See fatigue strength. { in 'dür·əns ,streŋkth }

energy beam [ENG] An intense beam of light, electrons, or other nuclear particles; used to cut, drill, form, weld, or otherwise process metals, ceramics, and other materials. { 'en·ər·jē ,bēm }

energy conversion efficiency [MECH ENG] The efficiency with which the energy of the working substance is converted into kinetic energy. { 'en·ər·jē kən'vər·zhən i,fish·ən·sē }

energy efficiency ratio [ELEC] A value that represents the relative electrical efficiency of air conditioners; it is the quotient obtained by dividing Btu-per-hour output by electrical-watts input during cooling. { 'en·ər·jē i'fish·ən·se ,rā·shō }

energy ellipsoid See momental ellipsoid. { ¦en·ər·jē i'lip,sȯid }

energy integral [MECH] A constant of integration resulting from integration of Newton's second law of motion in the case of a conservative force; equal to the sum of the kinetic energy of the particle and the potential energy of the force acting on it. { 'en·ər·jē 'in·tə·grəl }

enfleurage [CHEM ENG] Removal of the odoriferous components from flowers by placing them near an odorless mixture of lard and tallow; this mixture absorbs the perfume, which is subsequently extracted. { ¦än,flü¦räzh }

engaged column [CIV ENG] A column partially built into a wall, and not freestanding. { in'gājd 'käl·əm }

engine [MECH ENG] A machine in which power is applied to do work by the conversion of various forms of energy into mechanical force and motion. { 'en·jən }

engine balance [MECH ENG] Arrangement and construction of moving parts in reciprocating or rotating machines to reduce dynamic forces which may result in undesirable vibrations. { 'en·jən ,bal·əns }

engine block See cylinder block. { 'en·jən ,bläk }

engine cooling [MECH ENG] Controlling the temperature of internal combustion engine parts to prevent overheating and to maintain all operating dimensions, clearances, and alignment by a circulating coolant, oil, and a fan. { 'en·jən ,kül·iŋ }

engine cycle [THERMO] Any series of thermodynamic phases constituting a cycle for the conversion of heat into work; examples are the Otto cycle, Stirling cycle, and Diesel cycle. { 'en·jən ,sī·kəl }

engine cylinder [MECH ENG] A cylindrical chamber in an engine in which the energy of the working fluid, in the form of pressure and heat, is converted to mechanical force by performing work on the piston. Also known as cylinder. { 'en·jən ,sil·ən·dər }

engine displacement [MECH ENG] Volume displaced by each piston moving from bottom dead center to top dead center multiplied by the number of cylinders. { 'en·jən di,splās·mənt }

engine efficiency [MECH ENG] Ratio between the energy supplied to an engine to the energy output of the engine. { 'en·jən i'fish·ən·sē }

engineer [ENG] An individual who specializes in one of the branches of engineering. { ,en·jə'nir }

engineering economy [IND ENG] **1.** Application of engineering or mathematical analysis and synthesis to decision making in economics. **2.** The knowledge and techniques concerned with evaluating the worth of commodities and services relative to their cost. **3.** Analysis of the economics of engineering alternatives. { ,en·jə'nir·iŋ i'kän·ə·mē }

engineering geology [CIV ENG] The application of education and experience in geology and other geosciences to solve geological problems posed by civil engineering structures. { ,en·jə'nir·iŋ jē'äl·ə·jē }

engineer's chain [CIV ENG] A surveyor's measuring instrument consisting of 1-foot (30.48-centimeter) steel links joined together by rings, 100 feet (30.5 meters) or 50 feet (15.25 meters) long. Also known as chain. { ,en·jə'nirz ,chān }

engine inlet [MECH ENG] A place of entrance for engine fuel. { 'en·jən ,in·lət }

engine knock [MECH ENG] In spark ignition engines, the sound and other effects associated with ignition and rapid combustion of the last part of the charge to burn, before the flame front reaches it. Also known as combustion knock. { 'en·jən ,näk }

engine lathe [MECH ENG] A manually operated lathe equipped with a headstock of the back-geared, cone-driven type or of the geared-head type. { 'en·jən ,lāth }

engine performance [MECH ENG] Relationship between power output, revolutions per minute, fuel or fluid consumption, and ambient conditions in which an engine operates. { 'en·jən pər'fór·məns }

engine sludge [ENG] The insoluble products of degradation of lubricating oils and fuels formed during the operation of an internal combustion engine. { 'en·jən ,sləj }

Engler distillation test [CHEM ENG] A standard test for determination of the volatility characteristics of a gasoline by the measurement of the percent of gasoline distilled at various specific temperatures. { 'eŋ·glər dis·tə'lā·shən ,test }

Engler flask [CHEM ENG] A standardized flask of 100-milliliter volume used in the Engler distillation test. { 'eŋ·glər ,flask }

Engler viscometer [ENG] An instrument used in the measurement of the degree Engler, a measure of viscosity; the kinematic viscosity v in stokes for this instrument is obtained from the equation $v = 0.00147t - 3.74/t$, where t is the efflux time in seconds. { 'eŋ·glər vi'skäm·əd·ər }

English garden-wall bond [CIV ENG] A masonry bond in which there are three courses of stretchers to one of headers. { 'iŋ·glish ¦gärd·ən 'wól ,bänd }

enhancement [ELECTR] An increase in the density of charged carriers in a particular region of a semiconductor. { en'hans·mənt }

enhancement mode [ELECTR] Operation of a field-effect transistor in which no current flows when zero gate voltage is applied, and increasing the gate voltage increases the current. { en'hans·mənt ,mōd }

enhancement-mode high-electron-mobility transistor [ELECTR] A high-electron-mobility transistor in which application of a positive bias to the gate electrode is required for current to flow between the source and drain electrodes. Abbreviated E-HEMT. { en'hans·mənt ¦mōd 'hī i¦lek,trän mō¦bil·əd·ē tran'zis·tər }

enhancement-mode junction field-effect transistor [ELECTR] A type of gallium arsenide field-effect transistor in which the gate consists of the junction between the n-type gallium arsenide forming the conducting channel and p-type material implanted under a metal electrode. Abbreviate E-JFET. { en'hans·mənt ¦mōd 'jəŋk·shən 'fēld i,fekt tran'zis·tər }

enqueue [ENG] To place a data item in a queue. { en'kyü }

enriching column [CHEM ENG] The portion of a countercurrent contractor (liquid-liquid extraction or vapor-liquid distillation) above the feed point in which an upward-moving, product-rich

stream from the stripping column is further purified by countercurrent contact with a downward-flowing reflux stream from the overhead product-recovery vessel. { in'rich·iŋ ,käl·əm }

enrockment [CIV ENG] A grouping of large stones dropped into water to form a base, such as for supporting a pier. { in'räk·mənt }

entering angle [MECH ENG] The angle between the side cutting edge of a tool and the machined surface of the work; angle is 90° for a tool with 0° side-cutting edge angle effective. { 'ent·ə·riŋ ,aŋ·gəl }

enthalpy [THERMO] The sum of the internal energy of a system plus the product of the system's volume multiplied by the pressure exerted on the system by its surroundings. Also known as heat content; sensible heat; total heat. { en 'thal·pē }

enthalpy-entropy chart [THERMO] A graph of the enthalpy of a substance versus its entropy at various values of temperature, pressure, or specific volume; useful in making calculations about a machine or process in which this substance is the working medium. { en¦thal·pē 'en·trə·pē ,chärt }

enthalpy of vaporization See heat of vaporization. { en'thal·pē əv ,vā·pə·rə'zā·shən }

entrainer [CHEM ENG] An additive that forms an azeotrope with one component of a liquid mixture to aid in otherwise difficult separations by distillation, as in azeotropic distillation. { en'trān·ər }

entrainment [CHEM ENG] A process in which the liquid boils so violently that suspended droplets of liquid are carried in the escaping vapor. { en'trān·mənt }

entrance [CIV ENG] The seaward end of a channel, harbor, and so on. [ENG] A place of physical entering, such as a door or passage. { 'en·trəns }

entrance angle [ENG] In molding, the maximum angle, measured from the center line of the mandrel, at which molten material enters the land area of a die. { 'en·trəns ,aŋ·gəl }

entrance lock [CIV ENG] A lock between the tideway and an enclosed basin made necessary because the levels of the two bodies of water vary; by means of this lock, vessels can pass either way at all states of the tide. Also known as guard lock; tidal lock; tide lock. { 'en·trəns ,läk }

entropy [THERMO] Function of the state of a thermodynamic system whose change in any differential reversible process is equal to the heat absorbed by the system from its surroundings divided by the absolute temperature of the system. Also known as thermal charge. { 'en·trə·pē }

entry ballistics [MECH] That branch of ballistics which pertains to the entry of a missile, spacecraft, or other object from outer space into and through an atmosphere. { 'en·trē bə,lis·tiks }

entry point See entrance. { 'en·trē ,póint }

envelope [ENG] The glass or metal housing of

an electron tube or the glass housing of an incandescent lamp. { 'en·və‚lōp }

environment |ENG| The aggregate of all natural, operational, or other conditions that affect the operation of equipment or components. { in'vī·ərn·mənt *or* in'vī·rən·ment }

environmental cab |ENG| Operator's compartment in earthmovers equipped with tinted safety glass, soundproofing, air conditioning, and cleaning units. { in¦vī·ərn¦ment·əl 'kab }

environmental control |ENG| Modification and control of soil, water, and air environments of humans and other living organisms. { in¦vī·ərn¦mənt·əl kən'trōl }

environmental control system |ENG| A system used in a closed area, especially a spacecraft or submarine, to permit life to be sustained; the system provides the occupants with a suitably controlled atmosphere to permit them to live and work in the area. { in¦vī·ərn¦mənt·əl kən'trōl ‚sis·təm }

environmental engineering |ENG| The technology concerned with the reduction of pollution, contamination, and deterioration of the surroundings in which humans live. { in¦vī·ərn¦mənt·əl en·jə'nir·iŋ }

environmental impact analysis |IND ENG| Predetermination of the extent of pollution or environmental degradation which will be involved in a mining or processing project. { in¦vī·ərn¦mənt·əl 'im‚pakt ə‚nal·ə·səs }

environmental impact statement |ENG| A report of the potential effect of plans for land use in terms of the environmental, engineering, esthetic, and economic aspects of the proposed objective. { in¦vī·ərn¦mənt·əl 'im‚pakt ‚stāt·mənt }

environmental protection |ENG| The protection of humans and equipment against stresses of climate and other elements of the environment. { in¦vī·ərn¦ment·əl prə'tek·shən }

environmental range |ENG| The range of environment throughout which a system or portion thereof is capable of operation at not less than the specified level of reliability. { in¦vī·ərn¦mənt·əl 'rānj }

environmental stress cracking |MECH| The susceptibility of a material to crack or craze in the presence of surface-active agents or other factors. { in¦vī·ərn¦mənt·əl 'stres ‚krak·iŋ }

environmental test |ENG| A laboratory test conducted to determine the functional performance of a component or system under conditions that simulate the real environment in which the component or system is expected to operate. { in¦vī·ərn¦mənt·əl 'test }

environment simulator |ENG| Any machine or artificial device that simulates all or some of the attributes of an environment, such as the solar simulators with artificial suns used in testing spacecraft. { in¦vī·ərn¦mənt 'sim·yə‚lād·ər }

eolian anemometer |ENG| An anemometer which works on the principle that the pitch of the eolian tones made by air moving past an

obstacle is a function of the speed of the air. { ē'ōl·yən an·ə'mäm·əd·ər }

eon |MECH| A unit of time, equal to 10^9 years. { 'ē‚än }

Eötvös effect |MECH| An apparent decrease (or increase) in the weight of a body moving from west to east (or east to west) because of its greater (or smaller) centrifugal acceleration. { 'ət·vəsh i‚fekt }

Eötvös rule |THERMO| The rule that the rate of change of molar surface energy with temperature is a constant for all liquids; deviations are encountered in practice. { 'ət·vəsh ‚rül }

Eötvös torsion balance |ENG| An instrument which records the change in the acceleration of gravity over the horizontal distance between the ends of a beam; used to measure density variations of subsurface rocks. { 'ət·vəsh 'tòr·shən ‚bal·əns }

epicyclic gear |MECH ENG| A system of gears in which one or more gears travel around the inside or the outside of another gear whose axis is fixed. { ¦ep·ə¦sī·klik 'gir }

epicyclic train |MECH ENG| A combination of epicyclic gears, usually connected by an arm, in which some or all of the gears have a motion compounded of rotation about an axis and a translation or revolution of that axis. { ¦ep·ə¦sī·klik 'trān }

epitaxial diffused-junction transistor |ELECTR| A junction transistor produced by growing a thin, high-purity layer of semiconductor material on a heavily doped region of the same type. { ‚ep·ə'tak·sē·əl də¦fyüzd ¦jəŋk·shən tran'zis·tər }

epitaxial diffused-mesa transistor |ELECTR| A diffused-mesa transistor in which a thin, high-resistivity epitaxial layer is deposited on the substrate to serve as the collector. { ‚ep·ə'tak·sē·əl də¦fyüzd ¦mā·sə tran'zis·tər }

epitaxial transistor |ELECTR| Transistor with one or more epitaxial layers. { ‚ep·ə'tak·sē·əl tran'zis·tər }

Eppley pyrheliometer |ENG| A pyrheliometer of the thermoelectric type; radiation is allowed to fall on two concentric silver rings, the outer covered with magnesium oxide and the inner covered with lampblack; a system of thermocouples (thermopile) is used to measure the temperature difference between the rings; attachments are provided so that measurements of direct and diffuse solar radiation may be obtained. { 'ep·lē ¦pīr‚hē·lē'äm·əd·ər }

equal-arm balance |MECH| A simple balance in which the distances from the point of support of the balance-arm beam to the two pans at the end of the beam are equal. { ¦ē·kwal ¦ärm 'bal·əns }

equaling file |DES ENG| A slightly bulging double-cut file used in fine toolmaking. { 'ē·kwəl·iŋ ‚fīl }

equalizer |ELECTR| A network designed to compensate for an undesired amplitude-frequency or phase-frequency response of a system or component; usually a combination of coils, capacitors, and resistors. Also known as equalizing circuit. |MECH ENG| **1.** A bar to which one attaches a vehicle's whiffletrees to make the pull of draft animals equal. Also known as equalizing bar. **2.** A bar which joins a pair of axle springs on a railway locomotive or car for equalization of weight. Also known as equalizing bar. **3.** A device which distributes braking force among independent brakes of an automotive vehicle. Also known as equalizer brake. **4.** A machine which saws wooden stock to equal lengths. { 'ē·kwə‚līz·ər }

equalizing line |CHEM ENG| A pipe or tubing interconnection between two closed vessels, containers, or process systems to allow pressure equalization. { 'ē·kwə‚līz·iŋ ‚līn }

equalizing reservoir |CIV ENG| A reservoir located between a primary water supply and the consumer for the purpose of maintaining equilibrium between different portions of the distribution system. { 'ē·kwə‚līz·iŋ 'rez·əv‚wär }

equation of motion |MECH| **1.** Equation which specifies the coordinates of particles as functions of time. **2.** A differential equation, or one of several such equations, from which the coordinates of particles as functions of time can be obtained if the initial positions and velocities of the particles are known. { i'kwā·zhən əv 'mō·shən }

equation of piezotropy |THERMO| An equation obeyed by certain fluids which states that the time rate of change of the fluid's density equals the product of a function of the thermodynamic variables and the time rate of change of the pressure. { i'kwā·zhən əv pē·ə'zä·trə‚pē }

equatorial mounting |ENG| The mounting of an equatorial telescope; it has two perpendicular axes, the polar axis (parallel to the earth's axis) that turns on fixed bearings, and the declination axis, supported by the polar axis. { ‚e·kwə'tór·ē·əl 'maùnt·iŋ }

equatorial plane |MECH| A plane perpendicular to the axis of rotation of a rotating body and equidistant from the intersections of this axis with the body's surface, provided that the body is symmetric about the axis of rotation and is symmetric under reflection through this plane. { ‚e·kwə'tór·ē·əl 'plān }

equatorial telescope |ENG| An astronomical telescope that revolves about an axis parallel to the earth's axis and automatically keeps a star on which it has been fixed in its field of view. { ‚e·kwə'tór·ē·əl 'tel·ə‚skōp }

equilibrant |MECH| A single force which cancels the vector sum of a given system of forces acting on a rigid body and whose torque cancels the sum of the torques of the system. { i'kwil·ə·brənt }

equilibristat |ENG| A device for measuring the deviation from equilibrium of a railroad car as it goes around a curve. { ‚ē·kwə'lib·rə‚stat }

equilibrium |MECH| Condition in which a particle, or all the constituent particles of a body, are at rest or in unaccelerated motion in an inertial reference frame. Also known as static equilibrium. { ‚ē·kwə'lib·rē·əm }

equilibrium distillation See equilibrium flash vaporization. { ‚ē·kwə‚lib·rē·əm ‚dis·tə‚lā·shən }

equilibrium flash vaporization |CHEM ENG| Process in which a continuous liquid-mixture feed stream is partly vaporized in a column or vessel, with continuous withdrawal of vapor and liquid portions, the vapor and liquid in equilibrium. Also known as continuous equilibrium vaporization; equilibrium distillation; flash distillation; simple continuous distillation. { ‚ē·kwə'lib·rē·əm 'flash ‚vā·pə·rə'zā·shən }

equilibrium state |IND ENG| A state in which the numbers of customers or items waiting in a queue varies in such a way that the mean and distribution remain constant over a long period. { ‚ē·kwə'lib·rē·əm ‚stāt }

equipment |ENG| One or more assemblies capable of performing a complete function. { ə'kwip·mənt }

equipment chain |ENG| Group of equipments that are functionally in series; the failure of one or more of the equipments results in loss of the function. { ə'kwip·mənt ‚chān }

equipment replacement study |IND ENG| A cost analysis based on estimates of operating costs over a stated time for the old facility compared with the new facility. { ə'kwip·mənt ri'plās·mənt ‚stəd·ē }

equipollent |MECH| Of two systems of forces, having the same vector sum and the same total torque about an arbitrary point. { ‚e·kwə‚päl·ənt }

equipotential surface |ELEC| A surface on which the electric potential is the same at every point. |MECH| A surface which is always normal to the lines of force of a field and on which the potential is everywhere the same. { ‚e·kwə·pə'ten·chəl 'sər·fəs }

equivalent annual rate |IND ENG| A measure used in setting up a monthly rate on a comparable basis for each of the months regardless of their variation in working days, or for making the rate comparable with an annual rate regardless of the variation in working days during each month. { i'kwiv·ə·lənt ‚an·yə·wəl 'rāt }

equivalent bending moment |MECH| A bending moment which, acting alone, would produce in a circular shaft a normal stress of the same magnitude as the maximum normal stress produced by a given bending moment and a given twisting moment acting simultaneously. { i'kwiv·ə·lənt 'bend·iŋ ‚mō·mənt }

equivalent blackbody temperature |THERMO| For a surface, the temperature of a blackbody which emits the same amount of radiation per unit area as does the surface. { i'kwiv·ə·lənt 'blak‚bäd·ē ‚tem·prə·chər }

equivalent circuit |ELEC| A circuit whose behavior is identical to that of a more complex circuit or device over a stated range of operating conditions. { i'kwiv·ə·lənt 'sər·kət }

equivalent nitrogen pressure |MECH| The pressure that would be indicated by a device if the gas inside it were replaced by nitrogen of equivalent molecular density. { i'kwiv·ə·lənt 'nī·trə·jən ,presh·ər }

equivalent noise pressure [ENG ACOUS] In an electroacoustic transducer or sound reception system, the root-mean-square sound pressure of a sinusoidal plane progressive wave, which when propagated parallel to the primary axis of the transducer, produces an open-circuit signal voltage equivalent to the root-mean-square of the inherent open-circuit noise voltage of the transducer in a transmission band with a bandwidth of 1 hertz and centered on the frequency of the plane sound wave. Also known as inherent noise pressure. { i'kwiv·ə·lənt 'nóiz ,presh·ər }

equivalent orifice [MECH ENG] An expression of fan performance as the theoretical sharp-edge orifice area which would offer the same resistance to flow as the system resistance itself. { i'kwiv·ə·lənt 'ór·ə·fəs }

equivalent round [ENG] The diameter of a circle whose circumference is equal to the circumference of a pipe whose cross section is not a perfect circle. { i'kwiv·ə·lənt 'raúnd }

equivalent temperature [THERMO] A term used in British engineering for that temperature of a uniform enclosure in which, in still air, a sizable blackbody at 75°F (23.9°C) would lose heat at the same rate as in the environment. { i'kwiv·ə·lənt 'tem·prə·chər }

equivalent twisting moment [MECH] A twisting moment which, if acting alone, would produce in a circular shaft a shear stress of the same magnitude as the shear stress produced by a given twisting moment and a given bending moment acting simultaneously. { i'kwiv·ə·lənt 'twist·iŋ ,mō·mənt }

equivalent viscous damping [MECH] An assumed value of viscous damping used in analyzing a vibratory motion, such that the dissipation of energy per cycle at resonance is the same for the assumed or the actual damping force. { i'kwiv·ə·lənt ¦vis·kəs 'damp·iŋ }

equiviscous temperature [CHEM ENG] A measure of viscosity used in the tar industry, equal to the temperature in degrees Celsius at which the viscosity of tar is 50 seconds as measured in a standard tar efflux viscometer. Abbreviated EVT. { ¦e·kwə¦vis·kəs 'tem·prə·chər }

erection [CIV ENG] Positioning and fixing the frame of a structure. { i'rek·shən }

erection bolt [CIV ENG] A threaded rod with a head at one end, used to temporarily join parts of a structure during construction. { i'rek·shən ,bōlt }

erection stress [MECH] The internal forces exerted on a structural member during construction. { i'rek·shən ,stres }

erection tower [CIV ENG] A temporary framework built at a construction site for hoisting equipment. { i'rek·shən ,taú·ər }

ergograph [ENG] An instrument with a recording device used to measure work capacity of muscles. { 'ər·gə,graf }

ergometer [ENG] An instrument with a recording device used to measure work performed by muscles under control conditions. { ər'gäm·əd·ər }

ergonometrics [IND ENG] The application of various procedures for determining the time for an operator to perform a task satisfactorily, using the standard method in the usual environmental conditions, for example, time study or work sampling. Also known as work measurement. { ər,gän·ə'me,triks }

ergonomics [IND ENG] The study of human capability and psychology in relation to the working environment and the equipment operated by the worker. { ,ər·gə'näm·iks }

Ericsson cycle [THERMO] An ideal thermodynamic cycle consisting of two isobaric processes interspersed with processes which are, in effect, isothermal, but each of which consists of an infinite number of alternating isentropic and isobaric processes. { 'er·ik·sən ,sī·kəl }

error coefficient [CONT SYS] The steady-state value of the output of a control system, or of some derivative of the output, divided by the steady-state actuating signal. Also known as error constant. { 'er·ər ,kō·i'fish·ənt }

error constant See error coefficient. { 'er·ər ,kän·stənt }

error of closure [ENG] Also known as angular error of closure. **1.** The amount by which the measurement of the azimuth of the first line of a traverse, made after completing the circuit, fails to equal the initial measurement. **2.** The amount by which the sum of the angles measured around the horizon differs from 360° { 'er·ər əv 'klō·zhər }

error signal [CONT SYS] In an automatic control device, a signal whose magnitude and sign are used to correct the alignment between the controlling and the controlled elements. See error voltage. [ELECTR] A voltage that depends on the signal received from the target in a tracking system, having a polarity and magnitude dependent on the angle between the target and the center of the scanning beam. { 'er·ər ,sig·nəl }

escalation [IND ENG] Provision in actual or estimated costs for inflational increases in the costs of equipment, materials, labor, and so on, over those specified in an original contract. { ,es·kə'lā·shən }

escalator [MECH ENG] A continuously moving stairway and handrail. { 'es·kə,lād·ər }

escape hatch [ENG] A hatch which permits persons to escape from a compartment, such as the interior of a submarine or aircraft, when normal means of exiting are blocked. { ə'skāp ,hach }

escapement [MECH ENG] A ratchet device that permits motion in one direction slowly. { ə'skāp·mənt }

escutcheon [DES ENG] An ornamental shield, flange, or border used around a dial, window, control knob, or other panel-mounted part. Also known as escutcheon plate. { e'skəch·ən }

escutcheon plate See escutcheon.

esthesiometer [ENG] An instrument used to measure tactile sensibility by determining the distance by which two points pressed against the skin must be separated in order that they be felt as separate. Also spelled aesthesiometer. { es,thē·zē'äm·əd·ər }

estimated time [IND ENG] A predicted element or operation time. { 'es·tə,mād·əd 'tīm }

esu *See* electrostatic units.

etched circuit [ENG] A printed circuit formed by chemical or electrolytic removal of unwanted portions of a layer of conductive material bonded to an insulating base. { ¦echt 'sər·kət }

ethoxylation [CHEM ENG] A catalytic process which involves the direct addition of ethylene oxide to an alkyl phenol or to an aliphatic alcohol. { e,thäk·sə'lā·shən }

ethylene alkylation [CHEM ENG] A catalytic petroleum-refining process in which dry isobutane and ethylene react to form ethylene alkylate. { 'eth·ə,lēn ,al·kə'lā·shən }

EU *See* expected value.

eudiometer [ENG] An instrument for measuring changes in volume during the combustion of gases, consisting of a graduated tube that is closed at one end and has two wires sealed into it, between which a spark may be passed. { ,yü·dē'äm·əd·ər }

Euler angles [MECH] Three angular parameters that specify the orientation of a body with respect to reference axes. { 'ȯi·lər ,aŋ·gəlz }

Euler equation [MECH] Expression for the energy removed from a gas stream by a rotating blade system (as a gas turbine), independent of the blade system (as a radial- or axial-flow system). { 'ȯi·lər i,kwā·zhən }

Euler equations of motion [MECH] A set of three differential equations expressing relations between the force moments, angular velocities, and angular accelerations of a rotating rigid body. { 'ȯi·lər i¦kwā·zhənz əv 'mō·shən }

Euler force [MECH] The greatest load that a long, slender column can carry without buckling, according to the Euler formula for long columns. { 'ȯi·lər ,fȯrs }

Euler formula for long columns [MECH] A formula which gives the greatest axial load that a long, slender column can carry without buckling, in terms of its length, Young's modulus, and the moment of inertia about an axis along the center of the column. { 'ȯi·lər ¦fȯr·myə·lə fər ,lȯŋ 'käl·əmz }

Eulerian description *See* Euler method. { ȯi¦ler·ē·ən di'skrip·shən }

Euler method [MECH] A method of studying fluid motion and the mechanics of deformable bodies in which one considers volume elements at fixed locations in space, across which material flows; the Euler method is in contrast to the Lagrangian method. { 'ȯi·lər ,meth·əd }

Euler-Rodrigues parameter [MECH] One of four numbers which may be used to specify the orientation of a rigid body; they are components of a quaternion. { 'ȯi·lər rə'drē·gəs pə,ram·əd·ər }

EV *See* expected value.

evaporation gage *See* atmometer. { i,vap·ə'rā·shən ,gāj }

evaporation loss [CHEM ENG] The loss of a stored volatile liquid component or mixture by evaporation; controlled by temperature, pressure, and the presence or absence of vapor-recovery systems. { i,vap·ə'rā·shən ,lȯs }

evaporation pan [ENG] A type of atmometer consisting of a pan, used in the measurement of the evaporation of water into the atmosphere. { i,vap·ə'rā·shən ,pan }

evaporation tank [ENG] A tank used to measure the evaporation of water under controlled conditions. { i,vap·ə'rā·shən ,taŋk }

evaporative condenser [MECH ENG] An apparatus in which vapor is condensed within tubes that are cooled by the evaporation of water flowing over the outside of the tubes. { i'vap·ə,rād·iv kən'den·sər }

evaporative control system [MECH ENG] A motor vehicle system that prevents escape of gasoline vapors from the fuel tank or carburetor to the atmosphere while the engine is not operating. { i¦vap·ə,rād·iv kən'trōl ,sis·təm }

evaporative cooling [ENG] **1.** Lowering the temperature of a large mass of liquid by utilizing the latent heat of vaporization of a portion of the liquid. **2.** Cooling air by evaporating water into it. **3.** *See* vaporization cooling. { i'vap·ə,rād·iv 'kül·iŋ }

evaporative cooling tower *See* wet cooling tower. { i'vap·ə,rād·iv 'kül·iŋ ,taü·ər }

evaporator [CHEM ENG] A device used to vaporize part or all of the solvent from a solution; the valuable product is usually either a solid or concentrated solution of the solute. [MECH ENG] Any of many devices in which liquid is changed to the vapor state by the addition of heat, for example, distiller, still, dryer, water purifier, or refrigeration system element where evaporation proceeds at low pressure and consequent low temperature. { i'vap·ə,rād·ər }

evaporimeter *See* atmometer. { i,vap·ə'rim·əd·ər }

evaporite pond [IND ENG] Any containment area for brines or solution-mined effluents constructed to permit solar evaporation and harvesting of dewatered evaporite concentrates. { i'vap·ə,rīt ,pänd }

evapotranspirometer [ENG] An instrument which measures the rate of evapotranspiration; consists of a vegetation soil tank so designed that all water added to the tank and all water left after evapotranspiration can be measured. { i,vap·ō,tranz·pə'räm·əd·ər }

Evasé stack [CIV ENG] In tunnel engineering, an exhaust stack for air having a cross section that increases in the direction of airflow at a rate to regain pressure. { ā,vä¦zā ,stak }

even pitch [DES ENG] The pitch of a screw in which the number of threads per inch is a multiple (or submultiple) of the threads per inch of the lead screw of the lathe on which the screw is cut. { ¦ē·vən 'pich }

event [IND ENG] A specified accomplishment in a program at a particular time; appears as a node in a graphic representation of an endeavor with a specific objective (project). { i'vent }

event recorder |ENG| A recorder that plots on-off information against time, to indicate when events start, how long they last, and how often they recur. { i'vent ri‚kòrd·ər }

event tree |IND ENG| A graphical representation of the possible sequence of events that might occur following an event that initiates an accident. { i'vent ‚trē }

evolutionary operation |IND ENG| An iterative technique for optimizing a production process by systematically introducing small changes in the process and then observing and evaluating the results. { ¦ev·ə¦lü·shə‚ner·ē ‚äp·ə'rā·shən }

EVT See equiviscous temperature.

Ewing's hysteresis tester |ENG| An instrument for determining the hysteresis loss of a specimen of magnetic material by measuring the deflection of a horseshoe magnet when the specimen is rapidly rotated between the poles of the magnet and the magnet is allowed to rotate about an axis that is aligned with the axis of rotation of the specimen. { ¦yü·iŋz ‚his·tə'rē·səs ‚tes·tər }

excavation |CIV ENG| 1. The process of digging a hollow in the earth. 2. An uncovered cavity in the ground. { ‚ek·skə'vā·shən }

excavator |MECH ENG| A machine for digging and removing earth. { 'ek·skə‚vād·ər }

exception handling |CONT SYS| The actions taken by a control system when unpredictable conditions or situations arise in which the controller must respond quickly. { ek'sep·shən ‚hand·liŋ }

excess air |ENG| Amount of air in a combustion process greater than the amount theoretically required for complete oxidation. { ¦ek‚ses 'er }

excess coefficient |MECH ENG| The ratio $(A - R)/R$, where A is the amount of air admitted in the combustion of fuel and R is the amount required. { 'ek‚ses ‚kō·i‚fish·ənt }

exchange adsorption |CHEM ENG| Ion exchange process in which the fluid phase contains (or consists of) two adsorbable components which together entirely saturate the surfaces of the adsorbent. { iks'chānj ad'sórp·shən }

exchanger See heat exchanger. { iks'chānj·ər }

excitation |CONT SYS| The application of energy to one portion of a system or apparatus in a manner that enables another portion to carry out a specialized function; a generalization of the electricity and electronics definitions. |ELEC| The application of voltage to field coils to produce a magnetic field, as required for the operation of an excited-field loudspeaker or a generator. |ELECTR| 1. The signal voltage that is applied to the control electrode of an electron tube. Also known as drive. 2. Application of signal power to a transmitting antenna. { ‚ek ‚sī'tā·shən }

exergy |THERMO| The portion of the total energy of a system that is available for conversion to useful work; in particular, the quantity of work that can be performed by a fluid relative to a

reference condition, usually the surrounding ambient condition. { 'eks·ər·jē }

exhaust |MECH ENG| 1. The working substance discharged from an engine cylinder or turbine after performing work on the moving parts of the machine. 2. The phase of the engine cycle concerned with this discharge. 3. A duct for the escape of gases, fumes, and odors from an enclosure, sometimes equipped with an arrangement of fans. { ig'zòst }

exhaust deflecting ring |MECH ENG| A type of jetavator consisting of a ring so mounted at the end of a nozzle as to permit it to be rotated into the exhaust stream. { ig'zòst di‚flek·tiŋ ‚riŋ }

exhaust gas |MECH ENG| Spent gas leaving an internal combustion engine or gas turbine. { ig'zòst ‚gas }

exhaust-gas analyzer |ENG| An instrument that analyzes the gaseous products to determine the effectiveness of the combustion process. { ig'zòst ‚gas 'an·ə‚līz·ər }

exhaust head |ENG| A device placed on the end of an exhaust pipe to remove oil and water and to reduce noise. { ig'zòst ‚hed }

exhaustion region |ELECTR| A layer in a semiconductor, adjacent to its contact with a metal, in which there is almost complete ionization of atoms in the lattice and few charge carriers, resulting in a space-charge density. { ig'zòs·chən ‚rē·jən }

exhaust manifold |MECH ENG| A branched system of pipes to carry waste emissions away from the piston chambers of an internal combustion engine. { ig'zòst ‚man·ə‚fōld }

exhaust pipe |MECH ENG| The duct through which engine exhaust is discharged. { ig'zòst ‚pīp }

exhaust scrubber |ENG| A purifying device on internal combustion engines which removes noxious gases from engine exhaust. { ig'zòst ‚skrəb·ər }

exhaust stroke |MECH ENG| The stroke of an engine, pump, or compressor that expels the fluid from the cylinder. { ig'zòst ‚strōk }

exhaust suction stroke |MECH ENG| A stroke of an engine that simultaneously removes used fuel and introduces fresh fuel to the cylinder. { ig 'zòst 'sək·shən ‚strōk }

exhaust valve |MECH ENG| The valve on a cylinder in an internal combustion engine which controls the discharge of spent gas. { ig'zòst ‚valv }

exit |ENG| A door, passage, or place of egress. { 'eg·zət }

ex lighterage |IND ENG| Price quoted exclusive of lighterage fees. { ¦eks 'līd·ə·rij }

exotherm |CHEM ENG| The graphical plotting of heat rise and fall versus time for an exothermic reaction or process system. { 'ek·sə‚thərm }

expanded-flow bin |ENG| A bin formed by attaching a mass-flow hopper to the bottom of a funnel-flow bin. { ik¦spand·əd 'flō ‚bin }

expander flange |ENG| A type of butt-welded flange designed with a tapered bore so that various pipe sizes can be matched. { ik'span·dər ‚flaṅj }

expanding brake |MECH ENG| A brake that operates by moving outward against the inside rim of a drum or wheel. { ik'spand·iŋ 'brāk }

expansion |ELECTR| A process in which the effective gain of an amplifier is varied as a function of signal magnitude, the effective gain being greater for large signals than for small signals; the result is greater volume range in an audio amplifier and greater contrast range in facsimile. |MECH ENG| Increase in volume of working material with accompanying drop in pressure of a gaseous or vapor fluid, as in an internal combustion engine or steam engine cylinder. { ik 'span·shən }

expansion bolt |DES ENG| A bolt having an end which, when embedded into masonry or concrete, expands under a pull on the bolt, thereby providing anchorage. { ik'span·shən ,bōlt }

expansion chucking reamer |DES ENG| A machine reamer with an expansion screw at the end which increases the diameter. { ik'span·shən 'chək·iŋ ,rē·mər }

expansion coefficient See coefficient of cubical expansion. { ik'span·shən kō·ə'fish·ənt }

expansion cooling |MECH ENG| Cooling of a substance by having it undergo adiabatic expansion. { ik'span·shən ,kül·iŋ }

expansion engine |MECH ENG| Piston-cylinder device that cools compressed air via sudden expansion; used in production of pure gaseous oxygen via the Claude cycle. { ik'span·shən ,en·jən }

expansion fit |DES ENG| A condition of optimum clearance between certain mating parts in which the cold inner member is placed inside the warmer outer member and the temperature is allowed to equalize. { ik'span·shən ,fit }

expansion joint |CIV ENG| **1.** In masonry, a flexible bituminous fiber strip used to separate blocks or units of concrete to prevent cracking caused by thermally induced expansion and contraction. **2.** A union or gap between adjacent parts of a building, structure, or concrete work that permits the relative movement caused by temperature changes to occur without rupture or damage. |MECH ENG| **1.** A joint between parts of a structure or machine to avoid distortion when subjected to temperature change. **2.** A pipe coupling which, under temperature change, allows movement of a piping system without hazard to associated equipment. { ik 'span·shən ,jóint }

expansion loop |ENG| A complete loop installed in a pipeline to mitigate the effect of expansion or contraction of the line. { ik'span·shən 'lüp }

expansion opening |ENG| A chamber in line with a pipe or tunnel and of larger diameter than the conduit containing liquid or gas, to allow lowering of pressure within the conduit by expansion of the fluid. { ik'span·shən,ōp·ə·niŋ }

expansion ratio |MECH ENG| In a reciprocating piston engine, the ratio of cylinder volume with piston at bottom dead center to cylinder volume with piston at top dead center. { ik'span·shən ,rā·shō }

expansion reamer |ENG| A reamer whose diameter may be adjusted between limits by an expanding screw. { ik'span·shən ,rē·mər }

expansion rollers |CIV ENG| Rollers fitted to one support of a bridge or truss to allow for thermal expansion and contraction. { ik'span·shən ,rō·lərz }

expansion shield |DES ENG| An anchoring device that expands as it is driven into masonry or concrete, pressing against the sides of the hole. { ik'span·shən ,shēld }

expansion valve |MECH ENG| A valve in which fluid flows under falling pressure and increasing volume. { ik'span·shən ,valv }

expansive bit |DES ENG| A bit in which the cutting blade can be set at various sizes. { ek'span·siv ,bit }

expansivity See coefficient of cubical expansion. { ,ek,span'siv·əd·ē }

expected utility See expected value.

expected value |SYS ENG| In decision theory, a measure of the value or utility expected to result from a given strategy, equal to the sum over states of nature of the product of the probability of the state times the consequence or outcome of the strategy in terms of some value or utility parameter. Abbreviated EV. Also known as expected utility (EU). { ek'spek·təd 'val·yü }

expert control system |CONT SYS| A control system that uses expert systems to solve control problems. { ¦ek,spərt kən'trōl ,sis·təm }

expletive |ENG| Any material used as fill, for example, a piece of masonry used to fill a cavity. { 'ek·spləd·iv }

explicit programming |CONT SYS| Robotic programming that employs detailed and exact descriptions of the tasks to be performed. { ik'splis·ət 'prō,gram·iŋ }

exploding bridge wire |ENG| An initiator or system in which a very high energy electrical impulse is passed through a bridge wire, literally exploding the bridge wire and releasing thermal and shock energy capable of initiating a relatively insensitive explosive in contact with the bridge wire. { ik¦splōd·iŋ 'brij ,wīr }

explosion door |MECH ENG| A door in a furnace which is designed to open at a predetermined pressure. { ik'splō·zhən ,dór }

explosion method |THERMO| Method of measuring the specific heat of a gas at constant volume by enclosing the gas with an explosive mixture, whose heat of reaction is known, in a chamber closed with a corrugated steel membrane which acts as a manometer, and by deducing the maximum temperature reached on ignition of the mixture from the pressure change. { ik 'splō·zhən ,meth·əd }

explosion rupture disk device |MECH ENG| A protective device used where the pressure rise in the vessel occurs at a rapid rate. { ik¦splō·zhən 'rəp·chər ,disk di,vīs }

explosive-actuated device |ENG| Any of various devices actuated by means of explosive; includes devices actuated either by high explosives or low explosives, whereas propellant-actuated devices include only the latter. { ik'splō·sive ,ak·chə,wād·əd di,vīs }

explosive disintegration |ENG| Explosive shattering when pressure is suddenly released on a pressured, permeable material (wood, mineral, and such) containing gas or liquid; the rupture of wood by this process is used to manufacture Masonite. { ik'splō·siv di,sin·tə'grā·shən }

explosive echo ranging |ENG| Sonar in which a charge is exploded underwater to produce a shock wave that serves the same purpose as an ultrasonic pulse; the elapsed time for return of the reflected wave gives target range. { ik'splō·siv 'ek·ō ,rānj·iŋ }

explosive limits |CHEM ENG| The upper and lower limits of percentage composition of a combustible gas mixed with other gases or air within which the mixture explodes when ignited. { ik 'splō·siv 'lim·əts }

explosive rivet |ENG| A rivet holding a charge of explosive material; when the charge is set off, the rivet expands to fit tightly in the hole. { ik'splō·siv 'riv·ət }

exponential horn |ENG ACOUS| A horn whose cross-sectional area increases exponentially with axial distance. { ,ek·spə'nen·chəl 'hórn }

exponential smoothing |IND ENG| A mathematical-statistical method of forecasting used in industrial engineering which assumes that demand for the following period is some weighted average of the demands for the past periods. { ,ek·spə'nen·chəl 'smüth·iŋ }

exposure |BUILD| The distance from the butt of one shingle to the butt of the shingle above it, or the amount of a shingle that is seen. { ik 'spō·zhər }

exposure time |CIV ENG| The time period of interest for seismic hazard calculations such as the design lifetime of a building or the time over which the numbers of casualties should be estimated. { ik'spō·zhər ,tīm }

expression |CHEM ENG| Separation of liquid from a two-phase solid-liquid system by compression under conditions that permit liquid to escape while the solid is retained between the compressing surfaces. Also known as mechanical expression. { ik'spresh·ən }

expressway |CIV ENG| A limited-access, high-speed, divided highway having grade separations at points of intersection with other roads. Also known as limited-access highway. { ik'spres ,wā }

extended area |DES ENG| An engineering surface that has been extended areawise without increasing diameter, as by using pleats (as in filter cartridges) or fins (as in heat exchangers). { ik'stend·əd 'er·ē·ə }

extensibility |MECH| The amount to which a material can be stretched or distorted without breaking. { ik,sten·sə'bil·əd·ē }

extension bolt |DES ENG| A vertical bolt that can be slid into place by a long extension rod; used at the top of doors. { ik'sten·chən ,bōlt }

extension jamb |BUILD| A jamb that extends past the head of a door or window. { ik'sten·chən ,jam }

extension ladder |DES ENG| A ladder of two or more nesting sections which can be extended to almost the combined length of the sections. { ik'sten·chən ,lad·ər }

extension spring |DES ENG| A tightly coiled spring designed to resist a tensile force. { ik 'sten·chən ,spriŋ }

extensometer |ENG| **1.** A strainometer that measures the change in distance between two reference points separated 60–90 feet (20–30 meters) or more; used in studies of displacements due to seismic activities. **2.** An instrument designed to measure minute deformations of small objects subjected to stress. { ,ek,sten 'säm·əd·ər }

exterior ballistics |MECH| The science concerned with behavior of a projectile after it leaves the muzzle of the firing weapon. { ek'stir·ē·ər bə'lis·tiks }

external brake |MECH ENG| A brake that operates by contacting the outside of a brake drum. { ek'stərn·əl 'brāk }

external centerless grinding |MECH ENG| A process by which a metal workpiece is finished on its external surface by supporting the piece on a blade while it is advanced between a regulating wheel and grinding wheel. { ek'stərn·əl 'sen·tər·ləs ,grīnd·iŋ }

external combustion engine |MECH ENG| An engine in which the generation of heat is effected in a furnace or reactor outside the engine cylinder. { ek'stərn·əl kəm'bəs·chən ,en·jən }

external device |ENG| A piece of equipment that operates in conjunction with and under the control of a central system, such as a computer or control system, but is not part of the system itself. { ek'stərn·əl di'vīs }

external force |MECH| A force exerted on a system or on some of its components by an agency outside the system. { ek¦stərn·əl 'fórs }

external grinding |MECH ENG| Grinding the outer surface of a rotating piece of work. { ek¦stərn·əl 'grīnd·iŋ }

external header |MECH ENG| Manifold connecting sections of a cast iron boiler. { ek¦stərn·əl 'hed·ər }

externally fired boiler |MECH ENG| A boiler that has refractory or cooling tubes surrounding its furnace. { ek¦stərn·əl·ē ¦fīrd 'bóil·ər }

external-mix oil burner |ENG| A burner utilizing a jet stream of air to strike the liquid fuel after it has left the burner orifice. { ek¦stərn·əl ,miks 'óil,bərn·ər }

external sensor |CONT SYS| A device that senses information about the environment of a control system but is not part of the system itself. { ek'stərn·əl 'sen·sər }

external shoe brake |MECH ENG| A friction brake operated by the application of externally contracting elements. { ek¦stərn·əl 'shü ,brāk }

external thread | DES ENG| A screw thread cut on an outside surface. { ek¦stərn·əl 'thred }

external time |IND ENG| The time used to perform work by the operator outside the machine cycle, resulting in a loss of potential machine operating time. { ek¦stərn·əl 'tīm }

external work |THERMO| The work done by a system in expanding against forces exerted from outside. { ek¦stərn·əl 'wərk }

external working environment |IND ENG| The workplace environment that is external to the human body; ranges from air quality to specific features such as clothing or tool handles. { ek¦stirn·əl ¦wərk·iŋ in'vī·rən·mənt }

extraction column |CHEM ENG| Vertical-process vessel in which a desired product is separated from a liquid by countercurrent contact with a solvent in which the desired product is preferentially soluble. { ik'strak·shən ‚käl·əm }

extraction turbine |MECH ENG| A steam turbine equipped with openings through which partly expanded steam is bled at one or more stages. { ik'strak·shən 'tər‚bīn }

extractive distillation |CHEM ENG| A distillation process to separate components from eutectic mixtures; a solution of the mixture is cooled, causing one component to crystallize out and the other to remain in solution; used to separate p-xylene and m-xylene, using n-pentane as the solvent. { ik'strak·tiv ‚dis·tə'lā·shən }

extractor |CHEM ENG| An apparatus for solvent-contact with liquids or solids for removal of specified components. |ENG| **1.** A machine for extracting a substance by a solvent or by centrifugal force, squeezing, or other action. **2.** An instrument for removing an object. { ik 'strak·tər }

extra-high voltage |ELEC| A voltage above 345 kilovolts used for power transmission. Abbreviated ehv. { ¦ek·strə ¦hī 'vōl·tij }

extrinsic detector |ENG| A semiconductor detector of electromagnetic radiation that is doped with an electrical impurity and utilizes transitions of charge carriers from impurity states in the band gap to nearby energy bands. { ek ¦strinz·ik di'tek·tər }

extrinsic photoconductivity |ELECTR| Photoconductivity that occurs for photon energies smaller than the band gap and corresponds to optical excitation from an occupied imperfection level to the conduction band, or to an unoccupied imperfection level from the valence band,

of a material. { ek¦strinz·ik ‚fō·dō·kän·dək'tiv·əd·ē }

extrinsic photoemission |ELECTR| Photoemission by an alkali halide crystal in which electrons are ejected directly from negative ion vacancies, forming color centers. Also known as direct ionization. { ek¦strin·sik ‚fōd·ō·i'mish·ən }

extrinsic properties |ELECTR| The properties of a semiconductor as modified by impurities or imperfections within the crystal. { ek¦strinz·ik 'präp·ərd·ēz }

extrinsic semiconductor |ELECTR| A semiconductor whose electrical properties are dependent on impurities added to the semiconductor crystal, in contrast to an intrinsic semiconductor, whose properties are characteristic of an ideal pure crystal. { ek¦strinz·ik 'sem·i·kən‚dək·tər }

extrudate |ENG| Ductile metal, plastic, or other semisoft solid material that has been shaped into a continuous form (such as fiber, film, pipe, or wire coating) by forcing the semisolid material through a die opening of appropriate shape. { 'ek·strə‚dāt }

extruder |ENG| A device that forces ductile or semisoft solids through die openings of appropriate shape to produce a continuous film, strip, or tubing { ed'strüd·ər }

extrusion |ENG| A process in which a hot or cold semisoft solid material, such as metal or plastic, is forced through the orifice of a die to produce a continuously formed piece in the shape of the desired product. { ek'strü·zhən }

extrusion coating |ENG| A process of placing resin on a substrate by extruding a thin film of molten resin and pressing it onto or into the substrates, or both, without the use of adhesives. { ek'strü·zhən ‚kōd·iŋ }

exudation See sweating. { ‚ek·syə'dā·shən }

eyebar |DES ENG| A metal bar having a hole or eye through each enlarged end. { 'ī‚bär }

eyebolt |DES ENG| A bolt with a loop at one end. { 'ī‚bōlt }

eyelet |DES ENG| A small ring or barrel-shaped piece of metal inserted into a hole for reinforcement. { 'ī·lət }

eyeleting |ENG| Forming a lip around the rim of a hole. { 'ī·ləd·iŋ }

eye scanning |IND ENG| Scanning of the visual field by moving the eyeballs without rotation of the head. { 'ī ‚skan·iŋ }

eye screw |DES ENG| A screw with an open loop head. { 'ī ‚skrü }

F

F *See* farad.

fabrication |ENG| **1.** The manufacture of parts, usually structural or electromechanical parts. **2.** The assembly of parts into a structure. { ,fab·ri'kā·shən }

face |CIV ENG| **1.** The surface of the area that has been excavated in constructing a tunnel. **2.** In building construction, the exposed surface of a wall, masonry unit, or sheet of material. **3.** To install a surface layer of one material over another, such as laying brick on a wall built of concrete blocks. |DES ENG| The surface of a flange on a pipe that is fitted against another flange. |ELECTR| *See* faceplate. { fās }

face-discharge bit |MECH ENG| A liquid-coolant bit designed for drilling in soft formations and for use on a double-tube core barrel, the inner tube of which fits snugly into a recess cut into the inside wall of the bit directly above the inside reaming stones; the coolant flows through the bit and is ejected at the cutting face. Also known as bottom-discharge bit; face-ejection bit. { ¦fās 'dis,chärj ,bit }

faced wall |BUILD| A wall whose masonry facing and backing are of different materials. { ¦fāst 'wȯl }

face-ejection bit *See* face-discharge bit. { ¦fās ē'jek·shən ,bit }

face gear |DES ENG| A gear having teeth cut on the face. { 'fās ,gir }

face milling |MECH ENG| Milling flat surfaces perpendicular to the rotational axis of the cutting tool. { 'fās ,mil·iŋ }

face mold |ENG| A pattern for cutting forms out of sheets of wood, metal, or other material. { 'fās ,mōld }

face nailing |ENG| Nailing of facing wood to a base, leaving the nailheads exposed. { 'fās ,nāl·iŋ }

faceplate |ELECTR| The transparent or semitransparent glass front of a cathode-ray tube; through which the image is viewed or projected; the inner surface of the face is coated with fluorescent chemicals that emit light when hit by an electron beam. Also known as face. |ENG| **1.** A disk fixed perpendicularly to the spindle of a lathe and used for attachment of the workpiece. **2.** A protective plate used to cover holes in machines or other devices. **3.** In scuba or skin diving, a glass or plastic window positioned over the face to provide an air space between the diver's eyes and the water. { 'fās,plāt }

face shield |ENG| A detachable wraparound guard fitted to a worker's helmet to protect the face from flying particles. { 'fās ,shēld }

facework |CIV ENG| Ornamental or otherwise special material on the front side or outside of a wall. { 'fās,wərk }

facing |CIV ENG| A covering or casting of some material applied to the outer face of embankments, buildings, and other structures. |MECH ENG| Machining the end of a flat rotating surface by applying a tool perpendicular to the axis of rotation in a spiral planar path. { 'fās·iŋ }

facing-point lock |CIV ENG| A lock used on a railroad track, such as a switch track, which contains a plunger that engages a rod on the switch point to lock the device. { 'fās·iŋ ,pȯint ,läk }

facing wall |CIV ENG| Concrete lining against the earth face of an excavation; used instead of timber sheeting. { 'fās·iŋ ,wȯl }

factor comparison |IND ENG| A quantitative system of job evaluation in which jobs are given relative positions on a rating scale based on a comparison of factors composing the job with certain previously selected key jobs. { 'fak·tər kəm,par·ə·sən }

factor of safety |MECH| **1.** The ratio between the breaking load on a member, appliance, or hoisting rope and the safe permissible load on it. Also known as safety factor. **2.** *See* factor of stress intensity. { 'fak·tər əv 'sāf·tē }

factor of stress concentration |MECH| Any irregularity producing localized stress in a structural member subject to load. Also known as fatigue-strength reduction factor. { 'fak·tər əv 'stres ,käns·ən,trā·shən }

factor of stress intensity |MECH| The ratio of the maximum stress to which a structural member can be subjected, to the maximum stress to which it is likely to be subjected. Also known as factor of safety. { 'fak·tər əv 'stres in,ten·səd·ē }

factory |IND ENG| A building or group of buildings where goods are manufactured. { 'fak·trē }

Fahrenheit scale |THERMO| A temperature scale; the temperature in degrees Fahrenheit (°F) is the sum of 32 plus 9/5 the temperature in degrees Celsius; water at 1 atmosphere (101,325 pascals) pressure freezes very near 32°F and boils very near 212°F. { 'far·ən,hīt ,skāl }

Fahrenheit's hydrometer |ENG| A type of hydrometer which carries a pan at its upper end in which weights are placed; the relative density of a liquid is measured by determining the weights necessary to sink the instrument to a fixed mark, first in water and then in the liquid being studied. { 'far·ən‚hīts hī'dräm·əd·ər }

failed hole |ENG| A drill hole loaded with dynamite which did not explode. Also known as missed hole. { 'fāld 'hōl }

fail-safe system |ENG| A system designed so that failure of power, control circuits, structural members, or other components will not endanger people operating the system or other people in the vicinity. { 'fāl ‚sāf ‚sis·təm }

fail soft |ENG| A failure in the performance of a system component that neither results in immediate or major interruption of the system operation as a whole nor adversely affects the quality of its products. { fāl ‚sóft }

failure |ENG| A permanent change in the volume of a powder or the stresses within it. |MECH| Condition caused by collapse, break, or bending, so that a structure or structural element can no longer fulfill its purpose. { 'fāl·yər }

failure properties |ENG| The parameters that control the degree of the failure of a powder. { 'fāl·yər ‚präp·ərd·ēz }

failure rate |ENG| The probability of failure per unit of time of items in operation; sometimes estimated as a ratio of the number of failures to the accumulated operating time for the items. { 'fāl·yər ‚rāt }

faired cable |DES ENG| A trawling cable covered by streamlined surfaces to reduce hydrodynamic drag. { ‚ferd 'kā·bəl }

fairlead |MECH ENG| A group of pulleys or rollers used in conjunction with a winch or similar apparatus to permit the cable to be reeled from any direction. { 'fer‚lēd }

Fales-Stuart windmill |MECH ENG| A windmill developed for farm use from the two-blade airfoil propeller. Also known as Stuart windmill. { ‚fālz ‚stü·ərt 'wind‚mil }

Falk flexible coupling |MECH ENG| A spring coupling in which a continuous steel spring is threaded back and forth through axial slots in the periphery of two hubs on the shaft ends. { ‚fók ‚flek·sə·bəl 'kəp·liŋ }

fall |ENG| The minimum slope that is required to facilitate proper drainage of liquid inside a pipe. |MECH ENG| The rope or chain of a hoisting tackle. { fól }

fall block |MECH ENG| A pulley block that rises and falls with the load on a lifting tackle. { 'fól ‚bläk }

faller |MECH ENG| A machine part whose operation depends on a falling action. { 'fól·ər }

falling-ball viscometer See falling-sphere viscometer. { 'fól· iŋ ‚ból vi'skäm·əd·ər }

falling body |MECH| A body whose motion is accelerated toward the center of the earth by the force of gravity, other forces acting on it being negligible by comparison. { 'fól·iŋ 'bäd·ē }

falling-film cooler |ENG| Liquid cooling system

in which the cooling liquid flows down vertical tube exterior surfaces in a thin film, and hot process fluid flows upward through the tubes. { 'fól·iŋ ‚film ‚kül·ər }

falling-film evaporator |ENG| Liquid evaporator system with heated vertical tubes; liquid to be evaporated flows down the inside tube surfaces as a film, evaporating as it flows. { 'fól· iŋ ‚film i'vap·ə‚rād·ər }

falling-film molecular still See falling-film still. { 'fól·iŋ ‚film mə¦lek·yə·lər 'stil }

falling-film still |CHEM ENG| Special molecular distillation apparatus designed for high evaporative and separation efficiency. Also known as falling-film molecular still. { 'fól·iŋ ‚film 'stil }

falling-sphere viscometer |ENG| A viscometer which measures the speed of a spherical body falling with constant velocity in the fluid whose viscosity is to be determined. Also known as falling-ball viscometer. { 'fól·iŋ ‚sfir vi'skäm· əd·ər }

fallout shelter |CIV ENG| A structure that affords some protection against fallout radiation and other effects of nuclear explosion; maximum protection is in reinforced concrete shelters below the ground. Also known as radiation shelter. { 'fól‚aút ‚shel·tər }

false attic |BUILD| A section under a roof normally occupied by an attic, but which has no windows and does not enclose rooms. { ¦fóls 'ad·ik }

false bottom |CIV ENG| A temporary bottom installed in a caisson to add to its buoyancy. { ¦fóls 'bäd·əm }

false header |CIV ENG| A half brick used to complete a visible bond; it is not a header. { ¦fóls 'hed·ər }

falsework |CIV ENG| A temporary support used until the main structure is strong enough to support itself. { 'fóls‚wərk }

family mold |ENG| A multicavity injection mold where each cavity forms a component part of the finished product. { 'fam· lē ‚mōld }

fan |MECH ENG| **1.** A device, usually consisting of a rotating paddle wheel or an airscrew, with or without a casing, for producing currents in order to circulate, exhaust, or deliver large volumes of air or gas. **2.** A vane to keep the sails of a windmill facing the direction of the wind. { fan }

fan brake |MECH ENG| A fan used to provide a load for a driving mechanism. { 'fan ‚brāk }

fan cut |ENG| A cut in which holes of equal or increasing length are drilled in a pattern on a horizontal plane or in a selected stratum to break out a considerable part of the plane or stratum before the rest of the round is fired. { 'fan ‚kət }

fan drilling |ENG| **1.** Drilling boreholes in different vertical and horizontal directions from a single-drill setup. **2.** A radial pattern of drill holes from a setup. { 'fan ‚dril·iŋ }

fan efficiency |MECH ENG| The ratio obtained by dividing a fan's useful power output by the power input (the power supplied to the fan shaft); it is expressed as a percentage. { 'fan ‚fish·ən·sē }

fang bolt [DES ENG] A bolt having a triangular nut with sharp projections at its corners; used to attach metal pieces to wood. { 'faŋ ‚bōlt }

fan rating [MECH ENG] The head, quantity, power, and efficiency expected from a fan operating at peak efficiency. { 'fan ‚rād·iŋ }

fan ring [DES ENG] Circular metallic collar encircling (but spaced away from) the tips of the fan blade in process equipment, such as air-cooled heat exchangers; ring design is critical to the efficiency of fan performance. { 'fan ‚riŋ }

fan shaft [DES ENG] The spindle on which a fan impeller is mounted. { 'fan‚shaft }

fan shooting [ENG] Seismic exploration in which seismometers are placed in a fan-shaped array to detect anomalies in refracted-wave arrival times indicative of circular rock structures such as salt domes. { 'fan ‚shüd·iŋ }

fan static pressure [MECH ENG] The total pressure rise diminished by the velocity pressure in the fan outlet. { 'fan ‚stad·ik ‚presh·ər }

fan test [MECH ENG] Observations of the quantity, total pressure, and power of air circulated by a fan running at a known constant speed. { 'fan‚test }

fan total head [MECH ENG] The sum of the fan static head and the velocity head at the fan discharge corresponding to a given quantity of airflow. { 'fan ‚tōd·əl ‚hed }

fan total pressure [MECH ENG] The algebraic difference between the mean total pressure at the fan outlet and the mean total pressure at the fan inlet. { 'fan ‚tōd·əl ‚presh·ər }

fan truss [CIV ENG] A truss with struts arranged as radiating lines. { 'fan ‚trəs }

fan velocity pressure [MECH ENG] The velocity pressure corresponding to the average velocity at the fan outlet. { 'fan və'läs·əd·ē ‚presh·ər }

farad [ELEC] The unit of capacitance in the meter-kilogram-second system, equal to the capacitance of a capacitor which has a potential difference of 1 volt between its plates when the charge on one of its plates is 1 coulomb, there being an equal and opposite charge on the other plate. Symbolized F. { 'fa‚rad }

Faraday cage See Faraday shield. { 'far·ə‚dā ‚kāj }

Faraday cylinder [ELEC] **1.** A closed, or nearly closed, hollow conductor, usually grounded, within which apparatus is placed to shield it from electrical fields. **2.** A nearly closed, insulated, hollow conductor, usually shielded by a second grounded cylinder, used to collect and detect a beam of charged particles. { 'far·ə‚dā ‚sil·ən·dər }

Faraday screen See Faraday shield. { 'far·ə‚dā ‚skrēn }

Faraday shield [ELEC] Electrostatic shield composed of wire mesh or a series of parallel wires, usually connected at one end to another conductor which is grounded. Also known as Faraday cage; Faraday screen. { 'far·ə‚dā ‚shēld }

Faraday tube [ELEC] A tube of force for electric

displacement which is of such size that the integral over any surface across the tube of the component of electric displacement perpendicular to that surface is unity. { 'far·ə‚dā ‚tüb }

faradic current [ELEC] An intermittent and nonsymmetrical alternating current like that obtained from the secondary winding of an induction coil. Also spelled faradaic current. { fə'rad·ik ‚kə·rənt }

far-infrared maser [ENG] A gas maser that generates a beam having a wavelength well above 100 micrometers, and ranging up to the present lower wavelength limit of about 500 micrometers for microwave oscillators. { ‚fär in·frə'red 'mā·zər }

fascia [BUILD] A wide board fixed vertically on edge to the rafter ends or wall which carries the gutter around the eaves of a roof. { 'fā·shə }

fascine [CIV ENG] A cylindrical bundle of brushwood 1–3 feet (30–90 centimeters) in diameter and 10–20 feet (3–6 meters) long, used as a facing for seawalls on riverbanks, as a foundation mat, as a dam in an estuary, or to protect bridge, dike, and pier foundations from erosion. { fa'sēn }

fast coupling [MECH ENG] A flexible geared coupling that uses two interior hubs on the shafts with circumferential gear teeth surrounded by a casing having internal gear teeth to mesh and connect the two hubs. { ‚fast 'kəp·liŋ }

fast-delay detonation [ENG] The firing of blasts by means of a blasting timer or millisecond delay caps. { ‚fast di‚lā det·ən'ā·shən }

fastener [DES ENG] **1.** A device for joining two separate parts of an article or structure. **2.** A device for holding closed a door, gate, or similar structure. { 'fas·nər }

fastening [DES ENG] A spike, bolt, nut, or other device to connect rails to ties. { ‚fas·niŋ }

fast-joint [ENG] Pertaining to a joint with a permanently secured pin. { ‚fast ‚jóint }

fast pin [ENG] A pin that fastens immovably, particularly the pin in a fast joint. { ‚fast ‚pin }

fast-spiral drill See high-helix drill. { ‚fast ‚spī·rəl 'dril }

fatigue [ELECTR] The decrease of efficiency of a luminescent or light-sensitive material as a result of excitation. [MECH] Failure of a material by cracking resulting from repeated or cyclic stress. { fə'tēg }

fatigue allowance [IND ENG] An adjustment to normal time to compensate for production time lost due to exhaustion of the worker. { fə'tēg ə‚laů·əns }

fatigue factor [IND ENG] The element of physical and mental exhaustion in a time-motion study; the multiplier used to add the fatigue allowance to the normal time. { fə'tēg ‚fak·tər }

fatigue life [MECH] The number of applied repeated stress cycles a material can endure before failure. { fə'tēg ‚līf }

fatigue limit [MECH] The maximum stress that a material can endure for an infinite number of stress cycles without breaking. Also known as endurance limit. { fə'tēg ,lim·ət }

fatigue ratio |MECH| The ratio of the fatigue limit or fatigue strength to the static tensile strength. Also known as endurance ratio. { fə'tēg ,rā·shō }

fatigue strength |MECH| The maximum stress a material can endure for a given number of stress cycles without breaking. Also known as endurance strength. { fə'tēg ,streŋkth }

fatigue-strength reduction factor See factor of stress concentration. { fə'tēg ,streŋkth ri'dək·shən ,fak·tər }

fatigue test |ENG| Test to determine the range of alternating stress which a material can withstand without breaking. { fə'tēg ,test }

faucet |ENG| A fixture through which water is drawn from a pipe or vessel. { 'fȯs·ət }

Faugeron kiln |ENG| A coal-fired tunnel kiln for firing feldspathic porcelain; the distinctive feature is the separation of the tunnel into a series of chambers by division walls on the cars and drop arches in the roof. { 'fō·zhə,rän ,kil }

fault |ELEC| A defect, such as an open circuit, short circuit, or ground, in a circuit, component, or line. Also known as electrical fault; faulting. |ELECTR| Any physical condition that causes a component of a data-processing system to fail in performance. { fȯlt }

fault analysis |ENG| The detection and diagnosis of malfunctions in technical systems, in particular, by means of a scheme in which one or more computers monitor the technical equipment to signal any malfunction and designate the components responsible for it. { fȯlt ə,nal·ə·səs }

fault finder |ENG| Test set for locating trouble conditions in communications circuits or systems. { 'fȯlt ,fīnd·ər }

faulting See fault. { 'fȯl·tiŋ }

fault monitoring |SYS ENG| A procedure for systematically checking for errors and malfunctions in the software and hardware of a computer or control system. { 'fȯlt ,män·ə·triŋ }

fault tolerance |SYS ENG| The capability of a system to perform in accordance with design specifications even when undesired changes in the internal structure or external environment occur. { 'fȯlt ,täl·ə·rəns }

fault tree |IND ENG| A graphical representation of an undesired event caused by a combination of factors arising from equipment failure, human error, or environmental events. { 'fȯlt ,trē }

Faxen drag factor See drag factor. { 'fäk·sən 'drag ,fak·tər }

faying surface |ENG| The surfaces of materials in contact with each other and joined or about to be joined together. { 'fā·iŋ ,sər·fəs }

feasibility study |SYS ENG| **1.** A study of applicability or desirability of any management or procedural system from the standpoint of advantages versus disadvantages in any given case. **2.** A study to determine the time at which it would be practicable or desirable to install such a system when determined to be advantageous.

3. A study to determine whether a plan is capable of being accomplished successfully. { ,fēz·ə'bil·əd·ē ,stəd·ē }

feasibility test |SYS ENG| A test conducted to obtain data in support of a feasibility study or to demonstrate feasibility. { ,fēz·ə'bil·əd·ē ,test }

feasible method See interaction prediction method. { 'fēz·ə·bəl 'meth·əd }

feather |MECH ENG| To change the pitch on a propeller in order to reduce drag and prevent windmilling in case of engine failure. { 'feth·ər }

featheredge |CIV ENG| The thin edge of a gravel-surfaced road. |DES ENG| A wood tool with a level edge used to straighten angles in the finish coat of plaster. { 'feth·ər,ej }

feathering |MECH ENG| A pitch position in a controllable-pitch propeller; it is used in the event of engine failure to stop the windmilling action, and occurs when the blade angle is about 90° to the plane of rotation. Also known as full feathering. { 'feth·ə·riŋ }

feathering propeller |MECH ENG| A variable-pitch marine or airscrew propeller capable of increasing pitch beyond the normal high pitch value to the feathered position. { 'feth·ə·riŋ prə'pel·ər }

feather joint |ENG| A joint made by cutting a mating groove in each of the pieces to be joined and inserting a feather in the opening formed when the pieces are butted together. Also known as ploughed-and-tongued joint. { 'feth·ər ,jȯint }

feed |ELECTR| To supply a signal to the input of a circuit, transmission line, or antenna. |ENG| **1.** Process or act of supplying material to a processing unit for treatment. **2.** The material supplied to a processing unit for treatment. **3.** A device that moves stock or workpieces to, in, or from a die. |MECH ENG| Forward motion imparted to the cutters or drills of cutting or drilling machinery. { fēd }

feedback |ELECTR| The return of a portion of the output of a circuit or device to its input. { 'fēd,bak }

feedback branch |CONT SYS| A branch in a signal-flow graph that belongs to a feedback loop. { 'fēd,bak ,branch }

feedback circuit |ELECTR| A circuit that returns a portion of the output signal of an electronic circuit or control system to the input of the circuit or system. { 'fēd,bak ,sər·kət }

feedback compensation |CONT SYS| Improvement of the response of a feedback control system by placing a compensator in the feedback path, in contrast to cascade compensation. Also known as parallel compensation. { 'fēd ,bak ,käm·pən,sā·shən }

feedback control loop See feedback loop. { 'fēd ,bak kən'trōl ,lüp }

feedback control signal |CONT SYS| The portion of an output signal which is retransmitted as an input signal. { 'fēd,bak kən'trōl ,sig·nəl }

feedback control system |CONT SYS| A system in which the value of some output quantity is controlled by feeding back the value of the controlled quantity and using it to manipulate an input quantity so as to bring the value of the controlled quantity closer to a desired value. Also known as closed-loop control system. { 'fēd,bak kən'trōl ,sis·təm }

feedback loop [CONT SYS] A closed transmission path or loop that includes an active transducer and consists of a forward path, a feedback path, and one or more mixing points arranged to maintain a prescribed relationship between the loop input signal and the loop output signal. Also known as feedback control loop. { 'fēd ,bak ,lüp }

feedback regulator [CONT SYS] A feedback control system that tends to maintain a prescribed relationship between certain system signals and other predetermined quantities. { 'fēd,bak ,reg·yə,lād·ər }

feedback transfer function [CONT SYS] In a feedback control loop, the transfer function of the feedback path. { 'fēd,bak 'tranz·fər ,faŋk·shən }

feed-control valve [MECH ENG] A small valve, usually a needle valve, on the outlet of the hydraulic-feed cylinder on the swivel head of a diamond drill, used to control minutely the speed of the hydraulic piston travel and hence the rate at which the bit is made to penetrate the rock. { 'fēd kən,trōl ,valv }

feeder [ELEC] **1.** A transmission line used between a transmitter and an antenna. **2.** A conductor, or several conductors, connecting generating stations, substations, or feeding points in an electric power distribution system. **3.** A group of conductors in an interior wiring system which link a main distribution center with secondary or branch-circuit distribution centers. [MECH ENG] **1.** A conveyor adapted to control the rate of delivery of bulk materials, packages, or objects, or a control device which separates or assembles objects. **2.** A device for delivering materials to a processing unit. { 'fēd·ər }

feeder-breaker [MECH ENG] A unit that breaks and feeds ore or crushed rock to a materials-handling system at a required rate. { 'fēd·ər ¦brāk·ər }

feeder canal [CIV ENG] A canal serving to conduct water to a larger canal. { 'fēd·ər kə,nal }

feeder conveyor [MECH ENG] A short auxiliary conveyor designed to transport materials to another conveyor. Also known as stage loader. { 'fēd·ər kən,vā·ər }

feeder road [CIV ENG] A road that feeds traffic to a more important road. { 'fēd·ər ,rōd }

feedforward control [CONT SYS] Process control in which changes are detected at the process input and an anticipating correction signal is applied before process output is affected. { ¦fēd¦fór·wərd kən,trōl }

feeding zone [CONT SYS] The area on the planar surface of a conveyor or pallet where the center of an object to be manipulated by a robotic system is placed. { 'fēd·iŋ ,zōn }

feed nut [MECH ENG] The threaded sleeve fitting around the feed screw on a gear-feed drill swivel head, which is rotated by means of paired gears driven from the spindle or feed shaft. { 'fēd ,nət }

feed off [ENG] To lower the bit continuously or intermittently during a drilling operation by disengaging the drum brake. { ¦fēd 'óf }

feed pipe [MECH ENG] The pipe which conducts water to a boiler drum. { 'fēd ,pīp }

feed pitch [DES ENG] The distance between the centers of adjacent feed holes in punched paper tape. { 'fēd ,pich }

feed preparation unit [CHEM ENG] A processing unit (such as distillation or desulfurization units) providing feedstock for subsequent processing. { ¦fēd prep·ə'rā·shən ,yü·nət }

feed pressure [MECH ENG] Total weight or pressure, expressed in pounds or tons, applied to the drilling stem to make the drill bit cut and penetrate the geologic, rock, or ore formation. { 'fēd ,presh·ər }

feed pump [MECH ENG] A pump used to supply water to a steam boiler. { 'fēd ,pəmp }

feed rate *See* cutting speed. { 'fēd ,rāt }

feed ratio [MECH ENG] The number of revolutions a drill stem and bit must turn to advance the drill bit 1 inch when the stem is attached to and rotated by a screw- or gear-feed type of drill swivel head with a particular pair of the set of gears engaged. Also known as feed speed. { 'fēd ,rā·shō }

feed reel [ENG] The reel from which paper tape or magnetic tape is being fed. { 'fēd ,rēl }

feed screw [MECH ENG] The externally threaded drill-rod drive rod in a screw- or gear-feed swivel head on a diamond drill; also used on percussion drills, lathes, and other machinery. { 'fēd ,skrü }

feed shaft [MECH ENG] A short shaft or countershaft in a diamond-drill gear-feed swivel head which is rotated by the drill motor through gears or a fractional drive and by means of which the engaged pair of feed gears is driven. { 'fēd ,shaft }

feed speed *See* feed ratio. { 'fēd ,spēd }

feedstock [ENG] The raw material furnished to a machine or process. { 'fēd,stäk }

feed tank [ENG] A chamber that contains feedstock. { 'fēd ,taŋk }

feed travel [MECH ENG] The distance a drilling machine moves the steel shank in traveling from top to bottom of its feeding range. { 'fēd ,trav·əl }

feed tray [CHEM ENG] For a tray-type distillation column, that tray on which fresh feedstock is introduced into the system. { 'fēd ,trā }

feed trough [MECH ENG] A receptacle into which feedwater overflows from a boiler drum. { 'fēd ,tróf }

feedwater [MECH ENG] The water supplied to a boiler or still. { 'fēd,wód·ər }

feedwater heater [MECH ENG] An apparatus that utilizes steam extracted from an engine or turbine to heat boiler feedwater. { 'fēd,wȯd·ər ,hēd·ər }

feeler gage |MECH ENG| A tool with many blades of different thickness used to establish clearance between parts or for gapping spark plugs. { 'fēl·ər ,gāj }

feeler pin |MECH ENG| A pin that allows a duplicating machine to operate only when there is a supply of paper. { 'fēl·ər ,pin }

Fell system |CIV ENG| A method of traction intended for steep railroad slopes; a central rail is gripped between horizontal wheels on the locomotive. { 'fel ,sis·təm }

female connector |ELEC| A connector having one or more contacts set into recessed openings; jacks, sockets, and wall outlets are examples. { ¦fē,māl kə'nek·tər }

female fitting |DES ENG| In a paired pipe or an electrical or mechanical connection, the portion (fitting) that receives, contrasted to the male portion (fitting) that inserts. { ¦fē,māl 'fid·iŋ }

femitrons |ELECTR| Class of field-emission microwave devices. { 'fem·ə,tränz }

femtometer |MECH| A unit of length, equal to 10^{-15} meter; used particularly in measuring nuclear distances. Abbreviated fm. Also known as fermi. { 'fem·tō,mēd·ər }

fence |ENG| **1.** A line of data-acquisition or tracking stations used to monitor orbiting satellites. **2.** A line of radar or radio stations for detection of satellites or other objects in orbit. **3.** A line or network of early-warning radar stations. **4.** A concentric steel fence erected around a ground radar transmitting antenna to serve as an artificial horizon and suppress ground clutter that would otherwise drown out weak signals returning at a low angle from a target. **5.** An adjustable guide on a tool. { fens }

fender |CIV ENG| A timber, cluster of piles, or bag of rope placed along dock or bridge pier to prevent damage by docking ships or floating objects. |ENG| A cover over the upper part of a wheel of an automobile or other vehicle. { 'fen·der }

Fenske equation *See* Fenske-Underwood equation. { 'fen·skē i,kwā·zhən }

Fenske-Underwood equation |CHEM ENG| Equation in plate-to-plate distillation-column calculations relating the number of theoretical plates needed at total reflux to overall relative volatility and the liquid-vapor composition ratios on upper and lower plates. Also known as Fenske equation. { ¦fen·skē 'ən·dər,wu̇d i,kwā·zhən }

fermi *See* femtometer. { 'fer·mē }

ferrite device |ELEC| An electrical device whose principle of operation is based upon the use of ferrites in powdered, compressed, sintered form, making use of their ferrimagnetism and their high electrical resistivity, which makes eddy-current losses extremely low at high frequencies. { 'fe,rīt di,vīs }

ferrocyanide process |CHEM ENG| A regenerative chemical treatment for removal of mercaptans from petroleum fuels; uses caustic-sodium ferrocyanide reagent. { fe·rō'sī·ə,nīd ,präs·əs }

ferroelectric converter |ELEC| A converter that transforms thermal energy into electric energy by utilizing the change in the dielectric constant of a ferroelectric material when heated beyond its Curie temperature. { ¦fe·rō·i'lek·trik kən 'vərd·ər }

ferroelectric hysteresis |ELEC| The dependence of the polarization of ferroelectric materials not only on the applied electric field but also on their previous history; analogous to magnetic hysteresis in ferromagnetic materials. Also known as dielectric hysteresis; electric hysteresis. { fe·rō·i'lek·trik ,his·tə'rē·səs }

ferroelectric hysteresis loop |ELEC| Graph of polarization or electric displacement versus applied electric field of a material displaying ferroelectric hysteresis. { ¦fe·rō·i'lek·trik ,his·tə'rē· səs ,lüp }

ferrograph analyzer |ENG| An instrument used for ferrography; a pump delivers a small sample of the fluid to a microscope slide mounted above a magnet that generates a high-gradient magnetic field, causing particles to be deposited in a gradient of sizes along the slide. { 'fer·ə,graf 'an·ə,līz·ər }

ferrography |ENG| Wear analysis of machine bearing surfaces by collection of ferrous (or nonferrous) wear particles from lubricating oil in a ferrograph analyzer; the method can be applied to human joints by collecting fragments of cartilage, bone, or prosthetic materials from synovial fluid. { fe'räg· rə·fē }

ferromagnetics |ELECTR| The science that deals with the storage of binary information and the logical control of pulse sequences through the utilization of the magnetic polarization properties of materials. { ¦fe·rō·mag¦ned·iks }

ferrometer |ENG| An instrument used to make permeability and hysteresis tests of iron and steel. { fə'räm·əd·ər }

ferrule |DES ENG| **1.** A metal ring or cap attached to the end of a tool handle, post, or other device to strengthen and protect it. **2.** A bushing inserted in the end of a boiler flue to spread and tighten it. *See* stabilizer. { 'fer·əl }

FET *See* field-effect transistor.

fiber gyro *See* fiber-optic gyroscope. { 'fī·bər 'jī· rō }

fiber-optic current sensor |ENG| An instrument for measuring currents on high-voltage lines, in which the magnetic field associated with the current changes the phase of light traveling through an optical fiber, and the phase change is measured in an interferometer. { 'fī·bər ¦äp·tik 'kə· rənt ,sen·sər }

fiber-optic gyroscope |ENG| An instrument for measuring rotation rate, in which light from a laser or light-emitting diode is split into two beams which travel in opposite directions around a coil of optical fiber and recombine to generate interference fringes whose shift is a measure of the rotation rate of the coil. Also known as fiber gyro; laser/fiber-optics gyroscope. { 'fī·bər ¦äp·tik 'jī·rə,skōp }

fiber-optic hydrophone See interferometric hydrophone. { 'fī·bər ¦äp·tik 'hī·drə,fōn }

fiber-optic magnetometer [ENG] A magnetometer in which the deformation of a magnetostrictive body in the field causes phase changes in light traveling through an optical fiber wrapped around the body, and these phase changes are measured in an interferometer. { 'fī·bər ¦äp·tik ‚mag·na'täm·əd·ər }

fiber-optic sensor See optical-fiber sensor. { 'fī·bər ¦äp·tik 'sen·sər }

fiber-optic thermometer [ENG] A thermometer in which light from a mercury lamp is guided along an optical fiber to excite a tiny fluorescent crystal, whose light is in turn guided back along the fiber to an evaluation unit where the crystal temperature is determined from the ratios of the strengths of spectral lines in the fluorescent light or from the decay time of the fluorescence. { 'fī·bər ¦äp·tik thər'mäm·əd·ər }

fiber stress [MECH] **1.** The tensile or compressive stress on the fibers of a fiber metal or other fibrous material, especially when fiber orientation is parallel with the neutral axis. **2.** Local stress through a small area (a point or line) on a section where the stress is not uniform, as in a beam under bending load. { 'fī·bər ‚stres }

fibrous fracture [MECH] Failure of a material resulting from a ductile crack; broken surfaces are dull and silky. Also known as ductile fracture. { 'fī·brəs 'frak·chər }

fiducial temperature [THERMO] Any of the temperatures assigned to a number of reproducible equilibrium states on the International Practical Temperature Scale; standard instruments are calibrated at these temperatures. { fə'dü·shəl 'tem·prə·chər }

field [ELEC] That part of an electric motor or generator which produces the magnetic flux which reacts with the armature, producing the desired machine action. [ELECTR] One of the equal parts into which a frame is divided in interlaced scanning for television; includes one complete scanning operation from top to bottom of the picture and back again. { fēld }

field effect [ELECTR] The local change from the normal value that an electric field produces in the charge-carrier concentration of a semiconductor. { 'fēld i‚fekt }

field-effect capacitor [ELECTR] A capacitor in which the effective dielectric is a region of semiconductor material that has been depleted or inverted by the field effect. { 'fēld i‚fekt kə'pas·əd·ər }

field-effect device [ELECTR] A semiconductor device whose properties are determined largely by the effect of an electric field on a region within the semiconductor. { 'fēld i‚fekt di‚vīs }

field-effect diode [ELECTR] A semiconductor diode in which the charge carriers are of only one polarity. { 'fēld i‚fekt 'dī‚ōd }

field-effect phototransistor [ELECTR] A field-effect transistor that responds to modulated light as the input signal. { 'fēld i‚fekt ¦fōd·ō·tran 'zis·tər }

field-effect tetrode [ELECTR] Four-terminal device consisting of two independently terminated semiconducting channels so displaced that the conductance of each is modulated along its length by the voltage conditions in the other { 'fēld i‚fekt 'te‚trōd }

field-effect transistor [ELECTR] A transistor in which the resistance of the current path from source to drain is modulated by applying a transverse electric field between grid or gate electrodes; the electric field varies the thickness of the depletion layer between the gates, thereby reducing the conductance. Abbreviated FET. { 'fēld i‚fekt tran'zis·tər }

field-effect-transistor resistor [ELECTR] A field-effect transistor in which the gate is generally tied to the drain; the resultant structure is used as a resistance load for another transistor. { 'fēld i‚fekt tran¦zis·tər ri¦zis·tər }

field-effect varistor [ELECTR] A passive, two-terminal, nonlinear semiconductor device that maintains constant current over a wide voltage range. { 'fēld i‚fekt və'ris·tər }

field engineer [ENG] **1.** An engineer who is in charge of directing civil, mechanical, and electrical engineering activities in the production and transmission of petroleum and natural gas. **2.** An engineer who operates at a construction site. { 'fēld en·jə‚nir }

field excitation [MECH ENG] Control of the speed of a series motor in an electric or diesel-electric locomotive by changing the relation between the armature current and the field strength, either through a reduction in field current by shunting the field coils with resistance, or through the use of field taps. { 'fēld ‚ek·sī'tā·shən }

field-strength meter [ENG] A calibrated radio receiver used to measure the field strength of radiated electromagnetic energy from a radio transmitter. { 'fēld ‚streŋkth ‚mēd·ər }

FIFO See first-in, first-out. { 'fī‚fō }

fifteen-degrees calorie See calorie. { ¦fif·tēn di ¦grēz ¦kal·ə·rē }

fifth wheel [MECH ENG] A coupling device in the form of two horizontal disks that rotate on each other positioned between a tractor and a semitrailer so that they can change direction independently. { ¦fifth ¦wēl }

figure of merit [ELECTR] A performance rating that governs the choice of a device for a particular application; for example, the figure of merit of a magnetic amplifier is the ratio of usable power gain to the control time constant. { 'fig·yər əv 'mer·ət }

filament [ELEC] Metallic wire or ribbon which is heated in an incandescent lamp to produce light, by passing an electric current through the filament. [ELECTR] A cathode made of resistance wire or ribbon, through which an electric current is sent to produce the high temperature required for emission of electrons in a thermionic tube. Also known as directly heated cathode; filamentary cathode; filament-type cathode. { 'fil·ə·mənt }

filamentary cathode See filament. { ,fil·ə'ment·ə·rē }

filament-type cathode See filament. { 'fil·ə·mənt ,tīp 'kath,ōd }

filament winding |ELECTR| The secondary winding of a power transformer that furnishes alternating-current heater or filament voltage for one or more electron tubes. |ENG| A process for fabricating a composite structure in which continuous fiber reinforcement (glass, boron, silicon carbide), either previously impregnated with a matrix material or impregnated during winding, are wound under tension over a rotating core. { 'fil·ə·mənt ,wīnd·iŋ }

filar micrometer |DES ENG| An instrument used to measure small distances in the field of an eyepiece by using two parallel wires, one of which is fixed while the other is moved at right angles to its length by means of an accurately cut screw. Also known as bifilar micrometer. { 'fī·lər mī'kräm·əd·ər }

file |DES ENG| A steel bar or rod with cutting teeth on its surface; used as a smoothing or forming tool. { fīl }

file hardness |ENG| Hardness of a material as determined by testing with a file of standardized hardness; a material which cannot be cut with the file is considered as hard as or harder than the file. { 'fīl ,härd·nəs }

fill |CIV ENG| Earth used for embankments or as backfill. { fil }

filled-system thermometer |ENG| A thermometer which has a bourdon tube connected by a capillary tube to a hollow bulb; the deformation of the bourdon tube depends on the pressure of a gas (usually nitrogen or helium) or on the volume of a liquid filling the system. Also known as filled thermometer. { ¦fild ¦sis·təm thər'mäm·əd·ər }

filled thermometer See filled-system thermometer. { ¦fild thər'mäm·əd·ər }

fillet |BUILD| A flat molding that separates rounded or angular moldings. |DES ENG| A concave transition surface between two otherwise intersecting surfaces. |ENG| **1.** Any narrow, flat metal or wood member. **2.** A corner piece at the juncture of perpendicular surfaces to lessen the danger of cracks, as in core boxes for castings. { 'fil·ət }

fillet gage |DES ENG| A gage for measuring convex or concave surfaces. { 'fil·ət ,gāj }

fill factor |MECH ENG| The approximate load that the dipper of a shovel is carrying, expressed as a percentage of the rated capacity. { 'fil ,fak·tər }

filling |ENG| The loading of trucks with any material. { 'fil·iŋ }

fill-up work See internal work. { 'fil,əp ,wərk }

film |ELEC| The layer adjacent to the valve metal in an electrochemical valve, in which is located the high voltage drop when current flows in the direction of high impedance. { film }

film analysis |IND ENG| A systematic detailed analysis of work from a motion picture film, usually derived from a memomotion study. { ¦film ə'nal·ə·səs }

film boiling |THERMO| Boiling in which a continuous film of vapor forms at the hot surface of the container holding the boiling liquid, reducing heat transfer across the surface. { 'film ,bȯil·iŋ }

film coefficient |THERMO| For a fluid confined in a vessel, the rate of flow of heat out of the fluid, per unit area of vessel wall divided by the difference between the temperature in the interior of the fluid and the temperature at the surface of the wall. Also known as convection coefficient. { 'film ,kō·i,fish·ənt }

film condensation |THERMO| The formation of a continuous film of liquid on a wall in contact with a vapor, when the wall is cooled below the local vapor saturation temperature and the liquid wets the cold surface. { 'film ,kän·dən,sā·shən }

film cooling |THERMO| The cooling of a body or surface, such as the inner surface of a rocket combustion chamber, by maintaining a thin fluid layer over the affected area. { 'film ,kül·iŋ }

film platen |ENG| A device which holds film in the focal plane during exposure. { 'film ,plat·ən }

film resistor |ELEC| A fixed resistor in which the resistance element is a thin layer of conductive material on an insulated form; the conductive material does not contain binders or insulating material. { 'film ri,zis·tər }

film transport |MECH ENG| **1.** The mechanism for moving photographic film through the region where light strikes it in recording film tracks or sound tracks of motion pictures. **2.** The mechanism which moves the film print past the area where light passes through it in reproduction of picture and sound. { 'film ¦tranz,pȯrt }

film vault |ENG| A place for safekeeping of film. { 'film ,vȯlt }

filter See compensator. |ELECTR| Any transmission network used in electrical systems for the selective enhancement of a given class of input signals. Also known as electric filter; electricwave filter. |ENG| A porous article or material for separating suspended particulate matter from liquids by passing the liquid through the pores in the filter and sieving out the solids. |ENG ACOUS| A device employed to reject sound in a particular range of frequencies while passing sound in another range of frequencies. Also known as acoustic filter. { 'fil·tər }

filterability |ENG| The adaptability of a liquid-solid system to filtration; system is not filterable if it is too viscous to be forced through a filter medium, or if the solids are too small to be stopped by the filter medium. { ,fil·trə'bil·əd·ē }

filter bed |CIV ENG| A fill of pervious soil that

provides a site for a septic field. |ENG| A contact bed used for filtering purposes. { 'fil·tər ,bed }

filter cake See mud cake. { 'fil·tər ,kāk }

filter-cake washing |CHEM ENG| An operation performed at the end of a filtration, in which residual liquid impurities are washed out of the cake by the flow of another liquid through the cake. { 'fil·tər ,kāk ,wash·iŋ }

filter capacitor |ELEC| A capacitor used in a power-supply filter system to provide a low-reactance path for alternating currents and thereby suppress ripple currents, without affecting direct currents. { 'fil·tər kə,pas·əd·ər }

filtered-particle testing |ENG| A penetrant method of nondestructive testing by which cracks in porous objects (100 mesh or smaller) are indicated: a fluid containing suspended particles is sprayed on a test object; if a crack exists, particles are filtered out and concentrate at the surface as liquid flows into the crack. { |fil·tərd ¦pärd·ə·kəl ,test·iŋ }

filtering |ENG| The process of interpreting reported information on movements of aircraft, ships, and submarines in order to determine their probable true tracks and, where applicable, heights or depths. { 'fil·tə·riŋ }

filter leaf |CHEM ENG| The frame or structure in a filter press that holds the filter cloth or other filter medium; a number of leaves in series usually comprises a filter press. { 'fil·tər ,lēf }

filter photometer |ENG| A colorimeter in which the length of light is selected by the use of appropriate glass filters. { 'fil·tər fə'täm·əd·ər }

filter press |ENG| A metal frame on which iron plates are suspended and pressed together by a screw device; liquid to be filtered is pumped into canvas bags between the plates, and the screw is tightened so that pressure is furnished for filtration. { 'fil·tər ,pres }

filter pump |MECH ENG| An aspirator or vacuum pump which creates a negative pressure on the filtrate side of the filter to hasten the process of filtering. { 'fil·tər ,pəmp }

filter screen |ENG| A fine-pored medium through which a liquid will pass and on which solids deposit; the medium may be a metal sieve screen or a woven fabric of metal or of natural or synthetic fibers. { 'fil·tər ,skrēn }

filter thickener |ENG| Device that thickens a liquid-solid mixture by removing a portion of the liquid by filtration, rather than by settling. { |fil·tər 'thik·ə·nər }

filter-type respirator |ENG| A protective device which removes dispersoids from the air by physically trapping the particles on the fibrous material of the filter. { 'fil·tər ,tīp 'res·pə,rād·ər }

fin |DES ENG| A projecting flat plate or structure, as a cooling fin. |ENG| Material which remains in the holes of a molded part and which must be removed. { fin }

final boiling point See end point. { |fīn·əl 'bȯil·iŋ ,pȯint }

final filter See afterfilter. { |fīn·əl 'fil·tər }

financial life See venture life. { fə'nan·chəl ,līf }

find |IND ENG| The therblig representing the mental reaction which occurs on recognizing an object at the end of the elemental motion search; now seldom used. { fīnd }

finding circuit See lockout circuit. { 'fīnd·iŋ ,sər·kət }

fineblanking |ENG| A manufacturing process in which a part is fabricated to a shape very close to its final dimensions by use of high-precision tools that yield a final workpiece with smoothly sheared edges. { 'fīn,blaŋk·iŋ }

fin efficiency |ENG| In extended-surface heat-exchange equations, the ratio of the mean temperature difference from surface-to-fluid divided by the temperature difference from fin-to-fluid at the base or root of the fin. { 'fin ə,fish·ən·sē }

fine grinding |MECH ENG| Grinding performed in a mill rotating on a horizontal axis in which the material undergoes final size reduction, to −100 mesh. { |fīn 'grīnd·iŋ }

fineness modulus |ENG| A number denoting the fineness of a fine aggregate or other fine material such as sand or paint. { 'fīn·nəs 'mäj·ə·ləs }

finger bit |DES ENG| A steel rock-cutting bit having fingerlike, fixed or replaceable steel-cutting points. { 'fiŋ·gər ,bit }

finger gripper |CONT SYS| A robot component that uses two or more joints for grasping objects. { 'fiŋ·gər ,grip·ər }

fining |CHEM ENG| A process in which molten glass is cleared of bubbles, usually by the addition of chemical agents. { 'fīn·iŋ }

finished goods |IND ENG| Manufactured products in inventory ready for packaging, shipment, or sale. { |fin·isht 'gu̇dz }

finisher |CIV ENG| A construction machine used to smooth the freshly placed surface of a roadway, or to prepare the foundation for a pavement. { 'fin·ish·ər }

finish grinding |MECH ENG| The last action of a grinding operation to achieve a good finish and accurate dimensions. { 'fin·ish ,grīnd·iŋ }

finishing hardware |BUILD| Items, such as hinges, door pulls, and strike plates, made in attractive shapes and finishes, and usually visible on the completed structure. { 'fin·ish·iŋ ,härd,wer }

finishing nail |DES ENG| A wire nail with a small head that can easily be concealed. { 'fin·ish·iŋ ,nāl }

finish plate |DES ENG| A plate which covers and protects the cylinder setscrews; it is fastened to the underplate and forms part of the armored front for a mortise lock. { 'fin·ish ,plāt }

finish turning |MECH ENG| The operation of machining a surface to accurate size and producing a smooth finish. { 'fin·ish ,tərn·iŋ }

finite elasticity theory See finite strain theory. { |fī,nīt i,las'tis·əd·ē ,thē·ə·rē }

finite element method |ENG| An approximation method for studying continuous physical systems, used in structural mechanics, electrical field theory, and fluid mechanics; the system is broken into discrete elements interconnected at discrete node points. { ¦fi,nīt 'el·ə·mənt ‚meth·əd }

finite strain theory |MECH| A theory of elasticity, appropriate for high compressions, in which it is not assumed that strains are infinitesimally small. Also known as finite elasticity theory. { 'fī,nīt 'strān ‚thē·ə·rē }

Fink truss |CIV ENG| A symmetrical steel roof truss suitable for spans up to 50 feet (15 meters). { 'fiŋk ‚trəs }

finned surface |MECH ENG| A tubular heat-exchange surface with extended projections on one side. { ¦find 'sər·fəs }

fire |ENG| To blast with gunpowder or other explosives. { fīr }

firebox |MECH ENG| The furnace of a locomotive or similar type of fire-tube boiler. { 'fīr‚bäks }

fire bridge |ENG| A low wall separating the hearth and the grate in a reverberatory furnace. { 'fīr ‚brij }

fire crack |ENG| A crack resulting from thermal stress which propagates on the heated side of a shell or header in a boiler or a heat transfer surface. { 'fīr ‚krak }

firecracker |ENG| A cylindrically shaped item containing an explosive and a fuse; used to simulate the noise of an explosive charge. { 'fīr‚krak·ər }

fire cut |BUILD| An angular cut made at the end of a joist which will rest on a brick wall. { 'fīr ‚kət }

firedamp reforming process |CHEM ENG| A process in which methane (firedamp) is mixed with steam and passed over a nickel catalyst for conversion to a mixture of hydrogen and carbon monoxide; this mixture is blended with pure methane, and the result is a fuel of high calorific value. { 'fīr‚damp ri'for·miŋ ‚präs·əs }

fire-danger meter |ENG| A graphical aid used in fire-weather forecasting to calculate the degree of forest-fire danger (or burning index): commonly in the form of a circular slide rule, it relates numerical indices of the seasonal stage of foliage, the cumulative effect of past precipitation or lack thereof (buildup index), the measured fuel moisture, and the speed of the wind in the woods; the fuel moisture is determined by weighing a special type of wooden stick that has been exposed in the woods, its weight being proportional to its contained water; the calculated burning index falls on a scale of 1 to 100: 1 to 11 is no fire danger; 12 to 35 medium danger; 40 to 100 high danger. { 'fīr ‚dān·jər ‚mēd·ər }

fire detector |ENG| A temperature-sensing device designed to sound an alarm, to turn on a sprinkler system, or to activate some other fire preventive measure at the first signs of fire. { 'fīr di‚tek·tər }

fire door |ENG| **1.** The door or opening through which fuel is supplied to a furnace or stove. **2.** A door that can be closed to prevent the spreading of fire, as through a building or mine. { 'fīr ‚dor }

fired process equipment |ENG| Heaters, furnaces, reactors, incinerators, vaporizers, steam generators, boilers, and other process equipment for which the heat input is derived from fuel combustion (flames); can be direct-fired (flame in contact with the process stream) or indirect-fired (flame separated from the process fluid by a metallic wall). { ¦fīrd 'präs·əs i‚kwip·mənt }

fire escape |BUILD| An outside stairway usually made of steel and used to escape from a building in case of fire. { 'fīr ə‚skāp }

fire-exit bolt See panic exit device. { 'fīr ‚eg·zət ‚bōlt }

fire extinguisher |ENG| Any of various portable devices used to extinguish a fire by the ejection of a fire-inhibiting substance, such as water, carbon dioxide, gas, or chemical foam. { 'fīr ik ‚stiŋ·gwish·ər }

firefinder |ENG| An instrument consisting of a map and a sighting device; used in fire towers to locate forest fires. { 'fīr‚fīn·dər }

fire hook |ENG| **1.** A pole with a hooked metal head that is used in fire fighting to tear down walls or ceilings. Also known as pike pole. **2.** A hook used to rake a furnace fire. { 'fīr ‚húk }

fire hose |ENG| A collapsible, flameproof hose that can be attached to a hydrant, standpipe, or similar outlet to supply water to extinguish a fire. { 'fīr ‚hōz }

fire hydrant |CIV ENG| An outlet from a water main provided inside buildings or outdoors to which fire hoses can be connected. Also known as fire plug; hydrant. { 'fīr ‚hī·drənt }

fire line |ENG| A pipework system dedicated to providing water for extinguishing fires. { 'fīr ‚līn }

fire load |CIV ENG| The load of combustible material per square foot of floor space. { 'fīr ‚lōd }

fire partition |BUILD| A wall inside a building intended to retard fire. { 'fīr pər‚tish·ən }

fire plug See fire hydrant. { 'fīr ‚pləg }

fireproof |BUILD| Having noncombustible walls, stairways, and stress-bearing members, and having all steel and iron structural members which could be damaged by heat protected by refractory materials. { 'fīr‚prüf }

fire protection |CIV ENG| Measures for reducing injury and property loss by fire. { 'fīr prə‚tek·shən }

fire pump |MECH ENG| A pump for fire protection purposes usually driven by an independent, reliable prime mover and approved by the National Board of Fire Underwriters. { 'fīr ‚pəmp }

fire-resistant |CIV ENG| Of a structural element, able to resist combustion for a specified time under conditions of standard heat intensity without burning or failing structurally. { 'fīr ri‚zis·tənt }

fireroom |MECH ENG| That portion of a fossil fuel-burning plant which contains the furnace and associated equipment. { 'fīr‚rüm }

fire sprinkling system See sprinkler system. { ¦fīr 'spriŋk·liŋ ‚sis·təm }

fire standpipe |CIV ENG| A high, vertical pipe

or tank that holds water to assure a positive, relatively uniform pressure, particularly to provide fire protection to upper floors of tall buildings. { ¦fīr 'stan͵pīp }

fire stop |BUILD| An incombustible, horizontal or vertical barrier, as of brick across a hollow wall or across an open room, to stop the spread of fire. Also known as draught stop. { 'fīr ͵stäp }

fire tower |BUILD| A fireproof and smokeproof stairway compartment running the height of a building. { 'fīr ͵taů·ər }

fire-tube boiler |MECH ENG| A steam boiler in which hot gaseous products of combustion pass through tubes surrounded by boiler water. { 'fīr ͵tüb ͵bȯil·ər }

fire wall |CIV ENG| 1. A fire-resisting wall separating two parts of a building from the lowest floor to several feet above the roof to prevent the spread of fire. 2. A fire-resisting wall surrounding an oil storage tank to retain oil that may escape and to confine fire. { 'fīr ͵wȯl }

firing |ELECTR| 1. The gas ionization that initiates current flow in a gas-discharge tube. 2. Excitation of a magnetron or transmit-receive tube by a pulse. 3. The transition from the unsaturated to the saturated state of a saturable reactor. |ENG| 1. The act or process of adding fuel and air to a furnace. 2. Igniting an explosive mixture. 3. Treating a ceramic product with heat. { 'fīr·iŋ }

firing machine |ENG| An electric blasting machine. |MECH ENG| A mechanical stoker used to feed coal to a boiler furnace. { 'fīr·iŋ mə͵shēn }

firing mechanism |ENG| A mechanism for firing a primer; the primer may be for initiating the propelling charge, in which case the firing mechanism forms a part of the weapon; if the primer is for the purpose of initiating detonation of the main charge, the firing mechanism is a part of the ammunition item and performs the function of a fuse. { 'fīr·iŋ ͵mek·ə͵niz·əm }

firing pressure |MECH ENG| The highest pressure in an engine cylinder during combustion. { 'fīr·iŋ ͵presh·ər }

firing rate |MECH ENG| The rate at which fuel feed to a burner occurs, in terms of volume, heat units, or weight per unit time. { 'fīr·iŋ ͵rāt }

firmer chisel |DES ENG| A small hand chisel with a flat blade; used in woodworking. { 'fər·mər ͵chiz·əl }

firm-joint caliper |DES ENG| An outside or inside caliper whose legs are jointed together at the top with a nut and which must be opened and closed by hand pressure. { 'fərm ͵jȯint 'kal·ə·pər }

firmoviscosity |MECH| Property of a substance in which the stress is equal to the sum of a term proportional to the substance's deformation, and a term proportional to its rate of deformation. { ¦fər·mō·vis¦käs·əd·ē }

first arrival |ENG| In exploration refraction seismology, the first seismic event recorded on a seismogram; it is noteworthy in that only first

arrivals are considered in this usage. { ¦fərst ə'rī·vəl }

first cost |IND ENG| The sum of the initial expenditures involved in capitalizing a property; includes items such as transportation, installation, preparation for service, as well as other related costs. { ¦fərst 'kȯst }

first fire |ENG| The igniter used with pyrotechnic devices, consisting of first fire composition, loaded in direct contact with the main pyrotechnic charge; the ignition of the igniter or first fire is generally accomplished by fuse action. { ¦fərst 'fīr }

first-in, first-out |IND ENG| An inventory cost evaluation method which transfers costs of material to the product in chronological order. Abbreviated FIFO. { ¦fərst 'in ¦fərst 'aůt }

first law of motion *See* Newton's first law. { 'fərst ͵lȯ əv 'mō·shən }

first law of thermodynamics |THERMO| The law that heat is a form of energy, and the total amount of energy of all kinds in an isolated system is constant; it is an application of the principle of conservation of energy. { 'fərst ͵lȯ əv ͵thər·mō·dī'nam·iks }

first-level controller |CONT SYS| A controller that is associated with one of the subsystems into which a large-scale control system is partitioned by plant decomposition, and acts to satisfy local objectives and constraints. Also known as local controller. { ¦fərst ¦lev·əl kən 'trōl·ər }

first-order leveling |ENG| Spirit leveling of high precision and accuracy in which lines are run first forward to the objective point and then backward to the starting point. { ¦fərst ͵ȯrd·ər 'lev· ə·liŋ }

first-order transition |THERMO| A change in state of aggregation of a system accompanied by a discontinuous change in enthalpy, entropy, and volume at a single temperature and pressure. { ¦fərst ͵ȯrd·ər trans'zish·ən }

Fischer-Tropsch process |CHEM ENG| A catalytic process to synthesize hydrocarbons and their oxygen derivatives by the controlled reaction of hydrogen and carbon monoxide. { ¦fish· ər ¦träpsh ͵präs·əs }

fished joint |CIV ENG| A structural joint made with fish plates. { 'fisht ͵jȯint }

fishing |ENG| In drilling, the operation by which lost or damaged tools are secured and brought to the surface from the bottom of a well or drill hole. { 'fish·iŋ }

fishing space |CIV ENG| The space between base and head of a rail in which a joint bar is placed. { 'fish·iŋ ͵spās }

fishing tool |ENG| A device for retrieving objects from inaccessible locations. { 'fish·iŋ ͵tül }

fish ladder |CIV ENG| Contrivance that carries water around a dam through a series of stepped baffles or boxes and thus facilitates the migration of fish. Also known as fishway. { 'fish ͵lad·ər }

fish lead |ENG| A type of sounding lead used

without removal from the water between soundings. { 'fish ,led }

fish plate [CIV ENG] One of a pair of steel plates bolted to the sides of a rail or beam joint, to secure the joint. { 'fish ,plāt }

fish screen [CIV ENG] **1.** A screen set across a water intake canal or pipe to prevent fish from entering. **2.** Any similar barrier to prevent fish from entering or leaving a pond. { 'fish ,skrēn }

fishtail bit [DES ENG] A drilling bit shaped like the tail of a fish. { 'fish,tāl ,bit }

fishtail burner [ENG] A burner in which two jets of gas impinge on each other to form a flame shaped like a fish's tail. { 'fish,tāl ,bərn·ər }

fishway See fish ladder. { 'fish,wā }

fit [DES ENG] The dimensional relationship between mating parts, such as press, shrink, or sliding fit. { fit }

fitment [BUILD] A decorative or functional item or component in a room that is fixed in place but not actually built in. Also known as fitting. { 'fit·mənt }

fitter [ENG] One who maintains, repairs, and assembles machines in an engineering shop. { 'fid·ər }

fitting [BUILD] See fitment. [ENG] A small auxiliary part of standard dimensions used in the assembly of an engine, piping system, machine, or other apparatus. { 'fid·iŋ }

five-fourths power law [THERMO] The proposition that the rate of heat loss from a body by free convection is proportional to the five-fourths power of the difference between the temperature of the body and that of its surroundings. { ¦fīv ¦fȯrths 'pau̇·ər ,lȯ }

fixed-active tooling [CONT SYS] Stationary equipment in a robotic system, such as numerical control equipment, sensors, cameras, conveying systems and parts feeders, that is activated and controlled by signals. { 'fikst ¦ak·tiv 'tül·iŋ }

fixed arch [CIV ENG] A stiff arch having rotation prevented at its supports. { ¦fikst 'ärch }

fixed-bed hydroforming [CHEM ENG] A cyclic petroleum process that utilizes a fixed bed of molybdenum oxide catalyst deposited on activated alumina. { 'fikst ,bed 'hī·drə,fȯr·miŋ }

fixed-bed operation [CHEM ENG] An operation in which the additive material (catalyst, absorbent, filter media, ion-exchange resin) remains stationary in the chemical reactor. { 'fikst ,bed ,äp·ə'rā·shən }

fixed bias [ELECTR] A constant value of bias voltage, independent of signal strength. { ¦fikst 'bī·əs }

fixed bridge [CIV ENG] A bridge having permanent horizontal or vertical alignment. { ¦fikst 'brij }

fixed capacitor [ELEC] A capacitor having a definite capacitance value that cannot be adjusted. { ¦fikst kə'pas·əd·ər }

fixed-charge problem [IND ENG] A linear programming problem in which each variable has a fixed-charge coefficient in addition to the usual cost coefficient; the fixed charge (for example, a

setup time charge) is a nonlinear function and is incurred only when the variable appears in the solution with a positive level. { ¦fikst 'chärj ,präb·ləm }

fixed cost [IND ENG] A cost that remains unchanged during short-term changes in production level. Also known as overhead; overhead cost. { ¦fikst 'kȯst }

fixed-electrode method [ENG] A geophysical surveying method used in a self-potential system of prospecting in which one electrode remains stationary while the other is grounded at progressively greater distances from it. { ¦fikst i'lek,trōd ,meth·əd }

fixed end [MECH] An end of a structure, such as a beam, that is clamped in place so that both its position and orientation are fixed. { 'fikst ,end }

fixed-end beam [CIV ENG] A beam that is supported at both free ends and is restrained against rotation and vertical movement. Also known as built-in beam; encastré beam. { 'fikst ,end 'bēm }

fixed-end column [CIV ENG] A column with the end fixed so that it cannot rotate. { 'fikst ,end 'käl·əm }

fixed end moment See fixing moment. { 'fikst ,end 'mō·mənt }

fixed-feed grinding [MECH ENG] Feeding processed material to a grinding wheel, or vice versa, in predetermined increments or at a given rate. { 'fikst ,fēd 'grind·iŋ }

fixed inductor [ELEC] An inductor whose coils are wound in such a manner that the turns remain fixed in position with respect to each other, and which either has no magnetic core or has a core whose air gap and position within the coil are fixed. { ¦fikst in'dək·tər }

fixed linkage system [IND ENG] Linkage formed between the skeletal elements of a human and a fixed machine in a human-machine system. { ¦fikst 'liŋk·ij ,sis·təm }

fixed mooring berth [CIV ENG] A marine structure consisting of dolphins for securing a ship and a platform to support cargo-handling equipment. { ¦fikst 'mu̇r·iŋ ,bərth }

fixed-needle traverse [ENG] In surveying, a traverse with a compass fitted with a sight line which can be moved above a graduated horizontal circle, so that the azimuth angle can be read, as with a theodolite. { ¦fikst ,nēd·əl trə'vərs }

fixed-passive tooling [CONT SYS] Unpowered, accessory equipment in a robotic system, such as jigs, fixtures, and work-holding devices. { 'fikst ¦pas·iv 'tül·iŋ }

fixed point [ENG] A reproducible value, as for temperature, used to standardize measurements; derived from intrinsic properties of pure substances. { ¦fikst 'pȯint }

fixed resistor [ELEC] A resistor that has no provision for varying its resistance value. { ¦fikst ri'zis·tər }

fixed-sequence robot See fixed-stop robot. { 'fikst ¦sē·kwəns 'rō,bät }

fixed sonar [ENG] Sonar in which the receiving

transducer is not constantly rotated, in contrast to scanning sonar. { ¦fikst 'sō,när }

fixed-stop robot |CONT SYS| A robot in which the motion along each axis has a fixed limit, but the motion between these limits is not controlled and the robot cannot stop except at these limits. Also known as fixed-squence robot; limited-sequence robot; nonservo robot. { 'fikst ¦stäp 'rō,bät }

fixing moment |MECH| The bending moment at the end support of a beam necessary to fix it and prevent rotation. Also known as fixed end moment. { 'fik·siŋ ,mō·mənt }

fixity See continuity. { 'fik·səd·ē }

fixture |CIV ENG| An object permanently attached to a structure, such as a light or sink. |MECH ENG| A device used to hold and position a piece of work without guiding the cutting tool. { 'fiks·chər }

flag |ELECTR| A small metal tab that holds the getter during assembly of an electron tube. |ENG| **1.** A piece of fabric used as a symbol or as a signaling or marking device. **2.** A large sheet of metal or fabric used to shield television camera lenses from light when not in use. { flag }

flag alarm |ENG| A semaphore-type flag in the indicator of an instrument to serve as a signal, usually to warn that the indications are unreliable. { 'flag ə,lärm }

flag float |ENG| A pyrotechnic device that floats and burns upon the water, used for marking or signaling. { 'flag ,flōt }

flagman |CIV ENG| A range-pole carrier in a surveying party. { 'flag·mən }

flagpole |ENG| A single staff or pole rising from the ground and on which flags or other signals are displayed; on charts the term is used only when the pole is not attached to a building. { 'flag,pōl }

flagstaff |ENG| A pole or staff on which flags or other signals are displayed; on charts this term is used only when the pole is attached to a building. { 'flag,staf }

flair |CIV ENG| A gradual widening of the flangeway near the end of a guard line of a track or rail structure. { fler }

flaking |CHEM ENG| Continuous process operation to remove heat from material in the liquid state to cause its solidification. |ENG| **1.** Reducing or separating into flakes. **2.** See frosting. { 'flāk·iŋ }

flaking mill |MECH ENG| A machine for converting material to flakes. { 'flāk·iŋ ,mil }

flak jacket |ENG| A jacket or vest of heavy fabric containing metal, nylon, or ceramic plates, designed especially for protection against flak; usually covers the chest, abdomen, back, and genitals, leaving the arms and legs free. Also known as flak vest. { 'flak ,jak·ət }

flak vest See flak jacket. { 'flak ,vest }

flame arrester |ENG| An assembly of screens, perforated plates, or metal-gauze packing attached to the breather vent on a flammable-product storage tank. { 'flām ə,res·tər }

flame collector |ENG| A device used in atmospheric electrical measurements for the removal of induction charge on apparatus; based upon the principle that products of combustion are ionized and will consequently conduct electricity from charged bodies. { 'flām kə,lek·tər }

flame detector |MECH ENG| A sensing device which indicates whether or not a fuel is burning, or if ignition has been lost, by transmitting a signal to a control system. { 'flām di,tek·tər }

flame plate |ENG| One of the plates on a boiler firebox which are subjected to the maximum furnace temperature. { 'flām ,plāt }

flameproofing |CHEM ENG| The process of treating materials chemically so that they will not support combustion. { 'flām,prüf·iŋ }

flame retardant |CHEM ENG| A substance that can suppress, reduce, or delay the propagation of a flame through a polymer material; may be inserted chemically into the polymer molecule or blended in after polymerization. { 'flām ri ,tärd·ənt }

flame spraying |ENG| **1.** A method of applying a plastic coating onto a surface in which finely powdered fragments of the plastic, together with suitable fluxes, are projected through a cone of flame. **2.** Deposition of a conductor on a board in molten form, generally through a metal mask or stencil, by means of a spray gun that feeds wire into a gas flame and drives the molten particles against the work. { 'flām ,sprā·iŋ }

flamethrower |ENG| A device used to project ignited fuel from a nozzle so as to cause casualties to personnel or to destroy material such as weeds or insects. { 'flām,thrō·ər }

flame trap |ENG| A device that prevents a gas flame from entering the supply pipe. { 'flām ,trap }

flame treating |ENG| A method of rendering inert thermoplastic objects receptive to inks, lacquers, paints, or adhesives, in which the object is bathed in an open flame to promote oxidation of the surface. { 'flām ,trēd·iŋ }

flanged pipe |DES ENG| A pipe with flanges at the ends; can be bolted end to end to another pipe. { ¦flanjd 'pīp }

flange union |ENG| A pair of flanges that are screwed to the ends of pipes and then bolted or welded together to hold two pipes together. { 'flanj ,yün·yən }

flangeway |CIV ENG| Open way through a rail or track structure that provides a passageway for the flange of a wheel. { 'flanj,wā }

flanging |ENG| A forming process in which the edge of a metal part is bent over to make a flange at a sharp angle to the body of the part. { 'flanj·iŋ }

flank |CIV ENG| The outer edge of a carriageway |DES ENG| **1.** The end surface of a cutting tool, adjacent to the cutting edge. **2.** The side of a screw thread. { flaŋk }

flank angle |DES ENG| The angle made by the flank of a screw thread with a line perpendicular to the axis of the screw. { 'flaŋk ,aŋ·gəl }

flank wear |ENG| Loss of relief on the flank of a tool behind the cutting edge. { 'flaŋk ,wer }

flap gate |CIV ENG| A gate that opens or closes by rotation around hinges at the top of the gate. Also known as pivot leaf gate. { 'flap ,gāt }

flap hinge See backflap hinge. { 'flap ,hiŋ }

flap trap |ENG| In plumbing, a trap fitted with a hinged flap that permits flow in one direction only, thus preventing backflow. { 'flap ,trap }

flap valve |MECH ENG| A valve fitted with a hinged flap or disk that swings in one direction only. { 'flap ,valv }

flare |CHEM ENG| A device for disposing of combustible gases from refining or chemical processes by burning in the open, in contrast to combustion in a furnace or closed vessel or chamber. |DES ENG| An expansion at the end of a cylindrical body, as at the base of a rocket. |ELECTR| A radar screen target indication having an enlarged and distorted shape due to excessive brightness. |ENG| A pyrotechnic item designed to produce a single source of intense light for such purposes as target or airfield illumination. { fler }

flare chute |ENG| A flare attached to a parachute. { 'fler ,shüt }

flare factor |ENG ACOUS| Number expressing the degree of outward curvature of the horn of a loudspeaker. { 'fler ,fak·tər }

flare gas |CHEM ENG| Surplus gas that is disposed of by combustion in the open. { 'fler ,gas }

flare-type burner |ENG| A circular burner which discharges flame in the form of a cone. { 'fler ,tīp ,bərn·ər }

flash |ENG| In plastics or rubber molding or in metal casting, that portion of the charge which overflows from the mold cavity at the joint line. { flash }

flashback See backfire. { 'flash,bak }

flashback arrester |ENG| A device which prevents a flashback from passing the point where the arrester is installed in a torch, thereby preventing damage. { 'flash,bak ə,res·tər }

flashboard |CIV ENG| A relatively low, temporary barrier constructed of a series of boards along the top of a dam spillway to increase storage capacity. { 'flash,bórd }

flash boiler |MECH ENG| A boiler with hot tubes of small capacity; designed to immediately convert small amounts of water to superheated steam. { 'flash ,bóil·ər }

flash bomb |ENG| A bomb that illuminates the ground for night aerial photography. { 'flash ,bäm }

flash carbonization |CHEM ENG| A carbonization process in which coal is subjected to a very brief residence time in the reactor in order to produce the largest possible yield of tar. { 'flash ,kär·bə·nə'zā·shən }

flash chamber |CHEM ENG| A conventional oil-and-gas separator operated at low pressure, with the liquid from a higher-pressure vessel being flashed into it. Also known as flash trap; flash vessel. { 'flash ,chām·bər }

flash distillation See equilibrium flash vaporization. { ¦flash ,dis·tə'lā·shən }

flash drum |CHEM ENG| A facility, such as a tower, which receives the products of a preheater or heat exchanger to release pressure; volatile components are vaporized and separated for further fractionation. { 'flash ,drəm }

flash dry |CHEM ENG| The rapid evaporation of moisture from a porous or granular solid by a sudden reduction in pressure or by placing the material in an updraft of warm air. { 'flash ,drī }

flash groove |ENG| **1.** A groove in a casting die so that excess material can escape during casting. **2.** See cutoff. { 'flash ,grüv }

flashing |BUILD| A strip of sheet metal placed at the junction of exterior building surfaces to render the joint watertight. |CHEM ENG| Vaporization of volatile liquids by either heat or vacuum. |ENG| Burning brick in an intermittent air supply in order to impart irregular color to the bricks. { 'flash·iŋ }

flashing block See raggle. { 'flash·iŋ ,bläk }

flashing flow |CHEM ENG| The condition when a liquid at its boiling point flows through a heated conduit and is further heated to cause partial vaporization (flashing), with a resultant two-phase (vapor-liquid) flow. { 'flash·iŋ ,flō }

flashing ring |ENG| A ring around a pipe that holds it in place as it passes through a partition such as a floor or wall. { 'flash·iŋ ,riŋ }

flash line |ENG| A raised line on the surface of a molding where the mold faces joined. { 'flash ,līn }

flash mold |ENG| A mold which permits excess material to escape during closing. { 'flash ,mōld }

flashover |ELEC| An electric discharge around or over the surface of an insulator. |ENG| A condition occurring during a fire in a building in which the surfaces of everything within a compartment or room seem to burst into flame simultaneously. { 'flash,ō·vər }

flash process |CHEM ENG| Liquid-vapor system in which the composition remains constant, but the proportion of gas and liquid phases changes as pressure or temperature change. { 'flash ,präs·əs }

flash ridge |ENG| The part of a flash mold along which the excess material escapes before the mold is closed. { 'flash ,rij }

flash separation |CHEM ENG| Process for separation of gas (vapor) from liquid components under reduced pressure; the liquid and gas remain in contact as the gas evolves from the liquid. { ¦flash ,sep·ə'rā·shən }

flash steam |ENG| A mixture of steam and water that occurs when hot water under pressure moves to a region of lower pressure, such as in a flash boiler. { 'flash ,stēm }

flash tank |CHEM ENG| In a processing operation, a unit that is used to separate the liquid and gas phases. { 'flash ,taŋk }

flash trap See flash chamber. { 'flash ,trap }

flash vaporization |CHEM ENG| Rapid vaporization achieved by passing a volatile liquid through continuously heated coils. [ENG] A method used for withdrawing liquefied petroleum gas from storage in which liquid is first flashed into a vapor in an intermediate pressure system, and then a second stage regulator provides the low pressure required to use the gas in appliances. { ¦flash vā·pə·rə'zā·shən }

flash vessel See flash chamber. { 'flash ¦ves·əl }

flat [ENG] A nonglossy painted surface. { flat }

flatbed plotter [ENG] A graphics output device that draws by moving a pen in both horizontal and vertical directions over a sheet of paper; the overall size of the drawing is limited by the height and width of this bed. { 'flat,bed 'pläd·ər }

flatbed truck [ENG] A truck whose body is in the form of a platform. { 'flat,bed 'trək }

flat belt [DES ENG] A power transmission belt, in the form of leather belting, used where high-speed motion rather than power is the main concern. { 'flat ,belt }

flat-belt conveyor [MECH ENG] A conveyor belt in which the carrying run is supported by flat-belt idlers or pulleys. { 'flat ,belt kən,vā·ər }

flat-belt pulley [DES ENG] A smooth, flat-faced pulley made of cast iron, fabricated steel, wood, and paper and used with a flat-belt drive. { 'flat ,belt ,pul·ē }

flat-blade turbine [MECH ENG] An impeller with flat blades attached to the margin. { 'flat ,blād 'tər,bīn }

flat-bottom crown See flat-face bit. { 'flat ,bäd·əm 'kraun }

flatcar [ENG] A railroad car without fixed walls or a cover. { 'flat,kär }

flat chisel [DES ENG] A steel chisel used to obtain a flat and finished surface. { ¦flat 'chiz·əl }

flat crank [DES ENG] A crankshaft having one flat bearing journal. { 'flat ,kraŋk }

flat-crested weir [CIV ENG] A type of measuring weir whose crest is in the horizontal plane and whose length is great compared with the height of water passing over it. { 'flat ,krest·əd 'wer }

flat drill [DES ENG] A type of rotary drill constructed from a flat piece of material. { 'flat ,dril }

flat edge trimmer [MECH ENG] A machine designed to trim the notched edges of metal shells. { 'flat ,ej 'trim·ər }

flat-face bit [DES ENG] A diamond core bit whose face in cross section is square. Also known as flat-bottom crown; flat-nose bit; square-nose bit. { 'flat ,fās ,bit }

flat-flamed burner [ENG] A burner which emits a mixture of fuel and air in a flat stream through a rectangular nozzle. { 'flat ,flamd 'bərn·ər }

flat form tool [DES ENG] A tool having a square or rectangular cross section with the form along the end. { 'flat ,form ,tül }

flathead rivet [DES ENG] A small rivet with a flat manufactured head used for general-purpose riveting. { 'flat,hed 'riv·ət }

flat jack [CIV ENG] A hollow steel cushion which is made of two nearly flat disks welded around the edge and which can be inflated with oil or cement under controlled pressure; used at the

arch abutments and crowns to relieve the load on the formwork at the moment of striking the formwork. { 'flat ¦jak }

flat-nose bit See flat-face bit. { 'flat ,nōz ,bit }

flatpack [ELECTR] Semiconductor network encapsulated in a thin, rectangular package, with the necessary connecting leads projecting from the edges of the unit. { 'flat,pak }

flat-panel display See panel display. { 'flat ¦pan·əl di'splā }

flat-plate collector [ENG] A solar collector consisting of a shallow metal box covered by a transparent lid. { 'flat ¦plāt kə'lek·tər }

flat rope [DES ENG] A steel or fiber rope having a flat cross section and composed of a number of loosely twisted ropes placed side by side, the lay of the adjacent strands being in opposite directions to secure uniformity in wear and to prevent twisting during winding. { 'flat ¦rōp }

flat slab [CIV ENG] A flat plate of reinforced concrete designed to span in two directions. { ¦flat ¦slab }

flat spin [MECH] Motion of a projectile with a slow spin and a very large angle of yaw, happening most frequently in fin-stabilized projectiles with some spin-producing moment, when the period of revolution of the projectile coincides with the period of its oscillation; sometimes observed in bombs and in unstable spinning projectiles. { ¦flat 'spin }

flat spring See leaf spring. { ¦flat ¦spriŋ }

flat trajectory [MECH] A trajectory which is relatively flat, that is, described by a projectile of relatively high velocity. { ¦flat trə'jek·trē }

flat-turret lathe [MECH ENG] A lathe with a low, flat turret on a power-fed cross-sliding headstock. { 'flat ,tə·rət 'lath }

flat yard [CIV ENG] A switchyard in which railroad cars are moved by locomotives, not by gravity. { 'flat ,yärd }

fl dr See fluid dram.

fleam [DES ENG] The angle of bevel of the edge of the teeth of a saw with respect to the plane of the blade. { flēm }

fleet [MECH ENG] Sidewise movement of a rope or cable when winding on a drum. { flēt }

fleet angle [MECH ENG] In hoisting gear, the included angle between the rope, in its position of greatest travel across the drum, and a line drawn perpendicular to the drum shaft, passing through the center of the head sheave or lead sheave groove. { 'flēt ,aŋ·gəl }

Fleming cracking process [CHEM ENG] An obsolete liquid-phase thermal cracking process for heavy petroleum fractions; the charge was heated under pressure in a vertical shell still. { 'flem·iŋ 'krak·iŋ ,präs·əs }

Flemish bond [CIV ENG] A masonry bond consisting of alternating stretchers and headers in each course, laid with broken joints. { 'flem·ish 'bänd }

Flemish garden wall bond [CIV ENG] A masonry bond consisting of headers and stretchers in the ratio of one to three or four in each course, with joints broken to give a variety of patterns. { ¦flem·ish 'gärd·ən ¦wól ,bänd }

Flesh-Demag process [CHEM ENG] A gas-making process in which a cyclic water-gas apparatus is used for feeding and charring the coal charge and for gas generation, with periodic automatic removal of the resultant ash. { ¦flesh 'da·mäk ,präs·əs }

fleshing machine [ENG] A machine that removes flesh from hides in a tannery. { 'flesh· iŋ mə,shēn }

Fletcher radial burner [ENG] A burner with gas jets arranged radially. { 'flech·ər ¦rād·ē·əl 'bərn·ər }

Flettner windmill [MECH ENG] An inefficient windmill with four arms, each consisting of a rotating cylinder actuated by a Savonius rotor. { 'flet·nər 'wind,mil }

flexibility [MECH] The quality or state of being able to be flexed or bent repeatedly. { ,flek· sə'bil·əd·ē }

flexible circuit [ELECTR] A printed circuit made on a flexible plastic sheet that is usually die-cut to fit between large components. { ,flek·sə·bəl 'sər·kət }

flexible coupling [MECH ENG] A coupling used to connect two shafts and to accommodate their misalignment. { ,flek·sə·bəl 'kəp·liŋ }

flexible-joint pipe [ENG] Cast-iron pipe adapted to laying under water and capable of motion through several degrees without leakage. { ,flek·sə·bəl ,jóint 'pīp }

flexible manufacturing system [IND ENG] A form of computer-integrated manufacturing used to make small to moderate-sized batches of parts. { 'flek·sə·bəl ,man·yə·'fak·chə·riŋ ,sis·təm }

flexible mold [ENG] A coating mold made of flexible rubber or other elastomeric materials; used mainly for casting plastics. { ,flek·sə·bəl 'mōld }

flexible pavement [CIV ENG] A road or runway made of bituminous material which has little tensile strength and is therefore flexible. { ,flek·sə·bəl 'pāv·mənt }

flexible shaft [MECH ENG] **1.** A shaft that transmits rotary motion at any angle up to about 90° **2.** A shaft made of flexible material or of segments. **3.** A shaft whose bearings are designed to accommodate a small amount of misalignment. { ,flek·sə·bəl 'shaft }

flexicoking [CHEM ENG] A continuous coke-making process that has a gasification section in which coke can be gasified to produce refinery fuel gas, allowing the production of both gas and coke in line with market requirements. { 'flek· sə,kōk·iŋ }

flexometer [ENG] An instrument for measuring the flexibility of materials. { flek'säm·əd·ər }

flexural modulus [MECH] A measure of the resistance of a beam of specified material and cross section to bending, equal to the product of Young's modulus for the material and the square of the radius of gyration of the beam

about its neutral axis. { 'flek·shə·rəl 'mäj·ə· ləs }

flexural rigidity [MECH] The ratio of the sideward force applied to one end of a beam to the resulting displacement of this end, when the other end is clamped. { 'flek·shə·rəl ri'jid·əd· ē }

flexural strength [MECH] Strength of a material in blending, that is, resistance to fracture. { 'flek·shə·rəl 'streŋkth }

flexure [MECH] **1.** The deformation of any beam subjected to a load. **2.** Any deformation of an elastic body in which the points originally lying on any straight line are displaced to form a plane curve. { 'flek·shər }

flexure theory [MECH] Theory of the deformation of a prismatic beam having a length at least 10 times its depth and consisting of a material obeying Hooke's law, in response to stresses within the elastic limit. { 'flek·shər ,thē·ə·rē }

flight [CIV ENG] A series of stairs between landings or floors. [MECH ENG] Plain or shaped plates that are attached to the propelling mechanism of a flight conveyor. { flīt }

flight conveyor [MECH ENG] A conveyor in which paddles, attached to single or double strands of chain, drag or push pulverized or granulated solid materials along a trough. Also known as drag conveyor. { 'flīt kən,vā·ər }

flight feeder [MECH ENG] Short-length flight conveyor used to feed solids materials to a process vessel or other receptacle at a preset rate. { 'flīt ,fēd·ər }

flight recorder [ENG] Any instrument or device that records information about the performance of an aircraft in flight or about conditions encountered in flight, for future study and evaluation. { 'flīt ri,kórd·ər }

flinching [IND ENG] In inspection, failure to call a borderline defect a defect. { 'flin·chiŋ }

flint mill [MECH ENG] A mill employing pebbles to pulverize materials (for example, in cement manufacture). { 'flint ,mil }

flip chip [ELECTR] A tiny semiconductor die having terminations all on one side in the form of solder pads or bump contacts; after the surface of the chip has been passivated or otherwise treated, it is flipped over for attaching to a matching substrate. Also known as solder-ball flip chip. { 'flip ,chip }

flip-flop circuit See bistable multivibrator. { 'flip ,fläp ,sər·kət }

FLIR imager See forward-looking infrared imager. { 'flir ,im·ij·ər }

flitch beam See flitch girder. { 'flich ,bēm }

flitch girder [BUILD] A beam made of structural timbers bolted together with a steel plate between them. Also known as flitch beam; sandwich beam. { 'flich ,gərd·ər }

flitch plate [CIV ENG] The metal plate in a flitch beam or girder. { 'flich ,plāt }

float [DES ENG] A file which has a single set of parallel teeth. [ENG] **1.** A flat, rectangular piece of wood with a handle, used to apply and smooth coats of plaster. **2.** A mechanical device to finish the surface of freshly placed concrete paving. **3.** A marble-polishing block. **4.** Any structure that provides positive buoyancy such as a hollow, watertight unit that floats or rests on the surface of a fluid. **5.** *See* plummet. | IND ENG| *See* bank. { flōt }

float barograph |ENG| A type of siphon barograph in which the mechanically magnified motion of a float resting on the lower mercury surface is used to record atmospheric pressure on a rotating drum. { 'flōt 'bar·ə,graf }

float bowl |MECH ENG| A component of a carburetor that holds a small amount of liquid gasoline and serves as a constant-level reservoir of fuel that is metered into the passing flow of air. { 'flōt ,bōl }

float chamber |ENG| A vessel in which a float regulates the level of a liquid. { 'flōt ,chām·bər }

float control |ENG| Floating device used to transmit a liquid-level reading to a control apparatus, such as an on-off switch controlling liquid flow into and out of a storage tank. { 'flōt kən,trōl }

float-cut file |DES ENG| A coarse file used on soft materials. { 'flōt ,kət ,fīl }

float finish |CIV ENG| A rough concrete finish, obtained by using a wooden float for finishing. { 'flōt ,fin·ish }

float gage |ENG| Any one of several types of instruments in which the level of a liquid is determined from the height of a body floating on its surface, by using pullies, levers, or other mechanical devices. { 'flōt ,gāj }

floating |ELECTR| The condition wherein a device or circuit is not grounded and not tied to an established voltage supply. { 'flōd·iŋ }

floating action |ENG| Controller action in which there is a predetermined relation between the deviation and the speed of a final control element; a neutral zone, in which no motion of the final control element occurs, is often used. { 'flōd·iŋ ,ak·shən }

floating axle |MECH ENG| A live axle used to turn the wheels of an automotive vehicle; the weight of the vehicle is borne by housings at the ends of a fixed axle. { 'flōd·iŋ 'ak·səl }

floating block *See* traveling block. { 'flōd·iŋ 'bläk }

floating chase |ENG| A mold part that can move freely in a vertical plane, which fits over a lower member (such as a cavity or plug) and into which an upper plug can telescope. { 'flōd·iŋ 'chās }

floating control |ENG| Control device in which the speed of correction of the control element (such as a piston in a hydraulic relay) is proportional to the error signal. Also known as proportional-speed control. { 'flōd·iŋ kən'trōl }

floating crane |CIV ENG| A crane having a barge or scow for an undercarriage and moved by cables attached to anchors set some distance off the corners of the barge; used for water work and for work on waterfronts. { 'flōd·iŋ 'krān }

floating dock |CIV ENG| **1.** A form of dry dock for repairing ships; it can be partly submerged by controlled flooding to receive a vessel, then

raised by pumping out the water so that the vessel's bottom can be exposed. Also known as floating dry dock. **2.** A barge or flatboat which is used as a wharf. { 'flōd·iŋ 'däk }

floating dry dock *See* floating dock. { 'flōd·iŋ 'drī ,däk }

floating floor |BUILD| A floor constructed so that the wearing surface is separated from the supporting structure by an insulating layer of mineral wool, resilient quilt, or other material to provide insulation against impact sound. { 'flōd·iŋ 'flōr }

floating foundation |CIV ENG| **1.** A reinforced concrete slab that distributes the concentrated load from columns; used on soft soil. **2.** A foundation mat several meters below the ground surface when it is combined with external walls. { 'flōd·iŋ faun'dā·shən }

floating lever |MECH ENG| A horizontal brake lever with a movable fulcrum; used under railroad cars. { 'flōd·iŋ 'lēv·ər }

floating pan |ENG| An evaporation pan in which the evaporation is measured from water in a pan floating in a larger body of water. { 'flōd·iŋ 'pan }

floating platen |ENG| In a multidaylight press, a platen that is between the main head and the press table and can be moved independently of them. { 'flōd·iŋ 'plat·ən }

floating roof |ENG| A type of tank roof (steel, plastic, sheet, or microballoons) which floats upon the surface of the stored liquid; used to decrease the vapor space and reduce the potential for evaporation. { 'flōd·iŋ 'rüf }

floating scraper |MECH ENG| A balanced scraper blade that rests lightly on a drum filter; removes solids collected on the rotating drum surface by riding on the drum's surface contour. { 'flōd·iŋ 'skrā·pər }

floatless level control |ENG| Any nonfloat device for measurement and control of liquid levels in storage tanks or process vessels; includes use of manometers, capacitances, electroprobes, nuclear radiation, and sonics. { 'flōt·ləs 'lev·əl kən,trōl }

float level |MECH ENG| The position of the float in a carburetor at which the needle valve closes the fuel inlet to prevent entry of additional fuel. { 'flōt ,lev·əl }

float switch |ENG| A switch actuated by a float at the surface of a liquid. { 'flōt ,swich }

float-type rain gage |ENG| A class of rain gage in which the level of the collected rainwater is measured by the position of a float resting on the surface of the water; frequently used as a recording rain gage by connecting the float through a linkage to a pen which records on a clock-driven chart. { 'flōt ,tīp 'rān ,gāj }

float valve |ENG| A valve whose on-off action is controlled directly by the fall or rise of a float concurrent with the fall or rise of liquid level in a liquid-containing vessel. { 'flōt ,valv }

flood |ELECTR| To direct a large-area flow of electrons toward a storage assembly in a charge storage tube. |ENG| To cover or fill with fluid.

|MECH ENG| To supply an excess of fuel to a carburetor so that the level rises above the nozzle. { fləd }

flood control |CIV ENG| Use of levees, walls, reservoirs, floodways, and other means to protect land from water overflow. { 'fləd kən,trōl }

flood dam |CIV ENG| A dam for storing floodwater, or for supplying a flood of water. { 'fləd ,dam }

flooded system |ENG| A system filled with so much tracer gas that probe testing for leaks suffers from a loss of sensitivity. { ¦fləd·əd 'sis·təm }

floodgate |CIV ENG| **1.** A gate used to restrain a flow or, when opened, to allow a flood flow to pass. **2.** The lower gate of a lock. { 'fləd,gāt }

flooding |CHEM ENG| Condition in a liquid-vapor counterflow device (such as a distillation column) in which the rate of vapor rise is such as to prevent liquid downflow, causing a buildup of the liquid (flooding) within the device. { 'fləd·iŋ }

flood relief channel See bypass channel. { 'fləd ri,lēf ,chan·əl }

flood wall |CIV ENG| A levee or similar wall for the purpose of protecting the land from inundation by flood waters. { 'fləd ,wòl }

floodway See bypass channel. { 'fləd,wā }

floor |ENG| The bottom, horizontal surface of an enclosed space. { flòr }

floor beam |BUILD| A beam used in the framing of floors in buildings. |CIV ENG| A large beam used in a bridge floor at right angles to the direction of the roadway, to transfer loads to bridge supports. { 'flòr ,bēm }

floor collar |ENG| A relatively narrow upright structural part fitted around the periphery of a hole where a pipe passes through to prevent drainage water from entering the hole. { 'flòr ,käl·ər }

floor drain |CIV ENG| A pipe or channel to remove water from under a floor in contact with soil. { 'flòr ,drān }

floor framing |BUILD| Floor joists together with their strutting and supports. { 'flòr ,frām·iŋ }

flooring saw |DES ENG| A pointed saw with teeth on both edges; cuts its own entrance into a material. { 'flòr·iŋ ,sò }

floor light |BUILD| A window set in a floor that is adapted for walking on and admitting light to areas below. { 'flòr ,līt }

floor plate |BUILD| A flat board on a floor used to support wall studs. |ENG| A plate in a floor to which heavy work or machine tools can be bolted. { 'flòr ,plāt }

floor system |CIV ENG| The structural floor assembly between supporting beams or girders in buildings and bridges. { 'flòr ,sis·təm }

flotation |ENG| A process used to separate particulate solids by causing one group of particles to float; utilizes differences in surface chemical properties of the particles, some of which are entirely wetted by water, others are not; the process is primarily applied to treatment of minerals but can be applied to chemical and biological materials; in mining engineering it is referred to as froth flotation. { flō'tā·shən }

flotation collar |ENG| A buoyant bag carried by a spacecraft and designed so that it inflates and surrounds part of the outer surface if the spacecraft lands in the sea. { flō'tā·shən ,käl·ər }

flotsam |ENG| Floating articles, particularly those that are thrown overboard to lighten a vessel in distress. { 'flät·səm }

flow |ENG| A forward movement in a continuous stream or sequence of fluids or discrete objects or materials, as in a continuous chemical process or solids-conveying or production-line operations. { flō }

flow analysis |IND ENG| A detailed study of all aspects of the progressive travel by personnel or material from place to place during a particular operation or from one operation to another. { 'flō ə,nal·ə·səs }

flow brush |ENG| A hollow tool for the continuous application of a broad coat of an adhesive. { 'flō brəsh }

flow chart |ENG| A graphical representation of the progress of a system for the definition, analysis, or solution of a data-processing or manufacturing problem in which symbols are used to represent operations, data or material flow, and equipment, and lines and arrows represent interrelationships among the components. Also known as control diagram; flow diagram; flow sheet. { 'flō ,chärt }

flow-chart symbol |ENG| Any of the existing symbols normally used to represent operations, data or materials flow, or equipment in a data-processing problem or manufacturing-process description. { 'flō ,chärt ,sim·bəl }

flow coat |ENG| A coating formed by pouring a liquid material over the object and allowing it to flow over the surface and drain off. { 'flō ,kōt }

flow coefficient |MECH ENG| A dimensionless number used in studying the power required by fans, equal to the volumetric flow rate through the fan divided by the product of the rate of rotation of the fan and the cube of the impeller diameter. { ¦flō ,kō·i'fish·ənt }

flow control |ENG| Any system used to control the flow of gases, vapors, liquids, slurries, pastes, or solid particles through or along conduits or channels. { 'flō kən,trōl }

flow control valve |ENG| A valve whose flow opening is controlled by the rate of flow of the fluid through it; usually controlled by differential pressure across an orifice at the valve. Also known as rate-of-flow control valve. { 'flō kən,trōl ,valv }

flow curve |MECH| The stress-strain curve of a plastic material. { 'flō ,kərv }

flow diagram See flow chart. { 'flō ,dī·ə,gram }

flow direction |ENG| The antecedent-to-successor relation, indicated by arrows or other conventions, between operations on a flow chart. { 'flō də,rek·shən }

flow graph See signal-low graph. { 'flō ,graf }

flowing-temperature factor [THERMO] Calculation correction factor for gases flowing at temperatures other than that for which a flow equation is valid, that is, other than 60°F (15.5°C). { ¦flō·iŋ 'tem·prə·chər ‚fak·tər }

flow line [ENG] **1.** The connecting line or arrow between symbols on a flow chart or block diagram. **2.** Mark on a molded plastic or metal article made by the meeting of two input-flow fronts during molding. Also known as weld line; weld mark. { 'flō ‚līn }

flow measurement [ENG] The determination of the quantity of a fluid, either a liquid, a vapor, or a gas, that passes through a pipe, duct, or open channel. { 'flō ‚mezh·ər·mənt }

flowmeter [ENG] An instrument used to measure pressure, flow rate, and discharge rate of a liquid, vapor, or gas flowing in a pipe. Also known as fluid meter. { 'flō‚mēd·ər }

flow mixer [MECH ENG] Liquid-liquid mixing device in which the mixing action occurs as the liquids pass through it; includes jet nozzles and agitator vanes. Also known as line mixer. { 'flō ‚mik·sər }

flow nozzle [ENG] A flowmeter in a closed conduit, consisting of a short flared nozzle of reduced diameter inset into the inner diameter of a pipe; used to cause a temporary pressure drop in flowing fluid to determine flow rate via measurement of static pressures before and after the nozzle. { 'flō ‚näz·əl }

flow process [ENG] System in which fluids or solids are handled in continuous movement during chemical or physical processing or manufacturing. { 'flō ‚präs·əs }

flow-rating pressure [MECH ENG] The value of inlet static pressure at which the relieving capacity of a pressure-relief device is established. { 'flō ‚rād·iŋ ‚presh·ər }

flow reactor [CHEM ENG] A dynamic reactor system in which reactants flow continuously into the vessel and products are continuously removed, in contrast to a batch reactor. { 'flō rē‚ak·tər }

flow sheet See flow chart. { 'flō ‚shēt }

flow shop [IND ENG] A manufacturing facility in which machine tools and robots are employed in the same manner on all jobs. { 'flō ‚shäp }

flow soldering [ENG] Soldering of printed circuit boards by moving them over a flowing wave of molten solder in a solder bath; the process permits precise control of the depth of immersion in the molten solder and minimizes heating of the board. Also known as wave soldering. { 'flō ‚säd·ə·riŋ }

flow stress [MECH] The stress along one axis at a given value of strain that is required to produce plastic deformation. { 'flō ‚stres }

flow transmitter [ENG] A device used to measure the flow of liquids in pipelines and convert the results into proportional electric signals that can be transmitted to distant receivers or controllers. { 'flō tranz‚mid·ər }

flow valve [ENG] A valve that closes itself when the flow of a fluid exceeds a particular value. { 'flō ‚valv }

flow visualization [ENG] Method of making visible the disturbances that occur in fluid flow, using the fact that light passing through a flow field of varying density exhibits refraction and a relative phase shift among different rays. { ¦flō vizh·ə·lə'zā·shən }

fl oz See fluid ounce.

flue [ENG] A channel or passage for conveying combustion products from a furnace, boiler, or fireplace to or through a chimney. { flü }

flue exhauster [ENG] A device installed as part of a vent in order to provide a positive induced draft. { 'flü ig‚zós·tər }

flue gas [ENG] Gaseous combustion products from a furnace. { 'flü ‚gas }

flue gas analyzer [ENG] A device that monitors the composition of the flue gas of a boiler heating unit to determine if the mixture of air and fuel is at the proper ratio for maximum heat output. { 'flü ‚gas ‚an·ə‚līz·ər }

flue gas expander [MECH ENG] In a petroleum processing system, a turbine for recovering energy at the point where combustion gases are discharged under pressure to the atmosphere; the reduction in pressure drives the turbine impeller. { 'flü ‚gas ik'spand·ər }

fluid amplifier [ENG] An amplifier in which all amplification is achieved by interaction between jets of fluid, with no electronic circuit and usually no moving parts. { ¦flü·əd 'am·plə‚fī·ər }

fluid-bed process [CHEM ENG] A type of process based on the tendency of finely divided powders to behave in a fluidlike manner when supported and moved by a rising gas or vapor stream; used mainly for catalytic cracking of petroleum distillates. { ¦flü·əd ‚bed 'präs·əs }

fluid catalyst [CHEM ENG] Finely divided solid particles utilized as a catalyst in a fluid-bed process. { ¦flü·əd 'kad·əl‚ist }

fluid catalytic cracking [CHEM ENG] An oil refining process in which the gas-oil is cracked by a catalyst bed fluidized by using oil vapors. { ¦flü·əd ¦kad·əl¦id·ik 'krak·iŋ }

fluid clutch See fluid drive. { ¦flü·əd ¦kləch }

fluid coking [CHEM ENG] A thermal process utilizing the fluidized solids technique for continuous conversion of heavy, low-grade petroleum oils into petroleum coke and lighter hydrocarbon products. { ¦flü·əd ¦kōk·iŋ }

fluid-controlled valve [MECH ENG] A valve for which the valve operator is activated by a fluid energy, in contrast to electrical, pneumatic, or manual energy. { ¦flü·əd kən‚trōld 'valv }

fluid coupling [MECH ENG] A device for transmitting rotation between shafts by means of the acceleration and deceleration of a fluid such as oil. Also known as hydraulic coupling. { ¦flü·əd ¦kəp·liŋ }

fluid die [MECH ENG] A die for shaping parts by liquid pressure; a plunger forces the liquid against the part to be shaped, making the part conform to the shape of a die. { ¦flü·əd ¦dī }

fluid distributor [ENG] Device for the controlled

distribution of fluid feed to a process unit, such as a liquid-gas or liquid-solids contactor, reactor, mixer, burner, or heat exchanger; can be a simple perforated-pipe sparger, spray head, or such. { ¦flü·əd də'strib·yəd·ər }

fluid dram [MECH] Abbreviated fl dr. **1.** A unit of volume used in the United States for measurement of liquid substances, equal to 1/8 fluid ounce, or $3.6966911953125 \times 10^{-6}$ cubic meter. **2.** A unit of volume used in the United Kingdom for measurement of liquid substances and occasionally of solid substances, equal to 1/8 fluid ounce or $3.5516328125 \times 10^{-6}$ cubic meter. { ¦flü·əd 'dram }

fluid drive [MECH ENG] A power coupling operated on a hydraulic turbine principle in which the engine flywheel has a set of turbine blades which are connected directly to it and which are driven in oil, thereby turning another set of blades attached to the transmission gears of the automobile. Also known as fluid clutch; hydraulic clutch. { ¦flü·əd ¦drīv }

fluid end [MECH ENG] In a fluid pump, the section that contains parts which are directly involved in moving the fluid. { 'flü·əd ‚end }

fluid-energy mill [ENG] A size-reduction unit in which grinding is achieved by collision between the particles being ground and the energy supplied by a compressed fluid entering the grinding chamber at high speed. Also known as jet mill. { ¦flü·əd 'en·ər·jē ‚mil }

fluid-film bearing [MECH ENG] An antifriction bearing in which rubbing surfaces are kept apart by a film of lubricant such as oil. { ¦flü·əd 'film ‚ber·iŋ }

fluid hydroforming [CHEM ENG] A type of fluid catalytic cracking process used by petroleum refineries to upgrade low-octane-number stocks. { ¦flü·əd 'hī·drə‚fȯr·miŋ }

fluidic device [ENG] A device that operates by the interaction of streams of fluid. { flü¦id·ik di¦vīs }

fluidic flow sensor [ENG] A device for measuring the velocity of gas flows in which a jet of air or other selected gas is directed onto two adjacent small openings and is deflected by the flow of gas being measured so that the relative pressure on the two ports is a measure of gas velocity. Also known as deflected jet fluidic flowmeter. { flü¦id·ik 'flō ‚sen·sər }

fluidic oscillator meter [ENG] A flowmeter that measures the frequency with which a fluid entering the meter attaches to one of two opposite diverging side walls and then the other, because of the Coanda effect. { flü¦id·ik 'äs·ə‚lād·ər ‚mēd·ər }

fluidics [ENG] A control technology that employs fluid dynamic phenomena to perform sensing, control, information processing, and actuation functions without the use of moving mechanical parts. { flü'id·iks }

fluidic sensor [ENG] A proximity sensor that detects the presence of a nearby object from the back pressure created on an air jet when the

object blocks the jet's exit area. { flü'id·ik 'sen·sər }

fluidization [CHEM ENG] A roasting process in which finely divided solids are suspended in a rising current of air (or other fluid), producing a fluidized bed; used in the calcination of various minerals, in Fischer-Tropsch synthesis, and in the coal industry. { ‚flü·ə·də'zā·shən }

fluidized adsorption [CHEM ENG] Method of vapor- or gas-fractionation (separation via adsorption-desorption cycles) in a fluidized bed of adsorbent material. { ¦flü·ə‚dīzd ad'sȯrp·shən }

fluidized bed [ENG] A cushion of air or hot gas blown through the porous bottom slab of a container which can be used to float a powdered material as a means of drying, heating, quenching, or calcining the immersed components. { ¦flü·ə‚dīzd 'bed }

fluidized-bed coating [ENG] Method for plastic-coating of objects; the heated object is immersed into the fluidized bed of a thermoplastic resin that then fuses into a continuous uniform coating over the immersed object. { ¦flü·ə‚dīzd ¦bed 'kōd·iŋ }

fluidized-bed combustion [MECH ENG] A method of burning particulate fuel, such as coal, in which the amount of air required for combustion far exceeds that found in conventional burners; the fuel particles are continually fed into a bed of mineral ash in the proportions of 1 part fuel to 200 parts ash, while a flow of air passes up through the bed, causing it to act like a turbulent fluid. { ¦flü·ə‚dīzd ¦bed kəm'bəs·chən }

fluid logic [ENG] The simulation of logical operations by means of devices that employ fluid dynamic phenomena to control the interactions between sets of gases or liquids. { ¦flü·əd ¦läj·ik }

fluid mechanics [MECH] The science concerned with fluids, either at rest or in motion, and dealing with pressures, velocities, and accelerations in the fluid, including fluid deformation and compression or expansion. { ¦flü·əd mə'kan·iks }

fluid meter See flowmeter. { 'flü·əd ‚mēd·ər }

fluid ounce [MECH] Abbreviated fl oz. **1.** A unit of volume that is used in the United States for measurement of liquid substances, equal to 1/16 liquid pint, or 231/128 cubic inches, or $2.95735295625 \times 10^{-5}$ cubic meter. **2.** A unit of volume used in the United Kingdom for measurement of liquid substances, and occasionally of solid substances, equal to 1/20 pint or $2.84130625 \times 10^{-5}$ cubic meter. { ¦flü·əd 'aúns }

fluid stress [MECH] Stress associated with plastic deformation in a solid material. { ¦flü·əd 'stres }

fluid ton [MECH] A unit of volume equal to 32 cubic feet or approximately 0.90614 cubic meter; used for many hydrometallurgical, hydraulic, and other industrial purposes. { ¦flü·əd 'tən }

fluid transmission [MECH ENG] Automotive transmission with fluid drive. { ¦flü·əd tranz'mish·ən }

fluing [ENG] A forming process in which a

flange is formed around a hole in a sheet-metal part by pressing a cylindrical die through the hole. { 'flü·iŋ }

flume |ENG| **1.** An open channel constructed of steel, reinforced concrete, or wood and used to convey water to be utilized for power, to transport logs, and so on. **2.** To divert by a flume, as the waters of a stream, in order to lay bare the auriferous sand and gravel forming the bed. { flüm }

fluorescent lamp |ELECTR| A tubular discharge lamp in which ionization of mercury vapor produces radiation that activates the fluorescent coating on the inner surface of the glass. { flü¦res·ənt 'lamp }

fluorescent screen |ENG| A sheet of material coated with a fluorescent substance so as to emit visible light when struck by ionizing radiation such as x-rays or electron beams. { flü¦res·ənt 'skrēn }

fluoridation |ENG| The addition of the fluorine ion (F^-) to municipal water supplies in a final concentration of 0.8–1.6 parts per million to help prevent dental caries in children. { flür·ə'dā·shən }

fluorimeter See fluorometer. { flü'rim·əd·ər }

fluorologging |ENG| A well-logging technique in which well cuttings are examined under ultraviolet light for fluorescence radiation related to trace occurrences of oil. { 'flür·ō,läg·iŋ }

fluorometer |ENG| An instrument that measures the fluorescent radiation emitted by a sample which is exposed to monochromatic radiation, usually radiation from a mercury-arc lamp or a tungsten or molybdenum x-ray source that has passed through a filter; used in chemical analysis, or to determine the intensity of the radiation producing fluorescence. Also spelled fluorimeter. { flü'räm·əd·ər }

fluoroscope |ENG| A fluorescent screen designed for use with an x-ray tube to permit direct visual observation of x-ray shadow images of objects interposed between the x-ray tube and the screen. { 'flür·ə,skōp }

fluoroscopy |ENG| Use of a fluoroscope for x-ray examination. { flü'räs·kə·pē }

flush |ENG| Pertaining to separate surfaces that are on the same level. { fləsh }

flush bead See quirk bead. { 'fləsh ,bēd }

flush coat |CIV ENG| A coating of bituminous material, used to waterproof a surface. { 'fləsh ,kōt }

flush gate |CIV ENG| A gate for flushing a channel that lies below the gate of a dam. { 'fləsh ,gāt }

flushing |CIV ENG| The removal or reduction to a permissible level of dissolved or suspended contaminants in an estuary or harbor. |ENG| Removing lodged deposits of rock fragments and other debris by water flow at high velocity; used to clean water conduits and drilled boreholes. { 'fləsh·iŋ }

flushometer |ENG| A valve that discharges a fixed quantity of water when a handle is operated; used to flush toilets and urinals. { flə'shäm·əd·ər }

flush tank |CIV ENG| **1.** A tank in which water or sewage is retained for periodic release through a sewer. **2.** A small water-filled tank for flushing a water closet. { 'fləsh ,taŋk }

flush valve |ENG| A valve used for flushing toilets. { 'fləsh ,valv }

flute |DES ENG| A groove having a curved section, especially when parallel to the main axis, as on columns, drills, and other cylindrical or conical shaped pieces. { flüt }

fluted chucking reamer |DES ENG| A machine reamer with a straight or tapered shank and with straight or spiral flutes; the ends of the teeth are ground on a slight chamfer for end cutting. { 'flüd·əd 'chək·iŋ ,rēm·ər }

flute length |DES ENG| On a twist drill, the length measured from the outside corners of the cutting lips to the farthest point at the back end of the flutes. { 'flüt ,leŋkth }

fluting |MECH ENG| A machining operation whereby flutes are formed parallel to the main axis of cylindrical or conical parts. { 'flüd·iŋ }

flutter |ENG| The irregular alternating motion of the parts of a relief valve due to the application of pressure where no contact is made between the valve disk and the seat. { 'fləd·ər }

flutter valve |ENG| A valve that is operated by fluctuations in pressure of the material flowing over it; used in carburetors. { 'fləd·ər ,valv }

fluvarium |ENG| A large aquarium in which the tanks contain flowing stream water maintained by gravity, not pumps. { 'flü'ver·ē·əm }

flux gate |ENG| A detector that gives an electric signal whose magnitude and phase are proportional to the magnitude and direction of the external magnetic field acting along its axis; used to indicate the direction of the terrestrial magnetic field. { 'fləks ,gāt }

fluxmeter |ENG| An instrument for measuring magnetic flux. { 'fləks,mēd·ər }

fly |MECH ENG| A fan with two or more blades used in timepieces or light machinery to govern speed by air resistance. { flī }

fly ash |ENG| **1.** Fine particulate, essentially noncombustible refuse, carried in a gas stream from a furnace. **2.** Coal combustion residue. { 'flī ,ash }

fly cutter |MECH ENG| A cutting tool that revolves with the arbor of a lathe. { 'flī ,kəd·ər }

fly cutting |MECH ENG| Cutting with a milling cutter provided with only one tooth. { 'flī ,kəd·iŋ }

flying switch |ENG| Disconnection of railroad cars from a locomotive while they are moving and switching them to another track under their own momentum. { ¦flī·iŋ 'swich }

fly rock |ENG| The fragments of rock thrown and scattered during quarry or tunnel blasting. { 'flī ,räk }

flywheel |MECH ENG| A rotating element

attached to the shaft of a machine for the maintenance of uniform angular velocity and revolutions per minute. Also known as balance wheel. { 'flī,wēl }

fm *See* femtometer.

FM/AM multiplier |ELECTR| Multiplier in which the frequency deviation from the central frequency of a carrier is proportional to one variable, and its amplitude is proportional to the other variable; the frequency-amplitude-modulated carrier is then consecutively demodulated for frequency modulation (FM) and for amplitude modulation (AM); the final output is proportional to the product of the two variables. { 'ef,em 'ā,em 'məl·tə,plī·ər }

foam blanketing |ENG| A technique for fighting fire within an oil tank or similar facility by generating foam that forms a coating inside the tank, thus depriving the fire of air. { ,fōm 'blaŋ·kə·tiŋ }

foaming |ENG| Any of various processes by which air or gas is introduced into a liquid or solid to produce a foam material. { 'fōm·iŋ }

foam-in-place |ENG| The deposition of reactive foam ingredients onto the surface to be covered, allowing the foaming reaction to take place upon that surface, as with polyurethane foam; used in applying thermal insulation for homes and industrial equipment. { ¦fōm in 'plās }

focometer |ENG| An instrument for measuring focal lengths of optical systems. { fō'käm·əd·ər }

focused-current log |ENG| A resistivity log that is obtained by means of a multiple-electrode arrangement. { ¦fō·kəst ¦kə·rənt 'läg }

focusing collector |ENG| A solar collector that uses semicircular aluminum reflectors to focus sunlight onto copper pipes containing circulating water. { 'fō·kəs·iŋ kə'lek·tər }

foil decorating |ENG| The molding of paper, textile, or plastic foil, printed with compatible inks, into a plastic part so that the foil is visible below the surface of the part as a decoration. { ¦foil ¦dek·ə,rād·iŋ }

folded horn |ENG ACOUS| An acoustic horn in which the path from throat to mouth is folded or curled to give the longest possible path in a given volume. { ¦fōld·əd 'hórn }

folded-plate roof |BUILD| A roof constructed of flat plates, usually of reinforced concrete, joined at various angles. { ¦fōld·əd ¦plāt 'rüf }

folding door |ENG| A door in sections that can be folded back or can be moved apart by sliding. { 'fōld·iŋ ,dór }

Foley pits |ENG ACOUS| Open boxes that are used in ADR studios and contain various materials (such as water, sand, gravel, rice, and nails) for generating sound effects that could not be recorded well during filming or video recording. { 'fō·lē ,pits }

follower |ENG| A drill used for making all but the first part of a hole, the first part being made with a drill of larger gage. { 'fäl·ə·wər }

following error |CONT SYS| The difference between commanded and actual positions in contouring control. { 'fäl·ə·wiŋ ,er·ər }

food engineering |ENG| The technical discipline involved in food manufacturing and processing. { 'füd ,en·jə,nir·iŋ }

foot |MECH| The unit of length in the British systems of units, equal to exactly 0.3048 meter. Abbreviated ft. { fút }

footage |ENG| The extent or length of a material expressed in feet. { 'fúd·ij }

foot block |ENG| Flat pieces of wood placed under props in tunneling to give a broad base and thus prevent the superincumbent weight from pressing the props down. { 'fút ,bläk }

foot bridge |CIV ENG| A bridge structure used only for pedestrian traffic. { 'fút ,brij }

footguard |CIV ENG| A filler placed on the space between converging rails to prevent a foot from being wedged between the rails. { 'fút ,gärd }

footing |CIV ENG| The widened base or substructure forming the foundation for a wall or a column. { 'fúd·iŋ }

foot-pound |MECH| **1.** Unit of energy or work in the English gravitational system, equal to the work done by 1 pound of force when the point at which the force is applied is displaced 1 foot in the direction of the force; equal to approximately 1.355818 joule. Abbreviated ft-lb; ft-lbf. **2.** Unit of torque in the English gravitational system, equal to the torque produced by 1 pound of force acting at a perpendicular distance of 1 foot from an axis of rotation. Also known as pound-foot. Abbreviated lbf-ft. { 'fút ¦paúnd }

foot-poundal |MECH| **1.** A unit of energy or work in the English absolute system, equal to the work done by a force of magnitude 1 poundal when the point at which the force is applied is displaced 1 foot in the direction of the force; equal to approximately 0.04214011 joule. Abbreviated ft-pdl. **2.** A unit of torque in the English absolute system, equal to the torque produced by a force of magnitude 1 poundal acting at a perpendicular distance of 1 foot from the axis of rotation. Also known as poundal-foot. Abbreviated pdl-ft. { 'fút ¦paúnd·əl }

footprint |BUILD| A description of the exact size, shape, and location of a building's foundation as the foundation has been installed on a specific site. Also known as building footprint. { 'fút,print }

foot screw |ENG| **1.** One of the three screws connecting the tribach of a theodolite or other level with the plate screwed to the tripod head. **2.** An adjusting screw that serves also as a foot. { 'fút ,skrü }

foot section |MECH ENG| In both belt and chain conveyors that portion of the conveyor at the extreme opposite end from the delivery point { 'fút ,sek·shən }

footstock |MECH ENG| A device containing a center which supports the workpiece on a milling machine; usually used in conjunction with a dividing head. { 'fút,stäk }

foot valve |MECH ENG| A valve in the bottom

of the suction pipe of a pump which prevents backward flow of water. { 'fút ,valv }

Forbes bar |THERMO| A metal bar which has one end immersed in a crucible of molten metal and thermometers placed in holes at intervals along the bar; measurement of temperatures along the bar together with measurement of cooling of a short piece of the bar enables calculation of the thermal conductivity of the metal. { 'fòrbz ,bär }

force |MECH| That influence on a body which causes it to accelerate; quantitatively it is a vector, equal to the body's time rate of change of momentum. { fòrs }

force-balance meter |ENG| A flowmeter that measures a force, such as that associated with the air pressure in a small bellows, that is required to balance the net force created by the differential pressure, on opposite sides of a diaphragm or diaphragm capsule, generated by a differential-producing primary device. { 'fòrs ,bal·əns ,mēd·ər }

force compensation |ENG| On an analytical balance, the weight force of a load that is held in equilibrium by a force of equal size which acts in the opposite direction. { 'fòrs ,käm·pən,sā·shən }

force constant |MECH| The ratio of the force to the deformation of a system whose deformation is proportional to the applied force. { 'fòrs ,kän·stənt }

force-controlled motion commands |CONT SYS| Robot control in which motion information is provided by computer software but sensing of forces or feedback is used by the robot to adapt this information to the environment. { 'fòrs kən¦tròld 'mō·shən kə,manz }

forced-air heating |MECH ENG| A warm-air heating system in which positive air circulation is provided by means of a fan or a blower. { ¦fòrst ,er 'hēd·iŋ }

forced circulation |MECH ENG| The use of a pump or other fluid-movement device in conjunction with liquid-processing equipment to move the liquid through pipes and process vessels; contrasted to gravity or thermal circulation. { ¦fòrst ,sər·kyə'lā·shən }

forced-circulation boiler |MECH ENG| A once-through steam generator in which water is pumped through successive parts. { ¦fòrst ,sər·kyə'lā·shən ,bòil·ər }

forced convection |THERMO| Heat convection in which fluid motion is maintained by some external agency. { ¦fòrst kən'vek·shən }

forced draft |MECH ENG| Air under positive pressure produced by fans at the point where air or gases enter a unit, such as a combustion furnace. { ¦fòrst 'draft }

forced oscillation |MECH| An oscillation produced in a simple oscillator or equivalent mechanical system by an external periodic driving force. Also known as forced vibration. { ¦fòrst ,äs·ə'lā·shən }

forced ventilation |MECH ENG| A system of ventilation in which air is forced through ventilation ducts under pressure. { ¦fòrst ,vent·əl'ā·shən }

forced vibration See forced oscillation. { ¦fòrst vī'brā·shən }

force feedback |CONT SYS| A method of error detection in which the force exerted on the effector is sensed and fed back to the control, usually by mechanical, hydraulic, or electric transducers. { 'fòrs ¦fēd,bak }

force fit See press fit. { 'fòrs ,fit }

force gage |ENG| An instrument which measures the force exerted on an object. { 'fòrs ,gāj }

force main |CIV ENG| The discharge pipeline of a pumping station. { 'fòrs ,mān }

force plate |ENG| A plate that carries the plunger or force plug of a mold and the guide pins on bushings. { 'fòrs ,plāt }

force plug |ENG| A mold member that fits into the cavity block, exerting pressure on the molding compound. Also known as piston; plunger. { 'fòrs ,pləg }

force polygon |MECH| A closed polygon whose sides are vectors representing the forces acting on a body in equilibrium. { ¦fòrs 'päl·ə,gän }

forceps |DES ENG| A pincerlike instrument for grasping objects. { 'fòr·səps }

force pump |MECH ENG| A pump fitted with a solid plunger and a suction valve which draws and forces a liquid to a considerable height above the valve or puts the liquid under a considerable pressure. { 'fòrs ,pəmp }

force ratio See mechanical advantage. { 'fòrs ,rā·shō }

force-time |IND ENG| The product of an applied force and its time of application; used for quantitative determination of isometric work. { ¦fòrs ¦tīm }

fording depth |ENG| Maximum depth at which a particular vehicle can operate in water. { 'fòrd·iŋ ,depth }

forebay |CIV ENG| **1.** A small reservoir at the head of the pipeline that carries water to the consumer; it is the last free water surface of a distribution system. **2.** A reservoir feeding the penstocks of a hydro-power plant. { 'fòr,bā }

foreign-body locator |ENG| A device for locating foreign metallic bodies in tissue by means of suitable probes that generate a magnetic field; the presence of a magnetic body within this field is indicated by a meter or a sound signal. { 'fär·ən ¦bäd·ē 'lō,kād·ər }

foreign element |IND ENG| A work element which is not a part of the normal work cycle, either because it is accidental or because it occurs only occasionally. { 'fär·ən 'el·ə·mənt }

fore pump See backing pump. { 'fòr ,pəmp }

foresight |ENG| **1.** A sight or bearing on a new survey point, taken in a forward direction and made in order to determine its elevation. **2.** A sight on a previously established survey point, taken in order to close a circuit. **3.** A reading taken on a level rod to determine the elevation of the point on which the rod rests when read. Also known as minus sight. { 'fòr,sīt }

forest engineering |ENG| A branch of engineering concerned with the solution of forestry problems with regard to long-range environmental and economic effects. { 'fär·əst ,en·jə,nir·iŋ }

forklift |MECH ENG| A machine, usually powered by hydraulic means, consisting of two or more prongs which can be raised and lowered and are inserted under heavy materials or objects for hoisting and moving them. { 'fȯrk,lift }

forklift truck See fork truck. { 'fȯrk,lift ,trək }

fork pocket |MECH ENG| An opening in the base of a container or pallet for insertion of the prong of a forklift. { 'fȯrk ,pak·ət }

fork truck |MECH ENG| A vehicle equipped with a forklift. Also known as forklift truck. { 'fȯrk ,trək }

form |CIV ENG| Temporary boarding, sheeting, or pans of plywood, molded fiber glass, and so forth, used to give desired shape to poured concrete or the like. { fȯrm }

form clamp |CIV ENG| An adjustable metal clamp used to secure planks of wooden forms for concrete columns or beams. { 'fȯrm ,klamp }

form cutter See formed cutter. { 'fȯrm ,kəd·ər }

formed cutter |MECH| A cutting tool shaped to make surfaces with irregular geometry. Also known as form cutter. { ¦fȯrmd 'kəd·ər }

form factor |ELEC| **1.** The ratio of the effective value of a periodic function, such as an alternating current, to its average absolute value. **2.** A factor that takes the shape of a coil into account when computing its inductance. Also known as shape factor. |MECH| The theoretical stress concentration factor for a given shape, for a perfectly elastic material. { 'fȯrm ,fak·tər }

form grinding |MECH ENG| Grinding by use of a wheel whose cutting face is contoured to the reverse shape of the desired form. { 'fȯrm ,grīnd·iŋ }

forming |ELEC| Application of voltage to an electrolytic capacitor, electrolytic rectifier, or semiconductor device to produce a desired permanent change in electrical characteristics as a part of the manufacturing process. |MECH ENG| A process for shaping or molding sheets, rods, or other pieces of hot glass, ceramic ware, plastic, or metal by the application of pressure. { 'fȯrm·iŋ }

forming die |ENG| A die like a drawing die, but without a blank holder. { 'fȯrm·iŋ ,dī }

forming press |MECH ENG| A punch press for forming metal parts. { 'fȯrm·iŋ ,pres }

forming rolls |MECH ENG| Rolls contoured to give a desired shape to parts passing through them. { 'fȯrm·iŋ ,rōlz }

forming tool |DES ENG| A nonrotating tool that produces its inverse form on the workpiece. { 'fȯrm·iŋ ,tül }

form process chart |IND ENG| A graphic representation of the process flow of paperwork forms. Also known as forms analysis chart; functional forms analysis chart; information process analysis chart. { ¦fȯrm ¦präs·əs ,chärt }

forms analysis chart See form process chart. { ¦fȯrmz ə¦nal·ə·səs ,chärt }

form scabbing |CIV ENG| In placing of concrete using formwork, removal of the surface layer of concrete that adheres to the form when it is removed. { 'fȯrm ,skab·iŋ }

formwork |CIV ENG| A temporary wooden casing used to contain concrete during its placing and hardening. Also known as shuttering. { 'fȯrm,wərk }

fors See G; gram-force. { fȯrs }

Fortin barometer |ENG| A type of cistern barometer; provision is made to increase or decrease the volume of the cistern so that when a pressure change occurs, the level of the cistern can be maintained at the zero of the barometer scale (the ivory point). { 'fȯrd·ən bə'räm·əd·ər }

forward bias |ELECTR| A bias voltage that is applied to a pn-junction in the direction that causes a large current flow; used in some semiconductor diode circuits. { ¦fȯr·wərd 'bī·əs }

forward-looking infrared imager |ENG| An infrared imaging device which employs an optomechanical system to make a two-dimensional scan, and produces a visible image corresponding to the spatial distribution of infrared radiation. Abbreviated FLIR imager. Also known as framing imager. { 'fȯr·wərd ¦lük·iŋ ,in·frə,red 'im·ij·ər }

forward pass |ENG| In project management, scheduling from a known start date and calculating the finish date by proceeding from the first operation to the last. Also known as forward scheduling. { ¦fȯr·wərd 'pas }

forward path |CONT SYS| The transmission path from the loop actuating signal to the loop output signal in a feedback control loop. { 'fȯr·wərd ,path }

forward scheduling See forward pass. { ¦fȯr·wərd 'skej·əl·iŋ }

forward transfer function |CONT SYS| In a feedback control loop, the transfer function of the forward path. { ¦fȯr·wərd 'tranz·fər ,fəŋk·shən }

Foster's reactance theorem |CONT SYS| The theorem that the most general driving point impedance or admittance of a network, in which every mesh contains independent inductance and capacitance, is a meromorphic function whose poles and zeros are all simple and occur in conjugate pairs on the imaginary axis, and in which these poles and zeros alternate. { 'fȯs·tərz rē'ak·təns ,thir·əm }

Foucault pendulum |MECH| A swinging weight supported by a long wire, so that the wire's upper support restrains the wire only in the vertical direction, and the weight is set swinging with no lateral or circular motion; the plane of the pendulum gradually changes, demonstrating the rotation of the earth on its axis. { fü'kō 'pen·jə·ləm }

foul bottom |CIV ENG| A hard, uneven, rocky or obstructed bottom having poor holding qualities for anchors, or one having rocks or wreckage

that would endanger an anchored vessel. { ¦faúl
'bäd·əm }

fouling |CHEM ENG| Deposition on the surface
of a heat-transfer device of sediment in the form
of scale derived from burned particles of the
heated substance. { 'faúl·iŋ }

fouling factor |CHEM ENG| In heat transfer, the
lowering of clear-film transfer rates resulting
from corrosion, dirt, or roughness of the surface
of tube walls of heat exchangers. { 'faúl·iŋ
,fak·tər }

fouling plates |ENG| Metal plates submerged in
water to allow attachment of fouling organisms,
which are then analyzed to determine species,
growth rate, and growth pattern, as influenced
by environmental conditions and time. { 'faúl·
iŋ ,plāts }

fouling point |CIV ENG| **1.** The point at a switch
or turnout beyond which railroad cars must be
placed so as not to interfere with cars on the
main track. **2.** The location of insulated joints
in a turnout on signaled tracks. { 'faúl·iŋ
,póint }

foundation |CIV ENG| **1.** The ground that sup-
ports a building or other structure. **2.** The portion of a structure which transmits the building
load to the ground. { faún'dā·shən }

foundation engineering |CIV ENG| That branch
of engineering concerned with evaluating the
earth's ability to support a load and designing
substructures to transmit the load of superstruc-
tures to the earth. { faún'dā·shən ,en·jə,nir·iŋ }

foundation mat See raft foundation. { faún'dā·
shən ,mat }

foundry |ENG| A building where metal or glass
castings are produced. { 'faún·drē }

foundry engineering |ENG| The science and
practice of melting and casting glass or metal.
{ 'faún·drē ,en·jə,nir·iŋ }

four-ball tester |ENG| A machine designed to
measure the efficiency of lubricants by driving
one ball against three stationary balls clamped
together in a cup filled with the lubricant; performance is evaluated by measuring wear-scar diameters on the stationary balls. { ¦fór ¦ból
'tes·tər }

four-bar linkage |MECH ENG| A plane linkage
consisting of four links pinned tail to head in a
closed loop with lower, or closed, joints. { ¦fór
¦bär 'liŋk·ij }

Foucault process |ENG| A process for forming
sheet glass in which the molten glass is drawn
vertically upward. { für'kō ,präs·əs }

four-channel sound system See quadraphonic
sound system. { ¦fór ¦chan·əl 'saúnd ,sis·təm }

Fourdrinier machine |MECH ENG| A papermak-
ing machine; a paper web is formed on an endless wire screen; the screen passes through
presses and over dryers to the calenders and
reels. { ,for·drə'nir mə,shēn }

fourier See thermal ohm. { 'fúr·ē,ā }

Fourier analyzer |ENG| A digital spectrum ana-
lyzer that provides push-button or other switch

selection of averaging, coherence function, correlation, power spectrum, and other mathematical operations involved in calculating Fourier
transforms of time-varying signal voltages for
such applications as identification of underwater
sounds, vibration analysis, oil prospecting, and
brain-wave analysis. { ,fúr·ē,ā 'an·ə,līz·ər }

Fourier heat equation See Fourier law of heat conduction; heat equation. { ,fúr·ē,ā 'hēt i,kwā·
zhən }

Fourier law of heat conduction |THERMO| The
law that the rate of heat flow through a substance
is proportional to the area normal to the direction of flow and to the negative of the rate of
change of temperature with distance along the
direction of flow. Also known as Fourier heat
equation. { 'fúr·ē,ā ,ló əv 'hēt kən,dək·shən }

Fourier number |THERMO| A dimensionless
number used in the study of unsteady-state heat
transfer, equal to the product of the thermal conductivity and a characteristic time, divided by
the product of the density, the specific heat at
constant pressure, and the distance from the
midpoint of the body through which heat is passing to the surface. Symbolized N_{Foh}. { ,fúr·ē,ā
,nəm·bər }

four-pi counter |ENG| An instrument which
measures the radiation that a radioactive material emits in all directions. { ¦fór 'pī ,kaún·tər }

four-stroke cycle |MECH ENG| An internal com-
bustion engine cycle completed in four piston
strokes; includes a suction stroke, compression
stroke, expansion stroke, and exhaust stroke.
{ ¦fór ¦strōk 'sī·kəl }

four-track tape |ENG ACOUS| Magnetic tape on
which two tracks are recorded for each direction
of travel, to provide stereo sound reproduction
or to double the amount of source material that
can be recorded on a given length of 1/4-inch
(0.635-centimeter) tape. { 'fór ,trak 'tāp }

four-way reinforcing |CIV ENG| A system of re-
inforcing rods in concrete slab construction in
which the rods are placed parallel to two adjacent edges and to both diagonals of a rectangular
slab. { 'fór ,wā rē·ən'fórs·iŋ }

four-way valve |MECH ENG| A valve at the junc-
tion of four waterways which allows passage between any two adjacent waterways by means of
a movable element operated by a quarter turn.
{ 'fór ,wā 'valv }

four-wheel drive |MECH ENG| An arrangement
in which the drive shaft acts on all four wheels
of the automobile. { 'fór ¦wēl 'drīv }

fox lathe |MECH ENG| A lathe with chasing bar
and leaders for cutting threads; used for turning
brass. { 'fäks ,lāth }

fractionator |CHEM ENG| An apparatus used to
separate a mixture by fractionation, especially
by fractional distillation. { 'frak·shə,nād·ər }

fraction defective |IND ENG| The number of
units per 100 pieces which are defective in a lot;
expressed as a decimal. { 'frak·shən di'fek·tiv }

fracture strength See fracture stress. { 'frak·shər
,streŋkth }

fracture stress |MECH| The minimum tensile

stress that will cause fracture. Also known as fracture strength. { 'frak·shər ,stres }

fracture test |ENG| **1.** Macro- or microscopic examination of a fractured surface to determine characteristics such as grain pattern, composition, or the presence of defects. **2.** A test designed to evaluate fracture stress. { 'frak·shər ,test }

fracture wear |MECH| The wear on individual abrasive grains on the surface of a grinding wheel caused by fracture. { 'frak·shər ,wer }

Frahm frequency meter See vibrating-reed frequency metery. { främ 'frē·kwən·sē ,mēd·ər }

frame |BUILD| The skeleton structure of a building. Also known as framing. |ELECTR| **1.** One complete coverage of a television picture. **2.** A rectangular area representing the size of copy handled by a facsimile system. { främ }

framework |ENG| The load-carrying frame of a structure; may be of timber, steel, or concrete. { 'främ,wərk }

framing |BUILD| See frame. |ELECTR| **1.** Adjusting a television picture to a desired position on the screen of the picture tube. **2.** Adjusting a facsimile picture to a desired position in the direction of line progression. Also known as phasing. { 'främ·iŋ }

framing anchor |BUILD| A metal device for joining elements such as studs, joists, and rafters in light wood-frame construction. { 'främ·iŋ ,aŋk·ər }

framing imager See forward-looking infrared imageryy. { 'främ·iŋ ,im·ij·ər }

framing square |DES ENG| A graduated carpenter's square used for cutting off and making notches. { 'främ·iŋ ,skwer }

Francis turbine |MECH ENG| A reaction hydraulic turbine of relatively medium speed with radial flow of water in the runner. { 'fran·səs 'tər,bīn }

frangible |MECH| Breakable, fragile, or brittle. { 'fran·jə·bəl }

Franklin equation |ENG ACOUS| An equation for intensity of sound in a room as a function of time after shutting off the source, involving the volume and exposed surface area of the room, the speed of sound, and the mean sound-absorption coefficient. { 'fraŋk·lən i,kwā·zhən }

Frazer-Brace extraction method |CHEM ENG| A method used to extract oil from citrus fruit; utilizes a machine which has abrasive carborundum rolls to rasp the peel from the fruit under a water spray; the water-and-peel mixture is screened and settled to allow oil separation. { 'frā·zər 'brās ik'strak·shən ,meth·əd }

free ascent |ENG| Emergency ascent by a diver by floating to the surface through natural buoyancy or through assisted buoyancy with a life jacket. { ¦frē ə'sent }

freeboard |CHEM ENG| In a fluidized-bed reactor, the space between the top of the reaction bed and the top of the reactor. |CIV ENG| The height between normal water level and the crest of a dam or the top of a flume. |ENG| The vertical distance in a water tank between the maximum water level and the top of the tank. { 'frē,bȯrd }

free charge |ELEC| Electric charge which is not bound to a definite site in a solid, in contrast to the polarization charge. { ¦frē 'chärj }

free convection See natural convection. { ¦frē kən'vek·shən }

free diving |ENG| Diving with the use of scuba equipment to allow freedom and maneuverability. { ¦frē 'dīv·iŋ }

free-drop |ENG| To air-drop supplies or equipment without parachute. { 'frē ,dräp }

free energy |THERMO| **1.** The internal energy of a system minus the product of its temperature and its entropy. Also known as Helmholtz free energy; Helmholtz function; Helmholtz potential; thermodynamic potential at constant volume; work function. **2.** See Gibbs free energy. { ¦frē 'en·ər·jē }

free enthalpy See Gibbs free energy. { ¦frē 'en ,thal·pē }

free fall |MECH| The ideal falling motion of a body acted upon only by the pull of the earth's gravitational field. { 'frē ,fȯl }

free falling |MECH ENG| In ball milling, the peripheral speed at which part of the crop load breaks clear on the ascending side and falls clear to the toe of the charge. { 'frē ,fȯl·iŋ }

free-field room See anechoic chamber. { 'frē ,fēld ,rüm }

free fit |DES ENG| A fit between mating pieces where accuracy is not essential or where large variations in temperature may occur. { 'frē ,fit }

free flight |MECH| Unconstrained or unassisted flight. { 'frē ,flīt }

free-flight angle |MECH| The angle between the horizontal and a line in the direction of motion of a flying body, especially a rocket, at the beginning of free flight. { 'frē ,flīt ,aŋ·gəl }

free-flight trajectory |MECH| The path of a body in free fall. { 'frē ,flīt trə'jek·trē }

free float |IND ENG| The length of time, expressed as work units, that a specific activity may be delayed without delaying the start of another activity scheduled to follow immediately after. Also known as free slack. { ¦frē 'flōt }

free gyroscope |ENG| A gyroscope that uses the property of gyroscopic rigidity to sense changes in altitude of a machine, such as an airplane; the spinning wheel or rotor is isolated from the airplane by gimbals; when the plane changes from level flight, the gyro remains vertical and gives the pilot an artificial horizon reference. { ¦frē ¦jī·rə,skōp }

freehand grinding See offhand grinding. { ¦frē ,hand 'grind·iŋ }

free instruments |ENG| Instruments designed to initially sink to the ocean bottom, release their ballast, and then rise to the surface where they are retrieved with their acquired payload. { ¦frē 'in·strə·məns }

free joint |MECH ENG| A robotic articulation that has six degrees of freedom. { 'frē 'jȯint }

free-mass antenna |ENG| A detector of gravitational radiation that consists of suspended, almost inertial masses and a laser interferometer that detects their motions. { ,frē ,mas an'ten·ə }

free-piston engine |MECH ENG| A prime mover utilizing free-piston motion controlled by gas pressure in the cylinders. { 'frē ,pis·tən 'en·jən }

free-piston gage |ENG| An instrument for measuring high fluid pressures in which the pressure is applied to the face of a small piston that can move in a cylinder and the force needed to keep the piston stationary is determined. Also known as piston gage. { ¦frē ¦pis·tən 'gāj }

free port |CIV ENG| An isolated, enclosed, and policed port in or adjacent to a port of entry, without a resident population. { 'frē ,pòrt }

free slack See free float. { ¦frē 'slak }

free-swelling index |ENG| A test for measuring the free-swelling properties of coal; consists of heating 1 gram of pulverized coal in a silica crucible over a gas flame under prescribed conditions to form a coke button, the size and shape of which are then compared with a series of standard profiles numbered 1 to 9 in increasing order of swelling. { 'frē ,swel·iŋ 'in,deks }

free turbine |MECH ENG| In a turbine engine, a turbine wheel that drives the output shaft and is not connected to the shaft driving the compressor. { ¦frē 'tər·bən }

free vector |MECH| A vector whose direction in space is prescribed but whose point of application and line of application are not prescribed. { ¦frē 'vek·tər }

freeze |ENG| **1.** To permit drilling tools, casing, drivepipe, or drill rods to become lodged in a borehole by reason of caving walls or impaction of sand, mud, or drill cuttings, to the extent that they cannot be pulled out. Also known as bind seize. **2.** To burn in a bit. Also known as burn-in. **3.** The premature setting of cement, especially when cement slurry hardens before it can be ejected fully from pumps or drill rods during a borehole cementation operation. **4.** The act or process of drilling a borehole by utilizing a drill fluid chilled to minus 30–40°F, (minus 34–40°C) as a means of consolidating, by freezing, the borehole wall materials or core as the drill penetrates a water-saturated formation, such as sand or gravel. { frēz }

freeze drying |ENG| A method of drying materials, such as certain foods, that would be destroyed by the loss of volatile ingredients or by drying temperatures above the freezing point; the material is frozen under high vacuum so that ice or other frozen solvent will quickly sublime and a porous solid remain. { 'frēz ,drī·iŋ }

freezer |MECH ENG| An insulated unit, compartment, or room in which perishable foods are quick-frozen and stored. { 'frēz·ər }

freeze-up |MECH ENG| Abnormal operation of a refrigerating unit because ice has formed at the expansion device. { 'frēz,əp }

freezing microtome |ENG| A microtome used to cut frozen tissue. { ¦frēz·iŋ 'mī·krə,tōm }

freight car |ENG| A railroad car in or on which freight is transported. { 'frāt ,kär }

freighter |ENG| A ship or aircraft used mainly for carrying freight. { 'frād·ər }

freight ton See ton. { 'frāt ,tən }

french |MECH| A unit of length used to measure small diameters, especially those of fiber optic bundles, equal to 1/3 millimeter. { french }

french coupling |DES ENG| A coupling having both right- and left-handed threads. { ¦french 'kəp·liŋ }

French drain |CIV ENG| An underground passage for water, consisting of loose stones covered with earth. { ¦french 'drān }

frequency characteristic See frequency-response curve. { ¦frē·kwən·sē ,kar·ik·tə'ris·tik }

frequency compensation See compensation. { ¦frē·kwən·sē ,käm·pən'sā·shən }

frequency domain |CONT SYS| Pertaining to a method of analysis, particularly useful for fixed linear systems in which one does not deal with functions of time explicitly, but with their Laplace or Fourier transforms, which are functions of frequency. { 'frē·kwən·sē də,mān }

frequency locus |CONT SYS| The path followed by the frequency transfer function or its inverse, either in the complex plane or on a graph of amplitude against phase angle; used in determining zeros of the describing function. { 'frē·kwən·sē ,lō·kəs }

frequency meter |ENG| **1.** An instrument for measuring the frequency of an alternating current; the scale is usually graduated in hertz, kilohertz, and megahertz. **2.** A device calibrated to indicate frequency of a radio wave. { 'frē·kwən·sē ,mēd·ər }

frequency-modulated radar |ENG| Form of radar in which the radiated wave is frequency modulated, and the returning echo beats with the wave being radiated, thus enabling range to be measured. { 'frē·kwən·sē ,mäj·ə,lād·əd 'rā ,där }

frequency-modulation Doppler |ENG| Type of radar involving frequency modulation of both carrier and modulation on radial sweep. { 'frē·kwən·sē ,mäj·ə,lā·shən 'däp·lər }

frequency-modulation synthesis |ENG ACOUS| A method of synthesizing musical tones which, in its simplest form, is carried out using two digital oscillators, with the output of one adding to the frequency (or phase) control of the other. { ¦frē·kwən·sē ,mä·jə'lā·shən ,sin·thə·səs }

frequency response |ENG| A measure of the effectiveness with which a circuit, device, or system transmits the different frequencies applied to it; it is a phasor whose magnitude is the ratio of the magnitude of the output signal to that of a sine-wave input, and whose phase is that of the output with respect to the input. Also known as amplitude-frequency response; sine-wave response. { 'frē·kwən·sē ri,späns }

frequency-response curve |ENG| A graph showing the magnitude or the phase of the frequency response of a device or system as a function of frequency. Also known as frequency characteristic. { 'frē·kwən·sē ri,späns ,kərv }

frequency-response trajectory [CONT SYS] The path followed by the frequency-response phasor in the complex plane as the frequency is varied. { 'frē·kwən·sē ri,späns trə'jek·trē }

frequency spectrum [SYS ENG] In the analysis of a random function of time, such as the amplitude of noise in a system, the limit as T approaches infinity of $1/(2\pi T)$ times the ensemble average of the squared magnitude of the amplitude of the Fourier transform of the function from $-T$ to T. Also known as power-density spectrum; power spectrum; spectral density. { 'frē·kwən·sē ,spek·trəm }

frequency study See work sampling. { 'frē·kwən·sē ,stəd·ē }

frequency transformation [CONT SYS] A transformation used in synthesizing a band-pass network from a low-pass prototype, in which the frequency variable of the transfer function is replaced by a function of the frequency. Also known as low-pass band-pass transformation. { ¦frē·kwən·sē ,tranz·fər'mā·shən }

fretsaw [DES ENG] A narrow-bladed fine-toothed saw that is held under tension in a frame. { 'fret,só }

friction [MECH] A force which opposes the relative motion of two bodies whenever such motion exists or whenever there exist other forces which tend to produce such motion. { 'frik·shən }

frictional grip [MECH] The adhesion between the wheels of a locomotive and the rails of the railroad track. { ¦frik·shən·əl 'grip }

friction bearing [MECH ENG] A solid bearing that directly contacts and supports an axle end. { 'frik·shən ,ber·iŋ }

friction bonding [ENG] Soldering of a semiconductor chip to a substrate by vibrating the chip back and forth under pressure to create friction that breaks up oxide layers and helps alloy the mating terminals. { 'frik·shən ,bänd·iŋ }

friction brake [MECH ENG] A brake in which the resistance is provided by friction. { 'frik·shən ,brāk }

friction calendering [ENG] Process wherein an elastomeric compound is forced into the interstices of woven or cord fabrics while passing between calender rolls. { 'frik·shən ,kal·ən·driŋ }

friction catch [DES ENG] A catch consisting of a spring and plunger contained in a casing. { 'frik·shən ,kach }

friction clutch [MECH ENG] A clutch in which torque is transmitted by pressure of the clutch faces on each other. { 'frik·shən ,kləch }

friction coefficient See coefficient of friction. { 'frik·shən ,kō·i'fish·ənt }

friction damping [MECH] The conversion of the mechanical vibrational energy of solids into heat energy by causing one dry member to slide on another. { 'frik·shən ,damp·iŋ }

friction drive [MECH ENG] A drive that operates by the friction forces set up when one rotating

wheel is pressed against a second wheel. { 'frik·shən ,drīv }

friction fit [DES ENG] A perfect fit between two parts. { 'frik·shən ,fit }

friction force microscopy [ENG] The use of an atomic force microscope to measure the frictional forces on a surface. { ¦frik·shən ¦fors mī'krä·ska·pē }

friction gear [MECH ENG] Gearing in which motion is transmitted through friction between two surfaces in rolling contact. { 'frik·shən ,gir }

friction horsepower [MECH ENG] Power dissipated in a machine through friction. { 'frik·shən 'hórs,paů·ər }

friction loss [MECH] Mechanical energy lost because of mechanical friction between moving parts of a machine. { 'frik·shən ,lós }

friction pile [CIV ENG] A bearing pile surrounded by earth and supported entirely by friction; carries no load at its end. { 'frik·shən ,pīl }

friction saw [MECH ENG] A toothless circular saw used to cut materials by fusion due to frictional heat. { 'frik·shən ,só }

friction sawing [MECH ENG] A burning process to cut stock to length by using a blade saw operating at high speed; used especially for the structural parts of mild steel and stainless steel. { 'frik·shən ,só·iŋ }

friction shoe [ENG] An adjustable friction device that holds a window sash in any desired open position. { 'frik·shən ,shü }

friction torque [MECH] The torque which is produced by frictional forces and opposes rotational motion, such as that associated with journal or sleeve bearings in machines. { 'frik·shən ,tórk }

friction-tube viscometer [ENG] Device to determine liquid viscosity by measurement of pressure drop through a friction tube with the liquid in viscous flow; gives direct solution to Poiseuille's equation. { 'frik·shən ,tüb vi'skäm·əd·ər }

friction welding [ENG] A welding process for metals and thermoplastic materials in which two members are joined by rubbing the mating faces together under high pressure. { 'frik·shən ,weld·iŋ }

frigorie [THERMO] A unit of rate of extraction of heat used in refrigeration, equal to 1000 fifteen-degree calories per hour, or 1.16264 ± 0.00014 watts. { 'frig·ə·rē }

frigorimeter [ENG] A thermometer which measures low temperatures. { ,frig·ə'rim·əd·ər }

fringe howl [ENG ACOUS] Squeal or howl heard when some circuit in a receiver is on the verge of oscillation. { 'frinj ,haůl }

frit seal [ENG] A seal made by fusing together metallic powders with a glass binder, for such applications as hermetically sealing ceramic packages for integrated circuits. { 'frit ,sēl }

fritting [ENG] Fusing materials for glass by application of heat. { 'frid·iŋ }

frog [DES ENG] A hollow on one or both of the larger faces of a brick or block; reduces weight of the brick or block; may be filled with mortar. Also known as panel. [ENG] A device which permits the train or tram wheels on one rail of a track to cross the rail of an intersecting track. { fräg }

from-to tester [ENG] Test equipment which checks continuity or impedance between points. { ¦frəm ‚tü ‚test·ər }

front-end loader [MECH ENG] An excavator consisting of an articulated bucket mounted on a series of movable arms at the front of a crawler or rubber-tired tractor. { ¦frənt ¦end 'lōd·ər }

front-end volatility [CHEM ENG] The volatility of the lower-boiling fractions of gasoline, such as butanes. { ¦frənt ¦end väl·ə'til·əd·ē }

front slagging [ENG] Skimming slag from the mixture of slag and molten metal as it flows through a taphole. { 'frənt ‚slag·iŋ }

frosting [ENG] Decorating a scraped metal surface with a handscraper. Also known as flaking. { 'frȯst·iŋ }

frost-point hygrometer [ENG] An instrument for measuring the frost point of the atmosphere; air under test is passed continuously across a polished surface whose temperature is adjusted so that a thin deposit of frost is formed which is in equilibrium with the air. { 'frȯst ‚pȯint hī'gräm·əd·ər }

froth flotation [ENG] A process for recovery of particles of ore or other material, in which the particles adhere to bubbles and can be removed as part of the froth. { ¦frȯth flō'tā·shən }

frothing [ENG] The producing of relatively stable bubbles at an air-liquid interface as the result of agitation, aeration, ebulliation, or chemical reaction; it can be an undesired side effect, but in minerals beneficiation it is the basis of froth flotation. { frȯ·thiŋ }

ft-lb See foot-pound.

ft-lbf See foot-pound.

ft-pdl See foot-poundal.

fuel bed [MECH ENG] A layer of burning fuel, as on a furnace grate or a cupola. { 'fyül ‚bed }

fuel filter [ENG] A device, as in an internal combustion engine, that removes particles from the fuel. { 'fyül ‚fil·tər }

fuel injection [MECH ENG] The delivery of fuel to an internal combustion engine cylinder by pressure from a mechanical pump. { 'fyül in ‚iek·shən }

fuel injector [MECH ENG] A pump mechanism that sprays fuel into the cylinder of an internal combustion engine at the appropriate part of the cycle. { 'fyül in‚iek·tər }

fuel pump [MECH ENG] A pump for drawing fuel from a storage tank and delivering it to an engine or furnace. { 'fyül ‚pəmp }

fuel system [MECH ENG] A system which stores fuel for present use and delivers it as needed. { 'fyül ‚sis·təm }

fuel tank [MECH ENG] The operating, fuel-storage component of a fuel system. { 'fyül ‚taŋk }

fugacity [THERMO] A function used as an analog of the partial pressure in applying thermodynamics to real systems; at a constant temperature it is proportional to the exponential of the ratio of the chemical potential of a constituent of a system divided by the product of the gas

constant and the temperature, and it approaches the partial pressure as the total pressure of the gas approaches zero. { fyü'gas·əd·ē }

fugacity coefficient [THERMO] The ratio of the fugacity of a gas to its pressure. { fyü'gas·əd·ē ‚kō·ə‚fish·ənt }

fulchronograph [ENG] An instrument for recording lightning strokes, consisting of a rotating aluminum disk with several hundred steel fins on its rim; the fins are magnetized if they pass between two coils when these are carrying the surge current of a lightning stroke. { fül'krän·ə‚graf }

fulcrum [MECH] The rigid point of support about which a lever pivots. { 'fül·krəm }

fulgurator [ENG] An atomizer used to spray salt solutions into a flame for analysis. { 'fül·gə‚rād·ər }

full adder [ELECTR] A logic element which operates on two binary digits and a carry digit from a preceding stage, producing as output a sum digit and a new carry digit. Also known as three-input adder. { ¦fül 'ad·ər }

full-cell process [ENG] A process of preservative treatment of wood that uses a pressure vessel and first draws a vacuum on the charge of wood and then introduces the preservative without breaking the vacuum. Also known as Bethell process. { ¦fül ¦sel 'präs·əs }

full-face tunneling [CIV ENG] A system of tunneling in which the tunnel opening is enlarged to desired diameter before extension of the tunnel face. { ¦fül ‚fās 'tən·əl·iŋ }

full-gear [MECH ENG] The condition of a steam engine when the valve is operated to the maximum extent by the link motion. { ¦fül 'gir }

full-mill [BUILD] A type of construction in which all vertical apertures open onto shafts of brick or other fireproof material; used for fire retardance. { ¦fül 'mil }

full subtracter [ELECTR] A logic element which operates on three binary input signals representing a minuend, subtrahend, and borrow digit, producing as output a different digit and a new borrow digit. Also known as three-input subtracter. { ¦fül səb'trak·tər }

full-track vehicle [MECH ENG] A vehicle entirely supported, driven, and steered by an endless belt, or track, on each side; for example, a tank. { ¦fül ‚trak 've·ə·kəl }

full trailer [MECH ENG] A towed vehicle whose weight rests completely on its own wheels. { ¦fül 'trāl·ər }

fumble [IND ENG] An unintentional sensory-motor error that may be unavoidable. { 'fəm·bəl }

fumigating [ENG] The use of a chemical compound in a gaseous state to kill insects, nematodes, arachnids, rodents, weeds, and fungi in confined or inaccessible locations; also used to control weeds, nematodes, and insects in the field. { 'fyü·mə‚gād·iŋ }

funal See sthène. { 'fyün·əl }

functional analysis [SYS ENG] A part of the design process that addresses the activities that a system, software, or organization must perform to achieve its desired outputs, that is, the transformations necessary to turn available inputs into the desired outputs. { ¦'faŋk·shən·əl ə'nal·ə·səs }

functional analysis diagram [SYS ENG] A representation of functional analysis and, in particular, the transformations necessary to turn available inputs into the desired outputs, the flow of data or items between functions, the processing instructions that are available to guide the transformation, and the control logic that dictates the activation and termination of functions. { ¦faŋk·shən·əl ə'nal·ə·səs ‚dī·ə‚gram }

functional decomposition [CONT SYS] The partitioning of a large-scale control system into a nested set of generic control functions, namely the regulatory or direct control function, the optimizing control function, the adaptive control function, and the self-organizing function. { 'faŋk·shən·əl dē‚käm·pə'zish·ən }

functional design [SYS ENG] The aspect of system design concerned with the system's objectives and functions, rather than its specific components. { 'faŋk·shən·əl di'zīn }

functional forms analysis chart See form process chart. { 'faŋk·shən·əl 'fȯrmz ə‚nal·ə·səs ‚chärt }

function failure safety [ENG] The capability of an electronic-mass measuring instrument to withhold the release of an incorrect measurement when there is a function failure. { 'faŋk·shən ¦fālyər ‚sāf·tē }

fundamental interval [THERMO] 1. The value arbitrarily assigned to the difference in temperature between two fixed points (such as the ice point and steam point) on a temperature scale, in order to define the scale. 2. The difference between the values recorded by a thermometer at two fixed points; for example, the difference between the resistances recorded by a resistance thermometer at the ice point and steam point. { ¦fən·də¦men·təl 'int·ər·vəl }

fundamental motion See elemental motion. { ¦fən·də¦ment·əl 'mō·shən }

fungible [CHEM ENG] Pertaining to petroleum products whose characteristics are so similar they can be commingled. { 'fən·jə·bəl }

fungi-proofing [ENG] Application of a protective chemical coating that inhibits growth of fungi. { 'fən‚jī ‚prüf·iŋ }

funicular See funicular railroad. { fə'nik·yə·lər }

funicular polygon [MECH] 1. The figure formed by a light string hung between two points from which weights are suspended at various points. 2. A force diagram for such a string, in which the forces (weights and tensions) acting on points of the string from which weights are suspended are represented by a series of adjacent triangles. { fə'nik·yə·lər 'päl·ə‚gän }

funicular railroad [ENG] A railroad system used primarily to ascend and descend mountains; the weight of the descending train helps to move the ascending train up the mountain. Also known as funicular. { fə'nik·yə·lər 'rāl‚rōd }

funnel [DES ENG] A tube with one conical end

that sometimes holds a filter; the function is to direct flow of a liquid or, if a filter is present, to direct a flow that was filtered. { 'fən·əl }

funnel-flow bin [ENG] A bin in which solid flows toward the outlet in a channel that forms within stagnant material. { 'fən·əl ¦flō ‚bin }

furfural extraction [CHEM ENG] Process for the refining of lubricating oils and other organic materials by contact with furfural. { 'fər‚fa‚ral ik'strak·shən }

furlong [MECH] A unit of length, equal to 1/8 mile, 660 feet, or 201.168 meters. { 'fər‚lȯŋ }

furnace [ENG] An apparatus in which heat is liberated and transferred directly or indirectly to a solid or fluid mass for the purpose of effecting a physical or chemical change. { 'fər·nəs }

furnace lining [ENG] The interior part of a furnace in contact with a molten charge and hot gases; constructed of heat-resistant material. { 'fər·nəs ‚līn·iŋ }

furnish [CHEM ENG] In papermaking, the raw materials placed in a beater for producing paper pulp. { 'fər·nish }

furred ceiling [BUILD] A ceiling in which the furring units are attached directly to the structural units of the building. { ¦fərd 'sē·liŋ }

furring [BUILD] Thin strips of wood or metal fastened to joists, studs, ceilings, or inner walls of a building to provide a level surface or air space over which the finished surface can be applied. Also known as batten; furring strip. { 'fər·iŋ }

furring strip See furring. { 'fər·iŋ ‚strip }

furrow [ENG] A trench plowed in the ground. { 'fər‚ō }

fuse [ELEC] An expendable device for opening an electric circuit when the current therein becomes excessive, containing a section of conductor which melts when the current through it exceeds a rated value for a definite period of time. Also known as electric fuse. [ENG] Also spelled fuze. 1. A device with explosive components designed to initiate a train of fire or detonation in an item of ammunition by an action such as hydrostatic pressure, electrical energy, chemical energy, impact, or a combination of these. 2. A nonexplosive device designed to initiate an explosion in an item of ammunition by an action such as continuous or pulsating electromagnetic waves or acceleration. { fyüz }

fuse blasting cap [ENG] A small copper cylinder closed at one end and charged with a fulminate. { ¦fyüz 'blast·iŋ ‚kap }

fuse body [ENG] The part of a fuse contributing the major portion of the total weight, and which houses the majority of the functioning parts, and to which smaller parts are attached. { 'fyüz ‚bäd·ē }

fuse diode [ELECTR] A diode that opens under specified current surge conditions. { 'fyüz ‚dī‚ōd }

fused junction See alloy junction. { ¦füzd 'jəŋk·shən }

fused-junction diode See alloy-junction diode. { ¦fyüzd ¦jəŋk·shən 'dī‚ōd }

fused-junction transistor See alloy-junction transistor. { ¦fyüzd ¦jəŋk·shən tran'zis·tər }

fused semiconductor |ELECTR| Junction formed by recrystallization on a base crystal from a liquid phase of one or more components and the semiconductor. { ¦fyüzd 'sem·i·kən ‚dək·tər }

fuse gage |ENG| An instrument for slicing time fuses to length. { 'fyüz ‚gāj }

fusehead |ENG| That part of an electric detonator consisting of twin metal conductors, bridged by fine resistance wire, and surrounded by a bead of igniting compound which burns when the firing current is passed through the bridge wire. { 'fyüz‚hed }

fuse lighter |ENG| A device for facilitating the ignition of the powder core of a fuse. { 'fyüz ‚līd·ər }

fusibility |THERMO| The quality or degree of being capable of being liquefied by heat. { ‚fyü·zə'bil·əd·ē }

fusible plug See safety plug. { ¦fyü·zə·bəl 'pləg }

fusing disk |MECH ENG| A rapidly spinning disk that cuts metal by melting it. { 'fyüz·iŋ ‚disk }

fusion piercing |ENG| A method of producing vertical blastholes by virtually burning holes in rock. Also known as piercing. { 'fyü·zhən ‚pir·siŋ }

fusion-piercing drill |ENG| A machine designed to use the fusion-piercing mode of producing holes in rock. Also known as det drill; jet-piercing drill; Linde drill. { 'fyü·zhən ‚pirs·iŋ ‚dril }

fuzzy controller |CONT SYS| An automatic controller in which the relation between the state variables of the process under control and the action variables, whose values are computed from observations of the state variables, is given as a set of fuzzy implications or as a fuzzy relation. { ¦fəz·ē kən'trōl·ər }

fuzzy system |SYS ENG| A process that is too complex to be modeled by using conventional mathematical methods, and that gives rise to data that are, in general, soft, with no precise boundaries; examples are large-scale engineering complex systems, social systems, economic systems, management systems, medical diagnostic processes, and human perception. { ¦fəz·ē 'sis·təm }

G

g See gram.

G |ELEC| See conductance. |MECH| A unit of acceleration equal to the standard acceleration of gravity, 9.80665 meters per second per second, or approximately 32.1740 feet per second per second. Also known as fors; grav.

GaAs FET See gallium arsenide field-effect transistor. { 'gas,fet }

gabion |ENG| A bottomless basket of wickerwork or metal iron filled with earth or stones; used in building fieldworks or as revetments in mining. Also known as pannier. { 'gā·bē·ən }

gableboard See vergeboard. { 'gā·bəl,bórd }

Gabor trolley |ENG| A small three-wheel trolley with knife-edge wheels, used in constructing trajectories of charged particles in an electric field. { 'gä,bòr ,trä·lē }

gage Also spelled gauge. |CIV ENG| The distance between the inner faces of the rails of railway track; standard gage in the United States is 4 feet 8½ inches (1.44 meters). |DES ENG| **1.** A device for determining the relative shape or size of an object. **2.** The thickness of a metal sheet, a rod, or a wire. |ENG| The minimum sieve size through which most (95% or more) of an aggregate will pass. { 'gāj }

gage block |DES ENG| A chrome steel block having two flat, parallel surfaces with the parallel distance between them being the size marked on the block to a guaranteed accuracy of a few millionths of an inch; used as the standard of precise lineal measurement for most manufacturing processes. Also known as precision block; size block. { 'gāj ,bläk }

gage cock |ENG| A valve located on a water column of a boiler drum. { 'gāj ,käk }

gage glass |ENG| A glass, plastic, or metal tube, usually equipped with shutoff valves, that is connected by a suitable fitting to a tank or vessel, for the measurement of liquid level. { 'gāj,glas }

gage length |ENG| Original length of the portion of a specimen measured for strain, length changes, and other characteristics. { 'gāj ,leŋkth }

gage plate |CIV ENG| A plate inserted between the parallel rails of a railroad track to maintain the gage. { 'gāj ,plāt }

gage point |DES ENG| A point used to position

a part in a jig, fixture, or qualifying gage. { 'gāj ,pòint }

gage pressure |MECH ENG| The amount by which the total absolute pressure exceeds the ambient atmospheric pressure. { 'gāj ,presh·ər }

gaging hatch |ENG| An opening in a tank or other vessel through which measuring and sampling can be performed. { 'gāj·iŋ ,hach }

gaging tape |ENG| A metal measuring tape used to determine the depth of liquid in a tank. { 'gāj·iŋ ,tāp }

gain |ELECTR| The increase in signal power that is produced by an amplifier; usually given as the ratio of output to input voltage, current, or power, expressed in decibels. Also known as transmission gain. |ENG| A cavity in a piece of wood prepared by notching or mortising so that a hinge or other hardware or another piece of wood can be placed on the cavity. { gān }

gain asymptotes |CONT SYS| Asymptotes to a logarithmic graph of gain as a function of frequency. { 'gān 'as·əm,tōts }

gain-crossover frequency |CONT SYS| The frequency at which the magnitude of the loop ratio is unity. { ¦gān ¦kròs,ō·vər ,frē·kwən·sē }

gain margin |CONT SYS| The reciprocal of the magnitude of the loop ratio at the phase crossover frequency, frequently expressed in decibels. { 'gān ,mär·jən }

gain scheduling |CONT SYS| A method of eliminating influences of variations in the process dynamics of a control system by changing the parameters of the regulator as functions of auxiliary variables which correlate well with those dynamics. { 'gān ,skej·ə·liŋ }

gal |MECH| **1.** The unit of acceleration in the centimeter-gram-second system, equal to 1 centimeter per second squared; commonly used in geodetic measurement. Formerly known as galileo. Symbolized Gal. **2.** See gallon. { gal }

Gal See gal. { gal }

Galilean transformation |MECH| A mathematical transformation used to relate the space and time variables of two uniformly moving (inertial) reference systems in nonrelativistic kinematics. { ,gal·ə¦lē·ən ,tranz·fər'mā·shən }

galileo See gal. { ,gal·ə'lē·ō }

Galileo's law of inertia See Newton's first law. { ,gal·ə'lē·ōz ¦lò əv i'nər·shə }

Galitzin pendulum | MECH | A massive horizontal pendulum that is used to measure variations in the direction of the force of gravity with time, and thus serves as the basis of a seismograph. { gä¦lit·sən 'pen·jə·ləm }

galley | ENG | The kitchen of a ship, airplane, or trailer. { 'gal·ē }

gallium arsenide field-effect transistor | ELECTR | A field-effect transistor in which current between the ohmic source and drain contacts is carried by free electrons in a channel consisting of *n*-type gallium arsenide, and this current is modulated by a Schottky-barrier rectifying contact called the gate that varies the cross-sectional area of the channel. Abbreviated GaAs FET. { 'gal·ē·əm 'ärs·ən‚īd 'fēld i¦fekt tran'zis·tər }

gallon | MECH | Abbreviated gal. **1.** A unit of volume used in the United States for measurement of liquid substances, equal to 231 cubic inches, or to 3.785 411 784 × 10⁻³ cubic meter, or to 3.785 411 784 liters; equal to 128 fluid ounces. **2.** A unit of volume used in the United Kingdom for measurement of liquid and solid substances, usually the former; equal to 4.54609 × 10⁻³ cubic meter, or to 4.54609 liters; equal to 160 fluid ounces. Also known as imperial gallon. { 'gal·ən }

Galton whistle | ENG ACOUS | A short cylindrical pipe with an annular nozzle, which is set into resonant vibration in order to generate ultrasonic sound waves. { 'gȯl·tən ‚wis·əl }

galvanic | ELEC | Pertaining to electricity flowing as a result of chemical action. { gal'van·ik }

galvanic battery | ELEC | A galvanic cell, or two or more such cells electrically connected to produce energy. { gal'van·ik 'bad·ə·rē }

galvanic cell | ELEC | An electrolytic cell that is capable of producing electric energy by electrochemical action. { gal'van·ik 'sel }

galvanic couple | ELEC | A pair of unlike substances, such as metals, which generate a voltage when brought in contact with an electrolyte. { gal'van·ik 'kəp·əl }

galvanic current | ELEC | A steady direct current. { gal'van·ik 'kə·rənt }

galvanometer | ENG | An instrument for indicating or measuring a small electric current by means of a mechanical motion derived from electromagnetic or electrodynamic forces produced by the current. { ‚gal·və'näm·əd·ər }

galvanometer recorder | ENG ACOUS | A sound recorder in which the audio signal voltage is applied to a coil suspended in a magnetic field; the resulting movements of the coil cause a tiny attached mirror to move a reflected light beam back and forth across a slit in front of a moving photographic film. { ‚gal·və'näm·əd·ər ri'kȯrd·ər }

gambrel roof | BUILD | A roof with two sloping sides stepped at different angles on each side of the center ridge; the lower slope is steeper than the upper slope. { 'gam·brəl 'rüf }

gamma | MECH | A unit of mass equal to 10⁻⁶ gram or 10⁻⁹ kilogram. { 'gam·ə }

gamma camera | ENG | An instrument consisting of a large, thin scintillation crystal or array of photomultiplier tubes, a multichannel collimator, and circuitry to analyze the pulses produced by the photomultipliers; used to visualize the distribution of radioactive compounds in the human body. { 'gam·ə ‚kam·rə }

gamma counter | ENG | A device for detecting gamma radiation, primarily through the detection of fast electrons produced by the gamma rays; it either yields information about integrated intensity within a time interval or detects each photon separately. { 'gam·ə ‚kaȯnt·ər }

gamma logging | ENG | Obtaining, by means of a gamma-ray probe, a record of the intensities of gamma rays emitted by the rock strata penetrated by a borehole. { 'gam·ə ‚läg·iŋ }

gamma-ray altimeter | ENG | An altimeter, used at altitudes under several hundred feet, that measures the photon backscatter from the earth resulting from the transmission of photons to earth from a cobalt-60 gamma source in the plane. { 'gam·ə ‚rā al'tim·əd·ər }

gamma-ray detector | ENG | An instrument that registers the presence of gamma rays. { 'gam·ə ‚rā di'tek·tər }

gamma-ray level indicator | ENG | A level indicator in which the rising level of the liquid or other material reduces the amount of radiation passing from a gamma-ray source through the container to a Geiger counter or other radiation detector. { 'gam·ə ‚rā ¦lev·əl 'in·də‚kād·ər }

gamma-ray probe | ENG | A gamma-ray counter built into a watertight case small enough to be lowered into a borehole. { 'gam·ə ‚rā ‚prōb }

gamma-ray tracking | ENG | Use of three tracking stations, located at the three corners of a triangle centered on a missile about to be launched, to obtain accurate azimuthal tracking of a cobalt-60 gamma source in the tail. { 'gram·ə ‚rā 'trak·iŋ }

gamma-ray well logging | ENG | Measurement of gamma-ray intensity versus depth down the wellbore; used to identify rock strata, their position, and their thicknesses. { 'gam·ə ‚rā 'wel ‚läg·iŋ }

gammeter | ENG | A template fashioned of transparent material and marked with a calibrated scale; when positioned on a sensitometric curve it is used to determine the slope of the straight-line portion. { 'ga‚mēd·ər }

gang | ELEC | A mechanical connection of two or more circuit devices so that they can be varied at the same time. { gaŋ }

gang chart | IND ENG | A multiple-activity process chart used for groups of men on materials-handling operations. { 'gaŋ ‚chärt }

gang drill | MECH ENG | A set of drills operated together in the same machine; used in rock drilling. { 'gaŋ ‚dril }

gang milling | ENG | Rolling of material by means of a composite machine with numerous cutting blades. { 'gaŋ ‚mil·iŋ }

gang saw | MECH ENG | A steel frame in which

thin, parallel saws are arranged to operate simultaneously in cutting logs. { 'gaŋ ˌsȯ }

gantlet |CIV ENG| A stretch of overlapping railroad track, with one rail of one track being between the two rails of another track; used over narrow bridges and passes. { 'gȯnt·lət }

gantry |ENG| A frame erected on side supports so as to span an area and support and hoist machinery and heavy materials. { 'gan·trē }

gantry crane |MECH ENG| A bridgelike hoisting machine having fixed supports or arranged for running along tracks on ground level. { 'gan·trē ˌkrān }

gantry-type robot |CONT SYS| A continuous-path, Cartesian-coordinate robot constructed in a bridge shape that uses rails to move along a single horizontal axis or along either of two perpendicular horizontal axes. { 'gan·trē ˌtīp 'rō,bät }

Gantt chart |IND ENG| In production planning and control, a type of bar chart depicting the work planned and done in relation to time; each division of space represents both a time interval and the amount of work to be done during that interval. { 'gant ˌchärt }

Gantt task and bonus plan |IND ENG| A wage incentive plan in which high task efficiency is maintained by providing a percentage bonus as a reward for production in excess of standard. { 'gant ˌtask ən 'bō·nəs ˌplan }

gap |ELEC| The spacing between two electric contacts. { gap }

gap-filler radar |ENG| Radar used to fill gaps in radar coverage of other radar. { 'gap ˌfil·ər 'rā,där }

gap-framepress |MECH ENG| A punch press whose frame is open at bed level so that wide work or strip work can be inserted. { 'gap 'frām,pres }

gap lathe |MECH ENG| An engine lathe with a sliding bed providing enough space for turning large-diameter work. { 'gap ˌlāth }

gap scanning |ENG| In ultrasonic testing, a coupling technique in which a sound beam is projected through a short fluid column that flows through a nozzle on an ultrasonic search unit. { 'gap ˌskan·iŋ }

garnet hinge |DES ENG| A hinge with a vertical bar and horizontal strap. { 'gär·nət ˌhinj }

garret |BUILD| The part of a house just under the roof. { 'gar·ət }

garter spring |DES ENG| A closed ring formed of helically wound wire. { 'gärd·ər ˌspriŋ }

gas absorption operation |CHEM ENG| The recovery of solute gases present in gaseous mixtures of noncondensables; this recovery is generally achieved by contacting the gas stream with a liquid that offers specific or selective solubility for the solute gas to be recovered, or with an adsorbent (for example, synthetic or natural zeolite) that accepts only specific molecule sizes or shapes. { 'gas əb,sȯrp·shən ˌap·ə,rā·shən }

gas bag |ENG| A bag made of gas-impermeable

material and designed for insertion into a pipeline followed by inflation to halt the flow of gas. { 'gas ˌbag }

gas bearing |MECH ENG| A journal or thrust bearing lubricated with gas. Also known as gas-lubricated bearing. { 'gas ˌber·iŋ }

gas burner |ENG| A hole or a group of holes through which a combustible gas or gas-air mixture flows and burns. { 'gas ˌbər·nər }

gas cleaning |ENG| Removing ingredients, pollutants, or contaminants from domestic and industrial gases. { 'gas ˌklēn·iŋ }

gas-compression cycle |MECH ENG| A refrigeration cycle in which hot, compressed gas is cooled in a heat exchanger, then passes into a gas expander which provides an exhaust stream of cold gas to another heat exchanger that handles the sensible-heat refrigeration effect and exhausts the gas to the compressor. { 'gas kəm,presh·ən ˌsī·kəl }

gas compressor |MECH ENG| A machine that increases the pressure of a gas or vapor by increasing the gas density and delivering the fluid against the connected system resistance. { 'gas kəm,pres·ər }

gas constant |THERMO| The constant of proportionality appearing in the equation of state of an ideal gas, equal to the pressure of the gas times its molar volume divided by its temperature. Also known as gas-law constant; universal gas constant. { 'gas ˌkän·stənt }

gas cycle |THERMO| A sequence in which a gaseous fluid undergoes a series of thermodynamic phases, ultimately returning to its original state. { 'gas ˌsī·kəl }

gas cylinder |MECH ENG| The chamber in which a piston moves in a positive displacement engine or compressor. { 'gas ˌsil·ən·dər }

gas dehydrator |CHEM ENG| A device or system to remove moisture vapor from a gas stream, usually incorporates desiccant-type packed towers. { 'gas dē'hī,drād·ər }

gas-deviation factor See compressibility factor. { 'gas ˌdē·vē'ā·shən ˌfak·tər }

gas engine |MECH ENG| An internal combustion engine that uses gaseous fuel. { 'gas ˌen·jən }

gaseous conduction analyzer |ENG| A device to detect organic vapors in air by measuring the change in current that flows between a heated platinum anode and a concentric platinum cathode. { 'gash·əs kən'dək·shən 'an·ə,līz·ər }

gaseous diffusion |CHEM ENG| **1.** Pressure-induced free-molecular transfer of gas through microporous barriers as in the process of making fissionable fuel. **2.** Selective solubility diffusion of gas through nonporous polymers by absorption and solution of the gas in the polymer matrix. { 'gash·əs di'fyü·zhən }

gas etching |ENG| The removal of material from a semiconductor circuit by reaction with a gas that forms a volatile compound. { 'gas ˌech·iŋ }

GasFET |ENG| A gas sensor based on changes, upon exposure to hydrogen, in the surface part of the work function of a palladium component that serves as the gate contact of a metal oxide semiconductor field-effect transistor (MOSFET). { 'gas,fet }

gas-filled thermometer [ENG] A thermometer which uses a gas (usually nitrogen or hydrogen), that approximately follows the ideal gas law. { 'gas ,fild thər'mäm·əd·ər }

gas filter [CHEM ENG] A device used to remove liquid or solid particles from a flowing gas stream. { 'gas fil·tər }

gas furnace [ENG] An enclosure in which a gaseous fuel is burned. { 'gas ,fər·nəs }

gas generator [CHEM ENG] A chemical plant for producing gas from coal, for example, water gas. [MECH ENG] An apparatus that supplies a high-pressure gas flow to drive compressors, airscrews, and other machines. { 'gas ,jen·ə ,rād·ər }

gas heater [MECH ENG] A unit heater designed to supply heat by forced convection, using gas as a heat source. { 'gas ,hēd·ər }

gas holder [ENG] Gas storage container with vertically free top section that moves up or down to adjust to the volume of gas held. { 'gas ,hōl·dər }

gas hole [ENG] A cavity formed in a casting as a result of cavitation. { 'gas ,hōl }

gasification [CHEM ENG] Any chemical or heat process used to convert a substance to a gas; coal is converted by the Hygas process to a gaseous fuel. { 'gas·ə·fə'kā·shən }

gasifier [CHEM ENG] A unit for producing gas, particularly synthesis gas from coal. { 'gas· ə,fī·ər }

gas injection [MECH ENG] Injection of gaseous fuel into the cylinder of an internal combustion engine at the appropriate part of the cycle. { 'gas in,jek·shən }

gasket [ENG] A packing made of deformable material, usually in the form of a sheet or ring, used to make a pressure-tight joint between stationary parts. Also known as static seal. { 'gas·git }

gas law [THERMO] Any law relating the pressure, volume, and temperature of a gas. { 'gas ,lò }

gas-law constant See gas constant. { 'gas ,lò ,kän·stənt }

gas lift [CHEM ENG] Solids movement operation in which an upward-flowing gas stream in a closed conduit or vessel is used to lift and move powdered or granular solid material. { 'gas ,lift }

gas making [CHEM ENG] Making water gas or air gas by the action of steam and air upon hot coke. { 'gas ,māk·iŋ }

gas manometer [ENG] A gage for determining the difference in pressure of two gases, usually by measuring the difference in height of liquid columns in the two sides of a U-tube. { ¦gas mə'näm·əd·ər }

gas mask [ENG] A device to protect the eyes and respiratory tract from noxious gases, vapors, and aerosols, by removing contamination with a filter and a bed of adsorbent material. { 'gas ,mask }

gas meter [ENG] An instrument for measuring and recording the amount of gas flow through a pipe. { 'gas ,mēd·ər }

gasoline engine [MECH ENG] An internal combustion engine that uses a mixture of air and gasoline vapor as a fuel. { 'gas·ə,lēn 'en·jən }

gasoline pump [MECH ENG] A device that pumps and measures the gasoline supplied to a motor vehicle, as at a filling station. { 'gas· ə,lēn ,pəmp }

gasometer [ENG] A piece of equipment that holds and measures gas; may be used in analytical chemistry to measure the quantity of gas evolved in a reaction. { ga'säm·əd·ər }

gas packing [IND ENG] Packing a material such as food in an atmosphere consisting of an oxygen-free gas. { 'gas ,pak·iŋ }

gas pliers [DES ENG] Pliers for gripping round objects such as pipes, tubes, and circular rods. { 'gas ,plī·ərz }

gas producer [CHEM ENG] A device for complete gasification of coal by utilizing simultaneously the air and water-gas reactions. { 'gas prə,düs·ər }

gas reversion [CHEM ENG] A process which combines thermal cracking or reforming of naphtha with thermal polymerization or alkylation of hydrocarbon gases carried out in the same reaction zone. { 'gas ri'vər·zhən }

gas scrubbing [CHEM ENG] Removal of gaseous or liquid impurities from a gas by the action of a liquid; the gas is contacted with the liquid which removes the impurities by dissolving or by chemical combination. { 'gas ,skrəb·iŋ }

gas seal [ENG] A seal which prevents gas from leaking to or from a machine along a shaft. { 'gas ,sēl }

gassing [ELEC] The evolution of gas in the form of small bubbles in a storage battery when charging continues after the battery has been completely charged. [ENG] **1.** Absorption of gas by a material. **2.** Formation of gas pockets in a material. **3.** Evolution of gas from a material during a process or procedure. { 'gas·iŋ }

gas tank [ENG] A tank for storing gas or gasoline. { 'gas ,taŋk }

gas thermometer [ENG] A device to measure temperature by measuring the pressure exerted by a definite amount of gas enclosed in a constant volume; the gas (preferably hydrogen or helium) is enclosed in a glass or fused-quartz bulb connected to a mercury manometer. Also known as constant-volume gas thermometer. { ¦gas thər'mäm·əd·ər }

gas thermometry [ENG] Measurement of temperatures with a gas thermometer; used with helium down to about 1 K. { ¦gas thər'mäm· ə·trē }

gas trap [CIV ENG] A bend or chamber in a drain or sewer pipe that prevents sewer gas from escaping. { 'gas ,trap }

gas-treating system [CHEM ENG] A process system to remove nonhydrocarbon impurities (such as water vapor, hydrogen sulfide, or carbon dioxide) from wellhead gas. { 'gas ¦trēd·iŋ ¦sis·təm }

gas-tube boiler See waste-heat boiler. { 'gas ¦tüb 'bóilər }

gas turbine [MECH ENG] A heat engine that converts the energy of fuel into work by using compressed, hot gas as the working medium and that usually delivers its mechanical output power either as torque through a rotating shaft (industrial gas turbines) or as jet power in the form of velocity through an exhaust nozzle (aircraft jet engines). Also known as combustion turbine. { 'gas ¦tər·bən }

gas-turbine nozzle [MECH ENG] The component of a gas turbine in which the hot, high-pressure gas expands and accelerates to high velocity. { 'gas ¦tər·bən ¦näz·əl }

gas valve [ENG] An exhaust valve, held shut by rubber springs, used to discharge gas from the extreme top of a balloon. { 'gas ¦valv }

gas vent [ENG] A pipe or hole that allows gas to pass off. { 'gas ¦vent }

gate [CIV ENG] A movable barrier across an opening in a large barrier, a fence, or a wall. [ELECTR] **1.** A circuit having an output and a multiplicity of inputs and so designed that the output is energized only when a certain combination of pulses is present at the inputs. **2.** A circuit in which one signal, generally a square wave, serves to switch another signal on and off. **3.** One of the electrodes in a field-effect transistor. **4.** An output element of a cryotron. **5.** To control the passage of a pulse or signal. **6.** In radar, an electric waveform which is applied to the control point of a circuit to alter the mode of operation of the circuit at the time when the waveform is applied. Also known as gating waveform. [ENG] **1.** A device, such as a valve or door, for controlling the passage of materials through a pipe, channel, or other passageway. **2.** A device for positioning the film in a camera, printer, or projector. { gāt }

gate-array device [ELECTR] An integrated logic circuit that is manufactured by first fabricating a two-dimensional array of logic cells, each of which is equivalent to one or a few logic gates, and then adding final layers of metallization that determine the exact function of each cell and interconnect the cells to form a specific network when the customer orders the device. { 'gāt ə¦rā di¦vīs }

Gates crusher [MECH ENG] A gyratory crusher which has a cone or mantle that is moved eccentrically by the lower bearing sleeve. { 'gāts 'krəsh·ər }

gate valve [MECH ENG] A valve with a disk-shaped closing element that fits tightly over an opening through which water passes. { 'gāt ¦valv }

gathering iron [ENG] A rod used to collect molten glass for glassblowing. { 'gath·ə·riŋ ¦T·ərn }

gathering ring [ENG] A clay ring placed on molten glass to collect impurities and thus permit high-quality glass to be taken from the center. { 'gath·ə·riŋ ¦riŋ }

gating [ELECTR] The process of selecting those portions of a wave that exist during one or more selected time intervals or that have magnitudes between selected limits. [ENG] A network of connecting channels, including sprues, runners, gates, and cavities, which conduct molten metal to the mold. { 'gād·iŋ }

gating waveform See gate. { ¦gād·iŋ 'wāv¦fórm }

Gaussian weighing method [ENG] A method used to determine the accuracy of equal-arm balances and to test standard weights in which the sample is placed on one pan and the comparative weights on the other, and then the weights are interchanged in a second weighing. { 'gaús·ē·ən 'wā·iŋ ¦meth·əd }

gaussmeter [ENG] A magnetometer whose scale is graduated in gauss or kilogauss, and usually measures only the intensity, and not the direction, of the magnetic field. { 'gaús¦mēd·ər }

Gauss method of weighing See double weighing. { ¦gaús ¦meth·əd əv 'wā·iŋ }

Gauss' principle of least constraint [MECH] The principle that the motion of a system of interconnected material points subjected to any influence is such as to minimize the constraint on the system; here the constraint, during an infinitesimal period of time, is the sum over the points of the product of the mass of the point times the square of its deviation from the position it would have occupied at the end of the time period if it had not been connected to other points. { ¦gaús 'prin·sə·pəl əv ¦lēst kən'strānt }

Gay-Lussac's second law [THERMO] The law that the internal energy of an ideal gas is independent of its volume. { ¦gā·lü¦säks 'sek·ənd ¦ló }

Gay-Lussac tower [CHEM ENG] A component part in the chamber process for sulfuric acid production that absorbs nitrogen oxides to form nitrous vitriol. { ¦ga·lü¦säk 'taú·ər }

g-cal See calorie. { 'jē¦kal }

g-cm See gram-centimeter.

gear [DES ENG] A toothed machine element used to transmit motion between rotating shafts when the center distance of the shafts is not too large. [MECH ENG] **1.** A mechanism performing a specific function in a machine. **2.** An adjustment device of the transmission in a motor vehicle which determines mechanical advantage, relative speed, and direction of travel. { gir }

gear case [MECH ENG] An enclosure, usually filled with lubricating fluid, in which gears operate. { 'gir ¦kās }

gear cutter [MECH ENG] A machine or tool for cutting teeth in a gear. { 'gir ¦kəd·ər }

gear cutting [MECH ENG] The cutting or forming of a uniform series of toothlike projections on the surface of a workpiece. { 'gir ¦kəd·iŋ }

gear down [MECH ENG] To arrange gears so the driven part rotates at a slower speed than the driving part. { 'gir 'daún }

gear drive [MECH ENG] Transmission of motion

or torque from one shaft to another by means of direct contact between toothed wheels. { 'gir ,drīv }

geared turbine [MECH ENG] A turbine connected to a set of reduction gears. { ¦gird 'tər·bən }

gear forming [MECH ENG] A method of gear cutting in which the desired tooth shape is produced by a tool whose cutting profile matches the tooth form. { 'gir ,fȯr·miŋ }

gear generating [MECH ENG] A method of gear cutting in which the tooth is produced by the conjugate or total cutting action of the tool plus the rotation of the workpiece. { 'gir ,jen·ə ,rād·iŋ }

gear grinding [MECH ENG] A gear-cutting method in which gears are shaped by formed grinding wheels and by generation; primarily a finishing operation. { 'gir ,grīnd·iŋ }

gear hobber [MECH ENG] A machine that mills gear teeth; the rotational speed of the hob has a precise relationship to that of the work. { 'gir ,häb·ər }

gearing [MECH ENG] A set of gear wheels. { 'gir·iŋ }

gearing chain [MECH ENG] A continuous chain used to transmit motion from one toothed wheel, or sprocket, to another. { 'gir·iŋ ,chān }

gearless traction [MECH ENG] Direct drive, without reduction gears. { ¦gir·ləs 'trak·shən }

gear level [MECH ENG] To arrange gears so that the driven part and driving part turn at the same speed. { 'gir ,lev·əl }

gear loading [MECH ENG] The power transmitted or the contact force per unit length of a gear. { 'gir ,lōd·iŋ }

gear meter [ENG] A type of positive-displacement fluid quantity meter in which the rotating elements are two meshing gear wheels. { 'gir ,mēd·ər }

gearmotor [MECH ENG] A motor combined with a set of speed-reducing gears. { 'gir,mōd·ər }

gear pump [MECH ENG] A rotary pump in which two meshing gear wheels contrarotate so that the fluid is entrained on one side and discharged on the other. { 'gir ,pəmp }

gear ratio [MECH ENG] The ratio of the angular speed of the driving member of a gear train or similar mechanism to that of the driven member; specifically, the number of revolutions made by the engine per revolution of the rear wheels of an automobile. { 'gir ,rā·shō }

gear shaper [MECH ENG] A machine that makes gear teeth by means of a reciprocating cutter that rotates slowly with the work. { 'gir ,shāp·ər }

gear-shaving machine [MECH ENG] A finishing machine that removes excess metal from machined gears by the axial sliding motion of a straight-rack cutter or a circular gear cutter. { 'gir ,shāv·iŋ mə,shēn }

gearshift [MECH ENG] A device for engaging and disengaging gears. { 'gir,shift }

gear teeth [DES ENG] Projections on the circumference or face of a wheel which engage with complementary projections on another wheel to transmit force and motion. { 'gir ,tēth }

gear train [MECH ENG] A combination of two or more gears used to transmit motion between two rotating shafts or between a shaft and a slide. { 'gir ,trān }

gear up [MECH ENG] To arrange gears so that the driven part rotates faster than the driving part. { ¦gir 'əp }

gear wheel [MECH ENG] A wheel that meshes gear teeth with another part. { 'gir ,wēl }

geepound See slug. { 'jē,paünd }

Geiger-Müller probe [ENG] A Geiger-Müller counter in a watertight container, lowered into a borehole to log the intensity of the gamma rays emitted by radioactive substances in traversed rock. Also known as electronic logger; Geiger probe. { ¦gī·gər 'myül·ər ,prōb }

Geiger probe See Geiger-Müller probe. { 'gī·gər ,prōb }

Geissler pump [ENG] A type of air pump that uses the principle of the Torricellian vacuum, and in which the vacuum is produced by the flow of mercury back and forth between a vertically adjustable and a fixed reservoir. { 'gīs·lər ,pəmp }

gelatinize [ENG] To coat or treat with a solution of gelatin. { jə'lat·ən,īz }

gelation time [CHEM ENG] In the manufacture of a thermosetting resin, the time interval between the addition of the catalyst into a liquid adhesive system and the formation of a gel. { jə'lā·shən ,tīm }

GEM See air-cushion vehicle.

gender [ELEC] The classification of a connector as female or male. { 'jen·dər }

gender changer [ELEC] A small passive device that is placed between two connectors of the same gender to enable them to be joined. Also known as cable matcher. { 'jen·dər ,chān·jər }

generalized coordinates [MECH] A set of variables used to specify the position and orientation of a system, in principle defined in terms of Cartesian coordinates of the system's particles and of the time in some convenient manner; the number of such coordinates equals the number of degrees of freedom of the system Also known as Lagrangian coordinates. { 'jen·rə,līzd kō 'ȯrd·ən·əts }

generalized force [MECH] The generalized force corresponding to a generalized coordinate is the ratio of the virtual work done in an infinitesimal virtual displacement, which alters that coordinate and no other, to the change in the coordinate. { 'jen·rə,līzd 'fȯrs }

generalized momentum See conjugate momentum. { 'jen·rə,līzd mə'ment·əm }

generalized velocity [MECH] The derivative with respect to time of one of the generalized coordinates of a particle. Also known as Lagrangian generalized velocity. { 'jen·rə,līzd və 'läs·əd·ē }

general manager [IND ENG] The person of general authority who performs all reasonable tasks in conducting the usual and customary business

of the principal head or owner. { ¦jen·rəl 'man·
ə·jər }

generating magnetometer [ENG] A magnetometer in which a coil is rotated in the magnetic
field to be measured with the resulting generated
voltage being proportional to the strength of the
magnetic field. { 'jen·ə,rād·iŋ mag·nə'täm·əd·
ər }

generating plant See generating station. { 'jen·
ə,rād·iŋ ,plant }

generating station [MECH ENG] A stationary
plant containing apparatus for large-scale conversion of some form of energy (such as hydraulic, steam, chemical, or nuclear energy) into electrical energy. Also known as generating plant;
power station. { 'jen·ə,rād·iŋ ,stā·shən }

generation rate [ELECTR] In a semiconductor,
the time rate of creation of electron-hole pairs.
{ ,jen·ə'rā·shən ,rāt }

generator [ELEC] A machine that converts mechanical energy into electrical energy; in its commonest form, a large number of conductors are
mounted on an armature that is rotated in a
magnetic field produced by field coils. Also
known as dynamo; electric generator.
[ELECTR] **1.** A vacuum-tube oscillator or any
other nonrotating device that generates an alternating voltage at a desired frequency when energized with direct-current power or low-frequency
alternating-current power. **2.** A circuit that generates a desired repetitive or nonrepetitive
waveform, such as a pulse generator. { 'jen·
ə,rād·ər }

generator set [ENG] The aggregate of one or
more generators together with the equipment
and plant for producing the energy that drives
them. { 'jen·ə,rād·ər ,set }

geochemical prospecting [ENG] The use of
geochemical and biogeochemical principles and
data in the search for economic deposits of minerals, petroleum, and natural gases. { ¦jē·
ō¦kem·ə·kəl 'prä,spek·ting }

geochemical well logging [ENG] Well logging
dependent on geochemical analysis of the data.
{ ¦jē·ō¦kem·ə·kəl 'wel ,läg·iŋ }

geodetic survey [ENG] A survey in which the
figure and size of the earth are considered; it is
applicable for large areas and long lines and
is used for the precise location of basic points
suitable for controlling other surveys. { ¦jē·
ə¦ded·ik 'sər,vā }

geographical mile [MECH] The length of 1 minute of arc of the Equator, or 6087.08 feet (1855.34
meters), which approximates the length of the
nautical mile. { ¦jē·ə¦graf·ə·kəl 'mīl }

geologic thermometer See geothermometer. { ¦jē·
ə¦läj·ik thər'mäm·əd·ər }

geolograph [ENG] A device that records the
penetration rate of a bit during the drilling of a
well. { jē'äl·ə,graph }

geomagnetic electrokinetograph [ENG] An instrument that can be suspended from the side
of a ship to measure the direction and speed of
ocean currents while the ship is under way by
measuring the voltage induced in the moving

conductive seawater by the magnetic field of
the earth. { ¦jē·ō·mag¦ned·ik i¦lek·trə·kə'ned·ə
,graf }

geomembrane [CIV ENG] Any impermeable
membrane (usually made of synthetic polymers
in sheets) used with soils, rock, earth, or other
geotechnical material in order to block the migration of fluids. { ,jē·ō'mem,brān }

geometric construction [ENG] Construction
that employs only straightedge and compasses
or is carried out by drawing only straight lines
and circles. { ¦jē·ə¦me·trik kən'strək·shən }

geometric programming [SYS ENG] A nonlinear
programming technique in which the relative
contribution of each of the component costs is
first determined; only then are the variables in
the component costs determined. { ¦jē·ə¦me·
trik 'prō ,gram·iŋ }

geophysical engineering [ENG] A branch of engineering that applies scientific methods for locating mineral deposits. { ¦jē·ə¦fiz·ə·kəl ,en·
jə'nir·iŋ }

geophysical prospecting [ENG] Application of
quantitative concepts and principles of physics
and mathematics in geologic explorations to discover the character of and mineral resources in
underground rocks in the upper portions of the
earth's crust. { ¦jē·ə¦fiz·ə·kəl 'prä,spek·tiŋ }

geosynthetic [CIV ENG] Any synthetic material
used in geotechnical engineering, such as geotextiles and geomembranes. { ,jē·ō·sin'thed·
ik }

geotechnics [CIV ENG] The application of scientific methods and engineering principles to
civil engineering problems through acquiring, in-
terpreting, and using knowledge of materials of
the crust of the earth. { ¦jē·ō¦tek·niks }

geotechnology [ENG] Application of the methods of engineering and science to exploitation
of natural resources. { ¦jē·ō·tek'näl·ə·jē }

geotextiles [CIV ENG] Woven or nonwoven fabrics used with foundations, soils, rock, earth, or
other geotechnical material as an integral part
of a manufactured project, structure, or system.
Also known as civil engineering fabrics; erosion
control cloth; filter fabrics; support membranes.
{ ¦jē·ō¦tek,stīlz }

geothermal prospecting [ENG] Exploration for
sources of geothermal energy. { ¦jē·ō¦thər·məl
'prä,spek·tiŋ }

geothermal well logging [ENG] Measurement
of the change in temperature of the earth by
means of well logging. { ¦jē·ō¦thər·məl 'wel
,läg·iŋ }

geothermometer [ENG] A thermometer constructed to measure temperatures in boreholes
or deep-sea deposits. { ¦jē·ō·thər'mäm·əd·ər }

gerber beam [CIV ENG] A long, straight beam
that functions essentially as a cantilevered beam
by the insertion of two hinges in alternate spans.
{ 'gər·bər ,bēm }

get [IND ENG] A combination of two or more of
the elemental motions of search, select, grasp,
transport empty, and transport loaded; applied
to time-motion studies. { get }

getter-ion pump |ENG| A high-vacuum pump that employs chemically active metal layers which are continuously or intermittently deposited on the wall of the pump, and which chemisorb active gases while inert gases are "cleaned up" by ionizing them in an electric discharge and drawing the positive ions to the wall, where the neutralized ions are buried by fresh deposits of metal. Also known as sputter-ion pump. { ¦ged·ər ¦ī,än ‚pəmp }

getter sputtering |ELECTR| The deposition of high-purity thin films at ordinary vacuum levels by using a getter to remove contaminants remaining in the vacuum. { 'gəd·ər ‚spəd·ə·riŋ }

gewel hinge |DES ENG| A hinge consisting of a hook inserted in a loop. { 'jü·əl ‚hinj }

gf See gram-force.

Giaque's temperature scale |THERMO| The internationally accepted scale of absolute temperature, in which the triple point of water is defined to have a temperature of 273.16 K. { ¦zhyäks 'tem·prə·chər ‚skāl }

gib |ENG| A removable plate designed to hold other parts in place or act as a bearing or wear surface. { gib }

Gibbs apparatus |ENG| A compressed-oxygen breathing apparatus used by miners in the United States. { 'gibz ‚ap·ə'rad·əs }

Gibbs diaphragm cell |CHEM ENG| A type of electrolytic diaphragm cell for chlorine production, with graphite electrodes and a cylindrical shape. { 'gibz 'dī·ə‚fram ‚sel }

Gibbs free energy |THERMO| The thermodynamic function $G = H - TS$, where H is enthalpy, T absolute temperature, and S entropy. Also known as free energy; free enthalpy; Gibbs function. { 'gibz ¦frē 'en·ər·jē }

Gibbs function See Gibbs free energy. { 'gibz ‚fəŋk·shən }

Gibbs-Helmholtz equation |THERMO| 1. Either of two thermodynamic relations that are useful in calculating the internal energy U or enthalpy H of a system; they may be written $U = F - T(\partial F/\partial T)_V$ and $H = G - T(\partial G/\partial T)_P$ where F is the free energy, G is the Gibbs free energy, T is the absolute temperature, V is the volume, and P is the pressure. 2. Any of the similar equations for changes in thermodynamic potentials during an isothermal process. { 'gibz 'helm‚hōlts i‚kwā·zhən }

Glegy-Hardisty process |CHEM ENG| The production of sebacic acid from castor oil or its acids by reaction of the acid at a high temperature with caustic alkali. { 'gē·gē 'här·də·stē ‚präs·əs }

Giesler coal test |ENG| A plastometric method for estimating the coking properties of coals. { 'gēs·lər 'kōl ‚test }

Gilbrethian variables |IND ENG| A system of three sets of variables that are considered to be intrinsic to every task: variables involving the response of the worker to anatomic and psychological factors, environmental variables, and variables of motion; used in analyzing and designing work systems. { gil'breth·ē·ən 'ver·ē·ə·bəlz }

Gilbreth's micromotion study |IND ENG| A time and motion study based on the concept that all work is performed by using a relatively few basic operations in varying combinations and sequence; basic elements (therbligs) include grasp, search, move, reach, and hold. { 'gil·brəths ¦mī·krō¦mō·shən ‚stəd·ē }

gill |MECH| 1. A unit of volume used in the United States for the measurement of liquid substances, equal to 1/4 U.S. liquid pint, or to $1.1829411825 \times 10^{-4}$ cubic meter. 2. A unit of volume used in the United Kingdom for the measurement of liquid substances, and occasionally of solid substances, equal to 1/4 U.K. pint, or to approximately $1.420653125 \times 10^{-4}$ cubic meter. { gil }

Gilliland correlation |CHEM ENG| Approximation method for distillation-column calculations; correlates reflux ratio and number of plates for the column as functions of minimum reflux and minimum plates. { gə'lil·ənd ‚kä·rə‚lā·shən }

gill net |ENG| A net that entangles the gill covers of fish. { 'gil ‚net }

Gilmour heat-exchange method |ENG| Thermal design method for heat exchangers by solution of five unique equations containing a minimum number of variables and involving tubeside, shell-side, tube-wall, and dirt resistance. { 'gil·mór 'hēt iks‚chānj ‚meth·əd }

gimbal |ENG| 1. A device with two mutually perpendicular and intersecting axes of rotation, thus giving free angular movement in two directions, on which an engine or other object may be mounted. 2. In a gyro, a support which provides the spin axis with a degree of freedom. 3. To move a reaction engine about on a gimbal so as to obtain pitching and yawing correction moments. 4. To mount something on a gimbal. { 'gim·bəl }

gimbaled nozzle |MECH ENG| A nozzle supported on a gimbal. { 'gim·bəld 'näz·əl }

gimbal freedom |ENG| Of a gyro, the maximum angular displacement about the output axis of a gimbal. { 'gim·bəl ‚frē·dəm }

gimbal lock |ENG| A condition of a two-degree-of-freedom gyro wherein the alignment of the spin axis with an axis of freedom deprives the gyro of a degree-of-freedom and therefore its useful properties. { 'gim·bəl ‚läk }

gimlet |DES ENG| A small tool consisting of a threaded tip, grooved shank, and a cross handle; used for boring holes in wood. { 'gim·lət }

gimlet bit |DES ENG| A bit with a threaded point and spiral flute; used for drilling small holes in wood. { 'gim·lət ‚bit }

gin |MECH ENG| A hoisting machine in the form of a tripod with a windlass, pulleys, and ropes. { jin }

gin pole |MECH ENG| A hand-operated derrick which has a nearly vertical pole supported by guy ropes; the load is raised on a rope that passes through a pulley at the top and over a winch at the foot. Also known as guyed-mast derrick; pole derrick; standing derrick. { 'jin ‚pōl }

gin tackle |MECH ENG| A tackle made for use with a gin. { 'jin ˌtak·əl }

Girbotal process |CHEM ENG| A regenerative absorption process to remove carbon dioxide, hydrogen sulfide, and other acid impurities from natural gas, using mono-, di-, or triethanolamine as the reagent. { 'gər·bə,tȯl ,präs·əs }

girder |CIV ENG| A large beam made of metal or concrete, and sometimes of wood. { 'gər·dər }

girder clamp See beam clip. { 'gərd·ər ,klamp }

girder clip See beam clip. { 'gərd·ər ,klip }

girt |CIV ENG| **1.** A timber in the second-floor corner posts of a house to serve as a footing for roof rafters. **2.** A horizontal member to stiffen the framework of a building frame or trestle. |ENG| A brace member running horizontally between the legs of a drill tripod or derrick. { gərt }

gland |ENG| **1.** A device for preventing leakage at a machine joint, as where a shaft emerges from a vessel containing a pressurized fluid. **2.** A movable part used in a stuffing box to compress the packing. { gland }

glare filter |ENG| A screen that is placed over the face of a cathode-ray tube to reduce glare from ambient and overhead light. { 'gler ,fil·tər }

glassblowing |ENG| Shaping a mass of viscid glass by inflating it with air introduced through a tube. { 'glas,blō·iŋ }

glass cutter |ENG| A tool equipped with a steel wheel or a diamond point used to cut glass. { 'glas,kəd·ər }

glassed steel |CHEM ENG| Process piping or vessels lined with glass; a glass-steel composite has structural strength of steel and corrosion resistance of glass. { ¦glast ¦stēl }

glass furnace |ENG| A large, covered furnace or tank for melting large batches of glass, in which heat is supplied by a flame playing over the glass surface, and regenerative heating of combustion air and gas is usually employed. Also known as glass tank. { 'glas ,fər·nəs }

glass heat exchanger |ENG| Any heat exchanger in which glass replaces metal, such as shell-and-tube, cascade, double-pipe, bayonet, and coil exchangers. { ¦glas 'hēt iks,chān·jər }

glass pot |ENG| A crucible used for making small amounts of glass. { 'glas ,pät }

glass seal |ENG| An airtight seal made by molten glass. { 'glas ,sēl }

glass tank See glass furnace. { 'glas ,taŋk }

glass-tube manometer |ENG| A manometer for simple indication of difference of pressure, in contrast to the metallic-housed mercury manometer, used to record or control difference of pressure or fluid flow. { 'glas ,tüb mə'näm·əd·ər }

glaze |ENG| A glossy coating. Also known as enamel. { glāz }

glazed |MECH ENG| Pertaining to an abrasive surface that has become smooth and cannot abrade efficiently. { glāzd }

glazed frost See glaze. { ¦glāzd 'fròst }

glaze ice See glaze. { 'glāz ,īs }

glazier's point |ENG| A small piece of sheet metal, usually shaped like a triangle, used to hold a pane of glass in place. Also known as sprig. { 'glā·zərz ,pȯint }

glazing |ENG| **1.** Cutting and fitting panes of glass into frames. **2.** Smoothing the lead of a wiped pipe joint by passing a hot iron over it. { 'glāz·iŋ }

glazing bar See sash bar. { 'glāz·iŋ ,bär }

Gleason bevel gear system |DES ENG| The standard for bevel gear designs in the United States; employs a basic pressure angle of 20° with long and short addenda for ratios other than 1:1 to avoid undercut pinions and to increase strength. { 'glēs·ən ¦bev·əl ¦gir ,sis·təm }

globe valve |MECH ENG| A device for regulating flow in a pipeline, consisting of a movable disk-type element and a stationary ring seat in a generally spherical body. { 'glōb ,valv }

glory hole |CIV ENG| A funnel-shaped, fixed-crest spillway. |ENG| A furnace for resoftening or fire polishing glass during working, or an entrance in such a furnace. { 'glȯ·rē ,hōl }

glossimeter |ENG| An instrument, often photoelectric, for measuring the ratio of the light reflected from a surface in a definite direction to the total light reflected in all directions. Also known as glossmeter. { glä'sim·əd·ər }

glossmeter See glossimeter. { 'gläs,mēd·ər }

glost firing |CHEM ENG| The process of glazing and firing ceramic ware which has previously been fired at a higher temperature. { 'glȯst ,fīr·iŋ }

glove box |ENG| A sealed box with gloves attached and passing through openings into the box, so that workers can handle materials in the box; used to handle certain radioactive and biologically dangerous materials and to prevent contamination of materials and objects such as germfree rats or lunar rocks. { 'gləv ,bäks }

Glover tower |CHEM ENG| A tower in the lead chamber process for manufacturing sulfuric acid; in this tower the nitrogen oxide, sulfur dioxide, and air mixture is passed upward and sprayed with a sulfuric acid-nitrosyl sulfuric acid mixture. { 'gləv·ər ,taü·ər }

glow-discharge microphone |ENG ACOUS| Microphone in which the action of sound waves on the current forming a glow discharge between two electrodes causes corresponding variations in the current. { ¦glō ¦dis,chärj 'mī·krə,fōn }

glowing combustion |CHEM ENG| A reaction between oxygen or an oxidizer and the surface of a solid fuel so that there is emission of heat and light without a flame. Also known as surface burning. { ¦glō·iŋ kəm'bəs·chən }

glow plug |MECH ENG| A small electric heater, located inside a cylinder of a diesel engine, that preheats the air and aids the engine in starting. { 'glō ,pləg }

glue block See angle block. { 'glü ,bläk }

glue-joint ripsaw |MECH ENG| A heavy-gage ripsaw used on straight-line or self-feed rip machines; the cut is smooth enough to permit gluing of joints from the saw. { 'glü ,jȯint 'rip,sȯ }

glue-line heating |ENG| Dielectric heating in

which the electrodes are designed to give preferential heating to a thin film of glue or other relatively high-loss material located between layers of relatively low-loss material such as wood. { 'glü ˌlīn 'hēd·iŋ }

glug |MECH| A unit of mass, equal to the mass which is accelerated by 1 centimeter per second per second by a force of 1 gram-force, or to 980.665 grams. { gləg }

glycol dehydrator |CHEM ENG| Processing equipment for removing all or most of the water from a wet gas by contacting with glycol. { 'glī,kȯl dē'hī,drād·ər }

gm See gram.

gnomon |ENG| On a sundial, the inclined plate or pin that casts a shadow. Also known as style. { 'nō·mən }

goal coordination method |CONT SYS| A method for coordinating the subproblem solutions in plant decomposition, in which Lagrange multipliers enter into the subsystem cost functions as shadow prices, and these are adjusted by the second-level controller in an iterative procedure which culminates (if the method is applicable) in the satisfaction of the subsystem coupling relationships. Also known as interaction balance method; nonfeasible method. { 'gōl kȯ,ȯrd·ə·nā·shən ,meth·əd }

gobo |ENG| A panel used to shield a television camera lens from direct light. |ENG ACOUS| A sound-absorbing shield used with a microphone to block unwanted sounds. { 'gō,bō }

go-devil |ENG| 1. A device inserted in a pipe or hole for purposes such as cleaning or for detonating an explosive. 2. A sled for moving logs or cultivating. 3. A large rake for gathering hay. 4. A small railroad car used for transporting workers and materials. { 'gō ,dev·əl }

go gage |DES ENG| A test device that just fits a part if it has the proper dimensions (often used in pairs with a "no go" gage to establish maximum and minimum dimensions). { 'gō ,gāj }

goggles |ENG| Spectacle-like eye protectors having shields at the sides and short, projecting eye tubes. { 'gäg·əlz }

going |CIV ENG| On a staircase, the distance between the faces of two successive risers. { 'gō·iŋ }

Golay cell |ENG| A radiometer in which radiation absorbed in a gas chamber heats the gas, causing it to expand and deflect a diaphragm in accordance with the amount of radiation. { gə'lā ,sel }

goldbeater's-skin hygrometer |ENG| A hygrometer using goldbeater's skin as the sensitive element; variations in the physical dimensions of the skin caused by its hygroscopic character indicate relative atmospheric humidity. { 'gōl ,bēd·ərz ,skin hī'gräm·əd·ər }

gold doping |ELECTR| A technique for controlling the lifetime of minority carriers in a transistor; gold is diffused into the base and collector regions to reduce storage time in transistor circuits. { 'gōl ,dōp·iŋ }

gold point |THERMO| The temperature of the freezing point of gold at a pressure of 1 standard atmosphere (101,325 pascals); used to define the International Temperature Scale of 1940, on which it is assigned a value of 1337.33 K or 1064.18°C. { 'gōld ,pȯint }

Gold slide |ENG| A slide rule used on British ships to compute barometric corrections and reduction of pressure to sea level; it includes the effects of temperature, latitude, index correction, and barometric height above sea level. { 'gōld ,slīd }

golf ball |ENG| A printing element used on some typewriters and serial printers, consisting of a rotating, spherically shape, removable typehead that skims across the printed line while the typewriter or printer carriage does not move. { 'gälf ,bȯl }

gondola car |ENG| A flat-bottomed railroad car which has no top, fixed sides, and often removable ends, in which steel, rock, or heavy bulk commodities are transported. { 'gän·də·lə ,kär }

goniometer |ENG| 1. An instrument used to measure the angles between crystal faces. 2. An instrument which uses x-ray diffraction to measure the angular positions of the axes of a crystal. 3. Any instrument for measuring angles. { ,gō·nē'äm·əd·ər }

go/no-go detector |ENG| An instrument having only two operating states, such as a common fuse which is either intact or melted. { 'gō 'nō ,gō di,tek·tər }

go/no-go test |ENG| A test based on the measurement of one or more parameters but which can have only one of two possible results, to pass or reject the device under test. { 'gō 'nō ,gō ,test }

good oil See raffinate. { 'gùd ,ȯil }

gooseneck |DES ENG| 1. A pipe, bar, or other device having a curved or bent shape resembling that of the neck of a goose. 2. See water swivel. { 'güs,nek }

gopher hole |ENG| Horizontal T-shaped opening made in rock in preparation for blasting. Also known as coyote hole. { 'gō·fər ,hōl }

Gordon's formula |CIV ENG| An empirical formula which gives the collapsing load of a column in terms of its cross-sectional area, length, and least diameter. { 'gȯrd·ənz ,fȯr·myə·lə }

gore |CIV ENG| A small triangular parcel of land. { gȯr }

gouge |DES ENG| A curved chisel for wood, bone, stone, and so on. { gaùj }

gouging |ENG| The removal of material by electrical, mechanical, or manual means for the formation of a groove. { 'gaùj·iŋ }

governor |MECH ENG| A device, especially one actuated by the centrifugal force of whirling weights opposed by gravity or by springs, used to provide automatic control of speed or power of a prime mover. { 'gəv·ə·nər }

grab |ENG| An instrument for extricating broken boring tools from a borehole. { grab }

grabbing crane |MECH ENG| An excavator

made up of a crane carrying a large grab or bucket in the form of a pair of half scoops, hinged to dig into the earth as they are lifted. { 'grab·iŋ ,krān }

grab bucket |MECH ENG| A bucket with hinged jaws or teeth that is hung from cables on a crane or excavator and is used to dig and pick up materials. { 'grab ,bək·ət }

grab dredger |MECH ENG| Dredging equipment comprising a grab or grab bucket that is suspended from the jib head of a crane. Also known as grapple dredger. { 'grab ,drej·ər }

grabhook |DES ENG| A hook used for grabbing, as in lifting blocks of stone, in which case the hooks are used in pairs connected with a chain, and are so constructed that the tension of the chain causes them to adhere firmly to the rock. { 'grab,hůk }

grade |CIV ENG| **1.** To prepare a roadway or other land surface of uniform slope. **2.** A surface prepared for the support of rails, a road, or a conduit. **3.** The elevation of the finished surface of an engineering project. |ENG| The degree of strength of a high explosive. { grād }

gradeability |MECH ENG| The performance of earthmovers on various inclines, measured in percent grade. { ,grād·ə'bil·əd·ē }

grade beam |CIV ENG| A reinforced concrete beam placed directly on the ground to provide the foundation for the superstructure. { 'grād ,bēm }

grade crossing |CIV ENG| The intersection of roadways, railways, pedestrian walks, or combinations of these at grade. { 'grād ,krós·iŋ }

grade line |CIV ENG| A line or slope used as a longitudinal reference for a railroad or highway. { ¦grād 'līn }

grader |MECH ENG| A high bodied, wheeled vehicle with a leveling blade mounted between the front and rear wheels; used for fine-grading relatively loose and level earth. { 'grād·ər }

grade separation |CIV ENG| A grade crossing employing an underpass and overpass. { 'grād sep·ə,rā·shən }

grade slab |CIV ENG| A reinforced concrete slab placed directly on the ground to provide the foundation for the superstructure. { 'grād ,slab }

grade stake |CIV ENG| A stake used as an elevation reference. { 'grād ,stāk }

gradienter |ENG| An attachment placed on a surveyor's transit to measure angle of inclination in terms of the tangent of the angle. { 'grād·ē,en·tər }

gradient microphone |ENG ACOUS| A microphone whose electrical response corresponds to some function of the difference in pressure between two points in space. { 'grād·ē·ənt 'mī·krə,fōn }

grading |IND ENG| Segregating a product into a number of adjoining categories which often form a spectrum of quality. Also known as classification. { 'grād·iŋ }

gradiometer |ENG| Any instrument that measures the gradient of some physical quantity, such as certain types of magnetometers which are designed to measure the gradient of magnetic field, or the Eötvös torsion balance and related instruments which measure the gradient of gravitational field. { ,grād·ē'äm·əd·ər }

graduator |ENG| An evaporation unit in which liquid is forced to flow over large surfaces which are subjected to air currents. { 'graj·ə,wād·ər }

Graetz number |THERMO| A dimensionless number used in the study of streamline flow, equal to the mass flow rate of a fluid times its specific heat at constant pressure divided by the product of its thermal conductivity and a characteristic length. Also spelled Grätz number. Symbolized N_{Gz}. { 'grets ,nəm·bər }

Graham's pendulum |DES ENG| A type of compensated pendulum having a hollow bob containing mercury whose thermal expansion balances the thermal expansion of the pendulum rod. { ¦gramz 'pen·jə·ləm }

grain |MECH| A unit of mass in the United States and United Kingdom, common to the avoirdupois, apothecaries', and troy systems, equal to 1/7000 of a pound, or to 6.479891×10^{-5} kilogram. Abbreviated gr. { grān }

grainer process |CHEM ENG| A salt production method in which salt is produced by surface evaporation of brine in open-air flat pans. { 'grān·ər ,präs·əs }

graining |ENG| Simulating a grain such as wood or marble on a painted surface by applying a translucent stain, then working it into suitable patterns with tools such as special combs, brushes, and rags. { 'grān·iŋ }

grain spacing |DES ENG| Relative location of abrasive grains on the surface of a grinding wheel. { 'grān ,spās·iŋ }

gram |MECH| The unit of mass in the centimeter-gram-second system of units, equal to 0.001 kilogram. Abbreviated g; gm. { gram }

gram-calorie See calorie. { 'gram ¦kal·ə·rē }

gram-centimeter |MECH| A unit of energy in the centimeter-gram-second gravitational system, equal to the work done by a force of magnitude 1 gram force when the point at which the force is applied is displaced 1 centimeter in the direction of the force. Abbreviated g-cm. { 'gram 'sent·ə,mēd·ər }

gram-force |MECH| A unit of force in the centimeter-gram-second gravitational system, equal to the gravitational force on a 1-gram mass at a specified location. Abbreviated gf. Also known as fors; gram-weight; pond. { 'gram ,fórs }

gram-weight See gram-force. { 'gram¦wāt }

granular-bed separator |ENG| Vessel or chamber in which a bed of granular material is used to remove dust from a dust-laden gas as it passes through the bed. { 'gran·yə·lər ,bed 'sep·ə,rād·ər }

granularity |SYS ENG| The degree to which a system can be broken down into separate components, making it customizable and flexible. { ,gran·yə'lar·əd·ē }

graphical statics |MECH| A method of determining forces acting on a rigid body in equilibrium, in which forces are represented on a diagram by straight lines whose lengths are proportional to the magnitudes of the forces. (¦graf·ə·kəl 'stad·iks)

graphical symbol |ELEC| A true symbol, rather than a coarse picture, representing an element in an electrical diagram. (¦graf·ə·kəl 'sim·bəl)

graphic equalizer |ENG ACOUS| A device that allows the response of audio equipment to be modified independently in several frequency bands through the use of a bank of slide controls whose positions form a graph of the frequency response. (¦graf·ik 'ē·kwə,līz·zər)

graphic panel |CONT SYS| A master control panel which indicates the status of equipment and operations in a system, and their relationships. (¦graf·ik 'pan·əl)

graphic recording instrument |ENG| An instrument that makes a graphic record of one or more quantities as a function of another variable, usually time. ('graf·ik ri,kȯrd·iŋ ,in·strə·mənt)

graphite anode |CHEM ENG| One of the electrodes of graphite used in a mercury cell to produce chlorine by electrolysis. |ELECTR| 1. The rod of graphite which is inserted into the mercury-pool cathode of an ignitron to start current flow. 2. The collector of electrons in a beam power tube or other high-current tube. ('gra,fīt 'an,ōd)

grapnel |DES ENG| An implement with claws used to recover a lost core, drill fittings, and junk from a borehole or for other grappling operations. Also known as grapple. ('grap·nəl)

grapple See grapnel. ('grap·əl)

grapple dredger See grab dredger. ('grap·əl ,drej·ər)

grapple hook |DES ENG| An iron hook used on the end of a rope to snag lines, to hold one ship alongside another, or as a fishing tool. Also known as grappling iron. ('grap·əl ,hůk)

grappling iron See grapple hook. ('grap·liŋ ,ī·ərn)

grasp |IND ENG| A basic element (therblig) in time-motion study; a useful element that accomplishes work. (grasp)

grasshopper linkage |MECH ENG| A straight-line mechanism used in some early steam engines. ('gras,häp·ər ,liŋ·kij)

Grassot fluxmeter |ENG| A type of fluxmeter in which a light coil of wire is suspended in a magnetic field in such a way that it can rotate; the ends of the suspended coil are connected to a search coil of known area penetrated by the magnetic flux to be measured; the flux is determined from the rotation of the suspended coil when the search coil is moved. (,grä,sō 'fləks,mēd·ər)

grass-roots plant |CHEM ENG| A complete plant erected on a virgin site. ('gras ,rüts 'plant)

grate |ENG| A support for burning solid fuels; usually made of closely spaced bars to hold the burning fuel, while allowing combustion air to rise up to the fuel from beneath, and ashes to fall away from the burning fuel. (grāt)

Grätz number See Graetz number. ('grets ,nəm·bər)

grav See G. (grav)

gravel pump |MECH ENG| A centrifugal pump with renewable impellers and lining, used to pump a mixture of gravel and water. ('grav·əl ,pəmp)

gravel stop |BUILD| Metal flashing placed at the edge of a roof to prevent gravel from falling off. ('grav·əl ,stäp)

graveyard shift |IND ENG| The shift of workers that begins at or around midnight; the last shift of the day. ('grāv,yärd ,shift)

gravimeter |ENG| A highly sensitive weighing device used for relative measurement of the force of gravity by detecting small weight differences of a constant mass at different points on the earth. Also known as gravity meter. (grə'vim·əd·ər)

gravimetry |ENG| Measurement of gravitational force. (grə'vim·ə·trē)

graving dock |CIV ENG| A form of dry dock consisting of an artificial basin fitted with a gate or caisson, into which a vessel can be floated and the water pumped out to expose the vessel's bottom. ('grāv·iŋ ,däk)

gravitational constant |MECH| The constant of proportionality in Newton's law of gravitation, equal to the gravitational force between any two particles times the square of the distance between them, divided by the product of their masses. Also known as constant of gravitation. (,grav·ə'tā·shən·əl 'kän·stənt)

gravitational displacement |MECH| The gravitational field strength times the gravitational constant. Also known as gravitational flux density. (,grav·ə'tā·shən·əl dis'plās·mənt)

gravitational energy See gravitational potential energy. (,grav·ə'tā·shən·əl 'en·ər·jē)

gravitational field |MECH| The field in a region in space in which a test particle would experience a gravitational force; quantitatively, the gravitational force per unit mass on the particle at a particular point. (,grav·ə'tā·shən·əl 'fēld)

gravitational flux density See gravitational displacement. (,grav·ə'tā·shən·əl 'fləks ,den·səd·ē)

gravitational force |MECH| The force on a particle due to its gravitational attraction to other particles. (,grav·ə'tā·shən·əl 'fȯrs)

gravitational instability |MECH| Instability of a dynamic system in which gravity is the restoring force. (,grav·ə'tā·shən·əl ,in·stə'bil·əd·ē)

gravitational potential |MECH| The amount of work which must be done against gravitational forces to move a particle of unit mass to a specified position from a reference position, usually a point at infinity. (,grav·ə'tā·shən·əl pə'ten·chəl)

gravitational potential energy |MECH| The energy that a system of particles has by virtue of their positions, equal to the work that must be done against gravitational forces to assemble

252

the particles from some reference configuration, such as mutually infinite separation. Also known as gravitational energy. { ‚grav·ə'tā·shən·əl pə|ten·chəl 'en·ər·jē }

gravitational systems of units [MECH] Systems in which length, force, and time are regarded as fundamental, and the unit of force is the gravitational force on a standard body at a specified location on the earth's surface. { ‚grav·ə'tā·shən·əl ¦sis·təmz əv 'yü·nəts }

gravitometer See densimeter. { grav·ə'täm·əd·ər }

gravity [MECH] The gravitational attraction at the surface of a planet or other celestial body. { 'grav·əd·ē }

gravity bed [ENG] A moving body of solids in which particles (granules, pellets, beads, or briquets) flow downward by gravity through a vessel, while process fluid flows upward; the moving-bed technique is used in blast and shaft furnaces, petroleum catalytic cracking, pellet dryers, and coolers. { 'grav·əd·ē ‚bed }

gravity chute [ENG] A gravity conveyor in the form of an inclined plane, trough, or framework that depends on sliding friction to control the rate of descent. { 'grav·əd·ē ‚shüt }

gravity concentration [ENG] **1.** Any of various methods for separating a mixture of particles, such as minerals, based on the differences in density of the various species and on the resistance to relative motion exerted upon the particles by the fluid or semifluid medium in which separation takes place. **2.** The separation of liquid-liquid dispersions based on settling out of the dense phase by gravity. { 'grav·əd·ē ‚käns·ən'trā·shən }

gravity conveyor [ENG] Any unpowered conveyor such as a gravity chute or a roller conveyor, which uses the force of gravity to move materials over a downward path. { 'grav·əd·ē kən'vā·ər }

gravity corer [ENG] Any type of corer that achieves bottom penetration solely as a result of gravitational force acting upon its mass. { 'grav·əd·ē ‚kór·ər }

gravity dam [CIV ENG] A dam which depends on its weight for stability. { 'grav·əd·ē ‚dam }

gravity feed [ENG] Movement of materials from one location to another using the force of gravity. { 'grav·əd·ē ‚fēd }

gravity meter [ENG] **1.** U-tube-manometer type of device for direct reading of solution specific gravities in semimicro quantities. **2.** An electrical device for measuring variations in gravitation through different geologic formations; used in mineral exploration. **3.** See gravimeter. { 'grav·əd·ē ‚mēd·ər }

gravity prospecting [ENG] Identifying and mapping the distribution of rock masses of different specific gravity by means of a gravity meter. { 'grav·əd·ē 'präs‚pek·tiŋ }

gravity railroad [ENG] A cable railroad in which cars descend the slope by gravity and are hauled back up the slope by a stationary engine, or there may be two tracks with cars so connected that cars going down may help to raise the cars going up and thus conserve energy. { 'grav·əd·ē 'rāl‚rōd }

gravity segregation [ENG] Tendency of immiscible liquids or multicomponent granular mixtures to separate into distinct layers in accordance with their respective densities. { 'grav·əd·ē ‚seg·rə'gā·shən }

gravity separation [ENG] Separation of immiscible phases (gas-solid, liquid-solid, liquid-liquid, solid-solid) by allowing the denser phase to settle out under the influence of gravity; used in ore dressing and various industrial chemical processes. { 'grav·əd·ē ‚sep·ə'rā·shən }

gravity settling chamber [ENG] Chamber or vessel in which the velocity of heavy particles (solids or liquids) in a fluid stream is reduced to allow them to settle downward by gravity, as in the case of a dust-laden gas stream. { 'grav·əd·ē 'set·liŋ ‚chām·bər }

gravity station [ENG] The site of installation of gravimeters. { 'grav·əd·ē ‚stā·shən }

gravity survey [ENG] The measurement of the differences in gravity force at two or more points. { 'grav·əd·ē 'sər‚vā }

gravity vector [MECH] The force of gravity per unit mass at a given point. Symbolized **g**. { 'grav·əd·ē ‚vek·tər }

gravity wall [CIV ENG] A retaining wall which is kept upright by the force of its own weight. { 'grav·əd·ē ‚wòl }

gravity wheel conveyor [MECH ENG] A downward-sloping conveyor trough with closely spaced axle-mounted wheel units on which flat-bottomed containers or objects are conveyed from point to point by gravity pull. { 'grav·əd·ē ‚wel kən'vā·ər }

gravity yard See hump yard. { 'grav·əd·ē ‚yärd }

graybody [THERMO] An energy radiator which has a blackbody energy distribution, reduced by a constant factor, throughout the radiation spectrum or within a certain wavelength interval. Also known as nonselective radiator. { 'grā ‚bäd·ē }

Gray clay treating [CHEM ENG] A fixed-bed, vapor-phase treating process used to polymerize selectively unsaturated gum-forming constituents (diolefins); a fixed bed is used of 30- to 60-mesh fuller's earth. { 'grā 'klā ‚trēd·iŋ }

grease cup [ENG] A receptacle used to apply a solid or semifluid lubricant to a bearing; the receptacle is packed with grease and the cap forces the grease to the bearing. { 'grēs ‚kəp }

grease gun [ENG] A small hand-operated device that pumps grease under pressure into bearings. { 'grēs ‚gən }

grease seal [ENG] **1.** Type of seal used on floating pistons of some hydropneumatic recoil systems to prevent leakage past the piston of gas or oil; also used in cylinders of some hydropneumatic equilibrators. **2.** Seal used to retain grease in a case or housing, as on an axle shaft. { 'grēs‚sēl }

grease trap [CIV ENG] A trap in a drain or waste pipe to stop grease from entering a sewer system. { 'grēs ‚trap }

green design *See* industrial ecology. { ¦grēn di'zīn }

grid |DES ENG| A network of equally spaced lines forming squares, used for determining permissible locations of holes on a printed circuit board or a chassis. |ELEC| **1.** A metal plate with holes or ridges, used in a storage cell or battery as a conductor and a support for the active material. **2.** Any systematic network, such as of telephone lines or power lines. |ELECTR| An electrode located between the cathode and anode of an electron tube, which has one or more openings through which electrons or ions can pass, and serves to control the flow of electrons from cathode to anode. { grid }

grid nephoscope |ENG| A nephoscope constructed of a grid work of bars mounted horizontally on the end of a vertical column and rotating freely about the vertical axis; the observer rotates the grid and adjusts the position until some feature of the cloud appears to move along the major axis of the grid; the azimuth angle at which the grid is set is taken as the direction of the cloud motion. { 'grid 'nef·ə,skōp }

grid-rectification meter |ENG| A type of vacuum-tube voltmeter in which the grid and cathode of a tube act as a diode rectifier, and the rectified grid voltage, amplified by the tube, operates a meter in the plate circuit. { 'grid ,rek·tə·fə¦kā·shən ,mēd·ər }

Griffith's criterion |MECH| A criterion for the fracture of a brittle material under biaxial stress, based on the theory that the strength of such a material is limited by small cracks. { 'grif·əths krī,tir·ē·ən }

Griffiths' method |THERMO| A method of measuring the mechanical equivalent of heat in which the temperature rise of a known mass of water is compared with the electrical energy needed to produce this rise. { 'grif·əths ,meth·əd }

grillage |CIV ENG| A footing that consists of two or more tiers of closely spaced structural steel beams resting on a concrete block, each tier being at right angles to the one below. { grē'yazh }

grille |ENG| A grating or openwork barrier that is used to conceal or protect an opening in a floor, wall, or pavement. |ENG ACOUS| An arrangement of wood, metal, or plastic bars placed across the front of a loudspeaker in a cabinet for decorative and protective purposes. { gril }

grille cloth |ENG ACOUS| A loosely woven cloth stretched across the front of a loudspeaker to keep out dust and provide protection without appreciably impeding sound waves. { 'gril ,klôth }

grinder |MECH ENG| Any device or machine that grinds, such as a pulverizer or a grinding wheel. { 'grīn·dər }

grinding |ELECTR| **1.** A mechanical operation performed on silicon substrates of semiconductors to provide a smooth surface for epitaxial deposition or diffusion of impurities. **2.** A mechanical operation performed on quartz crystals to alter their physical size and hence their resonant frequencies. |MECH ENG| **1.** Reducing a material to relatively small particles. **2.** Removing material from a workpiece with a grinding wheel. { 'grīn·diŋ }

grinding aid |ENG| An additive to the charge in a ball mill or rod mill to accelerate the grinding process. { 'grīn·diŋ ,ād }

grinding burn |MECH ENG| Overheating a localized area of the work in grinding operations. { 'grīn·diŋ ,bərn }

grinding medium |ENG| Any material including balls and rods, used in a grinding mill. { 'grīn·diŋ ,mēd·ē·əm }

grinding mill |MECH ENG| A machine consisting of a rotating cylindrical drum, that reduces the size of particles of ore or other materials fed into it; three main types are ball, rod, and tube mills. { 'grīn·diŋ ,mil }

grinding pebbles |ENG| Pebbles, of chert or quartz, used for grinding in mills, where contamination with iron has to be avoided. { 'grīn·diŋ ,peb·əlz }

grinding ratio |MECH ENG| Ratio of the volume of ground material removed from the workpiece to the volume removed from the grinding wheel. { 'grīn·diŋ ,rā·shō }

grinding stress |MECH| Residual tensile or compressive stress, or a combination of both, on the surface of a material due to grinding. { 'grīn·diŋ ,stres }

grinding wheel |DES ENG| A wheel or disk having an abrasive material such as alumina or silicon carbide bonded to the surface. { 'grīn·diŋ ,wēl }

grindstone |ENG| A stone disk on a revolving axle, used for grinding, smoothing, and shaping. { 'grīnd,stōn }

gripper |CONT SYS| A component of a robot that grasps an object, generally through the use of suction cups, magnets, or articulated mechanisms. { 'grip·ər }

gripping zone |CONT SYS| The area in which the center of an object must be located in order for the object to be properly handled by the gripper of a robot. { 'grip·iŋ ,zōn }

grip vector |CONT SYS| A vector from a point on the wrist socket of a robot to the point where the end effector grasps an object; describes the orientation of the object in space. { 'grip ,vek·tər }

grit chamber |CIV ENG| A chamber designed to remove sand, gravel, or other heavy solids that have subsiding velocities or specific gravities substantially greater than those of the organic solids in waste water. { 'grit ,chām·bər }

grit size |DES ENG| Size of the abrasive particles on a grinding wheel. { 'grit ,sīz }

grizzly |ENG| **1.** A coarse screen used for rough sizing and separation of ore, gravel, or soil. **2.** A grating to protect chutes, manways, and winzes, in mines, or to prevent debris from entering a water inlet. { 'griz·lē }

grizzly crusher |MECH ENG| A machine with a series of parallel rods or bars for crushing rock and sorting particles by size. { 'griz·lē ¦krash·ər }

groin |CIV ENG| A barrier built out from a seashore or riverbank to protect the land from erosion and sand movements, among other functions. Also known as groyne; jetty; spur dike; wing dam. { gróin }

grommet |ENG| **1.** A metal washer or eyelet. **2.** A piece of fiber soaked in a packing material and used under bolt and nut heads to preserve tightness. { 'gräm·ət }

grommet nut |DES ENG| A blind nut with a round head; used with a screw to attach a hinge to a door. { 'gräm·ət ,nət }

groove |DES ENG| A long, narrow channel in a surface. { grüv }

grooved drum |DES ENG| Drum with a grooved surface to support and guide a rope. { 'grüvd ¦drəm }

groover |ENG| A tool for forming grooves in a slab of concrete not yet hardened. { 'grüv·ər }

grooving saw |MECH ENG| A circular saw for cutting grooves. { 'grüv·iŋ ,sȯ }

gross area |BUILD| Sum of the areas of all stories included within the outside face of the exterior walls of a building. { 'grōs ¦er·ē·ə }

gross rubber |CHEM ENG| In rubber manufacturing, the total weight of salable product, including elastomer, carbon black, extender oils, and other materials used in compounding the rubber. { 'grōs 'rəb·ər }

gross ton See ton. { ¦grōs ¦tən }

gross vehicle weight |IND ENG| A truck rating based on the combined weight of the vehicle and its load. Abbreviated gvw. { 'grōs 've·ə·kəl ,wāt }

gross weight |IND ENG| The weight of a vehicle or container when it is loaded with goods. Abbreviated gr wt. { ¦gros 'wāt }

ground |ELEC| **1.** A conducting path, intentional or accidental, between an electric circuit or equipment and the earth, or some conducting body serving in place of the earth. Abbreviated gnd. Also known as earth (British usage); earth connection. **2.** To connect electrical equipment to the earth or to some conducting body which serves in place of the earth. { graúnd }

ground anchor See anchor log. { 'graúnd ,aŋ·kər }

ground area |BUILD| The area of a building at ground level. { 'graúnd ,er·ē·ə }

ground block |CIV ENG| A pulley fastened to the anchor log which changes a horizontal pull to a vertical pull on a wire line. { 'graúnd ,bläk }

ground cable |ELEC| A heavy cable connected to earth for the purpose of grounding electric equipment. { 'graúnd ,kā·bəl }

ground check |ENG| **1.** A procedure followed prior to the release of a radiosonde in order to obtain the temperature and humidity corrections for the radiosonde system. **2.** Any instrumental check prior to the ground launch of an airborne experiment. Also known as base-line check. { 'graúnd ,chek }

ground-check chamber |ENG| A chamber that

is used to check the sensing elements of radiosonde equipment and that houses sources of heat and water vapor plus instruments for measuring temperature, humidity, and pressure, and in which air circulation is maintained by a motor-driven fan. { 'graúnd ,chek ,chām·bər }

ground circuit |ELEC| A telephone or telegraph circuit part of which passes through the ground. { 'graúnd ,sər·kət }

ground conductivity |ELEC| The effective conductivity of the ground, used in calculating the attenuation of radio waves. { 'graúnd ,kän·dək¦tiv·əd·ē }

ground control |CIV ENG| Supervision or direction of all airport surface traffic, except an aircraft landing or taking off. |ENG| The marking of survey, triangulation, or other key points or systems of points on the earth's surface so that they may be recognized in aerial photographs. { 'graúnd kən,trōl }

ground-controlled approach radar |ENG| A ground radar system providing information by which aircraft approaches may be directed by radio communications. Abbreviated GCA radar. { 'graúnd kən,trōld ə,prōch 'rā,där }

ground-controlled intercept radar |ENG| A radar system by means of which a controller may direct an aircraft to make an interception of another aircraft. Abbreviated GCI radar. { 'graúnd kən,trōld 'in·tər,sept ,rā,där }

ground controller |ENG| Aircraft controller stationed on the ground; a generic term, applied to the controller in ground-controlled approach, ground-controlled interception, and so on. { 'graúnd kən,trōl·ər }

ground current See earth current. { 'graúnd ,kə·rənt }

ground data equipment |ENG| Any device located on the ground that aids in obtaining space-position or tracking data (including computation function); reads out data telemetry, video, and so on, from payload instrumentation, or is capable of transmitting command and control signals to a satellite or space vehicle. { 'graúnd 'dad·ə i,kwip·mənt }

ground detector |ELEC| An instrument or equipment used for indicating the presence of a ground on an ungrounded system. Also known as ground indicator. { 'graúnd di,tek·tər }

ground dielectric constant |ELEC| Dielectric constant of the earth at a given location. { 'graúnd di·ə¦lek·trik 'kän·stənt }

grounded-anode amplifier See cathode follower. { ¦graúnd·əd 'an,ōd ,am·plə,fī·ər }

grounded-base amplifier |ELECTR| An amplifier that uses a transistor in a grounded-base connection. { ¦graúnd·əd 'bās ,am·plə,fī·ər }

grounded-base connection |ELECTR| A transistor circuit in which the base electrode is common to both the input and output circuits; the base need not be directly connected to circuit ground. Also known as common-base connection. { ¦graúnd·əd 'bās kə,nek·shən }

grounded-cathode amplifier |ELECTR| Electron-tube amplifier with a cathode at ground potential at the operating frequency, with input applied between control grid and ground, and with the output load connected between plate and ground. { ¦graùnd·əd 'kath‚ōd ‚am·plə‚fī·ər }

grounded-collector connection |ELECTR| A transistor circuit in which the collector electrode is common to both the input and output circuits; the collector need not be directly connected to circuit ground. Also known as common-collector connection. { ¦graùnd·əd kə'lek·tər kə‚nek·shən }

grounded-emitter amplifier |ELECTR| An amplifier that uses a transistor in a grounded-emitter connection. { ¦graùnd·əd i'mid·ər ‚am·plə ‚fī·ər }

grounded-emitter connection |ELECTR| A transistor circuit in which the emitter electrode is common to both the input and output circuits; the emitter need not be directly connected to circuit ground. Also known as common-emitter connection. { ¦graùnd·əd i'mid·ər kə‚nek·shən }

grounded-gate amplifier |ELECTR| Amplifier that uses thin-film transistors in which the gate electrode is connected to ground; the input signal is fed to the source electrode and the output is obtained from the drain electrode. { ¦graùnd·əd 'gāt ‚am·plə‚fī·ər }

grounded-grid amplifier |ELECTR| An electron-tube amplifier circuit in which the control grid is at ground potential at the operating frequency; the input signal is applied between cathode and ground, and the output load is connected between anode and ground. { ¦graùnd·əd 'grid ‚am·plə‚fī·ər }

grounded-grid-triode circuit |ELECTR| Circuit in which the input signal is applied to the cathode and the output is taken from the plate; the grid is at radio-frequency ground and serves as a screen between the input and output circuits. { ¦graùnd·əd ¦grid ¦trī‚ōd 'sər·kət }

grounded-grid-triode mixer |ELECTR| Triode in which the grid forms part of a grounded electrostatic screen between the anode and cathode, and is used as a mixer for centimeter wavelengths. { ¦graùnd·əd ¦grid ¦trī‚ōd 'mik·sər }

grounded-plate amplifier See cathode follower. { ¦graùnd·əd 'plāt ‚am·plə‚fī·ər }

grounded system |ELEC| Any conducting apparatus connected to ground. Also known as earthed system. { ¦graùnd·əd 'sis·təm }

ground-effect machine See air-cushion vehicle. { 'graùnd i‚fekt mə‚shēn }

ground electrode |ELEC| A conductor buried in the ground, used to maintain conductors connected to it at ground potential and dissipate current conducted to it into the earth, or to provide a return path for electric current in a direct-current power transmission system. Also known as earth electrode; grounding electrode. { 'graùnd i'lek‚trōd }

ground environment |ENG| **1.** Environment that surrounds and affects a system or piece of equipment that operates on the ground. **2.** System or part of a system, as of a guidance system, that functions on the ground; the aggregate of equipment, conditions, facilities, and personnel that go to make up a system, or part of a system, functioning on the ground. { 'graùnd in¦vī·ərn·mənt }

ground fault |ELEC| Accidental grounding of a conductor. { 'graùnd ‚fôlt }

ground fault interrupter |ELEC| A fast-acting circuit breaker that also senses very small ground fault currents such as might flow through the body of a person standing on damp ground while touching a hot alternating-current line wire. { 'graùnd ‚fôlt ‚int·ə‚rəp·tər }

ground instrumentation See spacecraft ground instrumentation. { 'graùnd ‚in·strə·mən'tā·shən }

ground joint |CIV ENG| A closely fitted masonry joint, usually set without mortar. |MECH ENG| A machined metal joint that makes a tight fit without packing or a gasket. { 'graùnd ‚jóint }

ground junction See grown junction. { 'graùnd ‚jəŋk·shən }

ground magnetic survey |ENG| A determination of the magnetic field at the surface of the earth by means of ground-based instruments. { 'graùnd mag¦ned·ik 'sər‚vā }

groundman |ENG| A person employed in digging or excavating. { 'graùnd‚man }

ground noise |ENG ACOUS| The residual system noise in the absence of the signal in recording and reproducing; usually caused by inhomogeneity in the recording and reproducing media, but may also include tube noise and noise generated in resistive elements in the amplifier system. { 'graùnd ‚nóiz }

ground-penetrating radar See ground-probing radar. { ¦graùnd ‚pen·ə¦trād·iŋ 'rā‚där }

ground potential |ELEC| Zero potential with respect to the ground or earth. { 'graùnd pə‚ten·chəl }

ground-probing radar |ENG| A nondestructive technique using electromagnetic waves to locate objects or interfaces buried beneath the earth's surface or located within a visually opaque structure. Also known as ground-penetrating radar; subsurface radar; surface-penetrating radar. { ¦graùnd ¦prōb·iŋ 'rā‚där }

ground protection |ELEC| Protection provided a circuit by a device which opens the circuit when a fault to ground occurs. { 'graùnd prə‚tek·shən }

ground resistance |ELEC| Opposition of the earth to the flow of current through it; its value depends on the nature and moisture content of the soil, on the material, composition, and nature of connections to the earth, and on the electrolytic action present. { 'graùnd ri‚zis·təns }

ground return |ELEC| Use of the earth as the return path for a transmission line. { 'graùnd ri‚tərn }

ground surveillance radar |ENG| **1.** A surveillance radar operated at a fixed point on the earth's surface for observation and control of the position of aircraft or other vehicles in the vicinity. **2.** A radar system capable of detecting objects on the ground from points on the ground. { 'graůnd sər,vā·ləns ,rā,där }

ground trace |ENG| The theoretical mark traced upon the surface of the earth by a flying object, missile, or satellite as it passes over the surface, the mark being made vertically from the object making the trace. { 'graůnd ,trās }

ground ways |CIV ENG| Supports, usually made of heavy timbers, which are placed on the ground on either side of the keel of a ship under construction, providing a track for launching, and supporting the sliding ways. Also known as standing ways. { 'graůnd ,wāz }

ground wire |CIV ENG| A small-gage, high-strength steel wire used to establish line and grade for air-blown mortar or concrete. Also known as alignment wire; screed wire. |ELEC| A conductor used to connect electric equipment to a ground rod or other grounded object. { 'graůnd ,wīr }

group bus |ELEC| A scheme of electrical connections for a generating station in which more than two feeder lines are supplied by two bus-selector circuit breakers which lead to a main bus and an auxiliary bus. { 'grüp ˌbəs }

group incentive |IND ENG| Any wage incentive applied to more than one employee who is engaged in group work characterized by interdependent relationship between operations with consequent physical proximity and unification of interest. { 'grüp in'sen·tiv }

group technology |IND ENG| A manufacturing system that uses a classification and coding scheme to group parts into families based on similar manufacturing requirements, and specifies parts characteristics, process plans, setups, and manufacturing sequences. { 'grüp tek'näl·ə·jē }

grouser |ENG| A temporary pile or a heavy, iron-shod pole driven into the bottom of a stream to hold a drilling or dredging boat or other floating object in position. Also known as spud. { 'graůs·ər }

grout curtain |ENG| A row of vertically drilled holes filled with grout under pressure to form the cutoff wall under a dam, or to form a barrier around an excavation through which water cannot seep or flow. { 'graůt ,kərt·ən }

grout hole |ENG| **1.** One of the holes in a grout curtain. **2.** Any hole into which grout is forced under pressure to consolidate the surrounding earth or rock. { 'graůt ,hōl }

grouting |ENG| The act or process of applying grout or of injecting grout into grout holes or crevices of a rock. { 'graůd·iŋ }

grout injector |ENG| A machine that mixes the dry ingredients for a grout with water and injects it, under pressure, into a grout hole. { 'graůt in,jek·tər }

grout pipe |ENG| A pipe that transports grout under pressure for injection into a grout hole or a rock formation. { 'graůt ,pīp }

grown-diffused transistor |ELECTR| A junction transistor in which the final junctions are formed by diffusion of impurities near a grown junction. { ¦grōn di¦fyüzd tran'zis·tər }

grown junction |ELECTR| A junction produced by changing the types and amounts of donor and acceptor impurities that are added during the growth of a semiconductor crystal from a melt. Also known as ground junction. { ¦grōn ¦jəŋk·shən }

grown-junction photocell |ELECTR| A photodiode consisting of a bar of semiconductor material having a *pn* junction at right angles to its length and an ohmic contact at each end of the bar. { ¦grōn ¦jəŋk·shən 'fōd·ō,sel }

grown-junction transistor |ELECTR| A junction transistor in which different impurities are placed in the melt in sequence as the silicon or germanium seed crystal is slowly withdrawn, to produce the alternate *pn* and *np* junctions. { ¦grōn ¦jəŋk·shən tran'zis·tər }

grubbing |CIV ENG| Clearing stumps and roots. { 'grəb·iŋ }

grub screw |DES ENG| A headless screw with a slot at one end to receive a screwdriver. { 'grəb ,skrü }

gr wt See gross weight.

g suit |ENG| A suit that exerts pressure on the abdomen and lower parts of the body to prevent or retard the collection of blood below the chest under positive acceleration. Also known as anti-g suit. { 'jē ,süt }

guard |ENG| A shield or other fixture designed to protect against injury. { gärd }

guard circle |DES ENG| The closed loop at the end of a grooved record. { 'gärd ,sər·kəl }

guard lock |CIV ENG| See entrance lock. |ENG| An auxiliary lock that must be opened before the key can be turned in a main lock. { 'gärd ,läk }

guardrail |CIV ENG| **1.** A handrail. **2.** A rail made of posts and a metal strip used on a road as a divider between lines of traffic in opposite directions or used as a safety barrier on curves. **3.** A rail fixed close to the outside of the inner rail on railway curves to hold the inner wheels of a railway car on the rail. Also known as check rail; safety rail; slide rail. { 'gärd ,rāl }

guard ring |ELEC| A ring-shaped auxiliary electrode surrounding one of the plates of a parallel-plate capacitor to reduce edge effects. |ELECTR| A ring-shaped auxiliary electrode used in an electron tube or other device to modify the electric field or reduce insulator leakage; in a counter tube or ionization chamber a guard ring may also serve to define the sensitive volume. |THERMO| A device used in heat flow experiments to ensure an even distribution of heat, consisting of a ring that surrounds the specimen and is made of a similar material. { 'gärd ,riŋ }

gudgeon |ENG| **1.** A pivot. **2.** A pin for fastening stone blocks. { gəj·ən }

Guggenheim process |CIV ENG| A method of chemical precipitation which employs ferric chloride and aeration to prepare sludge for filtration. { 'güg·ən·hīm ,präs·əs }

guidance site |ENG| Specific location of high-order geodetic accuracy containing equipment and structures necessary to provide guidance services or a given launch rate; it may be an integrated part of a launch site, or it may be a remote facility. { 'gīd·əns ,sīt }

guidance station equipment |ENG| The ground-based portion of the missile guidance system necessary to provide guidance during missile flight; it specifically includes the tracking radar, the rate measuring equipment, the data link equipment, and the computer, test, and maintenance equipment integral to these items. { 'gīd·əns ,stā·shən i,kwip·mənt }

guide bearing |MECH ENG| A plain bearing used to guide a machine element in its lengthwise motion, usually without rotation of the element. { 'gīd ,ber·iŋ }

guide idler |MECH ENG| An idler roll with its supporting structure mounted on a conveyor frame to guide the belt in a defined horizontal path, usually by contact with the edge of the belt. { 'gīd ,īd·lər }

guide key See home key. { 'gīd ,kē }

guideline |IND ENG| A document containing recommendations for methods that should be used to achieve a desired goal. { 'gīd,līn }

guidepath |ENG| The path over which an automated guided vehicle travels; often contains some means of communication with the guidance system, such as a guidewire. { 'gīd,path }

guide pin |ENG| A pin used to line up a tool or die with the work. { 'gīd ,pin }

guide post |CIV ENG| A post along a road that bears direction signs or guide boards. { 'gīd ,pōst }

guide rail |CIV ENG| A track or rail that serves to guide movement, as of a sliding door, window, or similar element. { 'gīd ,rāl }

guides |MECH ENG| **1.** Pulleys to lead a driving belt or rope in a new direction or to keep it from leaving its desired direction. **2.** Tracks that support and determine the path of a skip bucket and skip bucket bail. **3.** Tracks guiding the chain or buckets of a bucket elevator. **4.** The runway paralleling the path of the conveyor which limits the conveyor or parts of a conveyor to movement in a defined path. { gīdz }

guidewire |ENG| A wire embedded in the surface of the path traveled by an electromagnetically guided automated guided vehicle. { 'gīd,wīr }

guillotine shears |ENG| A cutting tool fitted with vertically mounted blades, the bottom blade being fixed in position and the top blade mounted on a movable ram. { 'gē·ə,tēn ,shirz }

Gukhman number |THERMO| A dimensionless number used in studying convective heat transfer in evaporation, equal to $(t_0 - t_m)/T_0$, where t_0 is the temperature of a hot gas stream, t_m is the temperature of a moist surface over which it is flowing, and T_0 is the absolute temperature of the gas stream. Symbolized Gu; N_{Gu}. { 'gúk·mən ,nəm·bər }

Guldberg-Waage group |CHEM ENG| A dimensionless number used in studying chemical reactions in blast furnaces; it is given by an equation relating volumes of reacting gases and reacting products. Symbolized N_{Gw}. { 'gúlt·berk 'väg·ə ,grüp }

gull-wing door |DES ENG| A door on an automotive vehicle that is hinged at the top, opens upward, and, in the open position, resembles an airplane gull wing. { 'gəl ,wiŋ 'dòr }

gum test |CHEM ENG| A standard American Society for Testing and Materials test to determine the amount of gums in gasolines. { 'gəm ,test }

gunbarrel |CHEM ENG| An atmospheric vessel used for treatment of waterflood waste water. { 'gən,bar·əl }

gun burner |ENG| A burner which sprays liquid fuel into a furnace for combustion. { 'gən ¦bər·nər }

gunite |CIV ENG| A mixture of cement, sand, and water that is sprayed on a surface for repairing portions of existing structures, lining reservoirs, and encasing steel for fireproofing. { 'gə,nīt }

gun-laying radar |ENG| Radar equipment specifically designed to determine range, azimuth, and elevation of a target and sometimes also to automatically aim and fire antiaircraft artillery or other guns. { 'gən ,lā·iŋ 'rā,där }

Gunn effect |ELECTR| Development of a rapidly fluctuating current in a small block of a semiconductor (perhaps n-type gallium arsenide) when a constant voltage above a critical value is applied to contacts on opposite faces. { 'gən i,fekt }

gunner's quadrant |ENG| Mechanical device having scales graduated in mils, with fine micrometer adjustments and leveling or cross-leveling vials; it is a separate, unattached instrument for hand placement on a reference surface. { ¦gən·ərz 'kwäd·rənt }

gun pendulum |ENG| A device used to determine the initial velocity of a projectile fired from a gun in which the gun is mounted as a pendulum and its excursion upon firing is measured. { 'gən ¦pen·jə·ləm }

gun reaction |MECH| The force exerted on the gun mount by the rearward movement of the gun resulting from the forward motion of the projectile and hot gases. Also known as recoil. { 'gən rē,ak·shən }

Gunter's chain |ENG| A chain 66 feet (20.1168 meters) long, consisting of 100 steel links, each 7.92 inches (20.1168 centimeters) long, joined by rings, which is used as the unit of length for surveying public lands in the United States. Also known as chain. { 'gən·tərz ¦chān }

gun-type burner |ENG| An oil burner that uses a nozzle to atomize the fuel. { 'gən,tīp ¦bər·nər }

gusset |CIV ENG| A plate that is used to strengthen truss joints. { 'gəs·ət }

gusset plate |CIV ENG| A rectangular or triangular steel plate that connects members of a truss. { 'gəs·ət ,plāt }

gust load |MECH| The wind load on an antenna due to gusts. { 'gəst ,lōd }

gustsonde |ENG| An instrument dropped from high altitude by a stable parachute, to measure the vertical component of turbulence aloft; consists of an accelerometer and radio telemetering equipment. { 'gəst,sänd }

gutter |BUILD| A trough along the edge of the eaves of a building to carry off rainwater. |CIV ENG| A shallow trench provided beside a canal, bordering a highway, or elsewhere, for surface drainage. { 'gəd·ər }

guttering |ENG| A process of quarrying stone in which channels, several inches wide, are cut by hand tools, and the stone block is detached from the bed by pinch bars. { 'gəd·ə·riŋ }

guy |ENG| A rope or wire securing a pole, derrick, or similar temporary structure in a vertical position. { gī }

guy derrick |MECH ENG| A derrick having a vertical pole supported by guy ropes to which a boom is attached by rope or cable suspension at the top and by a pivot at the foot. { 'gī ,der·ik }

gvw See gross vehicle weight.

gyratory breaker See gyratory crusher. { 'jī·rə,tór·ē 'brāk·ər }

gyratory crusher |MECH ENG| A primary breaking machine in the form of two cones, an outer fixed cone and a solid inner erect cone mounted on an eccentric bearing. Also known as gyratory breaker. { 'jī·rə,tór·ē 'krəsh·ər }

gyratory screen |MECH ENG| Boxlike machine with a series of horizontal screens nested in a vertical stack with downward-decreasing mesh-opening sizes; near-circular motion causes undersized material to sift down through each screen in succession. { 'jī·rə,tór·ē 'skrēn }

gyro See gyroscope. { 'jī·rō }

gyrodynamics |MECH| The study of rotating bodies, especially those subject to precession. { ,jī·rō·dī'nam·iks }

gyropendulum |MECH ENG| A gravity pendulum attached to a rapidly spinning gyro wheel. { ,jī·rō¦pen·jə·ləm }

gyrorepeater |ENG| That part of a remote indicating gyro compass system which repeats at a distance the indications of the master gyro compass system. { ¦jī·rō·ri'pēd·ər }

gyroscope |ENG| An instrument that maintains an angular reference direction by virtue of a rapidly spinning, heavy mass; all applications of the gyroscope depend on a special form of Newton's second law, which states that a massive, rapidly spinning body rigidly resists being disturbed and tends to react to a disturbing torque by precessing (rotating slowly) in a direction at right angles to the direction of torque. Also known as gyro. { 'jī·rə,skōp }

gyroscopic-clinograph method |ENG| A method used in borehole surveying which measures time, temperature, and temperature on 16-millimeter film while a gyroscope maintains the casing on a fixed bearing. { ,jī·rə'skäp·ik 'klīn·ə,graf ,meth·əd }

gyroscopic/Coriolis-type mass flowmeter |ENG| An instrument consisting of a C-shaped pipe and a T-shaped leaf-spring tuning fork which is excited by an electromagnetic forcer, resulting in an angular deflection of the pipe which is directly proportional to the mass-flow rate within the pipe. { ,jī·rə'skäp·ik ,kór·ē'ō·ləs ,tīp ¦mas 'flō ,mēd·ər }

gyroscopic couple |MECH ENG| The turning moment which opposes any change of the inclination of the axis of rotation of a gyroscope. { ,jī·rə'skäp·ik 'kəp·əl }

gyroscopic mass flowmeter |ENG| An instrument in which the torque on a rotating pipe of suitable shape, through which a fluid is made to flow, is measured to determine the mass flow through the pipe. { ,jī·rə'skäp·ik ¦mas 'flō ,mēd·ər }

gyroscopic precession |MECH| The turning of the axis of spin of a gyroscope as a result of an external torque acting on the gyroscope; the axis always turns toward the direction of the torque. { ,jī·rə'skäp·ik prē'sesh·ən }

gyroscopics |MECH| The branch of mechanics concerned with gyroscopes and their use in stabilization and control of ships, aircraft, projectiles, and other objects. { ,jī·rə'skäp·iks }

gyrostabilizer |ENG| A gyroscope used to stabilize ships and airplanes. { ¦jī·rō'stā·bə,līz·ər }

gyro wheel |MECH ENG| The rapidly spinning wheel in a gyroscope, which resists being disturbed. { 'jī·rō ,wēl }

H

ha *See* hectare.

Haber-Bosch process [CHEM ENG] Early nitrogen-fixation process for production of ammonia from hydrogen and nitrogen, catalyzed by iron; now replaced by more efficient ammonia synthesis processes. Also known as Haber process. { ¦hä·bər ¦bȯsh ¸prä·səs }

Haber process *See* Haber-Bosch process. { 'hä·bər ¸prä·səs }

hacking [ENG] The technique of roughening a surface by striking it with a tool. { 'hak·iŋ }

hacking knife [ENG] A tool for removing old putty from a window frame prior to reglazing. Also known as hacking-out tool. { 'hak·iŋ ¸nīf }

hacking-out tool *See* hacking knife. { 'hak·iŋ ¸aut ¸tül }

hacksaw [ENG] A hand or power tool consisting of a fine-toothed blade held in tension in a bow-shaped frame; used for cutting metal, wood, and other hard materials. { 'hak¸sȯ }

hair hygrometer [ENG] A hygrometer in which the sensing element is a bundle of human hair, which is held under slight tension by a spring and which expands and contracts with changes in the moisture of the surrounding air or gas. { 'her hī'gräm·əd·ər }

hairline *See* air line. { 'her¸līn }

hairpin tube [DES ENG] A boiler tube bent into a hairpin, or U, shape. { 'her¸pin ¸tüb }

half-adder [ELECTR] A logic element which operates on two binary digits (but no carry digits) from a preceding stage, producing as output a sum digit and a carry digit. { ¦haf ¦ad·ər }

half cycle [ENG] The time interval corresponding to half a cycle, or 180°, at the operating frequency of a circuit or device. { 'haf ¦sī·kəl }

half-dog setscrew [DES ENG] A setscrew with a short, blunt point. { 'haf ¸dȯg 'set¸skrü }

half nut [DES ENG] A nut split lengthwise so that it can be clamped around a screw. { 'haf ¸nət }

half-round file [DES ENG] A file that is flat on one side and convex on the other. { 'haf ¦raund ¸fīl }

half space [BUILD] A broad step between two half flights of a stair. { 'haf ¸spās }

half-subtracter [ELECTR] A logic element which operates on two digits from a preceding stage, producing as output a difference digit and a borrow digit. Also known as one-digit subtracter; two-input subtracter. { 'haf səb'trak·tər }

half-through arch [CIV ENG] A bridge arch having the roadway running through it at an elevation midway between the base and the crown. { 'haf ¸thrü 'ärch }

half-tide basin [CIV ENG] A lock of very large size and usually of irregular shape, the gates of which are kept open for several hours after high tide so that vessels may enter as long as there is sufficient depth over the sill; vessels remain in the half-tide basin until the ensuing flood tide, when they may pass through the gate to the inner harbor; if entry to the inner harbor is required before this time, water must be admitted to the half-tide basin from some external source. { 'haf ¸tīd ¸bās·ən }

half-timbered [BUILD] Pertaining to a timber frame building with brickwork, plaster, or wattle and daub filling the spaces between the timbers. { 'haf ¦tim·bərd }

half-track [MECH ENG] **1.** A chain-track drive system for a vehicle; consists of an endless metal belt on each side of the vehicle driven by one of two inside sprockets and running on bogie wheels; the revolving belt lays down on the ground a flexible track of cleated steel or hard-rubber plates; the front end of the vehicle is supported by a pair of wheels. **2.** A motor vehicle equipped with half-tracks. { 'haf ¸trak }

half-track tape recorder *See* double-track tape recorder. { 'haf ¸trak 'tāp ri¸kȯrd·ər }

Hall cyclic thermal reforming [CHEM ENG] A gas-making process that uses component parts of carbureted-water gas apparatus to generate high-Btu gas from feedstocks ranging from naphtha to Bunker C. { 'hȯl 'sī·klik ¦thər·məl rē'fȯr·miŋ }

Hall-effect gaussmeter [ENG] A gaussmeter that consists of a thin piece of silicon or other semiconductor material which is inserted between the poles of a magnet to measure the magnetic field strength by means of the Hall effect. { 'hȯl i¸fekt 'gaus¸med·ər }

Hall-plate device [ENG] A sensor that uses the Hall effect to measure magnetic field strength. { 'hȯl ¦plāt di¸vīs }

halo effect [IND ENG] A tendency when rating

a person in regard to a specific trait to be influenced by a general impression or by another trait of the person. { 'hā·lō i,fekt }

halophone |ENG| A device that records patterns in time in a manner analogous to the way that optical holograms record space. { 'hal·ə,fōn }

Halsey premium plan [IND ENG] A wage-incentive plan which sets a guaranteed daily rate to an employee and provides for predetermined compensation for superior performance. { 'hȯl·zē 'prē·mē·əm ,plan }

Hamiltonian function [MECH] A function of the generalized coordinates and momenta of a system, equal in value to the sum over the coordinates of the product of the generalized momentum corresponding to the coordinate, and the coordinate's time derivative, minus the Lagrangian of the system; it is numerically equal to the total energy if the Lagrangian does not depend on time explicitly; the equations of motion of the system are determined by the functional dependence of the Hamiltonian on the generalized coordinates and momenta. { ,ham·əl'tō·nē·ən ¦faŋk·shən }

Hamilton-Jacobi theory |MECH| A theory that provides a means for discussing the motion of a dynamic system in terms of a single partial differential equation of the first order, the Hamilton-Jacobi equation. { 'ham·əl·tən jə'kō·bē ,thē·ə·rē }

Hamilton's equations of motion |MECH| A set of first-order, highly symmetrical equations describing the motion of a classical dynamical system, namely $\dot{q}_i = \partial H/\partial p_i$, $\dot{p}_i = -\partial H/\partial q_i$; here q_i ($i = 1, 2, \ldots$) are generalized coordinates of the system, p_i is the momentum conjugate to q_i, and H is the Hamiltonian. Also known as canonical equations of motion. { 'ham·əl·tənz i¦kwā·zhənz əv 'mō·shən }

Hamilton's principle |MECH| A variational principle which states that the path of a conservative system in configuration space between two configurations is such that the integral of the Lagrangian function over time is a minimum or maximum relative to nearby paths between the same end points and taking the same time. { 'ham·əl·tənz ¦prin·sə·pəl }

hammer |DES ENG| **1.** A hand tool used for pounding and consisting of a solid metal head set crosswise on the end of a handle. **2.** An arm with a striking head for sounding a bell or gong. |MECH ENG| A power tool with a metal block or a drill for the head. { 'ham·ər }

hammer drill |MECH ENG| Any of three types of fast-cutting, compressed-air rock drills (drifter, sinker, and stoper) in which a hammer strikes rapid blows on a loosely held piston, and the bit remains against the rock in the bottom of the hole, rebounding slightly at each blow, but does not reciprocate. { 'ham·ər ,dril }

hammerhead |DES ENG| The striking part of a hammer. { 'ham·ər,hed }

hammerhead crane |MECH ENG| A crane with a horizontal jib that is counterbalanced. { 'ham·ər,hed ,krān }

hammer mill |MECH ENG| **1.** A type of impact mill or crusher in which materials are reduced in size by hammers revolving rapidly in a vertical plane within a steel casing. Also known as beater mill. **2.** A grinding machine which pulverizes feed and other products by several rows of thin hammers revolving at high speed. { 'ham·ər ,mil }

hammer milling |MECH ENG| Crushing or fracturing materials in a hammer mill. { 'ham·ər ,mil·iŋ }

hand See end effector. { hand }

hand auger |DES ENG| A hand tool resembling a large carpenters' bit or comprising a short cylindrical container with cutting lips attached to a rod; used to bore shallow holes in the soil to obtain samples of it and other relatively unconsolidated near-surface materials. { 'hand ¦ȯg·ər }

handbarrow |ENG| A flat, rectangular frame with handles at both ends, carried by two persons to transport objects. Also known as barrow. { 'hand,bar·ō }

hand brake |MECH ENG| A manually operated brake. { 'hand ,brāk }

handcar |MECH ENG| A small, four-wheeled, hand-pumped car used on railroad tracks to transport workers and equipment for construction or repair work; other cars for the same purpose are motor-operated. { 'hand,kär }

hand drill |DES ENG| A small, portable drilling machine which is operated by hand. { 'hand ,dril }

hand feed |ENG| A drill machine in which the rate at which the bit is made to penetrate the rock is controlled by a hand-operated ratchet and lever or a hand-turned wheel meshing with a screw mechanism. { 'hand ,fēd }

hand float |ENG| A wooden tool used to fill in and smooth a plaster surface in order to produce a level base coat or a textured finish coat. { 'hand ,flōt }

hand hammer drill |ENG| A hand-held rock drill. { 'hand 'ham·ər ,dril }

hand-held scanner |ENG| An image-reading device that is held and operated by a person. { ¦hand ,held 'skan·ər }

handhole |ENG| A shallow access hole large enough for a hand to be inserted for maintenance and repair of machinery or equipment. { 'hand,hōl }

hand lance |ENG| A hand-held pipe with a nozzle through which steam or air is discharged; used to remove soot deposits from the external surfaces of boiler tubes. { 'hand ,lans }

handle |MECH ENG| The arm connecting the bucket with the boom in a dipper shovel or hoe. { 'han·dəl }

hand lead |ENG| A light sounding lead (7–14 pounds or 3–6 kilograms) usually having a line not more than 25 fathoms (46 meters) in length. { 'hand ,led }

hand level |ENG| A hand-held surveyor's level,

basically a telescope with a bubble tube attached so that the position of the bubble can be seen when looking through the telescope. { 'hand ,lev·əl }

handling time |IND ENG| The time needed to transport parts or materials to or from a work area. { 'hand·liŋ ,tīm }

hand punch |DES ENG| A hand-held device for punching holes in paper or cards. { 'hand ,pənch }

handrail |ENG| A narrow rail to be grasped by a person for support. { 'hand,rāl }

handsaw |DES ENG| A saw operated by hand, with a backward and forward arm movement. { 'hand,só }

handset |DES ENG| A combination of a telephone-type receiver and transmitter, designed for holding in one hand. { 'hand,set }

handset bit |DES ENG| A bit in which the diamonds are manually set into holes that are drilled into a malleable-steel bit blank and shaped to fit the diamonds. { 'hand,set ,bit }

hand-tight |ENG| The extent of tightening of screwed fittings that can be accomplished without mechanical assistance. { 'hand ¦tīt }

hand time |IND ENG| The time necessary to complete a manual element. Also known as manual time. { 'hand ,tīm }

hand tool |ENG| Any implement used by hand. { 'hand ,tül }

hand truck |ENG| **1.** A manually operated, two-wheeled truck consisting of a rectangular frame with handles at the top and a plate at the bottom to slide under the load. **2.** Any of various small, manually operated, multiwheeled platform trucks for transporting materials. { 'hand ,trək }

hand winch |MECH ENG| A winch that is operated by hand. { 'hand ,winch }

hangar |CIV ENG| A building at an airport specially designed in height and width to enable aircraft to be stored or maintained in it. { 'haŋ·ər }

hanger |CIV ENG| An iron strap which lends support to a joist beam or pipe. { 'haŋ·ər }

hanger bolt |DES ENG| A bolt with a machine-screw thread on one end and a lag-screw thread on the other. { 'haŋ·ər ,bólt }

hangfire |ENG| Delay in the explosion of a charge. { 'haŋ,fīr }

hanging-drop atomizer |MECH ENG| An atomizing device used in gravitational atomization; functions by quasi-static emission of a drop from a wetted surface. Also known as pendant atomizer. { 'haŋ·iŋ ,dräp 'ad·ə,mīz·ər }

hanging load |MECH ENG| **1.** The weight that can be suspended on a hoist line or hook device in a drill tripod or derrick without causing the members of the derrick or tripod to buckle. **2.** The weight suspended or supported by a bearing. { 'haŋ·iŋ ¦lōd }

hanging scaffold |CIV ENG| A movable platform suspended by ropes and pulleys; used by workers for above-ground building construction and maintenance. { 'haŋ·iŋ ¦ska,fəld }

hang-up |ENG| A virtual leak resulting from the release of entrapped tracer gas from a leak detector vacuum system. { 'haŋ,əp }

HAP See hazardous air pollutants. { hap or ¦āch¦ā'pē }

harbor engineering |CIV ENG| Planning and design of facilities for ships to discharge or receive cargo and passengers. { 'här·bər ,en·jə'nir·iŋ }

harbor line |CIV ENG| The line beyond which wharves and other structures cannot be extended. { 'här·bər ,līn }

hard automation |IND ENG| Automation that makes use of specially designed equipment for production. { 'härd ,ód·ə'mā·shən }

hard beach |CIV ENG| A portion of a beach especially prepared with a hard surface extending into the water, employed for the purpose of loading or unloading directly into or from landing ships or landing craft. { 'härd ¦bēch }

hard goods See durable goods. { 'härd ,gúdz }

Hardgrove grindability index |ENG| The relative grindability of ores and minerals in comparison with standard coal, chosen as 100 grindability, as determined by a miniature ball-ring pulverizer. Also known as Hardgrove number. { 'här,grōv ,grīn·də'bil·əd·ē ,in,deks }

Hardgrove number See Hardgrove grindability index. { 'här,grōv ,nəm·bər }

hard hat |ENG| A safety hat usually having a metal crown; used by construction workers and miners. { 'härd ,hat }

Hardinge feeder-weigher |MECH ENG| A pivoted, short belt conveyor which controls the rate of material flow from a hopper by weight per cubic foot. { 'här·diŋ ¦fēd·ər ¦wā·ər }

Hardinge mill |MECH ENG| A tricone type of ball mill; the cones become steeper from the feed end toward the discharge end. { 'här·diŋ ,mil }

Hardinge thickener |ENG| A machine for removing the maximum amount of liquid from a mixture of liquid and finally divided solids by allowing the solids to settle out on the bottom as sludge while the liquid overflows at the top. { 'här·diŋ 'thik·ən·ər }

hard-laid |DES ENG| Pertaining to rope with strands twisted at a 45° angle. { 'härd ¦lād }

hardness |ENG| Property of an installation, facility, transmission link, or equipment that will prevent an unacceptable level of damage. { 'härd·nəs }

hardness number |ENG| A number representing the relative hardness of a mineral, metal, or other material as determined by any of more than 30 different hardness tests. { 'härd·nəs ,nəm·bər }

hardness test |ENG| A test to determine the relative hardness of a metal, mineral, or other material according to one of several scales, such as Brinell, Mohs, or Shore. { 'härd·nəs ,test }

hardstand |CIV ENG| **1.** A paved or stabilized area where vehicles or aircraft are parked. **2.** Open ground area having a prepared surface and used for storage of material. { 'härd,stand }

hard-surface |CIV ENG| To treat a ground surface in order to prevent muddiness. { 'härd ¦sər·fəs }

hardware |ENG| Items made of metal, such as tools, fittings, fasteners, and appliances. { 'härd,wer }

hard-wire |ELEC| To connect electric components with solid, metallic wires as opposed to radio links and the like. { 'härd |wīr }

hardwood bearing |MECH ENG| A fluid-film bearing made of lignum vitae which has a natural gum, or of hard maple which is impregnated with oil, grease, or wax. { 'härd,wúd |ber·iŋ }

Hardy plankton indicator |ENG| Metal-shrouded net sampler designed to collect specimens of plankton during normal passage of a ship. { 'härd·ē 'plaŋk·tən ,in·də,kād·ər }

Hare's hygrometer |ENG| A type of hydrometer in which the ratio of the densities of two liquids is determined by measuring the heights to which they rise in two vertical glass tubes, connected at their upper ends, when suction is applied. { 'herz hī'gräm·əd·ər }

Hargreaves process |CHEM ENG| A process for the manufacture of salt cake (sodium sulfate) by passing a mixture of sulfur dioxide and air through sodium chloride brine in a countercurrent manner. { 'här·grēvz ,prä·səs }

HARM See high-aspect-ratio micromachining. { |ach|a|är'em or härm }

harmonic drive |MECH ENG| A drive system that uses inner and outer gear bands to provide smooth motion. { här'män·ik 'drīv }

harmonic motion |MECH| A periodic motion that is a sinusoidal function of time, that is, motion along a line given by the equation $x = a \cos(kt + \theta)$, where t is the time parameter, and a, k, and θ are constants. Also known as harmonic vibration; simple harmonic motion (SHM). { här'män·ik 'mō·shən }

harmonic oscillator |ELECTR| See sinusoidal oscillator. |MECH| Any physical system that is bound to a position of stable equilibrium by a restoring force or torque proportional to the linear or angular displacement from this position. { här'män·ik 'äs·ə,lād·ər }

harmonic speed changer |MECH ENG| A mechanical-drive system used to transmit rotary, linear, or angular motion at high ratios and with positive motion. { här'män·ik 'spēd ,chān·jər }

harmonic synthesizer |MECH| A machine which combines elementary harmonic constituents into a single periodic function; a tide-predicting machine is an example. { här'män·ik 'sin·thə,sīz·ər }

harmonic vibration See harmonic motion. { här 'män·ik vī'brā·shən }

harness |ELEC| Wire and cables so arranged and tied together that they may be inserted and connected, or may be removed after disconnection, as a unit. { 'här·nəs }

harpoon |DES ENG| A barbed spear used to catch whales. { här'pün }

harpoon log |ENG| A log which consists essentially of a rotator and distance registering device combined in a single unit, and towed through the water; it has been largely replaced by the taffrail log; the two types of logs are similar except that the registering device of the taffrail log is located at the taffrail and only the rotator is in the water. { här'pün ,läg }

Harrison's gridiron pendulum |DES ENG| A type of compensated pendulum that has five iron rods and four brass rods arranged so that the effects of their thermal expansion cancel. { |har·i·sənz 'grid,ī·ərn ,pen·jə·ləm }

Hartford loop |MECH ENG| A condensate return arrangement for low-pressure, steam-heating systems featuring a steady water line in the boiler. { 'härt·fərd ,lüp }

Hartmann generator |ENG ACOUS| A device in which shock waves generated at the edges of a nozzle by a supersonic gas jet resonate with the opening of a small cylindrical pipe, placed opposite the nozzle, to produce powerful ultrasonic sound waves. { 'härt·mən ,jen·ə,rād·ər }

Hasche process |CHEM ENG| A thermal reforming process for hydrocarbon fuels; it is a noncatalytic regenerative method in which a mixture of hydrocarbon gas or vapor and air is passed through a regenerative mass that is progressively hotter in the direction of the gas flow; partial combustion occurs, liberating heat to crack the remaining hydrocarbons in a combustion zone. { 'häsh·ə ,prä·səs }

hasp |DES ENG| A two-piece fastening device having a loop on one piece and a hinged plate that fits over the loop on the other. { hasp }

hatch |ENG| A door or opening, especially on an airplane, spacecraft, or ship. { hach }

hatch beam |ENG| A heavy, portable beam which supports a hatch cover. { 'hach ,bēm }

hatch cover |ENG| A steel or wooden cover for a hatch. { 'hach ,kəv·ər }

hatchet |DES ENG| A small ax with a short handle and a hammerhead in addition to the cutting edge. { 'hach·ət }

haul |ENG| A single tow of a net or dredge. { hȯl }

hawk |ENG| A board with a handle underneath used by a workman to hold mortar. { hȯk }

Hayward grab bucket |MECH ENG| A clamshell type of grab bucket used for handling coal, sand, gravel, and other flowable materials. { 'hā·wərd 'grab ,bək·ət }

Hayward orange peel |MECH ENG| A grab bucket that operates like the clamshell type but has four blades pivoted to close. { 'hā·wərd 'ä·rənj ,pēl }

hazard |IND ENG| Any risk to which a worker is subject as a direct result (in whole or in part) of his being employed. { 'haz·ərd }

hazardous air pollutants |ENG| Chemicals that are known or suspected to cause cancer or other serious health effects, such as reproductive effects or birth defects, or adverse environmental effects. Listed hazardous air pollutants include benzene, found in gasoline; perchlorethlyene, emitted from some dry cleaning facilities; and methylene chloride, used as a solvent and paint stripper in industry; as well as dioxin, asbestos, toluene, and metals such as cadmium, mercury, chromium, and lead compounds. Also known as air toxics. Abbreviated HAP. { ,haz·ər·dəs 'er pə,lüt·əns }

hazemeter See transmissometer. { 'hāz,mēd·ər }

H beam |CIV ENG| A beam similar to the I beam but with longer flanges. Also known as wide-flange beam. { 'āch ,bēm }

H bit |DES ENG| A core bit manufactured and used in Canada having inside and outside diameters of 2.875 and 3.875 inches (73.025 and 98.425 millimeters), respectively; the matching reaming shell has an outside diameter of 3.906 inches (99.2124 millimeters). { 'āch ,bit }

head |BUILD| The upper part of the frame on a door or window. |ELECTR| The photoelectric unit that converts the sound track on motion picture film into corresponding audio signals in a motion picture projector. |ENG| **1.** The end section of a plastics blow-molding machine in which a hollow parison is formed from the melt. **2.** The section of a shell-and-tube heat exchanger from which fluid from the tube bundle is discharged. |ENG ACOUS| See cutter. { hed }

headache post |MECH ENG| A post installed on a cable-tool rig for supporting the end of the walking beam when the rig is not operating. { 'hed,āk ,pōst }

headbox |ENG| A device for controlling the flow of a suspension of solids into a machine. { 'hed,bäks }

header |BUILD| A framing beam positioned between trimmers and supported at each end by a tail beam. |CIV ENG| Brick or stone laid in a wall with its narrow end facing the wall. |ELEC| A mounting plate through which the insulated terminals or leads are brought out from a hermetically sealed relay, transformer, transistor, tube, or other device. |ENG| A pipe, conduit, or chamber which distributes fluid from a series of smaller pipes or conduits; an example is a manifold. |MECH ENG| A machine used for gathering or upsetting materials; used for screw, rivet, and bolt heads. { 'hed·ər }

header bond |CIV ENG| A masonry bond consisting of header courses exclusively. { 'hed·ər ,bänd }

header course |CIV ENG| A masonry course of bricks laid as headers. { 'hed·ər ,kórs }

header-type boiler See straight-tube boiler. { 'hed·ər ,tīp ,bóil·ər }

head gate |CIV ENG| **1.** A gate on the upstream side of a lock or conduit. **2.** A gate at the starting point of an irrigation ditch. { 'hed ,gāt }

heading |CIV ENG| In tunnel construction, one or more small tunnels excavated within a large tunnel cross section that will later be enlarged to full section. { 'hed·iŋ }

heading joint |BUILD| **1.** A joint between two pieces of timber which are joined in a straight line, end to end. **2.** A masonry joint formed between two stones in the same course. { 'hed·iŋ ,jóint }

head meter |ENG| A flowmeter that is dependent upon change of pressure head to operate. { 'hed ,mēd·ər }

head motion |MECH ENG| The vibrator on a reciprocating table concentrator which imparts motion to the deck. { 'hed ,mō·shən }

headphone |ENG ACOUS| An electroacoustic transducer designed to be held against an ear by a clamp passing over the head, for private listening to the audio output of a communications, radio, or television receiver or other source of audio-frequency signals. Also known as phone. { 'hed,fōn }

head pulley |MECH ENG| The pulley at the discharge end of a conveyor belt; may be either an idler or a drive pulley. { 'hed ,púl·ē }

head-pulley-drive conveyor |MECH ENG| A conveyor having the belt driven by the head pulley without a snub pulley. { 'hed ,púl·ē ¦drīv kən'vā·ər }

head scanning |IND ENG| Scanning of the visual field by using movement of both the head and the eyeballs. { 'hed ,skan·iŋ }

head section |ENG| That part of belt conveyor which consists of a drive pulley, a head pulley which may or may not be a drive pulley, belt idlers if included, and the necessary framing. { 'hed ,sek·shən }

headset |ENG ACOUS| A single headphone or a pair of headphones, with a clamping strap or wires holding them in position. { 'hed,set }

head shaft |MECH ENG| The shaft driven by a chain and mounted at the delivery end of a chain conveyor; it serves as the mount for a sprocket which drives the drag chain. { 'hed ,shaft }

headsill |BUILD| A horizontal beam at the top of the frame of a door or window. { 'hed,sil }

headstock |MECH ENG| **1.** The device on a lathe for carrying the revolving spindle. **2.** The movable head of certain measuring machines. **3.** The device on a cylindrical grinding machine for rotating the work. **4.** Also known as workhead. { 'hed ,stäk }

head up |ENG| To tighten bolts on a hatch cover or access hole plate to prevent leakage from or into an operating vessel. { 'hed ,əp }

headwall |CIV ENG| A retaining wall at the outlet of a drain or culvert. { 'hed,wól }

headworks |CIV ENG| Any device or structure at the head or diversion point of a waterway. { 'hed,wərks }

hearing aid |ENG ACOUS| A miniature, portable sound amplifier for persons with impaired hearing, consisting of a microphone, audio amplifier, earphone, and battery. { 'hir·iŋ ,ād }

heart bond |CIV ENG| A masonry bond in which two header stones meet in the middle of the wall, their joint being covered by another stone; no headers stretch across the wall. { 'härt ,bänd }

hearth |BUILD| **1.** The floor of a fireplace or brick oven. **2.** The projection in front of a fireplace, made of brick, stone, or cement. { härth }

heat |THERMO| Energy in transit due to a temperature difference between the source from which the energy is coming and a sink toward which the energy is going; other types of energy in transit are called work. { hēt }

heat balance |THERMO| The equilibrium which is known to exist when all sources of heat gain and loss for a given region or body are accounted for. { 'hēt ,bal·əns }

heat budget |THERMO| The statement of the total inflow and outflow of heat for a planet, spacecraft, biological organism, or other entity. { 'hēt ,bəj·ət }

heat capacity |THERMO| The quantity of heat required to raise a system one degree in temperature in a specified way, usually at constant pressure or constant volume. Also known as thermal capacity. { 'hēt kə,pas·əd·ē }

heat conduction |THERMO| The flow of thermal energy through a substance from a higher-to a lower-temperature region. { 'hēt kən,dək·shən }

heat conductivity See thermal conductivity. { 'hēt ,kän·dək'tiv·əd·ē }

heat content See enthalpy. { 'hēt ¦kän·tent }

heat convection |THERMO| The transfer of thermal energy by actual physical movement from one location to another of a substance in which thermal energy is stored. Also known as thermal convection. { 'hēt kən¦vek·shən }

heat cycle See thermodynamic cycle. { 'hēt ,sī·kəl }

heat death |THERMO| The condition of any isolated system when its entropy reaches a maximum, in which matter is totally disordered and at a uniform temperature, and no energy is available for doing work. { 'hēt ,deth }

heat distortion point |ENG| The temperature at which a standard test bar (American Society for Testing and Materials test) deflects 0.010 inch (0.254 millimeter) under a load of either 66 or 264 pounds per square inch (4.55 × 10⁵ or 18.20 × 10⁵ pascals), as specified. { 'hēt di,stȯr·shən ,pȯint }

heat energy See internal energy. { 'hēt ,en·ər·jē }

heat engine |MECH ENG| A machine that converts heat into work (mechanical energy). |THERMO| A thermodynamic system which undergoes a cyclic process during which a positive amount of work is done by the system; some heat flows into the system and a smaller amount flows out in each cycle. { 'hēt ,en·jən }

heat equation |THERMO| A parabolic second-order differential equation for the temperature of a substance in a region where no heat source exists: $\partial t/\partial \tau = (k/\rho c)(\partial^2 t/\partial x^2 + \partial^2 t/\partial y^2 + \partial t^2/\partial z^2)$, where x, y, and z are space coordinates, τ is the time, $t(x,y,z,\tau)$ is the temperature, k is the thermal conductivity of the body, ρ is its density, and c is its specific heat; this equation is fundamental to the study of heat flow in bodies. Also known as Fourier heat equation; heat flow equation. { 'hēt i,kwā·zhən }

heater |ELECTR| An electric heating element for supplying heat to an indirectly heated cathode in an electron tube. Also known as electron-tube heater. |ENG| A contrivance designed to give off heat. { 'hēd·ər }

heat exchange |CHEM ENG| A unit operation based on heat transfer which functions in the heating and cooling of fluids with or without phase change. { 'hēt iks,chānj }

heat exchanger |ENG| Any device, such as an automobile radiator, that transfers heat from one fluid to another or to the environment. Also known as exchanger. { 'hēt iks,chānj·ər }

heat flow |THERMO| Heat thought of as energy flowing from one substance to another; quantitatively, the amount of heat transferred in a unit time. Also known as heat transmission. { 'hēt ,flō }

heat flow equation See heat equation. { 'hēt ¦flō i,kwä·zhən }

heat flux |THERMO| The amount of heat transferred across a surface of unit area in a unit time. Also known as thermal flux. { 'hēt ,fləks }

heat gain |ENG| The increase of heat within a given space as a result of direct heating by solar radiation and of heat radiated by other sources such as lights, equipment, or people. { 'hēt ,gān }

heating chamber |ENG| The part of an injection mold in which cold plastic feed is changed into a hot melt. { 'hēd·iŋ ,chām·bər }

heating load |CIV ENG| The quantity of heat per unit time that must be provided to maintain the temperature in a building at a given level. { 'hēd·iŋ ,lōd }

heating plant |CIV ENG| The whole system for heating an enclosed space. Also known as heating system. { 'hēd·iŋ ,plant }

heating surface |ENG| The surface for the absorption and transfer of heat from one medium to another. { 'hēd·iŋ ,sər·fəs }

heating system See heating plant. { 'hēd·iŋ ,sis·təm }

heat-loss flowmeter |ENG| Any of various instruments that determine gas velocities or mass flows from the cooling effect of the flow on an electrical sensor such as a thermistor or resistor; a second sensor is used to compensate for the temperature of the fluid. Also known as thermal-loss meter. { 'hēt ,lós 'flō,mēd·ər }

heat of ablation |THERMO| A measure of the effective heat capacity of an ablating material, numerically the heating rate input divided by the mass loss rate which results from ablation. { 'hēt əv ə'blā·shən }

heat of adsorption |THERMO| The increase in enthalpy when 1 mole of a substance is adsorbed upon another at constant pressure. { 'hēt əv ad'sȯrp·shən }

heat of aggregation |THERMO| The increase in enthalpy when an aggregate of matter, such as a crystal, is formed at constant pressure. { 'hēt əv ,ag·rə'gā·shən }

heat of compression |THERMO| Heat generated when air is compressed. { 'hēt əv kəm 'presh·ən }

heat of condensation |THERMO| The increase in enthalpy accompanying the conversion of 1 mole of vapor into liquid at constant pressure and temperature. { 'hēt əv ,känd·ən'sā·shən }

heat of cooling |THERMO| Increase in enthalpy during cooling of a system at constant pressure.

resulting from an internal change such as an allotropic transformation. { 'hēt əv 'kül·iŋ }

heat of crystallization |THERMO| The increase in enthalpy when I mole of a substance is transformed into its crystalline state at constant pressure. { 'hēt əv ,krist·əl·ə'zā·shən }

heat of evaporation *See* heat of vaporization. { 'hēt əv i,vap·ə'rā·shən }

heat of fusion |THERMO| The increase in enthalpy accompanying the conversion of I mole, or a unit mass, of a solid to a liquid at its melting point at constant pressure and temperature. Also known as latent heat of fusion. { 'hēt əv 'fyü·zhən }

heat of mixing |THERMO| The difference between the enthalpy of a mixture and the sum of the enthalpies of its components at the same pressure and temperature. { 'hēt əv 'mik·siŋ }

heat of solidification |THERMO| The increase in enthalpy when I mole of a solid is formed from a liquid or, less commonly, a gas at constant pressure and temperature. { 'hēt əv sə,lid·ə·fə'kā·shən }

heat of sublimation |THERMO| The increase in enthalpy accompanying the conversion of I mole, or unit mass, of a solid to a vapor at constant pressure and temperature. Also known as latent heat of sublimation. { 'hēt əv ,səb·lə'mā·shən }

heat of transformation |THERMO| The increase in enthalpy of a substance when it undergoes some phase change at constant pressure and temperature. { 'hēt əv ,tranz·fər'mā·shən }

heat of vaporization |THERMO| The quantity of energy required to evaporate I mole, or a unit mass, of a liquid, at constant pressure and temperature. Also known as enthalpy of vaporization; heat of evaporation; latent heat of vaporization. { 'hēt əv ,vā·pə·rə'zā·shən }

heat of wetting |THERMO| **1.** The heat of adsorption of water on a substance. **2.** The additional heat required, above the heat of vaporization of free water, to evaporate water from a substance in which it has been absorbed. { 'hēt əv 'wed·iŋ }

heat pipe |ENG| A heat-transfer device consisting of a sealed metal tube with an inner lining of wicklike capillary material and a small amount of fluid in a partial vacuum; heat is absorbed at one end by vaporization of the fluid and is released at the other end by condensation of the vapor. { 'hēt ,pīp }

heat pump |MECH ENG| A device which transfers heat from a cooler reservoir to a hotter one, expending mechanical energy in the process, especially when the main purpose is to heat the hot reservoir rather than refrigerate the cold one. { 'hēt ,pəmp }

heat quantity |THERMO| A measured amount of heat; units are the small calorie, normal calorie, mean calorie, and large calorie. { 'hēt ¦kwän·əd·ē }

heat radiation |THERMO| The energy radiated by solids, liquids, and gases in the form of electromagnetic waves as a result of their temperature. Also known as thermal radiation. { 'hēt ,rād·ē'ā·shən }

heat rate |MECH ENG| An expression of the conversion efficiency of a thermal power plant or engine, as heat input per unit of work output; for example, Btu/kWh. { 'hēt ,rāt }

heat release |THERMO| The quantity of heat released by a furnace or other heating mechanism per second, divided by its volume. { 'hēt ri,lēs }

heat seal |ENG| A union between two thermoplastic surfaces by application of heat and pressure to the joint. { 'hēt ,sēl }

heatsink |ELEC| A mass of metal that is added to a device for the purpose of absorbing and dissipating heat; used with power transistors and many types of metallic rectifiers. Also known as dissipator. |THERMO| Any (gas, solid, or liquid) region where heat is absorbed. { 'hēt,siŋk }

heatsink cooling |ENG| Cooling a body or system by allowing heat to be absorbed from it by another body. { 'hēt,siŋk ¦kül·iŋ }

heat source |THERMO| Any device or natural body that supplies heat. { 'hēt ,sórs }

heat sterilization |ENG| An act of destroying all forms of life on and in bacteriological media, foods, hospital supplies, and other materials by means of moist or dry heat. { 'hēt ,ster·ə·lə'zā·shən }

heat transfer |THERMO| The movement of heat from one body to another (gas, liquid, solid, or combinations thereof) by means of radiation, convection, or conduction. { 'hēt ¦tranz·fər }

heat-transfer coefficient |THERMO| The amount of heat which passes through a unit area of a medium or system in a unit time when the temperature difference between the boundaries of the system is I degree. { 'hēt ¦tranz·fər ,kō·i'fish·ənt }

heat transmission *See* heat flow. { 'hēt tranz ,mish·ən }

heat transport |THERMO| Process by which heat is carried past a fixed point or across a fixed plane, as in a warm current. { 'hēt ¦tranz,pórt }

heat wheel |MECH ENG| In a ventilating system, a device to condition incoming air by causing it to approach thermal equilibrium with the exiting air; hot incoming air is cooled, and cold incoming air is warmed. { 'hēt ,wēl }

heavy-duty |ENG| Designed to withstand excessive strain. { ¦hev·ē ¦düd·ē }

heavy-duty car |MECH ENG| A railway motorcar weighing more than 1400 pounds (635 kilograms), propelled by an engine of 12–30 horsepower (8900–22,400 watts), and designed for hauling heavy equipment and for hump-yard service. { ¦hev·ē ¦düd·ē ¦kär }

heavy-duty tool block *See* open-side tool block. { ¦hev·ē ¦düd·ē ¦tül ,bläk }

heavy force fit |DES ENG| A fit for heavy steel parts or shrink fits in medium sections. { 'hev·ē 'fòrs ,fit }

heavy section car |MECH ENG| A railway motorcar weighing 1200–1400 pounds (544–635 kilograms) and propelled by an 8–12 horsepower (6000–8900 watts) engine. { 'hev·ē 'sek·shən ¦kär }

hectare |MECH| A unit of area in the metric system equal to 100 ares or 10,000 square meters. Abbreviated ha. { 'hek¦tar }

hectogram |MECH| A unit of mass equal to 100 grams. Abbreviated hg. { 'hek·tə¦gram }

hectoliter |MECH| A metric unit of volume equal to 100 liters or to 0.1 cubic meter. Abbreviated hl. { 'hek·tə¦lēd·ər }

hectometer |MECH| A unit of length equal to 100 meters. Abbreviated hm. { 'hek·tə¦mēd·ər }

heel *See* heel block. { hēl }

heel block |MECH ENG| A block or plate that is usually fixed on the die shoe to minimize deflection of a punch or cam. Also known as heel. { 'hēl ¦bläk }

heeling adjuster |ENG| A dip needle with a sliding weight that can be moved along one of its arms to balance the magnetic force; used to determine the correct position of a heeling magnet. Also known as heeling error instrument; vertical force instrument. { 'hēl·iŋ ə¦jəs·tər }

heeling error instrument *See* heeling adjuster. { 'hēl·iŋ ¦er·ər ¦in·strə·mənt }

heeling magnet |ENG| A permanent magnet placed vertically in a tube under the center of a marine magnetic compass, to correct for heeling error. { 'hēl·iŋ ¦mag·nət }

heel of a shot |ENG| **1.** In blasting, the front or face of a shot farthest from the charge. **2.** The distance between the mouth of the drill hole and the corner of the nearest free face. **3.** That portion of a drill hole which is filled with the tamping. { 'hēl əv ə 'shät }

heel plate |CIV ENG| A plate at the end of a truss. { 'hēl ¦plāt }

heel post |CIV ENG| A post to which are secured the hinges of a gate or door.

height equivalent of theoretical plate |CHEM ENG| In a packed fractionating column, a height of packing that makes a separation equivalent to that of a theoretical plate; used in sorption and distillation calculations. Abbreviated HETP. { 'hīt i¦kwiv·ə·lənt əv ¦thē·ə'red·ə·kəl 'plāt }

height finder |ENG| A radar equipment, used to determine height of aerial targets. { 'hīt ¦fīn·dər }

height finding |ENG| Determination of the height of an airborne object. { 'hīt ¦fīnd·iŋ }

height-finding radar |ENG| A radar set that measures and determines the height of an airborne object. { 'hīt ¦fīnd·iŋ 'rā¦där }

height gage |ENG| A gage used to measure heights by either a micrometer or a vernier scale. { 'hīt ¦gāj }

height of instrument |ENG| **1.** In survey leveling, the vertical height of the line of collimation of the instrument over the station above which it is centered, or above a specified datum level.

2. In spirit leveling, the vertical distance from datum to line of sight of the instrument. **3.** In stadia leveling the height of center of transit above the station stake. **4.** In differential leveling, the elevation of the line of sight of the telescope when the instrument is leveled. { 'hīt əv 'in·strə·mənt }

height of transfer unit |CHEM ENG| A dimensionless parameter used to calculate countercurrent sorption tower operations; it is proportional to the apparent resident time of the fluid. Abbreviated HTU. { 'hīt əv 'tranz·fər ¦yü·nət }

helical angle |MECH| In the study of torsion, the angular displacement of a longitudinal element, originally straight on the surface of an untwisted bar, which becomes helical after twisting. { 'hel·ə·kəl 'aŋ·gəl }

helical conveyor |MECH ENG| A conveyor for the transport of bulk materials which consists of a horizontal shaft with helical paddles or ribbons rotating inside a stationary tube. { 'hel·ə·kəl kən'vā·ər }

helical-fin section |CHEM ENG| Helical-shaped, extended-surface addition for the external surfaces of process-fluid tubes to increase heat-exchange efficiency; used for gas heating and cooling and in fuel oil residuum exchangers. { 'hel·ə·kəl 'fin ¦sek·shən }

helical-flow turbine |MECH ENG| A steam turbine in which the steam is directed tangentially and radially inward by nozzles against buckets milled in the wheel rim; the steam flows in a helical path, reentering the buckets one or more times. Also known as tangential helical-flow turbine. { 'hel·ə·kəl ¦flō 'tər·bən }

helical gear |MECH ENG| Gear wheels running on parallel axes, with teeth twisted oblique to the gear axis. { 'hel·ə·kəl 'gir }

helical milling |MECH ENG| Milling in which the work is simultaneously rotated and translated. { 'hel·ə·kəl 'mil·iŋ }

helical rake angle |DES ENG| The angle between the axis of a reamer and a plane tangent to its helical cutting edge; also applied to milling cutters. { 'hel·ə·kəl 'rāk ¦aŋ·gəl }

helical scanning |ELECTR| A method of recording on videotape and digital audio tape in which the tracks are recorded diagonally from top to bottom by wrapping the tape around the rotating-head drum in a helical path. |ENG| A method of radar scanning in which the antenna beam rotates continuously about the vertical axis while the elevation angle changes slowly from horizontal to vertical, so that a point on the radar beam describes a distorted helix. { 'hel·ə·kəl 'skan·iŋ }

helical-spline broach |MECH ENG| A broach used to produce internal helical splines having a straight-sided or involute form. { 'hel·ə·kəl 'splīn ¦brōch }

helical spring |DES ENG| A bar or wire of uniform cross section wound into a helix. { 'hel·ə·kəl 'spriŋ }

heliograph |ENG| An instrument that records the duration of sunshine and gives a qualitative measure of its amount by action of sun's rays on blueprint paper. { 'hē·lē·ə,graf }

heliostat |ENG| A clock-driven instrument mounting which automatically and continuously points in the direction of the sun; it is used with a pyrheliometer when continuous direct solar radiation measurements are required. { 'hē·lē·ə,stat }

heliotrope |ENG| An instrument that reflects the sun's rays over long distances; used in geodetic surveys. { 'hē·lē·ə,trōp }

helipad |CIV ENG| The launch and landing area of a heliport. Also known as pad. { 'hel·ə,pad }

heliport |CIV ENG| A place built for helicopter takeoffs and landings. { 'hel·ə,pòrt }

helium-oxygen diving |ENG| Diving operations employing a breathing mixture of helium and oxygen. { 'hē·lē·əm ¦äk·sə·jən 'dīv·iŋ }

helium refrigerator |MECH ENG| A refrigerator which uses liquid helium to cool substances to temperatures of 4 K or less. { 'hē·lē·əm ri'frij·ə,rād·ər }

helix angle |DES ENG| That angle formed by the helix of the thread at the pitch-diameter line and a line at right angles to the axis. { 'hē,liks ,aŋ·gəl }

helmet |ENG| A globe-shaped head covering made of copper and supplied with air pumped through a hose; attached to the breastplate of a diving suit for deep-sea diving. { 'hel·mət }

helmholtz |ELEC| A unit of dipole moment per unit area, equal to 1 Debye unit per square angstrom, or approximately 3.335 × 10^{-10} coulomb per meter. { 'helm,hōlts }

Helmholtz free energy See free energy. { 'helm ,hōlts ¦frē 'en·ər·jē }

Helmholtz function See free energy. { 'helm,hōlts ,fəŋk·shən }

Helmholtz potential See free energy. { 'helm,hōlts pə¦ten·chəl }

Helmholtz resonator |ENG ACOUS| An enclosure having a small opening consisting of a straight tube of such dimensions that the enclosure resonates at a single frequency determined by the geometry of the resonator. { 'helm,hōlts ¦rez·ən,ād·ər }

help-yourself system |IND ENG| A tool-crib system for temporary issue of tools employed in small shops; employees have access to tools in the crib and help themselves. { ¦help yùr'self ,sis·təm }

hemispherical pyrheliometer |ENG| An instrument for measuring the total solar energy from the sun and sky striking a horizontal surface, in which a thermopile measures the temperature difference between white and black portions of a thermally insulated target within a partially evacuated transparent sphere or hemisphere. { ,he·mē'sfir·ə·kəl ,pīr,hē·lē'äm·əd·ər }

hemming |MECH ENG| Forming of an edge by bending the metal back on itself. { 'hem·iŋ }

hemp-core cable See standard wire rope. { 'hemp ,kòr ,kā·bəl }

Hengstebeck approximation |CHEM ENG| A method of calculation to estimate the distribution of non-key components in distillation column products. { 'heŋ·stə·bek ə,präk·sə,mā·shən }

HEPA filter See high-efficiency particulate air filter. { 'hep·ə ,fil·tər }

hereditary mechanics |MECH| A field of mechanics in which quantities, such as stress, depend not only on other quantities, such as strain, at the same instant but also on integrals involving the values of such quantities at previous times. { hə'red·ə,ter·ē mi'kan·iks }

hermaphrodite caliper |DES ENG| A layout tool having one leg pointed and the other like that of an inside caliper; used to locate the center of irregularly shaped stock or to lay out a line parallel to an edge. { hər'maf·rə,dīt ¦kal·ə·pər }

hermetic seal |ENG| An airtight seal. { hər'med·ik 'sēl }

herpolhode |MECH| The curve traced out on the invariable plane by the point of contact between the plane and the inertia ellipsoid of a rotating rigid body not subject to external torque. { ¦hər·pəl'hōd }

herpolhode cone See space cone. { ¦hər·pəl'hōd ,kōn }

herringbone gear |MECH ENG| The equivalent of two helical gears of opposite hand placed side by side. { 'her·iŋ,bōn ,gir }

Herschel-type venturi tube |ENG| A type of venturi tube in which the converging and diverging sections are cones, the throat section is relatively short, the diverging cone is long, and the pressures preceding the inlet cone and in the throat are transferred through multiple openings into annular openings, called piezometer rings. { 'hər·shəl ,tīp ven'tùr·ē ,tüb }

Hertz's law |MECH| A law which gives the radius of contact between a sphere of elastic material and a surface in terms of the sphere's radius, the normal force exerted on the sphere, and Young's modulus for the material of the sphere. { 'hərt·səs ,lò }

heterodyne |ELECTR| To mix two alternating-current signals of different frequencies in a nonlinear device for the purpose of producing two new frequencies, the sum of and difference between the two original frequencies. { 'hed·ə·rə,dīn }

heterodyne detector |ELECTR| A detector in which an unmodulated carrier frequency is combined with the signal of a local oscillator having a slightly different frequency, to provide an audio-frequency beat signal that can be heard with a loudspeaker or headphones; used chiefly for code reception. { 'hed·ə·rə,dīn di'tek·tər }

heterodyne analyzer |ENG ACOUS| A type of constant-bandwidth analyzer in which the electric signal from a microphone beats with the signal from an oscillator, and one of the side bands produced by this modulation is then passed through a fixed filter and detected. { 'hed·ə·rə,dīn 'an·ə,liz·ər }

heterodyne frequency meter |ELECTR| A frequency meter in which a known frequency, which may be adjustable or fixed, is heterodyned with an unknown frequency to produce a zero beat or an audio-frequency signal whose value is measured by other means. Also known as heterodyne wavemeter. { 'hed·ə·rə,dīn 'frē·kwən·sē ,mēd·ər }

heterodyne measurement |ELECTR| A measurement carried out by a type of harmonic analyzer which employs a highly selective filter, at a frequency well above the highest frequency to be measured, and a heterodyning oscillator. { 'hed·ə·rə,dīn 'mezh·ər·mənt }

heterodyne modulator See mixer. { 'hed·ə·rə,dīn 'mäj·ə,lād·ər }

heterodyne oscillator |ELECTR| **1.** A separate variable-frequency oscillator used to produce the second frequency required in a heterodyne detector for code reception. **2.** See beat-frequency oscillator. { 'hed·ə·rə,dīn 'äs·ə,lād·ər }

heterodyne reception |ELECTR| Radio reception in which the incoming radio-frequency signal is combined with a locally generated rf signal of different frequency, followed by detection. Also known as beat reception. { 'hed·ə·rə,dīn ri'sep·shən }

heterodyne repeater |ELECTR| A radio repeater in which the received radio signals are converted to an intermediate frequency, amplified, and reconverted to a new frequency band for transmission over the next repeater section. { 'hed·ə·rə,dīn ri'pēd·ər }

heterodyne wavemeter See heterodyne frequency meter. { 'hed·ə·rə,dīn 'wāv,mēd·ər }

heterogeneous strain |MECH| A strain in which the components of the displacement of a point in the body cannot be expressed as linear functions of the original coordinates. { ,hed·ə·rə¦jē·nē·əs 'strān }

heterojunction |ELECTR| The boundary between two different semiconductor materials, usually with a negligible discontinuity in the crystal structure. { ¦hed·ə·rō'jəŋk·shən }

heterojunction bipolar transistor |ELECTR| A bipolar transistor that has two or more materials making up the emitter, base, and collector regions, giving it a much higher maximum frequency than a silicon bipolar transistor. Abbreviated HBT. { ¦hed·ə·rə,jəŋk·shən 'bī,pōl·ər tran,zis·tər }

heterojunction field-effect transistor See high-electron-mobility transistor. { ¦hed·ə·rə,jəŋk·shən 'fēld i,fekt tran,zis·tər }

heteromorphic transformation |THERMO| A change in the values of the thermodynamic variables of a system in which one or more of the component substances also undergo a change of state. { ,hed·ə·rə¦mȯr·fik ,tranz·fər'mā·shən }

HETP See height equivalent of theoretical plate.

hexagonal-head bolt |DES ENG| A standard wrench head bolt with a hexagonal head. { hek'sag·ə·nəl ,hed ,bōlt }

hexagonal nipple |DES ENG| A nipple for joining pipe with a hexagonal configuration around

the center of the exterior surface to permit tightening with a spanner. { ,hek¦sag·ən·əl 'nip·əl }

hexagonal nut |DES ENG| A plain nut in hexagon form. { hek'sag·ə·nəl 'nət }

hexapod |CONT SYS| A robot that uses six leglike appendages to stride over a surface. { 'hek·sə,päd }

hex nut |DES ENG| A nut in the shape of a hexagon. { 'heks ,nət }

HF alkylation |CHEM ENG| Petroleum refinery alkylation process in which olefins (C_3, C_4, C_5) are reacted with isobutane in the presence of hydrofluoric acid catalyst. { ¦ách¦ef ,al·kə'lā·shən }

hg See hectogram.

hierarchical control |CONT SYS| The organization of controllers in a large-scale system into two or more levels so that controllers in each level send control signals to controllers in the level below and feedback or sensing signals to controllers in the level above. Also known as control hierarchy. { ¦hī·ər¦är·kə·kəl kən'trōl }

hi-fi See high fidelity. { 'hī'fī }

Higbie model |CHEM ENG| Mass-transfer theory for packed absorption towers, stating that liquid flows across each packing piece in laminar flow and is mixed with other liquids meeting it at the points of discontinuity between packing elements. { 'hig·bē ,mäd·əl }

high-aspect-ratio micromachining |ENG| Microfabrication processes that produce tall microstructures with vertical sidewalls. Abbreviated HARM. { ,hī ¦as,pekt ,rā·shō ,mī·krō·mə'shēn·iŋ }

high-efficiency particulate air filter |MECH ENG| An air filter capable of reducing the concentration of solid particles (0.3 millimeter in diameter or larger) in the airstream by 99.97%. Also known as HEPA filter. { ,hī i¦fish·ən·sē pər,tik·yə·lət 'er ,fil·tər }

high-electron-mobility transistor |ELECTR| A type of field-effect transistor consisting of gallium arsenide and gallium aluminum arsenide, with a Schottky metal contact on the gallium aluminum arsenide layer and two ohmic contacts penetrating into the gallium arsenide layer, serving as the gate, source, and drain respectively. Abbreviated HEMT. Also known as heterojunction field-effect transistor (HFET); modulation-doped field-effect transistor (MODFET); selectively doped heterojunction transistor (SDHT); two-dimensional electron gas field-effect transistor (TEGFET). { 'hī i'lek,trän mō ¦bil·əd·ē tran,zis·tər }

higher pair |MECH ENG| A link in a mechanism in which the mating parts have surface (instead of line or point) contact. { 'hī·ər 'per }

high fidelity |ENG ACOUS| Audio reproduction that closely approximates the sound of the original performance. Also known as hi-fi. { ¦hī fi 'del·əd·ē }

high-frequency furnace |ENG| An induction furnace in which the heat is generated within the charge, within the walls of the containing crucible, or within both, by currents induced by high-frequency magnetic flux produced by a surrounding coil. Also known as coreless-type induction furnace; high-frequency heater. { 'hī ¦frē·kwən·sē 'fər·nəs }

high-frequency heater See high-frequency furnace. { 'hī ¦frē·kwən·sē 'hēd·ər }

high-frequency heating See electronic heating. { 'hī ¦frē·kwən·sē 'hēd·iŋ }

high-frequency resistance |ELEC| The total resistance offered by a device in an alternating-current circuit, including the direct-current resistance and the resistance due to eddy current, hysteresis, dielectric, and corona losses. Also known as alternating-current resistance; effective resistance; radio-frequency resistance. { 'hī ¦frē·kwən·sē ri'zis·təns }

high-frequency voltmeter |ELECTR| A voltmeter designed to measure currents alternating at high frequencies. { 'hī ¦frē·kwən·sē 'vōlt,mēd·ər }

high-front shovel |MECH ENG| A power shovel with a dipper stick mounted high on the boom for stripping and overburden removal. { 'hī ¦frənt 'shəv·əl }

high-gradient magnetic separation |ENG| A magnetic separation technique applicable to weakly paramagnetic compounds and to particle sizes down to the colloidal domain. { 'hī ,grād·ē·ənt mag'ned·ik ,sep·ə'rā·shən }

high hat |ENG| A very low tripod head resembling a formal top hat in shape. { 'hī ,hat }

high heat |THERMO| Heat absorbed by the cooling medium in a calorimeter when products of combustion are cooled to the initial atmospheric (ambient) temperature. { 'hī ¦hēt }

high-helix drill |DES ENG| A two-flute twist drill with a helix angle of 35–40°; used for drilling deep holes in metals, such as aluminum, copper, hard brass, and soft steel. Also known as fast-spiral drill. { 'hī ¦hē·liks ¦dril }

high-impedance voltmeter |ELEC| A voltage-measuring device with a high-impedance input to reduce load on the unit under test; a vacuum-tube voltmeter is one type. { 'hī im¦pēd·əns 'vōlt,mēd·ər }

high-intensity atomizer |MECH ENG| A type of atomizer used in electrostatic atomization, based on stress sufficient to overcome tensile strength of the liquid. { 'hī in,ten·səd·ē 'ad·ə,miz·ər }

high-K capacitor |ELEC| A capacitor whose dielectric material is a ferroelectric having a high dielectric constant, up to about 6000. { 'hī ,kā kə'pas·əd·ər }

high-lift truck |MECH ENG| A forklift truck with a fixed or telescoping mast to permit high elevation of a load. { 'hī ¦lift 'trək }

high-pass filter |ELECTR| A filter that transmits all frequencies above a given cutoff frequency and substantially attenuates all others. { 'hī ,pas 'fil·tər }

high-potting |ELEC| Testing with a high voltage, generally on a production line. { 'hī ¦päd·iŋ }

high-pressure gage glass |ENG| A gage glass

consisting of a metal tube with thick glass windows. { 'hī ¦presh·ər 'gāj ,glas }

high-pressure process |CHEM ENG| A chemical process operating at elevated pressure; for example, phenol manufacture at 330 atmospheres (1 atmosphere = 101,325 pascals), ethylene polymerization at 2000 atm, ammonia synthesis at 100–1000 atm, and synthetic-diamond manufacture up to 100,000 atm. { 'hī ¦presh·ər 'prä·səs }

high-pressure torch |ENG| A type of torch in which both acetylene and oxygen are delivered to the mixing chamber under pressure. { 'hī ¦presh·ər 'tórch }

high Q |ELECTR| A characteristic wherein a component has a high ratio of reactance to effective resistance, so that its Q factor is high. { 'hī 'kyü }

high-resistance voltmeter |ELEC| A voltmeter having a resistance considerably higher than 1000 ohms per volt, so that it draws little current from the circuit in which a measurement is made. { 'hī ri,zis·təns 'vōlt,mēd·ər }

high-resolution radar |ENG| A radar system which can discriminate between two close targets. { 'hī ,rez·ə,lü·shən 'rā,där }

high-rise building See tall building. { ¦hī ¦rīz 'bild·iŋ }

high-speed machine |MECH ENG| A diamond drill capable of rotating a drill string at a minimum of 2500 revolutions per minute, as contrasted with the normal maximum speed of 1600–1800 revolutions per minute attained by the average diamond drill. { 'hī,spēd mə'shēn }

high-technology robot |CONT SYS| A robot equipped with feedback, vision, real-time data acquisition, and powerful controllers. { 'hī tek'näl·ə·jē 'rō,bät }

high-temperature water boiler |MECH ENG| A boiler which provides hot water, under pressure, for space heating of large areas. { 'hī ,tem·prə·chər 'wód·ər ,bóil·ər }

high-tensile bolt |ENG| A bolt that is adjusted to a carefully controlled tension by means of a calibrated torsion wrench; used in place of a rivet. Also known as high-tension bolt. { 'hī ,ten·səl 'bōlt }

high tension See high voltage. { 'hī ¦ten·chən }

high-tension bolt See high-tensile bolt. { ¦hī ,ten·chən 'bōlt }

high-tension detonator |ENG| A detonator requiring an electric potential of about 50 volts for firing. { 'hī ,ten·chən 'det·ən,ād·ər }

high-tension separation See electrostatic separation. { 'hī ,ten·chən ,sep·ə'rā·shən }

high-test chain |ENG| Chain made from heat-treatable plain-carbon steel, usually with a carbon content of 0.15–0.20; used for load binding, tie-downs, and other applications where failure would be costly. { 'hī ,test 'chān }

high-vacuum insulation |CHEM ENG| High vacuum between the walls of double-wall vessels to serve as thermal insulation at ultralow (cryogenic) temperatures, such as in Dewar vessels. { 'hī ¦vak·yüm ,in·sə'lā·shən }

high voltage |ELEC| A voltage on the order of thousands of volts. Also known as high tension. { 'hī ¦vōl·tij }

highway |CIV ENG| A public road where traffic has the right to pass and to which owners of adjacent property have access. { 'hī,wā }

highway engineering |CIV ENG| A branch of civil engineering dealing with highway planning, location, design, and maintenance. { 'hī,wā ,en·jə'nir·iŋ }

Hildebrand function |THERMO| The heat of vaporization of a compound as a function of the molal concentration of the vapor; it is nearly the same for many compounds. { 'hil·də,brand ,faŋk·shən }

hill-climbing |MECH ENG| Adjustment, either continuous or periodic, of a self-regulating system to achieve optimum performance. { 'hil ,klim·iŋ }

Hindley screw |DES ENG| An endless screw or worm of hourglass shape that fits a part of the circumference of a worm wheel so as to increase the bearing area and thus diminish wear. Also known as hourglass screw; hourglass worm. { 'hind·lē ,skrü }

hindrance factor See drag factor. { 'hin·drəns ,fak·tər }

hinge |DES ENG| A pair of metal leaves forming a jointed device on which a swinging part turns. { hinj }

hinged arch |CIV ENG| A structure that can rotate at its supports or in the center or at both places. { 'hinjd ¦ärch }

hip |BUILD| 1. The external angle formed by the junction of two sloping roofs or the sides of a roof. 2. A rafter that is positioned at the junction of two sloping roofs or the sides of a roof. |CIV ENG| See hip joint. { hip }

HIP See hot isostatic pressing. { hip or ¦āch¦ī'pē }

hip joint |CIV ENG| The junction of an inclined head post and the top chord of a truss. Also known as hip. { 'hip ,jȯint }

hi pot |ELEC| High potential voltage applied across a conductor to test the insulation or applied to an etched circuit to burn out tenuous conducting paths that might later fail in service. { 'hī ,pät }

hip rafter |BUILD| A diagonal rafter extending from the plate to the ridge of a roof. { 'hip ,raf·tər }

hl See hectoliter.

hm See hectometer.

hob |DES ENG| A master model made from hardened steel which is used to press the shape of a plastics mold into a block of soft steel. |MECH ENG| A rotary cutting tool with its teeth arranged along a helical thread; used for generating gear teeth. { häb }

hobber See hobbing machine. { 'häb·ər }

hobbing |DES ENG| In plastics manufacturing, the act of creating multiple mold cavities by pressing a hob into soft metal cavity blanks. |MECH ENG| Cutting evenly spaced forms, such as gear teeth, on the periphery of cylindrical workpieces. { 'häb·iŋ }

hobbing machine |MECH ENG| A machine for cutting gear teeth in gear blanks or for cutting worm, spur, or helical gears. Also known as hobber. { 'häb·iŋ mə,shēn }

hobnail |DES ENG| A short, large-headed, sharp-pointed nail; used to attach soles to heavy shoes. { 'häb,nāl }

hobo connection |ENG| A parallel electrical connection used in blasting. { 'hō·bō kə,nek·shən }

hod |CIV ENG| A tray fitted with a handle by which it can be carried on the shoulder for transporting bricks or mortar. { häd }

Hodgson number |CHEM ENG| Method of predicting the metering error during pulsating gas flow when a surge tank is located between the pulsation source (pump or compressor) and the meter (orifice, nozzle, or venturi). { 'häj·sən ,nam·bər }

hoe |DES ENG| An implement consisting of a long handle with a thin, flat, straight-edged blade attached transversely to the end; used for cultivating and weeding. { hō }

hoe shovel |MECH ENG| A revolving shovel with a pull-type bucket rigidly attached to a stick hinged on the end of a live boom. { 'hō ¦shəv·əl }

Hoffmann electrometer |ENG| A variant of the quadrant electrometer that has two sections instead of four. { ¦häf·mən i,lek'träm·əd·ər }

hogging |ENG| Mechanical chipping of wood waste for fuel. { 'häg·iŋ }

hohlraum See blackbody. { 'hōl,raúm }

hoist |MECH ENG| 1. To move or lift something by a rope-and-pulley device. 2. A power unit for a hoisting machine, designed to lift from a position directly above the load and therefore mounted to facilitate mobile service. Also known as winding engine. { hȯist }

hoist back-out switch |MECH ENG| A protective switch that permits hoist operation only in the reverse direction in case of overwind. { 'hȯist ¦bak,aút ,swich }

hoist cable |MECH ENG| A fiber rope, wire rope, or chain by means of which force is exerted on the sheaves and pulleys of a hoisting machine. { 'hȯist ,kā·bəl }

hoist hook |DES ENG| A swivel hook attached to the end of a hoist cable for securing a load. { 'hȯist ,húk }

hoisting |MECH ENG| 1. Raising a load, especially by means of tackle. 2. Either of two power-shovel operations: the raising or lowering of the boom, or the lifting or dropping of the dipper stick in relation to the boom. { 'hȯist·iŋ }

hoisting drum See drum. { 'hȯist·iŋ ,drəm }

hoisting machine |MECH ENG| A mechanism for raising and lowering material with intermittent motion while holding the material freely suspended. { 'hȯist·iŋ mə,shēn }

hoisting power |MECH ENG| The capacity of the hoisting mechanism on a hoisting machine. { 'hȯist·iŋ ,paú·ər }

hoistman |ENG| One who operates steam or

electric hoisting machinery to lower and raise cages, skips, or instruments into a mine or an oil or gas well. Also known as hoist operator; winch operator. { 'hȯist·mən }

hoist operator *See* hoistman. { 'hȯist ˌäp·ə₁rād· ər }

hoist overspeed device [MECH ENG] A device used to prevent a hoist from operating at speeds greater than predetermined values by activating an emergency brake when the predetermined speed is exceeded. { 'hȯist ¦ō·vər₁spēd di₁vīs }

hoist overwind device [MECH ENG] A device which can activate an emergency brake when a hoisted load travels beyond a predetermined point into a danger zone. { 'hȯist ¦ō·vər₁wīnd di₁vīs }

hoist slack-brake switch [MECH ENG] A device that automatically cuts off power to the hoist motor and sets the brake if the links in the brake rigging require tightening or if the brakes require relining. { 'hȯist ¦slak ₁brāk ₁swich }

hoist tower [CIV ENG] A temporary shaft of scaffolding used to hoist materials for building construction. { 'hȯist ₁taú·ər }

hoistway [MECH ENG] A shaft for one or more elevators, lifts, or dumbwaiters. { 'hȯist₁wā }

hold [ELECTR] To maintain storage elements at equilibrium voltages in a charge storage tube by electron bombardment. [ENG] The interior of a ship or plane, especially the cargo compartment. [IND ENG] A therblig, or basic operation, in time-and-motion study in which the hand or other body member maintains an object in a fixed position and location. [MECH ENG] A machine motion that is halted by an operator or interlock until it is restarted. { hōld }

holdback [MECH ENG] A brake on an inclined-belt conveyor system which is automatically activated in the event of power failure, thus preventing the loaded belt from running downward. { 'hōl₁bak }

holddown groove [ENG] A groove in the side wall of the molding surface which assists in holding the molded plastic article in place when the mold opens. { 'hōl₁daún ₁grüv }

holdup [CHEM ENG] **1.** Volume of material held or contained in a process vessel or line. **2.** Liquid held up (suspended) in a vertical process vessel or line by rising gas or vapor streams. { 'hōl₁dəp }

hole conduction [ELECTR] Conduction occurring in a semiconductor when electrons move into holes under the influence of an applied voltage and thereby create new holes. { 'hōl kən¦dək·shən }

hole deviation [ENG] The change in the course or direction that a borehole follows. { 'hōl ₁dē·vē₁ā·shən }

hole injection [ELECTR] The production of holes in an *n*-type semiconductor when voltage is applied to a sharp metal point in contact with the surface of the material. { 'hōl in₁jek·shən }

hole mobility [ELECTR] A measure of the ability

of a hole to travel readily through a semiconductor, equal to the average drift velocity of holes divided by the electric field. { 'hōl mō₁bil·əd·ē }

hole saw *See* crown saw. { 'hōl ₁sȯ }

hole trap [ELECTR] A semiconductor impurity capable of releasing electrons to the conduction or valence bands, equivalent to trapping a hole. { 'hōl ₁trap }

holiday [ENG] An undesirable discontinuity or break in the anticorrosion protection on pipe or tubing. { 'häl·ə₁dā }

holiday detector [ENG] An electrical device used to determine the location of a gap or void in the anticorrosion coating of a metal surface. { 'häl·ə₁dā di₁tek·tər }

hollander [MECH ENG] An elongate tube with a central mid-feather and a cylindrical beater roll; formerly used for stock preparation in paper manufacture. { 'häl·ən·dər }

Holland formula [ENG] A formula used to calculate the height of a plume formed by pollutants emitted from a stack in terms of the diameter of the stack exit, the exit velocity and heat emission rate of the stack, and the mean wind speed. { 'häl·ənd ₁fȯr·myə·lə }

hollow-core construction [BUILD] Panel construction with wood faces bonded to a framed-core assembly of elements which support the facing at spaced intervals. { 'häl·ō ¦kȯr kən'strək·shən }

hollow drill [DES ENG] A drill rod or stem having an axial hole for the passage of water or compressed air to remove cuttings from a drill hole. Also known as hollow rod; hollow stem. { 'häl·ō 'dril }

hollow gravity dam [CIV ENG] A fixed gravity dam, usually of reinforced concrete, constructed of inclined slabs or arched sections supported by transverse buttresses. { 'häl·ō 'grav·əd·ē ₁dam }

hollow mill [MECH ENG] A milling cutter with three or more cutting edges that revolve around the cylindrical workpiece. { 'häl·ō ₁mil }

hollow reamer [ENG] A tool or bit used to correct the curvature in a crooked borehole. { 'häl·ō 'rēm·ər }

hollow rod *See* hollow drill. { 'häl·ō 'räd }

hollow-rod churn drill [MECH ENG] A churn drill with hollow rods instead of steel wire rope. { 'häl·ō ₁räd 'chərn ₁dril }

hollow-rod drilling [ENG] A modification of wash boring in which a check valve is introduced at the bit so that the churning action may be also used to pump the cuttings up the drill rods. { 'häl·ō ₁räd 'dril·iŋ }

hollow shafting [MECH ENG] Shafting made from hollowed-out rods or hollow tubing to minimize weight, allow internal support, or permit other shafting to operate through the interior. { 'häl·ō 'shaft·iŋ }

hollow stem *See* hollow drill. { 'häl·ō 'stem }

hollow wall [BUILD] A masonry wall provided with an air space between the inner and outer wythes. { 'häl·ō 'wȯl }

Holme mud sampler [ENG] A scooplike device

which can be lowered by cable to the ocean floor to collect sediment samples. { 'hōm 'mad ˌsam·plər }

holonomic constraints |MECH| An integrable set of differential equations which describe the restrictions on the motion of a system; a function relating several variables, in the form $f(x_1, . . ., x_n) = 0$, in optimization or physical problems. { ¦häl·ə¦näm·ik kən'sträns }

holonomic system |MECH| A system in which the constraints are such that the original coordinates can be expressed in terms of independent coordinates and possibly also the time. { ¦häl·ə¦näm·ik 'sis·təm }

holopulping process |CHEM ENG| A process for making paper pulp by alkaline oxidation of extremely thin wood chips at low temperature and pressure and then solubilization of the lignin fraction. { ¦häl·ō'pəl·piŋ ˌpräs·əs }

Holzer's method |MECH| A method of determining the shapes and frequencies of the torsional modes of vibration of a system, in which one imagines the system to consist of a number of flywheels on a massless flexible shaft and, starting with a trial frequency and motion for one flywheel, determines the torques and motions of successive flywheels. { 'hōt·sərz ˌmeth·əd }

home key |ENG| One of the eight keys on a keyboard on which the typist's fingers normally rest in the starting position for touch typing. Also known as guide key. { 'hōm ˌkē }

homenergic flow |THERMO| Fluid flow in which the sum of kinetic energy, potential energy, and enthalpy per unit mass is the same at all locations in the fluid and at all times. { 'häm·ə,nər·jik 'flō }

home row |ENG| The row on a keyboard that contains the home keys. { 'hōm ,rō }

home signal |CIV ENG| A signal at the beginning of a block of railroad track that indicates whether the block is clear. { 'hōm ¦sig·nəl }

homing device |ELECTR| A control device that automatically starts in the correct direction of motion or rotation to achieve a desired change, as in a remote-control tuning motor for a television receiver. |ENG| A device incorporated in a guided missile or the like to home it on a target. { 'hōm·iŋ di,vīs }

homing guidance |ENG| A guidance system in which a missile directs itself to a target by means of a self-contained mechanism that reacts to a particular characteristic of the target. { 'hōm·iŋ ˌgīd·əns }

homogeneous strain |MECH| A strain in which the components of the displacement of any point in the body are linear functions of the original coordinates. { ¦hō·mə,jē·nē·əs 'strän }

homogenizer |MECH ENG| A machine that blends or emulsifies a substance by forcing it through fine openings against a hard surface. { hō'mäj·ə,nīz·ər }

homojunction bipolar transistor |ELECTR| Any bipolar transistor that is composed entirely of one type of semiconductor. { ¦hō·mō,jəŋk·shən bī,pō·lər tran'zis·tər }

homologous motion |IND ENG| A motion produced by one set of muscles that can be substituted for an essentially similar motion performed by another set of muscles; the substitution is usually made in order to reduce the stress needed to perform a work task. { hə'mäl·ə·gəs 'mō·shən }

homomorphous transformation |THERMO| A change in the values of the thermodynamic variables of a system in which none of the component substances undergoes a change of state. { ˌhō·mə¦mòr·fəs ˌtranz·fər'mā·shən }

hone |MECH ENG| A machine for honing that consists of a holding device containing several oblong stones arranged in a circular pattern. { hōn }

honed-bore tube |DES ENG| Tubing manufactured to very close tolerances and having a very smooth surface in the bore. { ¦hōnd ¦bòr 'tüb }

honeycomb radiator |MECH ENG| A heat-exchange device utilizing many small cells, shaped like a bees' comb, for cooling circulating water in an automobile. { 'hən·ē,kōm 'rād·ē,ād·ər }

honeycomb wall |BUILD| A brick wall having openings created either by allowing gaps between stretchers or by omitting bricks and used to support floor joists and provide ventilation under floors. { 'hən·ē,kōm ,wòl }

honing |MECH ENG| The process of removing a relatively small amount of material from a cylindrical surface by means of abrasive stones to obtain a desired finish or extremely close dimensional tolerance. { 'hōn·iŋ }

honing gage |ENG| A device for keeping a chisel steady at the proper angle while it is sharpened on a flat stone. { 'hōn·iŋ ,gāj }

hood |DES ENG| An opaque shield placed above or around the screen of a cathode-ray tube to eliminate extraneous light. |ENG| **1.** Close-fitting, rubber head covering that leaves the face exposed; used in scuba diving. **2.** A protective covering, usually providing special ventilation to carry away objectionable fumes, dusts, and gases, in which dangerous chemical, biological, or radioactive materials can be safely handled. { hüd }

hood test |ENG| A leak detection method in which the vessel under test is enclosed by a metallic casing so that a dynamic leak test may be carried out on a large portion of the external surface. { 'hüd ,test }

hook |DES ENG| A piece of hard material, especially metal, formed into a curve for catching, holding, or pulling something. |ELECTR| A circuit phenomenon occurring in four-zone transistors, wherein hole or electron conduction can occur in opposite directions to produce voltage drops that encourage other types of conduction. { hùk }

hookah |ENG| An air supply device used in free diving, comprising a demand regulator worn by the diver and a hose extending to a compressed air supply at the surface. { 'hü·kə }

hook-and-eye hinge |DES ENG| A hinge consisting of a hook (usually attached to a gate post) over which an eye (usually attached to the gate) is placed. { ¦húk ən 'T ¸hinj }

hook bolt |DES ENG| A bolt with a hook or L band at one end and threads at the other to fit a nut. { 'húk ¸bōlt }

hook collector transistor |ELECTR| A transistor in which there are four layers of alternating n- and p-type semiconductor material and the two interior layers are thin compared to the diffusion length. Also known as hook transistor; pn hook transistor. { ¦húk kə'lek·tər tran¸zis·tər }

Hookean deformation |MECH| Deformation of a substance which is proportional to the force applied to it. { 'húk·ē·ən ¸dəf·ər'mā·shən }

Hookean solid |MECH| An ideal solid which obeys Hooke's law exactly for all values of stress, however large. { 'húk·ē·ən 'säl·əd }

Hooker diaphragm cell |CHEM ENG| A device used in industry for the electrolysis of brine (sodium chloride) to make chlorine and caustic soda (sodium hydroxide) or caustic potash (potassium hydroxide); saturated purified brine fed around the anode passes through the diaphragm to the cathode; chlorine is formed at the anode and hydrogen released at the cathode, leaving sodium hydroxide and residual sodium chloride in the cell liquor; the diaphragm prevents the products from mixing. { 'húk·ər 'dī·ə¸fram ¸sel }

Hooke's joint |MECH ENG| A simple universal joint; consists of two yokes attached to their respective shafts and connected by means of a spider. Also known as Cardan joint. { 'húks ¸jóint }

Hooke's law |MECH| The law that the stress of a solid is directly proportional to the strain applied to it. { 'húks ¸lò }

hook gage |ENG| An instrument used to measure changes in the level of the water in an evaporation pan; it consists of a pointed metal hook, mounted in the vertical, whose position with respect to its supporting member may be adjusted by means of a micrometer arrangement; the gage is placed on the still well, and a measurement is taken when the point of the hook just breaks above the surface of the water. { 'húk ¸gāj }

hook transistor See hook collector transistor. { 'húk tran¸zis·tər }

hookup |ELEC| An arrangement of circuits and apparatus for a particular purpose. { 'húk¸əp }

hook wrench |DES ENG| A wrench with a hook for turning a nut or bolt. { 'húk ¸rench }

hoop |CIV ENG| A ring-shaped binder placed around the main reinforcement in a reinforced concrete column. { 'húp }

hooped column |CIV ENG| A column of reinforced concrete with hoops around the main reinforcements. { 'húpt ¦kál·əm }

Hope's apparatus |THERMO| An apparatus consisting of a vessel containing water, a freezing mixture in a tray surrounding the vessel, and thermometers inserted in the water at points above and below the freezing mixture; used to show that the maximum density of water lies at about 4°C. { 'hōps ¸ap·ə¸rad·əs }

hopper |ENG| A funnel-shaped receptacle with an opening at the top for loading and a discharge opening at the bottom for bulk-delivering material such as grain or coal. { 'häp·ər }

hopper car |ENG| A freight car with a permanent roof and a hinged floor sloping to one or more hoppers for discharging contents by gravity. { 'häp·ər ¸kär }

hopper dryer |ENG| In extrusion and injection molding of plastics, a combined feeding and drying device in which hot air flows through the hopper. { 'häp·ər ¦drī·ər }

horizon sensor |ENG| A passive infrared device that detects the thermal discontinuity between the earth and space; used in establishing a stable vertical reference for control of the attitude or orientation of a missile or satellite in space. { hə'rīz·ən ¸sen·sər }

horizontal auger |MECH ENG| A rotary drill, usually powered by a gasoline engine, for making horizontal blasting holes in quarries and opencast pits. { ¸här·ə'zänt·əl 'òg·ər }

horizontal boiler |MECH ENG| A water-tube boiler having a main bank of straight tubes inclined toward the rear at an angle of 5 to 15° from the horizontal. { ¸här·ə'zänt·əl 'bòil·ər }

horizontal boring machine |MECH ENG| A boring machine adapted for work not conveniently revolved, for milling, slotting, drilling, tapping, boring, and reaming long holes and for making interchangeable parts that must be produced without jigs and fixtures. { ¸här·ə'zänt·əl 'bòr·iŋ ma¸shēn }

horizontal broaching machine |MECH ENG| A pull-type broaching machine having the broach mounted on the horizontal plane. { ¸här·ə'zänt·əl 'bròch·iŋ ma¸shēn }

horizontal circle |ENG| A graduated disk affixed to the base of a transit or theodolite which is used to measure horizontal angles. { ¸här·ə'zänt·əl 'sər·kəl }

horizontal crusher |MECH ENG| Rotary size reducer in which the crushing cone is supported on a horizontal shaft; needs less headroom than vertical models. { ¸här·ə'zänt·əl 'krəsh·ər }

horizontal drilling machine |MECH ENG| A drilling machine in which the drill bits extend in a horizontal direction. { ¸här·ə'zänt·əl 'dril·iŋ ma¸shēn }

horizontal engine |MECH ENG| An engine with horizontal stroke. { ¸här·ə'zänt·əl 'en·jən }

horizontal field balance |ENG| An instrument that measures the horizontal component of the magnetic field by means of the torque that the field component exerts on a vertical permanent magnet. { ¸här·ə'zänt·əl 'fēld ¸bal·əns }

horizontal firing |MECH ENG| The firing of fuel in a boiler furnace in which the burners discharge fuel and air into the furnace horizontally. { ¸här·ə'zänt·əl 'fīr·iŋ }

horizontal force instrument |ENG| An instrument used to make a comparison between the intensity of the horizontal component of the earth's magnetic field and the magnetic field at the compass location on board a craft; basically, it consists of a magnetized needle pivoted in a horizontal plane, as a dry-card compass; it settles in some position which indicates the direction of the resultant magnetic field; if the needle is started swinging, it damps down with a certain period of oscillation dependent upon the strength of the magnetic field. Also known as horizontal vibrating needle. { ,här·ə'zänt·əl ¦förs 'in·strə·mənt }

horizontal intensity variometer [ENG] Essentially a declination variometer with a larger, stiffer fiber than in the standard model; there is enough torsion in the fiber to cause the magnet to turn 90° out of the magnetic meridian; the magnet is aligned with the magnetic prime vertical to within 0.5° so it does not respond appreciably to changes in declination. Also known as H variometer. { ,här·ə'zänt·əl in'ten·səd·ē ,ver·ē'äm·əd·ər }

horizontal lathe [MECH ENG] A horizontally mounted lathe with which longitudinal and radial movements are applied to a workpiece that rotates. { ,här·ə'zänt·əl 'lāth }

horizontal magnetometer [ENG] A measuring instrument for ascertaining changes in the horizontal component of the magnetic field intensity. { ,här·ə'zänt·əl ,mag·nə'täm·əd·ər }

horizontal milling machine [MECH ENG] A knee-type milling machine with a horizontal spindle and a swiveling table for cutting helices. { ,här·ə'zänt·əl 'mil·iŋ mə,shēn }

horizontal pendulum [MECH] A pendulum that moves in a horizontal plane, such as a compass needle turning on its pivot. { ,här·ə'zänt·əl 'pen·jə·ləm }

horizontal return tubular boiler [MECH ENG] A fire-tube boiler having tubes within a cylindrical shell that are attached to the end closures; products of combustion are transported under the lower half of the shell and back through the tubes. { ,här·ə'zänt·əl ri'tərn ¦tü·byə·lər 'böil·ər }

horizontal scanning [ENG] In radar scanning, rotating the antenna in azimuth around the horizon or in a sector. Also known as searching lighting. { ,här·ə'zänt·əl 'skan·iŋ }

horizontal screen [MECH ENG] Shaking screen with horizontal plates. { ,här·ə'zänt·əl 'skrēn }

horizontal-tube evaporator [MECH ENG] A horizontally mounted tube-and-shell type of liquid evaporator, used most often for preparation of boiler feedwater. { ,här·ə'zänt·əl ¦tüb i'vap·ə,rād·ər }

horizontal vibrating needle See horizontal force instrument. { ,här·ə'zänt·əl ¦vī,brād·iŋ 'nēd·əl }

horn [BUILD] A section projecting from the end of one of the members of a right-angle wood framing joint. [ENG ACOUS] A tube whose cross-sectional area increases from one end to the other, used to radiate or receive sound waves and to intensify and direct them. Also known as acoustic horn. { hórn }

horn-loaded speaker [ENG ACOUS] A loudspeaker that has an acoustic horn between the

diaphragm and the air load. { ¦hórn ,lōd·əd 'spēk·ər }

horn loudspeaker [ENG ACOUS] A loudspeaker in which the radiating element is coupled to the air or another medium by means of a horn. { 'hórn 'laúd,spēk·ər }

horn socket [DES ENG] A cone-shaped fishing tool especially designed to recover lost collared drill rods, drill pipe, or tools in bored wells. { 'hórn ,säk·ət }

horsepower [MECH] The unit of power in the British engineering system, equal to 550 foot-pounds per second, approximately 745.7 watts. Abbreviated hp. { 'hórs¦paú·ər }

hose [DES ENG] Flexible tube used for conveying fluids. { hōz }

hose clamp [DES ENG] Band or brace to attach the raw end of a hose to a water outlet. { 'hōz ,klamp }

hose coupling [DES ENG] Device to interconnect two or more pieces of hose. { 'hōz ,kəp·liŋ }

hose fitting [DES ENG] Any attachment or accessory item for a hose. { 'hōz ,fid·iŋ }

hostile-environment machine [MECH ENG] A robot capable of operating in extreme conditions of temperature, vibration, moisture, pollution, or electromagnetic or nuclear radiation. { 'häs·təl in'vī·rən·mənt mə,shēn }

hot-air engine [MECH ENG] A heat engine in which air or other gases, such as hydrogen, helium, or nitrogen, are used as the working fluid, operating on cycles such as the Stirling or Ericsson. { 'hä¦der 'en·jən }

hot-air furnace [MECH ENG] An encased heating unit providing warm air to ducts for circulation by gravity convection or by fans. { 'hä¦der 'fər·nəs }

hot-air sterilization [ENG] A method of sterilization using dry heat for glassware and other heat-resistant materials which need to be dry after treatment; temperatures of 160–165°C are generated for at least 2 hours. { 'hä¦der ,ster·ə·lə'zā·shən }

hot-bulb [MECH ENG] Pertaining to an ignition method used in semidiesel engines in which the fuel mixture is ignited in a separate chamber kept above the ignition temperature by the heat of compression. { 'hät ,bəlb }

hot carrier [ELECTR] A carrier, which may be either an electron or a hole, that has relatively high energy with respect to the carriers normally found in majority-carrier devices such as thin-film transistors. { 'hät ¦kar·ē·ər }

hot-chamber die casting [ENG] A die-casting process in which a piston is driven through a reservoir of molten metal and thereby delivers a quantity of molten metal to the die cavity. { 'hät ¦chäm·bər 'dī ,kast·iŋ }

Hotchkiss drive [MECH ENG] An automobile rear suspension designed to take torque reactions through longitudinal leaf springs. { 'häch,kis ,drīv }

Hotchkiss superdip |ENG| A sensitive dip needle consisting of a freely rotating magnetic needle about a horizontal axis and a nonmagnetic bar with a counterweight at the end which is attached to the pivot point of the needle. { 'häch,kis 'sü·pər,dip }

hot-draw |ENG| To draw a material while it is hot. { 'hät ¦drȯ }

hot editing |CONT SYS| A method for detecting errors in the programming of a robot in which as many errors as possible are identified and resolved during testing, without setting the robotic program to its starting condition. { 'hät 'ed·əd·iŋ }

hot electron |ELECTR| An electron that is in excess of the thermal equilibrium number and, for metals, has an energy greater than the Fermi level; for semiconductors, the energy must be a definite amount above that of the edge of the conduction band. { 'hät i'lek,trän }

hot-electron transistor |ELECTR| A transistor in which electrons tunnel through a thin emitter-base barrier ballistically (that is, without scattering), traverse a very narrow base region, and cross a barrier at the base-collector interface whose height, controlled by the collector voltage, determines the fraction of electrons coming to the collector. { ¦häti 'lek,trän ,tran'zis·tər }

hot-gas welding |ENG| Joining of thermoplastic materials by softening first with a jet of hot air, then joining at the softened points. { 'hät ,gas 'weld·iŋ }

hot hole |ELECTR| A hole that can move at much greater velocity than normal holes in a semiconductor. { 'hät ,hōl }

hothouse |ENG| A greenhouse heated to grow plants out of season. { 'hät,haüs }

hot isostatic pressing |ENG| A process in which a ceramic or metal powder is consolidated by heating and compressing the powder equally from all directions inside a sealed flexible mold. Abbreviated HIP. { ¦hät ,ī·sō¦stad·ik 'pres·iŋ }

hot junction |ELECTR| The heated junction of a thermocouple. { 'hät 'jəŋk·shən }

hot patching |ENG| Repair of a hot refractory lining in a furnace, usually by spraying with a refractory slurry. { 'hät 'pach·iŋ }

hot pressing |ENG| **1.** Forming a metal-powder compact or a ceramic shape by applying pressure and heat simultaneously at temperatures high enough for sintering to occur. **2.** Fabrication of a composite material through joining the reinforcement and the matrix by means of heat and pressure, usually in a hydraulically actuated press. { 'hät 'pres·iŋ }

hot-runner mold |ENG| A plastics mold in which the runners are kept hot by insulation from the chilled cavities. { 'hät ,rən·ər 'mōld }

hot saw |MECH ENG| A power saw used to cut hot metal. { 'hät ,sȯ }

hot-solder coating |ENG| The application of a protective finish to a printed circuit board by dip soldering in a solder bath. { 'hät ¦säd·ər 'kōd·iŋ }

hot spot |CHEM ENG| An area or point within a reaction system at which the temperature is appreciably higher than in the bulk of the reactor; usually locates the reaction front. |ENG| An area in a pipeline that is subject to excessive corrosion. { 'hät ,spät }

hot spraying |ENG| A paint-spraying technique in which paint viscosity is reduced by heat rather than a solvent. { ¦hät ¦sprā·iŋ }

hot stamp |FNG| An impression on a forging made in a heated condition. { 'hät ,stamp }

hot strength See tensile strength. { 'hät ,streŋkth }

hot-water heating |MECH ENG| A heating system for a building in which the heat-conveying medium is hot water and the heat-emitting means are radiators, convectors, or panel coils. Also known as hydronic heating. { 'hät ,wȯd·ər 'hēd·iŋ }

hot well |MECH ENG| A chamber for collecting condensate, as in a steam condenser serving an engine or turbine. { 'hät ,wel }

hot-wire ammeter |ENG| An ammeter which measures alternating or direct current by sending it through a fine wire, causing the wire to heat and to expand or sag, deflecting a pointer. Also known as thermal ammeter. { 'hät ¦wīr 'a,med·ər }

hot-wire anemometer |ENG| An anemometer used in research on air turbulence and boundary layers; the resistance of an electrically heated fine wire placed in a gas stream is altered by cooling by an amount which depends on the fluid velocity. { 'hät ¦wīr ,an·ə'mäm·əd·ər }

hot-wire instrument |ENG| An instrument that depends for its operation on the expansion by heat of a wire carrying a current. { 'hät ¦wīr 'in·strə·mənt }

hot-wire microphone |ENG ACOUS| A velocity microphone that depends for its operation on the change in resistance of a hot wire as the wire is cooled by varying particle velocities in a sound wave. { 'hät ¦wīr 'mī·krə,fōn }

hot work |IND ENG| A task that requires working on, or in proximity to, exposed energized electrical equipment or wiring. { 'hät ,wərk }

Houdry butane dehydrogenation |CHEM ENG| A catalytic process for dehydrogenating light hydrocarbons from crude oil to their corresponding mono- or diolefins; chromia-alumina catalysts with inert material are used in pellet form. { 'hü·drē ¦byü,tān dē,hī·drə·jə'nā·shən }

Houdry fixed-bed catalytic cracking |CHEM ENG| A cyclic, regenerable process for cracking of petroleum distillates to produce high-octane gasoline from higher-boiling petroleum fractions; synthetic or natural bead catalysts of activated hydrosilicate of alumina may be used. Also known as Houdry process. { 'hü·drē ¦fixt ¦bed ,kad·əl¦id·ik 'krak·iŋ }

Houdry hydrocracking |CHEM ENG| A catalytic process combining cracking and desulfurization of crude petroleum oil in the presence of hydrogen; catalysts may be nickel oxide or nickel sulfide on silica alumina, and cobalt molybdate on alumina. { 'hü·drē ¦hī·drō¦krak·iŋ }

Houdry process See Houdry fixed-bed catalytic cracking. { 'hü·drē ˌprä·səs }

hour [MECH] A unit of time equal to 3600 seconds. Abbreviated h; hr. { aủr }

hourglass screw See Hindley screw. { 'aủr,glas ˌskrü }

hourglass worm See Hindley screw. { 'aủr,glas ˌwərm }

housed joint See dado joint. { 'haủzd ˌjȯint }

house drain [CIV ENG] Horizontal drain in a basement receiving waste from stacks. { 'haủs ˌdrān }

house sewer [CIV ENG] Connection between house drain and public sewer. { 'haủs ˌsü·ər }

housing [ENG] A case or enclosure to cover and protect a structure or a mechanical device. { 'haủ·ziŋ }

Houskeeper seal [ENG] A vacuum-tight seal made between copper and glass by bringing the copper to a flexible feather edge before fusing it to the glass; the copper then flexes as the glass shrinks during cooling. { 'haủs,kēp·ər ˌsēl }

hovercraft See air-cushion vehicle. { 'həv·ər,kraft }

Howell-Bunger valve See cone valve. { 'haủ·əl 'bəŋ·gər ˌvalv }

Howe truss [CIV ENG] A truss for spans up to 80 feet (24 meters) having both vertical and diagonal members; made of steel or timber or both. { 'haủ ˌtrəs }

howl [ENG ACOUS] Undesirable prolonged sound produced by a radio receiver or audio-frequency amplifier system because of either electric or acoustic feedback. { haủl }

Hoyer method of prestressing See pretensioning. { 'hȯi·yər ˌmeth·əd əv prē'stres·iŋ }

hp See horsepower.

H pile [CIV ENG] A steel pile that is H-shaped in section. { ¦āch ˌpīl }

hr See hour.

H rod [DES ENG] A drill rod having an outside diameter of 3-1/2 inches (8.89 centimeters). { 'āch ˌräd }

HTU See height of transfer unit.

hub [BUILD] The core section of a building from which corridors extend. [DES ENG] **1.** The cylindrical central part of a wheel, propeller, or fan. **2.** A piece in a lock that is turned by the knob spindle, causing the bolt to move. **3.** A short coupling that joins plumbing pipes. [ENG] In surveying, a stake that marks the position of a theodolite. { həb }

hubcap [DES ENG] A metal cap fastened or clamped to the end of an axle, as on motor vehicles. { 'həb,kap }

Huggenberger tensometer [ENG] A type of extensometer having a short gage length (10 to 20 millimeters) and employing a compound lever system that gives a magnification of about 1200. { 'həg·ən,bərg·ər ten'säm·əd·ər }

human engineering See human-factors engineering. { 'hyü·mən ˌen·jə'nir·iŋ }

human-factors engineering [ENG] The area of knowledge dealing with the capabilities and limitations of human performance in relation to design of machines, jobs, and other modifications of the human's physical environment. Also known as human engineering. { 'hyü·mən ¦fak·tərz ˌen·jə'nir·iŋ }

human-machine chart [IND ENG] A two-column, multiple-activity process chart listing the steps performed by an operator and the operations performed by a machine and showing the corresponding idle times for each. Also known as man-machine chart. { ¦yü·mən mə¦shēn 'chärt }

human-machine system [ENG] A system in which the functions of the worker and the machine are interrelated and necessary for the operation of the system. Also known as man-machine system. { ¦yü·mən mə¦shēn 'sis·təm }

hum-bucking coil [ENG ACOUS] A coil wound on the field coil of an excited-field loudspeaker and connected in series opposition with the voice coil, so that hum voltage induced in the voice coil is canceled by that induced in the hum-bucking coil. { 'həm ˌbək·iŋ ˌkȯil }

humidification [ENG] The process of increasing the water vapor content of a gas. { yü,mid·i·fə'kā·shən }

humidifier [MECH ENG] An apparatus for supplying moisture to the air and for maintaining desired humidity conditions. { yü'mid·ə,fī·ər }

humidistat [ENG] An instrument that measures and controls relative humidity. Also known as hydrostat. { yü'mid·ə,stat }

humidity element [ENG] The transducer of any hygrometer, that is, that part of a hygrometer that quantitatively senses atmospheric water vapor. { hyü'mid·əd·ē ,el·ə·mənt }

humidity strip [ENG] The humidity transducing element in a Diamond-Hinman radiosonde; it consists of a flat plastic strip bounded by electrodes on two sides and coated with a hygroscopic chemical compound such as lithium chloride; the electrical resistance of this coating is a function of the amount of moisture absorbed from the atmosphere and the temperature of the strip. Also known as electrolytic strip. { hyü'mid·əd·ē ,strip }

Humphrey gas pump [MECH ENG] A combined internal combustion engine and pump in which the metal piston has been replaced by a column of water. { 'həm·frē 'gas ,pəmp }

Humphries equation [THERMO] An equation which gives the ratio of specific heats at constant pressure and constant volume in moist air as a function of water vapor pressure. { 'həm·frēz i,kwā·zhən }

hump yard [CIV ENG] A switch yard in a railway system that has a hump or steep incline down which freight cars can coast to prescheduled locations. Also known as gravity yard. { 'həmp ,yärd }

hungry joint See starved joint. { 'həŋ·grē ¦jȯint }

hung shot [ENG] A shot whose explosion is delayed after detonation or ignition. { 'həŋ ¦shät }

hunting [CONT SYS] Undesirable oscillation of

an automatic control system, wherein the controlled variable swings on both sides of the desired value. {ELECTR} Operation of a selector in moving from terminal to terminal until one is found which is idle. {MECH ENG} Irregular engine speed resulting from instability of the governing device. { 'hənt·iŋ }

hunting circuit *See* lockout circuit. { 'hənt·iŋ ,sər·kət }

hunting tooth {DES ENG} An extra tooth on the larger of two gear wheels so that the total number of teeth will not be an integral multiple of the number on the smaller wheel. { 'hənt·iŋ ,tüth }

hurricane beacon {ENG} An air-launched balloon designed to be released in the eye of a tropical cyclone, to float within the eye at predetermined levels, and to transmit radio signals. { 'hər·ə,kān ,bē·kən }

hurricane lamp {ENG} An oil lamp with a glass chimney and perforated lid to protect the flame, or a candle with a glass chimney. { 'hər·ə,kān ,lamp }

hurricane tracking {ENG} Recording of the movement of individual hurricanes by means of airplane sightings and satellite photography. { 'hər·ə,kān ,trak·iŋ }

Huttig equation {THERMO} An equation which states that the ratio of the volume of gas adsorbed on the surface of a nonporous solid at a given pressure and temperature to the volume of gas required to cover the surface completely with a unimolecular layer equals $(1 + r) c'/ (1 + c')$, where r is the ratio of the equilibrium gas pressure to the saturated vapor pressure of the adsorbate at the temperature of adsorption, and c is the product of a constant and the exponential of $(q - q_l)/RT$, where q is the heat of adsorption into a first layer molecule, q_l is the heat of liquefaction of the adsorbate, T is the temperature, and R is the gas constant. { 'həd·ik i,kwā·zhən }

HVAC {CIV ENG} The abbreviation for heating, ventilation, and air conditioning systems, used in building design and construction. { ¦ach ¦vē¦a'sē *or* 'ach,vak }

H variometer *See* horizontal intensity variometer. { 'ach ,ver·ē'äm·əd·ər }

hybrid beam {ENG} A metal beam with flanges fabricated from a material that differs from that of the web plate and has a different minimum yield strength. { ¦hī·brəd ¦bēm }

hybrid inlet noise reduction {ENG ACOUS} A method of reducing the noise from the inlet of a jet engine, which involves the use of both high-Mach-number flows to retard or block the passage of sound waves and acoustic treatment of the walls of the inlet. { 'hī·brəd ¦in·lət 'nóiz ri,dək·shən }

hybrid integrated circuit {ELECTR} A circuit in which one or more discrete components are used in combination with integrated-circuit construction. { 'hī·brəd ¦int·ə,grād·əd 'sər·kət }

hybrid junction {ELECTR} A transformer, resistor, or waveguide circuit or device that has four pairs of terminals so arranged that a signal entering at one terminal pair divides and emerges from the two adjacent terminal pairs, but is unable to reach the opposite terminal pair. Also known as bridge hybrid. { 'hī·brəd 'jəŋk·shən }

hybrid microcircuit {ELECTR} Microcircuit in which thin-film, thick-film, or diffusion techniques are combined with separately attached semiconductor chips to form the circuit. { 'hī·brəd 'mī·krō,sər·kət }

hybrid thin-film circuit {ELECTR} Microcircuit formed by attaching discrete components and semiconductor devices to networks of passive components and conductors that have been vacuum-deposited on glazed ceramic, sapphire, or glass substrates. { 'hī·brəd ¦thin ,film 'sər·kət }

hydrant *See* fire hydrant. { 'hī·drənt }

hydraucone {DES ENG} A conical, spreading type of draft tube used on hydraulic turbine installations. { 'hī·dró,kōn }

hydraulic {ENG} Operated or effected by the action of water or other fluid of low viscosity. { hī'dró·lik }

hydraulic accumulator {MECH ENG} A hydraulic flywheel that stores potential energy by accumulating a quantity of pressurized hydraulic fluid in a suitable enclosed vessel. { hī'dró·lik ə'kyü·myə,lād·ər }

hydraulic actuator {MECH ENG} A cylinder or fluid motor that converts hydraulic power into useful mechanical work; mechanical motion produced may be linear, rotary, or oscillatory. { hī'dró·lik 'ak·chə,wād·ər }

hydraulic air compressor {MECH ENG} A device in which water falling down a pipe entrains air which is released at the bottom under compression to do useful work. { hī'dró·lik 'er kəm ,pres·ər }

hydraulic amplifier {CONT SYS} A device which increases the power of a signal in a hydraulic servomechanism or other system through the use of fixed and variable orifices. Also known as hydraulic intensifier. { hī'dró·lik 'am·plə,fī·ər }

hydraulic backhoe {MECH ENG} A backhoe operated by a hydraulic mechanism. { hī'dró·lik 'bak,hō }

hydraulic brake {MECH ENG} A brake in which the retarding force is applied through the action of a hydraulic press. { hī'dró·lik 'brāk }

hydraulic circuit {MECH ENG} A circuit whose operation is analogous to that of an electric circuit except that electric currents are replaced by currents of water or other fluids, as in a hydraulic control. { hī'dró·lik 'sər·kət }

hydraulic classification {ENG} Classification of particles in a tank by specific gravity, utilizing the action of rising water currents. { hī'dró·lik ,klas·ə·fə'kā·shən }

hydraulic classifier {MECH ENG} A classifier in which particles are sorted by specific gravity in a stream of hydraulic water that rises at a controlled rate; heavier particles gravitate down and are discharged at the bottom, while lighter ones

hydraulic clutch

are carried up and out. Also known as hydrosizer. { hī'drȯ·lik 'klas·ə,fī·ər }

hydraulic clutch See fluid drive. { hī'drȯ·lik 'kləch }

hydraulic conveyor |MECH ENG| A system for handling material, such as ash from a coal-fired furnace; refuse is flushed from a hopper or slag tank to a grinder which discharges to a pump for conveying to a disposal area or a dewatering bin. { hī'drȯ·lik kən'vā·ər }

hydraulic coupling See fluid coupling. { hī'drȯ·lik 'kəp·liŋ }

hydraulic cylinder |MECH ENG| The cylindrical chamber of a positive displacement pump. { hī'drȯ·lik 'sil·ən·dər }

hydraulic dredge |MECH ENG| A dredge consisting of a large suction pipe which is mounted on a hull and supported and moved about by a boom, a mechanical agitator or cutter head which churns up earth in front of the pipe, and centrifugal pumps mounted on a dredge which suck up water and loose solids. { hī'drȯ·lik 'drej }

hydraulic drill |MECH ENG| A rotary drill powered by hydrodynamic means and used to make shot-firing holes in coal or rock, or to make a well hole. { hī'drȯ·lik 'dril }

hydraulic drive |MECH ENG| A mechanism transmitting motion from one shaft to another, the velocity ratio of the shafts being controlled by hydrostatic or hydrodynamic means. { hī'drȯ·lik 'drīv }

hydraulic ejector |ENG| A pipe for removing excavated material from a pneumatic caisson. { hī'drȯ·lik i'jek·tər }

hydraulic elevator |MECH ENG| An elevator operated by water pressure. Also known as hydraulic lift. { hī'drȯ·lik 'el·ə,vād·ər }

hydraulic engineering |CIV ENG| A branch of civil engineering concerned with the design, erection, and construction of sewage disposal plants, waterworks, dams, water-operated power plants, and such. { hī'drȯ·lik 'en·jə'nir·iŋ }

hydraulic excavator digger |MECH ENG| An excavation machine which employs hydraulic pistons to actuate mechanical digging elements. { hī'drȯ·lik ,eks·kə'vād·ər 'dig·ər }

hydraulic intensifier See hydraulic amplifier. { hī'drȯ·lik in'ten·sə,fī·ər }

hydraulic jack |MECH ENG| A jack in which force is applied through the mechanism of a hydraulic press. { hī'drȯ·lik 'jak }

hydraulic jetting |ENG| Use of high-pressure water forced through nozzles to clean tube interiors and exteriors in heat exchangers and boilers. { hī'drȯ·lik 'jed·iŋ }

hydraulic lift See hydraulic elevator. { hī'drȯ·lik 'lift }

hydraulic machine |MECH ENG| A machine powered by a motor activated by the confined flow of a stream of liquid, such as oil or water under pressure. { hī'drȯ·lik mə'shēn }

hydraulic motor |MECH ENG| A motor activated by water or other liquid under pressure. { hī'drȯ·lik 'mōd·ər }

hydraulic nozzle |MECH ENG| An atomizing device in which fluid pressure is converted into fluid velocity. { hī'drȯ·lik 'näz·əl }

hydraulic packing |ENG| Packing material that resists the effects of water even under high pressure. { hī'drȯ·lik 'pak·iŋ }

hydraulic power system |MECH ENG| A power transmission system comprising machinery and auxiliary components which function to generate, transmit, control, and utilize hydraulic energy. { hī'drȯ·lik 'pau̇·ər ,sis·təm }

hydraulic press |MECH ENG| A combination of a large and a small cylinder connected by a pipe and filled with a fluid so that the fluid pressure created by a small force acting on the small-cylinder piston will result in a large force on the large piston. Also known as hydrostatic press. { hi'drȯ·lik 'pres }

hydraulic pump See hydraulic ram. { hi'drȯ·lik 'pəmp }

hydraulic ram |MECH ENG| A device for forcing running water to a higher level by using the kinetic energy of flow; the flow of water in the supply pipeline is periodically stopped so that a small portion of water is lifted by the velocity head of a larger portion. Also known as hydraulic pump. { hi'drȯ·lik 'ram }

hydraulic robot |CONT SYS| A robot that is powered by hydraulic actuators, usually controlled by servovalves and analog resolvers. { hī'drȯl·ik 'rō,bät }

hydraulic rope-geared elevator |MECH ENG| An elevator hoisted by a system of ropes and sheaves attached to a piston in a hydraulic cylinder. { hi'drȯ·lik 'rōp ,gird 'el·ə,vād·ər }

hydraulic scale |MECH ENG| An industrial scale in which the load applied to the load-cell piston is converted to hydraulic pressure. { hī'drȯ·lik 'skāl }

hydraulic separation |MECH ENG| Mechanical classification using a hydraulic classifier. { hī'drȯ·lik ,sep·ə'rā·shən }

hydraulic shovel |MECH ENG| A revolving shovel in which hydraulic rams or motors are substituted for drums and cables. { hi'drȯ·lik 'shəv·əl }

hydraulic sprayer |MECH ENG| A machine that sprays large quantities of insecticide or fungicide on crops. { hi'drȯ·lik 'sprā·ər }

hydraulic spraying See airless spraying. { hi'drȯ·lik 'sprā·iŋ }

hydraulic stacker |MECH ENG| A tiering machine whose carriage is raised or lowered by a hydraulic cylinder. { hī'drȯ·lik 'stak·ər }

hydraulic swivel head |MECH ENG| In a drill machine, a swivel head equipped with hydraulically actuated cylinders and pistons to exert pressure on and move the drill rod string longitudinally. { hi'drȯ·lik 'swiv·əl ,hed }

hydraulic transport |ENG| Movement of material by water. { hī'drȯ·lik 'tranz,pȯrt }

hydraulic turbine |MECH ENG| A machine which converts the energy of an elevated water supply into mechanical energy of a rotating shaft. { hī'drȯ·lik 'tər·bən }

hydrocarbon blending value |ENG| Octane number rating for a 20% blend of a hydrocarbon with a 60:40 mixture of isooctane:*n*-heptane, which has been recalculated for a hypothetical 100% concentration of the tested hydrocarbon. { ¦hī·drə'kär·bən 'blend·iŋ ˌval·yü }

hydroclone |CHEM ENG| A device for separating a solid-liquid mixture during an industrial process by using a conical vortex and centrifugal force. { 'hī·drəˌklōn }

hydrocracker |CHEM ENG| A high-pressure processing unit that cracks long hydrocarbon molecules under a high-hydrogen-content atmosphere. { 'hī·drō,krak·ər }

hydrocracking |CHEM ENG| A catalytic, high-pressure petroleum refinery process that is flexible enough to produce either high-octane gasoline or aviation jet fuel; the two main reactions are the adding of hydrogen to petroleum-derived molecules too massive and complex for gasoline and then the cracking of them to the required fuels; the catalyst is an acidic solid and a hydrogenating metal component. { 'hī·drō,krak·iŋ }

hydrocyclone |MECH ENG| A cyclone separator in which granular solids are removed from a stream of water and classified by centrifugal force. { ¦hī·drō¦sī,klōn }

hydrodealkylation |CHEM ENG| A petroleum refining operation in which heat and pressure are used to remove methyl groups or larger alkyl groups from hydrocarbons, or to change positions of these groups on the molecule; used to upgrade low-value products. { ˌhī·drō·dē,al·kə'lā·shən }

hydrodesulfurization |CHEM ENG| A catalytic process in which the petroleum feedstock is reacted with hydrogen to reduce the sulfur content in the oil. { ˌhī·drō·dēˌsəl·fə·rə'zā·shən }

hydrodynamic oscillator |ENG ACOUS| A transducer for generating sound waves in fluids, in which a continuous flow through an orifice is modulated by a reciprocating valve system controlled by acoustic feedback. { 'hī·drō·dī'nam·ik 'äs·ə,lād·ər }

hydroelectric generator |MECH ENG| An electric rotating machine that transforms mechanical power from a hydraulic turbine or water wheel into electric power. { ¦hī·drō·i'lek·trik 'jen·ə,rād·ər }

hydroelectricity |ELEC| Electric power produced by hydroelectric generators. Also known as hydropower. { ¦hī·drō·i,lek'tris·əd·ē }

hydroelectric plant |MECH ENG| A facility at which electric energy is produced by hydroelectric generators. Also known as hydroelectric power station. { ¦hī·drō·i'lek·trik 'plant }

hydroelectric power station See hydroelectric plant. { ¦hī·drō·i'lek·trik 'paü·ər ,stā·shən }

hydrofining |CHEM ENG| A fixed-bed catalytic process to desulfurize and hydrogenate a wide range of charge stocks, from gases through waxes; the catalyst comprises cobalt oxide and molybdenum oxide on an extruded alumina support and may be regenerated in place by air and steam or flue gas. { 'hī·drəˌfīn·iŋ }

hydroforming |CHEM ENG| A petroleum-refinery process in which naphthas are passed over a catalyst at elevated temperatures and moderate pressures in the presence of added hydrogen or hydrogen-containing gases, to form high-octane BTX aromatics for motor fuels or chemical manufacture. { ¦hī·drəˌfor·miŋ }

hydroformylation |CHEM ENG| The reaction of adding hydrogen and the —CHO group to the carbon atoms across a double bond to yield oxygenated derivatives; an example is in the oxo process where the term hydroformylation applies to those reactions brought about by treating olefins with a mixture of hydrogen and carbon monoxide in the presence of a cobalt catalyst. { ˌhī·drəˌfor·mə'lā·shən }

hydrogasification |CHEM ENG| A technique to manufacture synthetic pipeline gas from coal; pulverized coal is reacted with hot, raw, hydrogen-rich gas containing a substantial amount of steam at 1000 pounds per square inch gage (6.9 × 10⁶ pascals, gage) to form methane. { ˌhī·drəˌgas·ə·fə'kā·shən }

hydrogenation |CHEM ENG| Saturation of diolefin impurities in gasolines to form a stable product. { hī,dräj·ə'nā·shən }

hydrographic sextant |ENG| A surveying sextant similar to those used for celestial navigation but smaller and lighter, constructed so that the maximum angle that can be read is slightly greater than that on the navigating sextant; usually the angles can be read only to the nearest minute by means of a vernier; it is fitted with a telescope with a large object glass and field of view. Also known as sounding sextant; surveying sextant. { 'hī·drə'graf·ik 'seks·tənt }

hydrographic sonar |ENG| An echo sounder used in mapping ocean bottoms. { 'hī·drə'graf·ik 'sō,när }

hydrometer |ENG| A direct-reading instrument for indicating the density, specific gravity, or some similar characteristic of liquids. { hī'dräm·əd·ər }

hydrometrograph |ENG| An instrument that measures and records the rate of water discharge from a pipe or an orifice. { ˌhī·drə'me·trə,graf }

hydronic heating See hot-water heating. { hī'drän·ik 'hēd·iŋ }

hydrophone |ENG ACOUS| A device which receives underwater sound waves and converts them to electric waves. { 'hī·drə,fōn }

hydropneumatic |ENG| Operated by both water and air power. { ¦hī·drō·nü'mad·ik }

hydropneumatic recoil system |MECH ENG| A recoil mechanism that absorbs the energy of recoil by the forcing of oil through orifices and returns the gun to battery by compressed gas. { ¦hī·drō·nü'mad·ik 'rē,koil ,sis·təm }

hydropower See hydroelectricity. { 'hī·drə,paü·ər }

hydroseparator |MECH ENG| A separator in which solids in suspension are agitated by hydraulic pressure or stirring devices. { ¦hī·drō 'sep·ə,rād·ər }

281

hydrosizer

hydrosizer *See* hydraulic classifier. { 'hī·drə,sīz·ər }

hydrostat *See* humidistat. { 'hī·drə,stat }

hydrostatic balance [MECH] An equal-arm balance in which an object is weighed first in air and then in a beaker of water to determine its specific gravity. { ,hī·drə¦stad·ik i'kwä·zhən }

hydrostatic bearing [MECH ENG] A sleeve bearing in which high-pressure oil is pumped into the area between the shaft and the bearing so that the shaft is raised and supported by an oil film. { ,hī·drə'stad·ik 'ber·iŋ }

hydrostatic modulus *See* bulk modulus of elasticity. { ,hī·drə'stad·ik 'mäj·ə·ləs }

hydrostatic press *See* hydraulic press. { ,hī·drə'stad·ik 'pres }

hydrostatic pressing [ENG] Compacting ceramic or metal powders by packing them in a rubber bag which is subjected to pressure from a hydraulic press. { ,hī·drə'stad·ik 'pres·iŋ }

hydrostatic roller conveyor [MECH ENG] A portion of a roller conveyor that has rolls weighted with liquid to control the speed of the moving objects. { ,hī·drə'stad·ik rō·lər kən,vā·ər }

hydrostatic strength [MECH] The ability of a body to withstand hydrostatic stress. { ,hī·drə'stad·ik 'streŋkth }

hydrostatic stress [MECH] The condition in which there are equal compressive stresses or equal tensile stresses in all directions, and no shear stresses on any plane. { ,hī·drə'stad·ik 'stres }

hydrostatic test [ENG] Test of strength and leak-resistance of a vessel, pipe, or other hollow equipment by internal pressurization with a test liquid. { ,hī·drə'stad·ik 'test }

hydrothermal crystal growth [CHEM ENG] Formation of simple crystals of quartz at elevated temperatures and pressures in an autoclave with an alkaline solution. { ,hī·drə'thər·məl 'krist·əl ,grōth }

hydrotreating [CHEM ENG] Oil refinery catalytic process in which hydrogen is contacted with petroleum intermediate or product streams to remove impurities, such as oxygen, sulfur, nitrogen, or unsaturated hydrocarbons. { ¦hī·drō ¦trēd·iŋ }

hydrowire [ENG] A wire to which equipment is clamped so that it can be lowered over the side of the ship into the water. { 'hī·drō,wīr }

hygrodeik [ENG] A form of psychrometer with wet-bulb and dry-bulb thermometers mounted on opposite edges of a specially designed graph of the psychrometric tables, arranged so that the intersections of two curves determined by the wet-bulb and dry-bulb readings yield the relative humidity, dew-point, and absolute humidity. { 'hī·grə,dīk }

hygrogram [ENG] The record made by a hygrograph. { 'hī·grə,gram }

hygrograph [ENG] A recording hygrometer. { 'hī·grə,graf }

hygrometer [ENG] An instrument for giving a direct indication of the amount of moisture in the air or other gas, the indication usually being in terms of relative humidity as a percentage which the moisture present bears to the maximum amount of moisture that could be present at the location temperature without condensation taking place. { hī'gräm·əd·ər }

hygrometry [ENG] The study which treats of the measurement of the humidity of the atmosphere and other gases. { hī'gräm·ə·trē }

hygrothermograph [ENG] An instrument for recording temperature and humidity on a single chart. { ,hī·grə'thər·mə,graf }

hyl *See* metric-technical unit of mass.

hyperbaric chamber [ENG] A specially equipped pressure vessel used in medicine and physiological research to administer oxygen at elevated pressures. { ¦hī·pər¦bar·ik 'chäm·bər }

hyperbolic horn [ENG] Horn whose equivalent cross-sectional radius increases according to a hyperbolic law. { ¦hī·pər¦bäl·ik 'hórn }

hyperforming [CHEM ENG] A catalytic, petroleum-refinery hydrogenation process to improve naphtha octane number by removal of sulfur and nitrogen compounds; the catalyst is cobalt molybdate on a silica-alumina base. { 'hī·pər ,fór·miŋ }

hyperoid axle [MECH ENG] A type of rear-axle drive gear set which generally carries the pinion 1.5–2 inches (38–51 millimeters) or more below the centerline of the gear. { 'hī·pə,róid 'ak·səl }

hypersonic wind tunnel [ENG] A wind tunnel in which air flows at speeds roughly in the range from 5 to 15 times the speed of sound. { ¦hī·pər'sän·ik 'win ,tən·əl }

hypersorption [CHEM ENG] Process with recirculating bed of activated-carbon adsorbent for continuous recovery of ethylene from methane and other low-molecular-weight gases. { ¦hī·pər¦sórp·shən }

hyperspectral imaging system [ENG] An infrared imaging system that has more than 30 spectral channels with relatively fine spectral resolution, allowing imaging spectroscopy to be carried out. { ¦hī·pər,spek·trəl 'im·ij·iŋ ,sis·təm }

hypervelocity [MECH] **1.** Muzzle velocity of an artillery projectile of 3500 feet per second (1067 meters per second) or more. **2.** Muzzle velocity of a small-arms projectile of 5000 feet per second (1524 meters per second) or more. **3.** Muzzle velocity of a tank-cannon projectile in excess of 3350 feet per second (1021 meters per second). { ,hī·pər·və'läs·əd·ē }

hypervelocity wind tunnel [ENG] A wind tunnel in which higher airspeeds and temperatures can be attained than in a hypersonic wind tunnel. { ,hī·pər·və'läs·əd·ē 'win ,tən·əl }

hypochlorite sweetening [CHEM ENG] A petroleum refinery process to oxidize gasoline mercaptans by agitation with an aqueous, alkaline hypochlorite solution. { ,hī·pə'klór,īt 'swet·ən·iŋ }

hypoid gear |MECH ENG| Gear wheels connecting nonparallel, nonintersecting shafts, usually at right angles. { 'hī,pȯid ˈgir }

hypoid generator |MECH ENG| A gear-cutting machine for making hypoid gears. { 'hī,pȯid 'jen·ə,rād·ər }

hypsometer |ENG| **1.** An instrument for measuring atmospheric pressure to ascertain elevations by determining the boiling point of liquids. **2.** Any of several instruments for determining tree heights by triangulation. { hip'säm·əd·ər }

hypsometric |ENG| Pertaining to hypsometry. { ,hip·sə'me·trik }

hypsometry |ENG| The measuring of elevation with reference to sea level. { hip'säm·ə·trē }

hysteresimeter |ENG| A device for measuring hysteresis. { his,ter·ə'sim·əd·ər }

hysteresis |ELECTR| An oscillator effect wherein a given value of an operating parameter may result in multiple values of output power or frequency. { ,his·tə'rē·səs }

hysteresis clutch |MECH ENG| A clutch in which torque is produced by attraction between induced poles in a magnetized iron ring and the control field. { ,his·tə'rē·səs ,kləch }

hysteresis damping |MECH| Damping of a vibration due to energy lost through mechanical hysteresis. { ,his·tə'rē·səs 'dam·piŋ }

hysteretic damping |MECH| Damping of a vibrating system in which the retarding force is proportional to the velocity and inversely proportional to the frequency of the vibration. { ,his·tə'red·ik ˈdamp·iŋ }

283

I beam [CIV ENG] A rolled iron or steel joist having an I section, with short flanges. { 'ī ‚bēm }

IC See integrated circuit.

ice-accretion indicator [ENG] An instrument used to detect the occurrence of freezing precipitation, usually consisting of a strip of sheet aluminum about $1\frac{1}{2}$ inches (4 centimeters) wide, and is exposed horizontally, face up, in the free air a few meters above the ground. { 'īs ə‚krē·shən ‚ind·ə‚kād·ər }

ice apron [CIV ENG] A wedge-shaped structure which protects a bridge pier from floating ice. { 'īs ‚ā·prən }

ice buoy [ENG] A sturdy buoy, usually a metal spar, used to replace a more easily damaged buoy during a period when heavy ice is anticipated. { 'īs ‚bȯi }

ice calorimeter See Bunsen ice calorimeter. { 'īs ‚kal·ə'rim·əd·ər }

ice line [THERMO] A graph of the freezing point of water as a function of pressure. { 'īs ‚līn }

ice load [ENG] The weight of glaze deposited on an overhead wire in a power supply system; standard safety codes require allowance for 1/2-inch (12.7-millimeter) radial thickness in heavy loading districts and 1/4-inch (6.35-millimeter) in medium. { 'īs ‚lōd }

ice pick [DES ENG] A hand tool for chipping ice. { 'īs ‚pik }

ice tongs [DES ENG] Tongs for handling cubes or blocks of ice. { 'īs ‚täŋz }

icing-rate meter [ENG] An instrument for the measurement of the rate of ice accretion on an unheated body. { 'ī·siŋ ‚rāt ‚mēd·ər }

ID See inside diameter.

ideal gas [THERMO] Also known as perfect gas. **1.** A gas whose molecules are infinitely small and exert no force on each other. **2.** A gas that obeys Boyle's law (the product of the pressure and volume is constant at constant temperature) and Joule's law (the internal energy is a function of the temperature alone). { ī'dēl 'gas }

ideal gas law [THERMO] The equation of state of an ideal gas which is a good approximation to real gases at sufficiently high temperatures and low pressures; that is, PV = RT, where P is the pressure, V is the volume per mole of gas, T is the temperature, and R is the gas constant. { ī'dēl 'gas ‚lȯ }

ideal radiator See blackbody. { ī'dēl 'rād·ē‚ād·ər }

identification [CONT SYS] The procedures for deducing a system's transfer function from its response to a step-function input or to an impulse. { ī‚dent·ə·fə'kā·shən }

identification, friend or foe [ENG] A system using pulsed radio transmissions to which equipment carried by friendly forces automatically responds, by emitting a pulse code, thereby identifying themselves from enemy forces; a method of determining the friendly or unfriendly character of aircraft, ships, and army units by other aircraft, ships, or ground force units. Abbreviated IFF. { ī‚dent·ə·fə'kā·shən 'frend ər 'fō }

idle [MECH ENG] To run without a load. { 'īd·əl }

idler arm [MECH ENG] In an automotive steering system, a link that supports the tie rod and transmits steering motion to both wheels through the ends of the tie rod. { 'īd·lər ‚ärm }

idler gear [MECH ENG] A gear situated between a driving gear and a driven gear to transfer motion, without any change of direction or of gear ratio. { 'īd·lər ‚gir }

idler pulley [MECH ENG] A pulley used to guide and tighten the belt or chain of a conveyor system. { 'īd·lər ‚pu̇l·ē }

idler wheel [MECH ENG] **1.** A wheel used to transmit motion or to guide and support something. **2.** A roller with a rubber surface used to transfer power by frictional means in a sound-recording or sound-reproducing system. { 'īd·lər ‚wēl }

idle-stop solenoid [MECH ENG] An electrically operated plunger in a carburetor that provides a predetermined throttle setting at idle and closes the throttle completely when the ignition switch is turned off. Also known as antidieseling solenoid. { 'īd·əl ‚stäp 'sō·lə‚nȯid }

idle time [IND ENG] A period of time during a regular work cycle when a worker is not active because of waiting for materials or instruction. Also known as waiting time. { 'īd·əl ‚tīm }

idling jet [MECH ENG] A carburetor part that introduces gasoline during minimum load or speed of the engine. { 'īd·liŋ ‚jet }

idling system [MECH ENG] A system to obtain adequate metering forces at low airspeeds and small throttle openings in an automobile carburetor in the idling position. { 'īd·liŋ ‚sis·təm }

i-f *See* intermediate frequency.

i-f amplifier *See* intermediate-frequency amplifier. { ¦ī'ef 'am·plə‚fī·ər }

IFF *See* identification, friend or foe.

igniter [ENG] **1.** A device for igniting a fuel mixture. **2.** A charge, as of black powder, to facilitate ignition of a propelling or bursting charge. { ig'nīd·ər }

igniter cord [ENG] A cord which passes an intense flame along its length at a uniform rate to light safety fuses in succession. { ig'nīd·ər ‚kȯrd }

ignition delay *See* ignition lag. { ig'nish·ən di‚lā }

ignition lag [MECH ENG] In the internal combustion engine, the time interval between the passage of the spark and the inflammation of the air-fuel mixture. Also known as ignition delay. { ig'nish·ən ‚lag }

ignition quality [CHEM ENG] The property of a fuel that ignites when injected into the compressed-air charge in a diesel engine cylinder; measurement is given in terms of cetane number. { ig'nish·ən ‚kwäl·əd·ē }

ignition system [MECH ENG] The system in an internal combustion engine that initiates the chemical reaction between fuel and air in the cylinder charge by producing a spark. { ig'nish·ən ‚sis·təm }

ignorable coordinate *See* cyclic coordinate. { ig 'nȯr·ə·bəl kō'ȯrd·ən·ət }

I-head cylinder [MECH ENG] The internal combustion engine construction having both inlet and exhaust valves located in the cylinder head. { 'Ī‚hed ‚sil·ən·dər }

ihp *See* indicated horsepower.

I²L *See* integrated injection logic.

illumination design [ENG] Design of sources of lighting and of systems which distribute light in order to effect a comfortable and satisfactory environment for seeing. { ə‚lü·mə‚nā·shən di‚zīn }

image *See* electric image. { 'im·ij }

image force [ELEC] The electrostatic force on a charge in the neighborhood of a conductor, which may be thought of as the attraction to the charge's electric image. { 'im·ij ‚fȯrs }

image potential [ELEC] The potential set up by an electric image. { 'im·ij pə‚ten·chəl }

image table [CONT SYS] A data table that contains the status of all inputs, registers, and coils in a programmable controller. { 'im·ij ‚tā·bəl }

imaging radar [ENG] Radar carried on aircraft which forms images of the terrain. { 'im·i·jiŋ 'rā‚där }

Imhoff cone [CIV ENG] A graduated glass vessel for measuring settled solids in testing the composition of sewage. { 'im‚hȯf ‚kōn }

Imhoff tank [CIV ENG] A sewage treatment tank in which digestion and settlement take place in separate compartments, one below the other. { 'im‚hȯf ‚taŋk }

immersion coating [ENG] Applying material to the surface of a metal or ceramic by dipping into a liquid. { ə'mər·zhən ¦kōd·iŋ }

immersion scanning [ENG] Ultrasonic scanning in which the ultrasonic transducer and the object being scanned are both immersed in water or some other liquid that provides good coupling while the transducer is being moved around the object. { ə'mər·zhən ‚skan·iŋ }

immittance [ELEC] A term used to denote both impedance and admittance, as commonly applied to transmission lines, networks, and certain types of measuring instruments. { i'mit·əns }

impact [MECH] A forceful collision between two bodies which is sufficient to cause an appreciable change in the momentum of the system on which it acts. Also known as impulsive force. { 'im‚pakt }

impact area [ENG] An area with designated boundaries within which all objects that travel over a range are to make contact with the ground. { 'im‚pakt ‚er·ē·ə }

impact avalanche and transit time diode *See* IMPATT diode. { 'im‚pakt ¦av·ə‚lanch ən 'tran·zit ‚tīm 'dī‚ōd }

impact bar [ENG] Specimen used to test the relative susceptibility of a plastic material to fracture by shock. { 'im‚pakt ‚bär }

impact breaker [MECH ENG] A device that utilizes the energy from falling stones in addition to power from massive impellers for complete breaking up of stone. Also known as double impeller breaker. { 'im‚pakt ‚brāk·ər }

impact crusher [MECH ENG] A machine for crushing large chunks of solid materials by sharp blows imposed by rotating hammers, or steel plates or bars; some crushers accept lumps as large as 28 inches (about 70 centimeters) in diameter, reducing them to 1/4 inch (6 millimeters) and smaller. { 'im‚pakt ‚krəsh·ər }

impact energy [MECH] The energy necessary to fracture a material. Also known as impact strength. { 'im‚pakt ‚en·ər·jē }

impact force *See* set forward force. { 'im‚pakt ‚fȯrs }

impact grinding [MECH ENG] A technique used to break up particles by direct fall of crushing bodies on them. { 'im‚pakt ‚grīn·diŋ }

impact load [ENG] A force delivered by a blow, as opposed to a force applied gradually and maintained over a long period. { 'im‚pakt ‚lōd }

impact microphone [ENG ACOUS] An instrument that picks up the vibration of an object impinging upon another, used especially on space probes to record the impact of small meteoroids. { 'im‚pakt 'mī·krə‚fōn }

impact mill [MECH ENG] A unit that reduces the size of rocks and minerals by the action of rotating blades projecting the material against steel plates. { 'im‚pakt ‚mil }

impact-noise analyzer [ENG] An analyzer used with a sound-level meter to evaluate the characteristics of impact-type sounds and electric noise impulses that cannot be measured accurately with a noise meter alone. { 'im‚pakt ‚nȯiz 'an·ə‚līz·ər }

impactometer *See* impactor. (,im,pak'täm·əd· ər)

impactor |ENG| A general term for instruments which sample atmospheric suspensoids by impaction; such instruments consist of a housing which constrains the air flow past a sensitized sampling plate. Also known as impactometer. [MECH ENG] A machine or part whose operating principle is striking blows. (im'pak·tər)

impact roll [MECH ENG] An idler roll protected by a covering of a resilient material from the shock of the loading of material onto a conveyor belt, so as to reduce the damage to the belt. ('im,pakt ,rōl)

impact screen [MECH ENG] A screen designed to swing or rock forward when loaded and to stop abruptly by coming in contact with a stop. ('im,pakt ,skrēn)

impact strength |MECH| **1.** Ability of a material to resist shock loading. **2.** *See* impact energy. ('im,pakt ,streŋkth)

impact stress [MECH] Force per unit area imposed on a material by a suddenly applied force. ('im,pakt ,stres)

impact test [ENG] Determination of the degree of resistance of a material to breaking by impact, under bending, tension, and torsion loads; the energy absorbed is measured in breaking the material by a single blow. ('im,pakt ,test)

impact tube *See* pitot tube. ('im,pakt ,tüb)

impact velocity |MECH| The velocity of a projectile or missile at the instant of impact. Also known as striking velocity. ('im,pakt və'läs·əd·ē)

impact wrench [MECH ENG] A compressed-air or electrically operated wrench that gives a rapid succession of sudden torques. ('im,pakt ,rench)

IMPATT diode |ELECTR| A *pn* junction diode that has a depletion region adjacent to the junction, through which electrons and holes can drift, and is biased beyond the avalanche breakdown voltage. Derived from impact avalanche and transit time diode. ('im,pat ,dī,ōd)

Impedance *See* electrical impedance. (im'pēd· əns)

impedance bridge [ELEC] A device similar to a Wheatstone bridge, used to compare impedances which may contain inductance, capacitance, and resistance. (im'pēd·əns ,brij)

impedance coil [ELEC] A coil of wire designed to provide impedance in an electric circuit. (im'pēd·əns ,kòil)

impedance compensator |ELEC| Electric network designed to be associated with another network or a line with the purpose of giving the impedance of the combination a desired characteristic with frequency over a desired frequency range. (im'pēd·əns 'käm·pən,sād·ər)

impedance component [ELEC] **1.** Resistance or reactance. **2.** A device such as a resistor, inductor, or capacitor designed to provide impedance in an electric circuit. (im'pēd·əns kəm,pō· nənt)

impedance coupling [ELEC] Coupling of two

signal circuits with an impedance. (im'pēd· əns ,kəp·liŋ)

impedance drop |ELEC| The total voltage drop across a component or conductor of an alternating-current circuit, equal to the phasor sum of the resistance drop and the reactance drop. (im'pēd·əns ,dräp)

impedance magnetometer |ENG| An instrument for determining local variations in magnetic field by measuring the change in impedance of a high-permeability nickel-iron wire. (im'ped·əns ,mag·nə'täm·əd·ər)

impeller [MECH ENG] The rotating member of a turbine, blower, fan, axial or centrifugal pump, or mixing apparatus. Also known as rotor. (im'pel·ər)

impeller pump [MECH ENG] Any pump using a mechanical agency to provide continuous power to move liquids. (im'pel·ər ,pəmp)

imperfect gas *See* real gas. (im'pər·fikt 'gas)

imperial gallon *See* gallon. (im'pir·ē·əl 'gal·ən)

imperial pint *See* pint. (im'pir·ē·əl 'pīnt)

impersonal micrometer [ENG] An instrument consisting of a vertical wire that is mounted in the focal plane of a transit circle and can be moved across the field of view to follow a star, and instrumentation to record the position of the wire as a function of time; used to reduce systematic observational errors. (im'pərs·ən· əl mī'kräm·əd·ər)

impingement |ENG| Removal of liquid droplets from a flowing gas or vapor stream by causing it to collide with a baffle plate at high velocity, so that the droplets fall away from the stream. Also known as liquid knockout. (im'pinj· mənt)

impinger |ENG| A device used to sample dust in the air that draws in a measured volume of dusty air and directs it through a jet to impact on a wetted glass plate; the dust particles adhering to the plate are counted. (im'pin·jər)

implanted atom |ELECTR| An atom introduced into semiconductor material by ion implantation. (im'plant·əd 'ad·əm)

implicit programming [CONT SYS] Robotic programming that uses descriptions of the tasks at hand which are less exact than in explicit programming. (im'plis·ət 'prō,gram·iŋ)

imposed date |IND ENG| An assignment of a date to an activity that represents either the earliest or the latest date at which the activity can be either started or finished. (im'pōzd 'dāt)

imposed load |CIV ENG| Any load which a structure must sustain, other than the weight of the structure itself. (im'pōzd 'lōd)

impound |CIV ENG| To collect water for irrigation, flood control, or similar purpose. (im 'paùnd)

impounding reservoir |CIV ENG| A reservoir with outlets controlled by gates that release stored surface water as needed in a dry season; may also store water for domestic or industrial use or for flood control. Also known as storage reservoir. (im'paùnd·iŋ ,rez·əv,wär)

impregnate |ENG| To force a liquid substance

into the spaces of a porous solid in order to change its properties, as the impregnation of turquoise gems with plastic to improve color and durability, the impregnation of porous tungsten with a molten barium compound to manufacture a dispenser cathode, or the impregnation of wood with creosote to preserve its integrity against water damage. { im'preg,nāt }

impregnated bit [DES ENG] A sintered, powder-metal matrix bit with fragmented bort or whole diamonds of selected screen sizes uniformly distributed throughout the entire crown section. { im'preg,nād·əd 'bit }

impulse [MECH] The integral of a force over an interval of time. { 'im,pəls }

impulse modulation [CONT SYS] Modulation of a signal in which it is replaced by a series of impulses, equally spaced in time, whose strengths (integrals over time) are proportional to the amplitude of the signal at the time of the impulse. { 'im,pəls ,mäj·ə,lā·shən }

impulse response [CONT SYS] The response of a system to an impulse which differs from zero for an infinitesimal time, but whose integral over time is unity; this impulse may be represented mathematically by a Dirac delta function. { 'im,pəls ri,späns }

impulse sealing [ENG] Heat-sealing of plastic materials by applying a pulse of intense thermal energy to the sealing area for a very short time, followed immediately by cooling. { 'im,pəls ¦sēl·iŋ }

impulse tachometer [ENG] A tachometer in which each rotation of a shaft generates an electric pulse and the time rate of pulses is then measured; classified as capacitory-current, inductory, or interrupted direct-current tachometer. { 'im,pəls tə'käm·əd·ər }

impulse train [CONT SYS] An input consisting of an infinite series of unit impulses, equally separated in time. { 'im,pəls ,trān }

impulse turbine [MECH ENG] A prime mover in which fluid under pressure enters a stationary nozzle where its pressure (potential) energy is converted to velocity (kinetic) energy and absorbed by the rotor. { 'im,pəls ¦tər,bən }

impulse welding [ENG] A welding process in which two layers of thermoplastic film are heated and fused to form a welded seam by clamping them together in close contact with a shielded electric heating element. { 'im,pəls 'weld·iŋ }

impulsive force See impact. { im'pəl·siv 'fôrs }

impulsive stimulated thermal scattering [ENG] An optical, noncontacting method for characterizing the high-frequency acoustic behavior of surfaces, thin membrane, coatings, and multilayer assemblies, in which picosecond pulses of light from an excitation laser stimulate motions which are then detected with a continuous-wave probing laser. Abbreviated ISTS. Also known as transient grating photoacoustics. { im¦pəl·siv ¦stim·yə,lād·əd ¦thərm·əl 'skad·ər·iŋ }

in. See inch.

in-and-out bond [CIV ENG] Masonry bond composed of vertically alternating stretchers and headers. { ¦in ən ¦aút 'bänd }

inboard [ENG] Toward or close to the longitudinal axis of a ship or aircraft. { 'in,bôrd }

inbond [CIV ENG] Pertaining to bricks or stones laid as headers across a wall. { 'in,bänd }

incandescent lamp [ELEC] An electric lamp that produces light when a metallic filament is heated white-hot in a vacuum by passing an electric current through the filament. Also known as filament lamp; light bulb. { ,in·kən'des·ənt 'lamp }

incentive operator [IND ENG] An employee whose wage is based on the quantity or quality of output. { in'sen·tiv ,äp·ə,rād·ər }

incentive wage system See wage incentive plan. { in'sen·tiv 'wāj ,sis·təm }

inch [MECH] A unit of length in common use in the United States and the United Kingdom, equal to 1/12 foot or 2.54 centimeters. Abbreviated in. { inch }

inch of mercury [MECH] The pressure exerted by a 1-inch-high (2.54-centimeter) column of mercury that has a density of 13.5951 grams per cubic centimeter when the acceleration of gravity has the standard value of 9.80665 m/s^2 or approximately 32.17398 ft/s^2 equal to 3386.388640341 pascals; used as a unit in the measurement of atmospheric pressure. { 'inch əv 'mər·kyə·rē }

incidental element See irregular element. { ¦in·sa¦dent·əl 'el·ə·mənt }

incinerator [ENG] A furnace or other container in which materials are burned. { in'sin·ə,rād·ər }

inclined cableway [MECH ENG] A monocable arrangement in which the track cable has a slope sufficiently steep to allow the carrier to run down under its own weight. { in'klīnd 'kā·bəl,wā }

inclined drilling [ENG] The drilling of blastholes at an angle with the vertical. { in'klīnd 'dril·iŋ }

inclined plane [MECH] A plane surface at an angle to some force or reference line. { 'in ,klīnd 'plān }

inclined-tube manometer [ENG] A glass-tube manometer with the leg inclined from the vertical to extend the scale for more minute readings. { in'klīnd ,tüb mə'näm·əd·ər }

inclinometer [ENG] **1.** An instrument that measures the attitude of an aircraft with respect to the horizontal. **2.** An instrument for measuring the angle between the earth's magnetic field vector and the horizontal plane. Also known as dip circle. **3.** An apparatus used to ascertain the direction of the magnetic field of the earth with reference to the plane of the horizon. { ,in·klə'näm·əd·ər }

incompetent rock [ENG] Soft or fragmented rock in which an opening, such as a borehole or an underground working place, cannot be maintained unless artificially supported by casing, cementing, or timbering. { in'käm·pəd·ənt 'räk }

incomplete lubrication |MECH ENG| Lubrication that takes place when the load on the rubbing surfaces is carried partly by a fluid viscous film and partly by areas of boundary lubrication; friction is intermediate between that of fluid and boundary lubrication. { ‚in·kəm'plēt ‚lü·brə 'kā·shən }

incompressibility |MECH| Quality of a substance which maintains its original volume under increased pressure. { ¦in·kəm‚pres·ə'bil·əd·ē }

increaser |ENG| An adapter for connecting a small-diameter pipe to a larger-diameter pipe. { in'krēs·ər }

incremental cost |IND ENG| **1.** The difference between the costs and the revenues between two alternative procedures. **2.** The cost of the last unit produced at a given level of production. { ‚iŋ·krə¦ment·əl 'kȯst }

indented bolt |DES ENG| A type of anchor bolt that has indentations to hold better in cemented grout. { in'den·təd 'bōlt }

independent chuck |DES ENG| A chuck for holding work by means of four jaws, each of which is moved independently of the others. { ‚in·də'pen·dənt 'chək }

independent contractor |ENG| One who exercises independent control over the mode and method of operations to produce the results demanded by the contract. { ‚in·də'pen·dənt 'kän ‚trak·tər }

independent footing |CIV ENG| A footing that supports a concentrated load, such as a single column. { ‚in·də'pen·dənt 'fud·iŋ }

independent suspension |MECH ENG| In automobiles, a system of springs and guide links by which wheels are mounted independently on the chassis. { ‚in·də'pen·dənt sə'spen·chən }

independent wire-rope core |DES ENG| A core of steel in a wire rope made in accordance with the best practice and design, either bright (uncoated) galvanized or drawn galvanized wire. { ‚in·də'pen·dənt 'wīr ‚rōp ‚kȯr }

indeterminate truss |CIV ENG| A truss having redundant bars. { ‚in·də'tərm·ə·nət 'trəs }

index center |MECH ENG| One of two machine-tool centers used to hold work and to rotate it by a fixed amount. { 'in‚deks ‚sen·tər }

index chart |MECH ENG| **1.** A chart used in conjunction with an indexing or dividing head, which correlates the index plate, hole circle, and index crank motion with the desired angular subdivisions. **2.** A chart indicating the arrangement of levers in a machine to obtain desired output speed or fuel rate. { 'in‚deks ‚chärt }

index counter |ENG| A counter indicating revolutions of the tape supply reel, making it possible to index selections within a reel of tape. { 'in ‚deks ‚kaunt·ər }

index crank |MECH ENG| The crank handle of an index head used to turn the spindle. { 'in ‚deks ‚krank }

index error |ENG| An error caused by the misalignment of the vernier and the graduated circle (arc) of an instrument. { 'in‚deks ‚er·ər }

index head |MECH ENG| A headstock that can be affixed to the table of a milling machine, planer, or shaper; work may be mounted on it by a chuck or centers, for indexing. { 'in‚deks ‚hed }

indexing |MECH ENG| The process of providing discrete spaces, parts, or angles in a workpiece by using an index head. { 'in‚dek·siŋ }

indexing fixture |MECH ENG| A fixture that changes position with regular steplike movements. { 'in‚dek·siŋ ‚fiks·chər }

index of work tolerance |IND ENG| A measure of the period of time during which an individual can perform a given task with the required efficiency while maintaining appropriate levels of physiological and emotional well-being. { 'in ‚deks əv 'wərk ‚täl·ə·rəns }

index plate |DES ENG| A plate with circular graduations or holes arranged in circles, each circle with different spacing; used for indexing on machines. { 'in‚deks ‚plāt }

index thermometer |ENG| A thermometer in which steel index particles are carried by mercury in the capillary and adhere to the capillary wall in the high and low positions, thus indicating minimum and maximum inertial scales. { 'in ‚deks thər'mäm·əd·ər }

indicated horsepower |MECH ENG| The horsepower delivered by an engine as calculated from the average pressure of the working fluid in the cylinders and the displacement. Abbreviated ihp. { 'in·də‚kād·əd 'hȯrs‚pau·ər }

indicating gage |ENG| A gage consisting essentially of a case and mounting, a spindle carrying the contact point, an amplifying mechanism, a pointer, and a graduated dial; used to amplify and measure the displacement of a movable contact point. { 'in·də‚kād·iŋ ‚gāj }

indicating instrument |ENG| An instrument in which the present value of the quantity being measured is visually indicated. { 'in·də‚kād·iŋ ‚in·strə·mənt }

indication |ENG| In ultrasonic testing, determination of the presence of a flaw by detection of a reflected ultrasonic beam. { ‚in·də'kā·shən }

indicator |ELECTR| A cathode-ray tube or other device that presents information transmitted or relayed from some other source, as from a radar receiver. |ENG| An instrument for obtaining a diagram of the pressure-volume changes in a running positive-displacement engine, compressor, or pump cylinder during the working cycle. { 'in·də‚kād·ər }

indicator card |ENG| A chart on which an indicator diagram is produced by an instrument called an engine indicator which traces the real-performance cycle diagram as the machine is running. { 'in·də‚kād·ər ‚kärd }

indicator diagram |ENG| A pressure-volume diagram representing and measuring the work done by or on a fluid while performing the work cycle in a reciprocating engine, pump, or compressor cylinder. { 'in·də‚kād·ər 'dī·ə‚gram }

indicator unit |ENG| An instrument which detects the presence of an electrical quantity without necessarily measuring it. { 'in·də‚kād·ər ‚yü·nət }

indifferent stability See neutral stability. { in'dif·ərnt stə'bil·əd·ē }

indirect-arc furnace |ENG| A refractory-lined furnace in which the burden is heated indirectly by the radiant heat from an electric arc. { ‚in·də'rekt ‚ärk 'fər·nəs }

indirect cost |IND ENG| A cost that is not readily indentifiable with or chargeable to a specific product or service. { ‚in·də'rekt kóst }

indirect heater |ENG| A vessel containing equipment in which heat generated by a primary source is transferred to a fluid or solid which then serves as the heating medium. { ‚in·də'rekt 'hēd·ər }

indirect labor |IND ENG| Labor not directly engaged in the actual production of the product or performance of a service. { ‚in·də'rekt 'lā·bər }

indirect lighting |ENG| A system of lighting in which more than 90% of the light from luminaires is distributed upward toward the ceiling, from which it is diffusely reflected. { ‚in·də'rekt 'līd·iŋ }

indirect material |IND ENG| Any material used in the manufacture of a product which does not itself become a part of the product and whose cost is indirect. { ‚in·də'rekt mə'tir·ē·əl }

individual distributed numerical control |CONT SYS| A form of distributed numerical control involving only a few machines, each of which operates independently of the others and is unaffected by their failures. { ‚in·də'vij·ə·wəl di 'strib·yəd·əd nü'mer·ə·kəl kən'trōl }

induced dipole |ELEC| An electric dipole produced by application of an electric field. { in 'düst 'dī‚pōl }

induced draft |MECH ENG| A mechanical draft produced by suction stream jets or fans at the point where air or gases leave a unit. { in 'düst 'draft }

induced-draft cooling tower |MECH ENG| A structure for cooling water by circulating air where the load is on the suction side of the fan. { in'düst ‚draft 'kül·iŋ ‚taú·ər }

induced moment |ELEC| The average electric dipole moment per molecule which is produced by the action of an electric field on a dielectric substance. { in'düst 'mō·mənt }

inductance See coil. { in'dək·təns }

inductance coil See coil. { in'dək·təns ‚kóil }

induction See electrostatic induction. { in'dək·shən }

induction burner |ENG| Fuel-air burner into which the fuel is fed under pressure to entrain needed air into the combustion nozzle area. { in'dək·shən ‚bər·nər }

induction charging |ELEC| Production of electric charge on a body by means of electrostatic induction. { in'dək·shən ‚chär·jiŋ }

induction-electrical survey |ENG| Study of subterranean formations by combined induction and electrical logging. { in'dək·shən i‚lek·trə· kəl 'sər‚vā }

induction flowmeter |ENG| An instrument for measuring the flow of a conducting liquid passing through a tube, in which the tube is placed in a transverse magnetic field and the induced electromotive force between electrodes at opposite ends of a diameter of the tube perpendicular to the field is measured. { in'dək·shən 'flō ‚mēd·ər }

induction furnace |ENG| An electric furnace in which heat is produced in a metal charge by electromagnetic induction. { in'dək·shən ‚fər· nəs }

induction generator |ELEC| A nonsynchronous alternating-current generator whose construction is identical to that of an ac motor, and which is driven above synchronous speed by external sources of mechanical power. { in'dək·shən ‚jen·ə·rād·ər }

induction heating |ENG| Increasing the temperature in a material by induced electric current. Also known as eddy-current heating. { in'dək· shən ‚hēd·iŋ }

induction inclinometer See earth inductor. { in'dək·shən ‚in·klə'näm·əd·ər }

induction instrument |ENG| Meter that depends for its operation on the reaction between magnetic flux set up by current in fixed windings, and other currents set up by electromagnetic induction in conducting parts of the moving system. { in'dək·shən ‚in·strə·mənt }

induction log |ENG| An electric log of the conductivity of rock with depth obtained by lowering into an uncased borehole a generating coil that induces eddy currents on the rocks and these are detected by a receiver coil. { in'dək·shən ‚läg }

induction loudspeaker |ENG ACOUS| Loudspeaker in which the current which reacts with the steady magnetic field is induced in the moving member. { in'dək·shən ‚laúd‚spēk·ər }

induction motor |ELEC| An alternating-current motor in which a primary winding on one member (usually the stator) is connected to the power source, and a secondary winding on the other member (usually the rotor) carries only current induced by the magnetic field of the primary. { in'dək·shən ‚mōd·ər }

induction pump |MECH ENG| Any pump operated by electromagnetic induction. { in'dək· shən ‚pəmp }

induction salinometer |ENG| A device for measuring salinity by taking voltage readings of the current in seawater. { in'dək·shən ‚sal·ə'näm· əd·ər }

induction silencer |ENG| A device for reducing engine induction noise, which consists essentially of a low-pass acoustic filter with the inertance of the air-entrance tube and the acoustic compliance of the annular and central volumes providing acoustic filtering elements. { in'dək· shən ‚sī'lən·sər }

induction valve See inlet valve. { in'dək·shən ‚valv }

inductive charge |ELEC| The charge that exists

on an object as a result of its being near another charged object. { in'dək·tiv 'chärj }

inductive circuit [ELEC] A circuit containing a higher value of inductive reactance than capacitive reactance. { in'dək·tiv 'sər·kət }

inductive coupler [ELEC] A mutual inductance that provides electrical coupling between two circuits; used in radio equipment. { in'dək·tiv 'kəp·lər }

inductive coupling [ELEC] Coupling of two circuits by means of the mutual inductance provided by a transformer. Also known as transformer coupling. { in'dək·tiv 'kəp·liŋ }

inductive grounding [ELEC] Use of grounding connections containing an inductance in order to reduce the magnitude of short-circuit currents created by line-to-ground faults. { in'dək·tiv 'graúnd·iŋ }

inductive load [ELEC] A load that is predominantly inductive, so that the alternating load current lags behind the alternating voltage of the load. Also known as lagging load. { in'dək·tiv 'lōd }

inductive reactance [ELEC] Reactance due to the inductance of a coil or circuit. { in'dək·tiv rē'ak·təns }

inductive superconducting fault-current limiter See shielded-core superconducting fault-current limiter. { in'dək·tiv ,sü·pər·kən'dək·tiŋ 'fólt ,cər·ənt ,lim·əd·ər }

inductive susceptance [ELEC] In a circuit containing almost no resistance, the part of the susceptance due to inductance. { in'dək·tiv sə'sep·təns }

inductive waveform [ELEC] A graph or trace of the effect of current buildup across an inductive network; proportional to the exponential of the product of a negative constant and the time. { in'dək·tiv 'wāv,fórm }

inductor See coil. { in'dək·tər }

inductor microphone [ENG ACOUS] Moving-conductor microphone in which the moving element is in the form of a straight-line conductor. { in'dək·tər 'mī·krə,fōn }

inductor tachometer [ENG] A type of impulse tachometer in which the rotating member, consisting of a magnetic material, causes the magnetic flux threading a circuit containing a magnet and a pickup coil to rise and fall, producing pulses in the circuit which are rectified for a permanent-magnet, movable-coil instrument. { in'dək·tər tə'käm·əd·ər }

inductosyn [CONT SYS] A resolver whose output phase is proportional to the shaft angle. { in'dək·tə,sin }

inductrack [ENG] A magnetic levitation concept for trains and other moving objects that uses special arrays of permanent magnets to achieve levitation forces, and is inherently stable. { in'dək,trak }

industrial anthropometry [IND ENG] Application of the knowledge of physical anthropology to the design and construction of equipment for human use, such as automobiles. { in'dəs·trē·əl ¦an·thrə'päm·ə·trē }

industrial car [IND ENG] Any of various narrow-gage railcars used for indoor or outdoor handling of bulk and package materials. { in'dəs·trē·əl 'kär }

industrial cost control [IND ENG] A specific system or procedure used to keep manufacturing costs in line. Also known as cost control. { in'dəs·trē·əl 'kóst kən,trōl }

industrial ecology [IND ENG] The development and use of industrial processes that result in products based on simultaneous consideration of product functionality and competitiveness, natural-resource conservation, and environmental preservation. Also known as design for environment; green design. { in¦dəs·trē·əl ē'käl·ə·jē }

industrial engineering [ENG] A branch of engineering concerned with the design, improvement, and installation of integrated systems of people, materials, and equipment. Also known as management engineering. { in'dəs·trē·əl ,en·jə'nir·iŋ }

industrial mobilization [IND ENG] Transformation of industry and other productive facilities and contributory services from their peacetime activities to the fulfillment of the munitions program necessary to support a military effort. { in'dəs·trē·əl ,mō·bə·lə'zā·shən }

industrial railway [IND ENG] **1.** A usually short feeder line that is either owned or controlled and wholly operated by an industrial firm. **2.** Narrow-gage rail lines used on construction jobs or around industrial plants. { in'dəs·trē·əl 'rāl,wā }

industrial revolution [IND ENG] A widespread change in industrial or production methods, toward production by machine and away from manual labor. { in'dəs·trē·əl ,rēv·ə'lü·shən }

industrial security [IND ENG] The portion of internal security which refers to the protection of industrial installations, resources, utilities, materials, and classified information essential to protection from loss or damage. { in'dəs·trē·əl si'kyúr·əd·ē }

industrial truck [ENG] A manually propelled or powered wheeled vehicle for transporting materials over level or slightly inclined running surfaces in a manufacturing or warehousing facility. { in'dəs·trē·əl 'trək }

industrial waste [ENG] Worthless materials remaining from industrial operations. { in'dəs·trē·əl 'wāst }

inelastic [MECH] Not capable of sustaining a deformation without permanent change in size or shape. { ,in·ə'las·tik }

inelastic buckling [MECH] Sudden increase of deflection or twist in a column when compressive stress reaches the elastic limit but before elastic buckling develops. { ,in·ə'las·tik 'bək·liŋ }

inelastic collision [MECH] A collision in which the total kinetic energy of the colliding particles is not the same after the collision as before it. { ,in·ə'las·tik kə'lizh·ən }

inelastic stress [MECH] A force acting on a

solid which produces a deformation such that the original shape and size of the solid are not restored after removal of the force. { ,in·ə'las·tik 'stres }

inequality of Clausius See Clausius inequality. { ,in·i'kwäl·əd·ē əv 'klaú·zē·əs }

inert atmosphere |CHEM ENG| A nonreactive gas atmosphere, such as nitrogen, carbon dioxide, or helium; used to blanket reactive liquids in storage, to purge process lines and vessels of reactive gases and liquids, and to cover a reaction mix in a partially filled vessel. { i'nərt 'at·mə,sfir }

inert-gas blanketing |ENG| Purging the air from a unit of a heat exchanger by using an inert gas as the unit is being shut down. { i¦nərt ,gas 'blaŋ·kəd·iŋ }

inertia |MECH| That property of matter which manifests itself as a resistance to any change in the momentum of a body. { i'nər·shə }

inertia ellipsoid |MECH| An ellipsoid used in describing the motion of a rigid body; it is fixed in the body, and the distance from its center to its surface in any direction is inversely proportional to the square root of the moment of inertia about the corresponding axis. Also known as Poinsot ellipsoid. { i'nər·shə i'lip,sóid }

inertia governor |MECH ENG| A speed-control device utilizing suspended masses that respond to speed changes by reason of their inertia. { i'nər·shə ,gəv·ə·nər }

inertial coordinate system See inertial reference frame. { i'nər·shəl kō'órd·ən,at ,sis·təm }

inertial force |MECH| The fictitious force acting on a body as a result of using a noninertial frame of reference; examples are the centrifugal and Coriolis forces that appear in rotating coordinate systems. Also known as effective force. { i'nər·shəl 'fórs }

inertial mass |MECH| The mass of an object as determined by Newton's second law, in contrast to the mass as determined by the proportionality to the gravitational force. { i'nər·shəl 'mas }

inertial reference frame |MECH| A coordinate system in which a body moves with constant velocity as long as no force is acting on it. Also known as inertial coordinate system. { i'nər·shəl 'ref·rəns ,frām }

inertia matrix |MECH| A matrix **M** used to express the kinetic energy T of a mechanical system during small displacements from an equilibrium position, by means of the equation $T = \frac{1}{2} \dot{q}^T M \dot{q}$, where $\dot{q}$ is the vector whose components are the derivatives of the generalized coordinates of the system with respect to time, and $\dot{q}^T$ is the transpose of $\dot{q}$. { i'nər·shə ,mā·triks }

inertia starter |MECH ENG| A device utilizing inertial principles to start the rotator of an internal combustion engine. { i'nər·shə ¦stärd·ər }

inertia tensor |MECH| A tensor associated with a rigid body whose product with the body's rotation vector yields the body's angular momentum. { i'nər·shə ,ten·sər }

inert primer |ENG| A cylinder which enshrouds

a detonator but does not interfere with the detonation of the explosive charge. { i'nərt 'prī·mər }

inert retarder |CIV ENG| A braking device built into a railroad track and operating without an external source of power that reduces car speed by means of brake shoes applied to the lower sides of the wheels. { i¦nərt ri'tär·dər }

inextensional deformation |MECH| A bending of a surface that leaves unchanged the length of any line drawn on the surface and the curvature of the surface at each point. { in,ek'sten·chən·əl ,def·ər'mā·shən }

in-feed centerless grinding |MECH ENG| A metal-cutting process by which a cylindrical workpiece is ground to a prescribed surface smoothness and diameter by the insertion of the workpiece between a grinding wheel and a canted regulating wheel; the rotation of the regulating wheel controls the rotation and feed rate of the workpiece. { 'in,fēd ¦sen·tər,les 'grīnd·iŋ }

inferential flow meter |ENG| A flow meter in which the flow is determined by measurement of a phenomenon associated with the flow, such as a drop in static pressure at a restriction in a pipe, or the rotation of an impeller or rotor, rather than measurement of the actual mass flow. { ¦in·fə¦ren·chəl 'flō ,mēd·ər }

inferential liquid-level meter |ENG| A liquid-level meter in which the level of a liquid is determined by measurement of some phenomenon associated with this level, such as the buoyancy of a solid partly immersed in the liquid, the pressure at a certain level, the conductance of the liquid, or its absorption of gamma radiation, rather than by direct measurement. { ¦in·fə¦ren·chəl ¦lik·wəd 'lev·əl ,mēd·ər }

infiltration |ENG| Leakage of outdoor air into a building by natural forces, for example, by seepage through cracks or other openings. { ,in·fil 'trā·shən }

infiltration gallery |CIV ENG| A large, horizontal underground conduit of perforated or porous material with openings on the sides for collecting percolating water by infiltration. { ,in·fil 'trā·shən ,gal·rē }

infinite baffle |ENG ACOUS| A loudspeaker baffle which prevents interaction between the front and back radiation of the loudspeaker. { 'in·fə·nət 'baf·əl }

infinite-capacity loading |CONT SYS| The deliberate overloading of a robotic work center with excessive force or weight in order to determine the overload protection necessary to maintain proper load conditions. { 'in·fə·nət kə'pas·əd·ē ,lōd·iŋ }

inflatable gasket |DES ENG| A gasket whose seal is activated by inflation with compressed air. { in¦flād·ə·bəl 'gas·kət }

inflated |ENG| Filled or distended with air or gas. { in'flād·əd }

inflected arch See inverted arch. { in'flek·təd 'ärch }

influence diagram |SYS ENG| A graph-theoretic

representation of a decision, which may include four types of nodes (decision, chance, value, and deterministic), directed arcs between the nodes (which identify dependencies between them), a marginal or conditional probability distribution defined at each chance node, and a mathematical function associated with each of the other types of node. { 'in,flü·əns ,dī·ə,gram }

influence line |MECH| A graph of the shear, stress, bending moment, or other effect of a movable load on a structural member versus the position of the load. { 'in,flü·əns ,līn }

information process analysis chart See form process chart. { ,in·fər'mā·shən ¦prä·ses ə¦nal·ə·səs ,chärt }

information systems engineering |ENG| The discipline concerned with the design, development, testing, and maintenance of information systems. { ,in·fər¦mā·shən ¦sis·təmz ,en·jə'nir·iŋ }

infrared array |ENG| A collection of several thousand infrared detector elements arranged in a grid pattern and connected to readout electronics to display infrared images focused on the array by an astronomical telescope. { ¦in·frə¦red ə'rā }

infrared-emitting diode |ELECTR| A light-emitting diode that has maximum emission in the near-infrared region, typically at 0.9 micrometer for *pn* gallium arsenide. { ¦in·frə¦red i¦mid·iŋ 'dī,ōd }

infrared heating |ENG| Heating by means of infrared radiation. { ¦in·frə¦red 'hēd·iŋ }

infrared homing |ENG| Homing in which the target is tracked by means of its emitted infrared radiation. { ¦in·frə¦red 'hōm·iŋ }

infrared imaging device |ENG| Any device which converts an invisible infrared image into a visible image. { ¦in·frə¦red 'im·ə·jiŋ di,vīs }

infrared thermography |ENG| A method of measuring surface temperatures by observing the infrared emission from the surface. { ,in·frə¦red thər'mäg·rə·fē }

infrared thermometer |ENG| An instrument that focuses and detects the infrared radiation emitted by an object in order to determine its temperature. { ¦in·frə·'red thər'mäm·əd·ər }

Ingen-Hausz apparatus |THERMO| An apparatus for comparing the thermal conductivities of different conductors; specimens consisting of long wax-coated rods of equal length are placed with one end in a tank of boiling water covered with a radiation shield, and the lengths along the rods from which the wax melts are compared. { ¦iŋ·gən 'haús ,ap·ə,rad·əs }

inhabited building distance |ENG| The minimum distance permitted between an ammunition or explosive location and any building used for habitation or where people are accustomed to assemble, except operating buildings or magazines. { in'hab·əd·əd ¦bil·diŋ ,dis·təns }

inhaul cable |MECH ENG| In a cable excavator, the line that pulls the bucket to dig and bring in soil. Also known as digging line. { 'in,hól ,kā·bəl }

inherent damping |MECH ENG| A method of vibration damping which makes use of the mechanical hysteresis of such materials as rubber, felt, and cork. { in'hir·ənt 'dam·piŋ }

inherent noise pressure See equivalent noise pressure. { in'hir·ənt 'nóiz ,presh·ər }

inhibitor sweetening |CHEM ENG| Petroleum-refinery treating process to sweeten gasoline (convert mercaptans to disulfides) of low mercaptan content; uses a phenylenediamine inhibitor, air, and caustic. { in'hib·əd·ər ,swēt·ən·iŋ }

in-house |IND ENG| Pertaining to an operation produced or carried on within a plant or organization, rather than done elsewhere under contract. { ,in,haús }

initial boiling point |CHEM ENG| According to American Society for Testing and Materials petroleum-analysis distillation procedures, the recorded temperature when the first drop of distilled vapor is liquefied and falls from the end of the condenser. { i'nish·əl 'bóil·iŋ ,póint }

initial free space |MECH| In interior ballistics, the portion of the effective chamber capacity not displaced by propellant. { i'nish·əl ¦frē 'spās }

initial shot start pressure |MECH| In interior ballistics, the pressure required to start the motion of the projectile from its initial loaded position; in fixed ammunition, it includes pressure required to separate projectile and cartridge case and to start engraving the rotating band. { i'nish·əl 'shät ¦stärt ,presh·ər }

initial yaw |MECH| The yaw of a projectile the instant it leaves the muzzle of a gun. { i'nish·əl 'yó }

injection |ELECTR| **1.** The method of applying a signal to an electronic circuit or device. **2.** The process of introducing electrons or holes into a semiconductor so that their total number exceeds the number present at thermal equilibrium. |MECH ENG| The introduction of fuel, fuel and air, fuel and oxidizer, water, or other substance into an engine induction system or combustion chamber. { in'jek·shən }

injection blow molding |ENG| Plastics molding process in which a hollow-plastic tube is formed by injection molding. { in'jek·shən 'blō ,mól·diŋ }

injection carburetor |MECH ENG| A carburetor in which fuel is delivered under pressure into a heated part of the engine intake system. Also known as pressure carburetor. { in'jek·shən 'kär·bə,rād·ər }

injection efficiency |ELECTR| A measure of the efficiency of a semiconductor junction when a forward bias is applied, equal to the current of injected minority carriers divided by the total current across the junction. { in'jek·shən ə,fish·ən·sē }

injection electroluminescence |ELECTR| Radiation resulting from recombination of minority charge carriers injected in a *pn* or *pin* junction that is biased in the forward direction. Also known as Lossev effect; recombination electroluminescence. { in'jek·shən i¦lek·trō,lü·mə 'nes·əns }

injection locking [ELECTR] The capture or synchronization of a free-running oscillator by a weak injected signal at a frequency close to the natural oscillator frequency or to one of its subharmonics; used for frequency stabilization in IMPATT or magnetron microwave oscillators, gas-laser oscillators, and many other types of oscillators. { in'jek·shən ¦läk·iŋ }

injection luminescent diode [ELECTR] Gallium arsenide diode, operating in either the laser or the noncoherent mode, that can be used as a visible or near-infrared light source for triggering such devices as light-activated switches. { in 'jek·shən ¦lü·mə¦nes·ənt 'dī,ōd }

injection mold [ENG] A plastics mold into which the material to be formed is introduced from an exterior heating cylinder. { in'jek·shən ,mōld }

injection molding [ENG] Molding metal, plastic, or nonplastic ceramic shapes by injecting a measured quantity of the molten material into dies. { in'jek·shən 'mōl·diŋ }

injection pump [MECH ENG] A pump that forces a measured amount of fuel through a fuel line and atomizing nozzle in the combustion chamber of an internal combustion engine. { in'jek·shən 'pəmp }

injection ram [ENG] In injection molding, the ram that applies pressure to the feed plunger in the process of either injection or transfer molding. { in'jek·shən ,ram }

injection signal [ENG ACOUS] The sawtooth frequency-modulated signal which is added to the first detector circuit for mixing with the incoming target signal. { in'jek·shən ,sig·nəl }

injector [ELECTR] An electrode through which charge carriers (holes or electrons) are forced to enter the high-field region in a spacistor. [MECH ENG] **1.** An apparatus containing a nozzle in an actuating fluid which is accelerated and thus entrains a second fluid, so delivering the mixture against a pressure in excess of the actuating fluid. **2.** A plug with a valved nozzle through which fuel is metered to the combustion chambers in diesel- or full-injection engines. **3.** A jet through which feedwater is injected into a boiler, or fuel is injected into a combustion chamber. { in'jek·tər }

injector torch See low-pressure torch. { in'jek·tər ,tórch }

inkometer [ENG] An instrument for measuring adhesion of liquids by rotating drums in contact with the liquid. { iŋ'käm·əd·ər }

inlet [ENG] An entrance or orifice for the admission of fluid. { 'in,let }

inlet box [MECH ENG] A closure at the fan inlet or inlets in a boiler for attachment of the fan to the duct system. { 'in,let ,bäks }

inlet valve [MECH ENG] The valve through which a fluid is drawn into the cylinder of a positive-displacement engine, pump, or compressor. Also known as induction valve. { 'in ,let ,valv }

in line [ENG] **1.** Over the center of a borehole and parallel with its long axis. **2.** Of a drill motor, mounted so that its drive shaft and the drive rod in the drill swivel head are parallel, or mounted so that the shaft driving the drill-swivel-head bevel gear and the drill-motor drive shaft are centered in a direct line and parallel with each other. **3.** Having similar units mounted together in a line. { 'in ¦līn }

in-line assembly machine [IND ENG] An assembly machine that inserts components into a wiring board one at a time as the board is moved from station to station by a conveyor or other transport mechanism. { 'in ¦līn ə¦sem·blē mə,shēn }

in-line engine [MECH ENG] A multiple-cylinder engine with cylinders aligned in a row. { 'in ¦līn 'en·jən }

in-line equipment [ENG] **1.** A sequence of equipment or processing items mounted along the same vertical or horizontal plane. **2.** Equipment mounted within a process line, such as an in-line pump, pressure-drop flowmeter, or nozzle mixer. { 'in ¦līn i'kwip·mənt }

in-line linkage [MECH ENG] A power-steering linkage which has the control valve and actuator combined in a single assembly. { 'in ¦līn 'liŋ·kij }

innage [ENG] The volume or the measured height of liquid introduced into a tank or container. { 'in·ij }

inner barrel See inner tube. { ¦in·ər ¦bar·əl }

inner hearth See back hearth. { ¦in·ər 'härth }

inner tube [ENG] A rubber tube used inside a pneumatic tire casing to hold air under pressure. Also known as tube. { 'in·ər ,tüb }

in-phase component [ELEC] The component of the phasor representing an alternating current which is parallel to the phasor representing voltage. { 'in ,fāz kəm'pō·nənt }

in-place value [IND ENG] The site value of property, that is, the market value of equipment plus costs of transportation to the site and subsequent installation. { ¦in,plās 'val·yü }

input [ELECTR] **1.** The power or signal fed into an electrical or electronic device. **2.** The terminals to which the power or signal is applied. { 'in,pùt }

input/output relation [SYS ENG] The relation between two vectors whose components are the inputs (excitations, stimuli) of a system and the outputs (responses) respectively. { 'in,pùt 'aùt ,pùt ri,lā·shən }

insensitive time See dead time. { in'sen·sə·tiv ,tīm }

insert bit [DES ENG] A bit into which inset cutting points of various preshaped pieces of hard metal (usually a sintered tungsten carbide-cobalt powder alloy) are brazed or hand-peened into slots or holes cut or drilled into a blank bit. Also known as slug bit. { 'in,sərt ,bit }

inserted-tooth cutter [DES ENG] A milling cutter in which the teeth can be replaced. { in'sərd·əd ,tüth 'kəd·ər }

insertion meter [ENG] A type of flowmeter

which measures the rotation rate of a small propeller or turbine rotor mounted at right angles to the end of a support rod and inserted into the flowing stream or closed pipe. { in'sər·shən ‚mēd·ər }

inside caliper [DES ENG] A caliper that has two legs with feet that turn outward; used to measure inside dimensions, as the diameter of a hole. { 'in‚sīd 'kal·ə·pər }

inside diameter [DES ENG] The length of a line which passes through the center of a hollow cylindrical or spherical object, and whose end points lie on the inner surface of the object. Abbreviated ID. { 'in‚sīd dī'am·əd·ər }

inside face [DES ENG] That part of the bit crown nearest to or parallel with the inside wall of an annular or coring bit. { 'in‚sīd ¦fās }

inside gage [DES ENG] The inside diameter of a bit as measured between the cutting points, such as between inset diamonds on the inside-wall surface of a core bit. { 'in‚sīd ¦gāj }

inside micrometer [DES ENG] A micrometer caliper with the points turned outward for measuring the internal dimensions of an object. { 'in‚sīd mī'kräm·əd·ər }

inside work See internal work. { 'in‚sīd ‚wərk }

in situ foaming [ENG] Depositing of the ingredients of a foamable plastic onto the location where foaming is to take place; for example, in situ foam insulation on equipment or walls. { in 'si·chü 'fōm·iŋ }

inspect [IND ENG] To examine an object to determine whether it conforms to standards; may employ sight, hearing, touch, odor, or taste. { in'spekt }

inspection [IND ENG] The critical examination of a product to determine its conformance to applicable quality standards or specifications. { in'spek·shən }

inspection by variables [IND ENG] A quality-control inspection method in which the sampled articles are evaluated on the basis of quantitative criteria. { in‚spek·shən bī 'ver·ē·ə·bəlz }

instability [CONT SYS] A condition of a control system in which excessive positive feedback causes persistent, unwanted oscillations in the output of the system. { ‚in·stə'bil·əd·ē }

installation [ENG] Procedures for setting up equipment for use or service. { ‚in·stə'lā·shən }

instantaneous axis [MECH] The axis about which a rigid body is carrying out a pure rotation at a given instant in time. { ‚in·stən¦tā·nē·əs 'ak·səs }

instantaneous center [MECH] A point about which a rigid body is rotating at a given instant in time. Also known as instant center. { ‚in·stən¦tā·nē·əs 'sen·tər }

instantaneous cut [ENG] A cut that is set off by instantaneous detonators to be certain that all charges in the cut go off at the same time; the drilling and ignition are carried out so that all the holes break smaller top angles. { ‚in·stən¦tā·nē·əs 'kət }

instantaneous detonator [ENG] A type of detonator that does not have a delay period between the passage of the electric current through the detonator and its explosion. { ‚in·stən¦tā·nē·əs 'det·ən‚ād·ər }

instantaneous fuse [ENG] A fuse with an ignition rate of several thousand feet per minute; an example is PETN. { ‚in·stən¦tā·nē·əs 'fyüz }

instantaneous recording [ENG ACOUS] A recording intended for direct reproduction without further processing. { ‚in·stən¦tā·nē·əs ri'kȯrd·iŋ }

instantaneous recovery [MECH] The immediate reduction in the strain of a solid when a stress is removed or reduced, in contrast to creep recovery. { ‚in·stən¦tā·nē·əs ri'kəv·ə·rē }

instantaneous strain [MECH] The immediate deformation of a solid upon initial application of a stress, in contrast to creep strain. { ‚in·stən¦tā·nē·əs 'strān }

instant center See instantaneous center. { 'in·stənt 'sen·tər }

instruction card [IND ENG] A written description of the standard method used by a worker, to guide his activities. { in'strək·shən ‚kärd }

instrument [ENG] A device for measuring and sometimes also recording and controlling the value of a quantity under observation. { 'in·strə·mənt }

instrumental analysis [ENG] The use of an instrument to measure a component, to detect the completion of a quantitative reaction, or to detect a change in the properties of a system. { ‚in·strə'ment·əl ə'nal·ə·səs }

instrumentation [ENG] Designing, manufacturing, and utilizing physical instruments or instrument systems for detection, observation, measurement, automatic control, automatic computation, communication, or data processing. { ‚in·strə·men'tā·shən }

instrument correction [ENG] A correction of measurements made on a unit under test for either inaccuracy of the instrument or eroding effect of the instrument. { 'in·strə·mənt kə‚rek·shən }

instrument housing [ENG] A case or enclosure to cover and protect an instrument. { 'in·strə·mənt ‚haů·ziŋ }

instrument panel [ENG] A panel or board containing indicating meters. { 'in·strə·mənt ‚pan·əl }

instrument reading time [ENG] The time, after a change in a measured quantity, which it takes for the indication of an instrument to come and remain within a specified percentage of its final value. { 'in·strə·mənt 'rēd·iŋ ‚tīm }

instrument science [ENG] The systematically organized body of general concepts and principles underlying the design, analysis, and application of instruments and instrument systems. { 'in·strə·mənt ‚sī·əns }

instrument shelter [ENG] A boxlike structure designed to protect certain meteorological instruments from exposure to direct sunshine, precipitation, and condensation, while providing adequate ventilation. Also known as thermometer screen; thermometer shelter; thermoscreen. { 'in·strə·mənt ‚shel·tər }

instrument system |ENG| A system which integrates one or more instruments with auxiliary or associated devices for detection, observation, measurement, automatic control, automatic computation, communication, or data processing. { 'in·strə·mənt ‚sis·təm }

insulated |ELEC| Separated from other conducting surfaces by a nonconducting material. { 'in·sə‚lād·əd }

insulated-gate bipolar transistor |ELECTR| A power semiconductor device that combines low forward voltage drop, gate-controlled turnoff, and high switching speed. It structurally resembles a vertically diffused MOSFET, featuring a double diffusion of a p-type region and an n-type region, but differs from the MOSFET in the use of a $p+$ substrate layer (in the case of an n-channel device) for the drain. The effect is to change the transistor into a bipolar device, as this p-type region injects holes into the n-type drift region. Abbreviated IGBT. { 'in·sə‚lād·əd‚gāt bī̄‚pō·lər tran'zis·tər }

insulated-gate field-effect transistor *See* metal oxide semiconductor field-effect transistor. { 'in·sə‚lād·əd ‚gāt ‚fēld i‚fekt tran'zis·tər }

insulated-substrate monolithic circuit |ELECTR| Integrated circuit which may be either an all-diffused device or a compatible structure so constructed that the components within the silicon substrate are insulated from one another by a layer of silicon dioxide, instead of reverse-biased pn junctions used for isolation in other techniques. { 'in·sə‚lād·əd ‚səb‚strāt ‚män·ə‚lith·ik 'sər·kət }

insulating strength |ELEC| Measure of the ability of an insulating material to withstand electric stress without breakdown; it is defined as the voltage per unit thickness necessary to initiate a disruptive discharge; usually measured in volts per centimeter. { 'in·sə‚lād·iŋ ‚streŋkth }

insulation |BUILD| Material used in walls, ceilings, and floors to retard the passage of heat and sound. |ELEC| A material having high electrical resistivity and therefore suitable for separating adjacent conductors in an electric circuit or preventing possible future contact between conductors. Also known as electrical insulation. { ‚in·sə'lā·shən }

insulation resistance |ELEC| The electrical resistance between two conductors separated by an insulating material. { ‚in·sə'lā·shən ri‚zis·təns }

insulation sampler |ENG| A device for collecting deep water which prevents any significant conduction of heat from the water sample so that it maintains its original temperature as it is hauled to the surface. { ‚in·sə'lā·shən ‚sam·plər }

insulation testing set |ENG| An instrument for measuring insulation resistance, consisting of a high-range ohmmeter having a hand-driven direct-current generator as its voltage source. { ‚in·sə'lā·shən 'test·iŋ ‚set }

insulator |ELEC| A device having high electrical resistance and used for supporting or separating conductors to prevent undesired flow of current from them to other objects. Also known as electrical insulator. { 'in·sə‚lād·ər }

Intake |ENG| **1.** An entrance for air, water, fuel, or other fluid, or the amount of such fluid taken in. **2.** A main passage for air in a mine. { 'in‚tāk }

intake chamber |CIV ENG| A large chamber that gradually narrows to an intake tunnel; designed to avoid undesirable water currents. { 'in‚tāk ‚chām·bər }

intake gate |CIV ENG| A movable partition for opening or closing a water intake opening. { 'in‚tāk ‚gāt }

intake manifold |MECH ENG| A system of pipes which feeds fuel to the various cylinders of a multicylinder internal combustion engine. { 'in‚tāk ‚man·ə‚fōld }

intake stroke |MECH ENG| The fluid admission phase or travel of a reciprocating piston and cylinder mechanism as, for example, in an engine, pump, or compressor. { 'in‚tāk ‚strōk }

intake valve |MECH ENG| The valve which opens to allow air or an air-fuel mixture to enter an engine cylinder. { 'in‚tāk ‚valv }

integer programming |SYS ENG| A series of procedures used in operations research to find maxima or minima of a function subject to one or more constraints, including one which requires that the values of some or all of the variables be whole numbers. { 'int·ə·jər 'prō‚gram·iŋ }

integrable system |MECH| A dynamical system whose motion is governed by an integrable differential equation. { ‚int·i·grə·bəl ‚sis·təm }

integral action |CONT SYS| A control action in which the rate of change of the correcting force is proportional to the deviation. { 'int·ə·grəl ‚ak·shən }

integral compensation |CONT SYS| Use of a compensator whose output changes at a rate proportional to its input. { 'int·ə·grəl ‚käm·pən'sā·shən }

integral control |CONT SYS| Use of a control system in which the control signal changes at a rate proportional to the error signal. { 'int·ə·grəl kən‚trōl }

integral-furnace boiler |MECH ENG| A type of steam boiler which incorporates furnace water-cooling in the circulatory system. { 'int·ə·grəl ‚fər·nəs ‚bȯil·ər }

integral-mode controller |CONT SYS| A controller which produces a control signal proportional to the integral of the error signal. { 'int·ə·grəl ‚mōd kən‚trōl·ər }

integral network |CONT SYS| A compensating network which produces high gain at low input frequencies and low gain at high frequencies, and is therefore useful in achieving low steady-state errors. Also known as lagging network; lag network. { 'int·ə·grəl 'net‚wərk }

integral square error [CONT SYS] A measure of system performance formed by integrating the square of the system error over a fixed interval of time; this performance measure and its generalizations are frequently used in linear optimal control and estimation theory. { 'int·ə·grəl ¦skwer ¦er·ər }

integral-type flange [DES ENG] A flange which is forged or cast with, or butt-welded to, a nozzle neck, pressure vessel, or piping wall. { 'int·ə·grəl ¦tīp 'flanj }

integral waterproofing [ENG] Waterproofing concrete by adding the waterproofing material to the cement or to the mixing water. { 'int·ə·grəl 'wȯd·ər,prüf·iŋ }

integraph [ENG] A device used for completing a mathematical integration by graphical methods. { 'int·ə,graf }

integrated circuit [ELECTR] An interconnected array of active and passive elements integrated with a single semiconductor substrate or deposited on the substrate by a continuous series of compatible processes, and capable of performing at least one complete electronic circuit function. Abbreviated IC. Also known as integrated semiconductor. { 'int·ə,grād·əd 'sər·kət }

integrated electronics [ELECTR] A generic term for that portion of electronic art and technology in which the interdependence of material, device, circuit, and system-design consideration is especially significant; more specifically, that portion of the art dealing with integrated circuits. { 'in·tə,grād·əd i,lek'trän·iks }

integrated injection logic [ELECTR] Integrated-circuit logic that uses a simple and compact bipolar transistor gate structure which makes possible large-scale integration on silicon for logic arrays, memories, watch circuits, and various other analog and digital applications. Abbreviated I²L. Also known as merged-transistor logic. { 'in·tə,grād·əd in'jek·shən 'läj·ik }

integrated semiconductor See integrated circuit. { 'in·tə,grād·əd ¦sem·i·kən¦dək·tər }

integrated sensor [ENG] A very small device in which the sensing of some physical quantity is integrated with the functions of signal processing and information processing. { ¦in·tə,grād·əd 'sen·sər }

integrating accelerometer [ENG] A device whose output signals are proportional to the velocity of the vehicle or to the distance traveled (depending on the number of integrations) instead of acceleration. { 'in·tə,grād·əd ak,sel·ə'räm·əd·ər }

integrating frequency meter [ENG] An instrument that measures the total number of cycles through which the alternating voltage of an electric power system has passed in a given period of time, enabling this total to be compared with the number of cycles that would have elapsed if the prescribed frequency had been maintained. Also known as master frequency meter. { 'int·ə,grād·iŋ 'frē·kwən·sē ,mēd·ər }

integrating galvanometer [ENG] A modification of the d'Arsonval galvanometer which measures the integral of current over time; it is designed to be able to measure changes of flux in an exploring coil which last over periods of several minutes. { 'int·ə,grād·iŋ ,gal·və'näm·əd·ər }

integrating gyroscope [ENG] A gyroscope that senses the rate of angular displacement and measures and transmits the time integral of this rate. { 'int·ə,grād·iŋ 'jī·rə,skōp }

integrating meter [ENG] An instrument that totalizes electric energy or some other quantity consumed over a period of time. { 'int·ə,grād·iŋ 'mēd·ər }

integrating water sampler [ENG] A water sampling device comprising a cylinder with a free piston whose movement is regulated by the evacuation of a charge of fresh water. { 'int·ə,grād·iŋ 'wȯd·ər ,sam·plər }

integration [SYS ENG] The arrangement of components in a system so that they function together in an efficient and logical way. { ,int·ə'grā·shən }

intelligent agent [IND ENG] A computing hardware- or software-based system that operates without the direct intervention of humans or other agents; examples include robots, smart sensors, and Web-search software agents. { in¦tel·ə·jənt 'ā·jənt }

intelligent machine [ENG] Any machine that can accomplish its specific task in the presence of uncertainty and variability in its environment. { in'tel·ə·jənt mə'shēn }

intelligent manufacturing [IND ENG] **1.** The use of production process technology that can automatically adapt to changing environments and varying process requirements, with the capability of manufacturing various products with minimal supervision and assistance from operators. **2.** The development and implementation of artificial intelligence in manufacturing. { in¦tel·ə·jənt ,man·ə¦fak·chər·iŋ }

intelligent robot [CONT SYS] A robot that functions as an intelligent machine, that is, it can be programmed to take actions or make choices based on input from sensors. { in'tel·ə·jənt 'rō,bät }

intelligent sensor See smart sensor. { in¦tel·ə·jənt 'sen·sər }

intelligent transportation systems [CIV ENG] The application of advanced technologies to surface transportation problems, including traffic and transportation management, travel demand management, advanced public transportation management, electronic payment, commercial vehicle operations, emergency services management, and advanced vehicle control and safety systems. Previously known as intelligent vehicle highway systems. { in¦tel·ə·jənt ,tranz·pər'tā·shən ,sis·təmz }

intelligent vehicle highway systems See intelligent transportation systems. { in¦tel·ə·jənt ,vē·ə·kəl 'hī,wā ,sis·təmz }

interaction balance method See goal coordination method. { ¦in·tə¦rak·shən 'bal·əns ,meth·əd }

interaction prediction method [CONT SYS] A method for coordinating the subproblem solutions in plant decomposition, in which the interaction variables are specified by the second-level controller according to overall optimality conditions, and the subproblems are solved to satisfy local optimality conditions constrained by the specified values of the interaction variables. Also known as feasible method. { ¦in·tə¦rak·shən prə'dik·shən ,meth·əd }

interbase current [ELECTR] The current that flows from one base connection of a junction tetrode transistor to the other, through the base region. { 'in·tər,bās 'kə·rənt }

intercepting sewer [CIV ENG] A sewer that receives flow from transverse sewers and conducts the water to a treatment plant or disposal point. { in·tər'sep·tiŋ 'sü·ər }

interceptometer [ENG] A rain gage which is placed under trees or in foliage to determine the rainfall in that location; by comparing this catch with that from a rain gage set in the open, the amount of rainfall which has been intercepted by foliage is found. { ,in·tər,sep'täm·əd·ər }

interchange [CIV ENG] A junction of two or more highways at a number of separate levels so that traffic can pass from one highway to another without the crossing at grade of traffic streams. [ELEC] The current flowing into or out of a power system which is interconnected with one or more other power systems. { 'in·tər,chānj }

interchangeability [ENG] The ability to replace the components, parts, or equipment of one manufacturer with those of another, without losing function or suitability. { ,in·tər,chānj·ə'bil·əd·ē }

intercondenser [MECH ENG] A condenser between stages of a multistage steam jet pump. { ¦in·tər·kən'den·sər }

interconnection [ELEC] A link between power systems enabling them to draw on one another's reserves in time of need and to take advantage of energy cost differentials resulting from such factors as load diversity, seasonal conditions, time-zone differences, and shared investment in larger generating units. { ¦in·tər·kə'nek·shən }

intercooler [MECH ENG] A heat exchanger for cooling fluid between stages of a multistage compressor with consequent saving in power. { ¦in·tər¦kül·ər }

interface resistance [THERMO] 1. Impairment of heat flow caused by the imperfect contact between two materials at an interface. 2. Quantitatively, the temperature difference across the interface divided by the heat flux through it. { 'in·tər,fās ri'zis·təns }

interference fit [DES ENG] A fit wherein one of the mating parts of an assembly is forced into a space provided by the other part in such a way that the condition of maximum metal overlap is achieved. { ,in·tər'fir·əns ,fit }

interference time [IND ENG] Idle machine time

occurring when a machine operator, assigned to two or more semiautomatic machines, is unable to service a machine requiring attention. { ,in·ter'fir·əns ,tīm }

interferometric hydrophone [ENG] A hydrophone in which pressure changes act directly or indirectly to deform an optical fiber and thus produce a phase change in light from a laser or light-emitting diode; the phase change is detected in an interferometer. Also known as fiber-optic hydrophone. { ,in·tər¦fir·ə¦me·trik ,hī·drə,fōn }

interfit [ENG] The distance extended by the ends of one bit cone into the grooves of an adjacent one in a roller cone bit. Also known as intermesh. { 'in·tər,fit }

interior ballistics [MECH] The science concerned with the combustion of powder, development of pressure, and movement of a projectile in the bore of a gun. { in'tir·ē·ər bə'lis·tiks }

interlock [ENG] A switch or other device that prevents activation of a piece of equipment when a protective door is open or some other hazard exists. { 'in·tər,läk }

interlocking cutter [DES ENG] A milling cutter assembly consisting of two mating sections with uniform or alternate overlapping teeth. { ¦in·tər¦läk·iŋ 'kəd·ər }

intermediate frequency [ELECTR] The frequency produced by combining the received signal with that of the local oscillator in a superheterodyne receiver. Abbreviated i-f. { ,in·tər 'mēd·ē·ət 'frē·kwən·sē }

intermediate-frequency amplifier [ELECTR] The section of a superheterodyne receiver that amplifies signals after they have been converted to the fixed intermediate-frequency value by the frequency converter. Abbreviated i-f amplifier. { ,in·tər'mēd·ē·ət ¦frē·kwən·sē 'am·plə,fī·ər }

intermediate gear [MECH ENG] An idler gear interposed between a driver and driven gear. { ,in·tər'mēd·ē·ət ¦gir }

intermediate material [IND ENG] A manufactured product that requires additional processing before it becomes finished goods. { ,in·tər'mēd·ē·ət mə'tir·ē·əl }

intermesh See interfit. { ¦in·tər¦mesh }

intermittent current [ELEC] A unidirectional current that flows and ceases to flow at irregular or regular intervals. { ¦in·tər¦mit·ənt 'kə·rənt }

intermittent defect [ENG] A defect that is not continuously present. { ¦in·tər¦mit·ənt 'dē ,fekt }

intermittent-duty rating [ENG] An output rating based on operation of a device for specified intervals of time rather than continuous duty. Also known as intermittent rating. { ¦in·tər¦mit·ənt ¦düd·ē 'rād·iŋ }

intermittent firing [MECH ENG] Cyclic firing whereby fuel and air are burned in a furnace for frequent short time periods. { ¦in·tər¦mit·ənt 'fīr·iŋ }

intermittent operation [ENG] Condition in which a device operates normally for a time, then becomes defective for a time, with the process repeating itself at regular or irregular intervals. { ¦in·tər¦mit·ənt ‚äp·ə'rā·shən }

intermittent rating *See* intermittent-duty rating. { ¦in·tər¦mit·ənt 'rād·iŋ }

intermittent work [IND ENG] A type of task requiring moderate to highly demanding physical effort that is interrupted by short periods of rest or light work lasting a few seconds to a few minutes. { ¦in·tər¦mit·ənt 'wərk }

intermodulation [ELECTR] Modulation of the components of a complex wave by each other, producing new waves whose frequencies are equal to the sums and differences of integral multiples of the component frequencies of the original complex wave. { ‚in·tər‚mäj·ə'lā·shən }

internal biomechanical environment [IND ENG] A concept that is used in ergonomic design and considers that muscles, bones, and tissues are subject to the same Newtonian mechanical forces as are objects external to the body. { in¦tərn·əl ‚bī·ō·mi¦kan·ə·kəl in'vī·ərn·mənt }

internal brake [MECH ENG] A friction brake in which an internal shoe follows the inner surface of the rotating brake drum, wedging itself between the drum and the point at which it is anchored; used in motor vehicles. { in'tərn·əl 'brāk }

internal broaching [MECH ENG] The removal of material on internal surfaces, by means of a tool with teeth of progressively increasing size moving in a straight line or other prescribed path over the surface, other than for the origination of a hole. { in'tərn·əl 'brōch·iŋ }

internal combustion engine [MECH ENG] A prime mover in which the fuel is burned within the engine and the products of combustion serve as the thermodynamic fluid, as with gasoline and diesel engines. { in'tərn·əl kəm'bəs·chən ‚en·jən }

internal dielectric field *See* dielectric field. { in'tərn·əl ‚dī·ə'lek·trik 'fēld }

internal diffusion [CHEM ENG] The diffusion of liquid or gaseous reactants to the innermost pore depths of an adsorbent-base catalyst, necessary for full catalytic effect. { in'tərn·əl di'fyü·zhən }

internal energy [THERMO] A characteristic property of the state of a thermodynamic system, introduced in the first law of thermodynamics; it includes intrinsic energies of individual molecules, kinetic energies of internal motions, and contributions from interactions between molecules, but excludes the potential or kinetic energy of the system as a whole; it is sometimes erroneously referred to as heat energy. { in 'tərn·əl 'en·ər·jē }

internal floating-head exchanger [MECH ENG] Tube-and-shell heat exchanger in which the tube sheet (support for tubes) at one end of the tube bundle is free to move. { in'tərn·əl 'flōd·iŋ ¦hed iks'chānj·ər }

internal force [MECH] A force exerted by one part of a system on another. { in'tərn·əl 'fōrs }

internal friction [MECH] **1.** Conversion of mechanical strain energy to heat within a material subjected to fluctuating stress. **2.** In a powder, the friction that is developed by the particles sliding over each other; it is greater than the friction of the mass of solid that comprises the individual particles. { in'tərn·əl 'frik·shən }

internal furnace [MECH ENG] A boiler furnace having a firebox within a water-cooled heating surface. { in'tərn·əl 'fər·nəs }

internal gear [DES ENG] An annular gear having teeth on the inner surface of its rim. { in'tərn·əl 'gir }

internal grinder [MECH ENG] A machine designed for grinding the surfaces of holes. { in'tərn·əl 'grīn·dər }

internally fired boiler [MECH ENG] A fire-tube boiler containing an internal furnace which is water-cooled. { in'tərn·əl·ē ¦fīrd 'bȯil·ər }

internal mechanical environment [IND ENG] A concept that considers parts of the human body, such as muscles, bones, and tissues, in terms of how they are subject to Newtonian mechanics in their interaction with the external environment. { in¦tərn·əl mi¦kan·ə·kəl in'vī·rən·mənt }

internal mix atomizer [MECH ENG] A type of pneumatic atomizer in which gas and liquid are mixed prior to the gas expansion through the nozzle. { in'tərn·əl ¦miks 'ad·ə‚mīz·ər }

internal spring safety relief valve [ENG] A spring-loaded valve with a portion of the operating mechanism located inside the pressure vessel. { in'tərn·əl ¦spriŋ 'sāf·tē ri'lēf ‚valv }

internal stress [MECH] A stress system within a solid that is not dependent on external forces. Also known as residual stress. { in'tərn·əl 'stres }

internal thread [DES ENG] A screw thread cut on the inner surface of a hollow cylinder. { in'tərn·əl 'thred }

internal vibrator [MECH ENG] A vibrating device which is drawn vertically through placed concrete to achieve proper consolidation. { in'tərn·əl 'vī‚brād·ər }

internal work [IND ENG] Manual work done by a machine operator while the machine is automatically operating. Also known as fill-up work; inside work. [THERMO] The work done in separating the particles composing a system against their forces of mutual attraction. { in'tərn·əl 'wərk }

international ampere [ELEC] The current that, when flowing through a solution of silver nitrate in water, deposits silver at a rate of 0.001118 gram per second; it has been superseded by the ampere as a unit of current, and is equal to approximately 0.999850 ampere. { ¦in·tər¦nash·ən·əl 'am‚pir }

international ohm [ELEC] A unit of resistance, equal to that of a column of mercury of uniform cross section that has a length of 160.3 centimeters and a mass of 14.4521 grams at the temperature of melting ice; it has been superseded by the ohm, and is equal to 1.00049 ohms. { ¦in·tər¦nash·ən·əl 'ōm }

international practical temperature scale [THERMO] Temperature scale based on six points: the water triple point, the boiling points of oxygen, water, sulfur, and the solidification points of silver and gold; designated as °C, degrees Celsius, or t_{int}; replaced in 1990 by the international temperature scale. { 'in·tər¦nash· ən·əl ¦prak·tə·kəl 'tem·prə·chər ˌskāl }

international system of electrical units |ELEC| System of electrical units based on agreed fundamental units for the ohm, ampere, centimeter, and second, in use between 1893 and 1947, inclusive; in 1948, the Giorgi, or meter-kilogram-second-absolute system, was adopted for international use. { ¦in·tər¦nash·ən·əl ¦sistəm əv i¦lek·trə·kəl 'yü·nəts }

international table British thermal unit See British thermal unit. { ¦in·tər¦nash·ən·əl ¦tā·bəl ¦british 'thər·məl ˌyü·nət }

international table calorie See calorie. { ¦in·tər¦nash·ən·əl ¦tā·bəl 'kal·ə·rē }

international temperature scale [THERMO] A standard temperature scale, adopted in 1990, that approximates the thermodynamic scale, based on assigned temperature values of 17 thermodynamic equilibrium fixed points and prescribed thermometers for interpolation between them. Abbreviated ITS-90. { ¦in·tər¦nash·ən·əl 'tem·prə·chər ˌskāl }

international thread |DES ENG| A standardized metric system in which the pitch and diameter of the thread are related, with the thread having a rounded root and flat crest. { ¦in·tər¦nash·ən·əl 'thred }

international volt |ELEC| A unit of potential difference or electromotive force, equal to 1/1.01858 of the electromotive force of a Weston cell at 20°C; it was superseded by the volt, and is equal to 1.00034 volts. { ¦in·tər¦nash·ən·əl 'vōlt }

interrupted dc tachometer |ENG| A type of impulse tachometer in which the frequency of pulses generated by the interrupted direct current of an ignition-circuit primary of an internal combustion engine is used to measure the speed of the engine. { 'int·ə,rəp·təd ¦dē¦sē tə'käm·əd·ər }

interrupted screw |DES ENG| A screw with longitudinal grooves cut into the thread, and which locks quickly when inserted into a similar mating part. { 'int·ə,rəp·təd 'skrü }

interrupter |ELEC| An electric, electronic, or mechanical device that periodically interrupts the flow of a direct current so as to produce pulses. { 'int·ə,rəp·tər }

intersect |ENG| To find a position by the triangulation method. { ,in·tər'sekt }

intersection |CIV ENG| **1.** A point of junction or crossing of two or more roadways. **2.** A surveying method in which a plane table is used alternately at each end of a measured baseline. { ,in·tər'sek·shən }

intersection angle |CIV ENG| The angle of deflection at the intersection point between the straights of a railway or highway curve. { ,in·tər'sek·shən ,aŋ·gəl }

intersection point |CIV ENG| That point where

two straights or tangents to a railway or road curve would meet if extended. { ,in·tər'sek·shən ,pȯint }

interspace |BUILD| An air space. { 'in·tər ,spās }

interterminal switching |CIV ENG| The movement of railroad cars from one line to another within a switching area. { ¦in·tər'tər·mən·əl 'swich·iŋ }

intertube burner |MECH ENG| A burner which utilizes a nozzle that discharges between adjacent tubes. { 'in·tər,tüb ,bər·nər }

interval timer |ENG| A device which operates a set of contacts during a preset time interval and, at the end of the interval, returns the contacts to their normal positions. Also known as timer. { 'in·tər·vəl ,tīm·ər }

intraline distance |ENG| The minimum distance permitted between any two buildings within an explosives operating line; to protect buildings from propagation of explosions due to blast effect. { 'in·trə,līn 'dis·təns }

intrinsic-barrier diode |ELECTR| A *pin* diode, in which a thin region of intrinsic material separates the *p*-type region and the *n*-type region. { in'trin·sik ¦bar·ē·ər 'dī,ōd }

intrinsic-barrier transistor |ELECTR| A *pnip* or *npin* transistor, in which a thin region of intrinsic material separates the base and collector. { in 'trin·sik ¦bar·ē·ər tran'zis·tər }

intrinsic contact potential difference |ELEC| True potential difference between two perfectly clean metals in contact. { in'trin·sik ¦kän,takt pə¦ten·chəl 'dif·ərns }

intrinsic detector |ENG| A semiconductor detector of electromagnetic radiation that utilizes the generation of electron-hole pairs across the semiconductor band gap. { in'trin·sik di'tek·tər }

intrinsic electric strength |ELEC| The extremely high dielectric strength displayed by a substance at low temperatures. { in¦trin·sik i¦lek·trik ,streŋkth }

intrinsic layer |ELECTR| A layer of semiconductor material whose properties are essentially those of the pure undoped material. { in'trin·sik 'lā·ər }

intrusion grouting |ENG| A method of placing concrete by intruding the mortar component in position and then converting it into concrete as it is introduced into voids. { in'trü·zhən ,graúd·iŋ }

invariable line |MECH| A line which is parallel to the angular momentum vector of a body executing Poinsot motion, and which passes through the fixed point in the body about which there is no torque. { in'ver·ē·ə·bəl 'līn }

invariable plane |MECH| A plane which is perpendicular to the angular momentum vector of a rotating rigid body not subject to external torque, and which is always tangent to its inertia ellipsoid. { in'ver·ē·ə·bəl 'plān }

inventory |ENG| The amount of plastic in the heating cylinder or barrel in injection molding or extrusion. { 'in·vən,tȯr·ē }

inventory control |IND ENG| Systematic management of the balance on hand of inventory items, involving the supply, storage, distribution, and recording of items. { 'in·vən,tȯr·ē kən,trōl }

inverse cam |MECH ENG| A cam that acts as a follower instead of a driver. { 'in,vərs 'kam }

inverse current |ELECTR| The current resulting from an inverse voltage in a contact rectifier. { 'in,vərs 'kə·rənt }

inverse feedback See negative feedback. { 'in,vərs 'fēd,bak }

inverse problem |CONT SYS| The problem of determining, for a given feedback control law, the performance criteria for which it is optimal. { 'in,vərs 'präb·ləm }

inverse voltage |ELECTR| The voltage that exists across a rectifier tube or x-ray tube during the half cycle in which the anode is negative and current does not normally flow. { 'in,vərs 'vōl·tij }

inversion |ELEC| The solution of certain problems in electrostatics through the use of the transformation in Kelvin's inversion theorem. |MECH ENG| The conversion of basic four-bar linkages to special motion linkages, such as parallelogram linkage, slider-crank mechanism, and slow-motion mechanism by successively holding fast, as ground link, members of a specific linkage (as drag link). |THERMO| A reversal of the usual direction of a variation or process, such as the change in sign of the expansion coefficient of water at 4°C, or a change in sign in the Joule-Thomson coefficient at a certain temperature. { in'vər·zhən }

inversion temperature |ENG| The temperature to which one junction of a thermocouple must be raised in order to make the thermoelectric electromotive force in the circuit equal to zero, when the other junction of the thermocouple is held at a constant low temperature. |THERMO| The temperature at which the Joule-Thomson effect of a gas changes sign. { in'vər·zhən ,tem·prə·chər }

invert |CIV ENG| The floor or bottom of a conduit. { 'in,vərt }

inverted arch |CIV ENG| An arch with the crown downward, below the line of the springings; commonly used in tunnels and foundations. Also known as inflected arch. { in'vərd·əd 'ärch }

inverted engine |MECH ENG| An engine in which the cylinders are below the crankshaft. { in'vərd·əd 'en·jən }

inverted siphon |CIV ENG| A pressure pipeline crossing a depression or passing under a highway; sometimes called a sag line from its U-shape. { in'vərd·əd 'sī·fən }

inverter |ELEC| A device for converting direct current into alternating current; it may be electromechanical, as in a vibrator or synchronous inverter, or electronic, as in a thyratron inverter circuit. Also known as dc-to-ac converter; dc-to-ac inverter. |ELECTR| See phase inverter. { in'vərd·ər }

inverter circuit See NOT circuit. { in'vərd·ər ,sər·kət }

inverting amplifier |ELECTR| Amplifier whose output polarity is reversed as compared to its input; such an amplifier obtains its negative feedback by a connection from output to input, and with high gain is widely used as an operational amplifier. { in'vərd·iŋ 'am·plə,fī·ər }

inverting function |ELECTR| A logic device that inverts the input signal, so that the output is out of phase with the input. { in'vərd·iŋ ,faŋk·shən }

invert level |ENG| The level of the lowest portion at any given section of a liquid-carrying conduit, such as a drain or a sewer, and which determines the hydraulic gradient available for moving the contained liquid. { 'in,vərt ,lev·əl }

invisible hinge |DES ENG| A door hinge whose parts are not exposed when the door is closed. { in'viz·ə·bəl 'hinj }

involute gear tooth |DES ENG| A gear tooth whose profile is established by an involute curve outward from the base circle. { ¦in·və¦lüt 'gir ,tüth }

involute spline |DES ENG| A spline having the same general form as involute gear teeth, except that the teeth are one-half the depth and the pressure angle is 30°. { ¦in·və¦lüt 'splīn }

involute spline broach |MECH ENG| A broach that cuts multiple keys in the form of internal or external involute gear teeth. { ¦in·və¦lüt 'splīn ,brōch }

ion-beam mixing |ENG| A process in which bombardment of a solid with a beam of energetic ions causes the intermixing of atoms of two separate phases originally present in the near-surface region. { 'ī,än ¦bēm ,miks·iŋ }

ion-beam scanning |ELECTR| The process of analyzing the mass spectrum of an ion beam in a mass spectrometer either by changing the electric or magnetic fields of the mass spectrometer or by moving a probe. { 'ī,än ,bēm ,skan·iŋ }

ion-beam thinning See ion machining. { 'ī,än ,bēm ¦thin·iŋ }

ion fractionation |CHEM ENG| Separation of cations or anions from an ionic solution by use of a membrane permeable to the desired ion; equipment includes electrodialyzers and ion-fractionation stills. { 'ī,än ,frak·shə'nā·shən }

ionic membrane |CHEM ENG| Semipermeable membrane that conducts electricity; the application of an electric field to the membrane achieves an electrophoretic movement of ions through the membrane; used in electrodialysis. { ī'än·ik 'mem,brān }

ion implantation |ENG| A process of introducing impurities into the near-surface regions of solids by directing a beam of ions at the solid. { 'ī,än ,im,plan'tā·shən }

ionization spectrometer See Bragg spectrometer. { ,ī·ə·nə'zā·shən spek'träm·əd·ər }

ion machining |ENG| Use of a high-velocity ion beam to remove material from a surface. Also known as ion beam thinning, ion milling. { 'ī,än mə'shēn·iŋ }

ion microprobe mass spectrometer |ENG| A

type of secondary ion mass spectrometer in which primary ions are focused on a spot 1–2 micrometers in diameter, mass-charge separation of secondary ions is carried out by a double focusing mass spectrometer or spectrograph, and a magnified image of elemental or isotopic distributions on the sample surface is produced using synchronous scanning of the primary ion beam and an oscilloscope. { 'ī,än 'mī·krə,prōb ¦mas spek'träm·əd·ər }

ion migration |ELEC| Movement of ions produced in an electrolyte, semiconductor, and so on, by the application of an electric potential between electrodes. { 'ī,än mī'grā·shən }

ion milling See ion machining. { 'ī,än ,mil·iŋ }

ionogram |ENG| A record produced by an ionosonde, that is, a graph of the virtual height of the ionosphere plotted against frequency. { ī'än·ə,gram }

ionophone |ENG ACOUS| A high-frequency loudspeaker in which the audio-frequency signal modulates the radio-frequency supply to an arc maintained in a quartz tube, and the resulting modulated wave acts directly on ionized air to create sound waves. { ī'än·ə,fōn }

ionosonde |ENG| A radar system for determining the vertical height at which the ionosphere reflects signals back to earth at various frequencies; a pulsed vertical beam is swept periodically through a frequency range from 0.5 to 20 megahertz, and the variation of echo return time with frequency is photographically recorded. { ī'än·ə,sänd }

ion probe See secondary ion mass spectrometer. { 'ī,än ,prōb }

ion retardation |CHEM ENG| Sorbent extraction of strong electrolytes with an anion-exchange resin in which a cationic monomer has been polymerized, or vice versa. { 'ī,än ,rē·tär'dā·shən }

IR drop See resistance drop. { ¦ī¦är 'dräp }

iron count |CHEM ENG| An analytic determination of the iron compounds in a product stream; reflects the occurrence and the extent of corrosion. { 'ī·ərn ,kaůnt }

iron oxide process |CHEM ENG| A process by which a gas is passed through iron oxide and wood shavings to remove sulfides. { 'ī·ərn 'äk,sīd prä·səs }

irradiation |ENG| The exposure of a material, object, or patient to x-rays, gamma rays, ultraviolet rays, or other ionizing radiation. { i,rād·ē'ā·shən }

irregular element |IND ENG| An element whose frequency of occurrence is irregular but predictable. Also known as incidental element. { i'reg·yə·lər 'el·ə·mənt }

irreversible energy loss |THERMO| Energy transformation process in which the resultant condition lacks the driving potential needed to reverse the process; the measure of this loss is expressed by the entropy increase of the system. { ,i·ri'vər·sə·bəl 'en·ər·je ,lòs }

irreversible process |THERMO| A process which cannot be reversed by an infinitesimal

change in external conditions. { ,i·ri'vər·sə·bəl 'prä·səs }

irreversible thermodynamics See nonequilibrium thermodynamics. { ,i·ri'vər·sə·bəl ¦thər·mə·dī'nam·iks }

irrigation |CIV ENG| Artificial application of water to arable land for agricultural use. { ,ir·ə'gā·shən }

irrigation canal |CIV ENG| An artificial open channel for transporting water for crop irrigation. { ,ir·ə'gā·shən kə,nal }

irrigation pipe |CIV ENG| A conduit of connected pipes for transporting water for crop irrigation. { ,ir·ə'gā·shən ,pīp }

isenergic flow |THERMO| Fluid flow in which the sum of the kinetic energy, potential energy, and enthalpy of any part of the fluid does not change as that part is carried along with the fluid. { ¦ī·sə,nər·jik 'flō }

isenthalpic expansion |THERMO| Expansion which takes place without any change in enthalpy. { ¦īs·ən¦thal·mik ik'span·chən }

isenthalpic process |THERMO| A process that is carried out at constant enthalpy. { ,ī·sən¦thal·pik 'prä,ses }

isentrope |THERMO| A line of equal or constant entropy. { 'īs·ən,trōp }

isentropic |THERMO| Having constant entropy; at constant entropy. { ¦īs·ən¦träp·ik }

isentropic compression |THERMO| Compression which occurs without any change in entropy. { ¦īs·ən¦träp·ik kəm'presh·ən }

isentropic expansion |THERMO| Expansion which occurs without any change in entropy. { ¦īs·ən'träp·ik ik'span·chən }

isentropic flow |THERMO| Fluid flow in which the entropy of any part of the fluid does not change as that part is carried along with the fluid. { ¦īs·ən'träp·ik 'flō }

isentropic process |THERMO| A change that takes place without any increase or decrease in entropy, such as a process which is both reversible and adiabatic. { ¦īs·ən'träp·ik 'prä·ses }

island of automation |IND ENG| A single robotic system or other automatically operating machine that functions independently of any other machine or process. { 'ī·lənd əv ,òd·ə'mā·shən }

isobaric |THERMO| Of equal or constant pressure, with respect to either space or time. { ¦ī·sə¦bär·ik }

isobaric process |THERMO| A thermodynamic process of a gas in which the heat transfer to or from the gaseous system causes a volume change at constant pressure. { ¦ī·sə¦bär·ik 'prä·səs }

isochronism |MECH| The property of having a uniform rate of operation or periodicity, for example, of a pendulum or watch balance. { ī'sä·krə,niz·əm }

isochronous governor |MECH ENG| A governor that keeps the speed of a prime mover constant at all loads. Also known as astatic governor. { ī'sä·krə·nəs 'gəv·ər·nər }

isoconcentration |CHEM ENG| Constant concentration values. { |T·sō,käns·ən'trā·shən }

isoconcentration map |CHEM ENG| Map or diagram of a liquid or gas system's concentration with respect to a single component of the system, shown by constant-concentration contour lines. { |T·sō,käns·ən'trā·shən ,map }

isocracking |CHEM ENG| A hydrocracking process for conversion of hydrocarbons into more valuable, lower-boiling products; operates at relatively low temperatures and pressures in the presence of hydrogen and a catalyst. { |T·sō 'krak·iŋ }

isodynamic |MECH| Pertaining to equality of two or more forces or to constancy of a force. { |T·sō·dī'nam·ik }

isoelectric |ELEC| Pertaining to a constant electric potential. { |T·sō·i'lek·trik }

isoforming |CHEM ENG| A petroleum refinery process in which olefinic naphtha is contacted with an alumina catalyst at high temperature and low pressure to produce isomers of higher octane number. { 'T·sə,fór·miŋ }

isokinetic sampling |ENG| Any technique for collecting airborne particulate matter in which the collector is so designed that the airstream entering it has a velocity equal to that of the air passing around and outside the collector. { ,T· sə·ki¦ned·ik 'sam·pliŋ }

isolate |CHEM ENG| To separate two portions of a process system by means of valving or line blanks; used as safety measure during maintenance or repair, or to redirect process flows. |ELEC| To disconnect a circuit or piece of equipment from an electric supply system. { 'T· sə,lāt }

isolated footing |CIV ENG| A concrete slab or block under an individual load or column. { 'T· sə,lād·əd 'fúd·iŋ }

isolated system See closed system. { 'T·sə,lād·əd 'sis·təm }

isolation amplifier |ELECTR| An amplifier used to minimize the effects of a following circuit on the preceding circuit. { ,T·sə'lā·shən 'am· plə,fī·ər }

isolation diode |ELECTR| A diode used in a circuit to allow signals to pass in only one direction. { ,T·sə'lā·shən 'dī,ōd }

isolation test |ENG| A leak detection method which isolates the evacuated system from the pump, followed by observation of the rate of pressure rise. { ,T·sə'lā·shən ,test }

isolator |ELECTR| A passive attenuator in which the loss in one direction is much greater than that in the opposite direction; a ferrite isolator for waveguides is an example. |ENG| Any device that absorbs vibration or noise, or prevents its transmission. { 'T·sə,lād·ər }

isolith |ELECTR| Integrated circuit of components formed on a single silicon slice, but with the various components interconnected by beam leads and with circuit parts isolated by removal of the silicon between them. { 'T·sə,lith }

isometric process |THERMO| A constant-volume, frictionless thermodynamic process in which the system is confined by mechanically rigid boundaries. { |T·sə'me·trik 'prä·səs }

isostatics |MECH| In photoelasticity studies of stress analyses, those curves, the tangents to which represent the progressive change in principal-plane directions. Also known as stress trajectories. Also known as stress lines. { |T· sə'stad·iks }

isostatic surface |MECH| A surface in a three-dimensional elastic body such that at each point of the surface one of the principal planes of stress at that point is tangent to the surface. { |T·sə'stad·ik 'sər·fəs }

isoteniscope |ENG| An instrument for measuring the vapor pressure of a liquid, consisting of a U tube containing the liquid, one arm of which connects with a closed vessel containing the same liquid, while the other connects with a pressure gage where the pressure is adjusted until the levels in the arms of the U tube are equal. { ,T·sə'ten·ə,skōp }

isotherm |THERMO| A curve or formula showing the relationship between two variables, such as pressure and volume, when the temperature is held constant. Also known as isothermal. { 'T·sə,thərm }

isothermal See isotherm. { |T·sə¦thər·məl }

isothermal calorimeter |THERMO| A calorimeter in which the heat received by a reservoir, containing a liquid in equilibrium with its solid at the melting point or with its vapor at the boiling point, is determined by the change in volume of the liquid. { |T·sə¦thər·məl ,kal·ə'rim· əd·ər }

isothermal compression |THERMO| Compression at constant temperature. { |T·sə¦thər·məl kəm'presh·ən }

isothermal equilibrium |THERMO| The condition in which two or more systems are at the same temperature, so that no heat flows between them. { |T·sə¦thər·məl ,ē·kwə'lib·rē·əm }

isothermal expansion |THERMO| Expansion of a substance while its temperature is held constant. { |T·sə¦thər·məl ik'span·chən }

isothermal flow |THERMO| Flow of a gas in which its temperature does not change. { |T· sə¦thər·məl 'flō }

isothermal layer |THERMO| A layer of fluid, all points of which have the same temperature. { |T·sə¦thər·məl 'lā·ər }

isothermal magnetization |THERMO| Magnetization of a substance held at constant temperature; used in combination with adiabatic demagnetization to produce temperatures close to absolute zero. { |T·sə¦thər·məl ,mag·nə·tə'zā· shən }

isothermal process |THERMO| Any constant-temperature process, such as expansion or compression of a gas, accompanied by heat addition or removal from the system at a rate just adequate to maintain the constant temperature. { |T·sə¦thər·məl 'prä·səs }

isothermal transformation

isothermal transformation |THERMO| Any transformation of a substance which takes place at a constant temperature. { ¦ī·sə¦thər·məl ‚tranz·fər'mā·shən }

ISTS *See* impulsive stimulated thermal scattering.

IT calorie *See* calorie. { ¦ī'tē ‚kal·ə·rē }

ITS *See* intelligent transportation system.

ITS-90 *See* international temperature scale.

ivory point |ENG| A small pointer extending downward from the top of the cistern of a Fortin barometer; the level of the mercury in the cistern is adjusted so that it just comes in contact with the end of the pointer, thus setting the zero of the barometer scale. { 'īv·rē ¦pȯint }

J

J *See* joule.

jack |ELEC| A connecting device into which a plug can be inserted to make circuit connections; may also have contacts that open or close to perform switching functions when the plug is inserted or removed. |MECH ENG| A portable device for lifting heavy loads through a short distance, operated by a lever, a screw, or a hydraulic press. { jak }

jackbit |DES ENG| A drilling bit used to provide the cutting end in rock drilling; the bit is detachable and either screws on or is taper-fitted to a length of drill steel. Also known as ripbit. { 'jak,bit }

jack chain |DES ENG| **1.** A chain made of light wire, with links arranged in figure-eights with loops at right angles. **2.** A toothed endless chain for moving logs. { 'jak ,chān }

jacket |MECH ENG| The space around an engine cylinder through which a cooling liquid circulates. { 'jak·ət }

jacketed pipe |DES ENG| A double-walled pipe in which liquids that are too viscous for pipeline transport at normal temperatures flow through the inner pipe that is surrounded by a pipe circulating hot fluids. { ¦jak·əd·əd 'pīp }

jack ladder |ENG| A V-shaped trough holding a toothed endless chain, and used to move logs from pond to sawmill. { 'jak ,lad·ər }

jackleg |ENG| A supporting bar used with a jackhammer. { 'jak,leg }

jack plane |DES ENG| A general-purpose bench plane measuring over 1 foot (30 centimeters) in length. { 'jak ,plān }

jack rafter |BUILD| A short, secondary, or simulated rafter. { 'jak ,raf·tər }

jackscrew |MECH ENG| **1.** A jack operated by a screw mechanism. Also known as screw jack. **2.** The screw of such a jack. { 'jak,skrü }

jackshaft |MECH ENG| A countershaft, especially when used as an auxiliary shaft between two other shafts. { 'jak,shaft }

jack truss |BUILD| A minor truss in a hip roof where the roof has a reduced section. { 'jak ,trəs }

Jacobs taper |DES ENG| A machine tool used for mounting drill chucks in drilling machines. { 'jā·kəbz 'tā·pər }

Jaeger-Steinwehr method |THERMO| A refinement of the Griffiths method for determining the mechanical equivalent of heat, in which a large mass of water, efficiently stirred, is used, the temperature rise of the water is small, and the temperature of the surroundings is carefully controlled. { 'yā·gər 'shtīn·ver ,meth·əd }

jag bolt |DES ENG| An anchor bolt with barbs on a flaring shank. { 'jag,bōlt }

jalousie |BUILD| A window that consists of a number of long, narrow panels, each hinged at the top. { 'jal·ə·sē }

jamb |BUILD| The vertical member on the side of an opening, as a door or window. { jam }

jamb liner |BUILD| A small strip of wood applied to the edge of a window jamb to increase its width for use in thicker walls. { 'jam ,līn·ər }

jam nut *See* locknut. { 'jam ,nət }

Janecke coordinates |CHEM ENG| Use of a rectangular or Ponchon-type diagram to plot the solvent content of liquid-liquid equilibrium phases; used for solvent-extraction design calculations. { 'yä·nə·kē kō,órd·ən·əts }

jaw |ENG| A notched part that permits a railroad-car axle box to move vertically. { jó }

jawbreaker *See* jaw crusher. { 'jó,brāk·ər }

jaw clutch |MECH ENG| A clutch that provides positive connection of one shaft with another by means of interlocking faces; may be square or spiral; the most common type of positive clutch. { 'jó ,kləch }

jaw crusher |MECH ENG| A machine for breaking rock between two steel jaws, one fixed and the other swinging. Also known as jawbreaker. { 'jó ¦krəsh·ər }

J bolt |DES ENG| A J-shaped bolt, threaded on the long leg of the J. { 'jā ,bōlt }

J box *See* junction box. { 'jā ,bäks }

Jeans viscosity equation |THERMO| An equation which states that the viscosity of a gas is proportional to the temperature raised to a constant power, which is different for different gases. { 'jēnz vi'skäs·əd·ē i,kwā·zhən }

jeep |MECH ENG| A one-quarter-ton, four-wheel-drive utility vehicle in wide use in all United States military services. { jēp }

Jeremiassen crystallizer |CHEM ENG| Device used to grow solid crystals in a supersaturated liquid solution and to separate them from it. { ,yer·ə'mī·ə·sən 'krist·əl,īz·ər }

jerk |MECH| **1.** The rate of change of acceleration; it is the third derivative of position with respect to time. **2.** A unit of rate of change of acceleration, equal to 1 foot (30.48 centimeters) per second squared per second. { jərk }

jerk pump [MECH ENG] A pump that supplies a precise amount of fuel to the fuel injection valve of an internal combustion engine at the time the valve opens; used for fuel injection. { 'jərk ,pəmp }

jet bit [DES ENG] A modification of a drag bit or a roller bit that utilizes the hydraulic jet principle to increase drilling rate. { 'jet ¦bit }

jet compressor [MECH ENG] A device, utilizing an actuating nozzle and a combining tube, for the pumping of a compressible fluid. { ¦jet kəm¦pres·ər }

jet condenser [MECH ENG] A direct-contact steam condenser utilizing the aspirating effect of a jet for the removal of noncondensables. { ¦jet kən¦den·sər }

jet drilling [MECH ENG] A drilling method that utilizes a chopping bit, with a water jet run on a string of hollow drill rods, to chop through soils and wash the cuttings to the surface. Also known as wash boring. { ¦jet ¦dril·iŋ }

jet engine [MECH ENG] Any engine that ejects a jet or stream of gas or fluid, obtaining all or most of its thrust by reaction to the ejection. { ¦jet ¦en·jən }

jet hole [ENG] A borehole drilled by use of a directed, forceful stream of fluid or air. { 'jet ,hōl }

jet mill See fluid-energy mill. { 'jet ,mil }

jet mixer [MECH ENG] A type of flow mixer or line mixer, depending on impingement of one liquid on the other to produce mixing. { 'jet 'mik·sər }

jet molding [ENG] Molding method in which most of the heat is applied to the material to be molded as it passes through a nozzle or jet, rather than in a conventional heating cylinder. { 'jet ,mōl·diŋ }

jet nozzle [DES ENG] A nozzle, usually specially shaped, for producing a jet, such as the exhaust nozzle on a jet or rocket engine. { 'jet ¦näz·əl }

jet-piercing drill See fusion-piercing drill. { 'jet ¦pir·siŋ ,dril }

jet propulsion [ENG] Propulsion by means of a jet of fluid. { ¦jet prə¦pəl·shən }

jet pump [MECH ENG] A pump in which an accelerating jet entrains a second fluid to deliver it at elevated pressure. { 'jet ,pəmp }

jetsam [ENG] Articles that sink when thrown overboard, particularly those jettisoned for the purpose of lightening a vessel in distress. { 'jet·səm }

jet spinning [ENG] Production of plastic fibers in which a directed blast or jet of hot gas pulls the molten polymer from a die lip; similar to melt spinning. { ¦jet ¦spin·iŋ }

jetting [CIV ENG] A method of driving piles or well points into sand by using a jet of water to break the soil. [ENG] During molding of plastics, the turbulent flow of molten resin from an undersized gate or thin section into a thicker

mold section, as opposed to laminar, progressive flow. { 'jed·iŋ }

jettison [ENG] The throwing overboard of objects, especially to lighten a craft in distress. { 'jed·ə·sən }

jewel [ENG] **1.** A bearing usually made of synthetic corundum and used in precision timekeeping devices, gyros, and other instruments. **2.** A bearing lining of soft metal, used in railroad cars, for example. { jül }

J factor [THERMO] A dimensionless equation used for the calculation of free convection heat transmission through fluid films. { 'jā ,fak·tər }

JFET See junction field-effect transistor. { 'jā,fet }

jib boom [MECH ENG] An extension that is hinged to the upper end of a crane boom. { 'jib ,büm }

jib crane [MECH ENG] Any of various cranes having a projecting arm (jib). { 'jib ,krān }

jig [ENG] A machine for dyeing piece goods by moving the cloth at full width (open width) through the dye liquor on rollers. [MECH ENG] A device used to position and hold parts for machining operations and to guide the cutting tool. { jig }

jig back [MECH ENG] An aerial ropeway with a pair of containers that move in opposite directions and are loaded or stopped alternately at opposite stations but do not pass around the terminals. Also known as reversible tramway; to-and-fro ropeway. { 'jig ,bak }

jig borer [MECH ENG] A machine tool resembling a vertical milling machine designed for locating and drilling holes in jigs. { 'jig ,bȯr·ər }

jiggering [ENG] A mechanization of the ceramic-forming operation consisting of molding the outside of a piece by throwing plastic clay on a plaster of paris mold, placing the mold and clay on a rotating head, and forming the inner surface by forcing a template or jigger knife against the clay; method used in mass-producing dinnerware. { 'jig·ə·riŋ }

jig grinder [MECH ENG] A precision grinding machine used to locate and grind holes to size, especially in hardened steels and carbides. { 'jig ¦grīn·dər }

jigsaw [MECH ENG] A tool with a narrow blade suitable for cutting intricate curves and lines. { 'jig,sȯ }

jim crow [DES ENG] A device with a heavy buttress screw thread used for bending rails by hand. { 'jim 'krō }

JIT See just-in-time.

J-K flip-flop [ELECTR] A storage stage consisting only of transistors and resistors connected as flip-flops between input and output gates, and working with charge-storage transistors; gives a definite output even when both inputs are 1. { ¦jā¦kā 'flip,fläp }

job [IND ENG] **1.** The combination of duties, skills, knowledge, and responsibilities assigned to an individual employee. **2.** A work order. { jäb }

job analysis [IND ENG] A detailed study of the

work performed, the facilities required, the working conditions, and the skills required to complete a specific job. Also known as job study. { 'jäb ə,nal·ə·səs }

jobber's reamer [DES ENG] A machine reamer that is solid with straight or helical flutes and taper shanks. { 'jäb·ərz ‚rē·mər }

job breakdown [IND ENG] Separation of an operation into elements. Also known as operation breakdown. { 'jäb 'brāk,daůn }

job characteristic See job factor. { 'jäb ‚kar·ik·tə,ris·tik }

job class [IND ENG] A group of jobs involving a similar type of work, difficulty of performance, or range of pay. Also known as job family; job grade; labor grade. { 'jäb ‚klas }

job classification [IND ENG] Designating job classes on the basis of job factors or level of pay, or on the basis of job evaluation. { 'jäb ‚klas·ə·fə‚kā·shən }

job description [IND ENG] A detailed description of the essential activities required to perform a task. { 'jäb di‚skrip·shən }

job design [IND ENG] The arrangement of tasks over a work shift with the goal of achieving technological and organizational requirements as well as reducing sources of fatigue and human error. Also known as work design. { 'jäb di‚zīn }

job evaluation [IND ENG] Orderly qualitative appraisal of each job or position in an establishment either by a point system for the specific job characteristics or by comparison of job factors; used for establishing a job hierarchy and wage plans. { 'jäb i‚val·yə'wā·shən }

job factor [IND ENG] An essential job element which provides a basis for selecting and training employees and establishing the wage plan for the job. Also known as job characteristic. { 'jäb ‚fak·tər }

Jo block See Johansson block. { 'jō ‚bläk }

job plan [IND ENG] The organized approach to production management involving formal, step-by-step procedures. { 'jäb ‚plan }

job safety analysis [IND ENG] A method of studying a job by breaking it down into its components to determine any possible hazards it may involve and the qualifications needed by those who perform it. { 'jäb ‚sāf·tē ə,nal·ə·səs }

job schedule [CONT SYS] A control program that selects from a job queue the next job to be processed. { 'jäb ‚sked·yůl }

job shop [IND ENG] A manufacturing facility that generates a variety of products in relatively low numbers and in batch lots. { 'jäb ‚shäp }

job stream [CONT SYS] A collection of jobs in a job queue. { 'jäb ‚strēm }

job study See job analysis. { 'jäb ‚stəd·ē }

joggle [DES ENG] **1.** A flangelike offset on a flat piece of metal. **2.** A projection or notch on a sheet of building material to prevent protrusion. **3.** A dowel for joining blocks of masonry. { 'jäg·əl }

joggle joint [CIV ENG] In masonry or stonework, a joint between two blocks in which a projection on one fits into a recess in another. { 'jäg·əl ‚jóint }

joggle piece See joggle post. { 'jäg·əl ‚pēs }

joggle post [BUILD] **1.** A post constructed of two or more sections of lumber joined by joggles. **2.** A king post with notches or shoulders at its lower end that provide support for the feet of the struts. Also known as joggle piece. { 'jäg·əl ‚pōst }

Johansson block [DES ENG] A type of gage block ground to an accuracy of at least 1/100,000 inch (0.25 micrometer). Also known as Jo block. { jo'han·sən ‚bläk }

joint [ELEC] A juncture of two wires or other conductive paths for current. [ENG] The surface at which two or more mechanical or structural components are united. { jóint }

joint bar [CIV ENG] A rigid steel member used in pairs to join, hold, and align rail ends. { 'jóint ‚bär }

joint clearance [ENG] The distance between mating surfaces of a joint. { 'jóint ‚klir·əns }

jointed-arm robot [CONT SYS] A robot whose arm is constructed of rigid members connected by rotary joints. Also known as revolute-coordinate robot. { 'jóin·təd ‚ärm 'rō,bät }

jointer [ENG] **1.** Any tool used to prepare, make, or simulate joints, such as a plane for smoothing wood surfaces prior to joining them, or a hand tool for inscribing grooves in fresh cement. **2.** A file for making sawteeth the same height. **3.** An attachment to a plow that covers discarded material. **4.** A worker who makes joints, particularly a construction worker who cuts stone to proper fit. **5.** A pipe of random length made from two joined, relatively short lengths. { 'jóint·ər }

jointer gage [DES ENG] An attachment to a bench vise that holds a board at any angle desired for planing. { 'jóint·ər ‚gāj }

jointing [CIV ENG] Caulking of masonry joints. [ENG] A basic woodworking process for trueing or smoothing one surface of a workpiece by using a single peripheral cutting head in order to prepare the workpiece for further processing. { 'jóint·iŋ }

joint pole [ELEC] Pole used in common by two or more utility companies. { 'jóint 'pōl }

joint ring [DES ENG] A pipe-joint flange whose outside diameter is less than the diameter of the circle containing the connecting bolts and thus fits inside the bolts. { 'jóint ‚riŋ }

joint space [CONT SYS] The space defined by a vector whose components are the translational and angular displacements of each joint of a robotic link. { 'jóint ‚spās }

joist [CIV ENG] A steel or wood beam providing direct support for a floor. { 'jóist }

joist anchor See wall anchor. { 'jóist ‚aŋ·kər }

Jolly balance [ENG] A spring balance used to measure specific gravity of mineral specimens by weighing a specimen when in the air and when immersed in a liquid of known density. { 'jal·ē ‚bal·əns }

jolt molding [ENG] A process for shaping refractory blocks in which a mold containing prepared batch is jolted mechanically to consolidate the material. { 'jōlt ¦mōl·diŋ }

Joly steam calorimeter [ENG] **1.** A calorimeter in which the mass of steam that condenses on a specimen and a pan holding it is measured, as well as the mass of steam that condenses on an empty pan. **2.** See differential steam calorimeter. { ¦jäl·ē ¦stēm ,kal·ə'rim·əd·ər }

jordan [MECH ENG] A machine or engine used to refine paper pulp, consisting of a rotating cone, with cutters, that fits inside another cone, also with cutters. { 'jórd·ən }

Jordan sunshine recorder [ENG] A sunshine recorder in which the time scale is supplied by the motion of the sun; it consists of two opaque metal semicylinders mounted with their curved surfaces facing each other; each of the semicylinders has a short narrow slit in its flat side; sunlight entering one of the slits falls on light-sensitive paper (blueprint paper) which lines the curved side of the semicylinder. { 'jórd·ən 'sən,shīn ri,kórd·ər }

joule [MECH] The unit of energy or work in the meter-kilogram-second system of units, equal to the work done by a force of 1 newton magnitude when the point at which the force is applied is displaced 1 meter in the direction of the force. Symbolized J. Also known as newton-meter of energy. { jül or jaúl }

Joule and Playfairs' experiment [THERMO] An experiment in which the temperature of the maximum density of water is measured by taking the mean of the temperatures of water in two columns whose densities are determined to be equal from the absence of correction currents in a connecting trough. { ¦jül and 'plā,fārz ik,sper·ə·mənt }

Joule calorimeter [ENG] Any electrically heated calorimeter, such as that used in the Griffiths method. { ¦jül ,kal·ə'rim·əd·ər }

Joule cycle See Brayton cycle. { 'jül ,sī·kəl }

Joule equivalent [THERMO] The numerical relation between quantities of mechanical energy and heat; the present accepted value is 1 fifteen-degrees calorie equals 4.1855 ± 0.0005 joules. Also known as mechanical equivalent of heat. { 'jül i,kwiv·ə·lənt }

Joule experiment [THERMO] **1.** An experiment to detect intermolecular forces in a gas, in which one measures the heat absorbed when gas in a small vessel is allowed to expand into a second vessel which has been evacuated. **2.** An experiment to measure the mechanical equivalent of heat, in which falling weights cause paddles to rotate in a closed container of water whose temperature rise is measured by a thermometer. { 'jül,sper·ə·mənt }

Joule heat [ELEC] The heat which is evolved when current flows through a medium having electrical resistance, as given by Joule's law. { 'jül ,hēt }

Joule-Kelvin effect See Joule-Thomson effect. { 'jül 'kel·vən i,fekt }

Joule's law [ELEC] The law that when electricity flows through a substance, the rate of evolution of heat in watts equals the resistance of the substance in ohms times the square of the current in amperes. [THERMO] The law that at constant temperature the internal energy of a gas tends to a finite limit, independent of volume, as the pressure tends to zero. { 'jülz ,ló }

Joule-Thomson coefficient [THERMO] The ratio of the temperature change to the pressure change of a gas undergoing isenthalpic expansion. { 'jül 'täm·sən ,kō·ə,fish·ənt }

Joule-Thomson effect [THERMO] A change of temperature in a gas undergoing Joule-Thomson expansion. Also known as Joule-Kelvin effect. { 'jül 'täm·sən i,fekt }

Joule-Thomson expansion [THERMO] The adiabatic, irreversible expansion of a fluid flowing through a porous plug or partially opened valve. Also known as Joule-Thomson process. { 'jül 'täm·sən ik,span·chən }

Joule-Thomson inversion temperature [THERMO] A temperature at which the Joule-Thomson coefficient of a given gas changes sign. { ¦jül ¦täm·sən in'vər·zhən ,tem·prə·chər }

Joule-Thomson process See Joule-Thomson expansion. { 'jül 'täm·sən ,prä·səs }

journal [MECH ENG] That part of a shaft or crank which is supported by and turns in a bearing. { 'jərn·əl }

journal bearing [MECH ENG] A cylindrical bearing which supports a rotating cylindrical shaft. { 'jərn·əl ,ber·iŋ }

journal box [ENG] A metal housing for a journal bearing. { 'jərn·əl ,bäks }

journal friction [MECH ENG] Friction of the axle in a journal bearing arising mainly from viscous sliding friction between journal and lubricant. { 'jərn·əl ,frik·shən }

joystick [ENG] A two-axis displacement control operated by a lever or ball, for XY positioning of a device or an electron beam. { 'jói,stik }

jumbo See drill carriage. { 'jəm·bō }

jumper [ELEC] A short length of conductor used to make a connection between two points or terminals in a circuit or to provide a path around a break in a circuit. { 'jəm·pər }

jumper tube [MECH ENG] A short tube used to bypass the flow of fluid in a boiler or tubular heater. { 'jəmp ,tüb }

jump phenomenon [CONT SYS] A phenomenon occurring in a nonlinear system subjected to a sinusoidal input at constant frequency, in which the value of the amplitude of the forced oscillation can jump upward or downward as the input amplitude is varied through either of two fixed values, and the graph of the forced amplitude versus the input amplitude follows a hysteresis loop. { 'jəmp fə,näm·ə·nən }

jump resonance [CONT SYS] A jump discontinuity occurring in the frequency response of a nonlinear closed-loop control system with saturation in the loop. { 'jəmp ,rez·ən·əns }

junction [CIV ENG] A point of intersection of roads or highways, especially where one terminates. [ELEC] *See* major node. [ELECTR] A region of transition between two different semiconducting regions in a semiconductor device, such as a *pn* junction, or between a metal and a semiconductor. { 'jəŋk·shən }

junction box [ENG] A protective enclosure into which wires or cables are led and connected to form joints. Also known as J box. { 'jəŋk·shən ,bäks }

junction capacitance *See* barrier capacitance. { 'jəŋk·shən kə'pas·əd·əns }

junction capacitor [ELECTR] An integrated-circuit capacitor that uses the capacitance of a reverse-biased *pn* junction. { 'jəŋk·shən kə¦pas·əd·ər }

junction diode [ELECTR] A semiconductor diode in which the rectifying characteristics occur at an alloy, diffused, electrochemical, or grown junction between *n*-type and *p*-type semiconductor materials. Also known as junction rectifier. { 'jəŋk·shən ¦dī,ōd }

junction field-effect transistor [ELECTR] A field-effect transistor in which there is normally a channel of relatively low-conductivity semiconductor joining the source and drain, and this channel is reduced and eventually cut off by junction depletion regions, reducing the conductivity, when a voltage is applied between the gate electrodes. Abbreviated JFET. { 'jəŋk·shən 'fēld i,fekt tran¦zis·tər }

junction filter [ELECTR] A combination of a high-pass and a low-pass filter that is used to

separate frequency bands for transmission over separate paths. { 'jəŋk·shən ,fil·tər }

junction isolation [ELECTR] Electrical isolation of a component on an integrated circuit by surrounding it with a region of a conductivity type that forms a junction, and reverse-biasing the junction so it has extremely high resistance. { 'jəŋk·shən ,ī·sə'lā·shən }

junction phenomena [ELECTR] Phenomena which occur at the boundary between two semiconductor materials, or a semiconductor and a metal, such as the existence of an electrostatic potential in the absence of current flow, and large injection currents which may arise when external voltages are applied across the junction in one direction. { 'jəŋk·shən fə,näm·ə·nə }

junction pole [ELEC] Pole at the end of a transposition section of an open-wire line or the pole common to two adjacent transposition sections. { 'jəŋk·shən ,pōl }

junction rectifier *See* junction diode. { 'jəŋk·shən ¦rek·tə,fī·ər }

junction transistor [ELECTR] A transistor in which emitter and collector barriers are formed between semiconductor regions of opposite conductivity type. { 'jəŋk·shən tran¦zis·tər }

Junkers engine [MECH ENG] A double-opposed-piston, two-cycle internal combustion engine with intake and exhaust ports at opposite ends of the cylinder. { 'yùŋ·kərz 'en·jən }

just-in-time [IND ENG] A systems approach to developing and operating a manufacturing system so that the least amount of resources is expended in producing the final products. Abbreviated JIT. { ¦jəst in 'tīm }

just ton *See* ton. { 'jəst 'tən }

K

K *See* cathode.

Kalman filter |CONT SYS| A linear system in which the mean squared error between the desired output and the actual output is minimized when the input is a random signal generated by white noise. { 'kal·mən ,fil·tər }

kanban |IND ENG| An inventory control system for tracking the flow of in-process materials through the various operations of a just-in-time production process. Kanban means "card" or "ticket" in Japanese. { ¦kan¦ban }

Kapitza balance |ENG| A magnetic balance for measuring susceptibilities of materials in large magnetic fields that are applied for brief periods. { ka'pit·sə ,bal·əns }

Kapitza expander |CHEM ENG| Reciprocating-piston gas expander used for helium liquefaction; relies on close fit rather than packing or rings on the pistons. { 'kä·pit·sə ik¦span·dər }

Kaplan turbine |MECH ENG| A propeller-type hydraulic turbine in which the positions of the runner blades and the wicket gates are adjustable for load change with sustained efficiency. { 'kap·lən ,tər·bən }

Karrer method |CHEM ENG| An industrial method for the chemical synthesis of riboflavin. { 'kar·ər ,meth·əd }

Kata thermometer |ENG| An alcohol thermometer used to measure low velocities in air circulation, by heating the large bulb of the thermometer above 100°F (38°C) and noting the time it takes to cool from 100 to 95°F (38 to 35°C) or some other interval above ambient temperature, the time interval being a measure of the air current at that location. { 'kad·ə thər'mam·əd·ər }

Kater's reversible pendulum |MECH| A gravity pendulum designed to measure the acceleration of gravity and consisting of a body with two knife-edge supports on opposite sides of the center of mass. { 'kā·dərz ri¦vər·sə·bəl 'pen·jə·ləm }

katharometer |ENG| An instrument for detecting the presence of small quantities of gases in air by measuring the resulting change in thermal conductivity of the air. Also known as thermal conductivity cell. { ,kath·ə'räm·əd·ər }

Kauertz engine |MECH ENG| A type of cat-and-mouse rotary engine in which the pistons are vanes which are sections of a right circular cylinder; two pistons are attached to one rotor so

that they rotate with constant angular velocity, while the other two pistons are controlled by a gear-and-crank mechanism, so that angular velocity varies. { 'kaů·ərts ,en·jən }

kb *See* kilobar.

kcal *See* kilocalorie.

keel block |CIV ENG| A docking block used to support a ship's keel. { 'kēl ,bläk }

kellering |MECH ENG| Three-dimensional machining of a contoured surface by tracer-milling the die block or punch; the cutter path is controlled by a tracer that follows the contours on a die model. { 'kel·ə·riŋ }

Kellogg equation |THERMO| An equation of state for a gas, of the form

$$p = RT\rho + \sum_{n=2}^{\infty} |b_n T - a_n - (c_n//T^2)|\rho^n$$

where p is the pressure, T the absolute temperature, ρ the density, R the gas constant, and a_n, b_n, and c_n are constants. { 'kel¦äg i,kwā·zhən }

Kelly ball test |ENG| A test for the consistency of concrete using the penetration of a half sphere; a 1-inch (2.5-centimeter) penetration by the Kelly ball corresponds to about 2 inches (5 centimeters) of slump. { 'kel·ē 'bȯl ,test }

kelvin |ELEC| A name formerly given to the kilowatt-hour. Also known as thermal volt. |THERMO| A unit of absolute temperature equal to 1/273.16 of the absolute temperature of the triple point of water. Symbolized K. Formerly known as degree Kelvin. { 'kel·vən }

Kelvin absolute temperature scale |THERMO| A temperature scale in which the ratio of the temperatures of two reservoirs is equal to the ratio of the amount of heat absorbed from one of them by a heat engine operating in a Carnot cycle to the amount of heat rejected by this engine to the other reservoir; the temperature of the triple point of water is defined as 273.16 K. Also known as Kelvin temperature scale. { 'kel·vən ¦ab·sə,lüt 'tem·prə·chər ,skāl }

Kelvin body |MECH| An ideal body whose shearing (tangential) stress is the sum of a term proportional to its deformation and a term proportional to the rate of change of its deformation with time. Also known as Voigt body. { 'kel·vən ,bäd·ē }

Kelvin bridge |ELEC| A specialized version of the Wheatstone bridge network designed to eliminate, or greatly reduce, the effect of lead and contact resistance, and thus permit accurate measurement of low resistance. Also known as double bridge; Kelvin network; Thomson bridge. { 'kel·vən ,brij }

Kelvin equation |THERMO| An equation giving the increase in vapor pressure of a substance which accompanies an increase in curvature of its surface; the equation describes the greater rate of evaporation of a small liquid droplet as compared to that of a larger one, and the greater solubility of small solid particles as compared to that of larger particles. { 'kel·vən i,kwā·zhən }

Kelvin network See Kelvin bridge. { 'kel·vən ,net,wərk }

Kelvin scale |THERMO| The basic scale used for temperature definition; the triple point of water (comprising ice, liquid, and vapor) is defined as 273.16 K; given two reservoirs, a reversible heat engine is built operating in a cycle between them, and the ratio of their temperatures is defined to be equal to the ratio of the heats transferred. { 'kel·vən ,skāl }

Kelvin's statement of the second law of thermodynamics |THERMO| The statement that it is not possible that, at the end of a cycle of changes, heat has been extracted from a reservoir and an equal amount of work has been produced without producing some other effect. { 'kel·vənz 'stāt·mənt əv thə 'sek·ənd ,lȯ əv ,thər·mō·dī'nam·iks }

Kelvin temperature scale |THERMO| **1.** An International Temperature Scale which agrees with the Kelvin absolute temperature scale within the limits of experimental determination. **2.** See Kelvin absolute temperature scale. { 'kel·vən 'tem·prə·chər ,skāl }

Kennedy and Pancu circle |MECH| For a harmonic oscillator subject to hysteretic damping and subjected to a sinusoidally varying force, a plot of the in-phase and quadrature components of the displacement of the oscillator as the frequency of the applied vibration is varied. { 'ken·ə·dē ən 'pän·chü ,sər·kəl }

Kennedy key |DES ENG| A square taper key fitted into a keyway of square section and driven from opposite ends of the hub. { 'ken·ə·dē ,kē }

kerf |ENG| A cut made in wood, metal, or other material by a saw or cutting torch. { kərf }

Kern counter See dust counter. { 'kərn ,kaȯn·tər }

ketene lamp |CHEM ENG| An electrically heated Chromel filament by the means of which acetone is hydrolyzed to produce ketene. { 'kē,tēn ,lamp }

kettle reboiler |CHEM ENG| Tube-and-shell heat exchange device in which liquid is vaporized on the shell side from heat transferred from hot liquid flowing through the tubes; dome space allows liquid-vapor separation above the tube bundle. { 'ked·əl rē'bȯil·ər }

Kew barometer |ENG| A type of cistern barometer; no adjustment is made for the variation of the level of mercury in the cistern as pressure changes occur; rather, a uniformly contracting scale is used to determine the effective height of the mercury column. { 'kyü bə'räm·əd·ər }

key |BUILD| **1.** Plastering that is forced between laths to secure the rest of the plaster in place. **2.** The roughening on a surface to be glued or plastered to increase adhesiveness. |CIV ENG| A projecting portion that serves to prevent movement of parts at a construction joint. |DES ENG| **1.** An instrument that is inserted into a lock to operate the bolt. **2.** A device used to move in some manner in order to secure or tighten. **3.** One of the levers of a keyboard. **4.** See machine key. |ELEC| **1.** A hand-operated switch used for transmitting code signals. Also known as signaling key. **2.** A special lever-type switch used for opening or closing a circuit only as long as the handle is depressed. Also known as switching key. |ENG| The pieces of core causing a block in a core barrel, the removal of which allows the rest of the core in the barrel to slide out. { kē }

key activity |IND ENG| An activity that possesses major significance. Also known as milestone activity. { 'kē ak'tiv·əd·ē }

keyboard |ENG| A set of keys or control levers having a systematic arrangement and used to operate a machine or other piece of equipment such as a typewriter, typesetter, processing unit of a computer, or piano. { 'kē,bȯrd }

keyboard perforator |ENG| A typewriterlike device that prepares punched paper tape for communications or computing equipment. { 'kē ,bȯrd 'pər·fə,rād·ər }

Keyes equation |THERMO| An equation of state of a gas which is designed to correct the van der Waals equation for the effect of surrounding molecules on the term representing the volume of a molecule. { 'kēz i,kwā·zhən }

Keyes process |CHEM ENG| A distillation process used to obtain absolute alcohol; benzene is added to a constant-boiling 95% alcohol-water solution, and on distillation anhydrous alcohol leaves the bottom of the column. { 'kēz ,prä·səs }

key grasp See pinch grasp. { 'kē ,grasp }

keyhole |DES ENG| A hole or a slot for receiving a key. { 'kē,hōl }

keyhole saw |DES ENG| A fine compass saw with a blade 11–16 inches (28–41 centimeters) long. { 'kē,hōl ,sȯ }

keying |CIV ENG| Establishing a mechanical bond in a construction joint. |ELEC| The forming of signals, such as for telegraph transmission, by modulating a direct-current or other carrier between discrete values of some characteristic. { 'kē·iŋ }

key job |IND ENG| A job that has been evaluated and is considered representative of similar jobs in the same labor market and is used as a benchmark to evaluate the similar jobs and to establish non-key-job wages. { 'kē ,jäb }

key joint |CIV ENG| A mortar joint with a concave pointing. { 'kē ,jȯint }

key seat See keyway. { 'kē ,sēt }
keyseater |MECH ENG| A machine for milling beds or grooves in mechanical parts which receive keys. { 'kē,sēd·ər }
keyway |DES ENG| 1. An opening in a lock for passage of a flat metal key. 2. The pocket in the driven element to provide a driving surface for the key. 3. A groove or channel for a key in any mechanical part. Also known as key seat. |ENG| An interlocking channel or groove in a cement or wood joint to provide reinforcement. { 'kē,wā }
keyword spotting |ENG ACOUS| An approach to task-oriented speech understanding through detecting a limited number of keywords that would most likely express the intent of a speaker, rather than attempting to recognize every word in an utterance. { 'kē,wərd ,spät·iŋ }
kg See kilogram; kilogram force.
kg-cal See kilocalorie.
kgf See kilogram force.
kgf-m See meter-kilogram.
kg-wt See kilogram force.
kickback |MECH ENG| A backward thrust, such as the backward starting of an internal combustion engine as it is cranked, or the reverse push of a piece of work as it is fed to a rotary saw. { 'kik,bak }
kickdown |MECH ENG| 1. Shifting to lower gear in an automotive vehicle. 2. The device for shifting. { 'kik,daún }
kick over |MECH ENG| To start firing; applied to internal combustion engines. { 'kik ,ō·vər }
kickpipe |BUILD| A short pipe protecting an electrical cable at the point where it emerges from a floor. { 'kik,pīp }
kickplate |BUILD| A plate used on the bottom of doors and cabinets or on the risers of steps to protect them from shoe marks. Also known as toeplate. { 'kik,plāt }
Kick's law |ENG| The law that the energy needed to crush a solid material to a specified fraction of its original size is the same, regardless of the original size of the feed material. { 'kiks ,lò }
kick starter |MECH ENG| A mechanism for starting the operation of a motor by thrusting with the foot. { 'kik ,stärd·ər }
kick wheel |ENG| A potter's wheel worked by a foot pedal. { 'kik ,wēl }
kiln |ENG| A heated enclosure used for drying, burning, or firing materials such as ore or ceramics. { kil }
kilobar |MECH| A unit of pressure equal to 1000 bars (100 megapascals). Abbreviated kb. { 'kil·ə,bär }
kilocalorie |THERMO| A unit of heat energy equal to 1000 calories. Abbreviated kcal. Also known as kilogram-calorie (kg-cal); large calorie (Cal). { 'kil·ə,kal·ə·rē }
kilogram |MECH| 1. The unit of mass in the meter-kilogram-second system, equal to the mass of the international prototype kilogram stored at Sèvres, France. Abbreviated kg. 2. See kilogram force. { 'kil·ə,gram }
kilogram-calorie See kilocalorie. { 'kil·ə,gram 'kal·ə·rē }
kilogram force |MECH| A unit of force equal to the weight of a 1-kilogram mass at a point on the earth's surface where the acceleration of gravity is 9.80665 m/s². Abbreviated kgf. Also known as kilogram (kg), kilogram weight (kg-wt). { 'kil·ə,gram 'fòrs }
kilogram-meter See meter-kilogram. { 'kil·ə,gram 'mēd·ər }
kilogram weight See kilogram force. { 'kil·ə,gram 'wāt }
kiloliter |MECH| A unit of volume equal to 1000 liters or to 1 cubic meter. Abbreviated kl. { 'kil·ə,lēd·ər }
kilometer |MECH| A unit of length equal to 1000 meters. Abbreviated km. { 'kil·ə,mēd·ər }
kilowatt-hour |ELEC| A unit of energy or work equal to 1000 watt-hours. Abbreviated kWh; kW-hr. Also known as Board of Trade Unit. { 'kil·ə,wät ,aúr }
kinematically admissible motion |MECH| Any motion of a mechanical system which is geometrically compatible with the constraints. { ,kin·ə¦mad·ə·klē id¦mis·ə·bəl 'mō·shən }
kinematics |MECH| The study of the motion of a system of material particles without reference to the forces which act on the system. { ¦kin·ə¦mad·iks }
kinetic energy |MECH| The energy which a body possesses because of its motion; in classical mechanics, equal to one-half of the body's mass times the square of its speed. { kə'ned·ik 'en·ər·jē }
kinetic equilibrium See dynamic equilibrium. { kə'ned·ik ,ē·kwə'lib·rē·əm }
kinetic friction |MECH| The friction between two surfaces which are sliding over each other. { kə'ned·ik 'frik·shən }
kinetic momentum |MECH| The momentum which a particle possesses because of its motion; in classical mechanics, equal to the particle's mass times its velocity. { kə'ned·ik mə'men·təm }
kinetic potential See Lagrangian. { kə'ned·ik pə'ten·chəl }
kinetic reaction |MECH| The negative of the mass of a body multiplied by its acceleration. { kə'ned·ik rē'ak·shən }
kinetics |MECH| The dynamics of material bodies. { kə'ned·iks }
king closer |CIV ENG| In masonry work, a rectangular brick having one corner cut diagonally to half the end of the brick and used to fill an opening in a course larger than half a brick. Also known as beveled closer. { ¦kiŋ ¦klōz·ər }
kingpin |MECH ENG| The pin for articulation between an automobile stub axle and an axle-beam or steering head. Also known as swivel pin. { 'kiŋ,pin }
king post |BUILD| In a roof truss, the central vertical member against which the rafters abut and which supports the tie beam. { 'kiŋ ,pōst }

king post truss |BUILD| A wooden roof truss having two principal rafters held by a horizontal tie beam, a king post upright between tie beam and ridge, and usually two struts to the rafters from a thickening at the king post foot. { 'kiŋ ‚pōst ‚trəs }

kink |ENG| A tightened loop in a wire rope resulting in permanent deformation and damage to the wire. { kiŋk }

kip |MECH| A 1000-pound (453.6-kilogram) load. { kip }

Kirchhoff formula |THERMO| A formula for the dependence of vapor pressure p on temperature T, valid over limited temperature ranges; it may be written $\log p = A - (B/T) - C \log T$, where A, B, and C are constants. { 'kərk‚hŏf ‚fór·myə·lə }

Kirchhoff's current law |ELEC| The law that at any given instant the sum of the instantaneous values of all the currents flowing toward a point is equal to the sum of instantaneous values of all the currents flowing away from the point. Also known as Kirchhoff's first law. { 'kərk‚hŏfs 'kə·rənt ‚lȯ }

Kirchhoff's equations |THERMO| Equations which state that the partial derivative of the change of enthalpy (or of internal energy) during a reaction, with respect to temperature, at constant pressure (or volume) equals the change in heat capacity at constant pressure (or volume). { 'kərk‚hŏfs i‚kwā·zhənz }

Kirchhoff's first law See Kirchhoff's current law. { 'kərk‚hŏfs 'fərst ‚lȯ }

Kirchhoff's law |ELEC| Either of the two fundamental laws dealing with the relation of currents at a junction and voltages around closed loops in an electric network; they are known as Kirchhoff's current law and Kirchhoff's voltage law. |THERMO| The law that the ratio of the emissivity of a heat radiator to the absorptivity of the same radiator is the same for all bodies, depending on frequency and temperature alone, and is equal to the emissivity of a blackbody. Also known as Kirchhoff's principle. { 'kərk ‚hŏfs ‚lȯ }

Kirchhoff's principle See Kirchhoff's law. { 'kərk‚hŏfs ‚prin·sə·pəl }

Kirchhoff's second law See Kirchhoff's voltage law. { 'kərk‚hŏfs 'sek·ənd ‚lȯ }

Kirchhoff's voltage law |ELEC| The law that at each instant of time the algebraic sum of the voltage rises around a closed loop in a network is equal to the algebraic sum of the voltage drops, both being taken in the same direction around the loop. Also known as Kirchhoff's second law. { 'kərk‚hŏfs 'vōl·tij ‚lȯ }

Kirchhoff vapor pressure formula |THERMO| An approximate formula for the variation of vapor pressure p with temperature T, valid over a limited temperature range; it is $\ln p = A - B/T - C \ln T$, where A, B, and C are constants. { ‚kirch‚hŏf 'va·pər ‚pre·shər ‚fór·myə·lə }

Kirkwood-Brinkely's theory |MECH| In terminal ballistics, a theory formulating the scaling laws from which the effect of blast at high altitudes may be inferred, based upon observed results at ground level. { 'kərk‚wüd 'briŋk·lēz ‚thē·ə·rē }

kiss-roll coating |ENG| Procedure for coating a substrate web in which the coating roll carries a metered film of coating material; part of the film transfers to the web, part remains on the roll. { 'kis ‚rōl ‚kōd·iŋ }

kl See kiloliter.

klaxon |ENG ACOUS| A diaphragm horn sometimes operated by hand. { 'klak·sən }

klydonograph |ENG| A device attached to electric power lines for estimating certain electrical characteristics of lightning by means of the figures produced on photographic film by the lightning-produced surge carried over the lines; the size of the figure is a function of the potential and polarity of the lightning discharge. { klī'dän·ə‚graf }

km See kilometer.

knapping hammer |ENG| A steel hammer used for breaking and shaping stone. { 'nap·iŋ ‚ham·ər }

knee |MECH ENG| In a knee-and-column type of milling machine, the part which supports the saddle and table and which can move vertically on the column. { nē }

knee brace |BUILD| A stiffener between a column and a supported truss or beam to provide greater rigidity in a building frame under transverse loads. { 'nē ‚brās }

knee frequency See break frequency. { 'nē ‚frē·kwən·sē }

kneeler |CIV ENG| In masonry, a stone cut to provide a break in the horizontal-vertical pattern to begin the curve or angle of an arch or vault. { 'nēl·ər }

knee pad |ENG| A protective cushion, usually made of sponge rubber, that can be strapped to a worker's knee. { 'nē ‚pad }

knee rafter |BUILD| A brace placed diagonally between a principal rafter and a tie beam. { 'nē ‚raf·tər }

knee switch |ENG| A control mechanism operated with knee movements by a seated worker. { 'nē ‚swich }

knee tool |MECH ENG| A tool holder with a shape resembling a knee, such as the holder for simultaneous cutting and interval operations on a screw machine or turret lathe. { 'nē ‚tül }

knee wall |BUILD| A partition that forms a side wall or supports roof rafters under a pitched roof. { 'nē ‚wȯl }

knife |DES ENG| A sharp-edged blade for cutting. { nīf }

knife coating |ENG| Procedure for coating a continuous-web substrate in which coating thickness is controlled by the distance between the substrate and a movable knife or bar. { 'nīf ‚kōd·iŋ }

knife-edge |DES ENG| A sharp narrow edge resembling that of a knife, such as the fulcrum for a lever arm in a measuring instrument. { 'nīf ‚ej }

knife-edge bearing |MECH ENG| A balance beam or lever arm fulcrum in the form of a hardened steel wedge; used to minimize friction. { 'nīf ,ej ,ber·iŋ }

knife-edge cam follower |DES ENG| A cam follower having a sharp narrow edge or point like that of a knife; useful in developing cam profile relationships. { 'nīf ,ej 'kam ,fäl·ə·wər }

knife file |DES ENG| A tapered file with a thin triangular cross section resembling that of a knife. { 'nīf ,fīl }

knife switch |ELEC| An electric switch consisting of a metal blade hinged at one end to a stationary jaw, so that the blade can be pushed over to make contact between spring clips. { 'nīf ,swich }

knob |DES ENG| A component that is placed on a control shaft to facilitate manual rotation of the shaft; sometimes has a pointer or markings to indicate shaft position. { näb }

knocker See shell knocker. { 'näk·ər }

knock intensity |ENG| The intensity of knock (detonation) recorded when testing a motor gasoline for octane or knock rating. { 'näk in,ten·səd·ē }

knockmeter |ENG| A fuels-testing device used to measure the output of the detonation meter used in American Society for Testing and Materials knock-test ratings of motor fuels. { 'näk ,mēd·ər }

knock-off |MECH ENG| **1.** The automatic stopping of a machine when it is operating improperly. **2.** The device that causes automatic stopping. { 'näk,óf }

knock-off bit See detachable bit. { 'näk,óf ,bit }

knockout |ENG| A partially cutout piece in metal or plastic that can be forced out when a hole is needed. { 'näk,aút }

knockout pin See ejector pin. { 'näk,aút ,pin }

knockout vessel |CHEM ENG| A vessel, drum, or trap used to remove fluid droplets from flowing gases. { 'näk,aút ,ves·əl }

knock rating |ENG| Rating of gasolines according to knocking tendency. { 'näk,aút ,rād·iŋ }

known-good die |ELECTR| An unpackaged, fully tested integrated circuit chip. { ,nōn ¦gúd 'dī }

knuckle joint |DES ENG| A hinge joint between two rods in which an eye on one piece fits between two flat projections with eyes on the other piece and is retained by a round pin. { 'nək·əl ,jóint }

knuckle joint press |MECH ENG| A short-stroke press in which the slide is actuated by a crank attached to a knuckle joint hinge. { 'nək·əl ,jóint ,pres }

knuckle pin |DES ENG| The pin of a knuckle joint. { 'nək·əl ,pin }

knuckle post |MECH ENG| A post which acts as the pivot for the steering knuckle in an automobile. { 'nək·əl ,pōst }

Knudsen gage |ENG| An instrument for measuring very low pressures, which measures the force of a gas on a cold plate beside which there is an electrically heated plate. { kə'nüd·sən ,gāj }

Knudsen-Langmuir equation |CHEM ENG| Relationship of molecular distillation rate to vapor saturation pressure, solution temperature, and molecular weight during evaporation and no-recycle condensation. { kə'nüd·sən 'laŋ,myúr i,kwā·zhən }

Knudsen reversing water bottle |ENG| A type of frameless reversing bottle for collecting water samples; carries reversing thermometers. { kə'nüd·sən ri¦vərs·iŋ 'wód·ər ,bäd·əl }

Knudsen vacuum gage |ENG| Device to measure negative gas pressures; a rotatable vane is moved by the pressure of heated molecules, proportionately to the concentration of molecules in the system. { kə'nüd·sən 'vak·yəm ,gāj }

knurl |ENG| To provide a surface, usually a metal, with small ridges or knobs to ensure a firm grip or as a decorative feature. { nərl }

Kolosov-Muskhelishvili formulas |MECH| Formulas which express plane strain and plane stress in terms of two holomorphic functions of the complex variable $z = x + iy$, where x and y are plane coordinates. { ¦kōl·ə,sóf ,músh'kel·ish,vil·ē ,fór·myə·ləz }

konimeter |ENG| An air-sampling device used to measure dust as in a cement mill or a mine; a measured volume of air drawn through a jet impacts on a glycerin-jelly-coated glass surface; the particles are counted with a microscope. { kō'nim·əd·ər }

koniscope |ENG| An instrument which indicates the presence of dust particles in the atmosphere. Also spelled coniscope. { 'kän·ə ,skōp }

kraft process See sulfate pulping. { 'kraft ,prä·səs }

kraft pulping See sulfate pulping. { 'kraft ,pəlp·iŋ }

Kremser formula |CHEM ENG| Equation for calculating distillation-column material balances and equilibrium, assuming the ideal distribution law, that is, the concentrations in the two phases (vapor and liquid) are proportional to each other. { 'krem·zər ,fór·myə·lə }

Krigar-Menzel law |MECH| A generalization of the second Young-Helmholtz law which states that when a string is bowed at a point which is at a distance of p/q times the string's length from one of the ends, where p and q are relative primes, then the string moves back and forth with two constant velocities, one of which is $q - 1$ times as large as the other. { ¦krē·gər 'menz·əl ,ló }

kryptoclimate See cryptoclimate. { ¦krip·tō'klī·mət }

K truss |BUILD| A building truss in the form of a K due to the orientation of the vertical member and two oblique members in each panel. { 'kä ,trəs }

Kullenberg piston corer |MECH ENG| A piston-operated coring device used to obtain 2-inch-diameter (5-centimeter) core samples. { 'kəl·ən,bərg 'pis·tən ,kór·ər }

kWh *See* kilowatt-hour.

kW-hr *See* kilowatt-hour.

kyanize [CHEM ENG] To saturate wood with mercuric chloride as a decay preventive. { 'kī‧ə,nīz }

kymograph [IND ENG] A device used to measure extremely short work time intervals by using a system of transducers that are activated by an operator performing a job, with the impulses recorded as a function of time. { 'kī‧mə,graf }

L

l See liter.

L See liter.

labeled cargo [IND ENG] Cargo of a dangerous nature, such as explosives and flammable or corrosive liquids, which is designated by different-colored labels to indicate the requirements for special handling and storage. { 'lā·bəld ,kär·gō }

laboratory coordinate system [MECH] A reference frame attached to the laboratory of the observer, in contrast to the center-of-mass system. { 'lab·rə,tȯr·ē kō'ȯrd·ən,ət ,sis·təm }

labor cost [IND ENG] That part of the cost of goods and services attributable to wages, especially for direct labor. { 'lā·bər ,kȯst }

labor factor [IND ENG] The ratio of the number of hours required to perform a task under project conditions to the number of hours required to perform an identical task under standard conditions of work measurement. { 'lā·bər ,fak·tər }

labor relations [IND ENG] The management function that deals with a company's work force; usually the term is restricted to relations with organized labor. { 'lā·bər ri,lā·shənz }

labyrinth [ENG ACOUS] A loudspeaker enclosure having air chambers at the rear that absorb rearward-radiated acoustic energy, to prevent it from interfering with the desired forward-radiated energy. { 'lab·ə,rinth }

labyrinth seal [ENG] A minimum-leakage seal that offers resistance to fluid flow while providing radial or axial clearance; a labyrinth of circumferential knives or touch points provides for successive expansion of the fluid being piped; used for gas pipes, steam engines, and turbines. { 'lab·ə,rinth ,sēl }

lacing [CIV ENG] **1.** A lightweight metallic piece that is fixed diagonally to two channels or four angle sections, forming a composite strut. **2.** A course of brick, stone, or tiles in a wall of rubble to give strength. **3.** A course of upright bricks forming a bond between two or more arch rings. **4.** Distribution steel in a slab of reinforced concrete. **5.** A light timber fastened to pairs of struts or walings in the timbering of excavations (including mines). [ELEC] Tying insulated wires together to support each other and form a single neat cable, with separately laced branches. { 'lās·iŋ }

lactometer [ENG] A hydrometer used to measure the specific gravity of milk. { lak'täm·əd·ər }

ladder [ENG] A structure, often portable, for climbing up and down; consists of two parallel sides joined by a series of crosspieces that serve as footrests. { 'lad·ər }

ladder-bucket dredge See bucket-ladder dredge. { 'lad·ər ¦bək·ət ,drej }

ladder diagram [CONT SYS] A diagram used to program a programmable controller, in which power flows through a network of relay contacts arranged in horizontal rows called rungs between two vertical rails on the side of the diagram containing the symbolic power. { 'lad·ər ,dī·ə ,gram }

ladder ditcher See ladder trencher. { 'lad·ər ¦dich·ər }

ladder dredge See bucket-ladder dredge. { 'lad·ər ¦drej }

ladder drilling [MECH ENG] An arrangement of retractable drills with pneumatic powered legs mounted on banks of steel ladders connected to a holding frame; used in large-scale rock tunneling, with the advantage that many drills can be worked at the same time by a small labor force. { 'lad·ər ,dril·iŋ }

ladder jack [ENG] A scaffold support which hooks onto a ladder. { 'lad·ər ,jak }

ladder track [CIV ENG] A main track that joins successive body tracks in a railroad yard. { 'lad·ər ,trak }

ladder trencher [MECH ENG] A machine that digs trenches by means of a bucket-ladder excavator. Also known as ladder ditcher. { 'lad·ər ¦trench·ər }

ladle [DES ENG] A deep-bowled spoon with a long handle for dipping up, transporting, and pouring liquids. { 'lād·əl }

lag [CIV ENG] A flat piece of material, usually wood, used to wedge timber or steel supports against the ground and to make secure the space between supports. [ELECTR] A persistence of the electric charge image in a camera tube for a small number of frames. { lag }

lagan [ENG] A heavy object thrown overboard and buoyed to mark its location for future recovery. { 'lag·ən }

lag bolt See coach screw. { 'lag ,bōlt }

lagging [CIV ENG] **1.** Horizontal wooden strips fastened across an arch under construction to transfer weight to the centering form. **2.** Wooden members positioned vertically to prevent cave-ins in earthworking. { 'lag·iŋ }

lagging network *See* integral network. { 'lag·iŋ ˌnet·wərk }

lag-lead network *See* lead-lag network. { 'lag 'lēd ˌnet·wərk }

lag network *See* integral network. { 'lag ˌnet ˌwərk }

Lagrange bracket |MECH| Given two functions of coordinates and momenta in a system, their Lagrange bracket is an expression measuring how coordinates and momenta change jointly with respect to the two functions. { lə'gränj ˌbrak·ət }

Lagrange function *See* Lagrangian. { lə'gränj ˌfəŋk·shən }

Lagrange-Hamilton theory |MECH| The formalized study of continuous systems in terms of field variables where a Lagrangian density function and Hamiltonian density function are introduced to produce equations of motion. { lə'gränj 'ham·əl·tən ˌthē·ə·rē }

Lagrange's equations |MECH| Equations of motion of a mechanical system for which a classical (non-quantum-mechanical) description is suitable, and which relate the kinetic energy of the system to the generalized coordinates, the generalized forces, and the time. Also known as Lagrangian equations of motion. { lə'grän·jəz iˌkwā·zhənz }

Lagrangian |MECH| **1.** The difference between the kinetic energy and the potential energy of a system of particles, expressed as a function of generalized coordinates and velocities from which Lagrange's equations can be derived. Also known as kinetic potential; Lagrange function. **2.** For a dynamical system of fields, a function which plays the same role as the Lagrangian of a system of particles; its integral over a time interval is a maximum or a minimum with respect to infinitesimal variations of the fields, provided the initial and final fields are held fixed. { lə'grän·jē·ən }

Lagrangian coordinates *See* generalized coordinates. { lə'grän·jē·ən ko'ȯrd·ən·əts }

Lagrangian density |MECH| For a dynamical system of fields or continuous media, a function of the fields, of their time and space derivatives, and the coordinates and time, whose integral over space is the Lagrangian. { lə'grän·jē·ən 'den· səd·ē }

Lagrangian equations of motion *See* Lagrange's equations. { lə'grän·jē·ən iˌkwā·zhənz əv 'mō· shən }

Lagrangian function |MECH| The function which measures the difference between the kinetic and potential energy of a dynamical system. { lə'grän·jē·ən ˌfəŋk·shən }

Lagrangian generalized velocity *See* generalized velocity. { lə'grän·jē·ən ¦jen·rə,līzd və'läs·əd·ē }

lag screw *See* coach screw. { 'lag ˌskrü }

lally column |CIV ENG| A hollow and nearly circular steel column that supports girders or beams. { 'läl·ē ˌkäl·əm }

lambda |MECH| A unit of volume equal to 10^{-6} liter or 10^{-9} cubic meter. { 'lam·də }

lambda dispatch |IND ENG| The solution of the problem of finding the most economical use of generators to supply a given quantity of electric power, using the method of Lagrange multipliers, which are symbolized λ. { 'lam·də di ˌspach }

lambda point |THERMO| A temperature at which the specific heat of a substance has a sharply peaked maximum, observed in many second-order transitions. { 'lam·də ˌpȯint }

Lambert surface |THERMO| An ideal, perfectly diffusing surface for which the intensity of reflected radiation is independent of direction. { 'lam·bərt ˌsər·fəs }

Lamé constants |MECH| Two constants which relate stress to strain in an isotropic, elastic material. { lä'mā ˌkän·stəns }

lamella |CIV ENG| A thin member made of reinforced concrete, metal, or wood that is joined with similar members in an overlapping pattern to form an arch or a vault. { lə'mel·ə }

lamella arch |CIV ENG| An arch consisting basically of a series of intersecting skewed arches made up of relatively short straight members; two members are bolted, riveted, or welded to a third piece at its center. { lə'mel·ə ˌärch }

lamella roof |BUILD| A large span vault built of members connected in a diamond pattern. { lə'mel·ə ˌrüf }

laminated spring |DES ENG| A flat or curved spring made of thin superimposed plates and forming a cantilever or beam of uniform strength. { 'lam·ə,nād·əd 'spriŋ }

Lami's theorem |MECH| When three forces act on a particle in equilibrium, the magnitude of each is proportional to the sine of the angle between the other two. { la'mēz ˌthir·əm }

lamp |ENG| A device that produces light, such as an electric lamp. { lamp }

lamphouse |ENG| **1.** The light housing in a motion picture projector, located behind the projector head ordinarily consisting of a carbon arc lamp operating on direct current at about 60 volts, and a concave reflector behind the arc which collects the light and concentrates it on the film, and cooling devices. **2.** A box with a small hole containing an electric lamp and a concave mirror behind it, used as a concentrated source of light in a microscope, photographic enlarger, or other instrument. { 'lamp,haüs }

Lancashire boiler |MECH ENG| A cylindrical steam boiler consisting of two longitudinal furnace tubes which have internal grates at the front. { 'laŋ·kə·shir ˌbȯil·ər }

lance door |MECH ENG| The door to a boiler furnace through which a hand lance is inserted. { 'lans ˌdȯr }

Lanchester balancer |MECH ENG| A device for balancing four-cylinder engines; consists of two meshed gears with eccentric masses, driven by the crankshaft. { 'lan·chə·stər 'bal·ən·sər }

Lanchester's rule |MECH| The rule that a torque applied to a rotating body along an axis perpendicular to the rotation axis will produce precession in a direction such that, if the body is viewed along a line of sight coincident with the torque axis, then a point on the body's circumference, which initially crosses the line of sight, will appear to describe an ellipse whose sense is that of the torque. { 'lan,ches·tərz ,rülz }

land |DES ENG| The top surface of the tooth of a cutting tool, behind the cutting edge. |ELECTR| **1.** One of the regions between pits on a track on an optical disk. **2.** *See* terminal area. |ENG| **1.** In plastics molding equipment, the horizontal bearing surface of a semipositive or flash mold to allow excess material to escape; or the bearing surface along the top of the screw flight in a screw extruder; or the surface of an extrusion die that is parallel to the direction of melt flow. **2.** The surface between successive grooves of a diffraction grating or phonograph record. { land }

land accretion |CIV ENG| Gaining land in a wet area, such as a marsh or by the sea, by planting maritime plants to encourage silt deposition or by dumping dredged materials in the area. Also known as land reclamation. { 'land ə,krē·shən }

land drainage |CIV ENG| The removal of water from land to improve the soil as a medium for plant growth and a surface for land management operations. { land ,drān·ij }

landfill |CIV ENG| Disposal of solid waste by burying in layers of earth in low ground. { 'lan,fil }

landing |CIV ENG| A place where boats receive or discharge passengers, freight, and so on. { 'land·iŋ }

landing gear |MECH ENG| A pair of small wheels at the forward end of a semitrailer to support the vehicle when it is detached from the tractor. { 'land·iŋ ,gir }

landing stage |CIV ENG| A platform, usually floating and attached to the shore, for the discharge and embarkation of passengers, freight, and so on. { 'land·iŋ ,stāj }

landing tee *See* wind tee. { 'land·iŋ ,tee }

landmark |ENG| Any fixed natural or artificial monument or object used to designate a land boundary. { 'lan,märk }

land measure |MECH| **1.** Units of area used in measuring land. **2.** Any system for measuring land. { 'land ,mezh·ər }

land mile *See* mile. { 'lan ¦mīl }

land reclamation *See* land accretion. { 'land ,rek· lə'mā·shən }

landscape architecture |CIV ENG| The art of arranging and fitting land for human use and enjoyment. { 'lan,skāp 'är·kə,tek·chər }

landscape engineer |CIV ENG| A person who applies engineering principles and methods to planning, design, and construction of natural scenery arrangements on a tract of land. { 'lan ,skāp ,en·jə'nīr }

land surveyor |CIV ENG| A specialist who measures land and its natural features and any constructed features such as buildings or roads for

drawing to scale as plans or maps. { 'land sər,vā·ər }

land tie |CIV ENG| A rod or chain connecting an outside structure such as a retaining wall to a buried anchor plate. { 'land ,tī }

land-use classes |CIV ENG| Categories into which land areas can be grouped according to present or potential economic use. { 'land ,yüs ,klas·əz }

lane |CIV ENG| An established route, as an air lane, shipping lane, or highway traffic lane. { lān }

lang lay |DES ENG| A wire rope lay in which the wires of each strand are twisted in the same direction as the strands. { 'laŋ ,lā }

Langmuir diffusion pump |ENG| A type of diffusion pump in which the mercury vapor emerges from a nozzle, giving it motion in a direction away from the high-vacuum side of the pump. { ¦laŋ·myu̇r di'fyü·zhən ,pəmp }

lantern |ENG| A portable lamp. { 'lan·tərn }

lantern pinion |DES ENG| A pinion with bars (between parallel disks) instead of teeth. { 'lan· tərn ,pin·yən }

lantern ring |DES ENG| A ring or sleeve around a rotating shaft; an opening in the ring provides for forced feeding of oil or grease to bearing surfaces; particularly effective for pumps handling liquids. { 'lan·tərn ,riŋ }

lap |CIV ENG| The length by which a reinforcing bar must overlap the bar it will replace. { lap }

lapel microphone |ENG ACOUS| A small microphone that can be attached to a lapel or pocket on the clothing of the user, to permit free movement while speaking. { lə'pel ¦mī·krə,fōn }

lap joint |ENG| A simple joint between two members made by overlapping the ends and fastening them together with bolts, rivets, or welding. { 'lap ,jóint }

lapping |ELECTR| Moving a quartz, semiconductor, or other crystal slab over a flat plate on which a liquid abrasive has been poured, to obtain a flat polished surface or to reduce the thickness a carefully controlled amount. { 'lap·iŋ }

lap siding |BUILD| Beveled boards used for siding that are similar to clapboards but longer and wider. |CIV ENG| Two railroad sidings, the turnout of one overlapping that of the other. { 'lap ,sīd·iŋ }

Laray viscometer |ENG| An instrument designed to measure viscosity and other properties of ink. { lə'rā vi'skäm·əd·ər }

large calorie *See* kilocalorie. { 'lärj 'kal·ə·rē }

large dyne *See* newton. { 'lärj 'dīn }

large-scale integrated circuit |ELECTR| A very complex integrated circuit, which contains well over 100 interconnected individual devices, such as basic logic gates and transistors, placed on a single semiconductor chip. Abbreviated LSI circuit. Also known as chip circuit; multiplefunction chip. { 'lärj ¦skāl ,int·ə,grād·əd 'sər· kət }

large-systems control theory |CONT SYS| A branch of the theory of control systems concerned with the special problems that arise in the design of control algorithms (that is, control policies and strategies) for complex systems. { 'lärj ,sis·tǝmz kǝn'trōl ,thē·ǝ·rē }

Larson-Miller parameter |MECH| The effects of time and temperature on creep, being defined empirically as $P = T (C + \log t) \times 10^{-3}$, where T = test temperature in degrees Rankine (degrees Fahrenheit + 460) and t = test time in hours; the constant C depends upon the material but is frequently taken to be 20. { 'lärs·ǝn 'mil·ǝr pǝ'ram·ǝd·ǝr }

laryngophone |ENG ACOUS| A microphone designed to be placed against the throat of a speaker, to pick up voice vibrations directly without responding to background noise. { lǝ'riŋ·gǝ‚fōn }

LASCR See light-activated silicon controlled rectifier.

LASCS See light-activated silicon controlled switch.

laser amplifier |ELECTR| A laser which is used to increase the output of another laser. Also known as light amplifier. { 'lā·zǝr ¦am·plǝ‚fī·ǝr }

laser anemometer |ENG| An anemometer in which the wind being measured passes through two perpendicular laser beams, and the resulting change in velocity of one or both beams is measured. { 'lā·zǝr an·ǝ'mäm·ǝd·ǝr }

laser ceilometer |ENG| A ceilometer in which the time taken by a light pulse from a ground laser to travel straight up to a cloud ceiling and be reflected to a receiving photomultiplier is measured and converted into a cathode-ray display that indicates cloud-base height. { 'lā·zǝr sē'läm·ǝd·ǝr }

laser earthquake alarm |ENG| An early-warning system proposed for earthquakes, involving the use of two lasers with beams at right angles, positioned across a known geologic fault for continuous monitoring of distance across the fault. { 'lā·zǝr 'ǝrth‚kwāk ǝ‚lärm }

laser/fiber-optic gyroscope See fiber-optic gyroscope. { 'lā·zǝr 'fī·bǝr ¦äp·tik 'jī·rǝ‚skōp }

laser gyro |ENG| A gyro in which two laser beams travel in opposite directions over a ring-shaped path formed by three or more mirrors; rotation is thus measured without the use of a spinning mass. Also known as ring laser. { 'lā·zǝr ¦jī‚rō }

laser intrusion detector |ENG| A photoelectric intrusion detector in which a laser is a light source that produces an extremely narrow and essentially invisible beam around the perimeter of the area being guarded. { 'lā·zǝr in'trü·zhǝn di‚tek·tǝr }

laser ranging |ENG| A technique for determining the distance to a target by precise measurement of the time required for a laser pulse to travel from a transmitter to a reflector on the target and return to a detector. { 'lā·zǝr ‚rānj·iŋ }

laserscope |ENG| A pulsed high-power laser used with appropriate scanning and imaging devices to sense objects over the sea at night or in fog and provide three-dimensional images on a viewing screen. { 'lā·zǝr‚skōp }

laser scriber |ENG| A laser-cutting setup used in place of a diamond scriber for dicing thin slabs of silicon, gallium arsenide, and other semiconductor materials used in the production of semiconductor diodes, transistors, and integrated circuits; also used for scribing sapphire and ceramic substrates. { 'lā·zǝr ¦skrīb·ǝr }

laser seismometer |ENG| A laser interferometer system that detects seismic strains in the earth by measuring changes in distance between two granite piers located at opposite ends of an evacuated pipe through which a helium-neon or other laser beam makes a round trip; movements as small as 80 nanometers (one-eighth the wavelength of the 632.8-nanometer helium-neon laser radiation) can be detected. { 'lā·zǝr sīz 'mäm·ǝd·ǝr }

laser threshold |ELECTR| The minimum pumping energy required to initiate lasing action in a laser. { 'lā·zǝr ¦thresh·hōld }

laser tracking |ENG| Determination of the range and direction of a target by echoed coherent light. { 'lā·zǝr ¦trak·iŋ }

laser transit |ENG| A transit in which a laser is mounted over the sighting telescope to project a clearly visible narrow beam onto a small target at the survey site. { 'lā·zǝr ¦tranz·ǝt }

lashing |ENG| A rope, chain, or wire used for binding, fastening, or wrapping. { 'lash·iŋ }

lash-up |ENG| A model or test sample of equipment required in the testing of a new concept or idea which is in the embryo stage. { 'lash‚ǝp }

last in, first out |IND ENG| A method of determining the inventory costs by transferring the costs of material to the product in reverse chronological order. Abbreviated LIFO. { ‚last 'in ‚first 'aút }

latch |ELECTR| An electronic circuit that reverses and maintains its state each time that power is applied. |ENG| **1.** Any of various closing devices on a door that fit into a hook, notch, or cavity in the frame. **2.** In plastics fabrication, a device used to hold together the two members of a mold. { lach }

latch bolt |DES ENG| A self-acting spring bolt with a beveled head. { 'lach ‚bōlt }

latch-up phenomenon |ELECTR| In a bipolar or MOS integrated circuit, the generation of photocurrents by ionizing radiation which can provide a trigger signal for a parasitic *pnpn* circuit and possibly result in permanent damage or operational failure if the circuit remains in this state. { 'lach ‚ǝp fǝ‚näm·ǝ‚nän }

latent defect |IND ENG| A flaw or other imperfection in any article which is discovered after delivery; usually, latent defects are inherent weaknesses which normally are not detected by examination or routine tests, but which are present at time of manufacture and are aggravated by use. { 'lāt·ǝnt 'dē‚fekt }

latent heat |THERMO| The amount of heat absorbed or evolved by 1 mole, or a unit mass, of a substance during a change of state (such as fusion, sublimation or vaporization) at constant temperature and pressure. { 'lāt·ənt 'hēt }

latent heat of fusion See heat of fusion. { 'lāt·ənt ¦hēt əv 'fyü·zhən }

latent heat of sublimation See heat of sublimation. { 'lāt·ənt ¦hēt əv ‚səb·lə'mā·shən }

latent heat of vaporization See heat of vaporization. { 'lāt·ənt ¦hēt əv ‚vā·pə·rə'zā·shən }

latent load [MECH ENG] Cooling required to remove unwanted moisture from an air-conditioned space. { 'lāt·ənt ‚lōd }

lateral [ENG] In a gas distribution or transmission system, a pipe branching away from the central, primary part of the system. { 'lad·ə·rəl }

lateral compliance [ENG ACOUS] That characteristic of a stylus based on the force required to move it from side to side as it follows the grooves of a phonograph record. { 'lad·ə·rəl kəm'plī·əns }

lateral extensometer [ENG] An instrument used in photoelastic studies of the stresses on a plate; it measures the change in the thickness of the plate resulting from the stress at various points. { 'lad·ə·rəl ‚ek‚sten'säm·əd·ər }

lateral flow spillway See side-channel spillway. { 'lad·ə·rəl ‚flō 'spil‚wā }

lateral recording [ENG ACOUS] A type of disk recording in which the groove modulation is parallel to the surface of the recording medium so that the cutting stylus moves from side to side during recording. { 'lad·ə·rəl ri'kórd·iŋ }

lateral search See profiling. { 'lad·ə·rəl 'sərch }

lateral sewer [CIV ENG] A sewer discharging into a branch or other sewer and having no tributary sewer. { 'lad·ə·rəl 'sü·ər }

lateral support [CIV ENG] Horizontal propping applied to a column, wall, or pier across its smallest dimension. { 'lad·ə·rəl sə'pórt }

laterlog [ENG] A downhole resistivity measurement method wherein electric current is forced to flow radially through the formation in a sheet of predetermined thickness; used to measure the resistivity in hard-rock reservoirs as a method of determining subterranean structural features. { 'lad·ər‚läg }

lath [CIV ENG] **1.** A narrow strip of wood used in making a level base, as for plaster or tiles, or in constructing a light framework, as a trellis. **2.** A sheet of material used as a base for plaster. { lath }

lathe [MECH ENG] A machine for shaping a workpiece by gripping it in a holding device and rotating it under power against a suitable cutting tool for turning, boring, facing, or threading. { lāth }

lathing board See backup strip. { 'lath·iŋ ‚bórd }

latrine [ENG] A toilet facility, either fixed or of a portable nature, such as is maintained underground for use by miners. { lə‚trēn }

lattice [CIV ENG] A network of crisscrossed strips of metal or wood. { 'lad·əs }

lattice filter [ELECTR] An electric filter consisting of a lattice network whose branches have

L-C parallel-resonant circuits shunted by quartz crystals. { 'lad·əs ‚fil·tər }

lattice girder [CIV ENG] An open girder, beam, or column built from members joined and braced by intersecting diagonal bars. Also known as open-web girder. { 'lad·əs ‚gərd·ər }

lattice truss [CIV ENG] A truss that resembles latticework because of diagonal placement of members connecting the upper and lower chords. { 'lad·əs ‚trəs }

launching [CIV ENG] The act or process of floating a ship after only hull construction is completed; in some cases ships are not launched until after all construction is completed. { 'lón·chiŋ }

launching cradle [CIV ENG] A framework made of wood to support a vessel during launching from sliding ways. { 'lón·chiŋ ‚krād·əl }

launching ways [CIV ENG] Two (or more) sets of long, heavy timbers arranged longitudinally under the bottom of a ship during building and launching, with one set on each side, and sloping toward the water; the lower set, or ground ways, remain stationary and support the upper set, or sliding ways, which carry the weight of the ship after the shores and keel blocks are removed. { 'lón·chiŋ ‚wāz }

launder [ENG] An inclined channel or trough for the conveyance of a liquid, such as for water in mining and construction engineering or for molten metal. { 'lón·dər }

Lauson engine [ENG] Single-cylinder engine used in screening tests prior to the L-series lube oil tests (such as L-1 or L-2 tests). { 'laüz·ən ‚en·jən }

lawnmower [ELECTR] Type of radio-frequency preamplifier used with radar receivers. [ENG] A helix-type recorder mechanism. [MECH ENG] A machine for cutting grass on lawns. { 'lón‚mō·ər }

law of action and reaction See Newton's third law. { 'ló əv 'ak·shən ən 'rē‚ak·shən }

law of corresponding times [MECH] The principle that the times for corresponding motions of dynamically similar systems are proportional to L/V and also to $\sqrt{(L/G)}$, where L is a typical dimension of the system, V a typical velocity, and G a typical force per unit mass. { ¦ló əv ‚kär·ə¦spänd·iŋ 'tīmz }

law of electric charges [ELEC] The law that like charges repel, and unlike charges attract. { 'ló əv i¦lek·trik 'chärj·əz }

law of electrostatic attraction See Coulomb's law. { 'ló əv i¦lek·trə¦stad·ik ə'trak·shən }

law of gravitation See Newton's law of gravitation. { 'ló əv ‚grav·ə'tā·shən }

lay [DES ENG] The direction, length, or angle of twist of the strands in a rope or cable. { lā }

lay off [ENG] The process of fairing a ship's lines or an airplane's in a mold loft in order to make molds and templates for structural units. { 'lā ‚óf }

lay-up [ENG] Production of reinforced plastics by positioning the reinforcing material (such as glass fabric) in the mold prior to impregnation with resin. { 'lā,əp }

lazy jack [ENG] A device that accommodates changes in length of a pipeline or similar structure through the motion of two linked bell cranks. { 'lā·zē 'jak }

lb See pound.

lb ap See pound.

lb apoth See pound.

lbf See pound.

lbf-ft See foot-pound.

lb t See pound.

lb tr See pound.

LCA See life-cycle assessment.

LCD See liquid crystal display.

LCL See less-than-carload.

L/D ratio [ENG] Length to diameter ratio, a frequently used engineering relationship. { 'el'dē ,rā·shō }

leaching [CHEM ENG] The dissolving, by a liquid solvent, of soluble material from its mixture with an insoluble solid; leaching is an industrial separation operation based on mass transfer; examples are the washing of a soluble salt from the surface of an insoluble precipitate, and the extraction of sugar from sugarbeets. { 'lēch·iŋ }

lead [DES ENG] The distance that a screw will advance or move into a nut in one complete turn. |ELEC| A wire used to connect two points in a circuit. [ENG] A mass of lead attached to a line, as used for sounding at sea. { led }

lead angle [DES ENG] The angle that the tangent to a helix makes with the plane normal to the axis of the helix. { 'lēd ,aŋ·gəl }

lead-chamber process [CHEM ENG] A process for the preparation of impure or dilute (60–78) sulfuric acid; sulfur dioxide is oxidized by moist air with nitrogen oxide catalysts in a series of lead-lined chambers the Gay-Lussac tower and the Glover tower; used primarily in the manufacture of fertilizer. { 'led ,chām·bər ,prä·səs }

lead compensation [CONT SYS] A type of feedback compensation primarily employed for stabilization or for improving a system's transient response; it is generally characterized by a series compensation transfer function of the type

$$G_c(s) = K \frac{(s - z)}{(s - p)}$$

where $z < p$ and K is a constant. { 'lēd ,käm·pən'sā·shən }

lead curve [CIV ENG] The curve in a railroad turnout between the switch and the frog. { 'lēd ,kərv }

leader [BUILD] See downspout. [ENG] The unrecorded length of magnetic tape that enables the operator to thread the tape through the drive and onto the take-up reel without losing data or recorded music, speech, or such. [MECH ENG] In a hot-air heating system, a duct that conducts heated air to an outlet. { 'lēd·ər }

leader streamer See leader. { 'lēd·ər ,strēm·ər }

leading edge [DES ENG] The surfaces or inset cutting points on a bit that face in the same direction as the rotation of the bit. { 'lēd·iŋ 'ej }

lead-in groove [DES ENG] A blank spiral groove at the outside edge of a disk recording, generally of a pitch much greater than that of the recorded grooves, provided to bring the pickup stylus quickly to the first recorded groove. Also known as lead-in spiral. { 'lēd,in ,grüv }

leading truck [MECH ENG] A swiveling frame with wheels under the front end of a locomotive. { 'lēd·iŋ 'trək }

lead-in spiral See lead-in groove. { 'lēd,in 'spī·rəl }

lead joint [ENG] A pipe joint made by caulking with lead wool or molten lead. { 'led ,jóint }

lead-lag network [CONT SYS] Compensating network which combines the characteristics of the lag and lead networks, and in which the phase of a sinusoidal response lags a sinusoidal input at low frequencies and leads it at high frequencies. Also known as lag-lead network. { 'lēd 'lag 'net,wərk }

lead line See sounding line. { 'led ,līn }

lead lining [ENG] Lead sheeting used to line the inside surfaces of liquid-storage vessels and process equipment to prevent corrosion. { 'led 'līn·iŋ }

lead network See derivative network. { 'lēd ,net,wərk }

lead-out groove [DES ENG] A blank spiral groove at the end of a disk recording, generally of a pitch much greater than that of the recorded grooves, connected to either the locked or eccentric groove. Also known as throw-out spiral. { 'lēd,aút ,grüv }

lead-over groove [DES ENG] A groove cut between separate selections or sections on a disk recording to transfer the pickup stylus from one cut to the next. Also known as cross-over spiral. { 'lēd,ō·vər ,grüv }

lead rail [CIV ENG] In an ordinary rail switch, the turnout rail lying between the rails of the main track. { 'lēd ,rāl }

lead screw [MECH ENG] A threaded shaft used to convert rotation to longitudinal motion; in a lathe it moves the tool carriage when cutting threads; in a disk recorder it guides the cutter at a desired rate across the surface of an ungrooved disk. { 'lēd ,skrü }

lead susceptibility [CHEM ENG] The increase in octane number of gasoline imparted by the addition of a specified amount of TEL (tetraethyllead). { 'led sə,sep·tə'bil·əd·ē }

lead time [IND ENG] The time allowed or required to initiate and develop a piece of equipment that must be ready for use at a given time. { 'lēd ,tīm }

lead track [CIV ENG] A distance measured along a straight railroad track from a switch to a frog. { 'lēd ,trak }

lead wire [ENG] One of the heavy wires connecting a firing switch with the cap wires. { 'lēd ,wīr }

leaf [BUILD] **1.** A separately movable division of a folding or sliding door. **2.** One of a pair of doors or windows. **3.** One of the two halves of a cavity wall. { lēf }

leaf spring [DES ENG] A beam of cantilever design, firmly anchored at one end and with a large deflection under a load. Also known as flat spring. { 'lēf ,spriŋ }

league [MECH] A unit of length equal to 3 miles or 4828.032 meters. { lēg }

leakage [ENG] Undesired and gradual escape or entry of a quantity, such as loss of neutrons by diffusion from the core of a nuclear reactor, escape of electromagnetic radiation through joints in shielding, flow of electricity over or through an insulating material, and flow of magnetic lines of force beyond the working region. { 'lēk·ij }

leakage current [ELEC] **1.** Undesirable flow of current through or over the surface of an insulating material or insulator. **2.** The flow of direct current through a poor dielectric in a capacitor. [ELECTR] The alternating current that passes through a rectifier without being rectified. { 'lēk·ij ,kə·rənt }

leakage rate [ENG] Flow rate of all leaks from an evacuated vessel. { 'lēk·ij ,rāt }

leakage resistance [ELEC] The resistance of the path over which leakage current flows; it is normally high. { 'lēk·ij ri¦zis·təns }

leak detector [ENG] An instrument used for finding small holes or cracks in the walls of a vessel; the helium mass spectrometer is an example. { 'lēk di,tek·tər }

leak test pressure [MECH ENG] The inlet pressure used for a standard quantitative seat leakage test. { 'lēk ¦test ,presh·ər }

lean fuel mixture See lean mixture. { 'lēn 'fyül ,miks·chər }

leaning wheel grader [CIV ENG] A grader with skewed wheels to help cut or spread the soil. { 'lēn·iŋ ¦wēl 'grād·ər }

lean manufacturing [IND ENG] A production system consisting of manufacturing cells linked together with a functionally integrated system for inventory and production control that uses less of the key resources needed to make goods. { ¦lēn ,man·ə'fak·chər·iŋ }

lean manufacturing cells [IND ENG] Typically U-shaped manufacturing cells in which workers, cross-trained on all the related processes, move from machine to machine in counterclockwise loops. { ¦lēn ,man·ə'fak·chər·iŋ ,selz }

lean mixture [MECH ENG] A fuel-air mixture containing a low percentage of fuel and a high percentage of air, as compared with a normal or rich mixture. Also known as lean fuel mixture. { 'lēn ¦miks·chər }

lean-to [BUILD] A single-pitched roof whose summit is supported by the wall of a higher structure. { 'lēn,tü }

lear See lehr. { lir }

learning control [CONT SYS] A type of automatic control in which the nature of control parameters and algorithms is modified by the actual experience of the system. { 'lərn·iŋ kən ,trōl }

lease [IND ENG] **1.** Contract between landowner and another granting the latter the right to use the land, usually upon payment of an agreed rental, bonus, or royalty. **2.** A piece of land that is leased. { lēs }

least-action principle See principle of least action. { ¦lēst 'ak·shən ,prin·sə·pəl }

least-energy principle [MECH] The principle that the potential energy of a system in stable equilibrium is a minimum relative to that of nearby configurations. { ¦lēst 'en·ər·jē ,prin·sə·pəl }

least-work theory [MECH] A theory of statically indeterminate structures based on the fact that when a stress is applied to such a structure the individual parts of it are deflected so that the energy stored in the elastic members is minimized. { ¦lēst 'wərk ,thē·ə·rē }

LED See light-emitting diode.

LEDE room [ENG ACOUS] A control room in a sound-recording studio in which the rear wall is made reflective or diffusive, while the dead or sound-absorbent treatment is applied to the frontal sidewalls near the loudspeaker to prevent lateral reflections from mixing with direct signals from the loudspeaker. Derived from live-end-dead-end room. { 'lē,dē ,rüm }

ledge [BUILD] A horizontal timber on the back of a batten door or on a framed and braced door. [ENG] **1.** A raised edge or molding. **2.** A narrow shelf projecting from the side of a vertical structure. **3.** A horizontal timber that supports the put-logs of scaffolding. { lej }

ledged door See batten door. { 'lejd 'dór }

ledger [CIV ENG] A main horizontal member of formwork, supported on uprights and supporting the soffit of the formwork. [ENG] The horizontal support for a scaffold platform. { 'lej·ər }

Ledoux bell meter [ENG] A type of manometer used to measure the difference in pressure between two points generated by any one of several types of flow measurement devices such as a pitot tube; it is equipped with a shaped plug which makes the reading of the meter directly proportional to the flow rate. { lə'dü 'bel ,mēd·ər }

leer See lehr. { ler }

Lee's disk [THERMO] A device for determining the thermal conductivity of poor conductors in which a thin, cylindrical slice of the substance under study is sandwiched between two copper disks, a heating coil is placed between one of these disks and a third copper disk, and the temperatures of the three copper disks are measured. { 'lēz ,disk }

left-hand [DES ENG] Of drilling and cutting tools, screw threads, and other threaded devices, designed to rotate clockwise or cut to the left. { 'left ¦hand }

left-handed See left-laid. { 'left ¦hand·əd }

left-hand screw [DES ENG] A screw that advances when turned counterclockwise. { 'left ¦hand 'skrü }

left-laid [DES ENG] The lay of a wire or fiber rope or cable in which the individual wires or fibers in the strands are twisted to the right and the strands to the left. Also known as left-handed; regular-lay left twist. { 'left ¦lād }

leg [ENG] **1.** Anything that functionally or structurally resembles an animal leg. **2.** One of the branches of a forked or jointed object. **3.** One of the main upright members of a drill derrick or tripod. [MECH ENG] The case that encloses the vertical part of the belt carrying the buckets within a grain elevator. { leg }

leg wire [ENG] One of the two wires forming a part of an electric blasting cap or squib. { 'leg ,wīr }

lehr [ENG] A long oven in which glass is cooled and annealed after being formed. Also spelled lear; leer. { ler }

Leidenfrost point [THERMO] The lowest temperature at which a hot body submerged in a pool of boiling water is completely blanketed by a vapor film; there is a minimum in the heat flux from the body to the water at this temperature. { 'līd·ən,fróst ,póint }

Leidenfrost's phenomenon [THERMO] A phenomenon in which a liquid dropped on a surface that is above a critical temperature becomes insulated from the surface by a layer of vapor, and does not wet the surface as a result. { 'līd·ən,frósts fə,nam·ə,nän }

Lenard spiral [ENG] A type of magnetometer consisting of a spiral of bismuth wire and a Wheatstone bridge to measure changes in the resistance of the wire produced by magnetic fields and as a result of the transverse magnetoresistance of bismuth. { 'lā·närd ,spī·rəl }

length [MECH] Extension in space. { leŋkth }

lengthening joint [ENG] A joint between two members running in the same direction. { 'leŋk·thə,niŋ ,jóint }

length of lay [DES ENG] The distance measured along a line parallel to the axis of the rope in which the strand makes one complete turn about the axis of the rope, or the wires make a complete turn about the axis of the strand. { 'leŋkth əv 'lā }

length of shot [ENG] The depth of the shothole, in which powder is placed, or the size of the block of coal or rock to be loosened by a single blast, measured parallel with the hole. { 'leŋkth əv 'shät }

leo [MECH] A unit of acceleration, equal to 10 meters per second per second; it has rarely been employed. { 'lē·ō }

Leslie cube [THERMO] A metal box, with faces having different surface finishes, in which water is heated and next to which a thermopile is placed in order to compare the heat emission properties of different surfaces. { 'lez·lē ,kyüb }

Leslie effect [ENG ACOUS] A dynamic timbre-changing effect created by rotating one or more directional speakers inside a cabinet such that a mixture of Doppler-shifted reflections is generated in the output of an electronic instrument. { 'lez·lē i,fekt }

less-than-carload [IND ENG] Too light to fill a freight car and therefore not eligible for carload rate. Abbreviated LCL. { 'les thən 'kär,lōd }

letters patent See patent. { 'led·ərz 'pat·ənt }

levee [CIV ENG] **1.** A dike for confining a stream. **2.** A pier along a river. { 'lev·ē }

level [CIV ENG] **1.** A surveying instrument with a telescope and bubble tube used to take level sights over various distances, commonly 100 feet (30 meters). **2.** To make the earth surface horizontal. [DES ENG] A device consisting of a bubble tube that is used to find a horizontal line or plane. Also known as spirit level. [ELEC] A single bank of contacts, as on a stepping relay. [ELECTR] **1.** The difference between a quantity and an arbitrarily specified reference quantity, usually expressed as the logarithm of the ratio of the quantities. **2.** A charge value that can be stored in a given storage element of a charge storage tube and distinguished in the output from other charge values. { 'lev·əl }

leveled element time See normal element time. { 'lev·əld ,el·ə,ment 'tīm }

leveled time See normal time. { 'lev·əld ,tīm }

leveler [ENG] A back scraper, drag, or other form of device for smoothing land. { 'lev·ə·lər }

level indicator [ENG] An instrument that indicates liquid level. [ENG ACOUS] An indicator that shows the audio voltage level at which a recording is being made; may be a volume-unit meter, neon lamp, or cathode-ray tuning indicator. { 'lev·əl 'in·də,kād·ər }

leveling [ENG] Adjusting any device, such as a launcher, gun mount, or sighting equipment, so that all horizontal or vertical angles will be measured in the true horizontal and vertical planes. [IND ENG] A method of performance rating which seeks to rate the principal factors that cause the speed of motions rather than speed itself; it considers that the level at which the operator works is influenced by effort and skill. { 'lev·ə·liŋ }

leveling instrument [ENG] An instrument for establishing a horizontal line of sight, usually by means of a spirit level or a pendulum device. { 'lev·ə·liŋ ,in·strə·mənt }

leveling screw [ENG] An adjusting screw used to bring an instrument into level. { 'lev·ə·liŋ ,skrü }

level measurement [MECH] The determination of the linear vertical distance between a reference point or datum plane and the surface of a liquid or the top of a pile of divided solid. { 'lev·əl 'mezh·ər·mənt }

level point See point of fall. { 'lev·əl ,póint }

level rod [ENG] A straight rod or bar, with a flat face graduated in plainly visible linear units with zero at the bottom, used in measuring the vertical distance between a point of the earth's surface and the line of sight of a leveling instrument that has been adjusted to the horizontal position. { 'lev·əl ,räd }

level surface [ENG] A surface which is perpendicular to the plumb line at every point. { 'lev·əl 'sər·fəs }

level valve [MECH ENG] A valve operated by a

lever which travels through a maximum arc of 180°. { 'lev·əl ,valv }

Levenstein process [CHEM ENG] A process for the manufacture of mustard gas from ethene, $CH_2=CH_2$, and sulfur chloride, S_2Cl_2. { 'le·vən,stīn ,prä·səs }

lever [ENG] A rigid bar, pivoted about a fixed point (fulcrum), used to multiply force or motion; used for raising, prying, or dislodging an object. { 'lev·ər, lē·vər }

leverage [MECH] The multiplication of force or motion achieved by a lever. { 'lev·rij }

lever shears [DES ENG] A shears in which the input force at the handles is related to the output force at the cutting edges by the principle of the lever. Also known as alligator shears; crocodile shears. { 'lev·ər ,shirz }

levitated vehicle [MECH ENG] A train or other vehicle which travels at high speed at some distance above an electrically conducting track by means of levitation. { 'lev·ə,tād·əd 've·ə·kəl }

lewis [DES ENG] A device for hoisting heavy stones; employs a dovetailed tenon that fits into a mortise in the stone. { 'lü·əs }

lewis bolt [DES ENG] A bolt with an enlarged, tapered head that is inserted into masonry or stone and fixed with lead; used as a foundation bolt. { 'lü·əs ,bōlt }

Lewis-Matheson method [CHEM ENG] Trial-and-error calculation method for the design of multicomponent distillation columns, or for the determination of the separating ability of an existing column. { 'lü·əs 'math·ə·sən ,meth·əd }

L-head engine [MECH ENG] A type of four-stroke cycle internal combustion engine having both inlet and exhaust valves on one side of the engine block which are operated by pushrods actuated by a single camshaft. { 'el ,hed 'en·jən }

lie detector [ENG] An instrument that indicates or records one or more functional variables of a person's body while the person undergoes the emotional stress associated with a lie. Also known as polygraph; psychintegroammeter. { 'lī dī,tek·tər }

life-cycle assessment [SYS ENG] A methodology that identifies the environmental impacts associated with the life cycle of a material or product in a specific application, thus identifying opportunities for improvement in environmental performance. Abbreviated LCA. { 'līf ,sī·kəl ə,ses·mənt }

life-cycle cost [ENG] A measurement of the total cost of using equipment over the entire time of service of the equipment; includes initial, operating, and maintenance costs. { 'līf ,sī·kəl ,kȯst }

life expectancy [ENG] The predicted useful service life of an item of equipment. { 'līf ik'spek·tən·sē }

life preserver [ENG] A buoyant device that is used to prevent drowning by supporting a person in the water. { 'līf pri,zər·vər }

life support system [ENG] A system providing atmospheric control and monitoring, such as a breathing mixture supply system, air purification and filtering system, or carbon dioxide removal system; used in oceanographic submersibles and spacecraft. { 'līf sə,pȯrt ,sis·təm }

life test [CHEM ENG] In petroleum testing, an American Society for Testing and Materials oxidation test made on inhibited steam-turbine oils to determine their stability under oxidizing conditions. [ENG] A test in which a device is operated under conditions that simulate a normal lifetime of use, to obtain an estimate of service life. { 'līf ,test }

LIFO *See* last in, first out. { 'lī,fō }

lift *See* elevator. { lift }

lift bridge [CIV ENG] A drawbridge whose movable spans are raised vertically. { 'lift ,brij }

lifter flight [DES ENG] Spaced plates or projections on the inside surfaces of cylindrical rotating equipment (such as rotary dryers) to lift and shower the solid particles through the gas-drying stream during their passage through the dryer cylinder. { 'lif·tər ,flīt }

lifter roof [ENG] Gas storage tank in which the roof is raised by the incoming gas as the tank fills. { 'lif·tər ,rüf }

lifting block [MECH ENG] A combination of pulleys and ropes which allows heavy weights to be lifted with least effort. { 'lift·iŋ ,bläk }

lifting device [ENG] A device to manually open a pressure relief valve by decreasing the spring loading in order to determine if the valve is in working order. { 'lift·iŋ di,vīs }

lifting dog [ENG] **1.** A component part of the overshot assembly that grasps and lifts the inner tube or a wire-line core barrel. **2.** A clawlike hook for grasping cylindrical objects, such as drill rods or casing, while raising and lowering them. { 'lift·iŋ ,dȯg }

lifting magnet [ENG] A large circular, rectangular, or specially shaped magnet used for handling pig iron, scrap iron, castings, billets, rails, and other magnetic materials. { 'lift·iŋ ,mag·nət }

lifting task [IND ENG] A task that involves application of a moment to the vertebral column of the worker. { 'lift·iŋ ,task }

lift pump [MECH ENG] A pump for lifting fluid to the pump's own level. { 'lift ,pəmp }

lift-slab construction [CIV ENG] Pouring reinforced concrete roof and floor slabs at ground level, then lifting them into position after hardening. { 'lift ,slab kən,strək·shən }

lift truck [MECH ENG] A small hand- or power-operated dolly equipped with a platform or forklift. { 'lift ,trək }

lift valve [MECH ENG] A valve that moves perpendicularly to the plane of the valve seat. { 'lift ,valv }

ligament [ENG] The section of solid material in a tube sheet or shell between adjacent holes. { 'lig·ə·mənt }

light-activated silicon controlled rectifier [ELECTR] A silicon controlled rectifier having a glass window for incident light that takes the place of, or adds to the action of, an electric gate current in providing switching action. Abbreviated LASCR. Also known as photo-SCR; photothyristor. { 'līt ¦ak·tə,vād·əd ¦sil·ə·kən kən¦trōld 'rek·tə,fī·ər }

light-activated silicon controlled switch [ELECTR] A semiconductor device that has four layers of silicon alternately doped with acceptor and donor impurities, but with all four of the *p* and *n* layers made accessible by terminals; when a light beam hits the active light-sensitive surface, the photons generate electron-hole pairs that make the device turn on; removal of light does not reverse the phenomenon; the switch can be turned off only by removing or reversing its positive bias. Abbreviated LASCS. { 'līt ¦ak·tə,vād·əd ¦sil·ə·kən kən¦trōld 'swich }

light amplifier [ELECTR] **1.** Any electronic device which, when actuated by a light image, reproduces a similar image of enhanced brightness, and which is capable of operating at very low light levels without introducing spurious brightness variations (noise) into the reproduced image. Also known as image intensifier. **2.** *See* laser amplifier. { 'līt ,am·plə,fī·ər }

light-beam galvanometer *See* d'Arsonval galvanometer. { 'līt ,bēm ,gal·və'näm·əd·ər }

light-beam pickup [ENG ACOUS] A phonograph pickup in which a beam of light is a coupling element of the transducer. { 'līt ,bēm 'pik,əp }

light blasting [ENG] Loosening of shallow or small outcrops of rock and breaking boulders by explosives. { 'līt 'blast·iŋ }

light-emitting diode [ELECTR] A rectifying semiconductor device which converts electrical energy into electromagnetic radiation. The wavelength of the emitted radiation ranges from the near-ultraviolet to the near-infrared, that is, from about 400 to over 1500 nanometers. Abbreviated LED. { 'līt i,mid·iŋ 'dī,ōd }

lightening hole [CIV ENG] An opening cut into a strengthening member that decreases its weight without significantly altering its strength. { 'līt·niŋ ,hōl }

lighterage [IND ENG] **1.** Loading or unloading ships by means of a lighter. **2.** The fee charged for this operation. { 'līd·ə·rij }

lighting-off torch [ENG] A torch used to ignite a fuel oil burner; it consists of asbestos cloth wrapped around an iron rod and soaked with oil. { 'līd·iŋ ,óf ,tórch }

light-inspection car [MECH ENG] A railway motorcar weighing 400–600 pounds (180–270 kilograms) and having a capacity of 650–800 pounds (295–360 kilograms). { ¦līt in'spek·shən ,kär }

light meter [ENG] A small, portable device for measuring illumination; an exposure meter is a specific application, being calibrated to give photographic exposures. { 'līt ,mēd·ər }

light modulator [ELECTR] The combination of a source of light, an appropriate optical system, and a means for varying the resulting light beam to produce an optical sound track on motion picture film. { 'līt ,mäj·ə,lād·ər }

lightning arrester [ELEC] A protective device designed primarily for connection between a conductor of an electrical system and ground to limit the magnitude of transient overvoltages on equipment. Also known as arrester; surge arrester. { 'līt·niŋ ə,res·tər }

light section car [MECH ENG] A railway motorcar weighing 750–900 pounds (340–408 kilograms) and propelled by 4–6-horsepower (3000–4500-watt) engines. { 'līt 'sek·shən ,kär }

light-sensitive [ELECTR] Having photoconductive, photoemissive, or photovoltaic characteristics. Also known as photosensitive. { 'līt 'sen·səd·iv }

light-sensitive cell *See* photodetector. { 'līt ¦sen·səd·iv 'sel }

light-sensitive detector *See* photodetector. { 'līt ¦sen·səd·iv di'tek·tər }

light valve [ELECTR] **1.** A device whose light transmission can be made to vary in accordance with an externally applied electrical quantity, such as volatage, current, electric field, or magnetic field, or an electron beam. **2.** Any direct-view electronic display optimized for reflecting or transmitting an image with an independent collimated light source for projection purposes. { 'līt,valv }

Lilly controller [MECH ENG] A device on steam and electric winding engines that protects against overspeed, overwind, and other incidents injurious to workers and the engine. { 'lil·ē kən¦trōl·ər }

limb [DES ENG] **1.** The graduated margin of an arc or circle in an instrument for measuring angles, as that part of a marine sextant carrying the altitude scale. **2.** The graduated staff of a leveling rod. { limb }

lime kiln [CHEM ENG] Furnace-type apparatus, usually a long, tilted cylinder that is slowly rotated, used to heat calcium carbonate, $CaCO_3$, above 900°C to produce lime. { 'līm ,kil }

limelight [ENG] A light source once used in spotlights; it consisted of a block of lime heated to incandescence by means of an oxyhydrogen flame torch. { 'līm,līt }

limestone log [ENG] A log that employs an electrical resistivity element in the form of four symmetrically arranged current electrodes to give accurate readings in borehole surveying of hard formations. { 'līm,stōn 'läg }

liming [CHEM ENG] Soaking hides and skins in milk of lime and causing them to swell, to facilitate the removal of hair. { 'līm·iŋ }

limit control [MECH ENG] **1.** In boiler operation, usually a device, electrically controlled, that shuts down a burner at a prescribed operating point. **2.** In machine-tool operation, a sensing device which terminates motion of the workpiece or tool at prescribed points. { 'lim·ət kən,trōl }

limit dimensioning method [DES ENG] Method of dimensioning and tolerancing wherein the maximum and minimum permissible values for a dimension are stated specifically to indicate the size or location of the element in question. { 'lim·ət də,men·chən·iŋ ,meth·əd }

limited-access highway *See* expressway. { 'lim·əd·əd ¦ak·ses 'hī,wā }

limited-degree-of-freedom robot |CONT SYS|
Robot whose end effector can be positioned and
oriented in fewer than six degrees of freedom.
{ 'lim·əd·əd di'grē əv 'frē·dəm 'rō,bät }

limited integrator |ELECTR| A device used in an-
alog computers that has two input signals and
one output signal whose value is proportional
to the integral of one of the input signals with
respect to the other as long as this output signal
does not exceed specified limits. { 'lim·əd·əd
'int·ə,grād·ər }

limited-pressure cycle See mixed cycle. { 'lim·əd·
əd ¦presh·ər ,sī·kəl }

limited-rotation hydraulic actuator |MECH
ENG| A type of hydraulic actuator that produces
limited reciprocating rotary force and motion;
used for lifting, lowering, opening, closing, indexing, and transferring movements; examples
are the piston-rack actuator, single-vane actuator, and double-vane actuator. { 'lim·əd·əd
rō¦tā·shən hī¦drô·lik 'ak·chə,wād·ər }

limited-sequence robot See fixed-stop robot.
{ 'lim·əd·əd ¦sē·kwəns 'rō,bät }

limiter |ELECTR| An electronic circuit used to
prevent the amplitude of an electronic waveform
from exceeding a specified level while preserving
the shape of the waveform at amplitudes less
than the specified level. Also known as amplitude limiter; amplitude-limiting circuit; automatic peak limiter; clipper; clipping circuit; limiter circuit; peak limiter. { 'lim·əd·ər }

limit governor |MECH ENG| A mechanical gov-
ernor that takes over control from the main governor to shut the machine down when speed
reaches a predetermined excess above the allowable rate. Also known as topping governor.
{ 'lim·ət ,gəv·ər·nər }

limiting friction See static friction. { 'lim·əd·iŋ
,frik·shən }

limit lines |IND ENG| Lines on a chart designat-
ing specification limits. { 'lim·ət ,līnz }

limit-load design See ultimate-load design.
{ 'lim·ət ,lōd di,zīn }

limits |DES ENG| In dimensioning, the maxi-
mum and minimum values prescribed for a specific dimension; the limits may be of size if the
dimension concerned is a size dimension, or they
may be of location if the dimension concerned
is a location dimension. { 'lim·əts }

limit state |CIV ENG| The condition beyond
which a structure or a structural member is
deemed unsafe due to one or more loads or load
effects. { 'lim·ət ,stāt }

limit switch |ELEC| A switch designed to cut off
power automatically at or near the limit of travel
of a moving object controlled by electrical
means. { 'lim·ət ,swich }

limit velocity |MECH| In armor and projectile
testing, the lowest possible velocity at which any
one of the complete penetrations is obtained;
since the limit velocity is difficult to obtain, a
more easily obtainable value, designated as the
ballistic limit, is usually employed. { 'lim·ət
və'läs·əd·ē }

limnimeter |ENG| A type of tide gage for mea-
suring lake level variations. { lim'nim·əd·ər }

limnograph |ENG| A recording made on a limni-
meter. { 'lim·nə,graf }

Linde copper sweetening |CHEM ENG| A petro-
leum-refinery process to treat gasolines and distillates with a slurry of clay and cupric chloride
to remove mercaptans. { 'lin·də 'käp·ər ,swēt·
ən·iŋ }

Linde drill See fusion-piercing drill. { 'lin·də ,dril }

line-and-staff organization |IND ENG| A form of
organization structure which combines functional subunits with staff officers in line functions. { ¦līn ən ¦staf ,ȯr·gə·nə,zā·shən }

linear |CONT SYS| Having an output that varies
in direct proportion to the input. { 'lin·ē·ər }

linear actuator |MECH ENG| A device that con-
verts some kind of power, such as hydraulic or
electric power, into linear motion. { 'lin·ē·ər
'ak·chə,wād·ər }

linear control system |CONT SYS| A linear sys-
tem whose inputs are forced to change in a desired manner as time progresses. { 'lin·ē·ər
kən'trōl ,sis·təm }

linear expansity See coefficient of linear expansion.
{ 'lin·ē·ər ik'span·səd·ē }

linear feedback control |CONT SYS| Feedback
control in a linear system. { 'lin·ē·ər 'fēd,bak
kən,trōl }

linear integrated circuit |ELECTR| An integrated
circuit that provides linear amplification of signals. { 'lin·ē·ər ¦int·ə,grād·əd 'sər·kət }

linearization |CONT SYS| 1. The modification of
a system so that its outputs are approximately
linear functions of its inputs, in order to facilitate
analysis of the system. 2. The mathematical
approximation of a nonlinear system, whose de-
partures from linearity are small, by a linear system corresponding to small changes in the variables about their average values. { ,lin·ē·ər·
ə'zā·shən }

linear meter |ENG| A meter in which the deflec-
tion of the pointer is proportional to the quantity
measured. { 'lin·ē·ər 'mēd·ər }

linear momentum See momentum. { 'lin·ē·ər
mə'men·təm }

linear motion See rectilinear motion. { 'lin·ē·ər
'mō·shən }

linear-quadratic-Gaussian problem |CONT SYS|
An optimal-state regulator problem, containing
Gaussian noise in both the state and measurement equations, in which the expected value of
the quadratic performance index is to be minimized. Abbreviated LQG problem. { 'lin·ē·ər
kwə'drad·ik 'gaüs·ē·ən ,präb·ləm }

linear regulator problem |CONT SYS| A type of
optimal control problem in which the system to
be controlled is described by linear differential
equations and the performance index to be minimized is the integral of a quadratic function of
the system state and control functions. Also
known as optimal regulator problem; regulator
problem. { 'lin·ē·ər 'reg·yə,lād·ər ,präb·ləm }

linear scanning |ENG| Radar beam which
moves with constant angular velocity through

the scanning sector, which may be a complete 360° { 'lin·ē·ər 'skan·iŋ }

linear strain [MECH] The ratio of the change in the length of a body to its initial length. Also known as longitudinal strain. { 'lin·ē·ər ¦strān }

linear system [CONT SYS] A system in which the outputs are components of a vector which is equal to the value of a linear operator applied to a vector whose components are the inputs. { 'lin·ē·ər 'sis·təm }

linear system analysis [CONT SYS] The study of a system by means of a model consisting of a linear mapping between the system inputs (causes or excitations), applied at the input terminals, and the system outputs (effects or responses), measured or observed at the output terminals. { 'lin·ē·ər ¦sis·təm ə'nal·ə·səs }

linear velocity See velocity. { 'lin·ē·ər və'läs·əd·ē }

line clinometer [ENG] A clinometer designed to be inserted between rods at any point in a string of drill rods. { 'līn klī'näm·əd·ər }

line driver [ELECTR] An integrated circuit that acts as the interface between logic circuits and a two-wire transmission line. { 'līn ¦drīv·ər }

line functions [IND ENG] Organizational functions having direct authority and responsibility. { 'līn ¦faŋk·shənz }

line hydrophone [ENG ACOUS] A directional hydrophone consisting of one straight-line element, an array of suitably phased elements mounted in line, or the acoustic equivalent of such an array. { 'līn 'hī·drə,fōn }

line level [ENG] A small spirit level fitted with hooks at each end so that it can be hung on a horizontally stretched line. { 'līn ,lev·əl }

line loss [ELEC] Total of the various energy losses occurring in a transmission line. [ENG] The quantity of gas that is lost in a distribution system or pipeline. { 'līn ,lòs }

line lubricator See line oiler. { 'līn ,lü·brə,kād·ər }

line microphone [ENG ACOUS] A highly directional microphone consisting of a single straight-line element or an array of small parallel tubes of different lengths, with one end of each abutting a microphone element. Also known as machine-gun microphone. { 'līn ,mī·krə,fōn }

line mixer See flow mixer. { 'līn ,mik·sər }

line of action [MECH ENG] The locus of contact points as gear teeth profiles go through mesh. { 'līn əv 'ak·shən }

line of balance [IND ENG] A production planning system that schedules key events leading to completion of an assembly on the basis of the delivery date for the completed system. Abbreviated LOB. { 'līn əv 'bal·əns }

line of fall [MECH] The line tangent to the ballistic trajectory at the level point. { 'līn əv ¦fòl }

line of flight [MECH] The line of movement, or the intended line of movement, of an aircraft, guided missile, or projectile in the air. { 'līn əv ¦flīt }

line of impact [MECH] A line tangent to the trajectory of a missile at the point of impact. { 'līn əv 'im,pakt }

line-of-sight velocity See radial velocity. { 'līn əv 'sīt və'läs·əd·ē }

line of thrust [MECH] Locus of the points through which the resultant forces pass in an arch or retaining wall. { 'līn əv 'thròst }

line of tunnel [ENG] The width marked by the exterior lines or sides of a tunnel. { 'līn əv 'tən·əl }

line oiler [MECH ENG] An apparatus inserted in a line conducting air or steam to an air- or steam-activated machine that feeds small controllable amounts of lubricating oil into the air or steam. Also known as air-line lubricator; line lubricator. { 'līn ,òi·lər }

line pack [ENG] The actual amount of gas in a pipeline or distribution system. { 'līn ,pak }

liner [DES ENG] A replaceable tubular sleeve inside a hydraulic or pump-pressure cylinder in which the piston travels. [ENG] A string of casing in a borehole. { 'līn·ər }

liner bushing [DES ENG] A bushing, provided with or without a head, that is permanently installed in a jig to receive the renewable wearing bushings. Also known as master bushing. { 'līn·ər ,bùsh·iŋ }

liner rod See range rod. { 'līn ,räd }

liner plate cofferdam [CIV ENG] A cofferdam made from steel plates about 16 inches (41 centimeters) high and 3 feet (91 centimeters) long, and corrugated for added stiffness. { 'līn·ər ,plāt 'kòf·ər,dam }

line scanner [ENG] An infrared imaging device which utilizes the motion of a moving platform, such as an aircraft or satellite, to scan infrared radiation from the terrain. Also known as thermal mapper. { 'līn ,skan·ər }

line shafting [MECH ENG] One or more pieces of assembled shafting to transmit power from a central source to individual machines. { 'līn ,shaft·iŋ }

linesman [ENG] **1.** A worker who sets up and repairs communication and power lines. **2.** An assistant to a surveyor. { 'līnz·mən }

line space lever [MECH ENG] A lever on a typewriter used to move the carriage to a new line. { 'līn ¦spās ,lev·ər }

line voltage [ELEC] The voltage provided by a power line at the point of use. { 'līn ,vōl·tij }

lining bar [DES ENG] A crowbar with a pinch, wedge, or diamond point at its working end. { 'līn·iŋ ,bär }

lining pole See range rod. { 'līn·iŋ ,pōl }

link [CIV ENG] A standardized part of a surveyor's chain, which is 7.92 inches (20.1168 centimeters) in the Gunter's chain and 1 foot (30.48 centimeters) in the engineer's chain. [DES ENG] **1.** One of the rings of a chain. **2.** A connecting piece in the moving parts of a machine. { liŋk }

linkage [MECH ENG] A mechanism that transfers motion in a desired manner by using some combination of bar links, slides, pivots, and rotating members. { 'liŋ·kij }

link V belt [DES ENG] A V belt composed of a large number of rubberized-fabric links joined by metal fasteners. { 'liŋk 'vē ,belt }

lintel |BUILD| A horizontal member over an opening, such as a door or window, usually carrying the wall load. { 'lint·əl }

linter |MECH ENG| A machine for removing fuzz linters from ginned cottonseed. { 'lin·tər }

lip |CIV ENG| A parapet placed on the downstream margin of a millrace or apron in order to minimize scouring of the river bottom. |DES ENG| Cutting edge of a fluted drill formed by the intersection of the flute and the lip clearance angle, and extending from the chisel edge at the web to the circumference. { lip }

Lippmann electrometer See capillary electrometer. { 'lip·mən ‚i‚lek'träm·əd·ər }

liq pt See pint.

liquefier |ENG| Equipment or system used to liquefy gases; usually employs a combination of compression, heat exchange, and expansion operations. { 'lik·wə‚fī·ər }

liquid-column gage See U-tube manometer. { 'lik·wəd ¦käl·əm ‚gāj }

liquid compass |ENG| A compass in a bowl filled with liquid. { 'lik·wəd 'käm·pəs }

liquid-cooled dissipator See cold plate. { 'lik·wəd ¦küld 'dis·ə‚pād·ər }

liquid-cooled engine |MECH ENG| An internal combustion engine with a jacket cooling system in which liquid, usually water, is circulated to maintain acceptable operating temperatures of machine parts. { 'lik·wəd ¦küld 'en·jən }

liquid cooling |ENG| Use of circulating liquid to cool process equipment and hermetically sealed components such as transistors. { 'lik·wəd 'kül·iŋ }

liquid crystal display |ELECTR| A digital display that consists of two sheets of glass separated by a sealed-in, normally transparent, liquid crystal material; the outer surface of each glass sheet has a transparent conductive coating such as tin oxide or indium oxide, with the viewing-side coating etched into character-forming segments that have leads going to the edges of the display; a voltage applied between front and back electrode coatings disrupts the orderly arrangement of the molecules, darkening the liquid enough to form visible characters even though no light is generated. Abbreviated LCD. { 'lik·wəd 'krist·əl di'splā }

liquid extraction See solvent extraction. { 'lik·wəd ik'strak·shən }

liquid filter |CHEM ENG| A device for the removal of solids or coalesced droplets out of a liquid stream by use of a filter medium, such as a screen, cartridge, or granular bed. { 'lik·wəd 'fil·tər }

liquid-in-glass thermometer |ENG| A thermometer in which the thermally sensitive element is a liquid contained in a graduated glass envelope; the indication of such a thermometer depends upon the difference between the coefficients of thermal expansion of the liquid and the glass; mercury and alcohol are liquids commonly used in meteorological thermometers. { 'lik·wəd in ¦glas thər'mäm·əd·ər }

liquid-in-metal thermometer |ENG| A thermometer in which the thermally sensitive element is a liquid contained in a metal envelope, frequently in the form of a Bourdon tube. { 'lik·wəd in ¦med·əl thər'mäm·əd·ər }

liquid knockout See impingement. { 'lik·wəd 'nä‚kaut }

liquid level control |ENG| Regulation of the linear vertical distance between the surface of a liquid and some reference point. { 'lik·wəd 'lev·əl kən‚trōl }

liquid-liquid extraction |CHEM ENG| The removal of a soluble component from a liquid mixture by contact with a second liquid, immiscible with the carrier liquid in which the component is preferentially soluble. { 'lik·wəd 'lik·wəd ik 'strak·shən }

liquid measure |MECH| A system of units used to measure the volumes of liquid substances in the United States; the units are the fluid dram, fluid ounce, gill, pint, quart, and gallon. { 'lik·wəd ¦mezh·ər }

liquid penetrant test |ENG| A penetrant method of nondestructive testing used to locate defects open to the surface of nonporous materials; penetrating liquid is applied to the surface, and after 1–30 minutes excess liquid is removed, and a developer is applied to draw the penetrant out of defects, thus showing their location, shape, and size. { 'lik·wəd 'pen·ə·trənt ‚test }

liquid-phase hydrogenation |CHEM ENG| Hydrogen reaction with liquid-phase hydrogenatable material, such as unsaturated aliphatic or aromatic hydrocarbons. { 'lik·wəd ‚fāz ‚hī·drə·jə'nā·hshən }

liquid pint See pint. { 'lik·wəd 'pīnt }

liquid piston rotary compressor |MECH ENG| A rotary compressor in which a multiblade rotor revolves in a casing partly filled with liquid, for example, water. { 'lik·wəd ¦pis·tən ¦rōd·ə·rē kəm'pres·ər }

liquid seal |CHEM ENG| **1.** The depth of liquid above an opening from which gas or vapor issues, as for a riser in a distillation-column tray. **2.** Product drawoff in which a depth of liquid prevents the outflow of gas or vapor. { 'lik·wəd 'sēl }

liquid-sealed meter |ENG| A type of positive-displacement meter for gas flows consisting of a cylindrical chamber that is more than half filled with water and divided into four rotating compartments formed by trailing vanes; gas entering through the center shaft into one compartment after another forces rotation that allows the gas then to exhaust out the top as it is displaced by the water. Also known as drum meter. { 'lik·wəd ‚sēld 'mēd·ər }

liquid semiconductor |ELECTR| An amorphous material in solid or liquid state that possesses the properties of varying resistance induced by charge carrier injection. { 'lik·wəd 'sem·i·kən‚dək·tər }

liquid-sorbent dehumidifier |MECH ENG| A sorbent type of dehumidifier consisting of a main circulating fan, sorbent-air contactor, sorbent pump, and reactivator; dehumidification and reactivation are continuous operations, with a small part of the sorbent constantly bled off from the main circulating system and reactivated to the concentration required for the desired effluent dew point. { 'lik·wəd ¦sȯr·bənt ˌdē·yü'mid·ə͵fī·ər }

liquid sulfur dioxide-benzene process |CHEM ENG| A petroleum-refinery process using a mixed solvent (SO_2 and benzene) to dewax lubricating oils or improve their viscosity indices. { 'lik·wəd 'səl·fər dī'äk͵sīd ben'zēn ˌprä·səs }

liquidus line |THERMO| For a two-component system, a curve on a graph of temperature versus concentration which connects temperatures at which fusion is completed as the temperature is raised. { 'lik·wəd·əs ˌlīn }

liquor |CHEM ENG| **1.** Supernatant liquid decanted from a liquid-solids mixture in which the solids have settled. **2.** Liquid overflow from a liquid-liquid extraction unit. { 'lik·ər }

list |ENG| To lean to one side, or deviate from the vertical. { list }

listening station |ENG| A radio or radar receiving station that is continuously manned for various purposes, such as for radio direction finding or for gaining information about enemy electronic devices. { 'lis·ən·iŋ ˌstā·shən }

listing See lashing. { 'list·iŋ }

liter |MECH| A unit of volume or capacity, equal to 1 decimeter cubed, or 0.001 cubic meter, or 1000 cubic centimeters. Abbreviated l; L. { 'lēd·ər }

lithography |ELECTR| A technique used for integrated circuit fabrication in which a silicon slice is coated uniformly with a radiation-sensitive film, the resist, and an exposing source (such as light, x-rays, or an electron beam) illuminates selected areas of the surface through an intervening master template for a particular pattern. { lə'thäg·rə·fē }

live axle |MECH ENG| An axle to which wheels are rigidly fixed. { 'līv 'ak·səl }

live center |MECH ENG| A lathe center that fits into the headstock spindle. { 'līv 'sen·tər }

live-end-dead-end room See LEDE room. { ¦līv ˌend 'ded͵end ˌrüm }

live load |MECH| A moving load or a load of variable force acting upon a structure, in addition to its own weight. { 'līv 'lōd }

live load allowance |ENG| The permissible load that may be added to a completed building structure, including installations, equipment, and personnel. { 'līv ˌlōd ə͵laü·əns }

live-roller conveyor |MECH ENG| Conveying machine which moves objects over a series of rollers by the application of power to all or some of the rollers. { 'līv ¦rōl·ər kən͵vā·ər }

live steam |MECH ENG| Steam that is being delivered directly from a boiler under full pressure. { 'līv 'stēm }

Livingstone sphere |ENG| A clay atmometer in the form of a sphere; evaporation indicated by this instrument is supposed to be somewhat representative of that from plant growth. { 'liv·iŋ·stən ˌsfir }

livre |MECH| A unit of mass, used in France, equal to 0.5 kilogram. { 'lēv·rə }

lixiviate |CHEM ENG| To extract a soluble component from a solid mixture by washing or percolation processes. { lik'siv·ē͵āt }

lixuration See leaching. { ˌlik·syü'rā·shən }

Ljungström heater |MECH ENG| Continuous, regenerative, heat-transfer air heater (recuperator) made of slow-moving rotors packed with closely spaced metal plates or wires with a housing to confine the hot and cold gases to opposite sides. { 'yüŋ·strəm ˌhēd·ər }

Ljungström steam turbine |MECH ENG| A radial outward-flow turbine having two opposed rotation rotors. { 'yüŋ·strəm ¦stēm 'tər·bən }

load |ELEC| **1.** A device that consumes electric power. **2.** The amount of electric power that is drawn from a power line, generator, or other power source. **3.** The material to be heated by an induction heater or dielectric heater. Also known as work. |ELECTR| The device that receives the useful signal output of an amplifier, oscillator, or other signal source. |ENG| **1.** To place ammunition in a gun, bombs on an airplane, explosives in a missile or borehole, fuel in a fuel tank, cargo or passengers into a vehicle, and the like. **2.** The quantity of gas delivered or required at any particular point on a gas supply system; develops primarily at gas-consuming equipment. |MECH| **1.** The weight that is supported by a structure. **2.** Mechanical force that is applied to a body. **3.** The burden placed on any machine, measured by units such as horsepower, kilowatts, or tons. { lōd }

load-and-carry equipment |MECH ENG| Earthmoving equipment designed to load and transport material. { ¦lōd ən 'kar·ē i͵kwip·mənt }

load-carrying capacity |MECH ENG| The greatest weight that the end effector of a robot can manipulate without reducing its level of performance. { 'lōd ¦kar·ē·iŋ kə͵pas·əd·ē }

load chart |IND ENG| A graph showing the amount of work still to be performed by a factory producing unit such as a machine or assembly group. { 'lōd ˌchärt }

load compensation |CONT SYS| Compensation in which the compensator acts on the output signal after it has generated feedback signals. Also known as load stabilization. { 'lōd käm·pən'sā·shən }

load deflection |MECH ENG| The change in position of a body when a load is applied to it. { 'lōd di͵flek·shən }

load diagram |CIV ENG| A diagram showing the distribution and intensity of loads on a structure. { 'lōd ˌdī·ə͵gram }

loaded Q |ELEC| The Q factor of an impedance which is connected or coupled under working conditions. Also known as working Q. { 'lōd·əd kyü }

loaded wheel |ENG| A grinding wheel that is

dull as a result of becoming filled with particles from the material being ground. { 'lōd·əd 'wēl }

loader |MECH ENG| A machine such as a mechanical shovel used for loading bulk materials. { 'lōd·ər }

load factor |ELEC| The ratio of average electric load to peak load, usually calculated over a 1-hour period. |MECH| The ratio of load to the maximum rated load. { 'lōd ,fak·tər }

loading |CHEM ENG| Condition of vapor overcapacity in a liquid-vapor-contact tower, in which rising vapor lifts or holds falling liquid. |ELEC| The addition of inductance to a transmission line to improve its transmission characteristics throughout a given frequency band. Also known as electrical loading. |ENG| **1.** Buildup on a cutting tool of the material removed in cutting. **2.** Filling the pores of a grinding wheel with material removed in the grinding process. |ENG ACOUS| Placing material at the front or rear of a loudspeaker to change its acoustic impedance and thereby alter its radiation. { 'lōd·iŋ }

loading board |ENG| A device that holds preforms in positions corresponding to the multiple cavities in a compression mold, thus facilitating the simultaneous insertion of the preforms. { 'lōd·iŋ ,bȯrd }

loading density |ENG| The number of pounds of explosive per foot length of drill hole. { 'lōd·iŋ ,den·səd·ē }

loading head |MECH ENG| The part of a loader which gathers the bulk materials. { 'lōd·iŋ ,hed }

loading rack |ENG| The shelter and associated equipment for the withdrawal of liquid petroleum or a chemical product from a storage tank and loading it into a railroad tank car or tank truck. { 'lōd·iŋ ,rak }

loading space |ENG| Space in a compression mold for holding the plastic molding material before it is compressed. { 'lōd·iŋ ,spās }

loading station |MECH ENG| A device which receives material and puts it on a conveyor; may be one or more plates or a hopper. { 'lōd·iŋ ,stā·shən }

loading tray |ENG| A tray with a sliding bottom used to simultaneously load the plastic charge into the cavities of a multicavity mold. { 'lōd·iŋ ,trā }

loading weight |ENG| Weight of a powder put into a container. { 'lōd·iŋ ,wāt }

load limit |CIV ENG| The maximum weight that can be supported by a structure. |MECH ENG| The maximum recommended or permitted overall weight of a container or a cargo-carrying vehicle that is determined by combining the weight of the empty container or vehicle with the weight of the load. { 'lōd ,lim·ət }

load profile |ENG| A measure of the time distribution of a building's energy requirements, including the heating, cooling, and electrical loads. { 'lōd ,prō,fīl }

load stabilization See load compensation. { 'lōd ,stā·bə·lə,zā·shən }

load stress |MECH| Stress that results from a pressure or gravitational load. { 'lōd ,stres }

LOB See line of balance.

lobe |DES ENG| A projection on a cam wheel or a noncircular gear wheel. |ENG ACOUS| A portion of the directivity pattern of a transducer representing an area of increased emission or response. { lōb }

lobed impeller meter |ENG| A type of positive displacement meter in which a fluid stream is separated into discrete quantities by rotating, meshing impellers driven by interlocking gears. { lōbd im'pel·ər ,mēd·ər }

local buckling |MECH| Buckling of thin elements of a column section in a series of waves or wrinkles. { 'lō·kəl 'bək·liŋ }

local coefficient of heat transfer |THERMO| The heat transfer coefficient at a particular point on a surface, equal to the amount of heat transferred to an infinitesimal area of the surface at the point by a fluid passing over it, divided by the product of this area and the difference between the temperatures of the surface and the fluid. { 'lō·kəl ,kō·i'fish·ənt əv 'hēt ,tranz·fər }

local controller See first-level controller. { 'lō·kəl kən'trōl·ər }

localized vector |MECH| A vector whose line of application or point of application is prescribed, in addition to its direction. { 'lō·kə,līzd 'vek·tər }

local networking |CONT SYS| The system of communication linking together the components of a single robot. { 'lō·kəl 'net,wərk·iŋ }

local structural discontinuity |MECH| The effect of intensified stress on a small portion of a structure. { 'lō·kəl 'strək·chə·rəl dis,känt·ən'ü·əd·ē }

locating |MECH ENG| A function of tooling operations accomplished by designing and constructing the tooling device so as to bring together the proper contact points or surfaces between the workpiece and the tooling. { 'lō ,kād·iŋ }

locating hole |MECH ENG| A hole used to position the part in relation to a cutting tool or to other parts and gage points. { 'lō,kād·iŋ ,hōl }

locating surface |MECH ENG| A surface used to position an item being manufactured in a numerical control or robotic system for clamping. { 'lō,kād·iŋ ,sər·fəs }

location analysis |DES ENG| An initial step in the design of a robotic system consisting of a detailed study of all aspects of the placement of components such as work stations, buffers, and materials-handling equipment, as well as accessories, tools, and workpieces within a work station. { lō'kā·shən ə,nal·ə·səs }

location dimension |DES ENG| A dimension which specifies the position or distance relationship of one feature of an object with respect to another. { lō'kā·shən də'men·chən }

location fit |DES ENG| The characteristic wherein mechanical sizes of mating parts are such that, when assembled, the parts are accurately positioned in relation to each other. { lō'kā·shən ,fit }

locator |ENG| A radar or other device designed to detect and locate airborne aircraft. { 'lō ,kād·ər }

lock |CIV ENG| A chamber with gates on both ends connecting two sections of a canal or other waterway, to raise or lower the water level in each section. |DES ENG| A fastening device in which a releasable bolt is secured. |ELECTR| To fasten onto and automatically follow a target by means of a radar beam. |ENG| See air lock. { läk }

lock bolt |ENG| **1.** The bolt of a lock. **2.** A bolt equipped with a locking collar instead of a nut. **3.** A bolt for adjusting and securing parts of a machine. { 'läk ,bōlt }

lock chamber |CIV ENG| A compartment between lock gates in a canal. { 'läk ,chām·bər }

locked-coil rope |DES ENG| A completely smooth wire rope that resists wear, made of specially formed wires arranged in concentric layers about a central wire core. Also known as locked-wire rope. { 'läkt 'kȯil ,rōp }

locked groove |DES ENG| A blank and continuous groove placed at the end of the modulated grooves on a disk recording to prevent further travel of the pickup. Also known as concentric groove. { 'läkt 'grüv }

locked-wire rope See locked-coil rope. { 'läkt 'wīr ,rōp }

lock front |DES ENG| On a door lock or latch, the plate through which the latching or locking bolt (or bolts) projects. { 'läk ,frənt }

lock gate |CIV ENG| A movable barrier separating the water in an upper or lower section of waterway from that in the lock chamber. { 'läk ,gāt }

locking |ELECTR| Controlling the frequency of an oscillator by means of an applied signal of constant frequency. |ENG| Automatic following of a target by a radar antenna. { 'läk·iŋ }

locking fastener |DES ENG| A fastening used to prevent loosening of a threaded fastener in service, for example, a seating lock, spring stop nut, interference wedge, blind, or quick release. { 'läk·iŋ 'fas·nər }

lock joint |DES ENG| A joint made by interlocking the joined elements, with or without other fastening. { 'läk ,jȯint }

locknut |DES ENG| **1.** A nut screwed down firmly against another or against a washer to prevent loosening. Also known as jam nut. **2.** A nut that is self-locking when tightened. **3.** A nut fitted to the end of a pipe to secure it and prevent leakage. { 'läk,nət }

lockout circuit |ELECTR| A switching circuit which responds to concurrent inputs from a number of external circuits by responding to one, and only one, of these circuits at any time. Also known as finding circuit; hunting circuit. { 'läk ,aut ,sər·kət }

lock rail |BUILD| An intermediate horizontal structural member of a door, between the vertical stiles, at the height of the lock. { 'läk ,rāl }

lockset |ENG| **1.** A complete lock including the lock mechanism, keys, plates, and other parts.

2. A jig or template for making cuts in a door for holding a lock. { 'läk,set }

lock washer |DES ENG| A solid or split washer placed underneath a nut or screw that prevents loosening by exerting pressure. { 'läk ,wäsh·ər }

locomotive |MECH ENG| A self-propelling machine with flanged wheels, for moving loads on railroad tracks; utilizes fuel (for steam or internal combustion engines), compressed air, or electric energy. { ,lō·kə'mōd·iv }

locomotive boiler |MECH ENG| An internally fixed horizontal fire-tube boiler with integral furnace; the doubled furnace walls contain water which mixes with water in the boiler shell. { ,lō·kə'mōd·iv ¦bȯil·ər }

locomotive crane |MECH ENG| A crane mounted on a railroad flatcar or a special chassis with flanged wheels. Also known as rail crane. { ,lō·kə'mōd·iv ¦krān }

loft |BUILD| **1.** An upper part of a building. **2.** A work area in a factory or warehouse. { lȯft }

loft building |BUILD| A building with a large open floor area. { 'lȯft ,bild·iŋ }

log |ENG| The record of, or the act or process of recording, events or the type and characteristics of the rock penetrated in drilling a borehole as evidenced by the cuttings, core recovered, or information obtained from electronic devices. { läg }

logarithmic amplifier |ELECTR| An amplifier whose output signal is a logarithmic function of the input signal. { 'läg·ə,rith·mik 'am·plə,fī·ər }

logarithmic diode |ELECTR| A diode that has an accurate semilogarithmic relationship between current and voltage over wide and forward dynamic ranges. { 'läg·ə,rith·mik 'dī,ōd }

logarithmic multiplier |ELECTR| A multiplier in which each variable is applied to a logarithmic function generator, and the outputs are added together and applied to an exponential function generator, to obtain an output proportional to the product of two inputs. { 'läg·ə,rith·mik 'məl·tə,plī·ər }

logging |ENG| Continuous recording versus depth of some characteristic datum of the formations penetrated by a drill hole; for example, resistivity, spontaneous potential, conductivity, fluid content, radioactivity, or density. { 'läg· iŋ }

logic |ELECTR| **1.** The basic principles and applications of truth tables, interconnections of on/off circuit elements, and other factors involved in mathematical computation in a computer. **2.** General term for the various types of gates, flip-flops, and other on/off circuits used to perform problem-solving functions in a digital computer. { 'läj·ik }

logical gate See switching gate. { 'läj·ə·kəl 'gāt }

logic card |ELECTR| A small fiber chassis on which resistors, capacitors, transistors, magnetic cores, and diodes are mounted and interconnected in such a way as to perform some computer function; computers employing this type of construction may be repaired by removing

the faulty card and replacing it with a new card.
{ 'läj·ik ,kärd }

logic high |ELECTR| The electronic representation of the binary digit 1 in a digital circuit or device. { 'läj·ik 'hī }

logic level |ELECTR| One of the two voltages whose values have been arbitrarily chosen to represent the binary numbers 1 and 0 in a particular data-processing system. { 'läj·ik ,lev·əl }

logic low |ELECTR| The electronic representation of the binary digit 0 in a digital circuit or device. { 'läj·ik ,lō }

logic swing |ELECTR| The voltage difference between the logic levels used for 1 and 0; magnitude is chosen arbitrarily for a particular system and is usually well under 10 volts. { 'läj·ik ,swiŋ }

logic switch |ELECTR| A diode matrix or other switching arrangement that is capable of directing an input signal to one of several outputs. { 'läj·ik ,swich }

log line See current line. { 'läg ,līn }

log-mean temperature difference |THERMO| The log-mean temperature difference $T_{LM} = (T_2 - T_1)/\ln T_2/T_1$, where T_2 and T_1 are the absolute (K or °R) temperatures of the two extremes being averaged; used in heat transfer calculations in which one fluid is cooled or heated by a second held separate by pipes or process vessel walls. { 'läg ¦mēn 'tem·prə·chər ,dif·rəns }

long column |CIV ENG| A column so slender that bending is the primary deformation, generally having a slenderness ratio greater than 120–150. { 'lȯŋ ¦käl·əm }

Longhurst-Hardy plankton sampler |ENG| A nonquantitative metal-shrouded net for trapping plankton. { 'lȯŋ,hərst 'här·dē 'plaŋk·tən ,sam·plər }

longitudinal acceleration |MECH| The component of the linear acceleration of an aircraft, missile, or particle parallel to its longitudinal, or X, axis. { ,län·jə'tüd·ən·əl ak,sel·ə'rā·shən }

longitudinal baffle |CHEM ENG| Baffle sheets or plates within a process vessel (such as a heat exchanger) that are parallel to the long dimension of the vessel; used to direct fluid flow in the desired flow pattern. { ,län·jə'tüd·ən·əl 'baf·əl }

longitudinal drum boiler |MECH ENG| A boiler in which the axis of the horizontal drum is parallel to the tubes, both lying in the same plane. { ,län·jə'tüd·ən·əl 'drəm ,bȯil·ər }

longitudinal flow reactor |CHEM ENG| Theoretical reactor system in which there is no longitudinal mixing (back mixing) of reactants and products as they flow through the reactor, but in which there is complete radial (side-to-side) mixing. { ,län·jə'tüd·ən·əl 'flō rē,ak·tər }

longitudinal stability |ENG| The ability of a ship or aircraft to recover a horizontal position after a vertical motion of its ends about a horizontal axis perpendicular to the centerline. { ,län·jə'tüd·ən·əl stə'bil·əd·ē }

longitudinal strain See linear strain. { ,län·jə'tüd·ən·əl 'strān }

longitudinal vibration |MECH| A continuing periodic change in the displacement of elements of a rod-shaped object in the direction of the long axis of the rod. { ,län·jə'tüd·ən·əl vī'brā·shən }

long-nose pliers |DES ENG| Small pincer with long, tapered jaws. { 'lȯŋ ,nōz 'plī·ərz }

long-playing record |ENG ACOUS| A 10- or 12-inch (25.4- or 30.48-centimeter) phonograph record that operates at a speed of $33\frac{1}{3}$ rpm (revolutions per minute) and has closely spaced grooves, to give playing times up to about 30 minutes for one 12-inch side. Also known as LP record; microgroove record. { 'lȯŋ ,plā·iŋ 'rek·ərd }

long span |ENG| Span of open wire exceeding 250 feet (76 meters) in length. { 'lȯŋ ¦span }

long-span steel framing |BUILD| Framing system used when there is a greater clear distance between supports than can be spanned with rolled beams; girders, simple trusses, arches, rigid frames, and cantilever suspension spans are used in this system. { 'lȯŋ ¦span ¦stēl 'frām·iŋ }

long-term repeatability |CONT SYS| The close agreement of positional movements of a robotic system repeated under identical conditions over long periods of time. { 'lȯŋ ,tərm ri,pēd·ə'bil·əd·ē }

long ton See ton. { 'lȯŋ 'tən }

long-tube vertical evaporator |CHEM ENG| A liquid evaporator in which the material is force-fed into the bottom of a bundle ofong, vertical tubes; hot liquid on the outsides of the tubes transfers heat to the rising liquid feed, causing partial evaporation. { 'lȯŋ ,tüb ¦vərd·ə·kəl i'vap·ə,rād·ər }

look angle |ENG| The solid angle in which an instrument operates effectively, generally used to describe radars, optical instruments, and space radiation detectors. { 'lȯk ,aŋ·gəl }

look box |CHEM ENG| Box with glass windows built into distillation-column rundown lines (or other flow lines) so that the stream of condensate from the condenser can be watched. { 'lȯk ,bäks }

lookout |BUILD| A horizontal wood framing member that extends out from the studs to the end of rafters and overhangs a part of a roof, such as a gable. { 'lȯk,aȯt }

lookout station |ENG| A structure or place on shore at which personnel keep watch of events at sea or along the shore. { 'lȯk,aȯt ,stā·shən }

lookout tower |ENG| In marine operations, any tower surmounted by a small house in which a watch is habitually kept, as distinguished from an observation tower in which no watch is kept. { 'lȯk,aȯt ,taȯ·ər }

loop |ELEC| **1.** A closed path or circuit over which a signal can circulate, as in a feedback control system. **2.** Commercially, the portion of a connection from central office to subscriber in a telephone system. |ENG| **1.** A reel of motion picture film or magnetic tape whose ends are spliced together, so that it may be played repeatedly without interruption. **2.** A closed circuit of pipe in which materials and components may be placed to test them under different conditions of temperature, irradiation, and so forth. { lüp }

loop control *See* photoelectric loop control. { 'lüp kən,trōl }

loop filter |ELECTR| A low-pass filter, which may be a simple RC filter or may include an amplifier, and which passes the original modulating frequencies but removes the carrier-frequency components and harmonics from a frequency-modulated signal in a locked-oscillator detector. { 'lüp ,fil·tər }

loop gain |CONT SYS| The ratio of the magnitude of the primary feedback signal in a feedback control system to the magnitude of the actuating signal. |ELECTR| Total usable power gain of a carrier terminal or two-wire repeater; maximum usable gain is determined by, and may not exceed, the losses in the closed path. { 'lüp ,gān }

looping |ENG| Laying a parallel pipeline along another, or along just a section of it, to increase capacity. { 'lüp·iŋ }

loop ratio *See* loop transfer function. { 'lüp ,rā·shō }

loop seal |CHEM ENG| Antivapor seal for liquid drawoffs from process or storage vessels; liquid drawoff is made to flow through an immersed loop or beneath an obstruction, thus sealing off vapor flow. { 'lüp ,sēl }

loop strength *See* loop tenacity. { 'lüp ,streŋkth }

loop tenacity |ENG| A measure of the strength of a fibrous material determined by a test in which two linked loops of the material are pulled against each other to determine if the material will cut or crush itself. Also known a loop strength. { 'lüp tə,nas·əd·ē }

loop transfer function |CONT SYS| For a feedback control system, the ratio of the Laplace transform of the primary feedback signal to the Laplace transform of the actuating signal. Also known as loop ratio. { 'lüp 'tranz·fər ,fəŋk·shən }

loop transmittance |CONT SYS| **1.** The transmittance between the source and sink created by the splitting of a specified node in a signal flow graph. **2.** The transmittance between the source and sink created by the splitting of a node which has been inserted in a specified branch of a signal flow graph in such a way that the transmittance of the branch is unchanged. { 'lüp tranz,mit·əns }

loop tunnel |ENG| A tunnel which is looped or folded back on itself to gain grade in a tunnel location. { 'lüp ,tən·əl }

loose-detail mold |ENG| A plastics mold with parts that come out with the molded piece. { 'lüs ¦dē,tāl ,mōld }

loose fit |DES ENG| A fit with enough clearance to allow free play of the joined members. { 'lüs ,fit }

loose-joint butt |DES ENG| A knuckle hinge in which the pin on one half slides easily into a slot on the other half. { 'lüs ¦jöint 'bət }

loose pulley |MECH ENG| In belt-driven machinery, a pulley which turns freely on a shaft so that the belt can be shifted from the driving pulley to the loose pulley, thereby causing the machine to stop. { 'lüs 'púl·ē }

lopping shears |DES ENG| Long-handled shears used for pruning branches. { 'läp·iŋ ,shirz }

loss |ENG| Power that is dissipated in a device or system without doing useful work. Also known as internal loss. { lòs }

Lossev effect *See* injection electroluminescence. { ,lò,sef i,fekt }

loss factor |ELEC| The power factor of a material multiplied by its dielectric constant; determines the amount of heat generated in a material. { 'lòs ,fak·tər }

loss-in-weight feeder |MECH ENG| A device to apportion the output of granulated or powdered solids at a constant rate from a feed hopper; weight-measured decrease in hopper content actuates further opening of the discharge chute to compensate for flow loss as the hopper overburden decreases; used in the chemical, fertilizer, and plastics industries. { 'lòs in 'wāt ,fēd·ər }

loss-of-head gage |ENG| A gage on a rapid sand filter, which indicates loss of head for a filtering operation. { 'lòs əv 'hed ,gāj }

lost motion |MECH ENG| The delay between the movement of a driver and the movement of a follower. { 'lòst 'mō·shən }

lost time |ENG ACOUS| The period in a frequency-modulation sonar, just after flyback, during which the sound field must be reestablished; its duration equals travel time of the signal to and from the target. { 'lòst 'tīm }

lot |CIV ENG| A piece of land with fixed boundaries. |IND ENG| A quantity of material, such as propellant, the units of which were manufactured under identical conditions. Also known as lot batch. { lät }

lot batch *See* batch. { 'lät ,bach }

lot line |CIV ENG| The legal boundary line of a piece of property. { 'lät ,līn }

lot number |IND ENG| Identification number assigned to a particular quantity or lot of material from a single manufacturer. { 'lät ,nəm·bər }

lot plot method |IND ENG| A variables acceptance sampling plan based on the frequency plot of a random sample of 50 items taken from a lot. { 'lät ¦plät ,meth·əd }

lot tolerance percent defective |IND ENG| The percent of defectives in a lot which is considered bad and should be rejected for some specified fraction, usually 90, of the time. { 'lät ¦täl·ə·rəns pər¦sent di'fek·tiv }

loudness control |ENG ACOUS| A combination volume and tone control that boosts bass frequencies when the control is set for low volume, to compensate automatically for the reduced response of the ear to low frequencies at low volume levels. Also known as compensated volume control. { 'laúd·nəs kən,trōl }

loudspeaker |ENG ACOUS| A device that converts electrical signal energy into acoustical energy, which it radiates into a bounded space, such as a room, or into outdoor space. Also known as speaker. { 'laud,spēk·ər }

loudspeaker dividing network See crossover network. { 'laud,spēk·ər di'vīd·iŋ ,net,wərk }

loudspeaker voice coil See voice coil. { 'laud ,spēk·ər 'vȯis ,kȯil }

louver [BUILD] An opening in a wall or ceiling with slanted or sloping slats to allow sunlight and ventilation and exclude rain; may be fixed or adjustable, and may be at the opening of a ventilating duct. Also known as outlet ventilator. [ENG] Any arrangement of fixed or adjustable slatlike openings to provide ventilation. [ENG ACOUS] An arrangement of concentric or parallel slats or equivalent grille members used to conceal and protect a loudspeaker while allowing sound waves to pass. { 'lü·vər }

lowboy [MECH ENG] A trailer with low ground clearance for hauling construction equipment. { 'lō,bȯi }

Lowenhertz thread [DES ENG] A screw thread that differs from U.S. Standard form in that the angle between the flanks measured on an axial plane is 53°8'; height equals 0.75 times the pitch, and width of flats at top and bottom equals 0.125 times the pitch. { 'lō·ən,harts ,thred }

lower chord [CIV ENG] The bottom member of a truss. { 'lō·ər ¦kȯrd }

lower control limit [IND ENG] The horizontal line drawn on a control chart at a specified distance below the central line; points plotted below the lower control limit indicate that the process may be out of control. { ¦lō·ər kən'trōl ,lim·ət }

lower half-power frequency [ELECTR] The frequency on an amplifier response curve which is smaller than the frequency for peak response and at which the output voltage is $1\sqrt{2}$ of its midband or other refer { 'lō·ər 'haf ,paú·ər 'frē·kwən·sē }

lower heating value See low heat value. { 'lō·ər 'hēd·iŋ ,val·yü }

lower pair [MECH ENG] A link in a mechanism in which the mating parts have surface (instead of line or point) contact. { 'lō·ər 'pər }

lowest safe waterline [MECH ENG] The lowest water level in a boiler drum at which the burner may safely operate. { 'lō·əst ¦sāf 'wȯd·ər,līn }

low-frequency compensation [ELECTR] Compensation that serves to extend the frequency range of a broad-band amplifier to lower frequencies. { 'lō ,frē·kwən·sē ,käm·pə'sā·shən }

low-frequency current [ELEC] An alternating current having a frequency of less than about 300 kilohertz. { 'lō ,frē·kwən·sē 'kə·rənt }

low-frequency cutoff [ELECTR] A frequency below which the gain of a system or device decreases rapidly. { 'lō ,frē·kwən·sē 'kə,dȯf }

low-frequency gain [ELECTR] The gain of the voltage amplifier at frequencies less than those frequencies at which this gain is close to its maximum value. { 'lō ,frē·kwən·sē 'gān }

low-frequency impedance corrector [ELEC] Electric network designed to be connected to a basic network, or to a basic network and a building-out network, so that the combination will simulate, at low frequencies, the sending-end impedance, including dissipation, of a line. { 'lō ,frē·kwən·sē im'pēd·əns kə,rek·tər }

low-frequency induction furnace [ENG] An induction furnace in which current flow at the commercial power-line frequency is induced in the charge to be heated. { 'lō ,frē·kwən·sē in¦dək·shən ,fər·nəs }

low heat value [THERMO] The heat value of a combustion process assuming that none of the water vapor resulting from the process is condensed out, so that its latent heat is not available. Also known as lower heating value; net heating value. { 'lō 'hēt ,val·yü }

low-helix drill [DES ENG] A two-flute twist drill with a lower helix angle than a conventional drill. Also known as slow-spiral drill. { 'lō ,hē·liks ,dril }

low-impedance measurement [ELECTR] The measurement of an impedance which is small enough to necessitate use of indirect methods. { 'lō im,pēd·əns 'mezh·ər·mənt }

low-intensity atomizer [MECH ENG] A type of electrostatic atomizer operating on the principle that atomization is the result of Rayleigh instability, in which the presence of charge in the surface counteracts surface tension. { 'lō in ,ten·səd·ē 'ad·ə,mīz·ər }

low level [ELECTR] The less positive of the two logic levels or states in a digital logic system. { 'lō ,lev·əl }

low-level condenser [MECH ENG] A direct-contact water-cooled steam condenser that uses a pump to remove liquid from a vacuum space. { 'lō ,lev·əl kən'den·sər }

low-level logic circuit [ELECTR] A modification of a diode-transistor logic circuit in which a resistor and capacitor in parallel are replaced by a diode, with the result that a relatively small voltage swing is required at the base of the transistor to switch it on or off. Abbreviated LLL circuit. { 'lō ,lev·əl 'läj·ik ,sər·kət }

low-lift truck [MECH ENG] A hand or powered lift truck that raises the load sufficiently to make it mobile. { 'lō ,lift ,trək }

low-loss [ELEC] Having a small dissipation of electric or electromagnetic power. { 'lō ¦lȯs }

low-noise preamplifier [ELECTR] A low-noise amplifier placed in a system prior to the main amplifier, sometimes close to the source; used to establish a satisfactory noise figure at an early point in the system. { 'lō ,nȯiz prē'am·plə,fī·ər }

low-pass band-pass transformation See frequency transformation. { 'lō ,pas 'band ,pas ,tranz·fər,mā·shən }

low-pass filter [ELEC] A filter that transmits alternating currents below a given cutoff frequency and substantially attenuates all other currents. { 'lō ,pas 'fil·tər }

low-population zone |ENG| An area of low population density sometimes required around a nuclear installation; the number and density of residents is of concern in providing, with reasonable probability, that effective protection measures can be taken if a serious accident should occur. { 'lō ,päp·ə'lā·shən ,zōn }

low-pressure area |MECH ENG| The point in a bearing where the pressure is the least and the area or space for a lubricant is the greatest. { 'lō ¦presh·ər 'er·ē·ə }

low-pressure torch |ENG| A type of torch in which acetylene enters a mixing chamber, where it meets a jet of high-pressure oxygen; the amount of acetylene drawn into the flame is controlled by the velocity of this oxygen jet. Also known as injector torch. { 'lō ¦presh·ər 'tórch }

low-Q filter |ELECTR| A filter in which the energy dissipated in each cycle is a fairly large fraction of the energy stored in the filter. { 'lō ¦kyü 'fil·tər }

low-reactance grounding |ELEC| Use of grounding connections with a moderate amount of inductance to effect a moderate reduction in the short-circuit current created by a line-to-ground fault. { 'lō rē¦ak·təns 'graund·iŋ }

Lowry process |ENG| A system for wood preservation which uses atmospheric pressure at the start and then introduces preservative into the wood in a vacuum. { 'laú·rē ,prä·səs }

low-speed wind tunnel |ENG| A wind tunnel that has a speed up to 300 miles (480 kilometers) per hour and the essential features of most wind tunnels. { 'lō ,spēd 'win ,tən·əl }

low-technology robot |CONT SYS| The simplest type of robot, with only two or three degrees of freedom, and only the end points of motion specified, using fixed and adjustable stops. { 'lō tek¦näl·ə·jē 'rō,bät }

low-temperature carbonization |CHEM ENG| Low-temperature destructive distillation of coal to produce liquid products. { 'lō ,tem·prə·chər ,kär·bə·nə'zā·shən }

low-temperature hygrometry |ENG| The study that deals with the measurement of water vapor at low temperatures; the techniques used differ from those of conventional hygrometry because of the extremely small amounts of moisture present at low temperatures and the difficulties imposed by the increase of the time constants of the standard instruments when operated at these temperatures. { 'lō ,tem·prə·chər hī'gräm·ə·trē }

low-temperature separation |CHEM ENG| Liquid condensate recovery from wet gases at temperatures of 20 to −20°F (−6.7 to −28.9°C), the temperature range at which the gas-oil separator operates. { 'lō ,tem·prə·chər ,sep·ə'rā·shən }

low velocity |MECH| Muzzle velocity of an artillery projectile of 2499 feet (762 meters) per second or less. { 'lō və'läs·əd·ē }

low voltage |ELEC| **1.** Voltage which is small enough to be regarded as safe for indoor use, usually 120 volts in the United States. **2.** Voltage which is less than that needed for normal operation; a result of low voltage may be burnout of electric motors due to loss of electromotive force. { 'lō 'vōl·tij }

low-water fuel cutoff |MECH ENG| A float device which shuts off fuel supply and burner when boiler water level drops below the lowest safe waterline. { 'lō ,wód·ər ,fyül ,kə,dóf }

lozenge file |DES ENG| A small file with four sides and a lozenge-shaped cross section; used in forming dies. { 'läz·ənj ,fīl }

L pad |ENG ACOUS| A volume control having essentially the same impedance at all settings. { 'el ,pad }

LP record See long-playing record. { ¦el¦pē 'rek·ərd }

LQG problem See linear-quadratic-Gaussian problem. { ¦el¦kyü¦jē ,präb·ləm }

LSA diode |ELECTR| A microwave diode in which a space charge is developed in the semiconductor by the applied electric field and is dissipated during each cycle before it builds up appreciably, thereby limiting transit time and increasing the maximum frequency of oscillation. Derived from limited space-charge accumulation diode. { ¦el¦es¦a 'dī,ōd }

LSI circuit See large-scale integrated circuit. { ¦el¦es'Ī ,sər·kət }

L-1 test |ENG| A 480-hour engine test in a single-cylinder Caterpillar diesel engine to determine the detergency of heavy-duty lubricating oils. { ¦el 'wən ,test }

L-2 test |ENG| An engine test made in a single-cylinder Caterpillar diesel engine to determine the oiliness of an engine oil. Also known as scoring test. { ¦el 'tü ,test }

L-3 test |ENG| An engine test in a four-cylinder Caterpillar engine to determine stability of crankcase oil at high temperatures and under severe operating conditions. { ¦el 'thrē ,test }

L-4 test |ENG| An engine test in a six-cylinder spark-ignition Chevrolet engine to evaluate crankcase oil oxidation stability, bearing corrosion, and engine deposits. { ¦el 'fór ,test }

L-5 test |ENG| An engine test in a General Motors diesel engine to determine detergency, corrosiveness, ring sticking, and oxidation stability properties of lubricating oils. { ¦el 'fīv ,test }

LTPD See lot tolerance percent defective.

lubricator |ENG| A device for applying a lubricant. { 'lü·brə,kād·ər }

Luckiesh-Moss visibility meter |ENG| A type of photometer that consists of two variable-density filters (one for each eye) that are adjusted so that an object seen through them is just barely discernible; the reduction in visibility produced by the filters is read on a scale of relative visibility related to a standard task. { lü'kēsh 'mós ,viz·ə'bil·əd·ē ,mēd·ər }

Ludwig-Soret effect |THERMO| A phenomenon in which a temperature gradient in a mixture of substances gives rise to a concentration gradient. { ¦lüd,vik sə'rä i,fekt }

Luenberger observer |CONT SYS| A compensator driven by both the inputs and measurable outputs of a control system. { 'lün,bərg·ər əb'zər·vər }

lug |DES ENG| A projection or head on a metal part to serve as a cap, handle, support, or fitting connection. { ləg }

lug bolt |DES ENG| **1.** A bolt with a flat extension or hook instead of a head. **2.** A bolt designed for securing a lug. { 'ləg ˌbōlt }

lung-governed breathing apparatus |ENG| A breathing apparatus in which the oxygen that is supplied to the wearer is governed by the wearer's demand. { 'ləŋ ˌgəv·ərnd 'breth·iŋ ap·ə,rad·əs }

Lyapunov stability criterion |CONT SYS| A method of determining the stability of systems (usually nonlinear) by examining the sign-definitive properties of an associated Lyapunov function. { lē'ap·ə,nóf stə'bil·əd·ē krī,tir·ē·ən }

lyophilization |CHEM ENG| Rapid freezing of a material, especially biological specimens for preservation, at a very low temperature followed by rapid dehydration by sublimation in a high vacuum. { lī,äf·ə·lə'zā·shən }

lysimeter |ENG| An instrument for measuring the water percolating through soils and determining the materials dissolved by the water. { lī'sim·əd·ər }

M

m *See* meter.

macadam |CIV ENG| Uniformly graded stones consolidated by rolling to form a road surface; may be bound with water or cement, or coated with tar or bitumen. { mə'kad·əm }

maceration |CHEM ENG| The process of extracting fragrant oils from flower petals by immersing them in hot molten fat. { ,mas·ə'rā·shən }

machete |DES ENG| A knife with a broad blade 2 to 3 feet (60 to 90 centimeters) long. { mə'shed·ē *or* mə'ched·ē }

Mach indicator *See* Machmeter. { 'mäk ,in·də ,kād·ər }

machine |MECH ENG| A combination of rigid or resistant bodies having definite motions and capable of performing useful work. { mə'shēn }

machine attention time |IND ENG| Time during which a machine operator must observe the machine's functioning and be available for immediate servicing, while not actually operating or servicing the machine. Also known as service time. { mə'shēn ə'ten·chən ,tīm }

machine bolt |DES ENG| A heavy-weight bolt with a square, hexagonal, or flat head used in the automotive, aircraft, and machinery fields. { mə'shēn ,bōlt }

machine capability |IND ENG| A qualitative or quantitative statement of the performance potential of a specific item of power equipment. { mə'shēn ,kā·pə'bil·əd·ē }

machine controlled time |IND ENG| The time necessary for a machine to complete the automatic portion of a work cycle. Also known as independent machine time; machine movement; machine time. { mə'shēn kən'trōld 'tīm }

machine design |DES ENG| Application of science and invention to the development, specification, and construction of machines. { mə'shēn di,zīn }

machine drill |MECH ENG| Any mechanically driven diamond, rotary, or percussive drill. { mə'shēn ,dril }

machine element |DES ENG| Any of the elementary mechanical parts, such as gears, bearings, fasteners, screws, pipes, springs, and bolts used as essentially standardized components for most devices, apparatus, and machinery. *See* machine controlled time. { mə'shēn ,el·ə·mənt }

machine file |DES ENG| A file that can be clamped in the chuck of a power-driven machine. { mə'shēn ,fīl }

machine-gun microphone *See* line microphone. { mə'shēn ,gən 'mī·krə,fōn }

machine-hour |IND ENG| A unit representing the operation of one machine for 1 hour; used in the determination of costs and economics. { mə'shēn 'aúr }

machine idle time |IND ENG| Time during a work cycle when a machine is idle, awaiting completion of manual work. { mə'shēn 'īd·əl ,tīm }

machine interference |IND ENG| A situation in which two or more units of equipment simultaneously require service. { mə'shēn ,in·tər'fir·əns }

machine key |DES ENG| A piece inserted between a shaft and a hub to prevent relative rotation. Also known as key. { mə'shēn ,kē }

machine loading |IND ENG| 1. Feeding work into a machine. 2. Planning the amount of use of a unit of equipment during a given time period. { mə'shēn ,lōd·iŋ }

machine-paced operation |IND ENG| The proportion of an operation cycle during which the machine controls the speed of work progress. { mə'shēn 'pāst ,äp·ə'rā·shən }

machine rating |MECH ENG| The power that a machine can draw or deliver without overheating. { mə'shēn ,rād·iŋ }

machine run *See* run. { mə'shēn 'rən }

machinery |MECH ENG| A group of parts or machines arranged to perform a useful function. { mə'shēn·rē }

machine screw |DES ENG| A blunt-ended screw with a standardized thread and a head that may be flat, round, fillister, or oval, and may be slotted, or constructed for wrenching; used to fasten machine parts together. { mə'shēn ,skrü }

machine setting *See* mechanical setting. { mə'shēn ,sed·iŋ }

machine shop |ENG| A workshop in which work, metal or other material, is machined to specified size and assembled. { mə'shēn ,shäp }

machine shot capacity |ENG| In injection molding, the maximum weight of a given thermoplastic resin which can be displaced by a single stroke of the injection ram. { mə'shēn 'shät kə,pas·əd·ē }

machine taper |MECH ENG| A taper that provides a connection between a tool, arbor, or center and its mating part to ensure and maintain accurate alignment between the parts; permits easy separation of parts. { mə'shēn ,tā·pər }

machine-tight |ENG| The extent of the tightening of a screwed fitting that can be accomplished without damaging or stripping the thread. { mə'shēn ,tīt }

machine time See machine controlled time. { mə'shēn ,tīm }

machine tool |MECH ENG| A stationary power-driven machine for the shaping, cutting, turning, boring, drilling, grinding, or polishing of solid parts, especially metals. { mə'shēn ,tül }

machine utilization |ENG| The percentage of time that a machine is actually in use. { mə'shēn ,yüd·əl·ə'zā·shən }

machining |MECH ENG| Performing various cutting or grinding operations on a piece of work. { mə'shēn·iŋ }

machining center |MECH ENG| Manufacturing equipment that removes metal under computer numerical control by making use of several axes and a variety of tools and operations. { mə'shēn·iŋ ,sen·tər }

machinist's file |DES ENG| A type of double-cut file that removes metal fast and is used for rough metal filing. { mə'shē·nəsts ,fil }

Machmeter |ENG| An instrument that measures and indicates speed relative to the speed of sound, that is, indicates the Mach number. Also known as Mach indicator. { 'mäk,mēd·ər }

macroanalytical balance |ENG| A relatively large type of analytical balance that can weigh loads of up to 200 grams to the nearest 0.1 milligram. { 'mak·rō,an·ə'lid·ə·kəl 'bal·əns }

macroelement |IND ENG| An element of a work cycle whose time span is long enough to be observed and measured with a stopwatch. { ¦mak·rō'el·ə·mənt }

macromechanics See composite macromechanics. { ¦mak·rō·mə'kan·iks }

macrorheology |MECH| A branch of rheology in which materials are treated as homogeneous or quasi-homogeneous, and processes are treated as isothermal. { ¦mak·rō·rē'äl·ə·jē }

macroscopic anisotropy |ENG| Phenomenon in electrical downhole logging wherein electric current flows more easily along sedimentary strata beds than perpendicular to them. { ¦mak·rə¦skäp·ik ,an·ə'sä·trə,pē }

macroscopic property See thermodynamic property. { ¦mak·rə¦skäp·ik 'präp·ərd·ē }

macrotome |ENG| A device for making large anatomical sections. { 'mak·rə,tōm }

madistor |ELECTR| A cryogenic semiconductor device in which injection plasma can be steered or controlled by transverse magnetic fields, to give the action of a switch. { ma'dis·tər }

Madsen impedance meter |ENG| An instrument for measuring the acoustic impedance of normal and deaf ears, based on the principle of the Wheatstone bridge. { 'mad·zən im'pēd·əns ,mēd·ər }

MADT See microalloy diffused transistor.

MAG See maximum available gain.

magazine |ENG| **1.** A storage area for explosives. **2.** A building, compartment, or structure constructed and located for the storage of explosives or ammunition. { ¦mag·ə¦zēn }

magnesite wheel |ENG| A grinding wheel made with magnesium oxychloride as the bonding agent. { 'mag·nə,sīt ,wēl }

magnetic balance |ENG| **1.** A device for determining the repulsion or attraction between magnetic poles, in which one magnet is suspended and the forces needed to cancel the effects of bringing a pole of another magnet close to one end are measured. **2.** Any device for measuring the small forces involved in determining paramagnetic or diamagnetic susceptibility. { mag'ned·ik 'bal·əns }

magnetic bearing |MECH ENG| A device incorporating magnetic forces to cause a shaft to levitate and float in a magnetic field without any contact between the rotating and stationary elements. { mag'ned·ik 'ber·iŋ }

magnetic brake |MECH ENG| A friction brake under the control of an electromagnet. { mag'ned·ik 'brāk }

magnetic chuck |MECH ENG| A chuck in which the workpiece is held by magnetic force. { mag'ned·ik 'chək }

magnetic clutch See magnetic fluid clutch; magnetic friction clutch. { mag'ned·ik 'kləch }

magnetic cutter |ENG ACOUS| A cutter in which the mechanical displacements of the recording stylus are produced by the action of magnetic fields. { mag'ned·ik 'kəd·ər }

magnetic drag dynamometer See eddy-current brake. { mag'ned·ik ¦drag ,dī·nə'mäm·əd·ər }

magnetic drum See drum. { mag'ned·ik 'drəm }

magnetic drum storage See drum. { mag'ned·ik ¦drəm 'stór·ij }

magnetic earphone |ENG ACOUS| An earphone in which variations in electric current produce variations in a magnetic field, causing motion of a diaphragm. { mag'ned·ik 'ir,fōn }

magnetic element |ENG| That part of an instrument producing or influenced by magnetism. { mag'ned·ik 'el·ə·mənt }

magnetic field sensor |ENG| A proximity sensor that uses a combination of a reed switch and a magnet to detect the presence of a magnetic field. { mag'ned·ik feld ,sen·sər }

magnetic filter |CHEM ENG| Filtration device in which the filter screen is magnetized to trap and remove fine iron from liquids or liquid suspensions being filtered. { mag'ned·ik 'fil·tər }

magnetic fluid clutch |MECH ENG| A friction clutch that is engaged by magnetizing a liquid suspension of powdered iron located between pole pieces mounted on the input and output shafts. Also known as magnetic clutch. { mag'ned·ik ¦flü·əd 'kləch }

magnetic flux quantum |ELEC| A fundamental unit of magnetic flux, the total magnetic flux in a fluxoid in a type II superconductor, equal to $h/(2e)$, where h is Planck's constant and e is the magnitude of the electron charge, or approximately 2.07×10^{-15} weber. { mag,ned·ik 'fläks ,kwän·təm }

magnetic force microscopy |ENG| The use of an atomic force microscope to measure the gradient of a magnetic field acting on a tip made of a magnetic material, by monitoring the shift of the natural frequency of the cantilever due to the magnetic force as the tip is scanned over the sample. { mag¦ned·ik ¦fórs mī'krä·ska·pē }

magnetic friction clutch |MECH ENG| A friction clutch in which the pressure between the friction surfaces is produced by magnetic attraction. Also known as magnetic clutch. { mag'ned·ik 'frik·shən ,kləch }

magnetic hardness comparator |ENG| A device for checking the hardness of steel parts by placing a unit of known proper hardness within an induction coil; the unit to be tested is then placed within a similar induction coil, and the behavior of the induction coils compared; if the standard and test units have the same magnetic properties, the hardness of the two units is considered to be the same. { mag'ned·ik 'härd·nəs kəm,par·əd·ər }

magnetic head |ELECTR| The electromagnet used for reading, recording, or erasing signals on a magnetic disk, drum, or tape. Also known as magnetic read/write head. { mag'ned·ik 'hed }

magnetic induction gyroscope |ENG| A gyroscope without moving parts, in which alternating- and direct-current magnetic fields act on water doped with salts which exhibit nuclear paramagnetism. { mag'ned·ik in¦dək·shən 'jī·rə,skōp }

magnetic loudspeaker |ENG ACOUS| Loudspeaker in which acoustic waves are produced by mechanical forces resulting from magnetic reactions. Also known as magnetic speaker. { mag'ned·ik 'laúd,spēk·ər }

magnetic microphone |ENG ACOUS| A microphone consisting of a diaphragm acted upon by sound waves and connected to an armature which varies the reluctance in a magnetic field surrounded by a coil. Also known as reluctance microphone; variable-reluctance microphone. { mag'ned·ik 'mī·krə,fōn }

magnetic pickup See variable-reluctance pickup. { mag'ned·ik 'pi,kəp }

magnetic potentiometer |ENG| Instrument that measures magnetic potential differences. { mag'ned·ik pə,ten·chē'äm·əd·ər }

magnetic pressure transducer |ENG| A type of pressure transducer in which a change of pressure is converted into a change of magnetic reluctance or inductance when one part of a magnetic circuit is moved by a pressure-sensitive element, such as a bourdon tube, bellows, or diaphragm. { mag'ned·ik 'presh·ər tranz,dü·sər }

magnetic prospecting |ENG| Carrying out airborne or ground surveys of variations in the earth's magnetic field, using a magnetometer or other equipment, to locate magnetic deposits of

iron, nickel, or titanium, or nonmagnetic deposits which either contain magnetic gangue minerals or are associated with magnetic structures. { mag'ned·ik 'prä,spek·tiŋ }

magnetic pulley |ENG| Magnetized pulley device for a conveyor belt; removes tramp iron from dry products being moved by the belt. { mag'ned·ik 'púl·ē }

magnetic read/write head See magnetic head. { mag'ned·ik ¦rēd ¦rīt ,hed }

magnetic resonance imaging |ENG| A technique in which an object placed in a spatially varying magnetic field is subjected to a pulse of radio-frequency radiation, and the resulting nuclear magnetic resonance spectra are combined to give cross-sectional images. Abbreviated MRI. { mag'ned·ik 'rez·ən·əns 'im·ij·iŋ }

magnetic separator |ENG| A machine for separating magnetic from less magnetic or nonmagnetic materials by using strong magnetic fields; used for example, in tramp iron removal, or concentration and purification. { mag'ned·ik 'sep·ə,rād·ər }

magnetic sound track |ENG ACOUS| A magnetic tape, attached to a motion picture film, on which a sound recording is made. { mag'ned·ik 'saún ,trak }

magnetic source imaging |ENG| A method of mapping electric currents within an object, particularly currents associated with biological activity, by using an array of SQUID magnetometers to detect the resulting magnetic fields surrounding the object. Abbreviated MSI. { mag,ned·ik 'sórs ,im·ij·iŋ }

magnetic speaker See magnetic loudspeaker. { mag'ned·ik 'spēk·ər }

magnetic tunnel junction |ELECTR| A magnetic storage and switching device in which two magnetic layers are separated by an insulating barrier, typically aluminum oxide, that is only 1–2 nanometers thick, allowing an electronic current whose magnitude depends on the orientation of both magnetic layers to tunnel through the barrier when it is subject to a small electric bias. { mag¦ned·ik 'tən·əl ,jəŋk·shən }

magneto |ELEC| An alternating-current generator that uses one or more permanent magnets to produce its magnetic field; frequently used as a source of ignition energy on tractor, marine, industrial, and aviation engines. Also known as magnetoelectric generator. { mag'nēd·ō }

magneto anemometer |ENG| A cup anemometer with its shaft mechanically coupled to a magnet; both the frequency and amplitude of the voltage generated are proportional to the wind speed, and may be indicated or recorded by suitable electrical instruments. { mag'nēd·ō ,an·ə'mäm·əd·ər }

magnetocaloric effect |THERMO| The reversible change of temperature accompanying the change of magnetization of a ferromagnetic material. { mag¦nēd·ō·kə'lór·ik i,fekt }

magnetoelectronics |ELECTR| The use of electron spin (as opposed to charge) in electronic

341

devices. Also known as spin electronics; spintronics. (mag,ned·ō·i·lek'trän·iks)

magnetometer |ENG| An instrument for measuring the magnitude and sometimes also the direction of a magnetic field, such as the earth's magnetic field. (,mag·na'täm·ad·ar)

magnetooptic recording |ENG| An erasable data storage technology in which data are stored on a rotating disk in a thin magnetic layer that may be switched between two magnetization states by the combination of a magnetic field and a pulse of light from a diode laser. (mag,ned·ō,äp·tik ri'kòrd·iŋ)

magnetoresistance |ELECTR| The change in the electrical resistance of a material when it is subjected to an applied magnetic field; this property has widespread application in sensors and magnetic read heads. (mag¦nēd·ō·ri'zis·tans)

magnetoresistive memory |ELECTR| A random-access memory that uses the magnetic state of small ferromagnetic regions to store data, plus magnetoresistive devices to read the data, all integrated with silicon integrated-circuit electronics. (mag,ned·ō·ri,zis·tiv 'mem·rē)

magnetoresistor |ELECTR| Magnetic field-controlled variable resistor. (mag¦nēd·ō·ri'zis·tar)

magnetostrictive filter |ELECTR| Filter network which uses the magnetostrictive phenomena to form high-pass, low-pass, band-pass, or band-elimination filters; the impedance characteristic is the inverse of that of a crystal. (mag¦nēd·ō¦strik·tiv 'fil·tar)

magnetostrictive loudspeaker |ENG ACOUS| Loudspeaker in which the mechanical forces result from the deformation of a material having magnetostrictive properties. (mag¦nēd·ō¦strik·tiv 'laùd,spēk·ar)

magnetostrictive microphone |ENG ACOUS| Microphone which depends for its operation on the generation of an electromotive force by the deformation of a material having magnetostrictive properties. (mag¦nēd·ō¦strik·tiv 'mī·kra,fōn)

magnetostrictive oscillator |ELECTR| An oscillator whose frequency is controlled by a magnetostrictive element. (mag¦nēd·ō¦strik·tiv 'äs·a,lād·ar)

magnetovision |ENG| A method of measuring and displaying magnetic field distributions in which scanning results from a thin-film Permalloy magnetoresistive sensor are processed numerically and presented in the form of a color map on a video display unit. (mag'ned·a,vizh·an)

magnetron |ELECTR| One of a family of crossed-field microwave tubes, wherein electrons, generated from a heated cathode, move under the combined force of a radial electric field and an axial magnetic field in such a way as to produce microwave radiation in the frequency range 1–40 gigahertz; a pulsed microwave radiation source for radar, and continuous source for microwave cooking. ('mag·na,trän)

magnet wire |ELEC| The insulated copper or aluminum wire used in the coils of all types of electromagnetic machines and devices. ('mag·nat ,wīr)

magnistor |ELECTR| A device that utilizes the effects of magnetic fields on injection plasmas in semiconductors such as indium antimonide. (mag'nis·tar)

main |ELEC| **1.** One of the conductors extending from the service switch, generator bus, or converter bus to the main distribution center in interior wiring. **2.** *See* power transmission line. |ENG| A duct or pipe that supplies or drains ancillary branches. (mān)

main bearing |MECH ENG| One of the bearings that support the crankshaft in an internal combustion engine. (¦mān 'ber·iŋ)

main firing |ENG| The firing of a round of shots by means of current supplied by a transformer fed from a main power supply. ('mān 'fīr·iŋ)

main shaft |MECH ENG| The line of shafting receiving its power from the engine or motor and transmitting power to other parts. ('mān 'shaft)

maintainability |ENG| **1.** The ability of equipment to meet operational objectives with a minimum expenditure of maintenance effort under operational environmental conditions in which scheduled and unscheduled maintenance is performed. **2.** Quantitatively, the probability that an item will be restored to specified conditions within a given period of time when maintenance action is performed in accordance with prescribed procedures and resources. (mān,tā·na'bil·ad·ē)

maintenance |IND ENG| The upkeep of industrial facilities and equipment. ('mānt·an·ans)

maintenance engineering |IND ENG| The function of providing policy guidance for maintenance activities, and of exercising technical and management review of maintenance programs. ('mānt·an·ans ,en·ja'nir·iŋ)

maintenance kit |ENG| A collection of items not all having the same basic name, which are of a supplementary nature to a major component or equipment; the items within the collection may provide replacement parts and facilitate such functions as inspection, test repair, or preventive types of maintenance, for the specific purpose of restoring and improving the operational status of a component or equipment comparable to its original capacity and efficiency. ('mānt·an·ans ,kit)

maintenance vehicle |ENG| Vehicle used for carrying parts, equipment, and personnel for maintenance or evacuation of vehicles. ('mānt·an·ans ,vē·a·kal)

major assembly |ENG| A self-contained unit of individual identity; a completed assembly of component parts ready for operation, but utilized as a portion of, and intended for further installation in, an end item or major item. ('mā·jar a'sem·blē)

major defect |IND ENG| Defect which causes serious malfunctioning of a product. ('mā·jar 'dē,fekt)

major diameter |DES ENG| The largest diameter of a screw thread, measured at the crest for an external (male) thread and at the root for an internal (female) thread. { 'mā·jər dī'am·əd·ər }

majority carrier |ELECTR| The type of carrier, that is, electron or hole, that constitutes more than half the carriers in a semiconductor. { mə'jär·əd·ē 'kar·ē·ər }

majority emitter |ELECTR| Of a transistor, an electrode from which a flow of minority carriers enters the interelectrode region. { mə'jär·əd·ē i'mid·ər }

major repair |ENG| Repair work on items of material or equipment that need complete overhaul or substantial replacement of parts, or that require special tools. { 'mā·jər ri'per }

makeup air |ENG| The volume of air required to replace air exhausted from a given space. { 'māk,əp ,er }

makeup water |CHEM ENG| Water feed needed to replace that which is lost by evaporation or leakage in a closed-circuit, recycle operation. { 'mā,kəp ,wȯd·ər }

male connector |ELEC| An electrical connector with protruding contacts for joining with a female connector. { 'māl kə'nek·tər }

mallet |DES ENG| An implement with a barrel-shaped head made of wood, rubber, or other soft material; used for driving another tool, such as a chisel, or for striking a surface without causing damage. { 'mal·ət }

Mallory bonding |DES ENG| Hermetically sealing polished silicon chips to polished glass plates by placing the two pieces together, heating them to about 350°C (662°F), and applying approximately 8000 volts across the assembly. { 'mal·ə·rē ,bänd·iŋ }

management control system |IND ENG| Any one of the various systems used by a contractor to plan, control the cost, and schedule the work required to undertake and complete a project. { 'man·ij·mənt kən'trōl ,sis·təm }

management engineering See industrial engineering. { 'man·ij·mənt ,en·jə'nir·iŋ }

management game |IND ENG| A training exercise in which prospective decision makers act out managerial decision-making roles in a simulated environment. Also known as business game; operational game. { 'man·ij·mənt ,gām }

mandrel |ENG| The core around which continuous strands of impregnated reinforcement materials are wound to fabricate hollow objects made of composite materials. |MECH ENG| A shaft inserted through a hole in a component to support the work during machining. { 'man·drəl }

mandrel press |MECH ENG| A press for driving mandrels into holes. { 'man·drəl ,pres }

mangle gearing |MECH ENG| Gearing for producing reciprocating motion; a pinion rotating in a single direction drives a rack with teeth at the ends and on both sides. { 'maŋ·gəl ,gir·iŋ }

Manhattan Project |ENG| A United States project lasting from August 1942 to August 1946, which developed the atomic energy program,

with special reference to the atomic bomb. { man'hat·ən ,prä,jekt }

manhead See manhole. { 'man,hed }

manhole |ENG| An opening to provide access to a tank or boiler, to underground passages, or in a deck or bulkhead of a ship; usually covered with a cast iron or steel plate. Also known as access hole; manhead. { 'man,hōl }

man-hour |IND ENG| A unit of measure representing one person working for one hour. { 'man,aȯr }

manifold |ENG| The branch pipe arrangement which connects the valve parts of a multicylinder engine to a single carburetor or to a muffler. { 'man·ə,fōld }

manifolding |ENG| The gathering of multiple-line fluid inputs into a single intake chamber (intake manifold), or the division of a single fluid supply into several outlet streams (distribution manifold). { 'man·ə,fōld·iŋ }

manifold pressure |MECH ENG| The pressure in the intake manifold of an internal combustion engine. { 'man·ə,fōld ,presh·ər }

manikin |ENG| A correctly proportioned doll-like figure that is jointed and will assume any human position and hold it; useful in art to draw a human figure in action, or in medicine to show the relations of organs by means of movable parts. { 'man·ə·kən }

manipulative grasp See tripodal grasp. { mə'nip·yə·ləd·iv 'grasp }

manipulative skill |IND ENG| The ability of a worker to handle an object with the appropriate control and speed of movement required by a task. { mə'nip·yə·ləd·iv 'skil }

manipulators |CONT SYS| An armlike mechanism on a robotic system that consists of a series of segments, usually sliding or jointed which grasp and move objects with a number of degrees of freedom, under automatic control. See remote manipulator. { mə'nip·yə,läd·ərz }

man-machine chart See human-machine chart. { 'man mə;shēn 'chärt }

man-machine system See human-machine system. { 'man mə;shēn 'sis·təm }

manocryometer |THERMO| An instrument for measuring the change of a substance's melting point with change in pressure; the height of a mercury column in a U-shaped capillary supported by an equilibrium between liquid and solid in an adjoining bulb is measured, and the whole apparatus is in a thermostat. { ,man·ō ,krī'äm·əd·ər }

manometer |ENG| A double-leg liquid-column gage used to measure the difference between two fluid pressures. { mə'näm·əd·ər }

manometry |ENG| The use of manometers to measure gas and vapor pressures. { mə'näm·ə·trē }

manostat |ENG| Fluid-filled, upside-down manometer-type device used to control pressures within an enclosure, as for laboratory analytical distillation systems. { 'man·ə,stat }

M-A-N scavenging system |MECH ENG| A system for removing used oil and waste gases from a cylinder of an internal combustion engine in which the exhaust ports are located above the intake ports on the same side of the cylinder, so that gases circulate in a loop, leaving a dead spot in the center of the loop. { ¦em¦ā¦en 'skav·ənj·iŋ ˌsis·təm }

mantle |ENG| A lacelike hood or envelope (sack) of refractory material which, when positioned over a flame and heated to incandescence, gives light. { 'mant·əl }

manual control unit |CONT SYS| A portable, hand-held device that allows an operator to program and store instructions related to robot motions and positions. Also known as programming unit. { 'man·yə·wəl kən¦trōl ˌyü·nət }

manual element |IND ENG| A specific measurable subdivision of a work cycle or operation that is completed entirely by hand or with the use of tools. { 'man·yə·wəl 'el·ə·mənt }

manually controlled work See effort-controlled cycle. { 'man·yə·lē kən¦trōld 'wərk }

manual time See hand time. { 'man·yə·wəl 'tīm }

manual tracking |ENG| System of tracking a target in which all the power required is supplied manually through the tracking handwheels. { 'man·yə·wəl 'trak·iŋ }

manufacturer's part number |IND ENG| Identification number of symbol assigned by the manufacturer to a part, subassembly, or assembly. { ˌman·ə'fak·chər·ərz 'pärt ˌnəm·bər }

many-body problem |MECH| The problem of predicting the motions of three or more objects obeying Newton's laws of motion and attracting each other according to Newton's law of gravitation. Also known as n-body problem. { 'men·ē 'bäd·ē ˌpräb·ləm }

Marangoni effect |CHEM ENG| The effect that a disturbance of the liquid-liquid interface (due to interfacial tension) has on mass transfer in a liquid-liquid extraction system. { ˌmär·än'gō·nē i ˌfekt }

marbling |ENG| The use of antiquing techniques to achieve the appearance of marble in a paint film. { 'mär·bliŋ }

marginal cost |IND ENG| The extra cost incurred for an extra unit of output. { 'mär·jən·əl 'kȯst }

marginal product |IND ENG| The extra unit of output obtained by one extra unit of some factor, all other factors being held constant. { 'mär·jən·əl 'präd·əkt }

marginal revenue |IND ENG| The extra revenue achieved by selling an extra unit of output. { 'mär·jən·əl 'rev·ə,nü }

margin of safety |DES ENG| A design criterion, usually the ratio between the load that would cause failure of a member or structure and the load that is imposed upon it in service. { 'mär·jən əv 'sāf·tē }

Margoulis number See Stanton number. { mär 'gü·ləs ˌnəm·bər }

marigraph |ENG| A self-registering gage that records the heights of the tides. { 'mar·ə,graf }

marina |CIV ENG| A harbor facility for small boats, yachts, and so on, where supplies, repairs, and various services are available. { mə'rē·nə }

marine engineering |ENG| The design, construction, installation, operation, and maintenance of main power plants, as well as the associated auxiliary machinery and equipment, for the propulsion of ships. { mə'rēn ˌen·jə'nir·iŋ }

marine railway |CIV ENG| A type of dry dock consisting of a cradle of wood or steel with rollers on which the ship may be hauled out of the water along a fixed inclined track leading up the bank of a waterway. { mə'rēn 'rāl,wā }

marine terminal |CIV ENG| That part of a port or harbor with facilities for docking, cargo-handling, and storage. { mə'rēn 'tərm·ən·əl }

market analysis |IND ENG| The collection and evaluation of data concerned with the past, present, or future attributes of potential consumers for a product or service. { 'mar·kət ə,nal·ə·səs }

marmon clampband |DES ENG| A metal band that wraps around the circumference of a special cylindrical joint between two structures, holding the structures together. { 'mär·mən 'klamp ,ban }

Marvin sunshine recorder |ENG| A sunshine recorder in which the time scale is supplied by a chronograph, and consisting of two bulbs (one of which is blackened) that communicate through a glass tube of small diameter, which is partially filled with mercury and contains two electrical contacts; when the instrument is exposed to sunshine, the air in the blackened bulb is warmed more than that in the clear bulb; the warmed air expands and forces the mercury through the connecting tube to a point where the electrical contacts are shorted by the mercury; this completes the electrical circuit to the pen on the chronograph. { 'mär·vən 'sən,shīn ri,kȯrd·ər }

mask |DES ENG| A frame used in front of a television picture tube to conceal the rounded edges of the screen. |ELECTR| A thin sheet of metal or other material containing an open pattern, used to shield selected portions of a semiconductor or other surface during a deposition process. |ENG| A protective covering for the face or head in the form of a wire screen, a metal shield, or a respirator. { mask }

masking |ELECTR| **1.** Using a covering or coating on a semiconductor surface to provide a masked area for selective deposition or etching. **2.** A programmed procedure for eliminating radar coverage in areas where such transmissions may be of use to the enemy for navigation purposes, by weakening the beam in appropriate directions or by use of additional transmitters on the same frequency at suitable sites to interfere with homing; also used to suppress the beam in areas where it would interfere with television reception. |ENG| Preventing entrance of a tracer gas into a vessel by covering the leaks. { 'mask·iŋ }

masonry |CIV ENG| A construction of stone or similar materials such as concrete or brick. { 'mās·ən·rē }

masonry dam |CIV ENG| A dam constructed of

stone or concrete blocks set in mortar. { 'mās·ən·rē ,dam }

masonry drill |DES ENG| A drill tipped with cemented carbide for drilling in concrete or masonry. { 'mās·ən·rē ,dril }

masonry nail |DES ENG| Spiral-fluted nail designed to be driven into mortar joints in masonry. { 'mās·ən·rē ,nāl }

Mason's theorem |CONT SYS| A formula for the overall transmittance of a signal flow graph in terms of transmittances of various paths in the graph. { 'mās·ənz ,thir·əm }

mass |MECH| A quantitative measure of a body's resistance to being accelerated; equal to the inverse of the ratio of the body's acceleration to the acceleration of a standard mass under otherwise identical conditions. { mas }

mass burning rate |CHEM ENG| The loss in mass per unit time by materials burning under specified conditions. { 'mas 'bərn·iŋ ,rāt }

mass concrete |CIV ENG| Concrete set without structural reinforcement. { 'mas 'kän,krēt }

mass-distance |ENG| The mass carried by a vehicle multiplied by the distance it travels. { 'mas 'dis·təns }

mass flow |ENG| A pattern of powder flow occurring in hoppers that is characterized by the powder flowing at every point, including points adjacent to the hopper wall. { 'mas 'flō }

mass-flow bin |ENG| A bin whose hopper walls are sufficiently steep and smooth to cause flow of all the solid, without stagnant regions, whenever any solid is withdrawn. { 'mas ¦flō ,bin }

mass flowmeter |ENG| An instrument that measures the mass of fluid that flows through a pipe, duct, or open channel in a unit time. { 'mas ¦flō,mēd·ər }

mass-haul curve |CIV ENG| A curve showing the quantity of excavation in a cutting which is available for fill. { 'mas ¦hol ,kərv }

Massieu function |THERMO| The negative of the Helmholtz free energy divided by the temperature. { ma'syü ,fəŋk·shən }

mass law of sound insulation |CIV ENG| The rule stating that sound insulation for a single wall is determined almost wholly by its weight per unit area; doubling the weight of the partition increases the insulation by 5 decibels. { 'mas 'lo əv 'saund ,in·sə,lā·shən }

mass spectrograph |ENG| A mass spectroscope in which the ions fall on a photographic plate which after development shows the distribution of particle masses. { 'mas 'spek·trə ,graf }

mass spectrometer |ENG| A mass spectroscope in which a slit moves across the paths of particles with various masses, and an electrical detector behind it records the intensity distribution of masses. { 'mas spek'träm·əd·ər }

mass spectroscope |ENG| An instrument used for determining the masses of atoms or molecules, in which a beam of ions is sent through a combination of electric and magnetic fields so arranged that the ions are deflected according to their masses. { 'mas 'spek·trə,skōp }

mass units |MECH| Units of measurement having to do with masses of materials, such as pounds or grams. { 'mas ,yü·nəts }

mast |ENG| **1.** A vertical metal pole serving as an antenna or antenna support. **2.** A slender vertical pole which must be held in position by guy lines. **3.** A drill, derrick, or tripod mounted on a drill unit, which can be raised to operating position by mechanical means. **4.** A single pole, used as a drill derrick, supported in its upright or operating position by guys. |MECH ENG| A support member on certain industrial trucks, such as a forklift, that provides guideways for the vertical movement of the carriage. { mast }

master |ENG| **1.** A device which controls subsidiary devices. **2.** A precise workpiece through which duplicates are made. |ENG ACOUS| See master phonograph record. { 'mas·tər }

master arm |ENG| A component of a remote manipulator whose motions are automatically duplicated by a slave arm, sometimes with changes of scale in displacement or force. { 'mas·tər 'ärm }

master bushing See liner bushing. { 'mas·tər 'bush·iŋ }

master cylinder |MECH ENG| The container for the fluid and the piston, forming part of a device such as a hydraulic brake or clutch. { 'mas·tər 'sil·ən·dər }

master frequency meter See integrating frequency meter. { 'mas·tər 'frē·kwən·sē ,mēd·ər }

master gage |DES ENG| A locating device with fixed hole locations or part positions; locates in three dimensions and generally occupies the same space as the part it represents. { 'mas·tər 'gāj }

master layout |DES ENG| A permanent template record laid out in reference planes and used as a standard of reference in the development and coordination of other templates. { 'mas·tər 'lā,aut }

master mechanic |ENG| The supervisor, as at the mine, in charge of the maintenance and installation of equipment. { 'mas·tər mə'kan·ik }

master phonograph record |ENG ACOUS| The negative metal counterpart of a disk recording, produced by electroforming as one step in the production of phonograph records. Also known as master. { 'mas·tər 'fō·nə,graf ,rek·ərd }

master/slave manipulator |ENG| A mechanical, electromechanical, or hydromechanical device which reproduces the hand or arm motions of an operator, enabling the operator to perform manual motions while separated from the site of the work. { 'mas·tər 'slāv mə'nip·yə,lād·ər }

masticate |CHEM ENG| To process rubber on a machine to make it softer and more pliable before mixing with other substances. { 'mas·tə,kāt }

mat |CIV ENG| **1.** A steel or concrete footing under a post. **2.** Mesh reinforcement in a concrete slab. **3.** A heavy steel-mesh blanket used to suppress rock fragments during blasting. { mat }

match |ENG| **1.** A charge of gunpowder put in a paper several inches long and used for igniting explosives. **2.** A short flammable piece of wood, paper, or other material tipped with a combustible mixture that bursts into flame through friction. { mach }

matched edges |ENG| Die face edges machined at right angles to each other to provide for alignment of the dies in machining equipment. { 'macht 'ej·əz }

matched-metal molding |ENG| Forming of reinforced-plastic articles between two close-fitting metal molds mounted in a hydraulic press. { 'macht ¦med·əl ,mōld·iŋ }

material balance |CHEM ENG| A calculation to inventory material inputs versus outputs in a process system. { mə'tir·ē·əl 'bal·əns }

material particle |MECH| An object which has rest-mass and an observable position in space, but has no geometrical extension, being confined to a single point. Also known as particle. { mə'tir·ē·əl 'pärd·ə·kəl }

material requirements planning |IND ENG| A formal computerized approach to inventory planning, manufacturing scheduling, supplier scheduling, and overall corporate planning. Abbreviated MRP. { mə'tir·ē·əl ri'kwīr·məns ,plan·iŋ }

materials control |IND ENG| Inventory control of materials involved in manufacturing or assembly. { mə'tir·ē·əlz kən,trōl }

materials handling |ENG| The loading, moving, and unloading of materials. { mə'tir·ē·əlz ,hand·liŋ }

materials science |ENG| The study of the nature, behavior, and use of materials applied to science and technology. { mə'tir·ē·əlz ,sī·əns }

material well |CHEM ENG| In a plastics process, the space provided in a compression or transfer mold to allow for the bulk factor. { mə'tir·ē·əl ,wel }

mat foundation |CIV ENG| A large, thick, usually reinforced concrete mat which transfers loads from a number of columns, or columns and walls, to the underlying rock or soil. Also known as raft foundation. { mat faún'dā·shən }

Matheson joint |DES ENG| A wrought-pipe joint made by enlarging the end of one pipe length to receive the male end of the next length. { 'math·ə·sən ,jóint }

matrix |ELECTR| **1.** The section of a color television transmitter that transforms the red, green, and blue camera signals into color-difference signals and combines them with the chrominance subcarrier. Also known as color coder; color encoder; encoder. **2.** The section of a color television receiver that transforms the color-difference signals into the red, green, and blue signals needed to drive the color picture tube. Also known as color decoder; decoder. |ENG| A recessed mold in which something is formed or cast. { 'mā·triks }

matrix sound system |ENG ACOUS| A quadraphonic sound system in which the four input channels are combined into two channels by a coding process for recording or for stereo frequency-modulation broadcasting and decoded back into four channels for playback of recordings or for quadraphonic stereo reception. { 'mā·triks 'saúnd ,sis·təm }

matte feeder |IND ENG| A heavy-duty apron feeder composed of thick steel flights attached to a solid chain-link mat supported by closely spaced rollers. { 'mat ,fēd·ər }

Matthiessen sinker method |THERMO| A method of determining the thermal expansion coefficient of a liquid, in which the apparent weight of a sinker when immersed in the liquid is measured for two different temperatures of the liquid. { 'math·ə·sən 'siŋ·kər ,meth·əd }

mattock |DES ENG| A tool with the combined features of an adz, an ax, and a pick. { 'mad·ək }

mattress |CIV ENG| A woven mat, often of wire and cement blocks, used to prevent erosion of dikes, jetties, or river banks. { 'ma·trəs }

maul *See* rammer. { mól }

Maupertius' principle |MECH| The principle of least action is sufficient to determine the motion of a mechanical system. { mō'pər·shəs ,prin·sə·pəl }

max-flow min-cut theorem |IND ENG| In the analysis of networks, the concept that for any network with a single source and sink, the maximum feasible flow from source to sink is equal to the minimum cut value for any of the cuts of the network. { ,maks¦flō ,min'kət ,thir·əm }

maximal flow |IND ENG| Maximum total flow from the source to the sink in a connected network. { 'mak·sə·məl 'flō }

maximum allowable working pressure |MECH ENG| The maximum gage pressure in a pressure vessel at a designated temperature, used for the determination of the set pressure for relief valves. { 'mak·sə·məm ə¦laú·ə·bəl 'wərk·iŋ ,presh·ər }

maximum-and-minimum thermometer |ENG| A thermometer that automatically registers both the maximum and the minimum temperatures attained during an interval of time. { 'mak·sə·məm ən 'min·ə·məm thər'mäm·əd·ər }

maximum angle of inclination |MECH ENG| The maximum angle at which a conveyor may be inclined and still deliver an amount of bulk material within a given time. { 'mak·sə·məm 'aŋ·gəl əv ,in·klə'nā·shən }

maximum available gain |ELECTR| The theoretical maximum power gain available in a transistor stage; it is seldom achieved in practical circuits because it can be approached only when feedback is negligible. Abbreviated MAG. { 'mak·sə·məm ə¦vāl·ə·bəl 'gān }

maximum belt slope |MECH ENG| A slope beyond which the material on the belt of a conveyor tends to roll downhill. { 'mak·sə·məm 'belt ,slōp }

maximum belt tension |MECH ENG| The total of the starting and operating tensions in a conveyor. { 'mak·sə·məm 'belt ,ten·chən }

maximum continuous load |MECH ENG| The maximum load that a boiler can maintain for a designated length of time. { 'mak·sə·məm kən¦tin·yə·wəs 'lōd }

maximum gradability |MECH ENG| Steepest slope a vehicle can negotiate in low gear; usually expressed in precentage of slope, namely, the ratio between the vertical rise and the horizontal distance traveled; sometimes expressed by the angle between the slope and the horizontal. { 'mak·sə·məm ‚grād·ə'bil·əd·ē }

maximum ordinate |MECH| Difference in altitude between the origin and highest point of the trajectory of a projectile. { 'mak·sə·məm 'òrd·ən·ət }

maximum production life |MECH ENG| The length of time that a cutting tool performs at cutting conditions of maximum tool efficiency. { 'mak·sə·məm prə'dək·shən ‚līf }

maximum thermometer |ENG| A thermometer that registers the maximum temperature attained during an interval of time. { 'mak·sə·məm thər'mäm·əd·ər }

maximum working area |IND ENG| That portion of the working area that is readily accessible to the hands of a worker when in his normal operating position. { 'mak·sə·məm 'wərk·iŋ ‚er·ē·ə }

Maxwell equal-area rule |THERMO| At temperatures for which the theoretical isothermal of a substance, on a graph of pressure against volume, has a portion with positive slope (as occurs in a substance with liquid and gas phases obeying the van der Waals equation), a horizontal line drawn at the equilibrium vapor pressure and connecting two parts of the isothermal with negative slope has the property that the area between the horizontal and the part of the isothermal above it is equal to the area between the horizontal and the part of the isothermal below it. { 'mak‚swel ¦ē·kwəl 'er·ē·ə ‚rül }

Maxwell relation |THERMO| One of four equations for a system in thermal equilibrium, each of which equates two partial derivatives, involving the pressure, volume, temperature, and entropy of the system. { 'mak‚swel ri'lā·shən }

Maxwell's demon See demon of Maxwell. { 'mak ‚swel 'dē·mən }

Maxwell's stress functions |MECH| Three functions of position, ϕ_1, ϕ_2, and ϕ_3, in terms of which the elements of the stress tensor σ of a body may be expressed, if the body is in equilibrium and is not subjected to body forces; the elements of the stress tensor are given by $\sigma_{11} = \partial^2\phi_2/\partial x_3{}^2 + \partial^2\phi_3/\partial x_2{}^2$, $\sigma_{23} = -\partial^2\phi_1/\partial x_2\partial x_3$, and cyclic permutations of these equations. { 'mak ‚swel 'stres ‚fəŋk·shənz }

Maxwell's theorem |MECH| If a load applied at one point A of an elastic structure results in a given deflection at another point B, then the same load applied at B will result in the same deflection at A. { 'mak‚swel 'thir·əm }

mayer |THERMO| A unit of heat capacity equal to the heat capacity of a substance whose temperature is raised 1° Celsius by 1 joule. { 'mī·ər }

Mayer's formula |THERMO| A formula which

states that the difference between the specific heat of a gas at constant pressure and its specific heat at constant volume is equal to the gas constant divided by the molecular weight of the gas. { 'mī·ərz ‚fòr·myə·lə }

mb See millibar.

McCabe-Thiele diagram |CHEM ENG| Graphical method for calculation of the number of theoretical plates or contacting stages required for a given binary distillation operation. { mə'kāb 'tēl·ə ‚dī·ə‚gram }

M contour |CONT SYS| A line on a Nyquist diagram connecting points having the same magnitude of the primary feedback ratio. { 'em ‚kän·tùr }

M-design bit |DES ENG| A long-shank, box-threaded core bit made to fit M-design core barrels. { 'em di‚zīn ‚bit }

M-design core barrel |DES ENG| A double-tube core barrel in which a $2\frac{1}{2}°$-taper core lifter is carried inside a short tubular sleeve coupled to the bottom end of the inner tube, and the sleeve extends downward inside the bit shank to within a very short distance behind the face of the core bit. { 'em di‚zīn 'kòr ‚bar·əl }

meadow |ENG| Range of air-fuel ratio within which smooth combustion may be had. { 'med·ō }

mean-average boiling point |CHEM ENG| Pseudo boiling point for a hydrocarbon mixture; calculated from the American Society for Testing and Materials distillation curve's volumetric average boiling point. { ¦mēn ¦av·rij 'bòil·iŋ ‚pòint }

mean British thermal unit See British thermal unit. { 'mēn ¦brid·ish 'thər·məl ‚yü·nət }

mean calorie |THERMO| One-hundredth of the heat needed to raise 1 gram of water from 0 to 100°C. { 'mēn 'kal·ə·rē }

mean effective pressure |MECH ENG| A term commonly used in the evaluation for positive displacement machinery performance which expresses the average net pressure difference in pounds per square inch on the two sides of the piston in engines, pumps, and compressors. Abbreviated mep; mp. Also known as mean pressure. { 'mēn i¦fek·tiv 'presh·ər }

mean normal stress |MECH| In a system stressed multiaxially, the algebraic mean of the three principal stresses. { 'mēn ¦nòrm·əl 'stres }

mean pressure See mean effective pressure. { 'mēn 'presh·ər }

mean specific heat |THERMO| The average over a specified range of temperature of the specific heat of a substance. { 'mēn spə'sif·ik 'hēt }

mean-square-error criterion |CONT SYS| Evaluation of the performance of a control system by calculating the square root of the average over time of the square of the difference between the actual output and the output that is desired. { 'mēn 'skwer 'er·ər krī‚tir·ē·ən }

mean stress |MECH| **1.** The algebraic mean of

the maximum and minimum values of a periodically varying stress. **2.** *See* octahedral normal stress. { 'mēn 'stres }

mean temperature difference [CHEM ENG] In heat exchange calculations, a pseudo average temperature difference between the warmer and colder fluids at inlet and outlet conditions. { 'mēn 'tem·prə·chər ˌdif·rəns }

mean time to failure [ENG] A measure of reliability of a piece of equipment, giving the average time before the first failure. { 'mēn 'tīm tə 'fāl·yər }

mean time to repair [ENG] A measure of reliability of a piece of repairable equipment, giving the average time between repairs. { 'mēn 'tīm tə ri'per }

mean trajectory [MECH] The trajectory of a missile that passes through the center of impact or center of burst. { 'mēn trə'jek·trē }

measured daywork [IND ENG] Work done for an hourly wage on which specific productivity levels have been determined but which provides no incentive pay. { 'mezh·ərd 'dā,wərk }

measured drilling depth [ENG] The apparent depth of a borehole as measured along its longitudinal axis. { 'mezh·ərd 'dril·iŋ ,depth }

measured mile [CIV ENG] The distance of 1 mile (1609.344 meters), the units of which have been accurately measured and marked. { 'mezh·ərd 'mīl }

measured relieving capacity [DES ENG] The measured amounts of fluid which can be exhausted through a relief device at its rated operating pressure. { 'mezh·ərd ri'lēv·iŋ kə,pas·əd·ē }

measured work [IND ENG] Work, operations, or cycles for which a standard has been set. { ¦mezh·ərd 'wərk }

measurement ton *See* ton. { 'mezh·ər·mənt ,tən }

measuring machine [ENG] A device in which an astronomical photographic plate is viewed through a fixed low-power microscope with cross-hairs and which is mounted on a carriage that is moved by micrometer screws equipped with scales, in order to measure the relative positions of images on the plate. { 'mezh·ə·riŋ mə,shēn }

measuring tank [ENG] A tank that has been calibrated and fitted with devices to measure a volume of liquid and then release it. Also known as dump tank; metering tank. { 'mezh·ə·riŋ ,taŋk }

mechanical [ENG] Of, pertaining to, or concerned with machinery or tools. { mi'kan·ə·kəl }

mechanical advantage [MECH ENG] The ratio of the force produced by a machine such as a lever or pulley to the force applied to it. Also known as force ratio. { mi'kan·ə·kəl əd'van·tij }

mechanical analog [IND ENG] A mechanical model of a nonmechanical system that responds to an input with an output corresponding to the response of the real system. { mi'kan·i·kəl 'an·ə,läg }

mechanical analysis [MECH ENG] Mechanical separation of soil, sediment, or rock by sieving, screening, or other means to determine particle-size distribution. { mi'kan·ə·kəl ə'nal·ə·səs }

mechanical area [BUILD] The areas in a building that include equipment rooms, shafts, stacks, tunnels, and closets used for heating, ventilating, air conditioning, piping, communication, hoisting, conveying, and electrical services. { mi'kan·ə·kəl 'er·ē·ə }

mechanical bearing cursor *See* bearing cursor. { mi'kan·ə·kəl 'ber·iŋ ,kər·sər }

mechanical classification [MECH ENG] A sorting operation in which mixtures of particles of mixed sizes, and often of different specific gravities, are separated into fractions by the action of a stream of fluid, usually water. { mi'kan·ə·kəl ,klas·ə·fə'kā·shən }

mechanical classifier [MECH ENG] Any of various machines that are commonly used to classify mixtures of particles of different sizes, and sometimes of different specific gravities; the Dorr classifier is an example. { mi'kan·ə·kəl 'klas·ə,fī·ər }

mechanical comparator [ENG] A contact comparator in which movement is amplified usually by a rack, pinion, and pointer or by a parallelogram arrangement. { mi'kan·ə·kəl kəm'par·əd·ər }

mechanical damping [ENG ACOUS] Mechanical resistance which is generally associated with the moving parts of an electromechanically driven transducer such as a cutter or a reproducer. { mi'kan·ə·kəl 'damp·iŋ }

mechanical draft [MECH ENG] A draft that depends upon the use of fans or other mechanical devices; may be induced or forced. { mi'kan·ə·kəl 'draft }

mechanical-draft cooling tower [MECH ENG] Cooling tower that depends upon fans for introduction and circulation of its air supply. { mi'kan·ə·kəl ¦draft kül·iŋ ,taù·ər }

mechanical efficiency [MECH ENG] In an engine, the ratio of brake horsepower to indicated horsepower. { mi'kan·ə·kəl i'fish·ən·sē }

mechanical engineering [MECH ENG] The branch of engineering concerned with energy conversion, mechanics, and mechanisms and devices for diverse applications, ranging form automotive parts through nanomachines. { mi'kan·ə·kəl ,en·jə'nir·iŋ }

mechanical equivalent of heat [THERMO] The amount of mechanical energy equivalent to a unit of heat. { mi'kan·ə·kəl i'kwiv·ə·lənt əv 'hēt }

mechanical expression *See* expression. { mi'kan·ə·kəl ik'spresh·ən }

mechanical gripper [MECH ENG] A robot component that uses movable, fingerlike levers to grasp objects. { mi'kan·ə·kəl 'grip·ər }

mechanical hygrometer [ENG] A hygrometer in which an organic material, most commonly a bundle of human hair, which expands and contracts with changes in the moisture in the surrounding air or gas is held under slight tension by a spring, and a mechanical linkage actuates a pointer. { mi'kan·ə·kəl hī'gräm·əd·ər }

mechanical hysteresis |MECH| The dependence of the strain of a material not only on the instantaneous value of the stress but also on the previous history of the stress; for example, the elongation is less at a given value of tension when the tension is increasing than when it is decreasing. { mi'kan·ə·kəl ,his tə'rē·səs }

mechanical impedance |MECH| The complex ratio of a phasor representing a sinusoidally varying force applied to a system to a phasor representing the velocity of a point in the system. { mi'kan·ə·kəl im'pēd·əns }

mechanical lift dock [CIV ENG] A type of dry dock or marine elevator in which a vessel, after being placed on the keel and bilge blocks in the dock, is bodily lifted clear of the water so that work may be performed on the underwater body. { mi'kan·ə·kəl ¦lift 'däk }

mechanical linkage |MECH ENG| A set of rigid bodies, called links, joined together at pivots by means of pins or equivalent devices. { mi'kan·ə·kəl 'liŋ·kij }

mechanical loader |MECH ENG| A power machine for loading mineral, coal, or dirt. { mi 'kan·ə·kəl 'lōd·ər }

mechanical mucking |ENG| Loading of dirt or stone in tunnels or mines by machines. { mi 'kan·ə·kəl 'mək·iŋ }

mechanical ohm |MECH| A unit of mechanical resistance, reactance, and impedance, equal to a force of 1 dyne divided by a velocity of 1 centimeter per second. { mi'kan·ə·kəl 'ōm }

mechanical oscillograph See direct-writing recorder. { mi'kan·ə·kəl ä'sil·ə,graf }

mechanical patent [ENG] A patent granted for an inventive improvement in a process, manufacture, or machine. { mi'kan·ə·kəl 'pat·ənt }

mechanical press |MECH ENG| A press whose slide is operated by mechanical means. { mi 'kan·ə·kəl 'pres }

mechanical property |MECH| A property that involves a relationship between stress and strain or a reaction to an applied force. { mi'kan·ə· kəl 'präp·ərd·ē }

mechanical puddling See vibration puddling. { mi'kan·ə·kəl 'pəd·liŋ }

mechanical pulping |MECH ENG| Mechanical, rather than chemical, recovery of cellulose fibers from wood; unpurified, finely ground wood is made into newsprint, cheap Manila papers, and tissues. { mi'kan·ə·kəl 'pəlp·iŋ }

mechanical pump |MECH ENG| A pump through which fluid is conveyed by direct contact with a moving part of the pumping machinery. { mi'kan·ə·kəl 'pəmp }

mechanical reactance |MECH| The imaginary part of mechanical impedance. { mi'kan·ə·kəl rē'ak·təns }

mechanical refrigeration |MECH ENG| The removal of heat by utilizing a refrigerant subjected to cycles of refrigerating thermodynamics and employing a mechanical compressor. { mi'kan· ə·kəl ri,frij·ə'rā·shən }

mechanical resistance See resistance. { mi'kan· ə·kəl ri'zis·təns }

mechanical rotational impedance See rotational impedance. { mi'kan·ə·kəl rō'tā·shən·əl im 'pēd·əns }

mechanical rotational reactance See rotational reactance. { mi'kan·ə·kəl rō'tā·shən·əl rē'ak· təns }

mechanical rotational resistance See rotational resistance. { mi'kan·ə·kəl rō'tā·shən·əl ri'zis· təns }

mechanical scale |ENG| A weighing device that incorporates a number of levers with precisely located fulcrums to permit heavy objects to be balanced with counterweights or counterpoises. { mi'kan·ə·kəl 'skāl }

mechanical seal |MECH ENG| Mechanical assembly that forms a leakproof seal between flat, rotating surfaces to prevent high-pressure leakage. { mi'kan·ə·kəl 'sēl }

mechanical separation |MECH ENG| A group of industrial operations by means of which particles of solid or drops of liquid are removed from a gas or liquid, or are separated into individual fractions, or both, by gravity separation (settling), centrifugal action, and filtration. { mi 'kan·ə·kəl ,sep·ə'rā·shən }

mechanical setting |MECH ENG| Producing bits by setting diamonds in a bit mold into which a cast or powder metal is placed, thus embedding the diamonds and forming the bit crown; opposed to hand setting. Also known as cast setting; machine setting; sinter setting. { mi'kan· ə·kəl 'sed·iŋ }

mechanical shovel |MECH ENG| A loader limited to level or slightly graded drivages; when full, the shovel is swung over the machine, and the load is discharged into containers or vehicles behind. { mi'kan·ə·kəl 'shəv·əl }

mechanical splice |ENG| A splice made to terminate wire rope by pressing one or more metal sleeves over the rope junction. { mi'kan·ə·kəl 'splīs }

mechanical spring See spring. { mi'kan·ə·kəl 'spriŋ }

mechanical stage |ENG| A stage on a microscope provided with a mechanical device for positioning or changing the position of a slide. { mi'kan·ə·kəl 'stāj }

mechanical stepping motor |ELEC| A device in which a voltage pulse through a solenoid coil causes reciprocating motion by a solenoid plunger, and this is transformed into rotary motion through a definite angle by ratchet-and-pawl mechanisms or other mechanical linkages. { mi'kan·ə·kəl 'step·iŋ ,mōd·ər }

mechanical stoker See automatic stoker. { mi 'kan·ə·kəl 'stōk·ər }

mechanical torque converter |MECH ENG| A torque converter, such as a pair of gears, that transmits power with only incidental losses. { mi'kan·ə·kəl 'tȯrk kən,vərd·ər }

mechanical units |MECH| Units of length, time, and mass, and of physical quantities derivable from them. { mi'kan·ə·kəl ,yü·nəts }

mechanical vibration |MECH| The continuing motion, often repetitive and periodic, of parts of machines and structures. { mi'kan·ə·kəl vī'brā·shən }

mechanism |MECH ENG| That part of a machine which contains two or more pieces so arranged that the motion of one compels the motion of the others. { 'mek·ə,niz·əm }

mechanize |MECH ENG| 1. To substitute machinery for human or animal labor. 2. To produce or reproduce by machine. { 'mek·ə,nīz }

mechanized dew-point meter See dew-point recorder. { 'mek·ə,nīzd 'dü ,pȯint ,mēd·ər }

mechanomotive force |MECH| The root-mean-square value of a periodically varying force. { ¦mek·ə·nō¦mōd·iv ,fȯrs }

mechanooptical vibrometer |ENG| A vibrometer in which the motion given to a probe by a surface whose vibration amplitude is to be measured is used to rock a mirror; a light beam reflected from the mirror and focused onto a scale provides an indication of the vibration amplitude. { ¦mek·ə·nō¦äp·tə·kəl vī'bräm·əd·ər }

mechatronics |ENG| A branch of engineering that incorporates the ideas of mechanical and electronic engineering into a whole, and, in particular, covers those areas of engineering concerned with the increasing integration of mechanical, electronic, and software engineering into a production process. { ,mek·ə'trän·iks }

media migration |CHEM ENG| Carryover of fibers or other filter material by liquid effluent from a filter unit. { 'mē·dē·ə mī'grā·shən }

media mill See shot mill. { 'mēd·ē·ə ,mil }

median strip |CIV ENG| A paved or planted section dividing a highway into lanes according to direction of travel. { 'mē·dē·ən 'strip }

medical chemical engineering |CHEM ENG| The application of chemical engineering to medicine, frequently involving mass transport and separation processes, especially at the molecular level. { 'med·ə·kəl 'kem·ə·kəl ,en·jə'nir·iŋ }

medium |CHEM ENG| 1. The carrier in which a chemical reaction takes place. 2. Material of controlled pore size used to remove foreign particles or liquid droplets from fluid carriers. { 'mē·dē·əm }

medium-technology robot |CONT SYS| An automatically controlled machine that employs servomechanisms and microprocessor control units. { 'mē·dē·əm tek¦näl·ə·jē 'rō,bät }

megasecond |MECH| A unit of time, equal to 1,000,000 seconds. Abbreviated Ms; Msec. { 'meg·ə,sek·ənd }

megawatt |MECH| A unit of power, equal to 1,000,000 watts. Abbreviated MW. { 'meg·ə,wät }

megohm |ELEC| A unit of resistance, equal to 1,000,000 ohms. { 'me,gōm }

megohmmeter |ELEC| An instrument which is used for measuring the high resistance of electrical materials of the order of 20,000 megohms at 1000 volts; one direct-reading type employs a permanent magnet and a moving coil. { 'me ,gōmē,mēd·ər }

Melde's experiment |MECH| An experiment to study transverse vibrations in a long, horizontal thread when one end of the thread is attached to a prong of a vibrating tuning fork, while the other passes over a pulley and has weights suspended from it to control the tension in the thread. { 'mel·dēz ik,sper·ə·mənt }

meltback transistor |ELECTR| A junction transistor in which the junction is made by melting a properly doped semiconductor and allowing it to solidify again. { 'melt'back tran'zist·ər }

melter |ENG| A chamber used for melting. { 'melt·ər }

melt extractor |ENG| A device used to feed an injection mold, separating molten feed material from partially molten pellets. { 'melt ik,strak·tər }

melt fracture |MECH| Melt flow instability through a die during plastics molding, leading to helicular, rippled surface irregularities on the finished product. { 'melt ,frak·chər }

melt index |ENG| Number of grams of thermoplastic resin at 190°C that can be forced through a 0.0825-inch (2.0955-millimeter) orifice in 10 minutes by a 2160-gram force. { 'melt ,in,deks }

melting furnace |ENG| A furnace in which the frit for glass is melted. { 'melt·iŋ ,fər·nəs }

melting point |THERMO| 1. The temperature at which a solid of a pure substance changes to a liquid. Abbreviated mp. 2. For a solution of two or more components, the temperature at which the first trace of liquid appears as the solution is heated. { 'melt·iŋ ,pȯint }

melt instability |MECH| Instability of the plastic melt flow through a die. { 'melt ,in·stə'bil·əd·ē }

melt strength |MECH| Strength of a molten plastic. { 'melt ,streŋkth }

member |CIV ENG| A structural unit such as a wall, column, beam, or tie, or a combination of any of these. { 'mem·bər }

membrane |BUILD| In built-up roofing, a weather-resistant (flexible or semiflexible) covering consisting of alternate layers of felt and bitumen, fabricated in a continuous covering and surfaced with aggregate or asphaltic material. |CHEM ENG| 1. The medium through which the fluid stream is passed for purposes of filtration. 2. The ion-exchange medium used in dialysis, diffusion, osmosis and reverse osmosis, and electrophoresis. { 'mem,brān }

membrane analogy |MECH| A formal identity between the differential equation and boundary conditions for a stress function for torsion of an elastic prismatic bar, and those for the deflection of a uniformly stretched membrane with the same boundary as the cross section of the bar, subjected to a uniform pressure. { 'mem,brān ə,nal·ə·jē }

membrane curing See membrane waterproofing. { 'mem,brān ,kyur·iŋ }

membrane distillation |CHEM ENG| A separation method that uses a nonwetting, microporous membrane, with a liquid feed phase on one side and a condensing permeate phase on the other. Also known as membrane evaporation; thermopervaporation; transmembrane distillation. { 'mem,brān ,dis·tə'lā·shən }

membrane evaporation See membrane distillation. { 'mem,brān i,vap·ə'rā·shən }

membrane separation [CHEM ENG] The use of thin barriers (membranes) between miscible fluids for separating a mixture; a suitable driving force across the membrane, for example concentration or pressure differential, leads to preferential transport of one or more feed components. { 'mem,brān ,sep·ə'rā·shən }

membrane stress [MECH] Stress which is equivalent to the average stress across the cross section involved and normal to the reference plane. { 'mem,brān ,stres }

membrane waterproofing [CIV ENG] Curing concrete, especially in pavements, by spraying a liquid material over the surface to form a solid, impervious layer which holds the mixing water in the concrete. Also known as membrane curing. { 'mem,brān 'wȯd·ər,prüf·iŋ }

memomotion study [IND ENG] A technique of work measurement and methods analysis using a motion picture camera operated at less than normal camera speed. Also known as camera study; micromotion study. { ¦mem·ō¦mō·shən ,stəd·ē }

MEMS See micro-electro-mechanical system. { memz or ¦em¦ē¦em'es }

MEMS microphone [ENG ACOUS] A very small microphone, generally less than 1 millimeter, that can be incorporated directly onto an electronic chip and commonly uses a small thin membrane fabricated on the chip to detect sound. { ¦memz or ¦em¦ē¦em¦es 'mī·krə,fōn }

mep See mean effective pressure.

Mercer engine [MECH ENG] A revolving-block engine in which two opposing pistons operate in a single cylinder with two rollers attached to each piston; intake ports are uncovered when the pistons are closest together, and exhaust ports are uncovered when they are farthest apart. { 'mər·sər ,en·jən }

mercury barometer [ENG] An instrument which determines atmospheric pressure by measuring the height of a column of mercury which the atmosphere will support; the mercury is in a glass tube closed at one end and placed, open end down, in a well of mercury. Also known as Torricellian barometer. { 'mər·kyə·rē bə'räm·əd·ər }

mercury-cathode cell [CHEM ENG] Electrolytic cell used to manufacture chlorine and caustic soda from sodium chloride brine; includes Castner and DeNora cells. { 'mər·kyə·rē ¦kath,ōd ,sel }

mercury jet magnetometer [ENG] A type of magnetometer in which the magnetic field strength is determined by measuring the electromotive force between electrodes at opposite ends of a narrow pipe made of insulating material, through which mercury is forced to flow. { ¦mər·kyə·rē ,jet ,mag·nə'täm·əd·ər }

mercury manometer [ENG] A manometer in

which the instrument fluid is mercury; used to record or control difference of pressure or fluid flow. { 'mər·kyə·rē mə'näm·əd·ər }

mercury switch [ELEC] A switch that is closed by making a large globule of mercury move up to the contacts and bridge them; the mercury is usually moved by tilting the entire switch. { 'mər·kyə·rē ,swich }

mercury thermometer [ENG] A liquid-in-glass thermometer or a liquid-in-metal thermometer using mercury as the liquid. { 'mər·kyə·rē thər'mäm·əd·ər }

meridian circle See transit circle. { mə'rid·ē·ən ,sər·kəl }

meridian transit See transit circle. { mə'rid·ē·ən ,tran·zət }

merit [ELECTR] A performance rating that governs the choice of a device for a particular application; it must be qualified to indicate type of rating, as in gain-bandwidth merit or signal-to-noise merit. { 'mer·ət }

merit pay plan [IND ENG] Work performed for a set hourly wage that varies from one pay period to another as a function of the worker's productivity, but never declines below a guaranteed minimum wage. { 'mer·ət ¦pā ,plan }

Mersenne's law [MECH] The fundamental frequency of a vibrating string is proportional to the square root of the tension and inversely proportional both to the length and the square root of the mass per unit length. { mər'senz ,lȯ }

Merton nut [DES ENG] A nut whose threads are made of an elastic material such as cork, and are formed by compressing the material into a screw. { 'mərt·ən ,nət }

mesa device [ELECTR] Any device produced by diffusing the surface of a germanium or silicon wafer and then etching down all but selected areas, which then appear as physical plateaus or mesas. { 'mā·sə di,vīs }

mesa diode [ELECTR] A diode produced by diffusing the entire surface of a large germanium or silicon wafer and then delineating the individual diode areas by a photoresist-controlled etch that removes the entire diffused area except the island or mesa at each junction site. { 'mā·sə ,dī,ōd }

mesa transistor [ELECTR] A transistor in which a germanium or silicon wafer is etched down in steps so the base and emitter regions appear as physical plateaus above the collector region. { 'mā·sə tran'zis·tər }

MESFET See metal semiconductor field-effect transistor.

mesh [DES ENG] A size of screen or of particles passed by it in terms of the number of openings occurring per linear inch in each direction. Also known as mesh size. [ELEC] A set of branches forming a closed path in a network so that if any one branch is omitted from the set, the remaining branches of the set do not form a closed path. Also known as loop. [MECH ENG] Engagement or working contact of teeth of gears or of a gear and a rack. { mesh }

messenger [ENG] A small, cylindrical metal

weight that is attached around an oceanographic wire and sent down to activate the tripping mechanism on various oceanographic devices. { 'mes·ən·jər }

metabolic cost |IND ENG| The amount of energy consumed as the result of performing a given work task; usually expressed in calories. { ¦med·ə,bäl·ik 'kȯst }

metal lath |ENG| A mesh of metal used to provide a base for plaster. { 'med·əl 'lath }

metallic disk rectifier See metallic rectifier. { mə'tal·ik ¦disk 'rek·tə,fī·ər }

metallize |ENG| To coat or impregnate a metal or nonmetal surface with a metal, as by metal spraying or by vacuum evaporation. { 'med·əl,īz }

metallized slurry blasting |ENG| The breaking of rocks by using slurried explosive medium containing a powdered metal, such as powdered aluminum. { 'med·əl,īzd ¦slər·ē 'blast·iŋ }

metallurgical engineer |ENG| A person who specializes in metallurgical engineering. { ,med·əl'ər·jə·kəl ,en·jə'nir }

metallurgical engineering |ENG| Application of the principles of metallurgy to the engineering sciences. { ,med·əl'ər·jə·kəl ,en·jə'nir·iŋ }

metallurgical microscope |ENG| A microscope used in the study of metals, usually optical. { ,med·əl'ər·jə·kəl 'mī·krə,skōp }

metal oxide semiconductor field-effect transistor |ELECTR| A field-effect transistor having a gate that is insulated from the semiconductor substrate by a thin layer of silicon dioxide. Abbreviated MOSFET; MOST; MOS transistor. Formerly known as an insulated-gate field-effect transistor (IGFET). { 'med·əl ¦äk,sīd 'sem·i·kən,dək·tər 'fēld i,fekt tran'zis·tər }

metal oxide semiconductor integrated circuit |ELECTR| An integrated circuit using metal oxide semiconductor transistors; it can have a higher density of equivalent parts than a bipolar integrated circuit. { 'med·əl ¦äk,sīd 'sem·i·kən,dək·tər 'int·ə,grād·əd 'sər·kət }

metal rolling See rolling. { 'med·əl ,rōl·iŋ }

metal semiconductor field-effect transistor |ELECTR| A field-effect transistor that uses a thin film of gallium arsenide, with a Schottky barrier gate formed by depositing a layer of metal directly onto the surface of the film. Abbreviated MESFET. { 'med·əl 'sem·i·kən,dək·tər 'fēld i,fekt tran'zis·tər }

metal-slitting saw |MECH ENG| A milling cutter similar to a circular saw blade but sometimes with side teeth as well as teeth around the circumference; used for deep slotting and sinking in cuts. { 'med·əl ¦slid·iŋ 'sȯ }

metal spinning See spinning. { 'med·əl ,spin·iŋ }

metal spraying |ENG| Coating a surface with droplets of molten metal or alloy by using a compressed gas stream. { 'med·əl 'sprā·iŋ }

metarheology |MECH| A branch of rheology whose approach is intermediate between those of macrorheology and microrheology; certain processes that are not isothermal are taken into consideration, such as kinetic elasticity, surface tension, and rate processes. { ,med·ə·rē'äl·ə·jē }

meteorogram |ENG| A record obtained from a meteorograph. { ,med·ē'ȯr·ə,gram }

meteorograph |ENG| An instrument that measures and records meteorological data such as air pressure, temperature, and humidity. { ,med·ē'ȯr·ə,graf }

meteorological balloon |ENG| A balloon, usually of high-quality neoprene, polyethylene, or Mylar, used to lift radiosondes to high altitudes. { ,med·ē·ə·rə'läj·ə·kəl bə'lün }

meteorological instrumentation |ENG| Apparatus and equipment used to obtain quantitative information about the weather. { ,med·ē·ə·rə'läj·ə·kəl ,in·strə·mən'tā·shən }

meteorological rocket |ENG| Small rocket system used to extend observation of atmospheric character above feasible limits for balloon-borne observing and telemetering instruments. Also known as rocketsonde. { ,med·ē·ə·rə'läj·ə·kəl 'räk·ət }

meter |MECH| The international standard unit of length, equal to the length of the path traveled by light in vacuum during a time interval of 1/299,792,458 of a second. Abbreviated m. |ENG| A device for measuring the value of a quantity under observation; the term is usually applied to an indicating instrument alone. { 'mēd·ər }

meter bar |ENG| A metal bar for mounting a gas meter, having fittings at the ends for the inlet and outlet connections of the meter. { 'mēd·ər ,bär }

meter density |ENG| In an energy distribution system, the number of meters per unit area or per unit length. { 'mēd·ər ,den·səd·ē }

meter factor |ENG| A factor used with a meter to correct for ambient conditions, for example, the factor for a fluid-flow meter to compensate for such conditions as liquid temperature change and pressure shrinkage. { 'mēd·ər ,fak·tər }

metering pin See metering rod. { 'mēd·ə·riŋ ,pin }

metering pump |CHEM ENG| Plunger-type pump designed to control accurately small-scale fluid-flow rates; used to inject small quantities of materials into continuous-flow liquid streams. Also known as proportioning pump. { 'mēd·ə·riŋ ,pəmp }

metering rod |ENG| A device consisting of a long metallic pin of graduated diameters fitted to the main nozzle of a carburetor (on an internal combustion engine) or passage leading thereto in such a way that it measures or meters the amount of gasoline permitted to flow by it at various speeds. Also known as metering pin. { 'mēd·ə·riŋ ,räd }

metering screw |MECH ENG| An extrusion-type screw feeder or conveyor section used to feed pulverized or doughy material at a constant rate. { 'mēd·ə·riŋ ,skrü }

metering tank See measuring tank. { 'mēd·ə·riŋ ,taŋk }

metering valve |MECH ENG| In an automotive

hydraulic braking system, a valve that momentarily delays application of the front disk brakes until the rear drum brakes begin to act. { 'mēd·ə·riŋ ,valv }

meter-kilogram |MECH| **1.** A unit of energy or work in a meter-kilogram-second gravitational system, equal to the work done by a kilogram-force when the point at which the force is applied is displaced 1 meter in the direction of the force; equal to 9.80665 joules. Abbreviated m-kgf. Also known as meter kilogram-force. **2.** A unit of torque, equal to the torque produced by a kilogram-force acting at a perpendicular distance of 1 meter from the axis of rotation. Also known as kilogram-meter (kgf-m). { 'mēd·ər 'kil·ə ,gram }

meter kilogram-force See meter-kilogram. { 'mēd·ər 'kil·ə,gram 'fòrs }

meter-kilogram-second system |MECH| A metric system of units in which length, mass, and time are fundamental quantities, and the units of these quantities are the meter, the kilogram, and the second respectively. Abbreviated mks system. { 'mēd·ər 'kil·ə,gram 'sek·ənd ,sis·təm }

meter prover |ENG| A device that determines the accuracy of a gas meter; a quantity of air is collected over water or oil in a calibrated cylindrical bell, and then the bell is allowed to sink into the liquid, forcing the air through the meter; the calibrated measurement is then compared with the reading on the meter dial. { 'mēd·ər ,prü·vər }

meter-proving tank See calibrating tank. { 'mēd·ər ,prü·viŋ ,taŋk }

meter run |ENG| The length of straight, unobstructed fluid-flow conduit preceding an orifice or venturi meter. { 'mēd·ər ,rən }

meter sensitivity |ENG| The accuracy with which a meter can measure a voltage, current, resistance, or other quantity. { 'mēd·ər ,sen·sə'tiv·əd·ē }

meter stop |MECH ENG| A valve installed in a water service pipe for control of the flow of water to a building. { 'mēd·ər ,stäp }

meter-ton-second system |MECH| A modification of the meter-kilogram-second system in which the metric ton (1000 kilograms) replaces the kilogram as the unit of mass. { 'mēd·ər 'tən 'sek·ənd ,sis·təm }

meter wheel |ENG| A special block used to support the oceanographic wire paid out over the side of a ship; attached directly or connected by means of a speedometer cable to a gearbox which measures the length of wire. { 'mēd·ər ,wēl }

methanation |CHEM ENG| In coal gasification, the catalytic conversion of hydrogen and carbon monoxide to methane. { ,meth·ə'nā·shən }

method of joints |ENG| Determination of stresses for joints at which there are not more than two unknown forces by the methods of the stress polygon, resolution, or moments. { 'meth·əd əv 'jòins }

method of mixtures |THERMO| A method of determining the heat of fusion of a substance whose specific heat is known, in which a known amount of the solid is combined with a known amount of the liquid in a calorimeter, and the decrease in the liquid temperature during melting of the solid is measured. { 'meth·əd əv 'miks·chərz }

methods design |IND ENG| Design for a new, more efficient method of job performance. { 'meth·ədz di,zīn }

methods engineering |IND ENG| A technique used by management to improve working methods and reduce labor costs in all areas where human effort is required. { 'meth·ədz ,en·jə'nir·iŋ }

methods study |IND ENG| An analysis of the methods in use, of the means and potentials for their improvement, and of reducing costs. { 'meth·ədz ,stəd·ē }

metric centner |MECH| **1.** A unit of mass equal to 50 kilograms. **2.** A unit of mass equal to 100 kilograms. Also known as quintal. { 'me·trik 'sent·nər }

metric grain |MECH| A unit of mass, equal to 50 milligrams; used in commercial transactions in precious stones. { 'me·trik 'grān }

metric line See millimeter. { 'me·trik 'līn }

metric ounce See mounce. { 'me·trik 'aùns }

metric slug See metric-technical unit of mass. { 'me·trik 'sləg }

metric system |MECH| A system of units used in scientific work throughout the world and employed in general commercial transactions and engineering applications; its units of length, time, and mass are the meter, second, and kilogram respectively, or decimal multiples and submultiples thereof. { 'me·trik ,sis·təm }

metric-technical unit of mass |MECH| A unit of mass, equal to the mass which is accelerated by 1 meter per second per second by a force of 1 kilogram-force; it is equal to 9.80665 kilograms. Abbreviated TME. Also known as hyl; metric slug. { 'me·trik ‖tek·ni·kəl ‖yü·nət əv 'mas }

metric thread gearing |DES ENG| Gears that may be interchanged in change-gear systems to provide feeds suitable for cutting metric and module threads. { 'me·trik 'thred ,gir·iŋ }

metric ton See tonne. { 'me·trik 'tən }

mg See milligram.

mGal See milligal.

mho See siemens. { mō }

mi See mile.

MIC See microwave integrated circuit.

Michaelson actinograph |ENG| A pyrheliometer of the bimetallic type used to measure the intensity of direct solar radiation; the radiation is measured in terms of the angular deflection of a blackened bimetallic strip which is exposed to the direct solar beams. { 'mī·kəl·sən ak 'tin·ə,graf }

microacceleratometer |ENG| A MEMS device developed for the automotive industry to control air-bag inflation. { ,mī·krō·ik,sel·ə·rə'täm·əd·ər }

microactuator [ENG] A very small actuator, with physical dimensions in the submicrometer to millimeter range, generally batch-fabricated from silicon wafers. { ˌmī·krō'ak·chə͵wād·ər }

micro air vehicle [ENG] A very small airborne autonomous vehicle that can operate inside a building using primarily visual and other sensory information to navigate. { ¦mī·krō ˌvē·ə·kəl }

microalloy diffused transistor [ELECTR] A microalloy transistor in which the semiconductor wafer is first subjected to gaseous diffusion to produce a nonuniform base region. Abbreviated MADT. { ¦mī·krō'al͵ói də'fyüzd tran'zis·tər }

microalloy transistor [ELECTR] A transistor in which the emitter and collector electrodes are formed by etching depressions, then electroplating and alloying a thin film of the impurity metal to the semiconductor wafer, somewhat as in a surface-barrier transistor. { ¦mī·krō'al͵ói tran'zis·tər }

microangstrom [MECH] A unit of length equal to one-millionth of an angstrom, or 10^{-16} meter. Abbreviated μA. { ¦mī·krō'aŋ·strəm }

microbalance [ENG] A small, light type of analytical balance that can weigh loads of up to 0.1 gram to the nearest microgram. { ¦mī·krō'bal·əns }

microbar See barye. { 'mī·krə͵bär }

microbarogram [ENG] The record or trace made by a microbarograph. { ¦mī·krō'bar·ə͵gram }

microcalorimeter [ENG] A calorimeter for measuring very small amounts of heat, in which the heat source and a small heating coil are placed in identical vessels and the amount of current through the coil is varied until the temperatures of the vessels are identical, as indicated by thermocouples. { ˌmī·krō͵kal·ə'rim·əd·ər }

microcapacitor [ELECTR] Any very small capacitor used in microelectronics, usually consisting of a thin film of dielectric material sandwiched between electrodes. { ¦mī·krō·kə'pas·əd·ər }

microcapsule [CHEM ENG] A capsule with a plastic or waxlike coating having a diameter anywhere from well below 1 micrometer to over 2000 micrometers. { 'mī·krō͵kap·səl }

microcircuitry [ELECTR] Electronic circuit structures that are orders of magnitude smaller and lighter than circuit structures produced by the most compact combinations of discrete components. Also known as microelectronic circuitry; microminiature circuitry. { ¦mī·krō'sər·kə·trē }

microcontroller [ELECTR] A microcomputer, microprocessor, or other equipment used for precise process control in data handling, communication, and manufacturing. { ¦mī·krō·kən'trōl·ər }

microdiffusiometer [ENG] A type of diffusiometer in which diffusion is measured over microscopic distances, greatly reducing the time required for the measurement and the effects of vibration and temperature changes. { ˌmī·krō·də'fyüz·ər }

microelectrode [ENG] **1.** In biological research, an electrode with a microscopic tip dimension that may be placed adjacent to or inside a cell for the purpose of recording the electric potentials of single cells, passing electrical currents, or injecting electrically charged substances into the cell. **2.** In physical chemistry, a minute electrode used to perform electrolysis of small quantities of material. { ˌmī·krō·i'lek͵trōd }

micro-electro-mechanical system [ENG] A system in which micromechanisms are coupled with microelectronics, most commonly fabricated as microsensors or microactuators. Abbreviated MEMS. Also known as microsystem. { ¦mī·krōi͵lek·trə·mə'kan·ə·kəl ͵sis·təm }

microelectronic circuitry See microcircuitry. { ¦mī·krō·i͵lek'trän·ik 'sər·kə·trē }

microelectronics [ELECTR] The technology of constructing circuits and devices in extremely small packages by various techniques. Also known as microminiaturization; microsystem electronics. { ¦mī·krō·i͵lek'trän·iks }

microelement [ELECTR] Resistor, capacitor, transistor, diode, inductor, transformer, or other electronic element or combination of elements mounted on a ceramic wafer 0.025 centimeter thick and about 0.75 centimeter square; individual microelements are stacked, interconnected, and potted to form micromodules. [IND ENG] An element of a work cycle whose time span is too short to be observed by the unaided eye. { ¦mī·krō'el·ə·mənt }

microencapsulation [CHEM ENG] Enclosing of materials in capsules from well below 1 micrometer to over 2000 micrometers in diameter. { ¦mī·krō·in͵kap·sə'lā·shən }

microengineering [ENG] The design and production of small, three-dimensional objects, usually for manufacture in high volumes at low cost. { ˌmī·krō͵en·jə'nir·iŋ }

microfabrication [ENG] The technology of fabricating microsystems from silicon wafers, using standard semiconductor process technologies in combination with specially developed processes. { ¦mī·krō͵fab·rə'kā·shən }

microfiltration [CHEM ENG] A membrane separation process in which particles greater than about 20 nanometers in diameter are screened out of a liquid in which they are suspended. { ͵mi·krō·fil'trā·shən }

microfluoroscope [ENG] A fluoroscope in which a very fine-grained fluorescent screen is optically enlarged. { ¦mī·krō'flür·ə͵skōp }

microforge [ENG] In micromanipulation techniques, an optical-mechanical device for controlling the position of needles or pipets in the field of a low-power microscope by a simple micromanipulator. { 'mī·krə͵förj }

microgram [MECH] A unit of mass equal to one-millionth of a gram. Abbreviated μg. { 'mī·krə͵gram }

micrograph [ENG] An instrument for making very tiny writing or engraving. { 'mī·krə͵graf }

microgravity |MECH| A state of very weak gravity, such that the gravitational acceleration experienced by an observer inside the system in question is of the order of one-millionth of that on earth. { ,mī·krō'grav·əd·ē }

microgroove record See long-playing record. { 'mī·krə,grüv ,rek·ərd }

micro heat pipe |ENG| A very small heat pipe that has a diameter between about 100 micrometers and 2 millimeters (0.004 and 0.08 inch) and a triangular cross section or other cross section with sharp corners, and that uses the sharp corner regions instead of a wick to return the working fluid from the condenser to the evaporator; it has potential applications in the electronics (cooling circuit chips), medical, space, and aircraft industries. { ,mī·krō 'hēt ,pīp }

micromachining |ENG| The use of standard semiconductor process technologies in combination with specially developed processes to fabricate miniature mechanical devices and components on silicon and other materials. { 'mī·krō· mə,shēn·iŋ }

micromanipulator |ENG| A device for holding and moving fine instruments for the manipulation of microscopic specimens under a microscope. { ¦mī·krō·mə'nip·yə,lād·ər }

micromanometer |ENG| Any manometer that is designed to measure very small pressure differences. { ¦mī·krō·mə'näm·əd·ər }

micromechanical display |ENG| A video display based on an array of mirrors on a silicon chip that can be deflected by electrostatic forces. Abbreviated MMD. { ,mī·krō·mə,kan·i·kəl di'splā }

micromechanics |ENG| **1.** The design and fabrication of micromechanisms. **2.** See composite micromechanics. { ¦mī·krō·mə'kan·iks }

micromechanism |ENG| A mechanical component with submillimeter dimensions and corresponding tolerances of the order of 1 micrometer or less. { ¦mī·krō 'mek·ə,niz·əm }

micromechatronics |ENG| The branch of engineering concerned with micro-electro-mechanical systems. { ¦mī·krō,mek·ə'trän·iks }

micrometer |ENG| **1.** An instrument attached to a telescope or microscope for measuring small distances or angles. **2.** A caliper for making precise measurements; a spindle is moved by a screw thread so that it touches the object to be measured; the dimension can be read on a scale. Also known as micrometer caliper. |MECH| A unit of length equal to one-millionth of a meter. Abbreviated μm. Also known as micron (μ). { mī'kräm·əd·ər }

micrometer caliper See micrometer. { mī'kräm· əd·ər 'kal·ə·pər }

micrometer of mercury See micron. { mī'kräm· əd·ər əv 'mər·kyə·rē }

micromicrowatt See picowatt. { ¦mī·krō¦mī· krō'wät }

micromolding |ENG| An alternative technique to micromachining for fabricating microsystems, in which a sacrificial material serves as a mold

to which a deposited material conforms. { 'mī· krō,mōld·iŋ }

micromotion film |IND ENG| A record of a specific task made with motion picture film or video tape in which each component of the activity is recorded in an individual frame. { 'mī·krō,mō· shən ,film }

micromotion study See memomotion study. { ¦mī·krō¦mō shən 'stəd·ē }

micron |MECH| **1.** A unit of pressure equal to the pressure exerted by a column of mercury 1 micrometer high, having a density of 13.5951 grams per cubic centimeter, under the standard acceleration of gravity; equal to 0.133322387415 pascal; it differs from the millitorr by less than one part in seven million. Also known as micrometer of mercury. **2.** See micrometer. { 'mī ,krän }

micro-opto-electro-mechanical system |ENG| A microsystem that combines the functions of optical, mechanical, and electronic components in a single, very small package or assembly. Abbreviated MOEMS. { ¦mī·kro¦ äp·tō i¦lek·trō mə¦kan·ə·kəl 'sis·təm }

micro-opto-mechanical system |ENG| A microsystem that combines optical and mechanical functions without the use of electronic devices or signals. Abbreviated MOMS. { ¦mī·krō ¦op· to·mə¦kan·ə·kəl ,sis·təm }

microphone |ENG ACOUS| An electroacoustic device containing a transducer which is actuated by sound waves and delivers essentially equivalent electric waves. { 'mī·krə,fōn }

microphone transducer |ENG ACOUS| A device which converts variation in the position or velocity of some body into corresponding variations of some electrical quantity, in a microphone. { 'mī·krə,fōn tranz'dü·sər }

microphotometer |ENG| A photometer that provides highly accurate illumination measurements; in one form, the changes in illumination are picked up by a phototube and converted into current variations that are amplified by vacuum tubes. { ¦mī·krō·fə'täm·əd·ər }

micropipet |ENG| **1.** A pipet with capacity of 0.5 milliliter or less, to measure small volumes of liquids with a high degree of accuracy; types include lambda, straight-bore, and Lang-Levy. **2.** A pointed pipette used for microinjection. { ,mī·krō·pī'pet }

microporous barrier |CHEM ENG| A metallic or plastic membrane with micrometer-sized pores used for dialysis and other membrane-separation processes. { ¦mī·krō·pór·əs 'bar·ē·ər }

microprocessor |ELECTR| A single silicon chip on which the arithmetic and logic functions of a computer are placed. { ¦mī·krō'prä,ses·ər }

micropycnometer |ENG| A small-volume pycnometer with a capacity from 0.25 to 1.6 milliliters; weighing precision is 1 part in 10,000, or better. { ,mī·krō·pik'näm·əd·ər }

microreactor |CHEM ENG| A microsystem for chemical and biochemical reactions, including separation, fluid handling, and unit operations of chemical engineering, as well as analytical systems. Its small reaction volumes and high heat and mass transfer rates allow for precise adjustment of process conditions, short response times, and defined residence times, resulting in greater process control and higher yields and selectivity. { ¦mī·krō·rē'ak·tər }

micro-reciprocal-degree See mired. { ¦mī·krō ri'sip·rə·kəl di'grē }

microrheology [MECH] A branch of rheology in which the heterogeneous nature of dispersed systems is taken into account. { ¦mī·krō·rē'äl·ə·jē }

microsecond [MECH] A unit of time equal to one-millionth of a second. Abbreviated μs. { ¦mī·krə,sek·ənd }

microsensor [ENG] A submicrometer- to millimeter-size device that converts a nonelectrical physical or chemical quantity, such as pressure, acceleration, temperature, or gas concentration, into an electrical signal; it is generally able to offer better sensitivity, accuracy, dynamic range, and reliability, as well as lower power consumption, compared to larger counterparts. { 'mī·krō,sen·sər }

microsystem See micro-electro-mechanical system. { 'mī·krō,sis·təm }

microtome [ENG] An instrument for cutting thin sections of tissues or other materials for microscopical examination. { 'mī·krə,tōm }

microwatt [MECH] A unit of power equal to one-millionth of a watt. Abbreviated μW. { 'mī·krə,wät }

microwave early warning [ENG] High-power, long-range radar with a number of indicators, giving high resolution, and with a large traffic-handling capacity; used for early warning of missiles. { 'mī·krə,wāv ¦ər·lē 'wȯr·niŋ }

microwave impedance measurement [ENG] The determination of parameters, associated with microwave propagation in transmission lines or waveguides, which are generalizations of the impedance concept at lower frequencies and are derived from ratios of electric- or magnetic-field amplitudes. { 'mī·krə,wāv im'pēd·əns ,mezh·ər·mənt }

microwave integrated circuit [ELECTR] A microwave circuit that uses integrated-circuit production techniques involving such features as thin or thick films, substrates, dielectrics, conductors, resistors, and microstrip lines, to build passive assemblies on a dielectric. Abbreviated MIC. { 'mī·krə,wāv 'int·ə,grād·əd 'sər·kət }

microwave noise standard [ENG] An electrical noise generator of calculable intensity that is used to calibrate other noise sources by using comparison methods. { 'mī·krə,wāv 'nȯiz ,stan·dərd }

microwave oven [ENG] An oven that uses microwave heating for fast cooking of meat and other foods. { 'mī·krə,wāv 'əv·ən }

microwave solid-state device [ELECTR] A semiconductor device for the generation or amplification of electromagnetic energy at microwave frequencies. { 'mī·krə,wāv ¦säl·əd ¦stāt di'vīs }

middle-third rule [CIV ENG] The rule that no tension is developed in a wall or foundation if the resultant force lies within the middle third of the structure. { 'mid·əl ¦thərd ,rül }

midrange [ENG ACOUS] A loudspeaker designed to reproduce medium audio frequencies, generally used in conjunction with a crossover network, a tweeter, and a woofer. Also known as squawker. { 'mid,rānj }

Mie-Grüneisen equation [THERMO] An equation of state particularly useful at high pressure, which states that the volume of a system times the difference between the pressure and the pressure at absolute zero equals the product of a number which depends only on the volume times the difference between the internal energy and the internal energy at absolute zero. { 'mē 'grü,nīz·ən i,kwā·zhən }

migration See bleeding. { mī'grā·shən }

mil [MECH] **1.** A unit of length, equal to 0.001 inch, or to 2.54×10^{-5} meter. Also known as milli-inch; thou. **2.** See milliliter. { mil }

mile [MECH] A unit of length in common use in the United States, equal to 5280 feet, or 1609.344 meters. Abbreviated mi. Also known as land mile; statute mile. { mīl }

milepost [CIV ENG] **1.** A post placed a mile away from a similar post. **2.** A post indicating mileage from a given point. { 'mīl,pōst }

milestone activity See key activity. { 'mīl,stōn ,ak'tiv·əd·ē }

military engineering [ENG] Science, art, and practice involved in design and construction of defensive and offensive military works as well as construction and maintenance of transportation systems. { 'mil·i,ter·ē ,en·jə'nir·iŋ }

military geology [ENG] The application of the earth sciences to such military concerns as terrain analysis, water supply, foundations, and construction of roads and airfields. { 'mil·i,ter·ē jē'äl·ə·jē }

military technology [ENG] The technology needed to develop and support the armament used by the military. { 'mil·i,ter·ē tek'näl·ə·jē }

mill [IND ENG] **1.** A machine that manufactures paper, textiles, or other products by the continuous repetition of some simple process or action. **2.** A building that houses machinery for manufacturing processes. { mil }

mill building [CIV ENG] A steel-frame building in which roof trusses span columns in the outside wall; originally, this type of building housed milling machinery, as for wood or metal, hence the name. { 'mil ,bild·iŋ }

miller See milling machine. { 'mil·ər }

millibar [MECH] A unit of pressure equal to one-thousandth of a bar. Abbreviated mb. Also known as vac. { 'mil·ə,bär }

milliler See tonne. { mil'yā }

milligal [MECH] A unit of acceleration commonly used in geodetic measurements, equal to 10^{-3} galileo, or 10^{-5} meter per second per second. Abbreviated mGal. { 'mil·ə,gal }

milligram [MECH] A unit of mass equal to one-thousandth of a gram. Abbreviated mg. { 'mil·ə,gram }

millihg See millimeter of mercury.

milli-inch See mil.

milliliter [MECH] A unit of volume equal to 10^{-3} liter or 10^{-6} cubic meter. Abbreviated ml. Also known as mil. { 'mil·ə,lēd·ər }

millimeter [MECH] A unit of length equal to one-thousandth of a meter. Abbreviated mm. Also known as metric line; strich. { 'mil·ə,mēd·ər }

millimeter of mercury [MECH] A unit of pressure, equal to the pressure exerted by a column of mercury 1 millimeter high with a density of 13.5951 grams per cubic centimeter under the standard acceleration of gravity; equal to 133.322387415 pascals; it differs from the torr by less than 1 part in 7,000,000. Abbreviated mmHg. Also known as millihg. { 'mil·ə,mēd·ər əv 'mər·kyə·rē }

millimeter of water [MECH] A unit of pressure, equal to the pressure exerted by a column of water 1 millimeter high with a density of 1 gram per cubic centimeter under the standard acceleration of gravity; equal to 9.80665 pascals. Abbreviated mmH$_2$O. { 'mil·ə,mēd·ər əv 'wödər }

millimicron See nanometer. { 'mil·ə,mī·krón }

milling [MECH ENG] Mechanical treatment of materials to produce a powder, to change the size or shape of metal powder particles, or to coat one powder mixture with another. { 'mil·iŋ }

milling cutter [DES ENG] A rotary tool-steel cutting tool with peripheral teeth, used in a milling machine to remove material from the workpiece through the relative motion of workpiece and cutter. { 'mil·iŋ ,kəd·ər }

milling machine [MECH ENG] A machine for the removal of metal by feeding a workpiece through the periphery of a rotating circular cutter. Also known as miller. { 'mil·iŋ mə,shēn }

milling planer [MECH ENG] A planer that uses a rotary cutter rather than single-point tools. { 'mil·iŋ ,plān·ər }

millisecond [MECH] A unit of time equal to one-thousandth of a second. Abbreviated ms; msec. { 'mil·ə,sek·ənd }

millisecond delay cap [ENG] A delay cap with an extremely short (20–500 thousandths of a second) interval between passing of current and explosion. Also known as short-delay detonator. { 'mil·ə,sek·ənd di,lā ,kap }

milliwatt [MECH] A unit of power equal to one-thousandth of a watt. Abbreviated mW. { 'mil·ə,wät }

mill length See random length. { 'mil ,leŋkth }

millrace [CIV ENG] A canal filled with water that flows to and from a waterwheel acting as the power supply for a mill. { 'mil,rās }

millwright [ENG] **1.** A person who plans, builds, or sets up the machinery for a mill. **2.** A person who repairs milling machines. { 'mil,rīt }

min See minim. { min }

mine car [MECH ENG] An industrial car, usually of the four-wheel type, with a low body; the door is at one end, pivoted at the top with a latch at the bottom used for hauling bulk materials. { 'mīn ,kär }

mineral engineering See mining engineering. { 'min·rəl ,en·jə'nir·iŋ }

minim [MECH] A unit of volume in the apothecaries' measure; equals 1/60 fluidram (approximately 0.061612 cubic centimeter) or about 1 drop (of water). Abbreviated min. { 'min·əm }

minimal realization [CONT SYS] In linear system theory, a set of differential equations, of the smallest possible dimension, which have an input/output transfer function matrix equal to a given matrix function G(s). { 'min·ə·məl ,rē·ə·lə'zā·shən }

mini-maxi regret [CONT SYS] In decision theory, a criterion which selects that strategy which has the smallest maximum difference between its payoff and that of the best hindsight choice. { ¦min·ē ¦mak·sē ri'gret }

minimum metal condition [DES ENG] The condition corresponding to the removal of the greatest amount of material permissible in a machined part. { 'min·ə·məm 'med·əl kən,dish·ən }

minimum-phase system [CONT SYS] A linear system for which the poles and zeros of the transfer function all have negative or zero real parts. { 'min·ə·məm 'fāz ,sis·təm }

minimum reflux ratio [CHEM ENG] The smallest reflux ratio in a two-component liquid distillation system that will produce the desired overhead and bottom compositions. { 'min·ə·məm 'rē,fləks ,rā·shō }

minimum resolvable temperature difference [THERMO] The change in equivalent blackbody temperature that corresponds to a change in radiance which will produce a just barely resolvable change in the output of an infrared imaging device, taking into account the characteristics of the device, the display, and the observer. Abbreviated MRTD. { 'min·ə·məm ri'zäl·və·bəl 'tem·prə·chər ,dif·rəns }

minimum thermometer [ENG] A thermometer that automatically registers the lowest temperature attained during an interval of time. { 'min·ə·məm thər'mäm·əd·ər }

minimum turning circle [ENG] The diameter of the circle described by the outermost projection of a vehicle when the vehicle is making its shortest possible turn. { 'min·ə·məm 'tərn·iŋ ,sər·kəl }

minimum wetting rate [CHEM ENG] The smallest liquid-flow rate through a packed column that will thoroughly wet the column packing. { 'min·ə·məm 'wed·iŋ ,rāt }

mining engineering [ENG] Engineering concerned with the discovery, development, and exploitation of coal, ores, and minerals, as well as the cleaning, sizing, and dressing of the product. Also known as mineral engineering. { 'mīn·iŋ ,en·jə'nir·iŋ }

minor defect [IND ENG] A defect which reduces the effectiveness of the product, without causing serious malfunctioning. { 'mīn·ər di'fekt }

minor diameter [DES ENG] The diameter of a

cylinder bounding the root of an external thread or the crest of an internal thread. { 'mīn·ər dī'am·əd·ər }

minor loop [CONT SYS] A portion of a feedback control system that consists of a continuous network containing both forward elements and feedback elements. { 'mīn·ər ¦lüp }

minus angle See angle of depression. { 'mī·nəs 'aŋ·gəl }

minus sight See foresight. { 'mī·nəs ,sīt }

minute [MECH] A unit of time, equal to 60 seconds. { 'min·ət }

mired [THERMO] A unit used to measure the reciprocal of color temperature, equal to the reciprocal of a color temperature of 10^6 kelvins. Derived from micro-reciprocal-degree. { mīrd }

mirror-image programming [CONT SYS] Programming of a robot in which the x and y axes are reversed in all instructions, in order to create mirror images of workpieces. { 'mir·ər ¦im·ij 'prō,gram·iŋ }

mirror interferometer [ENG] An interferometer used in radio astronomy, in which the sea surface acts as a mirror to reflect radio waves up to a single antenna, where the reflected waves interfere with the waves arriving directly from the source. { 'mir·ər ,in·tər·fə'räm·əd·ər }

mirror nephoscope [ENG] A nephoscope in which the motion of a cloud is observed by its reflection in a mirror. Also known as cloud mirror; reflecting nephoscope. { 'mir·ər 'nef·ə ,skōp }

mirror scale [ENG] A scale with a mirror used to align the eye perpendicular to the scale and pointer when taking a reading; improves accuracy by eliminating parallax. { 'mir·ər ,skāl }

mirror transit circle [ENG] A development of the conventional transit circle in which light from a star is reflected into fixed horizontal telescopes pointing due north and south by a plane mirror that is mounted on a horizontal east-west axis and attached to a large circle with accurately calibrated markings to determine the mirror's position. { 'mir·ər 'tran·zit ,sər·kəl }

mismatch [ELEC] The condition in which the impedance of a source does not match or equal the impedance of the connected load or transmission line. { 'mis,mach }

missed hole See failed hole. { 'mist 'hōl }

missed round [ENG] A round in which all or part of the explosive has failed to detonate. { 'mist 'raünd }

missile attitude [MECH] The position of a missile as determined by the inclination of its axes (roll, pitch, and yaw) in relation to another object, as to the earth. { 'mis·əl ,ad·ə,tüd }

missile site radar [ENG] Phased array radar located at a missile launch area to provide a guidance link with interceptor missiles enroute to their targets. { 'mis·əl ¦sīt 'rā,där }

mist extractor [ENG] A device that removes liquid mist or droplets from a gas stream via impingement, flow-direction change, velocity change, centrifugal force, filters, or coalescing packs. { 'mist ik,strak·tər }

mistuning [MECH] The difference between the square of the natural frequency of vibration of a vibrating system, without the effect of damping, and the square of the frequency of an external, oscillating force. { mis'tün·iŋ }

miter bend [DES ENG] A pipe bend made by mitering (angle cutting) and joining pipe ends. { 'mīd·ər ,bend }

miter box [ENG] A troughlike device of metal or wood with vertical slots set at various angles in the upright sides, for guiding a handsaw in making a miter joint. { 'mīd·ər ,bäks }

miter gate [CIV ENG] Either of a pair of canal lock gates that swing out from the side walls and meet at an angle pointing toward the upper level. { 'mīd·ər ,gāt }

miter gear [DES ENG] A bevel gear whose bevels are in 1:1 ratio. { 'mīd·ər ,gir }

miter joint [DES ENG] A joint, usually perpendicular, in which the mating ends are beveled. { 'mīd·ər ,jóint }

miter saw [DES ENG] A hollow-ground saw in diameters from 6 to 16 inches (15.24 to 40.64 centimeters), used for cutting off and mitering on light stock such as moldings and cabinet work. { 'mīd·ər ,só }

miter valve [DES ENG] A valve in which a disk fits in a seat making a 45° angle with the axis of the valve. { 'mīd·ər ,valv }

mixed cycle [MECH ENG] An internal combustion engine cycle which combines the Otto cycle constant-volume combustion and the Diesel cycle constant-pressure combustion in high-speed compression-ignition engines. Also known as combination cycle; commercial Diesel cycle; limited-pressure cycle. { 'mikst 'sī·kəl }

mixed flow [CHEM ENG] Flow stream existing in two or more phases, such as gas, hydrocarbon, and water. Also known as mixed-phase flow. { 'mikst 'flō }

mixed-flow impeller [MECH ENG] An impeller for a pump or compressor which combines radial- and axial-flow principles. { 'mikst ¦flō im 'pel·ər }

mixed-phase flow See mixed flow. { 'mikst ¦fāz 'flō }

mixer-settler [CHEM ENG] Solvent-extraction system with alternating or combined arrangement of mixers and settlers; used for chemicals extraction, lubricating-oil refining, and uranium oxide recovery. Also known as mixer-settler extractor. { 'mik·sər 'set·lər }

mixer-settler extractor See mixer-settler. { 'mik·sər ¦set·lər ik'strak·tər }

mixing [CHEM ENG] The intermingling of different materials (liquid, gas, solid) to produce a homogeneous mixture. [ELECTR] Combining two or more signals, such as the outputs of several microphones. { 'mik·siŋ }

mixing chamber [ENG] The space in a welding torch in which the gases are mixed. { 'mik·siŋ ,chām·bər }

mixing valve [ENG] Multi-inlet valve used to mix two or more fluid intakes to give a mixed product of desired composition. { 'mik·siŋ ,valv }

m-kgf See meter-kilogram.

mks system See meter-kilogram-second system. { ¦em¦kā'es ,sis·təm }

ml See milliliter.

mm See millimeter.

MMD See micromechanical display.

M meter |ENG| A class of instruments which measure the liquid water content of the atmosphere. { 'em ,mēd·ər }

mmHg See millimeter of mercury.

mmH₂O See millimeter of water.

MMSCFD |CHEM ENG| Abbreviation for million standard cubic feet per day; usually refers to gas flow.

MMSCFH |CHEM ENG| Abbreviation for million standard cubic feet per hour; usually refers to gas flow.

MMSCFM |CHEM ENG| Abbreviation for million standard cubic feet per minute; usually refers to gas flow.

mobile crane |MECH ENG| **1.** A cable-controlled crane mounted on crawlers or rubber-tired carriers. **2.** A hydraulic-powered crane with a telescoping boom mounted on truck-type carriers or as self-propelled models. { 'mō·bəl 'krān }

mobile hoist |MECH ENG| A platform hoist mounted on a pair of pneumatic-tired road wheels, so it can be towed from one site to another. { 'mō·bəl 'hȯist }

mobile loader |MECH ENG| A self-propelling power machine for loading coal, mineral, or dirt. { 'mō·bəl 'lōd·ər }

mobile robot |CONT SYS| A robot mounted on a movable platform that transports it to the area where it carries out tasks. { 'mō·bəl 'rō,bät }

mobility |ENG| The ability of an analytical balance to react to small load changes; affected by friction and degree of looseness in the balance components. { mō'bil·əd·ē }

mobility threshhold |ENG| On an analytical balance, the smallest load change that will cause a noticeable change in the weight measurement. { mō'bil·əd·ē ¦thresh,hōld }

mockup |ENG| A model, often full-sized, of a piece of equipment, or installation, so devised as to expose its parts for study, training, or testing. { 'mäk,əp }

model basin |ENG| A large basin or tank of water where scale models of ships can be tested. Also known as model tank; towing tank. { 'mäd·əl 'bās·ən }

model-following problem |CONT SYS| The problem of determining a control that causes the response of a given system to be as close as possible to the response of a model system, given the same input. { 'mäd·əl ¦fäl·ə·wiŋ ,präb·ləm }

model reduction |CONT SYS| The process of discarding certain modes of motion while retaining others in the model used by an active control system, in order that the control system can compute control commands with sufficient rapidity. { 'mäd·əl ri'dək·shən }

model reference system |CONT SYS| An ideal system whose response is agreed to be optimum; computer simulation in which both the model system and the actual system are subjected to the same stimulus is carried out, and parameters of the actual system are adjusted to minimize the difference in the outputs of the model and the actual system. { 'mäd·əl 'ref·rəns ,sis·təm }

model tank See model basin. { 'mäd·əl ,taŋk }

modem |ELECTR| A combination modulator and demodulator at each end of a telephone line to convert binary digital information to audio tone signals suitable for transmission over the line, and vice versa. Also known as dataset. Derived from modulator-demodulator. { 'mō ,dem }

mode of oscillation See mode of vibration. { 'mōd əv ,äs·ə'lā·shən }

mode of vibration |MECH| A characteristic manner in which a system which does not dissipate energy and whose motions are restricted by boundary conditions can oscillate, having a characteristic pattern of motion and one of a discrete set of frequencies. Also known as mode of oscillation. { 'mōd əv vī'brā·shən }

modern control |CONT SYS| A control system that takes account of the dynamics of the processes involved and the limitations on measuring them, with the aim of approaching the condition of optimal control. { 'mäd·ərn kən'trōl }

MODFET See high-electron-mobility transistor. { 'mäd,fet }

modification |ENG| A major or minor change in the design of an item, effected in order to correct a deficiency, to facilitate production, or to improve operational effectiveness. { ,mäd·ə·fə 'kā·shən }

modification kit |ENG| A collection of items not all having the same basic name which are employed individually or conjunctively to alter the design of a component or equipment. { ,mäd·ə·fə'kā·shən ,kit }

MOD room |ENG ACOUS| A control room in a sound-recording studio in which the acoustic treatment comprises a uniform disposition of the sound-absorbent material all about the room. { 'mäd ,rüm }

modular structure |BUILD| A building that is constructed of preassembled or presized units of standard sizes; uses a 4-inch (10.16-centimeter) cubical module as a reference. |ELECTR| **1.** An assembly involving the use of integral multiples of a given length for the dimensions of electronic components and electronic equipment, as well as for spacings of holes in a chassis or printed wiring board. **2.** An assembly made from modules. { 'mäj·ə·lər 'strək·chər }

modulate |ELECTR| To vary the amplitude, frequency, or phase of a wave, or vary the velocity of the electrons in an electron beam in some characteristic manner. { 'mäj·ə,lāt }

modulation |MECH ENG| Regulation of the fuel-air mixture to a burner in response to fluctuations of load on a boiler. { ,mäj·ə'lā·shən }

modulation-doped field-effect transistor See high-electron-mobility transistor. { ˌmäj·ə'lā·shən ¦dōpt 'fēld i¦fekt tran'zis·tər }

modulation meter |ENG| Instrument for measuring the degree of modulation (modulation factor) of a modulated wave train, usually expressed in percent. { ˌmäj·ə'lā·shən ˌmēd·ər }

modulation transformer |ENG ACOUS| An audio-frequency transformer which matches impedances and transmits audio frequencies between one or more plates of an audio output stage and the grid or plate of a modulated amplifier. { ˌmäj·ə'lā·shən tranz¦fór·mər }

modulator |ELECTR| **1.** The transmitter stage that supplies the modulating signal to the modulated amplifier stage or that triggers the modulated amplifier stage to produce pulses at desired instants as in radar. **2.** A device that produces modulation by any means, such as by virtue of a nonlinear characteristic or by controlling some circuit quantity in accordance with the waveform of a modulating signal. **3.** One of the electrodes of a spacistor. { 'mäj·ə,lād·ər }

modulator-demodulator See modem. { 'mäj·ə,lād·ər dē'mäj·ə,lād·ər }

module |ELECTR| A packaged assembly of wired components, built in a standardized size and having standardized plug-in or solderable terminations. |ENG| A unit of size used as a basic component for standardizing the design and construction of buildings, building parts, and furniture. { 'mäj·ül }

modulus of compression See bulk modulus of elasticity. { 'mäj·ə·ləs əv kəm'presh·ən }

modulus of decay |MECH| The time required for the amplitude of oscillation of an underdamped harmonic oscillator to drop to 1/e of its initial value; the reciprocal of the damping factor. { ¦mäj·ə·ləs əv di'kā }

modulus of deformation |MECH| The modulus of elasticity of a material that deforms other than according to Hooke's law. { 'mäj·ə·ləs əv ˌdē,fór'mā·shən }

modulus of elasticity |MECH| The ratio of the increment of some specified form of stress to the increment of some specified form of strain, such as Young's modulus, the bulk modulus, or the shear modulus. Also known as coefficient of elasticity; elasticity modulus; elastic modulus. { 'mäj·ə·ləs əv i,las'tis·əd·ē }

modulus of elasticity in shear |MECH| A measure of a material's resistance to shearing stress, equal to the shearing stress divided by the resultant angle of deformation expressed in radians. Also known as coefficient of rigidity; modulus of rigidity; rigidity modulus; shear modulus. { 'mäj·ə·ləs əv i,las'tis·əd·ē in 'shir }

modulus of resilience |MECH| The maximum mechanical energy stored per unit volume of material when it is stressed to its elastic limit. { 'mäj·ə·ləs əv ri'zil·yəns }

modulus of rigidity See modulus of elasticity in shear. { 'mäj·ə·ləs əv ri'jid·əd·ē }

modulus of rupture in bending |MECH| The maximum stress per unit area that a specimen can withstand without breaking when it is bent, as calculated from the breaking load under the assumption that the specimen is elastic until rupture takes place. { 'mäj·ə·ləs əv 'rəp·chər in 'bend·iŋ }

modulus of rupture in torsion |MECH| The maximum stress per unit area that a specimen can withstand without breaking when its ends are twisted, as calculated from the breaking load under the assumption that the specimen is elastic until rupture takes place. { 'mäj·ə·ləs əv 'rəp·chər in 'tór·shən }

modulus of simple longitudinal extension See axial modulus. { ¦mäj·ə·ləs əv ¦sim·pəl ˌlän·jə¦tüd·ən·əl ik'sten·chən }

modulus of torsion See torsional modulus. { 'mäj·ə·ləs əv 'tór·shən }

modulus of volume elasticity See bulk modulus of elasticity. { 'mäj·ə·ləs əv 'väl·yəm i,las'tis·əd·ē }

MOEMS See micro-opto-electro-mechanical system. { 'mō,emz }

mohm |MECH| A unit of mechanical mobility, equal to the reciprocal of 1 mechanical ohm. { mōm }

Mohr cubic centimeter |CHEM ENG| A unit of volume used in saccharimetry, equal to the volume of 1 gram of water at a specified temperature, usually 17.5°C, in which case, it is equal to 1.00238 cubic centimeters. { 'mór 'kyü·bik 'sent·ə,mēd·ər }

Mohr liter |CHEM ENG| A unit of volume, equal to 1000 Mohr cubic centimeters. { 'mór 'lēd·ər }

Mohr's circle |MECH| A graphical construction making it possible to determine the stresses in a cross section if the principal stresses are known. { 'mórz 'sər·kəl }

moiré interferometry |ENG| An optical technique that measures the components of deformation of a specimen surface in the plane of the surface by superposing a reference grating and a diffraction grating that is applied to, and deforms with, the surface. { mó'rā ˌin·tər·fə'räm·ə·trē }

moist-heat sterilization |ENG| Sterilization with steam under pressure, as in an autoclave, pressure cooker, or retort; most bacteriological media are sterilized by autoclaving at 121°C, with 15 pounds (103 kilopascals) of pressure, for 20 minutes or more. { 'móist ¦hēt ˌster·ə·lə'zā·shən }

moist room |ENG| An enclosed space that is maintained at a specified temperature, usually 73°F (23°C), with the humidity maintained at 98% or above and that is used to cure and store test specimens of cementitious material. { 'móist ˌrüm }

moisture content |MECH| The quantity of water in a mass of soil, sewage, sludge, or screenings; expressed in percentage by weight of water in the mass. { 'móis·chər ˌkän·tent }

moisture gradient |ENG| The difference in moisture content between the surface and the inner portion of a section of wood. { 'móis·chər ˌgrād·ē·ənt }

moisture loss |MECH ENG| The difference in heat content between the moisture in the boiler exit gases and that of moisture at ambient air temperature. { 'mȯis·chər ,lȯs }

mold |ENG| **1.** A pattern or template used as a guide in construction. **2.** A cavity which imparts its form to a fluid or malleable substance. |ENG ACOUS| The metal part derived from the master by electroforming in reproducing disk recordings; has grooves similar to those of the recording. { mōld }

mold base |ENG| The assembly of all parts of an injection mold except the cavity, cores, and pins. { 'mōld ,bās }

molded-fabric bearing |DES ENG| A bearing composed of laminations of cotton or other fabric impregnated with a phenolic resin and molded under heat and pressure. { 'mōl·dəd ¦fab·rik 'ber·iŋ }

molded lines |ENG| Full-size lines of a ship or airplane which are laid out in a mold loft. { 'mōl·dəd 'līnz }

mold efficiency |ENG| In a multimold blow-molding system, the percentage of the total turnaround time actually required for the forming, cooling, and ejection of the formed objects. { 'mōld i,fish·ən·sē }

molding cycle |ENG| **1.** The time required for a complete sequence of molding operations. **2.** The combined operations required to produce a set of moldings. { 'mōl·diŋ ,sī·kəl }

molding pressure |ENG| Pressure needed to force softened plastic to fill a mold cavity. { 'mōl·diŋ ,presh·ər }

molding shrinkage |ENG| Difference in dimensions between the molding and the mold cavity, measured at normal room temperature. { 'mōl·diŋ ,shriŋk·ij }

molding time See curing time. { 'mōl·diŋ ,tīm }

mold loft |ENG| A large building with a smooth wooden floor where full-size lines of a ship or airplane are laid down and templates are constructed from them to lay off the steel for cutting. { 'mōld ,lȯft }

mold seam See seam. { 'mōld ,sēm }

mole |CIV ENG| A breakwater or berthing facility, extending from shore to deep water, with a core of stone or earth. |MECH ENG| A mechanical tunnel excavator. { mōl }

molecular circuit |ELECTR| A circuit in which the individual components are physically indistinguishable from each other. { mə'lek·yə·lər 'sər·kət }

molecular drag pump |ENG| A vacuum pump in which pumping is accomplished by imparting a high momentum to the gas molecules by impingement of a body rotating at very high speeds, as much as 16,000 revolutions per minute; such pumps achieve a vacuum as high as 10^{-6} torr. { mə'lek·yə·lər 'drag ,pəmp }

molecular engineering |ELECTR| The use of solid-state techniques to build, in extremely small volumes, the components necessary to provide the functional requirements of overall equipments, which when handled in more conventional ways are vastly bulkier. { mə'lek·yə·lər ,en·jə'nir·iŋ }

molecular gage |ENG| Any instrument, such as a rotating viscometer gage or a decrement gage, that uses the dependence of the viscosity of a gas on its pressure to measure pressures on the order of 1 pascal or less. Also known as viscosity gage; viscosity manometer. { mə'lek·yə·lər 'gāj }

molecular heat |THERMO| The heat capacity per mole of a substance. { mə'lek·yə·lər 'hēt }

molecular heat diffusion |THERMO| Transfer of heat through the motion of molecules. { mə'lek·yə·lər ¦hēt di,fyü·shən }

molecular pump |MECH ENG| A vacuum pump in which the molecules of the gas to be exhausted are carried away by the friction between them and a rapidly revolving disk or drum. { mə'lek·yə·lər 'pəmp }

mole drain |CIV ENG| A subsurface channel for water drainage; formed by pulling a solid object, usually a solid cylinder having a wedge-shaped point at one end, through the soil at the proper slope and depth. { 'mōl ,drān }

Mollier diagram |THERMO| Graph of enthalpy versus entropy of a vapor on which isobars, isothermals, and lines of equal dryness are plotted. { mȯl'yā ,dī·ə,gram }

Moll thermopile |ENG| A thermopile used in some types of radiation instruments; alternate junctions of series-connected manganan-constantan molybdenum, added as ferromolybdenum or calcium molybdenum; increases strength, toughness, and wear resistance. { mȯl 'thər·mə,pīl }

moment |MECH| Static moment of some quantity, except in the term "moment of inertia." { 'mō·mənt }

momental ellipsoid |MECH| An inertia ellipsoid whose size is specified to be such that the tip of the angular velocity vector of a freely rotating object, with origin at the center of the ellipsoid, always lies on the ellipsoid's surface. Also known as energy ellipsoid. { mō'ment·əl ə 'lip,sȯid }

moment diagram |MECH| A graph of the bending moment at a section of a beam versus the distance of the section along the beam. { 'mō·mənt ,dī·ə,gram }

moment of force See torque. { 'mō·mənt əv 'fȯrs }

moment of inertia |MECH| The sum of the products formed by multiplying the mass (or sometimes, the area) of each element of a figure by the square of its distance from a specified line. Also known as rotational inertia. { 'mō·mənt əv i'nər·shə }

moment of momentum See angular momentum. { 'mō·mənt əv mō'ment·əm }

moment sensor |ENG| A device that measures the force applied at a remote point in a robotic system. { 'mō·mənt ,sen·sər }

momentum |MECH| **1.** Also known as linear momentum; vector momentum. **2.** For a single nonrelativistic particle, the product of the mass and the velocity of a particle. **3.** For a single relativistic particle, $mv/(1 - v^2/c^2)^{1/2}$, where m is the rest-mass, $\mathbf{v}$ the velocity, and c the speed of light.   **4.** For a system of particles, the vector sum of the momenta (as in the first or second definition) of the particles. { mō'ment·əm }

momentum conservation *See* conservation of momentum. { mōm'ment·əm ‚kän·sər'vā·shən }

MOMS *See* micro-opto-mechanical system. { mämz *or* ‚em‚ō‚em'es }

monaural sound |ENG ACOUS| Sound produced by a system in which one or more microphones are connected to a single transducing channel which is coupled to one or two earphones worn by the listener. { män'ȯr·əl 'saȯnd }

monitor |ENG| **1.** An instrument used to measure continuously or at intervals a condition that must be kept within prescribed limits, such as radioactivity at some point in a nuclear reactor, a variable quantity in an automatic process control system, the transmissions in a communication channel or bank, or the position of an aircraft in flight.   **2.** To use meters or special techniques to measure such a condition.   **3.** A person who watches a monitor. { 'män·əd·ər }

monkey wrench |DES ENG| A wrench having one jaw fixed and the other adjustable, both of which are perpendicular to a straight handle. { 'maŋ·kē ‚rench }

monocable |MECH ENG| An aerial ropeway that uses one rope to both support and haul a load. { 'män·ō‚kā·bəl }

monochromatic emissivity |THERMO| The ratio of the energy radiated by a body in a very narrow band of wavelengths to the energy radiated by a blackbody in the same band at the same temperature. Also known as color emissivity. { ‚män·ə·krə'mad·ik ‚ē·mi'siv·əd·ē }

monochromatic temperature scale |THERMO| A temperature scale based upon the amount of power radiated from a blackbody at a single wavelength. { män·ə·krə'mad·ik 'tem·prə·chər ‚skāl }

monolithic |CIV ENG| Pertaining to concrete construction which is cast in one jointless piece. { ‚män·ə'lith·ik }

monophonic sound |ENG ACOUS| Sound produced by a system in which one or more microphones feed a single transducing channel which is coupled to one or more loudspeakers. { ‚män·ə‚fän·ik ‚saȯnd }

monopulse radar |ENG| Radar in which directional information is obtained with high precision by using a receiving antenna system having two or more partially overlapping lobes in the radiation patterns. { 'män·ə‚pəls 'rā‚där }

monorail |CIV ENG| A single rail used as a track; usually elevated, with cars straddling or hanging from it. { 'män·ə‚rāl }

monostat |ENG| Fluid-filled, upside-down manometer-type device used to control pressures within an enclosure, as for laboratory analytical distillation systems. { 'män·ə‚stat }

monostatic radar |ENG| Conventional radar, in which the transmitter and receiver are at the same location and share the same antenna; in contrast to bistatic radar. { ‚män·ə‚stad·ik 'rā‚där }

monument |ENG| A natural or artificial (but permanent) structure that marks the location on the ground of a corner or other survey point. { 'män·yə·mənt }

Moody formula |MECH ENG| A formula giving the efficiency e' of a field turbine, whose runner has diameter D', in terms of the efficiency e of a model turbine, whose runner has diameter D; $e' = 1 - (1 - e) (D/D')^{1/5}$ { 'müd·ē ‚fȯr·mya·lə }

Mooney unit |CHEM ENG| An arbitrary unit used to measure the plasticity of raw, or unvulcanized rubber; the plasticity in Mooney units is equal to the torque, measured on an arbitrary scale, on a disk in a vessel that contains rubber at a temperature of 100°C and rotates at two revolutions per minute. { 'mün·ē ‚yü·nət }

moor |ENG| Securing a ship or aircraft by attaching it to a fixed object or a mooring buoy with chains or lines, or with anchors or other devices. { mȯr }

mooring buoy |ENG| A buoy secured to the bottom by permanent moorings and provided with means for mooring a vessel by use of its anchor chain or mooring lines; in its usual form a mooring buoy is equipped with a ring. { 'mȯr·iŋ ‚bȯi }

Morera's stress functions |MECH| Three functions of position, ψ_1, ψ_2, and ψ_3, in terms of which the elements of the stress tensor σ of a body may be expressed, if the body is in equilibrium and is not subjected to body forces; the elements of the stress tensor are given by $\sigma_{11} = -2\partial^2\psi_1/\partial x_2\partial x_3$, $\sigma_{23} = \partial^2\psi_2/\partial x_1\partial x_2 + \partial^2\psi_3/\partial x_1\partial x_3$, and cyclic permutations of these equations. { mō'rer·əz 'stres ‚fəŋk·shənz }

Morgan equation |THERMO| A modification of the Ramsey-Shields equation, in which the expression for the molar surface energy is set equal to a quadratic function of the temperature rather than to a linear one. { 'mȯr·gən i‚kwā·zhən }

morning glory spillway *See* shaft spillway. { 'mȯrn·iŋ ‚glȯr·ē ‚spil‚wā }

Morse taper reamer |DES ENG| A machine reamer with a taper shank. { 'mȯrs 'tā·pər rēm·ər }

mortise |ENG| A groove or slot in a timber for holding a tenon. { 'mȯrd·əs }

mortise and tenon |DES ENG| A type of joint, principally used for wood, in which a hole, slot, or groove (mortise) in one member is fitted with a projection (tenon) from the second member. { 'mȯrd·əs ən 'ten·ən }

mortise lock |DES ENG| A lock designed to be installed in a mortise rather than on a door's surface. { 'mȯrd·əs ‚läk }

mortising machine |MECH ENG| A machine employing an auger and a chisel to produce a square or rectangular mortise in wood. { 'mȯrd·ə·siŋ mə‚shēn }

MOS-controlled thyristor |ELECTR| A type of thyristor in which there is a very thin metal oxide semiconductor (MOS) integrated circuit in the top surface of the high-power thyristor components, so that only a small gate current is needed to turn the entire device off or on.   Abbreviated MCT. { ¦em¦ō¦es kən,trōld thī'ris·tər }

MOSFET *See* metal oxide semiconductor field-effect transistor. { 'mós,fet }

MOST *See* metal oxide semiconductor field-effect transistor.

MOS transistor *See* metal oxide semiconductor field-effect transistor. { ¦em¦ō'es tran'zis·tər }

mother |ENG ACOUS| A mold derived by electroforming from a master; used to produce the stampers from which disk records are molded in large quantities. Also known as metal positive. { 'məth·ər }

mother liquor *See* discharge liquor. { 'məth·ər ,lik·ər }

motion |MECH| A continuous change of position of a body. { 'mō·shən }

motion analysis |IND ENG| Detailed study of the motions used in a work task or at a given work area. { 'mō·shən ə,nal·ə·səs }

motion cycle |IND ENG| The complete sequence of motions and activities required to complete one work cycle. { 'mō·shən ,sī·kəl }

motion economy |IND ENG| Simplification and reduction of body motions to simplify and reduce work content. { 'mō·shən i,kän·ə·mē }

motion picture projector |ENG| An optical and mechanical device capable of flashing pictures taken by a motion picture camera on a viewing screen at the same frequency the action was photographed, thus producing an image that appears to move. { 'mō·shən ¦pik·chər prə,jek·tər }

motions pathway |IND ENG| The locus of movement of an anatomical segment in moving from one point of the workplace to another; includes the elemental increments in such motions as reaching, changing position, examining, and holding. { 'mō·shənz 'path,wā }

motor |ELEC| A machine that converts electric energy into mechanical energy by utilizing forces produced by magnetic fields on current-carrying conductors. Also known as electric motor. { 'mōd·ər }

motorcycle |MECH ENG| An automotive vehicle, essentially a motorized bicycle, with two tandem and sometimes three rubber wheels. { 'mōd·ər,sī·kəl }

motor element |ENG ACOUS| That portion of an electroacoustic receiver which receives energy from the electric system and converts it into mechanical energy. { 'mōd·ər ,el·ə·mənt }

motor grader *See* autopatrol. { 'mōd·ər ,grād·ər }

motor meter |ENG| An integrating meter which has a rotor, one or more stators, a retarding element which makes the speed of the rotor proportional to the quantity (such as power or current) whose integral over time is being measured, and a register which counts the total number of revolutions of the rotor. { 'mōd·ər ,mēd·ər }

motor reducer |MECH ENG| Speed-reduction power transmission equipment in which the reducing gears are integral with drive motors. { 'mōd·ər ri,dü·sər }

motortruck |MECH ENG| An automotive vehicle which is used to transport freight. { 'mōd·ər,trək }

motor vehicle |MECH ENG| Any automotive vehicle that does not run on rails, and generally having rubber tires. { 'mōd·ər 'vē·ə·kəl }

mounce |MECH| A unit of mass, equal to 25 grams. Also known as metric ounce. { maúns }

mount |ENG| **1.** Structure supporting any apparatus, as a gun, searchlight, telescope, or surveying instrument. **2.** To fasten an apparatus in position, such as a gun on its support. { maúnt }

Mount Rose snow sampler |ENG| A particular pattern of snow sampler having an internal diameter of 1.485 inches (3.7719 centimeters), so that each inch of water in the sample weighs 1 ounce (28.3495 grams). { 'maúnt 'rōz 'snō ,sam·plər }

mouse trap |ENG| A cylindrical fishing tool having the open bottom end fitted with an inward opening valve. { 'maús ,trap }

mouth |ENG ACOUS| The end of a horn that has the larger cross-sectional area. { maúth }

movable-active tooling |MECH ENG| Any equipment in a robotic system that is able to move and that operates under power. { 'mü·və·bəl ¦ak·tiv 'tül·iŋ }

movable bridge |CIV ENG| A bridge in which either the horizontal or vertical alignment can be readily changed to permit the passage of traffic beneath it. Often called drawbridge (an anachronism). { 'müv·ə·bəl 'brij }

movable-passive tooling |MECH ENG| Equipment in a robotic system that moves but requires no power to operate, such as workpieces, clamps, and templates. { 'mü·və·bəl 'pas·iv 'tül·iŋ }

movable platen |ENG| The large platen at the back of an injection-molding machine to which the back half of the mold is fastened. { 'mü·və·bəl 'plat·ən }

movable-point crossing |CIV ENG| A small-angle rail crossing with two center frogs, each of which consists essentially of a knuckle rail and two opposed movable center points. { 'mü·və·bəl ¦póint 'krös·iŋ }

moving bed |CHEM ENG| Granulated solids in a process vessel that are circulated (moved) either mechanically or by gravity flow; used in catalytic and absorption processes. { 'müv·iŋ 'bed }

moving-bed catalytic cracking |CHEM ENG| Petroleum refining process for cracking (breaking) of long hydrocarbon molecules by use of heat, pressure, and a granular cracking catalyst that is continuously cycled between the reactor vessel and the catalyst regenerator. { 'müv·iŋ ¦bed ,kad·əl,id·ik 'krak·iŋ }

moving-coil galvanometer |ENG| Any galvanometer, such as the d'Arsonval galvanometer, in which the current to be measured is sent through a coil suspended or pivoted in a fixed magnetic field, and the current is determined by measuring the resulting motion of the coil. { 'müv·iŋ ¦kóil ,gal·və'näm·əd·ər }

moving-coil loudspeaker *See* dynamic loudspeaker. { 'müv·iŋ ¦kȯil 'laůd,spēk·ər }

moving-coil microphone *See* dynamic microphone. { 'müv·iŋ ¦kȯil 'mī·krə,fōn }

moving-coil voltmeter |ENG| A voltmeter in which the current, produced when the voltage to be measured is applied across a known resistance, is sent through coils pivoted in the magnetic field of permanent magnets, and the resulting torque on the coils is balanced by control springs so that the deflection of a pointer attached to the coils is proportional to the current. { 'müv·iŋ ¦kȯil 'vōlt,med·ər }

moving-coil wattmeter *See* electrodynamic wattmeter. { 'müv·iŋ ¦kȯil 'wät,mēd·ər }

moving-conductor loudspeaker |ENG ACOUS| A loudspeaker in which the mechanical forces result from reactions between a steady magnetic field and the magnetic field produced by current flow through a moving conductor. { 'müv·iŋ kən¦dək·tər 'laůd,spēk·ər }

moving constraint |MECH| A constraint that changes with time, as in the case of a system on a moving platform. { 'müv·iŋ kən'strānt }

moving-iron meter |ENG| A meter that depends on current in one or more fixed coils acting on one or more pieces of soft iron, at least one of which is movable. { 'müv·iŋ ¦ī·ərn 'mēd·ər }

moving-iron voltmeter |ENG| A voltmeter in which a field coil is connected to the voltage to be measured through a series resistor; current in the coil causes two vanes, one fixed and one attached to the shaft carrying the pointer, to be similarly magnetized; the resulting torque on the shaft is balanced by control springs. { 'müv·iŋ ¦ī·ərn 'vōlt,mēd·ər }

moving load |MECH| A load that can move, such as vehicles or pedestrians. { 'müv·iŋ 'lōd }

moving-magnet voltmeter |ENG| A voltmeter in which a permanent magnet aligns itself with the resultant magnetic field produced by the current in a field coil and another permanent control magnet. { 'müv·iŋ 'mag·nət 'vōlt,mēd·ər }

moving sidewalk |CIV ENG| A sidewalk constructed on the principle of an endless belt, on which pedestrians are moved. { 'müv·iŋ 'sīd,wȯk }

mp *See* mean effective pressure; melting point.

MRI *See* magnetic resonance imaging.

MRP *See* material requirements planning.

MRTD *See* minimum resolvable temperature difference.

ms *See* millisecond.

Ms *See* megasecond.

MSCFD |CHEM ENG| Abbreviation for thousand standard cubic feet per day; usually refers to gas flow.

MSCFH |CHEM ENG| Abbreviation for thousand standard cubic feet per hour; usually refers to gas flow.

MSCFM |CHEM ENG| Abbreviation for thousand standard cubic feet per minute; usually refers to gas flow.

msec *See* millisecond.

Msec *See* megasecond.

MSI *See* magnetic source imaging.

M synchronization |ENG| A linking arrangement between a camera lens and the flashbulb unit to allow a 15-millisecond delay of the shutter so that the bulb burns to its brightest point before the shutter opens. { 'em ,siŋ·krə·nə 'zā·shən }

MTTF *See* mean time to failure.

muck |CIV ENG| Rock or earth removed during excavation. { mək }

mucking |ENG| Clearing and loading broken rock and other excavated materials, as in tunnels or mines. { 'mək·iŋ }

mud *See* slime. { məd }

mud auger |DES ENG| A diamond-point bit with the wings of the point twisted in a shallow augerlike spiral. Also known as clay bit; diamond-point bit; mud bit. { 'məd ,ȯg·ər }

mud berth |CIV ENG| A berth where a vessel rests on the bottom at low water. { 'məd ,bərth }

mud bit *See* mud auger. { 'məd ,bit }

mud blasting |ENG| The detonation of sticks of explosive stuck on the side of a boulder with a mud covering, so that little of the explosive energy is used in breaking the boulder. { 'məd ,blast·iŋ }

mud cake |ENG| A caked layer of clay adhering to the walls of a well or borehole, formed where the water in the drilling mud filtered into a porous formation during rotary drilling. Also known as filter cake. { 'məd ,kāk }

mudcap |ENG| A quantity of wet mud, wet earth, or sand used to cover a charge of dynamite or other high explosive fired in contact with the surface of a rock in mud blasting. { 'məd ,kap }

mud pit *See* slushpit. { 'məd ,pit }

mudsill |CIV ENG| The lowest sill of a structure, usually embedded in the earth. { 'məd,sil }

mud still |ENG| An instrument used to separate oil, water, and other volatile materials in a mud sample by distillation, permitting determination of the quantities of oil, water, and total solid contents in the original sample. { 'məd ,stil }

mud sump |CHEM ENG| Upstream area in a process vessel where, because of a velocity drop, entrained solids drop out and are collected in a sump. { 'məd ,səmp }

mu factor |ELECTR| Ratio of the change in one electrode voltage to the change in another electrode voltage under the conditions that a specified current remains unchanged and that all other electrode voltages are maintained constant; a measure of the relative effect of the voltages on two electrodes upon the current in the circuit of any specified electrode. { 'myü ,fak·tər }

muffle furnace |ENG| A furnace with an externally heated chamber, the walls of which radiantly heat the contents of the chamber. { 'məf·əl ,fər·nəs }

muffler |ENG| A device to deaden the noise produced by escaping gases or vapors. { 'məf·lər }

mull |ENG| To mix thoroughly or grind. { məl }

muller |ENG| A foundry sand-mixing machine. { 'məl·ər }

mulling |ENG| The combining of clay, water, and sand, prior to molding, by compressing with a roller to ensure development of optimum sand properties by the adequate distribution of ingredients. { 'məl·iŋ }

mullion |BUILD| A vertical bar separating two windows in a multiple window { 'məl·yən }

multicellular horn |ENG ACOUS| A combination of individual horn loudspeakers having individual driver units or joined in groups to a common driver unit. Also known as cellular horn. { ¦məl·tē'sel·yə·lər 'hȯrn }

multichannel field-effect transistor |ELECTR| A field-effect transistor in which appropriate voltages are applied to the gate to control the space within the current flow channels. { ¦məl·tē'chan·əl 'fēld i¦fekt tran'zis·tər }

multichip microcircuit |ELECTR| Microcircuit in which discrete, miniature, active electronic elements (transistor or diode chips) and thin-film or diffused passive components or component clusters are interconnected by thermocompression bonds, alloying, soldering, welding, chemical deposition, or metallization. { 'məl·tē,chip 'mī·krō,sər·kət }

multicomponent distillation |CHEM ENG| The distillation separation of a single liquid feed stream containing three or more components into a single overhead product and a single bottoms product. { ¦məl·tē·kəm¦pō·nənt ,dist·əl'ā·shən }

multideck clarifiers |ENG| Extraction units which remove pollutants from recycled plant waste water. { 'məl·tə,dek 'klar·ə,fī·ərz }

multifuel burner |ENG| A burner which utilizes more than one fuel simultaneously for combustion. { 'məl·tē,fyül ,bər·nər }

multifunction array radar |ENG| Electronic scanning radar which will perform target detection and identification, tracking, discrimination, and some interceptor missile tracking on a large number of targets simultaneously and as a single unit. { ¦məl·tə'fəŋk·shən ə'rā 'rā,där }

multifuse igniter |ENG| A black powder cartridge that allows several fuses to be fired at the same time by lighting a single fuse. { 'məl·tə,fyüz ig'nīd·ər }

multilayer bit |DES ENG| A bit set with diamonds arranged in successive layers beneath the surface of the crown. { ¦məl·tē'lā·ər ,bit }

multilayer board |ELECTR| A printed wiring board that contains circuitry on internal layers throughout the cross section of the board as well as on the external layers. { ,məl·tē,lā·ər 'bȯrd }

multilevel control theory |CONT SYS| An approach to the control of large-scale systems based on decomposition of the complex overall control problem into simpler and more easily managed subproblems, and coordination of the subproblems so that overall system objectives and constraints are satisfied. { ¦məl·tə'lev·əl kən'trōl 'thē·ə·rē }

multimeter See volt-ohm-milliammeter. { 'məl·tə,mēd·ər or məl'tim·əd·ər }

multiphase flow |CHEM ENG| Mixture of two or more distinct phases (such as oil, water, and gas) flowing through a closed conduit. { 'məl·tə,fāz ,flō }

multiple-activity process chart |IND ENG| A chart showing the coordinated synchronous or simultaneous activities of a work system comprising one or more machines or individuals; separate, parallel columns indicate each machine's or person's activities as related to the other parts of the work system. { 'məl·tə·pəl ak¦tiv·əd·ē 'prä·səs ,chärt }

multiple-arch dam |CIV ENG| A dam composed of a series of arches inclined at about 45° and carried on parallel buttresses or piers. { 'məl·tə·pəl ¦ärch 'dam }

multiple cartridges |CHEM ENG| Filter medium made up of two or more filter cartridges, either fastened end to end or arranged side by side (in series or parallel flow respectively). { 'məl·tə·pəl 'kär·trə·jəz }

multiple connector |ENG| A flow chart symbol that indicates the merging of several flow lines into one line or the dispersal of a flow line into several lines. { 'məl·tə·pəl kə'nek·tər }

multiple-effect evaporation |CHEM ENG| Series-operation energy economizer system in which heat from the steam generated (evaporated liquid) in the first stage is used to evaporate additional liquid in the second stage (by reducing system pressure), and so on, up to 10 or more effects; commonly used in the pulp and paper industry. { 'məl·tə·pəl i¦fekt i,vap·ə'rā·shən }

multiple-effect evaporator |CHEM ENG| An evaporation system in which a series of evaporator bodies are connected so that the vapors from one body act as a heat source for the next body. { ¦məl·tə·pəl i¦fkt i'vap·ə,rād·ər }

multiple-factor incentive plan |IND ENG| A wage incentive plan based on productivity and other factors such as yield, material usage, and reduction of scrap. { 'məl·tə·pəl ¦fak·tər in'sen·tiv ,plan }

multiple firing |ENG| Electrically firing with delay blasting caps in a number of holes at one time. { 'məl·tə·pəl 'fīr·iŋ }

multiple-function chip See large-scale integrated circuit. { 'məl·tə·pəl ¦fəŋk·shən ,chip }

multiple-loop system |CONT SYS| A system whose block diagram has at least two closed paths, along each of which all arrows point in the same direction. { 'məl·tə·pəl ¦lüp ,sis·təm }

multiple midstop |MECH ENG| A peripheral device that allows a pick-and-place robot to swing and stop in several positions. { 'məl·tə·pəl 'mid,stäp }

multiple piece rate plan |IND ENG| A wage incentive plan wherein increasingly higher unit pay rates are given to the worker as his productivity increases. { 'məl·tə·pəl 'pēs ,rāt ,plan }

multiple-purpose tester See volt-ohm-milliammeter. { 'məl·tə·pəl ¦pər·pəs 'tes·tər }

multiple-row blasting |ENG| The drilling, charging, and firing of rows of vertical boreholes. { 'məl·tə·pəl ¦rō 'blast·iŋ }

multiple sampling |IND ENG| A plan for quality control in which a given number of samples from a group are inspected, and the group is either accepted, resampled, or rejected, depending on the number of failures found in the samples. { 'məl·tə·pəl 'sam·pliŋ }

multiple series |ENG| A method of wiring a large group of blasting charges by connecting small groups in series and connecting these series in parallel. Also known as parallel series. { 'məl·tə·pəl 'sir·ēz }

multiple shooting |ENG| The firing of an entire face at one time by means of connecting shot holes in a single series and shooting all holes at the same instant. { 'məl·tə·pəl 'shüd·iŋ }

multiple-slide press |MECH ENG| A press with individual adjustable slides built into the main slide or connected independently to the main shaft. { 'məl·tə·pəl ¦slīd 'pres }

multiple-strand conveyor |MECH ENG| A conveyor with two or more spaced strands of chain, belts, or cords as the supporting or propelling medium. { 'məl·tə·pəl ¦strand kən'vā·ər }

multiplex |ENG| Stereoscopic device to project aerial photographs onto surfaces so that the images may be viewed in three dimensions by using anaglyphic spectacles; used to prepare topographic maps. { 'məl·tə,pleks }

multiplexer |ELECTR| A device for combining two or more signals, as for multiplex, or for creating the composite color video signal from its components in color television. Also spelled multiplexor. { 'məl·tə,plek·sər }

multiplexor See multiplexer. { 'məl·tə,plek·sər }

multiple x-y recorder |ENG| Recorder that plots a number of independent charts simultaneously, each showing the relation of two variables, neither of which is time. { 'məl·tə·pəl ¦eks'wī ri,kórd·ər }

multiplication |ELECTR| An increase in current flow through a semiconductor because of increased carrier activity. { ,məl·tə·pli'kā·shən }

multiplier |ELEC| A resistor used in series with a voltmeter to increase the voltage range. Also known as multiplier resistor. |ELECTR| **1.** A device that has two or more inputs and an output that is a representation of the product of the quantities represented by the input signals;voltages are the quantities commonly multiplied. **2.** See electron multiplier; frequency multiplier. { 'məl·tə,plī·ər }

multiport burner |ENG| A burner having several nozzles which discharge fuel and air. { 'məl·tə,pórt 'bər·nər }

multiport network analyzer |ENG| A linear, passive microwave network having five or more ports which is used for measuring power and the complex reflection coefficient in a microwave circuit.

Also known as multiport reflectometer. { 'məl·tə,pórt ¦net,wərk 'an·ə,līz·ər }

multiport reflectometer See multiport network analyzer. { 'məl·tə,pórt ,rē,flek'täm·əd·ər }

multirole programmable device |CONT SYS| A device that contains a programmable memory to store data on positioning robots and sequencing their motion. { 'məl·tə,rōl prō¦gram·ə·bəl di'vīs }

multirope friction winder |MECH ENG| A winding system in which the drive to the winding ropes is the frictional resistance between the ropes and the driving sheaves. { 'məl·tə,rōp 'frik·shən ,wīn·dər }

multistage |ENG| Functioning or occurring in separate steps. { 'məl·tē,stāj }

multistage compressor |MECH ENG| A machine for compressing a gaseous fluid in a sequence of stages, with or without intercooling between stages. { 'məl·tē,stāj kəm'pres·ər }

multistage pump |MECH ENG| A pump in which the head is developed by multiple impellers operating in series. { 'məl·tē,stāj 'pəmp }

multistage queuing |IND ENG| A situation involving two or more sequential stages in a process, each of which involves waiting in line. { 'məl·tē,stāj 'kyü·iŋ }

multistatic radar |ENG| Radar in which successive antenna lobes are sequentially engaged to provide a tracking capability without physical movement of the antenna. { 'məl·tē,stad·ik 'rā,där }

multitrack recording system |ENG| Recording system which provides two or more recording paths on a medium, which may carry either related or unrelated recordings in common time relationship. { ¦məl·tē¦trak ri'kórd·iŋ ,sis·təm }

multivariable system |CONT SYS| A dynamical system in which the number of either inputs or outputs is greater than 1. { ¦məl·tē¦ver·ē·ə·bəl ,sis·təm }

municipal engineering |CIV ENG| Branch of engineering dealing with the form and functions of urban areas. { myü'nis·ə·pəl ,en·jə'nir·iŋ }

muntin See sash bar. { 'mənt·ən }

Murphree efficiency |CHEM ENG| In a plate-distillation column, the ratio of the actual change in vapor composition when the vapor passes through the liquid on a tray (plate) to the composition change of the vapor if it were in vapor-liquid equilibrium with the tray liquid. { 'mər·frē i'fish·ən·sē }

Muskhelishvili's method |MECH| A method of solving problems concerning the elastic deformation of a planar body that involves using methods from the theory of functions of a complex variable to calculate analytic functions which determine the plane strain of the body. { mə'skel·ish,vil·ēz ,meth·əd }

mW See milliwatt.

MW See megawatt.

myotome |ENG| An instrument used to divide a muscle. { 'mī·ə,tōm }

N

N *See* newton.

nail |DES ENG| A slender, usually pointed fastener with a head, designed for insertion by impact. |ENG| To drive nails in a manner that will position and hold two or more members, usually of wood, in a desired relationship. { 'nāl }

nail coat *See* devil float. { 'nāl ,kōt }

nailer |ENG| A wood strip or block which serves as a backing into which nails can be driven. { 'nāl·ər }

nailhead |DES ENG| Flat protuberance at the end of a nail opposite the point. { 'nāl,hed }

nail set |DES ENG| A small cylindrical steel tool, usually tapered at one end, that is used to drive a nail or a brad below or flush with a wood surface. Also known as punch. { 'nāl ,set }

NAND circuit |ELECTR| A logic circuit whose output signal is a logical 1 if any of its inputs is a logical 0, and whose output signal is a logical 0 if all of its inputs are logical 1. { 'nand ,sər·kət }

nanoelectronics |ELECTR| The technology of electronic devices whose dimensions range from atoms up to 100 nanometers. { ,nan·ō·i,lek 'trän·iks }

nanogram |MECH| One-billionth (10^{-9}) of a gram. Abbreviated ng. { 'nan·ə,gram }

nanometer |MECH| A unit of length equal to one-billionth of a meter, or 10^{-9} meter. Also known as millimicron (μm); nanon. { 'nan·ə,mēd·ər }

nanon *See* nanometer. { 'na,nän }

nanosecond |MECH| A unit of time equal to one-billionth of a second, or 10^{-9} second. { 'nan·ə,sek·ənd }

nanotechnology |ENG| **1.** Systems for transforming matter, energy, and information that are based on nanometer-scale components with precisely defined molecular features. **2.** Techniques that produce or measure features less than 100 nanometers in size. { ¦nan·ō·tek'näl·ə·jē }

Nansen bottle |ENG| A bottlelike water-sampling device with valves at both ends that is lowered into the water by wire; at the desired depth it is activated by a messenger which strikes the reversing mechanism and inverts the bottle, closing the valves and trapping the water sample inside. Also known as Petterson-Nansen water

bottle; reversing water bottle. { 'nan·sən ,bäd·əl }

narrow-band pyrometer |ENG| A pyrometer in which light from a source passes through a color filter, which passes only a limited band of wavelengths, before falling on a photoelectric detector. Also known as spectral pyrometer. { 'nar·ō ¦band pī'räm·əd·ər }

narrow gage |CIV ENG| A railway gage narrower than the standard gage of 4 feet $8\frac{1}{2}$ inches (143.51 centimeters). { 'nar·ō ¦gāj }

natural convection |THERMO| Convection in which fluid motion results entirely from the presence of a hot body in the fluid, causing temperature and hence density gradients to develop, so that the fluid moves under the influence of gravity. Also known as free convection. { 'nach·rəl kən'vek·shən }

natural-draft cooling tower |MECH ENG| A cooling tower that depends upon natural convection of air flowing upward and in contact with the water to be cooled. { 'nach·rəl ¦draft 'kül·iŋ ,taů·ər }

natural-gasoline plant |CHEM ENG| Compression, distillation, and absorption process facility used to remove natural gasoline (mostly butanes and heavier components) from natural gas. { 'nach·rəl ,gas·ə'lēn ,plant }

nautical chain |MECH| A unit of length equal to 15 feet or 4.572 meters. { 'nód·ə·kəl 'chān }

naval architecture |ENG| The study of the physical characteristics and the design and construction of buoyant structures, such as ships, boats, barges, submarines, and floats, which operate in water; includes the construction and operation of the power plant and other mechanical equipment of these structures. { 'nā·vəl 'är·kə,tek·chər }

Navier's equation |MECH| A vector partial differential equation for the displacement vector of an elastic solid in equilibrium and subjected to a body force. { nä'vyāz i,kwā·zhən }

navigation |ENG| The process of directing the movement of a craft so that it will reach its intended destination; subprocesses are position fixing, dead reckoning, pilotage, and homing. { ,nav·ə'gā·shən }

navigation dam |CIV ENG| A structure designed to raise the level of a stream to increase the depth for navigation purposes. { ,nav·ə'gā·shən ,dam }

n-body problem See many-body problem. { 'en ¦bad·ē ,präb·ləm }

n-channel |ELECTR| A conduction channel formed by electrons in an n-type semiconductor, as in an n-type field-effect transistor. { 'en ,chan·əl }

n-channel metal-oxide semiconductor See NMOS. { ¦en ,chan·əl ,med·əl ¦äk,sīd 'sem·i·kən,dək·tər }

neat line |CIV ENG| The line defining the limits of an aspect of construction, such as an excavation or a wall. Also known as net line. { 'nēt ,līn }

neck |ENG| The part of a furnace where the flame is contracted before reaching the stack. { nek }

neck-in |ENG| When coating by extrusion, the width difference between the extruded web leaving the die and that of the coating on the surface. { 'nek,in }

needle |DES ENG| **1.** A device made of steel pointed at one end with a hole at the other; used for sewing. **2.** A device made of steel with a hook at one end; used for knitting. |ENG| **1.** A piece of copper or brass about 1/2 inch (13 millimeters) in diameter and 3 or 4 feet (90 or 120 centimeters) long, pointed at one end, thrust into a charge of blasting powder in a borehole and then withdrawn, leaving a hole for the priming, fuse, or squib. Also known as pricker. **2.** A thin pointed indicator on an instrument dial. |ENG ACOUS| See stylus. { 'nēd·əl }

needle beam |CIV ENG| A temporary member thrust under a building or a foundation for use in underpinning. { 'nēd·əl ,bēm }

needle bearing |DES ENG| A roller-type bearing with long rollers of small diameter; the rollers are retained in a flanged cup, have no retainer, and bear directly on the shaft. { 'nēd·əl ,ber·iŋ }

needle blow |ENG| A blow-molding technique in which air is injected into the plastic article through a hollow needle inserted in the parison. { 'nēd·əl ,blō }

needle dam |CIV ENG| A barrier made of horizontal bars across a pass through a dam or of planks that can be removed in case of flooding. { 'nēd·əl ,dam }

needle file |DES ENG| A small file with an extended tang that serves as a needle. { 'nēd·əl ,fīl }

needle nozzle |MECH ENG| A streamlined hydraulic turbine nozzle with a movable element for converting the pressure and kinetic energy in the pipe leading from the reservoir to the turbine into a smooth jet of variable diameter and discharge but practically constant velocity. { 'nēd·əl ,näz·əl }

needle tubing |ENG| Stainless steel tubing with outside diameters from 0.014 to 0.203 inch (0.36 to 5.16 millimeters); used for surgical instruments and radon implanters. { 'nēd·əl ,tüb·iŋ }

needle valve |MECH ENG| A slender, pointed rod fitting in a hole or circular or conoidal seat; used in hydraulic turbines and hydroelectric systems. { 'nēd·əl ,valv }

needle weir |CIV ENG| A type of frame weir in which the wooden barrier is constructed of vertical square-section timbers placed side by side against the iron frames. { 'nēd·əl wer }

needling |CIV ENG| Underpinning the upper part of a building with horizontally placed timber or steel beams. { 'nēd·əl·iŋ }

negative acceleration |MECH| Acceleration in a direction opposite to the velocity, or in the direction of the negative axis of a coordinate system. { 'neg·əd·iv ik,sel·ə'rā·shən }

negative charge |ELEC| The type of charge which is possessed by electrons in ordinary matter, and which may be produced in a resin object by rubbing with wool. Also known as negative electricity. { 'neg·əd·iv 'chärj }

negative easement |CIV ENG| An easement that can be exercised to prevent the owner of a piece of land from using it in certain ways that he or she would otherwise be entitled to. { 'neg·əd·iv 'ēz·mənt }

negative electrode See cathode; negative plate. { 'neg·əd·iv i'lek,trōd }

negative feedback |CONT SYS| Feedback in which a portion of the output of a circuit, device, or machine is fed back 180° out of phase with the input signal, resulting in a decrease of amplification so as to stabilize the amplification with respect to time or frequency, and a reduction in distortion and noise. Also known as inverse feedback; reverse feedback; stabilized feedback. { 'neg·əd·iv 'fēd,bak }

negative g |MECH| In designating the direction of acceleration on a body, the opposite of positive g; for example, the effect of flying an outside loop in the upright seated position. { 'neg·əd·iv 'jē }

negative potential |ELEC| An electrostatic potential which is lower than that of the ground, or of some conductor or point in space that is arbitrarily assigned to have zero potential. { 'neg·əd·iv pə'ten·chəl }

negative rake |MECH ENG| The orientation of a cutting tool whose cutting edge lags the surface of the tooth face. { 'neg·əd·iv 'rāk }

negative temperature |THERMO| The property of a thermally isolated thermodynamic system whose elements are in thermodynamic equilibrium among themselves, whose allowed states have an upper limit on their possible energies, and whose high-energy states are more occupied than the low-energy ones. { 'neg·əd·iv 'tem·prə·chər }

negative terminal |ELEC| The terminal of a battery or other voltage source that has more electrons than normal; electrons flow from the negative terminal through the external circuit to the positive terminal. { 'neg·əd·iv 'tər·mən·əl }

negative work |IND ENG| Work that is performed with the assistance of gravity but the muscular effort required involves only control of the load. { 'neg·əd·iv 'wərk }

negotiated contract |IND ENG| A purchase or

sales agreement made by a United States government agency without normally employing techniques required by formal advertising. { nə'gō·shē,ăd·əd 'kăn,trakt }

Nelson diaphragm cell |CHEM ENG| Obsolete carbon-electrode type of electrolytic diaphragm cell once widely used to produce chlorine and caustic soda from brine. { 'nel·sən 'dī·ə,fram ,sel }

neohexane alkylation |CHEM ENG| A noncatalytic petroleum-refinery alkylation process that forms neohexane from a feed of ethylene and isobutane. { ¦nē·ō'hek,sān ,al·kə'lā·shən }

nepheloscope |ENG| An instrument for the production of clouds in the laboratory by condensation or expansion of moist air. { 'nef·ə·lə,skōp }

nephometer |ENG| A general term for instruments designed to measure the amount of cloudiness; an early type consists of a convex hemispherical mirror mapped into six parts; the amount of cloud coverage on the mirror is noted by the observer. { ne'fäm·əd·ər }

nephoscope |ENG| An instrument for determining the direction of cloud motion. { 'nef·ə,skōp }

Nernst approximation formula |THERMO| An equation for the equilibrium constant of a gas reaction based on the Nernst heat theorem and certain simplifying assumptions. { 'nernst ə,präk·sə'mā·shən ,fȯr·myə·lə }

Nernst heat theorem |THERMO| The theorem expressing that the rate of change of free energy of a homogeneous system with temperature, and also the rate of change of enthalpy with temperature, approaches zero as the temperature approaches absolute zero. { 'nernst 'hēt ,thir·əm }

Nernst-Lindemann calorimeter |ENG| A calorimeter for measuring specific heats at low temperatures, in which the heat reservoir consists of a metal of high thermal conductivity such as copper, to promote rapid temperature equalization; none of the material under study is more than a few millimeters from a metal surface, and the whole apparatus is placed in an evacuated vessel and heated by current through a platinum heating coil. { 'nernst 'lin·də·mən ,kal·ə'rim·əd·ər }

Nernst-Simon statement of the third law of thermodynamics |THERMO| The statement that the change in entropy which occurs when a homogeneous system undergoes an isothermal reversible process approaches zero as the temperature approaches absolute zero. { 'nernst 'sī·mən 'stāt·mənt əv tha 'thərd 'lȯ əv ,thər·mō·dī'nam·iks }

nesting |IND ENG| A production technique in which parts with similar patterns are manufactured together. { 'nest·iŋ }

net |ENG| **1.** Threads or cords tied together at regular intervals to form a mesh. **2.** A series of surveying or leveling stations that have been interconnected in such a manner that closed loops or circuits have been formed, or that are

arranged so as to provide a check on the consistency of the measured values. Also known as network. { net }

NETD See noise equivalent temperature difference.

net floor area |BUILD| Gross floor area of a building, excluding the area occupied by walls and partitions, the circulation area (where people walk), and the mechanical area (where there is mechanical equipment). { 'net 'flȯr ,er·ē·ə }

net flow area |DES ENG| The calculated net area which determines the flow after the complete bursting of a rupture disk. { 'net 'flō ,er·ē·ə }

net heating value See low heat value. { 'net 'hēd·iŋ ,val·yü }

net line See neat line. { 'net ,līn }

net load capacity |ENG| The weight of a material that can be handled, without failure, by a machine or process plus the weight of the container or device. { ¦net ,lȯd ,kə'pas·əd·ē }

net positive suction head |MECH ENG| The minimum suction head required for a pump to operate; depends on liquid characteristics, total liquid head, pump speed and capacity, and impeller design. Abbreviated NPSH. { 'net 'päz·əd·iv ¦sək·shən ,hed }

net radiometer |ENG| A Moll thermopile modified so that both sides are sensitive to radiation and the resulting electromotive force is proportional to the difference in intensities of radiation incident on the two sides; used to measure the difference in intensity between radiation entering and leaving the earth's surface. { ¦net ,rād·ē'äm·əd·ər }

net ton See ton. { 'net 'tən }

network |ELEC| A collection of electric elements, such as resistors, coils, capacitors, and sources of energy, connected together to form several interrelated circuits. Also known as electric network. See net. { 'net,wərk }

network analysis |ELEC| Derivation of the electrical properties of a network, from its configuration, element values, and driving forces. |IND ENG| An analytic technique used during project planning to determine the sequence of activities and their interrelationship within the network of activities that will be required by the project. Also known as network planning. { 'net,wərk ə'nal·ə·səs }

Neugebauer effect |ELEC| A small change in the polarization of an optically isotropic medium in an external electric field, related to the electrooptical Kerr effect. { 'nȯi·gə,baů·ər i,fekt }

Neumann-Kopp rule |THERMO| The rule that the heat capacity of 1 mole of a solid substance is approximately equal to the sum over the elements forming the substance of the heat capacity of a gram atom of the element times the number of atoms of the element in a molecule of the substance. { 'nȯi,män 'kȯp ,rül }

neuristor |ELECTR| A device that behaves like a nerve fiber in having attenuationless propagation of signals; one goal of research is development of a complete artificial nerve cell, containing many neuristors, that could duplicate the function of the human eye and brain in recognizing characters and other visual images. { nū'ris·tər }

neuromorphic engineering |ENG| Use of the functional principles of biological nervous systems to inspire the design and fabrication of artificial nervous systems, such as vision chips and roving robots. { ¦nù·rō¸mór·fik ¸en·jə'nir· iŋ }

neuronal interface |ENG| An artificial synapse capable of reversible chemical-to-electrical transduction processes between neural tissue and conventional solid-state electronic devices for applications such as aural, visual, and mechanical prostheses, as well as expanding human memory and intelligence. { nū¦rōn·əl 'in· tər¸fās }

neurotechnology |ENG| The application of microfabricated devices to achieve direct contact with the electrically active cells of the nervous system (neurons). { ¸nù·rō·tek'näl·ə·jē }

neutral |ELEC| Referring to the absence of a net electric charge. |MECH ENG| That setting in an automotive transmission in which all the gears are disengaged and the output shaft is disconnected from the drive wheels. { 'nü·trəl }

neutral atmosphere |ENG| An atmosphere which neither oxidizes nor reduces immersed materials. { 'nü·trəl 'at·mə¸sfir }

neutral axis |MECH| In a beam bent downward, the line of zero stress below which all fibers are in tension and above which they are in compression. { 'nü·trəl 'ak·səs }

neutral fiber |MECH| A line of zero stress in cross section of a bent beam, separating the region of compressive stress from that of tensile stress. { 'nü·trəl 'fī·bər }

neutrally buoyant float See swallow float. { 'nü· trə·lē ¦bòi·ənt 'flōt }

neutral stability |CONT SYS| Condition in which the natural motion of a system neither grows nor decays, but remains at its initial amplitude. { 'nü·trəl stə'bil·əd·ē }

neutral surface |MECH| A surface in a bent beam along which material is neither compressed nor extended. { 'nü·trəl 'sər·fəs }

neutron-gamma well logging |ENG| Neutron well logging in which the varying intensity of gamma rays produced artificially by neutron bombardment is recorded. { 'nü¸trän ¦gam·ə ¸wel ¸läg·iŋ }

neutron logging See neutron well logging. { 'nü ¸trän ¸läg·iŋ }

neutron shield |ENG| A shield that protects personnel from neutron irradiation. { 'nü¸trän ¸shēld }

neutron soil-moisture meter |ENG| An instrument for measuring the water content of soil and rocks as indicated by the scattering and absorption of neutrons emitted from a source, and resulting gamma radiation received by a detector, in a probe lowered into an access hole. { 'nü ¸trän 'sóil ¸mòis·chər ¸mēd·ər }

neutron well logging |ENG| Study of formation

fluid-content properties down a wellhole by neutron bombardment and detection of resultant radiation (neutrons or gamma rays). Also known as neutron logging. { 'nü¸trän ¦wel ¸läg·iŋ }

newel post |CIV ENG| **1.** A pillar at the end of an oblique retaining wall of a bridge. **2.** The post about which a circular staircase winds. **3.** A large post at the foot of a straight stairway or on a landing. { 'nü·əl ¸pōst }

newton |MECH| The unit of force in the meter-kilogram-second system, equal to the force which will impart an acceleration of 1 meter per second squared to the International Prototype Kilogram mass. Symbolized N. Formerly known as large dyne. { 'nüt·ən }

Newtonian attraction |MECH| The mutual attraction of any two particles in the universe, as given by Newton's law of gravitation. { nü'tō· nē·ən ə'trak·shən }

Newtonian mechanics |MECH| The system of mechanics based upon Newton's laws of motion in which mass and energy are considered as separate, conservative, mechanical properties, in contrast to their treatment in relativistic mechanics. { nü'tō·nē·ən mi'kan·iks }

Newtonian reference frame |MECH| One of a set of reference frames with constant relative velocity and within which Newton's laws hold; the frames have a common time, and coordinates are related by the Galilean transformation rule. { nü'tō·nē·ən 'ref·rəns ¸frām }

Newtonian velocity |MECH| The velocity of an object in a Newtonian reference frame, S, which can be determined from the velocity of the object in any other such frame, S', by taking the vector sum of the velocity of the object in S' and the velocity of the frame S' relative to S. { nü'tō· nē·ən və'läs·əd·ē }

newton-meter of energy See joule. { 'nüt·ən ¦mēd·ər əv 'en·ər·jē }

newton-meter of torque |MECH| The unit of torque in the meter-kilogram-second system, equal to the torque produced by 1 newton of force acting at a perpendicular distance of 1 meter from an axis of rotation. Abbreviated N-m. { 'nüt·ən ¸mēd·ər əv 'tòrk }

Newton's equations of motion |MECH| Newton's laws of motion expressed in the form of mathematical equations. { 'nüt·ənz i'kwā· zhənz əv 'mō·shən }

Newton's first law |MECH| The law that a particle not subjected to external forces remains at rest or moves with constant speed in a straight line. Also known as first law of motion; Galileo's law of inertia. { 'nüt·ənz 'fərst 'lò }

Newton's law of cooling |THERMO| The law that the rate of heat flow out of an object by both natural convection and radiation is proportional to the temperature difference between the object and its environment, and to the surface area of the object. { 'nüt·ənz 'lò əv 'kül·iŋ }

Newton's law of gravitation |MECH| The law that every two particles of matter in the universe attract each other with a force that acts along the line joining them, and has a magnitude proportional to the product of their masses and inversely proportional to the square of the distance between them. Also known as law of gravitation. { 'nüt·ənz 'lȯ əv ,grav·ə'tā·shən }

Newton's laws of motion [MECH] Three fundamental principles (called Newton's first, second, and third laws) which form the basis of classical, or Newtonian, mechanics, and have proved valid for all mechanical problems not involving speeds comparable with the speed of light and not involving atomic or subatomic particles. { 'nüt· ənz 'lȯz əv 'mō·shən }

Newton's second law [MECH] The law that the acceleration of a particle is directly proportional to the resultant external force acting on the particle and is inversely proportional to the mass of the particle. Also known as second law of motion. { 'nüt·ənz 'sek·ənd 'lȯ }

Newton's third law [MECH] The law that, if two particles interact, the force exerted by the first particle on the second particle (called the action force) is equal in magnitude and opposite in direction to the force exerted by the second particle on the first particle (called the reaction force). Also known as law of action and reaction; third law of motion. { 'nüt·ənz 'thərd 'lȯ }

ng See nanogram.

nib [ENG] A small projecting point. { nib }

nibbling [MECH ENG] Contour cutting of material by the action of a reciprocating punch that takes repeated small bites as the work is passed beneath it. { 'nib·liŋ }

Nichol's chart [CONT SYS] A plot of curves along which the magnitude M or argument α of the frequency control ratio is constant on a graph whose ordinate is the logarithm of the magnitude of the open-loop transfer function, and whose abscissa is the open-loop phase angle. { 'nik·əlz ,chärt }

Nicholson's hydrometer [ENG] A modification of Fahrenheit's hydrometer in which the lower end of the instrument carries a scale pan to permit the determination of the relative density of a solid. { 'nik·əl·sənz hī'dräm·əd·ər }

Nichols radiometer [ENG] An instrument, used to measure the pressure exerted by a beam of light, in which there are two small, silvered glass mirrors at the ends of a light rod that is suspended at the center from a fine quartz fiber within an evacuated enclosure. { 'nik·əlz ,rād·ē'äm·əd·ər }

nigre [CHEM ENG] Dark-colored layer formed between neat soap and lye during soap manufacture; contains more soap than lye, and a high concentration of salts and colored impurities. { 'nī·gər }

nine-light indicator [ENG] A remote indicator for wind speed and direction used in conjunction with a contact anemometer and a wind vane; the indicator consists of a center light, connected to the contact anemometer, surrounded by eight equally spaced lights which are individually connected to a set of similarly spaced electrical contacts on the wind vane; wind speed is determined

by counting the number of flashes of the center light during an interval of time; direction, indicated by the position of illuminated outer bulbs, is given to points of the compass. { 'nīn ¦līt 'in·də,kād·ər }

Nipher shield [ENG] A conically shaped, copper, rain-gage shield; used to prevent the formation of vertical wind eddies in the vicinity of the mouth of the gage, thereby making the rainfall catch a representative one. { 'nī·fər ,shēld }

nippers [DES ENG] Small pincers or pliers for cutting or gripping. { 'nip·ərz }

nipple [DES ENG] A short piece of tubing, usually with an internal or external thread at each end, used to couple pipes. Also known as bushing. { 'nip·əl }

nipple chaser [ENG] A member of a drilling crew who procures and delivers the tools and equipment necessary for an operation. { 'nip·əl ,chā·sər }

nitrogen fixation [CHEM ENG] Conversion of atmospheric nitrogen into compounds such as ammonia, calcium cyanamide, or nitrogen oxides by chemical or electric-arc processes. { 'nī·trə·jən ,fik¦sā·shən }

NLGI number [ENG] One of a series of numbers developed by the National Lubricating Grease Institute and used to classify the consistency range of lubricating greases; NLGI numbers are based on the American Society for Testing and Materials cone penetration number. { ¦en¦el¦jē'ī ,nəm·bər }

N-m See newton-meter of torque.

NMOS [ELECTR] Metal-oxide semiconductors that are made on p-type substrates, and whose active carriers are electrons that migrate between n-type source and drain contacts. Derived from n-channel metal-oxide semiconductor. { 'en,mȯs }

nn junction [ELECTR] In a semiconductor, a region of transition between two regions having different properties in n-type semiconducting material. { ¦en¦en ,jəŋk·shən }

no-bottom sounding [ENG] A sounding in the ocean in which the bottom is not reached. { 'nō 'bäd·əm ,saúnd·iŋ }

node [ELEC] See branch point. [ELECTR] A junction point within a network. [IND ENG] On a graphic presentation of a project, a symbol placed at the intersection of arrows that represent activities to identify the completion or start of an activity. { nōd }

nodulizing [ENG] Creation of spherical lumps from powders by working them together, coalescing them with binders, drying fluid-solid mixtures, heating, or chemical reaction. { 'näj·ə,līz·iŋ }

no-go gage [ENG] A limit gage designed not to fit a part being tested; usually employed with a go gage to set the acceptable maximum and minimum dimension limits of the part. { 'nō 'gō ,gāj }

noise [ELEC] Interfering and unwanted currents or voltages in an electrical device or system. { nȯiz }

noise-canceling microphone See close-talking microphone. { 'nȯiz ¦kans·liŋ 'mī·krə‚fōn }

noise equivalent temperature difference [THERMO] The change in equivalent blackbody temperature that corresponds to a change in radiance which will produce a signal-to-noise ratio of 1 in an infrared imaging device. Abbreviated NETD. { 'nȯiz i¦kwiv·ə·lənt 'tem·prə·chər ‚dif·rəns }

noise radial [ENG] The brightening of all range points on a particular plan position indicator bearing on a radar screen caused by noise reception from the indicated direction. { 'nȯiz 'rād·ē·əl }

noise reduction [ENG ACOUS] A process whereby the average transmission of the sound track of a motion picture print, averaged across the track, is decreased for signals of low level; since background noise introduced by the sound track is less at low transmission, this process reduces noise during soft passages. { 'nȯiz ri‚dək·shən }

noise-type flowmeter [ENG] A flowmeter that measures the noise generated in a selected frequency band. { 'nȯiz ¦tīp 'flō‚mēd·ər }

no-load current [ELEC] The current which flows in a network when the output is open-circuited. { 'nō ¦lōd 'kə·rənt }

no-load loss [ELEC] The power loss of a device that is operated at rated voltage and frequency but is not supplying power to a load. { 'nō ¦lōd 'lȯs }

no-load voltage See open-circuit voltage. { 'nō ¦lōd 'vōl·tij }

nominal bandwidth [ENG] The difference between the nominal upper and lower cutoff frequencies of an acoustic or electric filter. { 'näm·ə·nəl 'band‚width }

nominal pass-band center frequency [ENG] The geometric mean of the nominal upper and lower cutoff frequencies of an acoustic or electric filter. { 'näm·ə·nəl 'pas ‚band ¦sen·tər 'frē‚kwən·sē }

nominal size [DES ENG] Size used for purposes of general identification; the actual size of a part will be approximately the same as the nominal size but need not be exactly the same; for example, a rod may be referred to as 1/4 inch, although the actual dimension on the drawing is 0.2495 inch, and in this case 1/4 inch is the nominal size. { 'näm·ə·nəl 'sīz }

nonadiabatic See diabatic.

nonanticipatory system See causal system. { ¦nän·an'tis·ə·pə‚tȯr·ē ‚sis·təm }

nonbearing wall [CIV ENG] A wall that bears no vertical weight other than its own. { 'nän‚ber·iŋ 'wȯl }

nonblackbody [THERMO] A body that reflects some fraction of the radiation incident upon it; all real bodies are of this nature. { ¦nän'blak ( )‚bäd·ē }

noncontact sensor See proximity sensor. { ¦nän ( )'kän‚takt 'sen·sər }

noncontact thermometer See radiation pyrometer. { ¦nän'kän‚takt thər'mäm·əd·ər }

noncoring bit [ENG] A general type of bit made in many shapes which does not produce a core and with which all the rock cut in a borehole is ejected as sludge; used mostly for blasthole drilling and in the unmineralized zones in a borehole where a core sample is not wanted. Also known as borehole bit; plug bit. { 'nän‚kȯr·iŋ 'bit }

noncyclic element [IND ENG] An element of an operation or process that does not occur in every cycle but has a frequency of occurrence that is specified by the method. { ¦nän‚sī·klik 'el·ə·mənt }

nondestructive evaluation [IND ENG] A technique for probing and sensing material structure and properties without causing damage (as opposed to revealing flaws and defects). { ‚nän·di‚strək·tiv i‚val·yə'wā·shən }

nondestructive testing [ENG] A technique for revealing flaws and defects in a material or device without damaging or destroying the test sample; includes use of x-rays, ultrasonics, radiography, and magnetic flux. { ¦nän·di'strək·div 'test·iŋ }

nondissipative muffler See reactive muffler. { ¦nän'dis·ə‚pād·iv 'məf·lər }

nondurable goods [ENG] Products that are serviceable for a comparatively short time or are consumed or destroyed in a single usage. { ‚nän¦dùr·ə·bəl 'gùdz }

nonequilibrium thermodynamics [THERMO] A quantitative treatment of irreversible processes and of rates at which they occur. Also known as irreversible thermodynamics. { ¦nän‚ē·kwə'lib·rē·əm ‚thər·mō·dī'nam·iks }

nonexpendable [ENG] Pertaining to a supply item or piece of equipment that is not consumed, and does not lose its identity, in use, as a weapon, vehicle, machine, tool, piece of furniture, or instrument. { ¦nän·ik'spen·də·bəl }

nonfeasible method See goal coordination method. { ¦nän'fē·zə·bəl 'meth·əd }

nonflowing well [ENG] A well that yields water at the land surface only by means of a pump or other lifting device. { 'nän‚flō·iŋ 'wel }

nonholonomic system [MECH] A system of particles which is subjected to constraints of such a nature that the system cannot be described by independent coordinates; examples are a rolling hoop, or an ice skate which must point along its path. { ¦nän‚häl·ə'näm·ik 'sis·təm }

nonhoming [CONT SYS] Not returning to the starting or home position, as when the wipers of a stepping relay remain at the last-used set of contacts instead of returning to their home position. { ¦nän'hōm·iŋ }

nonintegrable system [MECH] A dynamical system whose motion is governed by an equation that is not an integrable differential equation. { ‚nän¦int·i·grə·bəl 'sis·təm }

noninteracting control [CONT SYS] A feedback control in a system with more than one input and more than one output, in which feedback transfer functions are selected so that each input influences only one output. (¦nän,in·tər'ak·tiŋ kən'trōl)

nonlinear circuit component |ELECTR| An electrical device for which a change in applied voltage does not produce a proportional change in current. Also known as nonlinear device; nonlinear element. ('nän,lin·ē·ər ¦sər·kət kəm ( )'pō·nənt)

nonlinear control system |CONT SYS| A control system that does not have the property of superposition, that is, one in which some or all of the outputs are not linear functions of the inputs. ('nän,lin·ē·ər kən'trōl ,sis·təmz)

nonlinear device See nonlinear circuit component. ('nän,lin·ē·ər di'vīs)

nonlinear distortion |ELECTR| Distortion in which the output of a system or component does not have the desired linear relation to the input. [ENG ACOUS] The ratio of the total root-mean-square (rms) harmonic distortion output of a microphone to the rms value of the fundamental component of the output. ('nän,lin·ē·ər di ( )'stór·shən)

nonlinear element See nonlinear circuit component. ('nän,lin·ē·ər 'el·ə·mənt)

nonlinear feedback control system |CONT SYS| Feedback control system in which the relationships between the pertinent measures of the system input and output signals cannot be adequately described by linear means. ('nän,lin·ē·ər 'fēd,bak kən'trōl ,sis·təm)

nonlinear vibration |MECH| A vibration whose amplitude is large enough so that the elastic restoring force on the vibrating object is not proportional to its displacement. ('nän,lin·ē·ər vī'brā·shən)

non-minimum-phase system |CONT SYS| A linear system whose transfer function has one or more poles or zeros with positive, nonzero real parts. (¦nän¦min·ə·məm 'fāz ,sis·təm)

nonpoint source |CIV ENG| A dispersed source of stormwater runoff; the water comes from land dedicated to uses such as agriculture, development, forest, and land fills and enters the surface water system as sheet flow at irregular rates. (,nän'póint ,sórs)

nonquantum mechanics |MECH| The classical mechanics of Newton and Einstein as opposed to the quantum mechanics of Heisenberg, Schrödinger, and Dirac; particles have definite position and velocity, and they move according to Newton's laws. (,nän¦kwän·təm mi'kan·iks)

nonreclosing pressure relief device |MECH ENG| A device which remains open after relieving pressure and must be reset before it can operate again. (¦nän·rē'klōz·iŋ 'presh·ər ri,lēf di,vīs)

nonrecording rain gage |ENG| A rain gage which indicates but does not record the amount of precipitation. (¦nän·ri'kórd·iŋ 'rān ,gāj)

nonrelativistic kinematics |MECH| The study of motions of systems of objects at speeds which are small compared to the speed of light, without reference to the forces which act on the system. (¦nän,rel·ə·tə'vis·tik ,kin·ə'mad·iks)

nonrelativistic mechanics |MECH| The study of the dynamics of systems in which all speeds are small compared to the speed of light. (¦nän ,rel·ə·tə'vis·tik mi'kan·iks)

nonreturn valve See check valve. (¦nän·ri'tərn ,valv)

nonselective radiator See graybody. (¦nän·si'lek·tiv 'rād·ē,ād·ər)

nonservo robot See fixed-stop robot. (¦nän'sər·vō 'rō,bät)

nonskid |CIV ENG| Pertaining to a surface that is roughened to reduce slipping, as a concrete floor treated with iron filings or carborundum powder, or indented while wet. (¦nän¦skid)

nonstranded rope |DES ENG| A wire rope with the wires in concentric sheaths instead of in strands, and in opposite directions in the different sheaths, giving the rope nonspinning properties. Also known as nonspinning rope. ('nän ,stran·dəd 'rōp)

nonwork unit |IND ENG| A time unit on a schedule during which work may not be performed on a given activity, for example, a weekend or a holiday. (¦nän'wərk ,yü·nət)

NOR circuit |ELECTR| A circuit in which output voltage appears only when signal is absent from all of its input terminals. ('nór ,sər·kət)

normal acceleration |MECH| **1.** The component of the linear acceleration of an aircraft or missile along its normal, or Z, axis. **2.** The usual or typical acceleration. ('nór·məl ak,sel·ə'rā·shən)

normal axis |MECH| The vertical axis of an aircraft or missile. ('nór·məl 'ak·səs)

normal barometer |ENG| A barometer of such accuracy that it can be used for the determination of pressure standards; an instrument such as a large-bore mercury barometer is usually used. ('nór·məl bə'räm·əd·ər)

normal coordinates |MECH| A set of coordinates for a coupled system such that the equations of motion each involve only one of these coordinates. ('nór·məl kō'órd·ən·əts)

normal effort |IND ENG| The effort expended by the average operator in performing manual work with average skill and application. ('nór·məl 'ef·ərt)

normal element time |IND ENG| The selected or average element time adjusted to obtain the element time used by an average qualified operator. Also known as base time; leveled element time. ('nór·məl ,el·ə¦ment 'tīm)

normal frequencies |MECH| The frequencies of the normal modes of vibration of a system. ('nór·məl 'frē·kwən,sēz)

normal impact |MECH| **1.** Impact on a plane perpendicular to the trajectory. **2.** Striking of a projectile against a surface that is perpendicular to the line of flight of the projectile. ('nór·məl 'im,pakt)

normal-incidence pyrheliometer |ENG| An instrument that measures the energy in the solar beam; it usually measures the radiation that strikes a target at the end of a tube equipped with a shutter and baffles to collimate the beam. { 'nȯr·məl ¦in·sad·əns ¦pīr,hē·lē'äm·əd·ər }

normal inspection [IND ENG] The number of items inspected as specified by the sampling inspection plan at the outset; if the quality of the product improves, the number of units to be inspected is reduced; if quality deteriorates, the number of units inspected is increased. { ¦nȯr·məl in'spek·shən }

normal mode of vibration [MECH] Vibration of a coupled system in which the value of one of the normal coordinates oscillates and the values of all the other coordinates remain stationary. { 'nȯrməl ¦mōd əv vī'brā·shən }

normal operation [MECH ENG] The operation of a boiler or pressure vessel at or below the conditions of coincident pressure and temperature for which the vessel has been designed. { 'nȯr·məl ,äp·ə'rā·shən }

normal pace [IND ENG] The manual pace achieved by normal effort. { 'nȯr·məl 'pās }

normal pitch [MECH ENG] The distance between working faces of two adjacent gear teeth, measured between the intersections of the line of action with the faces. { 'nȯr·məl 'pich }

normal-plate anemometer [ENG] A type of pressure-plate anemometer in which the plate, restrained by a stiff spring, is held perpendicular to the wind; the wind-activated motion of the plate is measured electrically; the natural frequency of this system can be made high enough so that resonance magnification does not occur. { 'nȯr·məl ¦plāt ,an·ə'mäm·əd·ər }

normal reaction [MECH] The force exerted by a surface on an object in contact with it which prevents the object from passing through the surface; the force is perpendicular to the surface, and is the only force that the surface exerts on the object in the absence of frictional forces. { 'nȯr·məl rē'ak·shən }

normal stress [MECH] The stress component at a point in a structure which is perpendicular to the reference plane. { 'nȯr·məl 'stres }

normal time [IND ENG] **1.** The time required by a trained worker to perform a task at a normal pace. **2.** The total of all the normal elemental times constituting a cycle or operation. Also known as base time; leveled time. { 'nȯr·məl 'tīm }

north-stabilized plan-position indicator [ENG] A heading-upward plan-position indicator; this term is deprecated because it may be confused with azimuth-stabilized plan-position indicator, a north-upward plan-position indicator. { 'nȯrth ¦stā·bə,līzd ¦plan pə¦zish·ən 'in·də,kād·ər }

north-upward plan position indicator [ENG] A plan position indicator on which north is maintained at the top of the indicator, regardless of the heading of the craft. { 'nȯrth 'əp·wərd ¦plan pə¦zish·ən 'in·də,kād·ər }

nose [ENG] The foremost point or section of a bomb, missile, or something similar. { nōz }

nose radius [MECH ENG] The radius measured

in the back rake or top rake plane of a cutting tool. { 'nōz ,rād·ē·əs }

nose sill [ENG] A short timber located under the end of the main sill of a standard rig front of a well. { 'nōz ,sil }

nosing [BUILD] Projection of a tread of a stair beyond the riser below it. [CIV ENG] A transverse, horizontal motion of a locomotive that exerts a lateral force on the track. { 'nōz·iŋ }

notch [ELECTR] Rectangular depression extending below the sweep line of the radar indicator in some types of equipment. [ENG] A V-shaped indentation or cut in a surface or edge. { näch }

notching [ELEC] Term indicating that a predetermined number of separate impulses are required to complete operation of a relay. [MECH ENG] Cutting out various shapes from the ends or edges of a workpiece. { 'näch·iŋ }

notching press [MECH ENG] A mechanical press for notching straight or rounded edges. { 'näch·iŋ ,pres }

NOT circuit [ELECTR] A logic circuit with one input and one output that inverts the input signal at the output; that is, the output signal is a logical 1 if the input signal is a logical 0, and vice versa. Also known as inverter circuit. { 'nät ,sər·kət }

nozzle [DES ENG] A tubelike device, usually streamlined, for accelerating and directing a fluid, whose pressure decreases as it leaves the nozzle. { 'näz·əl }

nozzle-contraction-area ratio [DES ENG] Ratio of the cross-sectional area for gas flow at the nozzle inlet to that at the throat. { 'näz·əl kən'trak·shən ¦er·ē·ə ,rā·shō }

nozzle efficiency [MECH ENG] The efficiency with which a nozzle converts potential energy into kinetic energy, commonly expressed as the ratio of the actual change in kinetic energy to the ideal change at the given pressure ratio. { 'näz·əl i,fish·ən·sē }

nozzle exit area [DES ENG] The cross-sectional area of a nozzle available for gas flow measured at the nozzle exit. { 'näz·əl ¦eg·zət ,er·ē·ə }

nozzle-expansion ratio [DES ENG] Ratio of the cross-sectional area for gas flow at the exit of a nozzle to the cross-sectional area available for gas flow at the throat. { 'näz·əl ik'pan·shən ,rā·shō }

nozzle-mix gas burner [ENG] A burner in which injection nozzles mix air and fuel gas at the burner tile. { 'näz·əl ,miks 'gas ,bər·nər }

nozzle throat [DES ENG] The portion of a nozzle with the smallest cross section. { 'näz·əl ,thrōt }

nozzle throat area [DES ENG] The area of the minimum cross section of a nozzle. { 'näz·əl ¦thrōt ,er·ē·ə }

npin transistor [ELECTR] An *npin* transistor which has a layer of high-purity germanium between the base and collector to extend the frequency range. { 'en,pin tran'zis·tər }

N-P-K [CHEM ENG] The code identifying the components in a fertilizer mixture: nitrogen (N), phosphorus pentoxide (P), and potassium oxide (K). Fertilizers are graded in the order N-P-K, with the numbers indicating the percentage of the total weight of each component. For example, 5-10-10 represents a mixture containing by weight 5% nitrogen, 10% phosphorus pentoxide, and 10% potassium oxide.

npnp diode See pnpn diode. { ¦en,pē¦en,pē 'dī,ōd }

npnp transistor |ELECTR| An npn-junction transistor having a transition or floating layer between p and n regions, to which no ohmic connection is made. Also known as pnpn transistor. { ¦en,pē¦en,pē tran'zis·tər }

npn semiconductor |ELECTR| Double junction formed by sandwiching a thin layer of p-type material between two layers of n-type material of a semiconductor. { 'en,pē'en 'sem·i·kən,dək·tər }

npn transistor |ELECTR| A junction transistor having a p-type base between an n-type emitter and an n-type collector; the emitter should then be negative with respect to the base, and the collector should be positive with respect to the base. { 'en,pē'en tran'zis·tər }

np semiconductor |ELECTR| Region of transition between n- and p-type material. { ¦en¦pē 'sem·i·kən,dək·tər }

NPSH See net positive suction head.

N rod bit |DES ENG| A Canadian standard noncoring bit having a set diameter of 2.940 inches (74.676 millimeters). { 'en 'räd ,bit }

n-type conduction |ELECTR| The electrical conduction associated with electrons, as opposed to holes, in a semiconductor. { 'en ,tīp kən ,dək·shən }

n-type germanium |ELECTR| Germanium to which more impurity atoms of donor type (with valence 5, such as antimony) than of acceptor type (with valence 3, such as indium) have been added, with the result that the conduction electron density exceeds the hole density. { 'en ,tīp jər'mā·nē·əm }

n-type semiconductor |ELECTR| An extrinsic semiconductor in which the conduction electron density exceeds the hole density. { 'en ,tīp 'sem·i·kən,dək·tər }

nuclear chemical engineering |CHEM ENG| The branch of chemical engineering that deals with the production and use of radioisotopes. { 'nü·klē·ər ¦kem·ə·kəl ,en·jə'nir·iŋ }

nuclear excavation |ENG| The use of nuclear explosions to remove earth for constructing harbors, canals, and other facilities. { 'nü·klē·ər ,ek·skə'vā·shən }

nuclear gyroscope |ENG| A gyroscope in which the conventional spinning mass is replaced by the spin of atomic nuclei and electrons; one version uses optically pumped mercury isotopes, and another uses nuclear magnetic resonance techniques. { 'nü·klē·ər 'jī·rə,skōp }

nuclear magnetic resonance flowmeter |ENG| A flowmeter in which nuclei of the flowing fluid are resonated by a radio-frequency field superimposed on an intense permanent magnetic field,

and a detector downstream measures the amount of decay of the resonance, thereby sensing fluid velocity. { 'nü·klē·ər mag'ned·ik 'rez·ən·əns 'flō,mēd·ər }

nuclear magnetic resonance gyroscope |ENG| A gyroscope that obtains information from the dynamic angular motion of atomic nuclei. { 'nü·klē·ər mag'ned·ik 'rez·ən·əns 'jī·rə,skōp }

nuclear magnetometer |ENG| Any magnetometer which is based on the interaction of a magnetic field with nuclear magnetic moments, such as the proton magnetometer. Also known as nuclear resonance magnetometer. { 'nü·klē·ər mag·nə'täm·əd·ər }

nuclear power plant |MECH ENG| A power plant in which nuclear energy is converted into heat for use in producing steam for turbines, which in turn drive generators that produce electric power. Also known as atomic power plant. { 'nü·klē·ər 'pau̇·ər ,plant }

nuclear resonance magnetometer See nuclear magnetometer. { 'nü·klē·ər 'rez·ən·əns ,mag·nə'täm·əd·ər }

nuclear snow gage |ENG| Any type of gage using a radioactive source and a detector to measure, by the absorption of radiation, the water-equivalent mass of a snowpack. { 'nü·klē·ər 'snō ,gāj }

nucleate boiling |CHEM ENG| Boiling in which bubble formation is at the liquid-solid interface rather than from external or mechanical devices; occurs in kettle-type and natural-circulation heaters or reboilers. { 'nü·klē,āt 'bȯil·iŋ }

nucleonics |ENG| The technology based on phenomena of the atomic nucleus such as radioactivity, fission, and fusion; includes nuclear reactors, various applications of radioisotopes and radiation, particle accelerators, and radiation-detection devices. { ,nü·klē'än·iks }

nucleus counter |ENG| An instrument which measures the number of condensation nuclei or ice nuclei per sample volume of air. { 'nü·klē·əs ,kau̇nt·ər }

null-balance recorder |ENG| An instrument in which a motor-driven slide wire in a measuring circuit is continuously adjusted so that the voltage or current to be measured will be balanced against the voltage or current from this circuit; a pen linked to the slide wire makes a graphical record of its position as a function of time. { 'nəl ¦bal·əns ri,kȯrd·ər }

null detector See null indicator. { 'nəl di,tek·tər }

null indicator |ENG| A galvanometer or other device that indicates when voltage or current is zero; used chiefly to determine when a bridge circuit is in balance. Also known as null detector. { 'nəl ,in·də,kād·ər }

null method |ENG| A method of measurement in which the measuring circuit is balanced to bring the pointer of the indicating instrument to zero, as in a Wheatstone bridge, and the settings of the balancing controls are then read. Also known as balance method; zero method. { 'nəl ,meth·əd }

Nusselt equation |THERMO| Dimensionless

equation used to calculate convection heat transfer for heating or cooling of fluids outside a bank of 10 or more rows of tubes to which the fluid flow is normal. { 'nús·əlt i,kwā·zhən }

Nusselt number [THERMO] A dimensionless number used in the study of forced convection which gives a measure of the ratio of the total heat transfer to conductive heat transfer, and is equal to the heat-transfer coefficient times a characteristic length divided by the thermal conductivity. Symbolized N_{Nu}. { 'nús·əlt ,nəm·bər }

nut [DES ENG] An internally threaded fastener for bolts and screws. { nət }

nutating antenna [ENG] An antenna system used in conical scan radar, in which a dipole or feed horn moves in a small circular orbit about the axis of a paraboloidal reflector without changing its polarization. { 'nü,tād·iŋ an 'ten·ə }

nutating-disk meter [ENG] An instrument for measuring flow of a liquid in which liquid passing through a chamber causes a disk to nutate, or roll back and forth, and the total number of rolls is mechanically counted. { 'nü,tād·iŋ ¦disk ¦mēd·ər }

nutation [MECH] A bobbing or nodding up-and-down motion of a spinning rigid body, such as a top, as it precesses about its vertical axis. { nü'tā·shən }

nutator [ENG] A mechanical or electrical device used to move a radar beam in a circular, conical, spiral, or other manner periodically to obtain greater air surveillance than could be obtained with a stationary beam. { 'nü,tād·ər }

Nyquist contour [CONT SYS] A directed closed path in the complex frequency plane used in constructing a Nyquist diagram, which runs upward, parallel to the whole length of the imaginary axis at an infinitesimal distance to the right of it, and returns from $+j_n$ to $-j_n$ along a semicircle of infinite radius in the right half-plane. { 'nī,kwist ,kän,túr }

Nyquist diagram [CONT SYS] A plot in the complex plane of the open-loop transfer function as the complex frequency is varied along the Nyquist contour; used to determine stability of a control system. { 'nī,kwist ,dī·ə,gram }

Nyquist stability criterion See Nyquist stability theorem. { 'nī,kwist stə'bil·əd·ē krī,tir·ē·ən }

Nyquist stability theorem [CONT SYS] The theorem that the net number of counterclockwise rotations about the origin of the complex plane carried out by the value of an analytic function of a complex variable, as its argument is varied around the Nyquist contour, is equal to the number of poles of the variable in the right half-plane minus the number of zeros in the right half-plane. Also known as Nyquist stability criterion. { 'nī,kwist stə'bil·əd·ē ,thir·əm }

Nyquist's theorem [ELECTR] The mean square noise voltage across a resistance in thermal equilibrium is four times the product of the resistance, Boltzmann's constant, the absolute temperature, and the frequency range within which the voltage is measured. { 'nī,kwists ,thir·əm }

nystagmogram [IND ENG] A recording of saccadic eye movements, that is, quick, rhythmic, and usually involuntary oscillations of the eyes. { ni'stag·mə,gram }

OBA *See* octave-band analyzer.

oblique valve |MECH ENG| A type of globe valve having an inclined orifice that serves to reduce the disruption of the flow pattern of the working fluid. { ə¦blēk 'valv }

obliterated corner |CIV ENG| In surveying, a corner for which visible evidence of the previous surveyor's work has disappeared, but whose original position can be established from other physical evidence and testimony. { ə'blid·ə,rād·əd 'kȯr·nər }

observability |CONT SYS| Property of a system for which observation of the output variables at all times is sufficient to determine the initial values of all the state variables. { əb,zər·və'bil·əd·ē }

observation spillover |CONT SYS| The part of the sensor output of an active control system caused by modes that have been omited from the control algorithm in the process of model reduction. { ,äb·zər'vā·shən 'spil,ō·vər }

observer |CONT SYS| A linear system B driven by the inputs and outputs of another linear system A which produces an output that converges to some linear function of the state of system A. Also known as state estimator; state observer. { əb'zər·vər }

obsolescence |ENG| Decreasing value of functional and physical assets or value of a product or facility from technological changes rather than deterioration. { ,äb·sə'les·əns }

obsolete |ENG| No longer satisfactory for the purpose for which obtained, due to improvements or revised requirements. { ,äb·sə'lēt }

occlusion |ENG| The retention of undissolved gas in a solid during solidification. { ə'klü·zhən }

occupational ecology |IND ENG| A discipline concerned with the interaction of workers with the environment, and with matching humans with the environment in the most ergonomically efficient way and with minimal disturbance of the environment. { ,ä·kyə'pā·shen·əl i'käl·ə·jē }

occupy |ENG| To set a surveying instrument over a point for the purpose of making observations or measurements. { 'äk·yə,pī }

ocean engineering |ENG| A subfield of engineering involved with the development of new equipment concepts and the methodical improvement of techniques which allow humans to operate successfully beneath the ocean surface in order to develop and utilize marine resources. { 'ō·shən ,en·jə'nir·iŋ }

oceanographic dredge |ENG| A device used aboard ship to bring up large samples of deposits and sediments from the ocean bottom. { ¦ō·shə·nə¦graf·ik 'drej }

oceanographic platform |ENG| A construction with a flat horizontal surface higher than the water, on which oceanographic equipment is suspended or installed. { ¦ō·shə·nə¦graf·ik 'plat ,fȯrm }

ocean thermal-energy conversion |MECH ENG| The conversion of energy arising from the temperature difference between warm surface water of oceans and cold deep-ocean current into electrical energy or other useful forms of energy. Abbreviated OTEC. { 'ō·shən 'thər·məl 'en·ər·jē kən,vər·zhən }

octahedral normal stress |MECH| The normal component of stress across the faces of a regular octahedron whose vertices lie on the principal axes of stress; it is equal in magnitude to the spherical stress across any surface. Also known as mean stress. { ¦äk·tə¦hē·drəl 'nȯr·məl ,stres }

octahedral shear stress |MECH| The tangential component of stress across the faces of a regular octahedron whose vertices lie on the principal axes of stress; it is a measure of the strength of the deviatoric stress. { ¦äk·tə¦hē·drəl 'shir ,stres }

octane number |ENG| A rating that indicates the tendency to knock when a fuel is used in a standard internal combustion engine under standard conditions; *n*-heptane is 0, isooctane is 100; different test methods yield other values variously known as research octane, motor octane, and road octane. { 'äk,tān ,nəm·bər }

octane requirement |MECH ENG| The fuel octane number needed for efficient operation (without knocking or spark retardation) of an internal combustion engine. { 'äk,tān ri,kwīr·mənt }

octane scale |ENG| Series of arbitrary numbers from 0 to 120.3 used to rate the octane number of a gasoline; *n*-heptane is 0 octane, isooctane is 100, and isooctane + 6 milliliters TEL (tetraethyllead) is 120.3. { 'äk,tān ,skāl }

octave-band analyzer |ENG ACOUS| A portable sound analyzer which amplifies a microphone signal, feeds it into one of several band-pass filters selected by a switch, and indicates the magnitude of sound in the corresponding frequency band on a logarithmic scale; all the bands except the highest and lowest span an octave in frequency. Abbreviated OBA. { 'äk·tiv ¦band 'an·ə,līz·ər }

octave-band filter |ENG ACOUS| A band-pass filter in which the upper cutoff frequency is twice the lower cutoff frequency. { 'äk·tiv ¦band 'fil·tər }

octoid |DES ENG| Pertaining to a gear tooth form used to generate the teeth in bevel gears; the octoid form closely resembles the involute form. { 'äk,tóid }

OD See outside diameter.

odd-leg caliper |DES ENG| A caliper in which the legs bend in the same direction instead of opposite directions. { 'äd ,leg 'kal·ə·pər }

odograph |ENG| An instrument installed in a vehicle to automatically plot on a map the course and distance traveled by the vehicle. { 'ō·də ,graf }

odometer |ENG| 1. An instrument for measuring distance traversed, as of a vehicle. 2. The indicating gage of such an instrument. 3. A wheel pulled by surveyors to measure distance traveled. { ō'däm·əd·ər }

odorize |CHEM ENG| To add an unpleasant odor as a safety measure to an odorless material such as fuel gas. { 'ō·də,rīz }

Oehman's survey instrument |ENG| A drill-hole surveying apparatus that makes a photographic record of the compass and clinometer readings. { 'ā·mənz 'sər,vā ,in·strə·mənt }

off |ENG| Designating the inoperative state of a device, or one of two possible conditions (the other being "on") in a circuit. { óf }

off-count mesh |DES ENG| A mesh in a wire cloth in which the count is not the same for both directions. { ¦óf ,kaůnt 'mesh }

offhand grinding |MECH ENG| Grinding operations performed with hand-held tools. Also known as freehand grinding. { ¦óf,hand 'grīnd·iŋ }

off-highway vehicle |MECH ENG| A bulk-handling machine, such as an earthmover or dump truck, that is designed to operate on steep or rough terrain and has a height and width that may exceed highway legal limits. { 'óf ,hī,wā 've·ə·kəl }

off-line |ENG| 1. A condition existing when the drive rod of the drill swivel head is not centered and parallel with the borehole being drilled. 2. A borehole that has deviated from its intended course. 3. A condition existing wherein any linear excavation (shaft, drift, borehole) deviates from a previously determined or intended survey line or course. |IND ENG| State in which an equipment or subsystem is in standby, maintenance, or mode of operation other than on-line. { 'óf¦līn }

off-road vehicle |MECH ENG| A conveyance designed to travel on unpaved roads, trails, beaches, or rough terrain rather than on public roads. { 'óf¦rōd 've·ə·kəl }

offset |BUILD| A horizontal ledge on the face of a wall or other member that is formed by diminishing the thickness of the wall at that point. Also known as setback. |CONT SYS| The steady-state difference between the desired control point and that actually obtained in a process control system. |ENG| 1. A short perpendicular distance measured to a traverse course or a surveyed line or principal line of measurement in order to locate a point with respect to a point on the course or line. 2. In seismic prospecting, the horizontal distance between a shothole and the line of profile, measured perpendicular to the line. 3. In seismic refraction prospecting, the horizontal displacement, measured from the detector, of a point for which a calculated depth is relevant. 4. In seismic reflection prospecting, the correction of a reflecting element from its position on a preliminary working profile to its actual position in space. |MECH| The value of strain between the initial linear portion of the stress-strain curve and a parallel line that intersects the stress-strain curve of an arbitrary value of strain; used as an index of yield stress; a value of 0.2% is common. { 'óf,set }

offset cab |ENG| Operator's cab positioned to one side of earthmoving equipment for greater visibility and safety. { 'óf,set 'kab }

offset cylinder |MECH ENG| A reciprocating part in which the crank rotates about a center off the centerline. { 'óf,set 'sil·ən·dər }

offset line |ENG| A secondary line established close to and roughly parallel with the primary survey line to which it is referenced by measured offsets. { 'óf,set 'līn }

offset screwdriver |DES ENG| A screwdriver with the blade set perpendicular to the shank for access to screws in otherwise awkward places. { 'óf,set 'skrü,drīv·ər }

offset voltage |ELECTR| The differential input voltage that must be applied to an operational amplifier to return the zero-frequency output voltage to zero volts, due to device mismatching at the input stage. { 'óf,set ,vōl·tij }

offset yield strength |MECH| That stress at which the strain surpasses by a specific amount (called the offset) an extension of the initial proportional portion of the stress-strain curve; usually expressed in pounds per square inch. { 'óf ,set 'yēld ,streŋkth }

offshore mooring |CIV ENG| An anchorage serving an area for which it is not considered feasible or cost-effective to construct a dock or provide a protected harbor, and providing equipment to which ships can attach mooring lines. { 'óf¦shór 'mür·iŋ }

off-site facility |CHEM ENG| In a chemical process plant, any supporting facility that is not a direct part of the reaction train, such as utilities, steam, and waste-treatment facilities. { 'ôf̩sĭt fə'sil·əd·ē }

off-the-shelf |IND ENG| Available for immediate shipment. { 'ôf t͟hə ¦shelf }

ohm |ELEC| The unit of electrical resistance in the rationalized meter-kilogram-second system of units, equal to the resistance through which a current of 1 ampere will flow when there is a potential difference of 1 volt across it. Symbolized Ω. { ŏm }

ohmic |ELEC| Pertaining to a substance or circuit component that obeys Ohm's law. { 'ŏ·mik }

ohmic dissipation |ELECTR| Loss of electric energy when a current flows through a resistance due to conversion into heat. Also known as ohmic loss. { 'ŏ·mik ‚dis·ə'pā·shən }

ohmic loss See ohmic dissipation. { 'ŏ·mik 'lŏs }

ohmmeter |ENG| An instrument for measuring electric resistance; scale may be graduated in ohms or megohms. { 'ŏ‚mēd·ər }

Ohm's law |ELEC| The law that the direct current flowing in an electric circuit is directly proportional to the voltage applied to the circuit; it is valid for metallic circuits and many circuits containing an electrolytic resistance. { 'ŏmz ‚lŏ }

ohms per volt |ENG| Sensitivity rating for measuring instruments, obtained by dividing the resistance of the instrument in ohms at a particular range by the full-scale voltage value at that range. { 'ŏmz pər 'vŏlt }

OHV engine See overhead-valve engine. { ¦ŏ ¦ăch'vē 'en·jən }

oil bath |ENG| **1.** Oil, in a container, within which a mechanism works or into which it dips. **2.** Oil in which a piece of apparatus is submerged. **3.** Oil that is poured on a cutting tool. { 'ôil ‚bath }

oil burner |ENG| Liquid-fuel burner device using a mixture of air and vaporized or atomized oil for combustion. { 'ôil ‚bər·nər }

oil cooler |MECH ENG| A small radiator used to cool the oil that lubricates an automotive engine. { 'ôil ‚kü·lər }

oil cup |ENG| A permanently mounted cup used to feed lubricant to a gear, usually with some means of regulating the flow. { 'ôil ‚kəp }

oil dilution valve |MECH ENG| A valve used to mix gasoline with engine oil to permit easier starting of the gasoline engine in cold weather. { 'ôil di‚lü·shən ‚valv }

oil filter |ENG| Cartridge-type filter used in automotive oil-lubrication systems to remove metal particles and products of heat decomposition from the circulating oil. { 'ôil ‚fil·tər }

oil fogging |ENG| Spraying a fine oil mist into the gas stream of a distribution system to alleviate the drying effects of gas on certain kinds of distribution and utilization equipment. { 'ôil 'fäg·iŋ }

oil furnace |MECH ENG| A combustion chamber in which oil is the heat-producing fuel. { 'ôil ‚fər·nəs }

oil-gas process |CHEM ENG| Process to manufacture high-caloric-value fuel gas by the destructive distillation of high-boiling petroleum oils. { 'ôil ‚gas ‚prä·səs }

oil groove |DES ENG| One of the grooves in a bearing which distribute and collect lubricating oil. { 'ôil ‚grüv }

oil hole |ENG| A small hole for injecting oil for a bearing. { 'ôil ‚hŏl }

oil-hole drill |DES ENG| A twist drill containing holes through which oil can be fed to the cutting edges. { 'ôil ‚hŏl ‚dril }

oiliness |ENG| The effect of a lubricant to reduce friction between two solid surfaces in contact; the effect is more than can be accounted for by viscosity alone. { 'ôi·lē·nəs }

oilless bearing |MECH ENG| A self-lubricating bearing containing solid or liquid lubricants in its material. { 'ôil·les 'ber·iŋ }

oil lift |MECH ENG| Hydrostatic lubrication of a journal bearing by using oil at high pressure in the area between the bottom of the journal and the bearing itself so that the shaft is raised and supported by an oil film whether it is rotating or not. { 'ôil ‚lift }

oil pump |MECH ENG| A pump of the gear, vane, or plunger type, usually an integral part of the automotive engine; it lifts oil from the sump to the upper level in the splash and circulating systems, and in forced-feed lubrication it pumps the oil to the tubes leading to the bearings and other parts. { 'ôil ‚pəmp }

oil reclaiming |ENG| **1.** A process in which oil is passed through a filter as it comes from equipment and then returned for reuse, in the same manner that crank case oil is cleaned by an engine filter. **2.** A method in which solids are removed from oil by treatment in settling tanks. { 'ôil ri'klām·iŋ }

oil ring |MECH ENG| **1.** A ring located at the lower part of a piston to prevent an excess amount of oil from being drawn up onto the piston during the suction stroke. **2.** A ring on a journal, dipping into an oil bath for lubrication. { 'ôil ‚riŋ }

oil seal |ENG| **1.** A device for preventing the entry or return of oil from a chamber. **2.** A device using oil as the sealing medium to prevent the passage of fluid from one chamber to another. { 'ôil ‚sēl }

Oldham coupling See slider coupling. { 'ŏl·dəm ‚kəp·liŋ }

oleometer |ENG| **1.** A device for measuring specific gravity of oils. **2.** An instrument for determining the proportion of oil in a substance. { ‚ŏ·lē'äm·əd·ər }

oleo strut |MECH ENG| A shock absorber consisting of a telescoping cylinder that forces oil into an air chamber, thereby compressing the air; used on aircraft landing gear. { 'ŏ·lē·ŏ ‚strət }

ombrometer See rain gage. { äm'bräm·əd·ər }

ombroscope |ENG| An instrument consisting of a heated, water-sensitive surface which indicates by mechanical or electrical techniques the occurrence of precipitation; the output of the instrument may be arranged to trip an alarm or to record on a time chart. { 'äm·brə,skōp }

omnibearing converter [ENG] An electromechanical device which combines an omnirange signal with heading information to furnish electrical signals for the operation of the pointer of a radio magnetic indicator. { 'äm·nə,ber·iŋ kən'vərd·ər }

omnibearing indicator [ENG] An instrument providing automatic and continuous indication of omnibearing. { 'äm·nə,ber·iŋ 'in·də,kād·ər }

omnibearing selector [ENG] A device capable of being set manually to any desired omnibearing, or its reciprocal, to control a course-line deviation indicator. Also known as radial selector. { 'äm·nə,ber·iŋ si'lek·tər }

omnidirectional hydrophone [ENG ACOUS] A hydrophone whose response is fundamentally independent of the incident sound wave's angle of arrival. { ,äm·nə·di'rek·shən·əl 'hī·drə,fōn }

omnigraph [ENG] An automatic acetylene cutter controlled by a mechanical pointer that traces a pattern; capable of cutting several duplicates simultaneously. { 'äm·nə,graf }

omnimeter [ENG] A theodolite with a microscope that can be used to observe vertical angular movement of the telescope. { äm'nim·əd·ər }

on [ENG] Designating the operating state of a device or one of two possible conditions (the other being "off") in a circuit. { ȯn }

on center [BUILD] The measurement made between the centers of two adjacent members. { 'ȯn 'sen·tər }

once-through boiler [MECH ENG] A boiler in which water flows, without recirculation, sequentially through the economizer, furnace wall, and evaporating and superheating tubes. { 'wəns ¦thrü ,bȯil·ər }

on composition See on grade. { ¦ȯn ,käm·pə 'zish·ən }

on-condition maintenance [IND ENG] Examination of those aspects of an installation that are predictive of pending failure, followed by performance of preventative maintenance activities before occurrence of total failure. { ¦ȯn kən¦dish·ən 'mānt·ən·əns }

one-digit subtracter See half-subtracter. { 'wən ,dij·ət səb'trak·tər }

one-hundred-percent premium plan [IND ENG] A wage incentive plan wherein each unit produced by an employee in excess of standard is compensated at the same rate paid for each unit of standard production. Also known as straight piecework system; straight proportional system. { 'wən ,hən·drəd pər¦sent 'prē·mē·əm ,plan }

one-shot molding [ENG] Production of urethane-plastic foam in which the isocyanate, polyol, and catalyst and other additives are mixed directly together and a foam is produced immediately. { 'wən ,shät 'mōld·iŋ }

one-sided acceptance sampling test [IND ENG] A test against a single specification only, in which permissible values in one direction are not limited. { 'wən ,sīd·əd ik'sep·təns 'samp·liŋ ,test }

one-way slab [CIV ENG] A concrete slab in which the reinforcing steel runs perpendicular to the supporting beams, that is, one way. { 'wən ,wā 'slab }

on grade [CIV ENG] **1.** At ground level. **2.** Supported directly on the ground. { ȯn 'grād }

onion diagram [SYS ENG] A schematic diagram of a system that is composed of concentric circles, with the innermost circle representing the core, and all the outer layers dependent on the core. { 'ən·yən ,dī·ə,gram }

on-off control [CONT SYS] A simple control system in which the device being controlled is either full on or full off, with no intermediate operating positions. Also known as on-off system. { 'ȯn ¦ȯf kən,trōl }

on-off system See on-off control. { 'ȯn 'ȯf ,sis·təm }

Onsager reciprocal relations [THERMO] A set of conditions which state that the matrix, whose elements express various fluxes of a system (such as diffusion and heat conduction) as linear functions of the various conjugate affinities (such as mass and temperature gradients) for systems close to equilibrium, is symmetric when certain definitions are chosen for these fluxes and affinities. { 'ȯn,säg·ər ri'sip·rə·kəl ri'lā·shənz }

on stream [CHEM ENG] Of a plant or process-operations unit, being in operation. { 'ȯn ,strēm }

on-stream factor [IND ENG] The ratio of the number of operating days to the number of calendar days per year. { 'ȯn ,strēm 'fak·tər }

on-stream time [CHEM ENG] In plant or process operations, the actual time that a unit is operating and producing product. { 'ȯn ,strēm ,tīm }

OPDAR [ENG] A laser system for measuring elevation angle, azimuth angle, and slant range of a missile during its firing period. Derived from optical direction and ranging. Also known as optical radar. { 'äp,där }

open [ELEC] **1.** Condition in which conductors are separated so that current cannot pass. **2.** Break or discontinuity in a circuit which can normally pass a current. { 'ō·pən }

open-belt drive [DES ENG] A belt drive having both shafts parallel and rotating in the same direction. { 'ō·pən ,belt ,drīv }

open berth [CIV ENG] An anchorage berth in an open roadstead. { 'ō·pən 'bərth }

open caisson [CIV ENG] A caisson in the form of a cylinder or shaft that is open at both ends; it is set in place, pumped dry, and filled with concrete. { 'ō·pən 'kā,sän }

open-center plan position indicator [ENG] A plan position indicator on which no signal is displayed within a set distance from the center. { 'ō·pən ,sen·tər 'plan pə,zish·ən 'in·də,kād·ər }

open circuit [ELEC] An electric circuit that has been broken, so that there is no complete path for current flow. { 'ō·pən 'sər·kət }

open-circuit grinding [MECH ENG] Grinding system in which material passes through the grinder without classification of product and without recycle of oversize lumps; in contrast to closed-circuit grinding. { 'ō·pən ¦sər·kət 'grīnd·iŋ }

open-circuit scuba |ENG| The simplest type of scuba equipment, in which all exhaled gas is discharged directly into the water and the utilization of gas is therefore equal to the mass exhaled. { 'ō·pən ¦sər·kət 'skü·bə }

opencut |CIV ENG| An open trench, such as across a hill. { 'ō·p·ən¦kət }

open cycle |THERMO| A thermodynamic cycle in which new mass enters the boundaries of the system and spent exhaust leaves it; the automotive engine and the gas turbine illustrate this process. { ¦ō·pən ¦sī·kəl }

open-cycle engine |MECH ENG| An engine in which the working fluid is discharged after one pass through boiler and engine. { ¦ō·pən ¦sī·kəl 'en·jən }

open-cycle gas turbine |MECH ENG| A gas turbine prime mover in which air is compressed in the compressor element, fuel is injected and burned in the combustor, and the hot products are expanded in the turbine element and exhausted to the atmosphere. { ¦ō·pən ¦sī·kəl 'gas 'tər·bən }

open-end wrench |DES ENG| A wrench consisting of fixed jaws at one or both ends of a handle. { ¦ō·pən ¦end 'rench }

open hole |ENG| **1.** A well or borehole, or a portion thereof, that has not been lined with steel tubing at the depth referred to. **2.** An unobstructed borehole. **3.** A borehole being drilled without cores. { 'ō·pən 'hōl }

opening die |MECH ENG| A die head for cutting screws that opens automatically to release the cut thread. { 'ōp·ə·niŋ ¸dī }

opening pressure |MECH ENG| The static inlet pressure at which discharge is initiated. { 'ōp·ə·niŋ ¸presh·ər }

open-loop control system |CONT SYS| A control system in which the system outputs are controlled by system inputs only, and no account is taken of actual system output. { ¦ō·pən ¦lüp kən'trōl ¸sis·təm }

open plan |BUILD| Arrangement of the interior of a building without distinct barriers such as partitions. { 'ō·pən ¸plan }

open shop |IND ENG| A shop in which employment is not restricted to members of a labor union. { 'ō·pən ¦shäp }

open-side planer |DES ENG| A planer constructed with one upright or housing to support the crossrail and tools. { 'ō·pən ¸sīd 'plān·ər }

open-side tool block |DES ENG| A toolholder on a cutting machine consisting of a T-slot clamp, a C-shaped block, and two or more tool clamping screws. Also known as heavy-duty tool block. { 'ō·pən ¸sīd 'tül ¸bläk }

open system |THERMO| A system across whose boundaries both matter and energy may pass. { 'ō·pən 'sis·təm }

open-timbered roof |BUILD| A roof in which the supporting timbers are left uncovered, forming part of the ceiling. { 'ō·pən ¦tim·bərd 'rüf }

open traverse |ENG| A surveying traverse in which the last leg, because of error, does not terminate at the origin of the first leg. { 'ō·pən 'tra¸vərs }

open valley |BUILD| A valley formed at the intersection of two roof surfaces and lined with either metal or a mineral-surfaced roofing material; the lining is exposed at the intersection. { 'ō·pən 'val·ē }

open-web girder See lattice girder. { 'ō·pən ¦web 'gərd·ər }

open well |CIV ENG| **1.** A well whose diameter is great enough (1 meter or more) for a person to descend to the water level. **2.** An artificial pond filling a large excavation in the zone of saturation up to the water table. { 'ō·pən 'wel }

operating line |CHEM ENG| In the graphical solution of equilibrium processes (such as distillation absorption extraction), the actual liquid-vapor relationship of a key component, in contrast to a true equilibrium relationship. { 'äp·ə¸rād·iŋ ¸līn }

operating pressure |ENG| The system pressure at which a process is operating. { 'äp·ə¸rād·iŋ ¸presh·ər }

operating stress |MECH| The stress to which a structural unit is subjected in service. { 'äp·ə¸rād·iŋ ¸stres }

operating water level |MECH ENG| The water level in a boiler drum which is normally maintained above the lowest safe level. { 'äp·ə¸rād·iŋ 'wȯd·ər ¸lev·əl }

operation |IND ENG| A job, usually performed in one location, and consisting of one or more work elements. { ¸äp·ə'rā·shən }

operational |ENG| Of equipment such as aircraft or vehicles, being in such a state of repair as to be immediately usable. { ¸äp·ə'rā·shən·əl }

operational game See management game. { ¸äp·ə'rā·shən·əl 'gām }

operational maintenance |ENG| The cleaning, servicing, preservation, lubrication, inspection, and adjustment of equipment; it includes that minor replacement of parts not requiring high technical skill, internal alignment, or special locative training. { ¸äp·ə'rā·shən·əl 'mānt·ən·əns }

operation analysis |IND ENG| An analysis of all procedures concerned with the design or improvement of production, the purpose of the operation, inspection standards, materials used and the manner of handling them, the setup, tool equipment, and working conditions and methods. { ¸äp·ə'rā·shən ə'nal·ə·səs }

operation analysis chart |IND ENG| A form that lists all the essential factors influencing the effectiveness of an operation. { ¸äp·ə'rā·shən ə'nal·ə·səs ¸chärt }

operation breakdown See job breakdown. { ¸äp·ə'rā·shən 'brāk¸daȯn }

operation process chart |IND ENG| A graphic representation that gives an overall view of an entire process, including the points at which materials are introduced, the sequence of inspections, and all operations not involved in material handling. { ,äp·ə'rā·shən 'prä·səs ,chärt }

operations sequence [CONT SYS] The logical series of procedures that constitute the task for a robot. { ,äp·ə'rā·shənz ,sē·kwəns }

operator [ENG] A person whose duties include the operation, adjustment, and maintenance of a piece of equipment. { 'äp·ə,rād·ər }

operator process chart [IND ENG] A chart of the time relationship of the movements made by the body members of a workman performing an operation. { 'äp·ə,rād·ər 'präs·əs ,chärt }

operator productivity [IND ENG] The ratio of standard hours to actual hours for a given task. { 'äp·ə,rād·ər ,präd·ək'tiv·əd·ē }

operator training [IND ENG] The process used to prepare the employee to make his expected contribution to his employer, usually involving the teaching of specialized skills. { 'äp·ə,rād·ər ,trān·iŋ }

operator utilization [IND ENG] The ratio of working time to total clock time; a ratio of 1.00 (or 100) indicates full utilization of the operator's work time. { 'äp·ə,rād·ər ,yüd·əl·ə'zā·shən }

opisometer [ENG] An instrument for measuring the length of curved lines, such as those on a map; a wheel on the instrument is traced over the line. { ,äp·ə'säm·əd·ər }

opposed engine [MECH ENG] A reciprocating engine having the pistons on opposite sides of the crankshaft, with the piston strokes on each side working in a direction opposite to the direction of the strokes on the other side. { ə'pōzd 'en·jən }

optical amplifier [ENG] An optoelectronic amplifier in which the electric input signal is converted to light, amplified as light, then converted back to an electric signal for the output. { 'äp·tə·kəl 'am·plə,fī·ər }

optical bench [ENG] A rigid horizontal bar or track for holding optical devices in experiments; it allows device positions to be changed and adjusted easily. { 'äp·tə·kəl 'bench }

optical comparator [ENG] Any comparator in which movement of a measuring plunger tilts a small mirror which reflects light in an optical system. Also known as visual comparator. { 'äp·tə·kəl kəm'par·əd·ər }

optical coupler See optoisolator. { 'äp·tə·kəl 'kəp·lər }

optical coupling [ELECTR] Coupling between two circuits by means of a light beam or light pipe having transducers at opposite ends, to isolate the circuits electrically. { 'äp·tə·kəl 'kəp·liŋ }

optical direction and ranging See OPDAR. { 'äp·tə·kəl di'rek·shən ən 'rānj·iŋ }

optical-fiber sensor [ENG] An instrument in which the physical quantity to be measured is made to modulate the intensity, spectrum, phase, or polarization of light from a light-emitting diode or laser diode traveling through an optical fiber; the modulated light is detected by a photodiode. Also known as fiber-optic sensor. { 'äp·tə·kəl ¦fī·bər 'sen·sər }

optical fluid-flow measurement [ENG] Any method of measuring the varying densities of a fluid in motion, such as schlieren, interferometer, or shadowgraph, which depends on the fact that light passing through a flow field of varying density is retarded differently through the field, resulting in refraction of the rays, and in a relative phase shift among different rays. { 'äp·tə·kəl 'flü·əd ¦flō ,mezh·ər·mənt }

optical gage [ENG] A gage that measures an image of an object, and does not touch the object itself. { 'äp·tə·kəl 'gāj }

optical indicator [ENG] An instrument which makes a plot of pressure in the cylinder of an engine as a function of piston (or volume) displacement, making use of magnification by optical systems and photographic recording; for example, the small motion of a pressure diaphragm may be transmitted to a mirror to deflect a beam of light. { 'äp·tə·kəl 'in·də,kād·ər }

optical isolator See optoisolator. { 'äp·tə·kəl 'ī·sə,lād·ər }

optical lantern [ENG] A device for projecting positive transparent pictures from glass or film onto a reflecting screen; it consists of a concentrated source of light, a condenser system, a holder (or changer) for the slide, a projection lens, and (usually) a blower for cooling the slide. Also known as slide projector. { 'äp·tə·kəl 'lan·tərn }

optical lithography [ELECTR] Lithography in which an integrated circuit pattern is first created on a glass plate or mask and is then transferred to the resist by one of a number of optical techniques by using visible or ultraviolet light. { 'äp·tə·kəl li'thäg·rə·fē }

optically coupled isolator See optoisolator.

optically pumped magnetometer [ENG] A type of magnetometer that measures total magnetic field intensity by observation of the precession frequency of magnetic atoms, usually gaseous rubidium, cesium, or helium, which are magnetized by irradiation with circularly polarized light of a suitable wavelength. { 'äp·tə·klē ¦pəmpt ,mag·nə'täm·əd·ər }

optical mask [ELECTR] A thin sheet of metal or other substance containing an open pattern, used to suitably expose to light a photoresistive substance overlaid on a semiconductor or other surface to form an integrated circuit. { 'äp·tə·kəl 'mask }

optical microphone [ENG ACOUS] A microphone in which the motion of a membrane is detected using a light beam reflected from it, either with the aid of an interferometer or by detecting the deflection of the beam. { ¦äp·tə·kəl 'mī·krə,fōn }

optical proximity sensor [ENG] A device that uses the principle of triangulation of reflected infrared or visible light to measure small distances in a robotic system. { 'äp·tə·kəl präk'sim·əd·ē ,sen·sər }

optical pyrometer [ENG] An instrument which

determines the temperature of a very hot surface from its incandescent brightness; the image of the surface is focused in the plane of an electrically heated wire, and current through the wire is adjusted until the wire blends into the image of the surface. Also known as disappearing filament pyrometer. { 'äp·tə·kəl pī'räm·əd·ər }

optical radar See OPDAR. { 'äp·tə·kəl 'rā,där }

optical rangefinder |ENG| An optical instrument for measuring distance, usually from its position to a target point, by measuring the angle between rays of light from the target, which enter the rangefinder through the windows spaced apart, the distance between the windows being termed the baselength of the rangefinder; the two types are coincidence and stereoscopic. { 'äp·tə·kəl 'rānj,fīnd·ər }

optical recording |ENG| Production of a record by focusing on photographic paper a beam of light whose position on the paper depends on the quantity to be measured, as in a light-beam galvanometer. { 'äp·tə·kəl ri'kòrd·iŋ }

optical reflectometer |ENG| An instrument which measures on surfaces the reflectivity of electromagnetic radiation at wavelengths in or near the visible region. { 'äp·tə·kəl ,rē,flek'täm·əd·ər }

optical relay |ELECTR| An optoisolator in which the output device is a light-sensitive switch that provides the same on and off operations as the contacts of a relay. { 'äp·tə·kəl 'rē,lā }

optical square |ENG| A surveyor's hand instrument used for laying of right angles; employs two mirrors at a 45° angle. { 'äp·tə·kəl 'skwer }

optical tracking |ENG| The determination of spatial positions of distant airplanes, missiles, and artificial satellites as a function of time, or the recording of engineering events, by precise time-correlated observations with various types of telescopes or ballistic cameras. { 'äp·tə·kəl 'trak·iŋ }

optician |ENG| A maker of optical instruments or lenses. { äp'tish·ən }

optimal control theory |CONT SYS| An extension of the calculus of variations for dynamic systems with one independent variable, usually time, in which control (input) variables are determined to maximize (or minimize) some measure of the performance (output) of a system while satisfying specified constraints. { 'äp·tə·məl kən'trōl ,thē·ə·rē }

optimal feedback control |CONT SYS| A subfield of optimal control theory in which the control variables are determined as functions of the current state of the system. { 'äp·tə·məl 'fēd,bak kən,trōl }

optimal programming |CONT SYS| A subfield of optimal control theory in which the control variables are determined as functions of time for a specified initial state of the system. { 'äp·tə·məl 'prō,gram·iŋ }

optimal regulator problem See linear regulator problem. { 'äp·tə·məl 'reg·yə,lād·ər ,präb·ləm }

optimal smoother |CONT SYS| An optimal filter

algorithm which generates the best estimate of a dynamical variable at a certain time based on all available data, both past and future. { 'äp·tə·məl 'smūth·ər }

optimization |SYS ENG| 1. Broadly, the efforts and processes of making a decision, a design, or a system as perfect, effective, or functional as possible. 2. Narrowly, the specific methodology, techniques, and procedures used to decide on the one specific solution in a defined set of possible alternatives that will best satisfy a selected criterion. Also known as system optimization. { ,äp·tə·mə'zā·shən }

optimizing control function |CONT SYS| That level in the functional decomposition of a large-scale control system which determines the necessary relationships among the variables of the system to achieve an optimal, or suboptimal, performance based on a given approximate model of the plant and its environment. { 'äp·tə,mīz·iŋ kən'trōl ,faŋk·shən }

optimum cure |CHEM ENG| The degree of vulcanization at which maximum desired property is reached. { 'äp·tə·məm 'kyúr }

optocoupler See optoisolator. { 'äp·tō'kəp·lər }

optoelectronic amplifier |ENG| An amplifier in which the input and output signals and the method of amplification may be either electronic or optical. { ¦äp·tō·i,lek'trän·ik 'am·plə,fī·ər }

optoelectronic integration |ELECTR| A technology that combines optical components with electronic components such as transistors on a single wafer to obtain highly functional circuits. { ,äp·tō,i·lek¦trän·ik ,in·tə'grā·shən }

optoelectronic isolator See optoisolator. { ¦äp·tō·i,lek'trän·ik 'ī·sə,lād·ər }

optoelectronics |ELECTR| 1. The branch of electronics that deals with solid-state and other electronic devices for generating, modulating, transmitting, and sensing electromagnetic radiation in the ultraviolet, visible-light, and infrared portions of the spectrum. 2. See photonics. { ¦äp·tō·i,lek'trän·iks }

optoelectronic shutter |ENG| A shutter that uses a Kerr cell to modulate a beam of light. { ¦äp·tō·i,lek'trän·ik 'shəd·ər }

optoisolator |ELECTR| A coupling device in which a light-emitting diode, energized by the input signal, is optically coupled to a photodetector such as a light-sensitive output diode, transistor, or silicon controlled rectifier. Also known as optical coupler; optical isolator; optically coupled isolator; optocoupler; optoelectronic isolator; photocoupler; photoisolator. { ¦äp·tō'ī·sə,lād·ər }

optophone |ENG ACOUS| A device with a photoelectric cell to convert ordinary printed letters into a series of sounds; used by the blind. { 'äp·tə,fōn }

orange-peel bucket |DES ENG| A type of grab bucket that is multileaved and generally round in configuration. { 'är·inj ,pēl ,bək·ət }

orbital angular momentum |MECH| The angular momentum associated with the motion of a particle about an origin, equal to the cross product of the position vector with the linear momentum. Also known as orbital momentum. { 'ȯr·bəd·əl 'aŋ·gyə·lər mə'men·təm }

orbital moment *See* orbital angular momentum. { 'ȯr·bəd·əl 'mō·mənt }

orbital momentum *See* orbital angular momentum. { 'ȯr·bəd·əl mə'men·təm }

orbital plane [MECH] The plane which contains the orbit of a body or particle in a central force field; it passes through the center of force. { 'ȯr·bəd·əl 'plān }

orbital sander [MECH ENG] An electric sander that moves the abrasive in an elliptical pattern. { 'ȯr·bəd·əl 'san·dər }

OR circuit *See* OR gate. { 'ȯr ‚sər·kət }

order of phase transition [THERMO] A phase transition in which there is a latent heat and an abrupt change in properties, such as in density, is a first-order transition; if there is not such a change, the order of the transition is one greater than the lowest derivative of such properties with respect to temperature which has a discontinuity. { 'ȯrd·ər əv 'fāz tran‚zish·ən }

order point [IND ENG] The inventory level at which a replenishment order must be placed. { 'ȯrd·ər ‚pȯint }

order quantity [IND ENG] The number of pieces ordered to replenish the inventory. { 'ȯrd·ər ‚kwän·əd·ē }

ordinary gear train [MECH ENG] A gear train in which all axes remain stationary relative to the frame. { 'ȯrd·ən‚er·ē 'gir ‚trān }

ordnance [ENG] Military materiel, such as combat weapons of all kinds, with ammunition and equipment for their use, vehicles, and repair tools and machinery. { 'ȯrd·nəns }

organic bonded wheel [DES ENG] A grinding wheel in which organic bonds are used to hold the abrasive grains. { ȯr'gan·ik ‚bän·dəd 'wēl }

organizational reengineering [SYS ENG] The study, capture, and modification of the internal mechanisms or functionality of existing system-management processes and practices in an organization in order to reconstitute them in a new form and with new features, often to take advantage of newly emerged organizational competitiveness requirements, but without changing the inherent purpose of the organization itself. Also known as systems management reengineering. { ‚ȯr·gə·nə‚zā·shən·əl ‚rē‚en·jə'nir·iŋ }

organization chart [IND ENG] Graphic representation of the interrelationships within an organization, depicting lines of authority and responsibility and provisions for control. { ‚ȯr·gə·nə'zā·shən ‚chärt }

OR gate [ELECTR] A multiple-input gate circuit whose output is energized when any one or more of the inputs is in a prescribed state; performs the function of the logical inclusive-or; used in digital computers. Also known as OR circuit. { 'ȯr ‚gāt }

orient [ENG] **1.** To place or set a map so that the map symbols are parallel with their corresponding ground features. **2.** To turn a transit so that the direction of the 0° line of its horizontal

circle is parallel to the direction it had in the preceding or initial setup, or parallel to a standard reference line. { 'ȯr·ē·ənt }

orientation [ENG] Establishment of the correct relationship in direction with reference to the points of the compass. { ‚ȯr·ē·ən'tā·shən }

orientation vector [MECH ENG] A vector whose direction indicates the orientation of a robot gripper. { ‚ȯr·ē·ən'tā·shən ‚vek·tər }

oriented core [ENG] A core that can be positioned on the surface in the same way that it was arranged in the borehole before extraction. { 'ȯr·ē‚ent·əd 'kȯr }

orifice meter [ENG] An instrument that measures fluid flow by recording differential pressure across a restriction placed in the flow stream and the static or actual pressure acting on the system. { 'ȯr·ə·fəs ‚med·ər }

orifice mixer [MECH ENG] Arrangement in which two or more liquids are pumped through an orifice constriction to cause turbulence and consequent mixing action. { 'ȯr·ə·fəs ‚mik·sər }

orifice plate [DES ENG] A disk, with a hole, placed in a pipeline to measure flow. { 'ȯr·ə·fəs ‚plāt }

original duration [IND ENG] The initial estimate of length of time required to complete a given activity. { ə‚rij·ən·əl də'rā·shən }

O ring [DES ENG] A flat ring made from synthetic rubber, used as an airtight seal or a seal against high pressures. { 'ō ‚riŋ }

orograph [ENG] A machine that records both distance and elevations as it is pushed across land surfaces; used in making topographic maps. { 'ȯr·ə‚graf }

orometer [ENG] A barometer with a scale that indicates elevation above sea level. { ȯ'räm·əd·ər }

orthometric correction [ENG] A systematic correction that must be applied to a measured difference in elevation since level surfaces at varying elevations are not absolutely parallel. { ‚ȯr·thə‚me·trik kə'rek·shən }

orthometric height [ENG] The distance above sea level measured along a plumb line. { ‚ȯr·thə‚me·trik 'hīt }

orthotropic [MECH] Having elastic properties such as those of timber, that is, with considerable variations of strength in two or more directions perpendicular to one another. { ‚ȯr·thə‚träp·ik }

orthotropic deck [CIV ENG] A bridge deck constructed typically of flat steel plate and longitudinal and transverse ribs; functions in carrying traffic and acting as top flanges of floor beams. { ‚ȯr·thə‚träp·ik 'dek }

oscillating conveyor [MECH ENG] A conveyor on which pulverized solids are moved by a pan or trough bed attached to a vibrator or oscillating mechanism. Also known as vibrating conveyor. { 'äs·ə‚lād·iŋ kən'vā·ər }

oscillating granulator [MECH ENG] Solids size-reducer in which particles are broken by a set of oscillating bars arranged in cylindrical form over a screen of suitable mesh.   { 'äs·ə‚lād·iŋ 'gran· yə‚lād·ər }

oscillating screen |MECH ENG| Solids separator in which the sifting screen oscillates at 300 to 400 revolutions per minute in a plane parallel to the screen.   { 'äs·ə‚lād·iŋ 'skrēn }

oscillation See cycling.   { ‚äs·ə'lā·shən }

oscillator |ELECTR| **1.** An electronic circuit that converts energy from a direct-current source to a periodically varying electric output.   **2.** The stage of a superheterodyne receiver that generates a radio-frequency signal of the correct frequency to mix with the incoming signal and produce the intermediate-frequency value of the receiver.   **3.** The stage of a transmitter that generates the carrier frequency of the station or some fraction of the carrier frequency.   { 'äs· ə‚lād·ər }

oscillatory circuit |ELEC| Circuit containing inductance or capacitance, or both, and resistance, connected so that a voltage impulse will produce an output current which periodically reverses or oscillates.   { 'äs·ə·lə‚tór·ē 'sər·kət }

oscillistor |ELECTR| A bar of semiconductor material, such as germanium, that will oscillate much like a quartz crystal when it is placed in a magnetic field and is carrying direct current that flows parallel to the magnetic field.   { 'äs· ə'lis·tər }

oscillogram |ENG| The permanent record produced by an oscillograph, or a photograph of the trace produced by an oscilloscope.   { ə'sil· ə‚gram }

oscillograph |ENG| A measurement device for determining waveform by recording the instantaneous values of a quantity such as voltage as a function of time.   { ə'sil·ə‚graf }

Ostwald process |CHEM ENG| An industrial preparation of nitric acid by the oxidation of ammonia; the oxidation takes place in successive stages to nitric oxide, nitrogen dioxide, and nitric acid; a catalyst of platinum gauze is used and high temperatures are needed.   { 'óst‚vält ‚prä·səs }

Ostwald's adsorption isotherm |THERMO| An equation stating that at a constant temperature the weight of material adsorbed on an adsorbent dispersed through a gas or solution, per unit weight of adsorbent, is proportional to the concentration of the adsorbent raised to some constant power.   { 'óst‚välts ad'sórp·shən 'ī·sə ‚thərm }

Ostwald viscometer |ENG| A viscometer in which liquid is drawn into the higher of two glass bulbs joined by a length of capillary tubing, and the time for its meniscus to fall between calibration marks above and below the upper bulb is compared with that for a liquid of known viscosity.   { 'óst‚vält vi'skäm·əd·ər }

OTEC See ocean thermal energy conversion.   { 'ō‚tek }

otter See paravane.   { 'äd·ər }

Otto cycle |THERMO| A thermodynamic cycle for the conversion of heat into work, consisting of two isentropic phases interspersed between two constant-volume phases.   Also known as spark-ignition combustion cycle.   { 'äd·ō ‚sī· kəl }

Otto engine |MECH ENG| An internal combustion engine that operates on the Otto cycle, where the phases of suction, compression, combustion, expansion, and exhaust occur sequentially in a four-stroke-cycle or two-stroke-cycle reciprocating mechanism.   { 'äd·ō ‚en·jən }

Otto-Lardillon method |MECH| A method of computing trajectories of missiles with low velocities (so that drag is proportional to the velocity squared) and quadrant angles of departure that may be high, in which exact solutions of the equations of motion are arrived at by numerical integration and are then tabulated.   { 'äd·ō ‚lär· dē'yón ‚meth·əd }

ounce |MECH| **1.** A unit of mass in avoirdupois measure equal to 1/16 pound or to approximately 0.0283495 kilogram.   Abbreviated oz.   **2.** A unit of mass in either troy or apothecaries' measure equal to 480 grains or exactly 0.0311034768 kilogram.   Also known as apothecaries' ounce or troy ounce (abbreviations are oz ap and oz t in the United States, and oz apoth and oz tr in the United Kingdom).   { 'aúns }

ouncedal |MECH| A unit of force equal to the force which will impart an acceleration of 1 foot per second per second to a mass of 1 ounce; equal to 0.0086409346485 newton.   { 'aún· sə‚dal }

outfall |CIV ENG| The point at which a sewer or drainage channel discharges to a body of water.   { 'aút‚fól }

outflow |CHEM ENG| Flow of fluid product out of a process facility.   { 'aút‚flō }

outgassing |ENG| The release of adsorbed or occluded gases or water vapor, usually by heating, as from a vacuum tube or other vacuum system.   { 'aút‚gas·iŋ }

outlet ventilator See louver.   { 'aút‚let 'vent· əl‚ād·ər }

output |ELECTR| **1.** The current, voltage, power, driving force, or information which a circuit or device delivers.   **2.** Terminals or other places where a circuit or device can deliver current, voltage, power, driving force, or information.   { 'aút‚pút }

output indicator |ENG| A meter or other device that is connected to a radio receiver to indicate variations in output signal strength for alignment and other purposes, without indicating the exact value of output.   { 'aút‚pút ‚in·də‚kād·ər }

output-limited |ENG| Restricted by the need to await completion of an output operation, as in process control or data processing.   { 'aút‚pút ‚lim·əd·əd }

output meter |ENG| An alternating-current voltmeter connected to the output of a receiver or amplifier to measure output signal strength in volume units or decibels.   { 'aút‚pút ‚mēd·ər }

output-meter adapter |ENG| Device that can be slipped over the plate prong of the output tube of a radio receiver to provide a conventional terminal to which an output meter can be connected during alignment. { 'aut‚put ‚mēd·ər ə‚dap·tər }

output power [ELEC] Power delivered by a system or transducer to its load. { 'aut‚put ‚pau·ər }

output shaft [MECH ENG] The shaft that transfers motion from the prime mover to the driven machines. { 'aut‚put ‚shaft }

output standard *See* standard time. { 'aut‚put ‚stan·dərd }

outrigger [ENG] A steel beam or lattice girder extending from a crane to provide stability by widening the base. { 'aut‚rig·ər }

outside caliper [DES ENG] A caliper having two curved legs which point toward each other; used for measuring outside dimensions of a workpiece. { 'aut‚sīd 'kal·ə·pər }

outside diameter [DES ENG] The outer diameter of a pipe, including the wall thickness; usually measured with calipers. Abbreviated OD. { 'aut‚sīd dī'am·əd·ər }

oven [ENG] A heated enclosure for baking, heating, or drying. { 'əv·ən }

overall plate efficiency [CHEM ENG] For a specified liquid-mixture separation in a fractionation (or distillation) tower, the ratio of actual to theoretical plates (or trays) required. { ¦ō·vər¦ól 'plāt i‚fish·ən·sē }

overarm [MECH ENG] One of the adjustable supports for the end of a milling-cutter arbor farthest from the machine spindle. { 'ō·vər‚ärm }

overbreak [CIV ENG] Rock excavated in excess of the neat lines of a tunnel or cutting. Also known as backbreak. { 'ō·vər‚brāk }

overcoating [ENG] Extruding a plastic web beyond the edge of the substrate web in extrusion coating. { 'ō·vər‚kōd·iŋ }

overcuring [CHEM ENG] A condition resulting from vulcanizing longer than necessary to achieve full development of physical strength; causes softness or brittleness and impaired age-resisting quality of the material. { 'ō·vər‚kyúr·iŋ }

overcurrent protection *See* overload protection. { ¦ō·vər¦kə·rənt prə'tek·shən }

overdrilling [ENG] The act or process of drilling a run or length of borehole greater than the core-capacity length of the core barrel, resulting in loss of the core. { ¦ō·vər'dril·iŋ }

overdrive [MECH ENG] An automobile engine device that lowers the gear ratio, thereby reducing fuel consumption. { 'ō·vər‚drīv }

overfall dam *See* overflow dam. { 'ō·vər‚fól ‚dam }

overfire draft [MECH ENG] The air pressure in a boiler furnace during occurrence of the main flame. { ¦ō·vər¦fīr 'draft }

overflow [CIV ENG] Any device or structure that conducts excess water or sewage from a conduit or container. { 'ō·vər‚flō }

overflow capacity [ENG] Capacity of a container measured to its top, or to the point of overflow. { 'ō·vər‚flō kə‚pas·əd·ē }

overflow channel [CIV ENG] An artificial waterway for conducting water away from an overflowing structure such as a reservoir or canal. { 'ō·vər‚flō ‚chan·əl }

overflow dam [CIV ENG] A dam built with a crest to allow the overflow of water. Also known as overfall dam; spillway dam. { 'ō·vər‚flō ‚dam }

overflow groove [ENG] Small groove on a plastics mold that allows material to flow freely, to prevent weld lines and low density in the finished product and to dispose of excess material. { 'ō·vər‚flō ‚grüv }

overflow pipe [ENG] Open pipe protruding above the surface of a liquid in a container, such as a distillation or absorption column or a toilet tank, to control the height of the liquid; excess liquid enters the pipe's open end and drains away. { 'ō·vər‚flō ‚pīp }

overgear [MECH ENG] A gear train in which the angular velocity ratio of the driven shaft to driving shaft is greater than unity, as when the propelling shaft of an automobile revolves faster than the engine shaft. { 'ō·vər‚gir }

overhang [BUILD] The distance measured horizontally that a roof projects beyond a wall. { 'ō·vər‚haŋ }

overhaul [ENG] A maintenance procedure for machinery involving disassembly, the inspecting, refinishing, adjusting, and replacing of parts, and reassembly and testing. { 'ō·vər‚hól }

overhead [CHEM ENG] Pertaining to fluid (gas or liquid) effluent from the top of a process vessel, such as a distillation column. *See* fixed cost. { 'ō·vər‚hed }

overhead camshaft [MECH ENG] A camshaft mounted above the cylinder head. { 'ō·vər‚hed 'kam‚shaft }

overhead cost *See* fixed cost. { 'ō·vər‚hed ‚kóst }

overhead shovel [MECH ENG] A tractor which digs with a shovel at its front end, swings the shovel rearward overhead, and dumps the shovel at its rear end. { 'ō·vər‚hed 'shəv·əl }

overhead traveling crane [MECH ENG] A hoisting machine with a bridgelike structure moved on wheels along overhead trackage which is usually fixed to the building structure. { 'ō·vər‚hed ¦trav·ə·liŋ 'krān }

overhead-valve engine [MECH ENG] A four-stroke-cycle internal combustion engine having its valves located in the cylinder head, operated by pushrods that actuate rocker arms. Abbreviated OHV engine. Also known as valve-in-head engine. { 'ō·vər‚hed ¦valv 'en·jən }

overlap radar [ENG] Radar located in one sector whose area of useful radar coverage includes a portion of another sector. { 'ō·vər‚lap 'rā‚där }

overlay [CIV ENG] A repair topping of asphalt or concrete placed on a worn roadway. [ENG] **1.** Nonwoven fibrous mat (glass or other fiber) used as the top layer in a cloth or mat lay-up to give smooth finish to plastic products or to minimize the fibrous pattern on the surface. Also known as surfacing mat. **2.** An ornamental covering, as of wood or metal. { 'ō·vər‚lā }

overload |CIV ENG| A load on a structure that is greater than that for which the structure was designed. |ELECTR| A load greater than that which a device is designed to handle; may cause overheating of power-handling components and distortion in signal circuits. { 'ō·vər,lōd }

overload capacity |ELEC| Current, voltage, or power level beyond which permanent damage occurs to the device considered. { 'ō·vər,lōd kə,pas· əd·ē }

overload level |ELEC| Level above which operation ceases to be satisfactory as a result of signal distortion, overheating, damage, and so forth. { 'ō·vər,lōd ,lev·əl }

overload protection |ELEC| Effect of a device operative on excessive current, but not necessarily on short circuit, to cause and maintain the interruption of current flow to the device governed. |MECH ENG| A safeguard against the application of excessive force against the wrist socket or end effector of a robot. Also known as overcurrent protection. { 'ō·vər,lōd prə,tek· shən }

overoccult |ENG| The action of a coronagraph that occults a region whose diameter is significantly greater than that of the photosphere and thereby cuts off the inner corona from observation, as may be necessary for a coronagraph aboard a spacecraft due to limitations on spacecraft control. { ,ō·vər·ə'kəlt }

overpass |CIV ENG| **1.** A grade separation in which traffic at the higher level is raised, and traffic at the lower level moves at approximately its original level. **2.** The upper level at such a grade separation. { 'ō·vər,pas }

overpotential See overvoltage. { ¦ō·vər·pə'ten· chəl }

override |CONT SYS| To cancel the influence of an automatic control by means of a manual control. { 'ō·və,rīd }

overriding process control |CONT SYS| Process control in which any one of several controllers associated with one control valve can be made to override another in accordance with a priority requirement of the process. { 'ō·və,rīd·iŋ 'prä· səs kən,trōl }

overrun |CIV ENG| A cleared area extending beyond the end of a runway. { 'ō·və,rən }

overrunning clutch |MECH ENG| A clutch that allows the driven shaft to turn freely only under certain conditions; for example, a clutch in an engine starter that allows the crank to turn freely when the engine attempts to run. { 'ō·və,rən· iŋ 'kləch }

oversail |BUILD| To project beyond the general face of a structure. { 'ō·vər,sāl }

overshoot |ENG| **1.** An initial transient response to a unidirectional change in input which exceeds the steady-state response. **2.** The maximum amount by which this transient response exceeds the steady-state response. { 'ō· vər,shüt }

overshot |ENG| **1.** A fishing tool for recovering lost drill pipe or casing. **2.** See bullet. { 'ō· vər,shät }

overshot wheel |MECH ENG| A horizontal-shaft waterwheel with buckets around the circumference; the weight of water pouring into the buckets from the top rotates the wheel. { 'ō· vər,shät ,wēl }

oversite concrete |BUILD| A layer of concrete that is installed below a slab or other type of floor surface. { ¦ō·vər,sīt 'kän,krēt }

overspeed governor |MECH ENG| A governor that stops the prime mover when speed is excessive. { 'ō·vər,spēd ,gəv·ə·nər }

overspin |MECH| In a spin-stabilized projectile, the overstability that results when the rate of spin is too great for the particular design of projectile, so that its nose does not turn downward as it passes the summit of the trajectory and follows the descending branch. Also known as overstabilization. { 'ō·vər,spin }

oversquare engine |MECH ENG| An engine with bore diameter greater than the stroke length. { 'ō·vər,skwer 'en·jən }

overstabilization See overspin. { ¦ō·vər,stā·bə· lə'zā·shən }

oversteer |MECH ENG| The tendency of an automotive vehicle to steer into a turn to a sharper degree than was intended by the driver; sometimes causes the vehicle's rear end to swing out. { 'ō·vər,stir }

overstressing |ENG| Cyclically stressing a material at a level higher than that used at the end of a fatigue test. { ¦ō·vər¦stres·iŋ }

overtone |MECH| One of the normal modes of vibration of a vibrating system whose frequency is greater than that of the fundamental mode. { 'ō·vər,tōn }

overtopping |CIV ENG| The flow of water over a dam or embankment. { 'ō·vər,täp·iŋ }

overturning |CIV ENG| Failure of a retaining wall caused by the soil pressure overcoming the stability of the structure. { ¦ō·vər¦tərn·iŋ }

overvoltage |ELEC| A voltage greater than that at which a device or circuit is designed to operate. Also known as overpotential. |ELECTR| The amount by which the applied voltage exceeds the Geiger threshold in a radiation counter tube. { ¦ō·vər¦vōl·tij }

overwind |ENG| To wind a spring, rope, or cable too tightly or too far. { ¦ō·vər¦wīnd }

Ovshinsky effect |ELECTR| The characteristic of a special thin-film solid-state switch that responds identically to both positive and negative polarities so that current can be made to flow in both directions equally. { ōv'shin·skē i,fekt }

oxidation pond |CIV ENG| A shallow lagoon or basin in which wastewater is purified by sedimentation and aerobic and anaerobic treatment. { ,äk·sə'dā·shən ,pänd }

oxide isolation |ELECTR| Isolation of the elements of an integrated circuit by forming a layer of silicon oxide around each element. { 'äk,sīd ,ī·sə'lā·shən }

oxide passivation |ELECTR| Passivation of a semiconductor surface by producing a layer of an insulating oxide on the surface. { 'äk,sīd ,pas·ə·vā·shən }

oxo process |CHEM ENG| Catalytic process for production of alcohols, aldehydes, and other oxygenated organic compounds by reaction of olefin vapors with carbon monoxide and hydrogen. { 'äk·sō ,prä·səs }

oxyacetylene cutting |ENG| The flame cutting of ferrous metals in which the preheating of the metal is accomplished with a flame produced by an oxyacetylene torch. Also known as acetylene cutting. { ¦äk·sē·ə'sed·əl,ēn ¦kəd·iŋ }

oxyacetylene torch |ENG| A torch that mixes acetylene and oxygen to produce a hot flame for the welding or cutting of metal. Also known as acetylene torch. { ¦äk·sē·ə'sed·əl,ēn ¦tȯrch }

oxyamination See ammoxidation. { ¦äk·sē,am·ə'nā·shən }

oxygen bomb calorimeter |ENG| Device to measure heat of combustion; the sample is burned with oxygen in a closed vessel, and the temperature rise is noted. { 'äk·sə·jən 'bäm ,kal·ə'rim·əd·ər }

oxygen cutting |ENG| Any of several types of cutting processes in which metal is removed with or without a flux by a chemical reaction of the base metal with oxygen at high temperatures. { 'äk·sə·jən ,kəd·iŋ }

oxygen-kerosine burner |ENG| Liquid-fuel device using a mixture of oxygen and vaporized or atomized kerosine for combustion. { 'äk·sə·jən 'ker·ə,sēn ,bərn·ər }

oxygen mask |ENG| A mask that covers the nose and mouth and is used to administer oxygen. { 'äk·sə·jən ,mask }

oxygen point |THERMO| The temperature at which liquid oxygen and its vapor are in equilibrium, that is, the boiling point of oxygen, at standard atmospheric pressure; it is taken as a fixed point on the International Practical Temperature Scale of 1968, at $-182.962°C$. { 'äk·sə·jən ,pȯint }

oxyl process |CHEM ENG| Modified Fischer-Tropsch process used to make alcohols, other oxygenated compounds, paraffins, and olefin hydrocarbons from carbon monoxide and hydrogen. { 'äk·səl ,prä·səs }

oz See ounce.

oz ap See ounce.

oz apoth See ounce.

ozone generator |ENG| Apparatus that converts oxygen, O_2, into ozone, O_3, by subjecting the oxygen to an electric-brush discharge. Also known as ozonizer. { 'ō,zōn ,jen·ə,rād·ər }

ozonizer See ozone generator. { 'ō,zō,nīz·ər }

oz t See ounce.

oz tr See ounce.

P

Pa *See* pascal.

pace rating *See* effort rating. { 'pās ,rād·iŋ }

pachimeter |ENG| An instrument for measuring the limit beyond which shear of a solid ceases to be elastic. { pə'kim·əd·ər }

Pachuca tank |CHEM ENG| Air-agitated, solid-liquid mixing vessel in which the air is injected into the bottom of a center draft tube; air and solids rise through the tube, with solids exiting the top of the tube and falling through the bulk of the liquid. { pə'chü·kə ,taŋk }

pachymeter |ENG| An instrument used to measure the thickness of a material, for example, a sheet of paper. { pə'kim·əd·ər }

pack |IND ENG| To provide protection for an article or group of articles against physical damage during shipment; packing is accomplished by placing articles in a shipping container, and blocking, bracing, and cushioning them when necessary, or by strapping the articles or containers on a pallet or skid. { pak }

packaged circuit *See* recap. { 'pak·ijd |sər·kət }

package freight |IND ENG| Freight shipped in lots insufficient to fill a complete car; billed by the unit instead of by the carload. { 'pak·ij ,frāt }

packaging |ELEC| The process of physically locating, connecting, and protecting devices or components. { 'pak·ə·jiŋ }

packaging density |ELECTR| The number of components per unit volume in a working system or subsystem. { 'pak·ə·jiŋ ,den·səd·ē }

packed bed |CHEM ENG| A fixed layer of small particles or objects arranged in a vessel to promote intimate contact between gases, vapors, liquids, solids, or various combinations thereof; used in catalysis, ion exchange, sand filtration, distillation, absorption, and mixing. { 'pakt 'bed }

packed tower |CHEM ENG| A fractionating or absorber tower filled with small objects (packing) to bring about intimate contact between rising fluid (vapor or liquid) and falling liquid. { 'pakt 'tau·ər }

packed tube |CHEM ENG| A pipe or tube filled with high-heat-capacity granular material; used to heat gases when tubes are externally heated. { 'pakt 'tüb }

packer |ENG| A device that is inserted into a

hole being grouted to prevent return of the grout around the injection pipe. { 'pak·ər }

packing |ENG| *See* stuffing. |ENG ACOUS| Excessive crowding of carbon particles in a carbon microphone, produced by excessive pressure or by fusion particles due to excessive current, and causing lowered resistance and sensitivity. { 'pak·iŋ }

packing density |ELECTR| The number of devices or gates per unit area of an integrated circuit. { 'pak·iŋ ,den·səd·ē }

packing ring *See* piston ring. { 'pak·iŋ ,riŋ }

pad |ELECTR| **1.** An arrangement of fixed resistors used to reduce the strength of a radio-frequency or audio-frequency signal by a desired fixed amount without introducing appreciable distortion. Also known as fixed attenuator. **2.** *See* terminal area. |ENG| **1.** A layer of material used as a cushion or for protection. **2.** A projection of excess metal on a casting forging, or welded part. **3.** An area within an airstrip or airway that is used for warming up the motors of an airplane before takeoff. **4.** A block of stone or masonry set on a wall to distribute a load that is concentrated at that portion of the wall. Also known as padstone. **5.** That portion of an airstrip or airway from which an airplane leaves the ground on takeoff or first touches the ground on landing. **6.** *See* helipad. { pad }

paddle |DES ENG| Any of various implements consisting of a shaft with a broad, flat blade or bladelike part at one or both ends. { 'pad·əl }

paddle wheel |MECH ENG| **1.** A device used to propel shallow-draft vessels, consisting of a wheel with paddles or floats on its circumference, the wheel rotating in a plane parallel to the ship's length. **2.** A wheel with paddles used to move leather in a processing vat. { 'pad·əl ,wēl }

padlock |DES ENG| An unmounted lock with a shackle that can be opened and closed; the shackle is usually passed through an eye, then closed to secure a hasp. { 'pad,läk }

pail |DES ENG| A cylindrical or slightly tapered container. { pāl }

pair |ELEC| Two like conductors employed to form an electric circuit. |MECH ENG| Two parts in a kinematic mechanism that mutually constrain relative motion; for example, a sliding pair composed of a piston and cylinder. { per }

pairing element |MECH ENG| Either of two machine parts connected to permit motion. { 'per·iŋ ,el·ə·mənt }

palladium barrier leak detector |ENG| A type of leak detector in which hydrogen is diffused through a barrier of hot palladium into an evacuated vacuum gage. { pə'lād·ē·əm 'bar·ē·ər 'lēk di,tek·tər }

pallet |BUILD| A flat piece of wood laid in a wall to which woodwork may be securely fastened. |ENG| **1.** A lever that regulates or drives a ratchet wheel. **2.** A hinged valve on a pipe organ. **3.** A tray or platform used in conjunction with a fork lift for lifting and moving materials. |MECH ENG| One of the disks or pistons in a chain pump. { 'pal·ət }

palletize |IND ENG| To package material for convenient handling on a pallet or lift truck. { 'pal·ə,tīz }

pall ring |CHEM ENG| A specially shaped steel ring used as packing for distillation columns. { 'pól ,riŋ }

palpable coordinate |MECH| A generalized coordinate that appears explicitly in the Lagrangian of a system. { 'pal·pə·bəl kō'órd·ən·ət }

pan bolt |DES ENG| A bolt with a head resembling an upside-down pan. { 'pan ,bólt }

pancake auger |DES ENG| An auger having one spiral web, 12 to 15 inches (30 to 38 centimeters) in diameter, attached to the bottom end of a slender central shaft; used as removable deadman to which a drill rig or guy line is anchored. { 'pan,kāk ‚óg·ər }

pancake engine |MECH ENG| A compact engine with cylinders arranged radially. { 'pan,kāk ¦en·jən }

pan conveyor |MECH ENG| A conveyor consisting of a series of pans. { 'pan kən,vā·ər }

pan crusher |MECH ENG| Solids-reduction device in which one or more grinding wheels or mullers revolve in a pan containing the material to be pulverized. { 'pan ,krəsh·ər }

pane |BUILD| A sheet of glass in a window or door. |DES ENG| One of the sides on a nut or on the head of a bolt. { pān }

panel |CIV ENG| **1.** One of the divisions of a lattice girder. **2.** A sheet of material held in a frame. **3.** A distinct, usually rectangular, raised or sunken part of a construction surface or a material. |DES ENG| See frog. |ENG| A metallic or nonmetallic sheet on which operating controls and dials of an electronic unit or other equipment are mounted. { 'pan·əl }

panel board |ELEC| See control board. |ENG| A drawing board with an adjustable outer frame that is forced over the drawing paper to hold and strain it. { 'pan·əl ,bórd }

panel coil See plate coil. { 'pan·əl ,kóil }

panel cooling |CIV ENG| A system in which the heat-absorbing units are in the ceiling, floor, or wall panels of the space which is to be cooled. { 'pan·əl ,kül·iŋ }

panel heating |CIV ENG| A system in which the heat-emitting units are in the ceiling, floor, or

wall panels of the space which is to be heated. { 'pan·əl ,hēd·iŋ }

panel length |CIV ENG| The distance between adjacent joints on a truss, measured along the upper or lower chord. { 'pan·əl ,leŋkth }

panel point |CIV ENG| The point in a framed structure where a vertical or diagonal member and a chord intersect. { 'pan·əl ,póint }

panel system |BUILD| A wall composed of factory-assembled units connected to the building frame and to each other by means of anchors. { 'pan·əl ,sis·təm }

panel wall |BUILD| A nonbearing partition between columns or piers. { 'pan·əl ,wól }

pan head |DES ENG| The head of a screw or rivet in the shape of a truncated cone. { 'pan ,hed }

panic exit device |ENG| A locking device installed on an exit door to release the latch when the crash bar is pushed. Also known as fire-exit bolt; panic hardware. { ¦pan·ik ¦eg·zit di,vīs }

panic hardware See panic exit device. { 'pan·ik ,härd,wer }

pannier See gabion. { 'pan·yər }

panoramic radar |ENG| Nonscanning radar which transmits signals over a wide beam in the direction of interest. { ¦pan·ə¦ram·ik 'rā,där }

pan tank See rundown tank. { 'pan ,taŋk }

pantograph |ENG| A device that sits on the top of an electric locomotive or cars in an electric train and picks up electricity from overhead wires to run the train. { 'pan·tə,graf }

pantography |ENG| System for transmitting and automatically recording radar data from an indicator to a remote point. { pan'täg·rə·fē }

pantometer |ENG| An instrument that measures all the angles necessary for determining distances and elevations. { pan'täm·əd·ər }

paper cutter |DES ENG| A hand-operated device to cut and trim paper, consisting of a cutting blade bolted at one end to a ruled board; when the blade is drawn flush with the board, which has a metal strip at the cutting edge, a shearing action takes place which cuts the paper cleanly and evenly. { 'pā·pər ,kəd·ər }

paper machine |MECH ENG| A synchronized series of mechanical devices for transforming a dilute suspension of cellulose fibers into a dry sheet of paper. { 'pā·pər mə,shēn }

paper mill |IND ENG| A building or complex of buildings housing paper machines. { 'pā·pər ,mil }

parabolic microphone |ENG ACOUS| A microphone used at the focal point of a parabolic sound reflector to give improved sensitivity and directivity, as required for picking up a band marching down a football field. { ¦par·ə¦bäl·ik 'mī·krə,fōn }

paraboloid |ENG| A reflecting surface which is a paraboloid of revolution and is used as a reflector for sound waves and microwave radiation. { pə'rab·ə,lóid }

parabomb |ENG| An equipment container with a parachute which is capable of opening automatically after a delayed drop. { 'par·ə,bäm }

paracentric [DES ENG] Pertaining to a key and keyway with longitudinal ribs and grooves that project beyond the center, as used in pin-tumbler cylinder locks to deter lockpicking. { 'par·ə¦sen·trik }

parachute flare [ENG] Pyrotechnic device attached to a parachute and designed to provide intense illumination for a short period; it may be discharged from aircraft or from the surface. { 'par·ə¦shüt ,fler }

parachute radiosonde See dropsonde. { 'par·ə¦shüt 'räd·ē¦sänd }

parachute weather buoy [ENG] A general-purpose automatic weather station which can be air-dropped; it is 10 feet (3 meters) long and 22 inches (56 centimeters) in diameter, and is designed to operate for 2 months on a 6-hourly schedule, transmitting station identification, wind speed, wind direction, barometric pressure, air temperature, and sea-water temperature. { 'par·ə¦shüt 'weth·ər ¦bȯi }

paracrate [ENG] Rigid equipment container for dropping equipment from an airplane by parachute. { 'par·ə¦krāt }

paraffin press [ENG] A filter press used during petroleum refining for the separation of paraffin oil and crystallizable paraffin wax from distillates. { 'par·ə·fən ¦pres }

parallel [ELEC] Connected to the same pair of terminals. Also known as multiple; shunt. { 'par·ə¦lel }

parallel axis theorem [MECH] A theorem which states that the moment of inertia of a body about any given axis is the moment of inertia about a parallel axis through the center of mass, plus the moment of inertia that the body would have about the given axis if all the mass of the body were located at the center of mass. Also known as Steiner's theorem. { 'par·ə¦lel ¦ak·səs ,thir·əm }

parallel baffle muffler [DES ENG] A muffler constructed of a series of ducts placed side by side in which the duct cross section is a narrow but long rectangle. { 'par·ə¦lel ¦baf·əl 'məf·lər }

parallel circuit [ELEC] An electric circuit in which the elements, branches (having elements in series), or components are connected between two points, with one of the two ends of each component connected to each point. { 'par·ə¦lel 'sər·kət }

parallel compensation See feedback compensation. { 'par·ə¦lel ¦käm·pən¦sā·shən }

parallel cut [ENG] A group of parallel holes, not all charged with explosive, to create the initial cavity to which the loaded holes break in blasting a development round. Also known as burn cut. { 'par·ə¦lel 'kət }

parallel drum [DES ENG] A cylindrical form of drum on which the haulage or winding rope is coiled. { 'par·ə¦lel 'drəm }

parallel firing [ENG] A method of connecting together a number of detonators which are to be fired electrically in one blast. { 'par·ə¦lel 'fīr·iŋ }

parallel flow [ELEC] Also known as loop flow. **1.** The flow of electric current from one point to another in an electric network over multiple paths, in accordance with Kirchhoff's laws. **2.** In particular, the flow of electric current through electric power systems over paths other than the contractual path. { 'par·ə¦lel 'flō }

parallel gripper [CONT SYS] A robot end effector made up of two jawlike components that grasp objects. { 'par·ə¦lel 'grip·ər }

parallel linkage [MECH ENG] An automotive steering system that has a short idler arm mounted parallel to the pitman arm. { 'par·ə¦lel 'liŋ·kij }

parallel-plate reactor [ENG] A type of plasma reactor in which a process gas is introduced into the space between two closely spaced parallel plane electrodes, and a plasma, generated by a radio-frequency excitation applied to the electrodes, acts directly on substrates placed on either electrode. { 'par·ə¦lel 'plāt rē¦ak·tər }

parallel reliability [SYS ENG] Property of a system composed of functionally parallel elements in such a way that if one of the elements fails, the parallel units will continue to carry out the system function. { 'par·ə¦lel ri'lī·ə,bil·əd·ē }

parallels [ENG] **1.** Spacers located between steam plate and press platen of the mold to prevent bending of the middle section. **2.** Spacers or pressure pods located between steam plates of a mold to regulate height and prevent crushing of mold parts. { 'par·ə¦lelz }

parallel series [ELEC] Circuit in which two or more parts are connected together in parallel to form parallel circuits, and in which these circuits are then connected together in series so that both methods of connection appear. [ENG] See multiple series. { 'par·ə¦lel 'sir·ēz }

parallel shot [ENG] In seismic prospecting, a test shot which is made with all the amplifiers connected in parallel and activated by a single geophone so that lead, lag, polarity, and phasing in the amplifier-to-oscillograph circuits can be checked. { 'par·ə¦lel 'shät }

parameter identification [SYS ENG] The problem of estimating the values of the parameters that govern a dynamical system from data on the observed behavior of the system. { pə'ram·əd·ər ī,dent·ə·fə'kā·shən }

parametric equalizer [ENG ACOUS] A device that allows control over the center frequencies, bandwidths, and amplitudes (parameters) of band-pass filters that determine the frequency response of audio equipment. { ¦par·ə¦me·trik ,ē·kwə'līz·ər }

parametric excitation [ENG] The method of exciting and maintaining oscillations in either an electrical or mechanical dynamic system, in which excitation results from a periodic variation in an energy storage element in a system such as a capacitor, inductor, or spring constant. { ¦par·ə¦me·trik ,ek·si'tā·shən }

parametrized voice response system [ENG ACOUS] A voice response system which first extracts informative parameters from human speech, such as natural resonant frequencies (formants) of the speaker's vocal tract and the fundamental frequency (pitch) of the voice, and which later reconstructs speech from such stored parameters. { pə'ram·ə₁trīzd 'vȯis ri₁späns ₁sis·təm }

parapack |ENG| A package or bundle with a parachute attached for dropping from an aircraft. { 'par·ə₁pak }

parasitic |ELECTR| An undesired and energy-wasting signal current, capacitance, or other parameter of an electronic circuit. { ¦par·ə¦sid·ik }

parasitic current |ELEC| An eddy current in a piece of electrical machinery; gives rise to energy losses. { ¦par·ə¦sid·ik 'kə·rənt }

paravane |ENG| A torpedo-shaped device with sawlike teeth along its forward end, towed with a wire rope underwater from either side of the bow of a ship to cut the cables of anchored mines. Also known as otter. { 'par·ə₁vān }

Pareto diagram |IND ENG| A histogram of defects or quality problems, classified by type and sorted in the order of descending frequency, that is used to focus on the major sources of problems. { pä're·tō ₁dī·ə₁gram }

Pareto's law |IND ENG| The principle that in most activities a small fraction (around 20%) of the total activity accounts for a large fraction (around 80%) of the result. Also known as rule of 80-20. { pə'rēd·ōz ₁lȯ }

parging |CIV ENG| A thin coating of mortar or plaster on a brick or stone surface. { 'pärj·iŋ }

paring |MECH ENG| A method of wood turning in which the piece is trimmed or reduced in size by cutting or shaving thin sections from the surface. { 'per·iŋ }

paring chisel |DES ENG| A long-handled chisel used to pare wood manually. { 'per·iŋ ₁chiz·əl }

paring gouge |DES ENG| A long, thin concave woodworker's gouge with the cutting edge beveled on the inside of the blade. { 'per·iŋ ₁gau̇j }

parison |ENG| A hollow plastic tube from which a bottle or other hollow object is blow-molded. { 'par·ə·sən }

parison swell |ENG| In blow molding, the ratio of the cross-sectional area of the parison to that of the die opening. { 'par·ə·sən ₁swel }

parking apron |CIV ENG| A hard-surfaced area used for parking aircraft. { 'pärk·iŋ ₁ā·prən }

parking brake |MECH ENG| In an automotive vehicle, a brake that functions independently of the service brake and is set after the vehicle has been brought to a stop. { 'pärk·iŋ ₁brāk }

parking lot |CIV ENG| An outdoor lot for parking automobiles. { 'pärk·iŋ ₁lät }

parkway |CIV ENG| A broad landscaped expressway which is not open to commercial vehicles. { 'pärk₁wā }

parquet flooring |BUILD| Wood flooring made of strips laid in a pattern to form designs. { pär'kā 'flȯr·iŋ }

Parshall flume |ENG| A calibrated device for measuring the flow of liquids in open conduits by measuring the upper and lower beads at a specified distance from an obstructing sill. { 'pär·shəl ₁flüm }

Parsons-stage steam turbine |MECH ENG| A

steam turbine having a reaction-type stage in which the pressure drop occurs partially across the stationary nozzles and partly across the rotating blades. { 'pär·sənz ¦stāj 'stēm 'tər·bən }

part |ENG| An element of a subassembly, not normally useful by itself and not amenable to further disassembly for maintenance purposes. { pärt }

part classification |IND ENG| A coding scheme employed in automated manfacturing processes that uses four or more digits to assign discrete products to families of parts. { pärt ₁klas·ə·fə₁kā·shən }

part detection |IND ENG| The recognition of parts and workpieces by a robot or a computer vision system. { pärt di₁tek·shən }

part family |IND ENG| In the group technology concept, a set of related parts that can be produced by the same sequence of machining operations because of similarity in shape and geometry or similarity in production operation processes. { 'pärt ₁fam·lē }

partial condensation |CHEM ENG| The cooling (or pressurization) of a saturated vapor until a part of it is condensed out as liquid. { 'pär·shəl ₁känd·ən'sā·shən }

particle See material particle. { 'pärd·ə·kəl }

particle dynamics |MECH| The study of the dependence of the motion of a single material particle on the external forces acting upon it, particularly electromagnetic and gravitational forces. { 'pärd·ə·kəl dī₁nam·iks }

particle energy |MECH| For a particle in a potential, the sum of the particle's kinetic energy and potential energy. { 'pärd·ə·kəl ₁en·ər·jē }

particle image velocimetry |ENG| A method of measuring local fluid velocities at thousands of locations in a fluid flow by optically observing large numbers of particles that are suspended in the fluid and move with it, using a photograph of the flow illuminated by two or more successive pulses of light or continuously for a known time interval. Also known as particle tracking velocimetry. { ₁pärd·i·kəl ₁im·ij ₁vel·ə'säm·ə·trē }

particle mechanics |MECH| The study of the motion of a single material particle. { 'pärd·ə·kəl mi₁kan·iks }

particle-size analysis |ENG| Determination of the proportion of particles of a specified size in a granular or powder sample. { 'pärd·ə·kəl ¦sīz ə₁nal·ə·səs }

particle-size distribution |ENG| The percentages of each fraction into which a granular or powder sample is classified, with respect to particle size, by number or weight. { 'pärd·ə·kəl ¦sīz ₁di·strə'byü·shən }

particle tracking velocimetry See particle image velocimetry. { ¦pärd·ə·kəl ¦trak·iŋ ₁vel·ə'sim·ə·trē }

particulate mass analyzer |ENG| A unit which measures dust concentrations in emissions from furnaces, kilns, cupolas, and scrubbers. { pär'tik·yə·lət 'mas 'an·ə₁līz·ər }

parting stop |BUILD| A thin strip of wood that

separates the sashes in a double-hung window. { 'pärd·iŋ ,stäp }

parting tool |DES ENG| A narrow-bladed hand tool with a V-shaped gouge used in woodworking for cutting grooves and in wood turning for cutting a piece in two. Also known as V-tool. { 'pärd·iŋ ,tül }

partition |BUILD| An interior wall having a height of one story or less, which divides a structure into sections. |IND ENG| A slotted sheet of paperboard that can be assembled with similar sheets to form cells for holding goods during shipment. { pär'tish·ən }

part programming |CONT SYS| The planning and specification of the sequence of steps or events in the operation of a numerically controlled machine tool. { 'pärt ,prō,gram·iŋ }

parts kit |ENG| A group of parts, not all having the same basic name, used for repair or replacement of the worn broken parts of an item; it may include instruction sheets and material, such as sandpaper, tape, cement, and gaskets. { 'pärts ,kit }

parts list |ENG| One or more printed sheets showing a manufacturer's parts or assemblies of an end item by illustration or a numerical listing of part numbers and names; it does not outline any assembly, maintenance, or operating instructions, and it may or may not have a price list cover sheet. { 'pärts ,list }

party wall |BUILD| A wall providing joint service between two buildings. { 'pärd·ē ,wȯl }

pascal |MECH| A unit of pressure equal to the pressure resulting from a force of 1 newton acting uniformly over an area of 1 square meter. Symbolized Pa. { pa'skal }

pass |MECH ENG| **1.** The number of times that combustion gases are exposed to heat transfer surfaces in boilers (that is, single-pass, double-pass, and so on). **2.** In metal rolling, the passage in one direction of metal deformed between rolls. **3.** In metal cutting, transit of a metal cutting tool past the workpiece with a fixed tool setting. { pas }

passband |ELECTR| A frequency band in which the attenuation of a filter is essentially zero. { 'pas,band }

pass-by |ENG| The double-track part of any single-track system of rail transport. { 'pas,bī }

passenger car |ENG| **1.** A railroad car in which passengers are carried. **2.** An automobile for carrying as many as nine passengers. { 'pas·ən·jər ,kär }

passing track |ENG| A sidetrack with switches at both ends. { 'pas·iŋ ,trak }

passivation |ELECTR| Growth of an oxide layer on the surface of a semiconductor to provide electrical stability by isolating the transistor surface from electrical and chemical conditions in the environment; this reduces reverse-current leakage, increases breakdown voltage, and raises power dissipation rating. { ,pas·ə'vā·shən }

passive accommodation |CONT SYS| The alteration in the positioning or motion of the end point of a robot manipulator that results from bending or deforming of the manipulator components in response to forces exerted on the robot. { 'pas·iv ə,käm·ə'dā·shən }

passive AND gate |ELECTR| *See* AND gate. |ENG| A fluidic device which achieves an output signal, by stream interaction, only when both of two control signals appear simultaneously. { 'pas·iv 'and ,gāt }

passive component *See* passive element. { 'pas·iv kəm'pō·nənt }

passive earth pressure |CIV ENG| The maximum value of lateral earth pressure exerted by soil on a structure, occurring when the soil is compressed sufficiently to cause its internal shearing resistance along a potential failure surface to be completely mobilized. { 'pas·iv 'ərth ,presh·ər }

passive element |ELEC| An element of an electric circuit that is not a source of energy, such as a resistor, inductor, or capacitor. Also known as passive component. { 'pas·iv 'el·ə·mənt }

passive method |CIV ENG| A construction method in permafrost areas in which the frozen ground near the structure is not disturbed or altered, and the foundations are provided with additional insulation to prevent thawing of the underlying ground. { 'pas·iv 'meth·əd }

passive radar |ENG| A technique for detecting objects at a distance by picking up the microwave electromagnetic energy that is both radiated and reflected by all bodies. { 'pas·iv 'rā,där }

passive radiator |ENG ACOUS| A loudspeaker driver with no voice-coil or magnet assemblies that is mounted in a box with a woofer and exhibits a resonance that can be used to improve the low-frequency response of the system. { ¦pas·iv 'rād·ē,ād·ər }

passive solar system |MECH ENG| A solar heating or cooling system that operates by using gravity, heat flows, or evaporation rather than mechanical devices to collect and transfer energy. { 'pas·iv 'sō·lər ,sis·təm }

passive sonar |ENG| Sonar that uses only underwater listening equipment, with no transmission of location-revealing pulses. { 'pas·iv 'sō,när }

passive transducer |ELECTR| A transducer containing no internal source of power. { 'pas·iv tranz'dü·sər }

paste mixer |ENG| Device for the blending together of solid particles and a liquid, with the final formation of a single paste phase. { 'pāst ,mik·sər }

paste-up *See* mechanical. { 'pāst,əp }

pasteurizer |ENG| An apparatus used for pasteurization of fluids. { 'pas·chə,rīz·ər }

patch |ELEC| A temporary connection between jacks or other terminations on a patch board. { pach }

patch bolt |DES ENG| A bolt with a countersunk head having a square knob that twists off when the bolt is screwed in tightly; used to repair boilers and steel ship hulls. { 'pach ,bōlt }

patent |IND ENG| A certificate of grant by a government of an exclusive right with respect to an invention for a limited period of time. Also known as letters patent. { 'pat·ənt }

path computation |CONT SYS| The calculations involved in specifying the trajectory followed by a robot. { 'path ˌkäm·pyəˌtā·shən }

pattern |ENG| A form designed and used as a model for making things. { 'pad·ərn }

pattern shooting |ENG| In seismic prospecting, firing of explosive charges arranged in geometric pattern. { 'pad·ərn ˌshüd·iŋ }

pavement |BUILD| A hard floor of concrete, brick, tiles, or other material. |CIV ENG| A paved surface. { 'pāv·mənt }

pavement light |CIV ENG| A window built into the surface of a pavement to admit daylight to a space below ground level. { 'pāv·mənt ˌlīt }

paver |MECH ENG| Any of several machines which, moving along the road, carry and lay paving material. { 'pāv·ər }

pawl |MECH ENG| The driving link or holding link of a ratchet mechanism, permits motion in one direction only. { pòl }

payback period |IND ENG| The amount of time required for achieving an amount in profits to offset the cost of a capital expenditure, such as the cost of investment in modifications in an industrial facility for the purpose of conserving energy. { 'pāˌbak ˌpir·ē·əd }

payout time |IND ENG| A measurement of profitability or liquidity of an investment, being the time required to recover the original investment in depreciable facilities from profit and depreciation; usually, but not always, calculated after income taxes. { 'pāˌaút ˌtīm }

p-channel metal-oxide semiconductor See PMOS. { ¦pē ˌchan·əl ˌmed·əl ¦äkˌsīd 'sem·i·kən,dək·tər }

p chart |IND ENG| A chart of the fraction defective, either observed in the sample or in some production period. { 'pē ˌchärt }

pdl-ft See foot-poundal.

PDM See precedence diagram method.

PDR See precision depth recorder.

peak load |ELEC| The maximum instantaneous load or the maximum average load over a designated interval of time. Also known as peak power. |ENG| The maximum quantity of a specified material to be carried by a conveyor per minute in a specified period of time. { 'pēk ˌlòd }

peak power See peak load. { 'pēk 'paú·ər }

Peaucellier linkage |MECH ENG| A mechanical linkage to convert circular motion exactly into straight-line motion. { pò'sel·yā ˌliŋ·kij }

pebble heater |CHEM ENG| Gas-heating device (for air, hydrogen, methane, and steam) in which heat is transferred to the gas via a countercurrent movement of preheated pebbles. { 'peb·əl ˌhēd·ər }

pebble mill |MECH ENG| A solids size-reduction device with a cylindrical or conical shell rotating on a horizontal axis, and with a grinding medium such as balls of flint, steel, or porcelain. { 'peb·əl ˌmil }

peck |MECH| Abbreviated pk. **1.** A unit of volume used in the United States for measurement

of solid substances, equal to 8 dry quarts, or 1/4 bushel, or 537.605 cubic inches, or 0.00880976754172 cubic meter. **2.** A unit of volume used in the United Kingdom for measurement of solid and liquid substances, although usually the former, equal to 2 gallons, or 0.00909218 cubic meter. { pek }

Peclet number |CHEM ENG| Dimensionless group used to determine the chemical reaction similitude for the scale-up from pilot-plant data to commercial-sized units; incorporates heat capacity, density, fluid velocity, and other pertinent physical parameters. { pə'klā ˌnəm·bər }

pedal |DES ENG| A lever operated by foot. { 'ped·əl }

pedestal |CIV ENG| **1.** The support for a column. **2.** A metal support carrying one end of a bridge truss or girder and transmitting any load to the top of a pier or abutment. |ELECTR| See blanking level. |ENG| A supporting part or the base of an upright structure, such as a radar antenna. { 'ped·əstˌəl }

pedestal design |MECH ENG| A robot design centered on the vertical axis of a central pedestal, in which the motion of any workpiece is confined to a spherical working envelope. { 'ped·əstˌəl diˌzīn }

pedestal flooring See raised flooring. { 'ped·əˌstəl ˌflór·iŋ }

pedestal pile |CIV ENG| A concrete pile with a bulbous enlargement at the bottom. { 'ped·əstˌəl ˌpīl }

pedometer |ENG| **1.** An instrument for measuring and weighing a newborn child. **2.** An instrument that registers the number of footsteps and distance covered in walking. { pə'däm·əd·ər }

peel-back |ENG| The separation of two bonded materials, one or both of which are flexible, by stripping or pulling the flexible material from the mating surface at a 90 or 180° angle to the plane in which it is adhered. { 'pēl ˌbak }

peel-off time |ENG| In seismic prospecting, the time correction applied to observed data to adjust them to a depressed reference datum. { 'pēl ˌòf ˌtīm }

peel test |ENG| A test to ascertain the adhesive strength of bonded strips of metals by peeling or pulling the metal strips back and recording the adherence values. { 'pēl ˌtest }

peen |DES ENG| The end of a hammer head with a hemispherical, wedge, or other shape; used to bend, indent, or cut. { pēn }

peepdoor |MECH ENG| A small door in a furnace with a glass opening through which combustion may be observed. { 'pēp,dòr }

peg |ENG| **1.** A small pointed or tapered piece, often cylindrical, used to pin down or fasten parts. **2.** A projection used to hang or support objects. { peg }

peg count meter |ENG| A meter or register that counts the number of trunks tested, the number of circuits passed busy, the number of test failures, or the number of repeat tests completed. { 'peg ˌkaúnt ˌmēd·ər }

PEL See permissible exposure limit. { pel }

pellet cooler |CHEM ENG| Gas-cooled, gravity-bed device for the cooling and drying of extruded pellets and briquets. { 'pel·ət ‚kül·ər }

pelleting |ENG| Method of accelerating solidification of cast explosive charges by blending precast pellets of the explosives into the molten charge. { 'pel·əd·iŋ }

pelletizer |CHEM ENG| A machine for cutting bulk plastic into pellets, suitable for use as feedstock, either from solidified polymer at the end of the manufacturing process or from the molten polymer as it emerges from the die. { 'pel·ə‚tīz·ər }

pellet mill |MECH ENG| Device for injecting particulate, granular or pasty feed into holes of a roller, then compacting the feed into a continuous solid rod to be cut off by a knife at the periphery of the roller. { ‚pel·ət ‚mil }

Pelton turbine *See* Pelton wheel. { 'pel·tən 'tər·bən }

Pelton wheel |MECH ENG| An impulse hydraulic turbine in which pressure of the water supply is converted into velocity by a few stationary nozzles, and the water jets then impinge on the buckets mounted on the rim of a wheel; usually limited to high head installations, exceeding 500 feet (150 meters). Also known as Pelton turbine. { 'pel·tən ‚wēl }

pen |ENG| **1.** A small place for confinement, storage, or protection. **2.** A device for writing with ink. { pen }

pencil |ENG| An implement for writing or making marks with a solid substance; the three basic kinds are graphite, carbon, and colored. { 'pen·səl }

pencil cave |ENG| A driller's term for hard, closely jointed shale that caves into a well in pencil-shaped fragments. { 'pen·səl ‚kāv }

pendant atomizer *See* hanging-drop atomizer. { 'pen·dənt 'ad·ə‚mīz·ər }

pendant post |BUILD| A post on a solid support and set against a wall to support a collar beam or other part of a roof. { 'pen·dənt ‚pōst }

pendulous gyroscope |MECH| A gyroscope whose axis of rotation is constrained by a suitable weight to remain horizontal; it is the basis of one type of gyrocompass. { 'pen·jə·ləs 'jī·rə‚skōp }

pendulum anemometer |ENG| A pressure-plate anemometer consisting of a plate which is free to swing about a horizontal axis in its own plane above its center of gravity; the angular deflection of the plate is a function of the wind speed; this instrument is not used for station measurements because of the false reading which results when the frequency of the wind gusts and the natural frequency of the swinging plate coincide. { 'pen·jə·ləm ‚an·ə'mäm·əd·ər }

pendulum level |ENG| A leveling instrument in which the line of sight is automatically kept horizontal by a built-in pendulum device (such as a horizontal arm and a plumb line at right angles to the arm). { 'pen·jə·ləm ‚lev·əl }

pendulum press |MECH ENG| A punch press actuated by a swinging treadle operated by the foot. { 'pen·jə·ləm ‚pres }

pendulum saw |MECH ENG| A circular saw that swings in a vertical arc for crosscuts. { 'pen·jə·ləm ‚sò }

pendulum scale |ENG| Weight-measurement device in which the load is balanced by the movement of one or more pendulums from vertical (zero weight) to horizontal (maximum weight). { 'pen·jə·ləm ‚skāl }

pendulum seismograph |ENG| A seismograph that measures the relative motion between the ground and a loosely coupled inertial mass; in some instruments, optical magnification is used whereas others exploit electromagnetic transducers, photocells, galvanometers, and electronic amplifiers to achieve higher magnification. { 'pen·jə·ləm 'sīz·mə‚graf }

penetration ballistics |MECH| A branch of terminal ballistics concerned with the motion and behavior of a missile during and after penetrating a target. { ‚pen·ə'trā·shən bə‚lis·tiks }

penetration depth |ELEC| In induction heating, the thickness of a layer, extending inward from a conductor's surface, whose resistance to direct current equals the resistance of the whole conductor to alternating current of a given frequency. |ENG| The greatest depth in an ultrasonic test piece at which indications can be measured. { ‚pen·ə'trā·shən ‚depth }

penetration number |ENG| The consistency of greases, waxes, petrolatum, and asphalt or other bituminous materials expressed as the distance that a standard needle penetrates the sample under specified American Society for Testing and Materials test conditions. { ‚pen·ə'trā·shən ‚nəm·bər }

penetration rate |MECH ENG| The actual rate of penetration of drilling tools. { ‚pen·ə'trā·shən ‚rāt }

penetration speed |MECH ENG| The speed at which a drill can cut through rock or other material. { ‚pen·ə'trā·shən ‚spēd }

penetration test |ENG| A test to determine the relative values of density of noncohesive sand or silt at the bottom of boreholes. { ‚pen·ə'trā·shən ‚test }

penetrometer |ENG| **1.** An instrument that measures the penetrating power of a beam of x-rays or other penetrating radiation. **2.** An instrument used to determine the consistency of a material by measurement of the depth to which a standard needle penetrates into it under standard conditions. { ‚pen·ə'träm·əd·ər }

Penex process |CHEM ENG| A continuous, nonregenerative petroleum-refinery process for isomerization of C_5 or C_6 fractions in the presence of hydrogen and a platinum catalyst. { 'pe‚neks ‚prä·səs }

Penning trap |ENG| A device for trapping electrons and isolating single electrons, consisting of a large, homogeneous magnetic field plus a superimposed weak parabolic electric potential created by a positive charge $+Q$ on a ring electrode and two negative charges $-Q/2$ each on two cap electrodes. { 'pen·iŋ ,trap }

Penning-trap mass spectrometer [ENG] A device for making highly accurate comparisons of the masses of charged atoms and molecules by comparing the cyclotron frequencies of single ions in a Penning trap. { ¦pen·iŋ,trap ,mas spek 'träm·əd·ər }

Pennsylvania truss [CIV ENG] A truss characterized by subdivided panels, curved top chords on through trusses, and curved bottom chords on deck spans; used on long bridge spans. { ¦pen· səl¦vā·nyə 'trəs }

pennyweight [MECH] A unit of mass equal to 1/20 troy ounce or to 1.55517384 grams; the term is employed in the United States and in England for the valuation of silver, gold, and jewels. Abbreviated dwt; pwt. { 'pen·ē,wāt }

pen recorder [ENG] A device in which the varying inputs (electrical, pneumatic, mechanical) are marked by a signal-controlled pen onto a continuous recorder chart (circular or roll chart). { 'pen ri,kòrd·ər }

Pensky-Martens closed tester [CHEM ENG] Device to determine the American Society for Testing and Materials flash point of fuel oils and cutback asphalts and other viscous materials and suspensions of solids. { 'pen·skē 'märt·ənz 'klōsd 'tes·tər }

penstock [CIV ENG] A valve or sluice gate for regulating water or sewage flow. [ENG] A closed water conduit controlled by valves and located between the intake and the turbine in a hydroelectric plant. { 'pen,stäk }

pentane lamp [ENG] A pentane-burning lamp formerly used as a standard for photometry. { 'pen,tān ,lamp }

penthouse [BUILD] **1.** An enclosed space built on a flat roof to cover a stairway, elevator, or other equipment. **2.** A dwelling built on top of the main roof. **3.** A sloping shed or roof attached to a wall or building. { 'pent,haùs }

percentage log [ENG] A sample log in which the percentage of each type of rock (except obvious cavings) present in each sample of cuttings is estimated and plotted. { pər'sen·tij ,läg }

percent compaction [ENG] The ratio, expressed as a percentage, of dry unit weight of a soil to maximum unit weight obtained in a laboratory compaction test. { pər'sent kəm'pak·shən }

percent defective [IND ENG] The ratio of defective pieces per lot or sample, expressed as a percentage. { pər'sent di'fek·div }

perch [MECH] Also known as pole; rod. **1.** A unit of length, equal to 5.5 yards, or 16.5 feet, or 5.0292 meters. **2.** A unit of area, equal to 30.25 square yards, or 272.25 square feet, or 25.29285264 square meters. { pərch }

percolation filtration [CHEM ENG] A continuous petroleum-refinery process in which lubricating oils and waxes are percolated through a clay bed to improve color, odor, and stability. { pər· kə'lā·shən fil,trā·shən }

percolation test [CIV ENG] A test to determine the suitability of a soil for the installation of a domestic sewage-disposal system, in which a hole is dug and filled with water and the rate of water-level decline is measured. { pər·kə'lā· shən ,test }

percussion bit [MECH ENG] A rock-drilling tool with chisellike cutting edges, which when driven by impacts against a rock surface drills a hole by a chipping action. { pər'kəsh·ən ,bit }

percussion drill [MECH ENG] A drilling machine usually using compressed air to drive a piston that delivers a series of impacts to the shank end of a drill rod or steel and attached bit. { pər'kəsh·ən ,dril }

percussion drilling [MECH ENG] A drilling method in which hammer blows are transmitted by the drill rods to the drill bit. { pər'kəsh·ən ,dril·iŋ }

perfect dielectric See ideal dielectric. { 'pər·fikt ,dī·ə'lek·trik }

perfect gas See ideal gas. { 'pər·fikt 'gas }

perfect lubrication [ENG] A complete, unbroken film of liquid formed over each of two metal surfaces moving relatively to one another with no contact. { 'pər·fikt ,lü·brə'kā·shən }

perforated-pipe distributor [CHEM ENG] Liquid distribution device consisting of a length of piping or tubing with holes at spaced intervals along the length; used in spray columns, liquid-vapor contactors, and spray driers. Also known as a sparger. { 'pər·fə,rād·əd ¦pīp di'strib·yəd·ər }

perforated plate [CHEM ENG] Flat plate with series of holes used to control fluid distribution, as in a perforated-plate (distillation) column. { 'pər·fə,rād·əd 'plāt }

perforated-plate column [CHEM ENG] Distillation column in which vapor-liquid contact is provided by perforated plates instead of bubble-cap trays. { 'pər·fə,rād·əd ¦plāt 'käl·əm }

perforated-plate distributor [CHEM ENG] **1.** A perforated plate or screen used to even out liquid-flow fluctuations through flow channels. **2.** A perforated plate as used in a distillation column or liquid-liquid extraction column. { 'pər·fə,rād·əd ¦plāt di'strib·yəd·ər }

perforated-plate extractor [CHEM ENG] A liquid-liquid extraction vessel in which perforated plates are used to bring about contact between the two or more liquid phases. { 'pər·fə,rād·əd ¦plāt ik'strak·tər }

performance bond [ENG] A bond that guarantees performance of a contract. { pər'fór· məns ,bänd }

performance characteristic [ENG] A characteristic of a piece of equipment, determined during its test or during its operation. { pər'fór·məns ,kar·ik·tə'ris·tik }

performance chart [ENG] A graph used in evaluating the performance of any device, for example, the performance of an electrical or electronic device, such as a graph of anode voltage versus anode current for a magnetron. { pər'fór· məns ,chärt }

performance curves |ENG| Graphical representations showing the abilities of rotating equipment at various operating conditions; for example, the performance curve for a compressor would include rotor speed for various intake and outlet pressures versus gas flow rate adjusted for temperature, density, viscosity, head, and other factors. { pər'fȯr·məns ‚kərvz }

performance data |ENG| Data on the manner in which a given substance or piece of equipment performs during actual use. { pər'fȯr·məns ‚dad·ə }

performance evaluation |IND ENG| The analysis in terms of initial objectives and estimates, and usually made on site, of accomplishments using an automatic data-processing system, to provide information on operating experience and to identify corrective actions required, if any. { pər'fȯr·məns i‚val·yə'wā·shən }

performance index |IND ENG| The ratio of standard hours to the hours of work actually used; a ratio exceeding 1.00 (or 100) indicates standard output is being exceeded. { pər'fȯr·məns ‚in‚deks }

performance measurement baseline |IND ENG| A time-phased budget plan developed for use in measuring contract performance; includes the budgets assigned to scheduled work elements and the related indirect budgets. { pər'fȯrm·əns ‚mezh·ər·mənt 'bās‚līn }

performance number |ENG| One of a series of numbers (constituting the PN, or performance-number, scale) used to convert fuel antiknock values in terms of a reference fuel into an index which is an indication of relative engine performance; used mostly to rate aviation gasolines with octane values greater than 100. { pər'fȯr·məns ‚nəm·bər }

performance rating See effort rating. { pər'fȯr·məns ‚rād·iŋ }

performance sampling |IND ENG| A technique in work measurement used to determine the leveling factor to be applied to an operator or a group of operators by short, randomly spaced observations of the performance index. { pər'fȯr·məns ‚sam·pliŋ }

peridynamic loudspeaker |ENG ACOUS| Box-type loudspeaker baffle designed to give good bass response by minimizing acoustic standing. { ‚per·ə·də'nam·ik 'laud‚spēk·ər }

periodic kiln |ENG| A kiln in which the cycle of setting ware in the kiln, heating up, "soaking" or holding at peak temperature for some time, cooling, and removing or "drawing" the ware is repeated for each batch. { ‚pir·ē‚äd·ik 'kil }

periodic motion |MECH| Any motion that repeats itself identically at regular intervals. { ‚pir·ē‚äd·ik 'mō·shən }

peripheral speed See cutting speed. { pə'rif·ə·rəl ‚spēd }

peristaltic pump |MECH ENG| A device for moving fluids by the action of multiple, equally spaced rollers, which rotate and compress a flexible tube. { ‚per·ə‚stal·tik 'pəmp }

permafrost drilling |ENG| Boreholes drilled in subsoil and rocks in which the contained water is permanently frozen. { 'pər·mə‚frȯst 'dril·iŋ }

permanent axis |MECH| The axis of the greatest moment of inertia of a rigid body, about which it can rotate in equilibrium. { 'pər·mə·nənt 'ak·səs }

permanent benchmark |ENG| A readily identifiable, relatively permanent, recoverable benchmark that is intended to maintain its elevation without change over a long period of time with reference to an adopted datum, and is located where disturbing influences are believed to be negligible. { 'pər·mə·nənt 'bench‚märk }

permanent gas |THERMO| A gas at a pressure and temperature far from its liquid state. { 'pər·mə·nənt 'gas }

permanent-magnet dynamic loudspeaker See permanent-magnet loudspeaker. { 'pər·mə·nənt ‚mag·nət dī‚nam·ik 'laud‚spēk·ər }

permanent-magnet loudspeaker |ENG ACOUS| A moving-conductor loudspeaker in which the steady magnetic field is produced by a permanent magnet. Also known as permanent-magnet dynamic loudspeaker. { 'pər·mə·nənt ‚mag·nət 'laud‚spēk·ər }

permanent-magnet moving-coil instrument |ENG| An ammeter or other electrical instrument in which a small coil of wire, supported on jeweled bearings between the poles of a permanent magnet, rotates when current is carried to it through spiral springs which also exert a restoring torque on the coil; the position of the coil is indicated by an attached pointer. { 'pər·mə·nənt ‚mag·nət ‚müv·iŋ ‚kȯil 'in·strə·mənt }

permanent-magnet moving-iron instrument |ENG| A meter that depends for its operation on a movable magnetic iron vane that aligns itself in the resultant magnetic field of a permanent magnet and adjacent current-carrying coil. { 'pər·mə·nənt ‚mag·nət ‚müv·iŋ ‚ī·ərn 'in·strə·mənt }

permanent set |MECH| Permanent plastic deformation of a structure or a test piece after removal of the applied load. Also known as set. { 'pər·mə·nənt 'set }

permanent stop |IND ENG| In a flexible manufacturing system a type of controlled stop where an automated guided vehicle will always halt, regardless of programming. { 'pər·mə·nənt 'stäp }

permeability number |ENG| A numbered value assigned to molding materials indicating the relative ease of passage of gases through them. { ‚pər·mē·ə'bil·əd·ē ‚nəm·bər }

permeameter |ENG| **1.** A laboratory device for measurement of permeability of materials, for example, soil or rocks; consists of a powder bed of known dimension and degree of packing through which the particles are forced; pressure drop and rate of flow are related to particle size, and pressure drop is related to surface area. **2.** A device for measuring the coefficient of permeability by measuring the flow of fluid through a sample across which there is a pressure drop produced by gravity. **3.** An instrument for measuring the magnetic flux or flux density produced in a test specimen of ferromagnetic material by a given magnetic intensity, to permit computation of the magnetic permeability of the material. { ,pər·mē'äm·əd·ər }

permeate |CHEM ENG| The clear fluid that passes through the membrane in a membrane filtration process. { 'pər·mē,āt }

permeator |CHEM ENG| A membrane assembly that performs an ion-exchange function, for example, desalting in a membrane water-desalting process. { 'pər·mē,ād·ər }

permissible exposure limit |IND ENG| The level of air contaminants that represents an acceptable exposure level as specified in standards set by a national government agency; generally expressed as 8-hour time-weighted average concentrations. Abbreviated PEL. { pərˌmis·ə·bəl ik'spō·zhər ,lim·ət }

permissible velocity |CIV ENG| The highest velocity at which water is permitted to pass through a structure or conduit without excessive damage. { pər'mis·ə·bəl və'läs·əd·ē }

permissive block system |CIV ENG| A block system in which a railroad train is permitted to enter a block section already occupied by a train. { pərˌmis·iv 'bläk ,sis·təm }

permissive stop |CIV ENG| A railway signal indicating the train must stop but can proceed slowly and cautiously after a specified interval, usually 1 minute. { pər'mis·iv 'stäp }

permittivity |ELEC| The dielectric constant multiplied by the permittivity of empty space, where the permittivity of empty space (ϵ_0) is a constant appearing in Coulomb's law, having the value of 1 in centimeter-gram-second electrostatic units, and of 8.854×10^{-12} farad/meter in rationalized meter-kilogram-second units. Symbolized ϵ. { ,pər·mə'tiv·əd·ē }

pernetti |ENG| 1. Small iron pins or tripods that support ware while it is being fired in a kiln. 2. The marks left on baked pottery by these supporting pins. { pər'ned·ē }

perpend |CIV ENG| A bondstone that extends completely through a masonry wall and is exposed on each side of the wall. { 'pər,pend }

perpendicular axis theorem |MECH| A theorem which states that the sum of the moments of inertia of a plane lamina about any two perpendicular axes in the plane of the lamina is equal to the moment of inertia about an axis through their intersection perpendicular to the lamina. { ,pər·pənˌdik·yə·lər 'ak·səs ,thir·əm }

Pers sunshine recorder |ENG| A type of sunshine recorder in which the time scale is supplied by the motion of the sun. { 'pərs 'sən,shīn ri,kord·ər }

PERT |SYS ENG| A management control tool for defining, integrating, and interrelating what must be done to accomplish a desired objective on time; a computer is used to compare current progress against planned objectives and give management the information needed for planning and decision making. Derived from program evaluation and review technique. { pərt }

peter out |ENG| To fail gradually in size, quantity, or quality; for example, a mine may be said to have petered out. { 'pēd·ər ˌaut }

Petersen grab |ENG| A bottom sampler consisting of two hinged semicylindrical buckets held apart by a cocking device which is released when the grab hits the ocean floor. { 'pēd·ər·sən ,grab }

petroleum engineering |ENG| The application of almost all types of engineering to the drilling for and production of oil, gas, and liquefiable hydrocarbons. { pə'trō·lē·əm ,en·jə'nir·iŋ }

petroleum isomerization process |CHEM ENG| A fixed-bed, vapor-phase petroleum-refinery process using a precious-metal catalyst and external hydrogen; feedstocks include natural gas, pentane, and hexane cuts; the product is high-octane blending stock. { pə'trō·lē·əm īˌsäm·ə·rə'zā·shən ,prä·səs }

petroleum processing |CHEM ENG| The recovery and processing of various usable fractions from the complex crude oils; usable fractions include gasoline, kerosine, diesel oil, fuel oil, and asphalt. Also known as petroleum refining. { pə'trō·lē·əm 'prä,ses·iŋ }

petroleum refining See petroleum processing. { pə'trō·lē·əm ri,fīn·iŋ }

Petterson-Nansen water bottle See Nansen bottle. { 'ped·ər·sən 'nan·sən 'wod·ər ,bäd·əl }

Pettit truss |CIV ENG| A bridge truss in which the panel is subdivided by a short diagonal and a short vertical member, both intersecting the main diagonal at its midpoint. { 'ped·ət ,trəs }

PGR See precision depth recorder.

pharmaceutical chemistry |CHEM ENG| The chemistry of drugs and of medicinal and pharmaceutical products. { ,fär·mə'süd·ə·kəl 'kem·ə·strē }

phase |THERMO| The type of state of a system, such as solid, liquid, or gas. { fāz }

phase advancer |ELEC| Phase modifier which supplies leading reactive volt-amperes to the system to which it is connected; may be either synchronous or asynchronous. { 'fāz id,van·sər }

phase-angle meter See phase meter. { 'fāz ˌaŋ·gəl ,mēd·ər }

phase-balance relay |ELEC| Relay which functions by reason of a difference between two quantities associated with different phases of a polyphase circuit. { 'fāz ˌbal·əns 'rē,lā }

phase-change material |ENG| A material which is used to store the latent heat absorbed in the material during a phase transition. { 'fāz ,chānj mə,tir·ē·əl }

phase-comparison relaying |ELEC| A method of detecting faults in an electric power system in which signals are transmitted from each of two terminals every half cycle so that a continuous signal is received at an intermediate point if there is no fault between the terminals, while a periodic signal is received if there is a fault. { 'fāz kəm,par·ə·sən 'rē,lā·iŋ }

phase conductor |ELEC| In a polyphase circuit,

398

any conductor other than the neutral conductor. { 'fāz kən,dək·tər }

phase converter |ELEC| A converter that changes the number of phases in an alternating-current power source without changing the frequency. { 'fāz kən,vərd·ər }

phase crossover |CONT SYS| A point on the plot of the loop ratio at which it has a phase angle of 180°. { 'fāz 'krós,ō·vər }

phase diagram |THERMO| **1.** A graph showing the pressures at which phase transitions between different states of a pure compound occur, as a function of temperature. **2.** A graph showing the temperatures at which transitions between different phases of a binary system occur, as a function of the relative concentrations of its components. { 'fāz ,dī·ə,gram }

phase factor See power factor. { 'fāz ,fak·tər }

phase integral See action. { 'fāz 'int·ə·grəl }

phase-locked system |ENG| A radar system, having a stable local oscillator, in which information regarding the target is gained by measuring the phase shift of the echo. { 'fāz 'läkt ,sis·təm }

phase margin |CONT SYS| The difference between 180° and the phase of the loop ratio of a stable system at the gain-crossover frequency. { 'fāz ,mär·jən }

phase meter |ENG| An instrument for the measurement of electrical phase angles. Also known as phase-angle meter. { 'fāz ,mēd·ər }

phase modifier |ELEC| Machine whose chief purpose is to supply leading or lagging reactive volt-amperes to the system to which it is connected; may be either synchronous or asynchronous. { 'fāz ,mäd·ə,fī·ər }

phase plane analysis |CONT SYS| A method of analyzing systems in which one plots the time derivative of the system's position (or some other quantity characterizing the system) as a function of position for various values of initial conditions. { 'fāz 'plān ə'nal·ə·səs }

phase portrait |CONT SYS| A graph showing the time derivative of a system's position (or some other quantity characterizing the system) as a function of position for various values of initial conditions. { 'fāz ,pòr·trət }

phase-rotation relay See phase-sequence relay. { 'fāz rō'tā·shən 'rē,lā }

phase-sequence relay |ELEC| Relay which functions according to the order in which the phase voltages successively reach their maximum positive values. Also known as phase-rotation relay. { 'fāz 'sē·kwəns 'rē,lā }

phase shift |ELECTR| The phase angle between the input and output signals of a network or system. { 'fāz ,shift }

phase-shift circuit |ELECTR| A network that provides a voltage component which is shifted in phase with respect to a reference voltage. { 'fāz 'shift ,sər·kət }

phase shifter |ELEC| A device used to change the phase relation between two alternating-current values. { 'fāz ,shif·tər }

phase-shifting transformer |ELEC| A transformer which produces a difference in phase

angle between two circuits. { 'fāz 'shif·tiŋ tranz ,fòr·mər }

phase splitter |ELEC| A circuit that takes a single input alternating voltage and produces two or more output alternating voltages that differ in phase from one another. { 'fāz ,splid·ər }

phase transformation |ELEC| A change of polyphase power from three-phase to six-phase, from three-phase to twelve-phase, and so forth, by use of transformers. { 'fāz ,tranz·fər,ma·shən }

phase transformer |ELEC| A transformer for changing a two-phase current to a three-phase current, or vice versa. { 'fāz tranz,fòr·mər }

phase undervoltage relay |ELEC| Relay which functions by reason of the reduction of one phase voltage in a polyphase circuit. { 'fāz 'ən·dər,vōl·tij 'rē,lā }

phasing See framing. { 'fāz·iŋ }

phenolate process |CHEM ENG| A process which employs sodium phenolate to remove hydrogen sulfide from gas. { 'fēn·əl,āt ,prä·səs }

phenol extraction |CHEM ENG| Petroleum-refinery solvent-extraction process using phenol as the solvent to remove aromatic, unsaturated and naphthenic constituents from lubricating-oil stocks. { 'fē,nól ik,strak·shən }

phenol process |CHEM ENG| A single-solvent petroleum-refining process in which phenol is the selective solvent. { 'fē,nól ,prä·səs }

Philips hot-air engine |MECH ENG| A compact hot-air engine that is a Philips Research Lab (Holland) design; it uses only one cylinder and piston, and operates at 3000 revolutions per minute, with hot-chamber temperature of 1200°F (650°C), maximum pressure of 50 atmospheres (5.07 megapascals), and mean effective pressure of 14 atmospheres (1.42 megapascals). { 'fil·əps 'hät 'er ,en·jən }

Phillips screw |DES ENG| A screw having in its head a recess in the shape of a cross; it is inserted or removed with a Phillips screwdriver that automatically centers itself in the screw. { 'fil·əps ,skrü }

phleger corer |ENG| A device for obtaining ocean bottom cores up to about 4 feet (1.2 meters) in length; consists of an upper tube, main body weight, and tailfin assembly with a check valve that prevents the flow of water into the upper section and a consequent washing out of the core sample while hoisting the corer. { 'flej· ər ,kòr·ər }

pH meter |ENG| An electronic voltmeter using a pH-responsive electrode that gives a direct conversion of voltage differences to differences of pH at the temperature of the measurement. { ,pē'āch ,mēd·ər }

phonation |ENG ACOUS| Production of speech sounds. { fō'nā·shən }

phone See headphone. { fōn }

phonemic synthesizer |ENG ACOUS| A voice response system in which each word is abstractly represented as a sequence of expected vowels and consonants, and speech is composed by juxtaposing the expected phonemic sequence for each word with the sequences for the preceding and following words. { fə'nē·mik 'sin·thə,sīz·ər }

phonograph |ENG ACOUS| An instrument for recording or reproducing acoustical signals, such as voice or music, by transmission of vibrations from or to a stylus that is in contact with a groove in a rotating disk. { 'fō·nə,graf }

phonograph cartridge See phonograph pickup. { 'fō·nə,graf ,kär·trij }

phonograph cutter See cutter. { 'fō·nə,graf ,kəd·ər }

phonograph needle See stylus. { 'fō·nə,graf ,nēd·əl }

phonograph pickup |ENG ACOUS| A pickup that converts variations in the grooves of a phonograph record into corresponding electric signals. Also known as cartridge; phonograph cartridge. { 'fō·nə,graf ,pik,əp }

phonograph record |ENG ACOUS| A shellac-composition or vinyl-plastic disk, usually 7 or 12 inches (18 or 30 centimeters) in diameter, on which sounds have been recorded as modulations in grooves. Also known as disk; disk recording. { 'fō·nə,graf ,rek·ərd }

phonon friction |MECH| Friction that arises when atoms of a surface are set into motion by the sliding action of atoms in an opposing surface, and the mechanical energy needed to slide one surface over the other is thereby converted to the energy of atomic lattice vibrations (phonons) and is eventually transformed into heat. { 'fō,nän ,frik·shən }

phonotelemeter |ENG ACOUS| A device consisting essentially of a stopwatch, for estimating the distance of guns in action by measuring the interval between the flash and the arrival of the sound waves from the discharge. { ¦fō·nō·tə'lem·əd·ər }

phosphate desulfurization |CHEM ENG| A continuous, regenerative petroleum-refinery process using a tripotassium phosphate solution to remove hydrogen sulfide from natural gas, refinery gas, or liquid hydrocarbons. { 'fä,sfāt dē,səl·fə·rə'zā·shən }

phosphoric acid polymerization |CHEM ENG| A petroleum-refinery process using phosphoric acid catalyst to convert propylene, butylene, or both, into high-octane gasoline or petrochemical polymers. { fä'sfór·ik 'as·əd pə,lim·ə·rə'zā·shən }

photoalidade |ENG| A photogrammetric instrument which has a telescopic alidade, a plateholder, and a hinged ruling arm and is mounted on a tripod frame; used for plotting lines of direction and measuring vertical angles to selected features appearing on oblique and terrestrial photographs. { ¦fōd·ō'al·ə,dād }

photocapacitative effect |ELEC| A change in the capacitance of a bulk semiconductor or semiconductor surface film upon exposure to light. { ,fōd·ō·kə'pas·ə,tā·tiv i,fekt }

photoclinometer |ENG| A directional surveying instrument which records photographically the direction and magnitude of well deviations from the vertical. { ¦fōd·ō·klə'näm·əd·ər }

photoconductive device |ELECTR| A photoelectric device which utilizes the photoinduced change in electrical conductivity to provide an electrical signal. { fōd·ō·kən'dək·tiv di'vīs }

photoconductive film |ELECTR| A film of material whose current-carrying ability is enhanced when illuminated. { fōd·ō·kən'dək·tiv 'film }

photoconductor diode See photodiode. { fōd·ō·kən'dək·tər 'dī,ōd }

photodetector |ELECTR| A detector that responds to radiant energy; examples include photoconductive cells, photodiodes, photoresistors, photoswitches, phototransistors, phototubes, and photovoltaic cells. Also known as light-sensitive cell; light-sensitive detector; light sensor photodevice; photodevice; photoelectric detector; photosensor. { ¦fōd·ō·di'tek·tər }

photodiffusion effect See Dember effect. { ¦fōd·ō·di'fyü·zhən i,fekt }

photodiode |ELECTR| A semiconductor diode in which the reverse current varies with illumination; examples include the alloy-junction photocell and the grown-junction photocell. Also known as photoconductor diode. { 'fōd·ō'dī,ōd }

photodraft |DES ENG| A photographic reproduction of a master layout or design on a specially prepared emulsion-coated piece of sheet metal; used as a master in a tool-construction department. { 'fōd·ō,draft }

photoecology |ENG| The application of air photography to ecology, integrated land resource studies, and forestry. { ¦fōd·ō·i'käl·ə·jē }

photoelectric |ELECTR| Pertaining to the electrical effects of light, such as the emission of electrons, generation of voltage, or a change in resistance when exposed to light. { ¦fōd·ō·i'lek·trik }

photoelectric absorption |ELECTR| Absorption of photons in one of the several photoelectric effects. { ¦fōd·ō·i'lek·trik əb'sórp·shən }

photoelectric cell See photocell. { ¦fōd·ō·i'lek·trik 'sel }

photoelectric colorimeter |ENG| A colorimeter that uses a phototube or photocell, a set of color filters, an amplifier, and an indicating meter for quantitative determination of color. { ¦fōd·ō·i'lek·trik ,kəl·ə'rim·əd·ər }

photoelectric constant |ELECTR| The ratio of the frequency of radiation causing emission of photoelectrons to the voltage corresponding to the energy absorbed by a photoelectron; equal to Planck's constant divided by the electron charge. { ¦fōd·ō·i'lek·trik 'kän·stənt }

photoelectric control |ELECTR| Control of a circuit or piece of equipment by changes in incident light. { ¦fōd·ō·i'lek·trik kən'trōl }

photoelectric densitometer |ENG| An electronic instrument used to measure the density or opacity of a film or other material; a beam of light is directed through the material, and the amount of light transmitted is measured with a photocell and meter. { ¦fōd·ō·i'lek·trik ˌden·sə'täm·əd·ər }

photoelectric detector See photodetector. { ¦fōd·ō·i'lek·trik di'tek·tər }

photoelectric device |ELECTR| A device which gives an electrical signal in response to visible, infrared, or ultraviolet radiation. { ¦fōd·ō·i'lek·trik di¦vīs }

photoelectric door opener |CONT SYS| A control system that employs a photocell or other photo device, used to open and close a power-operated door. { ¦fōd·ō·i'lek·trik 'dȯr ˌōp·ə·nər }

photoelectric effect See photoelectricity. { ¦fōd·ō·i'lek·trik iˌfekt }

photoelectric flame-failure detector |CONT SYS| A photoelectric control that cuts off fuel flow when the fuel-consuming flame is extinguished. { ¦fōd·ō·i'lek·trik 'flām ˌfāl·yər di¸tek·tər }

photoelectric fluorometer |ENG| Device using a photoelectric cell to measure fluorescence in a chemical sample that has been excited (one or more electrons have been raised to higher energy level) by ultraviolet or visible light; used for analysis of chemical mixtures. { ¦fōd·ō·i'lek·trik flü'räm·əd·ər }

photoelectricity |ELECTR| The liberation of an electric charge by electromagnetic radiation incident on a substance; includes photoemission, photoionization, photoconduction, the photovoltaic effect, and the Auger effect (an internal photoelectric process). Also known as photoelectric effect; photoelectric process. { ¦fōd·ō¸i¸lek'tris·əd·ē }

photoelectric liquid-level indicator |ENG| A level indicator in which rising liquid interrupts the light beam of a photoelectric control system; used in a tank or process vessel. { ¦fōd·ō·i'lek·trik ¦lik·wəd ¦lev·əl 'in·də¸kād·ər }

photoelectric loop control |CONT SYS| A photoelectric control system used as a position regulator for a loop of material passing from one strip-processing line to another that may travel at a different speed. Also known as loop control. { ¦fōd·ō·i'lek·trik 'lüp kənˌtrōl }

photoelectric photometer |ENG| A photometer that uses a photocell, phototransistor, or phototube to measure the intensity of light. Also known as electronic photometer. { ¦fōd·ō·i'lek·trik fə'täm·əd·ər }

photoelectric pyrometer |ENG| An instrument that measures high temperatures by using a photoelectric arrangement to measure the radiant energy given off by the heated object. { ¦fōd·ō·i'lek·trik pī'räm·əd·ər }

photoelectric reflectometer |ENG| A reflectometer that uses a photocell or phototube to measure the diffuse reflection of surfaces, powders, pastes, and opaque liquids. { ¦fōd·ō·i'lek·trik ˌre¸flek'täm·əd·ər }

photoelectric register control |CONT SYS| A register control using a light source, one or more phototubes, a suitable optical system, an amplifier, and a relay to actuate control equipment

when a change occurs in the amount of light reflected from a moving surface due to register marks, dark areas of a design, or surface defects. Also known as photoelectric scanner. { ¦fōd·ō·i'lek·trik 'rej·ə·stər kənˌtrōl }

photoelectric scanner See photoelectric register control. { ¦fōd·ō·i'lek·trik 'skan·ər }

photoelectric smoke-density control |CONT SYS| A photoelectric control system used to measure, indicate, and control the density of smoke in a flue or stack. { ¦fōd·ō·i'lek·trik 'smōk ˌden·səd·ē kənˌtrōl }

photoelectric sorter |CONT SYS| A photoelectric control system used to sort objects according to color, size, shape, or other light-changing characteristics. { ¦fōd·ō·i'lek·trik 'sȯrd·ər }

photoelectric transmissometer |ENG| A device to measure the runway visibility at an airport by measuring the degree to which a light beam falling on a photocell is obscured by clouds or fog. { ¦fōd·ō·i'lek·trik ˌtranz·mə'säm·əd·ər }

photoelectric turbidimeter |ENG| Device for measurement of solution turbidity by use of photocells to detect the loss of intensity of light beamed through the solution. { ¦fōd·ō·i'lek·trik ˌtər·bə'dim·əd·ər }

photoelectromotive force |ELECTR| Electromotive force caused by photovoltaic action. { ¦fōd·ō·i¸lek·trō'mōd·iv 'fȯrs }

photoelectron |ELECTR| An electron emitted by the photoelectric effect. { ¦fōd·ō·i'lekˌträn }

photoemission |ELECTR| The ejection of electrons from a solid (or less commonly, a liquid) by incident electromagnetic radiation. Also known as external photoelectric effect. { ¦fōd·ō·i'mish·ən }

photoemissive tube photometer |ENG| A photometer which uses a tube made of a photoemissive material; it is highly accurate, but requires electronic amplification, and is used mainly in laboratories. { ¦fōd·ō·i'mis·iv ¦tüb fə'täm·əd·ər }

photoemissivity |ELECTR| The property of a substance that emits electrons when struck by light. { ¦fōd·ō¸ē·mə'siv·əd·ē }

photofabrication |ELECTR| In manufacturing circuit boards and integrated circuits, a process in which the etching pattern is placed over the circuit board or semiconductor material, the board or chip is placed in a special solution, and the assembly is exposed to light. { ¸fōd·ō¸fab·rə'kā·shən }

photoflash bomb |ENG| A missile dropped from aircraft; it contains a photoflash mixture and a means for ignition at a distance above the ground, to produce a brilliant flash of short duration for photographic purposes. { 'fōd·ə¸flash ˌbäm }

photogoniometer |ENG| A goniometer that uses a phototube or photocell as a sensing device for studying x-ray spectra and x-ray diffraction effects in crystals. { ¦fōd·ō¸gō·nē'äm·əd·ər }

photogrammetry |ENG| **1.** The science of making accurate measurements and maps from aerial photographs. **2.** The practice of obtaining surveys by means of photography. { ‚fōd·ə'gram·ə·trē }

photographic barograph |ENG| A mercury barometer arranged so that the position of the upper or lower meniscus may be measured photographically. { ¦fōd·ə¦graf·ik 'bar·ə‚graf }

photographic interpretation *See* photointerpretation. { ¦fōd·ə¦graf·ik in‚tər·prə'tā·shən }

photographic surveying |ENG| Photographing of plumb bobs, clinometers, or magnetic needles in borehole surveying to provide an accurate permanent record. { ¦fōd·ə¦graf·ik sər'vā·iŋ }

photointerpretation |ENG| The science of identifying and describing objects in a photograph, such as deducing the topographic significance or the geologic structure of landforms on an aerial photograph. Also known as photographic interpretation. { ¦fōd·ō·in‚ter·prə'tā·shən }

photomask |ELECTR| A film or glass negative that has many high-resolution images, used in the production of semiconductor devices and integrated circuits. { 'fōd·ō‚mask }

photometer |ENG| An instrument used for making measurements of light or electromagnetic radiation, in the visible range. { fō'täm·əd·ər }

photon coupling |ELECTR| Coupling of two circuits by means of photons passing through a light pipe. { 'fō‚tän ‚kəp·liŋ }

photonegative |ELECTR| Having negative photoconductivity, hence decreasing in conductivity (increasing in resistance) under the action of light; selenium sometimes exhibits photonegativity. { ¦fōd·ō'neg·ə·tiv }

photonephelometer |ENG| A nephelometer that uses a photocell or phototube to measure the amount of light transmitted by a suspension of particles. { ¦fōd·ō‚nef·ə'läm·əd·ər }

photonics |ELECTR| The electronic technology involved with the practical generation, manipulation, analysis, transmission, and reception of electromagnetic energy in the visible, infrared, and ultraviolet portions of the light spectrum. It contributes to many fields, including astronomy, biomedicine, data communications and storage, fiber optics, imaging, optical computing, optoelectronics, sensing, and telecommunications. Also known as optoelectronics. { fō'tän·iks }

photopositive |ELECTR| Having positive photoconductivity, hence increasing in conductivity (decreasing in resistance) under the action of light; selenium ordinarily has photopositivity. { ¦fōd·ō'päz·əd·iv }

photoscanner |ENG| A scanner used to make a film record of gamma rays passing through tissue from an injected radioactive material. { 'fōd·ō‚skan·ər }

photosensitive *See* light-sensitive. { ¦fōd·ō'sen·səd·iv }

phototheodolite |ENG| A ground-surveying instrument used in terrestrial photogrammetry which combines the functions of a theodolite

and a camera mounted on the same tripod. { ¦fōd·ō·thē'äd·əl‚īt }

photothyristor *See* light-activated silicon controlled rectifier. { ¦fōd·ō·thī'ris·tər }

phototopography |ENG| The science of mapping and surveying in which details are plotted entirely from photographs taken at suitable ground stations. { ¦fōd·ō·tə'päg·rə·fē }

phototransistor |ELECTR| A junction transistor that may have only collector and emitter leads or also a base lead, with the base exposed to light through a tiny lens in the housing; collector current increases with light intensity, as a result of amplification of base current by the transistor structure. { ¦fōd·ō·tran'zis·tər }

phototriangulation |ENG| The extension of horizontal or vertical control points, or both, by photogrammetric methods, whereby the measurements of angles and distances on overlapping photographs are related into a spatial solution using the perspective principles of the photographs. { ¦fōd·ō·trī‚aŋ·gyə'lā·shən }

phototube current meter |ENG| A device for measuring the speed of water currents in which a perforated disk, which rotates with the current by means of a propeller, is placed in the path of a beam of light that is then reflected from a mirror onto a phototube. { 'fōd·ō‚tüb 'kə·rənt ‚mēd·ər }

photovoltaic |ELECTR| Capable of generating a voltage as a result of exposure to visible or other radiation. { ¦fōd·ō·vōl'tā·ik }

photovoltaic cell |ELECTR| A device that detects or measures electromagnetic radiation by generating a potential at a junction (barrier layer) between two types of material, upon absorption of radiant energy. Also known as barrier-layer cell; barrier-layer photocell; boundary-layer photocell; photronic photocell. { ¦fōd·ō·vōl'tā·ik ‚sel }

photovoltaic effect |ELECTR| The production of a voltage in a nonhomogeneous semiconductor, such as silicon, or at a junction between two types of material, by the absorption of light or other electromagnetic radiation. { ¦fōd·ō·vōl'tā·ik i‚fekt }

photovoltaic meter |ELECTR| An exposure cell in which a photovoltaic cell produces a current proportional to the light falling on the cell, and this current is measured by a sensitive microammeter. { ¦fōd·ō·vōl'tā·ik ‚mēd·ər }

physical compatibility |ENG| The ability of two or more materials, substances, or chemicals to be used together without ill effect. { 'fiz·ə·kəl kəm‚pad·ə'bil·əd·ē }

physical modeling synthesis |ENG ACOUS| A method of synthesizing the sounds of a musical instrument that uses computational algorithms that are based directly on the mathematical physics of the instrument. { ‚fiz·i·kəl 'mäd·əl·iŋ ‚sin·thə·səs }

physical realizability |CONT SYS| For a transfer function, the possibility of constructing a network with this transfer function. { 'fiz·ə·kəl ,rē·ə,l īz·ə'bil·əd·ē }

physical system See causal system. { 'fiz·ə·kəl 'sis·təm }

physical testing [ENG] Determination of physical properties of materials based on observation and measurement. { 'fiz·ə·kəl 'test·iŋ }

phytometer [ENG] A device for measuring transpiration, consisting of a vessel containing soil in which one or more plants are rooted and sealed so that water can escape only by transpiration from the plant. { fī'täm·əd·ər }

Picatinny test [ENG] An impact test used in the United States for evaluating the sensitivity of high explosives; a small sample of the explosive is placed in a depression in a steel die cup and capped by a thin brass cover, a cylindrical steel plug is placed in the center of the cover, and a 2-kilogram weight is dropped from varying heights on the plug; the reported sensitivity figure is the minimum height, in inches, at which at least 1 firing results from 10 trials. { pik·ə'tin·ē ,test }

Piche evaporimeter [ENG] A porous-paperwick atmometer. { 'pēsh i,vap·ə'rim·əd·ər }

pick [DES ENG] 1. The steel cutting points used on a coal-cutter chain. 2. A miner's steel or iron digging tool with sharp points at each end. [ENG] 1. To dress the sides of a shaft or other excavation. 2. To remove shale, dirt, and such from coal. { pik }

pick-and-place robot [CONT SYS] A simple robot, often with only two or three degrees of freedon and little or no trajectory control, whose sole function is to transfer items from one place to another. { ¦pik ən ¦plās 'rō,bät }

pickax [DES ENG] A pointed steel or iron tool mounted on a wooden handle and used for breaking earth and stone. { 'pik,aks }

pick hammer [DES ENG] A hammer with a point at one end of the head and a blunt surface at the other end. { 'pik ,ham·ər }

pick lacing [DES ENG] The pattern to which the picks are set in a cutter chain. { 'pik ,lās·iŋ }

pickling [CHEM ENG] A method of preparing hides for tanning by immersion in a salt solution with a pH of 2.5 or less. { 'pik·liŋ }

pickoff [ELECTR] A device used to convert mechanical motion into a proportional electric signal. [MECH ENG] A mechanical device for automatic removal of the finished part from a press die. { 'pik,óf }

pickup [ELEC] 1. A device that converts a sound, scene, measurable quantity, or other form of intelligence into corresponding electric signals, as in a microphone, phonograph pickup, or television camera. 2. The minimum current, voltage, power, or other value at which a relay will complete its intended function. 3. Interference from a nearby circuit or system. { 'pik,əp }

picoammeter [ENG] An ammeter whose scale is calibrated to indicate current values in picoamperes. { ,pē·kō'am,ēd·ər }

picosecond [MECH] A unit of time equal to 10^{-12} second, or one-millionth of a microsecond. Abbreviated ps; psec. { ,pē·kō'sek·ənd }

picowatt [MECH] A unit of power equal to 10^{-12} watt, or one-millionth of a microwatt. Abbreviated pW. { 'pē·kə,wät }

picture element [ELECTR] 1. That portion, in facsimile, of the subject copy which is seen by the scanner at any instant; it can be considered a square area having dimensions equal to the width of the scanning line. 2. In television, any segment of a scanning line, the dimension of which along the line is exactly equal to the nominal line width; the area which is being explored at any instant in the scanning process. Also known as critical area; elemental area; pixel; recording spot; scanning spot. { 'pik·chər ,el·ə·mənt }

picture window [BUILD] A large window framing an exterior view. { 'pik·chər ¦win·dō }

piece mark [ENG] Identification number for an individual part, subassembly, or assembly; shown on the drawing, but not necessarily on the part. { 'pēs,märk }

piece rate [IND ENG] Wages paid per unit of production. { 'pēs ,rāt }

piecewise linear system [CONT SYS] A system for which one can divide the range of values of input quantities into a finite number of intervals such that the output quantity is a linear function of the input quantity within each of these intervals. { 'pēs,wīz ¦lin·ē·ər ,sis·təm }

piecework [IND ENG] Work paid for in accordance with the amount done rather than the hours taken. { 'pēs,wərk }

pier [BUILD] A concrete block that supports the floor of a building. [CIV ENG] 1. A vertical, rectangular or circular support for concentrated loads from an arch or bridge superstructure. 2. A structure with a platform projecting from the shore into navigable waters for mooring vessels. { pir }

piercing See fusion piercing. { 'pirs·iŋ }

piercing gripper [CONT SYS] A robot component that first punctures a material such as cloth, rubber, or porous sheets, or soft plastic in order to lift and handle it. { 'pirs·iŋ ,grip·ər }

pier foundation See caisson foundation. { 'pir ,faün,dā·shən }

pierhead line [CIV ENG] The line in navigable waters beyond which construction is prohibited; open-pier construction may extend outward from the bulkhead line to the pierhead line. { 'pir,hed ,līn }

pièze [MECH] A unit of pressure equal to 1 sthène per square meter, or to 1000 pascals. Abbreviated pz. { pē'ez }

piezoelectric detector [ENG] A seismic detector constructed from a stack of piezoelectric crystals with an inertial mass mounted on top and intervening metal foil to collect the charges produced on the crystal faces when the crystals are strained. { pē¦ā·zō·ə'lek·trik di'tek·tər }

piezoelectric element [ELECTR] A piezoelectric crystal used in an electric circuit, for example, as a transducer to convert mechanical or acoustical

piezoelectric gage

signals to electric signals, or to control the frequency of a crystal oscillator. { pē¦ā·zō·ə'lek·trik 'el·ə·mənt }

piezoelectric gage [ENG] A pressure-measuring gage that uses a piezoelectric material to develop a voltage when subjected to pressure; used for measuring blast pressures resulting from explosions and pressures developed in guns. { pē¦ā·zō·ə'lek·trik 'gāj }

piezoelectric loudspeaker See crystal loudspeaker. { pē¦ā·zō·ə'lek·trik 'laùd,spēk·ər }

piezoelectric microphone See crystal microphone. { pē¦ā·zō·ə'lek·trik 'mī·krə,fōn }

piezoelectric oscillator See crystal oscillator. { pē¦ā·zō·ə'lek·trik 'äs·ə,lād·ər }

piezoelectric pickup See crystal pickup. { pē¦ā·zō·ə'lek·trik 'pik,əp }

piezoelectric resonator See crystal resonator. { pē¦ā·zō·ə'lek·trik 'rez·ən,ād·ər }

piezoelectric transducer [ELECTR] A piezoelectric crystal used as a transducer, either to convert mechanical or acoustical signals to electric signals, as in a microphone, or vice versa, as in ultrasonic metal inspection. { pē¦ā·zō·ə'lek·trik tranz'dü·sər }

piezojunction effect [ELECTR] A change in the current-voltage characteristic of a pn junction that is produced by a mechanical stress. { pē,ā·zō'jəŋk·shən i,fekt }

piezometer [ENG] 1. An instrument for measuring fluid pressure, such as a gage attached to a pipe containing a gas or liquid. 2. An instrument for measuring the compressibility of materials, such as a vessel that determines the change in volume of a substance in response to hydrostatic pressure. { ,pē·ə'zäm·əd·ər }

piezometer opening See pressure tap. { ,pē·ə'zäm·əd·ər ,ō·pən·iŋ }

piezoresistive microphone [ENG ACOUS] A microphone in which a piezoresistive material is deposited on the edges of a membrane, and variations in the resistance of this material resulting from motion of the membrane are sensed, typically in a Wheatstone bridge. { pē¦ā·zō·ri¦zis·tiv 'mī·krə,fōn }

piezoresistive sensor [ENG] A transducer which converts variations in mechanical stress into an electrical output; it consists of an element of piezoresistive material that is connected to a Wheatstone bridge circuit and is placed on a highly stressed part of a suitable mechanical structure, usually attached to a cantilever or other beam configuration. { pē¦ā·zō·ri¦zis·tiv 'sen·sər }

piezotransistor accelerometer [ENG] An accelerometer in which a seismic mass supported by a stylus transmits a concentrated force to the upper diode surface of a transistor and acceleration is determined from the resulting change in current across the pn junction of the transistor. { pē¦ā·zō·tran'zis·tər ak,sel·ə'räm·əd·ər }

pi filter [ELECTR] A filter that has a series element and two parallel elements connected in the shape of the Greek letter pi (π). { 'pī,fil·tər }

pig [ELECTR] 1. An ion source based on the

same principle as the Philips ionization gage. 2. See Philips ionization gage. [ENG] In-line scraper (brush, blade cutter, or swab) forced through pipelines by fluid pressure; used to remove scale, sand, water, and other foreign matter from the interior surfaces of the pipe. { pig }

pigtail [ELEC] A short, flexible wire, usually stranded or braided, used between a stationary terminal and a terminal having a limited range of motion, as in relay armatures. { 'pig,tāl }

pigtail splice [ELEC] A splice made by twisting together the bared ends of parallel conductors. { 'pig,tāl ¦splīs }

pike pole [ENG] 1. A pole with a sharp metal point in one end that is used to hold utility poles upright while they are being installed. 2. See fire hook. { 'pīk ,pōl }

pilaster [CIV ENG] A vertical rectangular architectural member that is structurally a pier and architecturally a column. { pə'las·tər }

pile [ENG] A long, heavy timber, steel, or reinforced concrete post that has been driven, jacked, jetted, or cast vertically into the ground to support a load. { pīl }

pile bent [CIV ENG] A row of timber or concrete bearing piles with a pile cap forming that part of a trestle which carries the adjacent ends of timber stringers or concrete slabs. { 'pīl ,bent }

pile cap [CIV ENG] A mass of reinforced concrete cast around the head of a group of piles to ensure that they act as a unit to support the imposed load. { 'pīl ,kap }

pile dike [CIV ENG] A dike consisting of a group of piles braced and lashed together along a riverbank. { 'pīl ,dīk }

pile driver [MECH ENG] A hoist and movable steel frame equipped to handle piles and drive them into the ground. { 'pīl ,drīv·ər }

pile extractor [MECH ENG] 1. A pile hammer which strikes the pile upward so as to loosen its grip and remove it from the ground. 2. A vibratory hammer which loosens the pile by high-frequency jarring. { 'pīl ik,strak·tər }

pile formula [MECH] An equation for the forces acting on a pile at equilibrium: $P = pA + tS + Sn \sin \phi$, where P is the load, A is the area of the pile point, p is the force per unit area on the point, S is the embedded surface of the pile, t is the force per unit area parallel to S, n is the force per unit area normal to S, and φ is the taper angle of the pile. { 'pīl ,fòr·myə·lə }

pile foundation [CIV ENG] A substructure supported on piles. { 'pīl faùn,dā·shən }

pile hammer [MECH ENG] The heavy weight of a pile driver that depends on gravity for its striking power and is used to drive piles into the ground. Also known as drop hammer. { 'pīl ,ham·ər }

pile shoe [CIV ENG] A cast-iron point on the foot of a timber or concrete driven pile to facilitate penetration of the ground. { 'pīl ,shü }

pillar [CIV ENG] A column for supporting part of a structure. { 'pil·ər }

pillar bolt [DES ENG] A bolt projecting from a part so as to support it. { 'pil·ər ,bōlt }

404

pillar crane [MECH ENG] A crane whose mechanism can be rotated about a fixed pillar. { 'pil·ər ,krān }

pillar press [MECH ENG] A punch press framed by two upright columns; the driving shaft passes through the columns, and the slide operates between them. { 'pil·ər ,pres }

pilot [DES ENG] A bullet-nosed cylindrical component used in a die that enters prepunched holes of a metal strip advancing through a series of operations to assure precise registration at each station. [MECH ENG] A cylindrical steel bar extending through, and about 8 inches (20 centimeters) beyond the face of, a reaming bit; it acts as a guide that follows the original unreamed part of the borehole and hence forces the reaming bit to follow, and be concentric with, the smaller-diameter, unreamed portion of the original borehole. { 'pī·lət }

pilot balloon [ENG] A small balloon whose ascent is followed by a theodolite in order to obtain data for the computation of the speed and direction of winds in the upper air. { 'pī·lət bə,lün }

pilot bit [DES ENG] A noncoring bit with a cylindrical diamond-set plug of somewhat smaller diameter than the bit proper, set in the center and projecting beyond the main face of the bit. { 'pī·lət ,bit }

pilot channel [CIV ENG] One of a series of cut-offs for converting a meandering stream into a straight channel of greater slope. { 'pī·lət ,chan·əl }

pilot drill [MECH ENG] A small drill to start a hole to ensure that a larger drill will run true to center. { 'pī·lət ,dril }

piloted ignition [ENG] The accidental initiation of combustion by means of contact of gaseous material with an external high-energy source, such as a flame, spark, electrical arc, or glowing wire. { ¦pī·ləd·əd ig'nish·ən }

pilot hole [DES ENG] In metal-forming operations, a prepunched hole in a metal strip into which the pilot component of the die enters in order to assure precise registration of the strip at each work station. [ENG] A small hole drilled ahead of a larger borehole. [MECH ENG] A hole drilled in a piece of wood to serve as a guide for a nail or a screw or for drilling a larger hole. { 'pī·lət ,hōl }

pilot lamp [ELEC] A small lamp used to indicate that a circuit is energized. Also known as pilot light. { 'pī·lət ,lamp }

pilot light [ELEC] See pilot lamp. [ENG] A small, constantly burning flame used to ignite a gas burner. { 'pī·lət ,līt }

pilot line operation [IND ENG] Minimum production of an item in order to preserve or develop the art of its production. { 'pī·lət 'līn ,äp·ə,rā·shən }

pilot materials [IND ENG] A minimum quantity of special materials, partially finished components, forgings, and castings, identified with specific production equipment and processes and required for the purpose of proofing, tooling,

and testing manufacturing processes to facilitate later reactivation. { 'pī·lət mə,tir·ē·əlz }

pilot model [IND ENG] An early production model of a product used to debug the manufacturing process. { 'pī·lət ,mäd·əl }

pilot plant [IND ENG] A small version of a planned industrial plant, built to gain experience in operating the final plant. { 'pī·lət ¦plant }

pilot reaming bit See reaming bit. { 'pī·lət 'rēm·iŋ ,bit }

pilot-scale chemical reaction [CHEM ENG] Small-scale chemical reaction used to test operating conditions and product yields; used as a pilot for design of large-scale reaction systems. { 'pī·lət ¦skāl ¦kem·ə·kəl rē'ak·shən }

pilot tunnel [ENG] A small tunnel or shaft excavated in advance of the main drivage in mining and tunnel building to gain information about the ground, create a free face, and thus simplify the blasting operations. { 'pī·lət ,tən·əl }

pilot wire regulator [CONT SYS] Automatic device for controlling adjustable gains or losses associated with transmission circuits to compensate for transmission changes caused by temperature variations, the control usually depending upon the resistance of a conductor or pilot wire having substantially the same temperature conditions as the conductors of the circuits being regulated. { 'pī·lət ¦wīr 'reg·yə,lād·ər }

pin [DES ENG] **1.** A cylindrical fastener made of wood, metal, or other material used to join two members or parts with freedom of angular movement at the joint. **2.** A short, pointed wire with a head used for fastening fabrics, paper, or similar materials. [ELECTR] A terminal on an electron tube, semiconductor, integrated circuit, plug, or connector. Also known as base pin; prong. { pin }

pinch [ENG] The closing-in of borehole walls before casing is emplaced, resulting from rock failure when drilling in formations having a low compressional strength. { pinch }

pinch bar [DES ENG] A pointed lever, used somewhat like a crowbar, to roll heavy wheels. { 'pinch ,bär }

pinch grasp [IND ENG] A grasp by the human hand that involves the thumb and the facing side of the index finger at the knuckle; used to apply a large force to a small object. Also known as key grasp. { 'pinch ,grasp }

pinch-off blades [ENG] In blow molding, the part that compresses the parison to seal it prior to blowing, and to allow easy cooling and removal of flash. { 'pinch,óf ,blādz }

pinch point [IND ENG] A point in a plant layout or on an automated guided vehicle such that the distance between the automated guided vehicle and the surrounding equipment and structures is so small that it represents a safety hazard to personnel. { 'pinch ,póint }

pinch-tube process [ENG] A plastics blow-molding process in which the extruder drops a tube between mold halves, and the tube is pinched off when the mold closes. { 'pinch ,tüb ,prä·səs }

pin diode |ELECTR| A diode consisting of a silicon wafer containing nearly equal *p*-type and *n*-type impurities, with additional *p*-type impurities diffused from one side and additional *n*-type impurities from the other side; this leaves a lightly doped intrinsic layer in the middle, to act as a dielectric barrier between the *n*-type and *p*-type regions. Also known as power diode. { 'pin 'dī,ōd }

pinger |ENG ACOUS| A battery-powered, low-energy source for an echo sounder. { 'piŋ·ər }

pinhole detector |ENG| A photoelectric device that detects extremely small holes and other defects in moving sheets of material. { 'pin,hōl di,tek·tər }

pinion |MECH ENG| The smaller of a pair of gear wheels or the smallest wheel of a gear train. { 'pin·yən }

pin joint |DES ENG| A joint made with a pin hinge which has a removable pin. { 'pin ,jóint }

pin junction |ELECTR| A semiconductor device having three regions: *p*-type impurity, intrinsic (electrically pure), and *n*-type impurity. { 'pin ,jəŋk·shən }

pinnate joint See feather joint. { 'pi,nāt ,jóint }

pinpoint gate |ENG| In plastics molding, an orifice of 0.030 inch (0.76 millimeter) or less in diameter through which molten resin enters a mold cavity. { 'pin,póint ,gāt }

pin rod |DES ENG| A rod designed to connect two parts so they act as one. { 'pin ,räd }

pint |MECH| Abbreviated pt. **1.** A unit of volume, used in the United States for measurement of liquid substances, equal to 1/8 U.S. gallon, or 231/8 cubic inches, or 4.73176473 × 10⁻⁴ cubic meter. Also known as liquid pint (liq pt). **2.** A unit of volume used in the United States for measurement of solid substances, equal to 1/64 U.S. bushel, or 107,521/3200 cubic inches, or approximately 5.50610 × 10⁻⁴ cubic meter. Also known as dry pint (dry pt). **3.** A unit of volume, used in the United Kingdom for measurement of liquid and solid substances, although usually the former, equal to 1/8 imperial gallon, or 5.6826125 × 10⁻⁴ cubic meter. Also known as imperial pint. { pīnt }

pintle |DES ENG| A vertical pivot pin, as on a rudder or a gun carriage. { 'pint·əl }

pintle chain |DES ENG| A chain with links held together by pivot pins; used with sprocket wheels. { 'pint·əl ,chān }

pin-type mill |MECH ENG| Solids pulverizer in which protruding pins on high-speed rotating disk provide the breaking energy. { 'pin ,tīp ,mil }

pipe |DES ENG| A tube made of metal, clay, plastic, wood, or concrete and used to conduct a fluid, gas, or finely divided solid. { pīp }

pipe bit |DES ENG| A bit designed for attachment to standard coupled pipe for use in socketing the pipe in bedrock. { 'pīp ,bit }

pipebox |ENG| In a pipework installation, a casing packed with loose insulation to enclose a set of pipes. { 'pīp,bäks }

pipe clamp |DES ENG| A device similar to a casing clamp, but used on a pipe to grasp it and facilitate hoisting or suspension. { 'pīp ,klamp }

pipe culvert |CIV ENG| A buried pipe for carrying a watercourse below ground level. { 'pīp ,kal·vərt }

pipe cutter |DES ENG| A hand tool consisting of a clamplike device with three cutting wheels which are forced inward by screw pressure to cut into a pipe as the tool is rotated around the pipe circumference. { 'pīp ,kəd·ər }

pipe elbow meter |ENG| A variable-head meter for measuring flow around the bend in a pipe. { 'pīp 'el·bō ,mēd·ər }

pipe fitter |ENG| A technician who fits, threads, installs, and repairs pipes in a pipework system. { 'pīp ,fid·ər }

pipe fitting |ENG| A piece, such as couplings, unions, nipples, tees, and elbows for connecting lengths of pipes. { 'pīp ,fid·iŋ }

pipe flow |ENG| Conveyance of fluids in closed conduits. { 'pīp ,flō }

pipe laying |ENG| The placing of pipe into position in a trench, as with buried pipelines for oil, water, or chemicals. { 'pīp ,lā·iŋ }

pipeline |ENG| A line of pipe connected to valves and other control devices, for conducting fluids, gases, or finely divided solids. { 'pīp,līn }

pipe pile |CIV ENG| A steel pipe 6–30 inches (15–76 centimeters) in diameter, usually filled with concrete and used for underpinning. { 'pīp ,pīl }

pipe run |ENG| The path followed by a piping system. { 'pīp ,rən }

pipe scale |ENG| Rust and corrosion products adhering to the inner surfaces of pipes; serve to decrease ability to transfer heat and to increase the pressure drop for flowing fluids. { 'pīp ,skāl }

pipe sleeve |ENG| A hollow, cylindrical insert placed in a form for a concrete wall at the position where a pipe is to penetrate in order to prevent flow of concrete into the opening. { 'pīp ,slēv }

pipe still |CHEM ENG| A petroleum-refinery still in which heat is applied to the oil while it is being pumped through a coil or pipe arranged in a firebox, the oil then running to a fractionator with continuous removal of overhead vapor and liquid bottoms. { 'pīp ,stil }

pipe tap |ENG| A small threaded hole or entry made into the wall of a pipe; used for sampling of pipe contents, or connection of control devices or pressure-drop-measurement devices. { 'pīp ,tap }

pipe tee |DES ENG| A T-shaped pipe fitting with two outlets, one at 90° to the connection to the main line. { 'pīp ,tē }

pipe thread |DES ENG| Most commonly, a 60° thread used on pipes and tubes, characterized by flat crests and roots and cut with 3/4-inch taper per foot (about 1.9 centimeters per 30 centimeters). Also known as taper pipe thread. { 'pīp ,thred }

pipe-thread protector *See* thread protector. { 'pīp ,thred prə,tek·tər }

pipe tongs |ENG| Heavy tongs that are hung on a cable and used for screwing pipe and tool joints. { 'pīp ,täŋz }

pipe train |ENG| In the extrusion of plastic pipe, the entire equipment assembly used to fabricate the pipe (such as the extruder, die, cooling bath, haul-off, and cutter). { 'pīp ,trān }

pipework *See* piping. { 'pīp,wərk }

pipe wrench |DES ENG| A tool designed to grip and turn a pipe or rod about its axis in one direction only. { 'pīp ,rench }

piping |ENG| A system of pipes provided to carry a fluid. Also known as pipework. { 'pīp·iŋ }

piston |ENG| *See* force plug. |MECH ENG| A sliding metal cylinder that reciprocates in a tubular housing, either moving against or moved by fluid pressure. { 'pis·tən }

piston blower |MECH ENG| A piston-operated, positive-displacement air compressor used for stationary, automobile, and marine duty. { 'pis·tən ¦blō·ər }

piston corer |MECH ENG| A steel tube which is driven into the sediment by a free fall and by lead attached to the upper end, and which is capable of recovering undistorted vertical sections of sediment. { 'pis·tən ¦kór·ər }

piston displacement |MECH ENG| The volume which a piston in a cylinder displaces in a single stroke, equal to the distance the piston travels times the internal cross section of the cylinder. { 'pis·tən di,splās·mənt }

piston drill |MECH ENG| A heavy percussion-type rock drill mounted either on a horizontal bar or on a short horizontal arm fastened to a vertical column; drills holes to 6 inches (15 centimeters) in diameter. Also known as reciprocating drill. { 'pis·tən ¦dril }

piston engine |MECH ENG| A type of engine characterized by reciprocating motion of pistons in a cylinder. Also known as displacement engine; reciprocating engine. { 'pis·tən ¦en·jən }

piston gage *See* free-piston gage. { 'pis·tən ,gāj }

piston head |MECH ENG| That part of a piston above the top ring. { 'pis·tən ,hed }

piston meter |ENG| A variable-area, constant-head fluid-flow meter in which the position of the piston, moved by the buoyant force of the liquid, indicates the flow rate. Also known as piston-type area meter. { 'pis·tən ¦mēd·ər }

pistonphone |ENG ACOUS| A small chamber equipped with a reciprocating piston having a measurable displacement and used to establish a known sound pressure in the chamber, as for testing microphones. { 'pis·tən,fōn }

piston pin |MECH ENG| A cylindrical pin that connects the connecting rod to the piston. Also known as wrist pin. { 'pis·tən ,pin }

piston pump |MECH ENG| A pump in which motion and pressure are applied to the fluid by a reciprocating piston in a cylinder. Also known as reciprocating pump. { 'pis·tən ¦pəmp }

piston ring |DES ENG| A sealing ring fitted around a piston and extending to the cylinder wall to prevent leakage. Also known as packing ring. { 'pis·tən ,riŋ }

piston rod |MECH ENG| The rod which is connected to the piston, and moves or is moved by the piston. { 'pis·tən ,räd }

piston skirt |MECH ENG| That part of a piston below the piston pin bore. { 'pis·tən ,skərt }

piston speed |MECH ENG| The total distance a piston travels in a given time; usually expressed in feet per minute. { 'pis·tən ,spēd }

piston-type area meter *See* piston meter. { 'pis·tən ,tīp 'er·ē·ə ,mēd·ər }

piston valve |MECH ENG| A cylindrical type of steam engine slide valve for admission and exhaust of steam. { 'pis·tən ¦valv }

piston viscometer |ENG| A device for the measurement of viscosity by the timed fall of a piston through the liquid being tested. { 'pis·tən vi'skäm·əd·ər }

pitch |DES ENG| The distance between similar elements arranged in a pattern or between two points of a mechanical part, as the distance between the peaks of two successive grooves on a disk recording or on a screw. |MECH| **1.** Of an aerospace vehicle, an angular displacement about an axis parallel to the lateral axis of the vehicle. **2.** The rising and falling motion of the bow of a ship or the tail of an airplane as the craft oscillates about a transverse axis. { pich }

pitch acceleration |MECH| The angular acceleration of an aircraft or missile about its lateral, or Y, axis. { 'pich ik,sel·ə,rā·shən }

pitch attitude |MECH| The attitude of an aircraft, rocket, or other flying vehicle, referred to the relationship between the longitudinal body axis and a chosen reference line or plane as seen from the side. { 'pich ,ad·ə,tüd }

pitch axis |MECH| A lateral axis through an aircraft, missile, or similar body, about which the body pitches. Also known as pitching axis. { 'pich ,ak·səs }

pitch circle |DES ENG| In toothed gears, an imaginary circle concentric with the gear axis which is defined at the thickest point on the teeth and along which the tooth pitch is measured. { 'pich ,sər·kəl }

pitch cone |DES ENG| A cone representing the pitch surface of a bevel gear. { 'pich ,kōn }

pitch cylinder |DES ENG| A cylinder representing the pitch surface of a spur gear. { 'pich ,sil·ən·dər }

pitch diameter |DES ENG| The diameter of the pitch circle of a gear. { 'pich dī,am·əd·ər }

pitched roof |BUILD| **1.** A roof that has one or more surfaces with a slope greater than 10°. **2.** A roof that has two slopes meeting at a central ridge. { ¦picht 'rüf }

pitching axis *See* pitch axis. { 'pich·iŋ ,ak·səs }

pitching moment |MECH| A moment about a lateral axis of an aircraft, rocket, or airfoil. { 'pich·iŋ ,mō·mənt }

pitch line *See* cam profile. { 'pich ,līn }

pitman |ENG| **1.** A worker in or near a pit, as in a quarry, mine, garage, or foundry. **2.** On a pumping unit, an arm connecting the crank with the walking beam for converting rotary motion to reciprocating motion. |MECH ENG| In an automotive steering system, the arm that is connected to the shaft of the steering gear sector and the tie rod, and swings back and forth as the steering wheel is turned. Also known as pitman arm. { 'pit·mən }

pitman arm See pitman. { 'pit·mən ‚ärm }

pitometer |ENG| Reversed pitot-tube-type flow-measurement device with one pressure opening facing upstream and the other facing downstream. { pə'täm·əd·ər }

pitometer log |ENG| A log consisting essentially of a pitot tube projecting into the water, and suitable registering devices. { pə'täm·əd·ər ‚läg }

pitot tube |ENG| An instrument that measures the stagnation pressure of a flowing fluid, consisting of an open tube pointing into the fluid and connected to a pressure-indicating device. Also known as impact tube. { pē'tō ‚tüb }

pitot-tube anemometer |ENG| A pressure-tube anemometer consisting of a pitot tube mounted on the windward end of a wind vane and a suitable manometer to measure the developed pressure, and calibrated in units of wind speed. { pē'tō ‚tüb ‚an·ə'mäm·əd·ər }

pitot-venturi flow element |ENG| Liquid-flow measurement device in which a pair of concentric venturi elements replaces the pitot-tube probe. { pē'tō ven'tür·ē ‚flō ‚el·ə·mənt }

pivot |MECH| A short, pointed shaft forming the center and fulcrum on which something turns, balances, or oscillates. { 'piv·ət }

pivot anchor |DES ENG| An anchor that permits a pipe to swivel around a fixed point. { 'piv·ət ‚aŋ·kər }

pivot bridge |CIV ENG| A bridge in which a span can open by pivoting about a vertical axis. { 'piv·ət 'brij }

pivot-bucket conveyor-elevator |MECH ENG| A bucket conveyor having overlapping pivoted buckets on long-pitch roller chains; buckets are always level except when tripped to discharge materials. { 'piv·ət ‚bək·ət kən¦vā·ər 'el·ə‚vād·ər }

pivoted window |BUILD| A window having a section which is pivoted near the center so that the top of the section swings in and the bottom swings out. { 'piv·əd·əd 'win·dō }

pixel |ELECTR| The smallest addressable element in an electronic display; a short form for picture element. Also known as pel. { pik'sel }

pk See peck.

plain concrete |CIV ENG| Concrete without reinforcement but often with light steel to reduce shrinkage and temperature cracking. { 'plān kän'krēt }

plain-laid |DES ENG| Pertaining to a rope whose strands are twisted together in a direction opposite to that of the twist in the strands. { 'plān‚lād }

plain milling cutter |DES ENG| A cylindrical milling cutter with teeth on the periphery only; used

for milling plain or flat surfaces. Also known as slab cutter. { 'plān ¦mil·iŋ ‚kəd·ər }

plain turning |MECH ENG| Lathe operations involved when machining a workpiece between centers. { 'plān ¦tərn·iŋ }

planar linkage |MECH ENG| A linkage that involves motion in only two dimensions. { 'plā·nər 'liŋ·kij }

planar process |ENG| A silicon-transistor manufacturing process in which a fractional-micrometer-thick oxide layer is grown on a silicon substrate; a series of etching and diffusion steps is then used to produce the transistor inside the silicon substrate. { 'plā·nər ‚prä·səs }

planchet |ENG| A small metal container or sample holder; usually used to hold radioactive materials that are being checked for the degree of radioactivity in a proportional counter or scintillation detector. { 'plan·chət }

Planck function |THERMO| The negative of the Gibbs free energy divided by the absolute temperature. { 'pläŋk ‚fəŋk·shən }

plane |DES ENG| A tool consisting of a smooth-soled stock from the face of which extends a wide-edged cutting blade for smoothing and shaping wood. |ELECTR| Screen of magnetic cores; planes are combined to form stacks. { plān }

plane correction |ENG| A correction applied to observed surveying data to reduce them to a common reference plane. { 'plān kə‚rek·shən }

plane lamina |MECH| A body whose mass is concentrated in a single plane. { 'plān 'lam·ə·nə }

plane of departure |MECH| Vertical plane containing the path of a projectile as it leaves the muzzle of the gun. { 'plān əv di'pär·chər }

plane of fire |MECH| Vertical plane containing the gun and the target, or containing a line of site. { 'plān əv 'fīr }

plane of maximum shear stress |MECH| Either of two planes that lie on opposite sides of and at angels of 45° to the maximum principal stress axis and that are parallel to the intermediate principal stress axis. { 'plān əv ¦mak·si·məm 'shir ‚stres }

plane of work |IND ENG| The plane in which most of a worker's motions occur in the performance of a task. { 'plān əv 'wərk }

plane of yaw |MECH| The plane determined by the tangent to the trajectory of a projectile in flight and the axis of the projectile. { 'plān əv 'yō }

plan equation |MECH ENG| The mathematical statement that horsepower = plan/33,000, where p = mean effective pressure (pounds per square inch), l = length of piston stroke (feet), a = net area of piston (square inches), and n = number of cycles completed per minute. { 'plan i‚kwā·zhən }

planer |MECH ENG| A machine for the shaping of long, flat, or flat contoured surfaces by reciprocating the workpiece under a stationary single-point tool or tools. { 'plān·ər }

plane strain |MECH| A deformation of a body in which the displacements of all points in the body are parallel to a given plane, and the values of these displacements do not depend on the distance perpendicular to the plane. { 'plān ¡strān }

plane stress |MECH| A state of stress in which two of the principal stresses are always parallel to a given plane and are constant in the normal direction. { 'plān ¡stres }

plane surveying |ENG| Measurement of areas on the assumption that the earth is flat. { 'plān sər¦vā·iŋ }

plane table |ENG| A surveying instrument consisting of a drawing board mounted on a tripod and fitted with a compass and a straight-edge ruler; used to graphically plot survey lines directly from field observations. { 'plān ¡tā·bəl }

planetary gear train |MECH ENG| An assembly of meshed gears consisting of a central gear, a coaxial internal or ring gear, and one or more intermediate pinions supported on a revolving carrier. { 'plan·ə,ter·ē 'gir ,trān }

planet carrier |MECH ENG| A fixed member in a planetary gear train that contains the shaft upon which the planet pinion rotates. { 'plan·ət ¡kar·ē·ər }

planet gear |MECH ENG| A pinion in a planetary gear train. { 'plan·ət ,gir }

planet pinion |MECH ENG| One of the gears in a planetary gear train that meshes with and revolves around the sun gear. { 'plan·ət ,pin·yən }

planimeter |ENG| A device used for measuring the area of any plane surface by tracing the boundary of the area. { plə'nim·əd·ər }

planing |ENG| Smoothing or shaping the surface of wood, metal, or plastic workpieces. { 'plān·iŋ }

planishing |MECH ENG| Smoothing the surface of a metal by a rapid series of overlapping, light hammerlike blows or by rolling in a planishing mill. { 'plan·ish·iŋ }

plankton net |ENG| A net for collecting plankton. { 'plaŋk·tən ,net }

planning horizon |IND ENG| In a materials-requirements planning system, the time from the present to some future date for which plans are being generated for acquisition of materials. { 'plan·iŋ hə,rīz·ən }

plant |IND ENG| The land, buildings, and equipment used in an industry. { plant }

plant decomposition |CONT SYS| The partitioning of a large-scale control system into subsystems along lines of weak interaction. { 'plant dē,käm·pə'zish·ən }

plant factor |ELEC| The ratio of the average power load of an electric power plant to its rated capacity. Also known as capacity factor. { 'plant ,fak·tər }

plant layout |IND ENG| The location of equipment and facilities in a manufacturing plant. { 'plant ,lā,aút }

plant protection |IND ENG| That portion of industrial security which concerns the safeguarding of industrial installations, resources, utilities, and materials by physical measures such as guards, fences, and lighting designation of restricted areas. { 'plant prə,tek·shən }

plasma-arc cutting |ENG| Metal cutting by melting a localized area with an arc followed by removal of metal by high-velocity, high-temperature ionized gas. { 'plaz·mə ¡ärk 'kəd·iŋ }

plasma processing |ENG| Methods and technologies that utilize a plasma to treat and manufacture materials, generally through etching, deposition, or chemical alteration at a surface inside or at the boundary of the plasma. { 'plaz·mə prä,ses·iŋ }

plasma-source ion implantation |ENG| A method of ion implantation in which the workpiece is placed in a plasma containing the appropriate ion species and is repetitively pulse-biased to a high negative potential so that positive plasma ions are accelerated to the surface and implant in the bulk material. Abbreviated PSII. { ¡plaz·mə ,sórs 'ī·ən ,im·plan,tā·shən }

plasma torch |ENG| A torch in which temperatures as high as 50,000°C are achieved by injecting a plasma gas tangentially into an electric arc formed between electrodes in a chamber; the resulting vortex of hot gases emerges at very high speed through a hole in the negative electrode, to form a jet for welding, spraying of molten metal, and cutting of hard rock or hard metals. { 'plaz·mə ,tórch }

plaster coat |BUILD| A thin layer of plaster lining walls in buildings. { 'plas·tər ¡kōt }

plaster ground |BUILD| A piece of wood used as a gage to control the thickness of a plaster coat placed on a wall; usually put around windows and doors and at the floor. { 'plas·tər ¡graúnd }

plaster shooting |ENG| A surface blasting method used when no rock drill is necessary or one is not available; consists of placing a charge of gelignite, primed with safety fuse and detonator, in close contact with the rock or boulder and covering it completely with stiff damp clay. { 'plas·tər 'shüd·iŋ }

plastic |MECH| Displaying, or associated with, plasticity. { 'plas·tik }

plasticate |ENG| To soften a material by heating or kneading. Also known as plastify. { 'plas·tə,kāt }

plastic bonding |ENG| The joining of plastics by heat, solvents, adhesives, pressure, or radio frequency. { 'plas·tik 'bänd·iŋ }

plastic collision |MECH| A collision in which one or both of the colliding bodies suffers plastic deformation and mechanical energy is dissipated. { ¡plas·tik kə'lizh·ən }

plastic deformation |MECH| Permanent change in shape or size of a solid body without fracture resulting from the application of sustained stress beyond the elastic limit. { 'plas·tik ,dē,fór'mā·shən }

409

plastic design See ultimate-load design. { 'plas·tik di'zīn }

plasticity [MECH] The property of a solid body whereby it undergoes a permanent change in shape or size when subjected to a stress exceeding a particular value, called the yield value. { plas'tis·əd·ē }

plasticize [ENG] To soften a material to make it plastic or moldable by adding a plasticizer or by using heat. { 'plas·tə,sīz }

plasticorder [ENG] Laboratory device used to predict the performance of a plastic material by measurement of temperature, viscosity, and shear-rate relationships. Also known as plastigraph. { 'plas·tə,kȯrd·ər }

plasticoviscosity [MECH] Plasticity in which the rate of deformation of a body subjected to stresses greater than the yield stress is a linear function of the stress. { ¦plas·tə·kō·vi'skäs·əd·ē }

plastify See plasticate. { 'plas·tə,fī }

plastigraph See plasticorder. { 'plas·tə,graf }

plastometer [ENG] Instrument used to determine the flow properties of a thermoplastic resin by forcing molten resin through a specified die opening or orifice at a given pressure and temperature. { pla'stäm·əd·ər }

plate [BUILD] **1.** A shoe or base member, such as of a partition or other kind of frame. **2.** The top horizontal member of a row of studs used in a frame wall. [DES ENG] A rolled, flat piece of metal of some arbitrary minimum thickness and width depending on the type of metal. [ELEC] **1.** One of the conducting surfaces in a capacitor. **2.** One of the electrodes in a storage battery. [ELECTR] See anode. { plāt }

plate anemometer See pressure-plate anemometer. { 'plāt ,an·ə'mäm·əd·ər }

plate bearing test [ENG] Former method to estimate the bearing capacity of a soil; a rigid steel plate about 1 foot (30 centimeters) square was placed on the foundation level and then loaded until the foundation failed, as evidenced by rapid sinking of the plate. { 'plāt 'ber·iŋ ,test }

plate-belt feeder See apron feeder. { 'plāt ,belt ,fēd·ər }

plate cam [MECH ENG] A flat, open cam that imparts a sliding motion. { 'plāt ,kam }

plate coil [MECH ENG] Heat-transfer device made from two metal sheets held together, one or both plates embossed to form passages between them for a heating or cooling medium to flow through. Also known as panel coil. { 'plāt ¦kȯil }

plate conveyor [MECH ENG] A conveyor with a series of steel plates as the carrying medium; each plate is a short trough, all slightly overlapped to form an articulated band, and attached to one center chain or to two side chains; the chains join rollers running on an angle-iron framework and transmit the drive from the driveheads, installed at intermediate points and sometimes also at the head or tail ends. { 'plāt kən,vā·ər }

plate cut [BUILD] The cut made in a rafter to rest on the plate. { 'plāt ,kət }

plated circuit [ELECTR] A printed circuit produced by electrodeposition of a conductive pattern on an insulating base. Also known as plated printed circuit. { 'plād·əd 'sər·kət }

plated printed circuit See plated circuit. { 'plād·əd 'print·əd 'sər·kət }

plate efficiency [CHEM ENG] The equilibrium produced by an actual plate of a distillation column or countercurrent tower extractor compared with that of a perfect plate, expressed as a ratio. [ELECTR] See anode efficiency. { 'plāt i,fish·ən·sē }

plate feeder See apron feeder. { 'plāt ¦fēd·ər }

plate-fin exchanger [MECH ENG] Heat-transfer device made up of a stack or layers, with each layer consisting of a corrugated fin between flat metal sheets sealed off on two sides by channels or bars to form passages for the flow of fluids. { 'plāt ,fin iks,chān·jər }

plate girder [CIV ENG] A riveted or welded steel girder having a deep vertical web plate with a pair of angles along each edge to act as compression and tension flanges. { 'plāt ,gərd·ər }

plate girder bridge [CIV ENG] A fixed bridge consisting, in its simplest form, of two flange plates welded to a web plate in the overall shape of an I. { 'plāt ¦gərd·ər ,brij }

plate modulus [MECH] The ratio of the stress component T_{xx} in an isotropic, elastic body obeying a generalized Hooke's law to the corresponding strain component S_{xx}, when the strain components S_{yy} and S_{zz} are 0; the sum of the Poisson ratio and twice the rigidity modulus. { 'plāt ,mäj·ə·ləs }

platen [ENG] **1.** A flat plate against which something rests or is pressed. **2.** The rubber-covered roller of a typewriter against which paper is pressed when struck by the typebars. [MECH ENG] A flat surface for exchanging heat in a boiler or heat exchanger which may have extended heat transfer surfaces. { 'plat·ən }

plate-shear test [ENG] A method used to get true shear data on a honeycomb core by bonding the core between two thick steel plates and subjecting the core to shear by displacing the plates relative to each other by loading in either tension or compression. { 'plāt ¦shir ,test }

plate tower [CHEM ENG] A distillation tower along the internal height of which is a series of transverse plates (bubble-cap or sieve) to force intimate contact between downward flowing liquid and upward flowing vapor. { 'plāt ,taú·ər }

plate-type exchanger [MECH ENG] Heat-exchange device similar to a plate-and-frame filter press; fluids flow between the frame-held plates, transferring heat between them. { 'plāt ,tīp iks,chān·jər }

plate vibrator [ENG] A mechanically operated tamper fitted with a flat base. { 'plāt vī'brād·ər }

platform balance [ENG] A weighing device with a flat plate mounted above a balanced beam. { 'plat,fȯrm ,bal·əns }

platform blowing [ENG] Special technique for

blow-molding large parts made of plastic without sagging of the part being formed. { 'plat ,fȯrm ,blō·iŋ }

platform conveyor |MECH ENG| A single- or double-strand conveyor with plates of steel or hardwood forming a continuous platform on which the loads are placed. { 'plat,fȯrm kən ,vā·ər }

platform framing |BUILD| A construction method in which each floor is framed independently by nailing the horizontal framing member to the top of the wall studs. { 'plat,fȯrm ,främ·iŋ }

platinum resistance thermometer |ENG| The basis of the International Practical Temperature Scale of 1968 from 259.35° to 630.74°C; used in industrial thermometers in the range 0 to 650°C; capable of high accuracy because platinum is noncorrosive, ductile, and nonvolatile, and can be obtained in a very pure state. Also known as Callendar's thermometer. { 'plat·ən·əm ri¦zis·təns thər'mäm·əd·ər }

play |MECH ENG| Free or unimpeded motion of an object, such as the motion between poorly fitted or worn parts of a mechanism. { plā }

playback |ENG ACOUS| Reproduction of a sound recording. { 'plā,bak }

playback robot |CONT SYS| A robot that repeats the same sequence of motions in all its operations, and is first instructed by an operator who puts it through this sequence. { 'plā,bak 'rō,bät }

play for position |IND ENG| The prepositioning of an object by a worker for a subsequent operation in the performance of a task. { 'plā fȯr pə'zish·ən }

pleated cartridge |DES ENG| A filter cartridge made into a convoluted form that resembles the folds of an accordion. { 'plēd·əd 'kär·trij }

plenum |ENG| A condition in which air pressure within an enclosed space is greater than that in the outside atmosphere. { 'plen·əm }

plenum blower assembly |MECH ENG| In an automotive air-conditioning system, the assembly through which air passes on its way to the evaporator or heater core. { ¦ple·nəm 'blō·ər ə,sem·blē }

plenum chamber |ENG| An enclosed space in which a plenum condition exists; air is forced into it for slow distribution through ducts. { 'plen·əm ,chām·bər }

plenum system |MECH ENG| A heating or air conditioning system in which air is forced through a plenum chamber for distribution to ducts. { 'plen·əm ,sis·təm }

pli |MECH| A unit of line density (mass per unit length) equal to 1 pound per inch, or approximately 17.8580 kilograms per meter. { plē }

pliers |DES ENG| A small instrument with two handles and two grasping jaws, usually long and roughened, working on a pivot; used for holding small objects and cutting, bending, and shaping wire. { 'plī·ərz }

plinth block See skirting block. { 'plinth ,bläk }

plot |CIV ENG| A measured piece of land. { plät }

plotter |ENG| A visual display or board on which a dependent variable is graphed by an automatically controlled pen or pencil as a function of one or more variables. { 'pläd·ər }

plotting board |ENG| The surface portion of a plotter, on which graphs are recorded. Also known as plotting table. { 'pläd·iŋ ,bȯrd }

plotting table See plotting board. { 'pläd·iŋ ,tā·bəl }

plough |ENG| A groove cut lengthwise with the grain in a piece of wood. { plaů }

ploughed-and-tongued joint See feather joint. { ¦plaůd ən 'təŋd ,jȯint }

plowshare |DES ENG| The pointed part of a moldboard plow, which penetrates and cuts the soil first. { 'plaů,sher }

plug |ELEC| The half of a connector that is normally movable and is generally attached to a cable or removable subassembly; inserted in a jack, outlet, receptacle, or socket. { pləg }

plug-and-feather hole |ENG| A hole drilled in quarries for the purpose of splitting a block of stone by the plug-and-feather method. { ¦pləg ən 'feth·ər ,hōl }

plug bit See noncoring bit. { 'pləg ,bit }

plug cock See plug valve. { 'pləg ,käk }

plug cutter |DES ENG| A device for boring out short dowels or plugs from wood that exactly match standard drill sizes. { 'pləg ,kəd·ər }

plug forming |ENG| Thermoforming process for plastics molding in which a plug or male mold is used to partially preform the part before forming is completed, using vacuum or pressure. { 'pləg ,fȯrm·iŋ }

plug gage |DES ENG| A steel gage that is used to test the dimension of a hole; may be straight or tapered, plain or threaded, and of any cross-sectional shape. { 'pləg ,gāj }

plugging |ELEC| Braking an electric motor by reversing its connections, so it tends to turn in the opposite direction; the circuit is opened automatically when the motor stops, so the motor does not actually reverse. |ENG| The formation of a barrier (plug) of solid material in a process flow system, such as a pipe or reactor. { 'pləg·iŋ }

plug meter |ENG| A variable-area flowmeter in which a tapered plug, located in an orifice and raised until the resulting opening is sufficient to handle the fluid flow, is used to measure the flow rate. { 'pləg ,mēd·ər }

plug valve |MECH ENG| A valve fitted with a plug that has a hole through which fluid flows and that is rotatable through 90° for operation in the open or closed position. Also known as plug cock. { 'pləg ,valv }

plumb |ENG| Pertaining to an object or structure in true vertical position as determined by a plumb bob. { pləm }

plumb bob |ENG| A weight suspended on a string to indicate the direction of the vertical. { 'pləm ,bäb }

plumb bond |CIV ENG| A masonry bond in

411

which corresponding joints (for example, on alternate courses) are aligned. { 'pləm ‚bänd }

plumbing |CIV ENG| The system of pipes and fixtures concerned with the introduction, distribution, and disposal of water in a building. { 'pləm·iŋ }

plumb line |ENG| The string on which a plumb bob hangs. { 'pləm ‚līn }

plummet |ENG| A loose-fitting metal plug in a tapered rotameter tube which moves upward (or downward) with an increase (or decrease) in fluid flow rate upward through the tube. Also known as float. { 'pləm·ət }

plunge |ENG| To set the horizontal cross hair of a theodolite in the direction of a grade when establishing a grade between two points of known level. { plənj }

plunge grinding |MECH ENG| Grinding in which the wheel moves radially toward the work. { 'plənj ‚grīnd·iŋ }

plunger |DES ENG| A wooden shaft with a large rubber suction cup at the end, used to clear plumbing traps and waste outlets. |ENG| See force plug. |MECH ENG| The long rod or piston of a reciprocating pump. { 'plən·jər }

plunger pump |MECH ENG| A reciprocating pump where the packing is on the stationary casing instead of the moving piston. { 'plən·jər ‚pəmp }

plunger-type instrument |ENG| Moving-iron instrument in which the pointer is attached to a long and specially shaped piece of iron that is drawn into or moved out of a coil carrying the current to be measured. { 'plən·jər ‚tīp 'in·strə·mənt }

pluviograph See recording rain gage. { 'plü·vē·ə‚graf }

pluviometer See rain gage. { ‚plü·vē'äm·əd·ər }

PMOS |ELECTR| Metal-oxide semiconductors that are made on *n*-type substrates, and whose active carriers are holes that migrate between *p*-type source and drain contacts. Derived from *p*-channel metal-oxide semiconductor. { 'pē ‚mós }

pneumatic |ENG| Pertaining to or operated by air or other gas. { nü'mad·ik }

pneumatic atomizer |MECH ENG| An atomizer that uses compressed air to produce drops in the diameter range of 5–100 micrometers. { nü'mad·ik 'ad·ə‚mīz·ər }

pneumatic caisson |CIV ENG| A caisson having a chamber filled with compressed air at a pressure equal to the pressure of the water outside. { nü'mad·ik 'kā‚sän }

pneumatic controller |MECH ENG| A device for the mechanical movement of another device (such as a valve stem) whose action is controlled by variations in pneumatic pressure connected to the controller. { nü'mad·ik kən'trōl·ər }

pneumatic control valve |MECH ENG| A valve in which the force of compressed air against a diaphragm is opposed by the force of a spring to control the area of the opening for a fluid stream. { nü'mad·ik kən'trōl ‚valv }

pneumatic conveyor |MECH ENG| A conveyor which transports dry, free-flowing, granular material in suspension, or a cylindrical carrier, within a pipe or duct by means of a high-velocity airstream or by pressure of vacuum generated by an air compressor. Also known as air conveyor. { nü'mad·ik kən'vā·ər }

pneumatic drill |MECH ENG| Compressed-air drill worked by reciprocating piston, hammer action, or turbo drive. { nü'mad·ik 'dril }

pneumatic drilling |MECH ENG| Drilling a hole when using air or gas in lieu of conventional drilling fluid as the circulating medium; an adaptation of rotary drilling. { nü'mad·ik 'dril·iŋ }

pneumatic hammer |MECH ENG| A hammer in which compressed air is utilized for producing the impacting blow. Also known as air hammer; jack hammer. { nü'mad·ik 'ham·ər }

pneumatic hoist See air hoist. { nü'mad·ik 'hóist }

pneumatic loudspeaker |ENG ACOUS| A loudspeaker in which the acoustic ouput results from controlled variation of an airstream. { nü'mad·ik 'laúd‚spēk·ər }

pneumatic riveter |MECH ENG| A riveting machine having a rapidly reciprocating piston driven by compressed air. { nü'mad·ik 'riv·əd·ər }

pneumatic servo See valve positioner.

pneumatic servomechanism |CONT SYS| A servomechanism in which power is supplied and transmission of signals is carried out through the medium of compressed air. { nü'mad·ik ‚sər·vō 'mek·ə‚niz·əm }

pneumatic telemetering |ENG| The transmission of a pressure impulse by means of pneumatic pressure through a length of small-bore tubing; used for remote transmission of signals from primary process-unit sensing elements for pressure, temperature, flow rate, and so on. { nü'mad·ik 'tel·ə‚mēd·ə·riŋ }

pneumatic test |ENG| Pressure testing of a process vessel by the use of air pressure. { nü'mad·ik 'test }

pneumatic weighing system |ENG| A system for weight measurement in which the load is detected by a nozzle and balanced by modulating the air pressure in an opposing capsule. { nü'mad·ik 'wā·iŋ ‚sis·təm }

pn hook transistor See hook collector transistor. { ‚pē¦en 'húk tran‚zis·tər }

pnip transistor |ELECTR| An intrinsic junction transistor in which the intrinsic region is sandwiched between the *n*-type base and the *p*-type collector. { ‚pē¦en‚ī‚pē tran‚zis·tər }

pn junction |ELECTR| The interface between two regions in a semiconductor crystal which have been treated so that one is a *p*-type semiconductor and the other is an *n*-type semiconductor; it contains a permanent dipole charge layer. { ‚pē¦en ‚jəŋk·shən }

pnpn diode |ELECTR| A semiconductor device consisting of four alternate layers of *p*-type and *n*-type semiconductor material, with terminal connections to the two outer layers. Also known as *npnp* diode. { ‚pē¦en¦pē¦en ‚dī‚ōd }

pnpn transistor See *npnp* transistor. { ¦pē¦en¦pē¦en tran,zis·tər }

pnp transistor │ELECTR│ A junction transistor having an *n*-type base between a *p*-type emitter and a *p*-type collector. { ¦pē¦en¦pē tran,zis·tər }

pocket │BUILD│ A recess in a wall designed to receive a folding or sliding door in the open position. │CIV ENG│ A recess made in masonry to receive the end of a beam. { 'päk·ət }

pod │DES ENG│ **1.** The socket for a bit in a brace. **2.** A straight groove in the barrel of a pod auger. { päd }

Podbielniak extractor │CHEM ENG│ A solvent-extraction device in which centrifugal action enhances liquid-liquid contact and increases resultant separation efficiency. { päd'bēl·nē,ak ik,strak·tər }

Pohlé air lift pump │MECH ENG│ A pistonless pump in which compressed air fills the annular space surrounding the uptake pipe and is free to enter the rising column at all points of its periphery. { pō'lā 'er ,lift ,pəmp }

poidometer │ENG│ An automatic weighing device for use on belt conveyors. { pói'däm·əd·ər }

Poincaré surface of section │MECH│ A method of displaying the character of a particular trajectory without examining its complete time development, in which the trajectory is sampled periodically, and the rate of change of a quantity under study is plotted against the value of that quantity at the beginning of each period. Also known as surface of section. { ,pwän,kä'rä 'sər·fəs əv 'sek·shən }

Poinsot ellipsoid See inertia ellipsoid. { pwän'sō ə'lip,sóid }

Poinsot motion │MECH│ The motion of a rigid body with a point fixed in space and with zero torque or moment acting on the body about the fixed point. { pwän'sō ,mō·shən }

Poinsot's central axis │MECH│ A line through a rigid body which is parallel to the vector sum **F** of a system of forces acting on the body, and which is located so that the system of forces is equivalent to the force **F** applied anywhere along the line, plus a couple whose torque is equal to the component of the total torque **T** exerted by the system in the direction **F**. { ¦pwän·sōz ¦sen·trəl 'ak·səs }

Poinsot's method │MECH│ A method of describing Poinsot motion, by means of a geometrical construction in which the inertia ellipsoid rolls on the invariable plane without slipping. { pwän'sōz ¦meth·əd }

point angle │DES ENG│ The angle at the point or edge of a cutting tool. { 'póint ,aŋ·gəl }

point-bearing pile See end-bearing pile. { 'póint ¦ber·iŋ ,pīl }

point-blank range │MECH│ Distance to a target that is so short that the trajectory of a bullet or projectile is practically a straight, rather than a curved, line. { 'póint¦blaŋk 'rānj }

point contact │ELECTR│ A contact between a specially prepared semiconductor surface and a metal point, usually maintained by mechanical

pressure but sometimes welded or bonded. { 'póint 'kän,takt }

point-contact diode │ELECTR│ A semiconductor rectifier that uses the barrier formed between a specially prepared semiconductor surface and a metal point to produce the rectifying action. { 'póint ¦kän,takt ,dī,ōd }

point-contact transistor │ELECTR│ A transistor having a base electrode and two or more point contacts located near each other on the surface of an *n*-type semiconductor. { 'póint ¦kän,takt tran,zis·tər }

pointer │ENG│ The needle-shaped rod that moves over the scale of a meter. { 'póint·ər }

pointing │CIV ENG│ **1.** Finishing a mortar joint. **2.** Pressing mortar into a raked joint. { 'póint·iŋ }

pointing trowel │ENG│ A tool used to apply pointing to the joints between bricks. { 'póint·iŋ ,traúl }

point initiation │ENG│ Application of the initial impulse from the detonator to a single point on the main charge surface; for a cylindrical charge this point is usually the center of one face. { 'póint i,nish·ē'ā·shən }

point-junction transistor │ELECTR│ Transistor having a base electrode and both point-contact and junction electrodes. { 'póint ,jəŋk·shən tran,zis·tər }

point of contraflexure │MECH│ A point at which the direction of bending changes. Also known as point of inflection. { 'póint əv ,kän·trə'flek·shər }

point of control │IND ENG│ Fraction defective in those lots that have a probability of .50 of acceptance according to a specific sampling acceptance plan. { 'póint əv kən'trōl }

point of fall │MECH│ The point in the curved path of a falling projectile that is level with the muzzle of the gun. Also known as level point. { 'póint əv 'fól }

point of frog │CIV ENG│ The place of intersection of the gage lines of the main track and a turnout. { 'póint əv 'fróg }

point of inflection See point of contraflexure. { 'póint əv in'flek·shən }

point of intersection │CIV ENG│ The point at which two straight sections or tangents to a road curve or rail curve meet when extended. { 'póint əv ,in·tər'sek·shən }

point of switch │CIV ENG│ That place in a track where a car passes from the main track to a turnout. { 'póint əv 'swich }

point of tangency │CIV ENG│ The point at which a road curve or railway curve becomes straight or changes its curvature. Also known as tangent point. { 'póint əv 'tan·jən·sē }

point source │CIV ENG│ A municipal or industrial wastewater discharge through a discrete pipe or channel. { 'póint ,sórs }

point system │IND ENG│ **1.** A system of job evaluation wherein job requirements are rated according to a scale of point values. **2.** A wage incentive plan based on points instead of man-minutes. { 'póint ,sis·təm }

point-to-point programming |CONT SYS| A method of programming a robot in which each major change in the robot's path of motion is recorded and stored for later use. { ¦pȯint tə ¦pȯint 'prō‚gram·iŋ }

poison |ELECTR| A material which reduces the emission of electrons from the surface of a cathode. { 'pȯiz·ən }

Poisson bracket |MECH| For any two dynamical variables, X and Y, the sum, over all degrees of freedom of the system, of $(\partial X/\partial q)(\partial Y/\partial p) - (\partial X/\partial p)(\partial Y/\partial q)$, where q is a generalized coordinate and p is the corresponding generalized momentum. { pwä'sōn ‚brak·ət }

Poisson number |MECH| The reciprocal of the Poisson ratio. { pwä'sōn ‚nəm·bər }

Poisson ratio |MECH| The ratio of the transverse contracting strain to the elongation strain when a rod is stretched by forces which are applied at its ends and which are parallel to the rod's axis. { pwä'sōn ‚rā·shō }

polarity effect |ELECTR| An effect for which the breakdown voltage across a vacuum separating two electrodes, one of which is pointed, is much higher when the pointed electrode is the anode. { pə'lar·əd·ē i‚fekt }

polarizability |ELEC| The electric dipole moment induced in a system, such as an atom or molecule, by an electric field of unit strength. { ‚pō·lə‚rīz·ə'bil·əd·ē }

polarization |ELEC| **1.** The process of producing a relative displacement of positive and negative bound charges in a body by applying an electric field. **2.** A vector quantity equal to the electric dipole moment per unit volume of a material. Also known as dielectric polarization; electric polarization. **3.** A chemical change occurring in dry cells during use, increasing the internal resistance of the cell and shortening its useful life. { ‚pō·lə·rə'zā·shən }

polarization charge See bound charge. { ‚pō·lə·rə'zā·shən ‚chärj }

polarized meter |ENG| A meter having a zero-center scale, with the direction of deflection of the pointer depending on the polarity of the voltage or the direction of the current being measured. { 'pō·lə‚rīzd 'mēd·ər }

polarized-vane ammeter |ENG| An ammeter of only moderate accuracy in which the current to be measured passes through a small coil, distorting the field of a circular permanent magnet, and an iron vane aligns itself with the axis of the distorted field, the deflection being roughly proportional to the current. { 'pō·lə‚rīzd ‚vān 'am‚ēd·ər }

polarizing pyrometer |ENG| A type of pyrometer, such as the Wanner optical pyrometer, in which monochromatic light from the source under investigation and light from a lamp with filament maintained at a constant but unknown temperature are both polarized and their intensities compared. { 'pō·lə‚rīz·iŋ pī'räm·əd·ər }

polar radiation pattern |ENG ACOUS| Diagram showing the strength of sound waves radiated from a loudspeaker in various directions in a given plane, or a similar response pattern for a microphone. { 'pō·lər ‚rād·ē'ā·shən ‚pad·ərn }

polar timing diagram |MECH ENG| A diagram of the events of an engine cycle relative to crankshaft position. { 'pō·lər 'tīm·iŋ ‚dī·ə‚gram }

polder |CIV ENG| Land reclaimed from the sea or other body of water by the construction of an embankment to restrain the water. { 'pōl·dər }

pole |ELEC| **1.** One of the electrodes in an electric cell. **2.** An output terminal on a switch; a double-pole switch has two output terminals. |MECH| **1.** A point at which an axis of rotation or of symmetry passes through the surface of a body. **2.** See perch. { pōl }

pole-dipole array |ENG| An electrode array used in a lateral search conducted during a resistivity or induced polarization survey, or in drill hole logging, in which one current electrode is placed at infinity while another current electrode and two potential electrodes in proximity are moved across the structure to be investigated. { 'pōl 'dī‚pōl ə‚rā }

pole lathe |MECH ENG| A simple lathe in which the work is rotated by a cord attached to a treadle. { 'pōl ‚lāth }

pole-pole array |ENG| An electrode array, used in lateral search or in logging, in which one current electrode and the other potential electrode are kept in proximity and traversed across the structure. { 'pōl 'pōl ə‚rā }

pole-positioning |CONT SYS| A design technique used in linear control theory in which many or all of a system's closed-loop poles are positioned as required, by proper choice of a linear state feedback law; if the system is controllable, all of the closed-loop poles can be arbitrarily positioned by this technique. { 'pōl pə‚zish·ən·iŋ }

polestar recorder |ENG| An instrument used to determine approximately the amount of cloudiness during the dark hours; consists of a fixed long-focus camera positioned so that Polaris is permanently within its field of view; the apparent motion of the star appears as a circular arc on the photograph and is interrupted as clouds come between the star and the camera. { 'pōl‚stär ri‚kȯrd·ər }

pole-zero configuration |CONT SYS| A plot of the poles and zeros of a transfer function in the complex plane; used to study the stability of a system, its natural motion, its frequency response, and its transient response. { 'pōl ¦zir·ō kən‚fig·yə'rā·shən }

polhode |MECH| For a rotating rigid body not subject to external torque, the closed curve traced out on the inertia ellipsoid by the intersection with this ellipsoid of an axis parallel to the angular velocity vector and through the center. { 'pä‚lōd }

polhode cone See body cone. { 'pä‚lōd ‚kōn }

poling |ELEC| Adjustment of polarity; specifically, in wire-line practice, the use of transpositions between transposition sections of open wire or between lengths of cable, to cause the residual cross-talk couplings in individual sections or lengths to oppose one another. { 'pōl·iŋ }

poling board |CIV ENG| A timber plank driven into soft soil to support the sides of an excavation. { 'pōl·iŋ ,bȯrd }

polishing |CHEM ENG| In petroleum refining, removal of final traces of impurities, as for a lubricant, by clay adsorption or mild hydrogen treating. |MECH ENG| Smoothing and brightening a surface such as a metal or a rock through the use of abrasive materials. { 'päl·ish·iŋ }

polishing roll |MECH ENG| A roll or series of rolls on a plastics mold; has highly polished chrome-plated surfaces; used to produce a smooth surface on a plastic sheet as it is extruded. { 'päl·ish·iŋ ,rōl }

polishing wheel |DES ENG| An abrasive wheel used for polishing. { 'päl·ish·iŋ ,wēl }

polyforming |CHEM ENG| A noncatalytic, petroleum-refinery process charging C_3 and C_4 gases with naphtha or gas oil at high temperature to produce high-quality gasoline and fuel oil; mostly replaced by catalytic reforming; the product is known as polyformdistillate. { 'päl·ē,fȯrm·iŋ }

polygraph See lie detector. { 'päl·i,graf }

polyimide |CHEM ENG| A group of polymers that contain a repeating imide group (−CON-HCO −). Aromatic polyimides are noted for their resistance to high temperatures, wear, and corrosion. { ¦päl·ē'ī,mīd }

polyliner |ENG| A perforated sleeve with longitudinal ribs that is used inside the cylinder of an injection-molding machine. { 'päl·i,līn·ər }

polyphase |ELEC| Having or utilizing two or more phases of an alternating-current power line. { 'päl·i,fāz }

polyphase circuit |ELEC| Group of alternating-current circuits (usually interconnected) which enter (or leave) a delimited region at more than two points of entry; they are intended to be so energized that, in the steady state, the alternating currents through the points of entry, and the alternating potential differences between them, all have exactly equal periods, but have differences in phase, and may have differences in waveform. { 'päl·i,fāz 'sər·kət }

polyphase meter |ENG| An instrument which measures some electrical quantity, such as power factor or power, in a polyphase circuit. { 'päl·i,fāz 'mēd·ər }

polyphase wattmeter |ENG| An instrument that measures electric power in a polyphase circuit. { 'päl·i,fāz 'wät,mēd·ər }

polysulfide treating |CHEM ENG| A petroleum-refinery process used to remove elemental sulfur from refinery liquids by contacting them with a nonregenerable solution of sodium polysulfide. { ¦päl·i'səl,fīd 'trēd·iŋ }

polytropic process |THERMO| An expansion or compression of a gas in which the quantity pV^n is held constant, where p and V are the pressure and volume of the gas, and n is some constant. { ¦päl·i¦träp·ik 'prä·səs }

PONA analysis |ENG| American Society for Testing and Materials analysis of paraffins (P), olefins (O), naphthenes (N), and aromatics (A) in gasolines. { 'pō·nə or ¦pē¦ō¦en'ā a,nal·ə·səs }

Ponchon-Savarit method |CHEM ENG| Graphical solution on an enthalpy-concentration diagram of liquid-vapor equilibrium values between trays of a distillation column. { ,pȯn ,shȯn ,sav·ə'rē ,meth·əd }

pond See gram-force. { pänd }

ponding |BUILD| An accumulation of water on a flat roof because of clogged or inadequate drains. |CIV ENG| **1.** The impoundment of stream water to form a pond. **2.** Covering the surface of newly poured concrete with a thin layer of water to promote curing. { 'pänd·iŋ }

pontoon bridge |CIV ENG| A fixed floating bridge supported by pontoons. { pän¦tün 'brij }

pontoon-tank roof |ENG| A type of floating tank roof, supported by buoyant floats on the liquid surface of a tank; the roof rises and falls with the liquid level in the tank; used to minimize vapor space above the liquid, thus reducing vapor losses during tank filling and emptying. { pänh¦tün ¦taŋk ,rüf }

pony truss |CIV ENG| A truss too low to permit overhead braces. { 'pō·nē ,trəs }

pool |CIV ENG| A body of water contained in a reservoir, by a dam, or by the gates of a lock. { pül }

Poole-Frenkel effect |ELEC| An increase in the electrical conductivity of insulators and semiconductors in strong electric fields. { ¦pül 'freŋ·kəl i,fekt }

pop action |MECH ENG| The action of a safety valve as it opens under steam pressure when the valve disk is lifted off its seat. { 'päp ,ak·shən }

Popov's stability criterion |CONT SYS| A frequency domain stability test for systems consisting of a linear component described by a transfer function preceded by a nonlinear component characterized by an input-output function), with a unity gain feedback loop surrounding the series connection. { pä'pȯfs stə'bil·əd·ē krī,tir·ē·ən }

poppet |CIV ENG| One of the timber and steel structures supporting the fore and aft ends of a ship for launching from sliding ways. |DES ENG| A spring-loaded ball engaging a notch; a ball latch. { 'päp·ət }

poppet valve |MECH ENG| A cam-operated or spring-loaded reciprocating-engine mushroom-type valve used for control of admission and exhaust of working fluid; the direction of movement is at right angles to the plane of its seat. { 'päp·ət ,valv }

popping pressure |MECH ENG| In compressible fluid service, the inlet pressure at which a safety valve disk opens. { 'päp·iŋ ,presh·ər }

population |ELECTR| The set of electronic components on a printed circuit board. { ,päp·yə'lā·shən }

porcupine boiler |MECH ENG| A boiler having dead end tubes projecting from a vertical shell. { 'pȯr·kyə,pīn ¦bȯil·ər }

pore diameter |DES ENG| The average or effective diameter of the openings in a membrane, screen, or other porous material. { 'pór dī,am·əd·ər }

porosimeter |ENG| Laboratory compressed-gas device used for measurement of the porosity of reservoir rocks. { ,pór·ə'sim·əd·ər }

porous bearing |DES ENG| A bearing made from sintered metal powder impregnated with oil by a vacuum treatment. { 'pór·əs 'ber·iŋ }

porous mold |ENG| A plastic-forming mold made from bonded or fused aggregates (such as powdered metal or coarse pellets) so that the resulting mass contains numerous open interstices through which air or liquids can pass. { 'pór·əs 'mōld }

porous wheel |DES ENG| A grinding wheel having a porous structure and a vitrified or resinoid bond. { 'pór·əs 'wēl }

port |ELEC| An entrance or exit for a network. |ENG| The side of a ship or airplane on the left of a person facing forward. |ENG ACOUS| An opening in a bass-reflex enclosure for a loudspeaker, designed and positioned to improve bass response. { pórt }

portable |ENG| Capable of being easily and conveniently transported. { 'pórd·ə·bəl }

portal |ENG| A redundant frame consisting of two uprights connected by a third member at the top. { 'pórd·əl }

portal crane |MECH ENG| A jib crane carried on a four-legged portal built to run on rails. { 'pórd·əl 'krān }

porthole |DES ENG| The opening or passageway connecting the inside of a bit or core barrel to the outside and through which the circulating medium is discharged. |ENG| A circular opening in the side of a ship or airplane, usually serving as a window and containing one or more panes of glass. { 'port,hōl }

port of entry |CIV ENG| A location for clearance of foreign goods and citizens through a customhouse. { 'pórt əv 'en·trē }

positional-error constant |CONT SYS| For a stable unity feedback system, the limit of the transfer function as its argument approaches zero. { pə'zish·ən·əl |er·ər ,kän·stənt }

positional servomechanism |CONT SYS| A feedback control system in which the mechanical position (as opposed to velocity) of some object is automatically maintained. { pə'zish·ən·əl |sər·vō'mek·ə,niz·əm }

position-analog unit |ENG| A device employed in machining operations to transmit analog information about the positions of machine parts to a servoamplifier which then compares it with input data. { pə'zish·ən |an·ə,läg ,yü·nət }

position-contouring system |CONT SYS| A numerical control system that exerts contouring control in two dimensions and position control in a third. { pə'zish·ən 'kän,túr·iŋ ,sis·təm }

position control |CONT SYS| A type of automatic control in which the input commands are the desired position of a body. { pə'zish·ən kən,trōl }

position indicator |ENG| An electromechanical dead-reckoning computer, either an air-position indicator or a ground-position indicator. { pə'zish·ən ,in·də,kād·ər }

positioning |MECH ENG| A tooling function concerned with manipulating the workpiece in relationship to the working tools. { pə'zish·ən·iŋ }

positioning action |CONT SYS| Automatic control action in which there is a predetermined relation between the value of a controlled variable and the position of a final control element. { pə'zish·ən·iŋ ,ak·shən }

positioning time |MECH ENG| The time required to move a machining tool from one coordinate position to the next. { pə'zish·ən·iŋ ,tīm }

position sensor |ENG| A device for measuring a position and converting this measurement into a form convenient for transmission. Also known as position transducer. { pə'zish·ən ,sen·sər }

position telemetering |ENG| A variation of voltage telemetering in which the system transmits the measurand by positioning a variable resistor or other component in a bridge circuit so as to produce relative magnitudes of electrical quantities or phase relationships. { pə'zish·ən |tel·ə'mēd·ə·riŋ }

position transducer See position sensor. { pə'zish·ən tranz,dü·sər }

positive |ELEC| Having fewer electrons than normal, and hence having ability to attract electrons. { 'päz·əd·iv }

positive acceleration |MECH| 1. Accelerating force in an upward sense or direction, such as from bottom to top, or from seat to head. 2. The acceleration in the direction that this force is applied. { 'päz·əd·iv ak,sel·ə'rā·shən }

positive charge |ELEC| The type of charge which is possessed by protons in ordinary matter, and which may be produced in a glass object by rubbing with silk. { 'päz·əd·iv 'chärj }

positive click adjustment |IND ENG| A means of adjusting dials or push buttons to incorporate audible clicks or their tactile counterparts at predetermined positions in order to provide appropriate motor-sensory feedback to the operator. { |päz·əd·iv |klik ə'jəz·mənt }

positive clutch |MECH ENG| A clutch designed to transmit torque without slip. { 'päz·əd·iv 'kləch }

positive-displacement compressor |MECH ENG| A compressor that confines successive volumes of fluid within a closed space in which the pressure of the fluid is increased as the volume of the closed space is decreased. { 'päz·əd·iv dis|pläs·mənt kəm,pres·ər }

positive-displacement meter |ENG| A fluid quantity meter that separates and captures definite volumes of the flowing stream one after another and passes them downstream, while counting the number of operations. { 'päz·əd·iv dis|pläs·mənt ,mēd·ər }

positive-displacement pump |MECH ENG| A

pump in which a measured quantity of liquid is entrapped in a space, its pressure is raised, and then it is delivered; for example, a reciprocating piston-cylinder or rotary-vane, gear, or lobe mechanism. { 'päz·əd·iv dis¦pläs·mənt ‚pəmp }

positive draft |MECH ENG| Pressure in the furnace or gas passages of a steam-generating unit which is greater than atmospheric pressure. { 'päz·əd·iv 'draft }

positive drive belt See timing belt. { 'päz·əd·iv 'drīv ‚belt }

positive electrode See anode. { 'päz·əd·iv i'lek‚trōd }

positive feedback |CONT SYS| Feedback in which a portion of the output of a circuit or device is fed back in phase with the input so as to increase the total amplification. Also known as reaction (British usage); regeneration; regenerative feedback; retroaction (British usage). { 'päz·əd·iv 'fēd‚bak }

positive mold |ENG| A plastics mold designed to trap all of the molding resin when the mold closes. { 'päz·əd·iv 'mōld }

positive motion |MECH ENG| Motion transferred from one machine part to another without slippage. { 'päz·əd·iv 'mō·shən }

positive temperature coefficient |THERMO| The condition wherein the resistance, length, or some other characteristic of a substance increases when temperature increases. { 'päz·əd·iv 'tem·prə·chər ‚kō·i‚fish·ənt }

positive terminal |ELEC| The terminal of a battery or other voltage source toward which electrons flow through the external circuit. { 'päz·əd·iv 'tərm·ən·əl }

positron camera |ENG| An instrument that uses photomultiplier tubes in combination with scintillation counters to detect oppositely directed gamma-ray pairs resulting from the annihilation with electrons of positrons emitted by short-lived radioisotopes used as tracers in the human body. { 'päz·ə‚trän ‚kam·rə }

post |CIV ENG| **1.** A vertical support such as a pillar, upright, or fence stake. **2.** A pole used as a boundary marker. { pōst }

post-and-beam construction |BUILD| A type of wall construction using posts instead of studs. { ‚pōst ən 'bēm kən‚strək·shən }

postauricular hearing aid |ENG ACOUS| A hearing aid that fits behind the ear and has a sound tip attached to plastic tubing that conducts sound through an ear mold to the ear canal. { ‚pōst·ȯ‚rik·yə·lər 'hēr·iŋ ‚ād }

post brake |MECH ENG| A brake occasionally fitted on a steam winder or haulage, and consisting of two upright posts mounted on either side of the drum that operate on brake paths bolted to the drum cheeks. { 'pōst ‚brāk }

postcure bonding |ENG| A method of postcuring at elevated temperatures of parts previously subjected to autoclave or press in order to obtain higher heat-resistant properties of the adhesive bond. { 'pōst‚kyür 'bänd·iŋ }

post drill |ENG| An auger or drill supported by a post. { 'pōst‚dril }

postemphasis See deemphasis. { ¦pōst'em·fə·səs }

postequalization See deemphasis. { ¦pōst‚ē·kwə·lə'zā·shən }

postforming |ENG| Forming, bonding, or shaping of heated, flexible thermoset laminates before the final thermoset reaction has occurred; upon cooling, the formed shape is held. { pōst 'fȯrm·iŋ }

posthole |CIV ENG| A hole bored in the ground to hold a fence post. { 'pōst‚hōl }

postsynchronizing studio See ADR studio. { ‚pōst‚siŋ·krə‚nīz·iŋ 'stüd·ē·ō }

posttensioning |ENG| Compressing of cast concrete beams or other structural members to impart the characteristics of prestressed concrete. { pōs'ten·shən·iŋ }

pot See potentiometer; pothole. { pät }

pot die forming |MECH ENG| Forming sheet or plate metal through a hollow die by the application of pressure which causes the workpiece to assume the contour of the die. { 'pät 'dī ‚fȯrm·iŋ }

potential See electric potential. { pə'ten·chəl }

potential difference |ELEC| Between any two points, the work which must be done against electric forces to move a unit charge from one point to the other. Abbreviated PD. { pə'ten·chəl ¦dif·rəns }

potential divider See voltage divider. { pə'ten·chəl di'vīd·ər }

potential drop |ELEC| The potential difference between two points in an electric circuit. { pə'ten·chəl ¦dräp }

potential energy |MECH| The capacity to do work that a body or system has by virtue of its position or configuration. { pə'ten·chəl 'en·ər·jē }

potential flow analyzer See electrolytic tank. { pə'ten·chəl ¦flō 'an·ə‚līz·ər }

potential gradient |ELEC| Difference in the values of the voltage per unit length along a conductor or through a dielectric. { pə'ten·chəl 'grād·ē·ənt }

potential temperature |THERMO| The temperature that would be reached by a compressible fluid if it were adiabatically compressed or expanded to a standard pressure, usually 1 bar. { pə'ten·chəl 'tem·prə·chər }

potential transformer See voltage transformer. { pə'ten·chəl tranz'fȯr·mər }

potential transformer phase angle |ELEC| Angle between the primary voltage vector and the secondary voltage vector reversed; this angle is conveniently considered as positive when the reversed, secondary voltage vector leads the primary voltage vector. { pə'ten·chəl tranz'fȯr·mər 'fāz ‚aŋ·gəl }

potentiometer |ELEC| A resistor having a continuously adjusted sliding contact that is generally mounted on a rotating shaft; used chiefly as a voltage divider. Also known as pot (slang). |ENG| A device for the measurement of an electromotive force by comparison with a known potential difference. { pə‚ten·chē'äm·əd·ər }

potentiometric controller |CONT SYS| A controller that operates on the null balance principle, in which an error signal is produced by balancing the sensor signal against a set-point voltage in the input circuit; the error signal is amplified for use in keeping the load at a desired temperature or other parameter. { pə¦ten·chē·ə¦me·trik kən'trōl·ər }

potentiostat |ENG| An automatic laboratory instrument that controls the potential of a working electrode to within certain limits during coulometric (electrochemical reaction) titrations. { pə'ten·chē·ə‚stat }

pot furnace |ENG| **1.** A furnace containing several pots in which glass is melted. **2.** A furnace in which the charge is contained in a pot or crucible. { 'pät ‚fər·nəs }

pothole |CIV ENG| A pot-shaped hole in a pavement surface. { 'pät‚hōl }

Potier diagram |ELEC| Vector diagram showing the voltage and current relations in an alternating-current generator. { pō'tyā ‚dī·ə‚gram }

pot life |CHEM ENG| See work life. |ENG| The period of time during which paint remains useful after its original package has been opened or after a catalyst or other additive has been incorporated. Also known as spreadable life; usable life. { 'pät ‚līf }

potometer |ENG| A device for measuring transpiration, consisting of a small vessel containing water and sealed so that the only escape of moisture is by transpiration from a leaf, twig, or small plant with its cut end inserted in the water. { pō'täm·əd·ər }

potomology |CIV ENG| The systematic study of the factors affecting river channels to provide the basis for predictions of the effects of proposed engineering works on channel characteristics. { ‚päd·ə'mäl·ə·jē }

pot plunger |ENG| A plunger used to force softened plastic molding material into the closed cavity of a transfer mold. { 'pät ‚plən·jər }

potter's wheel |ENG| A revolving horizontal disk that turns when a treadle is operated; used to shape clay by hand. { 'päd·ərz 'wēl }

potting |ELECTR| Process of filling a complete electronic assembly with a thermosetting compound for resistance to shock and vibration, and for exclusion of moisture and corrosive agents. { 'päd·iŋ }

pound |MECH| **1.** A unit of mass in the English absolute system of units, equal to 0.45359237 kilogram. Abbreviated lb. Also known as avoirdupois pound; pound mass. **2.** A unit of force in the English gravitational system of units, equal to the gravitational force experienced by a pound mass when the acceleration of gravity has its standard value of 9.80665 meters per second per second (approximately 32.1740 ft/s²) equal to 4.4482216152605 newtons. Abbreviated lb. Also spelled Pound (Lb). Also known as pound force (lbf). **3.** A unit of mass in the troy and apothecaries' systems, equal to 12 troy or apothecaries' ounces, or 5760 grains, or 5760/

7000 avoirdupois pound, or 0.3732417216 kilogram. Also known as apothecaries' pound (abbreviated lb ap in the United States or lb apoth in the United Kingdom); troy pound (abbreviated lb t in the United States, or lb tr or lb in the United Kingdom). { paùnd }

poundal |MECH| A unit of force in the British absolute system of units equal to the force which will impart an acceleration of 1 ft/s² to a pound mass, or to 0.138254954376 newton. { 'paùnd·əl }

poundal-foot See foot-poundal. { 'paùnd·əl 'fùt }

pound-foot See foot-pound. { 'paùnd 'fùt }

pound force See pound. { 'paùnd 'fòrs }

pound mass See pound. { 'paùnd 'mas }

pound per square foot |MECH| A unit of pressure equal to the pressure resulting from a force of 1 pound applied uniformly over an area of 1 square foot. Abbreviated psf. { 'paùnd pər ¦skwer 'fùt }

pound per square inch |MECH| A unit of pressure equal to the pressure resulting from a force of 1 pound applied uniformly over an area of 1 square inch. Abbreviated psi. { 'paùnd pər ¦skwer 'inch }

pounds per square inch absolute |MECH| The absolute, thermodynamic pressure, measured by the number of pounds-force exerted on an area of 1 square inch. Abbreviated lbf in.⁻² abs; psia. { 'paùns pər ¦skwer 'inch 'ab·sə‚lüt }

pounds per square inch differential |ENG| The difference in pressure between two points in a fluid-flow system, measured in pounds per square inch. Abbreviated psid. { 'paùns pər ¦skwer 'inch dif·ə'ren·chəl }

pounds per square inch gage |MECH| The gage pressure, measured by the number of pounds-force exerted on an area of 1 square inch. Abbreviated psig. { 'paùns pər ¦skwer 'inch 'gā }

pour test |ENG| The chilling of a liquid under specified test conditions to determine the American Society for Testing and Materials (ASTM) pour point. { 'pòr ‚test }

powder clutch |MECH ENG| A type of electromagnetic disk clutch in which the space between the clutch members is filled with dry, finely divided magnetic particles; application of a magnetic field coalesces the particles, creating friction forces between clutch members. { 'paùd·ər ‚kləch }

powder flowmeter |ENG| A device used to measure the flow rate of a metal powder. { 'paùd·ər 'flō‚mēd·ər }

powder house |CIV ENG| A magazine for the temporary storage of explosives. { 'paùd·ər ‚haùs }

powder keg |ENG| A small metal keg for black blasting powder. { 'paùd·ər ‚keg }

powder-moisture test |ENG| Determination of moisture in a propellant by drying under prescribed conditions; expressed as percentage by weight. { 'paùd·ər ‚mòis·chər ‚test }

powder molding |ENG| Generic term for plastics-molding techniques to produce objects of varying sizes and shapes by melting polyethylene powder, usually against the heated inside of a mold. { 'paud·ər ,mōld·iŋ }

powder train |ENG| **1.** Train, usually of compressed black powder, used to obtain time action in older fuse types. **2.** Train of explosives laid out for destruction by burning. { 'paud·ər ,trān }

power-actuated pressure relief valve |MECH ENG| A pressure relief valve connected to and controlled by a device which utilizes a separate energy source. { 'pau·ər ¦ak·chə,wād·əd 'presh·ər ri¦lēf ,valv }

power amplifier |ELECTR| The final stage in multistage amplifiers, such as audio amplifiers and radio transmitters, designed to deliver maximum power to the load, rather than maximum voltage gain, for a given percent of distortion. { 'pau·ər ¦am·plə,fī·ər }

power barker See barker. { 'pau·ər ,bärk·ər }

power brake |MECH ENG| An automotive brake with engine-intake-manifold vacuum used to amplify the atmospheric pressure on a piston operated by movement of the brake pedal. { 'pau·ər ,brāk }

power car |MECH ENG| **1.** A railroad car with equipment for furnishing heat and electric power to a train. **2.** A railroad car with controls, which can be operated by itself or as part of a train. { 'pau·ər ,kär }

power circuit |ELEC| The wires that carry current to electric motors and other devices that use electric power. { 'pau·ər ,sər·kət }

power component See active component. { 'pau·ər kəm,pō·nənt }

power control valve |MECH ENG| A safety relief device operated by a power-driven mechanism rather than by pressure. { 'pau·ər kən'trōl ,valv }

power cylinder |CONT SYS| A linear actuator consisting of a piston in a cylinder, driven by pneumatic or hydraulic fluid under high pressure. { 'pau·ər ,sil·ən·dər }

power dam |CIV ENG| A dam designed to raise the level of a stream to create or concentrate hydrostatic head for power purposes. { 'pau·ər ,dam }

power diode See pin diode. { 'pau·ər ,dī,ōd }

power drill |MECH ENG| A motor-driven drilling machine. { 'pau·ər ,dril }

power-driven |MECH ENG| Of a component or piece of equipment, moved, rotated, or operated by electrical or mechanical energy, as in a power-driven fan or power-driven turret. { 'pau·ər ,driv·ən }

power factor |ELEC| The ratio of the average (or active) power to the apparent power (root-mean-square voltage times rms current) of an alternating-current circuit. Abbreviated pf. Also known as phase factor. { 'pau·ər ,fak·tər }

power-factor meter |ENG| A direct-reading instrument for measuring power factor. { 'pau·ər ,fak·tər ,mēd·ər }

power-factor regulator |ELEC| Regulator which functions to maintain the power factor of a line or an apparatus at a predetermined value, or

to vary it according to a predetermined plan. { 'pau·ər ,fak·tər ,reg·yə,lād·ər }

power frequency |ELEC| The frequency at which electric power is generated and distributed; in most of the United States it is 60 hertz. { 'pau·ər ,frē·kwən·sē }

power generator |ELEC| A device for producing electric energy, such as an ordinary electric generator or a magnetohydrodynamic, thermionic, or thermoelectric power generator. { 'pau·ər ,jen·ə,rād·ər }

power grasp See power grip. { 'pau·ər ,grasp }

power grip |IND ENG| A basic grasp whereby the fingers are wrapped around an object and the thumb placed against it; used, for example, in certain hammering operations. Also known as power grasp. { 'pau·ər ,grip }

power level |ELEC| The ratio of the amount of power being transmitted past any point in an electric system to a reference power value; usually expressed in decibels. { 'pau·ər ,lev·əl }

power line |ELEC| Two or more wires conducting electric power from one location to another. Also known as electric power line. { 'pau·ər ,līn }

power-line carrier |ELEC| The use of transmission lines to transmit speech, metering indications, control impulses, and other signals from one station to another, without interfering with the lines' normal function of transmitting power. { 'pau·ər ,līn ,kar·ē·ər }

power-line filter See line filter. { 'pau·ər ,līn ,fil·tər }

power meter See electric power meter. { 'pau·ər ,mēd·ər }

power pack |ELECTR| Unit for converting power from an alternating- or direct-current supply into an alternating- or direct-current power at voltages suitable for supplying an electronic device. { 'pau·ər ,pak }

power package |MECH ENG| A complete engine and its accessories, designed as a single unit for quick installation or removal. { 'pau·ər ,pak·ij }

power plant |MECH ENG| Any unit that converts some form of energy into electrical energy, such as a hydroelectric or steam-generating station, a diesel-electric engine in a locomotive, or a nuclear power plant. Also known as electric power plant. { 'pau·ər ,plant }

power rating |ELEC| The power available at the output terminals of a component or piece of equipment that is operated according to the manufacturer's specifications. { 'pau·ər ,rād·iŋ }

power rectifier |ELEC| A device which converts alternating current to direct current and operates at high power loads. { 'pau·ər 'rek·tə,fī·ər }

power relay |ELEC| Relay that functions at a predetermined value of power; may be an overpower relay, an underpower relay, or a combination of both. { 'pau·ər 'rē,lā }

power resistor |ELEC| A resistor used in electric power systems, ranging in size from 5 watts to many kilowatts, and cooled by air convection, air blast, or water. { 'paủ·ǝr ri,zis·tǝr }

power saw [MECH ENG] A power-operated woodworking saw, such as a bench or circular saw. { 'paủ·ǝr ,sȯ }

power semiconductor [ELECTR] A semiconductor device capable of dissipating appreciable power (generally over 1 watt) in normal operation; may handle currents of thousands of amperes or voltages up into thousands of volts, at frequencies up to 10 kilohertz. { 'paủ·ǝr 'sem·i·kǝn,dǝk·tǝr }

power shovel [MECH ENG] A power-operated shovel that carries a short boom on which rides a movable dipper stick carrying an open-topped bucket; used to excavate and remove debris. { 'paủ·ǝr ,shǝv·ǝl }

power slips *See* automatic slips. { 'paủ·ǝr ,slips }

power station *See* generating station. { 'paủ·ǝr ,stā·shǝn }

power steering [MECH ENG] A steering control system for a propelled vehicle in which an auxiliary power source assists the driver by providing the major force required to direct the road wheels. { 'paủ·ǝr ,stir·iŋ }

power stroke [MECH ENG] The stroke in an engine during which pressure is applied to the piston by expanding steam or gases. { 'paủ·ǝr ,strȯk }

power supply circuit [ELEC] An electrical network used to convert alternating current to direct current. { 'paủ·ǝr sǝ,plī ,sǝr·kǝt }

power switch [ELEC] An electric switch which energizes or deenergizes an electric load; ranges from ordinary wall switches to load-break switches and disconnecting switches in power systems operating at voltages of hundreds of thousands of volts. { 'paủ·ǝr ,swich }

power train [MECH ENG] The part of a vehicle connecting the engine to propeller or driven axle; may include drive shaft, clutch, transmission, and differential gear. Also known as drive train. { 'paủ·ǝr ,trān }

power transformer [ELEC] An iron-core transformer having a primary winding that is connected to an alternating-current power line and one or more secondary windings that provide different alternating voltage values. { 'paủ·ǝr tranz,fȯr·mǝr }

power transistor [ELECTR] A junction transistor designed to handle high current and power; used chiefly in audio and switching circuits. { 'paủ·ǝr tran,zis·tǝr }

power transmission line [ELEC] The facility in an electric power system used to transfer large amounts of power from one location to a distant location; distinguished from a subtransmission or distribution line by higher voltage, greater power capability, and greater length. Also known as electric main; main (both British usages). { 'paủ·ǝr tranz'mish·ǝn ,līn }

power transmission tower [ELEC] A rigid steel tower supporting a high-voltage electric power transmission line, having a large enough spacing between conductors, and between conductors

and ground, to prevent corona discharge. { 'paủ·ǝr tranz'mish·ǝn ,taủ·ǝr }

power winding [ELEC] In a saturable reactor, a winding to which is supplied the power to be controlled; commonly the functions of the output and power windings are accomplished by the same winding, which is then termed the output winding. { 'paủ·ǝr ,wīnd·iŋ }

Poynting effect [MECH] The effect of torsion of a very long cylindrical rod on its length. { 'pȯin·tiŋ i,fekt }

Poynting's law [THERMO] A special case of the Clapeyron equation, in which the fluid is removed as fast as it forms, so that its volume may be ignored. { 'pȯint·iŋz ,lȯ }

pp junction [ELECTR] A region of transition between two regions having different properties in *p*-type semiconducting material. { ¦pē¦pē ,iǝŋk·shǝn }

practical entropy *See* virtual entropy. { 'prak·ti·kǝl 'en·trǝ·pē }

Prandtl number [THERMO] A dimensionless number used in the study of forced and free convection, equal to the dynamic viscosity times the specific heat at constant pressure divided by the thermal conductivity. Symbolized N_{Pr}. { 'pränt·ǝl ,nǝm·bǝr }

Pratt truss [CIV ENG] A truss having both vertical and diagonal members between the upper and lower chords, with the diagonals sloped toward the center. { 'prat ,trǝs }

preamplifier [ELECTR] An amplifier whose primary function is boosting the output of a low-level audio-frequency, radio-frequency, or microwave source to an intermediate level so that the signal may be further processed without appreciable degradation of the signal-to-noise ratio of the system. Also known as preliminary amplifier. { prē'am·plǝ,fī·ǝr }

preassembled [ENG] Assembled beforehand. { ¦prē·ǝ'sem·bǝld }

prebreaker [MECH ENG] Device used to break down large masses of solids prior to feeding them to a crushing or grinding device. { 'prē ¦brāk·ǝr }

precedence diagram method [IND ENG] A technique for constructing a network in which the activities are represented by symbols that are connected by lines to indicate the logical relationships between them. Abbreviated PDM. { ¦pres·ǝd·ǝns 'dī·ǝ,gram ,meth·ǝd }

precession [MECH] The angular velocity of the axis of spin of a spinning rigid body, which arises as a result of external torques acting on the body. { prē'sesh·ǝn }

precessional torque [MECH] A torque which causes a rotating body to precess. { prē¦sesh·ǝn·ǝl 'tȯrk }

prechlorination [CIV ENG] Chlorination of water before filtration. { ¦prē,klȯr·ǝ'nā·shǝn }

precipitation gage [ENG] Any device that measures the amount of precipitation; principally, a rain gage or snow gage. { prǝ,sip·ǝ'tā·shǝn ,gāj }

precipitator See electrostatic precipitator. { prə'sip·ə‚tād·ər }

precision block See gage block. { prə'sizh·ən ‚bläk }

precision depth recorder |ENG| A machine that plots sonar depth soundings on electrosensitive paper; can plot variations in depth over a range of 400 fathoms (730 meters) on a paper 18.85 inches (47.9 centimeters) wide. Abbreviated PDR. Also known as precision graphic recorder (PGR). { prə'sizh·ən 'depth ri‚kòrd·ər }

precision graphic recorder See precision depth recorder. { prə'sizh·ən 'graf·ik ri'kòrd·ər }

precision grinding |MECH ENG| Machine grinding to specified dimensions and low tolerances. { prə'sizh·ən ‚grīnd·iŋ }

precoat filter |ENG| A device designed to filter solid particles from a liquid-solid slurry after a precoat of builtup solid material (filter aid or filtered solid) has been applied to the inner surface of the filter medium. { 'prē‚kōt 'fil·tər }

precoating |ENG| The depositing of an inert material, such as filter aid, onto the filter medium prior to the filtration of suspended solids from a solid-liquid slurry. { ‚prē'kōd·iŋ }

precombustion chamber |MECH ENG| A small chamber before the main combustion space of a turbine or reciprocating engine in which combustion is initiated. { ‚prē·kəm‚bəs·chən ‚chäm·bər }

precooler |MECH ENG| A device for reducing the temperature of a working fluid before it is used by a machine. { prē'kül·ər }

preferential shop |IND ENG| An establishment in which preference is given to union members in hiring, layoffs, and dismissals, with the understanding that nonunion workers may be employed without being required to join the union when the union cannot supply workers. { ‚pref·ə'ren·chəl 'shäp }

prefilter |ENG| Filter used to remove gross solid contaminants before the liquid stream enters a separator-filter. { prē'fil·tər }

preform |ENG| **1.** A preshaped fibrous reinforcement. **2.** A compact mass of premixed plastic material that has been prepared for convenient handling and control of uniformity during the mold loading process. |ENG ACOUS| The small slab of record stock material that is loaded into a press to be formed into a disk recording. Also known as biscuit (deprecated usage). { prē'fórm }

preheater |MECH ENG| A device for preliminary heating of a material, substance, or fluid that will undergo further use or treatment by heating. { prē'hēd·ər }

preheat roll |ENG| In plastic-extrusion coating, the heated roll between the pressure roll and the unwind roll; used to heat the substrate before it is coated. { 'prē‚hēt ‚rōl }

preignition |MECH ENG| Ignition of the charge in the cylinder of an internal combustion engine before ignition by the spark. { ‚prē·ig'nish·ən }

preimpregnation |ENG| The mixing of a plastic resin with reinforcing material or substrate before molding takes place. { ‚prē·im‚preg'nā·shən }

preloading |ENG| For back-pressure-control gas valves, a weight or spring device to control the gas pressure at which the valve will open or close. { 'prē‚lōd·iŋ }

premix |ENG| In plastics molding, materials in which the resin, reinforcement, extenders, fillers, and so on have been premixed before molding. { 'prē‚miks }

premix gas burner |ENG| Fuel (gas or oil) burner in which fuel and air are premixed prior to ignition in the combustion chamber. { 'prē ‚miks 'gas ‚bər·nər }

preplastication |ENG| Premelting of injection-molding powders in a chamber separate from the injection cylinder. { ‚prē‚plas·tə'kā·shən }

prepolymer molding |ENG| A urethane-foam-producing system in which a portion of the polyol is prereacted with the isocyanate to form a liquid prepolymer with a pumpable viscosity; when combined with a second blend containing more polyol, catalyst, or blowing agent, the two components react and a foamed plastic results. { prē'päl·i·mər 'mōld·iŋ }

prepreg |ENG| A reinforced-plastics term for the reinforcing material that contains or is combined with the full complement of resin before the molding operation. { 'prē‚preg }

preprogrammed robot |CONT SYS| A robot that cannot adapt itself to the task it is carrying out, and must follow a built-in program. Also known as sequence robot. { ‚prē'prō‚gramd 'rō‚bät }

preset guidance |ENG| Guidance in which a predetermined path is set into the guidance mechanism of a craft, drone, or missile and is not altered after launching. { 'prē‚set 'gīd·əns }

preset tool |MECH ENG| A machine tool that is used to set an initial value of a parameter controlling another device. { 'prē‚set 'tül }

press |MECH ENG| Any of various machines by which pressure is applied to a workpiece, by which a material is cut or shaped under pressure, by which a substance is compressed, or by which liquid is expressed. { pres }

press bonding |ENG| A method of bonding structures or materials through the application of pressure by a platen press or other tool. { 'pres 'bänd·iŋ }

pressed loading |ENG| A loading operation in which bulk material, such as an explosive in granular form, is reduced in volume by the application of pressure. { 'prest 'lōd·iŋ }

press fit |ENG| An interference or force fit assembled through the use of a press. Also known as force fit. { 'pres ‚fit }

pressing |ENG ACOUS| A phonograph record produced in a record-molding press from a master or stamper. { 'pres·iŋ }

press polish |ENG| High-sheen finish on plastic sheet stock produced by contact with a smooth metal under heat and pressure. { 'pres ‚päl·ish }

press slide |MECH ENG| The reciprocating member of a power press on which the punch and upper die are fastened. { 'pres ˌslīd }

pressure |MECH| A type of stress which is exerted uniformly in all directions; its measure is the force exerted per unit area. { 'presh·ər }

pressure altimeter |ENG| A highly refined aneroid barometer that precisely measures the pressure of the air at the altitude an aircraft is flying, and converts the pressure measurement to an indication of height above sea level according to a standard pressure-altitude relationship. Also known as barometric altimeter. { 'presh·ər al 'tim·əd·ər }

pressure angle |MECH ENG| The angle that the line of force makes with a line at right angles to the center line of two gears at the pitch points. { 'presh·ər ˌaŋ·gəl }

pressure bag |ENG| A bag made of rubber, plastic, or other impermeable material that provides a flexible barrier between the pressure medium and the part being bonded. { 'presh·ər ˌbag }

pressure bar |MECH ENG| A bar that holds the edge of a metal sheet during press operations, such as punching, stamping, or forming, and prevents the sheet from buckling or becoming crimped. { 'presh·ər ˌbär }

pressure-base factor |CHEM ENG| Factor used in orifice pressure-drop calculations to allow for conditions where the pressure base used for calculating the orifice factor is not 14.73 pounds per square inch absolute (101.56 megapascals); calculated as $F_{pb} = 14.73$/pressure base (absolute). { 'presh·ər ˌbās ˌfak·tər }

pressure bulb |CIV ENG| The zone in a loaded soil mass bounded by an arbitrarily selected isobar of stress. { 'presh·ər ˌbəlb }

pressure carburetor See injection carburetor. { 'presh·ər ˌkär·bə'rād·ər }

pressure chamber |ENG| A chamber in which an artificial environment is established at low or high pressures to test equipment under simulated conditions of operation. { 'presh·ər ˌchām·bər }

pressure coefficient |THERMO| The ratio of the fractional change in pressure to the change in temperature under specified conditions, usually constant volume. { 'presh·ər ˌkō·i.fish·ənt }

pressure-containing member |MECH ENG| The part of a pressure-relieving device which is in direct contact with the pressurized medium in the vessel being protected. { 'presh·ər kən'tān·iŋ ˌmem·bər }

pressure control |ENG| Any device or system able to maintain, raise, or lower the pressure in a vessel or processing system as desired. { 'presh·ər kən,trōl }

pressure cooker |ENG| An autoclave designed for high-temperature cooking. { 'presh·ər ˌkük·ər }

pressure deflection |ENG| In a Bourdon or bellows-type pressure gage, the deflection or movement of the primary sensing element when pressured by the fluid being measured. { 'presh·ər di,flek·shən }

pressure-drop manometer |ENG| Manometer device (liquid-filled U tube) open at both ends, each end connected by tubing to a different location in a flow system (such as fluid- or gas-carrying pipe) to measure the drop in system pressure between the two points. { 'presh·ər ˌdräp ma'näm·əd·ər }

pressure dye test |ENG| A leak detection method in which a pressure vessel is filled with liquid dye and is pressurized under water to make possible leakage paths visible. { 'presh·ər 'dī ˌtest }

pressure elements |ENG| Those portions of a pressure-measurement gage which are moved or temporarily deformed by the gas or liquid of the system to which the gage is connected; the amount of movement or deformation is proportional to the pressure and is indicated by the position of a pointer or movable needle. { 'presh·ər ˌel·ə·məns }

pressure forming |ENG| A plastics thermoforming process using pressure to push the plastic sheet to be formed against the mold surface, as opposed to using vacuum to suck the sheet flat against the mold. { 'presh·ər ˌfȯrm·iŋ }

pressure gage |ENG| An instrument having metallic sensing element (as in a Bourdon pressure gage or aneroid barometer) or a piezoelectric crystal (as in a quartz pressure gage) to measure pressure. { 'presh·ər gāj }

pressure hydrophone |ENG ACOUS| A pressure microphone that responds to waterborne sound waves. { 'presh·ər 'hī·drə,fōn }

pressure measurement |ENG| Measurement of the internal forces of a process vessel, tank, or piping caused by pressurized gas or liquid; can be for a static or dynamic pressure, in English or metric units, either absolute (total) or gage (absolute minus atmospheric) pressure. { 'presh·ər ˌmezh·ər·mənt }

pressure microphone |ENG ACOUS| A microphone whose output varies with the instantaneous pressure produced by a sound wave acting on a diaphragm; examples are capacitor, carbon, crystal, and dynamic microphones. { 'presh·ər 'mī·krə·fōn }

pressure pad |ENG| A steel reinforcement in the face of a plastics mold to help the land absorb the closing pressure. |ENG ACOUS| A felt pad mounted on a spring arm, used to hold magnetic tape in close contact with the head on some tape recorders. { 'presh·ər ˌpad }

pressure pillow |ENG| A mechanical-hydraulic snow gage consisting of a circular rubber or metal pillow filled with a solution of antifreeze and water, and containing either a pressure transducer or a riser pipe to record increase in pressure of the snow. { 'presh·ər ˌpil·ō }

pressure plate |MECH ENG| The part of an automobile disk clutch that presses against the flywheel. { 'presh·ər ˌplāt }

pressure-plate anemometer |ENG| An anemometer which measures wind speed in terms of the drag which the wind exerts on a solid body; may be classified according to the means by which the wind drag is measured. Also known as plate anemometer. ('presh·ǝr ¦plāt ‚an·ǝ'mäm·ǝd·ǝr)

pressure process |CHEM ENG| Treatment of timber to prevent decay by forcing a preservative such as creosote and zinc chloride into the cells of the wood. ('presh·ǝr ‚prä·sǝs)

pressure rating |ENG| The operating (allowable) internal pressure of a vessel, tank, or piping used to hold or transport liquids or gases. ('presh·ǝr ‚rād·iŋ)

pressure-regulating valve |ENG| A valve that releases or holds process-system pressure (that is, opens or closes) either by preset spring tension or by actuation by a valve controller to assume any desired position between full open and full closed. ('presh·ǝr ¦reg·yǝ‚lād·iŋ ‚valv)

pressure regulator |ENG| Open-close device used on the vent of a closed, gas-pressured system to maintain the system pressure within a specified range. ('presh·ǝr 'reg·yǝ‚lād·ǝr)

pressure relief |ENG| A valve or other mechanical device (such as a rupture disk) that eliminates system overpressure by allowing the controlled or emergency escape of liquid or gas from a pressured system. ('presh·ǝr ri‚lēf)

pressure-relief device |MECH ENG| 1. In pressure vessels, a device designed to open in a controlled manner to prevent the internal pressure of a component or system from increasing beyond a specified value, that is, a safety valve. 2. A spring-loaded machine part which will yield, or deflect, when a predetermined force is exceeded. ('presh·ǝr ri‚lēf di‚vīs)

pressure-relief valve |MECH ENG| A valve which relieves pressure beyond a specified limit and recloses upon return to normal operating conditions. ('presh·ǝr ri‚lēf ‚valv)

pressure-retaining member |MECH ENG| That part of a pressure-relieving device loaded by the restrained pressurized fluid. ('presh·ǝr ri¦tān·iŋ ‚mem·bǝr)

pressure roll |ENG| In plastics-extrusion coating, the roll that with the chill roll applies pressure to the substrate and the molten extruded web. ('presh·ǝr ‚rōl)

pressure seal |ENG| A seal used to make pressure-proof the interface (contacting surfaces) between two parts that have frequent or continual relative rotational or translational motion. ('presh·ǝr ‚sēl)

pressure still |CHEM ENG| A continuous-flow, petroleum-refinery still in which heated oil (liquid and vapor) is kept under pressure so that it will crack (decompose into smaller molecules) to produce lower-boiling products (pressure distillate or pressure naphtha). ('presh·ǝr ‚stil)

pressure storage |ENG| The storage of a volatile liquid or liquefied gas under pressure to prevent evaporation. ('presh·ǝr ‚stȯr·ij)

pressure switch |ELEC| A switch that is actuated by a change in pressure of a gas or liquid. ('presh·ǝr ‚swich)

pressure system |ENG| Any system of pipes, vessels, tanks, reactors, and other equipment,

or interconnections thereof, operating with an internal pressure greater than atmospheric. ('presh·ǝr ‚sis·tǝm)

pressure tank |CHEM ENG| A pressurized tank into which timber is inserted for impregnation with preservative. |CIV ENG| An airtight water tank in which air is compressed to exert pressure on the water and which is used in connection with a water distribution system. ('presh·ǝr ‚taŋk)

pressure tap |ENG| A small perpendicular hole in the wall of a pressurized, fluid-containing pipe or vessel; used for connection of pressure-sensitive elements for the measurement of static pressures. Also known as piezometer opening; static pressure tap. ('presh·ǝr ‚tap)

pressure transducer |ENG| An instrument component that detects a fluid pressure and produces an electrical signal related to the pressure. Also known as electrical pressure transducer. ('presh·ǝr tranz‚dü·sǝr)

pressure-travel curve |MECH| Curve showing pressure plotted against the travel of the projectile within the bore of the weapon. ('presh·ǝr ¦trav·ǝl ‚kǝrv)

pressure treater |CHEM ENG| Any chemical treating device operated at higher-than-atmospheric pressure, as in the chemical and petroleum industries. ('presh·ǝr ‚trēd·ǝr)

pressure-tube anemometer |ENG| An anemometer which derives wind speed from measurements of the dynamic wind pressures; wind blowing into a tube develops a pressure greater than the static pressure, while wind blowing across a tube develops a pressure less than the static; this pressure difference, which is proportional to the square of the wind speed, is measured by a suitable manometer. ('presh·ǝr ¦tüb ‚an·ǝ'mäm·ǝd·ǝr)

pressure tunnel |CIV ENG| A waterway tunnel under pressure because the hydraulic gradient lies above the tunnel crown. ('presh·ǝr ‚tǝn·ǝl)

pressure vector |IND ENG| A stress on the human body produced at the interface between the operator and the equipment during the use of hand tools or other equipment, and described in terms of direction and magnitude. ('presh·ǝr ‚vek·tǝr)

pressure vessel |ENG| A metal container, generally cylindrical or spheroid, capable of withstanding bursting pressures. ('presh·ǝr ‚ves·ǝl)

pressurization |ENG| 1. Use of an inert gas or dry air, at several pounds above atmospheric pressure, inside the components of a radar system or in a sealed coaxial line, to prevent corrosion by keeping out moisture, and to minimize high-voltage breakdown at high altitudes. 2. The act of maintaining normal atmospheric pressure in a chamber subjected to high or low external pressure. (‚presh·ǝ·rǝ'zā·shǝn)

pressurize |ENG| To maintain normal atmospheric pressure in a chamber subjected to high or low external pressures. ('presh·ǝ‚rīz)

pressurized blast furnace |ENG| A blast furnace operated under pressure above the ambient; pressure is obtained by throttling the off-gas line, which permits a greater volume of air to be passed through the furnace at a lower velocity, and results in increase in smelting rate. { 'presh·ə,rīzd 'blast ,fər·nəs }

presswork |ENG| The entire range of bending and drawing operations in the cold forming of sheet metal products. { 'pres,wərk }

prestress |ENG| To apply a force to a structure to condition it to withstand its working load more effectively or with less deflection. { ¦prē'stres }

pretensioning |ENG| Process of precasting concrete beams with tensioned wires embedded in them. Also known as Hoyer method of prestressing. { prē'ten·shən·iŋ }

pretersonics See acoustoelectronics. { ¦prēd·ər¦sän·iks }

pretravel |CONT SYS| The distance or angle through which the actuator of a switch moves from the free position to the operating position. { 'prē,trav·əl }

preventive maintenance |ENG| A procedure of inspecting, testing, and reconditioning a system at regular intervals according to specific instructions, intended to prevent failures in service or to retard deterioration. { pri'ven·tiv 'mānt·ən·əns }

Prevost's theory |THERMO| A theory according to which a body is constantly exchanging heat with its surroundings, radiating an amount of energy which is independent of its surroundings, and increasing or decreasing its temperature depending on whether it absorbs more radiation than it emits, or vice versa. { 'prā·vōz ,thē·ə·rē }

Price meter |ENG| The ocean current meter in use in the United States: six conical cups, mounted around a vertical axis, rotate and cause a signal in a set of headphones with each rotation; tail vanes and a heavy weight stabilize the instrument. { 'prīs ,mēd·ər }

prick punch |DES ENG| A tool that has a sharp conical point ground to an angle of 30–60°C; used to make a slight indentation on a workpiece to locate the intersection of centerlines. { 'prik ,pənch }

prill |CHEM ENG| To form pellet-sized crystals or agglomerates of material by the action of upward-blowing air on falling hot solution; used in the manufacture of ammonium nitrate and urea fertilizers. { pril }

primary air |MECH ENG| That portion of the combustion air introduced with the fuel in a burner. { 'prī,mer·ē 'er }

primary breaker |MECH ENG| A machine which takes over the work of size reduction from blasting operations, crushing rock to maximum size of about 2-inch (5-centimeter) diameter; may be a gyratory crusher or jaw breaker. Also known as primary crusher. { 'prī,mer·ē 'brāk·ər }

primary creep |MECH| The initial high strain-rate region in a material subjected to sustained stress. { 'prī,mer·ē 'krēp }

primary crusher See primary breaker. { 'prī,mer·ē 'krəsh·ər }

primary detector See sensor. { 'prī,mer·ē di'tek·tər }

primary drilling |ENG| The process of drilling holes in a solid rock ledge in preparation for a blast by means of which the rock is thrown down. { 'prī,mer·ē 'dril·iŋ }

primary energy |ENG| Energy that exists in a naturally occurring form, such as coal, before being converted into an end-use form. { 'prī,mer·ē 'en·ər·jē }

primary excavation |ENG| Digging performed in undisturbed soil. { 'prī,mer·ē ,eks·kə'vā·shən }

primary instrument |ENG| A measuring instrument that can be calibrated without reference to another instrument. { 'prī,mer·ē 'in·strə·mənt }

primary measuring element |ENG| The portion of a measuring or sensing device that is in direct contact with the variables being measured (such as temperature, pressure, pH, or velocity). { 'prī,mer·ē 'mezh·ə·riŋ ,el·ə·mənt }

primary phase |THERMO| The only crystalline phase capable of existing in equilibrium with a given liquid. { 'prī,mer·ē 'fāz }

primary phase region |THERMO| On a phase diagram, the locus of all compositions having a common primary phase. { 'prī,mer·ē 'fāz ,rē·jən }

primary radar |ENG| Radar in which the incident beam is reflected from the target to form the return signal. Also known as primary surveillance radar (PSR). { 'prī,mer·ē 'rā,där }

primary sewage sludge |CIV ENG| A semiliquid waste resulting from sedimentation with no additional treatment. { 'prī,mer·ē 'sü·ij ,sləj }

primary stress |MECH| A normal or shear stress component in a solid material which results from an imposed loading and which is under a condition of equilibrium and is not self-limiting. { 'prī,mer·ē 'stres }

primary surveillance radar See primary radar. { 'prī,mer·ē sər'vā·ləns ,rā,där }

primary treatment |CIV ENG| Removal of floating solids and suspended solids, both fine and coarse, from raw sewage. { 'prī,mer·ē 'trēt·mənt }

prime |ENG| **1.** Main or primary, as in prime contractor. **2.** In blasting, to place a detonator in a cartridge or charge of explosive. **3.** To treat wood with a primer or penetrant primer. **4.** To add water to a pump to enable it to begin pumping. { prīm }

prime contractor |ENG| A contractor having a direct contract for an entire project; the contractor may in turn assign portions of the work to subcontractors. { 'prīm 'kän,trak·tər }

prime mover |MECH ENG| **1.** The component of a power plant that transforms energy from the thermal or the chemical to the mechanical form. **2.** A tractor or truck, usually with four-wheel drive, used for hauling tasks. { 'prīm 'müv·ər }

primer |ENG| In general, a small, sensitive initial explosive train component which on being actuated initiates functioning of the explosive train, and will not reliably initiate high explosive charge; classified according to the method of initiation, for example, percussion primer, electric primer, or friction primer. { 'prīm·ər }

primer cup |ENG| A small metal cup, into which the primer mixture is loaded. { 'prīm·ər ,kəp }

primer-detonator |ENG| A unit, in a metal housing, in which are assembled a primer, a detonator, and when indicated, an intervening delay charge. { 'prīm·ər 'det·ən,ād·ər }

primer leak |ENG| Defect in a cartridge which allows partial escape of the hot propelling gases in a primer, caused by faulty construction or an excessive charge. { 'prīm·ər ,lēk }

priming |MECH ENG| In a boiler, the excessive carryover of fine water particles along with the steam because of insufficient steam space, faulty boiler design, or faulty operating conditions. { 'prīm·iŋ }

priming pump |MECH ENG| A device on motor vehicles and tanks, providing a means of injecting a spray of fuel into the engine to facilitate starting. { 'prīm·iŋ ,pəmp }

primitive |CONT SYS| A basic operation of a robot, initialized by a single command statement in the program that controls the robot. { 'prim·əd·iv }

principal axis |ENG ACOUS| A reference direction for angular coordinates used in describing the directional characteristics of a transducer; it is usually an axis of structural symmetry or the direction of maximum response. |MECH| One of three perpendicular axes in a rigid body such that the products of inertia about any two of them vanish. { prin·sə·pəl 'ak·səs }

principal axis of strain |MECH| One of the three axes of a body that were mutually perpendicular before deformation. Also known as strain axis. { 'prin·sə·pəl 'ak·səs əv 'strān }

principal axis of stress |MECH| One of the three mutually perpendicular axes of a body that are perpendicular to the principal planes of stress. Also known as stress axis. { 'prin·sə·pəl 'ak·səs əv 'stres }

principal function |MECH| The integral of the Lagrangian of a system over time; it is involved in the statement of Hamilton's principle. { 'prin·sə·pəl 'faŋk·shən }

principal item |ENG| Item which, because of its major importance, requires detailed analysis and examination of all factors affecting its supply and demand, as well as an unusual degree of supervision; its selection is based upon such criteria as strategic importance, high monetary value, unusual complexity of issue, and procurement difficulties. { 'prin·sə·pəl 'īd·əm }

principal meridian |CIV ENG| One of the meridians established by the United States government as a reference for subdividing public land. { 'prin·sə·pəl mə'rid·ē·ən }

principal plane of stress |MECH| For a point in an elastic body, a plane at that point across

which the shearing stress vanishes. { 'prin·sə·pəl 'plān əv 'stres }

principal strain |MECH| The elongation or compression of one of the principal axes of strain relative to its original length. { 'prin·sə·pəl 'strān }

principal stress |MECH| A stress occurring at right angles to a principal plane of stress. { 'prin·sə·pəl 'stres }

principle of coincidence |ENG| The principle of operation of a vernier, according to which the fraction of the smallest division of the main scale is determined by the division of the vernier which is exactly in line with a division of the main scale. { 'prin·sə·pəl əv kō'in·sə·dəns }

principle of dynamical similarity |MECH| The principle that two physical systems which are geometrically and kinematically similar at a given instant, and physically similar in constitution, will retain this similarity at later corresponding instants if and only if the Froude number I for each independent type of force has identical values in the two systems. Also known as similarity principle. { 'prin·sə·pəl əv di¦nam·ə·kəl ,sim·ə'lar·əd·ē }

principle of inaccessibility See Carathéodory's principle. { 'prin·sə·pəl əv ,in·ak,ses·ə'bil·əd·ē }

principle of least action |MECH| The principle that, for a system whose total mechanical energy is conserved, the trajectory of the system in configuration space is that path which makes the value of the action stationary relative to nearby paths between the same configurations and for which the energy has the same constant value. Also known as least-action principle. { 'prin·sə·pəl əv ,lēst 'ak·shan }

principle of optimality |CONT SYS| A principle which states that for optimal systems, any portion of the optimal state trajectory is optimal between the states it joins. { 'prin·sə·pəl əv ,äp·tə'mal·əd·ē }

principle of reciprocity See reciprocity therorem. { 'prin·sə·pəl əv ,res·ə'präs·əd·ē }

principle of superposition |ELEC| **1.** The principle that the total electric field at a point due to the combined influence of a distribution of point charges is the vector sum of the electric field intensities which the individual point charges would produce at that point if each acted alone. **2.** The principle that, in a linear electrical network, the voltage or current in any element resulting from several sources acting together is the sum of the voltages or currents resulting from each source acting alone. Also known as superposition theorem. |MECH| The principle that when two or more forces act on a particle at the same time, the resultant force is the vector sum of the two. { 'prin·sə·pəl əv ,sü·pər·pə'zish·ən }

principle of virtual work |MECH| The principle that the total work done by all forces acting on a system in static equilibrium is zero for any infinitesimal displacement from equilibrium which is consistent with the constraints of the system. Also known as virtual work principle. { 'prin·sə·pəl əv 'vər·chə·wəl ,wərk }

printed circuit |ELECTR| A conductive pattern that may or may not include printed components, formed in a predetermined design on the surface of an insulating base in an accurately repeatable manner. { 'print·əd 'sər·kət }

printed circuit board |ELECTR| A flat board whose front contains slots for integrated circuit chips and connections for a variety of electronic components, and whose back is printed with electrically conductive pathways between the components. Also known as circuit board. { 'print·əd 'sər·kət ,bȯrd }

printed wiring board |ELECTR| A copper-clad dielectric material with conductors etched on the external or internal layers. { ¦print·əd 'wīr·iŋ ,bȯrd }

prior-art search [ENG] **1.** A search for prior art which may possibly anticipate an invention which is being considered for patentability. **2.** A similar search but for the purpose of determining what the status of existing technology is before going ahead with new research; it is done to avoid unwittingly retracing new steps taken by other workers in the field. { 'prī·ər ,ärt ,sərch }

prismatic astrolabe [ENG] A surveying instrument that makes use of a pan of mercury forming an artificial horizon, and a prism mounted in front of a horizontal telescope to determine the exact times at which stars reach a fixed altitude, and thereby to establish an astronomical position. { priz'mad·ik 'as·trə,lāb }

prismatic compass [ENG] A hand compass used by surveyors which is equipped with a prism that allows the compass to be read while the site is being taken. { priz'mad·ik 'kəm·pəs }

prism joint [MECH ENG] A robotic articulation that has only one degree of freedom, in sliding motion only. { 'priz·əm ,jȯint }

prism level [ENG] A surveyor's level with prisms that allow the levelman to view the level bubble without moving his eye from the telescope. { 'priz·əm 'lev·əl }

probe [ENG] A small tube containing the sensing element of electronic equipment, which can be lowered into a borehole to obtain measurements and data. { prōb }

probe gas [ENG] Tracer gas emitted from a small orifice for impingement on a restricted area being tested for leaks. { 'prōb ,gas }

probe-type liquid-level meter [ENG] Device to sense or measure the level of liquids in storage or process vessels by means of an immersed electrode or probe. { 'prōb ,tīp 'lik·wəd ¦lev·əl ,mēd·ər }

process [ENG] A system or series of continuous or regularly occurring actions taking place in a predetermined or planned manner to produce a desired result. { 'prä,ses }

process analyzer [CHEM ENG] An instrument for determining the chemical composition of the substances involved in a chemical process directly, or for measuring the physical parameters indicative of composition. { 'prä,ses ,an·ə,līz·ər }

process chart |IND ENG| A graphic representation of events occurring during a series of actions or operations. { 'prä,ses ,chärt }

process control [ENG] Manipulation of the conditions of a process to bring about a desired change in the output characteristics of the process. { 'prä,ses kən,trōl }

process control chart |IND ENG| A tabulated graphical arrangement of test results and other pertinent data for each production assembly unit, arranged in chronological sequence for the entire assembly. { 'prä,ses kən,trōl ,chärt }

process control engineering [ENG] A field of engineering dealing with ways and means by which conditions of continuous processes are automatically kept as close as possible to desired values or within a required range. { 'prä,ses kən,trōl ,en·jə,nir·iŋ }

process control system |CONT SYS| The automatic control of a continuous operation. { 'prä,ses kən,trōl ,sis·təm }

process dynamics [ENG] The dynamic response interrelationships between components (units) of a complex system, such as in a chemical process plant. { 'prä,ses dī,nam·iks }

process engineering [ENG] A service function of production engineering that involves selection of the processes to be used, determination of the sequence of all operations, and requisition of special tools to make a product. { 'prä,ses ,en·jə,nir·iŋ }

process furnace |CHEM ENG| Furnace used to heat process-stream materials (liquids, gases, or solids) in a chemical-plant operation; types are direct-fired, indirect-fired, and pebble heaters. { 'prä,ses ,fər·nəs }

process heater |CHEM ENG| Equipment for the heating of chemical process streams (gases, liquids, or solids); usually refers to furnaces, in contrast to heat exchangers. { 'prä,ses ,hēd·ər }

processing [ENG] The act of converting material from one form into another desired form. { 'prä,ses·iŋ }

process layout |IND ENG| In a processing plant, the layout of machines, equipment, and locations which groups the same or similar operations. { 'prä,ses ,lā,aut }

process monitoring |CHEM ENG| The observation of chemical process variables by means of pressure, temperature, flow, and other types of indicators; usually occurs in a central control room. { 'prä,ses ,män·ə·triŋ }

process piping [ENG] In an industrial facility, pipework whose function is to convey the materials used for the manufacturing processes. { 'prä,ses ,pīp·iŋ }

process planning |IND ENG| Determining the conditions necessary to convert material from one state to another. { 'prä,ses ,plan·iŋ }

process reengineering |SYS ENG| The study, capture, and modification of the internal mechanisms or functionality of an existing process or systems-engineering life cycle in order to reconstitute it in a new form and with new functional and nonfunctional features, often to take advantage of newly emerged or desired organizational or technological capabilities without changing the inherent purpose of the process that is being reengineered. { ˌprä'səs ˌrē,en·jə'nir·iŋ }

process sequencing |IND ENG| Specification of the appropriate order for the processes required to manufacture a part. { 'prä,ses ˌsē·kwəns·iŋ }

process time |IND ENG| **1.** Time needed for completion of the machine-controlled portion of a work cycle. **2.** Time required for completion of an entire process. { 'prä,ses ˌtīm }

process variable |CHEM ENG| Any of those varying operational and physical conditions associated with a chemical processing operation, such as temperature, pressure, flowrate, density, pH, viscosity, or chemical composition. { 'prä,ses ˌver·ē·ə·bəl }

producer's risk |IND ENG| The probability that in an acceptance sampling plan, material of an acceptable quality level will be rejected. { prə'dü·sərz ˌrisk }

product |CHEM ENG| *See* discharge liquor. |IND ENG| **1.** An item or goods made by an industrial firm. **2.** The total of such items or goods. { 'präd·əkt }

product design |DES ENG| The determination and specification of the parts of a product and their interrelationship so that they become a unified whole. { 'präd·əkt di,zīn }

production |ENG| Output, such as units made in a factory, oil from a well, or chemicals from a processing plant. { prə'dək·shən }

production control |IND ENG| The procedure for planning, routing, scheduling, dispatching, and expediting the flow of materials, parts, subassemblies, and assemblies within a plant, from the raw state to the finished product, in an orderly and efficient manner. { prə'dək·shən kən,trōl }

production engineering |IND ENG| The planning and control of the mechanical means of changing the shape, condition, and relationship of materials within industry toward greater effectiveness and value. { prə'dək·shən ˌen·jə'nir·iŋ }

production model |IND ENG| A model in its final mechanical and electrical form of final production design made by production tools, jigs, fixtures, and methods. { prə'dək·shən ˌmäd·əl }

production requirements |IND ENG| The sum of authorized stock levels and pipeline needs less stocks expected to become available, stock on hand, stocks due in, returned stocks, and stocks from salvage, reclamation, rebuild, and other sources. { prə'dək·shən ri,kwīr·məns }

production standard *See* standard time { prə'dək·shən ˌstan·dərd }

production track |ENG ACOUS| A sound track which is either prerecorded or recorded directly on the set, and which exists in the film at that time when the music breakdown for scoring is about to begin. { prə'dək·shən ˌtrak }

productive time |IND ENG| Time during which useful work is performed in an operation or process. { prə'dək·tiv 'tīm }

productivity |IND ENG| The ratio of output production to input effort; it is an indicator of the efficiency with which an enterprise converts its resources (inputs) into finished goods or services (outputs). { prä,dək'tiv·əd·ē }

product life-cycle |IND ENG| All the phases, from conception and scale-up, through production, growing use, maturity, and obsolescence of a product. { ¦präd·əkt 'līf,sī·kəl }

product line |IND ENG| **1.** The range of products offered by a firm. **2.** A group of basically similar products, differentiated only by such characteristics as color, style, or size. { 'prä·dəkt ˌlīn }

product of inertia |MECH| Relative to two rectangular axes, the sum of the products formed by multiplying the mass (or, sometimes, the area) of each element of a figure by the product of the coordinates corresponding to those axes. { 'prä·dəkt əv i'nər·shə }

product reengineering |SYS ENG| The study, capture, and modification of the internal mechanisms or functionality of an existing system or product in order to reconstitute it in a new form with new features, often to take advantage of newly emerged technologies without major change to the inherent functionality and purpose of the system. { ˌpräd·əkt ˌrē,en'jə'nir·iŋ }

product water |CHEM ENG| Fresh water that is produced by a desalination process; also known as converted water. { 'prä,dəkt ˌwód·ər }

profile die |ENG| A plastics extrusion die used to produce continuous shapes, but not tubes or sheets. { 'prō,fīl ˌdī }

profiled keyway |DES ENG| A keyway for a straight key formed by an end-milling cutter. Also known as end-milled keyway. { ¦prō,fīld 'kē,wā }

profiling |ENG| Electrical exploration wherein the transmitter and receiver are moved in unison across a structure to obtain a profile of mutual impedance between transmitter and receiver. Also known as lateral search. { 'prō,fīl·iŋ }

profiling machine |MECH ENG| A machine used for milling irregular profiles; the cutting tool is guided by the contour of a model. { 'prō,fīl·iŋ mə,shēn }

profilograph |ENG| An instrument for measuring and recording roughness of the surface over which it travels. { prō'fil·ə,graf }

profilometer |ENG| An instrument for measuring the roughness of a surface by means of a diamond-pointed tracer arm attached to a coil in an electric field; movement of the arm across the surface induces a current proportional to surface roughness. { ˌprō·fə'läm·əd·ər }

profit sharing |IND ENG| Sharing of company profits with the employees. { 'präf·ət ˌsher·iŋ }

program |IND ENG| An undertaking of significant scope that is enduring rather than occurring within a limited time span. { 'prō·grəm *or* 'prō,gram }

program control |CONT SYS| A control system

whose set point is automatically varied during definite time intervals in order to make the process variable vary in some prescribed manner. { 'prō·grəm kən,trōl }

program device |CONT SYS| In missile guidance, tha automatic device used to control time and sequence of events of a program. { 'prō·grəm di,vīs }

program evaluation and review technique See PERT. { 'prō·grəm i,val·yə'wā·shən ən ri'vyü tek,nēk }

program level |ENG ACOUS| The level of the program signal in an audio system, expressed in volume units. { 'prō·grəm ,lev·əl }

programmable controller |CONT SYS| A control device, normally used in industrial control applications, that employs the hardware architecture of a computer and a relay ladder diagram language. Also known as programmable logic controller. { prō'gram·ə·bəl kən'trōl·ər }

programmable counter |ELECTR| A counter that divides an input frequency by a number which can be programmed into decades of synchronous down counters; these decades, with additional decoding and control logic, give the equivalent of a divide-by-N counter system, where N can be made equal to any number. { prō'gram·ə·bəl 'kaúnt·ər }

programmable decade resistor |ELECTR| A decade box designed so that the value of its resistance can be remotely controlled by programming logic as required for the control of load, time constant, gain, and other parameters of circuits used in automatic test equipment and automatic controls. { prō'gram·ə·bəl 'de,kād ri,zis·tər }

programmable electronic system |SYS ENG| A system based on a computer and connected to sensors or actuators for the purpose of control, protection, or monitoring. { prō'gram·ə·bəl i'lek,trän·ik ,sis·təm }

programmable logic array See field-programmable logic array. { prō'gram·ə·bəl |läj·ik ə,rā }

programmable logic controller See programmable controller. { prō'gram·ə·bəl |läj·ik kən,trōl·ər }

programmed logic array |ELECTR| An array of AND/OR logic gates that provides logic functions for a given set of inputs programmed during manufacture and serves as a read-only memory. Abbreviated PLA. { 'prō,gramd |läj·ik ə,rā }

programmer |CONT SYS| A device used to control the motion of a missile in accordance with a predetermined plan. { 'prō,gram·ər }

programming |ENG| In a plastics process, extruding a parison whose thickness differs longitudinally in order to equalize wall thickness of the blown container. { 'prō,gram·iŋ }

programming panel |CONT SYS| A device used to edit a program or insert and monitor it in a programmable controller. { 'prō,gram·iŋ ,pan·əl }

programming unit See manual control unit. { 'prō,gram·iŋ ,yü·nət }

program scan |CONT SYS| The span of time during which a programmable controller processor

executes all the instructions of a given program. { 'prō·grəm ,skan }

progress chart |IND ENG| A graphical representation of the degree of completion of work in progress. { 'präg·rəs ,chärt }

progressive bonding |ENG| A method of curing a resin adhesive wherein heat and pressure are applied in successive steps. Also known as progressive gluing. { prə'gres·iv 'bänd·iŋ }

progressive gluing See progressive bonding. { prə'gres·iv 'glü·iŋ }

project |ENG| A specifically defined task within a research and development field, which is established to meet a single requirement, either stated or anticipated, for research data, an end item of material, a major component, or a technique. { 'prä,jekt }

projected-scale instrument |ENG| An indicating instrument in which a light beam projects an image of the scale on a screen. { prə'jek·təd ,skāl ,in·strə·mənt }

projected window |BUILD| A window having one or more rotatable sashes which swing either inward or outward. { prə'jek·təd 'win·dō }

project engineering |ENG| 1. The engineering design and supervision (coordination) aspects of building a manufacturing facility. 2. The engineering aspects of a specific project, such as development of a product or solution to a problem. { 'prä,jekt en·jə'nir·iŋ }

projection thermography |ENG| A method of measuring surface temperature in which thermal radiation from a surface is imaged by an optical system on a thin screen of luminescent material, and the pattern formed corresponds to the heat radiation of the surface. { prə'jek·shən thər 'mäg·rə·fē }

project life See economic life. { 'prä·jikt ,līf }

projector |ENG ACOUS| 1. A horn designed to project sound chiefly in one direction from a loudspeaker. 2. An underwater acoustic transmitter. { prə'jek·tər }

pronate |CONT SYS| To orient a robot toward a position in which the back or protected side of a manipulator faces up and is exposed. { 'prō,nāt }

prong See pin. { präŋ }

prony brake |MECH ENG| An absorption dynamometer that applies a friction load to the output shaft by means of wood blocks, a flexible band, or other friction surface. { 'prō·nē ,brāk }

proof |ENG| Reproduction of a die impression by means of a cast. { prüf }

proof load |ENG| A predetermined test load, greater than the service load, to which a specimen is subjected before acceptance for use. { 'prüf ,lōd }

proof resilience |MECH| The tensile strength necessary to stretch an elastomer from zero elongation to the breaking point, expressed in foot-pounds per cubic inch of original dimension. { 'prüf ri,zil·yəns }

proof stress |MECH| 1. The stress that causes a specified amount of permanent deformation in a material. 2. A specified stress to be applied to a member or structure in order to assess its ability to support service loads. { 'prüf ,stres }

propagated blast |ENG| A blast of a number of unprimed charges of explosives plus one hole primed, generally for the purpose of ditching, where each charge is detonated by the explosion of the adjacent one, the shock being transmitted through the wet soil. { 'präp·ə,gād·əd 'blast }

propane deasphalting |CHEM ENG| Petroleum-refinery solvent process using propane to remove and precipitate asphalt from petroleum stocks, such as for lubricating oils. { 'prō,pān dē'as,fȯld·iŋ }

propane decarbonizing |CHEM ENG| Petroleum-refinery solvent process using propane to recover catalytic-cracking feedstock from heavy-fuel residues; when butane or butane-propane solvent is used, the process is called solvent decarbonizing. { 'prō,pān dē'kär·bə,nīz·iŋ }

propane dewaxing |CHEM ENG| Petroleum-refinery solvent process using propane to remove waxes from lubricating oils to lower the lube-oil pour point. { 'prō,pān dē'waks·iŋ }

propane fractionation |CHEM ENG| Continuous, petroleum-refinery solvent process using liquid propane to segregate long-vacuum residue into two or more grades of lube-oil stock (such as heavy neutral stock or bright stock) and asphalt. { 'prō,pān ,frak·shə'nā·shən }

propellant-actuated device |ENG| A device that employs the energy supplied by the gases produced by burning propellants to accomplish or initiate a mechanical action other than propelling a projectile. { prə'pel·ənt ¦ak·chə,wād·əd di,vīs }

propeller |MECH ENG| A bladed device that rotates on a shaft to produce a useful thrust in the direction of the shaft axis. { prə'pel·ər }

propeller anemometer |ENG| A rotation anemometer which is encased in a strong glass outer shell that protects it against hydrostatic pressure. { prə'pel·ər ,an·ə'mäm·əd·ər }

propeller blade |DES ENG| One of two or more plates radiating out from the hub of a propeller and normally twisted to form part of a helical surface. { prə'pel·ər ,blād }

propeller boss |DES ENG| The central portion of the screw propeller which carries the blades, and forms the medium of attachment to the propeller shaft. Also known as propeller hub. { prə'pel·ər ,bȯs }

propeller efficiency |MECH ENG| The ratio of the thrust horsepower delivered by the propeller to the shaft horsepower as delivered by the engine to the propeller. { prə'pel·ər i,fish·ən·sē }

propeller fan |MECH ENG| An axial-flow blower, with or without a casing, using a propeller-type rotor to accelerate the fluid. { prə'pel·ər ¦fan }

propeller hub See propeller boss. { prə'pel·ər ,həb }

propeller meter |ENG| A quantity meter in which the flowing stream rotates a propellerlike device and revolutions are counted. { prə'pel·ər ¦mēd·ər }

propeller pump See axial-flow pump. { prə'pel·ər ¦pəmp }

propeller shaft |MECH ENG| A shaft, carrying a screw propeller at its end, that transmits power from an engine to the propeller. { prə'pel·ər ,shaft }

propeller slip angle |MECH ENG| The angle between the plane of the blade face and its direction of motion. { prə'pel·ər 'slip ,aŋ·gəl }

propeller tip speed |MECH ENG| The speed in feet per minute swept by the propeller tips. { prə'pel·ər 'tip ,spēd }

propeller turbine |MECH ENG| A form of reactive-type hydraulic turbine using an axial-flow propeller rotor. { prə'pel·ər ¦tər·bən }

propeller windmill |MECH ENG| A windmill that extracts wind power from horizontal air movements to rotate the blades of a propeller. { prə'pel·ər ¦win,mil }

proportional band |CONT SYS| The range of values of the controlled variable that will cause a controller to operate over its full range. { prə'pȯr·shən·əl 'band }

proportional control |CONT SYS| Control in which the amount of corrective action is proportional to the amount of error; used, for example, in chemical engineering to control pressure, flow rate, or temperature in a process system. { prə'pȯr·shən·əl kən'trōl }

proportional controller |CONT SYS| A controller whose output is proportional to the error signal. { prə'pȯr·shən·əl kən'trōl·ər }

proportional dividers |DES ENG| Dividers with two legs, pointed at both ends, and an adjustable pivot; distances measured by the points at one end can be marked off in proportion by the points at the other end. { prə'pȯr·shən·əl di'vīd·ərz }

proportional elastic limit |MECH| The greatest stress intensity for which stress is still proportional to strain. { prə'pȯr·shən·əl i'las·tik ,lim·ət }

proportional limit |MECH| The greatest stress a material can sustain without departure from linear proportionality of stress and strain. { prə'pȯr·shən·əl 'lim·ət }

proportional-plus-derivative control |CONT SYS| Control in which the control signal is a linear combination of the error signal and its derivative. { prə'pȯr·shən·əl ,pləs də'riv·əd·iv kən,trōl }

proportional-plus-integral control |CONT SYS| Control in which the control signal is a linear combination of the error signal and its integral. { prə'pȯr·shən·əl ,pləs 'int·ə·grəl kən,trōl }

proportional-plus-integral-plus-derivative control |CONT SYS| Control in which the control signal is a linear combination of the error signal, its integral, and its derivative. { prə'pȯr·shən·əl ,pləs 'int·ə·grəl ,pləs də'riv·əd·iv kən,trōl }

proportional-speed control See floating control. { prə'pȯr·shən·əl 'spēd kən,trōl }

proportioning probe |ENG| A leak-testing probe capable of changing the air-tracer gas ratio without changing the amount of flow it transmits to the testing device. { prə'pȯr·shən·iŋ ,prōb }

proportioning pump See metering pump. { prə'pȯr·shən·iŋ ‚pəmp }

propped cantilever |CIV ENG| A beam having one built-in support and one simple support. { 'präpt 'kant·əl‚ē·vər }

proprioceptor |CONT SYS| A device that senses the position of an arm or other computer-controlled articulated mechanism of a robot and provides feedback signals. { ‚prō·prē·ə'sep·tər }

propulsion |MECH| The process of causing a body to move by exerting a force against it. { prə'pəl·shən }

propulsion system |MECH ENG| For a vehicle moving in a fluid medium, such as an airplane or ship, a system that produces a required change in momentum in the vehicle by changing the velocity of the air or water passing through the propulsive device or engine; in the case of a rocket-propelled vehicle operating without a fluid medium, the required momentum change is produced by using up some of the propulsive device's own mass, called the propellant. { prə'pəl·shən ‚sis·təm }

protected thermometer |ENG| A reversing thermometer which is encased in a strong glass outer shell that protects it against hydrostatic pressure. { prə'tek·təd thər'mäm·əd·ər }

protective device See electric protective device. { prə'tek·tiv di'vīs }

protective finish |ENG| A coating applied to equipment to protect it from corrosion and wear; many substances, including metals, glass, and ceramics, are used. { prə'tek·tiv 'fin·ish }

protective grounding |ELEC| Grounding of the neutral conductor of a secondary power-distribution system, and of all metal enclosures for conductors, to protect persons from dangerous currents. { prə'tek·tiv 'graúnd·iŋ }

protective relay |ELEC| A relay whose principal function is to protect service from interruption or to prevent or limit damage to apparatus. { prə'tek·tiv 'rē‚lā }

prototype |ENG| A model suitable for use in complete evaluation of form, design, and performance. { 'prōd·ə‚tīp }

protractor |ENG| A semicircular instrument used to construct and measure angles formed by intersecting lines of a plane; the midpoint of the diameter of the semicircle is marked and serves as the vertex of angles constructed or measured. { 'prō‚trak·tər }

proving ring |DES ENG| A ring used for calibrating test machines; the diameter of the ring changes when a force is applied along a diameter. { 'prüv·iŋ ‚riŋ }

proximal |CONT SYS| Located close to the base or pedestal and away from the end effector of a robot. { 'präk·sə·məl }

proximate analysis |CHEM ENG| A technique that separates and identifies categories of compounds in a mixture; reported are moisture and ash content, the extracts of the mixture made with alcohol, petroleum ether, water, hydrochloric acid and resins, starches, reducing sugars, proteins, fats, esters, free acids, and so on; this type of analysis of solid fuels allows a prediction to be made as to how the fuel will behave in a furnace. { 'präk·sə·mət ə'nal·ə·səs }

proximity detector |ENG| A sensing device that produces an electrical signal when approached by an object or when approaching an object. { präk'sim·əd·ē di‚tek·tər }

proximity sensor |CONT SYS| Any device that measures short distances within a robotic system. Also known as noncontact sensor. { präk'sim·əd·ē 'sen·sər }

ps See picosecond.

psec See picosecond.

psf See pound per square foot.

psi See pound per square inch.

psia See pounds per square inch absolute.

psid See pounds per square inch differential.

psig See pounds per square inch gage.

PSII See plasma-source ion implantation.

psophometer |ENG| An instrument for measuring noise in electric circuits; when connected across a 600-ohm resistance in the circuit under study, the instrument gives a reading that by definition is equal to half of the psophometric electromotive force actually existing in the circuit. { sō'fäm·əd·ər }

psophometric electromotive force |ELECTR| The true noise voltage that exists in a circuit. { ‚säf·ə‚me·trik i‚lek·trə‚mōd·iv 'fȯrs }

psophometric voltage |ELECTR| The noise voltage as actually measured in a circuit under specified conditions. { ‚säf·ə‚me·trik 'vōl·tij }

PSR See primary radar.

psychogalvanometer |ENG| An instrument for testing mental reaction by determining how skin resistance changes when a voltage is applied to electrodes in contact with the skin. { ‚sī·kō‚gal·və'näm·əd·ər }

psychointegroammeter See lie detector. { ‚sī·kō‚in·tə·grō'am‚ēd·ər }

psychomotor performance |IND ENG| The degree of skill demonstrated by an operator in the completion of a task. { ‚sī·kə‚mōd·ər pər'fȯr·məns }

psychomotor task |IND ENG| An aspect of a job that requires the operator to use controlled movements of the body. { ‚sī·kə‚mōd·ər ‚task }

psychosomatograph |ENG| An instrument for recording muscular action currents or physical movements during tests of mental-physical coordination. { ‚sī·kō·sə'mad·ə‚graf }

psychromatic ratio |THERMO| Ratio of the heat-transfer coefficient to the product of the mass-transfer coefficient and humid heat for a gas-vapor system; used in calculation of humidity or saturation relationships. { ‚sī·krə'mad·ik 'rā·shō }

psychrometer |ENG| A device comprising two thermometers, one a dry bulb, the other a wet or wick-covered bulb, used in determining the moisture content or relative humidity of air or other gases. Also known as wet and dry bulb thermometer. { sī'kräm·əd·ər }

psychrometric calculator |ENG| A device for

quickly computing certain psychrometric data, usually the dew point and the relative humidity, from known values of the dry- and wet-bulb temperatures and the atmospheric pressure. { ¦sī·krə¦me·trik 'kal·kyə,lād·ər }

psychrometric chart |THERMO| A graph each point of which represents a specific condition of a gas-vapor system (such as air and water vapor) with regard to temperature (horizontal scale) and absolute humidity (vertical scale); other characteristics of the system, such as relative humidity, wet-bulb temperature, and latent heat of vaporization, are indicated by lines on the chart. { ¦sī·krə¦me·trik 'chärt }

psychrometric formula |THERMO| The semiempirical relation giving the vapor pressure in terms of the barometer and psychrometer readings. { ¦sī·krə¦me·trik 'fȯr·mya·lə }

psychrometric tables |THERMO| Tables prepared from the psychrometric formula and used to obtain vapor pressure, relative humidity, and dew point from values of wet-bulb and dry-bulb temperature. { ¦sī·krə¦me·trik 'tā·bəlz }

psychrometry |ENG| The science and techniques associated with measurements of the water vapor content of the air or other gases. { sī'käm·ə·trē }

pt *See* pint.

p-type conductivity |ELECTR| The conductivity associated with holes in a semiconductor, which are equivalent to positive charges. { 'pē ¦tīp ,kän,dək'tiv·əd·ē }

p-type crystal rectifier |ELECTR| Crystal rectifier in which forward current flows when the semiconductor is positive with respect to the metal. { 'pē ¦tīp 'krist·əl 'rek·tə,fī·ər }

p-type semiconductor |ELECTR| An extrinsic semiconductor in which the hole density exeeds the conduction electron density. { 'pē ¦tīp 'sem·i·kən,dək·tər }

p⁺-type semiconductor |ELECTR| A *p*-type semiconductor in which the excess mobile hole concentration is very large. { 'pē¦pləs ,tīp 'sem·i·kən,dək·tər }

p-type silicon |ELECTR| Silicon to which more impurity atoms of acceptor type (with valence of 3, such as boron) than of donor type (with valence of 5, such as phosphorus) have been added, with the result that the hole density exceeds the conduction electron density. { 'pē ¦tīp 'sil·ə,kän }

public address system *See* sound-reinforcement system. { 'pəb·lik ə'dres ,sis·təm }

public area |BUILD| The total nonrentable area of a building, such as public conveniences and rest rooms. { 'pəb·lik 'er·ē·ə }

public utility |IND ENG| A business organization considered by law to be vested with public interest and subject to public regulation. { 'pəb·lik yü'til·əd·ē }

public works |IND ENG| Government-owned and financed works and improvements for public enjoyment or use. { 'pəb·lik 'wərks }

puddle |ENG| To apply water in order to settle loose dirt. { pəd·əl }

puff |ELEC| *See* picofarad. |MECH ENG| A small explosion within a furnace due to combustion conditions. { pəf }

pug mill |MECH ENG| A machine for mixing and tempering a plastic material by the action of blades revolving in a drum or trough. { 'pəg ,mil }

puking |CHEM ENG| In a distillation column, the foaming and rising of liquid so that part of it is driven out of the vessel through the vapor line. { 'pyük·iŋ }

puller |MECH ENG| A lever-operated chain or wire-rope hoist for lifting or pulling at any angle, which has a reversible ratchet mechanism in the lever permitting short-stroke operation for both tensioning and relaxing, and which holds the loads with a Weston-type friction brake or a releasable ratchet. Also known as come-along. { 'pùl·ər }

pulley |DES ENG| A wheel with a flat, round, or grooved rim that rotates on a shaft and carries a flat belt, V-belt, rope, or chain to transmit motion and energy. { 'pùl·ē }

pulley lathe |MECH ENG| A lathe for turning pulleys. { 'pùl·ē ,lāth }

pulley stile |BUILD| The upright part of a window frame which holds the pulley and guides the sash. { 'pùl·ē ,stīl }

pulley top |MECH ENG| A top with a long shank used to tap setscrew holes in pulley hubs. { 'pùl·ē ,täp }

pull-in torque |MECH ENG| The largest steady torque with which a motor will attain normal speed after accelerating from a standstill. { 'pùl,in ,tȯrk }

pull-out torque |MECH ENG| Th largest torque under which a motor can operate without sharply losing speed. { 'pùl,aut ,tȯrk }

pullshovel *See* backhoe. { 'pùl,shəv·əl }

pull strength |MECH| A unit in tensile testing; the bond strength in pounds per square inch. { 'pùl ,streŋkth }

pulp *See* slime. { pəlp }

pulper |MECH ENG| A machine that converts materials to pulp, for example, one that reduces paper waste to pulp. { 'pəlp·ər }

pulping |ENG| Reducing wood to pulp. { 'pəlp·iŋ }

pulp molding |ENG| A plastics-industry process in which a resin-impregnated pulp material is preformed by application of a vacuum, after which it is oven-cured and molded. { 'pəlp ,mōld·iŋ }

pulsating flow |ENG| Irregular fluid flow in a piping system often resulting from the pressure variations of reciprocating compressors or pumps within the system. { 'pəl,sād·iŋ 'flō }

pulsation dampening |ENG| Device installed in a fluid piping system (gas or liquid) to eliminate or even out the fluid-flow pulsations caused by reciprocating compressors, pumps, and such. { pəl'sā·shən ,dam·pən·iŋ }

pulse altimeter |ENG| A device which is used to measure the distance of an aircraft above the ground by sending out radar signals in short pulses and measuring the time delay between the leading edge of the transmitted pulse and that of the pulse returned from the ground. { 'pəls al'tim·əd·ər }

pulse-amplitude discriminator |ENG| Electronic instrument used to investigate the amplitude distribution of the pulses produced in a nuclear detector. { 'pəls ¦am·plə,tüd di'skrim·ə,nād·ər }

pulse circuit |ELECTR| An active electrical network designed to respond to discrete pulses of current or voltage. { 'pəls ,sər·kət }

pulse column |CHEM ENG| Continuous-phase process column (such as liquid only or gas only) in which the flow-through is pulsating; used to increase mass-transfer rates, as in a liquid-liquid extraction operation. { 'pəls ,käl·əm }

pulse-compression radar |ENG| A radar system in which the transmitted signal is linearly frequency-modulated or otherwise spread out in time to reduce the peak power that must be handled by the transmitter; signal amplitude is kept constant; the receiver uses a linear filter to compress the signal and thereby reconstitute a short pulse for the radar display. { 'pəls kəm,presh·ən 'rā,där }

pulsed-bed sorption |CHEM ENG| Solid-liquid countercurrent adsorption process (such as an ion-exchange process) in which the granulated solids bed and the solution flow alternately, in opposite directions. { 'pəlst ¦bed 'sȯrp·shən }

pulsed fast neutron analysis |ENG| A technique for detecting contraband materials, in which a pulsed beam of high-energy neutrons is scanned up and down in a raster pattern while the object under inspection is conveyed through the beam; characteristic gamma rays emitted by materials in the object are detected in order to analyze and image these materials with the help of time-of-flight measurements. { ¦pəlst ,fast 'nü,trän ə,nal·əsəs }

pulsed-light ceilometer See pulsed-light cloud-height indicator. { 'pəlst 'līt sē'läm·əd·ər }

pulsed-light cloud-height indicator |ENG| An instrument used for the determination of cloud heights; it operates on the principle of pulse radar, employing visible light rather than radio waves. Also known as pulsed-light ceilometer. { 'pəlst 'līt 'klau̇d ,hīt 'in·də,kād·ər }

pulse-Doppler radar |ENG| Pulse radar that uses the Doppler effect to obtain information about the velocity of a target. { 'pəls ¦dap·lər 'rā,där }

pulsed oscillator |ELECTR| An oscillator that generates a carrier-frequency pulse or a train of carrier-frequency pulses as the result of self-generated or externally applied pulses. { 'pəlst 'äs·ə,lād·ər }

pulse dot soldering iron |ENG| A soldering iron that provides heat to the tip for a precisely controlled time interval, as required for making a good soldered joint without overheating adjacent parts. { 'pəls ,dät 'säd·ə·riŋ ,ī·ərn }

pulsed transfer function |CONT SYS| The ratio of the z-transform of the output of a system to

the z-transform of the input, when both input and output are trains of pulses. Also known as discrete transfer function; z-transfer function. { 'pəlst 'tranz·fər ,faŋk·shən }

pulsed video thermography |ENG| A method of nondestructive testing in which a source of heat is applied to an area of a specimen for a very short time duration, and an infrared detection system reveals anomalously hot or cold regions that then appear close to defects. { 'pəlst 'vid·ē·ō thər'mäg·rə·fē }

pulse generator |ELEC| See impulse generator. |ELECTR| A generator that produces repetitive pulses or signal-initiated pulses. { 'pəls ,jen·ə,rād·ər }

pulse height |ELECTR| The strength or amplitude of a pulse, measured in volts. { 'pəls ,hīt }

pulse integrator |ELECTR| An RC (resistance-capacitance) circuit which stretches in time duration a pulse applied to it. { 'pəls ,int·ə,grād·ər }

pulse-modulated radar |ENG| Form of radar in which the radiation consists of a series of discrete pulses. { 'pəls ,mäj·ə¦lād·əd 'rä,där }

pulse modulator |ELECTR| A device for carrying out the pulse modulation of a radio-frequency carrier signal. { 'pəls ,mäj·ə,lād·ər }

pulser |CHEM ENG| Device used to create a pulsating fluid flow through a process vessel, such as a liquid-liquid or vapor-liquid extraction tower; used to increase contact and mass transfer rates. |ELECTR| A generator used to produce high-voltage, short-duration pulses, as required by a pulsed microwave oscillator or a radar transmitter. { 'pəl·sər }

pulse radar |ENG| Radar in which the transmitter sends out high-power pulses that are spaced far apart in comparison with the duration of each pulse; the receiver is active for reception of echoes in the interval following each pulse. { 'pəls 'rā,där }

pulse repeater |ELECTR| Device used for receiving pulses from one circuit and transmitting corresponding pulses into another circuit; it may also change the frequencies and waveforms of the pulses and perform other functions. { 'pəls ri,pēd·ər }

pulse repetition frequency See pulse repetition rate. { 'pəls ,rep·ə¦tish·ən ,frē·kwən·sē }

pulse repetition rate |ELECTR| The number of times per second that a pulse is transmitted. Abbreviated PRR. Also known as pulse recurrence rate; pulse repetition frequency (PRF). { 'pəls ,rep·ə¦tish·ən ,rāt }

pulse scaler |ELECTR| A scaler that produces an output signal when a prescribed number of input pulses has been received. { 'pəls ,skāl·ər }

pulse shaper |ELECTR| A transducer used for changing one or more characteristics of a pulse, such as a pulse regenerator or pulse stretcher. { 'pəls ,shāp·ər }

pulse stretcher |ELECTR| A pulse shaper that produces an output pulse whose duration is greater than that of the input pulse and whose amplitude is proportional to the peak amplitude of the input pulse. { 'pəls ,strech·ər }

pulse synthesizer |ELECTR| A circuit used to supply pulses that are missing from a sequence due to interference or other causes. { 'pəls ,sin·thə,sīz·ər }

pulse-time-modulated radiosonde |ENG| A radiosonde which transmits the indications of the meteorological sensing elements in the form of pulses spaced in time; the meteorological data are evaluated from the intervals between the pulses. Also known as time-interval radiosonde. { 'pəls ¦tīm ¦mäj·ə,läd·əd 'rād·ē·ō ,sänd }

pulse tracking system |ENG| Tracking system which uses a high-energy, short-duration pulse radiated toward the target from which the velocity, direction, and range are determined by the characteristics of the reflected pulse. { 'pəls 'trak·iŋ ,sis·təm }

pulse transformer |ELECTR| A transformer capable of operating over a wide range of frequencies, used to transfer nonsinusoidal pulses without materially changing their waveforms. { 'pəls tranz,fór·mər }

pulse transmitter |ELECTR| A pulse-modulated transmitter whose peak-power-output capabilities are usually large with respect to the average-power-output rating. { 'pəls tranz,mid·ər }

pulse-width discriminator |ELECTR| Device that measures the pulse length of video signals and passes only those whose time duration falls into some predetermined design tolerance. { 'pəls ¦width di'skrim·ə,nād·ər }

pulsometer |MECH ENG| A simple, lightweight pump in which steam forces water out of one of two chambers alternately. { pəl'säm·əd·ər }

pultrusion |ENG| A process for producing continuous fibers for advanced composites which involves pulling reinforcements through tanks of thermoset resins, a preformer, and then a die, where the product is formed into its final shape. { půl'trü·zhən }

pulverization See comminution. { ,pəl·və·rə'zā·shən }

pulverizer |MECH ENG| Device for breaking down of solid lumps into a fine material by cleavage along crystal faces. { 'pəl·və,rīz·ər }

pump |ELECTR| Of a parametric device, the source of alternating-current power which causes the nonlinear reactor to behave as a time-varying reactance. |MECH ENG| A machine that draws a fluid into itself through an entrance port and forces the fluid out through an exhaust port. { pəmp }

pumpability test |ENG| Standard test to ascertain the lowest temperature at which a petroleum fuel oil may be pumped. { ,pəm·pə'bil·əd·ē ,test }

pumparound |CHEM ENG| A system or process vessel that moves liquid out of and back into the vessel at a new location; for example, in a bubble tower, the withdrawing of liquid from a plate or tray, followed by cooling, and returning

to another plate to induce condensation of vapors. { 'pəm·pə,raůnd }

pump bob |MECH ENG| A device such as a crank that converts rotary motion into reciprocating motion. { 'pəmp ,bäb }

pump-down time |ENG| The length of time required to evacuate a leak-tested vessel. { 'pəmp 'daůn ,tīm }

pumphouse |CIV ENG| A building in which are housed pumps that supply an irrigation system, a power plant, a factory, a reservoir, a farm, a home, and so on. { 'pəmp,haůs }

pumping loss |MECH ENG| Power consumed in purging a cylinder of exhaust gas and sucking in fresh air instead. { 'pəmp·iŋ ,lós }

pumping station |CIV ENG| A building in which two or more pumps operate to supply fluid flowing at adequate pressure to a distribution system. { 'pəmp·iŋ ,stā·shən }

punch |DES ENG| See nail set. |MECH ENG| A tool that forces metal into a die for extrusion or similar operations. { pənch }

punched-plate screen |ENG| Flat, perforated plate with round, square, hexagonal, or elongated openings; used for screening (size classification) of crushed or pulverized solids. { 'pəncht ¦plāt ,skrēn }

punching |ENG| **1.** A piece removed from a sheet of metal or other material by a punch press. **2.** A method of extrusion, cold heading, hot forging, or stamping in a machine for which mating die sections determine the shape or contour of the work. { 'pənch·iŋ }

punch press |MECH ENG| **1.** A press consisting of a frame in which slides or rams move up and down, of a bed to which the die shoe or bolster plate is attached, and of a source of power to move the slide. Also known as drop press. **2.** Any mechanical press. { 'pənch ,pres }

punch radius |DES ENG| The radius on the bottom end of the punch over which the metal sheet is bent in drawing. { 'pənch ,rād·ē·əs }

puncture-sealing tire |ENG| A tire whose interior surface is coated with a plastic material that is forced into a puncture by high-pressure air inside the tire and subsequently hardens to seal the puncture. { ¦pəŋk·chər ,sēl·iŋ 'tīr }

pure shear |MECH| A particular example of irrotational strain or flattening in which a body is elongated in one direction and shortened at right angles to it as a consequence of differential displacements on two sets of intersecting planes. { pyůr 'shir }

purge meter interlock |MECH ENG| A meter to maintain airflow through a boiler furnace at a specific level for a definite time interval; ensures that the proper air-fuel ratio is achieved prior to ignition. { 'pərj ¦mēd·ər 'in·tər,läk }

purging |ENG| Replacing the atmosphere in a container by an inert substance to prevent formation of explosive mixtures. { 'pərj·iŋ }

purify |ENG| To remove unwanted constituents from a substance. { 'pyůr·ə,fī }

purlin |BUILD| A horizontal roof beam, perpendicular to the trusses or rafters; supports the roofing material or the common rafters. { 'pər·lən }

purse seine |ENG| A net that can be dropped by two boats to encircle a school of fish, then pulled together at the bottom and raised, thereby catching the fish. { 'pərs ˌsān }

push-bar conveyor |MECH ENG| A type of chain conveyor in which two endless chains are cross-connected at intervals by push bars which propel the load along a stationary bed or trough of the conveyor. { 'pùsh ˌbar kən,vā·ər }

push bench |MECH ENG| A machine used for drawing tubes of moderately heavy gage by cupping metal sheet and applying pressure to the inside bottom of the cup to force it through a die. { 'pùsh ˌbench }

push fit |DES ENG| A hand-tight sliding fit between a shaft and a hole. { 'pùsh ˌfit }

push nipple |MECH ENG| A short length of pipe used to connect sections of cast iron boilers. { 'pùsh ˌnip·əl }

push-pull sound track |ENG ACOUS| A sound track having two recordings so arranged that the modulation in one is 180° out of phase with that in the other. { 'pùsh ˌpùl 'saùn ˌtrak }

push rod |MECH ENG| A rod, as in an internal combustion engine, which is actuated by the cam to open and close the valves. { 'pùsh ˌräd }

push-up |ENG| Concave bottom contour of a plastic container; allows an even bearing surface on the outer edge and prevents the container from rocking. { 'pùsh,əp }

putlog |CIV ENG| A crosspiece in a scaffold or formwork; supports the soffits and is supported by the ledgers. { 'pùt,läg }

putty knife |DES ENG| A knife with a broad flexible blade, used to apply and smooth putty. { 'pəd·ē ˌnīf }

pW See picowatt.

pwt See pennyweight.

pycnometer |ENG| A container whose volume is precisely known, used to determine the density of a liquid by filling the container with the liquid and then weighing it. Also spelled pyknometer. { pik'näm·əd·ər }

pyknometer See pycnometer. { pik'näm·əd·ər }

pylon |CIV ENG| 1. A massive structure, such as a truncated pyramid, on either side of an entrance. 2. A tower supporting a wire over a long span. 3. A tower or other structure marking a route for an airplane. { 'pī,län }

pyramidal horn |ENG| Horn whose sides form a pyramid. { ˌpir·əˌmid·əl 'hòrn }

pyranometer |ENG| An instrument used to measure the combined intensity of incoming direct solar radiation and diffuse sky radiation; compares heating produced by the radiation on blackened metal strips with that produced by an electric current. Also known as solarimeter. { ˌpir·ə'näm·əd·ər }

pyrgeometer |ENG| An instrument for measuring radiation from the surface of the earth into space. { ˌpīr·jē'äm·əd·ər }

pyrheliometer |ENG| An instrument for measuring the total intensity of direct solar radiation received at the earth. { ˌpir,hē·lē'äm·əd·ər }

pyrogenic distillation |CHEM ENG| A cracking process that runs at high temperatures, high pressures, or both, resulting in greater yields of the light hydrocarbon components of gasoline. { ˌpī·rōˌjen·ik ˌdist·əl'ā·shən }

pyroligneous |CHEM ENG| Referring to a substance obtained by the destructive distillation of wood. { ˌpī·rō'lig·nē·əs }

pyrometer |ENG| Any of a broad class of temperature-measuring devices; they were originally designed to measure high temperatures, but some are now used in any temperature range; includes radiation pyrometers, thermocouples, resistance pyrometers, and thermistors. { pī'räm·əd·ər }

pyrometry |THERMO| The science and technology of measuring high temperatures. { pī'räm·ə·trē }

pyrostat |ENG| 1. A sensing device that automatically actuates a warning or extinguishing mechanism in case of fire. 2. A high-temperature thermostat. { 'pī·rə,stat }

pyrotechnic pistol |ENG| A single-shot device designed specifically for projecting pyrotechnic signals. { ˌpī·rəˌtek·nik 'pis·təl }

pyrotechnics |ENG| Art and science of preparing and using fireworks. { ˌpī·rəˌtek·niks }

pz See pièze.

Q

Q |THERMO| A unit of heat energy, equal to 10^{18} British thermal units, or approximately 1.055×10^{21} joules.

Q meter |ENG| A direct-reading instrument which measures the Q of an electric circuit at radio frequencies by determining the ratio of inductance to resistance, and which has also been developed to measure many other quantities. Also known as quality-factor meter. { 'kyü ,mēd·ər }

Q multiplier |ELECTR| A filter that gives a sharp response peak or a deep rejection notch at a particular frequency, equivalent to boosting the Q of a tuned circuit at that frequency. { 'kyü 'məl·tə,plī·ər }

Q point See quiescent operating point. { 'kyü ,pȯint }

qr See quarter.

qr tr See quarter.

Q signal |ELECTR| The quadrature component of the chrominance signal in color television, having a bandwidth of 0 to 0.5 megahertz; it consists of $+0.48(R - Y)$ and $+0.41(B - Y)$, where Y is the luminance signal, R is the red camera signal, and B is the blue camera signal. { 'kyü ,sig·nəl }

qt See quart.

quad |ELEC| A series of four separately insulated conductors, generally twisted together in pairs. |ELECTR| A series-parallel combination of transistors; used to obtain increased reliability through double redundancy, because the failure of one transistor will not disable the entire circuit. |THERMO| A unit of heat energy, equal to 10^{15} British thermal units, or approximately 1.055×10^{18} joules. { kwäd }

quadrangle |CIV ENG| **1.** A four-cornered, four-sided courtyard, usually surrounded by buildings. **2.** The buildings surrounding such a courtyard. **3.** A four-cornered, four-sided building. { 'kwä,draŋ·gəl }

quadrant |ENG| **1.** An instrument for measuring altitudes, used, for example, in astronomy, surveying, and gunnery; employs a sight that can be moved through a graduated 90° arc. **2.** A lever that can move through a 90° arc. |MECH ENG| A device for converting horizontal reciprocating motion to vertical reciprocating motion. { 'kwä·drənt }

quadrant angle of fall |MECH| The vertical acute angle at the level point, between the horizontal and the line of fall of a projectile. { 'kwä·drənt 'aŋ·gəl əv 'fȯl }

quadrant electrometer |ENG| An instrument for measuring electric charge by the movement of a vane suspended on a wire between metal quadrants; the charge is introduced on the vane and quadrants in such a way that there is a proportional twist to the wire. { 'kwä·drənt i,lek'träm·əd·ər }

quadraphonic sound system |ENG ACOUS| A system for reproducing sound by means of four loudspeakers properly situated in the listening room, usually at the four corners of a square, with each loudspeaker being fed its own identifiable segment of the program signal. Also known as four-channel sound system. { 'kwä·drə¦fän·ik 'saȯnd }

quadratic performance index |CONT SYS| A measure of system performance which is, in general, the sum of a quadratic function of the system state at fixed times, and the integral of a quadratic function of the system state and control inputs. { kwä'drad·ik pər'fȯr·məns ,in ,deks }

quadricycle |MECH ENG| A four-wheeled human-powered land vehicle, usually propelled by the action of the rider's feet on the pedals. { 'kwäd·rə,sī·kəl }

quadrilateral See quadrangle. { ¦kwä·drə¦lad·ə·rəl }

quadruple thread |DES ENG| A multiple thread having four separate helices equally spaced around the circumference of the threaded member; the lead is equal to four times the pitch of the thread. { kwə'drüp·əl 'thred }

qualification test |ENG| A formally defined series of tests by which the functional, environmental, and reliability performance of a component or system may be evaluated in order to satisfy the engineer, contractor, or owner as to its satisfactory design and construction prior to final approval and acceptance. { ,kwäl·ə·fə'kā·shən ,test }

quality analysis |IND ENG| Examination of the quality goals of a product or service. { 'kwäl·əd·ē ə,nal·ə·səs }

quality assurance |IND ENG| A series of

planned or systematic actions required to provide adequate confidence that a product or service will satisfy given needs. { 'kwäl·əd·ē ə'shu̇r·əns }

quality control [IND ENG] The operational techniques and the activities that sustain the quality of a product or service in order to satisfy given requirements. It consists of quality planning, data collection, data analysis, and implementation, and is applicable to all phases of the product life cycle: design, development, manufacturing, delivery and installation, and operation and maintenance. { 'kwäl·əd·ē kən‚trōl }

quality-control chart [IND ENG] A control chart used to indicate and control the quality of a product. { 'kwäl·əd·ē kən‚trōl ‚chärt }

quality-factor meter See Q meter. { 'kwäl·əd·ē ‚fak·tər ‚mēd·ər }

quantity meter [ENG] A type of fluid meter used to measure volume of flow. { 'kwän·əd·ē ‚mēd·ər }

quantizer [ELECTR] A device that measures the magnitude of a time-varying quantity in multiples of some fixed unit, at a specified instant or specified repetition rate, and delivers a proportional response that is usually in pulse code or digital form. { kwän'tīz·ər }

quantum dot [ELECTR] A quantized electronic structure in which electrons are confined with respect to motion in all three dimensions. { ‚kwänt·əm 'dät }

quantum efficiency [ELECTR] The average number of electrons photoelectrically emitted from a photocathode per incident photon of a given wavelength in a phototube. { 'kwän·təm i‚fish·ən·sē }

quantum electronics [ELECTR] The branch of electronics associated with the various energy states of matter, motions within atoms or groups of atoms, and various phenomena in crystals; examples of practical applications include the atomic hydrogen maser and the cesium atomic-beam resonator. { 'kwän·təm ‚i‚lek'trän·iks }

quantum Hall effect [ELECTR] A phenomenon exhibited by certain semiconductor devices at low temperatures and high magnetic fields, whereby the Hall resistance becomes precisely equal to $(h/e^2)/n$, where h is Planck's constant, e is the electronic charge, and n is either an integer or a rational fraction. Also known as von Klitzing effect. { 'kwän·təm 'hȯl i‚fekt }

quantum well [ELECTR] A thin layer of material (typically between 1 and 10 nanometers thick) within which the potential energy of an electron is less than outside the layer, so that the motion of the electron perpendicular to the layer is quantized. { ‚kwän·təm 'wel }

quantum well injection transit-time diode [ELECTR] An active microwave diode that employs resonant tunneling through a gallium arsenide quantum well located between two aluminum gallium arsenide barriers to inject electrons into an undoped gallium arsenide drift region. Abbreviated QWITT diode. { ‚kwän·təm ‚wel in ‚jek·shən ‚tranz·it ‚tīm 'dī‚ōd }

quantum well infrared photodetector [ELECTR] A detector of infrared radiation composed of numerous alternating layers of controlled thickness of gallium arsenide and aluminum gallium arsenide; the spectral response of the device can be tailored within broad limits by adjusting the aluminum-to-gallium ratio and the thicknesses of the layers during growth. Abbreviated QWIP. { ‚kwänt·əm ‚wel ‚in·frə'red ‚fōd·ō·di'tek·tər }

quantum wire [ELECTR] A strip of conducting material about 10 nanometers or less in width and thickness that displays quantum-mechanical effects such as the Aharanov-Bohm effect and universal conductance fluctuations. { 'kwän·təm 'wīr }

quarantine anchorage [CIV ENG] An area where a vessel anchors when satisfying quarantine regulations. { 'kwär·ən‚tēn ‚aŋ·kər·ij }

quarry [ENG] An open or surface working or excavation for the extraction of building stone, ore, coal, gravel, or minerals. { 'kwär·ē }

quarry bar [ENG] A horizontal bar with legs at each end, used to carry machine drills. { 'kwär·ē ‚bär }

quarrying [ENG] The surface exploitation and removal of stone or mineral deposits from the earth's crust. { 'kwär·ē·iŋ }

quarrying machine [MECH ENG] Any machine used to drill holes or cut tunnels in native rock, such as a gang drill or tunneling machine; most commonly, a small locomotive bearing rock-drilling equipment operating on a track. { 'kwär·ē·iŋ mə‚shēn }

quarry sap See quarry water. { 'kwär·ē ‚sap }

quarry water [ENG] Subsurface water retained in freshly quarried rock. Also known as quarry sap. { 'kwär·ē ‚wȯd·ər }

quart [MECH] Abbreviated qt. **1.** A unit of volume used for measurement of liquid substances in the United States, equal to 2 pints, or 1/4 gallon, or 573/4 cubic inches, or 9.46352946 × 10^{-4} cubic meter. **2.** A unit of volume used for measurement of solid substances in the United States, equal to 2 dry pints, or 1/32 bushel, or 107,521/1600 cubic inches, or approximately 1.10122 × 10^{-3} cubic meter. **3.** A unit of volume used for measurement of both liquid and solid substances, although mainly the former, in the United Kingdom and Canada, equal to 2 U.K. pints, or 1/4 U.K. gallon, or approximately 1.1365225 × 10^{-3} cubic meter. { kwȯrt }

quarter [MECH] **1.** A unit of mass in use in the United States, equal to 1/4 short ton, or 500 pounds, or 226.796185 kilograms. **2.** A unit of mass used in troy measure, equal to 1/4 troy hundredweight, or 25 troy pounds, or 9.33104304 kilograms. Abbreviated qr tr. **3.** A unit of mass used in the United Kingdom, equal to 1/4 hundredweight, or 28 pounds, or 12.70058636 kilograms. Abbreviated qr. **4.** A unit of volume used in the United Kingdom for measurement of liquid and solid substances, equal to 8 bushels, or 64 gallons, or approximately 0.29094976 cubic meter. { 'kwȯrd·ər }

quartering machine [MECH ENG] A machine

that bores parallel holes simultaneously in such a way that the center lines of adjacent holes are 90° apart.   { 'kwȯrd·ə·riŋ mə,shēn }

quarter-turn drive   [MECH ENG]   A belt drive connecting pulleys whose axes are at right angles.   { 'kwȯrd·ər ,tərn 'drīv }

quartz crystal   [ELECTR]   A natural or artificially grown piezoelectric crystal composed of silicon dioxide, from which thin slabs or plates are carefully cut and ground to serve as a crystal plate.   { 'kwȯrts ¦krist·əl }

quartz-crystal filter   [ELECTR]   A filter which utilizes a quartz crystal; it has a small bandwidth, a high rate of cutoff, and a higher unloaded Q than can be obtained in an ordinary resonator.   { 'kwȯrts ,krist·əl 'fil·tər }

quartz-crystal resonator   [ELECTR]   A quartz plate whose natural frequency of vibration is used to control the frequency of an oscillator. Also known as quartz resonator.   { 'kwȯrts ,krist·əl 'rez·ən,ād·ər }

quartz fiber   [ENG]   An extremely fine and uniform quartz filament that may be used as a torsion thread or as an indicator in an electroscope or dosimeter.   { 'kwȯrts 'fī·bər }

quartz-fiber dosimeter   [ENG]   A dosimeter in which radiation dose is determined from the deflection of a quartz fiber that is initially charged, repelling it from its metal support, and has its charge reduced by ionizing radiation, causing a proportional reduction in its deflection.   { ¦kwȯrts ¦fī·bər dō'sim·əd·ər }

quartz-fiber manometer   See decrement gage.   { ¦kwȯrts ¦fī·bər mə'näm·əd·ər }

quartz horizontal magnetometer   [ENG]   A type of relative magnetometer used as a geomagnetic field instrument and as an observatory instrument for routine calibration of recording equipment.   { 'kwȯrts ,här·ə'zänt·əl ,mag·nə'täm·əd·ər }

quartz oscillator   [ELECTR]   An oscillator in which the frequency of the output is determined by the natural frequency of vibration of a quartz crystal.   { 'kwȯrts 'äs·ə,lād·ər }

quartz plate   See crystal plate.   { 'kwȯrts ¦plāt }

quartz pressure gage   [ENG]   A pressure gage that uses a highly stable quartz crystal resonator whose frequency changes directly with applied pressure.   { 'kwȯrts 'presh·ər ,gāj }

quartz resonator   See quartz-crystal resonator.   { 'kwȯrts 'rez·ən,ād·ər }

quartz resonator force transducer   [ENG]   A type of accelerometer which measures the change in the resonant frequency of a small quartz plate with a longitudinal slot, forming a double-ended tuning fork, when a longitudinal force associated with acceleration is applied to the plate.   { 'kwȯrts 'rez·ən,ād·ər 'fȯrs tranz,dü·sər }

quartz thermometer   [ENG]   A thermometer based on the sensitivity of the resonant frequency of a quartz crystal to changes in temperature.   { 'kwȯrts thər,mäm·əd·ər }

quasi-linear feedback control system   [CONT SYS]   Feedback control system in which the relationships between the pertinent measures of the system input and output signals are substantially linear despite the existence of nonlinear elements.   { ¦kwä·zē 'lin·ē·ər 'fēd,bak kən'trol ,sis·təm }

quasi-linear system   [CONT SYS]   A control system in which the relationships between the input and output signals are substantially linear despite the existence of nonlinear elements.   { ¦kwä·zē 'lin·ē·ər 'sis·təm }

quasi-particle detector   [ENG]   A detector of electromagnetic radiation at wavelengths close to 1 millimeter, based on the tunneling of single electrons (more precisely, quasi-particles) through a tunnel junction consisting of an oxide barrier between two superconductors, with a responsivity of one tunneling electron for each microwave photon absorbed.   { ¦kwä·zē ¦pard·ə·kəl di,tek·tər }

quasi-static process   See reversible process.   { ¦kwä·zē 'stad·ik 'prä·səs }

quay   [CIV ENG]   A solid embankment or structure parallel to a waterway; used for loading and unloading ships.   { kē }

queen closer   [CIV ENG]   In masonry work, a brick that has been cut in half along its length and is used at the end of a course.   { 'kwēn ,klōs·ər }

queen post   [CIV ENG]   Either of two vertical members, one on each side of the apex of a triangular truss.   { 'kwēn ,pōst }

quench bath   [ENG]   A liquid medium, such as oil, fused salt, or water, into which a material is plunged for heat-treatment purposes.   { 'kwench ,bath }

quenching   [ELECTR]   **1.** The process of terminating a discharge in a gas-filled radiation-counter tube by inhibiting reignition.   **2.** Reduction of the intensity of resonance radiation resulting from deexcitation of atoms, which would otherwise have emitted this radiation, in collisions with electrons or other atoms in a gas.   [ENG]   Shock cooling by immersing liquid or molten material into a cooling medium (liquid or gas); used in metallurgy, plastics forming, and petroleum refining.   [MECH ENG]   Rapid removal of excess heat from the combustion chamber of an automotive engine.   { 'kwench·iŋ }

quench-tank extrusion   [ENG]   Plastic-film or metal extrusion that is cooled in a quenching medium.   { 'kwench ,taŋk ik'strü·zhən }

quench temperature   [ENG]   The temperature of the medium used for quenching.   { 'kwench ,tem·prə·chər }

queue   See waiting line.   { kyü }

queueing   [ENG]   The movement of discrete units through channels, such as programs or data arriving at a computer, or movement on a highway of heavy traffic.   { 'kyü·iŋ }

quick-change gearbox   [MECH ENG]   A cluster of gears on a machine tool, the arrangement of which allows for the rapid change of gear ratios.   { 'kwik ¦chānj 'gir,bäks }

quickmatch   [ENG]   Fast-burning fuse made from a cord impregnated with black powder.   { 'kwik,mach }

quick return   [MECH ENG]   A device used in a

reciprocating machine to make the return stroke faster than the power stroke. { 'kwik ri'tərn }

quiescent |ELECTR| Pertaining to a circuit element which has no input signal, so that it does not perform its active function. |ENG| Pertaining to a body at rest, or inactive, such as an undisturbed liquid in a storage or process vessel. { kwē'es·ənt }

quill |DES ENG| A hollow shaft into which another shaft is inserted in mechanical devices. { kwil }

quill drive |MECH ENG| A drive in which the motor is mounted on a nonrotating hollow shaft surrounding the driving-wheel axle; pins on the armature mesh with spokes on the driving wheels, thereby transmitting motion to the wheels; used on electric locomotives. { 'kwil ‚drīv }

quill gear |MECH ENG| A gear mounted on a hollow shaft. { 'kwil ‚gir }

quintal See metric centner. { 'kwint·əl }

quirk |BUILD| 1. An indentation separating one element from another, as between moldings. 2. A V groove in the finish-coat plaster where it abuts the return on a door or window. { kwərk }

quirk bead |BUILD| 1. A bead with a quirk on one side only, as on the edge of a board. Also known as bead and quirk. 2. A bead that is flush with the adjoining surface and separated from it by a quirk on each side. Also known as bead and quirk; double-quirked bead; flush bead; recessed bead. 3. A bead located at a corner with quirks at either side at right angles to each other. Also known as bead and quirk; return bead. 4. A bead with a quirk on its face. Also known as bead and quirk. { kwərk ‚bēd }

Q unit |THERMO| A unit of energy, used in measuring the heat energy of fuel reserves, equal to 10^{18} British thermal units, or approximately 1.055×10^{21} joules. { 'kyü ‚yü·nət }

quoin |BUILD| One of the members forming an outside corner or exterior angle of a building, and differentiated from the wall by color, texture, size, or projection. { köin }

quoin post |CIV ENG| The vertical member at the jointed end of a gate in a navigation lock. { 'köin ‚pöst }

qwerty keyboard |ENG| A keyboard containing the standard arrangement of letters so named after the first letters on the top alphabetic row. { 'kwərd·ē kē‚börd }

R

rabbet |ENG| **1.** A groove cut into a part. **2.** A strip applied to a part as, for example, a stop or seal. **3.** A joint formed by fitting one member into a groove, channel, or recess in the face or edge of a second member. { 'rab·ət }

rabbet plane [DES ENG] A plane with the blade extending to the outer edge of one side that is open. { 'rab·ət ,plān }

rabbling |ENG| Stirring a molten charge, as of metal or ore. { 'rab·liŋ }

race [DES ENG] Either of the concentric pair of steel rings of a ball bearing or roller bearing. |ENG| A channel transporting water to or away from hydraulic machinery, as in a power house. { rās }

rack |CIV ENG| A fixed screen composed of parallel bars placed in a waterway to catch debris. |DES ENG| *See* relay rack. |ENG| A frame for holding or displaying articles. |MECH ENG| A bar containing teeth on one face for meshing with a gear. { rak }

rack and pinion |MECH ENG| A gear arrangement consisting of a toothed bar that meshes with a pinion. { 'rak ən 'pin·yən }

rack-and-pinion steering |MECH ENG| A steering system in which the rotation of pinion gear at the end of the steering column moves a toothed bar (the rack) left or right to transmit steering movements. { ¦rak ən ¦pin·yən 'stir·iŋ }

racking |CIV ENG| Setting back the end of each course of brick or stone from the end of the preceding course. { 'rak·iŋ }

rack railway |CIV ENG| A railway with a rack between the rails which engages a gear on the locomotive; used on steep grades. { 'rak 'rāl,wā }

radar |ENG| **1.** A system using beamed and reflected radio-frequency energy for detecting and locating objects, measuring distance or altitude, navigating, homing, bombing, and other purposes; in detecting and ranging, the time interval between transmission of the energy and reception of the reflected energy establishes the range of an object in the beam's path. Derived from radio detection and ranging. **2.** *See* radar set. { 'rā,där }

radar bombsight |ENG| An airborne radar set used to sight the target, solve the bombing problem, and drop bombs. { 'rā,där 'bäm,sīt }

radar command guidance |ENG| A missile guidance system in which radar equipment at the launching site determines the positions of both target and missile continuously, computes the missile course corrections required, and transmits these by radio to the missile as commands. { 'rā,där kə'mand ,gīd·əns }

radar contact |ENG| Recognition and identification of an echo on a radar screen; an aircraft is said to be on radar contact when its radar echo can be seen and identified on a PPI (plan-position indicator) display. { 'rā,där ,kän,takt }

radar coverage |ENG| The limits within which objects can be detected by one or more radar stations. { 'rā,där ,kəv·rij }

radar coverage indicator |ENG| Device that shows how far a given aircraft should be tracked by a radar station, and also provides a reference (detection) range for quality control; takes into account aircraft size, altitude, screening angle, site elevation, type radar, antenna radiation pattern, and antenna tilt. { 'rā,där ¦kəv·rij ,in·də,kād·ər }

radar dome |ENG| Weatherproof cover for a primary radiating element of a radar or radio device which is transparent to radio-frequency energy, and which permits active operation of the radiating element, including mechanical rotation or other movement as applicable. { 'rā,där ,dōm }

radar gun-layer |ENG| A radar device which tracks a target and aims a gun or guns automatically. { 'rā,där 'gən ,lā·ər }

radar homing |ENG| Homing in which a missile-borne radar locks onto a target and guides the missile to that target. { 'rā,där ,hōm·iŋ }

radar marker |ENG| A fixed facility which continuously emits a radar signal so that a bearing indication appears on a radar display. { 'rā,där ,mär·kər }

radar netting |ENG| The linking of several radars to a single center to provide integrated target information. { 'rā,där ,ned·iŋ }

radar netting station |ENG| A center which can receive data from radar tracking stations and exchange these data among other radar tracking stations, thus forming a radar netting system. { 'rā,där ¦ned·iŋ ,stā·shən }

radar picket |ENG| A ship or aircraft equipped with early-warning radar and operating at a distance from the area being protected, to extend the range of radar detection. { 'rā,där ,pik·ət }

radar prediction |ENG| A graphic portrayal of the estimated radar intensity, persistence, and shape of the cultural and natural features of a specific area. { 'rā,där pri'dik·shən }

radar range marker *See* distance marker. { 'rā,där 'rānj ,mär·kər }

radar relay |ENG| **1.** Equipment for relaying the radar video and appropriate synchronizing signal to a remote location. **2.** Process or system by which radar echoes and synchronization data are transmitted from a search radar installation to a receiver at a remote point. { 'rā,där 'rē,lā }

radar scanning |ENG| The process or action of directing a radar beam through a space search pattern for the purpose of locating a target. { 'rā,där ,skan·iŋ }

radarscope overlay |ENG| A transparent overlay placed on a radarscope for comparison and identification of radar returns. { 'rā,där,skōp 'ō·vər,lā }

radar set |ENG| A complete assembly of radar equipment for detecting and ranging, consisting essentially of a transmitter, antenna, receiver, and indicator. Also known as radar. { 'rā,där ,set }

radarsonde |ENG| **1.** An electronic system for automatically measuring and transmitting high-altitude meteorological data from a balloon, kite, or rocket by pulse-modulated radio waves when triggered by a radar signal. **2.** A system in which radar techniques are used to determine the range, elevation, and azimuth of a radar target carried aloft by a radiosonde. { 'rā,där,sänd }

radar station |ENG| The place, position, or location from which, or at which, a radar set transmits or receives signals. { 'rā,där ,stā·shən }

radar surveying |ENG| Surveying in which airborne radar is used to measure accurately the distance between two ground radio beacons positioned along a baseline; this eliminates the need for measuring distance along the baseline in inaccessible or extremely rough terrain. { 'rā ,där sər'vā·iŋ }

radar telescope |ENG| A large radar antenna and associated equipment used for radar astronomy. { 'rā,där 'tel·ə,skōp }

radar theodolite |ENG| A theodolite that uses radar to obtain azimuth, elevation, and slant range to a reflecting target, for surveying or other purposes. { 'rā,där thē'äd·əl,īt }

radar threshold limit |ENG| For a given radar and specified target, the point in space relative to the focal point of the antenna at which initial detection criteria can be satisfied. { 'rā,där 'thresh,hōld ,lim·ət }

radar tracking |ENG| Tracking a moving object by means of radar. { 'rā,där ,trak·iŋ }

radar tracking station |ENG| A radar facility which has the capability of tracking moving targets. { 'rā,där 'trak·iŋ ,stā·shən }

radar triangulation |ENG| A radar system of locating targets, usually aircraft, in which two or more separate radars are employed to measure range only; the target is located by automatic trigonometric solution of the triangle composed of a pair of radars and the target in which all three sides are known. { 'rā,där trī,aŋ·gyə'lā·shən }

radar wind system |ENG| Apparatus in which radar techniques are used to determine the range, elevation, and azimuth of a balloon-borne target, and hence to compute upper-air wind data. { 'rā,där |wind ,sis·təm }

radial acceleration *See* centripetal acceleration. { 'rād·ē·əl ak,sel·ə'rā·shən }

radial band pressure |MECH| The pressure which is exerted on the rotating band by the walls of the gun tube, and hence against the projectile wall at the band seat, as a result of the engraving of the band by the gun rifling. { 'rād·ē·əl |band ,presh·ər }

radial bearing |MECH ENG| A bearing with rolling contact in which the direction of action of the load transmitted is radial to the axis of the shaft. { 'rād·ē·əl 'ber·iŋ }

radial draw forming |MECH ENG| A metal-forming method in which tangential stretch and radial compression are applied gradually and simultaneously. { 'rād·ē·əl 'drò ,fórm·iŋ }

radial drill |MECH ENG| A drilling machine in which the drill spindle can be moved along a horizontal arm which itself can be rotated about a vertical pillar. { 'rād·ē·əl 'dril }

radial drilling |ENG| The drilling of several holes in one plane, all radiating from a common point. { 'rād·ē·əl 'dril·iŋ }

radial engine |MECH ENG| An engine characterized by radially arranged cylinders at equiangular intervals around the crankshaft. { 'rād·ē·əl 'en·jən }

radial-flow |ENG| Having the fluid working substance flowing along the radii of a rotating tank. { 'rād·ē·əl |flō }

radial-flow turbine |MECH ENG| A turbine in which the gases flow primarily in a radial direction. { 'rād·ē·əl |flō 'tər·bən }

radial force |MECH ENG| In machining, the force acting on the cutting tool in a direction opposite to depth of cut. { 'rād·ē·əl 'fórs }

radial gate *See* Tainter gate. { 'rād·ē·əl 'gāt }

radial heat flow |THERMO| Flow of heat between two coaxial cylinders maintained at different temperatures; used to measure thermal conductivities of gases. { 'rād·ē·əl 'hēt ,flō }

radial load |MECH ENG| The load perpendicular to the bearing axis. { 'rād·ē·əl 'lōd }

radial locating |MECH ENG| One of the three locating problems in tooling to maintain the desired relationship between the workpiece, the cutter, and the body of the machine tool; the other two locating problems are concentric and plane locating. { 'rād·ē·əl 'lō,kād·iŋ }

radial motion |MECH| Motion in which a body moves along a line connecting it with an observer or reference point; for example, the motion of stars which move toward or away from the earth without a change in apparent position. { 'rād·ē·əl 'mō·shən }

radial-ply |DES ENG| Pertaining to the construction of a tire in which the cords run straight across the tire, and an additional layered belt of fabric is placed around the circumference between the plies and the tread. { 'rād·ē·əl ¦plī }

radial-ply tire See radial tire. { ¦rād·ē·əl ¦plī 'tīr }

radial rake [MECH ENG] The angle between the cutter tooth face and a radial line passing through the cutting edge in a plane perpendicular to the cutter axis. { 'rād·ē·əl 'rāk }

radial road [CIV ENG] One of a group of roads leading outward from the center of a city in a pattern similar to spokes on a wheel. { 'rād·ē·əl ‚rōd }

radial saw [MECH ENG] A power saw that has a circular blade suspended from a transverse head mounted on a rotatable overarm. { 'rād·ē·əl 'sȯ }

radial selector See omnibearing selector. { 'rād·ē·əl si'lek·tər }

radial stress [MECH] Tangential stress at the periphery of an opening. { 'rād·ē·əl 'stres }

radial tire [ENG] A pneumatic tire constructed with a layer of fabric between the tread and the plies (cords), which run straight across the tire. Also known as radial-ply tire. { ¦rād·ē·əl 'tīr }

radial velocity [MECH] The component of the velocity of a body that is parallel to a line from an observer or reference point to the body; the radial velocities of stars are valuable in determining the structure and dynamics of the Galaxy. Also known as line-of-sight velocity. { 'rād·ē·əl və'läs·əd·ē }

radial wave equation [MECH] Solutions to wave equations with spherical symmetry can be found by separation of variables; the ordinary differential equation for the radial part of the wave function is called the radial wave equation. { 'rād·ē·əl ¦wāv i‚kwā·zhən }

radiant energy See radiation. { 'rād·ē·ənt 'en·ər·jē }

radiant-energy thermometer See radiation pyrometer. { 'rād·ē·ənt ¦en·ər·jē thər'mäm·əd·ər }

radiant heating [ENG] Any system of space heating in which the heat-producing means is a surface that emits heat to the surroundings by radiation rather than by conduction or convection. { 'rād·ē·ənt 'hēd·iŋ }

radiant superheater [MECH ENG] A superheater designed to transfer heat from the products of combustion to the steam primarily by radiation. { 'rād·ē·ənt 'sü·pər‚hēd·ər }

radiant-type boiler [MECH ENG] A water-tube boiler in which boiler tubes form the boundary of the furnace. { 'rād·ē·ənt ¦tīp 'bȯil·ər }

radiating power See emittance. { 'rād·ē‚ād·iŋ 'paů·ər }

radiation [ENG] A method of surveying in which points are located by knowledge of their distances and directions from a central point. { ‚rād·ē'ā·shən }

radiation correction See cooling correction. { ‚rād·ē'ā·shən kə‚rek·shən }

radiation hardening [ENG] Improving the ability of a device or piece of equipment to withstand

nuclear or other radiation; applies chiefly to dielectric and semiconductor materials. { ‚rād·ē'ā·shən 'härd·ən·iŋ }

radiation loss [MECH ENG] Boiler heat loss to the atmosphere by conduction, radiation, and convection. { ‚rād·ē'ā·shən ‚lȯs }

radiation noise See electromagnetic noise. { ‚rād·ē'ā·shən ‚nȯiz }

radiation oven [ENG] Heating chamber relying on tungsten-filament infrared lamps with reflectors to create temperatures up to 600°F (315°C); used to dry sheet and granular material and to bake surface coatings. { ‚rād·ē'ā·shən ‚av·ən }

radiation pyrometer [ENG] An instrument which measures the temperature of a hot object by focusing the thermal radiation emitted by the object and making some observation on it; examples include the total-radiation, optical, and ratio pyrometers. Also known as noncontact thermometer; radiant-energy thermometer; radiation thermometer. { ‚rād·ē'ā·shən pī'räm·əd·ər }

radiation shelter See fallout shelter. { ‚rād·ē'ā·shən ‚shel·tər }

radiation shield [ENG] A shield or wall of material interposed between a source of radiation and a radiation-sensitive body, such as a person, radiation-detection instrument, or photographic film, to protect the latter. { ‚rād·ē'ā·shən ‚shēld }

radiation thermometer See radiation pyrometer. { ‚rād·ē'ā·shən thər'mäm·əd·ər }

radiation vacuum gage [ENG] Vacuum (reduced-pressure) measurement device in which gas ionization from an alpha source of radiation varies measurably with changes in the density (molecular concentration) of the gas being measured. { ‚rād·ē'ā·shən 'vak·yəm ‚gāj }

radiation well logging See radioactive well logging. { ‚rād·ē'ā·shən 'wel ‚läg·iŋ }

radiator [ENG] Any of numerous devices, units, or surfaces that emit heat, mainly by radiation, to objects in the space in which they are installed. { 'rād·ē‚ād·ər }

radiator temperature drop [MECH ENG] In internal combustion engines, the difference in temperature of the coolant liquid entering and leaving the radiator. { 'rād·ē‚ād·ər 'tem·prə·chər ‚dräp }

radioacoustic position finding See radioacoustic ranging. { ¦rād·ē·ō·ə'küs·tik pə'zish·ən ‚find·iŋ }

radioacoustic ranging [ENG] A method for finding the position of a vessel at sea; a bomb is exploded in the water, and the sound of the explosion transmitted through water is picked up by the vessel and by shore stations, other vessels, or buoys whose positions are known; the received sounds are transmitted instantaneously by radio to the surveying vessel, and the elapsed times are proportional to the distances to the known positions. Abbreviated RAR. Also known as radioacoustic position finding; radioacoustic sound ranging. { ¦rād·ē·ō·ə'küs·tik 'ranj·iŋ }

radioacoustic sound ranging See radioacoustic ranging. { ¦rād·ē·ō·ə'küs·tik 'saůnd ˌrānj·iŋ }

radioactive heat [THERMO] Heat produced within a medium as a result of absorption of radiation from decay of radioisotopes in the medium, such as thorium-232, potassium-40, uranium-238, and uranium-235. { ¦rād·ē·ō'ak·tiv 'hēt }

radioactive snow gage [ENG] A device which automatically and continuously records the water equivalent of snow on a given surface as a function of time; a small sample of a radioactive salt is placed in the ground in a lead-shielded collimator which directs a beam of radioactive particles vertically upward; a Geiger-Müller counting system (located above the snow level) measures the amount of depletion of radiation caused by the presence of the snow. { ¦rād·ē·ō'ak·tiv 'snō ˌgāj }

radioactive well logging [ENG] The recording of the differences in radioactive content (natural or neutron-induced) of the various rock layers found down an oil well borehole; types include γ-ray, neutron, and photon logging. Also known as radiation well logging; radioactivity prospecting. { ¦rād·ē·ō'ak·tiv 'wel ˌläg·iŋ }

radioactivity log [ENG] Record of radioactive well logging. { ˌrād·ē·ō·ak'tiv·əd·ē ˌläg }

radioactivity prospecting See radioactive well logging. { ˌrād·ē·ō·ak'tiv·əd·ē 'prä,spekt·iŋ }

radio altimeter [ENG] An absolute altimeter that depends on the reflection of radio waves from the earth for the determination of altitude, as in a frequency-modulated radio altimeter and a radar altimeter. Also known as electronic altimeter; reflection altimeter. { 'rād·ē·ō al'tim·əd·ər }

radio atmometer [ENG] An instrument designed to measure the effect of sunlight upon evaporation from plant foliage; consists of a porous-clay atmometer whose surface has been blackened so that it absorbs radiant energy. { 'rād·ē·ō at'mäm·əd·ər }

radioautography See autoradiography. { ¦rād·ē·ō,ó'täg·rə·fē }

radio autopilot coupler [ENG] Equipment providing means by which an electrical navigational signal operates an automatic pilot. { 'rād·ē·ō 'ȯd·ō,pī·lət 'kəp·lər }

radio detection [ENG] The detection of the presence of an object by radiolocation without precise determination of its position. { 'rād·ē·ō di'tek·shən }

radio detection and location [ENG] Use of an electronic system to detect, locate, and predict future positions of earth satellites. { 'rād·ē·ō di'tek·shən ən lō'kā·shən }

radio detection and ranging See radar. { 'rād·ē·ō di'tek·shən ən 'rānj·iŋ }

radio Doppler [ENG] Direct determination of the radial component of the relative velocity of an object by an observed frequency change due to such velocity. { 'rād·ē·ō 'däp·lər }

radio echo observation [ENG] A method of determining the distance of objects in the atmosphere or outer space, in which a radar pulse is directed at the object and the time that elapses from transmission of the pulse to reception of a reflected pulse is measured. { 'rād·ē·ō ¦ekō ˌäb·zər'vā·shən }

radio engineering [ENG] The field of engineering that deals with the generation, transmission, and reception of radio waves and with the design, manufacture, and testing of associated equipment. { 'rād·ē·ō ˌen·jə'nir·iŋ }

radio-frequency current [ELEC] Alternating current having a frequency higher than 10,000 hertz. { 'rād·ē·ō ¦frē·kwən·sē ˌkə·rənt }

radio-frequency head [ENG] Unit consisting of a radar transmitter and part of a radar receiver, the two contained in a package for ready removal and installation. { 'rād·ē·ō ¦frē·kwən·sē 'hed }

radio-frequency heating See electronic heating. { 'rād·ē·ō ¦frē·kwən·sē 'hēd·iŋ }

radio-frequency preheating [ENG] Preheating of plastics-molding materials by radio frequencies of 10–100 megahertz per second to facilitate the molding operation or to reduce the molding-cycle time. Abbreviated rf preheating. { 'rād·ē·ō ¦frē·kwən·sē ¦prē'hēd·iŋ }

radio-frequency sensor [ENG] A device that uses radio signals to determine the position of objects to be manipulated by a robotic system. { 'rād·ē·ō ¦frē·kwən·sē ˌsen·sər }

radiogoniometry [ENG] Science of locating a radio transmitter by means of taking bearings on the radio waves emitted by such a transmitter. { ¦rād·ē·ō,gō·nē'äm·ə·trē }

radio-inertial guidance system [ENG] A command type of missile guidance system consisting essentially of a radar tracking unit; a computer that accepts missile position and velocity information from the tracking system and furnishes to the command link appropriate signals to steer the missile; the command link, which consists of a transmitter on the ground and an antenna and receiver on the missile; and an inertial system for partial guidance in case of radio guidance failure. { ¦rād·ē·ō i¦nər·shəl 'gīd·əns ˌsis·təm }

radio interferometer [ENG] Radiotelescope or radiometer employing a separated receiving antenna to measure angular distances as small as 1 second of arc; records the result of interference between separate radio waves from celestial radio sources. { 'rād·ē·ō ˌin·tər·fə'räm·əd·ər }

radiolocation [ENG] Determination of relative position of an object by means of equipment operating on the principle that propagation of radio waves is at a constant velocity and rectilinear. { ¦rād·ē·ō·lō'kā·shən }

radio mast [ENG] A tower, pole, or other structure for elevating an antenna. { 'rād·ē·ō 'mast }

radiometer [ELECTR] A receiver for detecting microwave thermal radiation and similar weak wide-band signals that resemble noise and are obscured by receiver noise; examples include the Dicke radiometer, subtraction-type radiometer, and two-receiver radiometer. Also known as microwave radiometer; radiometer-type receiver. |ENG| An instrument for measuring radiant energy; examples include the bolometer, microradiometer, and thermopile. { ,rād·ē'äm·əd·ər }

radiopasteurization |ENG| Pasteurization by surface treatment with low-energy irradiation. { ¦rād·ē·ō,pas·chúr·ə'zā·shən }

radio position finding |ENG| Process of locating a radio transmitter by plotting the intersection of its azimuth as determined by two or more radio direction finders. { 'rād·ē·ō pə'zish·ən ,fīnd·iŋ }

radio prospecting |ENG| Use of radio and electric equipment to locate mineral or oil deposits. { 'rād·ē·ō 'prä,spek·tiŋ }

radio shielding |ELEC| Metallic covering over all electric wiring and ignition apparatus, which is grounded at frequent intervals for the purpose of eliminating electric interference with radio communications. { 'rād·ē·ō ,shēld·iŋ }

radiosonde |ENG| A balloon-borne instrument for the simultaneous measurement and transmission of meteorological data; the instrument consists of transducers for the measurement of pressure, temperature, and humidity, a modulator for the conversion of the output of the transducers to a quantity which controls a property of the radio-frequency signal, a selector switch which determines the sequence in which the parameters are to be transmitted, and a transmitter which generates the radio-frequency carrier. { 'rād·ē·ō,sänd }

radiosonde-radio-wind system |ENG| An apparatus consisting of a standard radiosonde and radiosonde ground equipment to obtain upper-air data on pressure, temperature, and humidity, and a self-tracking radio direction finder to provide the elevation and azimuth angles of the radiosonde so that the wind vectors may be obtained. { 'rād·ē·ō,sänd ¦rād·ē·ō ,wind ,sis·təm }

radiosonde set |ENG| A complete set for automatically measuring and transmitting high-altitude meteorological data by radio from such carriers as a balloon or rocket. { 'rād·ē·ō,sänd ,set }

radio sonobuoy See sonobuoy. { 'rād·ē·ō 'sän·ə,bói }

radio telescope |ENG| An astronomical instrument used to measure the amount of radio energy coming from various directions in the sky, consisting of a highly directional antenna and associated electronic equipment. { 'rād·ē·ō 'tel·ə,skōp }

radio tracking |ENG| The process of keeping a radio or radar beam set on a target and determining the range of the target continuously. { 'rād·ē·ō 'trak·iŋ }

radius cutter |MECH ENG| A formed milling cutter with teeth ground to produce a radius on the workpiece. { 'rād·ē·əs ,kəd·ər }

radius of action |ENG| The maximum distance a ship, aircraft, or other vehicle can travel away from its base along a given course with normal load and return without refueling, but including the fuel required to perform those maneuvers made necessary by all safety and operating factors. { 'rād·ē·əs əv 'ak·shən }

radius of gyration |MECH| The square root of the ratio of the moment of inertia of a body about a given axis to its mass. { 'rād·ē·əs əv ji'rā·shən }

radius of protection |ENG| The radius of the circle within which a lightning discharge will not strike, due to the presence of an elevated lightning rod at the center. { 'rād·ē·əs əv prə'tek·shən }

radius rod |ENG| A rod which restricts movement of a part to a given arc. { 'rād·ē·əs ,räd }

raffinate |CHEM ENG| In solvent refining, that portion of the treated liquid mixture that remains undissolved and is not removed by the selective solvent. Also known as good oil to petroleum-refinery operators. { 'raf·ə,nāt }

raft |ENG| A quantity of timber or lumber secured together by means of ropes, chains, or rods and used for transportation by floating. { raft }

rafter |BUILD| A roof-supporting member immediately beneath the roofing material. { 'raf·tər }

rafter dam |CIV ENG| A dam made of horizontal timbers that meet in the center of the stream like rafters in a roof. { 'raf·tər ,dam }

raft foundation |CIV ENG| A continuous footing that supports an entire structure, such as a floor. Also known as foundation mat. { 'raft faún,dā·shən }

rag bolt See barb bolt. { 'rag ,bólt }

raggle |BUILD| **1.** A manufactured masonry unit, frequently made of terra cotta, having a slot or groove to receive a metal flashing. Also known as flashing block; raggle block. **2.** A groove cut into masonry to receive adjoining material. { 'rag·əl }

raggle block See raggle. { 'rag·əl ,bläk }

rail |ENG| **1.** A bar extending between posts or other supports as a barrier or guard. **2.** A steel bar resting on the crossties to provide track for railroad cars and other vehicles with flanged wheels. |MECH ENG| A high-pressure manifold in some fuel injection systems. { rāl }

rail anchor |CIV ENG| A device that prevents tracks from moving longitudinally and maintains the proper gap between sections of rail. { 'rāl ,aŋ·kər }

rail bender |ENG| A portable appliance for bending rails for track or for straightening bent or curved rails. { 'rāl ,ben·dər }

rail capacity |CIV ENG| The maximum number of trains which can be planned to move in both directions over a specified section of track in a 24-hour period. { 'rāl kə,pas·əd·ē }

rail clip |CIV ENG| **1.** A plate that holds a rail at its base. **2.** A device used to fasten a derrick or crane to the rails of a track to prevent tipping. **3.** A support on a track rail, used for holding a detector bar. { 'rāl ,klip }

rail crane See locomotive crane. { 'rāl ,krān }

railhead |CIV ENG| **1.** The topmost part of a rail, supporting the wheels of railway vehicles. **2.** A point at which railroad traffic originates and terminates. **3.** The temporary ends of a railroad line under construction. { 'rāl‚hed }

railing |CIV ENG| A barrier consisting of a rail and supports. |ELECTR| Radar pulse jamming at high recurrence rates (50 to 150 kilohertz); it results in an image on a radar indicator resembling fence railing. { 'rāl·iŋ }

rail joint |CIV ENG| A rigid connection of the ends of two sections of railway track. { 'rāl ‚jóint }

railroad |CIV ENG| A permanent line of rails forming a route for freight cars and passenger cars drawn by locomotives. { 'rāl‚rōd }

railroad engineering |CIV ENG| That part of transportation engineering involved in the planning, design, development, operation, construction, maintenance, use, or economics of facilities for transportation of goods and people in wheeled units of rolling stock running on, and guided by, rails normally supported on crossties and held to fixed alignment. Also known as railway engineering. { 'rāl‚rōd ‚en·jə'nir·iŋ }

railroad jack |MECH ENG| **1.** A hoist used for lifting locomotives. **2.** A portable jack for lifting heavy objects. **3.** A hydraulic jack, either powered or lever-operated. { 'rāl‚rōd ‚jak }

railway dry dock |CIV ENG| A railway dock consisting of tracks built on an incline on a strong foundation, and extending from a sufficient distance in shore to allow a vessel to be hauled out of the water. { 'rāl‚wa 'drī ‚däk }

railway end-loading ramp |CIV ENG| A sloping platform situated at the end of a track and rising to the level of the floor of the railcars (wagons). { 'rāl‚wā 'end ‚lōd·iŋ ‚ramp }

railway engineering See railroad engineering. { 'rāl‚wā ‚en·jə'nir·iŋ }

rain gage |ENG| An instrument designed to collect and measure the amount of rain that has fallen. Also known as ombrometer; pluviometer; udometer. { 'rān ‚gāj }

rain-gage shield |ENG| A device which surrounds a rain gage and acts to maintain horizontal flow in the vicinity of the funnel so that the catch will not be influenced by eddies generated near the gage. Also known as wind shield. { 'rān ‚gāj ‚shēld }

rain-intensity gage |ENG| An instrument which measures the instantaneous rate at which rain is falling on a given surface. Also known as rate-of-rainfall gage. { 'rān in'ten·səd·ē ‚gāj }

raised flooring |CIV ENG| A flooring system having removable panels supported on adjustable pedestals or stringers to allow convenient access to the space below. Also known as access flooring; elevated flooring; pedestal flooring. { 'rāzd 'flór·iŋ }

raising plate See wall plate. { 'rāz·iŋ ‚plāt }

Rajakaruna engine |MECH ENG| A rotary engine that uses a combustion chamber whose sides are pin-jointed together at their ends. { ‚rä·jä·kə'rün·ə ‚en·jən }

rake |BUILD| The exterior finish and trim applied parallel to the sloping end walls of a gabled roof. |DES ENG| A hand tool consisting of a long handle with a row of projecting prongs at one end; for example, the tool used for gathering leaves or grass on the ground. |ENG| The angle between an inclined plane and the vertical. |MECH ENG| The angle between the tooth face or a tangent to the tooth face of a cutting tool at a given point and a reference plane or line. { rāk }

rake blade |ENG| A blade on a bulldozer in the form of spaced tines that point down. { 'rāk ‚blād }

raked joint |CIV ENG| A mortar, or masonry, joint from which the mortar has been scraped out to about 3/4 inch (20 millimeters). { 'rākt 'jóint }

ram |MECH ENG| A plunger, weight, or other guided structure for exerting pressure or drawing something by impact. { ram }

ram effect |MECH ENG| The increased air pressure in a jet engine or in the manifold of a piston engine, due to ram. { 'ram i‚fekt }

rammer |ENG| An instrument for driving something, such as wood or stones, into another material with force. Also known as beetle; maul. { 'ram·ər }

ramming |ENG| Packing a powder metal or sand into a compact mass. { 'ram·iŋ }

ramp |ENG| **1.** A uniformly sloping platform, walkway, or driveway. **2.** A stairway which gives access to the main door of an airplane. { ramp }

ram penetrometer See ramsonde. { 'ram ‚pen·ə'träm·əd·ər }

ramping |ENG| In the production of parts fabricated from composite materials, a gradual and programmed sequence of changes in temperature or pressure that control curing and cooling. { 'ramp·iŋ }

RAMPS See resource allocation in multiproject scheduling. { ramps }

Ramsay-Shields-Eötvös equation |THERMO| An elaboration of the Eötvös rule which states that at temperatures not too near the critical temperature, the molar surface energy of a liquid is proportional to t_c-t-6 K, where t is the temperature and t_c is the critical temperature. { 'ram·zē 'shēlz 'öt·vósh i‚kwā·zhən }

Ramsay-Young method |THERMO| A method of measuring the vapor pressure of a liquid, in which a thermometer bulb is surrounded by cotton wool soaked in the liquid, and the pressure, measured by a manometer, is reduced until the thermometer reading is steady. { ‚ram·zē 'yəŋ ‚meth·əd }

Ramsay-Young rule |THERMO| An empirical relationship which states that the ratio of the absolute temperatures at which two chemically similar liquids have the same vapor pressure is independent of this vapor pressure. { 'ram·zē 'yəŋ ‚rül }

ramsonde |ENG| A cone-tipped metal rod or tube that is driven downward into snow to measure its hardness. Also known as ram penetrometer. { 'ram‚sänd }

ram travel |ENG| In injection or transfer molding, the distance moved by the injection ram when filling the mold. { 'ram ,trav·əl }

ram-type turret lathe |MECH ENG| A horizontal turret lathe in which the turret is mounted on a ram or slide which rides on a saddle. { 'ram ¦tīp 'tər·ət ,lath }

random length |ENG| One of a group of various lengths of pipe as delivered by the manufacturer, usually 13–23 feet (4–7 meters) long. Also known as mill length. { 'ran·dəm 'leŋkth }

random line |ENG| A trial surveying line that is directed as closely as circumstances permit toward a fixed terminal point that cannot be seen from the initial point. Also known as random traverse. { 'ran·dəm 'līn }

random-sampling voltmeter |ENG| A sampling voltmeter which takes samples of an input signal at random times instead of at a constant rate; the synchronizing portions of the instrument can then be simplified or eliminated. { 'ran·dəm ¦sam·pliŋ 'vōlt,mēd·ər }

random traverse See random line. { 'ran·dəm trə'vərs }

random vibration |MECH| A varying force acting on a mechanical system which may be considered to be the sum of a large number of irregularly timed small shocks; induced typically by aerodynamic turbulence, airborne noise from rocket jets, and transportation over road surfaces. { 'ran·dəm vī'brā·shən }

range |CIV ENG| Any series of contiguous townships of the U.S. Public Land Survey system. |CONT SYS| **1.** The maximum distance a robot's arm or wrist can travel. Also known as reach. **2.** The volume comprising the locations to which a robot's arm or wrist can travel. |ENG| **1.** The distance capability of an aircraft, missile, gun, radar, or radio transmitter. **2.** A line defined by two fixed landmarks, used for missile or vehicle testing and other test purposes. |MECH| The horizontal component of a projectile displacement at the instant it strikes the ground. { rānj }

range calibration |ENG| Adjustment of a radar set so that when on target the set will indicate the correct range. { 'rānj ,kal·ə,brā·shən }

range coding |ENG| Method of coding a radar transponder beacon response so that it appears as a series of illuminated bars on a radarscope; the coding provides identification. { 'rānj ,kōd·iŋ }

range corrector setting |ENG| Degree to which the range scale of a position-finding apparatus must be adjusted before use. { 'rānj kə¦rek·tər ,sed·iŋ }

range deviation |MECH| Distance by which a projectile strikes beyond, or short of, the target; the distance as measured along the gun-target line or along a line parallel to the gun-target line. { 'rānj ,dē·vē¦ā·shən }

range discrimination See distance resolution. { 'rānj di,skrim·ə¦nā·shən }

rangefinder |ELECTR| A device which determines the distance to an object by measuring the time it takes for a radio wave to travel to the object and return. See optical rangefinder. { 'rānj ,fīnd·ər }

range-height indicator |ENG| A scope which simultaneously indicates range and height of a radar target; this presentation is commonly used by height finders. { 'rānj 'hīt ,in·də,kād·ər }

range-imaging sensor |ENG| A robotic device that makes precise measurements, by using the principles of algebra, trigonometry, and geometry, of the distance from a robot's end effector to various parts of an object, in order to form an image of the object. { 'rānj ¦im·ij·iŋ ,sen·sər }

range marker See distance marker. { 'rānj ,mär·kər }

range pole See range rod. { 'rānj ,pōl }

range recorder |ENG| An item which makes a permanent representation of distance, expressed as range, versus time. |ENG ACOUS| A display used in sonar in which a stylus sweeps across a paper moving at a constant rate and chemically treated so that it is darkened by an electrical signal from the stylus; the stylus starts each sweep as a sound pulse is emitted so that the distance along the trace at which the echo signal appears is a measure of the range to the target. { 'rānj ri,kòrd·ər }

range resolution See distance resolution. { 'rānj ,rez·ə,lü·shən }

range rod |ENG| A long (6–8 feet or 1.8–2.4 meters) rod fitted with a sharp-pointed metal shoe and usually painted in 1-foot (30-centimeter) bands of alternate red and white; used for sighting points and lines in surveying or for showing the position of a ground point. Also known as line rod; lining pole; range pole; ranging rod; sight rod. { 'rānj ,räd }

range sensing |ENG| The precise measurement of the distance of a device from a robot's end effector. { 'rānj ,sens·iŋ }

range surveillance |ENG| Surveillance of a missile range by means of electronic and other equipment. { 'rānj sər,vā·ləns }

ranging rod See range rod. { 'rānj·iŋ ,räd }

rank |MECH ENG| The number of rotational joints belonging to a robot. { raŋk }

Rankine cycle |THERMO| An ideal thermodynamic cycle consisting of heat addition at constant pressure, isentropic expansion, heat rejection at constant pressure, and isentropic compression; used as an ideal standard for the performance of heat-engine and heat-pump installations operating with a condensable vapor as the working fluid, such as a steam power plant. Also known as steam cycle. { 'raŋ·kən ,sī·kəl }

Rankine efficiency |MECH ENG| The efficiency of an ideal engine operating on the Rankine cycle under specified conditions of steam temperature and pressure. { 'raŋ·kən i,fish·ən·sē }

Rankine-Hugoniot equations |THERMO| Equations, derived from the laws of conservation of mass, momentum, and energy, which relate the velocity of a shock wave and the pressure, density, and enthalpy of the transmitting fluid before and after the shock wave passes. { 'raŋ·kən yü'gō·nē·ō i,kwā·zhənz }

Rankine temperature scale |THERMO| A scale of absolute temperature; the temperature in degrees Rankine (°R) is equal to 9/5 of the temperature in kelvins and to the temperature in degrees Fahrenheit plus 459.67. { 'raŋ·kən 'tem·prə·chər ,skāl }

ranking method |IND ENG| A system of job evaluation wherein each job as a whole is given a rank with respect to all the other jobs, and no attempt is made to establish a measure of value. { 'raŋk·iŋ ,meth·əd }

Ranney well |CIV ENG| A well that has a center caisson with horizontal perforated pipes extending radially into an aquifer; particularly applicable to the development of thin aquifers at shallow depths. { 'ran·ē ,wel }

rapid prototyping |IND ENG| A modeling process used in product design in which a CAD drawing of a part is processed to create a file of the part in slices, and then a part is built by depositing layer (slice) upon layer of material; includes stereolithography, selective laser sintering, or fused deposition modeling. { ¦rap·əd 'prōd·ə,tīp·iŋ }

rapid sand filter |CIV ENG| A system for purifying water, which is forced through layers of sand and gravel under pressure. { 'rap·əd 'sand ,fil·tər }

rapid traverse |MECH ENG| A machine tool mechanism which rapidly repositions the workpiece while no cutting takes place. { 'rap·əd trə'vərs }

Raschig process |CHEM ENG| A method for production of phenol that begins with a first-stage chlorination of benzene, using an air-hydrochloric acid mixture. { 'rä·shik 'prä·səs }

Raschig ring |CHEM ENG| A type of packing in the shape of a short pipe; used in columns for absorption operations, and to a limited extent for distillation operations. { 'rä·shik ,riŋ }

RA size |ENG| One of a series of sizes to which untrimmed paper is manufactured; for reels of paper, the standard sizes in millimeters are 430, 610, 860, and 1220; for sheets of paper, the sizes are RA0, 860 × 1220; RA1, 610 × 860; RA2, 430 × 610; RA sizes correspond to A sizes when trimmed. { ¦är'ä ,sīz }

rasp |DES ENG| A metallic tool with a rough surface of small points used for shaping and finishing metal, plaster, stone, and wood; designed in a number of useful curved shapes. { rasp }

ratchet |DES ENG| A wheel, usually toothed, operating with a catch or a pawl so as to rotate in only a single direction. { 'rach·ət }

ratchet coupling |MECH ENG| A coupling between two shafts that uses a ratchet to allow the driven shaft to be turned in one direction only, and also to permit the driven shaft to overrun the driving shaft. { 'rach·ət ,kəp·liŋ }

ratchet jack |DES ENG| A jack operated by a ratchet mechanism. { 'rach·ət ,jak }

ratchet tool |DES ENG| A tool in which torque or force is applied in one direction only by means of a ratchet. { 'rach·ət ,tül }

rat distillate |CHEM ENG| A refinery designation for gasoline and other fuels as they come from the condenser, before undesirable substances are removed by further processing. { 'rat ,dist·əl·ət }

rate action See derivative action. { 'rāt ,ak·shən }

rate control |CONT SYS| A form of control in which the position of a controller determines the rate or velocity of motion of a controlled object. Also known as velocity control. { 'rāt kən,trōl }

rated capacity |MECH ENG| The maximum capacity for which a boiler is designed, measured in pounds of steam per hour delivered at specified conditions of pressure and temperature. { 'rād·əd kə'pas·əd·ē }

rated engine speed |MECH ENG| The rotative speed of an engine specified as the allowable maximum for continuous reliable performance. { 'rād·əd 'en·jən ,spēd }

rated flow |ENG| **1.** Normal operating flow rate at which a fluid product is passed through a vessel or piping system. **2.** Flow rate for which a vessel or process system is designed. { 'rād·əd 'flō }

rated horsepower |MECH ENG| The normal maximum, allowable, continuous power output of an engine, turbine motor, or other prime mover. { 'rād·əd 'hȯrs,paủ·ər }

rated load |MECH ENG| The maximum load a machine is designed to carry. { 'rād·əd 'lōd }

rated relieving capacity |DES ENG| The measured relieving capacity for which the pressure relief device is rated in accordance with the applicable code or standard. { 'rād·əd ri'lēv·iŋ kə,pas·əd·ē }

rate effect |ELECTR| The phenomenon of a *pnpn* device switching to a high-conduction mode when anode voltage is applied suddenly or when high-frequency transients exist. { 'rāt i,fekt }

rate feedback |ELECTR| The return of a signal, proportional to the rate of change of the output of a device, from the output to the input. { 'rāt 'fēd,bak }

rate-grown transistor |ELECTR| A junction transistor in which both impurities (such as gallium and antimony) are placed in the melt at the same time and the temperature is suddenly raised and lowered to produce the alternate *p*-type and *n*-type layers of rate-grown junctions. Also known as graded-junction transistor. { 'rāt ¦grōn tran 'zis·tər }

rate gyroscope |MECH ENG| A gyroscope that is suspended in just one gimbal whose bearings form its output axis and which is restrained by a spring; rotation of the gyroscope frame about an axis perpendicular to both spin and output axes produces precession of the gimbal within the bearings proportional to the rate of rotation. { 'rāt 'jī·rə,skōp }

rate integrating gyroscope |MECH ENG| A single-degree-of-freedom gyro having primarily viscous restraint of its spin axis about the output axis; an output signal is produced by gimbal angular displacement, relative to the base, which is proportional to the integral of the angular rate of the base about the input axis. { 'rāt ¦int·ə,grād·iŋ ,ī·rə,skōp }

rate of change of acceleration [MECH] Time rate of change of acceleration; this rate is a factor in the design of some items of ammunition that undergo large accelerations. { 'rāt əv 'chānj əv ik,sel·ə'rā·shən }

rate-of-flow control valve See flow control valve. { 'rāt əv 'flō kən¦trōl ,valv }

rate-of-rainfall gage See rain-intensity gage. { 'rāt əv 'rān,fól ,gāj }

rate of rise [ENG] The time rate of pressure increase during an isolation test for leaks. { 'rāt əv 'rīz }

rate response [ENG] Quantitative expression of the output rate of a control system as a function of its input signal. { 'rāt ri,späns }

rate servomechanism See velocity servomechanism. { 'rāt ¦sər·vō'mek·ə,niz·əm }

rating [ENG] A designation of an operating limit for a machine, apparatus, or device used under specified conditions. { 'rād·iŋ }

ratio control system [CONT SYS] Control system in which two process variables are kept at a fixed ratio, regardless of the variation of either of the variables, as when flow rates in two separate fluid conduits are held at a fixed ratio. { 'rā·shō kən'trōl ,sis·təm }

ratio delay study See work sampling. { 'rā·shō di'lā ,stəd·ē }

ratio meter [ENG] A meter that measures the quotient of two electrical quantities; the deflection of the meter pointer is proportional to the ratio of the currents flowing through two coils. { 'rā·shō ,mēd·ər }

ratio of expansion [MECH ENG] The ratio of the volume of steam in the cylinder of an engine when the piston is at the end of a stroke to that when the piston is in the cutoff position. { 'rā·shō əv ik'span·shən }

ratio of reduction [ENG] The ratio of the maximum size of the stone which will enter a crusher, to the size of its product. { 'rā·shō əv ri'dək·shən }

rattail file [DES ENG] A round tapering file used for smoothing or enlarging holes. { 'rat,tāl 'fīl }

Rauschelback rotor [ENG] A free-turning S-shaped propeller used to measure ocean currents; the number of rotations per unit time is proportional to the flow. { 'raúsh·əl,bak ,rōd·ər }

raw material [IND ENG] A crude, unprocessed or partially processed material used as feedstock for a processing operation; for example, crude petroleum, raw cotton, or steel scrap. Also known as crude material. { 'ró mə'tir·ē·əl }

raw sewage [CIV ENG] Untreated waste materials. { 'ró 'sü·ij }

raw sludge [CIV ENG] Sewage sludge preliminary to primary and secondary treatment processes. { 'ró 'sləj }

raw water [CIV ENG] Water that has not been purified. { 'ró 'wòd·ər }

Raykin fender [CIV ENG] Sandwich-type fender buffer to protect docks from the impact of mooring ships; made of a connected series of steel plates cemented to layers of rubber. { 'rā·kən ,fen·dər }

Rayleigh line [MECH] A straight line connecting points corresponding to the initial and final states on a graph of pressure versus specific volume for a substance subjected to a shock wave. { 'rā·lē ,līn }

Rayleigh number 2 [THERMO] A dimensionless number used in studying free convection, equal to the product of the Grashof number and the Prandtl number. Symbolized R'_2. { 'rā·lē ¦nam·bər 'tü }

Rayleigh number 3 [THERMO] A dimensionless number used in the study of combined free and forced convection in vertical tubes, equal to Rayleigh number 2 times the Nusselt number times the tube diameter divided by its entry length. Symbolized Ra_3. { 'rā·lē ¦nam·bər 'thrē }

Rayleigh's dissipation function [MECH] A function which enters into the equations of motion of a system undergoing small oscillations and represents frictional forces which are proportional to velocities; given by a positive definite quadratic form in the time derivatives of the coordinates. Also known as dissipation function. { 'rā·lē ,dis·ə'pā·shən ,faŋk·shən }

Rayleigh wave [MECH] A wave which propagates on the surface of a solid; particle trajectories are ellipses in planes normal to the surface and parallel to the direction of propagation. Also known as surface wave. { 'rā·lē ,wāv }

Raymond concrete pile [CIV ENG] A pile made by driving a thin steel shell into the ground with a tapered mandrel and filling it with concrete. { 'rā·mənd ¦kän¦krēt ,pīl }

R-C amplifier See resistance-capacitance coupled amplifier. { ¦är¦sē 'am·plə,fī·ər }

R-C coupled amplifier See resistance-capacitance coupled amplifier. { ¦är¦sē ¦kəp·əld 'am·plə,fī·ər }

R-C coupling See resistance coupling. { ¦är¦sē 'kəp·liŋ }

R-C oscillator See resistance-capacitance oscillator. { ¦är¦sē 'äs·ə,lād·ər }

RDC extractor See rotary-disk contactor. { ¦är ¦dē'sē ik'strak·tər }

reach [CIV ENG] A portion of a waterway between two locks or gages. [CONT SYS] See range. [ENG] The length of a channel, uniform with respect to discharge, depth, area, and slope. { rēch }

reach rod [MECH ENG] A rod motion in a link used to transmit motion from the reversing rod to the lifting shaft. { 'rēch ,räd }

reactance [ELEC] The imaginary part of the impedance of an alternating-current circuit. { rē'ak·təns }

reactance drop [ELEC] The component of the phasor representing the voltage drop across a component or conductor of an alternating-current circuit which is perpendicular to the current. { rē'ak·təns ,dräp }

447

reactance grounded [ELEC] Grounded through a reactance. { rē'ak·təns ,graún·dəd }

reaction [CONT SYS] *See* positive feedback. [MECH] The equal and opposite force which results when a force is exerted on a body, according to Newton's third law of motion. { rē'ak·shən }

reaction injection molding [ENG] A plastics fabrication process in which two streams of highly reactive, low-molecular-weight, low-viscosity resin systems are combined to form a solid material. { rē'ək·shən in'jek·shən 'mōl·diŋ }

reactions inventory [IND ENG] A summary of the various possible responses of an individual to a stimulus or group of stimuli. { rē'ak·shənz 'in·ven,tór·ē }

reaction turbine [MECH ENG] A power-generation prime mover utilizing the steady-flow principle of fluid acceleration, where nozzles are mounted on the moving element. { rē'ak·shən ,tər·bən }

reaction wheel [MECH ENG] A device capable of storing angular momentum which may be used in a space ship to provide torque to effect or maintain a given orientation. { rē'ak·shən ,wēl }

reaction zone [CHEM ENG] In a catalytic reactor vessel, the location or zone within the vessel where the bulk of the chemical reaction takes place. { rē'ak·shən ,zōn }

reactive [ELEC] Pertaining to either inductive or capacitance reactance; a reactive circuit has a high value of reactance in comparison with resistance. { rē'ak·tiv }

reactive ion etching [ELECTR] A directed chemical etching process used in integrated circuit fabrication in which chemically active ions are accelerated along electric field lines to meet a substrate perpendicular to its surface. { rē'ak·tiv 'ī,än ,ech·iŋ }

reactive muffler [ENG] A muffler that attenuates by reflecting sound back to the source. Also known as nondissipative muffler. { rē'ak·tiv 'məf·lər }

reactive volt-ampere meter *See* varmeter. { rē'ak·tiv 'vōlt 'am,pir ,mēd·ər }

reactor [CHEM ENG] Device or process vessel in which chemical reactions (catalyzed or noncatalyzed) take place during a chemical conversion type of process. [ELEC] A device that introduces either inductive or capacitive reactance into a circuit, such as a coil or capacitor. Also known as electric reactor. { rē'ak·tər }

read [ELECTR] To generate an output corresponding to the pattern stored in a charge storage tube. { rēd }

Read diode [ELECTR] A high-frequency semiconductor diode consisting of an avalanching *pn* junction, biased to fields of several hundred thousand volts per centimeter, at one end of a high-resistance carrier serving as a drift space for the charge carriers. { 'rēd ,dī,ōd }

readiness time [ENG] The length of time required to obtain a stabilized system ready to perform its intended function (readiness time includes warm-up time); the time is measured from the point when the system is unassembled or uninstalled to such time as it can be expected to perform as accurately as at any later time; maintenance time is excluded from readiness time. { 'rēd·i·nəs ,tīm }

reading [ENG] **1.** The indication shown by an instrument. **2.** Observation of the readings of one or more instruments. { 'rēd·iŋ }

reading point *See* breakpoint. { 'rēd·iŋ ,póint }

real gas [THERMO] A gas, as considered from the viewpoint in which deviations from the ideal gas law, resulting from interactions of gas molecules, are taken into account. Also known as imperfect gas. { 'rēl 'gas }

realizability [CONT SYS] Property of a transfer function that can be realized by a network that has only resistances, capacitances, inductances, and ideal transformers. { ,rē·ə,līz·ə'bil·əd·ē }

ream [ENG] To enlarge or clean out a hole. { rēm }

reamer [DES ENG] A tool used to enlarge, shape, smooth, or otherwise finish a hole. { 'rēm·ər }

reaming bit [DES ENG] A bit used to enlarge a borehole. Also known as broaching bit; pilot reaming bit. { 'rēm·iŋ ,bit }

rear response [ENG ACOUS] The maximum pressure within 60° of the rear of a transducer in decibels relative to the pressure on the acoustic axis. { 'rir ri,späns }

Réaumur temperature scale [THERMO] Temperature scale where water freezes at 0°R and boils at 80°R. { ¦rā·ō¦myúr 'tem·prə·chər ,skāl }

rebar [CIV ENG] A steel bar or rod used to reinforce concrete. { 'rē,bär }

reboiler [CHEM ENG] An auxiliary heating unit for a fractionating tower designed to supply additional heat to the lower portion of the tower; liquid withdrawn from the side or bottom of the tower is reheated by heat exchange, then reintroduced into the tower. { rē'bóil·ər }

rebound clip [DES ENG] A clip surrounding the back and one or two other leaves of a leaf spring, to distribute the load during rebounds. { 'rē ,baúnd ,klip }

rebound leaf [DES ENG] In a leaf spring, a leaf placed over the master leaf to limit the rebound and help carry the load imposed by it. { 'rē ,baúnd ,lēf }

rebreather [ENG] A closed-loop oxygen supply system consisting of gas supply and face mask. { rē'brēth·ər }

rebuild [ENG] To restore to a condition comparable to new by disassembling the item to determine the condition of each of its component parts, and reassembling it, using serviceable, rebuilt, or new assemblies, subassemblies, and parts. { rē'bild }

receiver [CHEM ENG] Vessel, container, or tank used to receive and collect liquid material from a process unit, such as the distillate receiver from the overhead condenser of a distillation column. [ELECTR] The complete equipment required for receiving modulated radio waves and converting them into the original intelligence, such as into sounds or pictures, or converting to desired useful information as in a radar receiver. |MECH ENG| An apparatus placed near the compressor to equalize the pulsations of the air as it comes from the compressor to cause a more uniform flow of air through the pipeline and to collect moisture and oil carried in the air. { ri'sēv·vər }

receiving gage |ENG| A fixed gage designed to inspect a number of dimensions and also their reaction to each other. { ri'sēv·iŋ ,gāj }

receiving house |CHEM ENG| A building where liquid streams from petroleum-refining-process condensers are observed through a look box, and samples are taken for testing, and also where products are diverted to storage tanks or to other processing units. { ri'sēv·iŋ ,haùs }

receiving station |MECH ENG| The location or device on conveyor systems where bulk material is loaded or otherwise received onto the conveyor. { ri'sēv·iŋ ,stā·shən }

receiving tank See rundown tank. { ri'sēv·iŋ ,taŋk }

recess |ENG| A surface groove or depression. { 'rē,ses }

recessed bead See quirk bead. { 'rē,sest ,bēd }

recessed tube wall |MECH ENG| A boiler furnace wall which has openings to partially expose waterwall tubes to the radiant combustion gases. { 'rē,sest 'tüb ,wòl }

recharge basin |CIV ENG| A basin constructed in sandy material to collect water, as from storm drains, for the purpose of replenishing groundwater supply. { 'rē,chärj ,bās·ən }

reciprocal impedance |ELEC| Two impedances Z_1 and Z_2 are said to be reciprocal impedances with respect to an impedance Z (invariably a resistance) if they are so related as to satisfy the equation $Z_1Z_2 = Z^2$ { ri'sip·rə·kəl im'pēd·əns }

reciprocal leveling |CIV ENG| A variant of straight differential leveling applied to long distances in which levels are taken on two points, and the average of the two elevation differences is the true difference. { ri'sip·rə·kəl 'lev·ə·liŋ }

reciprocal ohm See siemens. { ri'sip·rə·kəl 'ōm }

reciprocal ohm centimeter See roc. { ri'sip·rə·kəl 'ōm 'sent·i,mēd·ər }

reciprocal strain ellipsoid |MECH| In elastic theory, an ellipsoid of certain shape and orientation which under homogeneous strain is transformed into a set of orthogonal diameters of the sphere. { ri'sip·rə·kəl ¦strān i'lip,sòid }

reciprocating compressor |MECH ENG| A positive-displacement compressor having one or more cylinders, each fitted with a piston driven by a crankshaft through a connecting rod. { ri'sip·rə,kād·iŋ kəm'pres·ər }

reciprocating drill See piston drill. { ri'sip·rə,kād·iŋ 'dril }

reciprocating engine See piston engine. { ri'sip·rə,kād·iŋ 'en·jən }

reciprocating flight conveyor |MECH ENG| A reciprocating beam or beams with hinged flights that advance materials along a conveyor trough. { ri'sip·rə,kād·iŋ 'flīt kən,vā·ər }

reciprocating-plate column See reciprocating-plate extractor. { ri'sip·rə,kād·iŋ ¦plāt 'käl·əm }

reciprocating-plate extractor |CHEM ENG| A liquid-liquid contactor in which equally spaced perforated plates (as in a distillation column) move up and down rapidly over a short distance to cause liquid agitation and mixing. Also known as reciprocating-plate column. { ri'sip·rə,kād·iŋ ¦plāt ik'strak·tər }

reciprocating-plate feeder |MECH ENG| A back-and-forth shaking tray used to feed abrasive materials, such as pulverized coal, into process units. { ri'sip·rə,kād·iŋ ¦plāt 'fēd·ər }

reciprocating pump See piston pump. { ri'sip·rə,kād·iŋ 'pəmp }

reciprocating screen |MECH ENG| Horizontal solids-separation screen (sieve) oscillated back and forth by an eccentric gear; used for solids classification. { ri'sip·rə,kād·iŋ 'skrēn }

reciprocity calibration |ENG ACOUS| A measurement of the projector loss and hydrophone loss of a reversible transducer by means of the reciprocity theorem and comparisons with the known transmission loss of an electric network, without knowing the actual value of either the electric power or the acoustic power. { ,res·ə'präs·əd·ē ,kal·ə,brā·shən }

reciprocity theorem Also known as principle of reciprocity. |ELEC| 1. The electric potentials V_1 and V_2 produced at some arbitrary point, due to charge distributions having total charges of q_1 and q_2 respectively, are such that $q_1V_2 = q_2V_1$. 2. In an electric network consisting of linear passive impedances, the ratio of the electromotive force introduced in any branch to the current in any other branch is equal in magnitude and phase to the ratio that results if the positions of electromotive force and current are exchanged. |ENG ACOUS| The sensitivity of a reversible electroacoustic transducer when used as a microphone divided by the sensitivity when used as a source of sound is independent of the type and construction of the transducer. { ,res·ə'präs·əd·ē ,thir·əm }

recirculating-ball steering |MECH ENG| A steering system that transmits steering movements by means of steel balls placed between a worm gear and a nut. { rē'sər·kyə,lād·iŋ ¦bòl 'stir·iŋ }

recirculator |ENG| A self-contained underwater breathing apparatus that recirculates an oxygen supply (mix-gas or pure) to the diver until the oxygen is depleted. { rē'sər·kyə,lād·ər }

reclamation |CIV ENG| 1. The recovery of land or other natural resource that has been abandoned because of fire, water, or other cause. 2. Reclaiming dry land by irrigation. { ,rek·lə'mā·shən }

recoil See gun reaction. { 'rē,kòil }

reconditioning |ENG| Restoration of an object to a good condition. { ,rē·kən'dish·ən·iŋ }

reconnaissance |ENG| A mission to secure

data concerning the meteorological, hydrographic, or geographic characteristics of a particular area. { ri'kän·ə·səns }

reconnaissance survey |ENG| A preliminary survey, usually executed rapidly and at relatively low cost, prior to mapping in detail and with greater precision. { ri'kän·ə·səns ,sər,vā }

record changer |ENG ACOUS| A record player that plays a number of records automatically in succession. { 'rek·ərd ,chānj·ər }

recorder See recording instrument. { ri'kòrd·ər }

recording head |ELECTR| A magnetic head used only for recording. Also known as record head. See cutter. { ri'kòrd·iŋ ,hed }

recording instrument |ENG| An instrument that makes a graphic or acoustic record of one or more variable quantities. Also known as recorder. { ri'kòrd·iŋ ,in·strə·mənt }

recording optical tracking instrument |ENG| Optical system used for recording data in connection with missile flights. { ri'kòrd·iŋ ¦äp·tə·kəl 'trak·iŋ ,in·strə·mənt }

recording rain gage |ENG| A rain gage which automatically records the amount of precipitation collected, as a function of time. Also known as pluviograph. { ri'kòrd·iŋ 'rān ,gāj }

recording thermometer See thermograph. { ri'kòrd·iŋ thər,mäm·əd·ər }

record player |ENG ACOUS| A motor-driven turntable used with a phonograph pickup to obtain audio-frequency signals from a phonograph record. { 'rek·ərd ,plā·ər }

recovery |MECH| The return of a body to its original dimensions after it has been stressed, possibly over a considerable period of time. { ri'kəv·ə·rē }

recovery vehicle |MECH ENG| A special-purpose vehicle equipped with winch, hoist, or boom for recovery of vehicles. { ri'kəv·ə·rē ,vē·ə·kəl }

rectangular weir |CIV ENG| A weir with a rectangular notch at top for measurement of water flow in open channels; it is simple, easy to make, accurate, and popular. { rek'taŋ·gyə·lər 'wer }

rectification |CIV ENG| A new alignment to correct a deviation of a stream channel or bank. |ELEC| The process of converting an alternating current to a unidirectional current. { ,rek·tə·fə'kā·shən }

rectification distillation |CHEM ENG| A distillation technique in which a rectifying column is used. { ,rek·tə·fə'kā·shən ,dis·tə'lā·shən }

rectification factor |ELECTR| Quotient of the change in average current of an electrode by the change in amplitude of the alternating sinusoidal voltage applied to the same electrode, the direct voltages of this and other electrodes being maintained constant. { ,rek·tə·fə'kā·shən ,fak·tər }

rectifier |ELEC| A nonlinear circuit component that allows more current to flow in one direction than the other; ideally, it allows current to flow in one direction unimpeded but allows no current to flow in the other direction. { 'rek·tə,fī·ər }

rectifier filter |ELECTR| An electric filter used in smoothing out the voltage fluctuation of an electron tube rectifier, and generally placed between the rectifier's output and the load resistance. { 'rek·tə,fī·ər ,fil·tər }

rectifier instrument |ENG| Combination of an instrument sensitive to direct current and a rectifying device whereby alternating current (or voltages) may be rectified for measurement. { 'rek·tə,fī·ər ,in·strə·mənt }

rectifier rating |ELECTR| A performance rating for a semiconductor rectifier, usually on the basis of the root-mean-square value of sinusoidal voltage that it can withstand in the reverse direction and the average current density that it will pass in the forward direction. { 'rek·tə,fī·ər ,rād·iŋ }

rectifier stack |ELECTR| A dry-disk rectifier made up of layers or stacks of disks of individual rectifiers, as in a selenium rectifier or copper-oxide rectifier. { 'rek·tə,fī·ər ,stak }

rectifier transformer |ELECTR| Transformer whose secondary supplies energy to the main anodes of a rectifier. { 'rek·tə,fī·ər tranz'fòr·mər }

rectifying column |CHEM ENG| Portion of a distillation column above the feed tray in which rising vapor is enriched by interaction with a countercurrent falling stream of condensed vapor; contrasted to the stripping column section below the column feed tray. { 'rek·tə,fī·iŋ ,käl·əm }

rectilinear motion |MECH| A continuous change of position of a body so that every particle of the body follows a straight-line path. Also known as linear motion. { ¦rek·tə'lin·ē·ər 'mō·shən }

recuperative air heater |ENG| An air heater in which the heat-transferring metal parts are stationary and form a separating boundary between the heating and cooling fluids. { rē'küp·rəd·iv 'er ,hēd·ər }

recuperator |ENG| An apparatus in which heat is conducted from the combustion products to incoming cooler air through a system of thin-walled ducts. { rē'kü·pə,rād·ər }

recurring demand |IND ENG| A request made periodically or anticipated to be repetitive by an authorized requisitioner for material for consumption or use, or for stock replenishment. { ri'kər·iŋ di'mand }

recycle mixing |CHEM ENG| The mixing of a portion of a product stream (fluid or solid) from a processing unit with incoming raw feed. { rē'sī·kəl ,miks·iŋ }

recycle ratio |CHEM ENG| In a continuous chemical process, the ratio of recycle stock to fresh feed. { rē'sī·kəl ,rā·shō }

recycle stock |CHEM ENG| That portion of a feedstock that has passed through a processing unit and is recirculated (recycled) back through the process. { rē'sī·kəl ,stäk }

recycling |ELECTR| Returning to an original condition, as to 0 or 1 in a counting circuit. |ENG| The extraction and recovery of valuable materials from scrap or other discarded materials. { rē'sīk·liŋ }

Redler conveyor |MECH ENG| A conveyor in which material is dragged through a duct by skeletonized or U-shaped impellers which move the material in which they are submerged because the resistance to slip through the element is greater than the drag against the walls of the duct. { 'red·lər kən'vā·ər }

redox cell |ELEC| Cell designed to convert the energy of reactants to electrical energy; an intermediate reductant, in the form of liquid electrolyte, reacts at the anode in a conventional manner; it is then regenerated by reaction with a primary fuel. { 'rē,däks ,sel }

reduced frequency See Strouhal number. { ri 'düst 'frē·kwən·sē }

reduced inspection |IND ENG| The decrease in the number of items inspected from that specified in the original sampling plan because the quality of the item has consistently improved. { ri'düst in'spek·shən }

reduced mass |MECH| For a system of two particles with masses m_1 and m_2 exerting equal and opposite forces on each other and subject to no external forces, the reduced mass is the mass m such that the motion of either particle, with respect to the other as origin, is the same as the motion with respect to a fixed origin of a single particle with mass m acted on by the same force; it is given by $m = m_1m_2/(m_1 + m_2)$. { ri'düst 'mas }

reduced-order controller [CONT SYS] A control algorithm in which certain modes of the structure to be controlled are ignored, to enable control commands to be computed with sufficient rapidity. { ri'düst ¦ȯr·dər kən'trōl·ər }

reduced pressure |THERMO| The ratio of the pressure of a substance to its critical pressure. { ri'düst 'presh·ər }

reduced-pressure distillation See vacuum distillation. { ri'düst ¦presh·ər ,dis·tə'lā·shən }

reduced property See reduced value. { ri'düst 'präp·ərd·ē }

reduced temperature |THERMO| The ratio of the temperature of a substance to its critical temperature. { ri'düst 'tem·prə·chər }

reduced value |THERMO| The actual value of a quantity divided by the value of that quantity at the critical point. Also known as reduced property. { ri'düst 'val·yü }

reduced viscosity |ENG| In plastics processing, the ratio of the specific viscosity to concentration. { ri'düst vi'skäs·əd·ē }

reduced volume |THERMO| The ratio of the specific volume of a substance to its critical volume. { ri'düst 'väl·yəm }

reducer |DES ENG| A fitting having a larger size at one end than at the other and threaded inside, unless specifically flanged or for some special joint. { ri'dü·sər }

reducing coupling |ENG| A coupling used to connect a smaller pipe to a larger one. { ri'düs·iŋ ,kəp·liŋ }

reduction gear |MECH ENG| A gear train which lowers the output speed. { ri'dək·shən ,gir }

reduction ratio |ENG| Ratio of feed size to product size for a mill (crushing or grinding) operation; measured by lump and sieve sizes. { ri'dək·shən ,rā·shō }

reduction to sea level |ENG| The application of a correction to a measured horizontal length on the earth's surface, at any altitude, to reduce it to its projected or corresponding length at sea level. { ri'dək·shən tə 'sē ,lev·əl }

redundancy |MECH| A statically indeterminate structure. { ri'dən·dən·sē }

redundant system See duplexed system. { ri'dən·dənt ,sis·təm }

Redwood viscometer |ENG| A standard British-type viscometer in which the viscosity is determined by the time, in seconds, required for a certain quantity of liquid to pass out through the orifice under given conditions; used for determining viscosities of petroleum oils. { 'red ,wud vi'skäm·əd·ər }

reed |ENG| A thin bar of metal, wood, or cane that is clamped at one end and set into transverse elastic vibration, usually by wind pressure; used to generate sound in musical instruments, and as a frequency standard, as in a vibrating-reed frequency meter. { rēd }

reed frequency meter See vibrating-reed frequency meter. { 'rēd 'frē·kwən·sē ,mēd·ər }

reed horn |ENG ACOUS| A horn that produces sound by means of a steel reed vibrated by air under pressure. { 'rēd ,hȯrn }

reeding |ENG| Corrugating or serrating, as in coining or embossing. { 'rēd·iŋ }

reel |DES ENG| A revolving spool-shaped device used for storage of hose, rope, cable, wire, magnetic tape, and so on. { rēl }

reel and bead See bead and reel. { 'rēl ən 'bēd }

reengineering |SYS ENG| The application of technology and management science to the modification of existing systems, organizations, processes, and products in order to make them more effective, efficient, and responsive. { ,rē·en·jə'nir·iŋ }

reentrant |ENG| Having one or more sections directed inward, as in certain types of cavity resonators. { rē'en·trənt }

reference dimension |DES ENG| In dimensioning, a dimension without tolerance used for informational purposes only, and does not govern machining operations in any way; it is indicated on a drawing by writing the abbreviation REF directly following or under the dimension. { 'ref·rəns di,men·shən }

reference level |ENG| See datum plane. |ENG ACOUS| The level used as a basis of comparison when designating the level of an audio-frequency signal in decibels or volume units. Also known as reference signal level. { 'ref·rəns ,lev·əl }

reference lot |IND ENG| A lot of select components, used as a standard. { 'ref·rəns ,lät }

reference plane |ENG| See datum plane. |MECH ENG| The plane containing the axis and the cutting point of a cutter. { 'ref·rəns ,plān }

reference range |ENG| Range obtained from the radar coverage indicator for a given penetrating aircraft. { 'ref·rəns ,rānj }

reference seismometer |ENG| In seismic prospecting, a detector placed to record successive shots under similar conditions, to permit overall time comparisons. { 'ref·rəns sīz'mäm·əd·ər }

reference signal level See reference level. { 'ref·rəns 'sig·nəl ,lev·əl }

reference tone |ENG| Stable tone of known frequency continuously recorded on one track of multitrack signal recordings and intermittently recorded on signal track recordings by the collection equipment operators for subsequent use by the data analysts as a frequency reference. { 'ref·rəns ,tōn }

reference voltage |ELEC| An alternating-current voltage used for comparison, usually to identify an in-phase or out-of-phase condition in an ac circuit. { 'ref·rəns ,vōl·tij }

referencing |ENG| The process of measuring the horizontal (or slope) distances and directions from a survey station to nearby landmarks, reference marks, and other permanent objects which can be used in the recovery or relocation of the station. { 'ref·rən·siŋ }

refine |ENG| To free from impurities, as the separation of petroleum, ores, or chemical mixtures into their component parts. { ri'fīn }

refinery |CHEM ENG| System of process units used to convert crude petroleum into fuels, lubricants, and other petroleum-derived products. { ri'fīn·rē }

reflectance See reflection factor. { ri'flek·təns }

reflected signal indicator |ENG| Pen recorder which presents the radar signals within frequency gates; these recordings enable the operator to determine that an airborne object has penetrated the Doppler link and its direction of penetration. { ri'flek·təd ¦sig·nəl 'in·də,kād·ər }

reflecting nephoscope See mirror nephoscope. { ri'flek·tiŋ 'nef·ə,skōp }

reflecting sign |CIV ENG| A road sign painted with reflective paint so as to be easily visible in the light of a headlamp. { ri'flek·tiŋ 'sīn }

reflection altimeter See radio altimeter. { ri'flek·shən al'tim·əd·ər }

reflection factor |ELEC| Ratio of the load current that is delivered to a particular load when the impedances are mismatched to that delivered under conditions of matched impedances. Also known as mismatch factor; reflectance; transition factor. { ri'flek·shən ,fak·tər }

reflection goniometer |ENG| A goniometer that measures the angles between crystal faces by reflection of a parallel beam of light from successive crystal faces. { ri'flek·shən ,gō·nē'äm·əd·ər }

reflection loss |ELEC| **1.** Reciprocal of the ratio, expressed in decibels, of the scalar values of the volt-amperes delivered to the load to the volt-amperes that would be delivered to a load of the same impedance as the source. **2.** Apparent transmission loss of a line which results from a portion of the energy being reflected toward the

source due to a discontinuity in the transmission line. { ri'flek·shən ,lòs }

reflection profile |ENG| A seismic profile obtained by designing the spread geometry in such a manner as to enhance reflected energy. { ri'flek·shən ,prō,fīl }

reflection seismology See reflection shooting. { ri'flek·shən sīz'mäl·ə·jē }

reflection shooting |ENG| A procedure in seismic prospecting based on the measurement of the travel times of waves which, originating from an artificially produced disturbance, have been reflected to detectors from subsurface boundaries separating media of different elastic-wave velocities; used primarily for oil and gas exploration. Also known as reflection seismology. { ri'flek·shən ,shüd·iŋ }

reflection survey |ENG| Study of the presence, depth, and configuration of underground formations; a ground-level explosive charge (shot) generates vibratory energy (seismic rays) that strike formation interfaces and are reflected back to ground-level sensors. Also known as seismic survey. { ri'flek·shən ,sər,vā }

reflection x-ray microscopy |ENG| A technique for producing enlarged images in which a beam of x-rays is successively reflected at grazing incidence, from two crossed cylindrical surfaces; resolution is about 0.5–1 micrometer. { ri'flek·shən ¦eks,rā mi'kräs·kə·pē }

reflectometer |ENG| A photoelectric instrument for measuring the optical reflectance of a reflecting surface. { ,rē,flek'täm·əd·ər }

reflector microphone |ENG ACOUS| A highly directional microphone which has a surface that reflects the rays of impinging sound from a given direction to a common point at which a microphone is located, and the sound waves in the speech-frequency range are in phase at the microphone. { ri'flek·tər ,mī'kra,fōn }

reflex baffle |ENG ACOUS| A loudspeaker baffle in which a portion of the radiation from the rear of the diaphragm is propagated forward after controlled shift of phase or other modification, to increase the overall radiation in some portion of the audio-frequency spectrum. Also known as vented baffle. { 'rē,fleks ,baf·əl }

reflowing |ENG| Melting and resolidifying an electrodeposited or other type coating. { rē 'flō·iŋ }

reflux |CHEM ENG| In a chemical process, that part of the product stream that may be returned to the process to assist in giving increased conversion or recovery, as in distillation or liquid-liquid extraction. { 'rē,fləks }

reflux condenser |CHEM ENG| An auxiliary vessel for a distillation column that constantly condenses vapors and returns liquid to the column. { 'rē,fləks kən,den·sər }

reflux ratio |CHEM ENG| The quantity of liquid reflux per unit quantity of product removed from the process unit, such as a distillation tower or extraction column. { 'rē,fləks ,rā·shō }

reforming |CHEM ENG| The thermal or catalytic

conversion of petroleum naphtha into more volatile products of higher octane number; represents the total effect of numerous simultaneous reactions, such as cracking, polymerization, dehydrogenation and isomerization. { ¦rē'fórm·iŋ }

refracting angle See apical angle. { ri'frak·tiŋ ‚aŋ·gəl }

refraction process [ENG] Seismic (reflection) survey in which the distance between the explosive shot and the receivers (sensors) is large with respect to the depths to be mapped. { ri'frak·shən ‚prä·səs }

refraction profile [ENG] A seismic profile obtained by designing the spread geometry in such a manner as to enhance refracted energy. { ri 'frak·shən ‚prō‚fīl }

refraction shooting [ENG] A type of seismic shooting based on the measurement of seismic energy as a function of time after the shot and of distance from the shot, by determining the arrival times of seismic waves which have traveled nearly parallel to the bedding in high-velocity layers, in order to map the depth of such layers. { ri'frak·shən ‚shüd·iŋ }

refractometer [ENG] An instrument used to measure the index of refraction of a substance in any one of several ways, such as measurement of the refraction produced by a prism, measurement of the critical angle, observation of an interference pattern produced by passing light through the substance, and measurement of the substance's dielectric constant. { ‚rē‚frak'täm·əd·ər }

refractory-lined firebox boiler [MECH ENG] A horizontal fire-tube boiler with the front portion of the shell located over a refractory furnace; the rear of the shell contains the first-pass tubes, and the second-pass tubes are located in the upper part of the shell. { ri'frak·trē ¦līnd 'fīr‚bäks ‚bói·ər }

refrigerated truck [MECH ENG] An insulated truck equipped and used as a refrigerator to transport fresh perishable or frozen products. { ri'frij·ə‚rād·əd 'trək }

refrigeration [MECH ENG] The cooling of a space or substance below the environmental temperature. { ri‚frij·ə'rā·shən }

refrigeration condenser [MECH ENG] A vapor condenser in a refrigeration system, where the refrigerant is liquefied and discharges its heat to the environment. { ri‚frij·ə'rā·shən kən ‚den·sər }

refrigeration cycle [THERMO] A sequence of thermodynamic processes whereby heat is withdrawn from a cold body and expelled to a hot body. { ri‚frij·ə'rā·shən ‚sī·kəl }

refrigeration system [MECH ENG] A closed-flow system in which a refrigerant is compressed, condensed, and expanded to produce cooling at a lower temperature level and rejection of heat at a higher temperature level for the purpose of extracting heat from a controlled space. { ri‚frij·ə'rā·shən ‚sis·təm }

refrigerator [MECH ENG] An insulated, cooled compartment. { ri'frij·ə‚rād·ər }

refrigerator car [MECH ENG] An insulated freight car constructed and used as a refrigerator. { ri'frij·ə‚rād·ər ‚kär }

regelation [THERMO] Phenomenon in which ice (or any substance which expands upon freezing) melts under intense pressure and freezes again when this pressure is removed; accounts for phenomena such as the slippery nature of ice and the motion of glaciers. { ¦rē·jə'lā·shən }

regenerate [CHEM ENG] To clean of impurities and make reusable as in regeneration of a catalytic cracking catalyst by burning off carbon residue, regeneration of clay adsorbent by washing free of adherents, or regeneration of a filtration system by cleaning off the filter media. [ELECTR] **1.** To restore pulses to their original shape. **2.** To restore stored information to its original form in a storage tube in order to counteract fading and disturbances. { rē'jen·ə‚rāt }

regeneration [CONT SYS] See positive feedback. [ELECTR] Replacement or restoration of charges in a charge storage tube to overcome decay effects, including loss of charge by reading. { rē‚jen·ə'rā·shən }

regeneration system [MECH ENG] A system within a gas turbine that recovers waste heat from the turbine exhaust and uses it for the compression cycle. { rē‚jen·ə'rā·shən ‚sis·təm }

regenerative air heater [MECH ENG] An air heater in which the heat-transferring members are alternately exposed to heat-surrendering gases and to air. { rē'jen·rəd·iv 'er ‚hēd·ər }

regenerative cooling [ENG] A method of cooling gases in which compressed gas is cooled by allowing it to expand through a nozzle, and the cooled expanded gas then passes through a heat exchanger where it further cools the incoming compressed gas. { rē'jen·rəd·iv 'kül·iŋ }

regenerative cycle [MECH ENG] See bleeding cycle. [THERMO] An engine cycle in which low-grade heat that would ordinarily be lost is used to improve the cyclic efficiency. { rē'jen·rəd·iv ‚sī·kəl }

regenerative feedback See positive feedback. { rē'jen·rəd·iv 'fēd‚bak }

regenerative pump [MECH ENG] Rotating-vane device that uses a combination of mechanical impulse and centrifugal force to produce high liquid heads at low volumes. Also known as turbine pump. { rē'jen·rəd·iv 'pəmp }

regenerator [CHEM ENG] Device or system used to return a system or a component of it to full strength in a chemical process; examples are a furnace to burn carbon from a catalyst, a tower to wash impurities from clay, and a flush system to clean off the surface of filter media. [ELECTR] **1.** A circuit that repeatedly supplies current to a display or memory device to prevent data from decaying. **2.** See repeater. [MECH ENG] A device used with hot-air engines and gas-burning furnaces which transfers heat from effluent gases to incoming air or gas. { rē'jen·ə,rād·ər }

register [ENG] Also known as registration. **1.** The accurate matching or superimposition of two or more images, such as the three color images on the screen of a color television receiver, or the patterns on opposite sides of a printed circuit board, or the colors of a design on a printed sheet. **2.** The alignment of positions relative to a specified reference or coordinate, such as hole alignments in punched cards, or positioning of images in an optical character recognition device. [MECH ENG] The portion of a burner which directs the flow of air used in the combustion process. { 'rej·ə·star }

register circuit [ELECTR] A switching circuit with memory elements that can store from a few to millions of bits of coded information; when needed, the information can be taken from the circuit in the same code as the input, or in a different code. { 'rej·ə·star ,sər·kət }

register control [CONT SYS] Automatic control of the position of a printed design with respect to reference marks or some other part of the design, as in photoelectric register control. { 'rej·ə·star kən,trōl }

register mark [ENG] A mark or line printed or otherwise impressed on a web of material for use as a reference to maintain register. { 'rej·ə·star ,mark }

regular element [IND ENG] An element that occurs with a fixed frequency in each work cycle. Also known as repetitive element. { 'reg·yə·lər 'el·ə·mənt }

regular lay [DES ENG] The lay of a wire rope in which the wires in the strand are twisted in directions opposite to the direction of the strands. { 'reg·yə·lər 'lā }

regular-lay left twist See left-laid. { 'reg·yə·lər ¦lā 'left 'twist }

regulating reservoir [CIV ENG] A reservoir that regulates the flow in a water-distributing system. { 'reg·yə,lād·iŋ 'rez·əv,wär }

regulating system See automatic control system. { 'reg·yə,lād·iŋ ,sis·təm }

regulation [CONT SYS] The process of holding constant a quantity such as speed, temperature, voltage, or position by means of an electronic or other system that automatically corrects errors by feeding back into the system the condition being regulated; regulation thus is based on feedback, whereas control is not. [ELEC] The change in output voltage that occurs between no load and full load in a transformer, generator, or other source. [ELECTR] The difference between the maximum and minimum tube voltage drops within a specified range of anode current in a gas tube. { ,reg·yə'lā·shən }

regulator [CONT SYS] A device that maintains a desired quantity at a predetermined value or varies it according to a predetermined plan. { 'reg·yə,lād·ər }

regulator problem See linear regulator problem. { 'reg·yə,lād·ər ,präb·ləm }

regulatory control function [CONT SYS] That level in the functional decomposition of a large-scale control system which interfaces with the plant to implement the decisions of the optimizing controller inputted in the form of set points, desired trajectories, or targets. Also known as direct control function. { 'reg·yə·lə,tör·ē kən'trōl ,faŋk·shən }

rehabilitation engineering [ENG] The use of technology to make disabled persons as independent as possible by providing assistive devices to compensate for disability. { ,rē·ə,bil·ə'tā·shən ,en·jə,nir·iŋ }

reheating [THERMO] A process in which the gas or steam is reheated after a partial isentropic expansion to reduce moisture content. Also known as resuperheating. { rē'hēd·iŋ }

Reich process [CHEM ENG] Process to purify carbon dioxide produced during fermentation; organic impurities in the gas are oxidized and absorbed, then the gas is dehydrated. { 'rīk ,prä·səs }

Reid vapor pressure [ENG] A measure in a test bomb of the vapor pressure in pounds pressure of a sample of gasoline at 100°F (37.8°C). { 'rēd 'vā·pər ,presh·ər }

reinforced beam [CIV ENG] A concrete beam provided with steel bars for longitudinal tension reinforcement and sometimes compression reinforcement and reinforcement against diagonal tension. { ¦rē·ən'fórst 'bēm }

reinforced brickwork [CIV ENG] Brickwork strengthened by expanded metal, steel-wire mesh, hoop iron, or thin rods embedded in the bed joints. { ¦rē·ən'fórst 'brik,wərk }

reinforced column [CIV ENG] **1.** A long concrete column reinforced with longitudinal bars with ties or circular spirals. **2.** A composite column. **3.** A combination column. { ¦rē·ən'fórst 'käl·əm }

reinforced concrete [CIV ENG] Concrete containing reinforcing steel rods or wire mesh. { ¦rē·ən'fórst 'kän,krēt }

reinforcement [CIV ENG] Strengthening concrete, plaster, or mortar by embedding steel rods or wire mesh in it. { ,rē·ən'fórs·mənt }

reinforcing bars [CIV ENG] Steel rods that are embedded in building materials such as concrete for reinforcement. { ¦rē·ən'fórs·iŋ ,bärz }

rejection number [IND ENG] A predetermined number of defective items in a batch which, if not exceeded, requires acceptance of the batch. { ri'jek·shən ,nəm·bər }

rejector circuit See band-stop filter. { ri'jek·tər ,sər·kət }

relative compaction [ENG] The percentage ratio of the field density of soil to the maximum density as determined by standard compaction. { 'rel·əd·iv kəm'pak·shən }

relative density See specific gravity. { 'rel·əd·iv 'den·səd·ē }

relative-density bottle See specific-gravity bottle. { 'rel·əd·iv ¦den·səd·ē ,bäd·əl }

relative dielectric constant See dielectric constant. { 'rel·əd·iv ¦dī·i'lek·trik 'kän·stənt }

relative force [ENG] Ratio of the force of a test

propellant to the force of a standard propellant, measured at the same initial temperature and loading density in the same closed chamber. { 'rel·əd·iv 'fôrs }

relative gain array |CONT SYS| An analytical device used in process control multivariable applications, based on the comparison of single-loop control to multivariable control; expressed as an array (for all possible input-output pairs) of the ratios of a measure of the single-loop behavior between an input-output variable pair, to a related measure of the behavior of the same input-output pair under some idealization of multivariable control. { 'rel·əd·iv ¦gān ə,rā }

relative gravity instrument |ENG| Any device for measuring the differences in the gravity force or acceleration at two or more points. { 'rel·əd·iv 'grav·əd·ē ,in·strə·mənt }

relative interference effect |ENG ACOUS| Of a single-frequency electric wave in an electro-acoustic system, the ratio, usually expressed in decibels, of the amplitude of a wave of specified reference frequency to that of the wave in question when the two waves are equal in interference effects. { 'rel·əd·iv ,in·tər'fir·əns i,fekt }

relative ionospheric opacity meter See riometer. { 'rel·əd·iv ī¦än·ə¦sfir·ik ō'pas·əd·ē ,mēd·ər }

relative magnetometer |ENG| Any magnetometer which must be calibrated by measuring the intensity of a field whose strength is accurately determined by other means; opposed to absolute magnetometer. { 'rel·əd·iv ,mag·nə'täm·əd·ər }

relative momentum |MECH| The momentum of a body in a reference frame in which another specified body is fixed. { 'rel·əd·iv mə'men·təm }

relative motion |MECH| The continuous change of position of a body with respect to a second body or to a reference point that is fixed. Also known as apparent motion. { 'rel·əd·iv 'mō·shən }

relative permittivity See dielectric constant. { 'rel·əd·iv ,pər·mə'tiv·əd·ē }

relative pressure response |ENG ACOUS| The amount, in decibels, by which the acoustic pressure induced by a projector under some specified condition exceeds the pressure induced under a reference condition. { 'rel·əd·iv ¦presh·ər ri ,späns }

relative resistance |ELEC| The ratio of the resistance of a piece of a material to the resistance of a piece of specified material, such as annealed copper, having the same dimensions and temperature. { 'rel·əd·iv ri'zis·təns }

relative transmitting response |ENG ACOUS| In a sonar projector, the ratio of the transmitting response for a given bearing and frequency to the transmitting response for a specified bearing and frequency. { 'rel·əd·iv tranz'mid·iŋ ri ,späns }

relative velocity |MECH| The velocity of a body with respect to a second body; that is, its velocity in a reference frame where the second body is fixed. { 'rel·əd·iv və'läs·əd·ē }

relaxation |MECH| **1.** Relief of stress in a strained material due to creep. **2.** The lessening of elastic resistance in an elastic medium under an applied stress resulting in permanent deformation. { ,rē,lak'sā·shən }

relaxation circuit |ELECTR| Circuit arrangement, usually of vacuum tubes, reactances, and resistances, which has two states or conditions, one, both, or neither of which may be stable; the transient voltage produced by passing from one to the other, or the voltage in a state of rest, can be used in other circuits. { ,rē,lak'sā·shən ,sər·kət }

relaxation test |ENG| A creep test in which the decrease of stress with time is measured while the total strain (elastic and plastic) is maintained constant. { ,rē,lak'sā·shən ,test }

relay |ELEC| A device that is operated by a variation in the conditions in one electric circuit and serves to make or break one or more connections in the same or another electric circuit. Also known as electric relay. { 'rē,lā }

relay control system |CONT SYS| A control system in which the error signal must reach a certain value before the controller reacts to it, so that the control action is discontinuous in amplitude. { 'rē,lā kən'trōl ,sis·təm }

relay rack |DES ENG| A standardized steel rack designed to hold 19-inch (48.26-centimeter) panels of various heights, on which are mounted radio receivers, amplifiers, and other units of electronic equipment. Also known as rack. { 'rē,lā ,rak }

relay system |ELEC| Dial-switching equipment that does not use mechanical switches, but is made up principally of relays. { 'rē,lā ,sis·təm }

release |MECH ENG| A mechanical arrangment of parts for holding or freeing a device or mechanism as required. { ri'lēs }

release adiabat |MECH| A curve or locus of points which defines the succession of states through which a mass that has been shocked to a high-pressure state passes while monotonically returning to zero pressure. { ri'lēs 'ad·ē·ə,bat }

reliability |ENG| The probability that a component part, equipment, or system will satisfactorily perform its intended function under given circumstances, such as environmental conditions, limitations as to operating time, and frequency and thoroughness of maintenance for a specified period of time. { ri,lī·ə'bil·əd·ē }

relief |MECH ENG| **1.** A passage made by cutting away one side of a tailstock center so that the facing or parting tool may be advanced to or almost to the center of the work. **2.** Clearance provided around the cutting edge by removal of tool material. { ri'lēf }

relief angle |MECH ENG| The angle between a relieved surface and a tangential plane at a cutting edge. { ri'lēf ,aŋ·gəl }

relief frame |MECH ENG| A frame placed between the slide valve of a steam engine and the steam chest cover; reduces pressure on the valve and thereby reduces friction. { ri'lēf ,frām }

relief hole |ENG| Any of the holes fired after the

cut holes and before the lifter holes in breaking ground for tunneling or shaft sinking. { ri'lēf ,hōl }

relief valve See pressure-relief valve. { ri'lēf ,valv }

relief well [CIV ENG] A well that drains a pervious stratum, to relieve waterlogging at the surface. { ri'lēf ,wel }

relieving [MECH ENG] Treating an embossed metal surface with an abrasive to reveal the basemetal color on the elevations or highlights of the surface. { ri'lēv·iŋ }

relieving arch See discharging arch. { ri'lēv·iŋ ,ärch }

relieving platform [CIV ENG] A deck on the land side of a retaining wall to transfer loads vertically down to the wall. { ri'lēv·iŋ ,plat,fórm }

relish [ENG] The shoulder of a tenon, used in a mortise and tenon system. { 'rel·ish }

reluctance microphone See magnetic microphone. { ri'lək·təns ,mī·krə,fōn }

reluctance pickup See variable-reluctance pickup. { ri'lək·təns ,pik,əp }

reluctance pressure transducer [ENG] Pressure-measurement transducer in which pressure changes activate equivalent magnetic-property changes. { ri'lək·təns 'presh·ər tranz,dü·sər }

remaining velocity [MECH] Speed of a projectile at any point along its path of fire. { ri'mān·iŋ və'läs·əd·ē }

remedial operation [CHEM ENG] In a chemical process operation, the revision of operating conditions so as to correct the overall operation and bring the product into desired rote or specification limits. Also known as corrective operation. { ri'mēd·ē·əl ,äp·ə'rā·shən }

remote-access admittance [CONT SYS] A special piece of hardware, with built-in sensors and actuators, that is used by a robot to carry out the last stages of assembling several parts into a piece of equipment. { ri'mōt ,ak,ses ad'mit·əns }

remote-center compliance [MECH ENG] A compliant device that allows a part that is gripped by a robot or other automatic machinery to rotate about the tip of the robot end effector or to translate without rotation when it is pushed, thereby easing the mechanical assembly of parts. { ri'mōt ¦sen·tər kəm'plī·əns }

remote control [CONT SYS] Control of a quantity which is separated by an appreciable distance from the controlling quantity; examples include master-slave manipulators, telemetering, telephone, and television. { ri'mōt kən 'trōl }

remote manipulation [ENG] Use of mechanical equipment controlled from a distance to handle materials, such as radioactive materials. Also known as teleoperation. { ri'mōt mə,nip·yə 'lā·shən }

remote manipulator [ENG] A mechanical, electromechanical, or hydromechanical device that enables a person, directly controlling the device through handles or switches, to perform manual operations while separated from the site of the work. Also known as manipulator; teleoperator. { ri'mōt mə'nip·yə,lād·ər }

remote metering See telemetering. { ri'mōt 'mēd·ə·riŋ }

remote sensing [ELEC] Sensing, by a power supply, of voltage directly at the load, so that variations in the load lead drop do not affect load regulation. [ENG] The gathering and recording of information without actual contact with the object or area being investigated. { ri 'mōt 'sens·iŋ }

renewable energy source [ENG] A form of energy that is constantly and rapidly renewed by natural processes such as solar, ocean wave, and wind energy. { ri,nü·ə·bəl 'en·ər·jē ,sórs }

renewable resources [CHEM ENG] Agricultural materials used as feedstocks for industrial processes. { ri'nü·ə·bəl ri'sór·səs }

reorder cycle [IND ENG] The interval between successive reorder (procurement) actions. { re'ór·dər ,sī·kəl }

reorder point [IND ENG] An arbitrary level of stock on hand plus stock due in, at or below which routine requisitions for replenishment purposes are submitted in accordance with established requisitioning schedules. { re'ór·dər ,póint }

repair [ENG] To restore that which is unserviceable to a serviceable condition by replacement of parts, components, or assemblies. { ri'per }

repair cycle [ENG] The period that elapses from the time the item is removed in a reparable condition to the time it is returned to stock in a serviceable condition. { ri'per ,sī·kəl }

repair dock [CIV ENG] A graving dock or floating dry dock built primarily for ship repair. { ri'per ,däk }

repair forecast [ENG] The quantity of items estimated to be repaired or rebuilt for issue during a stated future period. { ri'per ,fór,kast }

repair kit [ENG] A group of parts and tools, not all having the same basic name, used for repair or replacement of the worn or broken parts of an item; it may include instruction sheets and material, such as sandpaper, tape, cement, gaskets, and the like. { ri'per ,kit }

repair parts list [ENG] List approved by designated authorities, indicating the total quantities of repair parts, tools, and equipment necessary for the maintenance of a specified number of end items for a definite period of time. { ri'per ¦pärts ,list }

repeatability [CONT SYS] The ability of a robot to reposition itself at a location to which it is directed or at which it is commanded to stop. { ri,pēd·ə'bil·əd·ē }

repeat accuracy [CONT SYS] The variations in the actual position of a robot manipulator from one cycle to the next when the manipulator is commanded to repeatedly return to the same point or position. { ri'pēt 'ak·yə·rə·sē }

repeated load [MECH] A force applied repeatedly, causing variation in the magnitude and sometimes in the sense, of the internal forces. { ri'pēd·əd 'lōd }

repeater |ELEC| See repeating coil. |ELECTR| **1.** An amplifier or other device that receives weak signals and delivers corresponding stronger signals with or without reshaping of waveforms; may be either a one-way or two-way repeater. Also known as regenerator. **2.** An indicator that shows the same information as is shown on a master indicator. Also known as remote indicator. { ri'pēd·ər }

repeater jammer |ELECTR| A jammer that intercepts an enemy radar signal and reradiates the signal after modifying it to incorporate erroneous data on azimuth, range, or number of targets. { ri'pēd·ər ,jam·ər }

repeating coil |ELEC| A transformer used to provide inductive coupling between two sections of a telephone line when a direct connection is undesirable. Also known as repeater. { ri 'pēd·iŋ ,kȯil }

repeating-coil bridge cord |ELEC| In telephony, a method of connecting the common office battery to the cord circuits by connecting the battery to the midpoints of a repeating coil, bridged across the cord circuit. { ri'pēd·iŋ ¦kȯil 'briჰ ,kȯrd }

repeller |ELECTR| An electrode whose primary function is to reverse the direction of an electron stream in an electron tube. Also known as reflector. { ri'pel·ər }

repetitive element See regular element. { rə'ped·əd·iv 'el·ə·mənt }

repetitive time method |IND ENG| A technique where the stopwatch is read and simultaneously returned to zero at each break point. Also known as snapback method. { ri'ped·əd·iv 'tīm ,meth·əd }

replacement bit See reset bit. { ri'plās·mənt ,bit }

replacement demand |ENG| A demand representing replacement of items consumed or worn out. { ri'plās·mənt di,mand }

replacement factor |ENG| The estimated percentage of equipment or repair parts in use that will require replacement during a given period. { ri'plās·mənt ,fak·tər }

replacement study |IND ENG| An economic analysis involving the comparison of an existing facility and a proposed replacement facility. { ri'plās·mənt ,stəd·ē }

replica |ENG| A thin plastic or inorganic film which is formed on a surface and then removed from it for study in an electron microscope. { 'rep·lə·kə }

replica master |MECH ENG| A robotlike machine whose motions are duplicated by another robot when the machine is moved by a human operator. { 'rep·lə·kə ,mas·tər }

Reppe process |CHEM ENG| A family of high-pressure, catalytic acetylene-reaction processes yielding (depending upon what the acetylene reacts with) butadiene, allyl alcohol, acrylonitrile, vinyl ethers and derivatives, acrylic acid esters, cyclooctatetraene, and resins. { 'rep·ə ,prä·səs }

reproducing stylus See stylus. { ¦rē·prə¦düs·iŋ ,stī·ləs }

reproducing system See sound-reproducing system. { ¦rē·prə¦düs·iŋ ,sis·təm }

repulsion |MECH| A force which tends to increase the distance between two bodies having like electric charges, or the force between atoms or molecules at very short distances which keeps them apart. Also known as repulsive force. { ri'pəl·shən }

repulsive force See repulsion. { ri'pəl·siv 'fȯrs }

required thickness |DES ENG| The thickness calculated by recognized formulas for boiler or pressure vessel construction before corrosion allowance is added. { ri'kwīrd 'thik·nəs }

requirements engineering |SYS ENG| The process of identifying and articulating needs for a new technology and applications. { ri¦kwīr·məns ,en·jə¦nir·iŋ }

rerailer |ENG| A small, lightweight Y-shaped device, used to retrack railroad cars and locomotives; as the car is pulled across the device, the derailed wheels are channeled back onto the tracks. Also known as retracker. { rē'rāl·ər }

rerun |CHEM ENG| To distill a liquid material that has already been distilled; usually implies taking a large proportion of the charge stock overhead. { 'rē,rən }

resaw |ENG| To cut lumber to boards of final thickness. { rē'sȯ }

resealing pressure |MECH ENG| The inlet pressure at which leakage stops after a pressure relief valve is closed. { rē'sēl·iŋ ,presh·ər }

research method |ENG| A standard test to determine the research octane number (or rating) of fuels for use in spark-ignition engines. { ri'sərch ,meth·əd }

research octane number |ENG| An expression for the antiknock rating of a motor gasoline as a guide to how vehicles will operate under mild conditions associated with low engine speeds. { ri'sərch 'äk,tān ,nəm·bər }

resection |ENG| **1.** A method in surveying by which the horizontal position of an occupied point is determined by drawing lines from the point to two or more points of known position. **2.** A method of determining a plane-table position by orienting along a previously drawn foresight line and drawing one or more rays through the foresight from previously located stations. { ri'sek·shən }

reservoir |CIV ENG| A pond or lake built for storage of water, usually by the construction of a dam across a river. { 'rez·əv,wär }

reset action |CONT SYS| Floating action in which the final control element is moved at a speed proportional to the extent of proportional-position action. { 'rē,set ,ak·shən }

reset bit |DES ENG| A diamond bit made by reusing diamonds salvaged from a used bit and setting them in the crown attached to a new bit blank. Also known as replacement bit. { 'rē ,set bit }

reset rate |ENG| The number of times per minute that the effect of the proportional-position action upon the final control element is repeated by the proportional-speed floating action. { 'rē ,set ,rāt }

residence time [CHEM ENG] The average length of time a particle of reactant spends within a process vessel or in contact with a catalyst. { 'rez·ə·dəns ,tīm }

residual mode [CONT SYS] A characteristic motion of a structure which is deliberately ignored in the control algorithm of an active control system for the structure in the process of model reduction. { rə'zij·ə·wəl ¦mōd }

residual stress See internal stress. { rə'zij·ə·wəl 'stres }

residue [CHEM ENG] **1.** The substance left after distilling off all but the heaviest components from crude oil in petroleum refinery operations. Also known as bottoms; residuum. **2.** Solids deposited onto the filter medium during filtration. Also known as cake; discharged solids. { 'rez·ə,dü }

residuum See residue. { rə'zij·ə·wəm }

resilience [MECH] **1.** Ability of a strained body, by virtue of high yield strength and low elastic modulus, to recover its size and form following deformation. **2.** The work done in deforming a body to some predetermined limit, such as its elastic limit or breaking point, divided by the body's volume. { rə'zil·yəns }

resin-in-pulp ion exchange [CHEM ENG] Combination of coarse anion-exchange resin with a slurry of finely ground uranium ore in an acid-leach liquor. { 'rez·ən in 'pəlp 'ī,än iks,chānj }

resinoid wheel [DES ENG] A grinding wheel bonded with a synthetic resin. { 'rez·ən,öid 'wēl }

resistance [ELEC] **1.** The opposition that a device or material offers to the flow of direct current, equal to the voltage drop across the element divided by the current through the element. Also known as electrical resistance. **2.** In an alternating-current circuit, the real part of the complex impedance. [MECH] In damped harmonic motion, the ratio of the frictional resistive force to the speed. Also known as damping coefficient; damping constant; mechanical resistance. { ri'zis·təns }

resistance bridge See Wheatstone bridge. { ri'zis·təns ,brij }

resistance-capacitance circuit [ELEC] A circuit which has a resistance and a capacitance in series, and in which inductance is negligible. Abbreviated R-C circuit. { ri'zis·təns kə'pas·əd·əns ,sər·kət }

resistance-capacitance coupled amplifier [ELECTR] An amplifier in which a capacitor provides a path for signal currents from one stage to the next, with resistors connected from each side of the capacitor to the power supply or to ground; it can amplify alternating-current signals but cannot handle small changes in direct currents. Also known as R-C amplifier; R-C coupled amplifier; resistance-coupled amplifier. { ri'zis·təns kə'pas·əd·əns ¦kəp·əld 'am·plə,fī·ər }

resistance-capacitance oscillator [ELECTR] Oscillator in which the frequency is determined by resistance and capacitance elements. Abbreviated R-C oscillator. { ri'zis·təns kə'pas·əd·əns 'äs·ə,lād·ər }

resistance-coupled amplifier See resistance-capacitance coupled amplifier. { ri'zis·təns ¦kəp·əld 'am·plə,fī·ər }

resistance coupling [ELECTR] Coupling in which resistors are used as the input and output impedances of the circuits being coupled; a coupling capacitor is generally used between the resistors to transfer the signal from one stage to the next. Also known as R-C coupling; resistance-capacitance coupling; resistive coupling. { ri'zis·təns ,kəp·liŋ }

resistance drop [ELEC] The voltage drop occurring between two points on a conductor due to the flow of current through the resistance of the conductor; multiplying the resistance in ohms by the current in amperes gives the voltage drop in volts. Also known as IR drop. { ri'zis·təns ,dräp }

resistance element [ELEC] An element of resistive material in the form of a grid, ribbon, or wire, used singly or built into groups to form a resistor for heating purposes, as in an electric soldering iron. { ri'zis·təns ,el·ə·mənt }

resistance furnace [ENG] An electric furnace in which the heat is developed by the passage of current through a suitable internal resistance that may be the charge itself, a resistor embedded in the charge, or a resistor surrounding the charge. Also known as electric resistance furnace. { ri'zis·təns ,fər·nəs }

resistance gage [ENG] An instrument for determining high pressures from the change in the electrical resistance of manganin or mercury produced by these pressures. { ri'zis·təns ,gāj }

resistance grounding [ELEC] Electrical grounding in which lines are connected to ground by a resistive (totally dissipative) impedance. { ri'zis·təns ,graůnd·iŋ }

resistance heating [ELEC] The generation of heat by electric conductors carrying current; degree of heating is proportional to the electrical resistance of the conductor; used in electrical home appliances, home or space heating, and heating ovens and furnaces. { ri'zis·təns ,hēd·iŋ }

resistance loss [ELEC] Power loss due to current flowing through resistance; its value in watts is equal to the resistance in ohms multiplied by the square of the current in amperes. { ri'zis·təns ,lös }

resistance magnetometer [ENG] A magnetometer that depends for its operation on variations in the electrical resistance of a material immersed in the magnetic field to be measured. { ri'zis·təns ,mag·nə'täm·əd·ər }

resistance material [ELEC] Material having sufficiently high resistance per unit length or volume to permit its use in the construction of resistors. { ri'zis·təns mə'tir·ē·əl }

resistance measurement [ELEC] The quantitative determination of that property of an electrically conductive material, component, or circuit called electrical resistance. { ri'zis·təns ,mezh·ər·mənt }

resistance meter [ENG] Any instrument which measures electrical resistance. Also known as electrical resistance meter. { ri'zis·təns ,mēd·ər }

resistance methanometer [ENG] A catalytic methanometer, with platinum used as the filament, which both heats the detecting element and acts as a resistance-type thermometer. { ri'zis·təns ,meth·ə'näm·əd·ər }

resistance pyrometer See resistance thermometer. { ri'zis·təns pī'räm·əd·ər }

resistance-rate flowmeter See resistive flowmeter. { ri'zis·təns ¦rāt 'flō,mēd·ər }

resistance thermometer [ENG] A thermometer in which the sensing element is a resistor whose resistance is an accurately known function of temperature. Also known as electrical resistance thermometer; resistance pyrometer. { ri 'zis·təns thər'mäm·əd·ər }

resisting moment [MECH] A moment produced by internal tensile and compressive forces that balances the external bending moment on a beam. { ri'zist·iŋ ,mō·mənt }

resistive coupling See resistance coupling. { ri 'zis·tiv 'kəp·liŋ }

resistive flowmeter [ENG] Liquid flow-rate measurement device in which flow rates are read electrically as the result of the rise or fall of a conductive differential-pressure manometer fluid in contact with a resistance-rod assembly. Also known as resistance-rate flowmeter. { ri'zis·tiv 'flō,mēd·ər }

resistive load [ELEC] A load whose total reactance is zero, so that the alternating current is in phase with the terminal voltage. Also known as nonreactive load. { ri'zis·tiv 'lōd }

resistivity See electrical resistivity. { ,rē,zis'tiv·əd·ē }

resistivity method [ENG] Any electrical exploration method in which current is introduced in the ground by two contact electrodes and potential differences are measured between two or more other electrodes. { ,rē,zis'tiv·əd·ē ,meth·əd }

resistor [ELEC] A device designed to have a definite amount of resistance; used in circuits to limit current flow or to provide a voltage drop. Also known as electrical resistor. { ri'zis·tər }

resistor bulb [ENG] A temperature-measurement device inside of which is a resistance winding; changes in temperature cause corresponding changes in resistance, varying the current in the winding. { ri'zis·tər ,bəlb }

resistor-capacitor-transistor logic [ELECTR] A resistor-transistor logic with the addition of capacitors that are used to enhance switching speed. { ri'zis·tər kə'pas·əd·ər tran'zis·tər ,läj·ik }

resistor-capacitor unit See rescap. { ri'zis·tər kə'pas·əd·ər ,yü·nət }

resistor color code [ELEC] Code adopted by the Electronic Industries Association to mark the values of resistance on resistors in a readily recognizable manner; the first color represents the first significant figure of the resistor value, the second color the second significant figure, and the third color represents the number of zeros following the first two figures; a fourth color is sometimes added to indicate the tolerance of the resistor. { ri'zis·tər 'kəl·ər ,kōd }

resistor core [ELEC] Insulating support on which a resistor element is wound or otherwise placed. { ri'zis·tər ,kór }

resistor element [ELEC] That portion of a resistor which possesses the property of electric resistance. { ri'zis·tər ,el·ə·mənt }

resistor furnace [ENG] An electric furnace in which heat is developed by the passage of current through distributed resistors (heating units) mounted apart from the charge. { ri'zis·tər ,fər·nəs }

resistor network [ELEC] An electrical network consisting entirely of resistances. { ri'zis·tər 'net,wərk }

resistor oven [ENG] Heating chamber relying on an electrical-resistance element to create temperatures of up to 800°F (430°C); used for drying and baking. { ri'zis·tər 'əv·ən }

resistor termination [ELECTR] A thick-film conductor pad overlapping and contacting a thick-film resistor area. { ri'zis·tər ,tər·mə'nā·shən }

resistor-transistor logic [ELECTR] One of the simplest logic circuits, having several resistors, a transistor, and a diode. Abbreviated RTL. { ri'zis·tər tran'zis·tər ,läj·ik }

resolution [CONT SYS] The smallest increment in distance that can be distinguished and acted upon by an automatic control system. [ELECTR] In television, the maximum number of lines that can be discerned on the screen at a distance equal to tube height; this ranges from 350 to 400 for most receivers. { ,rez·ə'lü·shən }

resolution in azimuth [ENG] The angle by which two targets must be separated in azimuth in order to be distinguished by a radar set when the targets are at the same range. { ,rez·ə'lü·shən in 'az·ə·məth }

resolution in range [ENG] Distance by which two targets must be separated in range in order to be distinguished by a radar set when the targets are on the same azimuth line. { ,rez·ə'lü·shən in 'rānj }

resolve motion-rate control [CONT SYS] A form of robotic control in which the controlled variables are the velocity vectors of the end points of a manipulator, and the angular velocities of the joints are determined to obtain the desired results. { ri'zolv 'mō·shən ¦rāt kən,trōl }

resolving power See resolution. { ri'zälv·iŋ ,paů·ər }

resolving time [ENG] Minimum time interval, between events, that can be detected; resolving time may refer to an electronic circuit, to a mechanical recording device, or to a counter tube. { ri'zälv·iŋ ,tīm }

resonance [ELEC] A phenomenon exhibited by an alternating-current circuit in which there are relatively large currents near certain frequencies, and a relatively unimpeded oscillation of energy from a potential to a kinetic form; a special case of the physics definition. { 'rez·ən·əns }

resonance method |ELEC| A method of determining the impedance of a circuit element, in which resonance frequency of a resonant circuit containing the element is measured. |ENG| In ultrasonic testing, a method of measuring the thickness of a metal by varying the frequency of the beam transmitted to excite a maximum amplitude of vibration. { 'rez·ən·əns ,meth·əd }

resonance vibration |MECH| Forced vibration in which the frequency of the disturbing force is very close to the natural frequency of the system, so that the amplitude of vibration is very large. { 'rez·ən·əns vī,brā·shən }

resonant capacitor |ELEC| A tubular capacitor that is wound to have inductance in series with its capacitance. { 'res·ən·ənt kə'pas·əd·ər }

resonant circuit |ELEC| A circuit that contains inductance, capacitance, and resistance of such values as to give resonance at an operating frequency. { 'res·ən·ənt 'sər·kət }

resonant coupling |ELEC| Coupling between two circuits that reaches a sharp peak at a certain frequency. { 'res·ən·ənt 'kəp·liŋ }

resonant gate transistor |ELECTR| Surface field-effect transistor incorporating a cantilevered beam which resonates at a specific frequency to provide high-Q-frequency discrimination. { 'res·ən·ənt 'gāt tran,zis·tər }

resonant-mass antenna |ENG| A detector of gravitational radiation, consisting of a mass of several tons of aluminum or other metal, in the shape of a cylinder or a truncated icosahedron, and attached electromechanical transducers that convert deformations of the mass to electronic signals. { ¦rez·ən·ənt ,mas an'ten·ə }

resonant resistance |ELEC| Resistance value to which a resonant circuit is equivalent. { 'res·ən·ənt ri'zis·təns }

resource allocation in multiproject scheduling |IND ENG| A system that employs network analysis as an aid in making the best assignment of resources which must be stretched over a number of projects. Abbreviated RAMPS. { 'rē,sórs ,al·ə'kā·shən in ¦məl·ti¦prä·jekt 'sked·jə·liŋ }

respirator |ENG| A device for maintaining artificial respiration to protect the respiratory tract against irritating and poisonous gases, fumes, smoke, and dusts, with or without equipment supplying oxygen or air; some types have a fitting which covers the nose and mouth. { 'res·pə,rād·ər }

respirometer |ENG| **1.** An instrument for studying respiration. **2.** A diver's helmet containing a compressed air supply for replenishing oxygen used by the diver. { ,res·pə'räm·əd·ər }

response |CONT SYS| A quantitative expression of the output of a device or system as a function of the input. Also known as system response. { ri'späns }

response characteristic |CONT SYS| The response as a function of an independent variable,

such as direction or frequency, often presented in graphical form. { ri'späns ,kar·ik·tə,ris·tik }

response time |CONT SYS| The time required for the output of a control system or element to reach a specified fraction of its new value after application of a step input or disturbance. |ELEC| The time it takes for the pointer of an electrical or electronic instrument to come to rest at a new value, after the quantity it measures has been abruptly changed. { ri'späns ,tīm }

restitution coefficient See coefficient of restitution. { ,res·tə'tü·shən ,kō·i,fish·ənt }

rest point |ENG| On a balance, the position of the pointer with respect to the pointer scale when the beam has ceased moving. { 'rest ,point }

rest potential |ELEC| Residual potential difference remaining between an electrode and an electrolyte after the electrode has become polarized. { 'rest pə,ten·chəl }

restraint of loads |ENG| The process of binding, lashing, and wedging items into one unit onto or into its transporter in a manner that will ensure immobility during transit. { ri'strānt əv 'lōdz }

restricted air cargo |IND ENG| Cargo which is not highly dangerous under normal conditions, but which possesses certain qualities which require extra precautions in packing and handling. { ri'strik·təd 'er ,kär·gō }

restricted gate |ENG| Small opening between runner and cavity in an injection or transfer mold which breaks cleanly when the piece is ejected. { ri'strik·təd 'gāt }

restricted job |IND ENG| A task whose performance time is governed by a machine, a process, another task, or the nature of the job itself, rather than being under the control of the worker. { ri'strik·təd 'jäb }

restricted work |IND ENG| Manual or machine work where the work pace is only partially under the control of the worker. { ri'strik·təd 'wərk }

resultant of forces |MECH| A system of at most a single force and a single couple whose external effects on a rigid body are identical with the effects of the several actual forces that act on that body. { ri'zəlt·ənt əv 'fórs·əz }

resultant rake |MECH ENG| The angle between the face of a cutting tooth and an axial plane through the tooth point measured in a plane at right angles to the cutting edge. { ri'zəlt·ənt 'rāk }

resuperheating See reheating. { rē¦sü·pər'hēd·iŋ }

resupply |IND ENG| The act of replenishing stocks in order to maintain required levels of supply. { ¦rē·sə'plī }

resuscitator |ENG| A device for supplying oxygen to and inducing breathing in asphyxiation victims. { ri'səs·ə,tād·ər }

retainer |ENG| A device that holds a mechanical component in place. { ri'tān·ər }

retainer plate |ENG| The plate on which removable mold parts (such as a cavity or ejector pin) are mounted during molding. { ri'tān·ər ,plāt }

retainer wall |ENG| A wall, usually earthen,

around a storage tank or an area of storage tanks (tank farm); used to hold (retain) liquid in place if one or more tanks begin to leak. { ri'tān·ər ,wȯl }

retaining ring |DES ENG| **1.** A shoulder inside a reaming shell that prevents the core lifter from entering the core barrel. **2.** A steel ring between the races of a ball bearing to maintain the correct distribution of the balls in the races. { ri'tān· iŋ ,riŋ }

retaining wall |CIV ENG| A wall designed to maintain differences in ground elevations by holding back a bank of material. { ri'tān·iŋ ,wȯl }

retard |CIV ENG| A permeable bank-protection structure, situated at and parallel to the toe of a slope and projecting into a stream channel, designed to check stream velocity and induce silting or accretion. { ri'tärd }

retarder |MECH ENG| **1.** A braking device used to control the speed of railroad cars moving along the classification tracks in a hump yard. **2.** A strip inserted in a tube of a fire-tube boiler to increase agitation of the hot gases flowing therein. { ri 'tärd·ər }

retarding basin |CIV ENG| A basin designed and operated to provide temporary storage and thus reduce the peak flood flows of a stream. { ri 'tärd·iŋ ,bās·ən }

retarding conveyor |MECH ENG| Any type of conveyor used to restrain the movement of bulk materials, packages, or objects where the incline is such that the conveyed material tends to propel the conveying medium. { ri'tärd·iŋ kən ,vā·ər }

retort |CHEM ENG| **1.** A closed refractory chamber in which coal is carbonized for manufacture of coal gas. **2.** A vessel for the distillation or decomposition of a substance. { ri'tȯrt }

retreater |ENG| A defective maximum thermometer of the liquid-in-glass type in which the mercury flows too freely through the constriction; such a thermometer will indicate a maximum temperature that is too low. { ri'trēd·ər }

retrievable inner barrel |ENG| The inner barrel assembly of a wire-line core barrel, designed for removing core from a borehole without pulling the rods. { ri'trēv·ə·bəl 'in·ər 'bar·əl }

retroaction See positive feedback. { ¦re·trō'ak· shən }

retrofit |ENG| A modification of equipment to incorporate changes made in later production of similar equipment; it may be done in the factory or field. Derived from retroactive refit. { 're· trō,fit }

retting |CHEM ENG| Soaking vegetable stalks to decompose the gummy material and release the fibers. { 'red·iŋ }

return |BUILD| The continuation of a molding, projection, member, cornice, or the like, in a different direction, usually at a right angle. See echo. { ri'tərn }

return bead See quirk bead. { ri'tərn ,bēd }

return bend |DES ENG| A pipe fitting, equal to two ells, used to connect parallel pipes so that

fluid flowing into one will return in the opposite direction through the other. { ri'tərn ,bend }

return connecting rod |MECH ENG| A connecting rod whose crankpin end is located on the same side of the crosshead as the cylinder. { ri'tərn kə'nek·tiŋ ,räd }

return difference |CONT SYS| The difference between 1 and the loop transmittance. { ri'tərn ,dif·rəns }

return-flow burner |MECH ENG| A mechanical oil atomizer in a boiler furnace which regulates the amount of oil to be burned by the portion of oil recirculated to the point of storage. { ri'tərn ¦flō ,bər·nər }

return idler |MECH ENG| The idler or roller beneath the cover plates on which the conveyor belt rides after the load which it was carrying has been dumped. { ri'tərn ,Īd·lər }

return wall |BUILD| An interior wall of about the same height as the outside wall of a building; distinct from a partition or a low wall. { ri'tərn ,wȯl }

return wire |ELEC| The ground wire, common wire, or negative wire of a direct-current power circuit. { ri'tərn ,wīr }

reveal |BUILD| **1.** The side of an opening for a door or window, doorway, or the like, between the doorframe or window frame and the outer surface of the wall. **2.** The distance from the face of a door to the face of the frame on the pivot side. { ri'vēl }

reverberatory furnace |ENG| A furnace in which heat is supplied by burning of fuel in a space between the charge and the low roof. { ri'vər· brə,tȯr·ē ¦fər·nəs }

reverse bias |ELECTR| A bias voltage applied to a diode or a semiconductor junction with polarity such that little or no current flows; the opposite of forward bias. { ri'vərs 'bī·əs }

reverse Brayton cycle |THERMO| A refrigeration cycle using air as the refrigerant but with all system pressures above the ambient. Also known as dense-air refrigeration cycle. { ri'vərs 'brāt·ən ,sī·kəl }

reverse Carnot cycle |THERMO| An ideal thermodynamic cycle consisting of the processes of the Carnot cycle reversed and in reverse order, namely, isentropic expansion, isothermal expansion, isentropic compression, and isothermal compression. { ri'vərs kär'nō ,sī·kəl }

reverse current |ELECTR| Small value of direct current that flows when a semiconductor diode has reverse bias. { ri'vərs 'kə·rənt }

reversed air-blast process |CHEM ENG| A gas-making process in which, after a short period of the ordinary blow, the air blast is reversed so as to enter the top of the superheater, and passes back to the top of the generator and down { ri'vərst 'er ,blast ,prä·səs }

reverse engineering |ENG| The analysis of a completed system in order to isolate and identify its individual components or building blocks. { ri'vərs ,en·jə'nir·iŋ }

reverse feedback See negative feedback. { ri'vərs 'fēd,bak }

reverse flange |ENG| A flange made by shrinking. { ri'vərs 'flanj }

reverse lay |DES ENG| The lay of a wire rope with strands alternating in a right and left lay. { ri'vərs 'lā }

reverse osmosis |CHEM ENG| A technique used in desalination and waste-water treatment; pressure is applied to the surface of a saline (or waste) solution, forcing pure water to pass from the solution through a membrane (hollow fibers of cellulose acetate or nylon) that will not pass sodium or chloride ions. { ri'vərs äs'mō·səs }

reverse pitch |MECH ENG| A pitch on a propeller blade producing thrust in the direction opposite to the normal one. { ri'vərs 'pich }

reverse-printout typewriter |ENG| An automatic typewriter that eliminates conventional carriage return by typing one line from left to right and the next line from right to left. { ri'vərs ¦print,aut 'tīp,rīd·ər }

reverse-roll coating |ENG| Substrate coating that is premetered between rolls and then wiped off on the web; amount of coating is controlled by the metering gap and the rotational speed of the roll. { ri'vərs ¦rōl 'kōd·iŋ }

reverse voltage |ELEC| In the case of two opposing voltages, voltage of that polarity which produces the smaller current. { ri'vərs 'vōl·tij }

reversible capacitance |ELECTR| Limit, as the amplitude of an applied sinusoidal capacitor voltage approaches zero, of the ratio of the amplitude of the resulting in-phase fundamental-frequency component of transferred charge to the amplitude of the applied voltage, for a given constant bias voltage superimposed on the sinusoidal voltage. { ri'vər·sə·bəl kə'pas·əd·əns }

reversible engine |THERMO| An ideal engine which carries out a cycle of reversible processes. { ri'vər·sə·bəl 'en·jən }

reversible path |THERMO| A path followed by a thermodynamic system such that its direction of motion can be reversed at any point by an infinitesimal change in external conditions; thus the system can be considered to be at equilibrium at all points along the path. { ri'vər·sə·bəl 'path }

reversible-pitch propeller |MECH ENG| A type of controllable-pitch propeller; of either controllable or constant speed, it has provisions for reducing the pitch to and beyond the zero value, to the negative pitch range. { ri'vər·sə·bəl ¦pich prə'pel·ər }

reversible process |THERMO| An ideal thermodynamic process which can be exactly reversed by making an indefinitely small change in the external conditions. Also known as quasistatic process. { ri'vər·sə·bəl 'prä·səs }

reversible steering gear |MECH ENG| A steering gear for a vehicle which permits road shock and wheel deflections to come through the system and be felt in the steering control. { ri'vər·sə·bəl 'stir·iŋ ,gir }

reversible tramway See jig back. { ri'vər·sə·bəl 'tram,wā }

reversible transit circle |ENG| A transit circle that can be lifted out of its bearings and rotated through 180°, enabling systematic errors in both orientations to be determined. { ri'vər·sə·bəl 'tran·zət ,sər·kəl }

reversing thermometer |ENG| A mercury-in-glass thermometer which records temperature upon being inverted and thereafter retains its reading until returned to the first position. { ri'vərs·iŋ thər'mäm·əd·ər }

reversing water bottle See Nansen bottle. { ri'vərs·iŋ 'wód·ər ,bäd·əl }

reversion |CHEM ENG| In rubber manufacture, a decrease in rubber modulus or viscosity caused by overworking. { ri'vər·zhən }

revetment |CIV ENG| A facing made on a soil or rock embankment to prevent scour by weather or water. { rə'vet·mənt }

revolute-coordinate robot See jointed-arm robot. { 'rev·ə,lüt kō¦ord·ən·ət 'rō,bät }

revolute joint |MECH ENG| A robotic articulation consisting of a pin with one degree of freedom. { 'rev·ə,lüt ,jóint }

revolution |MECH| The motion of a body around a closed orbit. { ,rev·ə'lü·shən }

revolution counter |ENG| An instrument for registering the number of revolutions of a rotating machine. Also known as revolution indicator. { ,rev·ə'lü·shən ,kaúnt·ər }

revolution indicator See revolution counter. { ,rev·ə'lü·shən ,in·də,kād·ər }

revolution per minute |MECH| A unit of angular velocity equal to the uniform angular velocity of a body which rotates through an angle of 360° (2π radians), so that every point in the body returns to its original position, in 1 minute. Abbreviated rpm. { ,rev·ə'lü·shən pər 'min·ət }

revolution per second |MECH| A unit of angular velocity equal to the uniform angular velocity of a body which rotates through an angle of 360° (2π radians), so that every point in the body returns to its original position, in 1 second. Abbreviated rps. { ,rev·ə'lü·shən pər 'sek·ənd }

revolving-block engine |MECH ENG| Any of various engines which combine reciprocating piston motion with rotational motion of the entire engine block. { ri'välv·iŋ ¦bläk 'en·jən }

revolving door |BUILD| A door consisting of four leaves that revolve together on a central vertical axis within a circular vestibule. { ri'välv·iŋ 'dór }

revolving shovel |MECH ENG| A digging machine, mounted on crawlers or on rubber tires, that has the machinery deck and attachment on a vertical pivot so that it can swing freely. { ri'välv·iŋ 'shəv·əl }

Reynier's isolator |ENG| A mechanical barrier made of steel that surrounds the area in which germ-free vertebrates and accessory equipment are housed; has electricity for light and power, an exit-entry opening with a steam barrier, a means for sterile air exchange, glass viewing port, and neoprene gloves which allow handling of the animals. { rān'yās 'īs·ə,lād·ər }

Reynolds analogy |CHEM ENG| Relationship

showing the similarity between the transfer of mass, heat, and momentum. { 'ren·əlz ə,nal·ə·jē }

rf preheating See radio-frequency preheating. { ¦ärʲef prē'hēd·iŋ }

rheogoniometry |MECH| Rheological tests to determine the various stress and shear actions on Newtonian and non-Newtonian fluids. { ¦rē·ə·gō·nē'äm·ə·trē }

rheology |MECH| The study of the deformation and flow of matter, especially non-Newtonian flow of liquids and plastic flow of solids. { rē'äl·ə·jē }

rheometer |ENG| An instrument for determining flow properties of solids by measuring relationships between stress, strain, and time. { rē'äm·əd·ər }

rheostat |ELEC| A resistor constructed so that its resistance value may be changed without interrupting the circuit to which it is connected. Also known as variable resistor. { 'rē·ə,stat }

rheostatic braking |ENG| A system of dynamic braking in which direct-current drive motors are used as generators and convert the kinetic energy of the motor rotor and connected load to electrical energy, which in turn is dissipated as heat in a braking rheostat connected to the armature. { ¦rē·ə¦stad·ik 'brāk·iŋ }

rheostriction See pinch effect. { 'rē·ə,strik·shən }

rheotaxial growth |ENG| A chemical vapor deposition technique for producing silicon diodes and transistors on a fluid layer having high surface mobility. { ¦rē·ə¦tak·sē·əl 'grōth }

RIAA curve |ENG ACOUS| 1. Recording Industry Association of America curve representing standard recording characteristics for long-play records. 2. The corresponding equalization curve for playback of long-play records { ¦är¦ī¦ā¦ā 'kərv }

rib arch |CIV ENG| An arch consisting of ribs placed side by side and extending from the springings on one end to those on the other end. { 'rib ,ärch }

ribbed-clamp coupling |DES ENG| A rigid coupling which is split longitudinally and bored to shaft diameter, with a shim separating the two halves. { 'ribd ¦klamp 'kəp·liŋ }

ribbon |BUILD| A horizontal piece of wood nailed to the face of studs; usually used to support the floor joists. { 'rib·ən }

ribbon conveyor |MECH ENG| A type of screw conveyor which has an open space between the shaft and a ribbon-shaped flight, used for wet or sticky materials which would otherwise build up on the spindle. { 'rib·ən kən'vā·ər }

ribbon microphone |ENG ACOUS| A microphone whose electric output results from the motion of a thin metal ribbon mounted between the poles of a permanent magnet and driven directly by sound waves; it is velocity-actuated if open to sound waves on both sides, and pressure-actuated if open to sound waves on only one side. { 'rib·ən 'mī·krə,fōn }

ribbon mixer |MECH ENG| Device for the mixing of particles, slurries, or pastes of solids by the

revolution of an elongated helicoid (spiral) ribbon of metal. { 'rib·ən 'mik·sər }

riblet |DES ENG| Any of the small, longitudinal striations, with spacing on the order of 0.002 inch or 50 micrometers, that are made on the surfaces of ships or aircraft to reduce the drag of turbulent flow. { 'rib·lət }

Richardson automatic scale |ENG| An automatic weighing and recording machine for flowable materials carried on a conveyor; weighs batches from 200 to 1000 pounds (90 to 450 kilograms). { 'rich·ard·sən ¦ȯd·ə¦mad·ik 'skāl }

riddle |DES ENG| A sieve used for sizing or for removing foreign material from foundry sand or other granular materials. { 'rid·əl }

ridge board |BUILD| A horizontal board placed on edge at the apex of the roof. { 'rij ,bȯrd }

ridge cap |BUILD| Wood or metal cap which is placed over the angle of the ridge. { 'rij ,kap }

ridge pole |BUILD| The horizontal supporting member placed along the ridge of a roof. { 'rij ,pōl }

riffler |DES ENG| A small, curved rasp or file for filing interior surfaces or enlarging holes. { 'rif·lər }

rifle |DES ENG| A drill core that has spiral grooves on its outside surface. |ENG| A borehole that is following a spiral course. { 'rī·fəl }

rifling |MECH ENG| The technique of cutting helical grooves inside a rifle barrel to impart a spinning motion to a projectile around its long axis. { 'rīf·liŋ }

rift saw |DES ENG| 1. A saw for cutting wood radially from the log. 2. A circular saw divided into toothed arms for sawing flooring strips from cants. { 'rift ,sȯ }

rig |MECH ENG| A tripod, derrick, or drill machine complete with auxiliary and accessory equipment needed to drill. { rig }

right-and-left-hand chart |IND ENG| A graphic symbolic representation of the motions made by one hand in relation to those made by the other hand. { ¦rīt ən ¦left ,hand 'chärt }

right-cut tool |DES ENG| A single-point lathe tool which has the cutting edge on the right side when viewed face up from the point end. { 'rīt ¦kət ,tül }

right-hand cutting tool |DES ENG| A cutter whose flutes twist in a clockwise direction. { 'rīt ¦hand 'kəd·iŋ ,tül }

right-handed |DES ENG| 1. Pertaining to screw threads that allow coupling only by turning in a clockwise direction. 2. See right-laid. { 'rīt ¦han·dəd }

right-hand screw |DES ENG| A screw that advances when turned clockwise. { 'rīt ¦hand 'skrü }

right-laid |DES ENG| Rope or cable construction in which strands are twisted counterclockwise. Also known as right-handed. { 'rīt ¦lād }

right lang lay |DES ENG| Rope or cable in which the individual wires or fibers and the strands are twisted to the right. { 'rīt ¦laŋ ,lā }

right-of-way |CIV ENG| 1. Areas of land used for a road and along the side of the roadway. 2. A thoroughfare or path established for public use. **3.** Land occupied and used by a railroad or a public utility. { ¦rīt əv ¦wā }

rigid body |MECH| An idealized extended solid whose size and shape are definitely fixed and remain unaltered when forces are applied. { 'rij·id 'bäd·ē }

rigid-body dynamics |MECH| The study of the motions of a rigid body under the influence of forces and torques. { 'rij·id ¦bäd·ē dī'nam·iks }

rigid coupling |MECH ENG| A mechanical fastening of shafts connected with the axes directly in line. { 'rij·id 'kəp·liŋ }

rigid frame |BUILD| A steel skeleton frame in which the end connections of all members are rigid so that the angles they make with each other do not change. { 'rij·id 'frām }

rigidity |MECH| The quality or state of resisting change in form. { ri'jid·əd·ē }

rigidity modulus *See* modulus of elasticity in shear. { ri'jid·əd·ē ¸mäj·ə·ləs }

rigidizer |ENG| A supporting structure providing ridigity to an instrument that might otherwise be subject to undesirable vibrations. { ¸ri·jə'dīz·ər }

rigid pavement |CIV ENG| A thick portland cement pavement on a gravel base and subbase, with steel reinforcement and often with transverse joints. { 'rij·əd 'pāv·mənt }

rim |DES ENG| **1.** The outer part of a wheel, usually connected to the hub by spokes. **2.** An outer edge or border, sometimes raised or projecting. { rim }

rim-bearing swing bridge |CIV ENG| A swing bridge that is supported by a cylindrical girder on rollers. { ¦rim ¸ber·iŋ 'swiŋ ¸brij }

rim clutch |MECH ENG| A frictional contact clutch having surface elements that apply pressure to the rim either externally or internally. { 'rim ¸kləch }

rim drive |ENG ACOUS| A phonograph or sound recorder drive in which a rubber-covered drive wheel is in contact with the inside of the rim of the turntable. { 'rim ¸drīv }

ring |DES ENG| A tie member or chain link; tension or compression applied through the center of the ring produces bending moment, shear, and normal force on radial sections. { riŋ }

ring-and-ball test |CHEM ENG| A test for determining the melting point of asphalt, waxes, and paraffins in which a small ring is fitted with a test sample upon which a small ball is then placed; the melting point is that temperature at which the sample softens sufficiently to allow the ball to fall through the ring. Also known as ball and ring method. { ¦riŋ ən ¦bȯl ¸test }

ring-and-circle shear |DES ENG| A rotary shear designed for cutting circles and rings where the edge of the metal sheet cannot be used as a start. { ¦riŋ ən ¦sər·kəl ¸shir }

ringbolt |DES ENG| An eyebolt with a ring passing through the eye. { 'riŋ¸bōlt }

ring crusher |MECH ENG| Solids-reduction device with a rotor having loose crushing rings held outwardly by centrifugal force, which crush the feed by impact with the surrounding shell. { 'riŋ ¸krəsh·ər }

Ringelmann chart |ENG| A chart used in making subjective estimates of the amount of solid matter emitted by smoke stacks; the observer compares the grayness of the smoke with a series of shade diagrams formed by horizontal and vertical black lines on a white background. { 'riŋ·gəl¸män ¸chärt }

ring gage |DES ENG| A cylindrical ring of steel whose inside diameter is finished to gage tolerance and is used for checking the external diameter of a cylindrical object. { 'riŋ ¸gāj }

ring gate |CIV ENG| A type of gate used to regulate and control the discharge of a morning-glory spillway; like a drum gate, it offers a minimum of interference to the passage of ice or drift over the gate and requires no external power for operation. |ENG| An annular opening through which plastics enter the cavity of an injection or transfer mold. { 'riŋ ¸gāt }

ring gear |MECH ENG| The ring-shaped gear in an automobile differential that is driven by the propeller shaft pinion and transmits power through the differential to the line axle. { 'riŋ ¸gir }

ringing |CONT SYS| An oscillatory transient occurring in the output of a system as a result of a sudden change in input. { 'riŋ·iŋ }

ringing circuit |ELECTR| A circuit which has a capacitance in parallel with a resistance and inductance, with the whole in parallel with a second resistance; it is highly underdamped and is supplied with a step or pulse input. { 'riŋ·iŋ ¸sər·kət }

ringing time |ENG| In an ultrasonic testing unit, the length of time that the vibrations in a piezoelectric crystal remain after the generation of ultrasonic waves ceases. { 'riŋ·iŋ ¸tīm }

ring jewel |DES ENG| A type of jewel used as a pivot bearing in a time-keeping device, gyro, or instrument. { 'riŋ ¸jül }

ring job |MECH ENG| Installation of new piston rings on a piston. { 'riŋ ¸jäb }

ring laser *See* laser gyro. { 'riŋ ¸lā·zər }

ring lifter *See* split-ring core lifter. { 'riŋ ¸lif·tər }

ringlock nail |DES ENG| A nail ringed with grooves to provide greater holding power. { 'riŋ ¸läk ¸nāl }

ring-oil |MECH ENG| To oil (a bearing) by conveying the oil to the point to be lubricated by means of a ring, which rests upon and turns with the journal, and dips into a reservoir containing the lubricant. { 'riŋ ¸ȯil }

ring road *See* beltway. { 'riŋ ¸rōd }

ring-roller mill |MECH ENG| A grinding mill in which material is fed past spring-loaded rollers that apply force against the sides of a revolving bowl. Also known as roller mill. { 'riŋ ¦rōl·ər ¸mil }

riometer |ENG| An instrument that measures changes in ionospheric absorption of electromagnetic waves by determining and recording the level of extraterrestrial cosmic radio noise.

Derived from relative ionospheric opacity meter. { rī'äm·əd·ər }

rip |ENG| To saw wood with the grain. { rip }

ripbit See detachable bit; jackbit. { 'rip,bit }

ripping bar |DES ENG| A steel bar with a chisel at one end and a curved claw for pulling nails at the other. Also known as claw bar; wrecking bar. { 'rip·iŋ ,bär }

ripping punch |DES ENG| A tool with a rectangular cutting edge, used in a punch press to crosscut metal plates. { 'rip·iŋ ,pənch }

ripple |ELEC| The alternating-current component in the output of a direct-current power supply, arising within the power supply from incomplete filtering or from commutator action in a dc generator. { 'rip·əl }

riprap |CIV ENG| A foundation or revetment in water or on soft ground made of irregularly placed stones or pieces of boulders; used chiefly for river and harbor work, for roadway filling, and on embankments. { 'rip,rap }

ripsaw |MECH ENG| A heavy-tooth power saw used for cutting wood with the grain. { 'rip,sò }

rise and run |CIV ENG| The pitch of an inclined surface or member, usually expressed as the ratio of the vertical rise to the horizontal span. { ¦rīz ən 'rən }

riser |CHEM ENG| That portion of a bubble-cap assembly in a distillation tower that channels the rising vapor and causes it to flow downward to pass through the liquid held on the bubble plate. |CIV ENG| **1.** A board placed vertically beneath the tread of a step in a staircase. **2.** A vertical steam, water, or gas pipe. { 'rīz·ər }

riser plate |CIV ENG| A plate used to support a tapering switch rail above the base of the rail; used with a railroad gage or tie plate to maintain minimum gage. { 'rīz·ər ,plāt }

rise time |CONT SYS| The time it takes for the output of a system to change from a specified small percentage (usually 5 or 10) of its steady-state increment to a specified large percentage (usually 90 or 95). |ELEC| The time for the pointer of an electrical instrument to make 90% of the change to its final value when electric power suddenly is applied from a source whose impedance is high enough that it does not affect damping. { 'rīz ,tīm }

rising hinge |BUILD| A hinge that raises a door slightly as it is opened. { 'rīz·iŋ 'hinj }

risk |ENG| The potential realization of undesirable consequences from hazards arising from a possible event. { risk }

risk analysis |ENG| The scientific study of risk. { 'risk ə,nal·ə·səs }

risk management |ENG| The overall systematic approach to analyzing risk and implementing risk controls. { 'risk ,man·ij·mənt }

Ritchie's experiment |THERMO| An experiment that uses a Leslie cube and a differential air thermometer to demonstrate that the emissivity of a surface is proportional to its absorptivity. { 'rich·ēz ik,sper·ə·mənt }

Rittinger's law |MECH ENG| The law that energy needed to reduce the size of a solid particle is directly proportional to the resultant increase in surface area. { 'rit·ən·jərz ,lò }

river engineering |CIV ENG| A branch of transportation engineering consisting of the physical measures which are taken to improve a river and its banks. { 'riv·ər ,en·jə,nir·iŋ }

river gage |ENG| A device for measuring the river stage; types in common use include the staff gage, the water-stage recorder, and wire-weight gage. Also known as stream gage. { 'riv·ər ,gāj }

rivet |DES ENG| A short rod with a head formed on one end; it is inserted through aligned holes in parts to be joined, and the protruding end is pressed or hammered to form a second head. { 'riv·ət }

riveting |ENG| The permanent joining of two or more machine parts or structural members, usually plates, by means of rivets. { 'riv·əd·iŋ }

riveting hammer |MECH ENG| A hammer used for driving rivets. { 'riv·əd·iŋ ,ham·ər }

rivet pitch |ENG| The center-to-center distance of adjacent rivets. { 'riv·ət ,pich }

road |CIV ENG| An open way for travel and transportation. { rōd }

roadbed |CIV ENG| The earth foundation of a highway or a railroad. { 'rōd,bed }

road capacity |CIV ENG| The maximum traffic flow obtainable on a given roadway, using all available lanes, usually expressed in vehicles per hour or vehicles per day. { 'rōd kə,pas·əd·ē }

road grade |CIV ENG| The level and gradient of a road, measured along its center way. { 'rōd ,grād }

road net |CIV ENG| The system of roads available within a particular locality or area. { 'rōd ,net }

road octane number |ENG| A numerical value for automotive antiknock properties of a gasoline; determined by operating a car over a stretch of level road or on a chassis dynamometer under conditions simulating those encountered on the highway. { ¦rōd 'äk,tān ,nam·bər }

road test |ENG| A motor-vehicle test conducted on the highway or on a chassis dynamometer to determine the performance of fuels or lubricants or the performance of the vehicle. { 'rōd ,test }

roadway |CIV ENG| The portion of the thoroughfare over which vehicular traffic passes. { 'rōd,wā }

roaster |ENG| Equipment for the heating of materials, such as in pyrite roasting; a furnace. { 'rōs·tər }

roasting regeneration |CHEM ENG| Regeneration of a processing (treating) clay by heating or burning it in contact with air to remove combustible impurities adsorbed onto the surface. { 'rōst·iŋ rē,jen·ə'rā·shən }

Roberts evaporator See short-tube vertical evaporator. { 'räb·ərts i'vap·ə,rād·ər }

Roberts' linkage |MECH ENG| A type of approximate straight-line mechanism which provided, early in the 19th century, a practical means of making straight metal guides for the slides in a metal planner. { 'räb·ərts ,liŋ·kij }

Robins-Messiter system |MECH ENG| A stacking conveyor system in which material arrives on a conveyor belt and is fed to one or two wing conveyors. { 'räb·ənz 'mes·ə·tər ˌsis·təm }

Robitzsch actinograph |ENG| A pyranometer whose design utilizes three bimetallic strips which are exposed horizontally at the center of a hemispherical glass bowl; the outer strips are white reflectors, and the center strip is a blackened absorber; the bimetals are joined in such a manner that the pen of the instrument deflects in proportion to the difference in temperature between the black and white strips, and is thus proportional to the intensity of the received radiation; this instrument must be calibrated periodically. { 'rō,bitsh ak'tin·ə,graf }

robot |CONT SYS| A mechanical device that can be programmed to perform a variety of tasks of manipulation and locomotion under automatic control. { 'rō,bät }

robotics |IND ENG| The study of problems associated with the design, application, and control and sensory systems of self-controlled devices. { rō'bäd·iks }

roc |ELEC| A unit of electrical conductivity equal to the conductivity of a material in which an electric field of 1 volt per centimeter gives rise to a current density of 1 ampere per square centimeter. Derived from reciprocal ohm centimeter. { räk }

Roche lobes |MECH| **1.** Regions of space surrounding two massive bodies revolving around each other under their mutual gravitational attraction, such that the gravitational attraction of each body dominates the lobe surrounding it. **2.** In particular, the effective potential energy (referred to a system of coordinates rotating with the bodies) is equal to a constant V_0 over the surface of the lobes, and if a particle is inside one of the lobes and if the sum of its effective potential energy and its kinetic energy is less than V_0, it will remain inside the lobe. { 'rōch ,lōbz }

rock bit |ENG| Any one of many different types of roller bits used on rotary-type drills for drilling large-size holes in soft to medium-hard rocks. { 'räk ,bit }

rockbolt |ENG| A bar, usually constructed of steel, which is inserted into predrilled holes in rock and secured for the purpose of ground control. { 'räk,bōlt }

rock bolting |ENG| A method of securing or strengthening closely jointed or highly fissured rocks in mine workings, tunnels, or rock abutments by inserting and firmly anchoring rock bolts oriented perpendicular to the rock face or mine opening. { 'räk ,bōlt·iŋ }

rock channeler |MECH ENG| A machine used in quarrying for cutting an artificial seam in a mass of stone. { 'räk ,chan·əl·ər }

rock drill |MECH ENG| A machine for boring relatively short holes in rock for blasting purposes; motive power may be compressed air, steam, or electricity. { 'räk ,dril }

rocker |CIV ENG| A support at the end of a truss or girder which permits rotation and horizontal movement to allow for expansion and contraction. { 'räk·ər }

rocker arm |MECH ENG| In an internal combustion engine, a lever that is pivoted near its center and operated by a pushrod at one end to raise and depress the valve stem at the other end. { 'räk·ər ,ärm }

rocker bearing |CIV ENG| A bridge support that is free to rotate but cannot move horizontally. { 'räk·ər ,ber·iŋ }

rocker bent |CIV ENG| A bent used on a bridge span; hinged at one or both ends to provide for the span's expansion and contraction. { 'räk·ər ,bent }

rocker cam |MECH ENG| A cam that moves with a rocking motion. { 'räk·ər ,kam }

rocker panel |ENG| The part of the paneling on a passenger vehicle located below the passenger compartment doorsill. { 'räk·ər ,pan·əl }

rocketsonde See meteorological rocket. { 'räk·ət,sänd }

rocket station |ENG| A life-saving station equipped with line-carrying rocket apparatus. { 'räk·ət ,stā·shən }

rock-fill |CIV ENG| Composed of large, loosely placed rocks. { 'räk ,fil }

rock-fill dam |CIV ENG| A dam constructed of loosely placed rock or stone. { 'räk ,fil ,dam }

rocking furnace |MECH ENG| A horizonal, cylindrical melting furnace that is rolled back and forth on a geared cradle. { 'räk·iŋ ,fər·nəs }

rocking pier |CIV ENG| A pier that is hinged to allow for longitudinal expansion or contraction of the bridge. { 'räk·iŋ ,pir }

rocking valve |MECH ENG| An engine valve in which a disk or cylinder turns in its seat to permit fluid flow. { 'räk·iŋ ,valv }

rock pedestal See pedestal. { 'räk ,ped·ə·stəl }

Rockwell hardness |ENG| A measure of hardness of a material as determined by the Rockwell hardness test. { 'räk,wel 'härd·nəs }

Rockwell hardness test |ENG| One of the arbitrarily defined measures of resistance of a material to indentation under static or dynamic load; depth of indentation of either a steel ball or a 120° conical diamond with rounded point, 1/16, 1/8, 1/4, or 1/2 inch (1.5875, 3.175, 6.35, 12.7 millimeters) in diameter, called a brale, under prescribed load is the basis for Rockwell hardness; 60, 100, 150 kilogram load is applied with a special machine, and depth of impression under initial minor load is indicated on a dial whose graduations represent hardness number. { 'räk ,wel 'härd·nəs ,test }

rod |DES ENG| **1.** A bar whose end is slotted, tapered, or screwed for the attachment of a drill bit. **2.** A thin, round bar of metal or wood. See perch. { räd }

rod bit |DES ENG| A bit designed to fit a reaming shell that is threaded to couple directly to a drill rod. { 'räd ,bit }

rod coupling |DES ENG| A double-pin-thread coupling used to connect two drill rods together. { 'räd ,kəp·liŋ }

rodding |ENG| An operation in which a rod is passed through a length of tubing such as a rifle or pipework to determine if the bore is clear. { 'räd·iŋ }

rod level |ENG| A spirit level attached to a level rod or stadia rod to ensure the vertical position of the rod prior to instrument reading. { 'räd ,lev·əl }

rod mill |MECH ENG| A pulverizer operated by the impact of heavy metal rods. { 'räd ,mil }

rod string |MECH ENG| Drill rods coupled to form the connecting link between the core barrel and bit in the borehole and the drill machine at the collar of the borehole. { 'räd ,striŋ }

rod stuffing box |ENG| An annular packing gland fitting between the drill rod and the casing at the borehole collar; allows the rod to rotate freely but prevents the escape of gas or liquid under pressure. { 'räd 'stəf·iŋ ,bäks }

roentgen current |ELEC| An electric current arising from the motion of polarization charges, as in the rotation of a dielectric in a charged capacitor. { 'rent·gən ,kər·ənt }

Rogowski coil |ENG| A device for measuring alternating current without making contact with the current-carrying conductor, which consists of an air-core coil placed around the conductor in a toroidal fashion so that the alternating magnetic field produced by the current induces a voltage in the coil. { rə'gäv·skē ,kȯil }

rolamite mechanism |MECH ENG| An elemental mechanism consisting of two rollers contained by two parallel planes and bounded by a fixed S-shaped band under tension. { rō·lə,mīt ,mek·ə,niz·əm }

roll |MECH| Rotational or oscillatory movement of an aircraft or similar body about a longitudinal axis through the body; it is called roll for any degree of such rotation. |MECH ENG| A cylinder mounted in bearings; used for such functions as shaping, crushing, moving, or printing work passing by it. { rōl }

roll acceleration |MECH| The angular acceleration of an aircraft or missile about its longitudinal or X axis. { 'rōl ik,sel·ə,rā·shən }

roll axis |MECH| A longitudinal axis through an aircraft, rocket, or similar body, about which the body rolls. { 'rōl ,ak·səs }

roll bar |DES ENG| A metal bar installed overhead on a roofless automotive vehicle in order to protect the occupants if the car rolls over. { 'rōl ,bär }

roll cage |DES ENG| A frame of metal bars that is installed in a racing car around the driver's seat to protect the driver in the event of an accident. { 'rōl ,kāj }

roll control |ENG| The exercise of control over a missile so as to make it roll to a programmed degree, usually just before pitchover. { 'rōl kən,trōl }

roll crusher |MECH ENG| A crusher having one or two toothed rollers to reduce the material. { 'rōl ,krəsh·ər }

rolled joint |ENG| A joint made by expanding a tube in a tube sheet hole by use of an expander. { 'rōld 'jȯint }

roller |DES ENG| A cylindrical device for transmitting motion and force by rotation. { 'rō·lər }

roller analyzer |ENG| Device for quantitative separation of fine particles (down to 5 micrometers) by use of the graduated lift of a variable-rate pneumatic stream. { 'rō·lər ,an·ə,līz·ər }

roller bearing |MECH ENG| A shaft bearing characterized by parallel or tapered steel rollers confined between outer and inner rings. { 'rō·lər ,ber·iŋ }

roller bit See cone rock bit. { 'rō·lər ,bit }

roller cam follower |MECH ENG| A follower consisting of a rotatable wheel at the end of the shaft. { 'rō·lər 'kam ,fäl·ə·wər }

roller chain |MECH ENG| A chain drive assembled from roller links and pin links. { 'rō·lər ,chān }

roller coating |ENG| The application of paints, lacquers, or other coatings onto raised designs or letters by means of a roller. { 'rō·lər ,kōd·iŋ }

roller cone bit |ENG| A drilling bit containing two to four cutters (cones) mounted on very rugged bearings. Also known as bit cone; rock bit. { 'rōl·ər ,kōn ,bit }

roller conveyor |MECH ENG| A gravity conveyor having a track of parallel tubular rollers set at a definite grade, usually on antifriction bearings, at fixed locations, over which package goods which are sufficiently rigid to prevent sagging between rollers are moved by gravity or propulsion. { 'rō·lər kən,vā·ər }

roller drying |CHEM ENG| A method used to dry milk for purposes other than human consumption; concentrated milk is fed between two heated and narrowly spaced stainless steel rollers, the adhering thin film of milk dries as the rollers turn and is scraped off the roller by a doctor blade. { 'rō·lər ,drī·iŋ }

roller gate |CIV ENG| A cylindrical, usually hollow crest gate that is raised and lowered by large toothed wheels running on sloping racks. { 'rō·lər ,gāt }

roller-hearth kiln |ENG| A type of tunnel kiln through which the ware is conveyed on ceramic rollers. { 'rō·lər ,härth ,kil }

roller leveling |MECH ENG| Leveling flat stock by passing it through a machine having a series of rolls whose axes are staggered about a mean parallel path by a decreasing amount. { 'rō·lər 'lev·ə·liŋ }

roller mill See ring-roller mill. { 'rō·lər ,mil }

roller pulverizer |MECH ENG| A pulverizer operated by the crushing action of rotating rollers. { 'rō·lər 'pəl·və,rīz·ər }

roller stamping die |MECH ENG| An engraved roller used for stamping designs and other markings on sheet metal. { 'rō·lər 'stamp·iŋ ,dī }

rolling |MECH| Motion of a body across a surface combined with rotational motion of the body so that the point on the body in contact with the surface is instantaneously at rest. { 'rōl·iŋ }

rolling contact |MECH| Contact between bodies

rolling-contact bearing

such that the relative velocity of the two contacting surfaces at the point of contact is zero. { 'rōl·iŋ 'kän,takt }

rolling-contact bearing |MECH ENG| A bearing composed of rolling elements interposed between an outer and inner ring. { 'rōl·iŋ 'kän,takt 'ber·iŋ }

rolling door [ENG] A door that moves up and down or from side to side by means of wheels moving along a track. { 'rōl·iŋ 'dȯr }

rolling friction |MECH| A force which opposes the motion of any body which is rolling over the surface of another. { 'rōl·iŋ 'frik·shən }

rolling lift bridge |CIV ENG| A bridge having on the shore end of the lifting portion a segmental bearing that rolls on a flat surface. { 'rōl·iŋ 'lift ,brij }

rolling radius |DES ENG| For an automotive vehicle, the distance from the center of an axle to the ground. { 'rōl·iŋ ,rād·ē·əs }

roll mill |MECH ENG| A series of rolls operating at different speeds for grinding and crushing. { 'rōl ,mil }

roll-off |ELECTR| Gradually increasing loss or attenuation with increase or decrease of frequency beyond the substantially flat portion of the amplitude-frequency response characteristic of a system or transducer. { 'rōl ,ȯf }

roll set |ENG| A series of paired convex and concave contoured rolls in a roll forming machine that progressively form a workpiece of uniform cross section. { 'rōl ,set }

roll straightening |ENG| Unbending of metal stock by passing it through staggered rolls in different planes. { 'rōl ,strāt·ən·iŋ }

roll threading |MECH ENG| Threading a metal workpiece by rolling it either between grooved circular rolls or between grooved straight lines. { 'rōl ,thred·iŋ }

rom |ELEC| A unit of electrical conductivity, equal to the conductivity of a material in which an electric field of 1 volt per meter gives rise to a current density of 1 ampere per square meter. Derived from reciprocal ohm meter. { räm }

rood |MECH| A unit of area, equal to 1/4 acre, or 10,890 square feet, or 1011.7141056 square meters. { rüd }

roof |BUILD| The cover of a building or similar structure. { rüf }

roof beam |BUILD| A load-bearing member in the roof structure. { 'rüf ,bēm }

roof drain |BUILD| A drain for receiving water that has collected on the surface of a roof and discharging it into a downspout. { 'rüf ,drān }

roofing nail |DES ENG| A nail used for attaching paper or shingle to roof boards; usually short with a barbed shank and a large flat head. { 'rüf·iŋ ,nāl }

roof truss |BUILD| A truss used in roof construction; it carries the weight of roof deck and framing and of wind loads on the upper chord; an example is a Fink truss. { 'rüf ,trəs }

room |BUILD| A partitioned-off area inside a building or dwelling. { rüm }

root |CIV ENG| The portion of a dam which penetrates into the ground where the dam joins the hillside. |DES ENG| The bottom of a screw thread. { rüt }

root circle |DES ENG| A hypothetical circle defined at the bottom of the tooth spaces of a gear. { 'rüt ,sər·kəl }

rooter |ENG| A heavy plowing device equipped with teeth and used for breaking up the ground surface; a towed scarifier. { 'rüd·ər }

root fillet |DES ENG| The rounded corner at the angle of a gear tooth flank and the bottom land. { 'rüt ,fil·ət }

root locus plot |CONT SYS| A plot in the complex plane of values at which the loop transfer function of a feedback control system is a negative number. { 'rüt ¦lō·kəs ,plät }

root-mean-square current See effective current. { 'rüt ,mēn 'skwer 'kə·rənt }

Roots blower |MECH ENG| A compressor in which a pair of hourglass-shaped members rotate within a casing to deliver large volumes of gas at relatively low pressure increments. { 'rüts ,blō·ər }

rope-and-button conveyor |MECH ENG| A conveyor consisting of an endless wire rope or cable with disks or buttons attached at intervals. { ¦rōp ən ¦bət·ən kən,vā·ər }

rope boring |ENG| A method similar to rod drilling except that rigid rods are replaced by a steel rope to which the boring tools are attached and allowed to fall by their own weight. { 'rōp ,bȯr·iŋ }

rope drive |MECH ENG| A system of ropes running in grooved pulleys or sheaves to transmit power over distances too great for belt drives. { 'rōp ,drīv }

rope sheave |DES ENG| A grooved wheel, usually made of cast steel or heat-treated alloy steel, used for rope drives. { 'rōp ,shēv }

rope socket |DES ENG| A drop-forged steel device, with a tapered hole, which can be fastened to the end of a wire cable or rope and to which a load may be attached. { 'rōp ,säk·ət }

ropewalk |ENG| A long walkway down which a worker carries and lays rope in a manufacturing plant. { 'rōp,wȯk }

ropeway |ENG| One or a pair of steel cables between several supporting towers which serve as tracks for transporting materials in mountainous areas or at sea. { 'rōp,wā }

rose bit |DES ENG| A hardened steel or alloy noncore bit with a serrated face to cut or mill out bits, casing, or other metal objects lost in the hole. { 'rōz ,bit }

rose chucking reamer |DES ENG| A machine reamer with a straight or tapered shank and a straight or spiral flute; cutting is done at the ends of the teeth only; produces a rough hole since there are few teeth. { 'rōz 'chək·iŋ ,rē·mər }

rose reamer |DES ENG| A reamer designed to cut on the beveled leading ends of the teeth rather than on the sides. { 'rōz 'rē·mər }

Rossby diagram |THERMO| A thermodynamic

diagram, named after its designer, with mixing ratio as abscissa and potential temperature as ordinate; lines of constant equivalent potential temperature are added. { 'ròs·bē ,dī·ə,gram }

Ross feeder |MECH ENG| A chute for conveying bulk materials by means of a screen of heavy endless chains hung on a sprocket shaft; rotation of the shaft causes materials to slide. { 'ròs 'fēd·ər }

Rossman drive |ENG| A method used to provide speed control of alternating-current motors; an induction motor stator is mounted on trunnion bearings and driven with an auxiliary motor, to provide the desired change in slip between the stator and rotor. { 'ròs·mən ,drīv }

rotameter |ENG| A variable-area, constant-head, rate-of-flow volume meter in which the fluid flows upward through a tapered tube, lifting a shaped weight to a position where upward fluid force just balances its weight. { rō'tam·əd·ər }

rotary |MECH ENG| **1.** A rotary machine, such as a rotary printing press or a rotary well-drilling machine. **2.** The turntable and its supporting and rotating assembly in a well-drilling machine. { 'rōd·ə·rē }

rotary abutment meter |ENG| A type of positive displacement meter in which two displacement rotating vanes interleave with cavities on an abutment rotor in such a way that the three elements are geared together. { 'rōd·ə·rē ə'bət·mənt ,mēd·ər }

rotary actuator |MECH ENG| A device that converts electric energy into controlled rotary force; usually consists of an electric motor, gear box, and limit switches. { 'rōd·ə·rē 'ak·chə,wād·ər }

rotary air heater |MECH ENG| A regenerative air heater in which heat-transferring members are moved alternately through the gas and air streams. { 'rōd·ə·rē 'er ,hēd·ər }

rotary annular extractor |MECH ENG| Vertical, cylindrical shell with an inner, rotating cylinder; liquids to be contacted flow countercurrently through the annular space between the rotor and shell; used for liquid-liquid extraction processes. { 'rōd·ə·rē 'an·yə·lər ik'strak·tər }

rotary atomizer |MECH ENG| A hydraulic atomizer having the pump and nozzle combined. { 'rōd·ə·rē 'ad·ə,mīz·ər }

rotary belt cleaner |MECH ENG| A series of blades symmetrically spaced about the axis of rotation and caused to scrape or beat against the conveyor belt for the purpose of cleaning. { 'rōd·ə·rē 'belt ,klē·nər }

rotary blower |MECH ENG| Positive-displacement, rotating-impeller, air-movement device; can be straight-lobe, screw, sliding-vane, or liquid-piston type. { 'rōd·ə·rē 'blō·ər }

rotary boring |MECH ENG| A system of boring in which rock penetration is achieved by the rotation of the hollow cutting tool. { 'rōd·ə·rē 'bór·iŋ }

rotary bucket |MECH ENG| A 12- to 96-inch-diameter (30- to 244-centimeter) posthole augerlike device, the bottom end of which is equipped with cutting teeth used to rotary-drill

large-diameter shallow holes to obtain samples of soil lying above the groundwater level. { 'rōd·ə·rē 'bək·ət }

rotary-combustion engine See Wankel engine. { rōd·ə·rē kəm'bəs·chən ,en·jən }

rotary compressor |MECH ENG| A positive-displacement machine in which compression of the fluid is effected directly by a rotor and without the usual piston, connecting rod, and crank mechanism of the reciprocating compressor. { 'rōd·ə·rē kəm'pres·ər }

rotary crane |MECH ENG| A crane consisting of a boom pivoted to a fixed or movable structure. { 'rōd·ə·rē 'krān }

rotary crusher |MECH ENG| Solids-reduction device in which a high-speed rotating cone on a vertical shaft forces solids against a surrounding shell. { 'rōd·ə·rē 'krəsh·ər }

rotary-cup oil burner |ENG| Oil burner that uses centrifugal force to spray fuel oil from a rotary fuel atomizing cup into the combustion chamber. { 'rōd·ə·rē 'kəp 'òil ,bər·nər }

rotary cutter |MECH ENG| Device used to cut tough or fibrous materials by the shear action between two sets of blades, one set on a rotating holder, the other stationary on the surrounding casing. { 'rōd·ə·rē 'kəd·ər }

rotary-disk contactor |CHEM ENG| Liquid-liquid contactor, having a vertical cylindrical shell with vertical rotating shaft upon which are mounted a spaced series of flat disks; spinning of the disks forces liquid into shell-mounted baffles, causing mixing; used for liquid-liquid extraction processes. Also known as RDC extractor. { 'rōd·ə·rē 'disk 'kän,tak·tər }

rotary drill |MECH ENG| Any of various drill machines that rotate a rigid, tubular string of rods to which is attached a rock cutting bit, such as an oil well drilling apparatus. { 'rōd·ə·rē 'dril }

rotary drilling |MECH ENG| The act or process of drilling a borehole by means of a rotary-drill machine, such as in drilling an oil well. { 'rōd·ə·rē 'dril·iŋ }

rotary dryer |MECH ENG| A cylindrical furnace slightly inclined to the horizontal and rotated on suitable bearings; moisture is removed by rising hot gases. { 'rōd·ə·rē 'drī·ər }

rotary engine |MECH ENG| A positive displacement engine (such as a steam or internal combustion type) in which the thermodynamic cycle is carried out in a mechanism that is entirely rotary and without the more customary structural elements of a reciprocating piston, connecting rods, and crankshaft. { 'rōd·ə·rē 'en·jən }

rotary excavator See bucket-wheel excavator. { 'rōd·ə·rē 'ek·skə,vād·ər }

rotary feeder |MECH ENG| Device in which a rotating element or vane discharges powder or granules at a predetermined rate. { 'rōd·ə·rē 'fēd·ər }

rotary filter See drum filter. { 'rōd·ə·rē 'fil·tər }

rotary furnace |MECH ENG| A heat-treating furnace of circular construction which rotates the workpiece around the axis of the furnace during heat treatment; workpieces are transported through the furnace along a circular path. { 'rōd·ə·rē 'fər·nəs }

rotary kiln |ENG| A long cylindrical kiln lined with refractory, inclined at a slight angle, and rotated at a slow speed. { 'rōd·ə·rē 'kil }

rotary-percussive drill |MECH ENG| Drilling machine which operates as a rotary machine by the action of repeated blows to the bit. { 'rōd·ə·rē pər'kəs·iv 'dril }

rotary pump |MECH ENG| A displacement pump that delivers a steady flow by the action of two members in rotational contact. { 'rōd·ə·rē 'pəmp }

rotary roughening |MECH ENG| A metal preparation technique in which the workpiece surface is roughened by a cutting tool. { 'rōd·ə·rē 'rəf·ə·niŋ }

rotary shear |MECH ENG| A sheet-metal cutting machine having two rotary-disk cutters mounted on parallel shafts and driven in unison. { 'rōd·ə·rē 'shir }

rotary shot drill |MECH ENG| A rotary drill used to drill blastholes. { 'rōd·ə·rē 'shät ,dril }

rotary swager |MECH ENG| A machine for reducing diameter or wall thickness of a bar or tube by delivering hammerlike blows to the surface of the work supported on a mandrel. { 'rōd·ə·rē 'swā·jər }

rotary table |MECH ENG| A milling machine attachment consisting of a round table with T-shaped slots and rotated by means of a handwheel actuating a worm and worm gear. { 'rōd·ə·rē 'tā·bəl }

rotary vacuum filter See drum filter. { 'rōd·ə·rē 'vak·yəm ,fil·tər }

rotary valve |MECH ENG| A valve for the admission or release of working fluid to or from an engine cylinder where the valve member is a ported piston that turns on its axis. { 'rōd·ə·rē 'valv }

rotary-vane meter |ENG| A type of positive-displacement rate-of-flow meter having spring-loaded vanes mounted on an eccentric drum in a circular cavity; each time the drum rotates, a fixed volume of fluid passes through the meter. { 'rōd·ə·rē ¦vān 'mēd·ər }

rotary voltmeter |ENG| Type of electrostatic voltmeter used for measuring high voltages. { 'rōd·ə·rē 'vōlt,mēd·ər }

rotating-beam ceilometer |ENG| An electronic, automatic-recording meteorological device which determines cloud height by means of triangulation. { 'rō,tād·iŋ ¦bēm sē'läm·əd·ər }

rotating-coil gaussmeter |ENG| An instrument for measuring low magnetic field strengths and flux densities by measuring the voltage induced in a search coil that is rotated in the field at constant speed. { ,rō,tād·iŋ ,kȯil 'gaȯs,mēd·ər }

rotating coordinate system |MECH| A coordinate system whose axes as seen in an inertial coordinate system are rotating. { 'rō,tād·iŋ kō'ȯrd·ən·ət ,sis·təm }

rotating-drum heat transfer |CHEM ENG|

Procedure for solidifying layers of solids onto the outside surface of an inside-cooled drum that is partly immersed in a melt of the solids material. { 'rō,tād·iŋ ¦drəm 'hēt ,tranz·fər }

rotating meter See velocity-type flowmeter. { 'rō,tād·iŋ 'mēd·ər }

rotating spreader |ENG| Plastics-molding injection device consisting of a finned torpedo that is rotated by a shaft extending through a tubular cross-section injection ram behind it. { 'rō,tād·iŋ 'spred·ər }

rotating viscometer vacuum gage |ENG| Vacuum (reduced-pressure) measurement device in which the torque on a spinning armature is proportional to the viscosity (and the pressure) of the rarefied gas being measured; sensitive for absolute pressures of 1 millimeter of mercury (133.32 pascals), down to a few tens of micrometers. { 'rō,tād·iŋ vi'skäm·əd·ər 'vak·yəm ,gāj }

rotation |MECH| Also known as rotational motion. **1.** Motion of a rigid body in which either one point is fixed, or all the points on a straight line are fixed. **2.** Angular displacement of a rigid body. **3.** The motion of a particle about a fixed point. { rō'tā·shən }

rotational casting |ENG| Method to make hollow plastic articles from plastisols and lattices using a hollow mold rotated in one or two planes; the hot mold fuses the plastisol into a gel, which is then chilled and the product stripped out. Also known as rotational molding. { rō'tā·shən·əl 'kast·iŋ }

rotational energy |MECH| The kinetic energy of a rigid body due to rotation. { rō'tā·shən·əl 'en·ər·jē }

rotational impedance |MECH| A complex quantity, equal to the phasor representing the alternating torque acting on a system divided by the phasor representing the resulting angular velocity in the direction of the torque at its point of application. Also known as mechanical rotational impedance. { rō'tā·shən·əl im'pēd·əns }

rotational inertia See moment of inertia. { rō'tā·shən·əl i'nər·shə }

rotational molding See rotational casting. { rō'tā·shən·əl 'mōld·iŋ }

rotational reactance |MECH| The imaginary part of the rotational impedance. Also known as mechanical rotational reactance. { rō'tā·shən·əl rē'ak·təns }

rotational resistance |MECH| The real part of rotational impedance; it is responsible for dissipation of energy. Also known as mechanical rotational resistance. { rō'tā·shən·əl ri'zis·təns }

rotational stability |MECH| Property of a body for which a small angular displacement sets up a restoring torque that tends to return the body to its original position. { rō'tā·shən·əl stə'bil·əd·ē }

rotational strain |MECH| Strain in which the orientation of the axes of strain is changed. { rō'tā·shən·əl 'strān }

rotational traverse |MECH ENG| The maximum angle through which a body can rotate with one point of the body remaining fixed at an axis or center. { rō'tā·shən·əl trə'vərs }

rotational viscometer See Couette viscometer. { rō'tā·shən·əl vi'skäm·əd·ər }

rotation anemometer |ENG| A type of anemometer in which the rotation of an element serves to measure the wind speed; rotation anemometers are divided into two classes: those in which the axis of rotation is horizontal, as exemplified by the windmill anemometer; and those in which the axis is vertical, such as the cup anemometer. { rō'tā·shən ,an·ə'mäm·əd·ər }

rotation coefficients |MECH| Factors employed in computing the effects on range and deflection which are caused by the rotation of the earth; they are published only in firing tables involving comparatively long ranges. { rō'tā·shən ,kō·i,fish·əns }

rotation firing |ENG| Setting off explosions so that each hole throws its burden toward the space made by the preceding explosions. { rō'tā·shən ,fīr·iŋ }

rotation moment See torque. { rō'tā·shən ,mō·mənt }

rotator |MECH| A rotating rigid body. { 'rō ,tād·ər }

rotor |ELEC| The rotating member of an electrical machine or device, such as the rotating armature of a motor or generator, or the rotating plates of a variable capacitor. |MECH ENG| See impeller. { 'rōd·ər }

rough-axed brick See axed brick. { 'rəf ¦akst 'brik }

roughcast |CIV ENG| A rough finish on a surface; in particular, a plaster made of lime and shells or pebbles, applied by throwing it against a wall with a trowel. { 'rəf,kast }

rough cut |ENG| A heavy cut (or cuts) made before the finish cut, the primary object of which is the rapid removal of material. { 'rəf ,kət }

rough grinding |MECH ENG| Preliminary grinding without regard to finish. { 'rəf 'grīnd·iŋ }

rough hardware |ENG| Utility items such as nails, sash balances, and studs, without attractive finished appearance. { 'rəf 'härd,wer }

roughing |ENG| The start of evacuation of a vacuum system under test for leaks. { 'rəf·iŋ }

roughing tool |ENG| A single-point cutting tool having a sharp or small-radius nose, used for deep cuts and rapid material removal from the workpiece. { 'rəf·iŋ ,tül }

rough machining |MECH ENG| Preliminary machining without regard to finish. { 'rəf mə 'shēn·iŋ }

roughness-width cutoff |MECH ENG| The maximum width of surface irregularities included in roughness height measurements. { 'rəf·nəs ¦width 'kəd,óf }

rough threading |ENG| **1.** Rapid removal of the bulk of the material in a threading operation. **2.** Roughening a surface prior to hot-metal spraying to enhance adhesions. { 'rəf 'thred·iŋ }

rough turning |MECH ENG| The removal of excess stock from a workpiece as rapidly and efficiently as possible. { 'rəf 'tərn·iŋ }

round |ENG| A series of shots fired either simultaneously or with delay periods between them. { raúnd }

round-face bit |DES ENG| Any bit with a rounded cutting face. { 'raúnd ¦fās 'bit }

round file |DES ENG| A file having a circular cross section. { 'raúnd 'fīl }

round-head bolt |DES ENG| A bolt having a rounded head at one end. { 'raúnd ¦hed ,bōlt }

round-head buttress dam |CIV ENG| A mass concrete dam built of parallel buttresses thickened at the upstream end until they meet. { 'raúnd ¦hed 'bə·trəs ,dam }

roundnose chisel |DES ENG| A chisel having a rounded cutting edge. { 'raúnd¦nōs 'chiz·əl }

roundnose tool |DES ENG| A large-radius-nose cutting tool generally used in finishing operations. { 'raúnd¦nōs 'tül }

round strand rope |DES ENG| A rope composed generally of six strands twisted together or laid to form the rope around a core of hemp, sisal, or manila, or, in a wire-cored rope, around a central strand composed of individual wires. { 'raúnd ¦strand 'rōp }

round trip |ENG| The combined operations of entering and leaving a hole during drilling operations. { 'raúnd ¦trip }

rout |MECH ENG| To gouge out, make a furrow, or otherwise machine a wood member. { raút }

route locking |CIV ENG| Electrically locking in position switches, movable point frogs, or derails on the route of a train, after the train has passed a proceed signal. { 'rüt ,läk·iŋ }

router |DES ENG| **1.** A chisel with a curved point for cleaning out features such as grooves and mortises on wood members. **2.** See router plane. |MECH ENG| A machine tool with a rapidly rotating vertical spindle and cutter for making furrows, mortises, and similar grooves. { 'raúd·ər }

router plane |DES ENG| A plane for cutting grooves and smoothing the bottom of grooves. Also known as router. { 'raúd·ər ,plān }

route survey |CIV ENG| A survey for the design and construction of linear works, such as roads and pipelines. { 'rüt ,sər,vā }

Routh's procedure |MECH| A procedure for modifying the Lagrangian of a system so that the modified function satisfies a modified form of Lagrange's equations in which ignorable coordinates are eliminated. { 'rüths prə,sē·jər }

Routh's rule of inertia |MECH| The moment of inertia of a body about an axis of symmetry equals $M(a^2 + b^2)/n$, where M is the body's mass, a and b are the lengths of the body's two other perpendicular semiaxes, and n equals 3, 4, or 5 depending on whether the body is a rectangular parallelepiped, elliptic cylinder, or ellipsoid, respectively. { 'raúths 'rül əv i'nər·shə }

routing |ENG| A manufacturing process in which wooden parts are fabricated in various configurations; in high-speed industrial applications, an overhead cutting tool drills into the workpiece and then cuts the desired interior shape. { 'rüd·iŋ }

rowlock course |CIV ENG| A course of bricks laid on their sides so that only their ends are visible. { 'rō,läk ,kȯrs }

rpm See revolution per minute.

rps See revolution per second.

RTL See resistor-transistor logic.

rubber belt |DES ENG| A conveyor belt that consists essentially of a rubber-covered fabric; fabric is cotton, or nylon or other synthetic fiber, with steel-wire reinforcement. { 'rəb·ər 'belt }

rubber blanket |ENG| A rubber sheet used as a functional die in rubber forming. { 'rəb·ər 'blaŋ·kət }

rubber-covered steel conveyor |DES ENG| A steel conveyor band with a cover of rubber bonded to the steel. { 'rəb·ər ¦kəv·ərd 'stēl kən,vā·ər }

rubber plating |ENG| The laying down of a rubber coating onto metals by electrodeposition or by ionic coagulation. { 'rəb·ər 'plād·iŋ }

rubber wheel |DES ENG| A grinding wheel made with rubber as the bonding agent. { 'rəb·ər 'wēl }

rubble |CIV ENG| **1.** Rough, broken stones and other debris resulting from the deterioration and destruction of a building. **2.** Rough stone or brick used in coarse masonry or to fill the space in a wall between the facing courses. { 'rəb·əl }

rubble-mound structure |CIV ENG| A mound of nonselectively formed and placed stones which are protected with a covering layer of selected stones or of specially shaped concrete armored elements. { 'rəb·əl ¦maʊnd ,strək·chər }

rubidium magnetometer See rubidium-vapor magnetometer. { rü'bid·ē·əm ,mag·nə'täm·əd·ər }

rubidium-vapor magnetometer |ENG| A highly sensitive magnetometer in which the spin precession principle is combined with optical pumping and monitoring for detecting and recording variations as small as 0.01 gamma (0.1 microoersted) in the total magnetic field intensity of the earth. Also known as rubidium magnetometer. { rü'bid·ē·əm ¦vā·pər ,mag·nə'täm·əd·ər }

rudder |ENG| **1.** A flat, usually foil-shaped movable control surface attached upright to the stern of a boat, ship, or aircraft, and used to steer the craft. **2.** See rudder angle. { 'rəd·ər }

rudder angle |ENG| The acute angle between a ship or plane's rudder and its fore-and-aft line. Also known as rudder. { 'rəd·ər ,aŋ·gəl }

rule-based control system See direct expert control system.

rule of 80-20 See Pareto's law. { ¦rül əv 'ād·ē ,twen·tē }

ruler |ENG| A graduated strip of wood, metal, or other material, used to measure lines or as a guide in drawing lines. { 'rül·ər }

rumble See turntable rumble. { 'rəm·bəl }

run |BUILD| **1.** The horizontal distance from the face of a wall to the ridge of the roof. **2.** The width of a single tread in a stairway. **3.** The horizontal distance traversed by a flight of steps. **4.** The runway or track for a window. |CHEM ENG| **1.** The amount of feedstock processed by a petroleum refinery unit during a given time; often used colloquially in relation to the type of stock being processed, as in crude run or naphtha run. **2.** A processing-cycle or batch-treatment operation. |ENG| A portion of pipe or fitting lying in a straight line in the same direction of flow as the pipe to which it is connected. { rən }

run a line of soundings |ENG| To obtain a series of soundings along a course line. { 'rən ə ¦līn əv 'saʊnd·iŋz }

runaway effect |ELECTR| The phenomenon whereby an increase in temperature causes an increase in a collector-terminal current in a transistor, which in turn results in a higher temperature and, ultimately, failure of the transistor; the effect limits the power output of the transistor. { 'rən·ə,wā i,fekt }

runback |CHEM ENG| A pipe through which all or part of a distillation column's overhead condensate can be run back into the column, instead of being drawn off as product. |ENG| **1.** To retract the drill feed mechanism to its starting position. **2.** To drill slowly downward toward the bottom of the hole when the drill string has been lifted off-bottom for rechucking. { 'rən ,bak }

rundown line |CHEM ENG| A line from a process unit that connects the look box in the receiving house with the tank in which the product is temporarily stored. { 'rən,daʊn ,līn }

rundown tank |CHEM ENG| A tank in which the product from a still, agitator, or other processing equipment is received, and from which the product is pumped to larger storage tanks. Also known as pan tank; receiving tank. { 'rən ,daʊn ,taŋk }

Runge vector |MECH| A vector which describes certain unchanging features of a nonrelativistic two-body interaction obeying an inverse-square law, either in classical or quantum mechanics; its constancy is a reflection of the symmetry inherent in the inverse-square interaction. { 'rəŋ·ə ,vek·tər }

run in |ENG| To lower the assembled drill rods and auxiliary equipment into a borehole. { 'rən 'in }

runner |ENG| In a plastics injection or transfer mold, the channel (usually circular) that connects the sprue with the gate to the mold cavity. { 'rən·ər }

running block See traveling block. { 'rən·iŋ ,bläk }

running bond |CIV ENG| A masonry bond involving the placing of each brick as a stretcher and overlapping the bricks in adjoining courses. { 'rən·iŋ ¦bänd }

running fit |DES ENG| The intentional difference in dimensions of mating mechanical parts that permits them to move relative to each other. { 'rən·iŋ ¦fit }

running gear |MECH ENG| The means employed to support a truck and its load and to provide rolling-friction contact with the running surface. { 'rən·iŋ ,gir }

running-in |ENG| The process of operating new

or repaired machinery or equipment in order to detect any faults and to ensure smooth, free operation of parts before delivery.   { 'rən·iŋ 'in }

run-on *See* dieseling.   { 'rən‚ȯn }

run-out time |IND ENG| Time required by machine tools after cutting time is finished before tool and material are completely free of interference and before the start of the next sequence of operation.   { 'rən‚aút ‚tīm }

run-time data |MECH ENG| Information obtained from sensors during a machine's regular operation and used to improve its performance.   { 'rən ‚tīm 'dad·ə }

runway |CIV ENG| A straight path, often hard-surfaced, within a landing strip, normally used for landing and takeoff of aircraft.   { 'rən‚wā }

Rüping process |ENG| A system for preservative treatment of wood by using positive initial pressure, followed by introduction of the preservative and release of air, creating a vacuum.   { 'rüp·iŋ ‚prä·səs }

rupture disk device |MECH ENG| A nonreclosing pressure relief device which relieves the inlet static pressure in a system through the bursting of a disk.   { 'rəp·chər ‚disk di‚vīs }

Rushton-Oldshue column |CHEM ENG| A mixing unit used for continuous pipeline blending in which two-phase contacting is desired; it is a column containing separation plates, baffles, and mixing impellers.   { 'rəsh·tən 'ȯl‚shü ‚käl·əm }

Russell movable-wall oven |CHEM ENG| An oven for coal carbonization which cokes a 400-pound (180-kilogram) charge in a horizontal, 12-inch-wide (30-centimeter) chamber, heated from both sides, but with one side floating and balanced against scales.   { 'rəs·əl ‚müv·ə·bəl ‚wȯl ‚əv·ən }

rust joint |ENG| A joint to which some oxidizing agent is applied either to cure a leak or to withstand high pressure.   { 'rəst ‚jȯint }

rust prevention |ENG| Surface protection of ferrous structures or equipment to prevent formation of iron oxide; can be by coatings, surface treatment, plating, chemicals, cathodic arrangements, or other means.   { 'rəst pri‚ven·chən }

R-value |ENG| An index of the ability of a substance or material to retard the flow of heat; higher numerical values correspond to higher insulating ability.   { 'är ‚val·yü }

Rzeppa joint |MECH ENG| A special application of the Bendix-Weiss universal joint in which four large balls are transmitting elements, while a center ball acts as a spacer; it transmits constant angular velocity through a single universal joint.   { 'zhep·ə ‚jȯint }

S

S *See* siemens.

Sabathé's cycle |MECH ENG| An internal combustion engine cycle in which part of the combustion is explosive and part at constant pressure. { 'sä·bəˌtāz ˌsī·kəl }

saber saw |MECH ENG| A portable saw consisting of an electric motor, a straight saw blade with reciprocating mechanism, a handle, baseplate, and other essential parts. { 'sā·bər ˌsȯ }

saccharimeter |ENG| An instrument for measuring the amount of sugar in a solution, often by determining the change in polarization produced by the solution. { ˌsak·ə'rim·əd·ər }

saccharometer |ENG| An instrument for measuring the amount of sugar in a solution, by determining either the specific gravity or the gases produced by fermentation. { ˌsak·ə'räm·əd·ər }

sacrificial compliant substrate *See* compliant substrate. { ˌsak·rəˈfish·əl kəmˈplī·ənt 'səbˌsträt }

saddle |DES ENG| A support shaped to fit the object being held. { 'sad·əl }

saddle-type turret lathe |MECH ENG| A turret lathe designed without a ram and with the turret mounted directly on a support (saddle) which slides on the bedways of the lathe. { 'sad·əl ˌtīp 'tər·ət ˌlath }

SAE number |ENG| A classification of motor, transmission, and differential lubricants to indicate viscosities, standardized by the Society of Automotive Engineers; SAE numbers do not connote quality of the lubricant. { ˌeˌsā'ē ˌnəm·bər }

safe load |MECH| The stress, usually expressed in tons per square foot, which a soil or foundation can safely support. { 'sāf ˌlōd }

safety |ENG| Methods and techniques of avoiding accident or disease. { 'sāf·tē }

safety belt |ENG| A strong strap or harness used to fasten a person to an object, such as the seat of an airplane or automobile. { 'sāf·tē ˌbelt }

safety bolt |CIV ENG| A bolt that can be opened from only one side of the door or gate it fastens. { 'sāf·tē ˌbōlt }

safety can |ENG| A cylindrical metal container used for temporary storage or handling of flammable liquids, such as gasoline, naphtha, and benzine, in buildings not provided with properly constructed storage rooms; these cans are also used to transport such liquids for filling and supply purposes within local areas. { 'sāf·tē ˌkan }

safety chuck |DES ENG| Any drill chuck on which the heads of the set screws do not protrude beyond the outer periphery of the chuck. { 'sāf·tē ˌchək }

safety engineer |IND ENG| A person who inspects all possible danger spots in a factory, mine, or other industrial building or plant. { 'sāf·tē ˌen·jə'nir }

safety engineering |IND ENG| The testing and evaluating of equipment and procedures to prevent accidents. { 'sāf·tē ˌen·jə'nir·iŋ }

safety factor |ELEC| The amount of load, above the normal operating rating, that a device can handle without failure. |MECH| *See* factor of safety. { 'sāf·tē ˌfak·tər }

safety flange |DES ENG| A type of flange with tapered sides designed to keep a wheel intact in the event of accidental breakage. { 'sāf·tē ˌflanj }

safety fuse |ENG| A train of black powder which is enclosed in cotton, jute yarn, and waterproofing compounds, and which burns at the rate of 2 feet (60 centimeters) per minute; it is used mainly for small-scale blasting. { 'sāf·tē ˌfyüz }

safety hoist |MECH ENG| A hoisting gear that does not continue running when tension is released. { 'sāf·tē ˌhȯist }

safety hook |DES ENG| A hoisting hook with a spring-loaded latch that prevents the load from accidentally slipping off the hook. { 'sāf·tē ˌhůk }

safety level of supply |IND ENG| The quantity of material, in addition to the operating level of supply, required to be on hand to permit continuous operations in the event of minor interruption of normal replenishment or unpredictable fluctuations in demand. { 'sāf·tē 'lev·əl əv sə'plī }

safety match |ENG| A match that can be ignited only when struck against a specially made friction surface. { 'sāf·tē ˌmach }

safety plug |ENG| A protective device used on a heated pressure vessel (for example, a steam boiler), and containing a fusible element that melts at a predetermined safe temperature to prevent the buildup of excessive pressure. Also known as fusible plug. { 'sāf·tē ˌpləg }

safety rail *See* guardrail. { 'sāf·tē ˌrāl }

safety relief valve See safety valve. { 'sāf·tē ri'lēf ˌvalv }

safety shoe |ENG| A special shoe without spark-producing nails or plates, worn by personnel working around explosives. { 'sāf·tē ˌshü }

safety stop [MECH ENG] **1.** On a hoisting apparatus, a device by which the load may be prevented from falling. **2.** An automatic device on a hoisting engine designed to prevent overwinding. { 'sāf·tē ˌstäp }

safety time |IND ENG| The difference between the time when a certain material will be required and the time when the material will actually be in stock. { 'sāf·tē ˌtīm }

safety valve [MECH ENG] A spring-loaded, pressure-actuated valve that allows steam to escape from a boiler at a pressure slightly above the safe working level of the boiler; fitted by law to all boilers. Also known as safety relief valve. { 'sāf·tē ˌvalv }

safe yield [CIV ENG] The maximum dependable draft that can be made continuously upon a source of water supply over a given period of time during which the probable driest period, and therefore period of greatest deficiency in water supply, is likely to occur. { 'sāf 'yēld }

Saint Venant's compatibility equations |MECH| Equations for the components e_{ij} of the strain tensor that follow from their integrability, namely, $(e_{ij})_{kl} + (e_{kl})_{ij} - (e_{ik})_{jl} - (e_{jl})_{ik} = 0$, where i, j, k, and l can take on any of the values x, y, and z, and subscripts outside the parentheses indicate partial differentiation. { ˌsän·və'nänz kəmˌpad·ə'bil·əd·ē ˌi̥kwā·shənz }

Saint Venant's principle |MECH| The principle that the strains that result from application, to a small part of a body's surface, of a system of forces that are statically equivalent to zero force and zero torque become negligible at distances which are large compared with the dimensions of the part. { ˌsän·və'nänz 'prin·sə·pəl }

salamander stove |ENG| A small portable stove used for temporary or emergency heat; for example, on construction sites or in greenhouses. { 'sal·əˌman·dər 'stōv }

salimeter |ENG| A hydrometer graduated to read directly the percentage of salt in a solution such as brine. { sə'lim·əd·ər }

salina See saltworks. { sə'lē·nə }

saline-water reclamation |CHEM ENG| Purification and removal of salts from brine or brackish water by ion exchange, crystallization, distillation, evaporation, and reverse osmosis. { 'sā ˌlēn 'wôd·ər ˌrek·lə‚mā·shən }

salinity-temperature-depth recorder |ENG| An instrument consisting of sensing elements usually lowered from a stationary ship, and a recorder on board which simultaneously records measurements of temperature, salinity, and depth. Also known as CTD recorder; STD recorder. { sə'lin·əd·ē 'tem·prə·chər 'depth ri ˌkôrd·ər }

salinometer |ENG| An instrument that measures water salinity by means of electrical conductivity or by a hydrometer calibrated to give percentage of salt directly. { ˌsal·ə'näm·əd·ər }

salt |ENG| To add an accelerator or retardant to cement. { sôlt }

salt-effect distillation |CHEM ENG| A process of extractive distillation in which a salt that is soluble in the liquid phase of the system being separated is used as a separating agent. { 'sôlt i‚fekt ˌdis·tə‚lā·shən }

saltern See salt garden; saltworks. { 'sôl·tərn }

salt garden |ENG| A large, shallow basin or pond where sea water is evaporated by solar heat. Also known as saltern. { 'sôlt ˌgärd·ən }

salt glaze |ENG| Glaze formed on the surface of stoneware by putting salt into the kiln during firing. { 'sôlt ˌglāz }

salt-gradient solar pond See solar pond. { 'sôlt ˌgrād·ē·ənt ˌsō·lər 'pänd }

salt grainer |CHEM ENG| Type of evaporative crystallizer in which the solution is kept hot, and supersaturation is developed by evaporation rather than by cooling. { 'sôlt ˌgrān·ər }

salting-out effect |CHEM ENG| The growth of crystals of a substance on heated, liquid-holding surfaces of a crystallizing evaporator as a result of the decrease in solubility of the substance with increase in temperature. { 'sôl·tiŋ ˌaut i‚fekt }

salt velocity meter |ENG| A rate-of-flow volume meter used to find the transit time of passage between two fixed points of a small quantity of salt or radioactive isotope in a flowing stream by measuring electrical conductivity or radiation level at those points. { 'sôlt və'läs·əd·ē ˌmēd· ər }

salt well |ENG| A bored or driven well from which brine is obtained. { 'sôlt ˌwel }

saltworks |ENG| A building or group of buildings where salt is produced commercially, as by extraction from sea water or from the brine of salt springs. Also known as salina; saltern. { 'sôlt‚wərks }

salvage procedure |ENG| The recovery, evacuation, and reclamation of damaged, discarded, condemned, or abandoned material, ships, craft, and floating equipment for reuse, repair, refabrication, or scrapping. { 'sal·vij ˌprə‚sē·jər }

salvage value |ENG| **1.** The cost that could be recovered from the sale of used equipment when removed or scrapped. **2.** The actual market value of a specific facility or equipment at a particular point in time. { 'sal·vij ˌval·yü }

sample-and-hold circuit |ELECTR| A circuit that measures an input signal at a series of definite times, and whose output remains constant at a value corresponding to the most recent measurement until the next measurement is made. { ˌsam·pəl ən 'hōld ˌsər·kət }

sampled-data control system |CONT SYS| A form of control system in which the signal appears at one or more points in the system as a sequence of pulses or numbers usually equally spaced in time. { 'sam·pəld ¦dad·ə kən'trōl ,sis·təm }

sample log |ENG| Record of core samples or drill cuttings; gives geological, visual, and hydrocarbon-content record versus depth of drilling. { 'sam·pəl ,läg }

sampler |CONT SYS| A device, used in sampled-data control systems, whose output is a series of impulses at regular intervals in time; the height of each impulse equals the value of the continuous input signal at the instant of the impulse. |ENG| A mechanical or other device designed to obtain small samples of materials for analysis; used in biology, chemistry, and geology. { 'sam·plər }

sample splitter |ENG| An instrument, generally constructed of acrylic resin, designed to subdivide a total sample of marine plankton while maintaining a quantitatively correct relationship between the various phyla in the sample. { 'sam·pəl ,splid·ər }

sampling |ENG| Process of obtaining a sequence of instantaneous values of a wave. { 'sam·pliŋ }

sampling bottle |ENG| A cylindrical container, usually closed at a chosen depth, to trap a water sample and transport it to the surface without introducing contamination. { 'sam·pliŋ ,bäd·əl }

sampling gate |ELECTR| A gate circuit that extracts information from the input waveform only when activated by a selector pulse. { 'sam·pliŋ ,gāt }

sampling interval |CONT SYS| The time between successive sampling pulses in a sampled-data control system. { 'sam·pliŋ ,in·tər·vəl }

sampling plan |IND ENG| A plan stating sample sizes and the criteria for accepting or rejecting items or taking another sample during inspection of a group of items. { 'sam·pliŋ ,plan }

sampling probe |ENG| A leak-testing probe which collects tracer gas from the test area of an object under pressure and feeds it to the leak detector at reduced pressure. { 'sam·pliŋ ,prōb }

sampling process |ENG| The process of obtaining a sequence of instantaneous values of some quantity that varies continuously with time. { 'sam·pliŋ ,prä·səs }

sampling rate |ENG| The rate at which measurements of physical quantities are made; for example, if it is desired to calculate the velocity of a missile and its position is measured each millisecond, then the sampling rate is 1000 measurements per second. { 'sam·pliŋ ,rāt }

sampling risk |IND ENG| In inspection procedure, the probability, under the sampling plan used, that acceptable material will be rejected or that unsatisfactory material will be accepted. { 'sam·pliŋ ,risk }

sampling synthesis |ENG ACOUS| Any method of synthesizing musical tones that is based on playing back digitally recorded sounds. { 'sam·pliŋ ,sin·thə·səs }

sampling time |ENG| The time between successive measurements of a physical quantity. { 'sam·pliŋ ,tīm }

sampling voltmeter |ENG| A special type of voltmeter that detects the instantaneous value of an input signal at prescribed times by means of an electronic switch connecting the signal to a memory capacitor; it is particularly effective in detecting high frequency signals (up to 12 gigahertz) or signals mixed with noise. { 'sam·pliŋ 'vōlt,mēd·ər }

samson post See king post. { 'sam·sən ,pōst }

sandbag |ENG| A bag filled with sand; used to build temporary protective walls. { 'san,bag }

sandblasting |ENG| Surface treatment in which steel grit, sand, or other abrasive material is blown against an object to produce a roughened surface or to remove dirt, rust, and scale. { 'san ,blast·iŋ }

sand drain |CIV ENG| A vertical boring through a clay or silty soil filled with sand or gravel to facilitate drainage. { 'san ,drān }

sander |MECH ENG| **1.** An electric machine used to sand the surface of wood, metal, or other material. **2.** A device attached to a locomotive or electric rail car which sands the rails to increase friction on the driving wheels. { 'san·dər }

sand filter |CIV ENG| A filter consisting of graded layers of sand and aggregate for purifying domestic water. { 'san ,fil·tər }

sand finish |ENG| A smooth finish on a plaster surface made by rubbing the sand or mortar coat. { 'san ,fin·ish }

sand heap analogy See sand hill analogy. { 'sand ,hēp ə,nal·ə·jē }

sand hill analogy |MECH| A formal identity between the differential equation and boundary conditions for a stress function for torsion of a perfectly plastic prismatic bar, and those for the height of the surface of a granular material, such as dry sand, which has a constant angle of rest. Also known as sand heap analogy. { 'sand ,hil ə,nal·ə·jē }

sandhog |ENG| A worker in compressed-air environments, as in driving tunnels by means of pneumatic caissons. { 'san,häg }

sanding |ENG| **1.** Covering or mixing with sand. **2.** Smoothing a surface with sandpaper or other abrasive paper or cloth. { 'sand·iŋ }

sand line |ENG| A wire line used to raise and lower a bailer or sand pump to remove cuttings from a borehole. { 'san ,līn }

sand mill |MECH ENG| Variation of a ball-type size-reduction mill in which grains of sand serve as grinding balls. { 'san ,mil }

sand pile |CIV ENG| A compacted filling of sand in a deep round hole formed by ramming the sand with a pile; used for foundations in soft soil. { 'san ,pīl }

sandpit |CIV ENG| An excavation dug in sand, especially as a source of sand for construction materials. { 'san,pit }

sand pump |MECH ENG| A pump, usually a centrifugal type, capable of handling sand- and

477

gravel-laden liquids without clogging or wearing unduly; used to extract mud and cuttings from a borehole. Also known as sludge pump. { 'san ,pəmp }

sand reel |MECH ENG| A drum, operated by a band wheel, for raising or lowering the sand pump or bailer during drilling operations. Also known as coring reel. { 'san ,rēl }

sand slinger |MECH ENG| A machine which delivers sand to and fills molds at high speed by centrifugal force. { 'san ,sliŋ·ər }

sand trap |ENG| A device in a conduit for trapping sand or soil particles carried by the water. { 'san ,trap }

sand wheel |MECH ENG| A wheel fitted with steel buckets around the circumference for lifting sand or sludge out of a sump to stack it at a higher level. { 'san ,wēl }

sandwich beam See flitch girder. { 'san,wich ,bēm }

sandwich construction |DES ENG| Composite construction of alloys, plastics, wood, or other materials consisting of a foam or honeycomb layer laminated and glued between two hard outer sheets. Also known as sandwich laminate. { 'san,wich kən,strək·shən }

sandwich heating |ENG| Method for heating both sides of a thermoplastic sheet simultaneously prior to forming or shaping. { 'san,wich ,hēd·iŋ }

sandwich laminate See sandwich construction. { 'san,wich 'lam·ə·nət }

sandwich molding See coinjection molding. { 'san,wich ,mōld·iŋ }

sanitary engineering |CIV ENG| A field of civil engineering concerned with works and projects for the protection and promotion of public health. { 'san·ə,ter·ē ,en·jə'nir·iŋ }

sanitary landfill |CIV ENG| The disposal of garbage by spreading it in layers covered with soil or ashes to a depth sufficient to control rats, flies, and odors. { 'san·ə,ter·ē 'lan,fil }

sanitary sewer |CIV ENG| A sewer which is restricted to carrying sewage and to which storm and surface waters are not admitted. { 'san·ə,ter·ē 'sü·ər }

sanitation |CIV ENG| The act or process of making healthy environmental conditions. { ,san·ə'tā·shən }

Sargent cycle |THERMO| An ideal thermodynamic cycle consisting of four reversible processes: adiabatic compression, heating at constant volume, adiabatic expansion, and isobaric cooling. { 'sär·jənt ,sī·kəl }

sarking |BUILD| A layer of boards or bituminous felt placed beneath tiles or other roofing to provide thermal insulation or to prevent ingress of water. { 'särk·iŋ }

SASAR See segmented aperture-synthetic aperture radar. { 'sā,sär }

sash |BUILD| A frame for window glass. { sash }

sash bar |BUILD| One of the strips of wood or metal that separate the panes of glass in a window. Also known as glazing bar; muntin; window bar. { 'sash ,bär }

sash cord |BUILD| A cord or chain used to attach a counterweight to the window sash. { 'sash ,kórd }

satellite and missile surveillance |ENG| The systematic observation of aerospace for the purpose of detecting, tracking, and characterizing objects, events, and phenomena associated with satellites and inflight missiles, friendly and enemy. { 'sad·əl,īt ən ¦mis·əl sər'vā·ləns }

saturable-core magnetometer |ENG| A magnetometer that depends for its operation on the changes in permeability of a ferromagnetic core as a function of the magnetic field to be measured. { 'sach·rə·bəl ¦kór ,mag·nə'täm·əd·ər }

saturated vapor |THERMO| A vapor whose temperature equals the temperature of boiling at the pressure existing on it. { 'sach·ə,rād·əd 'vā·pər }

saturation |ELECTR| 1. The condition that occurs when a transistor is driven so that it becomes biased in the forward direction (the collector becomes positive with respect to the base, for example, in a *pnp* type of transistor). 2. See anode saturation; temperature saturation. { ,sach·ə'rā·shən }

saturation specific humidity |THERMO| A thermodynamic function of state; the value of the specific humidity of saturated air at the given temperature and pressure. { ,sach·ə'rā·shən spə'sif·ik hyü'mid·əd·ē }

saturation vapor pressure |THERMO| The vapor pressure of a thermodynamic system, at a given temperature, wherein the vapor of a substance is in equilibrium with a plane surface of that substance's pure liquid or solid phase. { ,sach·ə'rā·shən 'vā·pər ,presh·ər }

saturator |ENG| A device, equipment, or person that saturates one material with another; examples are a tank in which vapors become saturated with ammonia from coal (in carbonization of coal), a humidifier, and the operator of a machine for impregnating roofing felt with asphalt. { ,sach·ə,rād·ər }

Saunders air-lift pump |MECH ENG| A device for raising water from a well by the introduction of compressed air below the water level in the well. { 'sòn·dərz 'er ,lift ,pəmp }

sauterelle |ENG| A device used by masons for tracing and forming angles. { ,sòd·ə'rel }

Savonius rotor |MECH ENG| A rotor composed of two offset semicylindrical elements rotating about a vertical axis. { sə'vō·nē·əs 'rōd·ər }

Savonius windmill |MECH ENG| A windmill composed of two semicylindrical offset cups rotating about a vertical axis. { sə'vō·nē·əs 'win,mil }

saw |DES ENG| 1. Any of various tools consisting of a thin, usually steel, blade with continuous cutting teeth on the edge. 2. Any similar device or tool, such as a rotating disc, in which a sharp continuous edge replaces the teeth. { sò }

saw bit | DES ENG| A bit having a cutting edge formed by teeth shaped like those in a handsaw. { 'sȯ ,bit }

saw gumming | MECH ENG| Grinding away the punch marks in the spaces between the teeth in saw manufacture. { 'sȯ ,gəm·iŋ }

sawhorse | ENG| A wooden rack used to support wood that is being sawed. { 'sȯ,hȯrs }

sawing | ENG| Cutting with a saw. { 'sȯ iŋ }

sawmill | IND ENG| A plant that houses sawing machines. | MECH ENG| A machine for cutting logs with a saw or a series of saws. { 'sȯ,mil }

sawtooth barrel See basket. { 'sȯ,tüth 'bar·əl }

sawtooth crusher | MECH ENG| Solids crusher in which feed is broken down between two saw-toothed shafts rotating at different speeds. { 'sȯ,tüth 'krəsh·ər }

sawtooth waveform | ELECTR| A waveform characterized by a slow rise time and a sharp fall, resembling a tooth of a saw. { 'sȯ,tüth 'wāv ,fȯrm }

sax | DES ENG| A tool for chopping away the edges of roof slates; it has a pick at one end for making nail holes. { saks }

Saybolt color | ENG| A color standard for petroleum products determined with a Saybolt chromometer. { 'sā,bȯlt ,kəl·ər }

Saybolt Furol viscosimeter | ENG| An instrument for measuring viscosity of very thick fluids, for example, heavy oils; similar to the Saybolt Universal viscosimeter, but with a larger-diameter tube so that the efflux time is about one-tenth that of the Universal instrument. { 'sā ,bȯlt 'fyu̇,rȯl ,vis·kə'sim·əd·ər }

Saybolt Universal viscosimeter | ENG| An instrument for measuring viscosity by the time it takes a fluid to flow through a calibrated tube; used for the lighter petroleum products and lubricating oils. { 'sā,bȯlt ,yü·nə'vər·səl ,vis·kə'sim·əd·ər }

scab | BUILD| A short, flat piece of lumber that is used to splice two pieces of wood set at right angles to each other. { skab }

SCADA See supervisory control and data acquisition. { 'skad·ə or ‚es¦sē¦a¦dē'ā }

scaffold | CIV ENG| A temporary or movable platform supported on the ground or suspended; used for working at considerable heights above the ground. { 'ska,fȯld }

scale | ENG| **1.** A series of markings used for reading the value of a quantity or setting. **2.** To change the magnitude of a variable in a uniform way, as by multiplying or dividing by a constant factor, or the ratio of the real thing's magnitude to the magnitude of the model or analog of the model. **3.** A weighing device. **4.** A ruler or other measuring stick. **5.** A dense deposit bonded on the surface of a tube in a heat exchanger or on the surface of an evaporating device. { skāl }

scale factor | ENG| The factor by which the reading of an instrument or the solution of a problem should be multiplied to give the true final value

when a corresponding scale factor is used initially to bring the magnitude within the range of the instrument or computer. { 'skāl ,fak·tər }

scaler | ELECTR| A circuit that produces an output pulse when a prescribed number of input pulses is received. Also known as counter; scaling circuit. { skāl·ər }

scale-up | DES ENG| Design process in which the data of an experimental-scale operation (model or pilot plant) is used for the design of a large (scaled-up) unit, usually of commercial size. | IND ENG| Transfer of a new process from a pilot plant operation to production at commercial levels. { 'skāl,əp }

scaling | ELECTR| Counting pulses with a scaler when the pulses occur too fast for direct counting by conventional means. | ENG| Removing scale (rust or salt) from a metal or other surface. | MECH| Expressing the terms in an equation of motion in powers of nondimensional quantities (such as a Reynolds number), so that terms of significant magnitude under conditions specified in the problem can be identified, and terms of insignificant magnitude can be dropped. { 'skāl·iŋ }

scaling circuit See scaler. { 'skāl·iŋ ,sər·kət }

scaling factor | ELECTR| The number of input pulses per output pulse of a scaling circuit. Also known as scaling ratio. | ENG| Factor used in heat-exchange calculations to allow for the loss in heat conductivity of a material because of the development of surface scale, as inside pipelines and heat-exchanger tubes. { 'skāl·iŋ ,fak·tər }

scaling ratio | ELECTR| See scaling factor. | ENG| The ratio of a certain property of a laboratory model to the same property in the natural prototype. { 'skāl·iŋ ,rā·shō }

scalpel | DES ENG| A small, straight, very sharp knife (or detachable blade for a knife), used for dissecting. { 'skal·pəl }

scan | ELECTR| The motion, usually periodic, given to the major lobe of an antenna; the process of directing the radio-frequency beam successively over all points in a given region of space. | ENG| **1.** To examine an area, a region in space, or a portion of the radio spectrum point by point in an ordered sequence; for example, conversion of a scene or image to an electric signal or use of radar to monitor an airspace for detection, navigation, or traffic control purposes. **2.** One complete circular, up-and-down, or left-to-right sweep of the radar, light, or other beam or device used in making a scan. { skan }

scanner | ENG| **1.** Any device that examines an area or region point by point in a continuous systematic manner, repeatedly sweeping across until the entire area or region is covered; for example, a flying-spot scanner. **2.** A device that automatically samples, measures, or checks a number of quantities or conditions in sequence, as in process control. { 'skan·ər }

scanning proton microprobe | ENG| An instrument used for determining the spatial distribution of trace elements in samples, in which a beam of energetic protons is focused on a narrow spot which is swept over the sample, and the characteristic x-rays emitted from the target are measured. { 'skan·iŋ 'prō,tän 'mī·krə,skōp }

scanning radiometer [ENG] An image-forming system consisting of a radiometer which, by the use of a plane mirror rotating at 45° to the optical axis, can see a circular path normal to the instrument. { 'skan·iŋ ,rād·ē'äm·əd·ər }

scanning sequence [ENG] The order in which the points in a region are scanned; for example, in television the picture is scanned horizontally from left to right and vertically from top to bottom. { 'skan·iŋ ,sēk·wəns }

scanning sonar [ENG] Sonar in which all targets of interest are shown simultaneously, as on a radar PPI (plan position indicator) display or sector display; the sound pulse may be transmitted in all directions simultaneously and picked up by a rotating receiving transducer, or transmitted and received in only one direction at a time by a scanning transducer. { 'skan·iŋ 'sō,när }

scantlings [BUILD] Sections of timber measuring less than 8 inches (20 centimeters) wide and from 2 to 6 inches (5 to 15 centimeters) thick; used for studding. { 'skant·liŋz }

scarf joint [DES ENG] A joint made by the cutting of overlapping mating parts so that the joint is not enlarged and the patterns are complementary, and securing them by glue, fasteners, welding, or other joining method. { 'skärf ,jöint }

scarifier [ENG] An implement or machine with downward projecting tines for breaking down a road surface 2 feet (60 centimeters) or less. { 'skär·ə,fī·ər }

scatterometer [ENG] A microwave sensor that is essentially a radar without ranging circuits, used to measure only the reflection or scattering coefficient while scanning the surface of the earth from an aircraft or a satellite. { ,skad·ə'räm·əd·ər }

scavenging [MECH ENG] Removal of spent gases from an internal combustion engine cylinder and replacement by a fresh charge or air. { 'skav·ən·jiŋ }

scenario-based design [SYS ENG] A family of techniques in which the use of a future system is concretely described at an early point in the development process, and narrative descriptions of the envisage usage episodes are then employed in a variety of ways to guide the development of the system. { sə¦ner·ē·ō ,bäst di'zīn }

scend [ENG] **1.** The upward motion of the bow and stern of a vessel associated with pitching. **2.** The lifting of the entire vessel by waves or swell. Also known as send. { send }

scheduling [IND ENG] A decision-making function that plays an important role in most manufacturing and service industries and often allows an organization to operate with a minimum of resources. Scheduling is applied in procurement and production, in transportation and distribution, and in information processing and communication. In manufacturing, the scheduling function coordinates the flow of parts and products through the system, and balances the workload on machines and personnel, departments, and the entire plant. { 'skej·əl·iŋ }

Scheffel engine [MECH ENG] A type of multirotor engine that uses nine approximately equal rotors turning in the same clockwise sense. { 'shef·əl ,en·jēn }

Scheibel column See Scheibel extractor. { 'shī·bəl ,käl·əm }

Scheibel extractor [CHEM ENG] Liquid-liquid contact vessel used in liquid-liquid extraction processes: a vertical cylinder with interspersed open spaces and wire-mesh packing along its height, with liquid agitators in the open spaces, or a vertical cylinder fully filled with wire-mesh packing. Also known as Scheibel column; Scheibel-York extractor; York-Scheibel column. { 'shī·bəl ik,strak·tər }

Scheibel-York extractor See Scheibel extractor. { 'shī·bəl 'yörk ik,strak·tər }

schematic circuit diagram See circuit diagram. { ski'mad·ik 'sər·kət ,dī·ə,gram }

Schleiermacher's method [THERMO] A method of determining the thermal conductivity of a gas, in which the gas is placed in a cylinder with an electrically heated wire along its axis, and the electric energy supplied to the wire and the temperatures of wire and cylinder are measured. { 'shlī·ər,mäk·ərz ,meth·əd }

Schlumberger dipmeter [ENG] An instrument that measures both the amount and direction of dip by readings taken in the borehole; it consists of a long, cylindrical body with two telescoping parts and three long, springy metal strips, arranged symmetrically round the body, which press outward and make contact with the walls of the hole. { 'shləm·bər,zhā 'dip,mēd·ər }

Schlumberger photoclinometer [ENG] An instrument that measures simultaneously the amount and direction of the deviation of a borehole; the sonde, designed to lie exactly parallel to the axis of the borehole, is fitted with a small camera on the axis of a graduated glass bowl, in which a steel ball rolls freely and a compass is mounted in gimbals; the camera is electrically operated from the surface and takes a photograph of the bowl, the steel ball marks the amount of deviation, and the position in relation to the image of the compass needle gives the direction of deviation. { 'shləm·bər,zhā ,fōd·ō·kli'näm·əd·ər }

Schmidt field balance [ENG] An instrument that operates as both a horizontal and vertical field balance and consists of a permanent magnet pivoted on a knife edge. { 'shmit 'fēld ,bal·əns }

Schneider recoil system [MECH ENG] A recoil system for artillery, employing the hydropneumatic principle without a floating piston. { 'shnī·dər 're̅,köil ,sis·təm }

Schoenherr-Hessberger process [CHEM ENG] A nitrogen-fixation process used in Norway; employs a very long (22 feet or 7 meters) alternating-current arc around which air moves in a helical path in a 746-kilowatt furnace. { 'shən·her 'hes,bərg·ər ,prä·səs }

Schoop process |ENG| A process for coating surfaces by spraying with high-velocity molten metal particles. { shöp ,prä·səs }

Schottky barrier |ELECTR| A transition region formed within a semiconductor surface to serve as a rectifying barrier at a junction with a layer of metal. { 'shät·kē ,bar·ē·ər }

Schottky barrier diode |ELECTR| A semiconductor diode formed by contact between a semiconductor layer and a metal coating; it has a nonlinear rectifying characteristic; hot carriers (electrons for n-type material or holes for p-type material) are emitted from the Schottky barrier of the semiconductor and move to the metal coating that is the diode base; since majority carriers predominate, there is essentially no injection or storage of minority carriers to limit switching speeds. Also known as hot-carrier diode; Schottky diode. { 'shät·kē ¦bar·ē·ər 'dī,ōd }

Schottky diode See Schottky barrier diode. { 'shät·kē 'dī,ōd }

Schottky-diode FET logic |ELECTR| A logic gate configuration used with gallium-arsenide field-effect transistors operating in the depletion mode, in which very small Schottky diodes at the gate input provide the logical OR function and the level shifting required to make the input and output voltage levels compatible. Abbreviated SDFL. { 'shät·kē ¦dī,ōd ¦ef¦e¦tē 'läj·ik }

Schottky noise See shot noise. { 'shät·kē ,nòiz }

Schottky transistor-transistor logic |ELECTR| A transistor-transistor logic circuit in which a Schottky diode with forward diode voltage is placed across the base-collector junction of the output transistor in order to improve the speed of the circuit. { 'shät·kē tran¦zis·tər tran¦zis·tər 'läj·ik }

Schuler pendulum |MECH| Any apparatus which swings, because of gravity, with a natural period of 84.4 minutes, that is, with the same period as a hypothetical simple pendulum whose length is the earth's radius; the pendulum arm remains vertical despite any motion of its pivot, and the apparatus is therefore useful in navigation. { 'shü·lər ,pen·jə·ləm }

Schuler tuning |ENG| The designing of gyroscopic devices so that their periods of oscillation will be about 84.4 minutes. { 'shü·lər ,tün·iŋ }

Schweydar mechanical detector |ENG| A seismic detector that senses and records refracted waves; a lead sphere is suspended by a flat spring, the sphere's motion is magnified by an aluminum cone that moves a bow around a spindle carrying a mirror, and this motion is then photographically recorded. { 'shwād·ər mi¦kan·i·kəl di'tek·tər }

scissor engine See cat-and-mouse engine. { 'siz·ər ,en·jən }

scissor jack |MECH ENG| A lifting jack driven by a horizontal screw; the linkages of the jack are parallelograms whose horizontal diagonals

are lengthened or shortened by the screw. { 'siz·ər ,jak }

scissors bridge |CIV ENG| A light metal bridge that can be folded and carried by a military tank. { 'siz·ərz ,brij }

scissors crossover |CIV ENG| A scissor-shaped junction between two parallel railway tracks. Also called double crossover. { 'siz·ərz 'krós,ō·vər }

scissors truss |BUILD| A roof truss in which the braces cross like scissors blades. { 'siz·ərz ,trəs }

sclerometer |ENG| An instrument used to determine the hardness of a material by measuring the pressure needed to scratch or indent a surface with a diamond point. { sklə'räm·əd·ər }

scleroscope |ENG| An instrument used to determine the hardness of a material by measuring the height to which a standard ball rebounds from its surface when dropped from a standard height. { 'skler·ə,skōp }

scoop |DES ENG| **1.** Any of various ladle-, shovel-, or bucketlike utensils or containers for moving liquid or loose materials. **2.** A funnel-shaped opening for channeling a fluid into a desired path. See ellipsoidal floodlight. |MECH ENG| A large shovel with a scoop-shaped blade. { sküp }

scoopfish See underway sampler. { 'süp,fish }

scope |ELECTR| See cathode-ray oscilloscope; radarscope. |ENG| The work that will actually be done on a project as documented by the terms in a contract. { skōp }

scorching |CHEM ENG| Premature vulcanization caused by heat during the processing of rubber. |ENG| **1.** Burning an exposed surface so as to change color, texture, or flavor without consuming. **2.** Destroying by fire. { 'skòrch·iŋ }

scorch time |CHEM ENG| In rubber manufacture, the time during which a rubber compound can be worked at a given temperature before curing begins. { 'skòrch ,tīm }

scoring |ENG| Scratching the surface of a material. { 'skòr·iŋ }

scoring test See L-2 test. { 'skòr·iŋ ,test }

scotch |DES ENG| See scutch. |ENG| A wooden stopblock or iron catch placed under a wheel or other curved object to prevent slipping or rolling. { skäch }

scotch boiler |MECH ENG| A fire-tube boiler with one or more cylindrical internal furnaces enveloped by a boiler shell equipped with five tubes in its upper part; heat is transferred to water partly in the furnace area and partly in passage of hot gases through the tubes. Also known as dry-back boiler; scotch marine boiler (marine usage). { 'skäch 'bòil·ər }

Scotch bond See American bond. { 'skäch 'bänd }

Scotch derrick See stiffleg derrick. { 'skäch 'der·ik }

scotch marine boiler See scotch boiler. { 'skäch mə¦rēn 'bòil·ər }

Scotch yoke |MECH ENG| A type of four-bar

linkage; it is employed to convert a steady rotation into a simple harmonic motion. { 'skäch 'yōk }

Scott connection |ELECTR| A type of transformer which transmits power from two-phase to three-phase systems, or vice versa. { 'skät ka,nek·shən }

Scott-Darcy process |CIV ENG| A chemical precipitation method used for fine solids removal in sewage plants; employs ferric chloride solution made by treating scrap iron with chlorine. { 'skät 'der·ē ,prä·səs }

scouring |ENG| Physical or chemical attack on process equipment surfaces, as in a furnace or fluid catalytic cracker. |MECH ENG| Mechanical finishing or cleaning of a hard surface by using an abrasive and low pressure. { 'skaúr·iŋ }

scouring basin |CIV ENG| A basin containing impounded water which is released at about low water in order to maintain the desired depth in the entrance channel. Also known as sluicing pond. { 'skaúr·iŋ ,bas·ən }

scout |ENG| An engineer who makes a preliminary examination of promising oil and mining claims and prospects. { skaút }

scrambler |ELECTR| A circuit that divides speech frequencies into several ranges by means of filters, then inverts and displaces the frequencies in each range so that the resulting reproduced sounds are unintelligible; the process is reversed at the receiving apparatus to restore intelligible speech. Also known as speech inverter; speech scrambler. { 'skram·blər }

scrap |ENG| Any solid material cutting or reject of a manufacturing operation, which may be suitable for recycling as feedstock to the primary operation; for example, scrap from plastic or glass molding or metalworking. { skrap }

scraped-surface exchanger |CHEM ENG| A liquid-liquid heat-exchange device that has a rotating element with spring-loaded scraper blades to wipe the process-fluid exchange surfaces clean of crystals or other foulants; used in paraffin-wax processing. { 'skrāpt ¦sər·fəs iks,chānjər }

scraper conveyor |MECH ENG| A type of flight conveyor in which the element (chain and flight) for moving materials rests on a trough. { 'skrāp·ər kən,vā·ər }

scraper hoist |MECH ENG| A drum hoist that operates the scraper of a scraper loader. { 'skrāp·ər ,hóist }

scraper loader |MECH ENG| A machine used for loading coal or rock by pulling a scoop through the material to an apron or ramp, where the load is discharged onto a car or conveyor. { 'skrāp·ər ,lōd·ər }

scraper ring |MECH ENG| A piston ring that scrapes oil from a cylinder wall to prevent it from being burned. { 'skrāp·ər ,riŋ }

scraper trap |ENG| A device for the insertion or recovery of pigs, or scrapers, that are used to clean the inside surfaces of pipelines. { 'skrāp·ər ,trap }

scratch coat |ENG| The first layer of plaster applied to a surface; the surface is scratched to improve the bond with the next coat. { 'skrach ,kōt }

scratch filter |ENG ACOUS| A low-pass filter circuit inserted in the circuit of a phonograph pickup to suppress higher audio frequencies and thereby minimize needle-scratch noise. { 'skrach ,fil·tər }

screed |BUILD| A long, narrow strip of plaster placed at intervals on a surface as a guide for the thickness of plaster to be applied. |CIV ENG| **1.** A straight-edged wood or metal template, fixed temporarily to a surface as a guide when plastering or concreting. **2.** An oscillating metal bar mounted on wheels and spanning a freshly placed road slab, used to strike off and smooth the surface. { skrēd }

screed wire See ground wire. { 'skrēd ,wīr }

screen |ELECTR| **1.** The surface on which a television, radar, x-ray, or cathode-ray oscilloscope image is made visible for viewing; it may be a fluorescent screen with a phosphor layer that converts the energy of an electron beam to visible light, or a translucent or opaque screen on which the optical image is projected. Also known as viewing screen. **2.** See screen grid. |ENG| **1.** A large sieve of suitably mounted wire cloth, grate bars, or perforated sheet iron used to sort rock, ore, or aggregate according to size. **2.** A covering to give physical protection from light, noise, heat, or flying particles. **3.** A filter medium for liquid-solid separation. { skrēn }

screen analysis |ENG| A method for finding the particle-size distribution of any loose, flowing, conglomerate material by measuring the percentage of particles that pass through a series of standard screens with holes of various sizes. { 'skrēn ə,nal·ə·səs }

screen deck |DES ENG| A surface provided with apertures of specified size, used for screening purposes. { 'skrēn ,dek }

screen dryer See traveling-screen dryer. { 'skrēn 'drī·ər }

screening |ENG| **1.** The separation of a mixture of grains of various sizes into two or more size-range portions by means of a porous or woven-mesh screening media. **2.** The removal of solid particles from a liquid-solid mixture by means of a screen. **3.** The material that has passed through a screen. |IND ENG| The elimination of defective pieces from a lot by inspection for specified defects. Also known as detailing. { 'skrēn·iŋ }

screen mesh |ENG| A wire network or cloth mounted on a frame for separating and classifying materials. { 'skrēn ,mesh }

screen overlay See glare filter. { ¦skrēn 'ō·vər,lā }

screen pipe |ENG| Perforated pipe with a straining device in the form of closely wound wire coils wrapped around it to admit well fluids while excluding sand. { 'skrēn ,pīp }

screw |DES ENG| **1.** A cylindrical body with a helical groove cut into its surface. **2.** A fastener with continuous ribs on a cylindrical or conical shank and a slotted, recessed, flat, or rounded head. Also known as screw fastener. { skrü }

screw blank See bolt blank. { 'skrü ,blaŋk }

screw compressor |MECH ENG| A rotary-element gas compressor in which compression is accomplished between two intermeshing, counterrotating screws. { 'skrü kəm'pres·ər }

screw conveyor |MECH ENG| A conveyor consisting of a helical screw that rotates upon a single shaft within a stationary trough or casing, and which can move bulk material along a horizontal, inclined, or vertical plane. Also known as auger conveyor; spiral conveyor; worm conveyor. { 'skrü kən'vā·ər }

screw displacement |MECH| A rotation of a rigid body about an axis accompanied by a translation of the body along the same axis. { 'skrü di,splās·mənt }

screw dowel |DES ENG| A metal dowel pin having a straight or tapered thread at one end. { 'skrü ,däül }

screwdriver |DES ENG| A tool for turning and driving screws in place; a thin, wedge-shaped or fluted end enters the slot or recess in the head of the screw. { 'skrü,drīv·ər }

screw elevator |MECH ENG| A type of screw conveyor for vertical delivery of pulverized materials. { 'skrü 'el·ə,vād·ər }

screw fastener See screw. { 'skrü ,fas·nər }

screwfeed |MECH ENG| A system or combination of gears, ratchets, and friction devices in the swivel head of a diamond drill, which controls the rate at which a bit penetrates a rock formation. { 'skrü,fēd }

screw feeder |MECH ENG| A mechanism for handling bulk (pulverized or granulated solids) materials, in which a rotating helicoid screw moves the material forward, toward and into a process unit. { 'skrü 'fēd·ər }

screw jack See jackscrew. { 'skrü 'jak }

screw machine |MECH ENG| A lathe for making relatively small, turned metal parts in large quantities. { 'skrü mə,shēn }

screw pile |CIV ENG| A pile having a wide helical blade at the foot which is twisted into position, for use in soft ground or other location requiring a large supporting surface. { 'skrü ,pīl }

screw plasticating injection molding |ENG| A plastic-molding technique in which plastic is converted from pellets to a viscous (plasticated) melt by an extruder screw that is an integral part of the molding machine. { 'skrü 'plas·ti,kād·iŋ in'jek·shən ,mōld·iŋ }

screw press |MECH ENG| A press having the slide operated by a screw mechanism. { 'skrü ,pres }

screw propeller |MECH ENG| A marine and airplane propeller consisting of a streamlined hub attached outboard to a rotating engine shaft on which are mounted two to six blades; the blades form helicoidal surfaces in such a way as to advance along the axis about which they revolve. { 'skrü prə,pel·ər }

screw pump |MECH ENG| A pump that raises water by means of helical impellers in the pump casing. { 'skrü ,pəmp }

screw rivet |DES ENG| A short rod threaded along the length of the shaft that is set without access to the point. { 'skrü ,riv·ət }

screw spike |DES ENG| A large nail with a helical thread on the upper portion of the shank; used to fasten railroad rails to the ties. { 'skrü ,spīk }

screwstock |MECH ENG| Free-machining bar, rod, or wire. { 'skrü,stäk }

screw thread |DES ENG| A helical ridge formed on a cylindrical core, as on fasteners and pipes. { 'skrü ,thred }

screw-thread gage |DES ENG| Any of several devices for determining the pitch, major, and minor diameters, and the lead, straightness, and thread angles of a screw thread. { 'skrü ,thred ,gāj }

screw-thread micrometer |DES ENG| A micrometer used to measure pitch diameter of a screw thread. { 'skrü ,thred mī'kräm·əd·ər }

scriber |DES ENG| A sharp-pointed tool used for drawing lines on metal workpieces. { 'skrī·bər }

scroll gear |DES ENG| A variable gear resembling a scroll with teeth on one face. { 'skrōl ,gir }

scroll saw |ENG| A saw with a narrow blade, used for cutting curves or irregular designs. { 'skrōl ,só }

scrubber |ENG| A device for the removal, or washing out, of entrained liquid droplets or dust, or for the removal of an undesired gas component from process gas streams. Also known as washer; wet collector. { 'skrəb·ər }

scrub plane |DES ENG| A narrow carpenter's plane with a blade that has a rough surface and a rounded cutting edge. { 'skrəb ,plān }

scuba diving |ENG| Any of various diving techniques using self-contained underwater breathing apparatus. { 'skü·bə ,dīv·iŋ }

scuffing |ENG| The dull mark, sometimes the result of abrasion, on the surface of glazed ceramic or glassware. { 'skəf·iŋ }

scuffle hoe |DES ENG| A hoe having two sharp edges so that it can be pushed and pulled. { 'skəf·əl ,hō }

scum chamber |CIV ENG| An enclosed compartment in an Imhoff tank, in which gas escapes from the scum which rises to the surface of sludge during sewage digestion. { 'skəm ,chām·bər }

scutch |DES ENG| A small, picklike tool which has flat cutting edges for trimming bricks. Also known as scotch. { skəch }

scuttle |BUILD| An opening in the ceiling to provide access to the attic or roof. { 'skəd·əl }

scythe |DES ENG| A tool with a long curved blade attached at a more or less right angle to a long handle with grips for both hands; used for cutting grass as well as grain and other crops. { sīth }

sea bank See seawall. { 'sē ,baŋk }

seadrome |CIV ENG| **1.** A designated area for

landing and takeoff of seaplanes. **2.** A platform at sea for landing and takeoff of land planes. { 'sē,drōm }

sea gate |CIV ENG| A gate which serves to protect a harbor or tidal basin from the sea, such as one of a pair of supplementary gates at the entrance to a tidal basin exposed to the sea. { 'sē ,gāt }

seal |ENG| **1.** Any device or system that creates a nonleaking union between two mechanical or process-system elements; for example, gaskets for pipe connection seals, mechanical seals for rotating members such as pump shafts, and liquid seals to prevent gas entry to or loss from a gas-liquid processing sequence. **2.** A tight, perfect closure or joint. { sēl }

Seale rope |DES ENG| A wire rope with six or eight strands, each having a large wire core covered by nine small wires, which, in turn, are covered by nine large wires. { 'sēl ,rōp }

sea-level datum |ENG| A determination of mean sea level that has been adopted as a standard datum for heights or elevations, based on tidal observations over many years at various tide stations along the coasts. { 'sē ¦lev·əl ,dad·əm }

seal off |ENG| To close off, as a tube or borehole, by using a cement or other sealant to eliminate ingress or egress. { 'sēl 'of }

seam |ENG| **1.** A mechanical or welded joint. **2.** A mark on ceramic or glassware where matching mold parts join. **3.** A line occurring on a molded or laminated piece of plastic material that differs in appearance from the rest of the surface and is caused by a parting of the mold. Also known as mold seam. { sēm }

sea marker |ENG| A patch of color on the ocean surface produced by releasing dye; used, for example, to attract the attention of the crew of a rescue airplane. { 'sē ,märkər }

seaport |CIV ENG| A harbor or town that has facilities for seagoing ships and is active in marine activities. { 'sē,pórt }

search |ENG| To explore a region in space with radar. { sərch }

search and rescue |ENG| The use of aircraft, surface craft, submarines, specialized rescue teams and equipment to search for and rescue personnel in distress on land or at sea. { 'sərch ən 'res,kyü }

searching control |ENG| A mechanism that changes the azimuth and elevation settings on a searchlight automatically and constantly, so that its beam is swept back and forth within certain limits. { 'sərch·iŋ kən,trōl }

searching lighting See horizontal scanning. { 'sərch·iŋ ,līd·iŋ }

searchlight-control radar |ENG| A ground-based radar used to direct searchlights at aircraft. { 'sərch,līt kən¦trōl ,rā,där }

searchlight-type sonar |ENG| A sonar system in which both transmission and reception are effected by the same narrow beam pattern. { 'sərch,līt ¦tīp 'sōnär }

search radar |ENG| A radar intended primarily to cover a large region of space and to display targets as soon as possible after they enter the region; used for early warning, in connection with ground-controlled approach and interception, and in air-traffic control. { 'sərch ,rā,där }

search unit |ENG| The portion of an ultrasonic testing system which incorporates sending and in some cases receiving transducers to scan the workpiece. { 'sərch ,yü·nət }

seasonal balancing |CHEM ENG| A seasonal adjustment of the front-end boiling range (volatility) of a motor gasoline to control engine starting characteristics by compensating for seasonal temperature changes. { 'sēz·ən·əl 'bal·əns·iŋ }

seasoning See curing. |ELECTR| Overcoming a temporary unsteadiness of a component that may appear when it is first installed. |ENG| Drying of wood either in the air or in a kiln. { 'sēz·ən·iŋ }

sea surveillance |ENG| The systematic observation of surface and subsurface sea areas by all available and practicable means primarily for the purpose of locating, identifying, and determining the movements of ships, submarines, and other vehicles, friendly and enemy, proceeding on or under the surface of seas and oceans. { 'sē sər,vā·ləns }

seat |MECH ENG| The fixed, pressure-containing portion of a valve which comes into contact with the moving portions of that valve. { sēt }

seating-lock locking fastener |DES ENG| A locking fastener that locks only when firmly seated and is therefore free-running on the bolt. { 'sēd·iŋ ¦läk 'läk·iŋ 'fas·nər }

sea van |IND ENG| Commercial or government-owned (or leased) shipping containers which are moved via ocean transportation; since wheels are not attached, they must be lifted on and off the ship. { 'sē ,van }

seawall |CIV ENG| A concrete, stone, or metal wall or embankment constructed along a shore to reduce wave erosion and encroachment by the sea. Also known as sea bank. { 'sē,wól }

seawater thermometer |ENG| A specially designed thermometer to measure the temperature of a sample of seawater; an instrument consisting of a mercury-in-glass thermometer protected by a perforated metal case. { 'sē,wód·ər thər'mäm·əd·ər }

Secchi disk |ENG| An opaque white disk used to measure the transparency or clarity of seawater by lowering the disk into the water horizontally and noting the greatest depth at which it can be visually detected. { 'sek·ē ,disk }

secondary air |MECH ENG| Combustion air introduced over the burner flame to enhance completeness of combustion. { 'sek·ən,der·ē 'er }

secondary creep |MECH| The change in shape of a substance under a minimum and almost constant differential stress, with the strain-time relationship a constant. Also known as steady-state creep. { 'sek·ən,der·ē 'krēp }

secondary crusher |MECH ENG| Any of a group of crushing and pulverizing machines used after the primary treatment to further reduce the particle size of shale or other rock. { 'sek·ən,der·ē 'krash·ər }

secondary grinding [MECH ENG] A further grinding of material previously reduced to sand size. { 'sek·ən,der·ē 'grīnd·iŋ }

secondary ion mass analyzer [ENG] A type of secondary ion mass spectrometer that provides general surface analysis and depth profiling capabilities. { 'sek·ən,der·ē 'ī,än ,'mas 'an·ə,līz·ər }

secondary ion mass spectrometer [ENG] An instrument for microscopic chemical analysis, in which a beam of primary ions with an energy in the range 5–20 kiloelectronvolts bombards a small spot on the surface of a sample, and positive and negative secondary ions sputtered from the surface are analyzed in a mass spectrometer. Abbreviated SIMS. Also known as ion microprobe; ion probe. { 'sek·ən,der·ē 'ī,än ,'mas spek'tram·əd·ər }

secondary port [CIV ENG] A port with one or more berths, normally at quays, which can accommodate oceangoing ships for discharge. { 'sek·ən,der·ē 'pȯrt }

secondary rescue facilities [ENG] Local airbase-ready aircraft, crash boats, and other air, surface, subsurface, and ground elements suitable for rescue missions, including government and privately operated units and facilities. { 'sek·ən,der·ē 'res,kyü fə,sil·əd·ēz }

secondary sewage sludge [CIV ENG] Sludge that includes activated sludge, mixed sludge, and chemically precipitated sludge. { 'sek·ən,der·ē 'sü·ij ,sləj }

secondary stress [MECH] A self-limiting normal or shear stress which is caused by the constraint of a structure and which is expected to cause minor distortions that would not result in a failure of the structure. { 'sek·ən,der·ē 'stres }

secondary tide station [ENG] A place at which tide observations are made over a short period to obtain data for a specific purpose. { 'sek·ən,der·ē 'tīd ,stā·shən }

second breakdown [ELECTR] Destructive breakdown in a transistor, wherein structural imperfections cause localized current concentrations and uncontrollable generation and multiplication of current carriers; reaction occurs so suddenly that the thermal time constant of the collector regions is exceeded, and the transistor is irreversibly damaged. { 'sek·ənd 'brāk,daùn }

second law of motion See Newton's second law. { 'sek·ənd 'lȯ əv 'mō·shən }

second law of thermodynamics [THERMO] A general statement of the idea that there is a preferred direction for any process; there are many equivalent statements of the law, the best known being those of Clausius and of Kelvin. { 'sek·ənd 'lȯ əv ,thər·mə·dī'nam·iks }

second-level controller [CONT SYS] A controller which influences the actions of first-level controllers, in a large-scale control system partitioned by plant decomposition, to compensate

for subsystem interactions so that overall objectives and constraints of the system are satisfied. Also known as coordinator. { 'sek·ənd ¦lev·əl kən'trōl·ər }

second-order leveling [ENG] Spirit leveling that has less stringent requirements than those of first-order leveling, in which lines between benchmarks established by first-order leveling are run in only one direction. { 'sek·ənd ¦ȯr·dər 'lev·ə·liŋ }

second-order transition [THERMO] A change of state through which the free energy of a substance and its first derivatives are continuous functions of temperature and pressure, or other corresponding variables. { 'sek·ənd ¦ȯr·dər tran'zish·ən }

section [CIV ENG] A piece of land usually 1 mile square (640 acres or approximately 2.58999 square kilometers) with boundaries conforming to meridians and parallels within established limits; 1 of 36 units of subdivision of a township in the U.S. Public Land survey system. { 'sek·shən }

sectional conveyor [MECH ENG] A belt conveyor that can be lengthened or shortened by the addition or the removal of interchangeable sections. { 'sek·shən·əl kən'vā·ər }

sectional core barrel [DES ENG] A core barrel whose length can be increased by coupling unit sections together. { 'sek·shən·əl 'kȯr ,bar·əl }

sectional header boiler [MECH ENG] A horizontal boiler in which tubes are assembled in sections into front and rear headers; the latter, in turn, are connected to the boiler drum by vertical tubes. { 'sek·shən·əl 'hed·ər ¦bȯil·ər }

section house [CIV ENG] A building near a railroad section for housing railroad workers, or for storing maintenance equipment for the section. { 'sek·shən ,haús }

section line [CIV ENG] A line representing the boundary of a section of land. { 'sek·shən ,līn }

section modulus [MECH] The ratio of the moment of inertia of the cross section of a beam undergoing flexure to the greatest distance of an element of the beam from the neutral axis. { 'sek·shən 'mäj·ə·ləs }

sector [CIV ENG] A clearly defined area or airspace designated for a particular purpose. { 'sek·tər }

sector gate [CIV ENG] A horizontal gate with a pie-slice cross section used to regulate the level of water at the crest of a dam; it is raised and lowered by a rack and pinion mechanism. { 'sek·tər ,gāt }

sector gear [DES ENG] **1.** A toothed device resembling a portion of a gear wheel containing the center bearing and a part of the rim with its teeth. **2.** A gear having such a device as its chief essential feature. [MECH ENG] A gear system employing such a gear as a principal part. { 'sek·tər ,gir }

secular [ENG] Of or pertaining to a long indefinite period of time. { 'sek·yə·lər }

sedimentation tank [ENG] A tank in which suspended matter is removed either by quiescent settlement or by continuous flow at high velocity and extended retention time to allow deposition. { ,sed·ə·mən'tā·shən ,taŋk }

sediment bulb |ENG| A bulb for holding sediment that settles from the liquid in a tank. { 'sed·ə·mənt ,bəlb }

sediment corer |ENG| A heavy coring tube which punches out a cylindrical sediment section from the ocean bottom. { 'sed·ə·mənt ,kȯr·ər }

sediment trap |ENG| A device for measuring the accumulation rate of sediment on the floor of a body of water. { 'sed·ə·mənt ,trap }

Seebeck coefficient |ELECTR| The ratio of the open-circuit voltage to the temperature difference between the hot and cold junctions of a circuit exhibiting the Seebeck effect. { 'zā,bek ,kō·i'fish·ənt }

Seebeck effect |ELECTR| The development of a voltage due to differences in temperature between two junctions of dissimilar metals in the same circuit. { 'zā,bek i,fekt }

Segas process |CHEM ENG| A process for the production of low-Btu gas by the catalytic method using a fixed bed catalyst, lime-bauxite mixture bonded with bentonite. { 'sē,gas ,prä·səs }

segmental gate *See* tainter gate. { seg'ment·əl 'gāt }

segmental meter |ENG| A variable head meter whose orifice plate has an opening in the shape of a half circle. { seg'ment·əl 'mēd·ər }

segmented aperture-synthetic aperture radar |ENG| An enhancement of synthetic aperture radar that overcomes restrictions on the effective length of the receiving antenna by using a receiving antenna array composed of a set of contiguous subarrays and employing signal processing to provide the proper phase corrections for each subarray. Abbreviated SASAR. { 'seg,ment·əd ¦ap·ə·chər sin'thed·ik ¦ap·ə·chər 'rā,där }

segment saw |MECH ENG| A saw consisting of steel segments attached around the edge of a flange and used for cutting veneer. { 'seg·mənt ,sȯ }

segregation |ENG| **1.** The keeping apart of process streams. **2.** In plastics molding, a close succession of parallel, relatively narrow, and sharply defined wavy lines of color on the surface of a plastic that differ in shade from surrounding areas and create the impression that the components have separated. { ,seg·rə'gā·shən }

seine net |ENG| A net used to catch fish by encirclement, usually by closure of the two ends and the bottom. { sān ,net }

seismic bracing |ENG| Reinforcement added to a structure to prevent collapse or deformation of building elements as a result of earthquakes. { ¦sīz·mik 'brās·iŋ }

seismic constant |CIV ENG| In building codes dealing with earthquake hazards, an arbitrarily set quantity of steady acceleration, in units of acceleration of gravity, that a building must withstand. { 'sīz·mik 'kän·stənt }

seismic detector |ENG| An instrument that receives seismic impulses. { 'sīz·mik di,tek·tər }

seismic exploration |ENG| The exploration for economic deposits by using seismic techniques, usually involving explosions, to map subsurface structures. { 'sīz·mik ,ek·splə'rā·shən }

seismic load |ENG| The force on a structure caused by acceleration induced on its mass by an earthquake. { ¦sīz·mik 'lōd }

seismic profiler |ENG| A continuous seismic reflection system used to study the structure beneath the sea floor to depths of 10,000 feet (3000 meters) or more, using a rotating drum to record reflections. { 'sīz·mik 'prō,fīl·ər }

seismic shooting |ENG| A method of geophysical prospecting in which elastic waves are produced in the earth by the firing of explosives. { 'sīz·mik 'shüd·iŋ }

seismic survey *See* reflection survey. { 'sīz·mik 'sər,vā }

seismochronograph |ENG| A chronograph for determining the time at which an earthquake shock appears. { ¦sīz·mə'krän·ə,graf }

seismogram |ENG| The record made by a seismograph. { 'sīz·mə,gram }

seismograph |ENG| An instrument that records vibrations in the earth, especially earthquakes. { 'sīz·mə,graf }

seismometer |ENG| An instrument that detects movements in the earth. { sīz'mäm·əd·ər }

seismoscope |ENG| An instrument for recording only the occurrence or time of occurrence (not the magnitude) of an earthquake. { 'sīz·mə,skōp }

seizing |ENG| Abrasive damage to a metal surface caused when the surface is rubbed by another metal surface. { 'sēz·iŋ }

selected time |IND ENG| An observed actual time value for an element, measured by time study, which is identified as being the most representative of the situation observed. { si'lek·təd 'tīm }

selective adsorbent |CHEM ENG| Material that will selectively adsorb (or reject) one or more specific components from a multicomponent mixture of gases or liquids; common adsorbents are silica gel, carbon and activated carbon, activated alumina, and synthetic or natural zeolites (molecular sieves). { si'lek·tiv ad'sȯr·bənt }

selective cracking |CHEM ENG| A refinery process in which recycled stock is distilled in equipment kept separate from that used for distillation of original stock. { si'lek·tiv 'krak·iŋ }

selectively doped heterojunction transistor *See* high-electron-mobility transistor. { si'lek·tiv·lē ¦dōpt ¦hed·ə·rō¦jəŋk·shən tran'zis·tər }

selective polymerization |CHEM ENG| The polymerization of a single type of molecule in a mixture of monomers; for example, the production of diisobutylene from a mixture of butylenes. { si'lek·tiv pə,lim·ə·rə'zā·shən }

selective solubility diffusion |CHEM ENG| The transmission of fluids through a nonporous, polymeric barrier (membrane) by an adsorption-solution-diffusion-desorption sequence. { si 'lek·tiv ,säl·yə'bil·əd·ē di,fyü·zhən }

selective solvent |CHEM ENG| A solvent that, at certain temperatures and ratios with other materials, preferentially dissolves more of one component of a liquid or solids mixture than of another, thereby permitting partial separation. { si'lek·tiv 'säl·vənt }

selective transmission |MECH ENG| A gear transmission with a single lever for changing from one gear ratio to another; used in automotive vehicles. { si'lek·tiv tranz·mish·ən }

selectivity diagram |CHEM ENG| A triangular plot of solubilities in a ternary liquid system; used to calculate the ability of a solvent to extract a component from a mixture (its selectivity) at various concentration combinations. { sə ,lek'tiv·əd·ē 'dī·ə,gram }

selector |CIV ENG| A device that automatically connects the appropriate railroad signal to control the track selected. |ELEC| An automatic or other device for making connections to any one of a number of circuits, such as a selector relay or selector switch. |ENG| **1.** A device for selecting objects or materials according to predetermined properties. **2.** A device for starting or stopping at predetermined positions. |MECH ENG| **1.** The part of the gearshift in an automotive transmission that selects the required gearshift bar. **2.** The lever with which a driver operates an automatic gearshift. { si 'lek·tər }

selenium cell |ELECTR| A photoconductive cell in which a thin film of selenium is used between suitable electrodes; the resistance of the cell decreases when the illumination is increased. { sə'lē·nē·əm ,sel }

selenium diode |ELECTR| A small area selenium rectifier which has characteristics similar to those of selenium rectifiers used in power systems. { sə'lē·nē·əm 'dī,ōd }

selenium rectifier |ELECTR| A metallic rectifier in which a thin layer of selenium is deposited on one side of an aluminum plate and a conductive metal coating is deposited on the selenium. { sə'lē·nē·əm 'rek·tə,fī·ər }

selenotrope |ENG| A device used in geodetic surveying for reflecting the moon's rays to a distant point, to aid in long-distance observations. { sə'lē·nə,trōp }

self-adapting system |SYS ENG| A system which has the ability to modify itself in response to changes in its environment. { ¦self ə¦dap·tiŋ 'sis·təm }

self-centering chuck |MECH ENG| A drill chuck that, when closed, automatically positions the drill rod in the center of the drive rod of a diamond-drill swivel head. { 'self ¦sen·tə·riŋ 'chək }

self-cleaning |ENG| Pertaining to any device that is designed to clean itself without disassembly, for example, a filter in which accumulated filter cake or sludge is removed by an internal scraper or by a blowdown or backwash action. { self 'klēn·iŋ }

self-contained breathing apparatus |ENG| A portable breathing unit which permits freedom of movement. { ¦self kən¦tānd 'breth·iŋ ,ap·ə,rad·əs }

self-contained range finder |ENG| Instrument used for measuring range by direct observation, without using a base line; the two types are the coincidence range finder and the stereoscopic range finder. { ¦self kən¦tānd 'rānj ,fīn·dər }

self-energizing brake |MECH ENG| A brake designed to reinforce the power applied to it, such as a hand brake. { ¦self ,en·ər¦jīz·iŋ 'brāk }

self-excited vibration See self-induced vibration. { ¦self ik'sīd·əd vī'brā·shən }

self-faced stone |CIV ENG| A type of stone used in masonry that splits along natural cleavage planes and does not have to be dressed. { ¦self ,fāst 'stōn }

self-healing dielectric breakdown |ELECTR| A dielectric breakdown in which the breakdown process itself causes the material to become insulating again. { ¦self ¦hēl·iŋ ,dī·ə¦lek·trik 'brāk,daún }

self-induced vibration |MECH| The vibration of a mechanical system resulting from conversion, within the system, of nonoscillatory excitation to oscillatory excitation. Also known as self-excited vibration. { ¦self in¦düst vī'brā·shən }

self-loading |MECH ENG| The capability of a powered industrial truck to pick up, transport, and deposit its load by using components that are part of its standard equipment, for example, a forklift. { 'self ¦lōd·iŋ }

self-locking nut |DES ENG| A nut having an inherent locking action, so that it cannot readily be loosened by vibration. { 'self ¦läk·iŋ 'nət }

self-locking screw |DES ENG| A screw that locks itself in place without requiring a separate nut or lock washer. { 'self ¦läk·iŋ 'skrü }

self-organizing function |CONT SYS| That level in the functional decomposition of a large-scale control system which modifies the modes of control action or the structure of the control system in response to changes in system objectives, contingency events, and so forth. { ¦self ¦òr·gə,nīz·iŋ 'fəŋk·shən }

self-organizing system |SYS ENG| A system that is able to affect or determine its own internal structure. { ¦self ¦òr·gə,nīz·iŋ 'sis·təm }

self-propelled |MECH ENG| Pertaining to a vehicle given motion by means of a self-contained motor. { ¦self prə¦peld }

self-sealing |ENG| A fluid container, such as a fuel tank or a tire, lined with a substance that allows it to close immediately over any small puncture or rupture. { 'self ¦sēl·iŋ }

self-starter |MECH ENG| An attachment for automatically starting an internal combustion engine. { 'self 'stär·dər }

self-tapping screw |DES ENG| A screw with a specially hardened thread that makes it possible for the screw to form its own internal thread in sheet metal and soft materials when driven into a hole. Also known as sheet-metal screw; tapping screw. { 'self ¦tap·iŋ 'skrü }

self-timer [ENG] A device that delays the tripping of a camera shutter so that the photographer can be included in the photograph. { 'self 'tīm·ər }

self-tuning regulator [CONT SYS] A type of adaptive control system composed of two loops, an inner loop which consists of the process and an ordinary linear feedback regulator, and an outer loop which is composed of a recursive parameter estimator and a design calculation, and which adjusts the parameters of the regulator. Abbreviated STR. { ¦self ¦tün·iŋ 'reg·yə,lād·ər }

sellers hob [MECH ENG] A hob that turns on the centers of a lathe, the work being fed to it by the lathe carriage. { 'sel·ərz 'häb }

Selwood engine [MECH ENG] A revolving-block engine in which two curved pistons opposed 180° run in toroidal tracks, forcing the entire engine block to rotate. { 'sel,wůd ,en·jən }

semiautomatic transmission [MECH ENG] An automobile transmission that assists the driver to shift from one gear to another. { ¦sem·ē,ȯd·ə'mad·ik tranz'mish·ən }

semibatch chemical reactor [CHEM ENG] A reactor in which a constant liquid volume is maintained without any overflow, and with the continuous addition of one reactant, usually a gas. { 'sem·i,bach 'kem·ə·kəl rē'ak·tər }

semichemical pulping [CHEM ENG] A method of producing wood-fiber products in which the wood chips are merely softened by chemical treatment (neutral sodium sulfite solution), while the remainder of the pulping action is supplied by a disk attrition mill or by some similar mechanical device for separating the fibers. { ¦sem·i'kem·ə·kəl 'pəlp·iŋ }

semiclosed-cycle gas turbine [MECH ENG] A heat engine in which a portion of the expanded gas is recirculated. { 'sem·i,klōzd,sī·kəl 'gas ,tər·bən }

semiconductive loading tube [ENG] A loading tube for blasthole explosives which dissipates static electric charges to prevent premature blasts. { ¦sem·i·kən¦dək·tiv 'lōd·iŋ ,tüb }

semiconductor device [ELECTR] Electronic device in which the characteristic distinguishing electronic conduction takes place within a semiconductor. { ¦sem·i·kən¦dək·tər di,vīs }

semiconductor diode [ELECTR] Also known as crystal diode; crystal rectifier; diode. **1.** A two-electrode semiconductor device that utilizes the rectifying properties of a pn junction or a point contact. **2.** More generally, any two-terminal electronic device that utilizes the properties of the semiconductor from which it is constructed. { ¦sem·i·kən¦dək·tər 'dī,ōd }

semiconductor-diode parametric amplifier [ELECTR] Parametric amplifier using one or more varactors. { ¦sem·i·kən¦dək·tər ¦dī,ōd ¦par·ə¦me·trik 'am·plə,fī·ər }

semiconductor doping See doping. { ¦sem·i·kən¦dək·tər 'dōp·iŋ }

semiconductor heterostructure [ELECTR] A

structure of two different semiconductors in junction contact having useful electrical or electrooptical characteristics not achievable in either conductor separately; used in certain types of lasers and solar cells. { ¦sem·i·kən¦dək·tər 'hed·ə·rō,strək·chər }

semiconductor junction [ELECTR] Region of transition between semiconducting regions of different electrical properties, usually between p-type and n-type material. { ¦sem·i·kən¦dək·tər ,jəŋk·shən }

semiconductor rectifier See metallic rectifier. { ¦sem·i·kən¦dək·tər 'rek·tə,fī·ər }

semiconductor thermocouple [ELECTR] A thermocouple made of a semiconductor, which offers the prospect of operation with high-temperature gradients, because semiconductors are good electrical conductors but poor heat conductors. { ¦sem·i·kən¦dək·tər 'thər·mə,kəp·əl }

semidiesel engine [MECH ENG] **1.** An internal combustion engine of a type resembling the diesel engine in using heavy oil as fuel but employing a lower compression pressure and spraying it under pressure, against a hot (uncooled) surface or spot, or igniting it by the precombustion or supercompression of a portion of the charge in a separate member or uncooled portion of the combustion chamber. **2.** A true diesel engine that uses a means other than compressed air for fuel injection. { ¦sem·i'dē·zəl 'en·jən }

semifloating axle [MECH ENG] A supporting member in motor vehicles which carries torque and wheel loads at its outer end. { ¦sem·i'flōd·iŋ 'ak·səl }

semilive skid [ENG] A platform having two fixed legs at one end and two wheels at the other; used for moving bulk materials. { ¦sem·i'līv 'skid }

semimember [CIV ENG] A part in a frame or truss that ceases to bear a load when the stress in it starts to reverse. { ¦sem·i'mem·bər }

semipositive mold [ENG] A plastics mold that allows a small amount of excess material to escape when it is closed. { ¦sem·i'päz·əd·iv 'mōld }

semitrailer [ENG] A cargo-carrying piece of equipment that has one or two axles at the rear; the load is carried on these axles and on the fifth wheel of the tractor that supplies motive power to the semitrailer. { ¦sem·i'trāl·ər }

sems [DES ENG] A preassembled screw and washer combination. { semz }

send See scend. { send }

sense [ENG] To determine the arrangement or position of a device or the value of a quantity. { sens }

sensible heat [THERMO] **1.** The heat absorbed or evolved by a substance during a change of temperature that is not accompanied by a change of state. **2.** See enthalpy. { 'sen·sə·bəl 'hēt }

sensible-heat factor [THERMO] The ratio of space sensible heat to space total heat; used for air-conditioning calculations. Abbreviated SHF. { 'sen·sə·bəl ¦hēt ˌfak·tər }

sensible-heat flow |THERMO| The heat given up or absorbed by a body upon being cooled or heated, as the result of the body's ability to hold heat; excludes latent heats of fusion and vaporization. { 'sen·sə·bəl ¦hēt 'flō }

sensing element *See* sensor. { 'sens·iŋ ˌel·ə·mənt }

sensitive altimeter |ENG| An aneroid altimeter constructed to respond to pressure changes (altitude changes) with a high degree of sensitivity; it contains two or more pointers to refer to different scales, calibrated in hundreds of feet, thousands of feet, and so on. { 'sen·sad·iv al'tim·əd·ər }

sensitivity |ELECTR| **1.** The minimum input signal required to produce a specified output signal, for a radio receiver or similar device. **2.** Of a camera tube, the signal current developed per unit incident radiation, that is, per watt per unit area. |ENG| **1.** A measure of the ease with which a substance can be caused to explode. **2.** A measure of the effect of a change in severity of engine-operating conditions on the antiknock performance of a fuel; expressed as the difference between research and motor octane numbers. Also known as spread. { ˌsen·sə'tiv·əd·ē }

sensitivity function |CONT SYS| The ratio of the fractional change in the system response of a feedback-compensated feedback control system to the fractional change in an open-loop parameter, for some specified parameter variation. { ˌsen·sə'tiv·əd·ē ˌfəŋk·shən }

sensitometer |ENG| An instrument for measuring the sensitivity of light-sensitive materials. { ˌsen·sə'täm·əd·ər }

sensor |ENG| The generic name for a device that senses either the absolute value or a change in a physical quantity such as temperature, pressure, flow rate, or pH, or the intensity of light, sound, or radio waves and converts that change into a useful input signal for an information-gathering system; a television camera is therefore a sensor, and a transducer is a special type of sensor. Also known as primary detector; sensing element. { 'sen·sər }

sensory control |CONT SYS| Control of a robot's actions on the basis of its sensor readings. { 'sen·sə·rē kən'trōl }

sensory controlled robot |CONT SYS| A robot whose programmed sequence of instructions can be modified by information about the environment received by the robot's sensors. { 'sen·sə·rē kən'trōld 'rō,bät }

separate sewage system |CIV ENG| A drainage system in which sewage and groundwater are carried in separate sewers. { 'sep·rət 'sü·ij ˌsis·təm }

separating power |CHEM ENG| The measure of the ability of a system (such as a rectifying system) to separate the components of a mixture, when the components have increasingly close boiling points. { 'sep·ə,rād·iŋ ˌpaů·ər }

separation |CHEM ENG| The separation of liquids or gases in a mixture, as by distillation or extraction. |ENG| **1.** The action segregating phases, such as gas-liquid, gas-solid, liquid-solid. **2.** The segregation of solid particles by size range, as in screening. |ENG ACOUS| The degree, expressed in decibels, to which left and right stereo channels are isolated from each other. { ˌsep·ə'rā·shən }

separation theorem |CONT SYS| A theorem in optimal control theory which states that the solution to the linear quadratic Gaussian problem separates into the optimal deterministic controller (that is, the optimal controller for the corresponding problem without noise) in which the state used is obtained as the output of an optimal state estimator. { ˌsep·ə'rā·shən ˌthir·əm }

separator |ELEC| A porous insulating sheet used between the plates of a storage battery. |ELECTR| A circuit that separates one type of signal from another by clipping, differentiating, or integrating action. |ENG| **1.** A machine for separating materials of different specific gravity by means of water or air. **2.** Any machine for separating materials, as the magnetic separator. |MECH ENG| *See* cage. { 'sep·ə,rād·ər }

separator-filter |ENG| A vessel that removes solids and entrained liquid from a liquid or gas stream, using a combination of a baffle or coalescer with a screening (filtering) element. { 'sep·ə,rād·ər 'fil·tər }

separatrix |CONT SYS| A curve in the phase plane of a control system representing the solution to the equations of motion of the system which would cause the system to move to an unstable point. { 'sep·ə,triks }

septic tank |CIV ENG| A settling tank in which settled sludge is in immediate contact with sewage flowing through the tank while solids are decomposed by anaerobic bacterial action. { 'sep·tik ,taŋk }

sequence |ENG| An orderly progression of items of information or of operations in accordance with some rule. { 'sē·kwəns }

sequencer |ENG| A mechanical or electronic device that may be set to initiate a series of events and to make the events follow in a given sequence. { 'sē·kwən·sər }

sequence robot *See* preprogrammed robot. { 'sē·kwəns ,rō,bät }

sequence-stressing loss |ENG| In posttensioning, the loss of elasticity in a stressed tendon that results from the shortening of the member as additional tendons are stressed. { 'sē·kwəns ,stres·iŋ ,lös }

sequencing |IND ENG| Designating the order of performance of tasks to assure optimal utilization of available production facilities. { 'sē·kwəns·iŋ }

sequential collation of range |ENG| Spherical, long-baseline, phase-comparison trajectory-measuring system using three or more ground stations, time-sharing a single transponder, to provide nonambiguous range measurements to determine the instantaneous position of a vehicle in flight. { si'kwen·chəl kə'lā·shən əv 'rānj }

sequential logic element [ELECTR] A circuit element having at least one input channel, at least one output channel, and at least one internal state variable, so designed and constructed that the output signals depend on the past and present states of the inputs. { si'kwen·chəl ¦läj·ik ‚el·ə·mənt }

sequential sampling [IND ENG] A sampling plan in which an undetermined number of samples are tested one by one, accumulating the results until a decision can be made. { si'kwen·chəl 'sam·pliŋ }

serial [IND ENG] An element or a group of elements within a series which is given a numerical or alphabetical designation for convenience in planning, scheduling, and control. { 'sir·ē·əl }

series [ELEC] An arrangement of circuit components end to end to form a single path for current. { 'sir·ēz }

series circuit [ELEC] A circuit in which all parts are connected end to end to provide a single path for current. { 'sir·ēz ‚sər·kət }

series compensation [CONT SYS] *See* cascade compensation. [ELEC] The insertion of variable, controlled, high-voltage series capacitors into transmission lines in order to modify the impedance structure of a transmission network so as to adjust the power-flow distribution on individual lines and thus increase the power flow across such compensated lines. { 'sir·ēz ‚käm·pən'sā·shən }

series connection [ELEC] A connection that forms a series circuit. { 'sir·ēz kə‚nek·shən }

series firing [ENG] The firing of detonators in a round of shots by passing the total supply current through each of the detonators. { 'sir·ēz 'fīr·iŋ }

series-parallel firing [ENG] The firing of detonators in a round of shots by dividing the total supply current into branches, each containing a certain number of detonators wired in series. { 'sir·ēz ¦par·ə‚lel ¦fīr·iŋ }

series production [IND ENG] The manufacture of a product or service by a group of operations sequenced so that all materials will be routed successively through each production state. Also known as batch production. { 'sir·ēz prə'dək·shən }

series reliability [SYS ENG] Property of a system composed of elements in such a way that failure of any one element causes a failure of the system. { 'sir·ēz ri‚lī·ə'bil·əd·ē }

series shots [ENG] The connecting and firing of a number of loaded holes one after the other. { 'sir·ēz ‚shäts }

serpentine cooler *See* cascade cooler. { 'sər·pən‚tēn 'kül·ər }

service [ENG] To perform services of maintenance, supply, repair, installation, distribution, and so on, for or upon an instrument, installation, vehicle, or territory. { 'sər·vəs }

serviceability [IND ENG] The reliability of equipment according to some objective criterion such as serviceability ratio, utilization ratio, or operating ratio. { ‚sər·və·sə'bil·əd·ē }

serviceability ratio [IND ENG] The ratio of up time to the sum of up time and down time. { ‚sər·və·sə'bil·əd·ē ‚rā·shō }

service agreement [ENG] A contract which agrees to provide mechanical maintenance of a machine for a fixed period of time at a stated charge. { 'sər·vəs ə‚grē·mənt }

service brake [MECH ENG] The brake used for ordinary driving in an automotive vehicle; usually foot-operated. { 'sər·vəs ‚brāk }

service dead load [ENG] The calculated dead load that will be supported by a member. { ¦sər·vəs 'ded ‚lōd }

service engineering [ENG] The function of determining the integrity of material and services in order to measure and maintain operational reliability, approve design changes, and assure their conformance with established specifications and standards. { 'sər·vəs ‚en·jə‚nir·iŋ }

service factor [ENG] For a chemical or a petroleum processing plant or its equipment, the measure of the continuity of an operation, computed by dividing the time on-stream (actual running time) by the total elapsed time. { 'sər·vəs ‚fak·tər }

service life [ENG] The length of time during which a machine, tool, or other apparatus or device can be operated or used economically or before breakdown. { 'sər·vəs ‚līf }

service pipe [CIV ENG] A pipe linking a building to a main pipe. { 'sər·vəs ‚pīp }

service road [CIV ENG] A small road parallel to the main road for convenient access to shops and houses. { 'sər·vəs ‚rōd }

service time *See* machine attention time. { 'sər·vəs ‚tīm }

service valve [ENG] In a pipework system, a valve that isolates a piece of equipment from the rest of the system. { 'sər·vəs ‚valv }

service wires [ELEC] The conductors that bring the electric power into a building. { 'sər·vəs ‚wīrz }

servicing [ENG] Replacement of consumable material or items needed to keep equipment in operating condition; does not include preventive or corrective maintenance. { 'sər·vəs·iŋ }

servo *See* servomotor. { 'sər·vō }

servoarm attachment [MECH ENG] A device that enhances the maximum distance over which the manipulator of a simple robot can travel. { 'sər·vō‚ärm ə‚tach·mənt }

servo brake [MECH ENG] **1.** A brake in which the motion of the vehicle is used to increase the pressure on one of the shoes. **2.** A brake in which the force applied by the operator is augmented by a power-driven mechanism. { 'sər·vō 'brāk }

servolink [CONT SYS] A power amplifier, usually mechanical, by which signals at a low power level are made to operate control surfaces requiring relatively large power inputs, for example, a relay and motor-driven actuator. { 'sər·vō‚liŋk }

servo loop See single-loop servomechanism. { 'sər·vō ‚lüp }

servomechanism |CONT SYS| An automatic feedback control system for mechanical motion; it applies only to those systems in which the controlled quantity or output is mechanical position, and so on). Also known as servo system. { ‚sər·vō'mek·ə‚niz·əm }

servomotor |CONT SYS| The electric, hydraulic, or other type of motor that serves as the final control element in a servomechanism; it receives power from the amplifier element and drives the load with a linear or rotary motion. Also known as servo. { 'sər·vō‚mōd·ər }

servonoise |ENG| Hunting action of the tracking servomechanism of a radar, which results from backlash and compliance in the gears, shafts, and structures of the mount. { 'sər·vō‚nóiz }

servo system See servomechanism. { 'sər·vō ‚sis·təm }

servovalve |MECH ENG| A transducer in which a low-energy signal controls a high-energy fluid flow so that the flow is proportional to the signal. { 'sər·vō‚valv }

set |ELECTR| The placement of a storage device in a prescribed state, for example, a binary storage cell in the high or 1 state. |ENG| **1.** A combination of units, assemblies, and parts connected or otherwise used together to perform an operational function, such as a radar set. **2.** In plastics processing, the conversion of a liquid resin or adhesive into a solid state by curing or evaporation of solvent or suspending medium, or by gelling. **3.** Saw teeth bent out of the plane of the saw body, resulting in a wide cut in the workpiece. |MECH| See permanent set. { set }

setback |BUILD| **1.** A withdrawal of the face of a building to a line toward the rear of the building line or the rear of the wall below in order to reduce obstruction of sunlight reaching the street or the lower stories of adjacent buildings. **2.** See offset. |CIV ENG| The distance that a section of a building is set back from the property line as required by local zoning codes. |MECH| The relative rearward movement of component parts in a projectile, missile, or fuse undergoing forward acceleration during its launching; these movements, and the setback force which causes them, are used to promote events which participate in the arming and eventual functioning of the fuse. { 'set‚bak }

setback force |MECH| The rearward force of inertia which is created by the forward acceleration of a projectile or missile during its launching phase; the forces are directly proportional to the acceleration and mass of the parts being accelerated. { 'set‚bak ‚fórs }

set bit |DES ENG| A bit insert with diamonds or other cutting media. { 'set ‚bit }

set casing |ENG| Introducing cement between the casing and the wall of the hole to seal off intermediate formations and prevent fluids from entering the hole. { 'set ‚kās·iŋ }

set forward |MECH| Relative forward movement of component parts which occurs in a projectile, missile, or bomb in flight when impact occurs; the effect is due to inertia and is opposite in direction to setback. { 'set 'fór·wərd }

set forward force |MECH| The forward force of inertia which is created by the deceleration of a projectile, missile, or bomb when impact occurs; the forces are directly proportional to the deceleration and mass of the parts being decelerated. Also known as impact force. { 'set 'fór·wərd ‚fórs }

set forward point |MECH| A point on the expected course of the target at which it is predicted the target will arrive at the end of the time of flight. { 'set 'fór·wərd ‚póint }

set hammer |DES ENG| **1.** A hammer used as a shaping tool by blacksmiths. **2.** A hollow-face tool used in setting rivets. { 'set ‚ham·ər }

setover |ENG| A device which helps move a lathe tailstock or headstock on its base so that a taper on a turned piece can be obtained. { 'set‚ō·vər }

set point |CONT SYS| The value selected to be maintained by an automatic controller. { 'set ‚póint }

set pressure |MECH ENG| The inlet pressure at which a relief valve begins to open as required by the code or standard applicable to the pressure vessel to be protected. { 'set ‚presh·ər }

set screw |DES ENG| A small headless machine screw, usually having a point at one end and a recessed hexagonal socket or a slot at the other end, used for such purposes as holding a knob or gear on a shaft. { 'set ‚skrü }

setting angle |MECH ENG| The angle, usually 90°, between the straight portion of the tool shank and the machined portion of the work. { 'sed·iŋ ‚aŋ·gəl }

setting circle |ENG| A coordinate scale on an optical pointing instrument, such as a telescope or surveyor's transit. { 'sed·iŋ ‚sər·kəl }

setting gage |ENG| A standard gage for testing a limit gage or setting an adjustable limit gage. { 'sed·iŋ ‚gāj }

setting temperature |ENG| The temperature at which a liquid resin or adhesive, or an assembly involving them, will set, that is, harden, gel, or cure. { 'sed·iŋ ‚tem·prə·chər }

setting time |ENG| The length of time that a resin or adhesive must be subjected to heat or pressure to cause them to set, that is, harden, gel, or cure. { 'sed·iŋ ‚tīm }

settleable solids test |CIV ENG| A test used in examination of sewage to help determine the sludge-producing characteristics of sewage; a measurement of the part of the suspended solids heavy enough to settle is made in an Imhoff cone. { 'sed·əl·ə·bəl 'säl·ədz ‚test }

settlement |CIV ENG| The gradual downward movement of an engineering structure, due to compression of the soil below the foundation. { 'sed·əl·mənt }

settler |ENG| A separator, such as a tub, pan, vat, or tank in which the partial separation of a mixture is made by density difference; used to separate solids from liquid or gas, immiscible liquid from liquid, or liquid from liquid. { 'set·lər }

settling |ENG| The gravity separation of heavy from light materials; for example, the settling out of dense solids or heavy liquid droplets from a liquid carrier, or the settling out of heavy solid grains from a mixture of solid grains of different densities. { 'set·liŋ }

settling basin |CIV ENG| An artificial trap designed to collect suspended stream sediment before discharge of the stream into a reservoir. |IND ENG| A sedimentation area designed to remove pollutants from factory effluents. { 'set·liŋ ,bās·ən }

settling chamber |ENG| A vessel in which solids or heavy liquid droplets settle out of a liquid carrier by gravity during processing or storage. { 'set·liŋ ,chām·bər }

settling reservoir |CIV ENG| A reservoir consisting of a series of basins connected in steps by long weirs; only the clear top layer of each basin is drawn off. { 'set·liŋ ,rez·əv,wär }

settling tank |ENG| A tank into which a two-phase mixture is fed and the entrained solids settle by gravity during storage. { 'set·liŋ ,taŋk }

settling time See correction time. { 'set·liŋ ,tīm }

settling velocity |MECH| The velocity reached by a particle as it falls through a fluid, dependent on its size and shape, and the difference between its specific gravity and that of the settling medium; used to sort particles by grain size. { 'set·liŋ və,läs·əd·ē }

setup |ELECTR| The ratio between the reference black level and the reference white level in television, both measured from the blanking level; usually expressed as a percentage. |IND ENG| The preparation of a facility or a machine for a specific work method, activity, or process. { 'sed,əp }

setup person |CONT SYS| A person who uses a teach pendant to instruct a robot in its motions. { 'sed,əp ,pər·sən }

setup time |CONT SYS| The total time needed to prepare a robot to carry out a task, including the time required to obtain the proper tools or end effectors and any work pieces. |IND ENG| In manufacturing operations, the time needed to perform tasks involved in starting up an operation. Also known as start-up time. { 'sed,əp ,tīm }

severity factor |CHEM ENG| A measure of the severeness or intensity of overall reaction conditions in a chemical reaction; for example, the temperature, pressure, or conversion in a catalytic cracker or reformer. { si'ver·əd·ē ,fak·tər }

sewage |CIV ENG| The fluid discharge from medical, domestic, and industrial sanitary appliances. Also known as sewerage. { 'sü·ij }

sewage disposal plant |CIV ENG| The land, building, and apparatus employed in the treatment of sewage by chemical precipitation or filtration, bacterial action, or some other method. { 'sü·ij di¦spōz·əl ,plant }

sewage sludge |CIV ENG| A semiliquid waste with a solid concentration in excess of 2500 parts per million, obtained from the purification of municipal sewage. Also known as sludge. { 'sü·ij ,sləj }

sewage system |CIV ENG| A drainage system for carrying surface water and sewage for disposal. { 'sü·ij ,sis·təm }

sewage treatment |CIV ENG| A process for the purification of mixtures of human and other domestic wastes; the process can be aerobic or anaerobic. { 'sü·ij ,trēt·mənt }

sewer |CIV ENG| An underground pipe or open channel in a sewage system for carrying water or sewage to a disposal area. { 'sü·ər }

sewerage See sewage. { 'sü·ə·rij }

sewing machine |MECH ENG| A mechanism that stitches cloth, leather, book pages, or other material by means of a double-pointed or eye-pointed needle. { 'sō·iŋ mə,shēn }

SFC See specific fuel consumption.

shackle |DES ENG| An open or closed link of various shapes with extended legs; each leg has a transverse hole to accommodate a pin, bolt, or the like, which may or may not be furnished. { 'shak·əl }

shackle bolt |DES ENG| A cylindrically shaped metal bar for connecting the ends of a shackle. { 'shak·əl ,bōlt }

shading coefficient |ENG| A ratio of the solar energy transmitted through a window to the incident solar energy; used to express the effectiveness of a shading device. { 'shād·iŋ ,kō·i,fish·ənt }

shading ring |ENG ACOUS| A heavy copper ring sometimes placed around the central pole of an electrodynamic loudspeaker to serve as a shorted turn that suppresses the hum voltage produced by the field coil. { 'shād·iŋ ,riŋ }

shadow photometer |ENG| A simple photometer in which a rod is placed in front of a screen and two light sources to be compared are adjusted in position until their shadows touch and are equal in intensity. { 'shad·ō fō'täm·əd·ər }

shaft |MECH ENG| A cylindrical piece of metal used to carry rotating machine parts, such as pulleys and gears, to transmit power or motion. { shaft }

shaft balancing |DES ENG| The process of redistributing the mass attached to a rotating body in order to reduce vibrations arising from centrifugal force. Also known as rotor balancing. { 'shaft ,bal·əns·iŋ }

shaft furnace |ENG| A vertical, refractory-lined cylinder in which a fixed bed (or descending column) of solids is maintained, and through which an ascending stream of hot gas is forced; for example, the pig-iron blast furnace and the phosphors-from-phosphate-rock furnace. { 'shaft ¦fər·nəs }

shaft hopper |MECH ENG| A hopper that feeds

shafts or tubes to grinders, threaders, screw machines, and tube benders. { 'shaft ¦hȧp·ər }

shaft horsepower |MECH ENG| The output power of an engine, motor, or other prime mover; or the input power to a compressor or pump. { 'shaft 'hȯrs,pau̇·ər }

shafting |MECH ENG| The cylindrical machine element used to transmit rotary motion and power from a driver to a driven element; for example, a steam turbine driving a ship's propeller. { 'shaft·iŋ }

shaft kiln |ENG| A kiln in which raw material fed into the top, moves down through hot gases flowing up from burners on either side at the bottom, and emerges as a product from the bottom; used for calcining operations. { 'shaft ¦kil }

shaft spillway |CIV ENG| A vertical shaft which has a funnel-shaped mouth and ends in an outlet tunnel, providing an overflow duct for a reservoir. Also known as morning glory spillway. { 'shaft 'spil,wā }

shakedown test |ENG| An equipment test made during the installation work. { 'shāk ,dau̇n ,test }

shake table See vibration machine. { 'shāk ,tā·bəl }

shake-table test |ENG| A laboratory test for vibration tolerance, in which the device to be tested is placed on a shake table. { 'shāk ¦tā·bəl ,test }

shaking-out |CHEM ENG| A procedure in which a sample of crude oil is centrifuged at high speed to separate its components; used to determine sediment and water content. { 'shāk·iŋ 'au̇t }

shaking screen |MECH ENG| A screen used in separating material into desired sizes; has an eccentric drive or an unbalanced rotating weight to produce shaking. { 'shāk·iŋ ,skrēn }

shank |DES ENG| **1.** The end of a tool which fits into a drawing holder, as on a drill. **2.** See bit blank. { shaŋk }

shank-type cutter |DES ENG| A cutter having a shank to fit into the machine tool spindle or adapter. { 'shaŋk ,tīp ,kəd·ər }

shape coding |DES ENG| The use of special shapes for control knobs, to permit recognition and sometimes also position monitoring by sense of touch. { 'shāp ,kōd·iŋ }

shaped-chamber manometer |ENG| A flow measurement device that measures differential pressure with a uniform flow-rate scale with a specially shaped chamber. { 'shāpt ¦chām·bər mə'näm·əd·ər }

shape factor |ELEC| See form factor. |ELECTR| The ratio of the 60-decibel bandwidth of a bandpass filter to the 3-decibel bandwidth. { 'shāp ,fak·tər }

shaper |MECH ENG| A machine tool for cutting flat-on-flat, contoured surfaces by reciprocating a single-point tool across the workpiece. { 'shā·pər }

shaping circuit See corrective network. { 'shāp·iŋ ,sər·kət }

shaping dies |MECH ENG| A set of dies for bending, pressing, or otherwise shaping a material to a desired form. { 'shāp·iŋ ,dīz }

shapometer |ENG| A device used to measure the shape of sedimentary particles. { shā'päm·əd·ər }

sharp-crested weir |CIV ENG| A weir in which the water flows over a thin, sharp edge. { 'shärp ¦kres·təd 'wer }

sharpen |ENG| To give a thin keen edge or a sharp acute point to. { 'shär·pən }

sharpening stone |ENG| A device such as a whetstone used for sharpening by hand. { 'shär·pə·niŋ ,stōn }

sharp iron |ENG| A tool used to open seams for caulking. { 'shärp 'ī·ərn }

sharp V thread |DES ENG| A screw thread having a sharp crest and root; the included angle is usually 60° { 'shärp 'vē ,thred }

shattering |MECH| The breaking up into highly irregular, angular blocks of a very hard material that has been subjected to severe stresses. { 'shad·ə·riŋ }

shave hook |DES ENG| A plumber's or metalworker's tool composed of a sharp-edged steel plate on a shank; used for scraping metal. { 'shāv ,hu̇k }

shaving |ENG ACOUS| Removing material from the surface of a disk recording medium to obtain a new recording surface. |MECH ENG| **1.** Cutting off a thin layer from the surface of a workpiece. **2.** Trimming uneven edges from stampings, forgings, and tubing. { 'shāv·iŋ }

shear |DES ENG| A cutting tool having two opposing blades between which a material is cut. |ENG| An apparatus for hoisting heavy loads consisting of two or more poles fastened together at their upper ends and spread apart at their lower ends, secured or steadied by a guy or guys, and provided with a tackle. Also known as shear legs. |MECH| See shear strain. { shir }

shear angle |MECH ENG| The angle made by the shear plane with the work surface. { 'shir ,aŋ·gəl }

shear cell |ENG| The component for holding the powder in an apparatus for making measurements of the failure properties of a sample of powder. { 'shir ,sel }

shear center See center of twist. { 'shir ,sen·tər }

shear diagram |MECH| A diagram in which the shear at every point along a beam is plotted as an ordinate. { 'shir ,dī·ə,gram }

shear fracture |MECH| A fracture resulting from shear stress. { 'shir ,frak·chər }

shearing |MECH ENG| Separation of material by the cutting action of shears. { 'shir·iŋ }

shearing die |MECH ENG| A die with a punch for shearing the work from the stock. { 'shir·iŋ ,dī }

shearing forces |MECH| Two forces that are equal in magnitude, opposite in direction, and act along two distinct parallel lines. { 'sher·iŋ ,fȯrs·əz }

shearing machine |MECH ENG| A machine for

cutting cloth or bars, sheets, or plates of metal or other material. { 'shir·iŋ mə,shēn }

shearing punch |MECH ENG| A punch that cuts material by shearing it, with minimal crushing effect. { 'shir·iŋ ,pənch }

shearing strain |MECH| The distortion that results from motion of material on opposite sides of a plane in opposite directions parallel to the plane. { 'shir·iŋ ,strān }

shearing stress |MECH| A stress in which the material on one side of a surface pushes on the material on the other side of the surface with a force which is parallel to the surface. Also known as shear stress; tangential stress. { 'shir·iŋ ,stres }

shearing tool |DES ENG| A cutting tool (for a lathe, for example) with a considerable angle between its face and a line perpendicular to the surface being cut. { 'shir·iŋ ,tül }

shear legs *See* shear. { 'shir ,legz }

shear mark |ENG| A crease on a piece of pressed glass; results when the piece is sheared off for pressing. { 'shir ,märk }

shear modulus *See* modulus of elasticity in shear. { 'shir ,mäj·ə·ləs }

shear pin |DES ENG| **1.** A pin or wire provided in a fuse design to hold parts in a fixed relationship until forces are exerted on one or more of the parts which cause shearing of the pin or wire; the shearing is usually accomplished by setback or set forward (impact) forces; the shear member may be augmented during transportation by an additional safety device. **2.** In a propellant-actuated device, a locking member which is released by shearing. **3.** In a power train, such as a winch, any pin, as through a gear and shaft, which is designed to fail at a predetermined force in order to protect a mechanism. { 'shir ,pin }

shear plane |MECH| A confined zone along which fracture occurs in metal cutting. { 'shir ,plān }

shear spinning |MECH ENG| A sheet-metal-forming process which forms parts with rotational symmetry over a mandrel with the use of a tool or roller in which deformation is carried out with a roller in such a manner that the diameter of the original blank does not change but the thickness of the part decreases by an amount dependent on the mandrel angle. { 'shir ,spin·iŋ }

shear strain |MECH| Also known as shear. **1.** A deformation of a solid body in which a plane in the body is displaced parallel to itself relative to parallel planes in the body; quantitatively, it is the displacement of any plane relative to a second plane, divided by the perpendicular distance between planes. **2.** The force causing such deformation. { 'shir ,strān }

shear strength |MECH| **1.** The maximum shear stress which a material can withstand without rupture. **2.** The ability of a material to withstand shear stress. { 'shir ,streŋkth }

shear stress *See* shearing stress. { 'shir ,stres }

shear test |ENG| Any of various tests to determine shear strength of soil samples. { 'shir ,test }

shear wave |MECH| A wave that causes an element of an elastic medium to change its shape without changing its volume. Also known as rotational wave. { 'shir ,wāv }

sheath |ELEC| A protective outside covering on a cable. |ELECTR| A space charge formed by ions near an electrode in a gas tube. { shēth }

sheathed explosive |ENG| A permitted explosive enveloped by a sheath containing a non-combustible powder which reduces the temperature of the resultant gases of the explosion and, therefore, reduces the risk of these hot gases causing a firedamp ignition. { 'shēthd ik'splō·siv }

sheave |DES ENG| A grooved wheel or pulley. { shēv }

sheepsfoot roller |DES ENG| A cylindrical steel drum to which knob-headed spikes are fastened; used for compacting earth. Also known as tamping roller. { 'shēps,fút ,rōl·ər }

sheepskin wheel |DES ENG| A polishing wheel made of sheepskin disks or wedges either quilted or glued together. { 'shēp,skin ,wēl }

sheet forming |ENG| The process of producing thin, flat sections of solid materials; for example, sheet metal, sheet plastic, or sheet glass. { 'shēt ,förm·iŋ }

sheet-metal screw *See* self-tapping screw. { 'shēt ¦med·əl ,skrü }

sheet piling |CIV ENG| Closely spaced piles of wood, steel, or concrete driven vertically into the ground to obstruct lateral movement of earth or water, and often to form an integral part of the permanent structure. { 'shēt ,pīl·iŋ }

sheet train |ENG| The entire assembly needed to produce plastic sheet; includes the extruder, die, polish rolls, conveyor, draw rolls, cutter, and stacker. { 'shēt ,trān }

Shelby tube |ENG| A thin-shelled tube used to take deep-soil samples; the tube is pushed into the undisturbed soil at the bottom of the casting of the borehole driven into the ground. { 'shel·bē ,tüb }

shelf angle |CIV ENG| A mild steel angle section, riveted or welded to the web of an I beam to support the formwork for hollow tiles or the floor or roof units, or to form a seat for precast concrete. { 'shelf ,aŋ·gəl }

shelf life |ENG| The time that elapses before stored food, chemicals, batteries, and other materials or devices become inoperative or unusable due to age or deterioration. { 'shelf ,līf }

shell |BUILD| A building without internal partitions or furnishings. |DES ENG| **1.** The case of a pulley block. **2.** A thin hollow cylinder. **3.** A hollow hemispherical structure. **4.** The outer wall of a vessel or tank. { shel }

shellac wheel |DES ENG| A grinding wheel having the abrasive bonded with shellac. { shə'lak ,wēl }

shell-and-tube exchanger |ENG| A device for the transfer of heat from a hot fluid to a cooler fluid; one fluid passes through a group (bundle) of tubes, the other passes around the tubes, through a surrounding shell. Also known as tubular exchanger. { ¦shel ən ¦tüb iks'chān·jər }

shell capacity [ENG] The amount of liquid that a tank car or tank truck will hold when the liquid just touches the underside of the top of the tank shell. { 'shel kə,pas·əd·ē }

shell clearance [DES ENG] The difference between the outside diameter of a bit or core barrel and the outside set or gage diameter of a reaming shell. { 'shel ,klir·əns }

shell innage [ENG] The depth of a liquid in a tank car or tank truck shell. { 'shel ,in·ij }

shell knocker [ENG] A device to strike the external surface of a horizontally rotating process vessel (for example, a kiln or a dryer) to loosen accumulations of solid materials from the inner walls or flights of the shell. Also known as knocker. { 'shel ,näk·ər }

shell outage [ENG] The unfilled portion of a tank car or tank truck shell; the distance from the underside of the top of the shell to the level of the liquid in the shell. { 'shel ,aúd·ij }

shell pump [MECH ENG] A simple pump for removing wet sand or mud; consists of a hollow cylinder with a ball or clack valve at the bottom. { 'shel ,pəmp }

shell reamer [DES ENG] A machine reamer consisting of two parts, the arbor and the replaceable reamer, with straight or spiral flutes; designed as a sizing or finishing reamer. { 'shel ,rēm·ər }

shell roof [BUILD] A roof made of a thin, curved, platelike structure, usually of concrete but lumber and steel are also used. { 'shel ,rüf }

shell still [CHEM ENG] A distillation device formerly used in petroleum refineries; oil was charged into a closed, cylindrical shell and heat was applied to the outside of the bottom by a firebox. { 'shel ,stil }

Shenstone effect [ELECTR] An increase in photoelectric emission of certain metals following passage of an electric current. { 'shen,stŏn i,fekt }

SHF *See* sensible-heat factor.

shield [ENG] An iron, steel, or wood framework used to support the ground ahead of the lining in tunneling and mining. { shēld }

shielded wire [ELEC] Insulated wire covered with a metal shield, usually of tinned braided copper wire. { 'shēl·dəd 'wīr }

shift [IND ENG] The number of hours or the part of any day worked. Also known as tour. [MECH ENG] To change the ratio of the driving to the driven gears to obtain the desired rotational speed or to avoid overloading and stalling an engine or a motor. { shift }

shift joint [BUILD] A vertical joint placed on a solid member of the course below. { 'shift ,jóint }

shift work [IND ENG] Work paid for by day wage. { 'shift ,wərk }

shim [ENG] **1.** In the manufacture of plywood, a long, narrow patch glued into the panel or cemented into the lumber core itself. **2.** A thin piece of material placed between two surfaces to obtain a proper fit, adjustment, or alignment. { shim }

shimmy [MECH] Excessive vibration of the front wheels of a wheeled vehicle causing a jerking motion of the steering wheel. { 'shim·ē }

shingle lap [DES ENG] A lap joint in which the two surfaces are tapered, with the thinner surface lapped over the thicker one. { 'shiŋ·gəl ,lap }

shingle nail [DES ENG] A nail about a half to a full gage thicker than a common nail of the same length. { 'shiŋ·gəl ,nāl }

ship auger [DES ENG] An auger consisting of a spiral body having a single cutting edge, with or without a screw; there is no spur at the outer end of the cutting edge. { 'ship ,óg·ər }

shipbuilding [CIV ENG] The construction of ships. { 'ship,bil·diŋ }

shipfitter [CIV ENG] A worker who builds the steel structure of a ship, including laying-off and fabricating the individual members, subassembly, and erection on the shipway. { 'ship,fid·ər }

ship motion [ENG] Translational and rotational motions of a ship in a wave system which cause the center of gravity to deviate from simple straight-line motion; these motions are heave, surge, sway, roll, pitch, and yaw. { 'ship ,mō·shən }

shipping and storage container [IND ENG] A reusable noncollapsible container of any configuration designed to provide protection for a specific item against impact, vibration, climatic conditions, and the like, during handling, shipment, and storage. { 'ship·iŋ ən 'stór·ij kən,tā·nər }

shipping document [IND ENG] A document listing the items in a shipment, and showing other supply and transportation information that is required by agencies concerned in the movement of material. { 'ship·iŋ ,däk·yə·mənt }

shipping time [ENG] The time elapsing between the shipment of material by the supplying activity and receipt of material by the requiring activity. { 'ship·iŋ ,tīm }

shipping ton *See* ton. { 'ship·iŋ ,tən }

shipway [CIV ENG] **1.** The ways on which a ship is constructed. **2.** The supports placed underneath a ship in dry dock. { 'ship,wā }

shipwright [CIV ENG] A worker whose responsibility is to ensure that the structure of a ship is straight and true and to the designed dimensions; the work starts with the laying down of the keel blocks and continues throughout the steelwork; applicable also to wood ship builders. { 'ship,rīt }

shipyard [CIV ENG] A facility adjacent to deep water where ships are constructed or repaired. { 'ship,yärd }

SHM *See* harmonic motion.

shock [MECH] A pulse or transient motion or force lasting thousandths to tenths of a second which is capable of exciting mechanical resonances; for example, a blast produced by explosives. { 'shäk }

shock absorber |MECH ENG| A spring, a dashpot, or a combination of the two, arranged to minimize the acceleration of the mass of a mechanism or portion thereof with respect to its frame or support. { 'shäk əb,zȯr·bər }

shock isolation |MECH ENG| The application of isolators to alleviate the effects of shock on a mechanical device or system. { 'shäk ‚ī·sə‚lā·shən }

Shockley diode |ELECTR| A *pnpn* silicon controlled switch having characteristics that permit operation as a unidirectional diode switch. { 'shäk·lē 'dī·ōd }

shock mount |MECH ENG| A mount used with sensitive equipment to reduce or prevent transmission of shock motion to the equipment. { 'shäk ‚maúnt }

shock resistance |ENG| The property which prevents cracking or general rupture when impacted. { 'shäk ri‚zis·təns }

shock test |ENG| The test to determine whether the armor sample will crack or spall under impact by kinetic energy or high-explosive projectiles. { 'shäk ‚test }

shock tunnel |ENG| A hypervelocity wind tunnel in which a shock wave generated in a shock tube ruptures a second diaphragm in the throat of a nozzle at the end of the tube, and gases emerge from the nozzle into a vacuum tank with Mach numbers of 6 to 25. { 'shäk ‚tən·əl }

shoe |ENG| In glassmaking, an open-ended crucible placed in a furnace for heating the blowing irons. |MECH ENG| **1.** A metal block used as a form or support in various bending operations. **2.** A replaceable piece used to break rock in certain crushing machines. **3.** *See* brake shoe. { shü }

shoe brake |MECH ENG| A type of brake in which friction is applied by a long shoe, extending over a large portion of the rotating drum; the shoe may be external or internal to the drum. { 'shü ‚brāk }

shoot |ENG| To detonate an explosive, used to break coal loose from a seam or in blasting operation or in a borehole. { shüt }

shooting board |ENG| **1.** A fixture used as a guide in planing boards; it is more accurate than a miter. **2.** A table and plane used for trimming printing plates. { 'shüd·iŋ ‚bȯrd }

shop fabrication |ENG| Making parts and materials in the shop rather than at the work site. { 'shäp ‚fab·rə‚kā·shən }

shop standards |ENG| Written criteria established to govern methods and procedures at an installation. { 'shäp ‚stan·dərdz }

shop supplies |ENG| Expendable items consumed in operation and maintenance (for example, waste, oils, solvents, tape, packing, flux, or welding rod). { 'shäp sə‚plīz }

shop weld |ENG| A weld made in the workshop prior to delivery to the construction site. { 'shäp ‚weld }

shore |ENG| Timber or other material used as a temporary prop for excavations or buildings; may be sloping, vertical, or horizontal. { shȯr }

Shore hardness |ENG| A method of rating the hardness of a metal or of a plastic or rubber material. { 'shȯr ‚härd·nəs }

shore protection |CIV ENG| Preventing erosion of the ground bordering a body of water. { 'shȯr prə‚tek·shən }

Shore scleroscope |ENG| A device used in rebound hardness testing of rubber, metal, and plastic; consists of a small, conical hammer fitted with a diamond point and acting in a glass tube. { 'shȯr 'skler·ə‚skōp }

shoring |ENG| Providing temporary support with shores to a building or an excavation. { 'shȯr·iŋ }

short |ELEC| *See* short circuit. |ENG| In plastics injection molding, the failure to fill the mold completely. Also known as short shot. { shȯrt }

short circuit |ELEC| A low-resistance connection across a voltage source or between both sides of a circuit or line, usually accidental and usually resulting in excessive current flow that may cause damage. Also known as short. { 'shȯrt 'sər·kət }

short-circuiting transfer |ENG| Transfer of melted material from a consumable electrode during short circuits. { 'shȯrt ¦sər·kəd·iŋ 'tranz·fər }

short column |CIV ENG| A column in which both compression and bending is significant, generally having a slenderness ratio between 30 and 120–150. { 'shȯrt 'käl·əm }

shortcoming |DES ENG| An imperfection or malfunction occurring during the life cycle of equipment, which should be reported and which must be corrected to increase efficiency and to render the equipment completely serviceable. { 'shȯrt‚kəm·iŋ }

short-delay blasting |ENG| A method of blasting by which explosive charges are detonated in a given sequence with short time intervals. { 'shȯrt di¦lā 'blas·tiŋ }

short-delay detonator *See* millisecond delay cap. { 'shȯrt di¦lā 'det·ən‚ād·ər }

short fuse |ENG| **1.** Any fuse that is cut too short. **2.** The practice of firing a blast, the fuse on the primer of which is not sufficiently long to reach from the top of the charge to the collar of the borehole; the primer, with fuse attached, is dropped into the charge while burning. { 'shȯrt 'fyüz }

short leg |ENG| One of the wires on an electric blasting cap, which has been shortened so that when placed in the borehole, the two splices or connections will not come opposite each other and make a short circuit. { 'shȯrt 'leg }

short-range radar |ENG| Radar whose maximum line-of-sight range, for a reflecting target having 1 square meter of area perpendicular to the beam, is between 50 and 150 miles (80 and 240 kilometers). { 'shȯrt ¦rānj 'rā‚där }

short residuum |CHEM ENG| A petroleum refinery term for residual oil from crude-oil distillation operations in which neutral oils are taken overhead with the distillate. { 'shȯrt ri'zij·ə·wəm }

shorts |ENG| Oversize particles held on a screen after sieving the fines through the screen. { 'shȯrts }

short shipment |ENG| Freight listed or manifested but not received. { 'shȯrt 'ship·mənt }

short stop |CHEM ENG| A substance added during a polymerization process to terminate the reaction. { 'shȯrt ˌstäp }

short supply |IND ENG| An item is in short supply when the total of stock on hand and anticipated receipts during a given period is less than the total estimated demand during that period. { 'shȯrt sə'plī }

short-term repeatability |CONT SYS| The close agreement of positional movements of a robotic system repeated under identical conditions over a short period of time and at the same location. { 'shȯrt ˌtərm ri,pēd·ə'bil·əd·ē }

short ton See ton. { 'shȯrt 'tən }

short-tube vertical evaporator |CHEM ENG| A liquid evaporation process unit with a vertical bundle of tubes 2–3 inches (5–8 centimeters) in diameter and 4–6 feet (1.2–1.8 meters) long; the heating fluid is inside the tubes, and the liquid to be evaporated is in the shell area outside the tubes; used mainly to evaporate cane-sugar juice. Also known as calandria evaporator; Roberts evaporator; standard evaporator. { 'shȯrt ˌtüb 'vərd·ə·kəl i'vap·ə,rād·ər }

shot |ENG| **1.** A charge of some kind of explosive. **2.** Small spherical particles of steel. **3.** Small steel balls used as the cutting agent of a shot drill. **4.** The firing of a blast. **5.** In plastics molding, the yield from one complete molding cycle, including scrap. { shät }

shot bit |DES ENG| A short length of heavy-wall steel tubing with diagonal slots cut in the flat-faced bottom edge. { 'shät ˌbit }

shot boring |ENG| The act or process of producing a borehole with a shot drill. { 'shät ˌbȯr·iŋ }

shot break |ENG| In seismic prospecting, the electrical impulse which records the instant of explosion. { 'shät ˌbrāk }

shot capacity |ENG| The maximum weight of molten resin that an accumulator can push out with one forward stroke of the ram during plastics forming operations. { 'shät kə,pas· əd·ē }

shotcreting |ENG| A process of conveying mortar or concrete through a hose at high velocity onto a surface; the material bonds tenaciously to a properly prepared concrete surface and to a number of other materials. { 'shät,krēd·iŋ }

shot depth |ENG| The distance from the surface to the charge. { 'shät ˌdepth }

shot drill See calyx drill. { 'shät ˌdril }

shot elevation |ENG| Elevation of the dynamite charge in the shot hole. { 'shät ˌel·ə,vā·shən }

shot feed |MECH ENG| A device to introduce chilled-steel shot, at a uniform rate and in the proper quantities, into the circulating fluid flowing downward through the rods or pipe connected to the core barrel and bit of a shot drill. { 'shät ˌfēd }

shothole |ENG| The borehole in which an explosive is placed for blasting. { 'shät,hōl }

shothole casing |ENG| A lightweight pipe, usually about 4 inches (10 centimeters) in diameter and 10 feet (3 meters) long, with threaded connections on both ends, used to prevent the shothole from caving and bridging. { 'shät,hōl ˌkās·iŋ }

shothole drill |MECH ENG| A rotary or churn drill for drilling shotholes. { 'shät,hōl ˌdril }

shot mill |ENG| A high-speed, continuous mill for deagglomerating, dispersing, and milling paints, inks, dyestuffs, adhesives, food, and pharmaceuticals; consists of a chamber with rotating disks that is filled with small steel or ceramic spheres (shot), and a pump to propel material through the mill. Also know as a media mill. { 'shät ˌmil }

shot point |ENG| The point at which an explosion (such as in seismic prospecting) originates, generating vibrations in the ground. { 'shät ˌpȯint }

shot rock |ENG| Blasted rock. { 'shät ˌräk }

shoulder |DES ENG| The portion of a shaft, a stepped object, or a flanged object that shows an increase of diameter. |ENG| A projection made on a piece of shaped wood, metal, or stone, where its width or thickness is suddenly changed. { 'shōl·dər }

shoulder harness |ENG| A harness in a vehicle that fastens over the shoulders to prevent a person's being thrown forward in the seat. { 'shōl·dər ,här·nəs }

shoulder screw |DES ENG| A screw with an unthreaded cylindrical section, or shoulder, between threads and screwhead; the shoulder is larger in diameter than the threaded section and provides an axis around which close-fitting moving parts operate. { 'shōl·dər ,skrü }

shovel |DES ENG| A hand tool having a flattened scoop at the end of a long handle for moving soil, aggregate, cement, or other similar material. |MECH ENG| A mechanical excavator. { 'shəv·əl }

shovel dozer See tractor loader. { 'shəv·əl ,dōz·ər }

shovel loader |MECH ENG| A loading machine mounted on wheels, with a bucket hinged to the chassis which scoops up loose material, elevates it, and discharges it behind the machine. { 'shəv·əl ,lōd·ər }

shrinkage |ENG| **1.** Contraction of a molded material, such as metal or resin, upon cooling. **2.** Contraction of a plastics casting upon polymerizing. { 'shriŋ·kij }

shrink fit |DES ENG| A tight interference fit between mating parts made by shrinking-on, that is, by heating the outer member to expand the bore for easy assembly and then cooling so that the outer member contracts. { 'shriŋk ˌfit }

shrink forming |DES ENG| Forming metal wherein the piece undergoes shrinkage during cooling following the application of heat, cold upset, or pressure. { 'shriŋk ˌfȯr·miŋ }

shrink ring |DES ENG| A heated ring placed on

an assembly of parts, which on subsequent cooling fixes them in position by contraction. { 'shriŋk ,riŋ }

shrink wrapping |ENG| A technique of packaging with plastics in which the strains in plastics film are released by raising the temperature of the film, causing it to shrink-fit over the object being packaged. { 'shriŋk ,rap·iŋ }

shroud |ENG| A protective covering, usually of metal plate or sheet. { shraúd }

shrouded propeller See ducted fan. { 'shraúd·əd prə'pel·ər }

shunt |CIV ENG| To shove or turn off to one side, as a car or train from one track to another. |ELEC| 1. A precision low-value resistor placed across the terminals of an ammeter to increase its range by allowing a known fraction of the circuit current to go around the meter. Also known as electric shunt. 2. To place one part in parallel with another. 3. See parallel. { shənt }

shunt valve |ENG| A valve that gives a fluid under pressure a more readily available escape route than the normal route. { shənt ,valv }

shut-down circuit |ENG| An electronic, electric, or pneumatic system designed to shut off and close down process systems or equipment; can be used for routine or emergency situations. { 'shət,daún ,sər·kət }

shut height |MECH ENG| The distance in a press between the bottom of the slide and the top of the bed, indicating the maximum die height that can be accommodated. { 'shət ,hīt }

shutoff head |MECH ENG| The pressure developed in a centrifugal or axial flow pump when there is zero flow through the system. { 'shət,óf ,hed }

shutter dam |CIV ENG| A dam consisting of a series of pieces that can be lowered or raised by revolving them about their horizontal axis. { 'shəd·ər ,dam }

shuttering See formwork. { 'shəd·ə·riŋ }

shuttle |MECH ENG| A back-and-forth motion of a machine which continues to face in one direction. { 'shəd·əl }

shuttle conveyor |MECH ENG| Any conveyor in a self-contained structure movable in a defined path parallel to the flow of the material. { 'shəd·əl kən,vā·ər }

shuttling |ENG| A movement involving two or more trips or partial trips by the same motor vehicles between two points. { 'shəd·əl·iŋ }

Siacci method |MECH| An accurate and useful method for calculation of trajectories of high-velocity missiles with low quadrant angles of departure; basic assumptions are that the atmospheric density anywhere on the trajectory is approximately constant, and the angle of departure is less than about 15° { sē'ä·chē ,meth·əd }

siamese blow |ENG| In the plastics industry, the blow molding of two or more parts of a product in a single blow, then cutting them apart. { 'sī·ə,mēz 'blō }

siamese connection |ENG| A Y-shaped standpipe installed close to the ground outside a building to provide two inlet connections for fire hoses to the standpipes and to the sprinkler system. { 'sī·ə,mēz kə'nek·shən }

SIC See dielectric constant.

sickle |DES ENG| A hand tool consisting of a hooked metal blade with a short handle, used for cutting grain or other agricultural products. { 'sik·əl }

side bar |ENG| A bar on which molding pins are carried; operated from outside the mold. { 'sīd ,bär }

side-channel spillway |CIV ENG| A dam spillway in which the initial and final flow are approximately perpendicular to each other. Also known as lateral flow spillway. { 'sīd ¦chan·əl 'spil ,wā }

side direction |MECH| In stress analysis, the direction perpendicular to the plane of symmetry of an object. { 'sīd di,rek·shən }

side draw pin |ENG| Projection used to core a hole in a molded article at an angle other than the line of mold closing; must be withdrawn before the article is ejected. { 'sīd 'drò ,pin }

side-facing tool |ENG| A single-point cutting tool having a nose angle of less than 60° and used for finishing the tailstock end of work being machined between centers or the face of a workpiece mounted in a chuck. { 'sīd ,fās·iŋ ,tül }

sidehill bit |DES ENG| A drill bit which is set off-center so that it cuts a hole of larger diameter than that of the bit. { 'sīd,hil ,bit }

side hook See bench hook. { 'sīd ,húk }

side-looking radar |ENG| A high-resolution airborne radar having antennas aimed to the right and left of the flight path; used to provide high-resolution strip maps with photographlike detail, to map unfriendly territory while flying along its perimeter, and to detect submarine snorkels against a background of sea clutter. { 'sīd ¦lúk·iŋ 'rā·där }

side milling |MECH ENG| Milling with a side-milling cutter to machine one vertical surface. { 'sīd ,mil·iŋ }

side-milling cutter |DES ENG| A milling cutter with teeth on one or both sides as well as around the periphery. { 'sīd ¦mil·iŋ 'kəd·ər }

side rake |MECH ENG| The angle between the tool face and a reference plane for a single-point turning tool. { 'sīd ,rāk }

side relief angle |DES ENG| The angle that the portion of the flanks of a cutting tool below the cutting edge makes with a plane normal to the base. { 'sīd ri'lēf ,aŋ·gəl }

side rod |MECH ENG| 1. A rod linking the crankpins of two adjoining driving wheels on the same side of a locomotive; distributes power from the main rod to the driving wheels. 2. One of the rods linking the piston-rod crossheads and the side levers of a side-lever engine. { 'sīd ,räd }

siderograph |ENG| An instrument that keeps the time of the Greenwich longitude; consists of a clock and a navigation instrument. { 'sid·ə·rə,graf }

side shot |ENG| A reading or measurement from a survey station to locate a point that is off the traverse or that is not intended to be used as a base for the extension of the survey. { 'sīd ,shät }

side slope [ENG] A test course used to determine lateral stability of a vehicle as well as steering, carburetion, and other functions. { 'sīd ,slōp }

sidestream [CHEM ENG] A liquid stream taken from an intermediate point of a liquids-processing unit, for example, a distillation or extraction tower. { 'sīd,strēm }

sidestream stripper [CHEM ENG] A device used to perform further distillation on a liquid stream (sidestream) from any one of the plates of a bubble tower, usually with the use of steam. { 'sīd,strēm 'strip·ər }

sidetrack [CIV ENG] **1.** To move railroad cars onto a siding. **2.** *See* siding. { 'sīd,trak }

sidetracking [ENG] The deliberate act or process of deflecting and drilling a borehole away from a normal, straight course. { 'sīd,trak·iŋ }

sidewalk [CIV ENG] **1.** A walkway for pedestrians on the side of a street or road. **2.** A foot pavement. { 'sīd,wók }

sidewall section [ENG ACOUS] A wall in a sound-recording studio with reversible panels or rotating columns that are sound-absorbent on one side and reflective on the other, used to vary the acoustic environment. { 'sīd,wòl ,sek·shən }

siding [CIV ENG] A short railroad track connected to the main track at one or more points and used to move railroad cars in order to free traffic on the main line or for temporary storage of cars. Also known as sidetrack. { 'sīd·iŋ }

siemens [ELEC] A unit of conductance, admittance, and susceptance, equal to the conductance between two points of a conductor such that a potential difference of 1 volt between these points produces a current of 1 ampere; the conductance of a conductor in siemens is the reciprocal of its resistance in ohms. Formerly known as mho (Ω); reciprocal ohm. Symbolized S. { 'sē·mənz }

sieve [ENG] **1.** A meshed or perforated device or sheet through which dry loose material is refined, liquid is strained, and soft solids are comminuted. **2.** A meshed sheet with apertures of uniform size used for sizing granular materials. { siv }

sieve analysis [ENG] The size distribution of solid particles on a series of standard sieves of decreasing size, expressed as a weight percent. Also known as sieve classification; sieving. { 'siv ə,nal·ə·səs }

sieve classification *See* sieve analysis. { 'siv ,klas·ə·fə,kā·shən }

sieve diameter [ENG] The size of a sieve opening through which a given particle will just pass. { 'siv dī,am·əd·ər }

sieve fraction [ENG] That portion of solid particles which pass through a standard sieve of given number and is retained by a finer sieve of a different number. { 'siv ,frak·shən }

sieve mesh [DES ENG] The standard opening in

sieve or screen, defined by four boundary wires (warp and woof); the laboratory mesh is square and is defined by the shortest distance between two parallel wires as regards aperture (quoted in micrometers or millimeters), and by the number of parallel wires per linear inch as regards mesh; 60-mesh equals 60 wires per inch. { 'siv ,mesh }

sieve plate [CHEM ENG] A distillation-tower tray that is perforated so that the vapor emerges vertically through the tray, passing through the liquid holdup on top of the tray; used as a replacement for bubble-cap trays in distillation. Also known as sieve tray. { 'siv ,plāt }

sieve shaker [CHEM ENG] A device used to shake a stacked column of standard sieve-test trays to cause solids to sift progressively from the top (large openings) to the bottom (small openings and a final pan), according to particle size. { 'siv ,shā·kər }

sieving *See* sieve analysis. { 'siv·iŋ }

sight-feed [ENG] Pertaining to piping in which the flowing liquid can be observed through a transparent tube or wall. { 'sīt,fēd }

sight glass [ENG] A glass tube or a glass-faced section on a process line or vessel; used for visual reading of liquid levels or of manometer pressures. { 'sīt ,glas }

sighting tube [ENG] A tube, usually ceramic, inserted into a hot chamber whose temperature is to be measured; an optical pyrometer is sighted into the tube to observe the interior end of the tube to give a temperature reading. { 'sīd·iŋ ,tüb }

sight rod *See* range rod. { 'sīt ,räd }

sigma-delta analog-to-digital converter [ELECTR] A converter that uses an analog circuit to generate a single-valued pulse stream in which the frequency of pulses is determined by the analog source, and then uses a digital circuit to repeatedly sum the number of these pulses over a fixed time interval, converting the pulses to numeric values. { ¦sig·mə ¦del·tə ,an·ə,läg tü ,dij·əd·əl kən,vərd·ər }

sigma-delta converter [ELECTR] A class of electronic systems containing both analog and digital subsystems whose most common application is the conversion of analog signals to digital form, and vice versa, using pulse density modulation to create a high-rate stream of single-amplitude pulses in either case. Also known as delta-sigma converter. { ,sig·mə ,del·tə kən 'vərd·ər }

sigma-delta digital-to-analog converter [ELECTR] A converter that uses a digital circuit to convert numeric values from a digital processor to a pulse stream and then uses an analog low-pass filter to produce an analog waveform. { ,sig·mə ¦del·tə ,dij·əd·əl tü ,an·ə,läg kən'vərd·ər }

sigma-delta modulator [ELECTR] The circuit used to generate a pulse stream in a sigma-delta converter. Also known as delta-sigma modulator. { ,sig·mə ,del·tə 'mäj·ə,lād·ər }

sigma function [THERMO] A property of a mixture of air and water vapor, equal to the difference between the enthalpy and the product of the specific humidity and the enthalpy of water (liquid) at the thermodynamic wet-bulb temperature; it is constant for constant barometric pressure and thermodynamic wet-bulb temperature. { 'sig·mə ,faŋk·shən }

signal correction [ENG] In seismic analysis, a correction to eliminate the time differences between reflection times, resulting from changes in the outgoing signal from shot to shot. { 'sig·nəl kə,rek·shən }

signal effect [ENG] In seismology, variation in arrival times of reflections recorded with identical filter settings, as a result of changes in the outgoing signal. { 'sig·nəl i,fekt }

signal flare [ENG] A pyrotechnic flare of distinct color and character used as a signal. { 'sig·nəl ,fler }

signal-flow graph [SYS ENG] An abbreviated block diagram in which small circles, called nodes, represent variables of the system, and the nodes are connected by lines, called branches, which represent one-way signal multipliers; an arrow on the line indicates direction of signal flow, and a letter near the arrow indicates the multiplication factor. Also known as flow graph. { 'sig·nəl ¦flō 'graf }

signal generator [ENG] An electronic test instrument that delivers a sinusoidal output at an accurately calibrated frequency that may be anywhere from the audio to the microwave range; the frequency and amplitude are adjustable over a wide range, and the output usually may be amplitude- or frequency-modulated. Also known as test oscillator. { 'sig·nəl ,jen·ə,rād·ər }

signaling key See key. { 'sig·nə·liŋ ,kē }

signal light [ENG] A signal, illumination, or any pyrotechnic light used as a sign. { 'sig·nəl ,līt }

signal-to-interference ratio [ELECTR] The relative magnitude of signal waves and waves which interfere with signal-wave reception. { 'sig·nəl tü ,in·tər'fir·əns ,rā·shō }

signal-to-noise ratio [ELECTR] The ratio of the amplitude of a desired signal at any point to the amplitude of noise signals at that same point; often expressed in decibels; the peak value is usually used for pulse noise, while the root-mean-square (rms) value is used for random noise. Abbreviated S/N; SNR. { 'sig·nəl tə 'nȯiz ,rā·shō }

signal tower [CIV ENG] A switch tower from which railroad signals are displayed or controlled. { 'sig·nəl ,tau̇·ər }

signal voltage [ELEC] Effective (root-mean-square) voltage value of a signal. { 'sig·nəl ,vōl·tij }

silent speed [ENG] The speed at which silent motion pictures are fed through a projector, equal to 16 frames per second (sound-film speed is 24 frames per second). { 'sī·lənt 'spēd }

silent stock support [MECH ENG] A flexible metal guide tube in which the stock tube of an automatic screw machine rotates; it is covered with a casing which deadens sound and prevents transfer of noise and vibration. { 'sī·lənt 'stäk sə,pȯrt }

silicate grinding wheel [DES ENG] A mild-acting grinding wheel where the abrasive grain is bonded with sodium silicate and fillers. { 'sil·ə·kət 'grīnd·iŋ ,wēl }

silicide resistor [ELECTR] A thin-film resistor that uses a silicide of molybdenum or chromium, deposited by direct-current sputtering in an integrated circuit when radiation hardness or high resistance values are required. { 'sil·ə,sīd ri'zis·tər }

silicon capacitor [ELECTR] A capacitor in which a pure silicon-crystal slab serves as the dielectric; when the crystal is grown to have a p zone, a depletion zone, and an n zone, the capacitance varies with the externally applied bias voltage, as in a varactor. { 'sil·ə·kən kə'pas·əd·ər }

silicon diode [ELECTR] A crystal diode that uses silicon as a semiconductor; used as a detector in ultra-high- and super-high-frequency circuits. Also known as silicon detector. { 'sil·ə·kən 'dī,ōd }

silicon homojunction See bipolar junction transistor. { ¦sil·ə·kən 'hä·mə,jəŋk·shən }

silicon-on-insulator [ELECTR] A semiconductor manufacturing technology in which thin films of single-crystalline silicon are grown over an electrically insulating substrate. { 'sil·ə·kən ȯn 'in·sə,lād·ər }

silicon-on-sapphire [ELECTR] A semiconductor manufacturing technology in which metal oxide semiconductor devices are constructed in a thin single-crystal silicon film grown on an electrically insulating synthetic sapphire substrate. Abbreviated SOS. { 'sil·ə·kən ȯn 'sa,fīr }

silicon rectifier [ELECTR] A metallic rectifier in which rectifying action is provided by an alloy junction formed in a high-purity silicon slab. { 'sil·ə·kən 'rek·tə,fī·ər }

silicon resistor [ELECTR] A resistor using silicon semiconductor material as a resistance element, to obtain a positive temperature coefficient of resistance that does not appreciably change with temperature; used as a temperature-sensing element. { 'sil·ə·kən ri'zis·tər }

silicon retina [ELECTR] An analog very large scale integrated circuit chip that performs operations which resemble some of the functions performed by the retina of the human eye. { ,sil·ə,kän 'ret·ən·ə }

silicon solar cell [ELECTR] A solar cell consisting of p and n silicon layers placed one above the other to form a pn junction at which radiant energy is converted into electricity. { 'sil·ə·kən 'sō·lər 'sel }

silicon transistor [ELECTR] A transistor in which silicon is used as the semiconducting material. { 'sil·ə·kən tran'zis·tər }

sill [BUILD] The lowest horizontal member of a framed partition or of a window or door frame. [CIV ENG] **1.** A timber laid across the foot of a trench or a heading under the side truss.

2. The horizontal overflow line of a dam spillway or other weir structure. **3.** A horizontal member on which a lift gate rests when closed. **4.** A low concrete or masonry dam in a small stream to retard bottom erosion. |CONT SYS| A type of robot articulation that has three degrees of freedom. { sil }

sill anchor |BUILD| A fastener projecting from a foundation wall or foundation slab to secure the sill to the foundation. { 'sil ,aŋ·kər }

silo |CIV ENG| A large vertical, cylindrical structure, made of reinforced concrete, steel, or timber, and used for storing grain, cement, or other materials. { 'sī·lō }

silting |CIV ENG| The filling up or raising of the bed of a body of water by depositing silt. { 'silt·iŋ }

silting index |ENG| The measurement of the tendency of a solids- or gel-carrying fluid to cause silting in close-tolerance devices, such as valves or other process-line flow constrictions. { 'silt·iŋ ,in,deks }

silver-disk pyrheliometer |ENG| An instrument used for the measurement of direct solar radiation; it consists of a silver disk located at the lower end of a diaphragmed tube which serves as the radiation receiver for a calorimeter; radiation falling on the silver disk is periodically intercepted by means of a shutter located in the tube, causing temperature fluctuations of the calorimeter which are proportional to the intensity of the radiation. { 'sil·vər 'disk ¦pīr,hē·lē'äm·əd·ər }

silvered mica capacitor |ELECTR| A mica capacitor in which a coating of silver is deposited directly on the mica sheets to serve in place of conducting metal foil. { 'sil·vərd ¦mī·kə kə'pas·ad·ər }

silver migration |ELEC| A process, causing reduction in insulation resistance and dielectric failure; silver, in contact with an insulator, at high humidity, and subjected to an electrical potential, is transported ionically from one location to another. { 'sil·vər mī'grā·shən }

similarity principle *See* principle of dynamical similarity. { ,sim·ə'lar·əd·ē ,prin·sə·pəl }

similitude |ENG| A likeness or resemblance; for example, the scale-up of a chemical process from a laboratory or pilot-plant scale to a commercial scale. { si'mil·ə,tüd }

simmer |ENG| The detectable leakage of fluid in a safety valve below the popping pressure. { 'sim·ər }

simo chart |IND ENG| A basic motion-time chart used to show the simultaneous nature of motions; commonly a therblig chart for two-hand work with motion symbols plotted vertically with respect to time, showing the therblig abbreviation and a brief description for each activity, and individual times values and body-member detail. Also known as simultaneous motion-cycle chart. { 'sī·mō ,chärt }

Simon's theory |ENG| A theory of drilling which includes the effects of drilling by percussion and by vibration with a rotary (oil well) bit, cable

tool, and pneumatic hammer; the rate of penetration of a chisel-shaped bit into brittle rock may be defined as follows: $R = NAf_v\sqrt{\pi}D$, where R equals the rate of advance of bit, N equals the number of wings of bit, f_v equals the number of impacts per unit time, D equals the diameter of the bit, and A equals the cross-sectional area of the crater at the periphery of the drill hole. { 'sī·mənz ,thē·ə·rē }

simple balance |ENG| An instrument for measuring weight in which a beam can rotate about a knife-edge or other point of support, the unknown weight is placed in one of two pans suspended from the ends of the beam and the known weights are placed in the other pan, and a small weight is slid along the beam until the beam is horizontal. { 'sim·pəl 'bal·ans }

simple continuous distillation *See* equilibrium flash vaporization. { 'sim·pəl kən'tin·yə·wəs ,dis·tə'lā·shən }

simple engine |MECH ENG| An engine (such as a steam engine) in which expansion occurs in a single phase, after which the working fluid is exhausted. { 'sim·pəl 'en·jən }

simple harmonic motion *See* harmonic motion. { 'sim·pəl här'män·ik 'mō·shən }

simple machine |MECH ENG| Any of several elementary machines, one or more being incorporated in every mechanical machine; usually, only the lever, wheel and axle, pulley (or block and tackle), inclined plane, and screw are included, although the gear drive and hydraulic press may also be considered simple machines. { 'sim·pəl mə'shēn }

simple pendulum |MECH| A device consisting of a small, massive body suspended by an inextensible object of negligible mass from a fixed horizontal axis about which the body and suspension are free to rotate. { 'sim·pəl 'pen·jə·ləm }

simplex concrete pile |CIV ENG| A molded-in-place pile made by using a hollow cylindrical mandrel which is filled with concrete after having been driven to the desired depth and raised a few feet at a time, the concrete flowing out at the bottom and filling the hole in the earth. { 'sim,pleks 'kän,krēt 'pīl }

simplex pump |MECH ENG| A pump with only one steam cylinder and one water cylinder. { 'sim,pleks ¦pəmp }

SIMS *See* secondary ion mass spectrometer. { simz }

simulate |ENG| To mimic some or all of the behavior of one system with a different, dissimilar system, particularly with computers, models, or other equipment. { 'sim·yə,lāt }

simulator |ENG| A computer or other piece of equipment that simulates a desired system or condition and shows the effects of various applied changes, such as a flight simulator. { 'sim·yə,lād·ər }

simultaneity |MECH| Two events have simultaneity, relative to an observer, if they take place at the same time according to a clock which is fixed relative to the observer. { ˌsī·məl·tə'nē·əd·ē }

simultaneous motion-cycle chart *See* simo chart. { ˌsī·məl'tā·nē·əs 'mō·shən ˌsī·kəl ˌchärt }

sine bar |DES ENG| A device consisting of a steel straight edge with two cylinders of equal diameter attached near the ends with their centers equidistant from the straightedge; used to measure angles accurately and to lay out work at a desired angle in relationship to a surface. { 'sīn ˌbär }

sine galvanometer |ENG| A type of magnetometer in which a small magnet is suspended in the center of a pair of Helmholtz coils, and the rest position of the magnet is measured when various known currents are sent through the coils. { 'sīn ˌgal·və'näm·əd·ər }

sine-wave response *See* frequency response. { 'sīn ˌwāv ri'späns }

singing |CONT SYS| An undesired, self-sustained oscillation in a system or component, at a frequency in or above the passband of the system or component; generally due to excessive positive feedback. { 'siŋ·iŋ }

singing margin |CONT SYS| The difference in level, usually expressed in decibels, between the singing point and the operating gain of a system or component. { 'siŋ·iŋ ˌmär·jən }

singing point |CONT SYS| The minimum value of gain of a system or component that will result in singing. { 'siŋ·iŋ ˌpóint }

single acting |MECH ENG| Acting in one direction only, as a single-acting plunger, or a single-acting engine (admitting the working fluid on one side of the piston only). { 'siŋ·gəl 'akt·iŋ }

single-action press |MECH ENG| A press having a single slide. { 'siŋ·gəl ˌak·shən 'pres }

single-axis gyroscope |ENG| A gyroscope suspended in just one gimbal whose bearings form its output axis; an example is a rate gyroscope. { 'siŋ·gəl ˌak·səs 'jī·rə,skōp }

single-block brake |MECH ENG| A friction brake consisting of a short block fitted to the contour of a wheel or drum and pressed up against the surface by means of a lever on a fulcrum; used on railroad cars. { 'siŋ·gəl ˌbläk 'brāk }

single-button carbon microphone |ENG ACOUS| Microphone having a carbon-filled buttonlike container on only one side of its flexible diaphragm. { 'siŋ·gəl ˌbət·ən 'kär·bən 'mī·krə,fōn }

single-cut file |DES ENG| A file with one set of parallel teeth, extending diagonally across the face of the file. { 'siŋ·gəl ˌkət 'fīl }

single-degree-of-freedom gyro |MECH| A gyro the spin reference axis of which is free to rotate about only one of the orthogonal axes, such as the input or output axis. { 'siŋ·gəl di¦grē əv ¦frē·dəm 'jī·rō }

single-edged push-pull amplifier circuit |ELECTR| Amplifier circuit having two transmission paths designed to operate in a complementary manner and connected to provide a single unbalanced output without the use of an output transformer. { 'siŋ·gəl ¦ejd ¦push ¦pul 'am·plə,fī·ər ,sər·kət }

single-effect evaporation |CHEM ENG| An evaporation process completed entirely in one vessel or by means of a single heating unit. { 'siŋ·gəl i¦fekt i,vap·ə'rā·shən }

single-electron transistor |ELECTR| A transistor whose dimensions are extremely small, in the nanometer range, causing it to exhibit characteristics that are sensitive to the transport and storage of single electrons. { ˌsiŋ·gəl i,lek·trän tran'zis·tər }

single-ended signal |ELECTR| A circuit signal that is the voltage difference between two nodes, one of which can be defined as being at ground or reference voltage. { ¦siŋ·gəl ¦en·dəd 'sig·nəl }

single-ended spread |ENG| A spread of geophones in which the shot point is located at one end of the arrangement. { ¦siŋ·gəl ¦end·əd 'spred }

single-hand drilling |ENG| A method of rock drilling in which the drill steel, which is held in the hand, is struck with a 4-pound (1.8-kilogram) hammer, the drill being turned between the blows. { 'siŋ·gəl ,han 'dril·iŋ }

single in-line package |ELECTR| A packaged resistor network or other assembly that has a single row of terminals or lead wires along one edge of the package. Abbreviated SIP. { 'siŋ·gəl 'in,līn 'pak·ij }

single-layer bit *See* surface-set bit. { 'siŋ·gəl ¦lā·ər 'bit }

single-loop feedback |CONT SYS| A system in which feedback may occur through only one electrical path. { 'siŋ·gəl ¦lüp 'fēd,bak }

single-loop servomechanism |CONT SYS| A servomechanism which has only one feedback loop. Also known as servo loop. { 'siŋ·gəl ¦lüp 'sər·vō,mek·ə,niz·əm }

single-phase |ELEC| Energized by a single alternating voltage. { 'siŋ·gəl 'fāz }

single-phase circuit |ELEC| Either an alternating-current circuit which has only two points of entry, or one which, having more than two points of entry, is intended to be so energized that the potential differences between all pairs of points of entry are either in phase or differ in phase by 180°. { 'siŋ·gəl ¦fāz 'sər·kət }

single-phase flow |CHEM ENG| The flow of a material, as a gas, single-phase liquid, or a solid, but not in any combination of the three. { 'siŋ·gəl ¦fāz 'flō }

single-phase meter |ENG| A type of power-factor meter that contains a fixed coil that carries the load current, and crossed coils that are connected to the load voltage; there is no spring to restrain the moving system, which takes a position to indicate the angle between the current and voltage. { 'siŋ·gəl ¦fāz 'mēd·ər }

single-phase motor |ELEC| A motor energized by a single alternating voltage. { 'siŋ·gəl ¦fāz 'mōd·ər }

single-piece milling |MECH ENG| A milling method whereby one part is held and milled in one machine cycle. { 'siŋ·gəl ¦pēs 'mil·iŋ }

single-point grounding |ELEC| Grounding system that attempts to confine all return currents to a network that serves as the circuit reference;

to be effective, no appreciable current is allowed to flow in the circuit reference, that is, the sum of the return currents is zero. { 'siŋ·gəl ¦pȯint 'graund·iŋ }

single-point tool |ENG| A cutting tool having one face and one continuous cutting edge. { 'siŋ·gəl ¦pȯint 'tül }

single-pole double-throw |ELEC| A three-terminal switch or relay contact arrangement that connects one terminal to either of two other terminals. Abbreviated SPDT. { 'siŋ·gəl 'pōl 'dəb·əl 'thrō }

single-pole single-throw |ELEC| A two-terminal switch or relay contact arrangement that opens or closes one circuit. Abbreviated SPST. { 'siŋ·gəl 'pōl 'siŋ·gəl 'thrō }

single sampling |IND ENG| A sampling inspection in which the lot is accepted or rejected on the basis of one sample. { 'siŋ·gəl 'sam·pliŋ }

single-shot blocking oscillator |ELECTR| Blocking oscillator modified to operate as a single-shot trigger circuit. { 'siŋ·gəl ¦shät 'bläk·iŋ 'äs·ə,lād·ər }

single-shot exploder |ENG| A magneto exploder operated by the twist action given by a half turn of the firing key. { 'siŋ·gəl ¦shät ik 'splōd·ər }

single-shot multivibrator See monostable multivibrator. { 'siŋ·gəl ¦shät ¦məl·ti'vī,brād·ər }

single-shot trigger circuit |ELECTR| Trigger circuit in which one triggering pulse initiates one complete cycle of conditions ending with a stable condition. Also known as single-trip trigger circuit. { 'siŋ·gəl ¦shät 'trig·ər ,sər·kət }

single-sided amplifier See single-end amplifier. { 'siŋ·gəl ¦sīd·əd 'am·plə,fī·ər }

single-sided board |ELECTR| A printed wiring board that contains all of the interconnect material on one of the external layers. { ,siŋ·gəl ,sīd·əd 'bȯrd }

single-stage compressor |MECH ENG| A machine that effects overall compression of a gas or vapor from suction to discharge conditions without any sequential multiplicity of elements, such as cylinders or rotors. { 'siŋ·gəl ¦stāj kəm'pres·ər }

single-stage pump |MECH ENG| A pump in which the head is developed by a single impeller. { 'siŋ·gəl ¦stāj 'pəmp }

single thread |DES ENG| A screw thread having a single helix in which the lead and pitch are equal. { 'siŋ·gəl 'thred }

single-throw switch |ELEC| A switch in which the same pair of contacts is always opened or closed. { 'siŋ·gəl ¦thrō 'swich }

single-trip trigger circuit See single-shot trigger circuit. { 'siŋ·gəl ¦trip 'trig·ər ,sər·kət }

single-tuned amplifier |ELECTR| An amplifier characterized by resonance at a single frequency. { 'siŋ·gəl ¦tünd 'am·plə,fī·ər }

single-unit semiconductor device |ELECTR| Semiconductor device having one set of electrodes associated with a single carrier stream. { 'siŋ·gəl ¦yü·nət 'sem·i·kən,dək·tər di,vīs }

singular arc |CONT SYS| In an optimal control problem, that portion of the optimal trajectory in which the Hamiltonian is not an explicit function of the control inputs, requiring higher-order necessary conditions to be applied in the process of solution. { 'siŋ·gyə·lər 'ärk }

sink-float separation process |ENG| A simple gravity process used in ore dressing that separates particles of different sizes or composition on the basis of differences in specific gravity. { 'siŋk 'flōt ,sep·ə'rā·shən ,prä·səs }

sinking fund |IND ENG| A fund established by periodically depositing funds at compound interest in order to accumulate a given sum at a given future time for some specific purpose. { 'siŋk·iŋ ,fənd }

sink mark |ENG| A shallow depression or dimple on the surface of an injection-molded plastic part due to collapsing of the surface following local internal shrinkage after the gate seals. { 'siŋk ,märk }

sinter setting See mechanical setting. { 'sin·tər ,sed·iŋ }

sinusoidal current See simple harmonic current. { ,sī·nə'sȯid·əl 'kə·rənt }

SIP See single in-line package. { sip }

siphon |ENG| A tube, pipe, or hose through which a liquid can be moved from a higher to a lower level by atmospheric pressure forcing it up the shorter leg while the weight of the liquid in the longer leg causes continuous downward flow. { 'sī·fən }

siphon barograph |ENG| A recording siphon barometer. { 'sī·fən 'bar·ə,graf }

siphon barometer |ENG| A J-shaped mercury barometer in which the stem of the J is capped and the cusp is open to the atmosphere. { 'sī·fən bə'räm·əd·ər }

siphon recorder |ENG| A recorder in which a small siphon discharges ink to make the record; used in submarine telegraphy. { 'sī·fən ri 'kȯrd·ər }

siphon spillway |CIV ENG| An enclosed spillway passing over the crest of a dam in which flow is maintained by atmospheric pressure. { 'sī·fən 'spil,wā }

siren |ENG ACOUS| An apparatus for generating sound by the mechanical interruption of the flow of fluid (usually air) by a perforated disk or cylinder. { 'sī·rən }

sister hook |DES ENG| **1.** Either of a pair of hooks which can be fitted together to form a closed ring. **2.** A pair of such hooks. { 'sis·tər ,hük }

site |ENG| Position of anything; for example, the position of a gun emplacement. { sīt }

six-axis system |MECH ENG| A robot that has six degrees of freedom, three rectangular and three rotational. { 'siks ¦ak·səs 'sis·təm }

six-phase circuit |ELEC| Combination of circuits energized by alternating electromotive forces which differ in phase by one-sixth of a cycle (60°). { 'siks ¦fāz 'sər·kət }

Six's thermometer |ENG| A combination maximum thermometer and minimum thermometer; the tube is shaped in the form of a U with a bulb at either end; one bulb is filled with creosote which expands or contracts with temperature variation, forcing before it a short column of mercury having iron indexes at either end; the indexes remain at the extreme positions reached by the mercury column, thus indicating the maximum and minimum temperatures; the indexes can be reset with the aid of a magnet. { 'sik·səz thər,mäm·əd·ər }

six-tenths factor |IND ENG| An empirical relationship between the cost and the size of a manufacturing facility; as size increases, cost increases by an exponent of six-tenths, that is $cost_1/cost_2 = (size_1/size_2)^{0.6}$ { 'siks ¦tenths ,fak·tər }

sixty degrees Fahrenheit British thermal unit See British thermal unit. { 'siks·tē di¦grēz 'far·ən,hīt 'brid·ish 'thər·məl ,yü·nət }

size analysis See particle-size analysis. { 'sīz ə,nal·ə·səs }

size block See gage block. { 'sīz ,bläk }

size classification See sizing. { 'sīz ,klas·ə·fə,kā·shən }

size dimension |DES ENG| In dimensioning, a specified value of a diameter, width, length, or other geometrical characteristic directly related to the size of an object. { 'sīz di,men·shən }

size enlargement |CHEM ENG| Making large particles out of small ones by crystallization, particle cementation, tableting, briquetting, agglomeration, flocculation, melting, casting, compaction and extrusion, and sintering or nodulizing. { 'sīz in,lärj·mənt }

size-frequency analysis See particle-size analysis. { 'sīz 'frē·kwən·sē ə,nal·ə·səs }

size reduction |MECH ENG| The breaking of large pieces of coal, ore, or stone by a primary breaker, or of small pieces by grinding equipment. { 'sīz ri,dək·shən }

sizing |ENG| **1.** Separating an aggregate of mixed particles into groups according to size, using a series of screens. Also known as size classification. **2.** See sizing treatment. |MECH ENG| A finishing operation to correct surfaces and shapes to meet specified dimensions and tolerances. { 'sīz·iŋ }

sizing screen |DES ENG| A mesh sheet with standard-size apertures used to separate granular material into classes according to size; the Tyler standard screen is an example. { 'sīz·iŋ ,skrēn }

sizing treatment |ENG| Also known as sizing; surface sizing. **1.** Application of material to a surface to fill pores and thus reduce the absorption of subsequently applied adhesive or coating; used for textiles, paper, and other porous materials. **2.** Surface-treatment applied to glass fibers used in reinforced plastics. { 'sīz·iŋ ,trēt·mənt }

Sk See Stefan number.

skeleton framing |BUILD| Framing in which steel framework supports all the gravity loading of the structure; this system is used for skyscrapers. { 'skel·ət·ən ,främ·iŋ }

skew |ELECTR| **1.** The deviation of a received

facsimile frame from rectangularity due to lack of synchronism between scanner and recorder; expressed numerically as the tangent of the angle of this deviation. **2.** The degree of nonsynchronism of supposedly parallel bits when bit-coded characters are read from magnetic tape. |MECH ENG| Gearing whose shafts are neither interesecting nor parallel. { skyü }

skewback |CIV ENG| The beveled or inclined support at each end of a segmental arch. { 'skyü ,bak }

skew bridge |CIV ENG| A bridge which spans a gap obliquely and is therefore longer than the width of the gap. { 'skyü ,brij }

skew chisel |ENG| A tool used for wood turning that has a straight cutting edge sharpened at an angle to the shank. { 'skyü ,chiz·əl }

skewed bridge |CIV ENG| A bridge for which the deck in plan is a parallelogram. { 'skyüd 'brij }

skew level gear |DES ENG| A level gear whose axes are not in the same place. { 'skyü 'lev·əl ,gir }

skid |ENG| **1.** A device attached to a chain and placed under a wheel to prevent its turning when descending a steep hill. **2.** A timber, bar, rail, or log placed under a heavy object when it is being moved over bare ground. **3.** A wood or metal platform support on wheels, legs, or runners used for handling and moving material. Also known as skid platform. |MECH ENG| A brake for a power machine. { skid }

skid-mounted |ENG| Equipment or processing systems mounted on a portable platform. { 'skid,maunt·əd }

skim coat |BUILD| A finish coat of plaster composed of lime putty and fine white sand. { 'skim ,kōt }

skimming plant |CHEM ENG| A petroleum refinery designed to remove and finish only the lighter constituents of crude oil, such as gasoline and kerosine; the heavy ends are sold as fuel oil or for further processing elsewhere. { 'skim·iŋ ,plant }

skin |BUILD| The exterior wall of a building. |ENG| In flexible bag molding, a protective covering for the mold; it may consist of a thin piece of plywood or a thin hardwood. { skin }

skin diving |ENG| Diving without breathing apparatus, using fins and faceplate only. { 'skin ,dīv·iŋ }

skintle |CIV ENG| To set bricks in an irregular fashion so that they are out of alignment with the face by 1/4 inch (6 millimeters) or more. { 'skint·əl }

skip See skip hoist. { skip }

skip distance |ENG| In angle-beam ultrasonic testing, the distance between the point of entry on the workpiece and the point of first reflection. { 'skip ,dis·təns }

skip hoist |MECH ENG| A basket, bucket, or open car mounted vertically or on an incline on wheels, rails, or shafts and hoisted by a cable; used to raise materials. Also known as skip. { 'skip ,hȯist }

skip logging |ENG| A phenomenon during

acoustical (sonic) logging in which the acoustical energy is attenuated by low-elasticity formations and lacks the energy to trip the second sonic receiver (skips a cycle). Also known as cycle skip. { 'skip ,läg·iŋ }

skip trajectory |MECH| A trajectory made up of ballistic phases alternating with skipping phases; one of the basic trajectories for the unpowered portion of the flight of a reentry vehicle or spacecraft reentering earth's atmosphere. { 'skip tra,jek·trē }

skirt See baseboard. { skərt }

skirting See baseboard. { 'skərd·iŋ }

skirting block |BUILD| Also known as base block; plinth block. **1.** A corner block where a base strip and vertical enframement meet. **2.** A concealed block to which a baseboard is anchored. { 'skərd·iŋ ,bläk }

skirt roof |BUILD| A false band of roofing projecting from between the stories of a building. { 'skərt ,rüf }

skiving |MECH ENG| **1.** Removal of material in thin layers or chips with a high degree of shear or slippage of the cutting tool. **2.** A machining operation in which the cut is made with a form tool with its face at an angle allowing the cutting edge to progress from one end of the work to the other as the tool feeds tangentially past ten rotating workpieces. { 'skīv·iŋ }

skull cracker |ENG| A heavy iron or steel ball that can be swung freely or dropped by a derrick to raze buildings or to compress bulky scrap. Also known as wrecking ball. { 'skəl ,krak·ər }

skylight |ENG| An opening in a roof or ship deck that is covered with glass or plastic and designed to admit daylight. { 'skī,līt }

skyscraper |BUILD| A very tall, multistory building. { 'skī,skrāp·ər }

slab |CIV ENG| That part of a reinforced concrete floor, roof, or platform which spans beams, columns, walls, or piers. |ELECTR| A relatively thick-cut crystal from which blanks are obtained by subsequent transverse cutting. |ENG| The outside piece cut from a log when sawing it into boards. { slab }

slabbing cutter |MECH ENG| A face-milling cutter used to make wide, rough cuts. { 'slab·iŋ ,kəd·ər }

slab cutter See plain milling cutter. { 'slab ,kəd·ər }

slabstone See slab. { 'slab,stōn }

slack |ENG| Looseness or play in a mechanism, as the play in the trigger of a small-arms weapon. { slak }

slackline cableway |MECH ENG| A machine, widely used in sand-and-gravel plants, employing an open-ended dragline bucket suspended from a carrier that runs upon a track cable, which can dig, elevate, and convey materials in one continuous operation. { 'slak,līn 'kā·bəl,wā }

slack time |ENG| For an activity in a PERT or critical-path-method network, the difference between the latest possible completion time of each activity which will not delay the completion

of the overall project, and the earliest possible completion time, based on all predecessor activities. { 'slak ,tīm }

slamming stile |BUILD| The vertical strip that a closed door abuts; it receives the bolt when the lock engages. { 'slam·iŋ ,stīl }

slant depth |DES ENG| The distance between the crest and root of a screw thread measured along the angle forming the flank of the thread. { 'slant ,depth }

slant drilling |ENG| The drilling of a borehole or well at an angle to the vertical. { 'slant ,dril·iŋ }

slat conveyor |MECH ENG| A conveyor consisting of horizontal slats on an endless chain. { 'slat kən,vā·ər }

slave |CONT SYS| A device whose motions are governed by instructions from another machine. { slāv }

slave arm |ENG| A component of a remote manipulator that automatically duplicates the motions of a master arm, sometimes with changes of scale in displacement or force. { 'slāv ,ärm }

sled |ENG| An item equipped with runners and a suitable body designed to transport loads over ice and snow. { sled }

sledgehammer |DES ENG| A large heavy hammer that is usually wielded with two hands; used for driving stakes or breaking stone. { 'slej ,ham·ər }

sleeper |CIV ENG| A timber, steel, or precast concrete beam placed under rails to hold them at the correct gage. { 'slēp·ər }

sleeve |ELEC| **1.** The cylindrical contact that is farthest from the tip of a phone plug. **2.** Insulating tubing used over wires or components. Also known as bushing; sleeving. |ENG| A cylindrical part designed to fit over another part. { slēv }

sleeve bearing |MECH ENG| A machine bearing in which the shaft turns and is lubricated by a sleeve. { 'slēv ,ber·iŋ }

sleeve burner |ENG| A type of oil burner for domestic heating. { 'slēv ,bər·nər }

sleeve coupling |DES ENG| A hollow cylinder which fits over the ends of two shafts or pipes, thereby joining them. { 'slēv ,kəp·liŋ }

sleeve joint |DES ENG| A device for joining the ends of two wires or cables together, constructed by forcing the ends of the wires or cables into both ends of a hollow sleeve. { 'slēv ,jóint }

sleeve valve |MECH ENG| An admission and exhaust valve on an internal-combustion engine consisting of one or two hollow sleeves that fit around the inside of the cylinder and move with the piston so that their openings align with the inlet and exhaust ports in the cylinder at proper stages in the cycle. { 'slēv ,valv }

slenderness ratio |CIV ENG| The ratio of the length of a column L to the radius of gyration r about the principal axes of the section. { 'slen·dər·nəs ,rā·shō }

slewing |ENG| Moving a radar antenna or a sonar transducer rapidly in a horizontal or vertical direction, or both. { 'slü·iŋ }

slewing mechanism |ENG| Device which permits rapid traverse or change in elevation of a weapon or instrument. { 'slü·iŋ ,mek·ə,niz·əm }

slew rate |CONT SYS| The maximum rate at which a system can follow a command. |ELECTR| The maximum rate at which the output voltage of an operational amplifier changes for a square-wave or step-signal input; usually specified in volts per microsecond. { 'slü ,rāt }

slice bar |ENG| A broad, flat steel blade used for chipping and scraping. { 'slīs ,bär }

slide |ENG| **1.** A sloping chute with a flat bed. **2.** A sliding mechanism. |MECH ENG| The main reciprocating member of a mechanical press, guided in a press frame, to which the punch or upper die is fastened. { slīd }

slide conveyor |ENG| A slanted gravity slide for the forward downward movement of flowable solids, slurries, liquids, or small objects. { 'slīd kən,vā·ər }

slide gate |CIV ENG| A crest gate which has high frictional resistance to opening because it slides on its bearings in opening and closing. { 'slīd ,gāt }

slide projector See optical lantern. { 'slīd prə,jek·tər }

slider |ELEC| Sliding type of movable contact. { 'slīd·ər }

slide rail See guardrail. { 'slīd ,rāl }

slider coupling |MECH ENG| A device for connecting shafts that are laterally misaligned. Also known as double-slider coupling; Oldham coupling. { 'slīd·ər ,kəp·liŋ }

slide rest |MECH ENG| An adjustable slide for holding a cutting tool, as on an engine lathe. { 'slīd ,rest }

slider support |ENG| A support designed to allow longitudinal movement of pipework in a horizontal plane. { 'slīd·ər sə'pórt }

slide-rule dial |ENG| A dial in which a pointer moves in a straight line over long straight scales resembling the scales of a slide rule. { 'slīd ,rül ,dīl }

slide valve |MECH ENG| A sliding mechanism to cover and uncover ports for the admission of fluid, as in some steam engines. { 'slīd ,valv }

sliding-block linkage |MECH ENG| A mechanism in which a crank and sliding block serve to convert rotary motion into translation, or vice versa. { 'slīd·iŋ ¦bläk 'liŋ·kij }

sliding-chain conveyor |MECH ENG| A conveying machine to handle cases, cans, pipes, or similar products on the plain or modified links of a set of parallel chains. { 'slīd·iŋ ¦chān kən'vā·ər }

sliding fit |DES ENG| A fit between two parts that slide together. { 'slīd·iŋ 'fit }

sliding form See slip form. { 'slīd·iŋ ,fórm }

sliding friction |MECH| Rubbing of bodies in sliding contact. { 'slīd·iŋ ,frik·shən }

sliding gear |DES ENG| A change gear in which speed changes are made by sliding gears along their axes, so as to place them in or out of mesh. { 'slīd·iŋ ,gir }

sliding-gear transmission |MECH ENG| A transmission system utilizing a pair of sliding gears. { 'slīd·iŋ ¦gir tranz'mish·ən }

sliding pair |MECH ENG| Two adjacent links, one of which is constrained to move in a particular path with respect to the other; the lower, or closed, pair is completely constrained by the design of the links of the pair. { 'slīd·iŋ 'per }

sliding-vane compressor |CHEM ENG| A rotary-element gas compressor in which spring-loaded sliding vanes (evenly spaced around a cylinder off-center in a surrounding chamber) pick up, compress, and discharge gas as the cylinder revolves. { 'slīd·iŋ ¦vān kəm'pres·ər }

sliding vector |MECH| A vector whose direction and line of application are prescribed, but whose point of application is not prescribed. { 'slīd·iŋ 'vek·tər }

sliding way |CIV ENG| One of the timbers which form the upper part of the cradle supporting a ship during its construction, and which slide over the ground ways with the ship when it is launched. { 'slīd·iŋ 'wā }

slime |ENG| Liquid slurry of very fine solids with slime- or mudlike appearance. Also known as mud; pulp; sludge. { slīm }

slim hole |ENG| A drill hole of the smallest practicable size, drilled with less-than normal-diameter tools, used primarily as a seismic shothole and for structure tests and sometimes for stratigraphic tests. { 'slīm ,hōl }

sling |ENG| A length of rope, wire rope, or chain used for attaching a load to a crane hook. { sliŋ }

sling psychrometer |ENG| A psychrometer in which the wet- and dry-bulb thermometers are mounted upon a frame connected to a handle at one end by means of a bearing or a length of chain; the psychrometer may be whirled in the air for the simultaneous measurement of wet- and dry-bulb temperatures. { 'sliŋ si'kräm·əd·ər }

sling thermometer |ENG| A thermometer mounted upon a frame connected to a handle at one end by means of a bearing or length of chain, so that the thermometer may be whirled by hand. { 'sliŋ thər'mäm·əd·ər }

slip |CIV ENG| A narrow body of water between two piers. |ELEC| **1.** The difference between synchronous and operating speeds of an induction machine. Also known as slip speed. **2.** Method of interconnecting multiple wiring between switching units by which trunk number 1 becomes the first choice for the first switch, trunk number 2 first choice for the second switch, trunk number 3 first choice for the third switch, and so on. |ELECTR| Distortion produced in the recorded facsimile image which is similar to that produced by skew but is caused by slippage in the mechanical drive system. { slip }

slip casting |ENG| A process in the manufacture of shaped refractories, cermets, and other materials in which the slip is poured into porous plaster molds. { 'slip ,kast·iŋ }

slip form |CIV ENG| A narrow section of formwork that can be easily removed as concrete placing progresses. { 'slip ,fòrm }

slip forming |ENG| A plastics-sheet forming technique in which some of the sheet is allowed to slip through the mechanically operated clamping rings during stretch-forming operations. { 'slip ,fòrm·iŋ }

slip friction clutch |MECH ENG| A friction clutch designed to slip when too much power is applied to it. { 'slip 'frik·shən ,kləch }

slip joint |CIV ENG| **1.** Contraction joint between two adjoining wall sections, or at the horizontal bearing of beams, slabs, or precast units, consisting of a vertical tongue fitted into a groove which allows independent movement of the two sections. **2.** A telescoping joint between two parts. |ENG| **1.** A method of laying-up plastic veneers in flexible-bag molding, wherein edges are beveled and allowed to overlap part or all of the scarfed area. **2.** A mechanical union that allows limited endwise movement of two solid items for example, pipe, rod, or duct with relation to each other. { 'slip ,jóint }

slippage |ENG| The leakage of fluid between the plunger and the bore of a pump piston. Also known as slippage loss. { 'slip·ij }

slippage loss |ENG| **1.** Unintentional movement between the faces of two solid objects. **2.** See slippage. { 'slip·ij ,lòs }

slipper brake |MECH ENG| **1.** A plate placed against a moving part to slow or stop it. **2.** A plate applied to the wheel of a vehicle or to the track roadway to slow or stop the vehicle. { 'slip·ər ,brāk }

slip plane |ENG| A plane visible by reflected light in a transparent material; caused by poor welding and shrinkage during cooling. { 'slip ,plān }

slip ratio |MECH ENG| For a screw propeller, relates the actual advance to the theoretic advance determined by pitch and spin. { 'slip ,rā·shō }

slips |ENG| A wedge-shaped steel collar fabricated in two sections, designed to hold a string of casing between various portions of the drilling operation. { 'slips }

slip speed See slip. { 'slip ,spēd }

slip tongue |ENG| A pole on a horse-drawn wagon that is fastened by slipping it between two plates connected to the forecarriage. { 'slip ,təŋ }

slipway |CIV ENG| The space in a shipyard where a foundation for launching ways and keel blocks exists and which is occupied by a ship while under construction. { 'slip,wā }

slit |DES ENG| A long, narrow opening through which radiation or streams of particles enter or leave certain instruments. { slit }

slitter |MECH ENG| A synchronized feeder-knife variation of a rotary cutter; used for precision cutting of sheet material, such as metal, rubber, plastics, or paper, into strips. { 'slid·ər }

slitting |MECH ENG| The passing of sheet or strip material (metal, plastic, paper, or cloth) through rotary knives. { 'slid·iŋ }

slop |CHEM ENG| A petroleum-refinery term for odds and ends of oil produced in the refinery; the slop must be rerun or further processed to make it suitable for use. Also known as slop oil. { släp }

slope conveyor |MECH ENG| A troughed belt conveyor used for transporting material on steep grades. { 'slōp kən,vā·ər }

slope course |ENG| A proving ground facility consisting of a large mound of earth with various sloping sides on which are roads having different grades; this slope course is used to measure the slope performance of military and other vehicles, including maximum speed on various grades, the most suitable gear for best performance, traction, and the holding ability of brakes. { 'slōp ,kòrs }

slope of fall |MECH| Ratio between the drop of a projectile and its horizontal movement; tangent of the angle of fall. { 'slōp əv 'fòl }

slop oil See slop. { 'släp ,óil }

slosh test |ENG| A test to determine the ability of the control system of a liquid-propelled missile to withstand or overcome the dynamic movement of the liquid within its fuel tanks. { 'släsh ,test }

slot |DES ENG| A narrow, vertical opening. |ELEC| One of the conductor-holding grooves in the face of the rotor or stator of an electric rotating machine. { slät }

slot distributor |ENG| A long, narrow discharge opening (slot) in a pipe or conduit; used for the extrusion of sheet material, such as plastics. { 'slät di'strib·yəd·ər }

slot dozing |ENG| A method of moving large quantities of material with a bulldozer using the same path for each trip so that the spillage from the sides of the blade builds up along each side; afterward all material pushed into the slot is retained in front of the blade. { 'slät ,dōz·iŋ }

slot extrusion |ENG| A method of extruding plastics-film sheet in which the molten thermoplastic compound is forced through a straight slot. { 'slät ik,strü·zhən }

slotted-head screw |DES ENG| A screw fastener with a single groove across the diameter of the head. { 'släd·əd ¦hed 'skrü }

slotted nut |DES ENG| A regular hexagon nut with slots cut across the flats of the hexagon so that a cotter pin or safety wire can hold it in place. { 'släd·əd 'nət }

slotter |MECH ENG| A machine tool used for making a mortise or shaping the sides of an aperture. { 'släd·ər }

slotting |MECH ENG| Cutting a mortise or a similar narrow aperture in a material using a machine with a vertically reciprocating tool. { 'släd·iŋ }

slotting machine |MECH ENG| A vertically reciprocating planing machine, used for making mortises and for shaping the sides of openings. { 'släd·iŋ mə,shēn }

slot washer |DES ENG| **1.** A lock washer with an indentation on its edge through which a nail or screw can be driven to hold it in place. **2.** A washer with a slot extending from its edge to the center hole to allow the washer to be removed without first removing the bolt. { 'slät ,wäsh·ər }

slough |ENG| The fragments of rocky material from the wall of a borehole. Also known as cavings. { slaù }

slow igniter cord |ENG| An igniter cord made with a central copper wire around which is extruded a plastic incendiary material with an iron wire embedded to give greater strength; the whole is enclosed in a thin extruded plastic coating. { 'slō ig'nīd·ər ,kórd }

slow match |ENG| A match or fuse that burns at a known slow rate; used for igniting explosive charges. { 'slō 'mach }

slow sand filter |CIV ENG| A bed of fine sand 20–48 inches (151–122 centimeters) deep through which water, being made suitable for human consumption and other purposes, is passed at a fairly low rate, 2,500,000 to 10,000,000 gallons per acre (23,000 to 94,000 cubic meters per hectare); an underdrain system of graded gravel and perforated pipes carries the water from the filters to the point of discharge. { 'slō 'sand ,fil·tər }

slow-spiral drill See low-helix drill. { 'slō ¦spī·rəl 'dril }

sludge |CHEM ENG| **1.** Residue left after acid treatment of petroleum oils. **2.** Any semisolid waste from a chemical process. |CIV ENG| See sewage sludge. |ENG| **1.** Mud from a drill hole in boring. **2.** Sediment in a steam boiler. **3.** A precipitate from petroleum oils or liquid fuels, for example, the insoluble degradation products formed during the operation of an internal combustion engine. **4.** An amorphous deposit that has accumulated on the surface of a tube in a heat exchanger or of an evaporating device, but is not bonded to the fouled surface. **5.** See slime. { sləj }

sludge bucket See calyx. { 'sləj ,bək·ət }

sludge coking |CHEM ENG| The recovery of sulfuric acid from dry acid sludge. { 'sləj ,kōk·iŋ }

sludge pit See slushpit. { 'sləj ,pit }

sludge pond See slushpit. { 'sləj ,pänd }

sludge pump See sand pump. { 'sləj ,pəmp }

sluff |ENG| The mud cake detached from the wall of a borehole. { sləf }

slug |MECH| A unit of mass in the British gravitational system of units, equal to the mass which experiences an acceleration of 1 foot per second per second when a force of 1 pound acts on it; equal to approximately 32.1740 pound mass or 14.5939 kilograms. Also known as geepound. { sləg }

slug bit See insert bit. { 'sləg ,bit }

sluice |CIV ENG| **1.** A passage fitted with a vertical sliding gate or valve to regulate the flow of water in a channel or lock. **2.** A body of water retained by a floodgate. **3.** A channel serving to drain surplus water. { slüs }

sluice gate |CIV ENG| The vertical slide gate of a sluice. { 'slüs ,gāt }

sluicing pond See scouring basin. { 'slüs·iŋ ,pänd }

slump test |ENG| Determining the consistency of concrete by filling a conical mold with a sample of concrete, then inverting it over a flat plate and removing the mold; the amount by which the concrete drops below the mold height is measured and this represents the slump. { 'sləmp ,test }

slurry bed reactor See ebullating-bed reactor. { 'slər·ē ,bed rē,ak·tər }

slurrying |ENG| The formation of a mud or a suspension from a liquid and nonsoluble solid particles. { 'slər·ē·iŋ }

slurry preforming |ENG| The preparation of reinforced plastics preforms by wet-processing techniques; similar to pulp molding. { 'slər·ē prē'fórm·iŋ }

slurry truck |ENG| A mobile unit that transports dry blasting ingredients, and mixes them in required proportions for introduction as explosive slurry into blastholes. { 'slər·ē ,trək }

slusher |ENG| A method for the application of vitreous enamel slip to ware by dashing it on the ware to cover all its parts, excess then being removed by shaking the ware. { 'sləsh·ər }

slush grouting |CIV ENG| Spreading a portland cement slurry over a surface that will subsequently be covered by concrete. { 'sləsh ,graùd·iŋ }

slush molding |ENG| A thermoplastic casting in which a liquid resin is poured into a hot, hollow mold where a viscous skin forms; excess slush is drained off, the mold is cooled, and the molded product is stripped out. { 'sləsh ,mōld·iŋ }

slushpit |ENG| An excavation or diked area to hold water, mud, sludge, and other discharged matter from an oil well. Also known as mud pit; sludge pit; sludge pond. { 'shəsh,pit }

small calorie See calorie. { 'smól 'kal·ə·rē }

small-diameter blasthole |ENG| A blast hole 1¹/₂ to 3 inches (3.8 to 7.6 centimeters) in diameter, in low-face quarries. { 'smól dī¦am·əd·ər 'blast,hōl }

small-lot storage |IND ENG| Generally, a quantity of less than one pallet stack, stacked to maximum storage height; thus, the term refers to a lot consisting of from one container to two or more pallet loads, but is not of sufficient quantity to form a complete pallet column. { 'smól ¦lät 'stór·ij }

small-scale hydropower |MECH ENG| The generation of electricity by using hydraulic turbines in which the installed capacity of the plant lies within the range from 5 kilowatts to 5 megawatts. { 'smól ,skāl 'hī·drə,paù·ər }

smart sensor |ENG| A microsensor integrated with signal-conditioning electronics such as analog-to-digital converters on a single silicon chip to form an integrated microelectromechanical component that can process information itself or communicate with an embedded microprocessor. Also known as intelligent sensor. { ,smärt 'sen·sər }

smart structures |ENG| Structures that are capable of sensing and reacting to their environment in a predictable and desired manner, through the integration of various elements, such as sensors, actuators, power sources, signal processors, and communications network. In addition to carrying mechanical loads, smart structures may alleviate vibration, reduce acoustic noise, monitor their own condition and environment, automatically perform precision alignments, or change their shape or mechanical properties on command. { ,smärt 'strək·chərz }

smart tool |CONT SYS| A robot end effector or fixed tool that uses sensors to measure the tool's position relative to reference markers or a workpiece or jig, and an actuator to adjust the tool's position with respect to the workpiece. { 'smärt ,tül }

Smithell's burner |ENG| Two concentric tubes that can be added to a bunsen burner to separate the inner and outer flame cones. { 'smith·əlz ,bər·nər }

Smith-McIntyre sampler |MECH ENG| A device for taking samples of sediment from the ocean bottom; the digging and hoisting mechanisms are independent: the digging bucket is forced into the sediment before the hoisting action occurs. { 'smith 'mak·ən,tīr ,sam·plər }

smoke |ENG| Dispersions of finely divided (0.01–5.0 micrometers) solids or liquids in a gaseous medium. { smōk }

smokebox |MECH ENG| A chamber external to a boiler for trapping the unburned products of combustion. { 'smōk,bäks }

smoke chamber |ENG| That area in a fireplace directly above the smoke shelf. { 'smōk ,chām·bər }

smoke detector |ENG| A photoelectric system for an alarm when smoke in a chimney or other location exceeds a predetermined density. { 'smōk di,tek·tər }

smoke point |ENG| The maximum flame height in millimeters at which kerosine will burn without smoking, tested under standard conditions; used as a measure of the burning cleanliness of jet fuel and kerosine. { 'smōk ,pöint }

smoke shelf |ENG| A horizontal surface directly behind the throat of a fireplace to prevent downdrafts. { 'smōk ,shelf }

smokestack |ENG| A chimney for the discharge of flue gases from a furnace operation such as in a steam boiler, powerhouse, heating plant, ship, locomotive, or foundry. { 'smōk,stak }

smoke test |ENG| A test used on kerosine to determine the highest point to which the flame can be turned before smoking occurs. { 'smōk ,test }

smoke washer |ENG| A device for removing particles from smoke by forcing it through a spray of water. { 'smōk ,wäsh·ər }

smooth blasting |ENG| Blasting to ensure even faces without cracks in the rock. { 'smüth 'blast·iŋ }

smooth drilling |ENG| Drilling in a rock formation in which a fast rotation of the drill stem, a

fast rate of penetration, and a high recovery of core can be achieved with vibration-free rotation of the drill stem. { 'smüth 'dril·iŋ }

smoothing |ENG| Making a level, or continuously even, surface. { 'smüth·iŋ }

smoothing mill |MECH ENG| A revolving stone wheel used to cut and bevel glass or stone. { 'smüth·iŋ ,mil }

smoothing plane |DES ENG| A finely set hand tool, usually 5.5–10 inches (14–25.4 centimeters) long, for finishing small areas on wood. { 'smüth·iŋ ,plān }

smother kiln |ENG| A kiln into which smoke can be introduced for blackening pottery. { 'smoth·ər ,kil }

smudging |ENG| A frost-preventive measure used in orchards; properly, it means the production of heavy smoke, supposed to prevent radiational cooling, but it is generally applied to both heating and smoke production. { 'sməj·iŋ }

S/N See signal-to-noise ratio.

snagging |MECH ENG| Removing surplus metal or large surface defects by using a grinding wheel. { 'snag·iŋ }

snake hole |ENG| **1.** A blasting hole bored directly under a boulder. **2.** A drill hole used in quarrying or bench blasting. { 'snāk ,hōl }

snaking |ENG| Towing a load with a long cable. { 'snāk·iŋ }

snap-back forming |ENG| A plastic-sheet-forming technique in which an extended, heated, plastic sheet is allowed to contract over a form shaped to the desired final contour. { 'snap ,bak ,förm·iŋ }

snapback method See repetitive time method. { 'snap,bak ,meth·əd }

snap fastener |DES ENG| A fastener consisting of a ball on one edge of an article that fits in a socket on an opposed edge, and used to hold edges together, such as those of a garment. { 'snap ,fas·ən·ər }

snap gage |DES ENG| A device with two flat, parallel surfaces spaced to control one limit of tolerance of an outside diameter or a length. { 'snap ,gāj }

snap hook See spring hook. { 'snap ,huk }

snap-off diode |ELECTR| Planar epitaxial passivated silicon diode that is processed so a charge is stored close to the junction when the diode is conducting; when reverse voltage is applied, the stored charge then forces the diode to snap off or switch rapidly to its blocking state. { 'snap,öf 'dī,ōd }

snapper |ENG| A device for collecting samples from the ocean bottom, and which closes to prevent the sample from dropping out as it is raised to the surface. { 'snap·ər }

snap ring |DES ENG| A form of spring used as a fastener; the ring is elastically deformed, put in place, and allowed to snap back toward its unstressed position into a groove or recess. { 'snap ,riŋ }

snatch block |DES ENG| A pulley frame or sheave with an eye through which lashing can be passed to fasten it to a scaffold or pole. { 'snach ,bläk }

snatch plate [ENG] A thick steel plate through which a hole about one-sixteenth of an inch larger than the outside diameter of the drill rod on which it is to be used is drilled; the plate is slipped over the drill rod and one edge is fastened to a securely anchored chain, and if rods must be pulled because high-pressure water is encountered, the eccentric pull of the chain causes the outside of the rods to be gripped and held against the pressure of water; the rod is moved a short distance out of the hole each time the plate is tapped. { 'snach ,plāt }

S-N diagram [ENG] In fatigue testing, a graphic representation of the relationship of stress S and the number of cycles N before failure of the material. { ¦es¦en 'dī·ə,gram }

snifter valve [ENG] A valve on a pump that allows air to enter or escape, and accumulated water to be released. { 'snif·tər ,valv }

snorkel [ENG] Any tube which supplies air for an underwater operation, whether it be for material or personnel. { 'snȯr·kəl }

snow bin [ENG] A box for measuring the amount of snowfall; a type of snow gage. { 'snō ,bin }

snow blower [MECH ENG] A machine that removes snow from a road surface or pavement using a screw-type blade to push the snow into the machine and from which it is ejected at some distance. { 'snō ,blō·ər }

snowbreak [CIV ENG] Any barrier designed to shelter an object or area from snow. { 'snō,brāk }

snow fence [CIV ENG] An open-slatted board fence usually 4 to 10 feet (1.2 to 3.0 meters) high, placed about 50 feet (15 meters) on the windward side of a railroad track or highway; the fence serves to disrupt the flow of the wind so that the snow is deposited close to the fence on the leeward side, leaving a comparatively clear, protected strip parallel to the fence and slightly farther downwind. { 'snō ,fens }

snow load [CIV ENG] The unit weight factor considered in the design of a flat or pitched roof for the probable amount of snow lying upon it. { 'snō ,lōd }

snow mat [ENG] A device used to mark the surface between old and new snow, consisting of a piece of white duck 28 inches (71 centimeters) square, having in each corner triangular pockets in which are inserted slats placed diagonally to keep the mat taut and flat. { 'snō ,mat }

snow-melting system [CIV ENG] A system of pipes containing a circulating nonfreezing liquid or electric-heating cables, embedded beneath the surface of a road, walkway, or other area to be protected from snow accumulation. { 'snō ,melt·iŋ ,sis·təm }

snow pillow [ENG] A device used to record the changing weight of the snow cover at a point, consisting of a fluid-filled bladder lying on the ground with a pressure transducer or a vertical pipe and float connected to it. { 'snō ,pil·ō }

snowplow [MECH ENG] A device for clearing away snow, as from a road or railway track. { 'snō,plaù }

snow resistograph [ENG] An instrument for recording a hardness profile of a snow cover by recording the force required to move a blade up through the snow. { 'snō ri'zis·tə,graf }

snow sampler [ENG] A hollow tube for collecting a sample of snow in place. Also known as snow tube. { 'snō ,sam·plər }

snow scale See snow stake. { 'snō ,skāl }

snowshed [CIV ENG] A structure to protect an exposed area as a road or rail line from snow. { 'snō,shed }

snow stake [ENG] A wood scale, calibrated in inches, used in regions of deep snow to measure its depth; it is bolted to a wood post or angle iron set in the ground. Also known as snow scale. { 'snō ,stāk }

snow tube See snow sampler. { 'snō ,tüb }

SNR See signal-to-noise ratio.

snubber [MECH ENG] A mechanical device consisting essentially of a drum, spring, and friction band, connected between axle and frame, in order to slow the recoil of the spring and reduce jolting. { 'snəb·ər }

Snyder sampler [ENG] A mechanical device for obtaining small representative quantities from a moving stream of pulverized or granulated solids; it consists of a cast-iron plate revolving in a vertical plane on a horizontal axis with an inclined sample spout; the material to be sampled comes to the sampler by way of an inclined chute whenever the sample spout comes in line with the moving stream. { 'snī·dər 'sam·plər }

soaking drum [CHEM ENG] A heated petroleum-refinery process vessel used in connection with petroleum thermal-cracking coils to furnish the residence time needed to complete the cracking reaction. { 'sōk·iŋ ,drəm }

soap bubble test [ENG] A leak test in which a soap solution is applied to the surface of the vessel under internal pressure test; soap bubbles form if the tracer gas leaks from the vessel. { 'sōp ¦bəb·əl ,test }

socket [ELEC] A device designed to provide electric connections and mechanical support for an electronic or electric component requiring convenient replacement. [ENG] A device designed to receive and grip the end of a tubular object, such as a tool or pipe. { 'säk·ət }

socket-head screw [DES ENG] A screw fastener with a geometric recess in the head into which an appropriate wrench is inserted for driving and turning, with consequent improved nontamperability. { 'säk·ət ¦hed ,skrü }

socket wrench [DES ENG] A wrench with a socket to fit the head of a bolt or a nut. { 'säk·ət ,rench }

soda-acid extinguisher [ENG] A fire-extinguisher from which water is expelled at a high rate by the generation of carbon dioxide, the result of mixing (when the extinguisher is tilted) of sulfuric acid and sodium bicarbonate. { 'sōd·ə 'as·əd ik'stiŋ·gwə·shər }

soda pulping process [CHEM ENG] The digestion of wood chips by caustic soda; used to manufacture pulp for paper products. { 'sōd·ə 'pəl·piŋ ,prä·səs }

sodar [ENG] Sound-wave transmitting and receiving equipment that is used to remotely measure the vertical turbulence structure and wind profile of the lower layer of the atmosphere by analyzing sound reflected in scattering by atmospheric turbulence. Derived from sonic detection and ranging. { 'sō,där }

sodium sulfite process [CHEM ENG] A process for the digestion of wood chips in a solution of magnesium, ammonium, or calcium disulfite containing free sulfur dioxide; used in papermaking. { 'sōd·ē·əm 'səl,fīt ,prä·səs }

soffit [CIV ENG] The underside of a horizontal structural member, such as a beam or a slab. { 'säf·ət }

soft automation [ENG] Automatic control, chiefly through the use of computer processing, with relatively little reliance on computer hardware. { 'sȯft ,ȯd·ə'mā·shən }

soft flow [ENG] The free-flowing characteristics of a plastic material under conventional molding conditions. { 'sȯft 'flō }

soft hammer [ENG] A hammer having a head made of a soft material, such as copper, lead, rawhide, or plastic; used to prevent damage to a finished surface. { 'sȯft 'ham·ər }

soft-iron ammeter [ENG] An ammeter in which current in a coil causes two pieces of magnetic material within the coil, one fixed and one attached to a pointer, to become similarly magnetized and to repel each other, moving the pointer; used for alternating-current measurement. { 'sȯft ¦ī·ərn 'am,ēd·ər }

soft missile base [CIV ENG] A missile-launching base that is not protected against a nuclear explosion. { 'sȯft 'mis·əl ,bās }

soft patch [ENG] A patch in a crack in a vessel such as a steam boiler consisting of a soft material inserted in the crack and covered by a metal plate bolted or riveted to the vessel. { 'sȯft 'pach }

soft-wired numerical control See computer numerical control. { 'sȯf ,wīrd nü'mer·ə·kəl kən'trōl }

soil line See soil pipe. { 'sȯil ,līn }

soil mechanics [ENG] The application of the laws of solid and fluid mechanics to soils and similar granular materials as a basis for design, construction, and maintenance of stable foundations and earth structures. { 'sȯil mi,kan·iks }

soil pipe [CIV ENG] A cast-iron or plastic pipe for carrying discharges from toilet fixtures from a building into the soil drain. Also known as soil line. { 'sȯil ,pīp }

soil stack [BUILD] The main vertical pipe into which flows the waste water from the soil pipes in a structure. { 'sȯil ,stak }

soil thermograph [ENG] A remote-recording thermograph whose sensing element may be buried at various depths in the earth. { 'sȯil 'thər·mə,graf }

soil thermometer [ENG] A thermometer used to

measure the temperature of the soil, usually the mercury-in-glass thermometer. Also known as earth thermometer. { 'sȯil thər,mäm·əd·ər }

soil vent See stack vent. { 'sȯil ,vent }

solar attachment [ENG] A device for determining the true meridian directly from the sun; used an an attachment on a surveyor's transit or compass. { 'sō·lər ə'tach·mənt }

solar battery [ELECTR] An array of solar cells, usually connected in parallel and series. { 'sō·lər 'bad·ə·rē }

solar cell [ELECTR] A pn-junction device which converts the radiant energy of sunlight directly and efficiently into electrical energy. { 'sō·lər 'sel }

solar chimney [ENG] A natural-draft drive device that uses solar radiation to provide upward momentum to a mass of air, thereby converting the thermal energy to kinetic energy, which can be extracted from the air with suitable wind machines. { ,sō·lər 'chim·nē }

solar collector [ENG] An installation designed to gather and accumulate energy in the form of solar radiation. { 'sō·lər kə'lek·tər }

solar distillation [CHEM ENG] A procedure in which the sun's heat is used to evaporate seawater in order to produce sodium chloride and other salts or potable water. { 'sō·lər ,dis·tə 'lā·shən }

solar engine [MECH ENG] An engine which converts thermal energy from the sun into electrical, mechanical, or refrigeration energy; may be used as a method of spacecraft propulsion, either directly by photon pressure on huge solar sails, or indirectly from solar cells or from a reflector-boiler combination used to heat a fluid. { 'sō·lər 'en·jən }

solar furnace [ENG] An image furnace in which high temperatures are produced by focusing solar radiation. { 'sō·lər 'fər·nəs }

solar heating [MECH ENG] The conversion of solar radiation into heat for technological, comfort-heating, and cooking purposes. { 'sō·lər 'hēd·iŋ }

solar heat storage [ENG] The storage of solar energy for later use; usually accomplished by the heating of water or fusing a salt, although sand and gravel have been used as storage media. { 'sō·lər 'hēt ,stȯr·ij }

solar house [BUILD] A house with large expanses of glass designed to catch solar radiation for heating. { ¦sō·lər ¦haus }

solarimeter [ENG] **1.** A type of pyranometer consisting of a Moll thermopile shielded from the wind by a bell glass. **2.** See pyranometer. { ,sō·lə'rim·əd·ər }

solar magnetograph [ENG] An instrument that utilizes the Zeeman effect to directly measure the strength and polarity of the complex patterns of magnetic fields at the sun's surface; comprises a telescope, a differential analyzer, a spectrograph, and a photoelectric or photographic means of differencing and recording. { 'sō·lər mag'ned·ə,graf }

solar pond [MECH ENG] A type of nonfocusing

solar collector consisting of a pool of salt water heated by the sun; used either directly as a source of heat or as a power source for an electric generator. Also known as salt-gradient solar pond. { 'sō·lər 'pänd }

solar power |MECH ENG| The conversion of the energy of the sun's radiation to useful work. { 'sō·lər 'paú·ər }

solar power satellite |ENG| A proposed collector of solar energy that would be placed in geostationary orbit where sunlight striking the satellite would be converted to electricity and then to microwaves, which would be beamed to earth. { ,sō·lər ,paú·ər 'sad·əl,īt }

solar sensor |ELECTR| A light-sensitive diode that sends a signal to the attitude-control system of a spacecraft when it senses the sun. Also known as sun sensor. { 'sō·lər 'sen·sər }

solar still |CHEM ENG| A device for evaporating seawater, in which water is confined in one or more shallow pools, over which is placed a roof-shaped transparent cover made of glass or plastic film; the sun's heat evaporates the water, leaving behind a residue of salt; the vapor from the evaporated water condenses on the surface of the cover and trickles down into gutters, which thus collect fresh water. { 'sō·lər 'stil }

solder-ball flip chip See flip chip. { ¦säd·ər ,bȯl 'flip ,chip }

soldering gun |ENG| A soldering iron shaped like a gun. { 'säd·ə·riŋ ,gən }

soldering iron |ENG| A rod of copper with a handle on one end and pointed or wedge-shaped at the other end, and used for applying heat in soldering. { 'säd·ə·riŋ ,ī·ərn }

soldering pencil |ENG| A small soldering iron, about the size and weight of a standard lead pencil, used for soldering or unsoldering joints on printed wiring boards. { 'säd·ə·riŋ ,pen·səl }

solder track |ELECTR| A conducting path on a printed circuit board that is formed by applying molten solder to the board. { 'säd·ər ,trak }

soldier course |CIV ENG| A course of bricks laid on their ends so that only their long sides are visible. { 'sōl·jər ,kȯrs }

sole |BUILD| The horizontal member beneath the studs in a framed building. |ELECTR| Electrode used in magnetrons and backward-wave oscillators to carry a current that generates a magnetic field in the direction wanted. { sōl }

solenoid brake |MECH ENG| A device that retards or arrests rotational motion by means of the magnetic resistance of a solenoid. { 'säl·ə,nȯid ,brāk }

solenoid valve |MECH ENG| A valve actuated by a solenoid, for controlling the flow of gases or liquids in pipes. { 'säl·ə,nȯid ,valv }

solepiece |CIV ENG| One of two steel plates, port and starboard, whose forward parts are bolted to the ground ways supporting a ship about to be launched, while their aft parts are attached to the sliding ways; at the start of the launch, they are cut simultaneously with burning torches to release the ship. Also known as soleplate. { 'sōl,pēs }

soleplate |BUILD| The plate on which stud bases butt in a stud partition. |CIV ENG| See solepiece. |ENG| 1. The supporting base of a machine. 2. A plate on which a bearing can be attached and, if necessary, adjusted slightly. { 'sōl,plāt }

solid box |MECH ENG| A solid, unadjustable ring bearing lined with babbitt metal, used on light machinery. { 'säl·əd 'bäks }

solid coupling |MECH ENG| A flanged-face or a compression-type coupling used to connect two shafts to make a permanent joint and usually designed to be capable of transmitting the full load capacity of the shaft; a solid coupling has no flexibility. { 'säl·əd 'kəp·liŋ }

solid cutter |DES ENG| A cutter made of a single piece of material. { 'säl·əd 'kəd·ər }

solid die |DES ENG| A one-piece screw-cutting tool with internal threads. { 'säl·əd 'dī }

solid drilling |ENG| In diamond drilling, using a bit that grinds the whole face, without preserving a core for sampling. { 'säl·əd 'dril·iŋ }

solid-electrolyte gas transducer |ENG| A device in which the concentration of a particular gas in a mixture is determined from the diffusion voltage across a heated solid electrolyte placed between this mixture and a reference gas. { 'säl·əd i'lek·trə,līt 'gas tranz,düs·ər }

solid injection system |MECH ENG| A fuel injection system for a diesel engine in which a pump forces fuel through a fuel line and an atomizing nozzle into the combustion chamber. { 'säl·əd in'jek·shən ,sis·təm }

solid logic technology |ELECTR| A method of computer construction that makes use of miniaturized modules, resulting in faster circuitry because of the reduced distances that current must travel. { 'säl·əd ¦läj·ik tek'näl·ə·jē }

solid shafting |MECH ENG| A solid round bar that supports a roller and wheel of a machine. { 'säl·əd 'shaft·iŋ }

solid shank tool |ENG| A cutting tool in which the shank and cutting edges are machined from one piece. { 'säl·əd ¦shaŋk 'tül }

solid state |ENG| Pertaining to a circuit, device, or system that depends on some combination of electrical, magnetic, and optical phenomena within a solid that is usually a crystalline semiconductor material. { 'säl·əd 'stāt }

solid-state circuit |ELECTR| Complete circuit formed from a single block of semiconductor material. { 'säl·əd ¦stāt 'sər·kət }

solid-state circuit breaker |ELECTR| A circuit breaker in which a Zener diode, silicon controlled rectifier, or solid-state device is connected to sense when load terminal voltage exceeds a safe value. { 'säl·əd ¦stāt 'sər·kət ,brāk·ər }

solid-state component |ELECTR| A component whose operation depends on the control of electrical or magnetic phenomena in solids, such as a transistor, crystal diode, or ferrite device. { 'säl·əd ¦stāt kəm'pō·nənt }

solid-state device |ELECTR| A device, other than a conductor, which uses magnetic, electrical, and other properties of solid materials, as opposed to vacuum or gaseous devices. { 'säl·əd ¦stāt di'vīs }

solid-state image sensor See charge-coupled image sensor. { 'säl·əd ¦stāt 'im·ij ˌsen·sər }

solid-state lamp See light-emitting diode. { 'säl·əd ¦stāt 'lamp }

solid-state power amplifier [ELECTR] An amplifier that uses field-effect transistors to provide useful amplification at gigahertz frequencies. { ˌsäl·əd ˌstāt 'paů·ər ˌam·plə¦fī·ər }

solid-state relay [ELECTR] A relay that uses only solid-state components, with no moving parts. Abbreviated SSR. { 'säl·əd ¦stāt 'rē¦lā }

solid-state switch [ELECTR] A microwave switch in which a semiconductor material serves as the switching element; a zero or negative potential applied to the control electrode will reverse-bias the switch and turn it off, and a slight positive voltage will turn it on. { 'säl·əd ¦stāt 'swich }

solid-state thyratron [ELECTR] A semiconductor device, such as a silicon controlled rectifier, that approximates the extremely fast switching speed and power-handling capability of a gaseous thyratron tube. { 'säl·əd ¦stāt 'thī·rəˌträn }

solid-web girder [CIV ENG] A beam, such as a box girder, having a web consisting of a plate or other solid section but not a lattice. { 'säl·əd ¦web 'gər·dər }

solution polymerization [CHEM ENG] A process for producing an addition polymer by heating the monomer, solvent, initiator, and catalyst together, with polymerization continuing as the solvent is removed. { səˈlü·shən pəˌlim·ə·rə'zā·shən }

solution process [CHEM ENG] An oil-refining process for separating mercaptans from gasoline by washing with a caustic solution containing organic compounds in which the mercaptans are soluble. { sə'lü·shən ˌprä·səs }

solutizer-air regenerative process [CHEM ENG] A petroleum refinery process that is identical to the solutizer-steam regeneration process, except for the regeneration step; the newer units use uncatalyzed air regeneration. { sə'lü·tīz·ər 'er rē¦jen·ə·rəd·iv ˌprä·səs }

solutizer-steam regenerative process [CHEM ENG] A petroleum refinery process used to extract mercaptans from gasoline or naphtha; uses solutizers (potassium isobutyrate or potassium alkyl phenolate) in strong potassium hydroxide solution as the selective solvent. { sə'lü·tīz·ər 'stēm rē¦jen·ə·rəd·iv ˌprä·səs }

solutizer-tannin process [CHEM ENG] A petroleum refinery process that is an early variation of the solutizer-air regenerative process for extraction of mercaptans from gasoline; uses tannin-catalyzed oxidation for the regeneration step. { sə'lü·tīz·ər 'tan·ən ˌprä·səs }

Solvay process [CHEM ENG] The process to make sodium carbonate and calcium chloride by treating sodium chloride with ammonia and carbon dioxide. { 'säl·vā ˌprä·səs }

solvent deasphalting [CHEM ENG] A petroleum refinery process used to remove asphaltic and resinous materials from reduced crude oils, lubricating oil stocks, gas oils, or middle distillates through the extractive or precipitant action of solvents. Also known as solvent deresining. { 'säl·vənt dē'as¦fólt·iŋ }

solvent deresining See solvent deasphalting. { 'säl·vənt di¦rez·ən·iŋ }

solvent dewaxing [CHEM ENG] A petroleum refinery process for solvent removal of wax from oils; the mixture of waxy oil and solvent is chilled, then filtered or centrifuged to remove the precipitated oil; the solvent is recovered for reuse. { 'säl·vənt di¦waks·iŋ }

solvent extraction [CHEM ENG] The separation of materials of different chemical types and solubilities by selective solvent action; that is, some materials are more soluble in one solvent than in another, hence there is a preferential extractive action; used to refine petroleum products, chemicals, vegetable oils, and vitamins. { 'säl·vənt ik¦strak·shən }

solvent molding [ENG] A process to form thermoplastic articles by dipping a mold into a solution or dispersion of the resin and drawing off (evaporating) the solvent to leave a plastic film adhering to the mold. { 'säl·vənt ˌmōld·iŋ }

solvent recovery [CHEM ENG] For reuse purposes, the catching and recovery of solvent vapors from vent lines, process vessels, or other sources of evaporative loss, usually with a solid adsorbent material. { 'säl·vənt ri¦kəv·ə·rē }

solvent-refined [CHEM ENG] Pertaining to any product material whose final quality and condition is in part the result of a solvent treatment during processing of the feedstock material. { 'säl·vənt ri¦fīnd }

solvent refining [CHEM ENG] The process of treating a mixed material with a solvent that preferentially dissolves and removes certain minor constituents (usually the undesired ones); common in the petroleum refining industry. { 'säl·vənt ri¦fīn·iŋ }

solvent welding [ENG] A technique for joining plastic pipework in which a mixture of solvent and cement is applied to the pipe end and to the socket, with the parts then being joined and allowed to set. { 'säl·vənt ˌweld·iŋ }

sonar [ENG] **1.** A system that uses underwater sound, at sonic or ultrasonic frequencies, to detect and locate objects in the sea, or for communication; the commonest type is echo-ranging sonar; other versions are passive sonar, scanning sonar, and searchlight sonar. Derived from sound navigation and ranging. **2.** See sonar set. { 'sō¦när }

sonar beacon [ENG ACOUS] An underwater beacon that transmits sonic or ultrasonic signals for the purpose of providing bearing information; it may have receiving facilities that permit triggering an external source. { 'sō¦när ˌbē·kən }

sonar boomer transducer [ENG ACOUS] A sonar transducer that generates a large pressure wave in the surrounding water when a capacitor bank discharges into a flat, epoxy-encapsulated coil, creating opposed magnetic fields from the coil and from eddy currents in an adjacent aluminum disk, which cause the disk to be driven away from the coils with great force. { 'sō,när 'büm·ər trans,dü·sər }

sonar capsule |ENG ACOUS| A capsule that reflects high-frequency sound waves; the sonar capsule, if attached to a reentry body, may be used to locate the reentry body. { 'sō,när ,kap·səl }

sonar dome |ENG| A streamlined, watertight enclosure that provides protection for a sonar transducer, sonar projector, or hydrophone and associated equipment, while offering minimum interference to sound transmission and reception. { 'sō,när ,dōm }

sonar projector |ENG ACOUS| An electromechanical device used under water to convert electrical energy to sound energy; a crystal or magnetostriction transducer is usually used for this purpose. { 'sō,när prə,jek·tər }

sonar set |ENG| A complete assembly of sonar equipment for detecting and ranging or for communication. Also known as sonar. { 'sō,när ,set }

sonar target |ENG ACOUS| An object which reflects a sufficient amount of a sonar signal to produce a detectable echo signal at the sonar equipment. { 'sō,när ,tär·gət }

sonar transducer |ENG ACOUS| A transducer used under water to convert electrical energy to sound energy and sound energy to electrical energy. { 'sō,när tranz,dü·sər }

sonar transmission |ENG ACOUS| The process by which underwater sound signals generated by a sonar set travel through the water. { 'sō,när tranz,mish·ən }

sonar window |ENG ACOUS| The portion of a sonar dome or sonar transducer that passes sound waves at sonar frequencies with little attenuation while providing mechanical protection for the transducer. { 'sō,när ,win·dō }

sonde |ENG| An instrument used to obtain weather data during ascent and descent through the atmosphere, in a form suitable for telemetering to a ground station by radio, as in a radiosonde. { sänd }

sonic altimeter |ENG| An instrument for determining the height of an aircraft above the earth by measuring the time taken for sound waves to travel from the aircraft to the surface of the earth and back to the aircraft again. { 'sän·ik al'tim·əd·ər }

sonic anemometer |ENG| An anemometer which measures wind speed by means of the properties of wind-borne sound waves; it operates on the principle that the propagation velocity of a sound wave in a moving medium is equal to the velocity of sound with respect to the medium plus the velocity of the medium. { 'sän·ik ,an·ə'mäm·əd·ər }

sonicate |ENG| To apply high-frequency sound waves to matter. { 'sän·ə,kāt }

sonicator |ENG ACOUS| An instrument for producing high-intensity ultrasound, consisting of a converter that transforms electrical energy into mechanical energy in the form of oscillation of piezoelectric transducers at a frequency of 20 kilohertz, and a titanium horn that focuses this oscillation and radiates energy into the liquid being treated through a tip. { 'sän·ə,kād·ər }

sonic chemical analyzer |ENG| A device to characterize the composition of a gas, liquid, or solid by the attenuation or change in the velocity of sound waves through a sample; the effect is related to molecular structure and intermolecular interactions. { 'sän·ik 'kem·ə·kəl 'an·ə,līz·ər }

sonic cleaning |ENG| Cleaning of contaminated materials by the action of intense sound in the liquid in which the material is immersed. { 'sän·ik 'klēn·iŋ }

sonic depth finder |ENG| A sonar-type instrument used to measure ocean depth and to locate underwater objects; a sound pulse is transmitted vertically downward by a piezoelectric or magnetostriction transducer mounted on the hull of the ship; the time required for the pulse to return after reflection is measured electronically. Also known as echo sounder. { 'sän·ik 'depth ,fīn·dər }

sonic detection and ranging See sodar. { ¦sän·ik di,tek·shən an 'rānj·iŋ }

sonic drilling |MECH ENG| The process of cutting or shaping materials with an abrasive slurry driven by a reciprocating tool attached to an audio-frequency electromechanical transducer and vibrating at sonic frequency. { 'sän·ik 'dril·iŋ }

sonic flaw detection |ENG| The process of locating imperfections in solid materials by observing internal reflections or a variation in transmission through the materials as a function of sound-path location. { 'sän·ik 'flō di,tek·shən }

sonic liquid-level meter |ENG| A meter that detects a liquid level by sonic-reflection techniques. { 'sän·ik 'lik·wəd ¦lev·əl ,mēd·ər }

sonic nucleation |CHEM ENG| In supersaturated solutions, the use of sonic or ultrasonic radiation to help bring about nucleation and corresponding crystallization of substances otherwise difficult to crystallize. { 'sän·ik ,nü·klē'ā·shən }

sonic sifter |MECH ENG| A high-speed vibrating apparatus used in particle size analysis. { 'sän·ik 'sif·tər }

sonic sounding |ENG| Determining the depth of the ocean bottom by measuring the time for an echo to return to a shipboard sound source. { 'sän·ik 'saúnd·iŋ }

sonic thermometer |ENG| A thermometer based upon the principle that the velocity of a sound wave is a function of the temperature of the medium through which it passes. { 'sän·ik thər'mäm·əd·ər }

sonic well logging |ENG| A well logging technique that uses a pulse-echo system to measure the distance between the instrument and a sound-reflecting surface; used to measure the size of cavities around brine wells, and capacities of underground liquefied petroleum gas storage chambers. { 'sän·ik 'wel ,läg·iŋ }

sonobuoy |ENG| An acoustic receiver and radio transmitter mounted in a buoy that can be dropped from an aircraft by parachute to pick up underwater sounds of a submarine and transmit them to the aircraft; to track a submarine, several buoys are dropped in a pattern that includes the known or suspected location of the submarine, with each buoy transmitting an identifiable signal; an electronic computer then determines the location of the submarine by comparison of the received signals and triangulation of the resulting time-delay data. Also known as radio sonobuoy. { 'sän·ə,bȯi }

sonograph |ENG| **1.** An instrument for recording sound or seismic vibrations. **2.** An instrument for converting sounds into seismic vibrations. { 'sän·ə,graf }

sonometer |ENG| **1.** In general, any device which consists of a thin metallic wire stretched over two bridges that are usually mounted on a soundboard and which is used to measure the vibration frequency, tension, density, or diameter of the wire, or to verify relations between these quantities. Also known as monochord. **2.** In particular, an instrument for measuring rock stress by means of a piano wire stretched between two bolts in the rock; any change of pitch after destressing is observed and used to indicate stress. { sə'näm·əd·ər }

sonoscan |ENG| A type of acoustic microscope in which an unfocused acoustic beam passes through the object and produces deformations in a liquid-solid interface that are sensed by a laser beam reflected from the surface. { 'sän·ə,skan }

soot blower |ENG| A system of steam or air jets used to maintain cleanliness, efficiency, and capacity of heat-transfer surfaces by the periodic removal of ash and slag from the heat-absorbing surfaces. { 'sut ,blō·ər }

sophisticated robot |CONT SYS| A robot that can be programmed and is controlled by a microprocessor. { sə'fis·tə,kād·əd 'rō,bät }

sorption pumping |ENG| A technique used to reduce the pressure of gas in an atmosphere; the gas is adsorbed on a granular sorbent material such as a molecular sieve in a metal container; when this sorbent-filled container is immersed in liquid nitrogen, the gas is sorbed. { 'sȯrp·shən ,pəmp·iŋ }

sound-field enhancement |ENG ACOUS| A system for enhancing the acoustical properties of both indoor and outdoor spaces, particularly for unamplified speech, song, and music; may consist of one or more microphones, systems for amplification and electronic signal processing, and one or more loudspeakers. { ¦saun ,fēld in 'hans·mənt }

sortie number |ENG| A reference used to identify the images taken by all the sensors during one air reconnaissance sortie. { 'sȯrd·ē ,nəm·bər }

sorting table |ENG| Any horizontal conveyor where operators, along its side, sort bulk material, packages, or objects from the conveyor. { 'sȯrd·iŋ ,tā·bəl }

sound analyzer |ENG| An instrument which measures the amount of sound energy in various frequency bands; it generally consists of a set of fixed electrical filters or a tunable electrical filter, along with associated amplifiers and a meter which indicates the filter output. { 'saund ,an·ə,līz·ər }

sound effects |ENG ACOUS| Mechanical devices or recordings used to provide lifelike imitations of various sounds. { 'saund i,feks }

sound film |ENG ACOUS| Motion picture film having a sound track along one side for reproduction of the sounds that are to accompany the film. { 'saund ,film }

sound filmstrip |ENG ACOUS| A filmstrip that has accompanying sound on a separate disk or tape, which is manually or automatically synchronized with projection of the pictures in the strip. { 'saund 'film,strip }

sound gate |ENG ACOUS| The gate through which film passes in a sound-film projector for conversion of the sound track into audio-frequency signals that can be amplified and reproduced. { 'saund ,gāt }

sound head |ENG ACOUS| **1.** The section of a sound motion picture projector that converts the photographic or magnetic sound track to audible sound signals. **2.** In a sonar system, the cylindrical container for the transmitting projector and the receiving hydrophone. { 'saund ,hed }

sounding |ENG| **1.** Determining the depth of a body of water by an echo sounder or sounding line. **2.** Measuring the depth of bedrock by driving a steel rod into the soil. **3.** Any penetration of the natural environment for scientific observation. { 'saund·iŋ }

sounding balloon |ENG| A small free balloon used for carrying radiosonde equipment aloft. { 'saund·iŋ bə,lün }

sounding lead |ENG| A lead used for determining the depth of water. { 'saund·iŋ ,led }

sounding line |ENG| The line attached to a sounding lead. Also known as lead line. { 'saund·iŋ ,līn }

sounding machine |ENG| An instrument for measuring the depth of water, consisting essentially of a reel of wire; to one end of this wire there is attached a weight which carries a device for measuring and recording the depth; a crank or motor reels in the wire. { 'saund·iŋ mə,shēn }

sounding pole |ENG| A pole or rod used for sounding in shallow water, and usually marked to indicate various depths. { 'saund·iŋ ,pōl }

sounding sextant See hydrographic sextant. { 'saund·iŋ ,sek·stənt }

sounding wire |ENG| A wire used with a sounding machine in determining depth of water. { 'saund·iŋ ,wīr }

sound-level meter |ENG| An instrument used

to measure noise and sound levels in a specified manner; the meter may be calibrated in decibels or volume units and includes a microphone, an amplifier, an output meter, and frequency-weighting networks. { 'saůnd ¦lev·əl ˌmēd·ər }

sound locator [ENG ACOUS] A device formerly used to detect aircraft in flight by sound, consisting of four horns, or sound collectors (two for azimuth detection and two for elevation), together with their associated mechanisms and controls, which enabled the listening operator to determine the position and angular velocity of an aircraft. { 'saůnd ˌlō‚kād·ər }

sound navigation and ranging See sonar. { 'saůnd ˌnav·ə'gā·shən ən 'rānj·iŋ }

sound-powered telephone [ENG ACOUS] A telephone operating entirely on current generated by the speaker's voice, with no external power supply; sound waves cause a diaphragm to move a coil back and forth between the poles of a powerful but small permanent magnet, generating the required audio-frequency voltage in the coil. { 'saůnd ¦paů·ərd 'tel·ə‚fŏn }

sound production [ENG ACOUS] Conversion of energy from mechanical or electrical into acoustical form, as in a siren or loudspeaker. { 'saůnd prə‚dək·shən }

soundproofing See damping. { 'saůnd‚prüf·iŋ }

sound ranging [ENG ACOUS] Determining the location of a gun or other sound source by measuring the travel time of the sound wave to microphones at three or more different known positions. { 'saůnd ˌrānj·iŋ }

sound reception [ENG ACOUS] Conversion of acoustical energy into another form, usually electrical, as in a microphone. { 'saůnd ri‚sep·shən }

sound recording [ENG ACOUS] The process of recording sound signals so they may be reproduced at any subsequent time, as on a phonograph disk, motion picture sound track, or magnetic tape. { 'saůnd ri‚kórd·iŋ }

sound-reinforcement system [ENG ACOUS] An electronic means for augmenting the sound output of a speaker, singer, or musical instrument in cases where it is either too weak to be heard above the general noise or too reverberant; basic elements of such a system are microphones, amplifiers, volume controls, and loudspeakers. Also known as public address system. { 'saůnd ˌrē·in'fórs·mənt ˌsis·təm }

sound-reproducing system [ENG ACOUS] A combination of transducing devices and associated equipment for picking up sound at one location and time and reproducing it at the same or some other location and at the same or some later time. Also known as audio system; reproducing system; sound system. { 'saůnd ˌrē·prə'düs·iŋ ˌsis·təm }

sound reproduction [ENG ACOUS] The use of a combination of transducing devices and associated equipment to pick up sound at one point and reproduce it either at the same point or at some other point, at the same time or at some subsequent time. { 'saůnd ˌrē·prə‚dək·shən }

sound spectrograph [ENG ACOUS] An instrument that records and analyzes the spectral composition of audible sound. { 'saůnd 'spek·trə‚graf }

sound speed [ENG] The speed of sound motion picture film, standardized at 24 frames per second (silent film speed is 18 frames per second). { 'saůnd ‚spēd }

soundstripe [ENG ACOUS] A longitudinal stripe of magnetic material placed on some motion picture films for recording a magnetic sound track. { 'saůnd‚strīp }

sound system See sound-reproducing system. { 'saůnd ‚sis·təm }

sound track [ENG ACOUS] A narrow band, usually along the margin of a sound film, that carries the sound record; it may be a variable-width or variable-density optical track or a magnetic track. { 'saůnd ‚trak }

sound transducer See electroacoustic transducer. { 'saůnd tranz‚düs·ər }

sound trap [ELECTR] A wave trap in a television receiver circuit that prevents sound signals from entering the picture channels. [ENG ACOUS] A pit between adjoining instrument sections in a sound-recording studio, generally filled with fiberglass panels, to absorb sound that would otherwise propagate from instruments in one section to microphones in adjacent sections. { 'saůnd ‚trap }

source [ELEC] The circuit or device that supplies signal power or electric energy or charge to a transducer or load circuit. [ELECTR] The terminal in a field-effect transistor from which majority carriers flow into the conducting channel in the semiconductor material. [THERMO] A device that supplies heat. { sórs }

source degeneration [ELECTR] The addition of a circuit element between a transistor source and ground, with several effects, including a reduction in gain. { ¦sórs di‚jen·ə'rā·shən }

source-follower amplifier See common-drain amplifier. { 'sórs 'fäl·ə·wər 'am·plə‚fī·ər }

space centrode [MECH] The path traced by the instantaneous center of a rotating body relative to an inertial frame of reference. { 'spās 'sen‚trōd }

space cloth [CHEM ENG] Woven cloth or wire used for solids screening, and for which the openings between the fibers or strands are designated in terms of space or clear opening. { 'spās ‚klóth }

space cone [MECH] The cone in space that is swept out by the instantaneous axis of a rigid body during Poinsot motion. Also known as herpolhode cone. { 'spās ‚kŏn }

spacecraft ground instrumentation [ENG] Instrumentation located on the earth for monitoring, tracking, and communicating with manned spacecraft, satellites, and space probes. Also known as ground instrumentation. { 'spās‚kraft 'graůnd ˌin·strə‚mən'tā·shən }

spacecraft tracking [ENG] The determination of the positions and velocities of spacecraft through radio and optical means. { 'spās,kraft ,trak·iŋ }

space detection and tracking system |ENG| System capable of detecting and tracking space vehicles from the earth, and reporting the orbital characteristics of these vehicles to a central control facility. Abbreviated SPADATS. { 'spās di¦tek·shən ən ¦trak·iŋ ,sis·təm }

spaced loading |ENG| Loading shot holes so that cartridges are separated by open spacers which do not prevent the concussion from one charge from reaching the next. { 'spāst 'lōd·iŋ }

space frame |BUILD| A three-dimensional steel building frame which is stable against wind loads. { 'spās ,frām }

space lattice |BUILD| A space frame built of lattice girders. { 'spās ,lad·əs }

space processing |ENG| The carrying out of various processes aboard orbiting spacecraft, utilizing the low-gravity, high-vacuum environment associated with these vehicles. { 'spās ,prä,ses·iŋ }

spacer |ENG| **1.** A piece of metal wire twisted at one end to form a guard to keep the explosive in a shothole in place and twisted at the other end to form a guard to hold the tamping in its place. **2.** A piece of wood doweling interposed between charges to extend the column of explosive. **3.** A device for holding two members at a given distance from each other. Also known as spacer block. **4.** The tapered section of a pug joining the barrel to the die; clay is compressed in this section before it issues through the die. { 'spās·ər }

spacer block See spacer. { 'spās·ər ,bläk }

space suit |ENG| A pressure suit for wear in space or at very low ambient pressures within the atmosphere, designed to permit the wearer to leave the protection of a pressurized cabin. { 'spās ,süt }

Space Tracking and Data Acquisition Network |ENG| A network of ground stations operated by the National Aeronautics and Space Administration, which tracks, commands, and receives telemetry for United States and foreign unmanned satellites. Abbreviated STADAN. { 'spās 'trak·iŋ ən ¦dad·ə ,ak·wə'zīsh·ən ,net ,wərk }

space velocity |CHEM ENG| The relationship between feed rate and reactor volume in a flow process; defined as the volume or weight of feed (measured at standard conditions) per unit time per unit volume of reactor (or per unit weight of catalyst). { 'spās və,läs·əd·ē }

spackling |ENG| The process of repairing a part of a plaster wall or mural by cleaning out the defective spot and then patching it with a plastering material. { 'spak·liŋ }

SPADATS See space detection and tracking system. { 'spā,dats }

spade |DES ENG| A shovellike implement with a flat oblong blade; used for turning soil by pushing against the blade with the foot. { spād }

spade bolt |DES ENG| A bolt having a spade-shaped flattened head with a transverse hole, used to fasten shielded coils, capacitors, and other components to a chassis. { 'spād ,bōlt }

spade drill |DES ENG| A drill consisting of three main parts: a cutting blade, a blade holder or shank, and a device, such as a screw, which fastens the blade to the holder; used for cutting holes over 1 inch (2.54 centimeters) in diameter. { 'spād ,dril }

spade lug |DES ENG| An open-ended flat termination for a wire lead, easily slipped under a terminal nut. { 'spād ,ləg }

spall |ENG| **1.** To reduce irregular stone blocks to an approximate size by chipping with a hammer. **2.** To break off thin chips from, and parallel to, the surface of a material, such as a metal or rock. { spȯl }

spalling hammer |ENG| A heavy axlike hammer with chisel edge, used for breaking and rough-dressing stone. { 'spȯl·iŋ ,ham·ər }

span |ENG| A structural dimension measured between certain extremities. { span }

spandrel |BUILD| The part of a wall between the sill of a window and the head of the window below it. { 'span·drəl }

spandrel beam |BUILD| In steel or concrete construction, the exterior beam that extends from column to column and marks the floor level between stories. { 'span·drəl ,bēm }

spandrel frame |BUILD| A triangular framing, as below a stair. { 'span·drəl ,frām }

spandrel wall |BUILD| A wall on the outer surface of a vault to fill the spandrels. { 'span·drəl ,wȯl }

spanner |DES ENG| A wrench with a semicircular head having a projection or hole at one end. |ENG| **1.** A horizontal brace. **2.** An artificial horizon attachment for a sextant. { 'span·ər }

spare part |ENG| In supply usage, any part, component, or subassembly kept in reserve for the maintenance and repair of major items of equipment. { 'spār ,pärt }

spare-parts list |ENG| List approved by designated authorities, indicating the total quantities of spare parts, tools, and equipment necessary for the maintenance of a specified number of major items for a definite period of time. { 'spār ¦pärts ,list }

sparger See perforated-pipe distributor. { 'spär·jər }

sparging |CHEM ENG| The process of forcing air through water to remove undesirable gases. { 'spärj·iŋ }

spark |ELEC| A short-duration electric discharge due to a sudden breakdown of air or some other dielectric material separating two terminals, accompanied by a momentary flash of light. Also known as electric spark; spark discharge; sparkover. { spärk }

spark arrester |ENG| **1.** An apparatus that prevents sparks from escaping from a chimney. **2.** A device that reduces or eliminates electric sparks at a point where a circuit is opened and closed. { 'spärk ə,res·tər }

spark-coil leak detector |ENG| A coil similar to a Tesla coil which detects leaks in a vacuum system by jumping a spark between the leak hole and the core of the coil. { 'spärk ¦kȯil 'lek di ‚tek·tər }

spark discharge *See* spark. { 'spärk 'dis‚chärj }

spark-ignition combustion cycle *See* Otto cycle. { 'spärk ig¦nish·ən kəm'bəs·chən ‚sī·kəl }

spark-ignition engine |MECH ENG| An internal combustion engine in which an electrical discharge ignites the explosive mixture of fuel and air. { 'spärk ig¦nish·ən ‚en·jən }

sparking potential *See* breakdown voltage. { 'spärk·iŋ pə‚ten·chəl }

sparking voltage *See* breakdown voltage. { 'spärk·iŋ ‚vōl·tij }

spark knock |MECH ENG| The knock produced in an internal combustion engine precedes the arrival of the piston at the top dead-center position. { 'spärk ‚näk }

spark lead |MECH ENG| The amount by which the spark precedes the arrival of the piston at its top (compression) dead-center position in the cylinder of an internal combustion engine. { 'spärk ‚lēd }

sparkover-initiated discharge machining |MECH ENG| An electromachining process in which a potential is impressed between the tool (cathode) and workpiece (anode) which are separated by a dielectric material; a heavy discharge current flows through the ionized path when the applied potential is sufficient to cause rupture of the dielectric. { 'spärk‚ō·vər i¦nish·ē‚ād·əd 'dis ‚chärj mə‚shēn·iŋ }

sparkproof |ENG| **1.** Treated with a material to prevent ignition or damage by sparks. **2.** Generating no sparks. { 'spärk‚prüf }

spark recorder |ENG| Recorder in which the recording paper passes through a spark gap formed by a metal plate underneath and a moving metal pointer above the paper; sparks from an induction coil pass through the paper periodically, burning small holes that form the record trace. { 'spärk ri'kȯrd·ər }

spatial linkage |MECH ENG| A linkage that involves motion in all three dimensions. { 'spā·shəl 'liŋ·kij }

spatter dash |CIV ENG| **1.** A finish put on stucco by dashing a mortar and sand mixture against it. **2.** Paint spattered on a different-colored ground coat. { 'spad·ər ‚dash }

speaker *See* loudspeaker. { 'spēk·ər }

speaker identification |ENG ACOUS| The use of automated equipment to find the identity of a talker, in a known population of talkers, using the speech input. { ‚spēk·ər ī‚dent·ə·tə'kā· shən }

speaker verification |ENG ACOUS| The use of automated equipment to authenticate a claimed speaker identity from a voice signal based on speaker-specific characteristics reflected in spoken words or sentences. Abbreviated SV. { ‚spēk·ər ‚ver·i·fə'kā·shən }

spear |DES ENG| A rodlike fishing tool having a barbed-hook end, used to recover rope, wire line, and other materials from a borehole. { spir }

special cargo |IND ENG| Cargo which requires special handling or protection, such as pyrotechnics, detonators, watches, and precision instruments. { 'spesh·əl 'kär·gō }

special-purpose item |ENG| In supply usage, any item designed to fill a special requirement, and having a limited application; for example, a wrench or other tool designed to be used for one particular model of a piece of machinery. { 'spesh·əl ¦pər·pəs 'īd·əm }

special-purpose vehicle |ENG| A vehicle having a special chassis, or a general-purpose chassis incorporating major modifications, designed to fill a specialized requirement; all tractors (except truck tractors) and tracklaying vehicles, regardless of design, size, or intended purpose, are classified as special-purpose vehicles. { 'spesh·əl ¦pər·pəs 'vē·ə·kəl }

specifications |ENG| An organized listing of basic requirements for materials of construction, product compositions, dimensions, or test conditions; a number of organizations publish standards (for example, American Society of Mechanical Engineers, American Petroleum Institute, and American Society for Testing and Materials), and many companies have their own specifications. Also known as specs. |IND ENG| A quantitative description of the required characteristics of a device, machine, structure, product, or process. { ‚spes·ə·fə'kā·shənz }

specific charge |ELEC| The ratio of a particle's charge to its mass. { spə'sif·ik 'chärj }

specific conductance *See* conductivity. { spə'sif· ik kən'dək·təns }

specific energy |THERMO| The internal energy of a substance per unit mass. { spə'sif·ik 'en· ər·jē }

specific fuel consumption |MECH ENG| The weight flow rate of fuel required to produce a unit of power or thrust, for example, pounds per horsepower-hour. Abbreviated SFC. Also known as specific propellant consumption. { spə'sif·ik 'fyül kən‚səm·shən }

specific gravity |MECH| The ratio of the density of a material to the density of some standard material, such as water at a specified temperature, for example, 4°C or 60°F, or (for gases) air at standard conditions of pressure and temperature. Abbreviated sp gr. Also known as relative density. { spə'sif·ik 'grav·əd·ē }

specific-gravity bottle |ENG| A small bottle or flask used to measure the specific gravities of liquids; the bottle is weighed when it is filled with the liquid whose specific gravity is to be determined, when filled with a reference liquid, and when empty. Also known as density bottle; relative-density bottle. { spə'sif·ik ¦grav·əd·ē ‚bäd·əl }

specific-gravity hydrometer |ENG| A hydrometer which indicates the specific gravity of a liquid, with reference to water at a particular temperature. { spə'sif·ik ¦grav·əd·ē hī'dräm·əd·ər }

specific heat |THERMO| **1.** The ratio of the amount of heat required to raise a mass of material 1 degree in temperature to the amount of heat required to raise an equal mass of a reference substance, usually water, 1 degree in temperature; both measurements are made at a reference temperature, usually at constant pressure or constant volume. **2.** The quantity of heat required to raise a unit mass of homogeneous material one degree in temperature in a specified way; it is assumed that during the process no phase or chemical change occurs. { spə'sif·ik 'hēt }

specific inductive capacity See dielectric constant. { spə'sif·ik in'dək·tiv kə'pas·əd·ē }

specific insulation resistance See volume resistivity. { spə'sif·ik ,in·sə'lā·shən ri,zis·təns }

specific propellant consumption See specific fuel consumption. { spə'sif·ik prə'pel·ənt kən,səm·shən }

specific resistance See electrical resistivity. { spə'sif·ik ri'zis·təns }

specific speed [MECH ENG] A number, N_s, used to predict the performance of centrifugal and axial pumps or hydraulic turbines: for pumps, $N_s = N \sqrt{Q}/H^{3/4}$; for turbines, $N_s = N \sqrt{P}/H^{5/4}$, where N_s is specific speed, N is the rotational speed in revolutions per minute, Q is the rate of flow in gallons per minute, H is head in feet, and P is shaft horsepower. { spə'sif·ik 'spēd }

specific surface [CHEM ENG] The surface area per unit weight or volume of a particulate solid; used in size-reduction (crushing and grinding) calculations. { spə'sif·ik 'sər·fəs }

specific volume [MECH] The volume of a substance per unit mass; it is the reciprocal of the density. Abbreviated sp vol. { spə'sif·ik 'väl·yəm }

specific weight [MECH] The weight per unit volume of a substance. { spə'sif·ik 'wāt }

specs See specifications. { speks }

spectral density See frequency spectrum. { 'spek·trəl 'den·səd·ē }

spectral emissivity [THERMO] The ratio of the radiation emitted by a surface at a specified wavelength to the radiation emitted by a perfect blackbody radiator at the same wavelength and temperature. { 'spek·trəl ,ē,mi'siv·əd·ē }

spectral hygrometer [ENG] A hygrometer which determines the amount of precipitable moisture in a given region of the atmosphere by measuring the attenuation of radiant energy caused by the absorption bands of water vapor; the instrument consists of a collimated energy source, separated by the region under investigation and a detector which is sensitive to those frequencies that correspond to the absorption bands of water vapor. { 'spek·trəl hī'gräm·əd·ər }

spectral pyrometer See narrow-band pyrometer. { 'spek·trəl pī'räm·əd·ər }

spectral response See spectral sensitivity. { 'spek·trəl ri'späns }

spectral sensitivity [ELECTR] Radiant sensitivity, considered as a function of wavelength. { 'spek·trəl ,sen·sə'tiv·əd·ē }

spectrum analyzer [ENG] Test instrument used to show the distribution of energy contained in the frequencies emitted by a pulse magnetron;

also used to measure the Q of resonant cavities and lines, and to measure the cold impedance of a magnetron. { 'spek·trəm 'an·ə,līz·ər }

speech amplifier [ENG ACOUS] An audio-frequency amplifier designed specifically for amplification of speech frequencies, as for public-address equipment and radiotelephone systems. { 'spēch ,am·plə,fī·ər }

speech clipper [ENG ACOUS] A clipper used to limit the peaks of speech-frequency signals, as required for increasing the average modulation percentage of a radiotelephone or amateur radio transmitter. { 'spēch ,klip·ər }

speech coil See voice coil. { 'spēch ,kȯil }

speech inverter See scrambler. { 'spēch in,vərd·ər }

speech recognition [ENG ACOUS] The process of analyzing an acoustic speech signal to identify the linguistic message that was intended, so that a machine can correctly respond to spoken commands. { 'spēch ,rek·ig'nish·ən }

speech scrambler See scrambler. { 'spēch ,skram·blər }

speed [MECH] The time rate of change of position of a body without regard to direction; in other words, the magnitude of the velocity vector. { spēd }

speed cone [MECH ENG] A cone-shaped pulley, or a pulley composed of a series of pulleys of increasing diameter forming a stepped cone. { 'spēd ,kōn }

speed lathe [MECH ENG] A light, pulley-driven lathe, usually without a carriage or back gears, used for work in which the tool is controlled by hand. { 'spēd ,lāth }

speedometer [ENG] An instrument that indicates the speed of travel of a vehicle in miles per hour, kilometers per hour, or knots. { spi'däm·əd·ər }

speed-payload tradeoff [MECH ENG] The relationship between the maximum speed with which a machine can move a workpiece and the maximum weight of the workpiece. { 'spēd 'pā ,lȯd 'trād,ȯf }

speed-power product [ELECTR] The product of the gate speed or propagation delay of an electronic circuit and its power dissipation. { 'spēd'paü·ər ,präd·əkt }

speed reducer [MECH ENG] A train of gears placed between a motor and the machinery which it will drive, to reduce the speed with which power is transmitted. { 'spēd ri,dü·sər }

speed-reliability tradeoff [MECH ENG] The relationship between the maximum speed at which a machine can move a workpiece and the reliability with which the machine's operations can be achieved to some degree of satisfaction. { 'spēd ri,lī·ə'bil·ədē 'trād,ȯf }

Sperry process [CHEM ENG] The electrolytic manufacture of basic lead carbonate (white lead) from desilverized lead that contains some bismuth; impure lead collects at the anode, and carbon dioxide is passed into the solution to convert the lead to carbonate. { 'sper·ē ,prä·səs }

sp gr *See* specific gravity.

spherical-coordinate robot |CONT SYS| A robot in which the degrees of freedom of the manipulator arm are defined primarily by spherical coordinates. { 'sfir·ə·kəl kō¦órd·ən·ət 'rō,bät }

spherical pendulum |MECH| A simple pendulum mounted on a pivot so that its motion is not confined to a plane; the bob moves over a spherical surface. { 'sfir·ə·kəl 'pen·jə·ləm }

spherical stress |MECH| The portion of the total stress that corresponds to an isotropic hydrostatic pressure; its stress tensor is the unit tensor multiplied by one-third the trace of the total stress tensor. { 'sfir·ə·kəl 'stres }

spherometer |ENG| A device used to measure the curvature of a spherical surface. { sfə'räm·əd·ər }

spider |ELEC| A structure on the shaft of an electric rotating machine that supports the core or poles of the rotor, consisting of a hub, spokes, and rim, or some similar arrangement. |ENG| **1.** The part of an ejector mechanism which operates ejector pins in a molding press. **2.** In extrusion, the membranes which support a mandrel within the head-die assembly. |ENG ACOUS| A highly flexible perforated or corrugated disk used to center the voice coil of a dynamic loudspeaker with respect to the pole piece without appreciably hindering in-and-out motion of the voice coil and its attached diaphragm. |MECH ENG| In a universal joint, a part with four projections that is pivoted between the forked ends of two shafts and transmits motion between the shafts. Also known as cross. { 'spīd·ər }

spike |DES ENG| A large nail, especially one longer than 3 inches (7.6 centimeters), and often of square section. { spīk }

spike microphone |ENG ACOUS| A device for clandestine aural surveillance in which the sensor is a spike driven into the wall of the target area and mechanically coupled to the diaphragm of a microphone on the other side of the wall. { 'spīk ,mī·krə,fōn }

spill |ENG| The accidental release of some material, such as nuclear material or oil, from a container. { spil }

spill box |CIV ENG| A device such as a flume that maintains a constant head on a measuring weir or orifice. { 'spil ,bäks }

spillway |CIV ENG| A passage in or about a dam or other hydraulic structure for escape of surplus water. { 'spil,wā }

spillway apron |CIV ENG| A concrete or timber floor at the bottom of a spillway to prevent soil erosion from heavy or turbulent flow. { 'spil,wā ,ā·prən }

spillway channel |CIV ENG| An outlet channel from a spillway. { 'spil,wā ,chan·əl }

spillway dam *See* overflow dam. { 'spil,wā ,dam }

spillway gate |CIV ENG| A gate for regulating the flow from a reservoir. { 'spil,wā ,gāt }

spin |MECH| Rotation of a body about its axis. { spin }

spincasting |ENG| A technique for manufacturing telescope mirrors in which molten glass is poured into a rotating mold and, as the glass cools and solidifies, the surface of the relatively thin mirror takes on a shape that is relatively close to the desired one, reducing substantially the need for grinding away excess glass. { 'spin ,kast·iŋ }

spin compensation |MECH| Overcoming or reducing the effect of projectile rotation in decreasing the penetrating capacity of the jet in shaped-charge ammunition. { 'spin ,käm·pən ,sā·shən }

spin-decelerating moment |MECH| A couple about the axis of the projectile, which diminishes spin. { 'spin di¦sel·ə,rād·iŋ 'mō·mənt }

spindle |DES ENG| A short, slender or tapered shaft. { 'spin·dəl }

spin electronics *See* magnetoelectronics. { 'spin ,i·lek,trän·iks }

spinner |ENG| **1.** Automatically rotatable radar antenna, together with directly associated equipment. **2.** Part of a mechanical scanner which rotates about an axis, generally restricted to cases where the speed of rotation is relatively high. { 'spin·ər }

spinneret |ENG| An extrusion die with many holes through which plastic melt is forced to form filaments. { ,spin·ə'ret }

spinning |ENG| The extrusion of a spinning solution (such as molten plastic) through a spinneret. |MECH ENG| Shaping and finishing sheet metal by rotating the workpiece over a mandrel and working it with a round-ended tool. Also known as metal spinning. { 'spin·iŋ }

spinning machine |MECH ENG| **1.** A machine that winds insulation on electric wire. **2.** A machine that shapes metal hollow ware. { 'spin·iŋ mə,shēn }

spin transistor *See* magnetic switch. { 'spin tran,zis·tər }

spintronics *See* magnetoelectronics. { spin'trän·iks }

spin valve *See* magnetic switch. { 'spin ,valv }

spin welding |ENG| Fusion of two objects (for example, plastics) by forcing them together while one of the pair is spinning; frictional heat melts the interface, spinning is stopped, and the bodies are held together until they are frozen in place (welded). { 'spin ,weld·iŋ }

spiral bevel gear |DES ENG| Bevel gear with curved, oblique teeth to provide gradual engagement and bring more teeth together at a given time than an equivalent straight bevel gear. { 'spī·rəl 'bev·əl ,gir }

spiral chute |DES ENG| A gravity chute in the form of a continuous helical trough spiraled around a column for conveying materials to a lower level. { 'spī·rəl 'shüt }

spiral conveyor *See* screw conveyor. { 'spī·rəl kən'vā·ər }

spiral flow tank |CIV ENG| An aeration tank of the activated sludge process into which air is diffused in a spiral helical movement guided by baffles and proper location of diffusers. { 'spī·rəl 'flō ‚taŋk }

spiral flow test |ENG| The determination of the flow properties of a thermoplastic resin by measuring the length and weight of resin flowing along the path of a spiral cavity. { 'spī·rəl 'flō ‚test }

spiral gage See spiral pressure gage. { 'spī·rəl 'gāj }

spiral gear |MECH ENG| A helical gear that transmits power from one shaft to another, nonparallel shaft. { 'spī·rəl ‚gir }

spiral-jaw clutch |MECH ENG| A modification of the square-jaw clutch permitting gradual meshing of the mating faces, which have a helical section. { 'spī·rəl ‚jó ‚klɔch }

spiral mold cooling |ENG| Cooling an injection mold by passing a liquid through a spiral cavity in the body of the mold. { 'spī·rəl ‚mōld 'kül·iŋ }

spiral pipe |DES ENG| Strong, lightweight steel pipe with a single continuous welded helical seam from end to end. { 'spī·rəl 'pīp }

spiral plate exchanger |CHEM ENG| A heat-transfer device made from a pair of plates rolled in a spiral to provide two relatively long, rectangular passages for heat-transfer between fluids in countercurrent flow. { 'spī·rəl ‚plāt iks 'chān·jər }

spiral pressure gage |ENG| A device for measurement of pressures; a hollow tube spiral receives the system pressure which deforms (unwinds) the spiral in direct relation to the pressure in the tube. Also known as spiral gage. { 'spī·rəl 'presh·ər ‚gāj }

spiral scanning |ENG| Scanning in which the direction of maximum radiation describes a portion of a spiral; the rotation is always in one direction; used with some types of radar antennas. { 'spī·rəl 'skan·iŋ }

spiral spring |DES ENG| A spring bar or wire wound in an Archimedes spiral in a plane; each end is fastened to the force-applying link of the mechanism. { 'spī·rəl 'spriŋ }

spiral thermometer |ENG| A temperature-measurement device consisting of a bimetal spiral that winds tighter or opens with changes in temperature. { 'spī·rəl thər'mäm·əd·ər }

spiral-tube heat exchanger |ENG| A countercurrent heat-exchange device made of a group of concentric spirally wound coils, generally connected by manifolds; used for cryogenic exchange in air-separation plants. { 'spī·rəl‚tüb 'hēt iks‚chān·jər }

spiral welded pipe |DES ENG| A steel pipe made of long strips of steel plate fitted together to form helical seams, which are welded. { 'spī·rəl ‚weld·əd 'pīp }

spirit level See level. { 'spir·ət ‚lev·əl }

spirit thermometer |ENG| A temperature-measurement device consisting of a closed capillary tube with a liquid (for example, alcohol) reservoir bulb at the bottom; as the bulb is heated, the liquid expands up into the capillary tubing, indicating the temperature of the bulb. { 'spir·ət thər'mäm·əd·ər }

spit |ENG| To light a fuse. { spit }

spitted fuse |ENG| A slow-burning fuse which has been cut open at the lighting end for ease of ignition. { 'spid·əd 'fyüz }

spitting rock |ENG| A rock mass under stress that breaks and ejects small fragments with considerable velocity. { 'spid·iŋ ‚räk }

splash block |BUILD| A small masonry block with a concave surface placed on the ground below a downspout at a sloping angle to carry roof drainage water away from a building and to prevent erosion of the soil. { 'splash ‚bläk }

splash lubrication |ENG| An engine-lubrication system in which the connecting-rod bearings dip into troughs of oil, splashing the oil onto the cylinder and piston rods. { 'splash ‚lü·brə‚kā·shən }

splay |ENG| A slanted or beveled surface making an oblique angle with another surface. { splā }

splayed arch |CIV ENG| An arch whose opening has a larger radius in front than at the back. { 'splād 'ärch }

splice |ELEC| A joint used to connect two lengths of conductor with good mechanical strength and good conductivity. |ENG| To unite two parts, such as rope or wire, to form a continuous length. { splīs }

splice plate |CIV ENG| A plate for joining the web plates or the flanges of girders. { 'splīs ‚plāt }

spline |DES ENG| One of a number of equally spaced keys cut integral with a shaft, or similarly, keyways in a hubbed part; the mated pair permits the transmission of rotation or translatory motion along the axis of the shaft. |ENG| A strip of wood, metal, or plastic. { splīn }

spline broach |MECH ENG| A broach for cutting straight-sided splines, or multiple keyways in holes. { 'splīn ‚brōch }

splined shaft |DES ENG| A shaft with longitudinal gearlike ridges along its interior or exterior surface. { 'splīnd 'shaft }

split barrel |DES ENG| A core barrel that is split lengthwise so that it can be taken apart and the sample removed. { 'split 'bar·əl }

split-barrel sampler |DES ENG| A drive-type soil sampler with a split barrel. { 'split ‚bar·əl 'sam·plər }

split bearing |DES ENG| A shaft bearing composed of two pieces bolted together. { 'split 'ber·iŋ }

split cavity |ENG| A cavity, such as in a mold, made in sections. { 'split 'kav·əd·ē }

split link |DES ENG| A metal link in the shape of a two-turn helix pressed together. { 'split 'liŋk }

splitnut |ENG| A nut cut axially into halves to allow for rapid engagement (closed) or disengagement (open). { 'split‚nət }

split pin |DES ENG| A pin with a split at one end so that it can be spread to hold it in place. { 'split 'pin }

split-ring core lifter |DES ENG| A hardened steel ring having an open slit, an outside taper, and an inside or outside serrated surface; in its expanded state it allows the core to pass through it freely, but when the drill string is lifted, the outside taper surface slides downward into the bevel of the bit or reaming shell, causing the ring to contract and grip tightly the core which it surrounds. Also known as core catcher; core gripper; core lifter; ring lifter; split-ring lifter; spring lifter. { 'split ¦riŋ 'kȯr ‚lif·tər }

split-ring lifter See split-ring core lifter. { 'split ¦riŋ 'lif·tər }

split-ring mold [ENG] A plastics mold in which a split-cavity block is assembled in a chase to permit the forming of undercuts in a molded piece. { 'split ¦riŋ 'mōld }

split-ring piston packing [MECH ENG] A metal ring mounted on a piston to prevent leakage along the cylinder wall. { 'split ¦riŋ 'pis·tən ‚pak·iŋ }

split shovel [DES ENG] A shovel containing parallel troughs separated by slots; used for sampling ground ore. { 'split 'shəv·əl }

split-stator variable capacitor [ELECTR] Variable capacitor having a rotor section that is common to two separate stator sections; used in grid and plate tank circuits of transmitters for balancing purposes. { 'split ¦stād·ər 'ver·ē·ə·bəl kə'pas·əd·ər }

splitter [CHEM ENG] A petroleum-refinery term for a fractionating tower that produces only an overhead and bottom stream. { 'splid·ər }

splitter vanes [ENG] A group of curved, parallel vanes located in a sharp (for example, miter) bend of a gas conduit; the vane shape and its location help guide the moving gas around the bend. { 'splid·ər ‚vānz }

split transducer [ENG] A directional transducer with electroacoustic transducing elements which are divided and arranged so that there is an electrical separation of each division. { 'split tranz'dü·sər }

SP logging See spontaneous-potential well logging. { ¦es¦pē 'läg·iŋ }

spoke [DES ENG] A bar or rod radiating from the center of a wheel. { spōk }

spokeshave [ENG] A small tool for planing convex or concave surfaces. { 'spōk‚shāv }

sponge [CHEM ENG] Wood shavings coated with iron oxide and used as a catalyst in processes for removing hydrogen sulfide from industrial gases. { spənj }

spongy [MECH ENG] Property of a robot whose end effector has high compliance, so that a small force applied to it results in a large motion. { 'spən·jē }

spontaneous combustion See autoignition. { spän'tā·nē·əs kəm'bəs·chən }

spontaneous-potential well logging [ENG] The recording of the natural electrochemical and electrokinetic potential between two electrodes, one above the other, lowered into a drill hole; used to detect permeable beds and their boundaries. Also known as SP logging. { spän'tā·nē·əs pə'ten·chəl ¦wel ‚läg·iŋ }

spontaneous process [THERMO] A thermodynamic process which takes place without the application of an external agency, because of the inherent properties of a system. { spän'tā·nē·əs 'prä·səs }

spool [MECH ENG] **1.** The drum of a hoist. **2.** The movable part of a slide-type hydraulic valve. { spül }

spool-type roller conveyor [MECH ENG] A type of roller conveyor in which the rolls are of conical or tapered shape with the diameter at the ends of the roll larger than that at the center. { 'spül ¦tīp 'rō·lər kən‚vā·ər }

spoon [DES ENG] A slender rod with a cup-shaped projection at right angles to the rod, used for scraping drillings out of a borehole. { spün }

spot check [IND ENG] A check or inspection of certain steps in an operation, process, or the like, of certain parts of a piece of equipment or of a representative lot of completed parts or articles; the steps or parts inspected would normally be only a small percentage of the total. { 'spät ‚chek }

spot drilling [MECH ENG] Drilling a small hole or indentation in the surface of a material to serve as a centering guide in later machining operations. { 'spät ‚dril·iŋ }

spot facing [MECH ENG] A finished circular surface around the top of a hole to seat a bolthead or washer, or to allow flush mounting of mating parts. { 'spät ‚fās·iŋ }

spot gluing [ENG] Applying heat to a glued assembly by dielectric heating to make the glue set in spots that are more or less regularly distributed. { 'spät ‚glü·iŋ }

spotting [ENG] Fitting one part of a die to another part by applying an oil color to the surface of the finished part and bringing this against the surface of the intended mating part, the high spots being marked by the transferred color. { 'späd·iŋ }

spouting [ENG] A term used in the feeding or ejection of powdered or granulated solids by means of vertical or slanted discharge spouts. { 'spaùd·iŋ }

sprag [ENG] A stake used as a brake for a vehicle by inserting it through the spokes of a wheel or digging it into the ground at an angle. { sprag }

sprag clutch [MECH ENG] A clutch designed to transmit power in one direction only. { 'sprag ‚kləch }

spray [ENG] A mechanically produced dispersion of liquid into a gas stream; as drops are large, the spray is unstable and the liquid will fall free of the gas stream when velocity decreases. { 'sprā }

spray chamber [MECH ENG] A compartment in an air conditioner where humidification is conducted. { 'sprā ‚chām·bər }

spray dryer [MECH ENG] A machine for drying an atomized mist by direct contact with hot gases. { 'sprā ¦drī·ər }

sprayed metal mold [ENG] A plastics mold made by spraying molten metal onto a master

522

form until a shell of predetermined thickness is achieved; the shell is then removed and backed up with plaster, cement, or casting resin; used primarily in plastic sheet forming. { 'sprād ¦med·əl ,mōld }

sprayer plate |ENG| A rotating flat-faced or dished metal plate used in an oil burner to enhance atomization. { 'sprā·ər ,plāt }

spray gun |MECH ENG| An apparatus shaped like a gun which delivers an atomized mist of liquid. { 'sprā ,gən }

spray nozzle |MECH ENG| A device in which a liquid is subdivided to form a stream (mist) of small drops. { 'sprā ,näz·əl }

spray painting |ENG| Applying a fine, even coat of paint by means of a spray nozzle. { 'sprā ,pānt·iŋ }

spray pond |ENG| An arrangement for cooling large quantities of water in open reservoirs or ponds; nozzles spray a portion of the water into the air for the evaporative cooling effect. { 'sprā ,pänd }

spray probe |ENG| A device which detects a jet spray of tracer gas in vacuum testing for leaks. { 'sprā ,prōb }

spray torch |ENG| In thermal spraying, a device used for the application of self-fluxing alloys; molten metal is propelled against the substrate by a stream of air and gas. { 'sprā ,tórch }

spray tower |CHEM ENG| A vertical column, at the top of which is a liquid spray device; used to contact liquids with gas streams for absorption, humidification, or drying. { 'sprā ,taů·ər }

spray-up |ENG| A term for a number of techniques in which a spray gun is used as the processing tool; for example, in reinforced plastics manufacture, fibrous glass and resin can simultaneously be spray-deposited into a mold or onto a form. { 'sprā,əp }

spread |ENG| The layout of geophone groups from which data from a single shot are recorded simultaneously. { spred }

spreadable life See pot life. { 'spred·ə·bəl ,līf }

spreader |CIV ENG| A wood or steel member inserted temporarily between form walls to keep them apart. |ELEC| An insulating crossarm used to hold apart the wires of a transmission line or multiple-wire antenna. |MECH ENG| **1.** A tool used in sharpening machine drill bits. **2.** A machine which spreads dumped material with its blades. { 'spred·ər }

spreader beam |ENG| A rigid beam hanging from a crane hook and fitted with a number of ropes at different points along its length; employed for such purposes as lifting reinforced concrete piles or large sheets of glass. { 'spred·ər ,bēm }

spreader stoker |MECH ENG| A coal-burning system where mechanical feeders and distributing devices form a thin fuel bed on a traveling grate, intermittent-cleaning dump grate, or reciprocating continuous-cleaning grate. { 'spred·ər ,stōk·ər }

spread footing |CIV ENG| A wide, shallow footing usually made of reinforced concrete. { 'spred ,fůd·iŋ }

spreading coefficient |THERMO| The work done in spreading one liquid over a unit area of another, equal to the surface tension of the stationary liquid, minus the surface tension of the spreading liquid, minus the interfacial tension between the liquids. { 'spred·iŋ ,kō·i,fish·ənt }

Sprengel pump |MECH ENG| An air pump that exhausts by trapping gases between drops of mercury in a tube. { 'spreŋ·gəl ,pəmp }

sprig |DES ENG| A small brad having no head. |ENG| See glazier's point. { sprig }

spring |ENG| To enlarge the bottom of a drill hole by small charges of a high explosive in order to make room for the full charge; to chamber a drill hole. |MECH ENG| An elastic, stressed, stored-energy machine element that, when released, will recover its basic form or position. Also known as mechanical spring. { spriŋ }

spring balance |ENG| An instrument which measures force by determining the extension of a helical spring. { 'spriŋ ,bal·əns }

spring bolt |DES ENG| A bolt which must be retracted by pressure and which is shot into place by a spring when the pressure is released. { 'spriŋ ,bōlt }

spring box mold |ENG| A compression mold with a spacing fork that is removed after partial compression. { 'spriŋ 'bäks ,mōld }

spring buffer |ENG| A buffer in the form of a spring that stores and dissipates the kinetic energy of an impact. { 'spriŋ ,bəf·ər }

spring calipers |ENG| Calipers in which tension against the adjusting nut is maintained by a circular spring. { 'spriŋ ,kal·ə·pərz }

spring clip |DES ENG| **1.** A U-shaped fastener used to attach a leaf spring to the axle of a vehicle. **2.** A clip that grips an inserted part under spring pressure; used for electrical connections. { 'spriŋ ,klip }

spring collet |DES ENG| A bushing that surrounds and holds the end of the work in a machine tool; the bushing is slotted and tapered, and when the collet is slipped over it, the slot tends to close and the bushing thereby grips the work. { 'spriŋ ¦käl·ət }

spring cotter |DES ENG| A cotter made of an elastic metal that has been bent double to form a split pin. { 'spriŋ ¦käd·ər }

spring coupling |MECH ENG| A flexible coupling with resilient parts. { 'spriŋ ,kəp·liŋ }

spring die |DES ENG| An adjustable die consisting of a hollow cylinder with internal cutting teeth, used for cutting screw threads. { 'spriŋ ¦dī }

spring faucet |ENG| A faucet that is kept closed by a spring; force must be exerted to open it, and it closes when the force is removed. { 'spriŋ ¦fȯs·ət }

spring gravimeter |ENG| An instrument for making relative measurements of gravity; the elongation s of the spring may be considered proportional to gravity g, $s = (1/k)g$, and the basic formula for relative measurements is $g_2 - g_1 = k(s_2 - s_1)$. { 'spriŋ grə'vim·əd·ər }

spring hammer |MECH ENG| A machine-driven hammer actuated by a compressed spring or by compressed air. { 'spriŋ ˌham·ər }

spring hinge |DES ENG| A hinge fitted with one or more springs. { 'spriŋ ˌhinj }

spring hook |DES ENG| A hook closed at the end by a spring snap. Also known as snap hook. { 'spriŋ ˌhúk }

spring-joint caliper |DES ENG| An outside or inside caliper having a heavy spring joining the legs together at the top; legs are opened and closed by a knurled nut. { 'spriŋ ˌjóint ˌkal·ə·pər }

spring lifter See split-ring core lifter. { 'spriŋ ˌlif·tər }

spring-load |ENG| To load or exert a force on an object by means of tension from a spring or by compression. { 'spriŋ ˌlōd }

spring-loaded meter |ENG| A variable-area flowmeter in which the force on an obstruction in a tapered tube created by the fluid flowing past the obstruction is balanced by the force of a spring to which the obstruction is attached, and the resulting differential pressure is used to determine the flow rate. { 'spriŋ ˌlōd·əd 'mēd·ər }

spring-loaded regulator |MECH ENG| A pressure-regulator valve for pressure vessels or flow systems; the regulator is preloaded by a calibrated spring to open (or close) at the upper (or lower) limit of a preset pressure range. { 'spriŋ ˌlōd·əd 'reg·yə,lād·ər }

spring modulus |MECH| The additional force necessary to deflect a spring an additional unit distance; if a certain spring has a modulus of 100 newtons per centimeter, a 100-newton weight will compress it 1 centimeter, a 200-newton weight 2 centimeters, and so on. { 'spriŋ 'mäj·ə·ləs }

spring pin |MECH ENG| An iron rod which is mounted between spring and axle on a locomotive, and which maintains a regulated pressure on the axle. { 'spriŋ ˌpin }

spring scale |ENG| A scale that utilizes the deflection of a spring to measure the load. { 'spriŋ ˌskāl }

spring shackle |ENG| A shackle for supporting the end of a spring, permitting the spring to vary in length as it deflects. { 'spriŋ ˌshak·əl }

spring stop-nut locking fastener |DES ENG| A locking fastener that functions by a spring action clamping down on the bolt. { 'spriŋ ˌstäp,nət 'läk·iŋ ,fas·nər }

spring switch |CIV ENG| A railroad switch that contains a spring to return it to the running position after it has been thrown over by trailing wheels moving on the diverging route. { 'spriŋ ,swich }

sprinkler system |ENG| A fire-protection system of pipes and outlets in a building, mine, or other enclosure for delivering a fire extinguishing liquid or gas, usually automatically by the action

of heat on the sprinkler head. Also known as fire sprinkling system. { 'spriŋk·lər ,sis·təm }

sprocket |DES ENG| A tooth on the periphery of a wheel or cylinder to engage in the links of a chain, the perforations of a motion picture film, or other similar device. { 'spräk·ət }

sprocket chain |MECH ENG| A continuous chain which meshes with the teeth of a sprocket and thus can transmit mechanical power from one sprocket to another. { 'spräk·ət ,chān }

sprocket hole |ENG| One of a series of perforations at the edge of a motion picture film, paper tape, or roll of continuous stationery, which are engaged by the teeth of a sprocket wheel to drive the material through some device. { 'spräk·ət ,hōl }

sprocket wheel |DES ENG| A wheel with teeth or cogs, used for a chain drive or to engage the blocks on a cable. { 'spräk·ət ,wēl }

sprue |ENG| **1.** A feed opening or vertical channel through which molten material, such as metal or plastic, is poured in an injection or transfer mold. **2.** A slug of material that solidifies in the channel. { sprü }

sprue bushing |ENG| A steel insert in an injection mold which contains the sprue hole and has a seat for the injection cylinder nozzle. { 'sprü ,bùsh·iŋ }

sprue gate |ENG| A passageway for the flow of molten resin from the nozzle to the mold cavity. { 'sprü ,gāt }

sprue puller |ENG| A pin with a Z-shaped slot to pull the sprue out of the sprue bushing in an injection mold. { 'sprü ,púl·ər }

sprung axle |MECH ENG| A supporting member for carrying the rear wheels of an automobile. { 'sprəŋ ,ak·səl }

sprung weight |MECH ENG| The weight of a vehicle which is carried by the springs, including the frame, radiator, engine, clutch, transmission, body, load, and so forth. { 'sprəŋ ˌwāt }

spud |DES ENG| **1.** A diamond-point drill bit. **2.** An offset type of fishing tool used to clear a space around tools stuck in a borehole. **3.** Any of various spade- or chisel-shaped tools or mechanical devices. **4.** See grouser. { spəd }

spur dike See groin. { 'spər ,gir }

spur gear |DES ENG| A toothed wheel with radial teeth parallel to the axis. { 'spər ,gir }

spur pile See batter pile. { 'spər ,pīl }

sputtering |ELECTR| Also known as cathode sputtering. **1.** The ejection of atoms or groups of atoms from the surface of the cathode of a vacuum tube as the result of heavy-ion impact. **2.** The use of this process to deposit a thin layer of metal on a glass, plastic, metal, or other surface in vacuum. { 'spəd·ə·riŋ }

sputter-ion pump See getter-ion pump. { 'spəd·ər ‡ī,än ,pəmp }

sp vol See specific volume.

sq See square.

square |MECH| Denotes a unit of area; if x is a unit of length, a square x is the area of a square whose sides have a length of $1x$; for example, a square meter, or a meter squared, is the area of a square whose sides have a length of 1 meter. Also known as monomino. Abbreviated sq. { skwer }

square-edged orifice |ENG| An orifice plate with straight-through edges for the hole through which fluid flows; used to measure fluid flow in fluid conduits by means of differential pressure drop across the orifice. { 'skwer ¦ejd 'ȯr·ə·fəs }

square engine |MECH ENG| An engine in which the stroke is equal to the cylinder bore. { 'skwer 'en·jən }

square-head bolt |DES ENG| A cylindrical threaded fastener with a square head. { 'skwer ¦hed ,bōlt }

square-jaw clutch |MECH ENG| A type of positive clutch consisting of two or more jaws of square section which mesh together when they are aligned. { 'skwer ¦jȯ ,kləch }

square joint See straight joint. { ¦skwer ¦jȯint }

square key |DES ENG| A machine key of square, usually uniform, but sometimes tapered, cross section. { 'skwer 'kē }

square mesh |DES ENG| A wire-cloth textile mesh count that is the same in both directions. { 'skwer 'mesh }

square-nose bit See flat-face bit. { 'skwer ¦nōz ,bit }

square thread |DES ENG| A screw thread having a square cross section; the width of the thread is equal to the pitch or distance between threads. { 'skwer 'thred }

square wave |ELEC| An oscillation the amplitude of which shows periodic discontinuities between two values, remaining constant between jumps. { 'skwer 'wāv }

square-wave amplifier |ELECTR| Resistance-coupled amplifier, the circuit constants of which are to amplify a square wave with the minimum amount of distortion. { 'skwer ¦wāv 'am·plə,fī·ər }

square-wave generator |ELECTR| A signal generator that generates a square-wave output voltage. { 'skwer ¦wāv 'jen·ə,rād·ər }

square-wave response |ELECTR| The response of a circuit or device when a square wave is applied to the input. { 'skwer ¦wāv ri,späns }

square wheel |DES ENG| A wheel with a flat spot on its rim. { 'skwer ¦wel }

squaring circuit |ELECTR| **1.** A circuit that reshapes a sine or other wave into a square wave. **2.** A circuit that contains nonlinear elements proportional to the square of the input voltage. { 'skwer·iŋ ,sər·kət }

squaring shear |MECH ENG| A machine tool consisting of one fixed cutting blade and another mounted on a reciprocating crosshead; used for cutting sheet metal or plate. { 'skwer·iŋ ,shir }

squawker See midrange. { 'skwȯk·ər }

squeegee |DES ENG| A device consisting of a handle with a blade of rubber or leather set transversely at one end and used for spreading, pushing, or wiping liquids off or across a surface. { 'skwē,jē }

squeeze |ENG| **1.** To inject a grout into a borehole under high pressure. **2.** The plastic movement of a soft rock in the walls of a borehole or mine working that reduced the diameter of the opening. { skwēz }

squeeze roll |MECH ENG| A roller designed to exert pressure on material passing between it and a similar roller. { 'skwēz ,rōl }

squib |ENG| A small tube filled with fine-grained black powder; upon the lighting and burning of the ignition match, the squib assumes a rocket effect and darts back into the hole to ignite the powder charge. { skwib }

SQUID See superconducting quantum interference device. { skwid }

squirt can |ENG| An oil can with a flexible bottom and a tapered spout; pressure applied to the bottom forces oil out the spout. { 'skwərt ,kan }

squirt gun |ENG| A device with a bulb and nozzle; when the bulb is pressed, liquid squirts from the nozzle. { 'skwərt ,gən }

SRA-size |ENG| One of a series of sizes to which untrimmed paper is manufactured; for reels of paper the standard sizes are 450, 640, 900, and 1280 millimeters; for sheets of paper the sizes are SRA0, 900 × 1280 millimeters; SRA1, 640 × 900 millimeters; and SRA2, 450 × 640 millimeters; SRA sizes correspond to A sizes when trimmed. { ¦es¦är'ā ,sīz }

stab |ENG| In a drilling operation, to insert the threaded end of a pipe joint into the collar of the joint already placed in the hole and to rotate it slowly to engage the threads before screwing up. { stab }

stability |CONT SYS| The property of a system for which any bounded input signal results in a bounded output signal. |ENG| The property of a body, such as an aircraft, rocket, or ship, to maintain its attitude or to resist displacement, and, if displaced, to develop forces and moments tending to restore the original condition. |MECH| See dynamic stability. { stə'bil·əd·ē }

stability criterion |CONT SYS| A condition which is necessary and sufficient for a system to be stable, such as the Nyquist criterion, or the condition that poles of the system's overall transmittance lie in the left half of the complex-frequency plane. { stə'bil·əd·ē krī,tir·ē·ən }

stability exchange principle |CONT SYS| In a linear system, which is either dynamically stable or unstable depending on the value of a parameter, the complex frequency varies with the parameter in such a way that its real and imaginary parts pass through zero simultaneously; the principle is often violated. { stə'bil·əd·ē iks'chānj ,prin·sə·pəl }

stability factor |ELECTR| A measure of a transistor amplifier's bias stability, equal to the rate of change of collector current with respect to reverse saturation current. { stə'bil·əd·ē ,fak·tər }

stability matrix See stiffness matrix. { stə'bil·əd·ē ,mā·triks }

525

stability test |ENG| Accelerated test to determine the probable suitability of an explosive material for long-term storage. { stə'bil·əd·ē ,test }

stabilization |CHEM ENG| A petroleum-refinery process for separating light gases from petroleum or gasoline, thus leaving a stable (less volatile) liquid so that it can be handled or stored with less change in composition. *See* compensation. |ELECTR| Feedback introduced into vacuum tube or transistor amplifier stages to reduce distortion by making the amplification substantially independent of electrode voltages and tube constants. |ENG| Maintenance of a desired orientation independent of the roll and pitch of a ship or aircraft. { ,stā·bə·lə'zā·shən }

stabilized feedback *See* negative feedback. { 'stā·bə,līzd 'fēd,bak }

stabilizer |CHEM ENG| The fractionation column in a petroleum refinery used to stabilize (remove fractions from) hydrocarbon mixtures. |ENG| **1.** A hardened, splined bushing, sometimes freely rotating, slightly larger than the outer diameter of a core barrel and mounted directly above the core barrel back head. Also known as ferrule; fluted coupling. **2.** A tool located near the bit in the drilling assembly to modify the deviation angle in a well by controlling the location of the contact point between the hole and the drill collars. { 'stā·bə,līz·ər }

stabilizer bar |MECH ENG| In an automotive vehicle, a shaft that interconnects the two lower suspension arms in order to reduce body roll when the vehicle is turning. Also known as sway bar. { 'stā·bə,līz·ər ,bär }

stable element |ENG| Any instrument or device, such as a gyroscope, used to stabilize a radar antenna, turret, or other piece of equipment mounted on an aircraft or ship. { 'stā·bəl 'el·ə·mənt }

stable vertical |ENG| Vertical alignment of any device or instrument maintained during motion of the mount. { 'stā·bəl 'vərd·ə·kəl }

stack |BUILD| The portion of a chimney rising above the roof. |CHEM ENG| In gas works, a row of benches containing retorts. |ELECTR| *See* pileup. |ENG| **1.** To stand and rack drill rods in a drill tripod or derrick. **2.** Any structure or part thereof that contains a flue or flues for the discharge of gases. **3.** One or more filter cartridges mounted on a single column. **4.** Tall, vertical conduit (such as smokestack, flue) for venting of combustion or evaporation products or gaseous process wastes. **5.** The exhaust pipe of an internal combustion engine. { stak }

stacked-beam radar |ENG| Three-dimensional radar system that derives elevation by emitting narrow beams stacked vertically to cover a vertical segment, azimuth information from horizontal scanning of the beam, and range information from echo-return time. { 'stakt ¦bēm 'rā,där }

stack effect |MECH ENG| The pressure difference between the confined hot gas in a chimney or stack and the cool outside air surrounding the outlet. { 'stak i,fekt }

stacker |MECH ENG| A machine for lifting merchandise on a platform or fork and arranging it in tiers; operated by hand, or electric or hydraulic mechanisms. { 'stak·ər }

stacker-reclaimer |MECH ENG| Equipment which transports and builds up material stockpiles, and recovers and transports material to processing plants. { 'stak·ər rē'klām·ər }

stack gas |ENG| Gas passed through a chimney. { 'stak ,gas }

stack pollutants |ENG| Smokestack emissions subject to Environmental Protection Agency standards regulations, including sulfur oxides, particulates, nitrogen oxides, hydrocarbons, carbon monoxide, and photochemical oxidants. { 'stak pə,lüt·əns }

stack vent |ENG| An extension to the atmosphere of a waste stack or a soil stack above the highest horizontal branch drain or fixture branch that is connected to the stack. Also known as soil vent; waste vent. { 'stak ,vent }

stactometer *See* stalagmometer. { stak'täm·əd·ər }

STADAN *See* Space Tracking and Data Acquisition Network. { 'stā,dan }

stadia |ENG| A surveying instrument consisting of a telescope with special horizontal parallel lines or wires, used in connection with a vertical graduated rod. { 'stād·ē·ə }

stadia hairs |ENG| Two horizontal lines in the reticule of a theodolite arranged symmetrically above and below the line of sight. Also known as stadia wires. { 'stād·ē·ə ,herz }

stadia rod |ENG| A graduated rod used with a stadia to measure the distance from the observation point to the rod by observation of the length of rod subtended by the distance between the stadia hairs. { 'stād·ē·ə ,räd }

stadia tables |ENG| Mathematical tables from which may be found, without computation, the horizontal and vertical components of a reading made with a transit and stadia rod. { 'stād·ē·ə ,tā·bəlz }

stadia wires *See* stadia hairs. { 'stād·ē·ə ,wīrz }

stadimeter |ENG| An instrument for determining the distance to an object, but its height must be known; the angle subtended by the object's bottom and top as measured at the observer's position is proportional to the object's height; the instrument is graduated directly in distance. { sta'dim·əd·ər }

staff bead |BUILD| **1.** A bead between a wooden frame and adjacent masonry. **2.** A molded or beaded angle of wood or metal set into the corner of plaster walls. { 'staf ,bēd }

staff gage |ENG| A graduated scale placed in a position so that the stage of a stream may be read directly therefrom; a type of river gage. { 'staf ,gāj }

stage loader *See* feeder conveyor. { 'stāj ,lōd·ər }

stagger-tooth cutter |MECH ENG| Side-milling cutter with successive teeth having alternating helix angles. { 'stag·ər ¦tüth ,kəd·ər }

stained glass |ENG| Glass colored by any of

several means and assembled to produce a vari-colored mosaic or representation. { 'stånd 'glas }

stair |CIV ENG| A series of steps between levels or from floor to floor in a building. { ster }

stairway |CIV ENG| One or more flights of stairs connected by landings. { 'ster,wā }

stairwell |BUILD| A vertical compartment that extends through a building to hold a stairway. { 'ster,wel }

stake |ELEC| An iron peg used as a power electrode to transfer current into the ground in electrical prospecting. |ENG| **1.** To fasten back or prop open with a piece of chain or otherwise the valves or clacks of a water barrel in order that the water may run back into the sump when necessary. **2.** A pointed piece of wood driven into the ground to mark a boundary, survey station, or elevation. { stāk }

staking |ENG| Joining two parts together by fitting a projection on one part against a mating feature in the other part and then causing plastic flow at the joint. { 'stāk·iŋ }

staking out |ENG| Driving stakes into the earth to indicate the foundation location of a structure to be built. { 'stāk·iŋ 'aút }

stalagmometer |ENG| An instrument for measuring the size of drops suspended from a capillary tube, used in the drop-weight method. Also known as stactometer; stalogometer. { ,stal·ig'mäm·əd·ər }

stall torque |MECH ENG| The amount of torque provided by a motor at close to zero speed. { 'stòl ,tòrk }

stalogometer See stalagmometer. { ,stal·ə'gäm·əd·ər }

stamper |ENG ACOUS| A negative, generally made of metal by electroforming, used for molding phonograph records. { 'stam·pər }

stamping |ELECTR| A transformer lamination that has been cut out of a strip or sheet of metal by a punch press. |MECH ENG| Almost any press operation including blanking, shearing, hot or cold forming, drawing, bending, and coining. { 'stam·piŋ }

stanchion |ENG| A structural steel member, usually larger than a strut, whose main function is to withstand axial compressive stresses. { 'stan·chən }

standard ballistic conditions |MECH| A set of ballistic conditions arbitrarily assumed as standard for the computation of firing tables. { 'stan·dərd bə'lis·tik kən'dish·ənz }

standard capacitor |ELEC| A capacitor constructed in such a manner that its capacitance value is not likely to vary with temperature and is known to a high degree of accuracy. Also known as capacitance standard. { 'stan·dərd kə'pas·əd·ər }

standard cell |ELEC| A primary cell whose voltage is accurately known and remains sufficiently constant for instrument calibration purposes; the Weston standard cell has a voltage of 1.018636 volts at 20°C. { 'stan·dərd 'sel }

standard elemental time |IND ENG| A standard time for individual work elements. { 'stan·dərd 'el·ə,ment·əl 'tīm }

standard evaporator See short-tube vertical evaporator. { 'stan·dərd i'vap·ə,rād·ər }

standard fit |DES ENG| A fit whose allowance and tolerance are standardized. { 'stan·dərd 'fit }

standard free-energy increase |THERMO| The increase in Gibbs free energy in a chemical reaction, when both the reactants and the products of the reaction are in their standard states. { 'stan·dərd 'frē ¦en·ər·jē 'in,krēs }

standard gage |CIV ENG| A railroad gage measuring 4 feet $8^1/_2$ inches (1.4351 meters). |DES ENG| A highly accurate gage used only as a standard for working gages. { 'stan·dərd 'gāj }

standard gravity |MECH| A value of the acceleration of gravity equal to 9.80665 meters per second per second. { 'stan·dərd 'grav·əd·ē }

standard heat of formation |THERMO| The heat needed to produce one mole of a compound from its elements in their standard state. { 'stan·dərd 'hēt əv fòr'mā·shən }

standard hole |DES ENG| A hole with zero allowance plus a specified tolerance; fit allowance is provided for by the shaft in the hole. { 'stan·dərd 'hōl }

standard hour |IND ENG| The quantity of output required of an operator to meet an hourly production quota. Also known as allowed hour. { 'stan·dərd 'aúr }

standard-hour plan |IND ENG| A wage incentive plan in which standard work times are expressed as standard hours and the worker is paid for standard hours instead of the actual work hours. { 'stan·dərd ¦aúr 'plan }

standardization |DES ENG| The adoption of generally accepted uniform procedures, dimensions, materials, or parts that directly affect the design of a product or a facility. |ENG| The process of establishing by common agreement engineering criteria, terms, principles, practices, materials, items, processes, and equipment parts and components. { ,stan·dər·də'zā·shən }

standardized product |DES ENG| A product that conforms to specifications resulting from the same technical requirements. { 'stan·dər,dīzd 'präd·əkt }

standard leak |ENG| Tracer gas allowed to enter a leak detector at a controlled rate in order to facilitate calibration and adjustment of the detector. { 'stan·dərd 'lēk }

standard load |DES ENG| A load which has been preplanned as to dimensions, weight, and balance, and designated by a number or some classification. { 'stan·dərd 'lòd }

standard noise temperature |ELECTR| The standard reference temperature for noise measurements, equal to 290 K. { 'stan·dərd 'nòiz ,tem·prə·chər }

standard output |IND ENG| The reciprocal of standard time. { 'stan·dərd 'aút,pùt }

standard performance |IND ENG| The performance of an individual or of a group on meeting standard output. { 'stan·dərd pər'fȯr·məns }

standard shaft |DES ENG| A shaft with zero allowance minus a specified tolerance. { 'stan·dərd 'shaft }

standard time |IND ENG| A unit time value for completion of a work task as determined by the proper application of the appropriate work-measurement techniques. Also known as direct labor standard; output standard; production standard; time standard. { 'stan·dərd 'tīm }

standard ton See ton. { 'stan·dərd 'tən }

standard trajectory |MECH| Path through the air that it is calculated a projectile will follow under given conditions of weather, position, and material, including the particular fuse, projectile, and propelling charge that are used; firing tables are based on standard trajectories. { 'stan·dərd trə'jek·trē }

standard wire rope |DES ENG| Wire rope made of six wire strands laid around a sisal core. Also known as hemp-core cable. { 'stan·dərd 'wīr 'rōp }

standby battery |ELEC| A storage battery held in reserve as an emergency power source in event of failure of regular power facilities at a radio station or other location. { 'stand͟¦bī ,bad·ə·rē }

standing ways See ground ways. { 'stand·iŋ 'wāz }

standpipe |ENG| **1.** A vertical pipe for holding a water supply for fire protection. **2.** A high tank or reservoir for holding water that is used to maintain a uniform pressure in a water-supply system. { 'stand,pīp }

standpipe system |ENG| A system that contains standpipes, pumps, siamese connections, piping, and equipment with hose outlets and is provided with an adequate supply of water for fire fighting. { 'stan,pīp ,sis·təm }

standstill feature |CONT SYS| A device which insures that false signals such as fluctuations in the power supply do not cause a controller to be altered. { 'stan,stil ,fē·chər }

Stanton number |THERMO| A dimensionless number used in the study of forced convection, equal to the heat-transfer coefficient of a fluid divided by the product of the specific heat at constant pressure, the fluid density, and the fluid velocity. Symbolized N_{St}. Also known as Margoulis number (M). { 'stant·ən ,nəm·bər }

staple |DES ENG| A U-shaped loop of wire with points at both ends; used as a fastener. { 'stā·pəl }

stapler |ENG| **1.** A device for inserting wire staples into paper or wood. **2.** A hammer for inserting staples. { 'stā·plər }

star drill |DES ENG| A tool with a star-shaped point, used for drilling in stone or masonry. { 'stär ,dril }

Stark number See Stefan number. { 'stärk ,nəm·bər }

starling |CIV ENG| A protective enclosure around the pier of a bridge that consists of piles driven close together and is often filled with gravel or stone to protect the pier by serving as a break to water, ice, or drift. { 'stär·liŋ }

starter |ELEC| **1.** A device used to start an electric motor and to accelerate the motor to normal speed. **2.** See engine starter. |ELECTR| An auxiliary control electrode used in a gas tube to establish sufficient ionization to reduce the anode breakdown voltage. Also known as trigger electrode. |ENG| A drill used for making the upper part of a hole, the remainder of the hole being made with a drill of smaller gage, known as a follower. { 'stär·dər }

starting barrel |ENG| A short (12 to 24 inches or 30 to 60 centimeters) core barrel used to begin coring operations when the distance between the drill chuck and the bottom of the hole or to the rock surface in which a borehole is to be collared is too short to permit use of a full 5- or 10-foot-long (1.5- or 3.0-meter) core barrel. { 'stärd·iŋ ,bar·əl }

starting friction See static friction. { 'stärd·iŋ ,frik·shən }

starting resistance |MECH ENG| The force needed to produce an oil film on the journal bearings of a train when it is at a standstill. { 'stärd·iŋ ri,zis·təns }

starting taper |DES ENG| A slight end taper on a reamer to aid in starting. { 'stärd·iŋ ,tā·pər }

start time |IND ENG| The calendar time at which the manufacturing work for a specific job begins on a machine or in a facility. { 'stärt ,tīm }

start-to-leak pressure |MECH ENG| The amount of inlet pressure at which the first bubble occurs at the outlet of a safety relief valve with a resilient disk when the valve is subjected to an air test under a water seal. { 'stärt tə 'lēk ,presh·ər }

start-up curve |IND ENG| A learning curve applied to a job for the purpose of adjusting work times that are longer than the standard because of the introduction of new jobs or new workers. { 'stärd,əp ,kərv }

starved joint |ENG| A glued joint containing insufficient or inadequate adhesive. Also known as hungry joint. { 'stärvd 'jȯint }

state |CONT SYS| A minimum set of numbers which contain enough information about a system's history to enable its future behavior to be computed. { stāt }

state equations |CONT SYS| Equations which express the state of a system and the output of a system at any time as a single valued function of the system's input at the same time and the state of the system at some fixed initial time. { 'stāt i,kwā·zhənz }

state estimator See observer. { 'stāt ,es·tə,mād·ər }

state feedback |CONT SYS| A class of feedback control laws in which the control inputs are explicit memoryless functions of the dynamical system state, that is, the control inputs at a given time t_a are determined by the values of the state variables at t_a and do not depend on the values of these variables at earlier times $t ≥ t_a$. { 'stāt 'fēd,bak }

state observer See observer. { 'stāt əb,zər·vər }

state of strain [MECH] A complete description, including the six components of strain, of the deformation within a homogeneously deformed volume. { 'stāt əv 'strān }

state of stress [MECH] A complete description, including the six components of stress, of a homogeneously stressed volume. { 'stāt əv 'stres }

state parameter See thermodynamic function of state. { 'stāt pə,ram·əd·ər }

state space [CONT SYS] The set of all possible values of the state vector of a system. { 'stāt ,spās }

state transition equation [CONT SYS] The equation satisfied by the $n \times n$ state transition matrix $\Phi(t,t_0)$: $\partial\Phi(t,t_0)/\partial t = A(t) \Phi(t,t_0)$, $\Phi(t_0,t_0) = I$; here I is the unit $n \times n$ matrix, and A(t) is the $n \times n$ matrix which appears in the vector differential equation $dx(t)/dt = A(t)x(t)$ for the n-component state vector $x(t)$. { 'stāt tran'zish·ən i,kwā·zhən }

state transition matrix [CONT SYS] A matrix $\Phi(t,t_0)$ whose product with the state vector x at an initial time t_0 gives the state vector at a later time t; that is, $x(t) = \Phi(t,t_0)x(t_0)$. { 'stāt tran'zish·ən ,mā·triks }

state variable [CONT SYS] One of a minimum set of numbers which contain enough information about a system's history to enable computation of its future behavior. See thermodynamic function of state. { 'stāt ,ver·ē·ə·bəl }

state vector [CONT SYS] A column vector whose components are the state variables of a system. { 'stāt ,vek·tər }

statically admissible loads [MECH] Any set of external loads and internal forces which fulfills conditions necessary to maintain the equilibrium of a mechanical system. { 'stad·ik·əl·ē əd'mis·ə·bəl 'lōdz }

static bed [CHEM ENG] A layer of solids in a process vessel (absorber, catalytic reactor, packed distillation column, or granular filter bed) in which the particles rest upon one another at essentially the settled bulk density of the solids phase; contrasted to moving-solids or fluidized-solids beds. { 'stad·ik 'bed }

static charge [ELEC] An electric charge accumulated on an object. { 'stad·ik 'chärj }

static discharger [ELEC] A rubber-covered cloth wick about 6 inches (15 centimeters) long, sometimes attached to the trailing edges of the surfaces of an aircraft to discharge static electricity in flight. { 'stad·ik 'dis,chär·jər }

static electricity [ELEC] **1.** The study of the effects of macroscopic charges, including the transfer of a static charge from one object to another by actual contact or by means of a spark that bridges an air gap between the objects. **2.** See electrostatics. { 'stad·ik ,i,lek'tris·əd·ē }

static equilibrium See equilibrium. { 'stad·ik ,ē·kwə'lib·rē·əm }

static friction [MECH] **1.** The force that resists the initiation of sliding motion of one body over the other with which it is in contact. **2.** The force required to move one of the bodies when they are at rest. Also known as limiting friction; starting friction. { 'stad·ik 'frik·shən }

static load [MECH] A nonvarying load; the basal pressure exerted by the weight of a mass at rest, such as the load imposed on a drill bit by the weight of the drill-stem equipment or the pressure exerted on the rocks around an underground opening by the weight of the superimposed rocks. Also known as dead load. { 'stad·ik 'lōd }

static moment [MECH] **1.** A scalar quantity (such as area or mass) multiplied by the perpendicular distance from a point connected with the quantity (such as the centroid of the area or the center of mass) to a reference axis. **2.** The magnitude of some vector (such as force, momentum, or a directed line segment) multiplied by the length of a perpendicular dropped from the line of action of the vector to a reference point. { 'stad·ik 'mō·mənt }

static-pressure tap See pressure tap. { 'stad·ik 'presh·ər ,tap }

static-pressure tube [ENG] A smooth tube with a rounded nose that has radial holes in the portion behind the nose and is used to measure the static pressure within the flow of a fluid. { 'stad·ik 'presh·ər ,tüb }

static reaction [MECH] The force exerted on a body by other bodies which are keeping it in equilibrium. { 'stad·ik rē'ak·shən }

statics [MECH] The branch of mechanics which treats of force and force systems abstracted from matter, and of forces which act on bodies in equilibrium. { 'stad·iks }

static seal See gasket. { 'stad·ik 'sēl }

static test [ENG] A measurement taken under conditions where neither the stimulus nor the environmental conditions fluctuate. { 'stad·ik 'test }

static tube [ENG] A device used to measure the static (not kinetic or total) pressure in a stream of fluid; consists of a perforated, tapered tube that is placed parallel to the flow, and has a branch tube that is connected to a manometer. { 'stad·ik ,tüb }

station [ELEC] An assembly line or assembly machine location at which a wiring board or chassis is stopped for insertion of one or more parts. [ELECTR] A location at which radio, television, radar, or other electric equipment is installed. [ENG] Any predetermined point or area on the seas or oceans which is patrolled by naval vessels. { stā·shən }

stationary cone classifier [MECH ENG] In a pulverizer directly feeding a coal furnace, a device which returns oversize coal to the pulverizing zone. { 'stā·shə,ner·ē ¦kōn 'klas·ə,fī·ər }

stationary engine [MECH ENG] A permanently placed engine, as in a power house, factory, or mine. { 'stā·shə,ner·ē 'en·jən }

station pole [CIV ENG] One of various rods used in surveying to mark stations, to sight points and lines; or to measure elevation with respect to the transit. { 'stā·shən ,pōl }

station roof [BUILD] **1.** A roof supported by a

single central post and having a shape that resembles an umbrella. Also known as umbrella roof. **2.** A long roof supported by a single row of posts and by cantilevers on one or both sides; typically used for railroad platforms. { 'stā·shən ,rüf }

statistical multiplexer |ELECTR| A device which combines several low-speed communications channels into a single high-speed channel, and which can manage more communications traffic than a standard multiplexer by analyzing traffic and choosing different transmission patterns. { stə'tis·tə·kəl 'məl·tə,plek·sər }

statistical quality control |IND ENG| The use of statistical techniques as a means of controlling the quality of a product or process. { stə'tis·tə·kəl 'kwäl·əd·ē kən,trōl }

stator |ELEC| The portion of a rotating machine that contains the stationary parts of the magnetic circuit and their associated windings. |MECH ENG| A stationary machine part in or about which a rotor turns. { 'stād·ər }

statoscope |ENG| **1.** A barometer that records small variations in atmospheric pressure. **2.** An instrument that indicates small changes in an aircraft's altitude. { 'stad·ə,skōp }

statute mile *See* mile. { 'stach·üt 'mīl }

stave |DES ENG| **1.** A rung of a ladder. **2.** Any of the narrow wooden strips or metal plates placed edge to edge to form the sides, top, or lining of a vessel or structure, such as a barrel. { stāv }

stay |ENG| In a structure, a tensile member which holds other members of the structure rigidly in position. { stā }

staybolt |DES ENG| A bolt with a thread along the entire length of the shaft; used to attach machine parts that are under pressure to separate. { 'stā,bōlt }

stayed-cable bridge |CIV ENG| A modified cantilever bridge consisting of girders or trusses cantilevered both ways from a central tower and supported by inclined cables attached to the tower at the top or sometimes at several levels. { 'stād ¦kā·bəl ,brij }

STD recorder *See* salinity-temperature-depth recorder. { ,es,tē'dē ri,kórd·ər }

steadiness |CONT SYS| Freedom of a robot arm or end effector from high-frequency vibrations and jerks. { 'sted·ē·nəs }

steady pin |ENG| **1.** A retaining device such as a dowel, pin, or key that prevents a pulley from turning on its axis. **2.** A guide pin used to lift a cope or pattern. { 'sted·ē ,pin }

steady rest |MECH ENG| A device that is used to support long, slender workpieces during turning or grinding and permits them to rotate without eccentric movement. { 'sted·ē ,rest }

steady-state conduction |THERMO| Heat conduction in which the temperature and heat flow at each point does not change with time. { 'sted·ē ¦stāt kən'dək·shən }

steady-state creep *See* secondary creep. { 'sted·ē ¦stāt 'krēp }

steady-state error |CONT SYS| The error that

remains after transient conditions have disappeared in a control system. { 'sted·ē ¦stāt 'er·ər }

steady-state flow |CHEM ENG| Fluid flow without any change in composition or phase equilibria relationships. { 'sted·ē ¦stāt 'flō }

steady-state vibration |MECH| Vibration in which the velocity of each particle in the system is a continuous periodic quantity. { 'sted·ē ¦stāt vī'brā·shən }

steam accumulator |MECH ENG| A pressure vessel in which water is heated by steam during off-peak demand periods and regenerated as steam when needed. { 'stēm ə'kyü·mə,lād·ər }

steam atomizing oil burner |ENG| A burner which has two supply lines, one for oil and the other for a jet of steam which assists in the atomization process. { 'stēm ¦ad·ə,mīz·iŋ 'óil ,bər·nər }

steam attemperation |MECH ENG| The control of the maximum temperature of superheated steam by water injection or submerged cooling. { 'stēm ə,tam·pə'rā·shən }

steam bending |ENG| Forming wooden members to a desired shape by pressure after first softening by heat and moisture. { 'stēm ,bend·iŋ }

steam boiler |MECH ENG| A pressurized system in which water is vaporized to steam by heat transferred from a source of higher temperature, usually the products of combustion from burning fuels. Also known as steam generator. { 'stēm ,bói·lər }

steam calorimeter |ENG| **1.** A calorimeter, such as the Joly or differential steam calorimeter, in which the mass of steam condensed on a body is used to calculate the amount of heat supplied. **2.** *See* throttling calorimeter. { 'stēm ,kal·ə'rim·əd·ər }

steam cock |ENG| A valve for the passage of steam. { 'stēm ,käk }

steam condenser |MECH ENG| A device to maintain vacuum conditions on the exhaust of a steam prime mover by transfer of heat to circulating water or air at the lowest ambient temperature. { 'stēm kən,den·sər }

steam cracking |CHEM ENG| High-temperature cracking of petroleum hydrocarbons in the presence of steam. { 'stēm 'krak·iŋ }

steam cure |ENG| To cure concrete or mortar in water vapor at an elevated temperature, at either atmospheric or high pressure. { 'stēm ,kyúr }

steam cycle *See* Rankine cycle. { 'stēm ,sī·kəl }

steam distillation |CHEM ENG| A distillation in which vaporization of the volatile constituents of a liquid mixture takes place at a lower temperature by the introduction of steam directly into the charge; steam used in this manner is known as open steam. Also known as steam stripping. { 'stēm ,dis·tə'lā·shən }

steam drive |MECH ENG| Any device which uses power generated by the pressure of expanding steam to move a machine or a machine part. { 'stēm ,drīv }

steam dryer |MECH ENG| A device for separating liquid from vapor in a steam supply system. { 'stēm ,drī·ər }

steam emulsion test |ENG| A test used for measuring the ability of oil and water to separate, especially for steam-turbine oil; after emulsification and separation, the time required for the emulsion to be reduced to 3 milliliters or less is recorded at 5-minute intervals. { 'stēm i'məl·shən ,test }

steam engine |MECH ENG| A thermodynamic device for the conversion of heat in steam into work, generally in the form of a positive displacement, piston and cylinder mechanism. { 'stēm ¦en·jən }

steam engine indicator |ENG| An instrument that plots the steam pressure in an engine cylinder as a function of piston displacement. { 'stēm ¦en·jən 'in·də,kād·ər }

steam gage |ENG| A device for measuring steam pressure. { 'stēm ,gāj }

steam-generating furnace See boiler furnace. { 'stēm ¦jen·ə,rād·iŋ ,fər·nəs }

steam generator See steam boiler. { 'stēm ¦jen·ə,rād·ər }

steam hammer |MECH ENG| A forging hammer in which the ram is raised, lowered, and operated by a steam cylinder. { 'stēm ,ham·ər }

steam-heated evaporator |MECH ENG| A structure using condensing steam as a heat source on one side of a heat-exchange surface to evaporate liquid from the other side. { 'stēm ¦hēd·əd i'vap·ə,rād·ər }

steam heating |MECH ENG| A system that used steam as the medium for a comfort or process heating operation. { 'stēm 'hēd·iŋ }

steam jacket |MECH ENG| A casing applied to the cylinders and heads of a steam engine, or other space, to keep the surfaces hot and dry. { 'stēm ,jak·ət }

steam jet |ENG| A blast of steam issuing from a nozzle. { 'stēm ,jet }

steam-jet cycle |MECH ENG| A refrigeration cycle in which water is used as the refrigerant; high-velocity steam jets provide a high vacuum in the evaporator, causing the water to boil at low temperature and at the same time compressing the flashed vapor up to the condenser pressure level. { 'stēm ¦jet ,sī·kəl }

steam-jet ejector |MECH ENG| A fluid acceleration vacuum pump or compressor using the high velocity of a steam jet for entrainment. { 'stēm ¦jet i'jek·tər }

steam line |THERMO| A graph of the boiling point of water as a function of pressure. { 'stēm ,līn }

steam locomotive |MECH ENG| A railway propulsion power plant using steam, generally in a reciprocating, noncondensing engine. { 'stēm ,lō·kə¦mōd·iv }

steam loop |ENG| Two vertical pipes connected by a horizontal one, used to condense boiler steam so that it can be returned to the boiler without a pump or injector. { 'stēm ,lüp }

steam molding |ENG| The use of steam, either directly on the material or indirectly on the mold surfaces, as a heat source to mold parts from preexpanded polystyrene beads. { 'stēm ,mōld·iŋ }

steam nozzle |MECH ENG| A streamlined flow structure in which heat energy of steam is converted to the kinetic form. { 'stēm ,näz·əl }

steam point |THERMO| The boiling point of pure water whose isotopic composition is the same as that of sea water at standard atmospheric pressure; it is assigned a value of 100°C on the International Practical Temperature Scale of 1968. { 'stēm ,pȯint }

steam pump |MECH ENG| A pump driven by steam acting on the coupled piston rod and plunger. { 'stēm ,pəmp }

steam purifier See steam separator. { 'stēm 'pyür·ə,fī·ər }

steam refining |CHEM ENG| A petroleum refinery distillation process, in which the only heat used comes from steam in open and closed coils near the bottom of the still; used to produce gasoline and naphthas where odor and color are of prime importance; where open steam is used, it is known as steam distillation. { 'stēm ri'fīn·iŋ }

steam reheater |MECH ENG| A steam boiler component in which heat is added to intermediate-pressure steam, which has given up some of its energy in expansion through the high-pressure turbine. { 'stēm rē,hēd·ər }

steam roller |MECH ENG| A road roller driven by a steam engine. { 'stēm ,rō·lər }

steam separator |MECH ENG| A device for separating a mixture of the liquid and vapor phases of water. Also known as steam purifier. { 'stēm 'sep·ə,rād·ər }

steam shovel |MECH ENG| A power shovel operated by steam. { 'stēm ,shəv·əl }

steam still |CHEM ENG| A still in which steam provides most of the heat; distillation requires a lower temperature than in standard equipment (except for a vacuum distillation unit). { 'stēm ,stil }

steam stripping See steam distillation. { 'stēm 'strip·iŋ }

steam superheater |MECH ENG| A boiler component in which sensible heat is added to the steam after it has been evaporated from the liquid phase. { 'stēm 'sü·pər,hēd·ər }

steam tracing |ENG| A steam-carrying heater (such as tubing or piping) next to or twisted around a process-fluid or instrument-air line; used to keep liquids from solidifying or condensing. { 'stēm ,trās·iŋ }

steam trap |MECH ENG| A device which drains and removes condensate automatically from steam lines. { 'stēm ,trap }

steam-tube dryer |MECH ENG| Rotary dryer with steam-heated tubes running the full length of the cylinder and rotating with the dryer shell. { 'stēm ¦tüb ,drī·ər }

steam turbine |MECH ENG| A prime mover for the conversion of heat energy of steam into work on a rotating shaft, utilizing fluid acceleration principles in jet and vane machinery. { 'stēm ‚tər·bən }

steam valve [ENG] A valve used to regulate the flow of steam. { 'stēm ‚valv }

steam washer [ENG] A device for removing contaminants, such as silica, from the steam produced in a boiler. { 'stēm ‚wäsh·ər }

steel-cable conveyor belt [DES ENG] A rubber conveyor belt in which the carcass is composed of a single plane of steel cables. { 'stēl ‚kā·bəl kən'vā·ər ‚belt }

steel-clad rope [DES ENG] A wire rope made from flat strips of steel wound helically around each of the six strands composing the rope. { 'stēl ‚klad 'rōp }

Steelflex coupling [MECH ENG] A flexible coupling made with two grooved steel hubs keyed to their respective shafts and connected by a specially tempered alloy-steel member called the grid. { 'stēl‚fleks 'kəp·liŋ }

steelyard [ENG] A weighing device with a counterbalanced arm supporting the load to be weighed on the short end. { 'stil·yärd }

steen [CIV ENG] To line an excavation such as a cellar or well with stone, cement, or similar material without the use of mortar. { stēn }

steering arm [MECH ENG] An arm that transmits turning motion from the steering wheel of an automotive vehicle to the drag link. { 'stir·iŋ ‚ärm }

steering brake [MECH ENG] Means of turning, stopping, or holding a tracked vehicle by braking the tracks individually. { 'stir·iŋ ‚brāk }

steering gear [MECH ENG] The mechanism, including gear train and linkage, for the directional control of a vehicle or ship. { 'stir·iŋ ‚gir }

steering wheel [MECH ENG] A hand-operated wheel for controlling the direction of the wheels of an automotive vehicle or of the rudder of a ship. { 'stir·iŋ ‚wēl }

Stefan number [THERMO] A dimensionless number used in the study of radiant heat transfer, equal to the Stefan-Boltzmann constant times the cube of the temperature times the thickness of a layer divided by the layer's thermal conductivity. Symbolized S*t*. Also known as Stark number (S*k*). { 'shte‚fän ‚nəm·bər }

Steiner's theorem *See* parallel axis theorem. { 'shtīn·ərz ‚thir·əm }

stem [ENG] **1.** The heavy iron rod acting as the connecting link between the bit and the balance of the string of tools on a churn rod. **2.** To insert packing or tamping material in a shothole. { stem }

stem correction [THERMO] A correction which must be made in reading a thermometer in which part of the stem, and the thermometric fluid within it, is at a temperature which differs from the temperature being measured. { 'stem kə‚rek·shən }

stemming rod [ENG] A nonmetallic rod used to push explosive cartridges into position in a shothole and to ram tight the stemming. { 'stem·iŋ ‚räd }

stem-winding [MECH ENG] Pertaining to a

timepiece that is wound by an internal mechanism turned by an external knob and stem (the winding button of a watch). { 'stem ‚wīnd·iŋ }

stenometer [ENG] An instrument for measuring distances; employs a telescope in which two target images a known distance apart are superimposed by turning a micrometer screw. { stə'näm·əd·ər }

step [ENG] A small offset on a piece of core or in a drill hole resulting from a sudden sidewise deviation of the bit as it enters a hard, tilted stratum or rock underlying a softer rock. { step }

step aeration [CIV ENG] An activated sludge process in which the settled sewage is introduced into the aeration tank at more than one point. { 'step e‚rā·shən }

step bearing [MECH ENG] A device which supports the bottom end of a vertical shaft. Also known as pivot bearing. { 'step ‚ber·iŋ }

step block [ENG] A metal block, usually of steel or cast iron, with integral stepped sections to allow application of clamps when securing a workpiece to a machine tool table. { 'step ‚bläk }

step-by-step system [CONT SYS] A control system in which the drive motor moves in discrete steps when the input element is moved continuously. { ‚step bī ‚step 'sis·təm }

step gage [DES ENG] **1.** A plug gage containing several cylindrical gages of increasing diameter mounted on the same axis. **2.** A gage consisting of a body in which a blade slides perpendicularly; used to measure steps and shoulders. { 'step ‚gāj }

stepped cone pulley [DES ENG] A one-piece pulley with several diameters to engage transmission belts and thereby provide different speed ratios. { 'stept ‚kōn 'pul·ē }

stepped footing [CIV ENG] A widening at the bottom of a wall consisting of a series of steps in the proportion of one horizontal to two vertical units. { 'stept 'fud·iŋ }

stepped gear wheel [DES ENG] A gear wheel containing two or more sets of teeth on the same rim, with adjacent sets slightly displaced to form a series of steps. { 'stept 'gir ‚wēl }

stepped screw [DES ENG] A screw from which sectors have been removed, the remaining screw surfaces forming steps. { 'stept 'skrü }

stepper motor [ELEC] A motor that rotates in short and essentially uniform angular movements rather than continuously; typical steps are 30, 45, and 90°; the angular steps are obtained electromagnetically rather than by the ratchet and pawl mechanisms of stepping relays. Also known as magnetic stepping motor; stepping motor; step-servo motor. { 'step·ər ‚mōd·ər }

stepping *See* zoning. { 'step·iŋ }

stepping motor *See* stepper motor. { 'step·iŋ ‚mōd·ər }

step pulley [MECH ENG] A series of pulleys of various diameters combined in a single concentric unit and used to vary the velocity ratio of shafts. Also known as cone pulley. { 'step ‚pul·ē }

step-recovery diode |ELECTR| A varactor in which forward voltage injects carriers across the junction, but before the carriers can combine, voltage reverses and carriers return to their origin in a group; the result is abrupt cessation of reverse current and a harmonic-rich waveform. { 'step ri¦kəv·rē 'dī,ōd }

step response |CONT SYS| The behavior of a system when its input signal is zero before a certain time and is equal to a constant nonzero value after this time. { 'step ri,späns }

step-up transformer |ELEC| Transformer in which the energy transfer is from a low-voltage winding to a high-voltage winding or windings. { 'step 'vōl·tij ,reg·yə,lād·ər }

step voltage regulator |ELEC| A type of voltage regulator used on distribution feeder lines; it provides increments or steps of voltage change. { 'step 'vōl·tij ,reg·yə,lād·ər }

stère |MECH| A unit of volume equal to 1 cubic meter; it is used mainly in France, and in measuring timber volumes. { stir }

stereo *See* stereophonic; stereo sound system. { 'ste·rē·ō }

stereo amplifier |ENG ACOUS| An audio-frequency amplifier having two or more channels, as required for use in a stereo sound system. { 'ste·rē·ō 'am·plə,fī·ər }

stereolithography |IND ENG| A three-dimensional printing process whereby a CAD drawing of a part is processed to create a file of the part in slices and the part is constructed one slice (or layer) at a time (from bottom to top) by depositing layer upon layer of material (usually a liquid resin that can be hardened using a scanning laser), used for rapid prototyping. { ,ster·ē·ō·li'thäg·rə·fē }

stereomicrometer |ENG| An instrument attached to an optical instrument (such as a telescope) to measure small angles. { ¦ster·ē·ə·mī'kräm·əd·ər }

stereophonic |ENG ACOUS| Pertaining to three-dimensional pickup or reproduction of sound, as achieved by using two or more separate audio channels. Also known as stereo. { ¦ster·ē·ə 'fän·ik }

stereophonics |ENG ACOUS| The study of reproducing or reinforcing sound in such a way as to produce the sensation that the sound is coming from sources whose spatial distribution is similar to that of the original sound sources. { ¦ster·ē·ə'fän·iks }

stereophonic sound system *See* stereo sound system. { ¦ster·ē·ə'fän·ik 'saúnd ,sis·təm }

stereo pickup |ENG ACOUS| A phonograph pickup designed for use with standard single-groove two-channel stereo records; the pickup cartridge has a single stylus that actuates two elements, one responding to stylus motion at 45° to the right of vertical and the other responding to stylus motion at 45° to the left of vertical. { 'ster·ē·ō 'pik,əp }

stereoplanigraph |ENG| An instrument for drawing topographic maps from observations of

stereoscopic aerial photographs with a stereocomparator. { ¦ster·ē·ə'plan·ə,graf }

stereo preamplifier |ENG ACOUS| An audio-frequency preamplifier having two channels, used in a stereo sound system. { 'ster·ē·ō ¦prē'am·plə,fī·ər }

stereo record |ENG ACOUS| A single-groove disk record having V-shaped grooves at 45° to the vertical; each groove wall has one of the two recorded channels. { 'ster·ē·ō 'rek·ərd }

stereo recorded tape |ENG ACOUS| Recorded magnetic tape having two separate recordings, one for each channel of a stereo sound system. { 'ster·ē·ō ri¦kòrd·əd 'tāp }

stereo sound system |ENG ACOUS| A sound reproducing system in which a stereo pickup, stereo tape recorder, stereo tuner, or stereo microphone system feeds two independent audio channels, each of which terminates in one or more loudspeakers arranged to give listeners the same audio perspective that they would get at the original sound source. Also known as stereo; stereophonic sound system. { 'ster·ē·ō 'saúnd ,sis·təm }

stereo tape recorder |ENG ACOUS| A magnetic-tape recorder having two stacked playback heads, used for reproduction of stereo recorded tape. { 'ster·ē·ō 'tāp ri,kòrd·ər }

stereo tuner |ENG ACOUS| A tuner having provisions for receiving both channels of a stereo broadcast. { 'ster·ē·ō 'tün·ər }

sterhydraulic |MECH ENG| Pertaining to a hydraulic press in which motion or pressure is produced by the introduction of a solid body into a cylinder filled with liquid. { ¦ster·hī'dròl·ik }

sterilizer |ENG| An apparatus for sterilizing by dry heat, steam, or water. { 'ster·ə,līz·ər }

sthène |MECH| The force which, when applied to a body whose mass is 1 metric ton, results in an acceleration of 1 meter per second per second; equal to 1000 newtons. Formerly known as funal. { sthēn }

stick |ENG| **1.** A rigid bar hinged to the boom of a dipper or pull shovel and fastened to the bucket. **2.** A long slender tool bonded with an abrasive for honing or sharpening tools and for dressing of wheels. { stik }

stick gage |ENG| A suitably divided vertical rod, or stick, anchored in an open vessel so that the magnitude of rise and fall of the liquid level may be observed directly. { 'stik ,gāj }

stick-slip friction |MECH| Friction between two surfaces that are alternately at rest and in motion with respect to each other. { 'stik ,slip ,frik·shən }

stiction |MECH| Friction that tends to prevent relative motion between two movable parts at their null position. { 'stik·shən }

stiffener |CIV ENG| A steel angle or plate attached to a slender beam to prevent its buckling by increasing its stiffness. { 'stif·nər }

stiffleg derrick |MECH ENG| A derrick consisting of a mast held in the vertical position by a fixed tripod of steel or timber legs. Also known as derrick crane; Scotch derrick. { 'stif ¦leg 'der·ik }

stiffness |MECH| The ratio of a steady force acting on a deformable elastic medium to the resulting displacement. { 'stif·nəs }

stiffness coefficient |MECH| The ratio of the force acting on a linear mechanical system, such as a spring, to its displacement from equilibrium. { 'stif·nəs ‚kō·i‚fish·ənt }

stiffness constant |MECH| Any one of the coefficients of the relations in the generalized Hooke's law used to express stress components as linear functions of the strain components. Also known as elastic constant. { 'stif·nəs ‚kän·stənt }

stiffness matrix [MECH] A matrix **K** used to express the potential energy V of a mechanical system during small displacements from an equilibrium position, by means of the equation $V = 1/2q^T\mathbf{K}q$, where q is the vector whose components are the generalized components of the system with respect to time and q^T is the transpose of q. Also known as stability matrix. { 'stif·nəs ‚mā·triks }

stigma |MECH| A unit of length used mainly in nuclear measurements, equal to 10^{-12} meter. Also known as bicron. { 'stig·mə }

stile |BUILD| The upright outside framing piece of a window or door. { stīl }

still |CHEM ENG| A device used to evaporate liquids; heat is applied to the liquid, and the resulting vapor is condensed to a liquid state. { stil }

stilling basin |ENG| A depressed area in a channel or reservoir that is deep enough to reduce the velocity of the flow. Also known as stilling box. { 'stil·iŋ ‚bas·ən }

stilling box See stilling basin. { 'stil·iŋ ‚bäks }

stimulus |CONT SYS| A signal that affects the controlled variable in a control system. { 'stim·yə·ləs }

Stirling cycle |THERMO| A regenerative thermodynamic power cycle using two isothermal and two constant volume phases. { 'stir·liŋ ‚sī·kəl }

Stirling engine |MECH ENG| An engine in which work is performed by the expansion of a gas at high temperature; heat for the expansion is supplied through the wall of the piston cylinder. { 'stir·liŋ ‚en·jən }

stirred-flow reactor |CHEM ENG| A reactor in which there is a device for achieving effective mixing, frequently in the form of a rapidly rotating basket holding the catalyst. { ¦stird 'flō rē‚ak·tər }

stirrup |CIV ENG| In concrete construction, a U-shaped bar which is anchored perpendicular to the longitudinal steel as reinforcement to resist shear. { 'stər·əp }

stitch bonding |ENG| A method of making wire connections between two or more points on an integrated circuit by using impulse welding or heat and pressure while feeding the connecting wire through a hole in the center of the welding electrode. { 'stich ‚bänd·iŋ }

stitching |ENG| Progressive welding of thermoplastic materials (resins) by successive applications of two small, mechanically operated, radio-frequency-heated electrodes; the mechanism is similar to that of a normal sewing machine. { 'stich·iŋ }

stitch rivet |ENG| One of a series of rivets joining the parallel elements of a structural member so that they act as a unit. { 'stich ‚riv·ət }

stochastic control theory |CONT SYS| A branch of control theory that aims at predicting and minimizing the magnitudes and limits of the random deviations of a control system through optimizing the design of the controller. { stō'kas·tik kən'trōl ‚thē·ə·rē }

stock |IND ENG| **1.** A product or material kept in storage until needed for use or transferred to some ultimate point for use, for example, crude oil tankage or paper-pulp feed. **2.** Designation of a particular material, such as bright stock or naphtha stock. { stäk }

stock accounting |IND ENG| The establishment and maintenance of formal records of material in stock reflecting such information as quantities, values, or condition. { 'stäk ə‚kaůnt·iŋ }

stock control |IND ENG| Process of maintaining inventory data on the quantity, location, and condition of supplies and equipment due in, on hand, and due out, to determine quantities of material and equipment available or required for issue and to facilitate distribution and management of material. { 'stäk kən‚trōl }

stock coordination |IND ENG| A supply management function exercised usually at department level which controls the assignment of material cognizance for items or categories of material to inventory managers. { 'stäk kō‚ȯrd·ən‚ā·shən }

stocking cutter |MECH ENG| **1.** A gear cutter having side rake or curved edges to rough out the gear-tooth spaces before they are formed by the regular gear cutter. **2.** A concave gear cutter ganged beside a regular gear cutter and used to finish the periphery of a gear blank by milling ahead of the regular cutter. { 'stäk·iŋ ¦kəd·ər }

stock number |IND ENG| Number assigned to an item, principally to identify that item for storage and issue purposes. { 'stäk ‚nəm·bər }

stockpile |ENG| A reserve stock of material, equipment, raw material, or other supplies. { 'stäk‚pīl }

stock rail |CIV ENG| The fixed rail in a track, against which the switch rail operates. { 'stäk ‚rāl }

stock record account |IND ENG| A basic record showing by item the receipt and issuance of property, the balances on hand, and such other identifying or stock control data as may be required by proper authority. { 'stäk ¦rek·ȯrd ə‚kaůnt }

Stodola method |MECH| A method of calculating the deflection of a uniform or nonuniform beam in free transverse vibration at a specified frequency, as a function of distance along the beam, in which one calculates a sequence of deflection curves each of which is the deflection resulting from the loading corresponding to the previous deflection, and these deflections converge to the solution. { 'stō·də·lə ,meth·əd }

stoker |MECH ENG| A mechanical means, as used in a furnace, for feeding coal, removing refuse, controlling air supply, and mixing with combustibles for efficient burning. { 'stō·kər }

Stokes number 2 |ENG| A dimensionless number used in the calibration of rotameters, equal to 1.042 $m_f g \rho (1 - \rho/\rho_f) R^3/\mu^2$, where ρ and μ are the density and dynamic viscosity of the fluid respectively, m_f and ρ_f are the mass and density of the float respectively, and R is the ratio of the radius of the tube to the radius of the float. Symbol St_2. { 'stōks ¦nəm·bər 'tü }

stone |MECH| A unit of mass in common use in the United Kingdom, equal to 14 pounds or 6.35029318 kilograms. { stōn }

stonework |CIV ENG| A structure or the part of a structure built of stone. { 'stōn,wərk }

Stoney gate |CIV ENG| A crest gate which moves along a series of rollers traveling vertically in grooves in masonry piers, independently of the gate and piers. { 'stō·nē ,gāt }

stop |CONT SYS| A bound or final position of a robot's movement. { stäp }

stop and stay See absolute stop. { 'stäp ən 'stā }

stop bead |BUILD| A molding on the pulley stile of a window frame; forms one side of the groove for the inner sash. { 'stäp ,bēd }

stop cock |ENG| A small valve for stopping or regulating the flow of a fluid through a pipe. { 'stäp ,käk }

stoplog |CIV ENG| A log, plank, or steel or concrete beam that fits into a groove or rack between walls or piers to prevent the flow of water through an opening in a dam, conduit, or other channel. { 'stäp,läg }

stop nut |DES ENG| **1.** An adjustable nut that restricts the travel of an adjusting screw. **2.** A nut with a compressible insert that binds it so that a lock washer is not needed. { 'stäp ,nət }

stopping capacitor See coupling capacitor. { 'stäp·iŋ kə,pas·əd·ər }

stop valve |ENG| A valve that can be opened or closed to regulate or stop the flow of fluid in a pipe. { 'stäp ,valv }

storage battery |ELEC| A connected group of two or more storage cells or a single storage cell. Also known as accumulator; accumulator battery; rechargeable battery; secondary battery. { 'stȯr·ij ,bad·ə·rē }

storage calorifier See cylinder. { 'stȯr·ij kə'lȯr·ə,fī·ər }

storage cell |ELEC| An electrolytic cell for generating electric energy, in which the cell after being discharged may be restored to a charged condition by sending a current through it in a direction opposite to that of the discharging current. Also known as secondary cell. { 'stȯr·ij ,sel }

storage reservoir See impounding reservoir. { 'stȯr·ij ,rez·əv,wär }

storage-retrieval machine |CONT SYS| A computer-controlled machine for an automated storage and retrieval system that operates on rails and moves material either vertically or horizontally between a storage compartment and a transfer station. { ¦stȯr·ij ri'trēv·əl mə,shēn }

stored-program numerical control See computer numerical control. { 'stȯrd ¦prō,gram nü'mer·ə·kəl kən,trōl }

storm cellar See cyclone cellar. { 'stȯrm ,sel·ər }

storm drain |CIV ENG| A drain which conducts storm surface, or wash water, or drainage after a heavy rain from a building to a storm or a combined sewer. Also known as storm sewer. { 'stȯrm ,drān }

storm sash See storm window. { 'stȯrm ,sash }

storm sewage |CIV ENG| Refuse liquids and waste carried by sewers during or following a period of heavy rainfall. { 'stȯrm ,sü·ij }

storm sewer See storm drain. { 'stȯrm ,sü·ər }

storm window |BUILD| A sash placed on the outside of an ordinary window to give added protection from the weather. Also known as storm sash. { 'stȯrm ,win·dō }

Storrow whirling hygrometer |ENG| A hygrometer in which the two thermometers are mounted side by side on a brass frame and fitted with a loose handle so that it can be whirled in the atmosphere to be tested; the instrument is whirled at some 200 revolutions per minute for about 1 minute and the readings on the wet- and dry-bulb thermometers are recorded; used in conjunction with Glaisher's or Marvin's hygrometrical tables. { 'stä·rō 'wərl·iŋ hī'gräm·əd·ər }

story |BUILD| The space between two floors or between a floor and the roof. { 'stȯr·ē }

story pole See story rod. { 'stȯr·ē ,pōl }

story rod |DES ENG| A pole cut to the exact specified height from finished floor to ceiling and used as a measuring device in the course of construction. Also known as story pole. { 'stȯr·ē ,räd }

stove |ENG| A chamber within which a fuel-air mixture is burned to provide heat, the heat itself being radiated outward from the chamber; used for space heating, process-fluid heating, and steel blast furnaces. { stōv }

stove bolt |DES ENG| A coarsely threaded bolt with a slotted head, which with a square nut is used to join metal parts. { 'stōv ,bōlt }

stovepipe |ENG| Large-diameter pipe made of sheet steel. { 'stōv,pīp }

stoving See baking. { 'stōv·iŋ }

STR See self-tuning regulator.

straddle milling |MECH ENG| Face milling of two parallel vertical surfaces of a workpiece simultaneously by using two side-milling cutters. { 'strad·əl ,mil·iŋ }

straddle truck |MECH ENG| A self-loading outrigger type of industrial truck that straddles the load before lifting it between the outrigger arms. { 'strad·əl ,trək }

straight beam |ENG| In ultrasonic testing, a longitudinal wave emitted from an ultrasonic search unit in a wavetrain which travels perpendicularly to the test surface. { 'strāt 'bēm }

straight bevel gear |DES ENG| A simple form of bevel gear having straight teeth which, if extended inward, would come together at the intersection of the shaft axes. { 'strāt 'bev·əl ‚gir }

straightedge |DES ENG| A strip of wood, plastic, or metal with one or more long edges made straight with a desired degree of accuracy. { 'strād‚ej }

straightening vanes |ENG| Horizontal vanes mounted on the inside of fluid conduits to reduce the swirling or turbulent flow ahead of the orifice or the venturi meters. { 'strāt·ən·iŋ ‚vānz }

straight filing |ENG| Filing by pushing a file in a straight line across the work. { 'strāt 'fīl·iŋ }

straight-flow turbine |MECH ENG| A horizontal-axis, low-head hydraulic turbine in which the upstream and downstream reservoirs are connected by a straight tube into which the runners are integrated, with the generator placed directly on the periphery of these runners. { 'strāt ‚flō 'tər·bən }

straight joint |BUILD| 1. A continuous joint formed by the ends of parallel floor boards or masonry units and oriented perpendicularly to their length. 2. A joint between two pieces of wood that are set edge to edge without tongues and grooves, dowels, or overlap to bind them. Also known as square joint. { 'strāt 'jȯint }

straight-line mechanism |MECH ENG| A linkage so proportioned and constrained that some point on it describes over part of its motion a straight or nearly straight line. { 'strāt ‚līn 'mek·ə‚niz·əm }

straight-line motion |CONT SYS| A method of moving a robot between via or way points in which the end effector moves only along segments of straight lines, stopping momentarily for any change in direction. { 'strāt ‚līn 'mō·shən }

straight piecework system See one-hundred-percent premium plan. { 'strāt 'pēs‚wərk ‚sis·təm }

straight proportional system See one-hundred-percent premium plan. { 'strāt prə'pȯr·shən·əl ‚sis·təm }

straight-run |CHEM ENG| Petroleum fractions derived from the straight distillation of crude oil without chemical reaction or molecular modification. Also known as virgin. { 'strāt 'rən }

straight-run distillation |CHEM ENG| Continuous nonreactive distillation of petroleum oil to separate it into products in the order of their boiling points. { 'strāt ‚rən ‚dis·tə'lā·shən }

straight strap clamp |DES ENG| A clamp made of flat stock with an elongated slot for convenient positioning; held in place by a T bolt and nut. { 'strāt ‚strap 'klamp }

straight-tube boiler |MECH ENG| A water-tube boiler in which all the tubes are devoid of curvature and therefore require suitable connecting devices to complete the circulatory system. Also known as header-type boiler. { 'strāt ‚tüb 'bȯi·lər }

straight turning |MECH ENG| Work turned in a lathe so that the diameter is constant over the length of the workpiece. { 'strāt 'tərn·iŋ }

straightway pump |MECH ENG| A pump with suction and discharge valves arranged to give a direct flow of fluid. { 'strāt‚wā ‚pəmp }

straight wheel |DES ENG| A grinding wheel whose sides or face are straight and not in any way changed from a cylindricalform. { 'strāt ‚wēl }

strain |MECH| Change in length of an object in some direction per unit undistorted length in some direction, not necessarily the same; the nine possible strains form a second-rank tensor. { strān }

strain axis See principal axis of strain. { 'strān ‚ak·səs }

strain ellipsoid |MECH| A mathematical representation of the strain of a homogeneous body by a strain that is the same at all points or of unequal stress at a particular point. Also known as deformation ellipsoid. { 'strān i'lip‚sȯid }

strain energy |MECH| The potential energy stored in a body by virtue of an elastic deformation, equal to the work that must be done to produce this deformation. { 'strān ‚en·ər·jē }

strainer |ENG| A porous or screen medium used ahead of equipment to filter out harmful solid objects and particles from a fluid stream; used for example, in river-water intakes for process plants or to remove decomposition products from the circulating fluid in a hydraulic system. { 'strān·ər }

strain foil |ENG| A strain gage produced from thin foil by photoetching techniques; may be applied to curved surfaces, has low transverse sensitivity, exhibits negligible hysteresis under cycling loads, and creeps little under sustained loads. { 'strān ‚fȯil }

strain gage |ENG| A device which uses the change of electrical resistance of a wire under strain to measure pressure. { 'strān ‚gāj }

strain-gage accelerometer |ENG| Any accelerometer whose operation depends on the fact that the resistance in a wire changes when it is strained; these devices are classified as bonded or unbonded. { 'strān ‚gāj ak‚sel·ə'räm·əd·ər }

strain-gage bridge |ENG| A bridge arrangement of four strain gages, cemented to a stressed part in such a way that two gages show increases in resistance and two show decreases when the part is stressed; the change in output voltage under stress is thus much higher than that for a single gage. { 'strān ‚gāj ‚brij }

straining beam |CIV ENG| A short piece of timber in a truss that holds the ends of struts or rafters. Also known as straining piece. { 'strān·iŋ ‚bēm }

straining piece See straining beam. { 'strān·iŋ ‚pēs }

strain rate |MECH| The time rate for the usual tensile test. { 'strān ‚rāt }

strain rosette |MECH| A pattern of intersecting lines on a surface along which linear strains are measured to find stresses at a point. { 'strān rō,zet }

strain seismograph |ENG| A seismograph that detects secular strains related to tectonic processes and tidal yielding of the solid earth; also detects strains associated with propagating seismic waves. { 'strān 'sīz·mə,graf }

strain seismometer |ENG| A seismometer that measures relative displacement of two points in order to detect deformation of the ground. { 'strān sīz'mäm·əd·ər }

strain tensor |MECH| A second-rank tensor whose components are the nine possible strains. { 'strān ,ten·sər }

strake |BUILD| A course of clapboarding on a house. |CIV ENG| A row of steel plates installed on a tall steel chimney. { strāk }

strand |ENG| **1.** One of a number of steel wires twisted together to form a wire rope or cable or an electrical conductor. **2.** A thread, yarn, string, rope, wire, or cable of specified length. **3.** One of the fibers or filaments twisted or laid together into yarn, thread, rope, or cordage. { strand }

strand burner |ENG| A device that determines the rate at which a propellant burns at various pressures by using a propellant strand. { 'strand ,bər·nər }

stranded caisson See box caisson. { 'stran·dəd 'kā,sän }

standing machine See closing machine. { 'strand·iŋ mə,shēn }

strap bolt |DES ENG| **1.** A bolt with a hook or flat extension instead of a head. **2.** A bolt with a flat center portion and which can be bent into a U shape. { 'strap ,bōlt }

strap hammer |MECH ENG| A heavy hammer controlled and operated by a belt drive in which the head is slung from a strap, usually of leather. { 'strap ,ham·ər }

strap hinge |DES ENG| A hinge fastened to a door and the adjacent wall by a long hinge. { 'strap ,hinj }

strapped wall See battened wall. { 'strapt ¦wȯl }

strategic material |IND ENG| A material needed for the industrial support of a war effort. { strə'tē·jik mə'tir·ē·əl }

stratified charge engine |MECH ENG| An internal combustion engine that uses a fuel charge consisting of two layers; a rich mixture is close to the spark plug, and combustion promotes ignition of a lean mixture in the remainder of the cylinder. { 'strad·ə,fīd 'chärj ,en·jən }

stray capacitance |ELECTR| Undesirable capacitance between circuit wires, between wires and the chassis, or between components and the chassis of electronic equipment. { 'strā kə'pas·əd·əns }

stray current |ELEC| **1.** A portion of a current that flows over a path other than the intended path, and may cause electrochemical corrosion of metals in contact with electrolytes. **2.** An undesirable current generated by discharge of static electricity; it commonly arises in loading and unloading petroleum fuels and some chemicals, and can initiate explosions. { 'strā ,kə·rənt }

stray line |ENG| An ungraduated portion of the line connected to a current pole, used so that the pole will acquire the speed of the current before a measurement is begun. { 'strā ¦līn }

stream day |CHEM ENG| Denoting a 24-hour actual operation of a processing unit, in contrast to the hours actually operated during a calendar (24-hour) day. { 'strēm 'dā }

stream gage See river gage. { 'strēm ,gāj }

streamlining |DES ENG| The contouring of a body to reduce its resistance to motion through a fluid. { 'strēm,līn·iŋ }

street |CIV ENG| A paved road for vehicular traffic in an urban area. { strēt }

street elbow |DES ENG| A pipe elbow with an internal thread at one end and an external thread at the other. { 'strēt ,el·bō }

stremmatograph |ENG| An instrument for measuring longitudinal stress in rails as trains pass over. { strə'mad·ə,graf }

strength |MECH| The stress at which material ruptures or fails. { streŋkth }

stress |MECH| The force acting across a unit area in a solid material resisting the separation, compacting, or sliding that tends to be induced by external forces. { stres }

stress amplitude |MECH ENG| One half the algebraic difference between the maximum and minimum stress values in one fatigue test cycle. { 'stres ,am·plə,tüd }

stress axis See principal axis of stress. { 'stres ,ak·səs }

stress concentration |MECH| A condition in which a stress distribution has high localized stresses; usually induced by an abrupt change in the shape of a member; in the vicinity of notches, holes, changes in diameter of a shaft, and so forth, maximum stress is several times greater than where there is no geometrical discontinuity. { 'stres ,kän·sən,trā·shən }

stress concentration factor |MECH| A theoretical factor K_t expressing the ratio of the greatest stress in the region of stress concentration to the corresponding nominal stress. { 'stres ,kän·sən¦trā·shən ,fak·tər }

stress crack |MECH| An external or internal crack in a solid body (metal or plastic) caused by tensile, compressive, or shear forces. { 'stres ,krak }

stress difference |MECH| The difference between the greatest and the least of the three principal stresses. { 'stres ,dif·rəns }

stressed skin construction |CIV ENG| A type of construction in which the outer skin and the framework interact, thus contributing to the flexural strength of the unit. { 'strest ¦skin kən 'strək·shən }

stress ellipsoid |MECH| A mathematical representation of the state of stress at a point that is defined by the minimum, intermediate, and maximum stresses and their intensities. { 'stres i'lip,sȯid }

stress equivalent |IND ENG| A quantitative expression that can be used to compare the physiological outputs generated by different types of work stress. { 'stres i,kwiv·ə·lənt }

stress function [MECH] A single function, such as the Airy stress function, or one of two or more functions, such as Maxwell's or Morera's stress functions, that uniquely define the stresses in an elastic body as a function of position. { 'stres ,faŋk·shən }

stress intensity |MECH| Stress at a point in a structure due to pressure resulting from combined tension (positive) stresses and compression (negative) stresses. { 'stres in,ten·səd·ē }

stress lines See isostatics. { 'stres ,līnz }

stress range |MECH| The algebraic difference between the maximum and minimum stress in one fatigue test cycle. { 'stres ,rānj }

stress ratio |MECH| The ratio of minimum to maximum stress in fatigue testing, considering tensile stresses as positive and compressive stresses as negative. { 'stres ,rā·shō }

stress sensor |CONT SYS| A contact sensor that responds to the forces produced by mechanical contact. { 'stres ,sen·sər }

stress-strain curve See deformation curve. { 'stres 'strān ,kərv }

stress tensor |MECH| A second-rank tensor whose components are stresses exerted across surfaces perpendicular to the coordinate directions. { 'stres ,ten·sər }

stress test |ENG| A test of equipment under extreme conditions, outside the range anticipated in normal operation. { 'stres ,test }

stress trajectories See isostatics. { 'stres trə,jek·trēz }

stress transmittal |IND ENG| Transfer of external force from a human-equipment interface to various points of the body. { 'stres tranz,mid·əl }

stretcher |CIV ENG| A brick or block that is laid with its length paralleling the wall. { 'strech·ər }

stretcher bond |CIV ENG| A bond that consists entirely of stretchers, with each vertical joint lying between the centers of the stretchers above and below. { 'strech·ər ,bänd }

stretch former |MECH ENG| A machine used to form materials, such as metals and plastics, by stretching over a mold. { 'strech ,fòr·mər }

stretch forming |MECH ENG| Shaping metals and plastics by applying tension to stretch the heated sheet or part, wrapping it around a die, and then cooling it. Also known as wrap forming. { 'strech ,fòrm·iŋ }

stretch out |IND ENG| A reduction in the delivery rate specified for a program without a reduction in the total quantity to be delivered. { 'strech ¦aùt }

strich See millimeter. { 'strich }

striding compass |ENG| A compass mounted on a theodolite for orientation. { 'strīd·iŋ ,käm·pəs }

strike-off board |ENG| A straight-edge board used to remove excess, freshly placed plaster,

concrete, or mortar from a surface. { 'strīk ,óf ,bórd }

strike plate |DES ENG| A metal plate or box which is set in a door jamb and is either pierced or recessed to receive the bolt or latch of a lock. { 'strīk ,plāt }

striking hammer |ENG| A hammer used to strike a rock drill. { 'strīk·iŋ ,ham·ər }

striking velocity See impact velocity. { 'strīk·iŋ və,läs·əd·ē }

string |ENG| A piece of pipe, casing, or other down-hole drilling equipment coupled together and lowered into a borehole. |MECH| A solid body whose length is many times as large as any of its cross-sectional dimensions, and which has no stiffness. { striŋ }

stringcourse |BUILD| A horizontal band of masonry, generally narrower than other courses and sometimes projecting, extending across the facade of a structure and in some instances encircling pillars or engaged columns. Also known as belt course. { 'striŋ,kòrs }

string electrometer |ENG| An electrometer in which a conducting fiber is stretched midway between two oppositely charged metal plates; the electrostatic field between the plates displaces the fiber laterally in proportion to the voltage between the plates. { 'striŋ ,i,lek'träm·əd·ər }

stringer |CIV ENG| **1.** A long horizontal member used to support a floor or to connect uprights in a frame. **2.** An inclined member supporting the treads and risers of a staircase. { 'striŋ·ər }

string galvanometer |ENG| A galvanometer consisting of a silver-plated quartz fiber under tension in a magnetic field, used to measure oscillating currents. Also known as Einthoven galvanometer. { 'striŋ ,gal·və'näm·əd·ər }

string milling |MECH ENG| A milling method in which parts are placed in a row and milled consecutively. { 'striŋ ,mil·iŋ }

strip |ENG| **1.** To remove insulation from a wire. **2.** To break or otherwise damage the threads of a nut or bolt. { strip }

strip-borer drill |MECH ENG| An electric or diesel skid- or caterpillar-mounted drill used at quarry or opencast sites to drill 3- to 6-inch-diameter (8- to 15-centimeter), horizontal blast holes up to 100 feet (30 meters) in length, without the use of flush water. { 'strip ,bòr·ər ,dril }

strip-chart recorder |ENG| A recorder in which one or more writing pens or other recording devices trace changes in a measured variable on the surface of a strip chart that is moved at constant speed by a time-clock motor. { 'strip ,chärt ri ,kòrd·ər }

stripper |CHEM ENG| An evaporative device for the removal of vapors from liquids; can be in a bubble-tray distillation tower, a vacuum vessel, or an evaporator; if it is a part of a distillation column below the feed tray, it is called the stripping section. |ENG| A hand or motorized tool used to remove insulation from wires. { 'strip·ər }

stripper plate |ENG| In plastics molding, a plate

that strips a molded article free of core pins or force plugs. { 'strip·ər ,plăt }

stripping |CHEM ENG| In petroleum refining, the removal (by flash evaporation or steam-induced vaporation) of the more volatile components from a cut or fraction; used to raise the flash point of kerosine, gas oil, or lubricating oil. { 'strip·iŋ }

strip printer |ENG| A device that prints computer, telegraph, or industrial output information along a narrow paper tape which resembles a ticker tape. { 'strip ,print·ər }

stroboscope |ENG| An instrument for making moving bodies visible intermittently, either by illuminating the object with brilliant flashes of light or by imposing an intermittent shutter between the viewer and the object; a high-speed vibration can be made visible by adjusting the strobe frequency close to the vibration frequency. { 'strō·bə¦skōp }

stroboscopic disk |ENG| A printed disk having a number of concentric rings each containing a different number of dark and light segments; when the disk is placed on a phonograph turntable or rotating shaft and illuminated at a known frequency by a flashing discharge tube, speed can be determined by noting which pattern appears to stand still or to rotate slowly. { 'sträb·ə¦skäp·ik 'disk }

stroboscopic tachometer |ENG| A stroboscope having a scale that reads in flashes per minute or in revolutions per minute; the speed of a rotating device is measured by directing the stroboscopic lamp on the device, adjusting the flashing rate until the device appears to be stationary, then reading the speed directly on the scale of the instrument. { ¦sträb·ə¦skäp·ik tə¦käm·əd·ər }

stroke |ELECTR| The penlike motion of a focused electron beam in cathode-ray-tube diplays. |MECH ENG| The linear movement, in either direction, of a reciprocating mechanical part. Also known as throw. { strōk }

stroke-bore ratio |MECH ENG| The ratio of the distance traveled by a piston in a cylinder to the diameter of the cylinder. { 'strōk 'bȯr ,rā·shō }

strongly typed language |CONT SYS| A high-level programming language in which the type of each variable must be declared at the beginning of the program, and the language itself then enforces rules concerning the manipulation of variables according to their types. { 'strȯŋ·lē ¦tīpt 'laŋ·gwij }

Strouhal number |MECH| A dimensionless number used in studying the vibrations of a body past which a fluid is flowing; it is equal to a characteristic dimension of the body times the frequency of vibrations divided by the fluid velocity relative to the body; for a taut wire perpendicular to the fluid flow, with the characteristic dimension taken as the diameter of the wire, it has a value between 0.185 and 0.2 Symbolized S_r. Also known as reduced frequency. { 'strü·əl ,nəm·bər }

struck joint |CIV ENG| A mortar joint in brick-work formed by pressing the trowel in at the lower edge, so that a recess is formed at the bottom of the joint; suitable only for interior work. { 'strək ,jȯint }

structural analysis |ENG| The determination of stresses and strains in a given structure. { 'strək·chə·rəl ə'nal·ə·səs }

structural connection |CIV ENG| A means of joining the individual members of a structure to form a complete assembly. { 'strək·chə·rəl kə'nek·shən }

structural deflections |MECH| The deformations or movements of a structure and its flexural members from their original positions. { 'strək·chə·rəl di'flek·shənz }

structural drill |MECH ENG| A highly mobile diamond- or rotary-drill rig complete with hydraulically controlled derrick mounted on a truck, designed primarily for rapidly drilling holes to determine the structure in subsurface strata or for use as a shallow, slim-hole producer or seismograph drill. { 'strək·chə·rəl 'dril }

structural drilling |ENG| Drilling done specifically to obtain detailed information delineating the location of folds, domes, faults, and other subsurface structural features indiscernible by studying strata exposed at the surface. { 'strək·chə·rəl 'dril·iŋ }

structural engineering |CIV ENG| A branch of civil engineering dealing with the design of structures such as buildings, dams, and bridges. { 'strək·chə·rəl ,en·jə'nir·iŋ }

structural frame |BUILD| The entire set of members of a building or structure required to transmit loads to the ground. { 'strək·chə·rəl 'frām }

structural riveting |ENG| Riveting structural members by using punched holes. { 'strək·chə·rəl 'riv·əd·iŋ }

structural wall See bearing wall. { 'strək·chə·rəl 'wȯl }

structure |CIV ENG| Something, as a bridge or a building, that is built or constructed and designed to sustain a load. { 'strək·chər }

structured analysis |SYS ENG| A method of breaking a large problem or process into smaller components to aid in understanding, and then identifying the components and their interrelationships and reassembling them. { 'strək·chərd ə'nal·ə·səs }

structure number |DES ENG| A number, generally from 0 to 15, indicating the spacing of abrasive grains in a grinding wheel relative to their grit size. { 'strək·chər ,nəm·bər }

strut |CIV ENG| A long structural member of timber or metal, or a bar designed to resist pressure in the direction of its length. |ENG| **1.** A brace or supporting piece. **2.** A diagonal brace between two legs of a drill tripod or derrick. { strət }

Stuart windmill See Fales-Stuart windmill. { 'stü·ərt 'win,mil }

stub |CIV ENG| A projection on a sewer pipe that provides an opening to accept a connection to another pipe or house sewer. { stəb }

stub axle |MECH ENG| An axle carrying only one wheel. { 'stəb ¦ak·səl }

stub mortise |ENG| A mortise which passes through only part of a timber. { 'stəb ,mȯrd·əs }

Stubs gage |DES ENG| A number system for denoting the thickness of steel wire and drills. { 'stəbz ,gāj }

stub switch |ENG| A pair of short switch rails, held only at or near one end and free to move at the other end; used in mining and to some extent on narrow-gage industrial tramways. { 'stəb ,swich }

stub tenon |ENG| A tenon that fits into a stub mortise. { 'stəb ;ten·ən }

stub tube |MECH ENG| A short tube welded to a boiler or pressure vessel to provide for the attachment of additional parts. { 'stəb ,tüb }

stud |BUILD| One of the vertical members in the walls of a framed building to which wallboards, lathing, or paneling is nailed or fastened. |DES ENG| **1.** A rivet, boss, or nail with a large, ornamental head. **2.** A short rod or bolt threaded at both ends without a head. { stəd }

stud driver |MECH ENG| A device, such as an impact wrench, for driving a hardened steel nail (stud) into concrete or other hard materials. { stəd ,drī·vər }

stud wall |BUILD| A wall formed with timbers; studs are usually spaced 12–16 inches (30–41 centimeters) on center. { 'stəb ,wȯl }

stuffing |ENG| A method of sealing the mechanical joint between two metal surfaces; packing (stuffing) material is inserted within the seal area container (the stuffing or packing box), and compressed to a liquid-proof seal by a threaded packing ring follower. Also known as packing. { 'stəf·iŋ }

stuffing box |ENG| A packed, pressure-tight joint for a rod that moves through a hole, to reduce or eliminate fluid leakage. { 'stəf·iŋ ,bäks }

stuffing nut |ENG| A nut for adjusting a stuffing box. { 'stəf·iŋ ,nət }

style *See* gnomon. { stīl }

stylus |ENG ACOUS| The portion of a phonograph pickup that follows the modulations of a record groove and transmits the resulting mechanical motions to the transducer element of the pickup for conversion to corresponding audio-frequency signals. Also known as needle; phonograph needle; reproducing stylus. { 'stī·ləs }

subaperture |ENG| Any subset of an array of transmitters of acoustic or electromagnetic radiation. { səb'ap·ə·chər }

subassembly |ELECTR| Two or more components combined into a unit for convenience in assembling or servicing equipment; an intermediate-frequency strip for a receiver is an example. |ENG| A structural unit, which, though manufactured separately, was designed for incorporation with other parts in the final assembly of a finished product. { ¦səb·ə'sem·blē }

subatmospheric heating system |MECH ENG| A system which regulates steam flow into the main throttle valve under automatic thermostatic control and maintains a fixed vacuum differential between supply and return by means of a differential controller and a vacuum pump. { ¦səb,at·mə'sfir·ik 'hēd·iŋ ,sis·təm }

subbottom depth recorder |ENG| A compact seismic instrument which can provide continuous soundings of strata beneath the ocean bottom utilizing the low-frequency output of an intense electrical spark discharge source in water. { ¦səb'bäd·əm 'depth ri,kȯrd·ər }

subcarrier oscillator |ELECTR| **1.** The crystal oscillator that operates at the chrominance subcarrier or burst frequency of 3.579545 megahertz in a color television receiver; this oscillator, synchronized in frequency and phase with the transmitter master oscillator, furnishes the continuous subcarrier frequency required for demodulators in the receiver. **2.** An oscillator used in a telemetering system to translate variations in an electrical quantity into variations of a frequency-modulated signal at a subcarrier frequency. { ¦səb'kar·ē·ər 'äs·ə,lād·ər }

subcomponent |DES ENG| A part of a component having characteristics of the component. { 'səb·kəm,pō·nənt }

subcontract |ENG| A contract made with a third party by one who has contracted to perform work or service for whole or part performance of that work or service. { ¦səb'kän,trakt }

subcontractor |ENG| A manufacturer or organization that receives a contract from a prime contractor for a portion of the work on a project. { ¦səb'kän,trak·tər }

subdrainage |CIV ENG| Natural or artificial removal of water from beneath a lined conduit. { ¦səb'drā·nij }

subdrilling |ENG| Refers to the breaking of the base in which boreholes are drilled 1 foot (0.3 meter) or several feet below the level of the quarry floor. { ¦səb'dril·iŋ }

subfloor |BUILD| The rough floor which rests on the floor joists and on which the finished floor is laid. Also known as blind floor; counterfloor. { 'səb,flȯr }

subgrade |CIV ENG| The soil or rock leveled off to support the foundation of a structure. { 'səb,grād }

sublimation |THERMO| The process by which solids are transformed directly to the vapor state or vice versa without passing through the liquid phase. { ,səb·lə'mā·shən }

sublimation cooling |THERMO| Cooling caused by the extraction of energy to produce sublimation. { ,səb·lə'mā·shən ¦kül·iŋ }

sublimation curve |THERMO| A graph of the vapor pressure of a solid as a function of temperature. { ,səb·lə'mā·shən ¦kərv }

sublimation energy |THERMO| The increase in internal energy when a unit mass, or 1 mole, of a solid is converted into a gas, at constant pressure and temperature. { ,səb·lə'mā·shən ¦en·ər·jē }

sublimation point |THERMO| The temperature at which the vapor pressure of the solid phase of a compound is equal to the total pressure of the gas phase in contact with it; analogous to the boiling point of a liquid. { ‚səb·lə'mā·shən ¦póint }

sublimation pressure |THERMO| The vapor pressure of a solid. { ‚səb·lə'mā·shən ¦presh·ər }

sublime |THERMO| To change from the solid to the gaseous state without passing through the liquid phase. { sə'blīm }

submarine blast |ENG| A charge of high explosives fired in boreholes drilled in the rock underwater for dislodging dangerous projections and for deepening channels. { ¦səb·mə'rēn 'blast }

submarine gate |ENG| An edge gate with the opening from the runner into the mold positioned below the printing line or mold surface. { ¦səb·mə'rēn 'gāt }

submarine oscillator |ENG ACOUS| A large, electrically operated diaphragm horn which produces a powerful sound for signaling through water. { ¦səb·mə'rēn 'äs·ə‚lād·ər }

submarine pipeline |ENG| A pipeline installed under water, resting on the bed of the waterway; frequently used for petroleum or natural gas transport across rivers, lakes, or bays. { ¦səb·mə'rēn 'pīp‚līn }

submarine sentry |ENG| A form of underwater kite towed at a predetermined constant depth in search of elevations of the bottom; the kite rises to the surface upon encountering an obstruction. { ¦səb·mə'rēn 'sen·trē }

submarine wave recorder |ENG| An instrument for measuring the changing water height above a hovering submarine by measuring the time required for sound emitted by an inverted echo sounder on the submarine to travel to the surface and return. { ¦səb·mə'rēn 'wāv ri‚kórd·ər }

submerged-combustion evaporator |ENG| A liquid-evaporation device in which heat is provided by combustion gases bubbling up through the liquid; the burner is submerged in the body of the liquid. { səb'mərjd kəm¦bəs·chən i'vap·ə‚rād·ər }

submerged-combustion heater |ENG| A combustion device in which fuel and combustion air are mixed and ignited below the surface of a liquid; used in heaters and evaporators where absorption of the combustion products will not be detrimental. { səb'mərjd kəm¦bəs·chən 'hēd·ər }

submerged weir |CIV ENG| A dam which, when in use, has the downstream water level at an elevation equal to or higher than the crest of the dam. { səb'mərjd 'wer }

submersible pump |MECH ENG| A pump and its electric motor together in a protective housing which permits the unit to operate under water. { səb'mər·sə·bəl 'pəmp }

suboptimization |SYS ENG| The process of fulfilling or optimizing some chosen objective which is an integral part of a broader objective; usually the broad objective and lower-level objective are different. { ‚səb‚äp·tə·mə'zā·shən }

subsidiary conduit |CIV ENG| Terminating branch of an underground conduit run extending from a manhole or handhole to a nearby building, handhole, or pole. { səb'sid·ē‚er·ē 'kän·dü·ət }

subsonic inlet |ENG| An entrance or orifice for the admission of fluid flowing at speeds less than the speed of sound in the fluid. { ¦səb'sän·ik 'in‚let }

subsonic nozzle |ENG| A nozzle through which a fluid flows at speed less than the speed of sound in the fluid. { ¦səb'sän·ik 'näz·əl }

substation |ELEC| See electric power substation. |ENG| An intermediate compression station to repressure a fluid being transported by pipeline over a long distance. { 'səb‚stā·shən }

substitution weighing |MECH| A method of weighing to allow for differences in lengths of the balance arms, in which the object to be weighed is first balanced against a counterpoise, and the known weights needed to balance the same counterpoise are then determined. Also known as counterpoise method. { ‚səb·stə'tü·shən ‚wā·iŋ }

substrate |ELECTR| The physical material on which a microcircuit is fabricated; used primarily for mechanical support and insulating purposes, as with ceramic, plastic, and glass substrates; however, semiconductor and ferrite substrates may also provide useful electrical functions. |ENG| Basic surface on which a material adheres, for example, paint or laminate. { 'səb‚strāt }

substructure |CIV ENG| The part of a structure which is below ground. { ¦səb'strək·chər }

subsurface radar See ground-probing radar. { ‚səb‚sər·fəs 'rā·dar }

subsurface waste disposal |ENG| A waste disposal method for manufacturing wastes in porous underground rock formations. { ¦səb'sər·fəs 'wāst di‚spōz·əl }

subsynchronous |ELEC| Operating at a frequency or speed that is related to a submultiple of the source frequency. { ¦səb'siŋ·krə·nəs }

subsystem |ENG| A major part of a system which itself has the characteristics of a system, usually consisting of several components. { 'səb‚sis·təm }

subtense bar |ENG| The horizontal bar of fixed length in the subtense technique of distance measurement method. { ¦səb'tens 'bär }

subtense technique |CIV ENG| A distance measuring technique in which the transit angle subtended by the subtense bar enables the computation of the transit-to-bar distance. { ¦səb 'tens tek'nēk }

subtracted time |IND ENG| In a continuous timing technique, the difference between two successive readings of a stopwatch. { səb¦trak·təd 'tīm }

subtractive synthesis |ENG ACOUS| A method of synthesizing musical tones, in which an electronic circuit produces a standard waveform (such as a sawtooth wave), which contains a very large number of harmonics at known relative amplitudes, and this circuit is followed by a variety of electric or electronic filters to convert the basic tone signals into the desired musical waveforms. { 'səb‚trak·tiv 'sin·thə·səs }

subtractor |ELECTR| A circuit whose output is determined by the differences in analog or digital input signals. { səb'trak·tər }

subway |CIV ENG| An underground passage. { 'səb‚wā }

subwoofer |ENG ACOUS| A loudspeaker designed to reproduce extremely low audio frequencies, extending into the infrasonic range, generally used in conjunction with a crossover network, a woofer, and a tweeter. { 'səb‚wúf·ər }

Sucksmith ring balance |ENG| A magnetic balance in which the specimen is rigidly suspended from a phosphor bronze ring carrying two mirrors that convert small deflections of the specimen in a nonuniform magnetic field into large deflections of a light beam; used chiefly to measure paramagnetic susceptibility. { ¦sək‚smith 'riŋ ‚bal·əns }

suction anemometer |ENG| An anemometer consisting of an inverted tube which is half-filled with water that measures the change in water level caused by the wind's force. { 'sək·shən ‚an·ə'mäm·əd·ər }

suction cup |ENG| A cup, often of flexible material such as rubber, in which a partial vacuum is created when it is inverted on a surface; the vacuum tends to hold the cup in place. { 'sək·shən ‚kəp }

suction-cutter dredger |MECH ENG| A dredger in which rotary blades dislodge the material to be excavated, which is then removed by suction as in a sand-pump dredger. { 'sək·shən ¦kəd·ər ‚drej·ər }

suction head See suction lift. { 'sək·shən ‚hed }

suction lift |MECH ENG| The head, in feet, that a pump must provide on the inlet side to raise the liquid from the supply well to the level of the pump. Also known as suction head. { 'sək·shən ‚lift }

suction line |ENG| A pipe or tubing feeding into the inlet of a fluid impelling device (for example, pump, compressor, or blower), consequently under suction. { 'sək·shən ‚līn }

suction pump |MECH ENG| A pump that raises water by the force of atmospheric pressure pushing it into a partial vacuum under the valved piston, which retreats on the upstroke. { 'sək·shən ‚pəmp }

suction stroke |MECH ENG| The piston stroke that draws a fresh charge into the cylinder of a pump, compressor, or internal combustion engine. { 'sək·shən ‚strōk }

Suhl effect |ELECTR| When a strong transverse magnetic field is applied to an n-type semiconducting filament, holes injected into the filament are deflected to the surface, where they may recombine rapidly with electrons or be withdrawn by a probe. { 'sül i‚fekt }

sulfate pulping |CHEM ENG| A wood-pulping process in which sodium sulfate is used in the caustic soda pulp-digestion liquor. Also known

as kraft process; kraft pulping. { 'səl‚fāt 'pəlp·iŋ }

sulfur hexameter |ENG| An instrument used to measure or to continuously monitor the amount of sulfur hexafluoride present in a waveguide or other device in which this gas is used as a dielectric. { 'səl·fər hek'sam·əd·ər }

sulfuric acid alkylation |CHEM ENG| A petroleum refinery alkylation process in which three-carbon, four-carbon, and five-carbon olefins combine with isobutane in the presence of a sulfuric acid catalyst to form high-octane, branched-chain hydrocarbons; used in motor gasoline. { ‚səl¦fyúr·ik 'as·əd ‚al·kə'lā·shən }

sullage |CIV ENG| Drainage or wastewater from a building, farmyard, or street. { 'səl·ij }

Sullivan angle compressor |MECH ENG| A two-stage compressor in which the low-pressure cylinder is horizontal and the high-pressure cylinder is vertical; a compact compressor driven by a belt, or directly connected to an electric motor or diesel engine. { 'səl·ə·vən 'aŋ·gəl kəm‚pres·ər }

Sulzer two-cycle engine |MECH ENG| An internal combustion engine utilizing the Sulzer Company system for the effective scavenging and charging of the two-cycle diesel engine. { 'səlt·sər 'tü ‚sī·kəl 'en·jən }

summing amplifier |ELECTR| An amplifier that delivers an output voltage which is proportional to the sum of two or more input voltages or currents. { 'səm·iŋ 'am·plə‚fī·ər }

sump |ENG| A pit or tank which receives and temporarily stores drainage at the lowest point of a circulating or drainage system. Also known as sump pit. { səmp }

sump fuse |ENG| A fuse used for underwater blasting. { 'səmp ‚fyüz }

sump pit See sump. { 'səmp ‚pit }

sump pump |MECH ENG| A small, single-stage vertical pump used to drain shallow pits or sumps. { 'səmp ‚pəmp }

sun-and-planet motion |MECH ENG| A train of two wheels moving epicyclically with a small wheel rotating a wheel on the central axis. { ¦sən ən ¦plan·ət 'mō·shən }

sun gear See central gear. { 'sən ‚gir }

sunk draft |BUILD| A recessed margin around a building stone that imparts a raised appearance to the stone. { 'səŋk ¦draft }

sunk face |BUILD| A building stone from whose face some material has been removed in order to impart the appearance of a sunk panel. { 'səŋk ¦fās }

sunk panel |BUILD| A panel that is recessed below the face of its framing or other surrounding surface. { 'səŋk ¦pan·əl }

sunshine integrator |ENG| An instrument for determining the duration of sunshine (daylight) in any locality. { 'sən‚shīn ‚int·ə‚grād·ər }

sunshine recorder |ENG| An instrument designed to record the duration of sunshine without regard to intensity at a given location; sunshine recorders may be classified in two groups according to the method by which the time scale is obtained: in one group the time scale is obtained from the motion of the sun in the manner of a sun dial, in the second group the time scale is supplied by a chronograph. { 'sən,shīn ri ,kȯrd·ər }

superabrasive [MECH ENG] A material having characteristically long life and high grinding productivity such as cubic boron nitride or polycrystalline diamond { 'sü·pər·ə,brā·siv }

supercalendering [ENG] A calendering process that uses both steam and high pressure to give calendered material, for example, paper, a high-density finish. { ¦sü·pər'kal·ən·driŋ }

supercardioid microphone [ENG ACOUS] A microphone whose response pattern resembles a cardioid but is exaggerated along the axis of maximum response, so that it is highly sensitive in one direction and insensitive in all others. Also known as superdirectional microphone. { ¦sü·pər,kärd·ē,ȯid 'mī·krə,fōn }

supercentrifuge [MECH ENG] A centrifuge built to operate at faster speeds than an ordinary centrifuge. { ¦sü·pər'sen·trə,fyüj }

supercharge method [ENG] A method for measuring the knock-limited power, under supercharge rich-mixture conditions, of fuels for use in spark-ignition aircraft engines. { ¦sü·pər ,chärj ,meth·əd }

supercharger [MECH ENG] An air pump or blower in the intake system of an internal combustion engine used to increase the weight of air charge and consequent power output from a given engine size. { 'sü·pər,chär·jər }

supercharging [MECH ENG] A method of introducing air for combustion into the cylinder of an internal combustion engine at a pressure in excess of that which can be obtained by natural aspiration { 'sü·pər,chärj·iŋ }

supercobalt drill [DES ENG] A drill made of 8% cobalt highspeed steel; used for drilling work-hardened stainless steels, silicon chrome, and certain chrome-nickel alloy steels. { ¦sü·pər'kō ,bȯlt ,dril }

supercompressibility factor See compressibility factor. { ¦sü·pər·kəm,pres·ə'bil·əd·ē ,fak·tər }

superconducting gyroscope See cryogenic gyroscope. { ¦sü·pər·kən'dəkt·iŋ 'jī·rə,skōp }

superconducting quantum interference device [ELECTR] A superconducting ring that couples with one or two Josephson junctions; applications include high-sensitivity magnetometers, near-magnetic-field antennas, and measurement of very small currents or voltages. Abbreviated SQUID. { ¦sü·pər·kən'dəkt·iŋ 'kwän·təm ,in·tər ¦fir·əns di,vīs }

supercooling [THERMO] Cooling of a substance below the temperature at which a change of state would ordinarily take place without such a change of state occurring, for example, the cooling of a liquid below its freezing point without freezing taking place; this results in a metastable state. { ¦sü·pər'kül·iŋ }

supercritical [THERMO] Property of a gas which is above its critical pressure and temperature. { ¦sü·pər'krid·ə·kəl }

supercritical fluid [THERMO] A fluid at a temperature and pressure above its critical point; also, a fluid above its critical temperature regardless of pressure. { ¦sü·pər¦krid·ə·kəl 'flü·əd }

supercritical-fluid extraction [CHEM ENG] A separation process that uses a supercritical fluid as the solvent. { ¦sü·pər¦krid·ə·kəl 'flü·əd ik 'strak·shən }

superdirectional microphone See supercardioid microphone. { ,sü·pər·di,rek·shən·əl 'mī·krə ,fōn }

superficial expansivity See coefficient of superficial expansion. { ¦sü·pər¦fish·əl ,ik,span'siv·əd·ē }

superheat [THERMO] Sensible heat in a gas above the amount needed to maintain the gas phase. { 'sü·pər,hēt }

superheated vapor [THERMO] A vapor that has been heated above its boiling point. { ¦sü· pər'hēd·əd 'vā·pər }

superheater [MECH ENG] A component of a steam-generating unit in which steam, after it has left the boiler drum, is heated above its saturation temperature. { ¦sü·pər'hēd·ər }

superheating [THERMO] Heating of a substance above the temperature at which a change of state would ordinarily take place without such a change of state occurring, for example, the heating of a liquid above its boiling point without boiling taking place; this results in a metastable state. { ¦sü·pər'hēd·iŋ }

superhighway [CIV ENG] A broad highway, such as an expressway, freeway, turnpike, for high-speed traffic. { ¦sü·pər'hī,wā }

superimposed back pressure [MECH ENG] The static pressure at the outlet of an operating pressure relief device, resulting from pressure in the discharge system. { ¦sü·pər·im'pōzd 'bak ,presh·ər }

superinsulation [CHEM ENG] A multilayer insulation for cryogenic systems, composed of many floating insulation shields in an evacuated double-wall annulus, closely spaced but thermally separated by a poor-conducting fiber. { ¦sü· pər,in·sə'lā·shən }

superlattice [ELECTR] A structure consisting of alternating layers of two different semiconductor materials, each several nanometers thick. { ¦sü· pər'lad·əs }

supernatant liquor [ENG] The liquid above settled solids, as in a gravity separator. { ¦sü· pər'nāt·ənt 'lik·ər }

superposition integral [CONT SYS] An integral which expresses the response of a linear system to some input in terms of the impulse response or step response of the system; it may be thought of as the summation of the responses to impulses or step functions occurring at various times. { ,sü·pər·pə'zish·ən 'int·ə·grəl }

superposition principle See principle of superposition. { ,sü·pər·pə'zish·ən 'prin·sə·pəl }

superposition theorem See principle of superposition. { ,sü·pər·pə'zish·ən 'thir·əm }

supersonic compressor [MECH ENG] A compressor in which a supersonic velocity is imparted to the fluid relative to the rotor blades, the stator blades, or both, producing oblique shock waves over the blades to obtain a high-pressure rise. { ¦sü·pər¦sän·ik kəm'pres·ər }

supersonic diffuser [MECH ENG] A diffuser designed to reduce the velocity and to increase the pressure of fluid moving at supersonic velocities. { ¦sü·pər¦sän·ik di'fyü·zər }

supersonic nozzle See convergent-divergent nozzle. { ¦sü·pər¦sän·ik 'näz·əl }

superstructure [CIV ENG] The part of a structure that is raised on the foundation. { 'sü·pər ‚strək·chər }

supertweeter [ENG ACOUS] A loudspeaker designed to reproduce extremely high audio frequencies, extending into the ultrasonic range, generally used in conjunction with a crossover network, a tweeter, and a woofer. { 'süp·ər ‚twēd·ər }

supervisory control [ENG] A control panel or room showing key readings or indicators (temperature, pressure, or flow rate) from an entire operating area, allowing visual supervision and control of the overall operation. { ¦sü·pər¦vīz·ə·rē kən'trōl }

supervisory control and data acquisition [ENG] A version of telemetry commonly used in wide-area industrial applications, such as electrical power generation and distribution and water distribution, which includes supervisory control of remote stations as well as data acquisition from those stations over a bidirectional communications link. Abbreviated SCADA. { ¦sü·pər¦vīz· ə·rē kən‚trōl ən 'dad·ə ‚ak·wə‚zish·ən }

supervisory controlled manipulation [ENG] A form of remote manipulation in which a computer enables the operator to teach the manipulator motion patterns to be remembered and repeated later. { ¦sü·pər¦vīz·ə·rē kən'trōld mə ‚nip·yə'lā·shən }

supervisory expert control system [CONT SYS] A control system in which an expert system is used to supervise a set of control, identification, and monitoring algorithms. { ¦sü·pər¦vīz· ə·rē ‚ek‚spərt kən'trōl ‚sis·təm }

supervoltage [ELEC] A voltage in the range of 500 to 2000 kilovolts, used for some x-ray tubes. { ¦sü·pər'vōl·tij }

supination [CONT SYS] The orientation and motion of a robot component with its front or unprotected side facing upward and exposed. { ‚sü· pə'nā·shən }

supplied-air respirator [ENG] An atmospheric-supplying device which provides the wearer with respirable air from a source outside the contaminated area; only those with manual or motor-operated blowers are approved for immediately harmful or oxygen-deficient atmospheres. { sə'plīd ¦er 'res·pə‚rād·ər }

supply chain management [IND ENG] An inventory process involving planning and processing orders; handling; transporting and storing all materials purchased, processed, or distributed; and managing inventories in a coordinated manner among all the players on the chain to fulfill customer orders as they arise rather than to build

up stock level to fulfill anticipated future demand. { sə'plī ‚chān ‚man·ij·mənt }

supply control [IND ENG] The process by which an item of supply is controlled within the supply system, including requisitioning receipt, storage, stock control, shipment, disposition, identification, and accounting. { sə'plī kən‚trōl }

supply voltage [ELEC] The voltage obtained from a power source for operation of a circuit or device. { sə'plī ‚vōl·tij }

support base [ENG] A place from which logistic support is provided for a group of launch complexes and their control center. { sə'pórt ‚bās }

supported end [MECH] An end of a structure, such as a beam, whose position is fixed but whose orientation may vary; for example, an end supported on a knife-edge. { sə'pórd·əd ‚end }

suppressed-zero instrument [ENG] An indicating or recording instrument in which the zero position is below the lower end of the scale markings. { sə'prest ¦zir·ō ‚in·strə·mənt }

suppression [ELECTR] Elimination of any component of an emission, as a particular frequency or group of frequencies in an audio-frequency of a radio-frequency signal. { sə'presh·ən }

suppressor [ELEC] 1. In general, a device used to reduce or eliminate noise or other signals that interfere with the operation of a communication system, usually at the noise source. 2. Specifically, a resistor used in series with a spark plug or distributor of an automobile engine or other internal combustion engine to suppress spark noise that might otherwise interfere with radio reception. See suppressor grid. { sə'pres·ər }

surcharge [CIV ENG] The load supported above the level of the top of a retaining wall. { 'sər‚chärj }

surcharged wall [CIV ENG] A retaining wall with an embankment on the top. { 'sər‚chärjd 'wól }

surface [ENG] The outer part (skin with a thickness of zero) of a body; can apply to structures, to micrometer-sized particles, or to extended-surface zeolites. { 'sər·fəs }

surface analyzer [ENG] An instrument that measures or records irregularities in a surface by moving the stylus of a crystal pickup or similar device over the surface, amplifying the resulting voltage, and feeding the output voltage to an indicator or recorder that shows the surface irregularities magnified as much as 50,000 times. { 'sər·fəs ‚an·ə‚līz·ər }

surface area [ENG] Measurement of the extent of the area (without allowance for thickness) covered by a surface. { 'sər·fəs ‚er·ē·ə }

surface barrier [ELECTR] A potential barrier formed at a surface of a semiconductor by the trapping of carriers at the surface. { 'sər·fəs ‚bar·ē·ər }

surface-barrier diode [ELECTR] A diode utilizing thin-surface layers, formed either by deposition of metal films or by surface diffusion, to serve as a rectifying junction. { 'sər·fəs ¦bar·ē· ər 'dī‚ōd }

surface-barrier transistor [ELECTR] A transistor in which the emitter and collector are formed on opposite sides of a semiconductor wafer, usually made of n-type germanium, by training two jets of electrolyte against its opposite surfaces to etch and then electroplate the surfaces. { 'sər·fəs ¦bar·ē·ər tran'zis·tər }

surface burning See glowing combustion. { 'sər·fəs ‚bərn·iŋ }

surface carburetor |MECH ENG| A carburetor in which air is passed over the surface of gasoline to charge it with fuel. { 'sər·fəs 'kär·bə‚rād·ər }

surface-charge transistor |ELECTR| An integrated-circuit transistor element based on controlling the transfer of stored electric charges along the surface of a semiconductor. { 'sər·fəs ¦chärj tran'zis·tər }

surface combustion |ENG| Combustion brought about near the surface of a heated refractory material by forcing a mixture of air and combustible gases through it or through a hole in it, or having the gas impinge directly upon it; used in muffles, crucibles, and certain types of boiler furnaces. { 'sər·fəs kəm‚bəs·chən }

surface condenser |MECH ENG| A heat-transfer device used to condense a vapor, usually steam under vacuum, by absorbing its latent heat in cooling fluid, ordinarily water. { 'sər·fəs kən‚den·sər }

surface-controlled avalanche transistor |ELECTR| Transistor in which avalanche breakdown voltage is controlled by an external field applied through surface-insulating layers, and which permits operation at frequencies up to the 10-gigahertz range. { 'sər·fəs kən¦trōld 'av·ə‚lanch tran‚zis·tər }

surface-effect ship |MECH ENG| A transportation device with fixed side walls, which is supported by low-pressure, low-velocity air and operates on water only. { 'sər·fəs i¦fekt ‚ship }

surface finish |ENG| The surface roughness of a component after final treatment, measured by a surface profile. { 'sər·fəs ‚fin·ish }

surface force |MECH| An external force which acts only on the surface of a body; an example is the force exerted by another object with which the body is in contact. { 'sər·fəs ‚fórs }

surface gage |DES ENG| **1.** A scribing tool in an adjustable stand, used to mark off castings and to test the flatness of surfaces. **2.** A gage for determining the distances of points on a surface from a reference plane. { 'sər·fəs ‚gāj }

surface grinder |MECH ENG| A grinding machine that produces a plane surface. { 'sər·fəs ‚grīn·dər }

surface ignition |ENG| The initiation of a flame in the combustion chamber of an automobile engine by any hot surface other than the spark discharge. { 'sər·fəs ig‚nish·ən }

surface leakage |ELEC| The passage of current over the surface of an insulator. { 'sər·fəs ‚lē·kij }

surface micromachining |ENG| A set of processes based upon deposition, patterning, and selective etching of thin films to form a free-standing microsensor on the surface of a silicon wafer. { 'sər·fəs ‚mī·krə·mə'shēn·iŋ }

surface-mount technology |ELECTR| The technique of mounting electronic circuit components and their electrical connections on the surface of a printed board, rather than through holes. { 'sər·fəs ¦maúnt tek'näl·ə·jē }

surface noise |ELECTR| The noise component in the electric output of a phonograph pickup due to irregularities in the contact surface of the groove. Also known as needle scratch. { 'sər·fəs ‚nóiz }

surface of section See Poincaré surface of section. { ¦sər·fəs əv 'sek·shən }

surface passivation |ELECTR| A method of coating the surface of a p-type wafer for a diffused junction transistor with an oxide compound, such as silicon oxide, to prevent penetration of the impurity in undesired regions. { 'sər·fəs ‚pas·ə'vā·shən }

surface-penetrating radar See ground-probing radar. { ‚sər·fəs ‚pen·ə‚trād·iŋ 'rā‚där }

surface planer See surfacer. { 'sər·fəs ‚plā·nər }

surface plate |DES ENG| A plate having a very accurate plane surface used for testing other surfaces or to provide a true surface for accurately measuring and locating testing fixtures. { 'sər·fəs ‚plāt }

surfacer |DES ENG| A machine that is used to dress or plane the surface of a material such as stone, metal, or wood. Also known as surface planer. { 'sər·fəs·ər }

surface resistivity |ELEC| The electric resistance of the surface of an insulator, measured between the opposite sides of a square on the surface; the value in ohms is independent of the size of the square and the thickness of the surface film. { 'sər·fəs ‚rē‚zis'tiv·əd·ē }

surface roughness |ENG| The closely spaced unevenness of a solid surface (pits and projections) that results in friction for solid-solid movement or for fluid flow across the solid surface. { 'sər·fəs ‚rəf·nəs }

surface-set bit |DES ENG| A bit containing a single layer of diamonds set so that the diamonds protrude on the surface of the crown. Also known as single-layer bit. { 'sər·fəs ‚set ‚bit }

surface sizing See sizing treatment. { 'sər·fəs ‚sīz·iŋ }

surface thermometer |ENG| A thermometer, mounted in a bucket, used to measure the temperature of the sea surface. { 'sər·fəs thər'mäm·əd·ər }

surface treating |ENG| Any method of treating a material (metal, polymer, or wood) so as to alter the surface, rendering it receptive to inks, paints, lacquers, adhesives, and various other treatments, or resistant to weather or chemical attack. { 'sər·fəs ‚trēd·iŋ }

surface vibrator |MECH ENG| A vibrating device used on the surface of a pavement or flat slab to consolidate the concrete. { 'sər·fəs ‚vī‚brād·ər }

surface waterproofing |ENG| Waterproofing concrete by painting a waterproofing liquid on the surface. { 'sər·fəs 'wód·ər ‚prüf·iŋ }

surface wave See Rayleigh wave. { 'sər·fəs ‚wāv }

surfacing mat See overlay. { 'sər·fə·siŋ ‚mat }
surge |ELEC| A momentary large increase in the current or voltage in an electric circuit. |ENG| **1.** An upheaval of fluid in a processing system, frequently causing a carryover (puking) of liquid through the vapor lines. **2.** The peak system pressure. **3.** An unstable pressure buildup in a plastic extruder leading to variable throughput and waviness of the hollow plastic tube. { sərj }
surge arrester |ELEC| A protective device designed primarily for connection between a conductor of an electrical system and ground to limit the magnitude of transient overvoltages on equipment. Also known as arrester; lightning arrester. { 'sərj ə‚res·tər }
surge current |ELEC| A short-duration, high-amperage electric current wave that may sweep through an electrical network, as a power transmission network, when some portion of it is strongly influenced by the electrical activity of a thunderstorm. { 'sərj ‚kə·rənt }
surge protector |ELEC| A device placed in an electrical circuit to prevent the passage of surges and spikes that could damage electronic equipment. { 'sərj prə‚tek·tər }
surge stress |MECH| The physical stress on process equipment or systems resulting from a sudden surge in fluid (gas or liquid) flow rate or pressure. { 'sərj ‚stres }
surge suppressor |ELECTR| A circuit that responds to the rate of change of a current or voltage to prevent a rise above a predetermined value; it may include resistors, capacitors, coils, gas tubes, and semiconducting disks. Also known as transient suppressor. { 'sərj sə‚pres·ər }
surge tank |ENG| **1.** A standpipe or storage reservoir at the downstream end of a closed aqueduct or feeder pipe, as for a water wheel, to absorb sudden rises of pressure and to furnish water quickly during a drop in pressure. Also known as surge drum. **2.** An open tank to which the top of a surge pipe is connected so as to avoid loss of water during a pressure surge. { 'sərj ‚taŋk }
surging |ENG| Motion of a ship that alternately moves forward and aft, usually when moored. { 'sərj·iŋ }
surveillance |ENG| Systematic observation of air, surface, or subsurface areas or volumes by visual, electronic, photographic, or other means, for intelligence or other purposes. { sər'vā·ləns }
survey |ENG| **1.** The process of determining accurately the position, extent, contour, and so on, of an area, usually for the purpose of preparing a chart. **2.** The information so obtained. { 'sər‚vā }
survey foot |MECH| A unit of length, used by the U.S. Coast and Geodetic Survey, equal to 12/39.37 meter, or approximately 1.000002 feet. { 'sər‚vā 'fút }
surveying altimeter |ENG| A barometric-type

instrument consisting of a pressure-sensitive element which contracts or expands in proportion to atmospheric pressure, connected through a linkage to a pointer; its dial is graduated in units of linear measurement (feet or meters) to indicate differences of elevation only. { sər'vā·iŋ al'tim·əd·ər }
surveying sextant See hydrographic sextant. { sər'vā·iŋ ‚seks·tənt }
surveyor's compass |ENG| An instrument used to measure horizontal angles in surveying. { sər'vā·ərz ‚käm·pəs }
surveyor's cross |ENG| An instrument for setting out right angles in surveying; consists of two bars at right angles with sights at each end. { sər'vā·ərz ‚krós }
surveyor's level |ENG| A telescope and spirit level mounted on a tripod, rotating vertically and having leveling screws for adjustment. { sər'vā·ərz ‚lev·əl }
surveyor's measure |ENG| A system of measurement used in surveying having the engineer's, or Gunter's, chain, as a unit. { sər'vā·ərz ‚mezh·ər }
survey traverse See traverse. { 'sər‚vā trə'vərs }
survivor curve |IND ENG| A curve showing the percentage of a group of machines or facilities surviving at a given age. { sər'vī·vər ‚kərv }
Surwell clinograph |ENG| A directional surveying instrument which records photographically the direction and magnitude of well deviations from the vertical; powered by batteries, it contains a box level gage (indicating vertical deviation), a gyroscopic compass (indicating azimuth direction) and a watch and a dial thermometer, so that a simultaneous record of amount and direction of deviation, temperature, and time can be made on 16-millimeter film. { 'sər‚wel 'klīn·ə‚graf }
susceptance |ELEC| The imaginary component of admittance. { sə'sep·təns }
susceptance standard |ELEC| Standard that introduces calibrated small values of shunt capacitance into 50-ohm coaxial transmission arrays. { sə'sep·təns ‚stan·dərd }
susceptibility See electric susceptibility. { sə‚sep·tə'bil·əd·ē }
susceptometer |ENG| An instrument that measures paramagnetic, diamagnetic, or ferromagnetic susceptibility. { ‚sə'sep'täm·əd·ər }
suspended acoustical ceiling |BUILD| An acoustical ceiling which is suspended from either the roof or a higher ceiling. { sə'spen·dəd ə'kü·stə·kəl 'sē·liŋ }
suspended ceiling |BUILD| The suspension of the furring members beneath the structural members of a ceiling. { sə'spen·dəd 'sē·liŋ }
suspended formwork |CIV ENG| Formwork suspended from supports for the floor being cast. { sə'spen·dəd 'fórm‚wərk }
suspended span |CIV ENG| A simple span supported from the free ends of cantilevers. { sə'spen·dəd 'span }
suspended transformation |THERMO| The cessation of change before true equilibrium is

reached, or the failure of a system to change immediately after a change in conditions, such as in supercooling and other forms of metastable equilibrium. { sə'spen·dəd ˌtranz·fər 'mā·shən }

suspended tray conveyor |MECH ENG| A vertical conveyor having pendant trays or other carriers on one or more endless chains. { sə'spen·dəd ˈtrā kən'vā·ər }

suspension |ENG| A fine wire or coil spring that supports the moving element of a meter. { sə'spen·shən }

suspension bridge |CIV ENG| A fixed bridge consisting of either a roadway or a truss suspended from two cables which pass over two towers and are anchored by backstays to a firm foundation. { sə'spen·shən ˌbrij }

suspension cable |ENG| A freely hanging cable; may carry mainly its own weight or a uniformly distributed load. { sə'spen·shən ˌkā·bəl }

suspension roof |BUILD| A roof that is supported by steel cables. { sə'spen·shən ˌrüf }

suspension system |MECH ENG| A system of springs, shock absorbers, and other devices supporting the upper part of a motor vehicle on its running gear. { sə'spen·shən ˌsis·təm }

sustainable development |ENG| Development of industrial and natural resources that meets the energy needs of the present without compromising the ability of future generations to meet their needs in a similar manner. { sə,stān·ə·bəl di'vel·əp·mənt }

sustained oscillation |CONT SYS| Continued oscillation due to insufficient attenuation in the feedback path. { sə'stānd ˌäs·ə'lā·shən }

Sutro weir |CIV ENG| A dam with at least one curved side and horizontal crest, so formed that the head above the crest is directly proportional to the discharge. { 'sü·trō ˌwer }

SV See speaker verification.

swage bolt |DES ENG| A bolt having indentations with which it can be gripped in masonry. { 'swāj ˌbōlt }

swallow buoy See swallow float. { 'swä·lō ˌbȯi }

swallow float |ENG| A tubular buoy used to measure current velocities; it can be adjusted to be neutrally buoyant and to drift at a selected density level while being tracked by shipboard listening devices. Also known as neutrally buoyant float; swallow buoy. { 'swä·lō ˌflōt }

swamp buggy |MECH ENG| A wheeled vehicle that runs on sand, on mud, or through shallow water; used especially in swamps. { 'swämp ˌbəg·ē }

swamping resistor |ELECTR| Resistor placed in the emitter lead of a transistor circuit to minimize the effects of temperature on the emitterbase junction resistance. { 'swämp·iŋ ri,zis·tər }

swarf |ENG| Chips, shavings, and other fine particles removed from the workpiece by grinding tools. { 'swȯrf }

swash-plate pump |MECH ENG| A rotary pump in which the angle between the drive shaft and

the plunger-carrying body is varied. { swäsh ˌplāt ˌpəmp }

sway bar See stabilizer bar. { 'swā ˌbär }

sway brace |CIV ENG| One or a pair of diagonal members designed to resist horizontal forces, such as wind. { 'swā ˌbrās }

sway frame |CIV ENG| A unit in the system of members of a bridge that provides bracing against side sway; consists of two diagonals, the verticals, the floor beam, and the bottom strut. { 'swā ˌfrām }

sweating |CHEM ENG| Separation of paraffin oil from low-melting petroleum wax obtained from paraffin wax in a chamber (sweater) by first cooling the mixture until it is a solid cake, then warming gradually to cause partial fusion of the mixture to allow drainage of liquid from the cake. Also known as exudation. { 'swed·iŋ }

sweetening |CHEM ENG| Improvement of a petroleum-product color and odor by converting sulfur compounds into disulfides with sodium plumbite (doctor treating), or by removing them by contacting the petroleum stream with alkalies or other sweetening agents. { 'swēt·ən·iŋ }

swing |ELEC| Variation in frequency or amplitude of an electrical quantity. |ENG| **1.** The arc or curve described by the point of a pick or mandril when being used. **2.** Rotation of the superstructure of a power shovel on the vertical shaft in the mounting. **3.** To rotate a revolving shovel on its base. { swiŋ }

swing bridge |CIV ENG| A movable bridge that pivots in a horizontal plane about a center pier. { 'swiŋ ˌbrij }

swing-frame grinder |MECH ENG| A grinding machine hanging by a chain so that it may swing in all directions for surface grinding heavy work. { 'swiŋ ˌfrām ˌgrīn·dər }

swinging load |ENG| The load in pressure equipment which changes at frequent intervals. { 'swiŋ·iŋ ˌlōd }

swing joint |DES ENG| A pipe joint in which the parts may be rotated relative to each other. { 'swiŋ ˌjȯint }

swing pipe |ENG| A discharge pipe whose intake end can be raised or lowered on a tank. { 'swiŋ ˌpīp }

swing shift |IND ENG| Working arrangement in a three-shift, continuously run plant with working hours changed at regular intervals; during a swing shift the morning shift becomes the afternoon shift, while the afternoon shift becomes the morning shift of the next day, with only an 8-hour break on the first day of change. { 'swiŋ ˌshift }

swirl flowmeter See vortex precession flowmeter. { 'swərl 'flō,mēd·ər }

Swiss pattern file |DES ENG| A type of fine file used for precision filing of jewelry, instrument parts, and dies. { 'swis ˌpad·ərn 'fīl }

switch |CIV ENG| **1.** A device for enabling a railway car to pass from one track to another. **2.** The junction of two tracks. |ELEC| A manual or mechanically actuated device for making, breaking, or changing the connections in an electric circuit. Also known as electric switch. Symbolized SW. { swich }

switch angle [CIV ENG] The angle between the switch and stock rails of a railroad track, measured at the point of juncture between the gage lines. { 'swich ,aŋ·gəl }

switchblade knife [DES ENG] A knife in which the blade is spring-loaded and swings open when released by a pushbutton. { 'swich,blād 'nīf }

switched capacitor [ELECTR] An integrated circuit element, consisting of a capacitor with two metal oxide semiconductor (MOS) switches, whose function is approximately equivalent to that of a resistor. { 'swicht kə'pas·əd·ər }

switch function [ELECTR] A circuit having a fixed number of inputs and outputs designed such that the output information is a function of the input information, each expressed in a certain code or signal configuration or pattern. { 'swich ,faŋk·shən }

switching [ELEC] Making, breaking, or changing the connections in an electrical circuit. { 'swich·iŋ }

switching circuit [ELEC] A constituent electric circuit of a switching or digital processing system which receives, stores, or manipulates information in coded form to accomplish the specified objectives of the system. { 'swich·iŋ ,sər·kət }

switching device [ENG] An electrical or mechanical device or mechanism, which can bring another device or circuit into an operating or nonoperating state. Also known as switching mechanism. { 'swich·iŋ di,vīs }

switching diode [ELECTR] A crystal diode that provides essentially the same function as a switch; below a specified applied voltage it has high resistance corresponding to an open switch, while above that voltage it suddenly changes to the low resistance of a closed switch. { 'swich·iŋ ,dī,ōd }

switching gate [ELECTR] An electronic circuit in which an output having constant amplitude is registered if a particular combination of input signals exists; examples are the OR, AND, NOT, and INHIBIT circuits. Also known as logical gate. { 'swich·iŋ ,gāt }

switching key See key. { 'swich·iŋ ,kē }

switching mechanism See switching device. { 'swich·iŋ ,mek·ə,niz·əm }

switching substation [ELEC] An electric power substation whose equipment is mainly for connections and interconnections, and does not include transformers. { 'swich·iŋ 'səb,stā·shən }

switching surface [CONT SYS] In feedback control systems employing bang-bang control laws, the surface in state space which separates a region of maximum control effort from one of minimum control effort. { 'swich·iŋ ,sər·fəs }

switching-through relay [ELEC] Control relay of a line-finder selector, connector, or other stepping switch, which extends the loop of a calling telephone through to the succeeding switch in a switch train. { 'swich·iŋ ¦thrü 'rē,lā }

switching time [ELECTR] **1.** The time interval between the reference time and the last instant at which the instantaneous voltage response of a magnetic cell reaches a stated fraction of its peak value. **2.** The time interval between the reference time and the first instant at which the instantaneous integrated voltage response of a magnetic cell reaches a stated fraction of its peak value. { 'swich·iŋ ,tīm }

switching transistor [ELECTR] A transistor designed for on/off switching operation. { 'swich·iŋ tran'zis·tər }

switching trunk [ELEC] Trunk from a long-distance office to a local exchange office used for completing a long-distance call. { 'swich·iŋ ,trəŋk }

switch jack [ELEC] Any of the devices that provide terminals for the control circuits of the switch. { 'swich ,jak }

swivel [DES ENG] A part that oscillates freely on a headed bolt or pin. { 'swiv·əl }

swivel block [DES ENG] A block with a swivel attached to its hook or shackle permitting it to revolve. { 'swiv·əl ,bläk }

swivel coupling [MECH ENG] A coupling that gives complete rotary freedom to a deflecting wedge-setting assembly. { 'swiv·əl ,kəp·liŋ }

swivel head [MECH ENG] The assembly of a spindle, chuck, feed nut, and feed gears on a diamond-drill machine that surrounds, rotates, and advances the drill rods and drilling stem; on a hydraulic-feed drill the feed gears are replaced by a hydraulically actuated piston assembly. { 'swiv·əl ,hed }

swivel hook [DES ENG] A hook with a swivel connection to its base or eye. { 'swiv·əl ,hủk }

swivel joint [DES ENG] A joint with a packed swivel that allows one part to move relative to the other. { 'swiv·əl ,jỏint }

swivel neck See water swivel. { 'swiv·əl ,nek }

swivel pin See kingpin. { 'swiv·əl ,pin }

swivel spindle [BUILD] A shaft in a door handle assembly designed with a center joint that permits one knob to remain fixed while the other is being turned. { 'swiv·əl ,spin·dəl }

symballophone [ENG] A double stethoscope for the comparison and lateralization of sounds; permits the use of the acute function of the two ears to compare intensity and varying quality of sounds arising in the body or mechanical devices. { sim'bȯl·ə,fōn }

symmetrical avalanche rectifier [ELECTR] Avalanche rectifier that can be triggered in either direction, after which it has a low impedance in the triggered direction. { sə'me·trə·kəl 'av·ə,lanch ,rek·tə,fī·ər }

symmetrical band-pass filter [ELECTR] A band-pass filter whose attenuation as a function of frequency is symmetrical about a frequency at the center of the pass band. { sə'me·trə·kəl 'band ,pas ,fil·tər }

symmetrical band-reject filter [ELECTR] A band-rejection filter whose attenuation as a function of frequency is symmetrical about a frequency at the center of the rejection band. { sə'me·trə·kəl 'band ri,jekt ,fil·tər }

symmetrical clipper |ELECTR| A clipper in which the upper and lower limits on the amplitude of the output signal are positive and negative values of equal magnitude. { sə'me·trə·kəl 'klip·ər }

symmetrical deflection |ELECTR| A type of electrostatic deflection in which voltages that are equal in magnitude and opposite in sign are applied to the two deflector plates. { sə'me·trə·kəl di'flek·shən }

symmetrical H attenuator |ELECTR| An H attenuator in which the impedance near the input terminals equals the corresponding impedance near the output terminals. { sə'me·trə·kəl 'āch ə,ten·yə'wād·ər }

symmetrical O attenuator |ELECTR| An O attenuator in which the impedance near the input terminals equals the corresponding impedance near the output terminals. { sə'me·trə·kəl 'ō ə,ten·yə,wād·ər }

symmetrical pi attenuator |ELECTR| A pi attenuator in which the impedance near the input terminals equals the corresponding impedance near the output terminals. { sə'me·trə·kəl 'pī ə,ten·yə,wād·ər }

symmetrical T attenuator |ELECTR| A T attenuator in which the impedance near the input terminals equals the corresponding impedance near the output terminals. { sə'me·trə·kəl 'tē ə,ten·yə,wād·ər }

symmetrical transducer |ELECTR| A transducer is symmetrical with respect to a specified pair of terminations when the interchange of that pair of terminations will not affect the transmission. { sə'me·trə·kəl tranz'dü·sər }

symmetry axis See axis of symmetry. { 'sim·ə,trē ,ak·səs }

sympathetic detonation |ENG| Explosion caused by the transmission of a detonation wave through any medium from another explosion. { ,sim·pə'thed·ik ,det·ən'ā·shən }

sync See synchronization. { siŋk }

synchro |ELEC| Any of several devices which are used for transmitting and receiving angular position or angular motion over wires, such as a synchro transmitter or synchro receiver. Also known as mag-slip (British usage); self-synchronous device; self-synchronous repeater; selsyn. { 'siŋ·krō }

synchromesh |MECH ENG| An automobile transmission device that minimizes clashing; acts as a friction clutch, bringing gears approximately to correct speed just before meshing. { 'siŋ·krō,mesh }

synchronization |ENG| The maintenance of one operation in step with another, as in keeping the electron beam of a television picture tube in step with the electron beam of the television camera tube at the transmitter. Also known as sync. { ,siŋ·krə·nə'zā·shən }

synchronization indicator |ENG| An indicator that presents visually the relationship between

two varying quantities or moving objects. { ,siŋ·krə·nə'zā·shən ,in·də,kād·ər }

synchronized shifting |MECH ENG| Changing speed gears, with the gears being brought to the same speed before the change can be made. { 'siŋ·krə,nīzd 'shift·iŋ }

synchronous |ENG| In step or in phase, as applied to two or more circuits, devices, or machines. { 'siŋ·krə·nəs }

synchronous belt See timing belt. { 'siŋ·krə·nəs 'belt }

synchronous gate |ELECTR| A time gate in which the output intervals are synchronized with an incoming signal. { 'siŋ·krə·nəs 'gāt }

synchroscope |ELECTR| A cathode-ray oscilloscope designed to show a short-duration pulse by using a fast sweep that is synchronized with the pulse signal to be observed. |ENG| An instrument for indicating whether two periodic quantities are synchronous; the indicator may be a rotating-pointer device or a cathode-ray oscilloscope providing a rotating pattern; the position of the rotating pointer is a measure of the instantaneous phase difference between the quantities. { 'siŋ·krə,skōp }

synchro-shutter |ENG| A camera shutter with a circuit that flashes a light the instant the shutter opens. { 'siŋ·krō ,shəd·ər }

syngas See synthesis gas. { 'sin,gas }

syntactic semigroup |SYS ENG| For a sequential machine, the set of all transformations performed by all input sequences. { sin'tak·tik 'sem·i,grüp }

synthesis See system design. { 'sin·thə·səs }

synthesis gas |CHEM ENG| A mixture of gases prepared as feedstock for a chemical reaction, for example, carbon monoxide and hydrogen to make hydrocarbons or organic chemicals, or hydrogen and nitrogen to make ammonia. Also known as syngas. { 'sin·thə·səs ,gas }

synthetic aperture |ENG| A method of increasing the ability of an imaging system, such as radar or acoustical holography, to resolve small details of an object, in which a receiver of large size (or aperture) is in effect synthesized by the motion of a smaller receiver and the proper correlation of the detected signals. { sin'thed·ik 'ap·ə·chər }

synthetic-aperture radar |ENG| A radar system in which an aircraft moving along a very straight path emits microwave pulses continuously at a frequency constant enough to be coherent for a period during which the aircraft may have traveled about 1 kilometer; all echoes returned during this period can then be processed as if a single antenna as long as the flight path had been used. { sin'thed·ik |ap·ə·chər 'rā,där }

synthetic data |IND ENG| Any production data applicable to a given situation that are not obtained by direct measurement. { sin'thed·ik 'dad·ə }

synthol process |CHEM ENG| A reaction of carbon monoxide and hydrogen with an iron and sodium carbonate catalyst; produces a mixture of higher alcohols, aldehydes, ketones, higher fatty acids, and aliphatic hydrocarbons, usable as a synthetic gasoline. { 'sin,thȯl ,prä·səs }

syntony |ELEC| Condition in which two oscillating circuits have the same resonant frequency. { 'sin·tə·nē }

system |ELECTR| A combination of two or more sets generally physically separated when in operation, and such other assemblies, subassemblies, and parts necessary to perform an operational function or functions. |ENG| A combination of several pieces of equipment integrated to perform a specific function; thus a fire control system may include a tracking radar, computer, and gun. { 'sis·təm }

system analysis |CONT SYS| The use of mathematics to determine how a set of interconnected components whose individual characteristics are known will behave in response to a given input or set of inputs. { 'sis·təm ə,nal·ə·səs }

systematic error |ENG| An error due to some known physical law by which it might be predicted; these errors produced by the same cause affect the mean in the same sense, and do not tend to balance each other but rather give a definite bias to the mean. { ,sis·tə'mad·ik 'er·ər }

system bandwidth |CONT SYS| The difference between the frequencies at which the gain of a system is $\sqrt{2}/2$ (that is, 0.707) times its peak value. { 'sis·təm 'band,width }

system design |CONT SYS| A technique of constructing a system that performs in a specified manner, making use of available components. Also known as synthesis. { 'sis·təm di,zīn }

system effectiveness |ENG| A measure of the extent to which a system may be expected to achieve a set of specific mission requirements expressed as a function of availability, dependability, and capability. { 'sis·təm i'fek·tiv·nəs }

system engineering See systems engineering. { 'sis·təm ,en·jə'nir·iŋ }

system life cycle |ENG| The continuum of phases through which a system passes from conception through disposition. { 'sis·təm 'līf ,sī·kəl }

system optimization See optimization. { 'sis·təm ,äp·tə·mə'zā·shən }

system reliability |ENG| The probability that a system will accurately perform its specified task under stated environmental conditions. { 'sis·təm ri,lī·ə'bil·əd·ē }

system safety |ENG| The optimum degree of safety within the constraints of operational effectiveness, time, and cost, attained through specific application of system safety engineering throughout all phases of a system. { 'sis·təm 'sāf·tē }

system safety engineering |ENG| An element of systems management involving the application of scientific and engineering principles for the timely identification of hazards, and initiation of those actions necessary to prevent or control hazards within the system. { 'sis·təm 'sāf·tē ,en·jə,nir·iŋ }

systems analysis |ENG| The analysis of an activity, procedure, method, technique, or business to determine what must be accomplished and how the necessary operations may best be accomplished. { 'sis·təmz ə,nal·ə·səs }

systems architecting |SYS ENG| The discipline that combines elements which, working together, create unique structural and behavioral capabilities in a system that none could produce alone. Also known as systems architecture. { 'sis·təmz 'är·kə,tek·tiŋ }

systems architecture See systems architecting. { 'sis·təmz ,är·kə,tek·chər }

systems engineering |ENG| The design of a complex interrelation of many elements (a system) to maximize an agreed-upon measure of system performance, taking into consideration all of the elements related in any way to the system, including utilization of worker power as well as the characteristics of each of the system's components. Also known as system engineering. { 'sis·təmz ,en·jə,nir·iŋ }

systems implementation test |ENG| The test program that exercises the complete system in its actual environment to determine its capabilities and limitations; this test also demonstrates that the system is functionally operative, and is compatible with the other subsystems and supporting elements required for its operational employment. { 'sis·təmz ,im·plə·mən'tā·shən ,test }

systems integration |SYS ENG| A discipline that combines processes and procedures from systems engineering, systems management, and product development for the purpose of developing large-scale complex systems that involve hardware and software and may be based on existing or legacy systems coupled with totally new requirements to add significant functionality. { 'sis·təmz ,in·tə'grā·shən }

systems-management reengineering See organizational reengineering. { 'sis·təmz ,man·ij·mənt ,rē,en·jə'nir·iŋ }

systems test |ENG| A test of an entire interconnected set of components for the purpose of determining proper functions and interconnections. { 'sis·təmz ,test }

Szechtman cell |CHEM ENG| An electrolytic process for manufacture of chlorine that is a variation of both the mercury cell and molten salt cell. { 'sekt·mən ,sel }

T

t *See* troy system.

tab-card cutter |DES ENG| A device for die-cutting card stock to uniform tabulating-card size. { 'tab ¦kärd ‚kəd·ər }

table |BUILD| A horizontal projection or molding on the exterior or interior face of a wall. |MECH ENG| That part of a grinding machine which directly or indirectly supports the work being ground. { 'tā·bəl }

tabled joint |CIV ENG| In cut stonework, a bed joint formed by a broad, shallow channel in the surface of one stone that fits a corresponding projection of the stone above or below. { 'tā·bəld ‚jóint }

tablespoonful |MECH| A unit of volume used particularly in cookery, equal to 4 fluid drams or 1/2 fluid ounce; in the United States this is equal to approximately 14.7868 cubic centimeters, in the United Kingdom to approximately 14.2065 cubic centimeters. Abbreviated tbsp. { 'tā·bəl¦spün‚fúl }

tableting |ENG| A punch-and-die procedure for the compaction of powdered or granular solids; used for pharmaceuticals, food products, fireworks, vitamins, and dyes. { 'tab·ləd·iŋ }

tabling |BUILD| Formation of a horizontal masonry joint by arranging building stones in a course so that they extend into the next course and thus prevent slippage. { 'tāb·liŋ }

tab stop |DES ENG| A column position to which the printing mechanism of a typewriter or computer printer advances upon receipt of a command. { 'tab ‚stäp }

tachometer |ENG| An instrument that measures the revolutions per minute or the angular speed of a rotating shaft. { tə'käm·əd·ər }

tack |DES ENG| A small, sharp-pointed nail with a broad flat head. { tak }

tack coat |CIV ENG| A thin layer of bitumen, road tar, or emulsion laid on a road to enhance adhesion of the course above it. { 'tak ‚kōt }

tackiness *See* tack. { 'tak·ē·nəs }

tackle |MECH ENG| Any arrangement of ropes and pulleys to gain a mechanical advantage. { 'tak·əl *or* 'tāk·əl (naval usage) }

tack range |ENG| The length of time during which an adhesive will remain in the tacky-dry condition after application to an adherent. { 'tak ‚rānj }

tactical aircraft shelter |CIV ENG| A shelter to house fighter-type aircraft and to provide protection to the aircraft from attack by conventional weapons, or damage from high winds or other elemental hazards. { 'tak·tə·kəl ¦er‚kraft ‚shel·tər }

tactical control radar |ENG| Antiaircraft artillery radar which has essentially the same inherent capabilities as the target acquisition radar (physically it may be the same type of set) but whose function is chiefly that of providing tactical information for the control of elements of the antiaircraft artillery defenses in battle. { 'tak·tə·kəl kən'trol ‚rā‚där }

tactical range recorder |ENG| A sonar device in surface ships used to plot the time-range coordinates of submarines and determine firing of depth charges. { 'tak·tə·kəl 'rānj ri‚kórd·ər }

tactile sensor |CONT SYS| A transducer, usually associated with a robot end effector, that is sensitive to touch; comprises stress and touch sensors. { 'tak·təl 'sen·sər }

taffrail log |ENG| A log consisting essentially of a rotator towed through the water by a braided log line attached to a distance-registering device usually secured at the taffrail, the railing at the stern. Also known as patent log. { 'taf‚rāl ‚läg }

Tag-Robinson colorimeter |ENG| A laboratory device used to determine the color shades of lubricating and other oils; the color, reported as a number, is determined by varying the thickness of a column of oil until its color matches that of a standard color glass. { 'tag 'räb·ən·sən ‚kə·lə'rim·əd·ər }

tailboard *See* tailgate. { 'tāl‚bórd }

tailgate |CIV ENG| The downstream gate of a canal lock. |ENG| A hinged gate at the rear of a vehicle that can be let down for convenience in loading. Also known as backboard. { 'tāl‚gāt }

tail house |CHEM ENG| An installation in a refinery containing a look box, facilities for sampling, and controls for diverting the products to storage tanks or to other locations in the refinery for further processing. { 'tāl ‚haús }

tailing |BUILD| The projecting portion of a stone or brick that has been set into a wall, for example, a cornice. { 'tāl·iŋ }

tailings |ENG| The lighter particles which pass over a sieve in milling, crushing, or purifying operations. { 'tāl·iŋz }

tail pulley

tail pulley |MECH ENG| A pulley at the tail of the belt conveyor opposite the normal discharge end; may be a drive pulley or an idler pulley. { 'tāl ˌpu̇l·ē }

tailrace |ENG| A channel for carrying water away from a turbine, waterwheel, or other industrial application. { 'tāl,rās }

tailstock |MECH ENG| A part of a lathe that holds the end of the work not being shaped, allowing it to rotate freely. { 'tāl,stäk }

tail warning radar |ENG| Radar installed in the tail of an aircraft to warn the pilot that an aircraft is approaching from the rear. { 'tāl 'wȯrn·iŋ ˌrā,där }

Tainter gate |CIV ENG| A spillway gate whose face is a section of a cylinder; rotates about a horizontal axis on the downstream end of the gate and can be closed under its own weight. Also known as radial gate. { 'tān·tər ˌgāt }

takeup |MECH ENG| A tensioning device in a belt-conveyor system for taking up slack of loose parts. { 'tāk,əp }

takeup pulley |MECH ENG| An adjustable idler pulley to accommodate changes in the length of a conveyor belt to maintain proper belt tension. { 'tāk,əp ˌpu̇l·ē }

takeup reel |ENG| The reel that accumulates magnetic tape after it is recorded or played by a tape recorder. { 'tāk,əp ˌrēl }

takt time |IND ENG| **1.** The rate of customer demand, calculated by dividing the available production time by the quantity the customer requires in that time. **2.** The reciprocal of the production rate. { 'tak ˌtīm }

talk-listen switch |ENG ACOUS| A switch provided on intercommunication units to permit using the loudspeaker as a microphone when desired. { 'tȯk 'lis·ən ˌswich }

tall building |CIV ENG| A structure that, because of its height, is affected by lateral forces due to wind or earthquake to the extent that the forces constitute an important element in structural design. Also known as high-rise building. { ˌtȯl ˌbil·diŋ }

tamp |ENG| To tightly pack a drilled hole with clay or other stemming material after the charge has been placed. { tamp }

tamper |CIV ENG| A ramming device for compacting a granular material such as soil, backfill, or unformed concrete; usually powered by a motor. { 'tam·pər }

tamping bag |ENG| A bag filled with stemming material such as sand for use in horizontal and upward sloping shotholes. { 'tamp·iŋ ˌbag }

tamping bar |ENG| A piece of wood for pushing explosive cartridges or forcing the stemming into shotholes. { 'tamp·iŋ ˌbär }

tamping plug |ENG| A plug of iron or wood used instead of tamping material to close up a loaded blasthole. { 'tamp·iŋ ˌpləg }

tamping roller See sheepsfoot roller. { 'tamp·iŋ 'rō·lər }

tampion |ENG| A cone-shaped hand tool usually fashioned of hardwood that is forced into

a lead pipe to increase its diameter. { 'tam·pē·ən }

tandem compensation See cascade compensation. { 'tan·dəm ˌkäm·pən'sā·shən }

tandem distributed numerical control |CONT SYS| A form of distributed numerical control involving a series of machines connected by a conveyor and automatic loading and unloading devices that are under control of the central computers. { 'tan·dəm di⸴strib·yəd·əd nü⸴mer·ə·kəl kən'trōl }

tandem-drive conveyor |MECH ENG| A conveyor having the conveyor belt in contact with two drive pulleys, both driven with the same motor. { 'tan·dəm ⸴drīv kən'vā·ər }

tandem roller |MECH ENG| A steam- or gasoline-driven road roller in which the weight is divided between heavy metal rolls, of dissimilar diameter, one behind the other. { 'tan·dəm 'rō·lər }

tang |ENG| **1.** The part of a file that fits into a handle. **2.** The end of a drill shank which allows transmission of torque from the drill press spindle to the body of the drill. { taŋ }

tangent galvanometer |ENG| A galvanometer in which a small compass is mounted horizontally in the center of a large vertical coil of wire; the current through the coil is proportional to the tangent of the angle of deflection of the compass needle from its normal position parallel to the magnetic field of the earth. { 'tan·jənt ˌgal·və'näm·əd·ər }

tangential acceleration |MECH| The component of linear acceleration tangent to the path of a particle moving in a circular path. { tan'jen·chəl ak,sel·ə'rā·shən }

tangential helical-flow turbine See helical-flow turbine. { tan'jen·chəl ˌhel·ə·kəl ˌflō 'tər·bən }

tangential stress See shearing stress. { tan'jen·chəl 'stres }

tangential velocity |MECH| **1.** The instantaneous linear velocity of a body moving in a circular path; its direction is tangential to the circular path at the point in question. **2.** The component of the velocity of a body that is perpendicular to a line from an observer or reference point to the body. { tan'jen·chəl və'läs·əd·ē }

tangent offset |ENG| In surveying, a method of plotting traverse lines; angles are laid out by linear measurement, using a constant times the natural tangent of the angle. { 'tan·jənt 'ȯf,set }

tangent point See point of tangency. { 'tan·jənt ˌpȯint }

tangent screw |ENG| A screw providing tangential movement along an arc, such as the screw which provides the final angular adjustment of a marine sextant during an observation. { 'tan·jənt ˌskrü }

tank |ELECTR| **1.** A unit of acoustic delay-line storage containing a set of channels, each forming a separate recirculation path. **2.** The heavy metal envelope of a large mercury-arc rectifier or other gas tube having a mercury-pool cathode. **3.** See tank circuit. |ENG| A large container for

552

holding, storing, or transporting a liquid. { taŋk }

tankage |ENG| Contents of a storage tank. { 'taŋ·kij }

tank balloon |ENG| An air- and vapor-tight flexible container fitted to the breather pipe of a gasoline storage tank to receive gasoline vapors; as the tank cools, the vapors return to the tank. { 'taŋk bə,lün }

tank bottom |CHEM ENG| The liquid material in a tank below the level of the outlet pipe; often a mixture of the stored liquid with rust and other sediment. { 'taŋk ,bäd·əm }

tank car |ENG| Railroad car onto which is mounted a cylindrical, horizontal tank designed for the transport of liquids, chemicals, gases, meltable solids, slurries, emulsions, or fluidizable solids. { 'taŋk ,kär }

tank gage |ENG| A device used to measure the contents of a liquid storage tank; can be manual or automatic. { 'taŋk ,gāj }

tank scale |ENG| A counterweighted suspension or platform weighing mechanism for tanks, hoppers, and similar solids or liquids containers. { 'taŋk ,skāl }

tank truck |ENG| A truck body onto which is mounted a cylindrical, horizontal tank, designed for the transport of liquids, chemicals, gases, meltable solids, slurries, emulsions, or fluidizable solids. { 'taŋk ,trək }

tanning |ENG| A process of preserving animal hides by chemical treatment (using vegetable tannins, metallic sulfates, and sulfurized phenol compounds, or syntans) to make them immune to bacterial attack, and subsequent treatment with fats and greases to make them pliable. { 'tan·iŋ }

tantalum nitride resistor |ELECTR| A thin-film resistor consisting of tantalum nitride deposited on a substrate, such as industrial sapphire. { 'tant·əl·əm 'nī,trīd ri'zis·tər }

tap |DES ENG| **1.** A plug of accurate thread, form, and dimensions on which cutting edges are formed; it is screwed into a hole to cut an internal thread. **2.** A threaded cone-shaped fishing tool. |ELEC| A connection made at some point other than the ends of a resistor or coil. |ENG| A small, threaded hole drilled into a pipe or process vessel; used as connection points for sampling devices, instruments, or controls. { tap }

tap bolt |DES ENG| A bolt with a head that can be screwed into a hole and held in place without a nut. Also known as tap screw. { 'tap ,bōlt }

tap crystal |ELECTR| Compound semiconductor that stores current when stimulated by light and then gives up energy as flashes of light when it is physically tapped. { 'tap ,krist·əl }

tap drill |MECH ENG| A drill used to make a hole of a precise size for tapping. { 'tap ,dril }

tape |ENG| A graduated steel ribbon used, instead of a chain, in surveying. { tāp }

tape-automated bonding |ELECTR| A semiconductor chip (die) assembly method, where the chips are connected to polyimide (tape) carriers, complete with circuitry for attachment to a printed circuit board. The chip-bonded tape carriers typically are supplied on a reel (like a roll of film) for automated circuit assembly processes. { ¦tap ,ód·ə,mād·əd 'bän·diŋ }

tape cartridge |ENG ACOUS| A cartridge that holds a length of magnetic tape in such a way that the cartridge can be slipped into a tape recorder and played without threading the tape; in stereophonic usage, usually refers to an eight-track continuous-loop cartridge, which is larger than a cassette. Also known as cartridge. { 'tāp ,kär·trij }

tape-controlled machine |MECH ENG| A machine tool whose movements are automatically controlled by means of a magnetic or punched tape. { 'tāp kən¦trōld mə,shēn }

tape correction |ENG| A quantity applied to a taped distance to eliminate or reduce errors due to the physical condition of the tape and the manner in which it is used. { 'tāp kə,rek·shən }

tape deck |ENG ACOUS| A tape-recording mechanism that is mounted on a motor board, including the tape transport, electronics, and controls, but no power amplifier or loudspeaker. { 'tāp ,dek }

tape drive See tape transport. |MECH ENG| A device that transmits power from an actuator to a remote mechanism by flexible tapes and pulleys. { 'tāp ,drīv }

tape-float liquid-level gage |ENG| A liquid-level measurement by a float connected by a flexible tape to a rotating member, in turn connected to an indicator mechanism. { 'tāp ¦flōt 'lik·wəd ¦lev·əl ,gāj }

tape gage |ENG| A box- or float-type tide gage which consists essentially of a float attached to a tape and counterpoise; the float operates in a vertical box or pipe which dampens out short-period wind waves while admitting the slower tidal movement; for the standard installation, the tape is graduated with numbers increasing toward the float and is arranged with pulleys and counterpoise to pass up and down over a fixed reading mark as the tide rises and falls. { 'tāp ,gāj }

tape loop |ENG ACOUS| A length of magnetic tape having the ends spliced together to form an endless loop; used in message repeater units and in some types of tape cartridges to eliminate the need for rewinding the tape. { 'tāp ,lüp }

tape player |ENG ACOUS| A machine designed only for playback of recorded magnetic tapes. { 'tāp ,plā·ər }

taper bit |DES ENG| A long, cone-shaped noncoring bit used in drilling blastholes and in wedging and reaming operations. { 'tā·pər ,bit }

tape recorder |ENG ACOUS| A device that records audio signals and other information on magnetic tape by selective magnetization of iron oxide particles that form a thin film on the tape; a recorder usually also includes provisions for playing back the recorded material. { 'tāp ri ,kórd·ər }

tape recording |ENG ACOUS| The record made on a magnetic tape by a tape recorder. { 'tāp ri,kȯrd·iŋ }

tapered core bit |DES ENG| A core bit having a conical diamond-inset crown surface tapering from a borehole size at the bit face to the next larger borehole size at its upper, shank, or reaming-shell end. { 'tā·pərd 'kȯr ,bit }

tapered joint |DES ENG| A firm, leakproof connection between two pieces of pipe having the thread formed with a slightly tapering diameter. { 'tā·pərd 'jȯint }

tapered thread |DES ENG| A screw thread cut on the surface of a tapered part; it may be either a pine or box thread, or a V-, Acme, or square-screw thread. { 'tā·pərd 'thred }

tapered wheel |DES ENG| A flat-face grinding wheel with greater thickness at the hub than at the face. { 'tā·pərd 'wēl }

taper gage |ENG| A precision gage that is used to check the accuracy of a standard taper. { 'tā·pər ,gāj }

taper key |DES ENG| A rectangular machine key that is slightly tapered along its length. { 'tā·pər ,kē }

taper pin |DES ENG| A small, tapered self-holding peg or nail used to connect parts together. { 'tā·pər ,pin }

taper pipe thread See pipe thread. { 'tā·pər 'pīp ,thred }

taper plug gage |DES ENG| An internal gage in the shape of a frustrum of a cone used to measure internal tapers. { 'tā·pər 'pləg ,gāj }

taper reamer |DES ENG| A reamer whose fluted portion tapers toward the front end. { 'tā·pər ,rē·mər }

taper ring gage |DES ENG| An external gage having a conical internal contour; used to measure external tapers. { 'tā·pər 'riŋ ,gāj }

taper-rolling bearing |MECH ENG| A roller bearing capable of sustaining end thrust by means of tapered rollers and coned races. { 'tā·pər 'rō·liŋ ,ber·iŋ }

taper shank |DES ENG| A cone-shaped part on a tool that fits into a tapered sleeve on a driving member. { 'tā·pər ,shaŋk }

taper tap |DES ENG| A threaded cone-shaped tool for cutting internal screw threads. { 'tā·pər ,tap }

taper washer |DES ENG| A type of washer designed to be used underneath nuts with tapered flanges to enable the bolt assembly to fit properly when tightened. { 'tā·pər ,wäsh·ər }

tape speed |ENG ACOUS| The speed at which magnetic tape moves past the recording head in a tape recorder; standard speeds are $^{15}/_{16}$, $1^7/_8$, $3^3/_4$, $7^1/_2$, 15, and 30 inches per second (2.38125, 4.7625, 9.525, 19.05, 38.1, and 76.2 centimeters per second); faster speeds give improved high-frequency response under given conditions. { 'tāp ,spēd }

tape transport |ENG ACOUS| The mechanism of a tape recorder that holds the tape reels, drives the tape past the heads, and controls various modes of operation. Also known as tape drive. { 'tāp ,tranz,pȯrt }

taping |ENG| The process of measuring distances with a surveyor's tape. { 'tāp·iŋ }

tappet |MECH ENG| A lever or oscillating member moved by a cam and intended to tap or touch another part, such as a push rod or valve system. { 'tap·ət }

tappet rod |MECH ENG| A rod carrying a tappet or tappets, as one for opening or closing the valves in a steam or an internal combustion engine. { 'tap·ət ,räd }

tapping |MECH ENG| Forming an internal screw thread in a hole or other part by means of a tap. { 'tap·iŋ }

tapping screw See self-tapping screw. { 'tap·iŋ ,skrü }

tap screw See tap bolt. { 'tap ,skrü }

tap wrench |ENG| A tool used to clamp taps during tapping operations. { 'tap ,rench }

tare |MECH| The weight of an empty vehicle or container; subtracted from gross weight to ascertain net weight. { ter }

target |ELECTR| **1.** In an x-ray tube, the anode or anticathode which emits x-rays when bombarded with electrons. **2.** In a television camera tube, the storage surface that is scanned by an electron beam to generate an output signal current corresponding to the charge-density pattern stored there. **3.** In a cathode-ray tuning indicator tube, one of the electrodes that is coated with a material that fluoresces under electron bombardment. |ENG| **1.** The sliding weight on a leveling rod used in surveying to enable the staffman to read the line of collimation. **2.** The point that a borehole or an exploratory work is intended to reach. **3.** In radar and sonar, any object capable of reflecting the transmitted beam. { 'tär·gət }

target acquisition radar |ENG| An antiaircraft artillery radar, normally of lesser range capabilities but of greater inherent accuracy than that of surveillance radar, whose normal function is to acquire aerial targets either by independent search or on direction of the surveillance radar, and to transfer these targets to tracking radars. { 'tär·gət ,ak·wə̩zish·ən 'rā,där }

target-type flowmeter |ENG| A fluid-flow measurement device with a small circular target suspended centrally in the flow conduit; the target transmits force to a force-balance transmitter by means of a pivoted bar. { 'tär·gət ‖tīp 'flō ,mēd·ər }

tariff |IND ENG| A government-imposed duty on imported or exported goods. { 'tar·əf }

tarring |ENG| The coating of piles for permanent underground work with prepared acid-free tar. { 'tär·iŋ }

task analysis |IND ENG| A process for determining in detail the specific behaviors required of the personnel involved in a human-machine system. { 'task ə,nal·ə·səs }

task element |IND ENG| The smallest logically

definable set of perceptions, decisions, and responses required of a human being in the performance of a task. { 'task ,el·ə·mənt }

taut-band ammeter [ENG] A modification of the permanent-magnet movable-coil ammeter in which the jeweled bearings and control springs are replaced by a taut metallic band rigidly held at the ends; the coil is firmly attached to the band, and restoring torque is supplied by twisting of the band. { 'tȯt ¦band 'am,ēd·ər }

taut-line cableway [MECH ENG] A cableway whose operation is limited to the distance between two towers, usually 3000 feet (914 meters) apart, has only one carrier, and the traction cable is reeved at the carrier so that loads can be raised and lowered; the towers can be mounted on trucks or crawlers, and the machine shifted across a wide area. { 'tȯt ¦līn 'kā·bəl,wā }

tawing [ENG] A tanning process in which alum is used as a partial tannage, supplementing or replacing chrome. { 'tȯ·iŋ }

taxi channel [CIV ENG] A defined path, on a water airport, intended for the use of taxiing aircraft. { 'tak·sē ,chan·əl }

taxiway [CIV ENG] A specially prepared or designated path on an airport for taxiing aircraft. { 'tak·sē,wā }

T beam [CIV ENG] A metal beam or bar with a T-shaped cross section. { 'tē ,bēm }

T bolt [DES ENG] A bolt with a T-shaped head, made to fit into a T-shaped slot in a drill swivel head or in the bed of a machine. { 'tē ,bȯlt }

tbsp See tablespoonful.

teach [CONT SYS] To program a robot by guiding it through its motions, which are then recorded and stored in its computer. { tēch }

teach box See teach pendant. { 'tēch ,bäks }

teach-by-doing [CONT SYS] A method of programming a robot in which the operator guides the robot through its intended motions by holding it and performing the work. { ¦tēch ·bī 'dü·iŋ }

teach-by-driving [CONT SYS] Programming a robot by using a teach pendant. { ¦tēch ·bī 'drīv·iŋ }

teach gun See teach pendant. { 'tēch ,gən }

teaching interface [CONT SYS] The devices and hardware that are used to instruct robots and other machinery how to operate, and to specify their motions. { 'tēch·iŋ 'in·tər,fās }

teach mode [CONT SYS] The mode of operation in which a robot is instructed in its motions, usually by guiding it through these motions using a teach pendant. { 'tēch ,mōd }

teach pendant [CONT SYS] A hand-held device used to instruct a robot, specifying the character and types of motions it is to undertake. Also known as teach box; teach gun. { 'tēch ,pendənt }

tear down [ENG] **1.** To disassemble a drilling rig preparatory to moving it to another drill site. **2.** To disassemble a machine or change the jigs and fixtures. { 'ter 'daȯn }

tear-down time [IND ENG] The downtime of a machine following a given work order which usually involves removing parts such as jigs and fixtures and which must be completely finished before setting up for the next order. { 'ter ¦daȯn ,tīm }

tear strength [MECH] The force needed to initiate or to continue tearing a sheet or fabric. { 'ter ,streŋkth }

teaspoonful [MECH] A unit of volume used particularly in cookery and pharmacy, equal to 1¹/₃ fluid drams, or 1/3 tablespoonful; in the United States this is equal to approximately 4.9289 cubic centimeters, in the United Kingdom to approximately 4.7355 cubic centimeters. Abbreviated tsp; tspn. { 'tē,spün,fül }

technical atmosphere [MECH] A unit of pressure in the metric technical system equal to one kilogram-force per square centimeter. Abbreviated at. { 'tek·nə·kəl 'at·mə,sfir }

technical characteristics [ENG] Those characteristics of equipment which pertain primarily to the engineering principles involved in producing equipment possessing desired characteristics, for example, for electronic equipment; technical characteristics include such items as circuitry, and types and arrangement of components. { 'tek·nə·kəl ,kar·ik·tə'ris·tiks }

technical evaluation [ENG] The study and investigation to determine the technical suitability of material, equipment, or a system. { 'tek·nə·kəl i,val·yə'wā·shən }

technical information [ENG] Information, including scientific information, which relates to research, development, engineering, testing, evaluation, production, operation, use, and maintenance of equipment. { 'tek·nə·kəl ,in·fər'mā·shən }

technical inspection [ENG] Inspection of equipment to determine whether it is serviceable for continued use or needs repairs. { 'tek·nə·kəl in'spek·shən }

technical maintenance [ENG] A category of maintenance that includes the replacement of unserviceable major parts, assemblies, or subassemblies, and the precision adjustment, testing, and alignment of internal components. { 'tek·nə·kəl 'mānt·ən·əns }

technical manual [ENG] A publication containing detailed information on technical procedures, including instructions on the operation, handling, maintenance, and repair of equipment. { 'tek·nə·kəl ,man·yə·wəl }

technical representative [IND ENG] A person who represents one or more manufacturers in an area and who gives technical advice on the application, installation, operation, and maintenance of their products, in addition to selling the products. { 'tek·nə·kəl ¦rep·ri¦zent·əd·iv }

technical specifications [ENG] A detailed description of technical requirements stated in terms suitable to form the basis for the actual design, development, and production processes of an item having the qualities specified in the operational characteristics. { 'tek·nə·kəl ,spes·ə·fə'kā·shənz }

tectonics |CIV ENG| **1.** The science and art of construction with regard to use and design. **2.** Design relating to crustal deformations of the earth. { tek'tän·iks }

tectonometer [ENG] An apparatus, including a microammeter, used on the surface to obtain knowledge of the structure of the underlying rocks. { ,tek·tə'näm·əd·ər }

tee |ENG| Shaped like the letter T. { tē }

tee joint |ENG| A joint in which members meet at right angles, forming a T. { 'tē ,jóint }

telechir |CONT SYS| A handlike remote manipulator. { 'tel·ə,kir }

telechirics |CONT SYS| The use of teleoperators or remote manipulators. { |tel·ə|kir·iks }

telegraph buoy |ENG| A buoy used to mark the position of a submarine telegraph cable. { 'tel·ə,graf ,bói }

telemeteorograph [ENG] Any meteorological instrument, such as a radiosonde, in which the recording instrument is located at some distance from the measuring apparatus; for example, a meteorological telemeter. { |tel·ə,mēd·ē'ór·ə,graf }

telemeteorography [ENG] The science of the design, construction, and operation of various types of telemeteorographs. { |tel·ə,mēd·ē·ə'räg·rə·fe }

telemeter [ENG] **1.** The complete measuring, transmitting, and receiving apparatus for indicating or recording the value of a quantity at a distance. Also known as telemetering system. **2.** To transmit the value of a measured quantity to a remote point. { 'tel·ə,mēd·ər }

telemetering [ENG] Transmitting the readings of instruments to a remote location by means of wires, radio waves, or other means. Also known as remote metering; telemetry. { ,tel·ə'mēd·ə·riŋ }

telemetering system See telemeter. { ,tel·ə'mēd·ə·riŋ ,sis·təm }

telemetering wave buoy [ENG] A buoy assembly that transmits a radio signal that varies in frequency proportional to the vertical acceleration experienced by the buoy, thereby conveying information about the buoy's vertical motion as it rides the waves. { ,tel·ə'mēd·ə·riŋ 'wāv ,bói }

telemetry See telemetering. { tə'lem·ə·trē }

teleoperation [ENG] **1.** The real-time control of remotely located machines that act as the eyes and hands of a person located elsewhere; it has been used in undersea and lunar exploration, mining, and microsurgery. **2.** Operation from a remote location. Also known as remote manipulation. { ,tel·ē,äp·ə'rā·shən }

teleoperator See remote manipulator. { ,tel·ē,äp·ə,rād·ər }

telephone See telephone set. { 'tel·ə,fōn }

telephone dial [ENG] A switch operated by a finger wheel, used to make and break a pair of contacts the required number of times for setting up a telephone circuit to the party being called. { 'tel·ə,fōn ,dīl }

telephone receiver |ENG ACOUS| The portion of

a telephone set that converts the audio-frequency current variations of a telephone line into sound waves, by the motion of a diaphragm activated by a magnet whose field is varied by the electrical impulses that come over the telephone wire. { 'tel·ə,fōn ri,sē·vər }

telephone set |ENG ACOUS| An assembly including a telephone transmitter, a telephone receiver, and associated switching and signaling devices. Also known as telephone. { 'tel·ə,fōn ,set }

telephone transmitter |ENG ACOUS| The microphone used in a telephone set to convert speech into audio-frequency electric signals. { 'tel·ə,fōn tranz,mid·ər }

telephotometer [ENG] A photometer that measures the received intensity of a distant light source. { |tel·ə·fə'täm·əd·ər }

telepresence |CONT SYS| The quality of sensory feedback from a teleoperator or telerobot to a human operator such that the operator feels present at the remote site. { |tel·ə'prez·əns }

telepsychrometer [ENG] A psychrometer in which the wet- and dry-bulb thermal elements are located at a distance from the indicating elements. { |tel·ə·sī'kräm·əd·ər }

telerecording bathythermometer [ENG] A device which transmits measurements of sea water depth and temperature over a wire to a ship, where a graph of temperature versus depth is recorded. { 'tel·ə·ri,kórd·iŋ |bath·i·thər'mäm·əd·ər }

telerobot |CONT SYS| A type of teleoperator that embodies features of a robot and is programmed for communication with a human operator in a high-level language but can revert to direct control in the event of unplanned contingencies. { ,tel·ə'rō,bät }

telescope [ENG] Any device that collects radiation, which may be in the form of electromagnetic or particle radiation, from a limited direction in space. { 'tel·ə,skōp }

telescopic alidade |ENG| An alidade used with a plane table, consisting of a telescope mounted on a straightedge ruler, fitted with a level bubble, scale, and vernier to measure angles, and calibrated to measure distances. { |tel·ə|skäp·ik 'al·ə,dād }

telescopic derrick [ENG] A drill derrick divided into two or more sections, with the uppermost sections nesting successively into the lower sections. { |tel·ə|skäp·ik 'de,rik }

telescopic tripod [ENG] A drill or surveyor's tripod each leg of which is a series of two or more closely fitted nesting tubes, which can be locked rigidly together in an extended position to form a long leg or nested one within the other for easy transport. { |tel·ə|skäp·ik 'trī,päd }

telescoping gage |DES ENG| An adjustable internal gage with a telescoping plunger that expands under spring tension in the hole to be measured; it is locked into position to allow measurement after being withdrawn from the hole. { |tel·ə|skōp·iŋ 'gāj }

telescoping valve |MECH ENG| A valve, with

sliding, telescoping members, to regulate water flow in a pipe line with minimum disturbance to stream lines. { |tel·ə|skōp·iŋ 'valv }

telethermometer |ENG| A temperature-measuring system in which the heat-sensitive element is located at a distance from the indicating element. { |tel·ə·thər'mäm·əd·ər }

telethermoscope |ENG| A temperature telemeter, frequently used in a weather station to indicate the temperature at the instrument shelter located outside. { |tel·ə'thər·mə,skōp }

telethesis |ENG| A robotic manipulation aid for the physically disabled that may be located remote from the body. There are two forms, operated by voice command, or operated through a body-powered prosthesis or a joystick. { tə'le·th·ə·səs }

televiewer |ENG| An acoustic camera that provides an ultrasonic image of the borehole wall during borehole logging. { 'tel·ə,vyü·ər }

television film scanner |ENG| A motion picture projector adapted for use with a television camera tube to televise 24-frame-per-second motion picture film at the 30-frame-per-second rate required for television. { 'tel·ə,vizh·ən 'film ,skan·ər }

television tower |ENG| A tall metal structure used as a television transmitting antenna, or used with another such structure to support a television transmitting antenna wire. { 'tel·ə,vizh·ən ,taù·ər }

telford pavement |CIV ENG| A road pavement having a firm foundation of large stones and stone fragments, and a smooth hard-rolled surface of small stones. { 'tel·fərd ,pāv·mənt }

Tellerette |CHEM ENG| A type of inert packing with the appearance of a circular-wound spiral, used to create a large surface area to increase contact between falling liquid and rising vapor; used in gas-absorption operations. { 'tel·ə,rīt }

telltale |ENG| A marker on the outside of a tank that indicates on an exterior scale the amount of fluid inside the tank. { 'tel,tāl }

telltale float |CIV ENG| A water-level indicator in a reservoir. { 'tel,tāl |flōt }

tellurometer |ENG| A microwave instrument used in surveying to measure distance; the time for a radio wave to travel from one observation point to the other and return is measured and converted into distance by phase comparison, much as in radar. { ,tel·yə'räm·əd·ər }

telpher |MECH ENG| An electric hoist hanging from and driven by a wheeled cab rolling on a single overhead rail or a rope. { 'tel·fər }

Telsmith breaker |MECH ENG| A type of gyratory crusher, often used for primary crushing; consists of a spindle mounted in a long eccentric sleeve which rotates to impart a gyratory motion to the crushing head, but gives a parallel stroke, that is, the axis of the spindle describes a cylinder rather than a cone, as in the suspended spindle gyratory. { 'tel,smith ,brā·kər }

TEMA standard |CHEM ENG| Shell-and-tube heat-exchange standard designed to supplement the American Society of Mechanical Engineers code for unfired pressure vessels. { 'tē·mə ,stan·dərd }

temper |ENG| **1.** To moisten and mix clay, plaster, or mortar to the proper consistency for use. **2.** See anneal. { 'tem·pər }

temperature |THERMO| A property of an object which determines the direction of heat flow when the object is placed in thermal contact with another object; heat flows from a region of higher temperature to one of lower temperature; it is measured either by an empirical temperature scale, based on some convenient property of a material or instrument, or by a scale of absolute temperature, for example, the Kelvin scale. { 'tem·prə·chər }

temperature-actuated pressure relief valve |MECH ENG| A pressure relief valve which operates when subjected to increased external or internal temperature. { 'tem·prə·chər |ak·chə,wād·əd 'presh·ər ri|lēf ,valv }

temperature bath |THERMO| A relatively large volume of a homogeneous substance held at constant temperature, so that an object placed in thermal contact with it is maintained at the same temperature. { 'tem·prə·chər ,bath }

temperature-chlorinity-depth recorder |ENG| An instrument in which an underwater unit suspended from a cable records temperature, chlorinity, and depth sequentially on a single-pen strip recorder, each quantity being recorded for several seconds at a time. { 'tem·prə·chər klō'rin·əd·ē 'depth ri,kórd·ər }

temperature color scale |THERMO| The relation between an incandescent substance's temperature and the color of the light it emits. { 'tem·prə·chər 'kəl·ər ,skāl }

temperature-compensated Zener diode |ELECTR| Positive-temperature-coefficient reversed-bias Zener diode (*pn* junction) connected in series with one or more negative-temperature forward-biased diodes within a single package. { 'tem·prə·chər |käm·pən,sād·əd 'zē·nər 'dī,ōd }

temperature compensation |ELECTR| The process of making some characteristic of a circuit or device independent of changes in ambient temperature. { 'tem·prə·chər ,käm·pən,sā·shən }

temperature control |ENG| A control used to maintain the temperature of an oven, furnace, or other enclosed space within desired limits. { 'tem·prə·chər kən,trōl }

temperature error |ENG| That instrument error due to nonstandard temperature of the instrument. { 'tem·prə·chər ,er·ər }

temperature gradient |THERMO| For a given point, a vector whose direction is perpendicular to an isothermal surface at the point, and whose magnitude equals the rate of change of temperature in this direction. { 'tem·prə·chər ,grād·ē·ənt }

temperature profile recorder |ENG| A portable instrument for measuring temperature as a function of depth in shallow water, particularly in lakes, in which a thermistor element transmits data over an electrical cable to a recording drum and depth is measured by the amount of wire paid out. { 'tem·prə·chər ¦prō¸fīl ri¸kórd·ər }

temperature scale [THERMO] An assignment of numbers to temperatures in a continuous manner, such that the resulting function is single valued; it is either an empirical temperature scale, based on some convenient property of a substance or object, or it measures the absolute temperature. { 'tem·prə·chər ¸skāl }

temperature sensor [ENG] A device designed to respond to temperature stimulation. { 'tem·prə·chər ¸sen·sər }

temperature transducer [ENG] A device in an automatic temperature-control system that converts the temperature into some other quantity such as mechanical movement, pressure, or electric voltage; this signal is processed in a controller, and is applied to an actuator which controls the heat of the system. { 'tem·prə·chər tranz ¸dü·sər }

tempering air [ENG] Low-temperature air added to a heated airstream to regulate the stream temperature. { 'tem·pə·riŋ ¸er }

template [ENG] **1.** A two-dimensional representation of a machine or other equipment used for building layout design. **2.** A guide or a pattern used in manufacturing items. Also spelled templet. { 'tem·plət }

temporal decomposition [CONT SYS] The partitioning of the control or decision-making problem associated with a large-scale control system into subproblems based on the different time scales relevant to the associated action functions. { 'tem·prəl ¸dē¸käm·pə'zish·ən }

temporary structures [CIV ENG] Structures used to facilitate the construction of buildings, bridges, tunnels, and other above- and belowground facilities by providing access, support, and protection for the facility as well as assuring the safety of the workers and the public. { ¦tem·pə¸rer·ē 'strək·chərz }

Ten Broecke chart [THERMO] A graphical plot of heat transfer and temperature differences used to calculate the thermal efficiency of a countercurrent cool-fluid-warm-fluid heat-exchange system. { 'ten ¸brü·kə ¸chärt }

tender [MECH ENG] A vehicle that is attached to a locomotive and carries supplies of fuel and water. { 'ten·dər }

tendon [CIV ENG] A steel bar or wire that is tensioned, anchored to formed concrete, and allowed to regain its initial length to induce compressive stress in the concrete before use. { 'ten·dən }

tenon [ENG] A tonguelike projection from the end of a framing member which is made to fit into a mortise. { 'ten·ən }

tenon saw [ENG] A precision saw that has a metal strip for stiffening along its back. { 'ten·ən ¸só }

tensile bar [ENG] A molded, cast, or machined specimen of specified cross-sectional dimensions used to determine the tensile properties of a material by use of a calibrated pull test. Also known as tensile specimen; test specimen. { 'ten·səl ¸bär }

tensile modulus [MECH] The tangent or secant modulus of elasticity of a material in tension. { 'ten·səl ¸mäj·ə·ləs }

tensile specimen See tensile bar. { 'ten·səl ¸spes·ə·mən }

tensile strength [MECH] The maximum stress a material subjected to a stretching load can withstand without tearing. Also known as hot strength. { 'ten·səl ¸streŋkth }

tensile stress [MECH] Stress developed by a material bearing a tensile load. { 'ten·səl ¸stres }

tensile test [ENG] A test in which a specimen is subjected to increasing longitudinal pulling stress until fracture occurs. { 'ten·səl ¸test }

tensimeter [ENG] A device for measuring differences in the vapor pressures of two liquids in which the liquids are placed in sealed, evacuated bulbs connected by a differential manometer. { ten'sim·əd·ər }

tensiometry [ENG] A discipline concerned with the measurement of tension or tensile strength. { ¸ten·sē'äm·ə·trē }

tension [MECH] **1.** The condition of a string, wire, or rod that is stretched between two points. **2.** The force exerted by the stretched object on a support. [MECH ENG] A device on a textile manufacturing machine or a sewing machine that regulates the tautness and the movement of the thread or the fabric. Also known as tension device. { 'ten·chən }

tension device See tension. { 'ten·chən di¸vīs }

tension member [CIV ENG] A structural member subject to tensile stress. { 'ten·chən ¸mem·bər }

tension pulley [MECH ENG] A pulley around which an endless rope passes mounted on a trolley or other movable bearing so that the slack of the rope can be readily taken up by the pull of the weights. { 'ten·chən ¸púl·ē }

tension rod [DES ENG] A rod held in place by tension devices at the ends, such as a rod for a clothes closet. [ENG] A rod in a truss or other structure that connects opposite parts in order to prevent their spreading. { 'ten·chən ¸räd }

tensometer [ENG] A portable machine that is used to measure the tensile strength and other mechanical properties of materials. { ten'säm·əd·ər }

tenthmeter See angstrom. { 'tenth¸mēd·ər }

terahertz technology [ENG] The generation, detection, and application (such as in communications and imaging) of electromagnetic radiation roughly in the frequency range from 0.05 to 20 terahertz, corresponding to wavelengths from 6 millimeters down to 15 micrometers. { ¸ter·ə¸harts tek'näl·ə·jē }

teraohmmeter [ENG] An ohmmeter having a teraohm range for measuring extremely high insulation resistance values. { ¦ter·ə'ōm¸mēd·ər }

terminal [ELEC] **1.** A screw, soldering lug, or other point to which electric connections can be made. Also known as electric terminal. **2.** The equipment at the end of a microwave relay system or other communication channel. **3.** One of the electric input or output points of a circuit or component. { 'tər·mən·əl }

terminal area |ELECTR| The enlarged portion of conductor material surrounding a hole for a lead on a printed circuit. Also known as land; pad. { 'tər·mən·əl |er·ē·ə }

terminal clearance capacity |ENG| The amount of cargo or personnel that can be moved through and out of a terminal on a daily basis. { 'tər·mən·əl 'klir·əns kə,pas·əd·ē }

terminal operations |ENG| The reception, processing, and staging of passengers; the receipt, transit storage, and marshaling of cargo; the loading and unloading of ships or aircraft; and the manifesting and forwarding of cargo and passengers to destination. { 'tər·mən·əl ,äp·ə'rā·shənz }

terminal pressure |ENG| A pressure drop across a unit when the maximum allowable pressure drop is reached, as for a filter press. { 'tər·mən·əl |presh·ər }

terminal throw velocity |ENG| The velocity at which a stream of air exiting a diffuser impinges on an object or surface. { |tər·mən·əl 'thrō və,läs·əd·ē }

terminal unit |MECH ENG| In an air-conditioning system, a unit at the end of a branch duct through which air is transferred or delivered to the conditioned space. { 'tər·mən·əl ,yü·nət }

terminating |ELEC| Closing of the circuit at either end of a line or transducer by connecting some device thereto; terminating does not imply any special condition such as the elimination of reflection. { 'tər·mə,nād·iŋ }

termite shield |BUILD| A strip of metal, usually galvanized iron, bent down at the edges and placed between the foundation of a house and a timber floor, around pipes, and other places where termites can pass. { 'tər,mīt ,shēld }

terrace |BUILD| **1.** A flat roof. **2.** A colonnaded promenade. **3.** An open platform extending from a building, usually at ground level. { 'ter·əs }

terrain-clearance indicator See absolute altimeter. { tə'rān |klir·əns ,in·də,kād·ər }

terrain profile recorder See airborne profile recorder. { tə'rān |prō,fīl ri,kȯrd·ər }

terrain sensing |ENG| The gathering and recording of information about terrain surfaces without actual contact with the object or area being investigated; in particular, the use of photography, radar, and infrared sensing in airplanes and artificial satellites. { tə'rān ,sens·iŋ }

tertiary air |MECH ENG| Combustion air added to primary and secondary air. { 'tər,shē,er·ē 'er }

tertiary sewage treatment |CIV ENG| A process for purification of wastewater in which nitrates and phosphates, as well as fine particles, are removed; the process follows removal of raw sludge and biological treatment. Also known

as advanced sewage treatment. { |tər·shē,er·ē |sü·ij ,trēt·mənt }

test |IND ENG| A procedure in which the performance of a product is measured under various conditions. { test }

testboard |ELEC| Switchboard equipped with testing apparatus, arranged so that connections can be made from it to telephone lines or central-office equipment for testing purposes. { 'test,bȯrd }

test chamber |ENG| A place, section, or room having special characteristics where a person or object is subjected to experimental procedures, as an altitude chamber. { 'test ,chām·bər }

test oscillator See signal generator. { 'test ,äs·ə,lād·ər }

test pile |CIV ENG| A pile equipped with a platform on which a load of sand or pig iron is placed in order to determine the load a pile can support (usually twice the working load) without settling. { 'test ,pīl }

test pit |CIV ENG| An open excavation used to obtain soil samples in foundation studies. { 'test ,pit }

test point |ELEC| A terminal or plug-in connector provided in a circuit to facilitate monitoring, calibration, or trouble-shooting. { 'test ,pȯint }

test specimen See tensile bar. { 'test ,spes·ə·mən }

tetrode junction transistor See double-base junction transistor. { 'te,trōd 'jəŋk·shən tran,zis·tər }

tetrode transistor |ELECTR| A four-electrode transistor, such as a tetrode point-contact transistor or double-base junction transistor. { 'te,trōd tran'zis·tər }

Texas tower |ENG| A radar tower built in the sea offshore, to serve as part of an early-warning radar network. { 'tek·səs 'tau·ər }

text-to-speech synthesizer |ENG ACOUS| A voice response system that provides an automatic means to take a specification of any English text at the input and generate a natural and intelligible acoustic speech signal at the output by using complex sets of rules for predicting the needed phonemic states directly from the input message and dictionary pronunciations. { |tekst tə |spēch 'sin·thə,sīz·ər }

th See thermie.

thaw house |ENG| A small building that is designed for thawing frozen dynamite and which is capacious enough for a supply of thawed dynamite for a day's work. { 'thȯ ,haus }

thawing |ENG| Warming dynamite, to reduce risk of premature explosion. { 'thȯ·iŋ }

theoretical air |ENG| The amount of air that is theoretically required for complete combustion. { ,thē·ə'red·ə·kəl 'er }

theoretical cutoff frequency |ELEC| Of an electric structure, a frequency at which, disregarding the effects of dissipation, the attenuation constant changes from zero to a positive value or vice versa. { ,thē·ə'red·ə·kəl 'kəd,ȯf ,frē·kwən·sē }

theoretical plate |CHEM ENG| A distillation column plate or tray that produces perfect distillation (that is, produces the same difference in composition as that existing between a liquid mixture and the vapor in equilibrium with it); the packed-column equivalent of a theoretical plate is the HETP, or height (of packing) equivalent to a theoretical plate. { ,thē·ə'red·ə·kəl 'plāt }

theoretical relieving capacity |MECH ENG| The capacity of a theoretically perfect nozzle calculated in volumetric or gravimetric units. { ,thē·ə'red·ə·kəl ri'lēv·iŋ kə,pas·əd·ē }

Therberg system |IND ENG| A system of categorizing hand movements that is used in the standard motion-and-time analysis technique. { 'thər,bərg ,sis·təm }

therblig See elemental motion. { 'thər,blig }

therblig chart |IND ENG| An operation chart with the suboperations divided into basic motions, all designated with appropriate symbols. { 'thər,blig ,chärt }

therm |THERMO| A unit of heat energy, equal to 100,000 international table British thermal units, or approximately 1.055×10^8 joules. { thərm }

thermactor See air-injection system. { 'thər,mak·tər }

thermal |THERMO| Of or concerning heat. { 'thər·məl }

thermal ammeter See hot-wire ammeter. { 'thər·məl 'am,ēd·ər }

thermal-arrest calorimeter |ENG| A vacuum device for measurement of heats of fusion; a sample is frozen under vacuum and allowed to melt as the calorimeter warms to room temperature. { 'thər·məl ə¦rest ,kal·ə'rim·əd·ər }

thermal barrier See thermal break. { 'thər·məl 'bar·ē·ər }

thermal break |BUILD| A component that is a poor conductor of heat and is placed in an assembly containing highly conducting materials in order to reduce or prevent the flow of heat. Also known as thermal barrier. { ¦thər·məl ¦brāk }

thermal bulb |ENG| A device for measurement of temperature; the liquid in a bulb expands with increasing temperature, pressuring a spiral Bourdon-type tube element and causing it to deform (unwind) in direct relation to the temperature in the bulb. { 'thər·məl ¦bəlb }

thermal capacitance |THERMO| The ratio of the entropy added to a body to the resulting rise in temperature. { 'thər·məl kə'pas·əd·əns }

thermal capacity See heat capacity. { 'thər·məl kə'pas·əd·ē }

thermal compressor |MECH ENG| A steam-jet ejector designed to compress steam at pressures above atmospheric. { 'thər·məl kəm'pres·ər }

thermal conductance |THERMO| The amount of heat transmitted by a material divided by the difference in temperature of the surfaces of the material. Also known as conductance. { 'thər·məl kən'dək·təns }

thermal conductimetry |THERMO| Measurement of thermal conductivities. { 'thər·məl ,kän,dək'tim·ə·trē }

thermal conductivity |THERMO| The heat flow across a surface per unit area per unit time, divided by the negative of the rate of change of temperature with distance in a direction perpendicular to the surface. Also known as coefficient of conductivity; heat conductivity. { 'thər·məl ,kan,dək'tiv·əd·ē }

thermal conductivity cell See katharometer. { 'thər·məl ,kän,dək'tiv·əd·ē ,sel }

thermal conductivity gage |ENG| A pressure measurement device for high-vacuum systems; an electrically heated wire is exposed to the gas under pressure, the thermal conductivity of which changes with changes in the system pressure. { 'thər·məl ,kän,dək'tiv·əd·ē ,gāj }

thermal conductor |THERMO| A substance with a relatively high thermal conductivity. { 'thər·məl kən'dək·tər }

thermal convection See heat convection. { 'thər·məl kən'vek·shən }

thermal converter |ELECTR| A device that converts heat energy directly into electric energy by using the Seebeck effect; it is composed of at least two dissimilar materials, one junction of which is in contact with a heat source and the other junction of which is in contact with a heat sink. Also known as thermocouple converter; thermoelectric generator; thermoelectric power generator; thermoelement. |ENG| An instrument used with external resistors for ac current and voltage measurements over wide ranges, consisting of a conductor heated by an electric current, with one or more hot junctions of a thermocouple attached to it, so that the output emf responds to the temperature rise, and hence the current. { 'thər·məl kən'vərd·ər }

thermal coulomb |THERMO| A unit of entropy equal to 1 joule per kelvin. { 'thər·məl 'kü,läm }

thermal cracking |CHEM ENG| A petroleum refining process that decomposes, rearranges, or combines hydrocarbon molecules by the application of heat, without the aid of catalysts. { 'thər·məl 'krak·iŋ }

thermal detector See bolometer. { 'thər·məl di 'tek·tər }

thermal diffusivity See diffusivity. { 'thər·məl ,di·fyü·siv·əd·ē }

thermal drift |ELECTR| Drift caused by internal heating of equipment during normal operation or by changes in external ambient temperature. { 'thər·məl 'drift }

thermal drilling |MECH ENG| A machining method in which holes are drilled in a workpiece by heat generated from the friction of a rotating tool. { ¦thər·məl ¦dril·iŋ }

thermal efficiency |CHEM ENG| In a tube-and-shell heat-exchange system, the ratio of the actual temperature range of the tube-side fluid (inlet versus outlet temperature) to the maximum possible temperature range. See efficiency. { 'thər·məl i'fish·ən·sē }

thermal effusion See thermal transpiration. { 'thər·məl e'fyü·zhən }

thermal emissivity See emissivity. { 'thər·məl ‚ē·mi'siv·əd·ē }

thermal environment [IND ENG] Those aspects of the workplace that include local temperature, humidity, and air velocity as well as the presence of radiating surfaces. { 'thərm·əl in'vī·rən·mənt }

thermal equilibrium [THERMO] Property of a system all parts of which have attained a uniform temperature which is the same as that of the system's surroundings. { 'thər·məl ‚ē·kwə'lib·rē·əm }

thermal farad [THERMO] A unit of thermal capacitance equal to the thermal capacitance of a body for which an increase in entropy of 1 joule per kelvin results in a temperature rise of 1 kelvin. { 'thər·məl 'far‚ad }

thermal flame safeguard [MECH ENG] A thermocouple located in the pilot flame of a burner; if the pilot flame is extinguished, an elective circuit is interrupted and the fuel supply is shut off. { 'thər·məl 'flām ‚saf‚gärd }

thermal flux See heat flux. { 'thər·məl 'fləks }

thermal henry [THERMO] A unit of thermal inductance equal to the product of a temperature difference of 1 kelvin and a time of 1 second divided by a rate of flow of entropy of 1 watt per kelvin. { 'thər·məl 'hen·rē }

thermal hysteresis [THERMO] A phenomenon sometimes observed in the behavior of a temperature-dependent property of a body; it is said to occur if the behavior of such a property is different when the body is heated through a given temperature range from when it is cooled through the same temperature range. { 'thər·məl ‚his·tə'rē·səs }

thermal inductance [THERMO] The product of temperature difference and time divided by entropy flow. { 'thər·məl in'dək·təns }

thermal instrument [ENG] An instrument that depends on the heating effect of an electric current, such as a thermocouple or hot-wire instrument. { 'thər·məl 'in·strə·mənt }

thermal-liquid system [CHEM ENG] A system with a special liquid that acts as a heat sink or heat source (for example, steam, hot water, mercury, Dowtherm, molten salts, or mineral oils); used for process heating and cooling. { 'thər·məl ‚lik·wəd ‚sis·təm }

thermal-loss meter See heat-loss flowmeter. { 'thər·məl ‚los ‚mēd·ər }

thermal mapper See line scanner. { 'thər·məl 'map·ər }

thermal microphone [ENG ACOUS] Microphone depending for its action on the variation in the resistance of an electrically heated conductor that is being alternately increased and decreased in temperature by sound waves. { 'thər·məl 'mī·krə‚fōn }

thermal neutron analysis [ENG] A technique for detecting explosives, in which the object under inspection is conveyed through a cloud of thermal neutrons (generated by slowing down fast neutrons in multiple collisions in a moderator surrounding the source), and the characteristic high-energy gamma rays that are then emitted by the objects are used in analysis and imaging. { ‚thər·məl 'nü‚trän ə‚nal·ə·səs }

thermal ohm [THERMO] A unit of thermal resistance equal to the thermal resistance for which a temperature difference of 1 kelvin produces a flow of entropy of 1 watt per kelvin. Also known as fourier. { 'thər·məl 'ōm }

thermal polymerization [CHEM ENG] A thermal, petroleum refining process used to convert light hydrocarbon gases into liquid fuels; paraffinic hydrocarbons are cracked to produce olefinic material which is concurrently polymerized by heat and pressure to form liquids, the product being known as polymer gasoline. { 'thər·məl pə‚lim·ə·rə'zā·shən }

thermal potential difference [THERMO] The difference between the thermodynamic temperatures of two points. { 'thər·məl pə‚ten·chəl 'dif·rəns }

thermal power plant [ENG] A facility to produce electric energy from thermal energy released by combustion of a fuel or consumption of a fissionable material. { 'thər·məl 'paù·ər ‚plant }

thermal probe [ENG] An instrument which measures the heat flow from ocean bottom sediment. [MECH ENG] A calorimeter in a boiler furnace which measures heat absorption rates. { 'thər·məl 'prōb }

thermal process [CHEM ENG] Any process that utilizes heat, without the aid of a catalyst, to accomplish chemical change; for example, thermal cracking, thermal reforming, or thermal polymerization. { 'thər·məl 'prä·səs }

thermal radiation See heat radiation. { 'thər·məl ‚rād·ē'ā·shən }

thermal reactor [CHEM ENG] A device, system, or vessel in which chemical reactions take place because of heat (no catalysis); for example, thermal cracking, thermal reforming, or thermal polymerization. { 'thər·məl rē'ak·tər }

thermal reforming [CHEM ENG] A petroleum refining process using heat (but no catalyst) to effect molecular rearrangement of a low-octane naphtha to form high-octane motor gasoline. { 'thər·məl ri'förm·iŋ }

thermal relief [ENG] A valve or other device that is preset to open when pressure becomes excessive due to increased temperature of the system. { 'thər·məl ri'lēf }

thermal resistance [ELECTR] See effective thermal resistance. [THERMO] A measure of a body's ability to prevent heat from flowing through it, equal to the difference between the temperatures of opposite faces of the body divided by the rate of heat flow. Also known as heat resistance. { 'thər·məl ri'zis·təns }

thermal resistivity [THERMO] The reciprocal of the thermal conductivity. { 'thər·məl rē‚zis'tiv·əd·ē }

thermal shock [MECH] Stress produced in a body or in a material as a result of undergoing

561

a sudden change in temperature. { 'thǝr·mǝl 'shäk }

thermal soakback |ENG| A phenomenon whereby, due to the lag in propagation of temperature changes through insulating materials, the maximum temperature of a thermally protected structure may be reached a certain time after the protective coating has reached its maximum temperature. { ,thǝr·mǝl 'sōk,bak }

thermal stress |MECH| Mechanical stress induced in a body when some or all of its parts are not free to expand or contract in response to changes in temperature. { 'thǝr·mǝl 'stres }

thermal stress cracking |MECH| Crazing or cracking of materials (plastics or metals) by overexposure to elevated temperatures and sudden temperature changes or large temperature differentials. { 'thǝr·mǝl ¦stres 'krak·iŋ }

thermal telephone receiver |ENG ACOUS| A thermophone used as a telephone receiver. { 'thǝr·mǝl 'tel·ǝ,fōn ri,sē·vǝr }

thermal transducer |ENG| Any device which converts energy from some form other than heat energy into heat energy; an example is the absorbing film used in the thermal pulse method. { 'thǝr·mǝl tranz'dü·sǝr }

thermal transpiration |THERMO| The formation of a pressure gradient in gas inside a tube when there is a temperature gradient in the gas and when the mean free path of molecules in the gas is a significant fraction of the tube diameter. Also known as thermal effusion. { 'thǝr·mǝl ,tranz·pǝ'rā·shǝn }

thermal value |THERMO| Heat produced by combustion, usually expressed in calories per gram or British thermal units per pound. { 'thǝr·mǝl ,val·yü }

thermal valve |MECH ENG| A valve controlled by an element made of material that exhibits a significant change in properties in response to a change in temperature. { 'thǝr·mǝl 'valv }

thermal volt See kelvin. { 'thǝr·mǝl 'vōlt }

thermal wattmeter |ENG| A wattmeter in which thermocouples are used to measure the heating produced when a current is passed through a resistance. { 'thǝr·mǝl 'wät,mēd·ǝr }

thermic boring |ENG| Boring holes into concrete by means of a high temperature, produced by a steel lance packed with steel wool which is ignited and kept burning by oxyacetylene or other gas. { 'thǝr·mik 'bòr·iŋ }

thermie |THERMO| A unit of heat energy equal to the heat energy needed to raise 1 tonne of water from 14.5°C to 15.5°C at a constant pressure of 1 standard atmosphere; equal to 10^6 fifteen-degrees calories or $(4.1855 \pm 0.0005) \times 10^6$ joules. Abbreviated th. { 'thǝr·mē }

thermion |ELECTR| A charged particle, either negative or positive, emitted by a heated body, as by the hot cathode of a thermionic tube. { ¦thǝrm'Ī,än }

thermionic |ELECTR| Pertaining to the emission of electrons as a result of heat. { ,thǝr·mē'än·ik }

thermionic emission |ELECTR| **1.** The outflow of electrons into vacuum from a heated electric conductor. Also known as Edison effect; Richardson effect. **2.** More broadly, the liberation of electrons or ions from a substance as a result of heat. { ,thǝr·mē'än·ik i'mish·ǝn }

thermistor |ELECTR| A resistive circuit component, having a high negative temperature coefficient of resistance, so that its resistance decreases as the temperature increases; it is a stable, compact, and rugged two-terminal ceramiclike semiconductor bead, rod, or disk. Derived from thermal resistor. { thǝr'mis·tǝr }

thermoacoustic engine |ENG| A heat engine that harnesses the combination of the pressure oscillations of a sound wave with the accompanying adiabatic temperature oscillations. { ¦thǝr·mō·ǝ¦kü·stik 'en·jǝn }

thermoacoustic refrigerator |ENG| A device that uses acoustic power to pump heat from a region of low temperature to a region of ambient temperature. { ,thǝr·mō·ǝ,kü·stik ri'frij·ǝ,rād·ǝr }

thermoacoustic-Stirling engine |ENG| A device in which the thermodynamic cycle of a Stirling engine is accomplished in a traveling-wave acoustic network, and acoustic power is produced from heat. { ,thǝr·mō·ǝ,kü·stik¦stǝr·liŋ 'en·jǝn }

thermoammeter |ENG| An ammeter that is actuated by the voltage generated in a thermocouple through which is sent the current to be measured; used chiefly for measuring radio-frequency currents. Also known as electrothermal ammeter; thermocouple ammeter. { ¦thǝr·mō'am ,ēd·ǝr }

thermochemical calorie See calorie. { ¦thǝr·mō 'kem·ǝ·kǝl 'kal·ǝ·rē }

thermocompression bonding |ENG| Use of a combination of heat and pressure to make connections, as when attaching beads to integrated-circuit chips; examples include wedge bonding and ball bonding. { ¦thǝr·mō·kǝm'presh·ǝn 'bänd·iŋ }

thermocompression evaporator |MECH ENG| A system to reduce the energy requirements for evaporation by compressing the vapor from a single-effect evaporator so that the vapor can be used as the heating medium in the same evaporator. { ¦thǝr·mō·kǝm'presh·ǝn i'vap·ǝ ,rād·ǝr }

thermocouple |ENG| A device consisting basically of two dissimilar conductors joined together at their ends; the thermoelectric voltage developed between the two junctions is proportional to the temperature difference between the junctions, so the device can be used to measure the temperature of one of the junctions when the other is held at a fixed, known temperature, or to convert radiant energy into electric energy. { 'thǝr·mǝ,kǝp·ǝl }

thermocouple ammeter See thermoammeter. { 'thǝr·mǝ,kǝp·ǝl 'am,ēd·ǝr }

thermocouple pyrometer See thermoelectric pyrometer. { 'thǝr·mǝ,kǝp·ǝl pī'räm·ǝd·ǝr }

thermocouple vacuum gage |ENG| A vacuum

gage that depends for its operation on the thermal conduction of the gas present; pressure is measured as a function of the voltage of a thermocouple whose measuring junction is in thermal contact with a heater that carries a constant current; ordinarily, used over a pressure range of 10^{-1} to 10^{-3} millimeter of mercury. { 'thər·mə₁kəp·əl 'vak·yəm ₁gāj }

thermodynamic cycle |THERMO| A procedure or arrangement in which some material goes through a cyclic process and one form of energy, such as heat at an elevated temperature from combustion of a fuel, is in part converted to another form, such as mechanical energy of a shaft, the remainder being rejected to a lower temperature sink. Also known as heat cycle. { ¦thər·mō·dī'nam·ik 'sī·kəl }

thermodynamic efficiency |IND ENG| An index for rating the effort required by a worker performing a task in terms of the ratio of work performed to the energy consumed. { ¦thər·mō·dī'nam·ik i'fish·ən·sē }

thermodynamic equation of state |THERMO| An equation that relates the reversible change in energy of a thermodynamic system to the pressure, volume, and temperature. { ¦thər·mō·dī'nam·ik i'kwā·zhən əv 'stāt }

thermodynamic equilibrium |THERMO| Property of a system which is in mechanical, chemical, and thermal equilibrium. { ¦thər·mō·dī'nam·ik ₁ē·kwə'lib·rē·əm }

thermodynamic function of state |THERMO| Any of the quantities defining the thermodynamic state of a substance in thermodynamic equilibrium; for a perfect gas, the pressure, temperature, and density are the fundamental thermodynamic variables, any two of which are, by the equation of state, sufficient to specify the state. Also known as state parameter; state variable; thermodynamic variable. { ¦thər·mō·dī'nam·ik 'fəŋk·shən əv 'stāt }

thermodynamic potential |THERMO| One of several extensive quantities which are determined by the instantaneous state of a thermodynamic system, independent of its previous history, and which are at a minimum when the system is in thermodynamic equilibrium under specified conditions. { ¦thər·mō·dī'nam·ik pə'ten·chəl }

thermodynamic potential at constant volume See free energy. { ¦thər·mō·dī'nam·ik pe¦ten·chəl at 'kän·stənt 'väl·yəm }

thermodynamic principles |THERMO| Laws governing the conversion of energy from one form to another. { ¦thər·mō·dī'nam·ik 'prin·sə·pəlz }

thermodynamic probability |THERMO| Under specified conditions, the number of equally likely states in which a substance may exist; the thermodynamic probability Ω is related to the entropy S by $S = k \ln \Omega$, where k is Boltzmann's constant. { ¦thər·mō·dī'nam·ik ₁präb·ə'bil·ad·ē }

thermodynamic process |THERMO| A change of any property of an aggregation of matter and energy, accompanied by thermal effects. { ¦thər·mō·dī'nam·ik 'prä·səs }

thermodynamic property |THERMO| A quantity which is either an attribute of an entire system or is a function of position which is continuous and does not vary rapidly over microscopic distances, except possibly for abrupt changes at boundaries between phases of the system; examples are temperature, pressure, volume, concentration, surface tension, and viscosity. Also known as macroscopic property. { ¦thər·mō·dī'nam·ik 'präp·ərd·ē }

thermodynamic system |THERMO| A part of the physical world as described by its thermodynamic properties. { ¦thər·mō·dī'nam·ik 'sis·təm }

thermodynamic temperature scale |THERMO| Any temperature scale in which the ratio of the temperatures of two reservoirs is equal to the ratio of the amount of heat absorbed from one of them by a heat engine operating in a Carnot cycle to the amount of heat rejected by this engine to the other reservoir; the Kelvin scale and the Rankine scale are examples of this type. { ¦thər·mō·dī'nam·ik 'tem·prə·chər ₁skāl }

thermodynamic variable See thermodynamic function of state. { ¦thər·mō·dī'nam·ik 'ver·ē·ə·bəl }

thermoelectric converter |ELECTR| A converter that changes solar or other heat energy to electric energy; used as a power source on spacecraft. { ¦thər·mō·i'lek·trik kən'vərd·ər }

thermoelectric cooler |ENG| An electronic heat pump based on the Peltier effect, involving the absorption of heat when current is sent through a junction of two dissimilar metals; it can be mounted within the housing of a device to prevent overheating or to maintain a constant temperature. { ¦thər·mō·i'lek·trik 'kü·lər }

thermoelectric cooling |ENG| Cooling of a chamber based on the Peltier effect; an electric current is sent through a thermocouple whose cold junction is thermally coupled to the cooled chamber, while the hot junction dissipates heat to the surroundings. Also known as thermoelectric refrigeration. { ¦thər·mō·i'lek·trik 'kül·iŋ }

thermoelectric generator See thermal converter. { ¦thər·mō·i'lek·trik 'jen·ə₁rād·ər }

thermoelectric heating |ENG| Heating based on the Peltier effect, involving a device which is in principle the same as that used in thermoelectric cooling except that the current is reversed. { ¦thər·mō·i'lek·trik 'hēd·iŋ }

thermoelectric junction See thermojunction. { ¦thər·mō·i'lek·trik 'jəŋk·shən }

thermoelectric laws |ENG| Basic relationships used in the design and application of thermocouples for temperature measurement; for example, the law of the homogeneous circuit, the law of intermediate metals, and the law of successive or intermediate temperatures. { ¦thər·mō·i'lek·trik 'lōz }

thermoelectric material |ELECTR| A material that can be used to convert thermal energy into electric energy or provide refrigeration directly from electric energy; good thermoelectric materials include lead telluride, germanium telluride, bismuth telluride, and cesium sulfide. { ¦thər·mō·i'lek·trik mə'tir·ē·əl }

thermoelectric pyrometer [ENG] An instrument which uses one or more thermocouples to measure high temperatures, usually in the range between 800 and 2400°F (425 and 1315°C). Also known as thermocouple pyrometer. { ¦thər·mō·i'lek·trik pī'räm·əd·ər }

thermoelectric refrigeration See thermoelectric cooling. { ¦thər·mō·i'lek·trik ri,frij·ə'rā·shən }

thermoelectric thermometer [ENG] A type of electrical thermometer consisting of two thermocouples which are series-connected with a potentiometer and a constant-temperature bath; one couple, called the reference junction, is placed in a constant-temperature bath, while the other is used as the measuring junction. { ¦thər·mō·i'lek·trik thər'mäm·əd·ər }

thermoelectromotive force [ELEC] Voltage developed due to differences in temperature between parts of a circuit containing two or more different metals. { ¦thər·mō·i¦lek·trə¦mōd·iv 'fórs }

thermoforming [ENG] Forming of thermoplastic sheet by heating it and then pulling it down onto a mold surface to shape it. { 'thər·mə,fórm·iŋ }

thermogalvanometer [ENG] Instrument for measuring small high-frequency currents by their heating effect, generally consisting of a direct-current galvanometer connected to a thermocouple that is heated by a filament carrying the current to be measured. { ¦thər·mō·gal·və'näm·əd·ər }

thermograd probe [ENG] An instrument that makes a record of temperature versus depth as it is lowered to the ocean floor, and measures heat flow through the ocean floor. { 'thər·mə,grad 'prōb }

thermogram [ENG] The recording made by a thermograph. { 'thər·mə,gram }

thermograph [ENG] An instrument that senses, measures, and records the temperature of the atmosphere. Also known as recording thermometer. { 'thər·mə,graf }

thermograph correction card [ENG] A table for quick and accurate correction of the reading of a thermograph to that of the more accurate dry-bulb thermometer at the same time and place. { 'thər·mə,graf kə'rek·shən ,kärd }

thermography [ENG] A method of measuring surface temperature by using luminescent materials: the two main types are contact thermography and projection thermography. { thər'mäg·rə·fē }

thermogravitational column [CHEM ENG] A device in which thermal diffusion results from the countercurrent flow of hot and cold material, thus increasing the separation of materials in a solution by the formation of a concentration gradient (difference). Also known as Clausius-Dickel column. { ¦thər·mō,grav·ə'tā·shən·əl 'käl·əm }

thermointegrator [ENG] An apparatus, used in studying soil temperatures, for measuring the total supply of heat during a given period; it consists of a long nickel coil (inserted into the soil by an attached rod) forming a 100-ohm resistance thermometer and a 6-volt battery, the current used being recorded on a galvanometer; a mercury thermometer can be used. { ¦thər·mō'int·ə,grād·ər }

thermojunction [ELECTR] One of the surfaces of contact between the two conductors of a thermocouple. Also known as thermoelectric junction. { ¦thər·mō'jəŋk·shən }

thermometer [ENG] An instrument that measures temperature. { thər'mäm·əd·ər }

thermometer anemometer [ENG] An anemometer consisting of two thermometers, one with an electric heating element connected to the bulb; the heated bulb cools in an airstream, and the difference in temperature as registered by the heated and unheated thermometers can be translated into air velocity by a conversion chart. { thər'mäm·əd·ər ,an·ə'mäm·əd·ər }

thermometer-bulb liquid-level meter [ENG] Detection of liquid level by temperature measurement changes using an immersed bulb-type thermometer. { thər'mäm·əd·ər ¦bəlb 'lik·wəd ¦lev·əl ,mēd·ər }

thermometer frame [ENG] A frame designed to hold two or more reversing thermometers; such a frame is often attached directly to a Nansen bottle. { thər'mäm·əd·ər ,frām }

thermometer screen See instrument shelter. { thər'mäm·əd·ər ,skrēn }

thermometer shelter See instrument shelter. { thər'mäm·əd·ər ,shel·tər }

thermometer support [ENG] A device used to hold liquid-in-glass maximum and minimum thermometers in the proper recording position inside an instrument shelter, and to permit them to be read and reset. { thər'mäm·əd·ər sə,pórt }

thermometric conductivity See diffusivity. { ¦thər·mə¦me·trik ,kän,dək,tiv·əd·ē }

thermometric fluid [THERMO] A fluid that has properties, such as a large and uniform thermal expansion coefficient, good thermal conductivity, and chemical stability, that make it suitable for use in a thermometer. { ,thər·mə¦me·trik 'flü·əd }

thermometric property [THERMO] A physical property that changes in a known way with temperature, and can therefore be used to measure temperature. { ¦thər·mə¦me·trik 'präp·ərd·ē }

thermometry [THERMO] The science and technology of measuring temperature, and the establishment of standards of temperature measurement. { thər'mäm·ə·trē }

thermomigration [ELECTR] A technique for doping semiconductors in which exact amounts of known impurities are made to migrate from the cool side of a wafer of pure semiconductor material to the hotter side when the wafer is heated in an oven. { ¦thər·mō·mī'grā·shən }

thermo-pervaporation See membrane distillation. { ¦thər·mō·pər,vap·ə'rā·shən }

thermophone |ENG ACOUS| An electroacoustic transducer in which sound waves having an accurately known strength are produced by the expansion and contraction of the air adjacent to a strip of conducting material, whose temperature varies in response to a current input that is the sum of a steady current and a sinusoidal current; used chiefly for calibrating microphones. { 'thǝr·mǝ,fōn }

thermophoresis |THERMO| The movement of particles in a thermal gradient from high to low temperatures. { ,thǝr·mǝ·fǝ'rē·sǝs }

thermopile |ENG| An array of thermocouples connected either in series to give higher voltage output or in parallel to give higher current output, used for measuring temperature or radiant energy or for converting radiant energy into electric power. { 'thǝr·mǝ,pīl }

thermoregulator |ENG| A high-accuracy or high-sensitivity thermostat; one type consists of a mercury-in-glass thermometer with sealed-in electrodes, in which the rising and falling column of mercury makes and breaks an electric circuit. { ¦thǝr·mō'reg·yǝ,lād·ǝr }

thermorelay See thermostat. { ¦thǝr·mō'rē,lā }

thermoscreen See instrument shelter. { 'thǝr·mǝ,skrēn }

thermosiphon |MECH ENG| A closed system of tubes connected to a water-cooled engine which permit natural circulation and cooling of the liquid by utilizing the difference in density of the hot and cool portions. { ¦thǝr·mō'sī·fǝn }

thermosiphon reboiler |CHEM ENG| A liquid reheater (as for distillation-column bottoms) in which natural circulation of the boiling liquid is obtained by maintaining a sufficient liquid head. { ¦thǝr·mō'sī·fǝn ¦rē'bȯi·lǝr }

thermostat |ENG| An instrument which measures changes in temperature and directly or indirectly controls sources of heating and cooling to maintain a desired temperature. Also known as thermorelay. { 'thǝr·mǝ,stat }

thermostatic switch |ELEC| A temperature-operated switch that receives its operating energy by thermal conduction or convection from the device being controlled or operated. { ¦thǝr·mǝ¦stad·ik 'swich }

thermoswitch See thermal switch. { 'thǝr·mǝ,swich }

thermovoltmeter |ENG| A voltmeter in which a current from the voltage source is passed through a resistor and a fine vacuum-enclosed platinum heater wire; a thermocouple, attached to the midpoint of the heater, generates a voltage of a few millivolts, and this voltage is measured by a direct-current millivoltmeter. { ¦thǝr·mō'vōlt,mēd·ǝr }

thetagram |THERMO| A thermodynamic diagram with coordinates of pressure and temperature, both on a linear scale. { 'thād·ǝ,gram }

thickener |ENG| A nonfilter device for the removal of liquid from a liquid-solids slurry to give a dewatered (thickened) solids product; can be by gravity settling or centrifugation. { 'thik·ǝ·nǝr }

thickening |CHEM ENG| The concentration of the solids in a suspension in order to recover a fraction with a higher concentration of solids than in the original suspension. { 'thik·ǝ·niŋ }

thick-film capacitor |ELEC| A capacitor in a thick-film circuit, made by successive screen-printing and firing processes. { 'thik ¦film kǝ'pas·ǝd·ǝr }

thick-film circuit |ELECTR| A microcircuit in which passive components, of a ceramic-metal composition, are formed on a ceramic substrate by successive screen-printing and firing processes, and discrete active elements are attached separately. { 'thik ¦film 'sǝr·kǝt }

thick-film hybrid |ELECTR| An assembly consisting of a thick-film circuit pattern with mounting positions for the insertion of conventional silicon devices. { ,thik ,film 'hī·brǝd }

thick-film resistor |ELEC| Fixed resistor whose resistance element is a film well over 0.001 inch (25 micrometers) thick. { 'thik ¦film ri'zis·tǝr }

thick-film sensor |ENG| A thick-film circuit that is fabricated from suitable materials to measure a physical quantity such as mechanical stress or temperature or to perform a chemical sensing application such as the measurement of gas or liquid composition, acidity, or humidity. { ,thik ,film 'sen·sǝr }

thickness gage |ENG| A gage for measuring the thickness of a sheet of material, the thickness of an object, or the thickness of a coating; examples include penetration-type and backscattering radioactive thickness gages and ultrasonic thickness gages. { 'thik·nǝs ,gāj }

Thiele coordinates |CHEM ENG| A graphical method for calculating the solvent-free composition of two components being separated by solvent extraction. { 'tēl·ǝ kō,ȯrd·ǝn·ǝts }

Thiele-Geddes method |CHEM ENG| A method for the prediction of the product distribution from a multicomponent distillation system. { 'tēl·ǝ 'ged·ǝs ,meth·ǝd }

thin film |ELECTR| A film a few molecules thick deposited on a glass, ceramic, or semiconductor substrate to form a capacitor, resistor, coil, cryotron, or other circuit component. { 'thin 'film }

thin-film capacitor |ELEC| A capacitor that can be constructed by evaporation of conductor and dielectric films in sequence on a substrate; silicon monoxide is generally used as the dielectric. { 'thin ¦film kǝ'pas·ǝd·ǝr }

thin-film circuit |ELECTR| A circuit in which the passive components and conductors are produced as films on a substrate by evaporation or sputtering; active components may be similarly produced or mounted separately. { 'thin ¦film 'sǝr·kǝt }

thin-film field-emitter cathode |ELECTR| A sharply pointed microminiature electron field emitter with an integral low-voltage extraction gate. { ¦thin ,film ¦fēld i,mid·ǝr 'kath,ōd }

thin-film integrated circuit |ELECTR| An integrated circuit consisting entirely of thin films deposited in a patterned relationship on a substrate. { 'thin ¦film 'int·ǝ,grād·ǝd 'sǝr·kǝt }

565

thin-film material |ELECTR| A material that can be deposited as a thin film in a desired pattern by a variety of chemical, mechanical, or high-vacuum evaporation techniques. { 'thin ¦film mə'tir·ē·əl }

thin-film resistor |ELEC| A fixed resistor whose resistance element is a metal, alloy, carbon, or other film having a thickness of about 0.000001 inch (25 nanometers). { 'thin ¦film ri'zis·tər }

thin-film semiconductor |ELECTR| Semiconductor produced by the deposition of an appropriate single-crystal layer on a suitable insulator. { 'thin ¦film 'sem·i·kən,dək·tər }

thin-film transistor |ELECTR| A field-effect transistor constructed entirely by thin-film techniques, for use in thin-film circuits. Abbreviated TFT. { 'thin ¦film tran'zis·tər }

thin-plate orifice |ENG| A thin-metal orifice sheet used in fluid-flow measurement in fluid conduits by means of differential pressure drop across the orifice. { 'thin ¦plāt 'ȯr·ə·fəs }

third law of motion See Newton's third law. { 'thərd 'lȯ əv 'mō·shən }

third law of thermodynamics |THERMO| The entropy of all perfect crystalline solids is zero at absolute zero temperature. { 'thərd 'lȯ əv ¦thər·mō·də'nam·iks }

third rail |CIV ENG| The electrified metal rail which carries current to the motor of an electric locomotive or other railway car. { 'thərd 'rāl }

13.0 temperature See annealing point. { ¦thər,tēn 'tem·prə·chər }

Thoma cavitation coefficient |MECH ENG| The equation for measuring cavitation in a hydraulic turbine installation, relating vapor pressure, barometric pressure, runner setting, tail water, and head. { 'tō·mə ,kav·ə'tā·shən ,kō·i,fish·ənt }

Thomas meter |ENG| An instrument used to determine the rate of flow of a gas by measuring the rise in the gas temperature produced by a known amount of heat. { 'täm·əs ,mēd·ər }

Thomson bridge See Kelvin bridge. { 'täm·sən ,brij }

thoroughfare |CIV ENG| **1.** An important, unobstructed public street or highway. **2.** A street going through from one street to another. **3.** An inland waterway for passage of ships usually not between two bodies of water. { 'thər·ə,fer }

thou See mil.

thread |DES ENG| A continuous helical rib, as on a screw or pipe. { thred }

thread contour |DES ENG| The shape of thread design as observed in a cross section along the major axis, for example, square or round. { 'thred ,kän,tùr }

thread cutter |MECH ENG| A tool used to cut screw threads on a pipe, screw, or bolt. { 'thred ,kəd·ər }

thread gage |DES ENG| A design gage used to measure screw threads. { 'thred ,gāj }

threading die |MECH ENG| A die which may be solid, adjustable, or spring adjustable, or a self-opening die head, used to produce an external thread on a part. { 'thred·iŋ ,dī }

threading machine |MECH ENG| A tool used to cut or form threads inside or outside a cylinder or cone. { 'thred·iŋ mə,shēn }

thread plug |ENG| Mold part which shapes an internal thread onto a molded article; must be unscrewed from the finished piece. { 'thred ,pləg }

thread plug gage |DES ENG| A thread gage used to measure female screw threads. { 'thred ,pləg ,gāj }

thread protector |ENG| A short-threaded ring to screw onto a pipe or into a coupling to protect the threads while the pipe is being handled or transported. Also known as pipe-thread protector. { 'thred prə,tek·tər }

thread rating |ENG| The maximum internal working pressure allowable for threaded pipe or tubing joints; important for pressure systems, chemical processes, and oil-well systems. { 'thred ,rād·iŋ }

thread ring gage |DES ENG| A thread gage used to measure male screw threads. { 'thred 'riŋ ,gāj }

three-body problem |MECH| The problem of predicting the motions of three objects obeying Newton's laws of motion and attracting each other according to Newton's law of gravitation. { 'thrē ¦bäd·ē ,präb·ləm }

three-dimensional braiding See through-the-thickness braiding. { ¦thrē di¦men·chən·əl 'brād·iŋ }

three-dimensional sound See virtual acoustics. { ¦thrē də,men·shən·əl 'saúnd }

three-input adder See full adder. { 'thrē ¦in,pút 'ad·ər }

three-input subtracter See full subtracter. { 'thrē ¦in,pút səb'trak·tər }

three-jaw chuck |DES ENG| A drill chuck having three serrated-face movable jaws that can grip and hold fast an inserted drill rod. { 'thrē ¦jȯ 'chək }

three-junction transistor |ELECTR| A *pnpn* transistor having three junctions and four regions of alternating conductivity; the emitter connection may be made to the *p* region at the left, the base connection to the adjacent *n* region, and the collector connection to the *n* region at the right, while the remaining *p* region is allowed to float. { 'thrē ¦jəŋk·shən tran'zis·tər }

three-layer diode |ELECTR| A junction diode with three conductivity regions. { 'thrē ¦lā·ər 'dī,ōd }

three-phase circuit |ELEC| A circuit energized by alternating-current voltages that differ in phase by one-third of a cycle or 120° { 'thrē ¦fāz 'sər·kət }

three-point problem |ENG| The problem of locating the horizontal position of a point of observation from the two observed horizontal angles subtended by three known sides of a triangle. { 'thrē ¦pȯint 'präb·ləm }

three-way switch |ELEC| An electric switch with

three terminals used to control a circuit from two different points. { 'thrē ¦wā 'swich }

threshold |BUILD| A piece of stone, wood, or metal that lies under an outside door. |ELECTR| In a modulation system, the smallest value of carrier-to-noise ratio at the input to the demodulator for all values above which a small percentage change in the input carrier-to-noise ratio produces a substantially equal or smaller percentage change in the output signal-to-noise ratio. |ENG| The least value of a current, voltage, or other quantity that produces the minimum detectable response in an instrument or system. { 'thresh,hōld }

threshold frequency |ELECTR| The frequency of incident radiant energy below which there is no photoemissive effect. { 'thresh,hōld ,frē·kwən·sē }

threshold speed |ENG| The minimum speed of current at which a particular current meter will measure at its rated reliability. { 'thresh,hōld ,spēd }

threshold treatment |CHEM ENG| The process of stopping a precipitation-type reaction at the threshold of precipitate formation; used in water-treatment reactions. { 'thresh,hōld ,trēt·mənt }

threshold value |CONT SYS| The minimum input that produces a corrective action in an automatic control system. { 'thresh,hōld ,val·yü }

threshold voltage |ELECTR| **1.** In general, the voltage at which a particular characteristic of an electronic device first appears. **2.** The voltage at which conduction of current begins in a *pn* junction. **3.** The voltage at which channel formation occurs in a metal oxide semiconductor field-effect transistor. **4.** The voltage at which a solid-state lamp begins to emit light. { 'thresh,hōld ,vōl·tij }

throat |DES ENG| The narrowest portion of a constricted duct, as in a diffuser or a venturi tube; specifically, a nozzle throat. |ENG| **1.** The smaller end of a horn or tapered waveguide. **2.** The area in a fireplace that forms the passageway from the firebox to the smoke chamber. { thrōt }

throatable |DES ENG| Of a nozzle, designed to allow a change in the velocity of the exhaust stream by changing the size and shape of the throat of the nozzle. { 'thrōd·ə·bəl }

throat microphone |ENG ACOUS| A contact microphone that is strapped to the throat of a speaker and reacts directly to throat vibrations rather than to the sound waves they produce. { 'thrōt 'mī·krə,fōn }

throttle See throttle valve. { 'thräd·əl }

throttle valve |MECH ENG| A choking device to regulate flow of a liquid, for example, in a pipeline, to an engine or turbine, from a pump or compressor. Also known as throttle. { 'thräd·əl 'valv }

throttling |CONT SYS| Control by means of intermediate steps between full on and full off. |THERMO| An adiabatic, irreversible process in

which a gas expands by passing from one chamber to another chamber which is at a lower pressure than the first chamber. { 'thräd·əl·iŋ }

throttling calorimeter |ENG| An instrument utilizing the principle of constant enthalpy expansion for the measurement of the moisture content of steam; steam drawn from a steampipe through sampling nozzles enters the calorimeter through a throttling orifice and moves into a well-insulated expansion chamber in which its temperature is measured. Also known as steam calorimeter. { 'thräd·əl·iŋ ,kal·ə'rim·əd·ər }

through arch |CIV ENG| An arch bridge from which the roadway is suspended as distinct from one which carries the roadway on top. { 'thrü ,ärch }

through bridge |CIV ENG| A bridge that carries the deck within the height of the superstructure. { 'thrü ,brij }

through-feed centerless grinding |MECH ENG| A metal cutting process by which the external surface of a cylindrical workpiece of uniform diameter is ground by passing the workpiece between a grinding and regulating wheel. { 'thrü ¦fēd 'sen·tər·ləs 'grīnd·iŋ }

throughput |CHEM ENG| The volume of feedstock charged to a process equipment unit during a specified time. { 'thrü,pút }

throughstone See bond header. { 'thrü,stōn }

through street |CIV ENG| A street at which all cross traffic is required to stop before crossing or entering. Also known as throughway. { 'thrü ,strēt }

through-the-thickness braiding |ENG| A technique for preparing composite materials in which fibers are intertwined continuously, producing three-dimensional seamless patterns that resist growth of cracks and delamination in the finished parts. Also known as three-dimensional braiding. { ¦thrü thə ¦thik·nəs 'brād·iŋ }

through transmission |ENG| An ultrasonic testing method in which mechanical vibrations are transmitted into one end of the workpiece and received at the other end. { 'thrü tranz,mish·ən }

throughway See expressway; through street. { 'thrü,wā }

throw |ENG| The scattering of fragments in a blasting operation. |MECH ENG| The maximum diameter of the circle moved by a rotary part. { thrō }

throwout |MECH ENG| In automotive vehicles, the mechanism or assemblage of mechanisms by which the driven and driving plates of a clutch are separated. { 'thrō,aút }

throw-out spiral See lead-out groove. { 'thrō,aút ,spī·rəl }

thrust |MECH| **1.** The force exerted in any direction by a fluid jet or by a powered screw. **2.** Force applied to an object to move it in a desired direction. |MECH ENG| The weight or pressure applied to a bit to make it cut. { thrəst }

thrust bearing |MECH ENG| A bearing which

sustains axial loads and prevents axial movement of a loaded shaft. { 'thrəst ,ber·iŋ }

thrust load |MECH ENG| A load or pressure parallel to or in the direction of the shaft of a vehicle. { 'thrəst ,lōd }

thrust meter |ENG| An instrument for measuring static thrust, especially of a jet engine or rocket. { 'thrəst ,mēd·ər }

thrust yoke |MECH ENG| The part connecting the piston rods of the feed mechanism on a hydraulically driven diamond-drill swivel head to the thrust block, which forms the connecting link between the yoke and the drive rod, by means of which link the longitudinal movements of the feed mechanism are transmitted to the swivel-head drive rod. Also known as back end. { 'thrəst ,yōk }

thumbscrew |DES ENG| A screw with a head flattened in the same axis as the shaft so that it can be gripped and turned by the thumb and forefinger. { 'thəm,skrü }

thump |ENG ACOUS| Low-frequency transient disturbance in a system or transducer characterized audibly by the vocal imitation of the word. { thəmp }

thurm |ENG| To work wood across the grain with a saw and chisel in order to produce an effect similar to turning the piece on a lathe. { thərm }

tidal lock *See* entrance lock. { 'tīd·əl 'läk }

tidal quay |CIV ENG| A quay in an open harbor or basin with sufficient depth to enable ships lying alongside to remain afloat at any state of the tide. { 'tīd·əl 'kē }

tide gage |ENG| A device for measuring the height of a tide; may be observed visually or may consist of an elaborate recording instrument. { 'tīd ,gāj }

tide gate |CIV ENG| **1.** A restricted passage through which water runs with great speed due to tidal action. **2.** An opening through which water may flow freely when the tide sets in one direction, but which closes automatically and prevents the water from flowing in the other direction when the direction of flow is reversed. { 'tīd ,gāt }

tide indicator |ENG| That part of a tide gage which indicates the height of tide at any time; the indicator may be in the immediate vicinity of the tidal water or at some distance from it. { 'tīd 'in·də,kād·ər }

tide lock *See* entrance lock. { 'tīd ,läk }

tide machine |ENG| An instrument that computes, sometimes for years in advance, the times and heights of high and low waters at a reference station by mechanically summing the harmonic constituents of which the tide is composed. { 'tīd mə,shēn }

tide pole |ENG| A graduated spar used for measuring the rise and fall of the tide. Also known as tide staff. { 'tīd ,pōl }

tide staff *See* tide pole. { 'tīd ,staf }

tie |CIV ENG| One of the transverse supports to which railroad rails are fastened to keep them to line, gage, and grade. |ELEC| **1.** Electrical connection or strap. **2.** *See* tie wire. |ENG| A

beam, post, rod, or angle to hold two pieces together; a tension member in a construction. { tī }

tie bar |CIV ENG| **1.** A bar used as a tie rod. **2.** A rod connecting two switch rails on a railway to hold them to gage. { 'tī ,bär }

tied arch |CIV ENG| An arch having the horizontal reaction component provided by a tie between the skewbacks of the arch ends. { 'tīd 'ärch }

tied concrete column |CIV ENG| A concrete column reinforced with longitudinal bars and horizontal ties. { 'tīd 'kän,krēt 'käl·əm }

tie-down diagram |ENG| A drawing indicating the prescribed method of securing a particular item of cargo within a specific type of vehicle. { 'tī,daún ,dī·ə,gram }

tie-down point |ENG| An attachment point provided on or within a vehicle. { 'tī,daún ,póint }

tie-down point pattern |ENG| The pattern of tie-down points within a vehicle. { 'tī,daún ¦póint 'pad·ərn }

tie plate |CIV ENG| A metal plate between a rail and a tie to hold the rail in place and reduce wear on the tie. |MECH ENG| A plate used in a furnace to connect tie rods. { 'tī ,plāt }

tier building |CIV ENG| A multistory skeleton frame building. { 'tir ,bil·diŋ }

tie rod |CIV ENG| A structural member used as a brace to take tensile loads. |ENG| A round or square iron rod passing through or over a furnace and connected with buckstays to assist in binding the furnace together. |MECH ENG| A rod used as a mechanical or structural support between elements of a machine. { 'tī ,räd }

TIGA *See* truncated icosahedral gravitational-wave antenna. { ¦tē¦¦iˈə *or* 'tī·gə }

tight |ENG| **1.** Unbroken, crack-free, and solid rock in which a naked hole will stand without caving. **2.** A borehole made impermeable to water by cementation or casing. |MECH ENG| **1.** Inadequate clearance or the barest minimum of clearance between working parts. **2.** The absence of leaks in a pressure system. { tīt }

tight fit |DES ENG| A fit between mating parts with slight negative allowance, requiring light to moderate force to assemble. { 'tīt 'fit }

tilting dozer |MECH ENG| A bulldozer whose blade can be pivoted on a horizontal center pin to cut low on either side. { 'tilt·iŋ 'dō·zər }

tilting idlers |MECH ENG| An arrangement of idler rollers in which the top set is mounted on vertical arms which pivot on spindles set low down on the frame of the roller stool. { 'tilt·iŋ 'īd·lərz }

tilting mixer |MECH ENG| A small-batch mixer consisting of a rotating drum which can be tilted to discharge the contents; used for concrete or mortar. { 'tilt·iŋ 'mik·sər }

tilting-type boxcar unloader |CIV ENG| A mechanism that is used to unload material such as grain from a boxcar; the car, with its door open, is held by end clamps on the specialized piece of track and tilted 15% from the vertical and then tilted endwise 40% to the horizontal to discharge the material at one end of the car, and 40% in the opposite direction to discharge the material from the opposite end. { 'tilt·iŋ ¦tip 'bäks,kär ən'lōd·ər }

tiltmeter |ENG| An instrument used to measure small changes in the tilt of the earth's surface, usually in relation to a liquid-level surface or to the rest position of a pendulum. { 'tilt,mēd·ər }

tilt/rotate code |ENG| A code that instructs a "golf ball" printing element which angle of tilt and rotation is needed to print a given character. { 'tilt'rō,tāt ,kōd }

tilt slab construction See tilt-up construction. { 'tilt ,slab kən,strək·shən }

tilt-up construction |BUILD| A method for constructing concrete wall panels by casting them horizontally adjacent to their final positions and then tilting them into vertical positions after the concrete has cured. Also known as tilt slab construction. { 'tilt,əp kən,strək·shən }

timber connector |ENG| A metal fastener that has a series of sharp teeth digging into the wood and is tightened with bolts to join sections of timber in heavy construction. { 'tim·bər kə,nek·tər }

time and material contract |IND ENG| A contract providing for the procurement of supplies or services on the basis of direct labor hours at specified fixed hourly rates (which rates include direct and indirect labor, overhead, and profit), and material at cost. { ¦tīm ən mə'tir·ē·əl ,kän,trakt }

time and motion study |IND ENG| Observation, analysis, and measurement of the steps in the performance of a job to determine a standard time for each performance. Also known as time-motion study. { ¦tīm ən 'mō·shən ,stəd·ē }

time break |ENG| A distinctive mark shown on an exploration seismogram to indicate the exact detonation time of an explosive energy source. { 'tīm ,brāk }

time-change component |ENG| A component which because of design limitations or safety is specified to be rebuilt or overhauled after a specified period of operation (for example, an engine or propeller of an airplane). { 'tīm ¦chānj kəm,pō·nənt }

time-controlled system See clock control system. { 'tīm kən¦trōld ,sis·təm }

time formula |IND ENG| A formula to determine the standard time of an operation as a function of one or more variables in the operation. { 'tīm ,fȯr·myə·lə }

time fuse |ENG| A fuse which contains a graduated time element to regulate the time interval after which the fuse will function. { 'tīm ,fyüz }

time-interval radiosonde See pulse-time-modulated radiosonde. { 'tīm ,in·tər·vəl 'rād·ē·ō,sänd }

time-invariant system |CONT SYS| A system in which all quantities governing the system's behavior remain constant with time, so that the

system's response to a given input does not depend on the time it is applied. { 'tīm in,ver·ē·ənt ,sis·təm }

time-motion study See time and motion study. { 'tīm 'mō·shən ,stəd·ē }

time of flight |MECH| Elapsed time in seconds from the instant a projectile or other missile leaves a gun or launcher until the instant it strikes or bursts. { 'tīm əv 'flīt }

time-of-flight spectrometer |ENG| Any instrument in which the speed of a particle is determined directly by measuring the time it takes to travel a measured distance. { ¦tīm əv ¦flīt spek 'träm·əd·ər }

timeout |CONT SYS| A test of the reliability of robotic software in which the robot is halted if a portion of software does not function properly until the problem is corrected. { 'tīm,aut }

time phasing |IND ENG| Production scheduling of components for product assembly so that each component is available at the correct time. { 'tīm ,fāz·iŋ }

timer |ELECTR| A circuit used in radar and in electronic navigation systems to start pulse transmission and synchronize it with other actions, such as the start of a cathode-ray sweep. |ENG| **1.** A device for automatically starting or stopping a machine or other device. **2.** See interval timer. |MECH ENG| A device that controls timing of the ignition spark of an internal combustion engine at the correct time. { 'tīm·ər }

time-sharing |IND ENG| Division of the time required for observation, decision making, and responding by an operator among the activities or tasks that must be performed almost simultaneously. { 'tīm ,sher·iŋ }

time standard See standard time. { 'tīm ,stan·dərd }

time study |IND ENG| A work measurement technique, generally using a stopwatch or other timing device, to record the actual elapsed time for performance of a task, adjusted for any observed variance from normal effort or pace, unavoidable or machine delays, rest periods, and personal needs. { 'tīm ,stəd·ē }

time switch |ENG| A clock-controlled switch used to open or close a circuit at one or more predetermined times. { 'tīm ,swich }

time system |CONT SYS| A system of clocks and control devices, with or without a master timepiece, to indicate time at various remote locations. { 'tīm ,sis·təm }

time-varying system |CONT SYS| A system in which certain quantities governing the system's behavior change with time, so that the system will respond differently to the same input at different times. { 'tīm ¦ver·ē·iŋ ,sis·təm }

timing |MECH ENG| Adjustment in the relative position of the valves and crankshaft of an automobile engine in order to produce the largest effective output of power. { 'tīm·iŋ }

timing belt |DES ENG| A power transmission belt with evenly spaced teeth on the bottom side which mesh with grooves cut on the periphery of the pulley to produce a positive, no-slip, constant-speed drive. Also known as cogged belt; synchronous belt. |MECH ENG| A positive drive belt that has axial cogs molded on the underside of the belt which fit into grooves on the pulley; prevents slip, and makes accurate timing possible; combines the advantages of belt drives with those of chains and gears. Also known as positive drive belt. { 'tīm·iŋ ‚belt }

timing-belt pulley |MECH ENG| A pulley that is similar to an uncrowned flat-belt pulley, except that the grooves for the belt's teeth are cut in the pulley's face parallel to the axis. { 'tīm·iŋ ¦belt ‚pùl·ē }

timing gears |MECH ENG| The gear train of reciprocating engine mechanisms for relating camshaft speed to crankshaft speed. { 'tīm·iŋ ‚girz }

timing motor |ELEC| A motor which operates from an alternating-current power system synchronously with the alternating-current frequency, used in timing and clock mechanisms. Also known as clock motor. { 'tīm·iŋ ‚mōd·ər }

Timken film strength |ENG| A test used on a gear lubricant to determine the amount of pressure the film of oil can withstand before rupturing. { 'tim·kən 'film ‚streŋkth }

Timken wear test |ENG| A test used on a gear lubricant to determine its abrasive effect on gear metals. { 'tim·kən 'wer ‚test }

tingle |BUILD| A support used in masonry to reduce sagging in a long layer of bricks. |DES ENG| **1.** A small nail. **2.** A flexible metal clip used to hold a sheet of material such as glass or metal. |ENG| A patch designed to cover a hole in a boat. { 'tiŋ·gəl }

tinner's rivet |DES ENG| A special-purpose rivet that has a flat head, used in sheet metal work. { 'tin·ərz ‚riv·ət }

tip |DES ENG| A piece of material secured to and differing from a cutter tooth or blade. |ELEC| The contacting part at the end of a phone plug. |ELECTR| A small protuberance on the envelope of an electron tube, resulting from the closing of the envelope after evacuation. { tip }

tipped bit |DES ENG| A drill bit in which the cutting edge is made of especially hard material. { 'tipt 'bit }

tipped solid cutters |DES ENG| Cutters made of one material and having tips or cutting edges of another material bonded in place. { 'tipt 'säl·əd 'kəd·ərz }

tipping-bucket rain gage |ENG| A type of recording rain gage; the precipitation collected by the receiver empties into one side of a chamber which is partitioned transversely at its center and is balanced bistably upon a horizontal axis; when a predetermined amount of water has been collected, the chamber tips, spilling out the water and placing the other half of the chamber under the receiver; each tip of the bucket is recorded on a chronograph, and the record obtained indicates the amount and rate of rainfall. { 'tip·iŋ ‚bək·ət 'rān ‚gāj }

tire |ENG| A continuous metal ring, or pneumatic rubber and fabric cushion, encircling and fitting the rim of a wheel. { tīr }

tire iron |DES ENG| A single metal bar having bladelike ends of various shapes to insert between the rim and the bead of a pneumatic tire to remove or replace the tire. { 'tīr ‚ī·ərn }

tirrill burner |ENG| A modification of the bunsen burner which allows greater flexibility in the adjustment of the air-gas mixture. { 'tir·əl ‚bər·nər }

T junction |ELECTR| A network of waveguides with three waveguide terminals arranged in the form of a letter T; in a rectangular waveguide a symmetrical T junction is arranged by having either all three broadsides in one plane or two broadsides in one plane and the third in a perpendicular plane. { 'tē ‚jəŋk·shən }

T²L *See* transistor-transistor logic.

TME *See* metric-technical unit of mass.

to-and-fro ropeway *See* jig back. { ¦tü ən ¦frō 'rōp‚wā }

toe |CIV ENG| The part of a base of a dam or retaining wall on the side opposite to the retained material. { tō }

toeboard |BUILD| A board placed around a platform or on a sloping roof to prevent personnel or materials from falling off. |ENG| A support or reinforcement that forms the lowest vertical face of a cabinet or similar installation, at toe level, and is frequently recessed. { 'tō‚bōrd }

toe cut |ENG| In underground blasting, the cut obtained by the use of toe holes. { 'tō ‚kət }

toe hole |ENG| A blasting hole, usually drilled horizontally or at a slight inclination into the base of a bank, bench, or slope of a quarry or open-pit mine. { 'tō ‚hōl }

toe-in |MECH ENG| The degree (usually expressed in fractions of an inch) to which the forward part of the front wheels of an automobile are closer together than the rear part, measured at hub height with the wheels in the normal "straight ahead" position of the steering gear. { 'tō ‚in }

toenailing |ENG| The technique of driving a nail at an angle to join two pieces of lumber. { 'tō¦nāl·iŋ }

toe-out |MECH ENG| The outward inclination of the wheels of an automobile at the front on turns due to setting the steering arms at an angle. { 'tō ‚aùt }

toeplate *See* kickplate. { 'tō‚plāt }

toe-to-toe drilling |ENG| The drilling of vertical large-diameter blasting holes in quarries and opencast pits. { ¦tō tə ¦tō 'dril·iŋ }

toe wall |CIV ENG| A low wall constructed at the bottom of an embankment to prevent slippage or spreading of the soil. { 'tō ‚wól }

toggle |ELECTR| To switch over to an alternate state, as in a flip-flop. |MECH ENG| A form of jointed mechanism for the amplification of forces. { 'täg·əl }

toggle bolt |DES ENG| A bolt having a nut with a pair of pivotal wings that close against a spring; wings open after emergence through a hole or passage in a thin or hollow wall to fasten the unit securely. { 'täg·əl ,bȯlt }

toggle press |MECH ENG| A mechanical press in which a toggle mechanism actuates the slide. { 'täg·əl ,pres }

toggle switch |ELEC| A small switch that is operated by manipulation of a projecting lever that is combined with a spring to provide a snap action for opening or closing a circuit quickly. |ELECTR| An electronically operated circuit that holds either of two states until changed. { 'täg· əl ,swich }

tolerance |DES ENG| The permissible variations in the dimensions of machine parts. |ENG| A permissible deviation from a specified value, expressed in actual values or more often as a percentage of the nominal value. { 'täl·ə·rəns }

tolerance chart |DES ENG| A chart indicating graphically the sequence in which dimensions must be produced on a part so that the finished product will meet the prescribed tolerance limits. { 'täl·ə·rəns ,chärt }

tolerance limits |DES ENG| The extreme values (upper and lower) that are permitted by the tolerance. { 'täl·ə·rəns ,lim·əts }

tolerance unit |DES ENG| A unit of length used to express the degree of tolerance allowed in fitting cylinders into cylindrical holes, equal, in micrometers, to 0.45 $D^{1/3}$ + 0.001 D, where D is the cylinder diameter in millimeters. { 'täl·ə· rəns ,yü·nət }

ton |IND ENG| A unit of volume of sea freight, equal to 40 cubic feet or approximately 1.1327 cubic meters. Also known as freight ton; measurement ton; shipping ton. |MECH| **1.** A unit of weight in common use in the United States, equal to 2000 pounds or 907.18474 kilogram-force. Also known as just ton; net ton; short ton. **2.** A unit of mass in common use in the United Kingdom equal to 2240 pounds, or to 1016.0469088 kilogram-force. Also known as gross ton; long ton. **3.** A unit of weight in troy measure, equal to 2000 troy pounds, or to 746.4834432 kilogram-force. **4.** See tonne. |MECH ENG| A unit of refrigerating capacity, that is, of rate of heat flow, equal to the rate of extraction of latent heat when one short ton of ice of specific latent heat 144 international table British thermal units per pound is produced from water at the same temperature in 24 hours; equal to 200 British thermal units per minute, or to approximately 3516.85 watts. Also known as standard ton. { tən }

tondal |MECH| A unit of force equal to the force which will impart an acceleration of 1 foot per second to a mass of 1 long ton; equal to approximately 309.6911 newtons. { 'tänd·əl }

tongs |DES ENG| Any of various devices for holding, handling, or lifting materials and consisting of two legs joined eccentrically by a pivot or spring. { taŋz }

tongue and groove |DES ENG| A joint in which a projecting rib on the edge of one board fits into a groove in the edge of another board. { 'təŋ ən 'grüv }

ton-mile |CIV ENG| In railroading, a standard measure of traffic, based on the rate of carriage per mile of each passenger or ton of freight. { tən 'mīl }

tonne |MECH| A unit of mass in the metric system, equal to 1000 kilograms or to approximately 2204.62 pound mass. Also known as metric ton; millier; ton; tonneau. { tən }

tonneau See tonne. { tə'nō }

tool |ENG| Any device, instrument, or machine for the performance of an operation, for example, a hammer, saw, lathe, twist drill, drill press, grinder, planer, or screwdriver. |IND ENG| To equip a factory or industry for production by designing, making, and integrating machines, machine tools, and special dies, jigs, and instruments, so as to achieve manufacture and assembly of products on a volume basis at minimum cost. { tül }

tool bit |ENG| A piece of high-strength metal, usually steel, ground to make single-point cutting tools for metal-cutting operations. { 'tül ,bit }

toolbox |ENG| A box to hold tools. { 'tül ,bäks }

tool-center point |CONT SYS| The location on the end effector or tool of a robot manipulator whose position and orientation define the coordinates of the controlled object. { 'tül 'sen· tər ,pȯint }

tool changer |MECH ENG| In program-controlled machines and robotics, a mechanism that allows the use of multiple tools. { 'tül ,chānj· ər }

tool-check system |IND ENG| A system for temporary issue of tools in which the employee is issued a number of small metal checks stamped with the same number; a check is surrendered for each tool obtained from the crib. { 'tül ,chek ,sis·təm }

tool design |DES ENG| The division of mechanical design concerned with the design of tools. { 'tül di,zīn }

tool-dresser |MECH ENG| A tool-stone-grade diamond inset in a metal shank and used to trim or form the face of a grinding wheel. { 'tül ,dres·ər }

tool extractor |ENG| An implement for grasping and withdrawing drilling tools when broken, detached, or lost in a borehole. { 'tül ik,strak·tər }

tool-function controller |CONT SYS| A unit that selects and controls tools for machining operations; it may be internal or external to the main controller. { 'tül ¦fəŋk·shən kən'trōl·ər }

toolhead |MECH ENG| The adjustable tool-carrying part of a machine tool. { 'tül,hed }

tooling |MECH ENG| Tools or end effectors with which a robot performs the actual work on a workpiece. { 'tül·iŋ }

tool joint |ENG| A coupling element for a drill pipe; designed to support the weight of the drill stem and the strain of frequent use, and to provide a leakproof seal. { 'tül ,jȯint }

tool-length compensation |CONT SYS| Programming of machining operations so that all tools are positioned correctly in advance for any tasks to be carried out. { 'tül ¦leŋkth ¡kämpən'sā·shən }

toolmaker's vise *See* universal vise. { 'tül,mäk·ərz ¡vīs }

tool offset [MECH ENG] The adjustment of tool positions in machines to compensate for their wear, finishing, or displacement from an axis. { 'tül 'óf,set }

tool post [MECH ENG] A device to clamp and position a tool holder on a machine tool. { 'tül ¡pōst }

tooth [DES ENG] **1.** One of the regular projections on the edge or face of a gear wheel. **2.** An angular projection on a tool or other implement, such as a rake, saw, or comb. { 'tüth }

tooth point [DES ENG] The chamfered cutting edge of the blade of a face mill. { 'tüth ¡póint }

top [MECH] A rigid body, one point of which is held fixed in an inertial reference frame, and which usually has an axis of symmetry passing through this point; its motion is usually studied when it is spinning rapidly about the axis of symmetry. { täp }

top dead center [MECH ENG] The dead-center position of an engine piston and its crankshaft arm when at the top or outer end of its stroke. { ¦täp 'ded 'sen·tər }

top-down design [IND ENG] A design methodology that proceeds from the highest level to the lowest and from the general to the particular, and that provides a formal mechanism for breaking complex process designs into functional descriptions, reviewing progress, and allowing modifications. { 'täp ¦daún di'zīn }

topographic survey [ENG] A survey that determines ground relief and location of natural and man-made features thereon. { ¦täp·ə¦graf·ik 'sər,vā }

topping [CHEM ENG] The distillation of crude petroleum to remove the light fractions only; the unrefined distillate is called tops. [CIV ENG] A layer of mortar placed over concrete to form a finishing surface on a floor, driveway, sidewalk, or curb. { 'täp·iŋ }

topping governor *See* limit governor. { 'täp·iŋ ¡gəv·ə·nər }

topping joint [CIV ENG] In concrete finishing, a small space or break set at regular intervals, particularly over expansion joints, to allow for contraction and expansion of the topping layer. { 'täp·iŋ ¡jóint }

top plate [BUILD] **1.** The top horizontal member of a building frame to which the rafters are fastened. **2.** The horizontal member of a building frame at the top of the partition studs. { 'täp ¡plāt }

topple [MECH] In gyroscopes for marine or aeronautical use, the condition of a sudden upset gyroscope or a gyroscope platform evidenced by a sudden and rapid precession of the spin axis due to large torque disturbances such as the spin axis striking the mechanical stops. Also known as tumble. { 'täp·əl }

topple axis [MECH] Of a gyroscope, the horizontal axis, perpendicular to the horizontal spin axis, around which topple occurs. Also known as tumble axis. { 'täp·əl ¡ak·səs }

top rail [BUILD] The uppermost horizontal member of a unit of framing, such as a door or a sash. { 'täp ¡rāl }

top steam [CHEM ENG] Steam admitted near the top of a shell still to purge the still, and to prevent a vacuum from forming when pumping out the liquid contents. { 'täp ¡stēm }

tor *See* pascal. { tór }

torch [BUILD] To apply lime mortar under the top edges of roof tiles or slates. [ENG] A gas burner used for brazing, cutting, or welding. { tórch }

tornado cellar *See* cyclone cellar. { tór'nād·ō ¡sel·ər }

toromatic transmission [MECH ENG] A semiautomatic transmission; it contains a compound planetary gear train with a torque converter. { ¦tòr·ə¦mad·ik tranz'mish·ən }

torpedo [ENG] An encased explosive charge slid, lowered, or dropped into a borehole and exploded to clear the hole of obstructions or to open communications with an oil or water supply. Also known as bullet. { tòr'pēd·ō }

torque [MECH] **1.** For a single force, the cross product of a vector from some reference point to the point of application of the force with the force itself. Also known as moment of force; rotation moment. **2.** For several forces, the vector sum of the torques (first definition) associated with each of the forces. { tòrk }

torque arm [MECH ENG] In automotive vehicles, an arm to take the torque of the rear axle. { 'tòrk ¡ärm }

torque-coil magnetometer [ENG] A magnetometer that depends for its operation on the torque developed by a known current in a coil that can turn in the field to be measured. { 'tòrk ¡kóil ¡mag·nə'täm·əd·ər }

torque converter [MECH ENG] A device for changing the torque speed or mechanical advantage between an input shaft and an output shaft. { 'tòrk kən,vərd·ər }

torque-load characteristic [ENG] For electric motors, the armature torque developed versus the load on the motor at constant speed. { 'tòrk ¦lōd ¡kar·ik·tə,ris·tik }

torquemeter [ENG] An instrument to measure torque. { 'tòrk,mēd·ər }

torque reaction [MECH ENG] On a shaft-driven vehicle, the reaction between the bevel pinion with its shaft (which is supported in the rear axle housing) and the bevel ring gear (which is fastened to the differential housing) that tends to rotate the axle housing around the axle instead of rotating the axle shafts alone. { 'tòrk rē,ak·shən }

torque ripple *See* cog. { 'tòrk ¡rip·əl }

torque-tube flowmeter [ENG] A liquid-flow measurement device in which a flexible torque tube transmits bellows motion (caused by differential pressure from the liquid flow through the pipe) to the recording pen arm. { 'tȯrk ¦tüb 'flō ‚mēd·ər }

torque-type viscometer |ENG| A device that measures liquid viscosity by the torque needed to rotate a vertical paddle submerged in the liquid; used for both Newtonian and non-Newtonian liquids and for suspensions. { 'tȯrk ¦tīp vi'skäm·əd·ər }

torque-winding diagram |MECH ENG| A diagram showing how the winding load on a winch drum varies and is used to decide the method of balancing needed; made by plotting the turning moment in pounds per foot on the vertical axis against time, or revolutions or depth on the horizontal axis. { 'tȯrk ¦wīnd·iŋ ‚dī·ə‚gram }

torque wrench |ENG| **1.** A hand or power tool used to turn a nut on a bolt that can be adjusted to deliver a predetermined amount of force to the bolt when tightening the nut. **2.** A wrench that measures torque while being turned. { 'tȯrk ‚rench }

torr |MECH| A unit of pressure, equal to 1/760 atmosphere; it differs from 1 millimeter of mercury by less than one part in seven million; approximately equal to 133.3224 pascals. { tȯr }

Torricellian barometer See mercury barometer. { ¦tȯr·ə¦chel·ē·ən bə'räm·əd·ər }

torsel |BUILD| A section of wood, stone, or steel that supports one end of a beam or joist and distributes the load. { 'tȯr·səl }

torsiometer |MECH ENG| An instrument which measures power transmitted by a rotating shaft; consists of angular scales mounted around the shaft from which twist of the loaded shaft is determined. Also known as torsionmeter. { ‚tȯr·shē'äm·əd·ər }

torsion |MECH| A twisting deformation of a solid body about an axis in which lines that were initially parallel to the axis become helices. { 'tȯr·shən }

torsional angle |MECH| The total relative rotation of the ends of a straight cylindrical bar when subjected to a torque. { 'tȯr·shən·əl 'aŋ·gəl }

torsional compliance |MECH| The reciprocal of the torsional rigidity. { ¦tȯr·shə·nəl kəm'plī·əns }

torsional hysteresis |MECH| Dependence of the torques in a twisted wire or rod not only on the present torsion of the object but on its previous history of torsion. { ¦tȯr·shə·nəl ‚his·tə'rē·səs }

torsional modulus |MECH| The ratio of the torsional rigidity of a bar to its length. Also known as modulus of torsion. { 'tȯr·shən·əl 'mäj·ə·ləs }

torsional pendulum |MECH| A device consisting of a disk or other body of large moment of inertia mounted on one end of a torsionally flexible elastic rod whose other end is held fixed; if the disk is twisted and released, it will undergo simple harmonic motion, provided the torque in the rod is proportional to the angle of twist. Also known as torsion pendulum. { 'tȯr·shən·əl 'pen·jə·ləm }

torsional rigidity |MECH| The ratio of the torque

applied about the centroidal axis of a bar at one end of the bar to the resulting torsional angle, when the other end is held fixed. { 'tȯr·shən·əl ri'jid·əd·ē }

torsional vibration |MECH| A periodic motion of a shaft in which the shaft is twisted about its axis first in one direction and then in the other; this motion may be superimposed on rotational or other motion. { 'tȯr·shən·əl vī'brā·shən }

torsion balance |ENG| An instrument, consisting essentially of a straight vertical torsion wire whose upper end is fixed while a horizontal beam is suspended from the lower end; used to measure minute gravitational, electrostatic, or magnetic forces. { 'tȯr·shən ‚bal·əns }

torsion bar |MECH ENG| A spring flexed by twisting about its axis; found in the spring suspension of truck and passenger car wheels, in production machines where space limitations are critical, and in high-speed mechanisms where inertia forces must be minimized. { 'tȯr·shən ‚bär }

torsion damper |MECH ENG| A damper used on automobile internal combustion engines to reduce torsional vibration. { 'tȯr·shən ‚dam·pər }

torsion function |MECH| A harmonic function, $\phi(x,y) = w/\tau$, expressing the warping of a cylinder undergoing torsion, where the x, y, and z coordinates are chosen so that the axis of torsion lies along the z axis, w is the z component of the displacement, and τ is the torsion angle. Also known as warping function. { 'tȯr·shən ‚fəŋk·shən }

torsion galvanometer |ENG| A galvanometer in which the force between the fixed and moving systems is measured by the angle through which the supporting head of the moving system must be rotated to bring the moving system back to its zero position. { 'tȯr·shən ‚gal·və'näm·əd·ər }

torsion hygrometer |ENG| A hygrometer in which the rotation of the hygrometric element is a function of the humidity; such hygrometers are constructed by taking a substance whose length is a function of the humidity and twisting or spiraling it under tension in such a manner that a change in length will cause a further rotation of the element. { 'tȯr·shən hī'gräm·əd·ər }

torsionmeter See torsiometer. { 'tȯr·shən‚mēd·ər }

torsion pendulum See torsional pendulum. { 'tȯr·shən 'pen·jə·ləm }

torsion-string galvanometer |ENG| A sensitive galvanometer in which the moving system is suspended by two parallel fibers that tend to twist around each other. { 'tȯr·shən ¦striŋ ‚gal·və'näm·əd·ər }

total air |ENG| The actual quantity of air supplied for combustion of fuel in a boiler, expressed as a percentage of theoretical air. { 'tōd·əl 'er }

total coincidence |MECH ENG| The condition in which all the joints of a robot become locked in position. { 'tōd·əl kō'in·səd·əns }

total heat See enthalpy. { 'tōd·əl 'hēt }

total pressure |MECH| The gross load applied on a given surface. { 'tōd·əl 'presh·ər }

total quality management |SYS ENG| A philosophy and set of guiding concepts that provides a comprehensive means of improving total organization performance and quality by examining each process through which work is done in a systematic, integrated, consistent, organization-wide manner. Abbreviated TQM. { ¦tōd·əl 'kwäl·əd·ē ,man·ij·mənt }

total radiation pyrometer |ENG| A pyrometer which focuses heat radiation emitted by a hot object on a detector (usually a thermopile or other thermal type detector), and which responds to a broad band of radiation, limited only by absorption of the focusing lens, or window and mirror. { 'tōd·əl 'rād·ē¦ā·shən pī'räm·əd·ər }

touch feedback |ENG| A type of force feedback in which servos provide the manipulator fingers with a sense of resistance when an object is grasped, so that the operator does not crush the object. { 'təch ,fēd,bak }

touch sensor |CONT SYS| A device such as a small, force-sensitive switch that uses contact to generate feedback in robotic systems. { 'təch ,sen·sər }

toughness |MECH| A property of a material capable of absorbing energy by plastic deformation; intermediate between softness and brittleness. { 'təf·nəs }

tow |ENG| **1.** To haul by a rope or chain, for example, to haul a disabled ship by another vessel or an automotive vehicle by another vehicle. **2.** To propel by pushing, as a tugboat piloting a ship. { tō }

towbar |ENG| An element which connects to a vehicle that is not equipped with an integral drawbar, for the purpose of towing or moving the vehicle. { 'tō,bär }

towed load |MECH| The weight of a carriage, trailer, or other equipment towed by a prime mover. { 'tōd 'lōd }

tower |CHEM ENG| A vertical, cylindrical vessel used in chemical and petroleum processing to increase the degree of separation of liquid mixtures by distillation or extraction. Also known as column. |ENG| A concrete, metal, or timber structure that is relatively high for its length and width, and used for various purposes, including the support of electric power transmission lines, radio and television antennas, and rockets and missiles prior to launching. { 'taù·ər }

tower bolt See barrel bolt. { 'taù·ər ,bōlt }

tower crane |CIV ENG| A crane mounted on top of a tower which is sometimes incorporated in the frame of a building. { 'taù·ər ,krān }

towing tank See model basin. { 'tō·iŋ ,taŋk }

Townsend avalanche See avalanche. { 'taùn·zənd ,av·ə,lanch }

TPR See airborne profile recorder.

TQM See total quality management.

trace |ELECTR| The visible path of a moving spot on the screen of a cathode-ray tube. Also known as line. |ENG| The record made by a recording device, such as a seismometer or electrocardiograph. { trās }

trace heating |ENG| Heating the layer between insulation and pipes in an insulated pipework system to reduce viscosity and thereby facilitate flow of the liquid. { ¦trās ¦hēd·iŋ }

tracer |ENG| A thread of contrasting color woven into the insulation of a wire for identification purposes. { 'trā·sər }

tracer gas |ENG| In vacuum testing for leaks, a gas emitting through a leak in a pressure system and subsequently conducted into the detector. { 'trā·sər ,gas }

tracer milling |MECH ENG| Cutting a duplicate of a three-dimensional form by using a mastic form to direct the tracer-controlled cutter. { 'trā·sər ,mil·iŋ }

tracing distortion |ENG ACOUS| The nonlinear distortion introduced in the reproduction of a mechanical recording because the curve traced by the motion of the reproducing stylus is not an exact replica of the modulated groove. { 'trās·iŋ di,stòr·shən }

track |DES ENG| As applied to a pattern of setting diamonds in a bit crown, an arrangement of diamonds in concentric circular rows in the bit crown, with the diamonds in a specific row following in the track cut by a preceding diamond. |ELECTR| **1.** A path for recording one channel of information on a magnetic tape, drum, or other magnetic recording medium; the location of the track is determined by the recording equipment rather than by the medium. **2.** The trace of a moving target on a plan-position-indicator radar screen or an equivalent plot. |ENG| **1.** The groove cut in a rock by a diamond inset in the crown of a bit. **2.** A pair of parallel metal rails for a railway, railroad, tramway, or for any wheeled vehicle. |MECH ENG| **1.** The slide or rack on which a diamond-drill swivel head can be moved to positions above and clear of the collar of a borehole. **2.** A crawler mechanism for earth-moving equipment. Also known as crawler track. { trak }

track cable |ENG| Steel wire rope, usually a locked-coil rope which supports the wheels of the carriers of a cableway. { 'trak ,kā·bəl }

track gage |CIV ENG| The width between the rails of a railroad track; in the United States the standard gage is 4 feet 8½ inches. { 'trak ,gāj }

track hopper |ENG| A hopper-shaped receiver mounted beside or below railroad tracks, into which railroad boxcars or bottom-dump cars are discharged; used for solid materials. { 'trak ,häp·ər }

tracking |ELEC| A leakage or fault path created across the surface of an insulating material when a high-voltage current slowly but steadily forms a carbonized path. |ELECTR| The condition in which all tuned circuits in a receiver accurately follow the frequency indicated by the tuning dial over the entire tuning range. |ENG| **1.** A motion given to the major lobe of a radar or radio antenna such that some preassigned moving target in space is always within the major lobe.

2. The process of following the movements of an object; may be accomplished by keeping the reticle of an optical system or a radar beam on the object, by plotting its bearing and distance at frequent intervals, or by a combination of techniques. |ENG ACOUS| **1.** The following of a groove by a phonograph needle. **2.** Maintaining the same ratio of loudness in the two channels of a stereophonic sound system at all settings of the ganged volume control. { 'trak·iŋ }

tracking error |ENG ACOUS| Deviation of the vibration axis of a phonograph pickup from tangency with a groove; true tangency is possible for only one groove when the pickup arm is pivoted; the longer the pickup arm, the less is the tracking error. { 'trak·iŋ ,er·ər }

tracking jitter |ENG| Minor variations in the pointing of an automatic tracking radar. { 'trak·iŋ ,jid·ər }

tracking network |ENG| A group of tracking stations whose operations are coordinated in tracking objects through the atmosphere or space. { 'trak·iŋ ,net,wərk }

tracking problem |CONT SYS| The problem of determining a control law which when applied to a dynamical system causes its output to track a given function; the performance index is in many cases taken to be of the integral square error variety. { 'trak·iŋ ,präb·ləm }

tracking radar |ENG| Radar used to monitor the flight and obtain geophysical data from space probes, satellites, and high-altitude rockets. { 'trak·iŋ ,rā,där }

tracking station |ENG| A radio, radar, or other station set up to track an object moving through the atmosphere or space. { 'trak·iŋ ,stā·shən }

tracking system |ENG| Apparatus, such as tracking radar, used in following and recording the position of objects in the sky. { 'trak·iŋ ,sis·təm }

trackshifter |ENG| A machine or appliance used to shift a railway track laterally. { 'trak ,shif·tər }

traction |MECH| Pulling friction of a moving body on the surface on which it moves. { 'trak·shən }

traction-control system |MECH ENG| An acceleration sensor-control system which, when a driving tire has no traction, slows the wheel movement by braking or reduces the engine speed and torque if braking alone will not prevent wheel spin. { 'trak·shən kən'trōl ,sis·təm }

traction meter |ENG| A load-sensing device placed between a locomotive and the car immediately behind it to measure pulling force exerted by the locomotive. { 'trak·shən ,mēd·ər }

traction tube |ENG| A device for measuring the minimum water velocities capable of moving various sizes of sand grains; it consists of a horizontal glass tube half-filled with sand. { 'trak·shən ,tüb }

tractor |MECH ENG| **1.** An automotive vehicle having four wheels or a caterpillar tread used for pulling agricultural or construction implements. **2.** The front pulling section of a semitrailer. Also known as truck-tractor. { 'trak·tər }

tractor drill |MECH ENG| A drill having a crawler mounting to support the feed-guide bar on an extendable arm. { 'trak·tər ,dril }

tractor gate |CIV ENG| A type of outlet control gate used to release water from a reservoir; there are two types, roller and wheel. { 'trak·tər ,gāt }

tractor loader |MECH ENG| A tractor equipped with a tipping bucket which can be used to dig and elevate soil and rock fragments to dump at truck height. Also known as shovel dozer; tractor shovel. { 'trak·tər ,lōd·ər }

tractor shovel See tractor loader. { 'trak·tər ,shəv·əl }

traffic |ENG| The passage or flow of vehicles, pedestrians, ships, or planes along defined routes such as highways, sidewalks, sea lanes, or air lanes. { 'traf·ik }

trafficability |CIV ENG| Capability of terrain to bear traffic, or the extent to which the terrain will permit continued movement of any or all types of traffic. { ,traf·ə·kə'bil·ad·ē }

traffic control |ENG| Control of the movement of vehicles, such as airplanes, trains, and automobiles, and the regulatory mechanisms and systems used to exert or enforce control. { 'traf·ik kən,trōl }

traffic cop |CONT SYS| The portion of a programmable controller's executive program concerned with input/output. { 'traf·ik ,käp }

traffic density |CIV ENG| The average number of vehicles that occupy 1 mile or 1 kilometer of road space, expressed in vehicles per mile or per kilometer. { 'traf·ik ,den·səd·ē }

traffic engineering |CIV ENG| The determination of the required capacity and layout of highway and street facilities that can safely and economically serve vehicular movement between given points. { 'traf·ik ,en·jə,nir·iŋ }

traffic flow |CIV ENG| The total number of vehicles passing a given point in a given time, expressed as vehicles per hour. { 'traf·ik ,flō }

traffic noise |ENG| The general disturbance in sonar transmissions which is due to ships but is not associated with a specific vessel. { 'traf·ik ,nȯiz }

traffic recorder |ENG| A mechanical counter or recorder used to determine traffic movements (hourly variations and total daily volumes of traffic at a point) on an existing route; the air-impulse counter, magnetic detector, photoelectric counter, and radar detector are used. { 'traf·ik ri,kȯrd·ər }

traffic signal |CIV ENG| With the exception of traffic signs, any power-operated device for regulating, directing, or warning motorists or pedestrians. { 'traf·ik ,sig·nəl }

T rail |CIV ENG| A rail shaped like a T in cross section due to a wide head, web, and flanged base. { 'tē ,rāl }

trailer |ELECTR| A bright streak at the right of a dark area or dark line in a television picture, or a dark area or streak at the right of a bright part; usually due to insufficient gain at low video frequencies. |MECH ENG| The section of a semitrailer that is pulled by the tractor. { 'trā·lər }

trail formation |ENG| Vehicles proceeding one behind the other at designated intervals. Also known as column formation. { 'trāl fȯr,mā·shən }

trailing edge |ELECTR| The major portion of the decay of a pulse. { 'trāl·iŋ 'ej }

train |ENG| To aim or direct a radar antenna in azimuth. { trān }

training aid |ENG| Any item which is developed or procured primarily to assist in training and the process of learning. { 'trān·iŋ ,ād }

training data |CONT SYS| Data entered into a robot's computer at the beginning of an operation. { 'trān·iŋ ,dad·ə }

training wall |CIV ENG| A wall built along the bank of a river or estuary parallel to the direction of flow to direct and confine the flow. { 'trān·iŋ ¦wȯl }

train shed |CIV ENG| **1.** A structure to protect trains from weather. **2.** The part of a railroad station that covers the tracks. { 'trān ,shed }

trajectory |MECH| The curve described by an object moving through space, as of a meteor through the atmosphere, a planet around the sun, a projectile fired from a gun, or a rocket in flight. { trə'jek·trē }

trajectory control |CONT SYS| A type of continuous-path control in which a robot's path is calculated based on mathematical models of joint acceleration, arm loads, and actuating signals. { trə'jek·trē kən,trōl }

trajectory-measuring system |ENG| A system used to provide information on the spatial position of an object at discrete time intervals throughout a portion of the trajectory or flight path. { trə'jek·trē ¦mezh·ə·riŋ ,sis·təm }

trammel |ENG| A device consisting of a bar, each of whose ends is constrained to move along one of two perpendicular lines; used in drawing ellipses and in the Rowland mounting. { 'tram·əl }

tramway |MECH ENG| An overhead rail, rope, or cable on which wheeled cars run to convey a load. { 'tram,wā }

transceiver |ELECTR| A radio transmitter and receiver combined in one unit and having switching arrangements such as to permit both transmitting and receiving. Also known as transmitter-receiver. { tran'sē·vər }

transcription |ENG ACOUS| A recording of a complete radio program, made especially for broadcast purposes. Also known as electrical transcription. { tranz'krip·shən }

transducer |ENG| Any device or element which converts an input signal into an output signal of a different form; examples include the microphone, phonograph pickup, loudspeaker, barometer, photoelectric cell, automobile horn, doorbell, and underwater sound transducer. { tranz'dü·sər }

transfer caliper |DES ENG| A caliper having one leg which can be opened (or closed) to remove the instrument from the piece being measured;

used to measure inside recesses or over projections. { 'tranz·fər ,kal·ə·pər }

transfer case |MECH ENG| In a vehicle with more than one driving axle, a housing fitted with gears that distribute the driving power among the axles. { 'tranz·fər ,kās }

transfer chamber |ENG| In plastics processing, a vessel in which thermosetting plastic is softened by heat and pressure before being placed in a closed mold for final curing. { 'tranz·fər ,chām·bər }

transfer chute |ENG| A chute used at a transfer point in a conveyor system; the chute is designed with a curved base or some other feature so that the load be discharged in a centralized stream and in the same direction as the receiving conveyor. { 'tranz·fər ,shüt }

transfer constant |ENG| A transducer rating, equal to one-half the natural logarithm of the complex ratio of the product of the voltage and current entering a transducer to that leaving the transducer when the latter is terminated in its image impedance; alternatively, the product may be that of force and velocity or pressure and volume velocity; the real part of the transfer constant is the image attenuation constant, and the imaginary part is the image phase constant. Also known as transfer factor. { 'tranz·fər ,kän·stənt }

transfer factor See transfer constant. { 'tranz·fər ,fak·tər }

transfer function |CONT SYS| The mathematical relationship between the output of a control system and its input; for a linear system, it is the Laplace transform of the output divided by the Laplace transform of the input under conditions of zero initial-energy storage. { 'tranz·fər ,fəŋk·shən }

transfer grille |ENG| In an air-conditioning system, a grille that permits air to flow from one space to another; may be one of a pair if installed on opposite sides of a wall or door. { 'tranz·fər ,gril }

transfer machine |MECH ENG| **1.** Equipment that moves parts from one production location in a factory to another. **2.** A device that holds a workpiece and moves it automatically through the stages of a manufacturing process. { 'tranz·fər mə,shēn }

transfer matrix |CONT SYS| The generalization of the concept of a transfer function to a multivariable system; it is the matrix whose product with the vector representing the input variables yields the vector representing the output variables. { 'tranz·fər ,mā·triks }

transfer-matrix method |MECH| A method of analyzing vibrations of complex systems, in which the system is approximated by a finite number of elements connected in a chainlike manner, and matrices are constructed which can be used to determine the configuration and forces acting on one element in terms of those on another. { 'tranz·fər ,mā·triks ,meth·əd }

transfer molding |ENG| Molding of thermosetting materials in which the plastic is softened by heat and pressure in a transfer chamber, then forced at high pressure through suitable sprues, runners, and gates into a closed mold for final curing. { 'tranz·fər ˌmōld·iŋ }

transfer ratio |ENG| From one point to another in a transducer at a specified frequency, the complex ratio of the generalized force or velocity at the second point to the generalized force or velocity applied at the first point; the generalized force or velocity includes not only mechanical quantities, but also other analogous quantities such as acoustical and electrical; the electrical quantities are usually electromotive force and current. { 'tranz·fər ˌrā·shō }

transfer register |ENG| A transfer grille fitted with a mechanism for controlling the volume of airflow. { 'tranz·fər ˌrej·ə·stər }

transfer robot |CONT SYS| A fixed-sequence robot that moves parts from one location to another. { 'tranz·fər 'rō,bät }

transfer unit |CHEM ENG| The relationship between the overall rate coefficient (for whatever transfer operation is being calculated), column volume, and fluid volumetric flow rate in fixed-bed sorption operations. { 'tranz·fər ˌyü·nət }

transformation |ELEC| For two networks which are equivalent as far as conditions at the terminals are concerned, a set of equations giving the admittances or impedances of the branches of one circuit in terms of the admittances or impedances of the other. { ˌtranz·fər'mā·shən }

transformer loss |ELEC| Ratio of the signal power that an ideal transformer of the same impedance ratio would deliver to the load impedance, to the signal power that the actual transformer delivers to the load impedance; this ratio is usually expressed in decibels. { tranz'för·mər ˌlós }

transformer substation |ELEC| An electric power substation whose equipment includes transformers. { tranz'för·mər 'səb,stā·shən }

transient grating photoacoustics See impulsive stimulated thermal scattering. { ˌtranch·ənt !grād·iŋ ˌfōd·ō·ə'kü·stiks }

transillumination |ENG| 1. Indirect lighting on a console panel that uses edge and backlighting techniques on clear, fluorescent, or layered plastic materials. 2. Transmission of light through sections of material in order to enhance inspection for deviations in quality. { ˌtranz·ə,lü·mə'nā·shən }

transistance |ELECTR| The characteristic that makes possible the control of voltages or currents so as to accomplish gain or switching action in a circuit; examples of transistance occur in transistors, diodes, and saturable reactors. { tran'zis·təns }

transistor |ELECTR| An active component of an electronic circuit consisting of a small block of semiconducting material to which at least three electrical contacts are made, usually two closely spaced rectifying contacts and one ohmic (nonrectifying) contact; it may be used as an amplifier, detector, or switch. { tran'zis·tər }

transistor amplifier |ELECTR| An amplifier in which one or more transistors provide amplification comparable to that of electron tubes. { tran'zis·tər ˌam·plə,fī·ər }

transistor biasing |ELECTR| Maintaining a direct-current voltage between the base and some other element of a transistor. { tran'zis·tər ˌbī·əs·iŋ }

transistor characteristics |ELECTR| The values of the impedances and gains of a transistor. { tran'zis·tər ˌkar·ik·tə,ris·tiks }

transistor chip |ELECTR| An unencapsulated transistor of very small size used in microcircuits. { tran'zis·tər ˌchip }

transistor circuit |ELECTR| An electric circuit in which a transistor is connected. { tran'zis·tər ˌsər·kət }

transistor gain |ELECTR| The increase in signal power produced by a transistor. { tran'zis·tər ˌgān }

transistor input resistance |ELECTR| The resistance across the input terminals of a transistor stage. Also known as input resistance. { tran'zis·tər 'in,pút ri,zis·təns }

transistor-transistor logic |ELECTR| A logic circuit containing two transistors, for driving large output capacitances at high speed. Abbreviated T²L; TTL. { tran'zis·tər tran'zis·tər 'läj·ik }

transit |ENG| 1. A surveying instrument with the telescope mounted so that it can measure horizontal and vertical angles. Also known as transit theodolite. 2. To reverse the direction of the telescope of a transit by rotating 180° about its horizontal axis. Also known as plunge. { 'trans·ət }

transit circle |ENG| A type of astronomical transit instrument having a micrometer eyepiece that has an extra pair of moving wires perpendicular to the vertical set to measure the zenith distance or declination of the celestial object in conjunction with readings taken from a large, accurately calibrated circle attached to the horizontal axis. Also known as meridian circle; meridian transit. { 'trans·ət ˌsər·kəl }

transit declinometer |ENG| A type of declinometer; a surveyor's transit, built to exacting specifications with respect to freedom from traces of magnetic impurities and quality of the compass needle, has a 17-power telescope for sighting on a mark and for making solar and stellar observations to determine true directions. { 'trans·ət ˌdek·lə'näm·əd·ər }

transition |THERMO| A change of a substance from one of the three states of matter to another. { tran'zish·ən }

transitional fit |DES ENG| A fit with varying clearances due to specified tolerances on the shaft and sleeve or hole. { tran'zish·ən·əl 'fit }

transition curve See easement curve. { tran'zish·ən ˌkərv }

transition factor See reflection factor. { tran'zish·ən ˌfak·tər }

transition frequency |ENG ACOUS| The frequency corresponding to the intersection of the asymptotes to the constant-amplitude and constant-velocity portions of the frequency-response curve for a disk recording; this curve is plotted with output-voltage ratio in decibels as the ordinate, and the logarithm of the frequency as the abscissa. Also known as cross-over frequency; turnover frequency. { tran'zish·ən ˌfrē·kwən·sē }

transition loss [ELEC] At a junction between a source and a load, the ratio of the available power to the power delivered to the load. { tran'zish·ən ˌlós }

transition point [THERMO] Either the temperature at which a substance changes from one state of aggregation to another (a first-order transition), or the temperature of culmination of a gradual change, such as the lambda point, or Curie point (a second-order transition). Also known as transition temperature. { tran'zish·ən ˌpóint }

transition temperature See transition point. { tran'zish·ən ˌtem·prə·chər }

transit survey [ENG] A ground surveying method in which a transit instrument is set up at a control point and oriented, and directions and distances to observed points are recorded. { 'trans·ət 'sər,vā }

transit theodolite See transit. { 'trans·ət thē'äd·əl,īt }

translation [MECH] The linear movement of a point in space without any rotation. { tran'slā·shən }

translational motion [MECH] Motion of a rigid body in such a way that any line which is imagined rigidly attached to the body remains parallel to its original direction. { tran'slā·shən·əl 'mō·shən }

transmembrane distillation See membrane distillation. { ˌtranzˌmem,brān ,dis·tə'lā·shən }

transmissibility [MECH] A measure of the ability of a system either to amplify or to suppress an input vibration, equal to the ratio of the response amplitude of the system in steady-state forced vibration to the excitation amplitude; the ratio may be in forces, displacements, velocities, or accelerations. { tranzˌmis·ə'bil·əd·ē }

transmission [ELECTR] **1.** The process of transferring a signal, message, picture, or other form of intelligence from one location to another location by means of wire lines, radio, light beams, infrared beams, or other communication systems. **2.** A message, signal, or other form of intelligence that is being transmitted. [MECH ENG] The gearing system by which power is transmitted from the engine to the live axle in an automobile. Also known as gearbox. { tranz'mish·ən }

transmission access [ELEC] The use of electric power lines and other power transmitting facilities by parties other than the owners of the lines. Also known as common carriage. { tranz'mish·ən 'ak,ses }

transmission dynamometer [ENG] A device for measuring torque and power (without loss) between a propulsion power plant and the driven

mechanism, for example, wheels or propellers. { tranz'mish·ən ˌdī·nə'mäm·əd·ər }

transmission line [ELEC] A system of conductors, such as wires, waveguides, or coaxial cables, suitable for conducting electric power or signals efficiently between two or more terminals. { tranz'mish·ən ˌlīn }

transmission-line admittance [ELEC] The complex ratio of the current flowing in a transmission line to the voltage across the line, where the current and voltage are expressed in phasor notation. { tranz'mish·ən ˌlīn ad,mit·əns }

transmission-line attenuation [ELEC] The decrease in power of a transmission-line signal from one point to another, expressed as a ratio or in decibels. { tranz'mish·ən ˌlīn ə,ten·yə,wā·shən }

transmission-line cable [ELEC] The coaxial cable, waveguide, or microstrip which forms a transmission line; a number of standard types have been designated, specified by size and materials. { tranz'mish·ən ˌlīn ,kā·bəl }

transmission-line constants See transmission-line parameters. { tranz'mish·ən ˌlīn ,kän·stəns }

transmission-line current [ELEC] The amount of electrical charge which passes a given point in a transmission line per unit time. { tranz'mish·ən ˌlīn ,kə·rənt }

transmission-line efficiency [ELEC] The ratio of the power of a transmission-line signal at one end of the line to that at the other end where the signal is generated. { tranz'mish·ən ˌlīn i,fish·ən·sē }

transmission-line impedance [ELEC] The complex ratio of the voltage across a transmission line to the current flowing in the line, where voltage and current are expressed in phasor notation. { tranz'mish·ən ˌlīn im,pēd·əns }

transmission-line parameters [ELEC] The quantities which are necessary to specify the impedance per unit length of a transmission line, and the admittance per unit length between various conductors of the line. Also known as linear electrical parameters; line parameters; transmission line constants. { tranz'mish·ən ˌlīn pə,ram·əd·ərz }

transmission-line power [ELEC] The amount of energy carried past a point in a transmission line per unit time. { tranz'mish·ən ˌlīn ,pau·ər }

transmission-line reflection coefficient [ELEC] The ratio of the voltage reflected from the load at the end of a transmission line to the direct voltage. { tranz'mish·ən ˌlīn ri'flek·shən ,kō·i,fish·ənt }

transmission-line theory [ELEC] The application of electrical and electromagnetic theory to the behavior of transmission lines. { tranz'mish·ən ˌlīn ,thē·ə·rē }

transmission-line transducer loss [ELEC] The ratio of the power delivered by a transmission line to a load to that produced at the generator, expressed in decibels; equal to the sum of the attenuation of the line and the mismatch loss. { tranz'mish·ən ˌlīn trans'dü·sər,lós }

transmission-line voltage [ELEC] The work that

would be required to transport a unit electrical charge between two specified conductors of a transmission line at a given instant. { tranz 'mish·ən ⎪līn ‚vōl·tij }

transmission substation ⎪ELEC⎪ An electric power substation associated with high voltage levels. { tranz'mish·ən 'səb‚stā·shən }

transmission tower ⎪ENG⎪ A concrete, metal, or timber structure used to carry a transmission line. { tranz'mish·ən ‚taú·ər }

transmissometer ⎪ENG⎪ An instrument for measuring the extinction coefficient of the atmosphere and for the determination of visual range. Also known as hazemeter; transmittance meter. { ‚tranz·mə'säm·əd·ər }

transmittance meter See transmissometer. { tranz 'mid·əns ‚mēd·ər }

transmitter See synchro transmitter. { tranz'mid·ər }

transmitter noise See frying noise. { tranz'mid·ər ‚nóiz }

transobuoy ⎪ENG⎪ A free-floating or moored automatic weather station developed for the purpose of providing weather reports from the open oceans; it transmits barometric pressure, air temperature, sea-water temperature, and wind speed and direction. { 'tran·sə‚bói }

transom ⎪BUILD⎪ A window above a door. { 'tran·səm }

transonic wind tunnel ⎪ENG⎪ A type of high-speed wind tunnel capable of testing the effects of airflow past an object at speeds near the speed of sound, Mach 0.7 to 1.4; sonic speed occurs where the cross section of the tunnel is at a minimum, that is, where the test object is located. { tran'sän·ik 'wind ‚tən·əl }

transosonde ⎪ENG⎪ The flight of a constant-level balloon, whose trajectory is determined by tracking with radio-direction-finding equipment; thus, it is a form of upper-air, quasi-horizontal sounding. { 'tran·zə‚sänd }

transponder set ⎪ELECTR⎪ A complete electronic set which is designed to receive an interrogation signal, and which retransmits coded signals that can be interpreted by the interrogating station; it may also utilize the received signal for actuation of additional equipment such as local indicators or servo amplifiers. { tranz'pän·dər ‚set }

transport ⎪ENG⎪ Conveyance equipment such as vehicular transport, hydraulic transport, and conveyor-belt setups. { trans'pórt (verb), 'tranz ‚pórt (noun) }

transportation emergency ⎪ENG⎪ A situation which is created by a shortage of normal transportation capability and of a magnitude sufficient to frustrate movement requirements, and which requires extraordinary action by the designated authority to ensure continued movement. { ‚tranz·pər'tā·shən i‚mər·jən·sē }

transportation engineering ⎪ENG⎪ That branch of engineering relating to the movement of goods and people; major types of transportation are highway, water, rail, subway, air, and pipeline. { ‚tranz·pər'tā·shən ‚en·jə‚nir·iŋ }

transportation lag See distance/velocity lag. { ‚tranz·pər'tā·shən ‚lag }

transportation priorities ⎪ENG⎪ Indicators assigned to eligible traffic which establish its movement precedence; appropriate priority systems apply to the movement of traffic by sea and air. { ‚tranz·pər'tā·shən prī‚är·əd·ēz }

transportation problem ⎪IND ENG⎪ A programming problem that is concerned with the optimal pattern of the distribution of goods from several points of origin to several different destinations, with the specified requirements at each destination. { ‚tranz·pər'tā·shən ‚präb·ləm }

transport capacity ⎪ENG⎪ The number of persons or the tonnage (or volume) of equipment which can be carried by a vehicle under given conditions. { 'tranz‚pórt kə‚pas·əd·ē }

transport case ⎪ENG⎪ A moistureproof nonconductive wood, plastic, or fabric container used to transport safely small quantities of dynamite sticks to and from blasting sites. { 'tranz‚pórt ‚kās }

transporter crane ⎪MECH ENG⎪ A long lattice girder supported by two lattice towers which may be either fixed or moved along rails laid at right angles to the girder; a crab with a hoist suspended from it travels along the girder. { trans'pórd·ər ‚krān }

transport lag See distance/velocity lag. { 'tranz ‚pórt ‚lag }

transport network ⎪ENG⎪ The complete system of the routes pertaining to all means of transport available in a particular area, made up of the network particular to each means of transport. { 'tranz‚pórt ‚net‚wərk }

transport vehicle ⎪MECH ENG⎪ Vehicle primarily intended for personnel and cargo carrying. { 'tranz‚pórt ‚vē·ə·kəl }

transverse baffle See cross-flow baffle. { trans'vərs ‚baf·əl }

transverse magnetization ⎪ENG ACOUS⎪ Magnetization of a magnetic recording medium in a direction perpendicular to the line of travel and parallel to the greatest cross-sectional dimension. { trans'vərs ‚mag·nəd·ə'zā·shən }

transverse stability ⎪ENG⎪ The ability of a ship or aircraft to recover an upright position after waves or wind roll it to one side. { trans'vərs stə'bil·əd·ē }

transverse vibration ⎪MECH⎪ Vibration of a rod in which elements of the rod move at right angles to the axis of the rod. { trans'vərs vī'brā·shən }

trap ⎪CIV ENG⎪ A bend or dip in a soil drain which is always full of water, providing a water seal to prevent odors from entering the building. ⎪ELECTR⎪ **1.** A tuned circuit used in the radio-frequency or intermediate-frequency section of a receiver to reject undesired frequencies; traps in television receiver video circuits keep the sound signal out of the picture channel. Also known as rejector. **2.** See wave trap. ⎪ENG⎪ A sealed passage such as a U-shaped bend in a pipe or pump that prevents the return flow of liquid or gas. ⎪MECH ENG⎪ A device which reduces the effect of the vapor pressure of oil or mercury on the high-vacuum side of a diffusion pump. { trap }

TRAPATT diode |ELECTR| A *pn* junction diode, similar to the IMPATT diode, but characterized by the formation of a trapped space-charge plasma within the junction region; used in the generation and amplification of microwave power. Derived from trapped plasma avalanche transit time diode. { 'tra,pat ,dī,ōd }

trapdoor |BUILD| **1.** A hinged, sliding, or lifting door to cover an opening in a roof, ceiling, or floor. **2.** An undocumented entry point into a computer program, which is generally inserted by a programmer to allow discreet access to the program. { 'trap,dór }

trapezoidal excavator |MECH ENG| A digging machine which removes earth in a trapezoidal cross-section pattern for canals and ditches. { ¦trap·ə¦zóid·əl 'eks·kə,vād·ər }

trapped-air process |ENG| A procedure for the blow-mold forming of closed plastic objects; the bottom pinch is conventional and, after blowing, sliding pinchers close off the top to form a sealed-air, inflated product. { 'trapt ¦er 'prä·səs }

trapped fuel |ENG| The fuel in an engine or fuel system that is not in the fuel tanks. { 'trapt 'fyül }

trap seal |CIV ENG| The vertical distance between the crown weir and the top of the dip of the trap in a plumbing system. { 'trap ,sēl }

trash screen |CIV ENG| A screen placed in a waterway to prevent the passage of trash. { 'trash ,skrēn }

Trauzl test |ENG| A test to determine the relative disruptive power of explosives, in which a standard quantity of explosive (10 grams) is placed in a cavity in a lead block and exploded; the resulting volume of cavity in the block is compared with the volume produced under the same conditions by a standard explosive, usually trinitrotoluene (TNT). { 'traút·səl ,test }

trave |BUILD| **1.** A division or bay (as in a ceiling) made by or appearing to be made by crossbeams. **2.** *See* crossbeam. { 'trāv }

travel |MECH ENG| The vertical distance of the path of an elevator or escalator as measured from the bottom terminal landing to the top terminal landing. { 'trav·əl }

travel chart |IND ENG| A tabulation of the various distances traveled by personnel or material between points in a manufacturing facility. { 'trav·əl ,chärt }

travel envelope |IND ENG| The clearance in space required by an automated guided vehicle when the vehicle is carrying a load with the maximum permissible dimensions. { ¦trav·əl ¦en·və,lōp }

traveling block |MECH ENG| The movable unit, consisting of sheaves, frame, clevis, and hook, connected to, and hoisted or lowered with, the load in a block-and-tackle system. Also known as floating block; running block. { 'trav·əl·iŋ 'bläk }

traveling detector |ENG| Radio-frequency probe which incorporates a detector used to measure the standing-wave ratio in a slotted-line section. { 'trav·əl·iŋ di'tek·tər }

traveling gantry crane |ENG| A type of hoisting machine with a bridgelike structure spanning the area over which it operates and running along tracks at ground level. { 'trav·əl·iŋ 'gan·trē ,krān }

traveling-grate stoker |MECH ENG| A type of furnace stoker; coal feeds by gravity into a hopper located on top of one end of a moving (traveling) grate; as the grate passes under the hopper, it carries a bed of fresh coal toward the furnace. { 'trav·əl·iŋ ¦grāt 'stō·kər }

traveling-screen dryer |CHEM ENG| A moving screen belt on which damp material is conveyed through a heated drying zone. Also known as screen dryer. { 'trav·əl·iŋ ¦skrēn 'drī·ər }

traveling-wave tube |ELECTR| An electron tube in which a stream of electrons interacts continuously or repeatedly with a guided electromagnetic wave moving substantially in synchronism with it, in such a way that there is a net transfer of energy from the stream to the wave; the tube is used as an amplifier or oscillator at frequencies in the microwave region. { 'trav·əl·iŋ ¦wāv ,tüb }

traverse |ENG| **1.** A survey consisting of a set of connecting lines of known length, meeting each other at measured angles. Also known as survey traverse. **2.** Movement to right or left on a pivot or mount, as of a gun, launcher, or radar antenna. { tra'vərs }

traverse adjustment *See* balancing a survey. { tra'vərs ə,jəs·mənt }

traversing mechanism |ENG| Mechanism by which a gun or other device can be turned in a horizontal plane. { tra'vərs·iŋ ,mek·ə,niz·əm }

trawl |ENG| A baglike net whose mouth is kept open by boards or by a leading diving vane or depressor at the foot of the opening and a spreader bar at the top; towed by a ship at specified depths for catching forms of marine life. { tról }

tray elevator |MECH ENG| A device for lifting drums, barrels, or boxes; a parallel pair of vertical-mounted continuous chains turn over upper and lower drive gears, and spaced trays on the chains cradle and lift the objects to be moved. { 'trā ,el·ə,vād·ər }

tray tower |CHEM ENG| A vertical process tower for liquid-vapor contacting (as in distillation, absorption, stripping, evaporation, spray drying, dehumidification, humidification, flashing, rectification, dephlegmation), along the height of which is a series of trays designed to cause intimate contact between the falling liquid and the rising vapor. { 'trā ,taú·ər }

tread |CIV ENG| **1.** The horizontal part of a step in a staircase. **2.** The distance between two successive risers in a staircase. |ENG| The part of a wheel or tire that bears on the road or rail. { tred }

treater [CHEM ENG] A vessel or system for the contacting of a process stream with reagent (treating) chemicals; for example, acid treating or caustic treating. { 'trēd·ər }

treating [CHEM ENG] Usually, the contacting of a fluid stream (for example, water, sewage, petroleum products, or mixed gases) with chemicals to improve the fluid properties by removing, sequestering, or converting undesirable impurities. { 'trēd·iŋ }

tremolo circuit [ENG ACOUS] A device which imparts a simple periodic amplitude modulation on the sound produced by an electronic instrument. { 'trem·ə·lō ‚sər·kət }

tremie [ENG] An apparatus for placing concrete underwater, consisting of a large metal tube with a hopper at the top end and a valve arrangement at the bottom, submerged end. { 'trem·ē }

trench duct [CIV ENG] A metal-lined trough set into a concrete floor with removable cover plates that are level with the top of the floor; used to house electrical connections. { 'trench ‚dəkt }

trencher See trench excavator. { 'trench·ər }

trench excavator [MECH ENG] A digging machine, usually on crawler tracks, and having either a movable wheel or a continuous chain on which buckets are mounted. Also known as bucket-ladder excavator; ditcher; trencher; trenching machine. { 'trench 'ek·ska‚vād·ər }

trenching machine See trench excavator. { 'trench·iŋ ma‚shēn }

trench shield [CIV ENG] A movable shoring system consisting of steel plates and braces that are bolted or welded together; used to support the walls of a trench while work is in progress. { 'trench ‚shēld }

trennschaukel apparatus [ENG] An instrument for determining the thermal diffusion factors of gases and gas mixtures, consisting of 20 suitably interconnected tubes whose top ends are maintained at the same temperature and whose bottom ends are maintained at the same temperature, with the temperature of the top ends greater than that of the bottom ends. { 'tren ‚shaù·kəl ‚ap·ə‚rad·əs }

trepanning tool [MECH ENG] A cutting tool in the form of a circular tube, having teeth on the end; the workpiece or tube, or both, are rotated and the tube is fed axially into the workpiece, leaving behind a narrow grooved surface in the workpiece. { trə'pan·iŋ ‚tül }

Tresca criterion [MECH] The assumption that plastic deformation of a material begins when the difference between the maximum and minimum principal stresses equals twice the yield stress in shear. { 'tres·kə krī‚tir·ē·ən }

trestle [CIV ENG] A series of short bridge spans supported by a braced tower. [ENG] **1.** A movable support usually with legs that spread diagonally. **2.** A braced structure of timber, reinforced concrete, or steel spanning a land depression to carry a road or railroad. { 'tres·əl }

trestle bent [CIV ENG] A transverse frame that supports the ends of the stringers in adjoining spans of a trestle. { 'tres·əl ‚bent }

trial batch [ENG] A batch of concrete mixed to determine the water-cement ratio that will produce the required slump and compressive strength; from a trial batch, one can also compute the yield, cement factor, and required quantities of each material. { ‚trīl 'bach }

trial shots [ENG] The experimental shots and rounds fired in a sinking pit, tunnel, opencast, or quarry to determine the best drill-hole pattern to use. { 'trīl ‚shäts }

triangle equation See angle equation. { 'trī‚aŋ·gəl i‚kwā·zhən }

triangle of forces [MECH] A triangle, two of whose sides represent forces acting on a particle, while the third represents the combined effect of these forces. { 'trī‚aŋ·gəl əv 'fȯr·səs }

triangular-notch weir [CIV ENG] A measuring weir with a V-shaped notch for measuring small flows. Also known as V-notch weir. { trī'aŋ·gyə·lər ‚näch 'wer }

triangulation [ENG] A surveying method for measuring a large area of land by establishing a base line from which a network of triangles is built up; in a series, each triangle has at least one side common with each adjacent triangle. { trī‚aŋ·gyə'lā·shən }

triangulation mark [ENG] A bronze disk set in the ground to identify a point whose latitude and longitude have been determined by triangulation. { trī‚aŋ·gyə'lā·shən ‚märk }

tribometer [ENG] A device for measuring coefficients of friction, consisting of a loaded sled subject to a measurable force. { trī'bäm·əd·ər }

trickle charge [ELEC] A continuous charge of a storage battery at a low rate to maintain the battery in a fully charged condition. { 'trik·əl ‚chärj }

trickle cooler See cascade cooler. { 'trik·əl ‚kü·lər }

trickle drain [CIV ENG] A drain that is set vertically in water, such as a pond, with its top open and level with the normal water surface in order to carry off excess water. { 'trik·əl ‚drān }

trickle hydrodesulfurization [CHEM ENG] A fixed-bed, petroleum refining process for desulfurization of middle distillates and gas oils; catalyst is cobalt molybdenum on alumina. { 'trik·əl ‚hī·drō·dē‚səl·fə·rə'zā·shən }

trickling filter [CIV ENG] A bed of broken rock or other coarse aggregate onto which sewage or industrial waste is sprayed intermittently and allowed to trickle through, leaving organic matter on the surface of the rocks, where it is oxidized and removed by biological growths. { 'trik·liŋ ‚fil·tər }

tricone bit [ENG] A rock bit with three toothed, conical cutters, each of which is mounted on friction-reducing bearings. { 'trī‚kōn 'bit }

trifilter hydrophotometer [ENG] An instrument that uses red, green, and blue filters to measure the transparency of the water at three wavelengths. { 'trī‚fil·tər ‚hī·drō·fə'täm·əd·ər }

trigger bolt See auxiliary dead latch. { 'trig·ər ‚bōlt }

trigger pull |MECH| Resistance offered by the trigger of a rifle or other weapon; force which must be exerted to pull the trigger. { 'trig·ər ‚pul }

trigonometric leveling |ENG| A method of determining the difference of elevation between two points, by using the principles of triangulation and trigonometric calculations. { ¦trig·ə·nə¦me·trik 'lev·əl·iŋ }

trilateration |ENG| The measurement of a series of distances between points on the surface of the earth, for the purpose of establishing relative positions of the points in surveying. { trī‚lad·ə'rā·shən }

trim |ELECTR| Fine adjustment of capacitance, inductance, or resistance of a component during manufacture or after installation in a circuit. { trim }

trimmer |BUILD| One of the single or double joists or rafters that go around an opening in the framing type of construction. { 'trim·ər }

trimmer conveyor |MECH ENG| A self-contained, lightweight portable conveyor, usually of the belt type, for use in unloading and delivering bulk materials from trucks to domestic storage places, and for trimming bulk materials in bins or piles. { 'tim·ər kən‚vā·ər }

triode transistor |ELECTR| A transistor that has three terminals. { 'trī‚ōd tran'zis·tər }

trip |ENG| To release a lever or set free a mechanism. { trip }

trip hammer |MECH ENG| A large power hammer whose head is tripped and falls by cam or lever action. { 'trip ‚ham·ər }

triple thread |DES ENG| A multiple screw thread having three threads or starts equally spaced around the periphery; the lead is three times the pitch. { 'trip·əl 'thred }

triplex chain block |MECH ENG| A geared hoist using an epicyclic train. { 'trip‚leks 'chān ‚bläk }

tripod |DES ENG| An adjustable, collapsible three-legged support, as for a camera or surveying instrument. { 'trī‚päd }

tripodal grasp |IND ENG| A basic grasp whereby an object is held by the thumb, index finger, and middle finger, to provide delicate rotational control. Also known as manipulative grasp. { ¦trī‚pōd·əl 'grasp }

tripod drill |MECH ENG| A reciprocating rock drill mounted on three legs and driven by steam or compressed air; the drill steel is removed and a longer drill inserted about every 2 feet (61 centimeters). { 'trī‚päd ‚dril }

tripper |CIV ENG| A device activated by a passing train to work a signal or switch or to apply brakes. |MECH ENG| A device that snubs a conveyor belt causing the load to be discharged. { 'trip·ər }

trip spear |ENG| A fishing tool intended to recover lost casing; if the casing is found to be immovable, the hold is broken by operating the trip release. { 'trip ‚spir }

trisistor |ELECTR| Fast-switching semiconductor consisting of an alloyed junction *pnp* device

in which the collector is capable of electron injection into the base; characteristics resemble those of a thyratron electron tube, and switching time is in the nanosecond range. { tri'zis·tər }

tristate logic |ELECTR| A form of transistor-transistor logic in which the output stages or input and output stages can assume three states; two are the normal low-impedance 1 and 0 states, and the third is a high-impedance state that allows many tristate devices to time-share bus lines. { 'trī‚stāt 'läj·ik }

trolley |MECH ENG| **1.** A wheeled car running on an overhead track, rail, or ropeway. **2.** An electric streetcar. { 'träl·ē }

trolley locomotive |MECH ENG| A locomotive operated by electricity drawn from overhead conductors by means of a trolley pole. { 'träl·ē ‚lōk·ə'mōd·iv }

tropical finish |ENG| A finish that is applied to electronic equipment to resist the high relative humidity, fungus, and insects encountered in tropical climates. { 'träp·ə·kəl 'fin·ish }

tropicalize |ENG| To prepare electronic equipment for use in a tropical climate by applying a coating that resists moisture and fungi. { 'träp·ə·kə‚līz }

tropometer |ENG| An instrument for measuring the angle through which one end of a bar is twisted in determining the strength of a material in torsion. { trə'päm·əd·ər }

troughed belt conveyor |MECH ENG| A belt conveyor with the conveyor belt edges elevated on the carrying run to form a trough by conforming to the shape of the troughed carrying idlers or other supporting surface. { 'tróft 'belt kən‚vā·ər }

troughed roller conveyor |MECH ENG| A roller conveyor having two rows of rolls set at an angle to form a trough over which objects are conveyed. { 'tróft 'rō·lər kən‚vā·ər }

troughing idler |MECH ENG| A belt idler having two or more rolls arranged to turn up the edges of the belt so as to form the belt into a trough. { 'tróf·iŋ ‚īd·lər }

troughing rolls |MECH ENG| The rolls of a troughing idler that are so mounted on an incline as to elevate each edge of the belt into a trough. { 'tróf·iŋ ‚rōlz }

Trouton's rule |THERMO| The rule that, for a nonassociated liquid, the latent heat of vaporization in calories is equal to approximately 22 times the normal boiling point on the Kelvin scale. { 'traut·ənz ‚rül }

trowel |DES ENG| Any of various hand tools consisting of a wide, flat or curved blade with a short wooden handle; used by gardeners, plasterers, and bricklayers. { 'traul }

troweling machine |MECH ENG| A motorized device used to spread concrete by operating orbiting steel trowels on radial arms rotated on a vertical shaft. { 'trawl·iŋ mə‚shēn }

troy ounce *See* ounce. { 'trói 'auns }

troy pound *See* pound. { 'trói 'paund }

troy system |MECH| A system of mass units used primarily to measure gold and silver; the

582

ounce is the same as that in the apothecaries' system, being equal to 480 grains or 31.1034768 grams. Abbreviated t. Also known as troy weight. { 'tròi ,sis·tǝm }

troy weight See troy system. { 'tròi ,wāt }

truck |MECH ENG| A self-propelled wheeled vehicle, designed primarily to transport goods and heavy equipment; it may be used to tow trailers or other mobile equipment. { trǝk }

truck crane |MECH ENG| A crane carried on the bed of a motortruck. { 'trǝk ,krān }

truck-mounted drill rig |MECH ENG| A drilling rig mounted on a lorry or caterpillar tracks. { 'trǝk ¦maúnt·ǝd 'dril ,rig }

truck-tractor See tractor. { 'trǝk ¦trak·tǝr }

true-boiling-point analysis |CHEM ENG| A standard laboratory technique used to predict the refining qualities of crude petroleum; gives distillation cuts for gasoline, kerosine, distillate (diesel) fuel, cracking, and lube distillate stocks. Also known as true-boiling-point distillation. { trü 'bóil·iɳ ¦póint ǝ,nal·ǝ·sǝs }

true-boiling-point distillation See true-boiling-point analysis. { trü 'bóil·iɳ ¦póint ,dis·tǝ,lā·shǝn }

true rake |MECH ENG| The angle, measured in degrees, between a plane containing a tooth face and the axial plane through the tooth point in the direction of chip flow. { 'trü 'rāk }

truing |MECH ENG| **1.** Cutting a grinding wheel to make its surface run concentric with the axis. **2.** Aligning a wheel to be concentric and in one plane. { 'trü·iɳ }

truncate |CONT SYS| To stop a robotic process before it has been completed. { 'trǝɳ,kāt }

truncated icosahedral gravitational-wave antenna |ENG| A resonant-mass antenna for detecting gravitational radiation in which the shape of the mass is a truncated icosahedron, which is much more efficient for this purpose than a cylinder. Abbreviated TIGA. { ¦trǝɳ,kād·ǝd ī,käs·ǝ¦he·drǝl ,grav·ǝ¦tā·shǝn·ǝl 'wāv an,ten·ǝ }

truncation error |ENG| The error resulting from the analysis of a partial set of data in place of a complete or infinite set. { trǝɳ'kā·shǝn ,er·ǝr }

trunk buoy |ENG| A mooring buoy having a pendant extending through an opening in the buoy, with the ship's anchor chain or mooring line being secured to this pendant. { 'trǝɳk ,bói }

trunk sewer |CIV ENG| A sewer receiving sewage from many tributaries serving a large territory. { 'trǝɳk ,sü·ǝr }

trunnion |DES ENG| **1.** Either of two opposite pivots, journals, or gudgeons, usually cylindrical and horizontal, projecting one from each side of a piece of ordnance, the cylinder of an oscillating engine, a molding flask, or a converter, and supported by bearings to provide a means of swiveling or turning. **2.** A pin or pivot usually mounted on bearings for rotating or tilting something. |ENG| A tubular section of steel welded to the side of a pipe in order to help support the pipe. { 'trǝn·yǝn }

truss |CIV ENG| A frame, generally of steel, timber, concrete, or a light alloy, built from members in tension and compression. { trǝs }

truss bridge |CIV ENG| A fixed bridge consisting of members vertically arranged in a triangular pattern. { 'trǝs ,brij }

trussed beam |CIV ENG| A beam stiffened by a steel tie rod to reduce its deflection. { 'trǝst 'bēm }

trussed rafter |BUILD| A triangulated beam in a trussed roof. { 'trǝst 'raf·tǝr }

truss rod |CIV ENG| A rod attached to the ends of a trussed beam which transmits the strain due to downward pressure. { 'trǝs ,räd }

try square |ENG| An instrument consisting of two straightedges secured at right angles to each other, used for laying off right angles and testing whether work is square. { 'trī ,skwer }

Tschudi engine |MECH ENG| A cat-and-mouse engine in which the pistons, which are sections of a torus, travel around a toroidal cylinder; motion of the pistons is controlled by two cams which bear against rollers attached to the rotors. { 'chü·dē ,en·jǝn }

tsi |MECH| A unit of force equal to 1 ton-force per square inch; equal to approximately 1.54444 × 10⁷ pascals. { sī or ,tē,es'Ī }

T slot |DES ENG| A recessed slot, in the form of an inverted T, in the table of a machine tool, to receive the square head of a T-slot bolt. { 'tē ,slät }

tsp See teaspoonful.

tspn See teaspoonful.

TTL See transistor-transistor logic.

tube |ELECTR| See electron tube. |ENG| **1.** A long cylindrical body with a hollow center used especially to convey fluid. **2.** See inner tube. { 'tüb }

tube bank |MECH ENG| An array of tubes designed to be used as a heat exchanger. { 'tüb ,baɳk }

tube bundle |ENG| In a shell-and-tube heat exchanger, an assembly of parallel tubes that is tied together with tie rods. { 'tüb ,bǝn·dǝl }

tube cleaner |MECH ENG| A device equipped with cutters or brushes used to clean tubes in heat transfer equipment. { 'tüb ,klēn·ǝr }

tube door |MECH ENG| A door in a boiler furnace wall which facilitates the removal or installation of tubes. { 'tüb ,dór }

tube hole |ENG| A hole in a tube sheet through which a tube is passed prior to sealing. { 'tüb ,hōl }

tubeless tire |ENG| A tire that does not require an inner tube to hold air. { ¦tüb·lǝs 'tīr }

tube mill |MECH ENG| A revolving cylinder used for fine pulverization of ore, rock, and other such materials; the material, mixed with water, is fed into the chamber from one end, and passes out the other end as slime. { 'tüb ,mil }

tube plug |ENG| A solid plug inserted into the end of a tube in a tube sheet. { 'tüb ,plǝg }

tube seat |ENG| The surface of the tube hole in a tube sheet which contacts the tube. { 'tüb ,sēt }

tube sheet |ENG| A mounting plate for elements of a larger item of equipment; for example, filter cartridges, or tubes for heat exchangers, coolers, or boilers. { 'tüb ,shēt }

tube shield |ENG| A shield designed to be placed around an electron tube. { 'tüb ,shēld }

tube socket |ENG| A socket designed to accommodate electrically and mechanically the terminals of an electron tube. { 'tüb ,säk·ət }

tube-still heater |CHEM ENG| A firebox containing a pipe coil through which oil for a tube still (pipe still) is pumped. { 'tüb ¦stil ,hēd·ər }

tube turbining |MECH ENG| Cleaning tubes by passing a power-driven rotary device through them. { 'tüb ,tər·bən·iŋ }

tube voltmeter See vacuum-tube voltmeter. { 'tüb 'vōlt,mēd·ər }

tubing |ENG| Material in the form of a tube, most often seamless. { 'tüb·iŋ }

tubular exchanger See shell-and-tube exchanger. { 'tü·byə·lər iks'chānj·ər }

tuck-and-pat pointing See tuck pointing. { ¦tək ən ¦pat ,póint·iŋ }

tuck joint pointing See tuck pointing. { 'tək ,jóint ,póint·iŋ }

tuck pointing |BUILD| The finishing of old masonry joints in which the joints are first cleaned out and then filled with fine mortar which projects slightly or has a fillet of putty or lime. Also known as tuck-and-pat pointing; tuck joint pointing. { 'tək ,póint·iŋ }

Tukon tester |ENG| A device that uses a diamond (Knoop) indenter applying average loads of 1 to 2000 grams to determine microhardness of a metal. { 'tü,kän ,tes·tər }

tumble See topple. { 'təm·bəl }

tumble axis See topple axis. { 'təm·bəl ,ak·səs }

tumbler |ENG| **1.** A device in a lock cylinder that must be moved to a particular position, as by a key, before the bolt can be thrown. **2.** A device or mechanism in which objects are tumbled. { 'təm·blər }

tumbler feeder See drum feeder. { 'təm·blər ,fēd·ər }

tumbler gears |MECH ENG| Idler gears interposed between spindle and stud gears in a lathe gear train; used to reverse rotation of lead screw or feed rod. { 'təm·blər ,girz }

tumbling |ENG| A surface-finishing operation for small articles in which irregularities are removed or surfaces are polished by tumbling them together in a barrel, along with wooden pegs, sawdust, and polishing compounds. |MECH ENG| Loss of control in a two-frame free gyroscope, occurring when both frames of reference become coplanar. { 'təm·bliŋ }

tumbling mill |MECH ENG| A grinding and pulverizing machine consisting of a shell or drum rotating on a horizontal axis. { 'təm·bliŋ ,mil }

tune |ELECTR| To adjust for resonance at a desired frequency. { tün }

tuned amplifier |ELECTR| An amplifier in which the load is a tuned circuit; load impedance and amplifier gain then vary with frequency. { ¦tünd 'am·plə,fī·ər }

tuned-anode oscillator |ELECTR| A vacuum-tube oscillator whose frequency is determined by a tank circuit in the anode circuit, coupled to the grid to provide the required feedback. Also known as tuned-plate oscillator. { ¦tünd 'an,ōd ,äs·ə,lād·ər }

tuned circuit |ELECTR| A circuit whose components can be adjusted to make the circuit responsive to a particular frequency in a tuning range. Also known as tuning circuit. { ¦tünd 'sər·kət }

tuned filter |ELECTR| Filter that uses one or more tuned circuits to attenuate or pass signals at the resonant frequency. { ¦tünd 'fil·tər }

tuned-reed frequency meter See vibrating-reed frequency meter. { ¦tünd 'rēd 'frē·kwən·sē ,mēd·ər }

tuner |ELECTR| The portion of a receiver that contains circuits which can be tuned to accept the carrier frequency of the alternating current supplied to the primary, thereby causing the secondary voltage to build up to higher values than would otherwise be obtained. { 'tü·nər }

tuning fork |ENG| A U-shaped bar for hard steel, fused quartz, or other elastic material that vibrates at a definite natural frequency when struck or when set in motion by electromagnetic means; used as a frequency standard. { 'tün·iŋ ,fórk }

tunnel |ENG| A long, narrow, horizontal or nearly horizontal underground passage that is open to the atmosphere at both ends; used for aqueducts and sewers, carrying railroad and vehicular traffic, various underground installations, and mining. { 'tən·əl }

tunnel blasting |ENG| A method of heavy blasting in which a heading is driven into the rock and afterward filled with explosives in large quantities, similar to a borehole, on a large scale, except that the heading is usually divided in two parts on the same level at right angles to the first heading, forming in plan a T, the ends of which are filled with explosives and the intermediate parts filled with inert material like an ordinary borehole. { 'tən·əl ,blast·iŋ }

tunnel borer |MECH ENG| Any boring machine for making a tunnel; often a ram armed with cutting faces operated by compressed air. { 'tən·əl ,bór·ər }

tunnel carriage |MECH ENG| A machine used for rapid tunneling, consisting of a combined drill carriage and manifold for water and air so that immediately the carriage is at the face, drilling may commence with no lost time for connecting up or waiting for drill steels; the air is supplied at pressures of 95 to 100 pounds per square inch (655,000 to 689,000 pascals). { 'tən·əl ,kar·ij }

tunnel diode |ELECTR| A heavily doped junction diode that has a negative resistance at very low voltage in the forward bias direction, due to quantum-mechanical tunneling, and a short circuit in the negative bias direction. Also known as Esaki tunnel diode. { 'tən·əl ,dī,ōd }

tunnel junction |ELECTR| A two-terminal electronic device having an extremely thin potential barrier to electron flow, so that the transport characteristic (the current-voltage curve) is primarily governed by the quantum-mechanical tunneling process which permits electrons to penetrate the barrier. { 'tən·əl ˌjäŋk·shən }

tunnel liner |CIV ENG| Any of various materials, especially timber, concrete, and cast iron, applied to the inner surface of a vehicular or railroad tunnel. { 'tən·əl ˌlīn·ər }

tunnel resistor |ELECTR| Resistor in which a thin layer of metal is plated across a tunneling junction, to give the combined characteristics of a tunnel diode and an ordinary resistor. { 'tən·əl ri,zis·tər }

tunnel triode |ELECTR| Transistorlike device in which the emitter-base junction is a tunnel diode and the collector-base junction is a conventional diode. { 'tən·əl ˌtrī,ōd }

turbine |MECH ENG| A fluid acceleration machine for generating rotary mechanical power from the energy in a stream of fluid. { 'tər·bən }

turbine propulsion |MECH ENG| Propulsion of a vehicle or vessel by means of a steam or gas turbine. { 'tər·bən prə,pəl·shən }

turbine pump See regenerative pump. { 'tər·bən ˌpəmp }

turbining |MECH ENG| The removal of scale or other foreign material from the internal surface of a metallic cylinder. { 'tər·bən·iŋ }

turboblower |MECH ENG| A centrifugal or axial-flow compressor. { 'tər·bō,blō·ər }

turbogrid plate |CHEM ENG| A tray for distillation columns that consists of a flat grid of parallel slots extending over the entire cross-sectional area of the column; the liquid level on each tray is maintained by a dynamic balance between down-flowing liquid and up-flowing vapor. { 'tər·bō,grid 'plāt }

turbopump |MECH ENG| A pump that is powered by a turbine. { 'tər·bō,pəmp }

turboshaft |MECH ENG| A gas turbine engine that is similar to a turboprop but operates through a transmission system to power a device such as a helicopter rotor or pump. { 'tər·bō,shaft }

turbosupercharger |MECH ENG| A centrifugal air compressor, gas-turbine driven, usually used to increase induction system pressure in an internal combustion reciprocating engine. { ˈtər·bō'sü·pər,chär·jər }

turbulent burner |ENG| An atomizing burner which mixes fuel and air to produce agitated flow. { 'tər·byə·lənt 'bər·nər }

turbulization |ENG| In a heat-transfer process involving the interaction of a solid, heat-conducting, and impermeable surface with a surrounding fluid, destruction of the boundary layer in order to intensify the convective heat transfer. { ˌtər·bə·lə'zā·shən }

turn |ELEC| One complete loop of wire. { 'tərn }

turnaround |CHEM ENG| In petroleum refining, the shutdown of a unit after a normal run for maintenance and repair work, then putting the unit back into operation. |ENG| The length of time between arriving at a point and departing from that point; it is used in this sense for the turnaround of vehicles, ships in ports, and aircraft. { 'tərn·ə,raùnd }

turnaround cycle |ENG| A term used in conjunction with vehicles, ships, and aircraft, and comprising the following: loading time at home, time to and from destination, unloading and loading time at destination, unloading time at home, planned maintenance time, and, where applicable, time awaiting facilities. { 'tərn·ə,raùnd ,sī·kəl }

turnbuckle |DES ENG| A sleeve with a thread at one end and a swivel at the other, or with threads of opposite hands at each end so that by turning the sleeve connected rods or wire rope will be drawn together and tightened. { 'tərn,bək·əl }

turning |MECH ENG| Shaping a member on a lathe. { 'tərn·iŋ }

turning bar See chimney bar. { 'tərn·iŋ ,bär }

turning basin |CIV ENG| An open area at the end of a canal or narrow waterway to allow boats to turn around. { 'tərn·iŋ ,bās·ən }

turning-block linkage |MECH ENG| A variation of the sliding-block mechanical linkage in which the short link is fixed and the frame is free to rotate. Also known as the Wentworth quick-return motion. { 'tərn·iŋ ˌbläk ,liŋ·kij }

turning center |MECH ENG| A numerically controlled lathe that sometimes functions together with a robot in boring and other machining work. { 'tərn·iŋ ,sen·tər }

turning table |ENG| In plastics molding, a rotating table or wheel carrying various molds in a multimold, single-parison blow-molding operation. { 'tərn·iŋ ,tā·bəl }

turnkey contract |ENG| A contract in which an independent agent undertakes to furnish for a fixed price all materials and labor, and to do all the work needed to complete a project. { 'tərn,kē 'kän,trakt }

turnout |ENG| **1.** A contrivance consisting of a switch, a frog, and two guardrails for passing from one track to another. **2.** The branching off of one rail track from another. **3.** A siding. { 'tərn,aùt }

turnover cartridge |ENG ACOUS| A phonograph pickup having two styli and a pivoted mounting that places in playing position the correct stylus for a particular record speed. { 'tərn,ō·vər ,kär·trij }

turnover frequency See transition frequency. { 'tərn,ō·vər ,frē·kwən·sē }

turnover number |CHEM ENG| In an industrial catalytic process, a value that indicates the amount of feed or substrate converted per a measured amount of catalyst. { 'tərn,ō·vər ,nəm·bər }

turnover rate |CHEM ENG| In an industrial catalytic process, a value corresponding to the turnover number per specified unit of time. { 'tərn,ō·vər ,rāt }

turnpike |CIV ENG| A toll expressway. { 'tərn ,pīk }

turns ratio |ELEC| The ratio of the number of turns in a secondary winding of a transformer to the number of turns in the primary winding. { 'tərnz ˌrā·shō }

turnstile |ENG| A barrier that rotates about a vertical axis and usually is arranged to allow the passage of only one person at a time through an opening. { 'tərn,stīl }

turntable |ENG ACOUS| The rotating platform on which a disk record is placed for recording or playback. { 'tərn,tā·bəl }

turntable rumble |ENG ACOUS| Low-frequency vibration that is mechanically transmitted to a recording or reproducing turntable and superimposed on the reproduction. Also known as rumble. { 'tərn,tā·bəl ,rəm·bəl }

turret lathe |MECH ENG| A semiautomatic lathe differing from the engine lathe in having the tailstock replaced with a multisided, indexing tool holder or turret designed to hold several tools. { 'tə·rət ,lāth }

turret robot |CONT SYS| A tower-shaped robot whose manipulator makes circular motions about the robot's base. { 'tər·ət 'rō,bät }

Twaddell scale |ENG| A scale for specific gravity of solutions that is the first two digits to the right of the decimal point multiplied by two; for example, a specific gravity of 1.4202 is equal to 84.04°Tw. { 'twä'del ,skāl }

tweeter |ENG ACOUS| A loudspeaker designed to handle only the higher audio frequencies, usually those well above 3000 hertz; generally used in conjunction with a crossover network and a woofer. { 'twēd·ər }

twin-cable ropeway |MECH ENG| An aerial ropeway which has parallel track cables with carriers running in opposite directions; both rows of carriers are pulled by the same traction rope. { 'twin ˌkāb·əl 'rōp,wā }

twin-geared press |MECH ENG| A crank press having the drive gears attached to both ends of the crankshaft. { 'twin ˌgird 'pres }

twist |DES ENG| In a fiber, rope, yarn, or cord, the turns about its axis per unit length; usually expressed as TPI (turns per inch). { twist }

twist drill |DES ENG| A tool having one or more helical grooves, extending from the point to the smooth part of the shank, for ejecting cuttings and admitting a coolant. { 'twist ,dril }

two-body problem |MECH| The problem of predicting the motions of two objects obeying Newton's laws of motion and exerting forces on each other according to some specified law such as Newton's law of gravitation, given their masses and their positions and velocities at some initial time. { 'tü ˌbäd·ē 'präb·ləm }

two-cycle engine |MECH ENG| A reciprocating internal combustion engine that requires two piston strokes or one revolution to complete a cycle. { 'tü ˌsī·kəl 'en·jən }

two-degrees-of-freedom gyro |MECH| A gyro whose spin axis is free to rotate about two orthogonal axes, not counting the spin axis. { 'tü ˌdi¦grēz əv ¦frē·dəm ¦jī·rō }

two-level mold |ENG| Placement of one cavity of a plastics mold above another instead of alongside it; reduces clamping force needed. { 'tü ¦lev·əl 'mōld }

two-lip end mill |MECH ENG| An end-milling cutter having two cutting edges and straight or helical flutes. { 'tü ¦lip 'end ,mil }

two-phase alternating-current circuit |ELEC| A circuit in which there are two alternating currents on separate wires, the two currents being 90° out of phase. { 'tü ¦fāz 'öl·tər,nād·iŋ kə·rənt ,sər·kət }

two-phase current |ELEC| Current delivered through two pairs of wires at a phase difference of one-quarter cycle (90°) between the current in the two pairs. { 'tü ¦fāz 'kə·rənt }

two-point press |MECH ENG| A mechanical press in which the slide is actuated at two points. { 'tü ¦pöint 'pres }

two-port system |CONT SYS| A system which has only one input or excitation and only one response or output. { 'tü ¦pört 'sis·təm }

two-sided sampling plans |IND ENG| Any sampling plan whereby the acceptability of material is determined against upper and lower limits. { 'tü ¦sīd·əd 'sam·pliŋ ,planz }

two-step grooving system |ENG| A method of spooling a drum in which the wire rope, controlled by grooves, moves parallel to the drum flanges for one-half the circumference and then crosses over to start the next wrap. Also known as counterbalance system. { 'tü ¦step 'grüv·iŋ ,sis·təm }

two-stroke cycle |MECH ENG| An internal combustion engine cycle completed in two strokes of the piston. { 'tü ¦strök 'sī·kəl }

two-tone diaphone |ENG ACOUS| A diaphone producing blasts of two tones, the second tone being of a lower pitch than the first tone. { 'tü ¦tōn 'dī·ə,fōn }

two-way slab |CIV ENG| A concrete slab supported by beams along all four edges and reinforced with steel bars arranged perpendicularly. { 'tü ¦wā 'slab }

two-way valve |MECH ENG| A mechanical device that controls the flow of fluid by allowing flow in either of two directions. { 'tü ¦wā 'valv }

two-wire circuit |ELEC| A metallic circuit formed by two conductors insulated from each other; in contrast with a four-wire circuit, it uses only one line or channel for transmission of electric waves in both directions. { 'tü ¦wīr 'sər·kət }

tyfon See typhon. { 'tī,fän }

Tyler screen |CHEM ENG| A screen standard for the openings in screen-type mediums based on meshes per linear inch; convertible to the U.S. Sieve Series. { 'tī·lər ,skrēn }

Tyler Standard screen scale |ENG| A scale for classifying particles in which the particle size in micrometers is correlated with the meshes per inch of a screen. { 'ti·lər 'stan·dərd 'skrēn ,skāl }

Tyndallization [ENG] Heat sterilization by steaming the food or medium for a few minutes at atmospheric pressure on three or four successive occasions, separated by 12- to 18-hour intervals of incubation at a temperature favorable for bacterial growth. { ‚tind·əl·ə'zā·shən }

type I assembly [ELECTR] An assembly consisting entirely of surface-mounted electronic components, on either one or both sides of a printed board. { 'tīp ¦wən ə'sem·blē }

type II assembly [ELECTR] An assembly of both surface-mounted and leaded electronic components, in which the surface-mounted components are on both sides of the printed board. { 'tīp ¦tü ə'sem·blē }

type III assembly [ELECTR] An assembly of both surface-mounted and leaded electronic components, in which the surface-mounted components are only on the bottom side of the printed board. { 'tīp ¦thrē ə'sem·blē }

typhon [ENG ACOUS] A diaphragm horn which operates under the influence of compressed air or steam. Also spelled tyfon. { 'tī‚fän }

U

U-bend die [MECH ENG] A die with a square or rectangular cross section which provides two edges over which metal can be drawn. { 'yü ,bend ,dī }

U blades [DES ENG] Curved bulldozer blades designed to increase moving capacity of tractor equipment. { 'yü ,blädz }

U bolt [DES ENG] A U-shaped bolt with threads at the ends of both arms to receive nuts. { 'yü ,bōlt }

udometer See rain gage. { yü'däm·əd·ər }

UJT See unijunction transistor.

ullage [ENG] The amount that a container, such as a fuel tank, lacks of being full. { 'əl·ij }

ultimate bearing capacity [CIV ENG] The average load per unit area that will cause failure by rupture of a supporting soil mass. { 'əl·tə·mət 'ber·iŋ kə,pas·əd·ē }

ultimate load See breaking load. { 'əl·tə·mət ,lōd }

ultimate-load design [DES ENG] Design of a beam that is proportioned to carry at ultimate capacity the design load multiplied by a safety factor. Also known as limit-load design, plastic design; ultimate-strength design. { 'əl·tə·mət ¦lōd di,zīn }

ultimate set [ENG] The ratio of the length of a specimen plate or bar before testing to the length at the moment of fracture; usually expressed as a percentage. { 'əl·tə·mət 'set }

ultimate strength [MECH] The tensile stress, per unit of the original surface area, at which a body will fracture, or continue to deform under a decreasing load. { 'əl·tə·mət 'streŋkth }

ultimate-strength design See ultimate-load design. { 'əl·tə·mət ¦streŋkth di,zīn }

ultracentrifuge [ENG] A laboratory instrument which develops centrifugal fields of more than 100,000 times gravity, used for the quantitative measurement of sedimentation velocity or sedimentation equilibrium, or for the separation of solutes in liquid solutions to study high polymers, particularly proteins, nucleic acids, viruses, and other macromolecules of biological origin. { ,əl·trə'sen·trə,fyüj }

ultrafiltration [CHEM ENG] Separation of colloidal or very fine solid materials by filtration through microporous or semipermeable mediums. { ¦əl·trə·fil'trā·shən }

ultramicrobalance [ENG] A differential weighing device with accuracies better than 1 microgram; used for analytical weighings in microanalysis. { ¦əl·trə'mī·krō,bal·əns }

ultramicrotome [ENG] A microtome which uses a glass or diamond knife, allowing sections of cells to be cut 300 nanometers in thickness. { ¦əl·trə'mī·krə,tōm }

ultrasonic atomizer [MECH ENG] An atomizer in which liquid is fed to, or caused to flow over, a surface which vibrates at an ultrasonic frequency; uniform drops may be produced at low feed rates. { ¦əl·trə'sän·ik 'ad·ə,mīz·ər }

ultrasonic cleaning [ENG] A method used to clean debris and swarf from surfaces by immersion in a solvent in which ultrasonic vibrations are excited. { ¦əl·trə'sän·ik 'klēn·iŋ }

ultrasonic delay line [ENG ACOUS] A delay line in which use is made of the propagation time of sound through a medium such as fused quartz, barium titanate, or mercury to obtain a time delay of a signal. Also known as ultrasonic storage cell. { ¦əl·trə'sän·ik di'lā ,līn }

ultrasonic depth finder [ENG] A direct-reading instrument which employs frequencies above the audible range to determine the depth of water; it measures the time interval between the emission of an ultrasonic signal and the return of its echo from the bottom. { ¦əl·trə'sän·ik 'depth ,fīn·dər }

ultrasonic drill [MECH ENG] A drill in which a magnetostrictive transducer is attached to a tapered cone serving as a velocity transformer; with an appropriate tool at the end of the transformer, practically any shape of hole can be drilled in hard, brittle materials such as tungsten carbide and gems. { ¦əl·trə'sän·ik 'dril }

ultrasonic drilling [MECH ENG] A vibration drilling method in which ultrasonic vibrations are generated by the compression and extension of a core of electrostrictive or magnetostrictive material in a rapidly alternating electric or magnetic field. { ¦əl·trə'sän·ik 'dril·iŋ }

ultrasonic flaw detector [ENG ACOUS] An ultrasonic generator and detector used together, much as in radar, to determine the distance to a wave-reflecting internal crack or other flaw in a solid object. { ¦əl·trə'sän·ik 'flò di,tek·tər }

ultrasonic generator [ENG ACOUS] A generator

consisting of an oscillator driving an electroacoustic transducer, used to produce acoustic waves above about 20 kilohertz. { |əl·trə'sän·ik 'jen·ə,rād·ər }

ultrasonic imaging device |ENG ACOUS| An imaging device in which a wave is generated by a transducer external to the body; the reflected wave is detected by the same transducer. { |əl·trə'sän·ik 'im·ij·iŋ di,vīs }

ultrasonic inspectoscope |ENG ACOUS| An instrument that transmits sound waves, at frequencies between 500 kilohertz and 15 megahertz, into a metal casting or other solid piece and determines the presence of flaws by reflections or by an interruption of the sound-wave transmission through the piece. { |əl·trə'sän·ik in'spek·tə,skōp }

ultrasonic leak detector |ENG| An instrument which detects ultrasonic energy resulting from the transition from laminar to turbulent flow of a gas passing through an orifice. { |əl·trə'sän·ik 'lēk di,tek·tər }

ultrasonic machining |MECH ENG| The removal of material by abrasive bombardment and crushing in which a ball-ended tool of soft alloy steel is made to vibrate at a frequency of about 20,000 hertz and an amplitude of 0.001–0.003 inch (0.0254–0.0762 millimeter) while a fine abrasive of silicon carbide, aluminum oxide, or boron carbide is carried by a liquid between tool and work. { |əl·trə'sän·ik mə'shēn·iŋ }

ultrasonic sealing |ENG| A method for sealing plastic film by localized heat developed by vibratory mechanical pressure at ultrasonic frequencies. { |əl·trə'sän·ik 'sēl·iŋ }

ultrasonic storage cell See ultrasonic delay line. { |əl·trə'sän·ik 'stòr·ij ,sel }

ultrasonic testing |ENG| A nondestructive test method that employs high-frequency mechanical vibration energy to detect and locate structural discontinuities or differences and to measure thickness of a variety of materials. { |əl·trə'sän·ik 'test·iŋ }

ultrasonic thickness gage |ENG| A thickness gage in which the time of travel of an ultrasonic beam through a sheet of material is used as a measure of the thickness of the material. { |əl·trə'sän·ik 'thik·nəs ,gāj }

ultrasonic transducer |ENG ACOUS| A transducer that converts alternating-current energy above 20 kilohertz to mechanical vibrations of the same frequency; it is generally either magnetostrictive or piezoelectric. { |əl·trə'sän·ik tranz 'dü·sər }

ultrasonic transmitter |ENG ACOUS| A device used to track seals, fish, and other aquatic animals: the device is fastened to the outside of the animal or fed to it, and has a loudspeaker which is made to vibrate at an ultrasonic frequency, propagating ultrasonic waves through the water to a special microphone or hydrophone. { |əl·trə'sän·ik tranz'mid·ər }

ultrasonoscope |ENG| An instrument that displays an echosonogram on an oscilloscope; usually has auxiliary output to a chart-recording instrument. { |əl·trə'sän·ə,skōp }

umbrella roof See station roof. { əm'brel·ə ,rüf }

unavailable energy |THERMO| That part of the energy which, when an irreversible process takes place, is initially in a form completely available for work and is converted to a form completely unavailable for work. { |ən·ə|vāl·ə·bəl 'en·ər· jē }

unavoidable delay |IND ENG| Any delay in a task, the occurrence of which is outside the control or responsibility of the worker. { |ən·ə'vòid· ə·bəl di'lā }

unavoidable-delay allowance |IND ENG| An adjustment of standard time to allow for unavoidable delays in a task. { |ən·ə'vòid·ə·bəl di'lā ə,lau·əns }

unbonded member |CIV ENG| A posttensioned member that is made of prestressed concrete and has the tensioning force applied only against the end anchorages. { ən|bänd·əd 'mem·bər }

unbonded strain gage |ENG| A type of strain gage that consists of a grid of fine wires strung under slight tension between a stationary frame and a movable armature; pressure applied to the bellows or to the diaphragm sensing element moves the armature with respect to the frame, increasing tension in one half of the filaments and decreasing tension in the rest. { |ən'bän· dəd 'strān ,gāj }

uncage |ENG| To release the caging mechanism of a gyroscope, that is, the mechanism that erects the gyroscope or locks it in position. { |ən'kāj }

uncharged demolition target |ENG| A demolition target which has been prepared to receive the demolition agent, the necessary quantities of which have been calculated, packaged, and stored in a safe place. { |ən'chärjd ,dem·ə'lish· ən ,tär·gət }

unconfined explosion |ENG| Explosion occurring in the open air where the (atmospheric) pressure is constant. { |ən·kən'fīnd ik'splō· zhən }

uncouple |ENG| To unscrew or disengage. { |ən'kəp·əl }

underbody |ENG| The lower portion or underside of the body of a vehicle or airplane. { 'ən· dər,bäd·ē }

undercut |ELECTR| Undesirable lateral etching by chemicals in the fabrication of semiconductor devices. |ENG| Underside recess either cut or molded into an object so as to leave a topside lip or protuberance. { 'ən·dər,kət }

undercutting |CHEM ENG| In distillation, the technique of taking the products coming off the distillation tower at a temperature below the desired ultimate boiling point range to prevent contaminating the products with the compound that would distill just beyond the ultimate boiling point range. { |ən·dər|kəd·iŋ }

underdrain |CIV ENG| A subsurface drain with

holes into which water flows when the water table reaches the drain level. { 'ən·dər,drān }

underdrive press [MECH ENG] A mechanical press having the driving mechanism located within or under the bed. { 'ən·dər,drīv 'pres }

underfeed stoker [ENG] A coal-burning system in which green coal is fed from beneath the burning fuel bed. { 'ən·dər,fēd 'stō·kər }

underfloor raceway [BUILD] A raceway for electric wires which runs beneath the floor. { 'ən·dər,flōr 'rās,wā }

underground [ENG] Situated, done, or operating beneath the surface of the ground. { ¦ən·dər¦graúnd }

underhung crane [MECH ENG] An overhead traveling crane in which the end trucks carry the bridge suspended below the rails. { 'ən·dər,haŋ 'krān }

underpinning [CIV ENG] **1.** Permanent supports replacing or reinforcing the older supports beneath a wall or a column. **2.** Braced props temporarily supporting a structure. { 'ən·dər,pin·iŋ }

underplate [DES ENG] An unfinished plate which forms part of an armored front for a mortise lock, and which is fastened to the case. { 'ən·dər,plāt }

underream [ENG] To enlarge a drill hole below the casing. { ¦ən·dər¦rēm }

undershoot [CONT SYS] The amount by which a system's response to an abrupt change in input falls short of that desired. { 'ən·dər,shüt }

undershoot wheel [MECH ENG] A water wheel operated by the impact of flowing water against blades attached around the periphery of the wheel, the blades being partly or totally submerged in the moving stream of water. { 'ən·dər,shät ,wēl }

undersize [ENG] That part of a crushed material (for example, ore) which passes through a screen. { 'ən·dər,sīz }

underspin [MECH] Property of a projectile having insufficient rate of spin to give proper stabilization. { 'ən·dər,spin }

underwater sound projector [ENG ACOUS] A transducer used to produce sound waves in water. { ¦ən·dər¦wód·ər 'saúnd prə,jek·tər }

underwater transducer [ENG ACOUS] A device used for the generation or reception of underwater sounds. { ¦ən·dər¦wód·ər tranz'dü·sər }

underway bottom sampler *See* underway sampler. { ¦ən·dər¦wā 'bäd·əm ,sam·plər }

underway sampler [ENG] A device for collecting samples of sediment on the ocean bottom, consisting of a cup in a hollow tube; on striking the bottom, the cup scoops up a small sample which is forced into the tube which is then closed with a lid, and the device is hoisted to the surface. Also known as scoopfish; underway bottom sampler. { ¦ən·dər¦wā 'sam·plər }

Underwood chart [CHEM ENG] A graphical solution of mass balances for a single equilibrium stage in the calculation of a solvent-extraction operation. { 'ən·dər,wúd ,chärt }

Underwood distillation method [CHEM ENG] A

method for calculation of liquid separations from binary distillation systems operated at partial reflux. { 'ən·dər,wúd ,dis·tə'lā·shən ,meth·əd }

undisturbed [ENG] Pertaining to a sample of material, as of soil, subjected to so little disturbance that it is suitable for determinations of strength, consolidation, permeability characteristics, and other properties of the material in place. { ¦ən·di'stərbd }

unfinished bolt [DES ENG] One of three degrees of finish in which standard hexagon wrench-head bolts and nuts are available; only the thread is finished. { ¦ən'fin·isht 'bōlt }

unfired pressure vessel [CHEM ENG] A pressure vessel that is not in direct contact with a heating flame. { ¦ən'fīrd 'presh·ər ,ves·əl }

uniaxial stress [MECH] A state of stress in which two of the three principal stresses are zero. { ¦yü·nē'ak·sē·əl 'stres }

unidirectional hydrophone [ENG ACOUS] A hydrophone mainly sensitive to sound that is incident from a single solid angle of one hemisphere or less. { ¦yü·nə·də'rek·shən·əl 'hī·drə,fōn }

unidirectional microphone [ENG ACOUS] A microphonethat is responsive predominantly to sound incident from one hemisphere, without picking up sounds from the sides or rear. { ¦yü·nə·də'rek·shən·əl 'mī·krə,fōn }

unified screw thread [DES ENG] Three series of threads: coarse (UNC), fine (UNF), and extra fine (UNEF); a 1/4-inch-diameter (0.006-millimeter) thread in the UNC series has 20 threads per inch, while in the UNF series it has 28. { 'yü·nə,fīd 'skrü ,thred }

unifilar suspension [ENG] The suspension of a body from a single thread, wire, or strip. { ¦yü·nə'fil·ər sə'spen·chən }

uniflow engine [MECH ENG] A steam engine in which steam enters the cylinder through valves at one end and escapes through openings uncovered by the piston as it completes its stroke. { 'yü·nə,flō 'en·jən }

uniform circular motion [MECH] Circular motion in which the angular velocity remains constant. { 'yü·nə,fōrm 'sər·kyə·lər 'mō·shən }

uniform click track [ENG ACOUS] A click track with regularly spaced clicks. { 'yü·nə,fōrm 'klik ,trak }

uniform load [MECH] A load distributed uniformly over a portion or over the entire length of a beam; measured in pounds per foot. { 'yü·nə,fōrm 'lōd }

uniform mat [CIV ENG] A type of foundation mat, consisting of a reinforced concrete slab of constant thickness, supporting walls, and columns; it is thick, rigid, and strong. { 'yü·nə,fōrm 'mat }

unijunction transistor [ELECTR] An *n*-type bar of semiconductor with a *p*-type alloy region on one side; connections are made to base contacts at either end of the bar and to the *p*-region. Abbreviated UJT. Formely known as double-base diode; double-base junction diode. { 'yü·nə ,jəŋk·shən tran'zis·tər }

unilateral conductivity |ELECTR| Conductivity in only one direction, as in a perfect rectifier. { ¦yü·nə'lad·ə·rəl ˌkän·dək'tiv·əd·ē }

unilateral tolerance method |DES ENG| Method of dimensioning and tolerancing wherein the tolerance is taken as plus or minus from an explicitly stated dimension; the dimension represents the size or location which is nearest the critical condition (that is maximum material condition), and the tolerance is applied either in a plus or minus direction, but not in both directions, in such a way that the permissible variation in size or location is away from the critical condition. { ¦yü·nə'lad·ə·rəl 'täl·ə·rəns ˌmeth·əd }

union |DES ENG| A screwed or flanged pipe coupling usually in the form of a ring fitting around the outside of the joint. { 'yün·yən }

union joint |DES ENG| A threaded assembly used for the joining of ends of lengths of installed pipe or tubing where rotation of neither length is feasible. { 'yün·yən ˌjóint }

union shop |IND ENG| An establishment in which union membership is not a requirement for original employment but becomes mandatory after a specified period of time. { 'yün·yən 'shäp }

unipolar |ELEC| Having but one pole, polarity, or direction; when applied to amplifiers or power supplies, it means that the output can vary in only one polarity from zero and, therefore, must always contain a direct-current component. { ¦yü·nə'pō·lər }

unipolar transistor |ELECTR| A transistor that utilizes charge carriers of only one polarity, such as a field-effect transistor. { ¦yü·nə'pō·lər tran'zis·tər }

unit |ENG| An assembly or device capable of independent operation, such as a radio receiver, cathode-ray oscilloscope, or computer subassembly that performs some inclusive operation or function. { 'yü·nət }

unitary air conditioner |MECH ENG| A small self-contained electrical unit enclosing a motor-driven refrigeration compressor, evaporative cooling coil, air-cooled condenser, filters, fans, and controls. { 'yü·nəˌter·ē 'er kən,dish·ən·ər }

unit assembly |IND ENG| Assemblage of machine parts which constitutes a complete auxiliary part of an end item, and which performs a specific auxiliary function, and which may be removed from the parent item without itself being disassembled. { 'yü·nət ə'sem·blē }

unit charge See statcoulomb. { 'yü·nət 'chärj }

unit construction |BUILD| An assembly comprising two or more walls, plus floor and ceiling construction, ready for shipping to a building site. { 'yü·nət kən'strək·shən }

unit cost |IND ENG| Cost allocated to a specified unit of a product; computed as the cost over a period of time divided by the number of units produced. { 'yü·nət 'kóst }

United States standard dry seal thread |DES ENG| A modified pipe thread used for pressure-tight connections that are to be assembled without lubricant or sealer in refrigeration pipes, automotive and aircraft fuel-line fittings, and gas and chemical shells. { yə'nīd·əd 'stāts 'standərd 'drī ¦sēl ˌthred }

unit heater |MECH ENG| A heater consisting of a fan for circulating air over a heat-exchange surface, all enclosed in a common casing. { 'yü·nət 'hēd·ər }

unitized body |ENG| An automotive body that has the body and frame in one unit; side members are designed on the principle of a bridge truss to gain stiffness, and sheet metal of the body is stressed so that it carries some of the load. { 'yü·nəˌtīzd 'bäd·ē }

unitized cargo |IND ENG| Grouped cargo carried aboard a ship in pallets, containers, wheeled vehicles, and barges or lighters. { 'yü·nəˌtīzd 'kär·gō }

unitized load |IND ENG| A single item or a number of items packaged, packed, or arranged in a specified manner and capable of being handled as a unit; unitization may be accomplished by placing the item or items in a container or by banding them securely together. Also known as unit load. { 'yü·nəˌtīzd 'lōd }

unitized tooling |DES ENG| A die having its upper and lower members incorporated into a self-contained unit arranged to maintain the die members in alignment. { ¦yü·nəˌtīzd 'tül·iŋ }

unit load See unitized load. { 'yü·nət 'lōd }

unit mold |ENG| A simple plastics mold composed of a simple cavity without further mold devices; used to produce sample containers having shapes difficult to blow-mold. { 'yü·nət 'mōld }

unit of issue |IND ENG| In reference to special storage, the quantity of an item, such as each number, dozen, gallon, pair, pound, ream, set, or yard. { 'yü·nət əv 'ish·ü }

unit operations |CHEM ENG| The basic physical operations of chemical engineering in a chemical process plant, that is, distillation, fluid transport, heat and mass transfer, evaporation, extraction, drying, crystallization, filtration, mixing, size separation, crushing and grinding, and conveying. { 'yü·nət ˌäp·ə'rā·shənz }

unit process |CHEM ENG| In chemical manufacturing, a process that involves chemical conversion. { 'yü·nət ˌprä,ses }

unit procurement cost |IND ENG| The net basic cost paid or estimated to be paid for a unit of a particular item including, where applicable, the cost of government-furnished property and the cost of manufacturing operations performed at government-owned facilities. { 'yü·nət prə'kyúr·mənt ˌkóst }

unit strain |MECH| **1.** For tensile strain, the elongation per unit length. **2.** For compressive strain, the shortening per unit length. **3.** For shear strain, the change in angle between two lines originally perpendicular to each other. { 'yü·nət 'strān }

unit stress |MECH| The load per unit of area. { 'yü·nət 'stres }

unity power factor |ELEC| Power factor of 1.0, obtained when current and voltage are in phase, as in a circuit containing only resistance or in a reactive circuit at resonance. { 'yü·nəd·ē 'paů·ər ,fak·tər }

univariant system |THERMO| A system which has only one degree of freedom according to the phase rule. { ¦yü·nə¦ver·ē·ənt 'sis·təm }

universal chuck |ENG| A self-centering chuck whose jaws move in unison when a scroll plate is rotated. { ¦yü·nə¦vər·səl 'chək }

universal dividing head |MECH ENG| An accessory fixture on a milling machine that rotates the workpiece to specified angles between machining steps. { ¦yü·nə¦vər·səl di'vīd·iŋ ,hed }

universal gas constant See gas constant. { ¦yü·nə¦vər·səl 'gas ,kän·stənt }

universal grinding machine |MECH ENG| A grinding machine having a swivel table and headstock, and a wheel head that can be rotated on its base. { ¦yü·nə¦vər·səl 'grīnd·iŋ mə,shēn }

universal gripper |CONT SYS| A versatile robot component that can grasp most kinds of objects. { ¦yü·nə¦vər·səl 'grip·ər }

universal instrument See altazimuth. { ¦yü·nə¦vər·səl 'inz·trə·mənt }

universal joint |MECH ENG| A linkage that transmits rotation between two shafts whose axes are coplanar but not coinciding. { ¦yü·nə¦vər·səl 'jóint }

universal motor |ELEC| A motor that may be operated at approximately the same speed and output on either direct current or single-phase alternating current. Also known as ac/dc motor. { ¦yü·nə¦vər·səl 'mōd·ər }

universal output transformer |ENG ACOUS| An output transformer having a number of taps on its winding, to permit its use between the audio-frequency output stage and the loudspeaker of practically any radio receiver by proper choice of connections. { ¦yü·nə¦vər·səl 'aůt,půt tranz ,fór·mər }

universal robot |CONT SYS| A robot whose end effector would be flexible enough to perform any desired task. { ¦yü·nə¦vər·səl 'rō,bät }

universal vise |ENG| A vise which has two or three swivel settings so that the workpiece can be set at a compound angle. Also known as toolmaker's vise. { ¦yü·nə¦vər·səl 'vīs }

unloaded Q |ELECTR| The Q of a system when there is no external coupling to it. { ¦ən'lōd· əd 'kyü }

unloader |MECH ENG| A power device for removing bulk materials from railway freight cars or highway trucks; in the case of railway cars, the car structure may aid the unloader; a transitional device between interplant transportation means and intraplant handling equipment. { ¦ən'lōd· ər }

unloading |CHEM ENG| **1.** The release downstream of a trapped contaminant. **2.** A filter medium failure and release of system pressure.

3. The depressuring or emptying of a process unit. { ¦ən'lōd·iŋ }

unloading conveyor |MECH ENG| Any of several types of portable conveyors adapted for unloading bulk materials, packages, or objects from conveyances. { ¦ən'lōd·iŋ kən'vā·ər }

unprotected reversing thermometer |ENG| A reversing thermometer for sea-water temperature which is not protected against hydrostatic pressure. { ¦ən·prə'tek·təd ri'vərs·iŋ thər'mäm· əd·ər }

unrestricted element |IND ENG| An element of an operation that is entirely under the control of a worker. { ¦ən·ri'strik·təd 'el·ə·mənt }

unscheduled maintenance |IND ENG| Those unpredictable maintenance requirements that had not been previously planned or programmed but require prompt attention and must be added to, integrated with, or substituted for previously scheduled workloads. { ¦ən'skej·əld 'mānt·ən· əns }

unscrambler |IND ENG| A part of a feeding and packaging line that aids in arranging cartons for the filling machines; there are rotary, straight-line, and walking-beam types. { ¦ən'skram· blər }

Unsin engine |MECH ENG| A type of rotary engine in which the trochoidal rotors of eccentric-rotor engines are replaced with two circular rotors, one of which has a single gear tooth upon which gas pressure acts, and the second rotor has a slot which accepts the gear tooth. { 'ən· sən ,en·jən }

unsprung axle |MECH ENG| A rear axle in an automobile in which the housing carries the right and left rear-axle shafts and the wheels are mounted at the outer end of each shaft. { ¦ən'sprəŋ 'ak·səl }

unsprung weight |MECH ENG| The weight of the various parts of a vehicle that are not carried on the springs, such as wheels, axles, and brakes. { ¦ən'sprəŋ 'wāt }

unwater |ENG| To remove or draw off water; to drain. { ¦ən'wód·ər }

unwind |MECH ENG| To reverse the direction of rotation of a threaded device. { ¦ən'wīnd }

up |ENG| Fully in operation. { əp }

up-converter |ELECTR| Type of parametric amplifier which is characterized by the frequency of the output signal being greater than the frequency of the input signal. { 'əp kən,vərd·ər }

up-Doppler |ENG ACOUS| The sonar situation wherein the target is moving toward the transducer, so the frequency of the echo is greater than the frequency of the reverberations received immediately after the end of the outgoing ping; opposite of down-Doppler. { 'əp ,däp·lər }

updraft carburetor |MECH ENG| For a gasoline engine, a fuel-air mixing device in which both the fuel jet and the airflow are upward. { 'əp,draft 'kär·bə,rād·ər }

updraft furnace |MECH ENG| A furnace in which volumes of air are supplied from below the fuel bed or supply. { 'əp,draft 'fər·nəs }

uplift pressure |CIV ENG| Pressure in an upward direction against the bottom of a structure, as a dam, a road slab, or a basement floor. { 'əp,lift ,presh·ər }

upmilling |MECH ENG| Milling a workpiece by rotating the cutter against the direction of feed of the workpiece. { 'əp,mil·iŋ }

upper console temperature See console temperature. { 'əp·ər 'kän·sə,lüt 'tem·prə·chər }

upper control limit |IND ENG| A horizontal line on a control chart at a specified distance above the central line; if all the plotted points fall between the upper and lower control lines, the process is said to be in control. { 'əp·ər kən'trōl ,lim·ət }

upper critical solution temperature See consolute temperature. { 'əp·ər ¦krid·ə·kəl sə¦lü·shən 'tem·prə·chər }

upright |CIV ENG| A vertical structural member, post, or stake. { 'əp,rīt }

upset |ENG| To increase the diameter of a rock drill by blunting the end. { əp'set }

upstand |BUILD| That section of a roof covering that turns up against a vertical surface. Also known as upturn. { 'əp,stand }

upstream |CHEM ENG| That portion of a process stream that has not yet entered the system or unit under consideration; for example, upstream to a refinery or to a distillation column. { 'əp¦strēm }

upstream face |CIV ENG| The side of a dam nearer the source of water. { 'əp¦strēm 'fās }

uptake |ENG| A large pipe for exhaust gases from a boiler furnace that runs upward to a chimney or smokestack. { 'əp,tāk }

up time |IND ENG| A period during which value is being added to a product by a machine or a process. { 'əp ,tīm }

upturn See upstand. { 'əp,tərn }

urbanization |CIV ENG| The state of being or becoming a community with urban characteristics. { ,ər·bə·nə'zā·shən }

urban renewal |CIV ENG| Redevelopment and revitalization of a deteriorated urban community. { 'ər·bən ri'nü·əl }

urea dewaxing |CHEM ENG| A continuous, petroleum refinery process used to produce low-pour-point oils; urea forms a filterable solid complex (adduct) with the straight-chain wax paraffins in the stock. { yü'rē·ə dē'waks·iŋ }

usability |IND ENG| The characteristics which enter into a product's design and are related to its quality and reliability that enable users to perform tasks quickly and error free, as well as reduce the time and mental effort to learn or operate the product. Also known as ease of use; user friendliness. { ,yüz·ə'bil·əd·ē }

usable life See pot life. { ¦yüz·ə·bəl 'līf }

user friendliness See usability. { 'yü·zər 'frend·lē·nəs }

U-shaped abutment |CIV ENG| A bridge abutment with wings perpendicular to the face which act as counterforts; a very stable abutment, often used for architectural effect. { 'yü ¦shāpt ə'bət·mənt }

utilidor |CIV ENG| An insulated, heated conduit built below the ground surface or supported above the ground surface to protect the contained water, steam, sewage, and fire lines from freezing. { yü'til·ə,dòr }

utility |ENG| One of the nonprocess (support) facilities for a manufacturing plant; usually considered as facilities for steam, cooling water, deionized water, electric power, refrigeration, compressed and instrument air, and effluent treatment. { yü'til·əd·ē }

U-tube heat exchanger |CHEM ENG| A heat-exchanger system consisting of a bundle of U tubes (hairpin tubes) surrounded by a shell (outer vessel); one fluid flows through the tubes, and the other fluid flows through the shell, around the tubes. { 'yü ¦tüb 'hēt iks,chān·jər }

U-tube manometer |ENG| A manometer consisting of a U-shaped glass tube partly filled with a liquid of known specific gravity; when the legs of the manometer are connected to separate sources of pressure, the liquid rises in one leg and drops in the other; the difference between the levels is proportional to the difference in pressures and inversely proportional to the liquid's specific gravity. Also known as liquid-column gage. { 'yü ¦tüb mə'näm·əd·ər }

U-value |ENG| A measure of heat transmission through a building part or a given thickness of insulating material, expressed as the number of British thermal units that will flow in 1 hour through 1 square foot of the structure or material from air to air with a temperature differential of 1°F. { 'yü ,väl·yü }

V

V *See* electric potential; volt.

VA *See* volt-ampere.

vac *See* millibar.

vacuum brake |MECH ENG| A form of air brake which operates by maintaining low pressure in the actuating cylinder; braking action is produced by opening one side of the cylinder to the atmosphere so that atmospheric pressure, aided in some designs by gravity, applies the brake. { 'vak·yəm ˌbrāk }

vacuum breaker |ENG| A device used to relieve a vacuum formed in a water supply line to prevent backflow. Also known as backflow preventer. { 'vak·yəm ˌbrāk·ər }

vacuum cleaner |MECH ENG| An electrically powered mechanical appliance for the dry removal of dust and loose dirt from rugs, fabrics, and other surfaces. { 'vak·yəm ˌklē·nər }

vacuum concrete |CIV ENG| Concrete poured into a framework that is fitted with a vacuum mat to remove water not required for setting of the cement; in this framework, concrete attains its 28-day strength in 10 days and has a 25% higher crushing strength. { 'vak·yəm 'kän,krēt }

vacuum crystallizer |CHEM ENG| Crystallizer in which a warm saturated solution is fed to a lagged, closed vessel maintained under vacuum; the solution evaporates and cools adiabatically, resulting in crystallization. { 'vak·yəm 'krist·əl,īz·ər }

vacuum distillation |CHEM ENG| Liquid distillation under reduced (less than atmospheric) pressure; used to lower boiling temperatures and lessen the risk of thermal degradation during distillation. Also known as reduced-pressure distillation. { 'vak·yəm ˌdis·tə'lā·shən }

vacuum drying |ENG| The removal of liquid from a solid material in a vacuum system; used to lower temperatures needed for evaporation to avoid heat damage to sensitive material. { 'vak·yəm 'drī·iŋ }

vacuum evaporation |ENG| Deposition of thin films of metal or other materials on a substrate, usually through openings in a mask, by evaporation from a boiling source in a hard vacuum. { 'vak·yəm iˌvap·ə'rā·shən }

vacuum evaporator |ENG| A vacuum device used to evaporate metals and spectrographic carbon to coat (replicate) a specimen for electron spectroscopic analysis or for electron microscopy. { 'vak·yəm iˈvap·əˌrād·ər }

vacuum filter |ENG| A filter device into which a liquid-solid slurry is fed to the high-pressure side of a filter medium, with liquid pulled through to the low-pressure side of the medium and a cake of solids forming on the outside of the medium. { 'vak·yəm ˌfil·tər }

vacuum filtration |ENG| The separation of solids from liquids by passing the mixture through a vacuum filter. { 'vak·yəm fil'trā·shən }

vacuum flashing |CHEM ENG| The heating of a liquid that, upon release to a lower pressure (vacuum), undergoes considerable vaporization (flashing). Also known as flash vaporization. { 'vak·yəm 'flash·iŋ }

vacuum forming |ENG| Plastic-sheet forming in which the sheet is clamped to a stationary frame, then heated and drawn down into a mold by vacuum. { 'vak·yəm 'fòrm·iŋ }

vacuum freeze dryer |ENG| A type of indirect batch dryer used to dry materials that would be destroyed by the loss of volatile ingredients or by drying temperatures above the freezing point. { 'vak·yəm 'frēz ˌdrī·ər }

vacuum gage |ENG| A device that indicates the absolute gas pressure in a vacuum system. { 'vak·yəm ˌgāj }

vacuum gripper |CONT SYS| A robot component that uses a suction cup connected to a vacuum source to lift and handle objects. { 'vak·yəm 'grip·ər }

vacuum heating |MECH ENG| A two-pipe steam heating system in which a vacuum pump is used to maintain a suction in the return piping, thus creating a positive return flow of air and condensate. { 'vak·yəm 'hēd·iŋ }

vacuum mat |CIV ENG| A rigid flat metal screen faced by a linen filter, the back of which is kept under partial vacuum; used to suck out surplus air and water from poured concrete to produce a dense, well-shrunk concrete. { 'vak·yəm ˌmat }

vacuum measurement |ENG| The determination of a fluid pressure less in magnitude than the pressure of the atmosphere. { 'vak·yəm 'mezh·ər·mənt }

vacuum pan salt |CHEM ENG| A salt made from salt brine boiled at reduced pressure in a triple-effect evaporator. { 'vak·yəm ˌpan ˌsȯlt }

vacuum pencil |ENG| A pencillike length of tubing connected to a small vacuum pump, for picking up semiconductor slices or chips during fabrication of solid-state devices. { 'vak·yəm ‚pen·səl }

vacuum pump |MECH ENG| A compressor for exhausting air and noncondensable gases from a space that is to be maintained at subatmospheric pressure. { 'vak·yəm ‚pəmp }

vacuum relief valve |ENG| A pressure relief device which is designed to allow fluid to enter a pressure vessel in order to avoid extreme internal vacuum. { 'vak·yəm ri'lēf ‚valv }

vacuum shelf dryer |ENG| A type of indirect batch dryer which generally consists of a vacuum-tight cubical or cylindrical chamber of cast-iron or steel plate, heated supporting shelves inside the chamber, a vacuum source, and a condenser; used extensively for drying pharmaceuticals, temperature-sensitive or easily oxidizable materials, and small batches of high-cost products where any product loss must be avoided. { 'vak·yəm 'shelf‚drī·ər }

vacuum support |MECH ENG| That portion of a rupture disk device which prevents deformation of the disk resulting from vacuum or rapid pressure change. { 'vak·yəm sə‚pórt }

vacuum-tube voltmeter |ENG| Any of several types of instrument in which vacuum tubes, acting as amplifiers or rectifiers, are used in circuits for the measurement of alternating-current or direct-current voltage. Abbreviated VTVM. Also known as tube voltmeter. { 'vak·yəm ‚tüb 'vōlt‚mēd·ər }

vacuum-type insulation |CHEM ENG| Highly reflective double-wall structure with high vacuum between the walls; used as insulation for cryogenic systems; Dewar flasks have vacuum-type insulation. { 'vak·yəm ‚tīp in·sə'lā·shən }

VAD See vapor-phase axial deposition. { vad or ‚vē‚ā'dē }

valley |BUILD| An inside angle formed where two sloping sides intersect. { 'val·ē }

valley rafter |BUILD| A part of the roof frame that extends diagonally from an inside corner plate to the ridge board at the intersection of two roof surfaces. { 'val·ē ‚raf·tər }

valley roof |BUILD| A pitched roof with one or more valleys. { 'val·ē ‚rüf }

value analysis See value engineering. { 'val·ē ‚nal·ə·səs }

value control See value engineering. { 'val·yü kən‚trōl }

value engineering |IND ENG| The systematic application of recognized techniques which identify the function of a product or service, and provide the necessary function reliably at lowest overall cost. Also known as value analysis; value control. { 'val·yü ‚en·jə‚nir·iŋ }

value theory |SYS ENG| A concept normally associated with decision theory; it strives to evaluate relative utilities of simple and mixed parameters which can be used to describe outcomes. { 'val·yü ‚thē·ə·rē }

valve See electron tube. |MECH ENG| A device

used to regulate the flow of fluids in piping systems and machinery. { valv }

valve follower |MECH ENG| A linkage between the cam and the push rod of a valve train. { 'valv ‚fäl·ə·wər }

valve guide |MECH ENG| A channel which supports the stem of a poppet valve for maintenance of alignment. { 'valv ‚gīd }

valve head |MECH ENG| The disk part of a poppet valve that gives a tight closure on the valve seat. { 'valv ‚hed }

valve-in-head engine See overhead-valve engine. { ‚valv in ‚hed 'en·jən }

valve lifter |MECH ENG| A device for opening the valve of a cylinder as in an internal combustion engine. { 'valv ‚lif·tər }

valve positioner |CONT SYS| A pneumatic servomechanism which is used as a component in process control systems to improve operating characteristics of valves by reducing hysteresis. Also known as pneumatic servo. { 'valv pə‚zish·ə·nər }

valve seat |DES ENG| The circular metal ring on which the valve head of a poppet valve rests when closed. { 'valv ‚sēt }

valve stem |MECH ENG| The rod by means of which the disk or plug is moved to open and close a valve. { 'valv ‚stem }

valve train |MECH ENG| The valves and valve-operating mechanism for the control of fluid flow to and from a piston-cylinder machine, for example, steam, diesel, or gasoline engine. { 'valv ‚trān }

van der Waals surface tension formula |THERMO| An empirical formula for the dependence of the surface tension on temperature: $\gamma = K p_c^{2/3} T_c^{1/3} (1 - T/T_c)^n$, where γ is the surface tension, T is the temperature, T_c and p_c are the critical temperature and pressure, K is a constant, and n is a constant equal to approximately 1.23. { 'van dər ‚wólz 'sər·fəs ‚ten·chən ‚fór·myə·lə }

Van Dorn sampler |ENG| A sediment sampler that consists of a Plexiglas cylinder closed at both ends by rubber force cups; in the armed position the cups are pulled outside the cylinder and restrained by a releasing mechanism, and after the sample is taken, a length of surgical rubber tubing connecting the cups is sufficiently prestressed to permit the force cups to retain the sample in the cylinder. { van 'dórn ‚sam·plər }

vane |MECH ENG| A flat or curved surface exposed to a flow of fluid so as to be forced to move or to rotate about an axis, to rechannel the flow, or to act as the impeller; for example, in a steam turbine, propeller fan, or hydraulic turbine. { vān }

vane anemometer |ENG| A portable instrument used to measure low wind speeds and airspeeds in large ducts; consists of a number of vanes radiating from a common shaft and set to rotate when facing the wind. { 'vān an·ə'mäm·əd·ər }

vane motor rotary actuator |MECH ENG| A type of rotary motor actuator which consists of a rotor with several spring-loaded sliding vanes in an elliptical chamber; hydraulic fluid enters the chamber and forces the vanes before it as it moves to the outlets. { 'vān ¦mŏd·ər 'rŏd·ə·rē 'ak·chə,wād·ər }

vane-type instrument |ENG| A measuring instrument utilizing the force of repulsion between fixed and movable magnetized iron vanes, or the force existing between a coil and a pivoted vane-shaped piece of soft iron, to move the indicating pointer. { vān ,tīp ,in·strə·mənt }

vapor |THERMO| A gas at a temperature below the critical temperature, so that it can be liquefied by compression, without lowering the temperature. { 'vā·pər }

vapor barrier |CIV ENG| A layer of material applied to the inner (warm) surface of a concrete wall or floor to prevent absorption and condensation of moisture. { 'vā·pər ,bar·ē·ər }

vapor-compression cycle |MECH ENG| A refrigeration cycle in which refrigerant is circulated through a machine which allows for successive boiling (or vaporization) of liquid refrigerant as it passes through an expansion valve, thereby producing a cooling effect in its surroundings, followed by compression of vapor to liquid. { 'vā·pər kəm'presh·ən ,sī·kəl }

vapor cycle |THERMO| A thermodynamic cycle, operating as a heat engine or a heat pump, during which the working substance is in, or passes through, the vapor state. { 'vā·pər ,sī·kəl }

vapor degreasing |ENG| A type of cleaning procedure for metals to remove grease, oils, and lightly attached solids; a solvent such as trichloroethylene is boiled, and its vapors are condensed on the metal surfaces. { 'vā·pər dē 'grēs·iŋ }

vapor-filled thermometer |ENG| A gas- or vapor-filled temperature measurement device that moves or distorts in response to temperature-induced pressure changes from the expansion or contraction of the sealed, vapor-containing chamber. { 'vā·pər ¦fild thər'mäm·əd·ər }

vaporimeter |ENG| An instrument used to measure a substance's vapor pressure, especially that of an alcoholic liquid, in order to determine its alcohol content. { ,vap·ə'rim·əd·ər }

vaporization See volatilization. { ,vā·pə·rə'zā·shən }

vaporization coefficient |THERMO| The ratio of the rate of vaporization of a solid or liquid at a given temperature and corresponding vapor pressure to the rate of vaporization that would be necessary to produce the same vapor pressure at this temperature if every vapor molecule striking the solid or liquid were absorbed there. { ,vā·pə·rə'zā·shən ,kō·ə·fish·ənt }

vaporization cooling |ENG| Cooling by volatilization of a nonflammable liquid having a low boiling point and high dielectric strength; the liquid is flowed or sprayed on hot electronic equipment in an enclosure where it vaporizes, carrying the heat to the enclosure walls, radiators, or heat exchanger. Also known as evaporative cooling. { ,vā·pə·rə'zā·shən ,kül·iŋ }

vaporizer |CHEM ENG| A process vessel in which a liquid is heated until it vaporizes; heat

can be indirect (steam or heat-transfer fluid) or direct (hot gases or submerged combustion). { 'vā·pə,rīz·ər }

vapor-liquid separation |CHEM ENG| The removal of liquid droplets from a flowing stream of gas or vapor; accomplished by impingement, cyclonic action, and absorption or adsorption operations. { 'vā·pər 'lik·wəd ,sep·ə'rā·shən }

vapor-phase axial deposition |ENG| A method of fabricating graded-index optical fibers in which fine glass particles of silicon dioxide and germanium dioxide are synthesized and deposited on a rotating seed rod, and the synthesized porous preform is then pulled up and passes through a hot zone, undergoing dehydration and sintering, to become a porous preform. Abbreviated VAD. { 'vā·pər ¦fāz 'ak·sē·əl ,dep·ə'zish·ən }

vapor-phase reactor |CHEM ENG| A heavy steel vessel for carrying out chemical reactions on an industrial scale where efficient control over a vapor phase is needed, for example, in an oxidation process. { 'vā·pər ¦fāz rē'ak·tər }

vapor pressure |THERMO| For a liquid or solid, the pressure of the vapor in equilibrium with the liquid or solid. { 'vā·pər ,presh·ər }

vapor-pressure thermometer |ENG| A thermometer in which the vapor pressure of a homogeneous substance is measured and from which the temperature can be determined; used mostly for low-temperature measurements. { 'vā·pər ¦presh·ər thər'mäm·əd·ər }

vapor rate |CHEM ENG| In distillation, the upward flow rate of vapor through a distillation column. { 'vā·pər ,rāt }

vapor-recovery unit |ENG| **1.** A device or system to catch vaporized materials (usually fuels or solvents) as they are vented. **2.** In petroleum refining, a process unit to which gases and vaporized gasoline from various processing operations are charged, separated, and recovered for further use. { 'vā·pər ri'kəv·ə·rē ,yü·nət }

vara |CIV ENG| A surveyors' unit of length equal to 33¹/₃ inches (84.7 centimeters). { 'vär·ə }

varactor |ELECTR| A semiconductor device characterized by a voltage-sensitive capacitance that resides in the space-charge region at the surface of a semiconductor bounded by an insulating layer. Also known as varactor diode; variable-capacitance diode; varicap; voltage-variable capacitor. { va'rak·tər }

varactor diode See varactor. { va'rak·tər 'dī,ōd }

varactor tuning |ELECTR| A method of tuning in which varactor diodes are used to vary the capacitance of a tuned circuit. { va'rak·tər 'tün·iŋ }

var hour meter |ENG| An instrument that measures and registers the integral of reactive power over time in the circuit to which it is connected. { 'var ¦aủr ,mēd·ər }

variable-area meter |ENG| A flowmeter that works on the principle of a variable restrictor in the flowing stream being forced by the fluid to a position to allow the required flow-through. { 'ver·ē·ə·bəl ¦er·ē·ə 'mēd·ər }

variable-area track [ENG ACOUS] A sound track divided laterally into opaque and transparent areas; a sharp line of demarcation between these areas corresponds to the waveform of the recorded signal. { 'ver·ē·ə·bəl ¦er·ē·ə 'trak }

variable attenuator [ELECTR] An attenuator for reducing the strength of an alternating-current signal either continuously or in steps, without causing appreciable signal distortion, by maintaining a substantially constant impedance match. { 'ver·ē·ə·bəl ə'ten·yə,wād·ər }

variable-capacitance diode *See* varactor. { 'ver·ē·ə·bəl kə¦pas·əd·əns 'dī,ōd }

variable capacitor [ELEC] A capacitor whose capacitance can be varied continuously by moving one set of metal plates with respect to another. { 'ver·ē·ə·bəl kə'pas·əd·ər }

variable click track [ENG ACOUS] A click track with irregularly spaced clicks. { 'ver·ē·ə·bəl 'klik ,trak }

variable costs [IND ENG] Costs which vary directly with the number of units produced; direct labor and material are examples. { 'ver·ē·ə·bəl 'kôsts }

variable-density sound track [ENG ACOUS] A constant-width sound track in which the average light transmission varies along the longitudinal axis in proportion to some characteristic of the applied signal. { 'ver·ē·ə·bəl ¦den·səd·ē 'saún ,trak }

variable-depth sonar [ENG] Sonar in which the projector and receiving transducer are mounted in a watertight pod that can be lowered below a vessel to an optimum depth for minimizing thermal effects when detecting underwater targets. { 'ver·ē·ə·bəl ¦depth 'sō,när }

variable element [IND ENG] **1.** An element with a time that varies significantly from cycle to cycle as a function of one or more variables occurring within the job. **2.** An element that is common to two different jobs but whose time varies because of differences between the two jobs. { ¦var·ē·ə·bəl 'el·ə·mənt }

variable force [MECH] A force whose direction or magnitude or both change with time. { 'ver·ē·ə·bəl 'fôrs }

variable-inductance accelerometer [ENG] An accelerometer consisting of a differential transformer with three coils and a mass which passes through the coils and is suspended from springs; the center coil is excited from an external alternating-current power source, and two end coils connected in series opposition are used to produce an ac output which is proportional to the displacement of the mass. { 'ver·ē·ə·bəl in¦dək·təns ik,sel·ə'räm·əd·ər }

variable-pitch propeller [ENG] A controllable-pitch propeller whose blade angle may be adjusted to any angle between the low and high pitch limits. { 'ver·ē·ə·bəl ¦pich prə'pel·ər }

variable radio-frequency radiosonde [ENG] A radiosonde whose carrier frequency is modulated by the magnitude of the meteorological variables being sensed. { 'ver·ē·ə·bəl 'rād·ē·ō ¦frē·kwən·sē 'rād·ē·ō,sänd }

variable-reluctance microphone *See* magnetic microphone. { 'ver·ē·ə·bəl ri¦lək·təns 'mī·krə,fōn }

variable-reluctance pickup [ENG ACOUS] A phonograph pickup that depends for its operation on variations in the reluctance of a magnetic circuit due to the movements of an iron stylus assembly that is a part of the magnetic circuit. Also known as magnetic cartridge; magnetic pickup; reluctance pickup. { 'ver·ē·ə·bəl ri¦lək·təns 'pik,əp }

variable-resistance accelerometer [ENG] Any accelerometer which operates on the principle that electrical resistance of any conductor is a function of its dimensions; when the dimensions of the conductor are varied mechanically, as constant current flows through it, the voltage across it varies as a function of this mechanical excitation; examples include the strain-gage accelerometer, and an accelerometer making use of a slide-wire potentiometer. { 'ver·ē·ə·bəl ri¦zis·təns ik,sel·ə'räm·əd·ər }

variable resistor *See* rheostat. { 'ver·ē·ə·bəl ri'zis·tər }

variable-sequence robot [CONT SYS] A robot controlled by instructions that can be modified. { 'ver·ē·ə·bəl ¦sē·kwəns 'rō,bät }

variable-speed drive [MECH ENG] A mechanism transmitting motion from one shaft to another that allows the velocity ratio of the shafts to be varied continuously. { 'ver·ē·ə·bəl ¦spēd 'drīv }

variable-volume air system [MECH ENG] An air-conditioning system in which the volume of air delivered to each controlled zone is varied automatically from a preset minimum to a maximum value, depending on the load in each zone. { ¦ver·ē·ə·bəl ¦väl·yəm 'er ,sis·təm }

varicap *See* varactor. { 'var·ə,kap }

variety [SYS ENG] The logarithm (usually to base 2) of the number of discriminations that an observer or a sensing system can make relative to a system. { və'rī·əd·ē }

Varignon's theorem [MECH] The theorem that the moment of a force is the algebraic sum of the moments of its vector components acting at a common point on the line of action of the force. { var·ən'yōnz ,thir·əm }

variograph [ENG] A recording variometer. { 'ver·ē·ə,graf }

variometer [ENG] A geomagnetic device for detecting and indicating changes in one of the components of the terrestrial magnetic field vector, usually magnetic declination, the horizontal intensity component, or the vertical intensity component. { ,ver·ē'äm·əd·ər }

varistor [ELECTR] A two-electrode semiconductor device having a voltage-dependent nonlinear resistance; its resistance drops as the applied voltage is increased. Also known as voltage-dependent resistor. { və'ris·tər }

varmeter [ENG] An instrument for measuring reactive power in vars. Also known as reactive volt-ampere meter. { 'vär,mēd·ər }

V belt [DES ENG] An endless power-transmission belt with a trapezoidal cross section which runs in a pulley with a V-shaped groove; it transmits higher torque at less width and tension than a flat belt. |MECH ENG| A belt, usually endless, with a trapezoidal cross section which runs in a pulley with a V-shaped groove, with the top surface of the belt approximately flush with the top of the pulley. { 've ,belt }

V-bend die |MECH ENG| A die with a triangular cross-sectional opening to provide two edges over which bending is accomplished. { 've ¦bend 'dī }

V block |ENG| A square or rectangular steel block having a 90° V groove through the center, and sometimes provided with clamps to secure round workpieces. { 've ,bläk }

V-bucket carrier |MECH ENG| A conveyor consisting of two strands of roller chain separated by V-shaped steel buckets; used for elevating and conveying nonabrasive materials, such as coal. { 've ¦bək·ət ,kar·ē·ər }

V cut |ENG| In mining and tunneling, a cut where the material blasted out in plan is like the letter V; usually consists of six or eight holes drilled into the face, half of which form an acute angle with the other half. { 've ,kət }

vectopluviometer |ENG| A rain gage or array of rain gages designed to measure the inclination and direction of falling rain; vectopluviometers may be constructed in the fashion of a wind vane so that the receiver always faces the wind, or they may consist of four or more receivers arranged to point in cardinal directions. { ¦vek·tō,plü·vē'äm·əd·ər }

vector impedance meter |ENG| An instrument that not only determines the ratio between voltage and current, to give the magnitude of impedance, but also determines the phase difference between these quantities, to give the phase angle of impedance. { 'vek·tər im'pēd·əns ,mēd·ər }

vector momentum See momentum. { 'vek·tər mə'men·təm }

vector power |ELEC| Vector quantity equal in magnitude to the square root of the sum of the squares of the active power and the reactive power. { 'vek·tər ,pau̇·ər }

vector-power factor |ELEC| Ratio of the active power to the vector power; it is the same as power factor in the case of simple sinusoidal quantities. { 'vek·tər ¦pau̇·ər ,fak·tər }

vector voltmeter |ENG| A two-channel high-frequency sampling voltmeter that measures phase as well as voltage of two input signals of the same frequency. { 'vek·tər 'vōlt,mēd·ər }

vee path |ENG| In ultrasonic testing, the path of an angle beam from an ultrasonic search unit in which the waves are reflected off the opposite surface of the test piece and returned to the examination surface in a manner which has the appearance of the letter V. { 've ,path }

vegetable tanning |ENG| Leather tanning using plant extracts, such as tannic acid. { 'vej·tə·bəl 'tan·iŋ }

vehicle |MECH ENG| A self-propelled wheeled machine that transports people or goods on or off roads; automobiles and trucks are examples. { 've·ə·kəl }

velocimeter |ENG| An instrument for measuring the speed of sound in water; two transducers transmit acoustic pulses back and forth over a path of fixed length, each transducer immediately initiating a pulse upon receiving the previous one; the number of pulses occurring in a unit time is measured. { ,vel·ə'sim·əd·ər }

velocity |MECH| **1.** The time rate of change of position of a body; it is a vector quantity having direction as well as magnitude. Also known as linear velocity. **2.** The speed at which the detonating wave passes through a column of explosives, expressed in meters or feet per second. { və'läs·əd·ē }

velocity analysis |MECH| A graphical technique for the determination of the velocities of the parts of a mechanical device, especially those of a plane mechanism with rigid component links. { və'läs·əd·ē ə,nal·ə·səs }

velocity constant |CONT SYS| The ratio of the rate of change of the input command signal to the steady-state error, in a control system where these two quantities are proportional. { və'läs·əd·ē ,kän·stənt }

velocity control See rate control. { və'läs·əd·ē kən,trōl }

velocity error |CONT SYS| The difference between the rate of change of the actual position of a control system component and the rate of change of the desired position. { və'läs·əd·ē ,er·ər }

velocity-head tachometer |ENG| A type of tachometer in which the device whose speed is to be measured drives a pump or blower, producing a fluid flow, which is converted to a pressure. { və'läs·əd·ē ¦hed tə'käm·əd·ər }

velocity hydrophone |ENG ACOUS| A hydrophone in which the electric output essentially matches the instantaneous particle velocity in the impressed sound wave. { və'läs·əd·ē 'hī·drə,fōn }

velocity microphone |ENG ACOUS| A microphone whose electric output depends on the velocity of the air particles that form a sound wave; examples are a hot-wire microphone and a ribbon microphone. { və'läs·əd·ē 'mī·krə,fōn }

velocity pressure See wind pressure. { və'läs·əd·ē ,presh·ər }

velocity ratio |MECH ENG| The ratio of the velocity given to the effort or input of a machine to the velocity acquired by the load or output. { və'läs·əd·ē ,rā·shō }

velocity servomechanism |CONT SYS| A servomechanism in which the feedback-measuring device generates a signal representing a measured value of the velocity of the output shaft. Also known as rate servomechanism. { və'läs·əd·ē 'sər·vō,mek·ə,niz·əm }

velocity-type flowmeter |ENG| A turbine-type fluid-flow measurement device in which the fluid flow actuates the movement of a wheel or turbine-type impeller, giving a volume-time reading. Also known as current meter; rotating meter. { və'läs·əd·ē ¦tīp 'flō,mēd·ər }

veneered construction |BUILD| A type of construction in which the framework is faced with a thin external layer of material, such as marble. { və¦nird kən'strək·shən }

vent |ENG| **1.** A small passage made with a needle through stemming, for admitting a squib to enable the charge to be lighted. **2.** A hole, extending up through the bearing at the top of the core-barrel inner tube, which allows the water and air in the upper part of the inner tube to escape into the borehole. **3.** A small hole in the upper end of a core-barrel inner tube that allows water and air in the inner tube to escape into the annular space between the inner and outer barrels. **4.** An opening provided for the discharge of pressure or the release of pressure from tanks, vessels, reactors, processing equipment, and so on. **5.** A pipe for providing airflow to or from a drainage system or for circulating air within the system to protect trap seals from siphonage and back pressure. { vent }

vented baffle See reflex baffle. { 'ven·təd 'baf·əl }

ventilation |ENG| Provision for the movement, circulation, and quality control of air in an enclosed space. { ,vent·əl'ā·shən }

ventilator |ENG| A device with an adjustable aperture for regulating the flow of fresh or stagnant air. |MECH ENG| A mechanical apparatus for producing a current of air, as a blowing or exhaust fan. { 'vent·əl,ād·ər }

vent stack |BUILD| The portion of a soil stack above the highest fixture. { 'vent ,stak }

venture life |IND ENG| The period of time during which expenditures and reimbursements involving a given venture occur. Also known as financial life. { 'ven·chər ,līf }

venturi flume |ENG| An open flume with a constricted flow which causes a drop in the hydraulic grade line; used in flow measurement. { ven 'tùr·ē ,flüm }

venturi meter |ENG| An instrument for efficiently measuring fluid flow rate in a piping system; a nozzle section increases velocity and is followed by an expanding section for recovery of kinetic energy. { ven'tùr·ē ,mēd·ər }

venturi scrubber |CHEM ENG| A gas-cleaning device in which liquid injected at the throat of a venturi is used to scrub dust and mist from the gas flowing through the venturi. { ven'tùr·ē 'skrəb·ər }

venturi tube |ENG| A constriction that is placed in a pipe and causes a drop in pressure as fluid flows through it, consisting essentially of a short straight pipe section or throat between two tapered sections; it can be used to measure fluid flow rate (a venturi meter), or to draw fuel into the main flow stream, as in a carburetor. { ven'tùr·ē ,tüb }

verbal information verification |ENG ACOUS| A method of talker authentication that involves checking the content of a spoken password or

pass-phrase, such as a personal identification number, a social security number, or a mother's maiden name. Abbreviated VIV. { ,vər·bəl ,in·fər¦mā·shən ,ver·i·fə'kā·shən }

verge |BUILD| The edge of a sloping roof which projects over a gable. { vərj }

vergeboard |BUILD| One of the boards utilized as the finish of the eaves on the gable end of a structure. Also known as bargeboard; gableboard. { 'vərj,bord }

verglas See glaze. { 'vər'glä }

vernier |ENG| A short, auxiliary scale which slides along the main instrument scale to permit accurate fractional reading of the least main division of the main scale. { 'vər·nē·ər }

vernier caliper |ENG| A caliper rule with an attached vernier scale. { 'vər·nē·ər 'kal·ə·pər }

vernier dial |ENG| A tuning dial in which each complete rotation of the control knob causes only a fraction of a revolution of the main shaft, permitting fine and accurate adjustment. { 'vər·nē·ər 'dīl }

vertical band saw |MECH ENG| A band saw whose blade operates in the vertical plane; ideal for contour cutting. { 'vərd·ə·kəl 'band ,sö }

vertical boiler |MECH ENG| A fire-tube boiler having vertical tubes between top head and tube sheet, connected to the top of an internal furnace. { 'vərd·ə·kəl 'bói·lər }

vertical boring mill |MECH ENG| A large type of boring machine in which a rotating workpiece is fastened to a horizontal table, which resembles a four-jaw independent chuck with extra radial T slots, and the tool has a traverse motion. { 'vərd·ə·kəl 'bòr·iŋ ,mil }

vertical broaching machine |MECH ENG| A broaching machine having the broach mounted in the vertical plane. { 'vərd·ə·kəl 'bröch·iŋ mə,shēn }

vertical compliance |ENG ACOUS| The ability of a stylus to move freely in a vertical direction while in the groove of a phonograph record. { 'vərd·ə·kəl kəm'plī·əns }

vertical conveyor |MECH ENG| A materials-handling machine designed to move or transport bulk materials or packages upward or downward. { 'vərd·ə·kəl kən'vā·ər }

vertical-current recorder |ENG| An instrument which records the vertical electric current in the atmosphere. { 'vərd·ə·kəl ¦kə·rənt ri,kòrd·ər }

vertical curve |CIV ENG| A curve inserted between two lengths of a road or railway which are at different slopes. { 'vərd·ə·kəl 'kərv }

vertical drop |MECH| The drop of an object in trajectory or along a plumb line, measured vertically from its line of departure to the object. { 'vərd·ə·kəl 'dräp }

vertical-face breakwater |CIV ENG| A breakwater whose mound of rubble does not rise above the water, but is surmounted by a vertical-face superstructure of masonry or concrete; may be built without mound rubble, provided sea bed is firm. { 'vərd·ə·kəl ¦fās 'brāk,wòd·ər }

vertical field balance |ENG| An instrument that

measures the vertical component of the magnetic field by means of the torque that the field component exerts on a horizontal permanent magnet. { 'vərd·ə·kəl 'fēld ,bal·əns }

vertical firing [MECH ENG] The discharge of fuel and air perpendicular to the burner in a furnace. { 'vərd·ə·kəl 'fīr·iŋ }

vertical force instrument *See* heeling adjuster. { 'vərd·ə·kəl ¦fȯrs 'in·strə·mənt }

vertical guide idlers [MECH ENG] Idler rollers about 3 inches (8 centimeters) in diameter so placed as to make contact with the edge of the belt conveyor should it run too much to one side. { 'vərd·ə·kəl ¦gīd 'īd·lərz }

vertical intensity variometer [ENG] A variometer employing a large permanent magnet and equipped with very fine steel knife-edges or pivots resting on agate planes or saddles and balanced so that its magnetic axis is horizontal. Also known as Z variometer. { 'vərd·ə·kəl ¦ten·səd·ē ,ver·ē'äm·əd·ər }

vertical-lift bridge [CIV ENG] A movable bridge with a span that rises on towers, lifted by steel ropes. { 'vərd·ə·kəl ¦lift 'brij }

vertical-lift gate [CIV ENG] A dam spillway gate of which the movable parts are raised and lowered vertically to regulate water flow. { 'vərd·ə·kəl ¦lift 'gāt }

vertical metal oxide semiconductor technology [ELECTR] For semiconductor devices, a technology that involves essentially the formation of four diffused layers in silicon and etching of a V-shaped groove to a precisely controlled depth in the layers, followed by deposition of metal over silicon dioxide in the groove to form the gate electrode. Abbreviated VMOS technology. { 'vərd·ə·kəl ¦med,əl ¦äk,sīd ¦sem·i·kən,dək·tər tek'näl·ə·jē }

vertical obstacle sonar [ENG] An active sonar used to determine heights of objects in the path of a submersible vehicle; its beam sweeps along a vertical plane, about 30° above and below the direction of the vehicle's motion. Abbreviated VOS. { 'vərd·ə·kəl ¦äb·stə·kəl 'sō,när }

vertical recording [ELECTR] Magnetic recording in which bits are magnetized in directions perpendicular to the surface of the recording medium, allowing the bits to be smaller. Also known as perpendicular recording. [ENG ACOUS] A type of disk recording in which the groove modulation is perpendicular to the surface of the recording medium, so the cutting stylus moves up and down rather than from side to side during recording. Also known as hill-and-dale recording. { 'vərd·ə·kəl ri'kȯrd·iŋ }

vertical scale [DES ENG] The ratio of the vertical dimensions of a laboratory model to those of the natural prototype; usually exaggerated in relation to the horizontal scale. { 'vərd·ə·kəl 'skāl }

vertical seismograph [ENG] An instrument that records the vertical component of the ground motion during an earthquake. { 'vərd·ə·kəl 'sīz·mə,graf }

vertical traverse [MECH ENG] The angle

through which a robot's arm can swing up and down, typically 30° { 'vərd·ə·kəl trə'vərs }

vertical turbine pump *See* deep-well pump. { 'vərd·ə·kəl 'tər·bən ,pəmp }

vertical turret lathe [DES ENG] Similar in principle to the horizontal turret lathe but capable of handling heavier, bulkier workpieces; it is constructed with a rotary, horizontal worktable whose diameter (30–74 inches, or 76–188 centimeters) normally designates the capacity of the machine; a crossrail mounted above the worktable carries a turret, which indexes in a vertical plane with tools that may be fed either across or downward. { 'vərd·ə·kəl 'tə·rət ,la̱th }

very high frequency oscillator [ELECTR] An oscillator whose frequency lies in the range from a few to several hundred megahertz; it uses distributed, rather than lumped, impedances, such as parallel wire transmission lines or coaxial cables. { ¦ver·ē ¦hī 'frē·kwən·sē 'äs·ə,lad·ər }

very high frequency tuner [ELECTR] A tuner in a television receiver for reception of stations transmitting in the very high frequency band; it generally has 12 discrete positions corresponding to channels 2–13. { ¦ver·ē ¦hī 'frē·kwən·sē 'tün·ər }

very large scale integrated circuit [ELECTR] A complex integrated circuit that contains between 20,000 and 1,000,000 transistors. Abbreviated VLSI circuit. { ¦ver·ē ¦lärj ¦skāl 'int·ə,grād·əd 'sər·kət }

vessel [ENG] A container or structural envelope in which materials are processed, treated, or stored; for example, pressure vessels, reactor vessels, agitator vessels, and storage vessels (tanks). { 'ves·əl }

vestibule [BUILD] A hall or chamber between the outer door and the interior, or rooms, of a building. { 'ves·tə,byül }

vestibule school [IND ENG] A school organized by an industrial concern to train new employees in specific tasks or prepare employees for promotion. { 'ves·tə,byül ,skül }

vestibule training [IND ENG] A procedure used in operator training in which the training location is separate from the main productive areas of the plant; includes student carrels, lecture rooms, and in many instances the same type of equipment that the trainee will use in the work station. { 'ves·tə,byül ,trān·iŋ }

VGC *See* viscosity-gravity constant.

V guide [MECH ENG] A V-shaped groove serving to guide a wedge-shaped sliding machine element. { 'vē ,gīd }

VI *See* viscosity index.

via [ELECTR] A pathway that is etched to allow electrical contact between different layers of a semiconductor device. { 'vē·ə *or* 'vī·ə }

viaduct [CIV ENG] A bridge structure supported on high towers with short masonry or reinforced concrete arched spans. { 'vī·ə,dəkt }

via point [CONT SYS] A point located midway between the starting and stopping positions of a robot tool tip, through which the tool tip passes without stopping. Also known as way point. { 'vē·ə ,póint }

vibrating conveyor See oscillating conveyor. { 'vī,brād·iŋ kən'vā·ər }

vibrating coring tube [ENG] A sediment corer made to vibrate in order to eliminate the resistance of compacted ocean floor sediments, sands, and gravel. { 'vī,brād·iŋ 'kór·iŋ ,tüb }

vibrating feeder [MECH ENG] A feeder for bulk materials (pulverized or granulated solids), which are moved by the vibration of a slightly slanted, flat vibrating surface. { 'vī,brād·iŋ 'fēd·ər }

vibrating grizzlies [MECH ENG] Bar grizzlies mounted on eccentrics so that the entire assembly is given a forward and backward movement at a speed of some 100 strokes a minute. { 'vī,brād·iŋ 'griz·lēz }

vibrating needle [ENG] A magnetic needle used in compass adjustment to find the relative intensity of the horizontal components of the earth's magnetic field and the magnetic field at the compass location. { 'vī,brād·iŋ 'nēd·əl }

vibrating pebble mill [MECH ENG] A size-reduction device in which feed is ground by the action of vibrating, moving pebbles. { 'vī,brād·iŋ 'peb·əl ,mil }

vibrating-reed electrometer [ENG] An instrument using a vibrating capacitor to measure a small charge, often in combination with an ionization chamber. { 'vī,brād·iŋ ¦rēd ,i,lek'träm·əd·ər }

vibrating-reed frequency meter [ENG] A frequency meter consisting of steel reeds having different and known natural frequencies, all excited by an electromagnet carrying the alternating current whose frequency is to be measured. Also known as Frahm frequency meter; reed frequency meter; tuned-reed frequency meter. { 'vī,brād·iŋ ¦rēd 'frē·kwən·sē ,mēd·ər }

vibrating-reed magnetometer [ENG] An instrument that measures magnetic fields by noting their effect on the vibration of reeds excited by an alternating magnetic field. { 'vī,brād·iŋ ¦rēd ,mag·nə'täm·əd·ər }

vibrating-reed tachometer [ENG] A tachometer consisting of a group of reeds of different lengths, each having a specific natural frequency of vibration; observation of the vibrating reed when in contact with a moving mechanical device indicates the frequency of vibration for the device. { 'vī,brād·iŋ ¦rēd tə'käm·əd·ər }

vibrating screen [MECH ENG] A sizing screen which is vibrated by solenoid or magnetostriction, or mechanically by eccentrics or unbalanced spinning weights. { 'vī,brād·iŋ 'skrēn }

vibrating screen classifier [MECH ENG] A classifier whose screening surface is hung by rods and springs, and moves by means of electric vibrators. { 'vī,brād·iŋ ¦skrēn 'klas·ə,fī·ər }

vibrating wire transducer [ENG] A device for measuring ocean depth, consisting of a very fine tungsten wire stretched in a magnetic field so that it vibrates at a frequency that depends on the tension in the wire, and thereby on pressure and depth. { 'vī,brād·iŋ ¦wīr tranz'dü·sər }

vibration [MECH] A continuing periodic change in a displacement with respect to a fixed reference. { vī'brā·shən }

vibration damping [MECH ENG] The processes and techniques used for converting the mechanical vibrational energy of solids into heat energy. { vī'brā·shən 'damp·iŋ }

vibration drilling [MECH ENG] Drilling in which a frequency of vibration in the range of 100 to 20,000 hertz is used to fracture rock. { vī'brā·shən 'dril·iŋ }

vibration galvanometer [ENG] An alternating-current galvanometer in which the natural oscillation frequency of the moving element is equal to the frequency of the current being measured. { vī'brā·shən ,gal·və'näm·əd·ər }

vibration isolation [ENG] The isolation, in structures, of those vibrations or motions that are classified as mechanical vibration; involves the control of the supporting structure, the placement and arrangement of isolators, and control of the internal construction of the equipment to be protected. { vī'brā·shən ,ī·sə'lā·shən }

vibration limit [CIV ENG] The amount of time during which fresh concrete remains mobile when subjected to vibration. { vī'brā·shən ,lim·ət }

vibration machine [MECH ENG] A device for subjecting a system to controlled and reproducible mechanical vibration. Also known as shake table. { vī'brā·shən mə,shēn }

vibration magnetometer [ENG] An instrument that measures the period of vibration of a magnetic needle to determine the horizontal magnetic field strength at the needle. { vī'brā·shən ,mag·nə'täm·əd·ər }

vibration meter See vibrometer. { vī'brā·shən ,mēd·ər }

vibration puddling [CIV ENG] A technique used to achieve proper consolidation of concrete; vibrating machines may be drawn vertically through the cement, or used on the surface, or placed against the form holding the concrete in place. Also known as mechanical puddling. { vī'brā·shən 'pəd·liŋ }

vibration separation [MECH ENG] Classification or separation of grains of solids in which separation through a screen is expedited by vibration or oscillatory movement of the screening mediums. { vī'brā·shən ,sep·ə'rā·shən }

vibration suppression [MECH ENG] The prevention of undesirable vibration, either through passive means such as damping or through active techniques involving feedback control. { vī'brā·shən sə,presh·ən }

vibrator [ELEC] An electromechanical device used primarily to convert direct current to alternating current but also used as a synchronous rectifier; it contains a vibrating reed which has a set of contacts that alternately hit stationary contacts attached to the frame, reversing the direction of current flow; the reed is activated when a soft-iron slug at its tip is attracted to the pole piece of a driving coil. {MECH ENG} An instrument which produces mechanical oscillations. { 'vī,brād·ər }

vibratory centrifuge {MECH ENG} A high-speed rotating device to remove moisture from pulverized coal or other solids. { 'vī·brə,tür·ē 'sen·trə,fyüj }

vibratory equipment {MECH ENG} Reciprocating or oscillating devices which move, shake, dump, compact, settle, tamp, pack, screen, or feed solids or slurries in process. { 'vī·brə,tór·ē i'kwip·mənt }

vibratory hammer {MECH ENG} A type of pile hammer which uses electrically activated eccentric cams to vibrate piles into place. { 'vī·brə,tór·ē 'ham·ər }

vibroenergy separator {MECH ENG} A screen-type device for classification or separation of grains of solids by a combination of gyratory motion and auxiliary vibration caused by balls bouncing against the lower surface of the screen cloth. { ¦vī·brō'en·ər·jē 'sep·ə,rād·ər }

vibrograph {ENG} An instrument that provides a complete oscillographic record of a mechanical vibration; in one form a moving stylus records the motion being measured on a moving paper or film. { 'vī·brə,graf }

vibrometer {ENG} An instrument designed to measure the amplitude of a vibration. Also known as vibration meter. { vī'bräm·əd·ər }

Vicat needle {ENG} An apparatus used to determine the setting time of cement by measuring the pressure of a special needle against the cement surface. { vē'kä ,nēd·əl }

Victaulic coupling {DES ENG} A development in which a groove is cut around each end of pipe instead of the usual threads; two ends of pipe are then lined up and a rubber ring is fitted around the joint; two semicircular bands, forming a sleeve, are placed around the ring and are drawn together with two bolts, which have a ridge on both edges to fit into the groove of the pipe; as the bolts are tightened, the rubber ring is compressed, making a watertight joint, while the ridges fitting in the grooves make it strong mechanically. { vik'tól·ik 'kəp·liŋ }

videomagnetograph {ENG} A sensitive and accurate device for measuring the strength and sign of solar magnetic fields, using the signal that results when successive images in right- and left-circularly polarized light are subtracted; the images are taken in the wing of a spectral line, using a birefringent filter. { ,vid·ē·ō·mag'ned·ə,graf }

virgin See straight-run. { 'vər·jən }

virial coefficients {THERMO} For a given temperature T, one of the coefficients in the expansion of P/RT in inverse powers of the molar volume, where P is the pressure and R is the gas constant. { 'vir·ē·əl ,kō·i'fish·əns }

Virmel engine {MECH ENG} A cat-and-mouse engine that employs vanelike pistons whose motion is controlled by a gear-and-crank system; each set of pistons stops and restarts when a

chamber reaches the spark plug. { vər'mel ,en·jən }

virtual acoustics {ENG ACOUS} Digitally processing sounds so that they appear to come from particular locations in three-dimensional space, with the goal of simulating the complex acoustic field experienced by a listener within a natural environment. Also known as auralization; three-dimensional sound. { ,vər·chə·wəl ə'küs·tiks }

virtual displacement {MECH} **1.** Any change in the positions of the particles forming a mechanical system. **2.** An infinitesimal change in the positions of the particles forming a mechanical system, which is consistent with the geometrical constraints on the system. { 'vər·chə·wəl di 'splās·mənt }

virtual entropy {THERMO} The entropy of a system, excluding that due to nuclear spin. Also known as practical entropy. { 'vər·chə·wəl 'en·trə·pē }

virtual leak {ENG} The semblance of the vacuum system leak caused by a gradual desorptive release of gas at a rate which cannot be accurately predicted. { 'vər·chə·wəl 'lēk }

virtual manufacturing {IND ENG} The modeling of manufacturing systems using audiovisual or other sensory features to simulate or design an actual manufacturing environment, or the prototyping and manufacture of a proposed product mainly through effective use of computers, used to predict potential problems and inefficiencies in product functionality and manufacturability before real manufacturing occurs. { ,vər·chə·wəl ,man·ə'fak·chər·iŋ }

virtual PPI reflectoscope {ENG} A device for superimposing a virtual image of a chart on a plan position indicator (PPI) pattern; the chart is usually prepared with white lines on a black background to the scale of the plan position indicator range scale. { 'vər·chə·wəl ¦pē¦pē'ī ri'flek·tə ,skōp }

virtual work {MECH} The work done on a system during any displacement which is consistent with the constraints on the system. { 'vər·chə·wəl 'wərk }

virtual work principle See principle of virtual work. { 'vər·chə·wəl ¦wərk ,prin·sə·pəl }

visbreaking See viscosity breaking. { 'vis,brāk·iŋ }

viscoelasticity {MECH} Property of a material which is viscous but which also exhibits certain elastic properties such as the ability to store energy of deformation, and in which the application of a stress gives rise to a strain that approaches its equilibrium value slowly. { ¦vis·kō,i,las'tis·əd·ē }

viscoelastic theory {MECH} The theory which attempts to specify the relationship between stress and strain in a material displaying viscoelasticity. { ¦vis·kō·i¦las·tik 'thē·ə·rē }

viscometer {ENG} An instrument designed to measure the viscosity of a fluid. { vi'skäm·əd·ər }

viscometer gage {ENG} A vacuum gage in

viscometry

which the gas pressure is determined from the viscosity of the gas. { vi'skäm·əd·ər ,gāj }

viscometry |ENG| A branch of rheology; the study of the behavior of fluids under conditions of internal shear; the technology of measuring viscosities of fluids. { vi'skäm·ə·trē }

viscose process |CHEM ENG| A process for the manufacture of rayon by treating cellulose with caustic soda, and with carbon disulfide to form cellulose xanthate, which is then dissolved in a weak caustic solution to form the viscose; fibers are used as silk substitutes. { 'vis,kōs ,prä·səs }

viscosity blending chart |CHEM ENG| A graphical means for estimating the viscosity at a given temperature of a blend of petroleum products. { vi'skäs·əd·ē 'blend·iŋ ,chärt }

viscosity breaking |CHEM ENG| A petroleum refinery process used to lower or break the viscosity of high-viscosity residuum by thermal cracking of molecules at relatively low temperatures. Also known as visbreaking. { vi'skäs·əd·ē 'brāk·iŋ }

viscosity conversion table |CHEM ENG| A table or chart with which kinematic viscosity, in centistokes, can be converted to Saybolt viscosity, in seconds, at the same temperature. { vi'skäs·əd·ē kən'vər·zhən ,tā·bəl }

viscosity gage See molecular gage. { vi'skäs·əd·ē ,gāj }

viscosity-gravity constant |CHEM ENG| An index of the chemical composition of crude oil; defined as the general relation between specific gravity and Saybolt Universal viscosity; the constant is low for paraffinic crude oils, high for naphthenic crude oils. Abbreviated VGC. { vi'skäs·əd·ē 'grav·əd·ē ,kän·stənt }

viscosity index |CHEM ENG| An arbitrary scale used to show the magnitude of viscosity changes in lubricating oils with changes in temperature. Abbreviated VI. { vi'skäs·əd·ē ,in,deks }

viscosity manometer See molecular gage. { vi'skäs·əd·ē mə'näm·əd·ē }

viscosity-temperature chart |CHEM ENG| A chart with which the kinematic or Saybolt viscosity of a petroleum oil at any temperature within a limited range may be ascertained, provided viscosities at two temperatures are known. { vi'skäs·əd·ē 'tem·prə·chər ,chärt }

viscous damping |MECH ENG| A method of converting mechanical vibrational energy of a body into heat energy, in which a piston is attached to the body and is arranged to move through liquid or air in a cylinder or bellows that is attached to a support. { 'vis·kəs 'damp·iŋ }

viscous-drag gas-density meter |ENG| A device to measure gas-mixture densities; driven impellers in sample and standard chambers create measurable turbulences (drags) against respective nonrotating impellers. { 'vis·kəs 'drag ¦gas ¦den·səd·ē ,mēd·ər }

viscous fillers |MECH ENG| A packaging machine that fills viscous product into cartons; there are two basic types, straight-line and rotary plunger; the former operates intermittently on a given number of containers, while the latter fills and discharges containers continuously. { 'vis·kəs 'fil·ərz }

viscous filter |ENG| An air-cleaning filter having a surface coated with a viscous liquid to trap particulates in the airstream. { ¦vis·kəs ¦fil·tər }

viscous impingement filter |ENG| A filter made up of a relatively loosely arranged medium, such that the airstream is forced to change direction frequently as it passes through the filter medium; the medium usually consists of spun-glass fibers, metal screens, or layers of crimped expanded metal whose surfaces are coated with a tacky oil. { 'vis·kəs im'pinj·mənt ,fil·tər }

viscous lubrication See complete lubrication. { 'vis·kəs ,lü·brə'kā·shən }

vise |DES ENG| A tool consisting of two jaws for holding a workpiece; opened and closed by a screw, lever, or cam mechanism. { vīs }

visibility meter |ENG| An instrument for making direct measurements of visual range in the atmosphere or of the physical characteristics of the atmosphere which determine the visual range. { ,viz·ə'bil·əd·ē ,mēd·ər }

vision light |BUILD| A viewing window set in a fire door, usually glazed with wire glass. { 'vizh·ən ,līt }

visual comparator See optical comparator. { 'vizh·ə·wəl kəm'par·əd·ər }

visual servoing |CONT SYS| The use of a solid-state camera on the end effector of a robot to provide feedback. { 'vizh·ə·wəl 'sər·vō·iŋ }

vitrification |ENG| Heat treatment of a material such as a ceramic to produce a glazed surface. { ,vi·trə·fə'kā·shən }

vitrified wheel |DES ENG| A grinding wheel with a glassy or porcelanic bond. { 'vi·trə,fīd 'wēl }

VIV See verbal information verification.

vixen file |DES ENG| A flat file with curved teeth; used for filing soft metals. { 'vik·sən ,fīl }

V jewels |DES ENG| Jewel bearings used in conjunction with a conical pivot, the bearing surface being a small radius located at the apex of a conical recess; found primarily in electric measuring instruments. { 'vē ,jülz }

VLSI circuit See very large scale integrated circuit. { ¦vē¦el¦es¦ī 'sər·kət }

VMOS technology See vertical metal oxide semiconductor technology. { 'vē,mòs tek,näl·ə·jē }

V-notch weir See triangular-notch weir. { 'vē ¦näch 'wer }

VOC See volatile organic compounds.

voice coil |ENG ACOUS| The coil that is attached to the diaphragm of a moving-coil loudspeaker and moves through the air gap between the pole pieces due to interaction of the fixed magnetic field with that associated with the audio-frequency current flowing through the voice coil. Also known as loudspeaker voice coil; speech coil (British usage). { 'vòis ,kòil }

voice print |ENG ACOUS| A voice spectrograph that has individually distinctive patterns of voice characteristics that can be used to identify one person's voice from other voice patterns. { 'vòis ,print }

voice response |ENG ACOUS| The process of

generating an acoustic speech signal that communicates an intended message, such that a machine can respond to a request for information by talking to a human user. Also known as speech synthesis. { 'vȯis ri,späns }

void channels [ENG] The open passages of a porous or packed medium through which liquid or gas can flow. { 'vȯid ,chan·əlz }

Voigt body See Kelvin body. { 'fȯit ,bäd·ē }

Voigt notation [MECH] A notation employed in the theory of elasticity in which elastic constants and elastic moduli are labeled by replacing the pairs of letters xx, yy, zz, yz, zx, and xy by the number 1, 2, 3, 4, 5, and 6 respectively. { 'fȯit nō,tā·shən }

volatile organic compounds [ENG] Organic chemicals that produce vapors readily at room temperature and normal atmospheric pressure, including gasoline and solvents such as toluene, xylene, and tetrachloroethylene. They form photochemical oxidants (including ground-level ozone) that affect health, damage materials, and cause crop and forest losses. Many are also hazardous air pollutants. Abbreviated VOC. { ¦väl·ə·təl ȯr,gan·ik 'käm,paůnz }

volatility [THERMO] The quality of having a low boiling point or subliming temperature at ordinary pressure or, equivalently, of having a high vapor pressure at ordinary temperatures. { ,väl·ə'til·əd·ē }

volatilization [THERMO] The conversion of a chemical substance from a liquid or solid state to a gaseous or vapor state by the application of heat, by reducing pressure, or by a combination of these processes. Also known as vaporization. { ,väl·əd·əl·ə'zā·shən }

volley [ENG] A round of holes fired at any one time. { 'väl·ē }

volt [ELEC] The unit of potential difference or electromotive force in the meter-kilogram-second system, equal to the potential difference between two points for which 1 coulomb of electricity will do 1 joule of work in going from one point to the other. Symbolized V. { vōlt }

Volta effect See contact potential difference. { 'vōl·tə i,fekt }

voltage [ELEC] Potential difference or electromotive force measured in volts. { 'vōl·tij }

voltage amplification [ELECTR] The ratio of the magnitude of the voltage across a specified load impedance to the magnitude of the input voltage of the amplifier or other transducer feeding that load; often expressed in decibels by multiplying the common logarithm of the ratio by 20. { 'vōl·tij ,am·plə·fə'kā·shən }

voltage amplifier [ELECTR] An amplifier designed primarily to build up the voltage of a signal, without supplying appreciable power. { 'vōl·tij 'am·plə,fī·ər }

voltage coefficient [ELEC] For a resistor whose resistance varies with voltage, the ratio of the fractional change in resistance to the change in voltage. { 'vōl·tij ,kō·i,fish·ənt }

voltage-current dual [ELEC] A pair of circuits in which the elements of one circuit are replaced by their dual elements in the other circuit according to the duality principle; for example, currents are replaced by voltages, capacitances by resistances. { 'vōl·tij 'kə·rənt 'dül }

voltage-dependent resistor See varistor. { 'vōl·tij di¦pen·dənt ri'zis·tər }

voltage drop [ELEC] The voltage developed across a component or conductor by the flow of current through the resistance or impedance of that component or conductor. { 'vōl·tij ,dräp }

voltage gain [ELECTR] The difference between the output signal voltage level in decibels and the input signal voltage level in decibels; this value is equal to 20 times the common logarithm of the ratio of the output voltage to the input voltage. { 'vōl·tij ,gān }

voltage generator [ELECTR] A two-terminal circuit element in which the terminal voltage is independent of the current through the element. { 'vōl·tij ,jen·ə,rād·ər }

voltage gradient [ELEC] The voltage per unit length along a resistor or other conductive path. { 'vōl·tij ,grād·ē·ənt }

voltage level [ELEC] At any point in a transmission system, the ratio of the voltage existing at that point to an arbitrary value of voltage used as a reference. { 'vōl·tij ,lev·əl }

voltage measurement [ELEC] Determination of the difference in electrostatic potential between two points. { 'vōl·tij ,mezh·ər·mənt }

voltage multiplier [ELEC] See instrument multiplier. [ELECTR] A rectifier circuit capable of supplying a direct-current output voltage that is two or more times the peak value of the alternating-current voltage. { 'vōl·tij ,məl·tə,plī·ər }

voltage-multiplier circuit [ELEC] A rectifier circuit capable of supplying a direct-current output voltage that is two or more times the peak value of the alternating-current input voltage; useful for high-voltage, low-current supplies. { 'vōl·tij ¦məl·tə,plī·ər ,sər·kət }

voltage phasor [ELEC] A line whose length represents the magnitude of a sinusoidally varying voltage and whose angle with the positive x-axis represents its phase. { 'vōl·tij ,fā·zər }

voltage quadrupler [ELECTR] A rectifier circuit, containing four diodes, which supplies a direct-current output voltage which is four times the peak value of the alternating-current input voltage. { 'vōl·tij kwä,drüp·lər }

voltage rating [ELEC] The maximum sustained voltage that can safely be applied to an electric device without risking the possibility of electric breakdown. Also known as working voltage. { 'vōl·tij ,rād·iŋ }

voltage ratio [ELEC] The root-mean-square primary terminal voltage of a transformer divided by the root-mean-square secondary terminal voltage under a specified load. { 'vōl·tij ,rā·shō }

voltage regulation [ELEC] The ratio of the difference between no-load and full-load output voltage of a device to the full-load output voltage, expressed as a percentage. { 'vōl·tij ,reg·yə,lā·shən }

voltage regulator |ELECTR| A device that maintains the terminal voltage of a generator or other voltage source within required limits despite variations in input voltage or load. Also known as automatic voltage regulator; voltage stabilizer. { 'vōl·tij ‚reg·yə‚lād·ər }

voltage-regulator diode |ELECTR| A diode that maintains an essentially constant direct voltage in a circuit despite changes in line voltage or load. { 'vōl·tij ‚reg·yə‚lād·ər ‚dī‚ōd }

voltage stabilizer See voltage regulator. { 'vōl·tij ‚stā·bə‚līz·ər }

voltage transformer |ELEC| An instrument transformer whose primary winding is connected in parallel with a circuit in which the voltage is to be measured or controlled. Also known as potential transformer. { 'vōl·tij tranz‚fór·mər }

voltage-variable capacitor See varactor. { 'vōl·tij ¦ver·ē·əbəl kə'pas·əd·ər }

voltaic cell |ELEC| A primary cell consisting of two dissimilar metal electrodes in a solution that acts chemically on one or both of them to produce a voltage. { vōl'tā·ik 'sel }

voltammeter |ELEC| An instrument that may be used either as a voltmeter or ammeter. { väl 'tam·əd·ər }

volt-ampere |ELEC| The unit of apparent power in the International System; it is equal to the apparent power in a circuit when the product of the root-mean-square value of the voltage, expressed in volts, and the root-mean-square value of the current, expressed in amperes, equals 1. Abbreviated VA. { 'vōlt 'am‚pir }

volt-ampere hour |ELEC| A unit for expressing the integral of apparent power over time, equal to the product of 1 volt-ampere and 1 hour, or to 3600 joules. { 'vōlt 'am‚pir 'aúr }

volt-ampere-hour reactive See var hour. { 'vōlt 'am‚pir 'aúr rē'ak·tiv }

volt-ampere reactive |ELEC| The unit of reactive power in the International System; it is equal to the reactive power in a circuit carrying a sinusoidal current when the product of the root-mean-square value of the voltage, expressed in volts, by the root-mean-square value of the current, expressed in amperes, and by the sine of the phase angle between the voltage and the current, equals 1. Abbreviated var. Also known as reactive volt-ampere. { 'vōlt 'am‚pir rē'ak·tiv }

voltmeter |ENG| An instrument for the measurement of potential difference between two points, in volts or in related smaller or larger units. { 'vōlt‚mēd·ər }

voltmeter-ammeter |ENG| A voltmeter and an ammeter combined in a single case but having separate terminals. { 'vōlt‚mēd·ər 'am‚ēd·ər }

volt-ohm-milliammeter |ENG| A test instrument having a number of different ranges for measuring voltage, current, and resistance. Also known as circuit analyzer; multimeter; multiple-purpose tester. { 'vōlt 'ōm ¦mil·ē'am ‚ēd·ər }

volume |ENG ACOUS| The magnitude of a complex audio-frequency current as measured in volume units on a standard volume indicator. { 'väl·yəm }

volume compressor |ENG ACOUS| An audio-frequency circuit that limits the volume range of a radio program at the transmitter, to permit using a higher average percent modulation without risk of overmodulation; also used when making disk recordings, to permit a closer groove spacing without overcutting. Also known as automatic volume compressor. { 'väl·yəm kəm ‚pres·ər }

volume control |ENG ACOUS| A potentiometer used to vary the loudness of a reproduced sound by varying the audio-frequency signal voltage at the input of the audio amplifier. { 'väl·yəm kən‚trōl }

volume control system |ENG ACOUS| An electronic system that regulates the signal amplification or limits the output of a circuit, such as a volume compressor or a volume expander. { 'väl·yəm kən‚trōl ‚sis·təm }

volume expander |ENG ACOUS| An audio-frequency control circuit sometimes used to increase the volume range of a radio program or recording by making weak sounds weaker and loud sounds louder; the expander counteracts volume compression at the transmitter or recording studio. Also known as automatic volume expander. { 'väl·yəm ik‚span·dər }

volume indicator |ENG ACOUS| A standardized instrument for indicating the volume of a complex electric wave such as that corresponding to speech or music; the reading in volume units is equal to the number of decibels above a reference level which is realized when the instrument is connected across a 600-ohm resistor that is dissipating a power of 1 milliwatt at 100 hertz. Also known as volume unit meter. { 'väl·yəm ‚in·də‚kād·ər }

volume meter |ENG| Any flowmeter in which the actual flow is determined by the measurement of a phenomenon associated with the flow. { 'väl·yəm ‚mēd·ər }

volumenometer |ENG| An instrument for determining the volume of a body by measuring the pressure in a closed air space when the specimen is present and when it is absent. { ‚väl‚yü·mə'näm·əd·ər }

volume range |ELEC| In a transmission system, the difference, expressed in decibels, between the maximum and minimum volumes that can be satisfactorily handled by the system. |ENG ACOUS| The difference, expressed in decibels, between the maximum and minimum volumes of a complex audio-frequency signal occurring over a specified period of time. { 'väl·yəm ‚rānj }

volume resistivity |ELEC| Electrical resistance between opposite faces of a 1-centimeter cube of insulating material, commonly expressed in ohm-centimeters. Also known as specific insulation resistance. { 'väl·yəm ‚rē‚zis'tiv·əd·ē }

volumeter |ENG| Any instrument for measuring

volumes of gases, liquids, or solids. { 'väl·yə,mēd·ər }

volumetric efficiency |MECH ENG| In describing an engine or gas compressor, the ratio of volume of working substance actually admitted, measured at a specified temperature and pressure, to the full piston displacement volume; for a liquid-fuel engine, such as a diesel engine, volumetric efficiency is the ratio of the volume of air drawn into a cylinder to the piston displacement. { ¦väl·yə¦me·trik i'fish·ən·sē }

volumetric radar |ENG| Radar capable of producing three-dimensional position data on a multiplicity of targets. { ¦väl·yə¦me·trik 'rā,där }

volumetric strain |MECH| One measure of deformation; the change of volume per unit of volume. { ¦väl·yə¦me·trik 'strān }

volume unit |ENG ACOUS| A unit for expressing the audio-frequency power level of a complex electric wave, such as that corresponding to speech or music; the power level in volume units is equal to the number of decibels above a reference level of 1 milliwatt as measured with a standard volume indicator. Abbreviated VU. { 'väl·yəm ,yü·nət }

volume unit meter See volume indicator. { 'väl·yəm ,yü·nət ,mēd·ər }

volute |DES ENG| A spiral casing for a centrifugal pump or a fan designed so that speed will be converted to pressure without shock. { və'lüt }

volute pump |MECH ENG| A centrifugal pump housed in a spiral casing. { və'lüt 'pəmp }

von Arx current meter |ENG| A type of current-measuring device using electromagnetic induction to determine speed and, in some models, direction of deep-sea currents. { fön 'ärks 'kə·rənt ,mēd·ər }

von Mises yield criterion |MECH| The assumption that plastic deformation of a material begins when the sum of the squares of the principal components of the deviatoric stress reaches a certain critical value. { fön ¦mēz·əz 'yēld ,krī,tir·ē·ən }

Vorce diaphragm cell |CHEM ENG| A cylindrical cell with graphite anodes and asbestos-covered cathode, used in the electrolytic process for the manufacture of chlorine. { 'vörs 'dī·ə,fram ,sel }

vortex amplifier |ENG| A fluidic device in which the supply flow is introduced at the circumference of a shallow cylindrical chamber; the vortex field developed can substantially reduce or throttle flow; used in fluidic diodes, throttles, pressure amplifiers, and a rate sensor. { 'vör,teks 'am·plə,fī·ər }

vortex burner |ENG| Combustion device in which the combustion air is fed tangentially into the burner, creating a spin (vortex) to mix it with the fuel as it is injected. { 'vör,teks 'bər·nər }

vortex cage meter |ENG| In flow measurement, a type of quantity meter which exerts only a slight retardation on the flowing fluid; the elements rotate at a speed that is linear with fluid velocity; revolutions are counted either by coupling to a local mounted counter or by a proximity detector for remote transmission. { 'vör,teks 'kāj ,mēd·ər }

vortex precession flowmeter |ENG| An instrument for measuring gas flows from the rate of precession of vortices generated by a fixed set of radial vanes placed in the flow. Also known as swirl flowmeter. { 'vör,teks prē'sesh·ən 'flō ,mēd·ər }

vortex-shedding meter |ENG| A flowmeter in which fluid velocity is determined from the frequency at which vortices are generated by an obstruction in the flow. { 'vör,teks ¦shed·iŋ ,mēd·ər }

vortex thermometer |ENG| A thermometer, used in aircraft, which automatically corrects for adiabatic and frictional temperature rises by imparting a rotary motion to the air passing the thermal sensing element. { 'vör,teks thər'mäm·əd·ər }

VOS See vertical obstacle sonar.

V-tool See parting tool. { 've,tül }

VTVM See vacuum-tube voltmeter.

v-type engine |MECH ENG| An engine in which the cylinders are arranged in two rows set at an angle to each other, with the crankshaft running through the point of a V. { 've ,tīp ,en·jən }

vulcanization |CHEM ENG| A chemical reaction of sulfur (or other vulcanizing agent) with rubber or plastic to cause cross-linking of the polymer chains; it increases strength and resiliency of the polymer. Also known as cure. { ,vəl·kə·nə 'zā·shən }

607

W

Wacker process |CHEM ENG| A process for the oxidation of ethylene to acetaldehyde by oxygen in the presence of palladium chloride and cupric chloride. { 'wak·ər ,prä·səs }

wafer |ELECTR| A thin semiconductor slice on which matrices of microcircuits can be fabricated, or which can be cut into individual dice for fabricating single transistors and diodes. |ENG| A flat element for a process unit, as in a series of stacked filter elements. { 'wā·fər }

wage curve |IND ENG| A graphic representation of the relationship between wage rates and point values for key jobs. { 'wāj ,kərv }

wage incentive plan |IND ENG| A wage system which provides additional pay for qualitative and quantitative performance which exceeds standard or normal levels. Also known as incentive wage system. { 'wāj in'sen·tiv ,plan }

wagon drill |MECH ENG| **1.** A vertically mounted, pneumatic, percussive-type rock drill supported on a three- or four-wheeled wagon. **2.** A wheel-mounted diamond drill machine. { 'wag·ən ,dril }

wainscot |BUILD| A decorative or protective panel installed over the lower portion of an interior partition or wall. { 'wānz·kət }

waist |ENG| The center portion of a vessel or container that has a smaller cross section than the adjacent areas. { wāst }

wait |CONT SYS| Cessation of motion of a robot manipulator, under computer control, until further notice. { wāt }

waiting line |IND ENG| A line formed by units waiting for service. Also known as queue. { 'wād·iŋ ,līn }

wale See waler. { wāl }

waler |CIV ENG| A horizontal reinforcement utilized to keep newly poured concrete forms from bulging outward. Also spelled whaler. Also known as wale. { 'wā·lər }

walking beam |MECH ENG| A lever that oscillates on a pivot and transmits power in a manner producing a reciprocating or reversible motion; used in rock drilling and oil well pumping. { 'wók·iŋ ,bēm }

walking dragline |MECH ENG| A large-capacity dragline built with moving feet; disks 20 feet (6 meters) in diameter support the excavator while working. { 'wók·iŋ 'drag,līn }

walking machine |MECH ENG| A machine designed to carry its operator over various types of terrain; the operator sits on a platform carried on four mechanical legs, and movements of his arms control the front legs of the machine while movements of his legs control the rear legs of the machine. { 'wók·iŋ mə·shēn }

walkthrough method |CONT SYS| The instruction of a robot by taking it through its sequences of motions, so that these actions are stored in its memory and recalled when necessary. { 'wók¦thrü ,meth·əd }

wall |ENG| A vertical structure or member forming an enclosure or defining a space. { wól }

wall anchor |BUILD| A steel strap fastened to the end of every second or third common joist and built into the brickwork of a wall to provide lateral support. Also known as joist anchor. { 'wól ,aŋ·kər }

wall box |BUILD| **1.** A frame or box set into a wall to receive a beam or joist. Also known as beam box; wall frame. **2.** A frame set into a wall to provide a sealed space for pipework to pass through. |ELEC| A metal box set into a wall to hold switches, receptacles, or similar electrical wiring components. { 'wól ,bäks }

wall coping |CIV ENG| The covering course on top of a brick or stone wall. { 'wól ,kōp·iŋ }

wall crane |MECH ENG| A jib crane mounted on a wall. { 'wól ,krān }

Walley engine |MECH ENG| A multirotor engine employing four approximately elliptical rotors that turn in the same clockwise sense, leading to excessively high rubbing velocities. { 'wäl·ē ,en·jən }

wall frame See wall box. { 'wól ,frām }

wall furnace |MECH ENG| A self-contained vented furnace that is permanently attached to a wall and provides heated air directly to the surrounding space. { 'wól ,fər·nəs }

wall grille |BUILD| A perforated plate or a framed structure composed of rods or bars that is used to cover a wall opening to restrict vision but allow movement of air. { 'wól ,gril }

wall guard |BUILD| A protective strip of resilient material applied to the surface of a wall (especially along a corridor) several feet off the floor to prevent damage by vehicles used within a building. { 'wól ,gärd }

wall hanger |BUILD| A bracket installed in a masonry wall to support the end of a horizontal member. { 'wȯl ,haŋ·ər }

wall off |ENG| To seal cracks or crevices in the wall of a borehole with cement, mud cake, compacted cuttings, or casing. { 'wȯl 'ȯf }

wall plate |BUILD| A piece of timber laid flat along the tip of the wall; it supports the rafters. Also known as raising plate. { 'wȯl ,plāt }

wall ratio |DES ENG| Ratio of the outside radius of a gun, a tube, or jacket to the inside radius; or ratio of the corresponding diameters. { 'wȯl ,rā·shō }

wall spacer |CIV ENG| A metal tie that holds a form for poured concrete in position until the concrete has set. { 'wȯl ,spās·ər }

wall superheat |THERMO| The difference between the temperature of a surface and the saturation temperature (boiling point at the ambient pressure) of an adjacent liquid that is heated by the surface. { ¦wȯl 'sü·pər,hēt }

wall tie |BUILD| A rigid, corrosion-resistant metal tie fitted into the bed joints across the cavity of a cavity wall. { 'wȯl ,tī }

Walter engine |MECH ENG| A multirotor rotary engine that uses two different-sized elliptical rotors. { 'wȯl·tər ,en·jən }

Wankel engine |MECH ENG| An eccentric-rotor-type internal combustion engine with only two primary moving parts, the rotor and the eccentric shaft; the rotor moves in one direction around the trochoidal chamber containing peripheral intake and exhaust ports. Also known as rotary-combustion engine. { 'väŋ·kəl ,en·jən }

Wanner optical pyrometer |ENG| A type of polarizing pyrometer in which beams from the source under investigation and a comparison lamp are polarized at right angles and then passed through a Nicol prism and a red filter; the source temperature is determined from the angle through which the Nicol prism must be rotated in order to equalize the intensities of the resulting patches of light. { ¦wän·ər ¦äp·tə·kəl pī'räm·əd·ər }

Ward-Leonard speed-control system |CONT SYS| A system for controlling the speed of a direct-current motor in which the armature voltage of a separately excited direct-current motor is controlled by a motor-generator set. { 'wȯrd 'len·ərd 'spēd kən¦trōl ,sis·təm }

warehouse |IND ENG| A building used for storing merchandise and commodities. { 'wer ,haùs }

warm-air heating |MECH ENG| Heating by circulating warm air; system contains a direct-fired furnace surrounded by a bonnet through which air circulates to be heated. { 'wȯrm ¦er 'hēd·iŋ }

warm-up time |ENG| A span of time between the first application of power to a system and the moment when the system can function fully. { 'wȯrm,əp ,tīm }

warning pipe |ENG| An overflow pipe with a conspicuous outlet permitting prompt observation of discharge. { 'wȯrn·iŋ ,pīp }

warpage |MECH| The action, process, or result of twisting or turning out of shape. { 'wȯr·pij }

warping function See torsion function. { 'wȯrp·iŋ ,faŋk·shən }

Warren truss |CIV ENG| A truss having only sloping members between the top and bottom horizontal members. { 'wär·ən ,trəs }

wash |BUILD| Any member that serves to carry water away from a section of a structure. |ENG| **1.** To clean cuttings or other fragmental rock materials out of a borehole by the jetting and buoyant action of a copious flow of water or a mud-laden liquid. **2.** The erosion of core or drill string equipment by the action of a rapidly flowing stream of water or mud-laden drill-circulation liquid. { wäsh }

washboard course |ENG| A test course for vehicles consisting of a series of waves or convolutions having arbitrary amplitude and frequency; a common type is the so-called sine-wave course. { 'wäsh,bȯrd ,kȯrs }

wash boring See jet drilling. { 'wäsh ,bȯr·iŋ }

wash coat |ENG| A sealer consisting of a very thin, semitransparent coat of paint. { 'wäsh ,kōt }

washer |DES ENG| A flattened, ring-shaped device used to improve the tightness of a screw fastener. |ENG| **1.** A device for removing dirt and soluble impurities from pulp and paper stock. **2.** A system for washing photographic materials to remove soluble products of developing or fixing. **3.** A power-driven machine for washing clothes and household linens. Also known as washing machine. **4.** See scrubber. { 'wäsh·ər }

washing |CHEM ENG| In a process operation, cleaning of a solids bed (settler) or cake (filter) with a liquid in which the solid is not soluble. { 'wäsh·iŋ }

washing machine See washer. { 'wäsh·iŋ mə,shēn }

washout |ENG| **1.** An overlarge well bore caused by the solvent and erosional action of drilling fluid. **2.** A fluid-cut opening resulting from leaking fluid. { 'wäsh,aùt }

wash water |CHEM ENG| Water contacted with process streams (liquid or gas), packed beds, or filter cakes to flush or dissolve out impurities. { 'wäsh ,wȯd·ər }

waste |ENG| **1.** Rubbish from a building. **2.** Dirty water from mining, industrial, and domestic use. **3.** The amount of excavated material exceeding fill. { wāst }

waste heat |ENG| Sensible heat in gases not subject to combustion and used for processes downstream in a system. { 'wāst 'hēt }

waste-heat boiler |CHEM ENG| A heat-retrieval unit using hot by-product gas or oil from chemical processes; used to produce steam in a boiler-type system. Also known as gas-tube boiler. { 'wāst ¦hēt 'bȯi·lər }

waste lubrication |ENG| A method in which a lubricant is delivered to a bearing surface by the wicking action of cloth waste or yarn. { 'wāst ,lü·brə¦kā·shən }

waste pipe |CIV ENG| A pipe to carry waste water from a basin, bath, or sink in a building. { 'wāst ,pīp }

waste vent *See* stack vent. { 'wāst ,vent }

watchdog timer |CONT SYS| In a flexible manufacturing system, a safety device in the form of a control interface on an automated guided vehicle that shuts down part or all of the system under certain conditions. { 'wäch,dóg ,tīm·ər }

water bar |BUILD| A strip of material attached to the sill of a window or external door to prevent penetration by water. Also known as weather bar. { 'wòd·ər ,bär }

water brake |ENG| An absorption dynamometer for measuring power output of an engine shaft; the mechanical energy is converted to heat in a centrifugal pump, with a free casing where turning moment is measured. { 'wòd·ər ,brāk }

water calorimeter |ENG| A calorimeter that measures radio-frequency power in terms of the rise in temperature of water in which the r-f energy is absorbed. { 'wòd·ər ,kal·ə'rim·əd·ər }

water column |MECH ENG| A tubular column located at the steam and water space of a boiler to which protective devices such as gage cocks, water gage, and level alarms are attached. { 'wòd·ər ,käl·əm }

water-cooled condenser |MECH ENG| A steam condenser which is for the maintenance of vacuum, and in which water is the heat-receiving fluid. { 'wòd·ər ¦küld kən'den·sər }

water-cooled furnace |MECH ENG| A fuel-fired furnace containing tubes in which water is circulated to limit heat loss to the surroundings, control furnace temperature, and generate steam. { 'wòd·ər ¦küld 'fər·nəs }

water cooling |ELECTR| Cooling the electrodes of an electron tube by circulating water through or around them. |ENG| Cooling in which the primary coolant is water. { 'wòd·ər ,kül·iŋ }

water demineralizing |CHEM ENG| The removal of minerals (for example, compounds of Ca, Mg, and Na) from water by chemical, ion-exchange, or distillation procedures. { 'wòd·ər dē'min·rə,līz·iŋ }

water-flow pyrheliometer |ENG| An absolute pyrheliometer, in which the radiation-sensing element is a blackened, water calorimeter; it consists of a cylinder, blackened on the interior, and surrounded by a special chamber through which water flows at a constant rate; the temperatures of the incoming and outgoing water, which are monitored continuously by thermometers, are used to compute the intensity of the radiation. { 'wòd·ər ¦flō ¦pir,hē·lē'äm·əd·ər }

water gage |ENG| A gage glass with attached fittings which indicates water level in a vessel. { 'wòd·ər ,gāj }

water-gas reaction |CHEM ENG| A method used to prepare carbon monoxide by passing steam over hot coke or coal at 600–1000°C. { 'wòd·ər ¦gas rē,ak·shən }

water heater |MECH ENG| A tank for heating and storing hot water for domestic use. { 'wòd·ər ,hēd·ər }

water jacket |ENG| A casing for circulation of cooling water. { 'wòd·ər ,jak·ət }

water-jet cutting |ENG| A machining method that uses a jet of pressurized water containing abrasive powder for cutting steel and other dense materials. { 'wòd·ər ,jet ,kəd·iŋ }

water joint |CIV ENG| A joint in a stone pavement containing stones that are set slightly higher to prevent water from settling in the joint. { 'wòd·ər ,jòint }

water leg |ENG| The vertical area of a vessel or accessory to a vessel for the collection of water. Also known as sump. { 'wòd·ər ,leg }

water main |CIV ENG| The water pipe in a street from which water is delivered to individual service pipes supplying domestic property. { 'wòd·ər ,mān }

water meter |ENG| An instrument for measuring the amount of water passing a specified point in a piping system. { 'wòd·ər ,mēd·ər }

water path |ENG| In ultrasonic testing, distance from an ultrasonic search unit to the test piece in an immersion or water column examination. { 'wòd·ər ,path }

waterpower |MECH| Power, usually electric, generated from an elevated water supply by the use of hydraulic turbines. { 'wòd·ər,paủ·ər }

waterproof |ENG| Impervious to water. { 'wòd·ər,prüf }

water purification |CIV ENG| Any of several processes in which undesirable impurities in water are removed or neutralized; for example, chlorination, filtration, primary treatment, ion exchange, and distillation. { 'wòd·ər ,pyúr·ə·fə'kā·shən }

water right |ENG| The right to use water for mining, agricultural, or other purposes. { 'wòd·ər ,rīt }

water sample |ENG| A portion of water brought up from a depth to determine its composition. { 'wòd·ər ,sam·pəl }

water scrubber |CHEM ENG| A device or system in which gases are contacted with water (either by spray or bubbling through) to wash out traces of water-soluble components of the gas stream. { 'wòd·ər ,skrəb·ər }

water seal |ENG| A seal formed by water to prevent the passage of gas. { 'wòd·ər ,sēl }

water-sealed holder |ENG| A low-pressure gas holder which consists of cylindrical sections or lifts telescoping into a pit or tank filled with water; the inside section is closed in on top. { 'wòd·ər ¦sēld 'hōl·dər }

waterspout |ENG| A pipe or orifice through which water is discharged or by which it is conveyed. { 'wòd·ər,spaủt }

water-supply engineering |CIV ENG| A branch of civil engineering concerned with the development of sources of supply, transmission, distribution, and treatment of water. { 'wòd·ər sə,plī ,en·jə'nir·iŋ }

water swivel |DES ENG| A device connecting the water hose to the drill-rod string and designed to permit the drill string to be rotated in the borehole while water is pumped into it to create the circulation needed to cool the bit and remove the cuttings produced. Also known as gooseneck; swivel neck. { 'wȯd·ər ,swiv·əl }

water table |BUILD| A ledge or slight projection of the masonry or wood construction on the exterior of a foundation wall, or just above it, to protect the foundation by directing rainwater away from the wall. Also known as canting strip. { 'wȯd·ər ,tā·bəl }

water tower |CIV ENG| A tower or standpipe for storing water in areas where ordinary water pressure is inadequate for distribution to consumers. { 'wȯd·ər ,taú·ər }

water treatment |CIV ENG| Purification of water to make it suitable for drinking or for any other use. { 'wȯd·ər ,trēt·mənt }

water-tube boiler |MECH ENG| A steam boiler in which water circulates within tubes and heat is applied from outside the tubes to generate steam. { 'wȯd·ər ¦tüb ,bȯi·lər }

water tunnel |CIV ENG| A tunnel to transport water in a water-supply system. { 'wȯd·ər ,tən·əl }

waterwall |MECH ENG| The side of a boiler furnace consisting of water-carrying tubes which absorb radiant heat and thereby prevent excessively high furnace temperatures. { 'wȯd·ər,wȯl }

waterway |CIV ENG| A channel for the escape or passage of water. { 'wȯd·ər,wā }

water well |CIV ENG| A well sunk to extract water from a zone of saturation. { 'wȯd·ər ,wel }

waterwheel |MECH ENG| A vertical wheel on a horizontal shaft that is made to revolve by the action or weight of water on or in containers attached to the rim. { 'wȯd·ər,wēl }

waterworks |CIV ENG| The whole system of supply and treatment utilized in acquisition and distribution of water to consumers. { 'wȯd·ər ,wərks }

Watson factor See characterization factor. { 'wät·sən ,fak·tər }

watt-hour |ELEC| A unit of energy used in electrical measurements, equal to the energy converted or consumed at a rate of 1 watt during a period of 1 hour, or to 3600 joules. Abbreviated Wh. { 'wät ¦aúr }

watt-hour meter |ENG| A meter that measures and registers the integral, with respect to time, of the active power of the circuit in which it is connected; the unit of measurement is usually the kilowatt-hour. { 'wät ¦aúr ,mēd·ər }

wattmeter |ENG| An instrument that measures electric power in watts ordinarily. { 'wät,mēd·ər }

Watt's law |THERMO| A law which states that the sum of the latent heat of steam at any temperature of generation and the heat required to raise water from 0°C to that temperature is constant; it has been shown to be substantially in error. { 'wäts ,lȯ }

wave filter |ELEC| A transducer for separating waves on the basis of their frequency; it introduces relatively small insertion loss to waves in one or more frequency bands and relatively large insertion loss to waves of other frequencies. { 'wāv ,fil·tər }

wave gage |ENG| A device for measuring the height and period of waves. { 'wāv ,gāj }

wave gait |MECH ENG| A mode of motion of a mobile robot with several legs in which its components have a wavy motion. { 'wāv ,gāt }

waveguide junction See junction. { 'wāv,gīd ¦jəŋk·shən }

waveguide synthesis |ENG ACOUS| A method of synthesizing the sounds of a string or wind instrument that simulates traveling waves on a string or inside a bore or horn using digital delay lines. { ,wāv,gīd 'sin·thə·səs }

wavemeter |ENG| A device for measuring the geometrical spacing between successive surfaces of equal phase in an electromagnetic wave. { 'wāv,mēd·ər }

wave microphone |ENG ACOUS| Any microphone whose directivity depends upon some type of wave interference, such as a line microphone or a reflector microphone. { 'wāv 'mī·krə,fōn }

wave motor |MECH ENG| A motor that depends on the lifting power of sea waves to develop its usable energy. { 'wāv ,mōd·ər }

wave noise |ELECTR| Noise in the electric current of a detector that results from fluctuations in the intensity of electromagnetic radiation falling on the detector. { 'wāv ,nȯiz }

wave polarization See polarization. { 'wāv ,pō·lə·rə,zā·shən }

wave shaper |ENG| Of explosives, an insert or core of inert material or of explosives having different detonation rates, used for changing the shape of the detonation wave. { 'wāv ,shāp·ər }

wave-shaping circuit |ELECTR| An electronic circuit used to create or modify a specified time-varying electrical quantity, usually voltage or current, using combinations of electronic devices, such as vacuum tubes or transistors, and circuit elements, including resistors, capacitors, and inductors. { 'wāv ¦shāp·iŋ ,sər·kət }

wave soldering See flow soldering. { 'wāv ,säd·ə·riŋ }

wave tail |ELECTR| Part of a signal-wave envelope (in time or distance) between the steady-state value (or crest) and the end of the envelope. { 'wāv ,tāl }

wave trap |CIV ENG| A device used to reduce the size of waves from sea or swell entering a harbor before they penetrate as far as the quayage; usually in the form of diverging breakwaters, or small projecting breakwaters situated close within the entrance. |ELECTR| A resonant circuit connected to the antenna system of a receiver to suppress signals at a particular frequency, such as that of a powerful local station that is interfering with reception of other stations. Also known as trap. { 'wāv ,trap }

wax fractionation |CHEM ENG| A continuous solvent-recovery/crystallization petroleum-refinery process for the production of waxes with low oil content from wax concentrates; for example,

MEK (methyl ethyl ketone) deoiling. { 'waks ,frak·shə'nā·shən }

wax manufacturing |CHEM ENG| A petroleum refinery process similar to wax fractionation for the manufacture of oil-free waxes by chilling and crystallization from a solvent. { 'waks ,man· ə'fak·chə·riŋ }

wax master See wax original. { 'waks 'mas·tər }

wax original |ENG ACOUS| An original recording made on a wax surface and used to make a master. Also known as wax master. { 'waks ə'rij· ən·əl }

way point See via point. { 'wā ,pȯint }

ways |CIV ENG| **1.** The tracks and sliding timbers used in launching a vessel. **2.** The building slip or space upon which the sliding timbers or ways, supporting a vessel to be launched, travel. |MECH ENG| Bearing surfaces used to guide and support moving parts of machine tools; may be flat, V-shaped, or dovetailed. { wāz }

wear |ENG| Deterioration of a surface due to material removal caused by relative motion between it and another part. { wer }

wearing course |CIV ENG| The top layer of surfacing on a road. { 'wer·iŋ ,kȯrs }

weather bar See water bar. { 'weth·ər ,bär }

weathered joint See weather-struck joint. { ¦weth· ərd ¦jȯint }

weather observation radar See weather radar. { 'weth·ər ,äb·zər,vā·shən 'rā,där }

weatherometer |ENG| A device used to subject articles and finishes to accelerated weathering conditions; for example, a rich ultraviolet source, water spray, or salt water. { ,weth·ə'räm·əd·ər }

weatherproof |ENG| Able to withstand exposure to weather without damage. { 'weth·ər ,prüf }

weather radar |ENG| Generally, any radar which is suitable or can be used for the detection of precipitation or clouds. Also known as weather observation radar. { 'weth·ər 'rā,där }

weather resistance |ENG| The ability of a material, paint, film, or the like to withstand the effects of wind, rain, or sun and to retain its appearance and integrity. { 'weth·ər ri,zis·təns }

weather strip |BUILD| A piece of material, such as wood or rubber, applied to the joints of a window or door to stop drafts. { 'weth·ər ,strip }

weather-struck joint |CIV ENG| A horizontal joint in a course of masonry in which the mortar at the upper edge has been pressed in, forming a convex surface that sheds water. Also known as weathered joint. { ¦weth·ər ¦strək ,jȯint }

web |CIV ENG| The vertical strip connecting the upper and lower flanges of a rail or girder. |MECH ENG| For twist drills and reamers, the central portion of the tool body that joins the loads. { web }

web angle See chisel-edge angle. { 'web ,aŋ·gəl }

Weber number 3 |CHEM ENG| A dimensionless number used in interfacial area determination in distillation equipment, equal to the surface tension divided by the product of the liquid density, the acceleration of gravity, and the depth of liquid on the tray under consideration. Symbolized N_{We3}. { 'vā·bər ¦nəm·bər 'thrē }

web plate |ENG| A steel plate that forms the web of a beam, girder, or truss. { 'web ,plāt }

wedge |DES ENG| A piece of resistant material whose two major surfaces make an acute angle. |ENG| In ultrasonic testing, a device which directs waves of ultrasonic energy into the test piece at an angle { wej }

wedge bit |DES ENG| A tapered-nose noncoring bit, used to ream out the borehole alongside the steel deflecting wedge in hole-deflection operations. Also known as bull-nose bit; wedge reaming bit; wedging bit. { 'wej ,bit }

wedge bonding |ENG| A type of thermocompression bonding in which a wedge-shaped tool is used to press a small section of the lead wire onto the bonding pad of an integrated circuit. { 'wej ,bänd·iŋ }

wedge core lifter |MECH ENG| A core-gripping device consisting of a series of three or more serrated-face, tapered wedges contained in slotted and tapered recesses cut into the inner surface of a lifter case or sleeve; the case is threaded to the inner tube of a core barrel, and as the core enters the inner tube, it lifts the wedges up along the case taper; when the barrel is raised, the wedges are pulled tight, gripping the core. { 'wej ¦kȯr ,lif·tər }

wedge photometer |ENG| A photometer in which the luminous flux density of light from two sources is made equal by pushing into the beam from the brighter source a wedge of absorbing material; the wedge has a scale indicating how much it reduces the flux density, so that the luminous intensities of the sources may be compared. { 'wej fə'täm·əd·ər }

wedge reaming bit See wedge bit. { 'wej 'rēm· iŋ ,bit }

wedging |ENG| **1.** A method used in quarrying to obtain large, regular blocks of building stones; a row of holes is drilled, either by hand or by pneumatic drills, close to each other so that a longitudinal crevice is formed into which a gently sloping steel wedge is driven, and the block of stone can be detached without shattering. **2.** The act of changing the course of a borehole by using a deflecting wedge. **3.** The lodging of two or more wedge-shaped pieces of core inside a core barrel, and therefore blocking it. **4.** The material, moss, or wood used to render the shaft lining tight. { 'wej·iŋ }

wedging bit See wedge bit. { 'wej·iŋ ,bit }

weep hole |CIV ENG| A hole in a wood sill, retaining wall, or other structure to allow accumulated water to escape. { 'wēp ,hōl }

weighing rain gage |ENG| A type of recording rain gage, consisting of a receiver in the shape of a funnel which empties into a bucket mounted upon a weighing mechanism; the weight of the catch is recorded, on a clock-driven chart, as inches of precipitation; used at climatological stations. { 'wā·iŋ 'rān gāj }

weight |MECH| **1.** The gravitational force with

which the earth attracts a body. **2.** By extension, the gravitational force with which a star, planet, or satellite attracts a nearby body. { wāt }

weight barometer |ENG| A mercury barometer which measures atmospheric pressure by weighing the mercury in the column or the cistern. { 'wāt bə,räm·əd·ər }

weighting |ENG| The artificial adjustment of measurements to account for factors that, in the normal use of the device, would otherwise be different from conditions during the measurements. { 'wād·iŋ }

weighting network |ENG ACOUS| One of three or more circuits in a sound-level meter designed to adjust its response; the A and B weighting networks provide responses approximating the 40- and 70-phon equal loudness contours, respectively, and the C weighting network provides a flat response up to 8000 hertz. { 'wād·iŋ ,net ,wərk }

weightlessness |MECH| A condition in which no acceleration, whether of gravity or other force, can be detected by an observer within the system in question. Also known as zero gravity. { 'wāt·ləs·nəs }

weight-loaded regulator |ENG| A pressure-regulator valve for pressure vessels or flow systems; the regulator is preloaded by counterbalancing weights to open (or close) at the upper (or lower) limit of a preset pressure range. { 'wāt ¦lōd·əd 'reg·yə,lād·ər }

weight thermometer |ENG| A glass vessel for determining the thermal expansion coefficient of a liquid by measuring the mass of liquid needed to fill the vessel at two different temperatures. { 'wāt ,thər,mäm·əd·ər }

weir |CIV ENG| A dam in a waterway over which water flows, serving to regulate water level or measure flow. { wer }

weld gage |ENG| A device used to check the shape and size of welds. { 'weld ,gāj }

welding tip |ENG| A replaceable nozzle for a gas torch used in welding. { 'weld·iŋ ,tip }

welding torch |ENG| A gas-mixing and burning tool for the welding of metal. { 'weld·iŋ ,tórch }

weld-interval timer |ENG| A device used to control weld interval. { 'weld ¦in·tər·vəl ,tīm·ər }

weld line See flow line. { 'weld ,līn }

weld mark See flow line. { 'weld ,märk }

weldment |ENG| An assembly or structure whose component parts are joined by welding. { 'weld·mənt }

well |BUILD| An open shaft in a building, extending vertically through floors to accommodate stairs or an elevator. |ENG| A hole dug into the earth to reach a supply of water, oil, brine, or gas. { wel }

well core |ENG| A sample of rock penetrated in a well or other borehole obtained by use of a hollow bit that cuts a circular channel around a central column or core. { 'wel ,kòr }

well drill |MECH ENG| A drill, usually a churn drill, used to drill water wells. { 'wel ,dril }

wellhead |CIV ENG| The top of a well. { 'wel ,hed }

well logging |ENG| The technique of analyzing and recording the character of a formation penetrated by a drill hole in petroleum exploration and exploitation work. { 'wel ,läg·iŋ }

wellpoint |CIV ENG| A component of a wellpoint system consisting of a perforated pipe about 4 feet (1.2 meters) long and about 2 inches (5 centimeters) in diameter, equipped with a ball valve, a screen, and a jetting tip. { 'wel,póint }

wellpoint system |CIV ENG| A method of keeping an excavated area dry by intercepting the flow of groundwater with pipe wells located around the excavation area. { 'wel,póint ,sis·təm }

well-regulated system |CONT SYS| A system with a regulator whose action, together with that of the environment, prevents any disturbance from permanently driving the system from a state in which it is stable, that is, a state in which it retains its structure and survives. { 'wel ¦reg·yə,lād·əd ,sis·təm }

well shooting |ENG| The firing of a charge of nitroglycerin, or other high explosive, in the bottom of a well for the purpose of increasing the flow of water, oil, or gas. { 'wel ,shüd·iŋ }

well-type manometer |ENG| A type of double-leg, glass-tube manometer; one leg has a relatively small diameter, and the second leg is a reservoir; the level of the liquid in the reservoir does not change appreciably with change of pressure; a mercury barometer is a common example. { 'wel ¦tīp mə'näm·əd·ər }

welt |BUILD| **1.** In sheet-metal roofing, a seam consisting of two joined sheets of metal whose edges have been folded over each other and fastened down flat. **2.** A strip of wood fastened over a flush seam or joint for added strength. |ENG| A strip that has been fastened to the edges of plates that form a butt joint in a steam boiler. { welt }

Wentworth quick-return motion See turning-block linkage. { 'went,wərth 'kwik ri¦tərn ,mō·shən }

Weston standard cell |ELEC| A standard cell used as a highly accurate voltage source for calibrating purposes; the positive electrode is mercury, the negative electrode is cadmium, and the electrolyte is a saturated cadmium sulfate solution; the Weston standard cell has a voltage of 1.018636 volts at 20°C. { 'wes·tən 'stan·dərd 'sel }

Westphal balance |ENG| A direct-reading instrument for determining the densities of solids and liquids; a plummet of known mass and volume is immersed in the liquid whose density is to be measured or, alternatively, a sample of the solid whose density is to be measured is immersed in a liquid of known density, and the loss in weight is measured, using a balance with movable weights. { 'west,fól ,bal·əns }

wet and dry bulb thermometer See psychrometer. { ¦wet ən ¦drī ,bəlb thər'mäm·əd·ər }

wet blasting |ENG| Shot firing in wet holes. { 'wet 'blast·iŋ }

wet-bulb thermometer |ENG| A thermometer having the bulb covered with a cloth, usually muslin or cambric, saturated with water. { 'wet ¦bəlb thər'mäm·əd·ər }

wet cell |ELEC| A primary cell in which there is a substantial amount of free electrolyte in liquid form. { 'wet ,sel }

wet classifier |ENG| A device for the separation of solid particles in a mixture of solids and liquid into fractions, according to particle size or density by methods other than screening; operates by the difference in the settling rate between coarse and fine or heavy and light particles in a tank-confined liquid. { 'wet 'klas·ə,fī·ər }

wet collector See scrubber. { 'wet kə'lek·tər }

wet cooling tower |MECH ENG| A structure in which water is cooled by atomization into a stream of air; heat is lost through evaporation. Also known as evaporative cooling tower. { 'wet 'kül·iŋ ,taù·ər }

wet drill |MECH ENG| A percussive drill with a water feed either through the machine or by means of a water swivel, to suppress the dust produced when drilling. { 'wet ¦dril }

wet engine |MECH ENG| An engine with its oil, liquid coolant (if any), and trapped fuel inside. { 'wet 'en·jən }

wet grinding |MECH ENG| 1. The milling of materials in water or other liquid. 2. The practice of applying a coolant to the work and the wheel to facilitate the grinding process. { 'wet ¦grīnd·iŋ }

wet hole |ENG| A borehole that traverses a water-bearing formation from which the flow of water is great enough to keep the hole almost full of water. { 'wet ,hōl }

wet mill |MECH ENG| 1. A grinder in which the solid material to be ground is mixed with liquid. 2. A mill in which the grinding energy is developed by a fast-flowing liquid stream; for example, a jet pulverizer. { 'wet 'mil }

wet scrubber |ENG| A device designed to clean a gas stream by bringing it into contact with a liquid. { 'wet 'skrəb·ər }

wet sleeve |MECH ENG| A cylinder liner which is exposed to the coolant over 70% or more of its surface. { 'wet 'slēv }

wet slip |CIV ENG| An opening between two wharves or piers where dock trials are usually conducted, and the final fitting out is done. { 'wet 'slip }

wetted-wall column |CHEM ENG| A vertical column that operates with the inner walls wetted by the liquid being processed; used in theoretical studies of mass transfer rates and in analytical distillations; an example is a spinning-band column. { 'wed·əd ¦wol 'käl·əm }

wet-test meter |ENG| A device to measure gas flow by counting the revolutions of a shaft upon which water-sealed, gas-carrying cups of fixed capacity are mounted. { 'wet ¦test ,mēd·ər }

wetting |ELECTR| The coating of a contact surface with an adherent film of mercury. { 'wed·iŋ }

wetting agent |CHEM ENG| A substance that increases the rate at which a liquid spreads across a surface when it is added to the liquid in small amounts. { 'wed·iŋ ,ā·jənt }

wet well |MECH ENG| A chamber which is used for collecting liquid, and to which the suction pipe of a pump is attached. { 'wet ,wel }

whaler See waler. { 'wāl·ər }

wharf |CIV ENG| A structure of open construction built parallel to the shoreline; used by vessels to receive and discharge passengers and cargo. { 'worf }

Wheatstone bridge |ELEC| A four-arm bridge circuit, all arms of which are predominantly resistive; used to measure the electrical resistance of an unknown resistor by comparing it with a known standard resistance. Also known as resistance bridge; Wheatstone network. { 'wēt ,stōn 'brij }

wheel |DES ENG| A circular frame with a hub at the center for attachment to an axle, about which it may revolve and bear a load. { 'wēl }

wheelbarrow |ENG| A small, hand-pushed vehicle with a single wheel and axle between the front ends of two shafts that support a boxlike body and serve as handles at the rear. Also known as barrow. { 'wēl,bar·ō }

wheel base |DES ENG| The distance in the direction of travel from front to rear wheels of a vehicle, measured between centers of ground contact under each wheel. { 'wēl ,bās }

wheel dresser |ENG| A tool for cleaning, resharpening, and restoring the mechanical accuracy of the cutting faces of grinding wheels. { 'wēl ,dres·ər }

wheeled crane |MECH ENG| A self-propelled crane that rides on a rubber-tired chassis with power for transportation provided by the same engine that is used for hoisting. { 'wēld 'krān }

wheel load capacity |CIV ENG| The capacity of airfield runways, taxiways, parking areas, or roadways to bear the pressures exerted by aircraft or vehicles in a gross weight static configuration. { 'wēl 'lōd kə,pas·əd·ē }

wheel sleeve |DES ENG| A flange used as an adapter on precision grinding machines where the hole in the wheel is larger than the machine arbor. { 'wēl ,slēv }

white coat |BUILD| The finishing coat in plastering. { 'wīt ,kōt }

Whitworth screw thread |DES ENG| A British screw thread standardized to form and dimension. { 'wit,wərth 'skrü ,thred }

whr See watt-hour.

wicket dam |CIV ENG| A movable dam consisting of a number of rectangular panels of wood or iron hinged to a sill and propped vertically; the prop is hinged and can be tripped to drop the wickets flat on the sill. { 'wik·ət ,dam }

wicking |ENG| The flow of solder under the insulation of covered wire. { 'wik·iŋ }

wide band |ELECTR| Property of a tuner, amplifier, or other device that can pass a broad range of frequencies. { 'wīd ¦band }

wide-flange beam See H beam. { ¦wīd ¦flanj 'bēm }

Wiese formula |ENG| An empirical relationship for motor fuel antiknock values above 100 in relation to performance numbers; basis for the American Society for Testing and Materials scale, in which octane numbers above 100 are related to increments of tetraethyllead added to isooctane. { 'vē·zə ,fȯr·myə·lə }

Wild fence |ENG| A wooden enclosure about 16 feet (4.8 meters) square and 8 feet (2.4 meters) high with a precipitation gage in its center; the function of the fence is to minimize eddies around the gage, and thus ensure a catch which will be representative of the actual rainfall or snowfall. { 'wīld ,fens }

Willans line |MECH ENG| The line (nearly straight) on a graph showing steam consumption (pounds per hour) versus power output (kilowatt or horsepower) for a steam engine or turbine; frequently extended to show total fuel consumed (pounds per hour) for gas turbines, internal combustion engines, and complete power plants. { 'wil·ənz ,līn }

winch |MECH ENG| A machine having a drum on which to coil a rope, cable, or chain for hauling, pulling, or hoisting. { winch }

winch operator See hoistman. { 'winch ,äp·ə,rād·ər }

windage |MECH| 1. The deflection of a bullet or other projectile due to wind. 2. The correction made for such deflection. { 'win·dij }

windage loss |ENG| In a ventilating or air-conditioning system, the decrease in the water content of the circulating air due to the loss of entrained droplets of water; expressed as a percentage of the rate of circulation. { 'win·dij ,lȯs }

wind box |ENG| A plenum chamber that supplies air for combustion to a stoker, gas burner, or oil burner. { 'wind ,bäks }

windbreak |ENG| Any device designed to obstruct wind flow and intended for protection against any ill effects of wind. { 'win,brāk }

wind cone |ENG| A tapered fabric sleeve, shaped like a truncated cone and pivoted at its larger end on a standard, for the purpose of indicating wind direction; since the air enters the fixed end, the small end of the cone points away from the wind. Also known as wind sleeve; wind sock. { 'win ,kōn }

wind correction |ENG| Any adjustment which must be made to allow for the effect of wind; especially, the adjustments to correct for the effect on a projectile in flight, on sound received by sound ranging instruments, and on an aircraft flown by dead reckoning navigation. { 'win kə'rek·shən }

wind deflection |MECH| Deflection caused by the influence of wind on the course of a projectile in flight. { 'win di,flek·shən }

wind-direction indicator |ENG| A device to indicate the direction from which the wind blows; an example is a weather vane. { 'win də¦rek·shən ,in·də,kād·ər }

winder |BUILD| A step, generally wedge-shaped, with a tread that is wider at one end than the other; often used in spiral staircases. { 'wīn·dər }

wind guard |CIV ENG| A building component that protects the building or some part of it against the wind, for example, a chimney cap. { 'win,gärd }

winding |ELEC| 1. One or more turns of wire forming a continuous coil for a transformer, relay, rotating machine, or other electric device. 2. A conductive path, usually of wire, that is inductively coupled to a magnetic storage core or cell. { 'wīnd·iŋ }

winding engine See hoist. { 'wīnd·iŋ ,en·jən }

windmill |MECH ENG| Any of various mechanisms, such as a mill, pump, or electric generator, operated by the force of wind against vanes or sails radiating about a horizontal shaft. { 'win,mil }

windmill anemometer |ENG| A rotation anemometer in which the axis of rotation is horizontal; the instrument has either flat vanes (as in the air meter) or helicoidal vanes (as in the propeller anemometer); the relation between wind speed and angular rotation is almost linear. { 'win,mil ,an·ə'mäm·əd·ər }

windmilling |MECH ENG| The rotation of a propeller from the force of the air when the engine is not operating. { 'win,mil·iŋ }

window |BUILD| An opening in the wall of a building or the body of a vehicle to admit light and usually to permit vision through a transparent or translucent material, usually glass. |ELECTR| A material having minimum absorption and minimum reflection of radiant energy, sealed into the vacuum envelope of a microwave or other electron tube to permit passage of the desired radiation through the envelope to the output device. { 'win·dō }

window bar |BUILD| 1. A bar for securing a casement window or window shutters. 2. A bar that prevents ingress or egress through a window. 3. See sash bar. { 'win,dō ,bär }

wind power |MECH ENG| The extraction of kinetic energy from the wind and conversion of it into a useful type of energy: thermal, mechanical, or electrical. { 'win ,pau̇·ər }

wind pressure |MECH| The total force exerted upon a structure by wind. Also known as velocity pressure. { 'win ,presh·ər }

windshield |ENG| A transparent glass screen that protects the passengers and compartment of a vehicle from wind and rain. { 'win,shēld }

wind shield See rain-gage shield. { 'win ,shēld }

wind sleeve See wind cone. { 'win ,slēv }

wind sock See wind cone. { 'win ,säk }

wind tee |ENG| A weather vane shaped like the letter T or like an airplane, situated on an airport or landing field to indicate the wind direction. Also known as landing tee. { 'win ,tē }

wind tunnel |ENG| A duct in which the effects of airflow past objects can be determined. { 'win ,tən·əl }

wind-tunnel instrumentation |ENG| Measuring devices used in wind-tunnel tests; in addition to conventional laboratory instruments for fluid flow, thermometry, and mechanical measurements, there are sensing devices capable of precision measurement in the small-scale environment of the test setup. { 'win ˌtən·əl ˌin·strə·mən'tā·shən }

windup |MECH ENG| The twisting of a shaft under a torsional load, usually resulting in vibration and other undesirable effects as the shaft relaxes. { 'wīn,dəp }

wind vane |ENG| An instrument used to indicate wind direction, consisting basically of an asymmetrically shaped object mounted at its center of gravity about a vertical axis; the end which offers the greater resistance to the motion of air moves to the downwind position; the direction of the wind is determined by reference to an attached oriented compass rose. { 'win ˌvān }

wing dam See groin. { 'wiŋ ˌdam }

wingless abutment |CIV ENG| A straight-sided bridge abutment designed to resist pressure in back and provide a bridge seat. { 'wiŋ·ləs ə'bət·mənt }

wing nut |DES ENG| An internally threaded fastener with wings to permit it to be tightened or loosened by finger pressure only. Also known as butterfly nut. { 'wiŋ ˌnət }

wing screw |DES ENG| A screw with a wing-shaped head that can be turned manually. { 'wiŋ ˌskrü }

winterization |ENG| The preparation of equipment for operation in conditions of winter weather; this applies to preparation not only for cold temperatures, but also for snow, ice, and strong winds. { ˌwin·tə·rə'zā·shən }

wire |ELEC| A single bare or insulated metallic conductor having solid, stranded, or tinsel construction, designed to carry current in an electric circuit. Also known as electric wire. { wīr }

wire bonding |ELEC| Lead-covered tie used to connect two cable sheaths until a splice is permanently closed and covered. |ELECTR| **1.** A method of connecting integrated-circuit chips to their substrate, using ultrasonic energy to weld very fine wires mechanically from metallized terminal pads along the periphery of the chip to corresponding bonding pads on the substrate. **2.** The attachment of very fine aluminum or gold wire (by thermal compression or ultrasonic welding) from metallized terminal pads along the periphery of an integrated circuit chip to corresponding bonding pads on the surface of the package leads. { 'wīr ˌbänd·iŋ }

wire cloth |DES ENG| Screen composed of wire crimped or woven into a pattern of squares or rectangles. { 'wīr ˌklȯth }

wire comb |ENG| A tool for roughening a base coat of plaster in order to improve bonding of the next coat. Also known as wire scratcher. { ˈwīr ˈkōm }

wire drag |ENG| An apparatus for surveying rocky underwater areas where normal sounding methods are insufficient to ensure the discovery of all existing submerged obstructions, small shoals, or rocks above a given depth or for determining the least depth of an area; it consists

essentially of a buoyed wire towed at the desired depth by two launches. { 'wīr 'drag }

wire-fabric reinforcing |CIV ENG| Reinforcing concrete or mortar with a welded wire fabric. { 'wīr ˈfab·rik ˌrē·ən'fȯrs·iŋ }

wire flame spray gun |ENG| A device which utilizes the heat from a gas flame and material in the form of wire or rod to perform a flame-spraying operation. { 'wīr ˈflām 'sprā ˌgən }

wire fusing current |ELEC| The electric current which will cause a wire to melt. { 'wīr ˈfyüz·iŋ ˌkə·rənt }

wire gage |DES ENG| **1.** A gage for measuring the diameter of wire or thickness of sheet metal. **2.** A standard series of sizes arbitrarily indicated by numbers, to which the diameter of wire or the thickness of sheet metal is usually made, and which is used in describing the size or thickness. { 'wīr ˌgāj }

wire lath |ENG| A netting formed of welded wire, usually with a paper backing, and used as a base for plaster. { ˈwīr 'lath }

wire line |DES ENG| **1.** Any cable or rope made of steel wires twisted together to form the strands. **2.** A steel wire rope 5/16 inch (7.94 millimeters) or less in diameter. |ELECTR| One or more current-conducting wires or cables, used for communication, control, or telemetry. { 'wīr ˌlīn }

wire nail |DES ENG| A nail made of wire and having a circular cross section. { 'wīr ˌnāl }

wire recorder |ENG ACOUS| A magnetic recorder that utilizes a round stainless steel wire about 0.004 inch (0.01 centimeter) in diameter instead of magnetic tape. { 'wīr ri,kȯrd·ər }

wire recording |ENG ACOUS| Magnetic recording by use of a magnetized wire. { 'wīr ri ˌkȯrd·iŋ }

wire rope |ENG| A rope formed of twisted strands of wire. { 'wīr ˌrōp }

wire saw |MECH ENG| A machine employing one- or three-strand wire cable, up to 16,000 feet (4900 meters) long, running over a pulley as a belt; used in quarries to cut rock by abrasion. { 'wīr 'sȯ }

wire scratcher See wire comb. { 'wīr ˌskrach·ər }

wiresonde |ENG| An atmospheric sounding instrument which is supported by a captive balloon and used to obtain temperature and humidity data from the ground level to a height of a few kilometers; height is determined by means of a sensitive altimeter, or from the amount of cable released and the angle which the cable makes with the ground, and the information is telemetered to the ground through a wire cable. { 'wīr,sänd }

wire stripper |ENG| A hand-operated tool or special machine designed to cut and remove the insulation for a predetermined distance from the end of an insulated wire, without damaging the solid or stranded wire inside. { 'wīr ˌstrip·ər }

wire tack |DES ENG| A tack made from wire stock. { 'wīr ˌtak }

wire train |ENG| An assembly that normally consists of an extruder, a crosshead and die, a means of cooling, and feed and take-up spools for the wire; used to coat wire with resin. { 'wīr ,trān }

wireway |ENG| A trough which is lined with sheet metal and has hinged covers, designed to house electrical conductors or cables. { 'wīr,wā }

wire weight gage |ENG| A river gage in which a weight suspended on a wire is lowered to the water surface from a bridge or other overhead structure to measure the distance from a point of known elevation on the bridge to the water surface; the distance is usually measured by counting the number of revolutions of a drum required to lower the weight, and a counter is provided which reads the water stage directly. { 'wīr 'wāt ,gāj }

wiring |ELEC| The installation and utilization of a system of wire for conduction of electricity. Also known as electric wiring. |ENG| A forming process in which the edge of a sheet-metal part is rolled over a wire to produce a tubular rim containing the wire. { 'wīr·iŋ }

wiring diagram See circuit diagram. { 'wīr·iŋ ,dī·ə,gram }

wiring harness |ELEC| An array of insulated conductors bound together by lacing cord, metal bands, or other binding, in an arrangement suitable for use only in specific equipment for which the harness was designed; it may include terminations. { 'wīr·iŋ ,här·nəs }

Wobbe index |THERMO| A measure of the amount of heat released by a gas burner with a constant orifice, equal to the gross calorific value of the gas in British thermal units per cubic foot at standard temperature and pressure divided by the square root of the specific gravity of the gas. { 'wä·bə ,in,deks }

wobble friction |ENG| A force that occurs in prestressed concrete when the prestressing tendon deviates from its specified profile. { 'wäb·əl ,frik·shən }

wobble wheel roller |MECH ENG| A roller with freely suspended pneumatic tires used in soil stabilization. { 'wäb·əl ¦wēl ,rō·lər }

Wollaston wire |ENG| An extremely fine platinum wire, produced by enclosing a platinum wire in a silver sheath, drawing them together, and using acid to dissolve away the silver; used in electroscopes, microfuses, and hot-wire instruments. { 'wul·ə·stən ,wīr }

wood-carving tools |DES ENG| The tools normally used in wood carving; they consist of adzes, chisels, gouges, files, and rasps, all of which vary in size and shape. { 'wud ¦kärv·iŋ ,tülz }

Woodruff key |DES ENG| A self-aligning machine key made by a side-milling cutter in the form of a segment of a disk. { 'wu·drəf ,kē }

wood screw |DES ENG| A threaded fastener with a pointed shank, a slotted or recessed head, and a sharp tapered thread of relatively coarse pitch for use only in wood. { 'wud ,skrü }

woodstave pipe |DES ENG| A pipe made of narrow strips of wood placed side by side and banded with wire, metal collars, and inserted joints, used largely for municipal water supply, outfall sewers, and mining irrigation. { 'wud ,stāv ,pīp }

woofer |ENG ACOUS| A large loudspeaker designed to reproduce low audio frequencies at relatively high power levels; usually used in combination with a crossover network and a high-frequency loudspeaker called a tweeter. { 'wuf·ər }

word concatenation system |ENG ACOUS| The simplest form of voice response system, which retrieves previously spoken versions of words or phrases and carefully forms them into a sequence without pauses, to approximate normally spoken word sequences. { 'wórd kən,kat·ən'ā·shən ,sis·təm }

work |ELEC| See load. |IND ENG| The physical or mental effort expended in the performance of a task. |MECH| The transference of energy that occurs when a force is applied to a body that is moving in such a way that the force has a component in the direction of the body's motion; it is equal to the line integral of the force over the path taken by the body. { wərk }

work breakdown structure |IND ENG| A hierarchy designed to organize, define, and display all the work that must be performed in order to accomplish the objectives of a project. { ¦wərk ¦brāk,daún ,strək·chər }

work cycle |IND ENG| A sequence of tasks, operations, and processes, or a pattern of manual motions, elements, and activities that is repeated for each unit of work. { 'wərk ,sī·kəl }

work design See job design. { 'wərk di,zīn }

worked penetration |ENG| Penetration of a sample of lubricating grease immediately after it has been brought to a specified temperature and subjected to strokes in a standard grease worker. { 'wərkt ,pen·ə'trā·shən }

work element |IND ENG| In planning a manufacturing process, a single task that cannot be subdivided. { 'wərk ,el·ə·mənt }

work function See free energy. { 'wərk ,fəŋk·shən }

workhead See headstock. { 'wərk,hed }

working area |IND ENG| A portion of the workplace in which a worker moves about while fulfilling work tasks. { 'wərk·iŋ ,er·ē·ə }

working envelope |MECH ENG| The surface bounding the maximum extent and reach of a robot's wrist, excluding the tool tip. Also known as working profile. { 'wərk·iŋ 'en·və,lōp }

working life See work life. { ,wərk·iŋ ,līf }

working load |ENG| The maximum load that any structural member is designed to support. { 'wərk·iŋ ,lōd }

working pressure |ENG| The allowable operating pressure in a pressurized vessel or conduit, usually calculated by ASME (American Society of Mechanical Engineers) or API (American Petroleum Institute) codes. { 'wərk·iŋ ,presh·ər }

working profile See working envelope. { 'wərk·iŋ 'prō,fīl }

working Q See loaded Q. { 'wərk·iŋ 'kyü }

working space-volume |MECH ENG| The volume enclosed by a robot's working envelope. { 'wərk·iŋ 'spās 'väl·yəm }

working voltage See voltage rating. { 'wərk·iŋ ,vōl·tij }

work-kinetic energy theorem |MECH| The theorem that the change in the kinetic energy of a particle during a displacement is equal to the work done by the resultant force on the particle during this displacement. { 'wərk ki'ned·ik ¦en·ər·jē ,thir·əm }

work life |CHEM ENG| The period of time a resin or an adhesive will remain usable after it is mixed with a catalyst and other ingredients. Also known as pot life; working life. { 'wərk ,līf }

work measurement |IND ENG| Determination of the difficulty of a given task by using both physiologic and biomechanical parameters to evaluate compatibility of available motions with motions required to perform the task. **2.** See ergonometrics. { 'wərk ,mezh·ər·mənt }

work of adhesion See adhesional work. { 'wərk əv ad'hē·zhən }

work package |IND ENG| The amount of work required to complete a given job that falls within the responsibility of a single unit of the organization handling the project. { 'wərk ,pak·ij }

work physiology |IND ENG| An aspect of industrial engineering that takes into account metabolic cost, measurement and prevention of work strain, and other ergonomic factors in the design of tasks and workplaces. { 'wərk ,fiz·ē,äl·ə·jē }

workpiece |IND ENG| An object that is being manufactured. { 'wərk,pēs }

workpiece program |CONT SYS| A program that directs the machining of a component under numerical or computer control. { 'wərk,pēs ,prō ,gram }

work sampling |IND ENG| A technique to measure work activity as related to delays consisting of intermittent observations of actual work and delays. Also known as activity sampling; frequency study; ratio delay study. { 'wərk ,sam·pliŋ }

work standardization |IND ENG| The establishment of uniformity of working conditions, tools, equipment, technical procedures, administrative procedures, workplace arrangements, motion sequences, materials, quality requirements, and similar factors which affect the performance of work. { 'wərk ,stan·dər·də'zā·shən }

work station |IND ENG| A workplace that is included in a production system or on a piece of equipment at which an individual worker may spend only a portion of a working shift. { 'wərk ,stā·shən }

work station independence |CONT SYS| Property of a numerical control or robot program which does not depend on the nature of the work station. { 'wərk,stā·shən ,in·də'pen·dəns }

work stress |IND ENG| Any external force that acts on the body of a worker during the performance of a task. { 'wərk ,stres }

work task |IND ENG| A specified amount of work, set of responsibilities, or occupation assigned to an individual or to a group. { 'wərk ,task }

work tolerance |IND ENG| A time period during which a worker can effectively perform a task without a rest period while maintaining acceptable levels of physiological and emotional well-being. { 'wərk ,täl·ə·rəns }

work unit |IND ENG| An amount of work or the result of an amount of work that is treated as an integer (a single piece of information) when work is being characterized quantitatively. { 'wərk ,yü·nət }

world coordinates |CONT SYS| A robotic coordinate system that is fixed with respect to the Earth. { 'wərld kō'órd·ən·əts }

world modeling |CONT SYS| Robot programming that allows the system to perform complex tasks, based on stored data. { 'wərld 'mäd·əl·iŋ }

worm |DES ENG| A shank having at least one complete tooth (thread) around the pitch surface; the driver of a worm gear. { wərm }

worm conveyor See screw conveyor. { 'wərm kən'vā·ər }

worm gear |DES ENG| A gear with teeth cut on an angle to be driven by a worm; used to connect nonparallel, nonintersecting shafts. { 'wərm ,gir }

worm wheel |DES ENG| A gear wheel with curved teeth that meshes with a worm. { 'wərm ,wēl }

wow |ENG ACOUS| A low-frequency flutter; when caused by an off-center hole in a disk record, occurs once per revolution of the turntable. { waú }

wrap-around grasp |IND ENG| A basic grasp whereby an object is held against the palm by the fingers wrapped around it, with the thumb opposing the index finger. { 'rap·ə,raúnd ,grasp }

wrap forming See stretch forming. { 'rap ,fórm·iŋ }

wrapper sheet |MECH ENG| **1.** The outer plate enclosing the firebox in a fire-tube boiler. **2.** The thinner sheet of a boiler drum having two sheets. { 'rap·ər ,shēt }

wrecking ball See skull cracker. { 'rek·iŋ ,bòl }

wrecking bar See ripping bar. { 'rek·iŋ ,bär }

wrecking strip |CIV ENG| A small section that is fitted into a form for poured concrete and is easily removed before the main panels to facilitate disassembly of the main components of the form. { 'rek·iŋ ,strip }

wrench |ENG| A manual or power tool with adapted or adjustable jaws or sockets either at the end or between the ends of a lever for holding or turning a bolt, pipe, or other object. |MECH| The combination of a couple and a force which is parallel to the torque exerted by the couple. { rench }

wrench-head bolt |DES ENG| A bolt with a square or hexagonal head designed to be gripped between the jaws of a wrench. { 'rench ¦hed ,bōlt }

wringing fit |DES ENG| A fit of zero-to-negative allowance. { 'riŋ·iŋ 'fit }

wrist |MECH ENG| A set of rotary joints to which the end effector of a robot is attached. Also known as wrist socket. { rist }

wrist pin *See* piston pin. { 'ris ,pin }

write head |ELECTR| Device that stores digital information as coded electrical pulses on a magnetic drum, disk, or tape. { 'rīt ,hed }

W-truss |CIV ENG| A truss having upper and lower chords joined by web members that form a shape resembling the letter W. { 'dəb·əl,yü ,trəs }

Wulf electrometer |ENG| **1.** A variant of the string electrometer in which charged metal plates are replaced by charged knife-edges. **2.** An electrometer in which two conducting fibers are placed side by side, and their separation upon charging is measured. { ¦wúlf i,lek'träm·əd·ər }

Wulff process |CHEM ENG| A chemical process to make acetylene and ethylene by cracking a hydrocarbon gas (for example, butane) with high-temperature steam in a regenerative furnace. { 'wúlf ,prä·səs }

Wurster process *See* air-suspension encapsulation. { 'wər·stər ,prä·səs }

wye |ELEC| Polyphase circuit whose phase differences are 120° and which when drawn resembles the letter Y. |ENG| A pipe branching off a straight main run at an angle of 45°. Also known as Y; yoke. { wī }

wye branch *See* Y branch. { 'wī ,branch }

wye fitting *See* Y fitting. { 'wī ,fid·iŋ }

wye level *See* Y level. { 'wī ,lev·əl }

X

X engine |MECH ENG| An in-line engine with the cylinder banks so arranged around the crankshaft that they resemble the letter X when the engine is viewed from the end. { 'eks ,en·jən }

X frame |DES ENG| An automotive frame which either has side rails bent in at the center of the vehicle, making the overall form that of an X, or has an X-shaped member which joins the side rails with diagonals for added strength and resistance to torsional stresses. { 'eks ,frām }

x-ray diffractometer |ENG| An instrument used in x-ray analysis to measure the intensities of the diffracted beams at different angles. { 'eks ,rā ,di,frak'täm·əd·ər }

x-ray goniometer |ENG| A scale designed to measure the angle between the incident and refracted beams in x-ray diffraction analysis. { 'eks ,rā ,gō·nē'äm·əd·ər }

x-ray machine |ENG| The x-ray tube, power supply, and associated equipment required for producing x-ray photographs. { 'eks ,rā mə,shēn }

x-ray microscope |ENG| **1.** A device in which an ultra-fine-focus x-ray tube or electron gun produces an electron beam focused to an extremely small image on a transmission-type x-ray target that serves as a vacuum seal; the magnification is by projection; specimens being examined can thus be in air, as also can the photographic film that records the magnified image. **2.** Any of several instruments which utilize x-radiation for chemical analysis and for magnification of 100–1000 diameters; it is based on contact or projection microradiography, reflection x-ray microscopy, or x-ray image spectrography. { 'eks ,rā 'mī·krə,skōp }

x-ray monochromator |ENG| An instrument in which x-rays are diffracted from a crystal to produce a beam having a narrow range of wavelengths. { 'eks ,rā ¦män·ə¦krō,mād·ər }

x-ray telescope |ENG| An instrument designed to detect x-rays emanating from a source outside the earth's atmosphere and to resolve the x-rays into an image; they are carried to high altitudes by balloons, rockets, or space vehicles; although several types of x-ray detector, involving gas counters, scintillation counters, and collimators, have been used, only one, making use of the phenomenon of total external reflection of x-rays from a surface at grazing incidence, is strictly an x-ray telescope. { 'eks ,rā 'tel·ə,skōp }

x-ray thickness gage |ENG| A thickness gage used for measuring and indicating the thickness of moving cold-rolled sheet steel during the rolling process without making contact with the sheet; an x-ray beam directed through the sheet is absorbed in proportion to the thickness of the material and its atomic number. { 'eks ,rā 'thik·nəs ,gāj }

XY recorder |ENG| A recorder that traces on a chart the relation of two variables, neither of which is time. { ¦eks¦wī ri'kȯrd·ər }

Y

yard |CIV ENG| A facility for building and repairing ships. |MECH| A unit of length in common use in the United States and United Kingdom, equal to 0.9144 meter, or 3 feet. Abbreviated yd. { 'yärd }

yardage |MECH| An amount expressed in yards. { 'yärd·ij }

yard crane *See* crane truck. { 'yärd ‚krān }

yard drain |CIV ENG| A drain for clearing an open area of surface water. { 'yärd ‚drān }

yard lumber |BUILD| A category of lumber up to 5 inches (12.5 centimeters) thick. { 'yärd ‚ləm·bər }

yard maintenance |ENG| A category of maintenance that includes the complete rebuilding of parts, subassemblies, or components. { 'yärd ‚maint·ən·əns }

yaw |MECH| **1.** The rotational or oscillatory movement of a ship, aircraft, rocket, or the like about a vertical axis. Also known as yawing. **2.** The amount of this movement, that is, the angle of yaw. **3.** To rotate or oscillate about a vertical axis. { yò }

yaw acceleration |MECH| The angular acceleration of an aircraft or missile about its normal or Z axis. { 'yò ak‚sel·ə'rā·shən }

yaw axis |MECH| A vertical axis through an aircraft, rocket, or similar body, about which the body yaws; it may be a body, wind, or stability axis. Also known as yawing axis. { 'yò ‚ak·səs }

yawing *See* yaw. { 'yò·iŋ }

yawing axis *See* yaw axis. { 'yò·iŋ ‚ak·səs }

yaw simulator |CONT SYS| A test instrument used to derive and thereby permit study of probable aerodynamic behavior in controlled flight under specific initial conditions; certain components of the missile guidance system, such as the receiver or servo loop, are connected into the simulator circuitry; also, certain aerodynamic parameters of the specific missile must be known and set into the simulator; applicable to the yaw plane. { 'yò ‚sim·yə‚läd·ər }

Y branch |ENG| A Y-shaped branch in a piping system. Also known as wye branch. { 'wī ‚branch }

yd *See* yard.

Y fitting |CIV ENG| A pipe fitting with one end

subdivided to form two openings, usually at a 45° angle to the run of the pipe. Also known as wye fitting. { 'wī ‚fid·iŋ }

yield |ENG| Product of a reaction or process as in chemical reactions or food processing. |MECH| That stress in a material at which plastic deformation occurs. { yēld }

yield factor |IND ENG| The ratio of the amount of material that results from an industrial process to the amount of material that went into it. { 'yēld ‚fak·tər }

yield point |MECH| The lowest stress at which strain increases without increase in stress. { 'yēld ‚pòint }

yield rate |IND ENG| The amount of satisfactory material available after the completion of a given manufacturing process expressed as a percentage of the total amount produced. { 'yēld ‚rāt }

yield strength |MECH| The stress at which a material exhibits a specified deviation from proportionality of stress and strain. { 'yēld ‚streŋkth }

yield stress |MECH| The lowest stress at which extension of the tensile test piece increases without increase in load. { 'yēld ‚stres }

yield temperature |ENG| The temperature at which a fusible plug device melts and is dislodged by its holder and thus relieves pressure in a pressure vessel; it is caused by the melting of the fusible material, which is then forced from its holder. { 'yēld ‚tem·prə·chər }

yig device |ELECTR| A filter, oscillator, parametric amplifier, or other device that uses an yttrium-iron-garnet crystal in combination with a variable magnetic field to achieve wide-band tuning in microwave circuits. Derived from yttrium-iron-garnet device. { 'yig di‚vīs }

Y level |ENG| A surveyor's level with Y-shaped rests to support the telescope. Also known as wye level. { 'wī ‚lev·əl }

yoke |DES ENG| A clamp or similar device to embrace and hold two other parts. |ELECTR| *See* deflection yoke. |ENG| **1.** A bar of wood used to join the necks of draft animals for working together. **2.** *See* wye. |MECH ENG| A slotted crosshead used instead of a connecting rod in some steam engines. { yōk }

York-Scheibel column *See* Scheibel extractor. { 'yòrk 'shī·bəl ‚käl·əm }

Young-Helmholtz laws |MECH| Two laws de-

Young's modulus

scribing the motion of bowed strings; the first states that no overtone with a node at the point of excitation can be present; the second states that when the string is bowed at a distance of $1/n$ times the string's length from one of the ends, where n is an integer, the string moves back and forth with two constant velocities, one of which has the same direction as that of the bow and is equal to it, while the other has the opposite direction and is $n - 1$ times as large. { ¦yȯŋ 'helm‚hōlts ‚lȯz }

Young's modulus [MECH] The ratio of a simple tension stress applied to a material to the re-sulting strain parallel to the tension. Also known as modulus of elasticity { 'yȯŋz ‚mäj‧ə‧ləs }

y parameter [ELECTR] One of a set of four transistor equivalent-circuit parameters, used especially with field-effect transistors, that conveniently specify performance for small voltage and current in an equivalent circuit; the equivalent circuit is a current source with shunt impedance at both input and output. { 'wī pə‚ram‧əd‧ər }

yttrium-iron-garnet device See yig device. { ¦i‧trē‧əm ¦ī‧ərn ¦gär‧nət di‚vīs }

Z

zee |CIV ENG| A metal member whose cross section has a modified Z shape; the internal angles are slightly less than 90° { zē }

Zener breakdown |ELECTR| Nondestructive breakdown in a semiconductor, occurring when the electric field across the barrier region becomes high enough to produce a form of field emission that suddenly increases the number of carriers in this region. Also known as Zener effect. { 'zē·nər 'brāk,daún }

Zener diode |ELECTR| A semiconductor breakdown diode, usually constructed of silicon, in which reverse-voltage breakdown is based on the Zener effect. { 'zē·nər 'dī,ōd }

Zener diode voltage regulator *See* diode voltage regulator. { 'zē·nər 'dī,ōd 'vōl·tij ,reg·yə,lād·ər }

Zener effect *See* Zener breakdown.

zero adjuster |ENG| A device for adjusting the pointer position of an instrument or meter to read zero when the measured quantity is zero. { 'zir·ō ə,jəs·tər }

zero bevel gear |DES ENG| A special form of bevel gear having curved teeth with a zero-degree spiral angle. { 'zir·ō ¦bev·əl 'gir }

zero bias |ELECTR| The condition in which the control grid and cathode of an electron tube are at the same direct-current voltage. { 'zir·ō 'bī·əs }

zero defects |IND ENG| A program for improving product quality to the point of perfection, so there will be no failures due to defects in construction. { 'zir·ō 'dē,feks }

zero gravity *See* weightlessness. { 'zir·ō 'grav·əd·ē }

zero level |ENG ACOUS| Reference level used for comparing sound or signal intensities; in audio-frequency work, a power of 0.006 watt is generally used as zero level; in sound, the threshold of hearing is generally assumed as the zero level. { 'zir·ō ,lev·əl }

zero method *See* null method. { 'zir·ō ,meth·əd }

zero-order hold |CONT SYS| A device which converts a sampled output into an output which is held constant between samples at the last sampled value. { 'zir·ō ¦ord·ər 'hōld }

zeroth law of thermodynamics |THERMO| A law that if two systems are separately found to be in thermal equilibrium with a third system, the first two systems are in thermal equilibrium with each other, that is, all three systems are at the same temperature. { ¦zir,ōth ,ló əv ,thər·mō·dī'nam·iks }

Ziegler process |CHEM ENG| A process for the low-pressure linear polymerization of ethylene and stereospecific polymerization of propylene; the product is a high-density polymer or elastomer. { 'zē·glər ,prä·səs }

zigzag rule |ENG| A folding ruler having pivoted sections that lock when the ruler is opened. { 'zig,zag ,rül }

zipper |ENG| A generic name for slide fasteners in which two sets of interlocking teeth of the same design provide sturdy and continuous closure for adjacent pieces of textile, leather, and other materials. { 'zip·ər }

zipper conveyor |MECH ENG| A type of conveyor belt with zipperlike teeth that mesh to form a closed tube; used to handle fragile materials. { 'zip·ər kən,vā·ər }

zirconium oxide-based oxygen transducer |ENG| A device in which the concentration of oxygen in a mixture of gases is determined from the diffusion voltage across a heated, suitably doped zirconium oxide material placed between this mixture and a reference gas. { zər¦kōn·ē·əm ¦äk,sīd ,bāst ¦äks·ə·jən tranz'düs·ər }

zone |MECH ENG| 1. In a heating or air-conditioning system, one or more spaces whose temperature is regulated by a single control. 2. A subdivision of a sprinkler, water-supply, or standpipe system. { zōn }

zone control |ENG| The zoning of a process or building, and the independent heating or temperature controls for each zone. { 'zōn kən,tról }

zone heat |CIV ENG| A central heating system arranged to allow different temperatures to be maintained at the same time in two or more areas of a building. { 'zōn ,hēt }

zone melting crystallization |CHEM ENG| A method for purification of crystalline solids; the sample, packed in a narrow column, is heated so that a molten zone passes down through the sample, carrying impurities with it. { 'zōn ¦mel·tiŋ ,krist·əl·ə'zā·shən }

zone-position indicator |ENG| Auxiliary radar set for indicating the general position of an object to another radar set with a narrower field. { 'zōn pə¦zish·ən 'in·də,kād·ər }

zoning |CIV ENG| Designation and reservation under a master plan of land use for light and heavy industry, dwellings, offices, and other buildings; use is enforced by restrictions on types of buildings in each zone. { 'zōn·iŋ }

zoom |ENG| To enlarge or reduce the size of an image in an optical system or electronic display. { züm }

Z parameter |ELECTR| One of a set of four transistor equivalent-circuit parameters; they are the inverse of the Y parameters. { 'zē pə,ram·əd·ər }

z-transfer function See pulsed transfer function. { 'zē 'tranz·fər ,fəŋk·shən }

Z variometer See vertical intensity variometer. { 'zē ,ver·ē'äm·əd·ər }

Zyglo method |ENG| A procedure for visualizing incipient cracks caused by fatigue failure, in which the part is immersed in a special activated penetrating oil and viewed under black light. { 'zī·glō ,meth·əd }

Appendix

Equivalents of commonly used units for the U.S. Customary System and the metric system

1 inch = 2.5 centimeters (25 millimeters)	1 centimeter = 0.4 inch	1 inch = 0.083 foot
1 foot = 0.3 meter (30 centimeters)	1 meter = 3.3 feet	1 foot = 0.33 yard (12 inches)
1 yard = 0.9 meter	1 meter = 1.1 yards	1 yard = 3 feet (36 inches)
1 mile = 1.6 kilometers	1 kilometer = 0.62 mile	1 mile = 5280 feet (1760 yards)

1 acre = 0.4 hectare	1 hectare = 2.47 acres	
1 acre = 4047 square meters	1 square meter = 0.00025 acre	

1 gallon = 3.8 liters	1 liter = 1.06 quarts = 0.26 gallon	1 quart = 0.25 gallon (32 ounces; 2 pints)
1 fluid ounce = 29.6 milliliters	1 milliliter = 0.034 fluid ounce	1 pint = 0.125 gallon (16 ounces)
32 fluid ounces = 946.4 milliliters		1 gallon = 4 quarts (8 pints)

1 quart = 0.95 liter	1 gram = 0.035 ounce	1 ounce = 0.0625 pound
1 ounce = 28.35 grams	1 kilogram = 2.2 pounds	1 pound = 16 ounces
1 pound = 0.45 kilogram	1 kilogram = 1.1×10^{-3} ton	1 ton = 2000 pounds
1 ton = 907.18 kilograms		

$°F = (1.8 \times °C) + 32$

$°C = (°F - 32) \div 1.8$

Appendix

Conversion factors for the U.S. Customary System, metric system, and International System

A. Units of length

Units	cm	m	in.	ft	yd	mi
1 cm =	1	0.01	0.3937008	0.03280840	0.01093613	6.213712×10^{-6}
1 m =	100.	1	39.37008	3.280840	1.093613	6.213712×10^{-4}
1 in. =	2.54	0.0254	1	0.08333333.	0.02777777.	1.578283×10^{-5}
1 ft =	30.48	0.3048	12.	1	0.3333333.	$1.893939... \times 10^{-4}$
1 yd =	91.44	0.9144	36.	3.	1	$5.681818... \times 10^{-4}$
1 mi =	1.609344×10^{5}	1.609344×10^{3}	6.336×10^{4}	5280.	1760.	1

B. Units of area

Units	cm^2	m^2	in.^{2}	ft^2	yd^2	mi^2
1 cm^2 =	1	10^{-4}	0.1550003	1.076391×10^{-3}	1.195990×10^{-4}	3.861022×10^{-11}
1 m^2 =	10^{4}	1	1550.003	10.76391	1.195990	3.861022×10^{-7}
1 in.^{2} =	6.4516	6.4516×10^{-4}	1	$6.944444. \times 10^{-3}$	7.716049×10^{-4}	2.490977×10^{-10}
1 ft^2 =	929.0304	0.09290304	144.	1	0.1111111...	3.587007×10^{-8}
1 yd^2 =	8361.273	0.8361273	1296.	9.	1	3.228306×10^{-7}
1 mi^2 =	2.589988×10^{10}	2.589988×10^{6}	4.014490×10^{9}	2.78784×10^{7}	3.0976×10^{6}	1

C. Units of volume

Units	m^3	cm^3	liter	$in.^3$	ft^3	qt	gal
1 m^3 =	1	10^6	10^3	6.102374×10^4	35.31467×10^{-3}	1.056688	264.1721
1 cm^3 =	10^{-6}	1	10^{-3}	0.06102374	3.531467×10^{-5}	1.056688×10^{-3}	2.641721×10^{-4}
1 liter =	10^{-3}	1000.	1	61.02374	0.03531467	1.056688	0.2641721
1 $in.^3$ =	1.638706×10^{-5}	16.38706	0.01638706	1	5.787037×10^{-4}	0.01731602	4.329004×10^{-3}
1 ft^3 =	2.831685×10^{-2}	28316.85	28.31685	1728.	1	2.992208	7.480520
1 qt =	9.463529×10^{-4}	946.3529	0.9463529	57.75	0.03342014	1	0.25
1 gal (U.S.) =	3.785412×10^{-3}	3785.412	3.785412	231.	0.1336806	4.	1

D. Units of mass

Units	g	kg	oz	lb	metric ton	ton
1 g =	1	10^{-3}	0.03527396	2.204623×10^{-3}	10^{-6}	1.102311×10^{-6}
1 kg =	1000.	1	35.27396	2.204623	10^{-3}	1.102311×10^{-3}
1 oz (avdp) =	28.34952	0.02834952	1	0.0625	2.834952×10^{-5}	3.125×10^{-5}
1 lb (avdp) =	453.5924	0.4535924	16.	1	4.535924×10^{-4}	$5. \times 10^{-4}$
1 metric ton =	10^8	1000.	35273.96	2204.623	1	1.102311
1 ton =	907184.7	907.1847	32000.	2000.	0.9071847	1

Conversion factors for the U.S. Customary System, metric system, and International System (cont.)

E. Units of density

Units	$g \cdot cm^{-3}$	$g \cdot L^{-1}, kg \cdot m^{-3}$	$oz \cdot in.^{-3}$	$lb \cdot in.^{-3}$	$lb \cdot ft^{-3}$	$lb \cdot gal^{-1}$
$1 \, g \cdot cm^{-3}$ = 1	1000.	0.5780365	0.03612728	62.42795	8.345403	
$1 \, g \cdot L^{-1}, kg \cdot m^{-3}$ = 10^{-3}	1	5.780365×10^{-4}	3.612728×10^{-5}	0.06242795	8.345403×10^{-3}	
$1 \, oz \cdot in.^{-3}$ = 1.729994	1729.994	1	0.0625	108.	14.4375	
$1 \, lb \cdot in.^{-3}$ = 27.67991	27679.91	16.	1	1728.	231.	
$1 \, lb \cdot ft^{-3}$ = 0.01601847	16.01847	9.259259×10^{-3}	5.787037×10^{-4}	1	0.1336806	
$1 \, lb \cdot gal^{-1}$ = 0.1198264	119.8264	4.749536×10^{-3}	4.329004×10^{-3}	7.480519	1	

F. Units of pressure

Units	$Pa, N \cdot m^{-2}$	$dyn \cdot cm^{-2}$	bar	atm	$kgf \cdot cm^{-2}$	$mmHg$ (torr)	$in. Hg$	$lbf \cdot in.^{-2}$
$1 \, Pa, 1 \, N \cdot m^{-2}$ = 1	10	10^{-5}	9.869233×10^{-6}	1.019716×10^{-5}	7.500617×10^{-3}	2.952999×10^{-4}	1.450377×10^{-4}	
$1 \, dyn \cdot cm^{-2}$ = 0.1	1	10^{-6}	9.869233×10^{-7}	1.019716×10^{-6}	7.500617×10^{-4}	2.952999×10^{-5}	1.450377×10^{-5}	
$1 \, bar$ = 10^5	10^6	1	0.9869233	1.019716	750.0617	29.52999	14.50377	
$1 \, atm$ = 101325	1013250	1.01325	1	1.033227	760.	29.92126	14.69595	
$1 \, kgf \cdot cm^{-2}$ = 98066.5	980665	0.980665	0.9678411	1	735.5592	28.95903	14.22334	
$1 \, mmHg$ (torr) = 133.3224	1333.224	1.333224×10^3	1.315789×10^{-3}	1.359510×10^{-3}	1	0.03937008	0.01933678	
$1 \, in. Hg$ = 3386.388	33863.88	0.03386388	0.03342105	0.03453155	25.4	1	0.4911541	
$1 \, lbf \cdot in.^{-2}$ = 6894.757	68947.57	0.06894757	0.06804596	0.07030696	51.71493	2.036021	1	

G. Units of energy

Units	g mass (energy equiv)	J	eV	cal	cal_IT	Btu_IT	kWh	hp-h	ft-lbf	ft³·lbf·in⁻²	liter-atm
1 g mass = 1 (energy equiv)	1	8.987552×10^{13}	5.609589×10^{32}	2.148076×10^{13}	2.146640×10^{13}	8.518555×10^{10}	2.496542×10^{7}	3.347918×10^{7}	6.628878×10^{13}	4.603388×10^{11}	8.870024×10^{11}
1 J =	1.112650×10^{-14}	1	6.241510×10^{18}	0.2390057	0.2388459	9.478172×10^{-4}	2.777777×10^{-7}	3.725062×10^{-7}	0.7375622	5.121960×10^{-3}	9.869233×10^{-3}
1 eV =	1.782662×10^{-33}	1.602176×10^{-19}	1	3.829293×10^{-20}	3.826733×10^{-20}	1.518570×10^{-22}	4.450490×10^{-26}	5.968206×10^{-26}	1.181705×10^{-19}	8.216283×10^{-22}	1.581225×10^{-21}
1 cal =	4.655328×10^{-14}	4.184	2.611448×10^{19}	1	0.9993312	3.965667×10^{-3}	1.1622222×10^{-6}	1.558562×10^{-6}	3.085960	2.143028×10^{-2}	0.04129287
1 cal_IT =	4.658443×10^{-14}	4.1868	2.613195×10^{19}	1.000669	1	3.968321×10^{-3}	1.163×10^{-6}	1.559609×10^{-6}	3.088025	2.144462×10^{-2}	0.04132050
1 Btu_IT =	1.173908×10^{-11}	1055.056	6.585141×10^{21}	252.1644	251.9958	1	2.930711×10^{-4}	3.930148×10^{-4}	778.1693	5.403953	10.41259
1 kWh =	4.005540×10^{-8}	3600000.	2.246944×10^{25}	860420.7	859845.2	3412.142	1	1.341022	2655224.	18349.06	35529.24
1 hp-h =	2.986931×10^{-8}	2384519.	1.675545×10^{25}	641615.6	641186.5	2544.33	0.7456998	1	1980000.	13750.	26494.15
1 ft-lbf =	1.508551×10^{-14}	1.355818	8.462351×10^{18}	0.3240483	0.3238315	1.285067×10^{-3}	3.766161×10^{-7}	5.050505×10^{-7}	1	6.944444×10^{-3}	0.01338088
1 ft³ lbf · in⁻² =	2.172313×10^{-12}	195.2378	1.218579×10^{21}	46.66295	46.63174	0.1850497	5.423272×10^{-5}	7.272727×10^{-5}	144.	1	1.926847
1 liter-atm =	1.127393×10^{-12}	101.325	6.324210×10^{20}	24.21726	24.20106	0.09603757	2.814583×10^{-5}	3.774419×10^{-5}	74.73349	0.5183825	1

Appendix

Special constants

$\pi = 3.14159\ 26535\ 89793\ 23846\ 2643\ ..$

$e = 2.71828\ 18284\ 59045\ 23536\ 0287\ ... = \lim\limits_{n \to \infty} \left(1 + \dfrac{1}{n}\right)^n$

= natural base of logarithms

$\sqrt{2} = 1.41421\ 35623\ 73095\ 0488\ .$

$\sqrt{3} = 1.73205\ 08075\ 68877\ 2935\ ..$

$\sqrt{5} = 2.23606\ 79774\ 99789\ 6964\ ..$

$\sqrt[3]{2} = 1.25992\ 1050\ .$

$\sqrt[3]{3} = 1.44224\ 9570\ ...$

$\sqrt[5]{2} = 1.14869\ 8355\ ...$

$\sqrt[5]{3} = 1.24573\ 0940\ .$

$e^{\pi} = 23.14069\ 26327\ 79269\ 006\ .$

$\pi^e = 22.45915\ 77183\ 61045\ 47342\ 715\ ..$

$e^e = 15.15426\ 22414\ 79264\ 190\ .$

$\log_{10} 2 = 0.30102\ 99956\ 63981\ 19521\ 37389\ .$

$\log_{10} 3 = 0.47712\ 12547\ 19662\ 43729\ 50279$

$\log_{10} e = 0.43429\ 44819\ 03251\ 82765\ ..$

$\log_{10} \pi = 0.49714\ 98726\ 94133\ 85435\ 12683\ ..$

$\log_e 10 = \ln 10 = 2.30258\ 50929\ 94045\ 68401\ 7991\ ..$

$\log_e 2 = \ln 2 = 0.69314\ 71805\ 59945\ 30941\ 7232\ .$

$\log_e 3 = \ln 3 = 1.09861\ 22886\ 68109\ 69139\ 5245\ .$

$\gamma = 0.57721\ 56649\ 01532\ 86060\ 6512\qquad$ = Euler's constant

$\qquad = \lim\limits_{n \to \infty} \left(1 + \dfrac{1}{2} + \dfrac{1}{3} + \cdots + \dfrac{1}{n} - \ln n\right)$

$e^{\gamma} = 1.78107\ 24179\ 90197\ 9852\ .$

$\sqrt{e} = 1.64872\ 12707\ 00128\ 1468\ .$

$\sqrt{\pi} = \Gamma(\tfrac{1}{2}) = 1.77245\ 38509\ 05516\ 02729\ 8167\ ..$

where Γ is the gamma function

$\Gamma(\tfrac{1}{3}) = 2.67893\ 85347\ 07748\ ..$

$\Gamma(\tfrac{1}{4}) = 3.62560\ 99082\ 21908$

1 radian $= 180°/\pi = 57.29577\ 95130\ 8232\ .\ °$

$1° = \pi/180$ radians $= 0.01745\ 32925\ 19943\ 29576\ 92\ .$ radians

---

SOURCE: Murray R. Spiegel and John Liu, *Mathematical Handbook of Formulas and Tables*, 2d ed., Schaum's Outline Series, McGraw-Hill, 1999.

Electrical and magnetic units

Quantity	Unit and symbol	Derivation
SI base units		
Mass	kilogram, kg	
Time	second, s	
Length	meter, m	
Electric current	ampere, A	
Thermodynamic temperature	kelvin, K	
Luminous intensity	candela, cd	
Amount of substance	mole, mol	
Derived units		
Potential difference, emf	vol, V	$W \cdot A^{-1} = m^2 \cdot kg \cdot s^{-3} \cdot A^{-1}$
Resistance	ohm, Ω	$V \cdot A^{-1} = m^2 \cdot kg \cdot s^{-3} \cdot A^{-2}$
Electric charge	coulomb, C	$s \cdot A$
Capacitance	farad, F	$C \cdot V^{-1} = m^{-2} \cdot kg^{-1} \cdot s^4 \cdot A^2$
Conductance	siemens, S	$A \cdot V^{-1} = m^{-2} \cdot kg^{-1} \cdot s^3 \cdot A^2$
Magnetic flux	weber, Wb	$V \cdot s = m^2 \cdot kg \cdot s^{-2} \cdot A^{-1}$
Inductance	henry, H	$Wb \cdot A^{-1} = m^2 \cdot kg \cdot s^{-2} \cdot A^{-2}$
Magnetic flux density	tesla, T	$Wb \cdot m^{-2} = kg \cdot s^{-2} \cdot A^{-1}$
Magnetic field strength	ampere per meter	$m^{-1} \cdot A$
Current density	ampere per square meter	$m^{-2} \cdot A$
Electric field strength	volt per meter	$V \cdot m^{-1} = m \cdot kg \cdot s^{-3} \cdot A^{-1}$
Permittivity	farad per meter	$F \cdot m^{-1} = m^{-3} \cdot kg^{-1} \cdot s^4 \cdot A^2$
Permeability	henry per meter	$H \cdot m^{-1} = m \cdot kg \cdot s^{-2} \cdot A^{-2}$

Dimensional formulas of common quantities

Quantity	Definition	Dimensional formula
Mass	Fundamental	M
Length	Fundamental	L
Time	Fundamental	T
Velocity	Distance/time	LT^{-1}
Acceleration	Velocity/time	LT^{-2}
Force	Mass × acceleration	MLT^{-2}
Momentum	Mass × velocity	MLT^{-1}
Energy	Force × distance	ML^2T^{-2}
Angle	Arc/radius	I
Angular velocity	Angle/time	T^{-1}
Angular acceleration	Angular velocity/time	T^{-2}
Torque	Force × lever arm	ML^2T^{-2}
Angular momentum	Momentum × lever arm	ML^2T^{-1}
Moment of inertia	Mass × radius squared	ML^2
Area	Length squared	L^2
Volume	Length cubed	L^3
Density	Mass/volume	ML^{-3}
Pressure	Force/area	$ML^{-1}T^{-2}$
Action	Energy × time	ML^2T^{-1}
Viscosity	Force per unit area per unit velocity gradient	$ML^{-1}T^{-1}$

Appendix

Internal energy and generalized work

Type of energy	Intensive factor	Extensive factor	Element of work
Mechanical			
Expansion	Pressure (P)	Volume (V)	$-PdV$
Stretching	Surface tension (γ)	Area (A)	γdA
Extension	Tensile stretch (F)	Length (l)	Fdl
Thermal	Temperature (T)	Entropy (S)	TdS
Chemical	Chemical potential (gm)	Moles (n)	μdn
Electrical	Electric potential (E)	Charge (Q)	EdQ
Gravitational	Gravitational field strength (mg)	Height (h)	$mgdh$
Polarization			
Electrostatic	Electric field strength (E)	Total electric polarization (P)	EdP
Magnetic	Magnetic field strength (H)	Total magetic polarization (M)	HdM

General rules of integration*

$$\int a\, dx = ax$$

$$\int af(x)\, dx = a\int f(x)\, dx$$

$$\int (u \pm v \pm w \pm \cdots)\, dx = \int u\, dx \pm \int v\, dx \pm \int w\, dx \pm \cdots$$

$$\int u\, dv = uv - \int v\, du \quad \text{[integration by parts]}$$

$$\int f(ax)\, dx = \frac{1}{a}\int f(u)\, du$$

$$\int F\{f(x)\}\, dx = \int \frac{F(u)}{f'(x)}\, du \qquad \text{where } u = f(x)$$

$$\int u^n\, du = \frac{u^{n+1}}{n+1}, \quad n \neq -1 \qquad \text{[for } n = -1\text{]}$$

$$\int \frac{du}{u} = \ln u \quad \text{if } u > 0 \text{ or } \ln(-u) \text{ if } u < 0$$
$$= \ln|u|$$

$$\int e^u\, du = e^u$$

$$\int a^u\, du = \int e^{u\ln a}\, du = \frac{e^{u\ln a}}{\ln a} = \frac{a^u}{\ln a}, \qquad a > 0,\ a \neq 1$$

$$\int \sin u\, du = -\cos u$$

$$\int \cos u\, du = \sin u$$

$$\int \tan u\, du = \ln \sec u = -\ln \cos u$$

$$\int \cot u\, du = \ln \sin u$$

$$\int \sec u\, du = \ln(\sec u + \tan u) = \ln \tan\left(\frac{u}{2} + \frac{\pi}{4}\right)$$

$$\int \csc u\, du = \ln(\csc u - \cot u) = \ln \tan\frac{u}{2}$$

$$\int \sec^2 u\, du = \tan u$$

$$\int \csc^2 u\, du = -\cot u$$

$$\int \tan^2 u\, du = \tan u - u \qquad \int \cot^2 u\, du = -\cot u - u$$

$$\int \sin^2 u\, du = \frac{u}{2} - \frac{\sin 2u}{4} = \frac{1}{2}(u - \sin u \cos u)$$

$$\int \cos^2 u\, du = \frac{u}{2} + \frac{\sin 2u}{4} = \frac{1}{2}(u + \sin u \cos u)$$

$$\int \sec u \tan u\, du = \sec u$$

$$\int \csc u \cot u\, du = -\csc u$$

$$\int \sinh u\, du = \cosh u$$

$$\int \cosh u\, du = \sinh u$$

$$\int \tanh u\, du = \ln \cosh u$$

$$\int \coth u\, du = \ln \sinh u$$

Appendix

General rules of integration* (cont.)

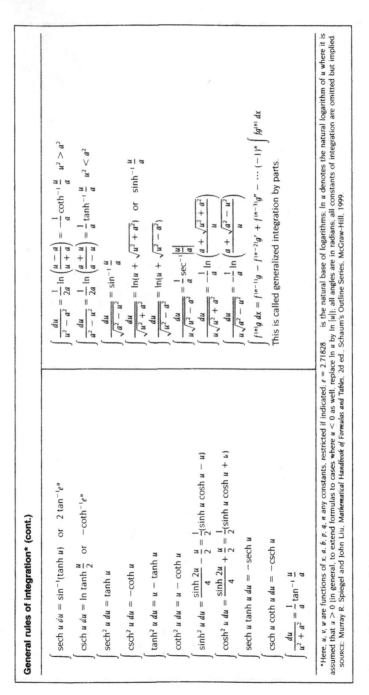

$$\int \operatorname{sech} u \, du = \sin^{-1}(\tanh u) \quad \text{or} \quad 2 \tan^{-1} e^u$$

$$\int \operatorname{csch} u \, du = \ln \tanh \frac{u}{2} \quad \text{or} \quad -\coth^{-1} e^u$$

$$\int \operatorname{sech}^2 u \, du = \tanh u$$

$$\int \operatorname{csch}^2 u \, du = -\coth u$$

$$\int \tanh^2 u \, du = u - \tanh u$$

$$\int \coth^2 u \, du = u - \coth u$$

$$\int \sinh^2 u \, du = \frac{\sinh 2u}{4} - \frac{u}{2} = \frac{1}{2}(\sinh u \cosh u - u)$$

$$\int \cosh^2 u \, du = \frac{\sinh 2u}{4} + \frac{u}{2} = \frac{1}{2}(\sinh u \cosh u + u)$$

$$\int \operatorname{sech} u \tanh u \, du = -\operatorname{sech} u$$

$$\int \operatorname{csch} u \coth u \, du = -\operatorname{csch} u$$

$$\int \frac{du}{u^2 + a^2} = \frac{1}{a} \tan^{-1} \frac{u}{a}$$

$$\int \frac{du}{u^2 - a^2} = \frac{1}{2a} \ln \left(\frac{u-a}{u+a} \right) = -\frac{1}{a} \coth^{-1} \frac{u}{a} \quad u^2 > a^2$$

$$\int \frac{du}{a^2 - u^2} = \frac{1}{2a} \ln \left(\frac{a+u}{a-u} \right) = \frac{1}{a} \tanh^{-1} \frac{u}{a} \quad u^2 < a^2$$

$$\int \frac{du}{\sqrt{a^2 - u^2}} = \sin^{-1} \frac{u}{a}$$

$$\int \frac{du}{\sqrt{u^2 + a^2}} = \ln(u + \sqrt{u^2 + a^2}) \quad \text{or} \quad \sinh^{-1} \frac{u}{a}$$

$$\int \frac{du}{\sqrt{u^2 - a^2}} = \ln(u + \sqrt{u^2 - a^2})$$

$$\int \frac{du}{u\sqrt{u^2 - a^2}} = \frac{1}{a} \sec^{-1} \left| \frac{u}{a} \right|$$

$$\int \frac{du}{u\sqrt{u^2 + a^2}} = -\frac{1}{a} \ln \left(\frac{a + \sqrt{u^2 + a^2}}{u} \right)$$

$$\int \frac{du}{u\sqrt{a^2 - u^2}} = -\frac{1}{a} \ln \left(\frac{a + \sqrt{a^2 - u^2}}{u} \right)$$

$$\int f^{(n)} g \, dx = f^{(n-1)} g - f^{(n-2)} g' + f^{(n-3)} g'' - \cdots (-1)^n \int f g^{(n)} \, dx$$

This is called generalized integration by parts.

*Here, u, v, w are functions of x; a, b, p, q, n any constants, restricted if indicated; $e = 2.71828$ is the natural base of logarithms; $\ln u$ denotes the natural logarithm of u where it is assumed that $u > 0$ (in general, to extend formulas to cases where $u < 0$ as well, replace $\ln u$ by $\ln |u|$); all angles are in radians; all constants of integration are omitted but implied.
SOURCE: Murray R. Spiegel and John Liu, *Mathematical Handbook of Formulas and Tables*, 2d ed., Schaum's Outline Series, McGraw-Hill, 1999.

Schematic electronic symbiols*

Symbol		Symbol	
Ammeter		Coaxial cable	
Amplifier, general		Crystal, piezoelectric	
Amplifier, inverting			
Amplifier, operational		Delay line	
and.gate		Diac	
Antenna, balanced		Diode, field-effect	
Antenna, general			
Antenna, loop		Diode, general	
		Diode, Gunn	
Antenna, loop, multiturn		Diode, light-emitting	
Battery			
Capacitor, feedthrough		Diode, photosensitive	
Capacitor, fixed			
Capacitor, variable		Diode, PIN	
Capacitor, variable, split-rotor		Diode, Schottky	
		Diode, tunnel	
Capacitor, variable, split-stator		Diode, varactor	
Cathode, electron-tube, cold		Diode, Zener	
Cathode, electron-tube, directly heated		Directional coupler	
Cathode, electron-tube indirectly heated		Directional wattmeter	
Cavity resonator		Exclusive-OR gate	
Cell, electrochemical		Female contact, general	
Circuit breaker		Ferrite bead	

*From S. Gibilisco, *The Illustrated Dictionary of Electronics*, 8th ed., McGraw-Hill, 2001.

Appendix

Filament, electron-tube		Inductor, powdered-iron core	
		Inductor, powdered-iron core, bifilar	
Fuse		Inductor, powdered-iron core, tapped	
Galvanometer			
		Inductor, powdered-iron core, variable	
			or
Grid, electron-tube			
Ground, chassis		Integrated, circuit, general	
		Jack, coaxial or photo	
Ground, earth			
		Jack, phone, two-conductor	
Headset			
		Jack, phone, three-conductor	
Handset, double			
		Key, telegraph	
Headset, single		Lamp, incandescent	
Headset, stereo		Lamp, neon	
Inductor, air core		Male contact, general	
Inductor, air core, bifilar		Meter, general	
Inductor, air core, tapped		Microammeter	
Inductor, air core, variable		Microphone	
		Microphone, directional	
Inductor, iron core			
Inductor, iron core, bifilar		Milliammeter	
Inductor, iron core, tapped		NAND gate	
		Negative voltage connection	
Inductor iron core, variable		NOR gate	

NOT gate		Rectifier, gas-filled	
Optoisolator			
		Rectifier, high-vacuum	
OR gate			
Outlet, two-wire, nonpolarized		Rectifier, semiconductor	
Outlet, two-wire, polarized		Rectifier, silicon-controlled	
Outlet, three-wire		Relay, double-pole, double-throw	
Outlet, 234-V			
Plate, electron-tube		Relay, double-pole, single-throw	
Plug, two-wire, nonpolarized			
Plug, two-wire, polarized		Relay, single-pole, double-throw	
Plug, three-wire		Relay, single-pole, single-throw	
Plug, 234-V		Resistor, fixed	
Plug, coaxial or phono		Resistor, preset	
		Resistor, tapped	
Plug, phone, two-conductor		Resonator	
Plug, phone, three-conductor		Rheostat	
Positive voltage connection		Saturable reactor	
Potentiometer		Signal generator	
Probe, radio-frequency		Solar battery	

Appendix

Solar cell

Source, constant-current

Source, constant-voltage

Speaker

Switch, double-pole, double-throw

Switch, double-pole, rotary

Switch, double-pole, single-throw

Switch, momentary-contact

Switch, silicon-controlled

Switch, single-pole, rotary

Switch, single-pole, double-throw

Switch, single-pole, single-throw

Terminals, general, balanced

Terminals, general, unbalanced

Test point

Thermocouple

Transformer, air core

Transformer, air core, step-down

Transformer, air core, step-up

Transformer, air core, tapped primary

Transformer, air core, tapped secondary

Transformer, iron core

Transformer, iron core, step-down

Transformer, iron core, step-up

Transformer, iron core, tapped primary

Transformer, iron core, tapped secondary

Transformer, powdered-iron core

Transformer, powdered-iron core, step-down

Transformer, powdered-iron core, step-up

Transformer, powdered-iron core, tapped primary

Transformer, powdered-iron core, tapped secondary

Transistor, bipolar, *NPN*

Transistor, bipolar, *PNP*

Transistor, field-effect, *N*-channel

Transistor, field-effect, *P*-channel

Transistor, MOS field-effect, *N*-channel

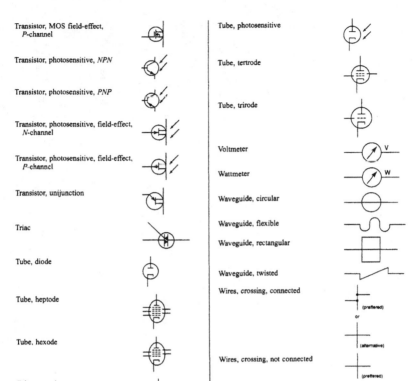

Transistor, MOS field-effect, P-channel	Tube, photosensitive
Transistor, photosensitive, *NPN*	Tube, tertrode
Transistor, photosensitive, *PNP*	Tube, trirode
Transistor, photosensitive, field-effect, N-channel	Voltmeter
Transistor, photosensitive, field-effect, P-channel	Wattmeter
Transistor, unijunction	Waveguide, circular
Triac	Waveguide, flexible
	Waveguide, rectangular
Tube, diode	Waveguide, twisted
Tube, heptode	Wires, crossing, connected
	(preferred)
	or
	(alternative)
Tube, hexode	Wires, crossing, not connected
	(preferred)
	or
Tube, pentode	(alternative)